Regression Modeling

Methods, Theory, and Computation with SAS

T0251863

Regression Modeling

Methods, Theory, and Computation with SAS

Michael Panik
University of Hartford
Connecticut, U. S. A.

CRC Press
Taylor & Francis Group
Boca Raton London New York

CRC Press is an imprint of the
Taylor & Francis Group an **informa** business
A CHAPMAN & HALL BOOK

Chapman & Hall/CRC
Taylor & Francis Group
6000 Broken Sound Parkway NW, Suite 300
Boca Raton, FL 33487-2742

© 2009 by Taylor & Francis Group, LLC
Chapman & Hall/CRC is an imprint of Taylor & Francis Group, an Informa business

No claim to original U.S. Government works
Printed in the United States of America on acid-free paper
10 9 8 7 6 5 4 3 2 1

International Standard Book Number-13: 978-1-4200-9197-7 (Hardcover)

This book contains information obtained from authentic and highly regarded sources. Reasonable efforts have been made to publish reliable data and information, but the author and publisher cannot assume responsibility for the validity of all materials or the consequences of their use. The authors and publishers have attempted to trace the copyright holders of all material reproduced in this publication and apologize to copyright holders if permission to publish in this form has not been obtained. If any copyright material has not been acknowledged please write and let us know so we may rectify in any future reprint.

Except as permitted under U.S. Copyright Law, no part of this book may be reprinted, reproduced, transmitted, or utilized in any form by any electronic, mechanical, or other means, now known or hereafter invented, including photocopying, microfilming, and recording, or in any information storage or retrieval system, without written permission from the publishers.

For permission to photocopy or use material electronically from this work, please access www.copyright.com (http://www.copyright.com/) or contact the Copyright Clearance Center, Inc. (CCC), 222 Rosewood Drive, Danvers, MA 01923, 978-750-8400. CCC is a not-for-profit organization that provides licenses and registration for a variety of users. For organizations that have been granted a photocopy license by the CCC, a separate system of payment has been arranged.

Trademark Notice: Product or corporate names may be trademarks or registered trademarks, and are used only for identification and explanation without intent to infringe.

Library of Congress Cataloging-in-Publication Data

Panik, Michael J.
Regression modeling : methods, theory, and computation with SAS / Michael Panik.
p. cm.
Includes bibliographical references and index.
ISBN 978-1-4200-9197-7 (alk. paper)
1. Regression analysis—Data processing. 2. SAS (Computer program language) I. Title.

QA278.2.P27 2009
519.5'36—dc22 2008045944

Visit the Taylor & Francis Web site at
http://www.taylorandfrancis.com

and the CRC Press Web site at
http://www.crcpress.com

To Paula

Contents

Preface

This book is designed to introduce the reader to the richness and diversity of regression techniques and is particularly well suited for use in a second course in statistics at the undergraduate or first-year graduate level; it is ideal for students of applied mathematics/statistics, the sciences, economics, and engineering who routinely use regression analysis for decision making and problem solving. It can also be used by researchers who need a hands-on reference work on regression methodology and modeling. The only prerequisite is a course in basic statistics and in calculus. A knowledge of matrices and linear algebra is not absolutely essential but, at certain points in the presentation, can be very helpful. For convenience, a review of the essentials of random variables, probability distributions, and classical statistical inference is provided in Chapter 1.

Although there are many varieties of regression analyses from which to choose, one is most often exposed to ordinary least squares (OLS), but OLS is only a part of the regression story. This text fully explores OLS and then offers many alternative regression methodologies. Specifically, the regression routines presented herein include the following: OLS (along with the method of maximum likelihood); nonparametric regression; logistic regression (including Poisson regression); Bayesian regression; robust regression; fuzzy regression; random coefficients regression; L_1 and q-quantile regression; regression in a spatial domain; ridge regression; semiparametric regression; nonlinear least squares; and time-series regression issues.

This book has many strengths and important features. It is highly readable, and the material is quite accessible to those enrolled in applied statistics courses or engaged in self-study. In this regard, an objective of this work is to make students aware of the power and diversity of regression techniques without overwhelming them with calculations.

For most of the regression methods presented, SAS® procedure code is included for the convenience of the reader. Although the various sets of SAS code will enable students and practitioners alike to immediately perform their own regression runs, the code given is not all-encompassing and is by no means a substitute for reading the SAS Manuals; it is only intended to give the reader a jumpstart in solving regression problems. Hence, this is not the type of book that only offers theory and proofs; with a modicum of study and effort, one can "hit the ground running," so to speak, and readily generate some fairly sophisticated regression results. Once a regression technique is explained, SAS handles the "how to" portion of the presentation. This is imperative because one can first study a regression method and then, for the most part, directly apply it because numerous example problems are included, with the SAS results explained in considerable detail.

The presentation of each regression technique is fairly streamlined and designed to offer the reader an unencumbered look into its operation in that proofs and derivations are only supplied in chapter appendices. Those readers who simply want to know "how to" can ignore the appendices; for those readers who want a more technical treatment, the appendices are a "must read." Moreover, to facilitate the understanding and appreciation of the various regression methods, only the bivariate case is covered in Chapters 2 through 11. In these chapters, most of the SAS programs can easily be extended to handle additional explanatory variables once multiple regression (Chapter 12) is covered.

Over the years, I have benefited tremendously from seemingly endless conversations about statistical techniques/methodologies with my colleagues in the Department of Economics, Finance, and Insurance. In particular, Bharat Kolluri, Rao Singamsetti, and Mahmoud Wahab have been particularly helpful. I hope I finally got it right. Additionally, Ed Gullason, Jay Stewart, and Mahmoud Wahab have been most generous with their expertise and insights concerning the application(s) of SAS code, and Randy Gordon has been extremely helpful with computer support.

I am most grateful to my graduate assistants: Caglar Uzun, who expertly typed the chapters of the main text; Goknur Gurz, who steadfastly typed the chapter exercises and some of the minor revisions; and Muhammed Cenk, who helped put some of the finishing touches to the final manuscript. I am also appreciative of the efforts of Alice Schoenrock, our Office Coordinator, who typed many of the ancillaries and monitored a sizeable portion of the activities involved in the securing of a complete draft of this work.

A special note of thanks is extended to Provost Elizabeth Ivey for providing me with the opportunity to devote a concentrated and uninterrupted period of time to the completion of this project. Without her support, this book would have taken much longer to complete.

Additional offerings of appreciation are extended to David Grubbs, Editor, *Mathematics and Statistics*, and to Jennifer Ahringer, Production Coordinator, both at Taylor & Francis/CRC Press. Their professionalism and suggestions/insights were most helpful. Finally, Eve Malakoff-Klein, Project Manager, Cadmus Communications, helped to secure a smooth production process.

Author

Michael J. Panik received his PhD in economics from Boston College. As a doctoral student, he held a NASA Fellowship for three years and developed specializations in the areas of statistics, econometrics, mathematical economics, and microeconomics.

Dr. Panik held a lectureship at Boston College and then taught for many years in the Department of Economics at the University of Hartford, where he is now Professor Emeritus. In 2000, Dr. Panik received the University's James E. and Francis W. Bent Medal for Scholarly Creativity. He is currently a member of the Visiting Faculty at Trinity College, Department of Mathematics, and Wesleyan University, Department of Economics.

Dr. Panik is the author of numerous articles in professional journals, has served as a consultant in the area of health care research and to the state of Connecticut, and has written four other books in the fields of optimization, convex analysis, linear programming, and statistics. His current research pursuits deal with growth curve modeling and estimation and analytical microeconomics.

1

Review of Fundamentals of Statistics

Probability

According to the *classical view of probability*, if a *random experiment* (e.g., tossing a fair coin a large number of times under essentially unchanged conditions) has n equiprobable outcomes, and, if n_A of these outcomes constitute or favor event A, then the probability of event A, denoted $P(A)$, is simply n_A/n. Here $P(A) = n_A/n$ is termed the *theoretical relative frequency* of event A.

Suppose P is a *probability measure*, a set function from the *sample space* S (the collection of all possible outcomes of a random experiment) to the unit interval [0,1]. Then for events $A, B \subseteq S$, the following probability axioms hold:

A.1. $P(A)$ is defined for every $A \subseteq S$;

A.2. $P(A) \geq 0$ for all $A \subseteq S$;

A.3. $P(S) = 1$; and

A.4. If $A, B \subseteq S$ with $A \cap B = \phi$ (A and B are *disjoint* or *mutually exclusive events*), then $P(A \cup B) = P(A) + P(B)$.

Given $A \subseteq S$, a set of *corollaries to the axiom system* are as follows:

C.1. The probability that event A does not occur is one minus the probability that it does occur or $P(\bar{A}) = 1 - P(A)$;

C.2. $0 \leq P(A) \leq 1$; and

C.3. $P(\phi) = 0$.

We next have the *general addition rule for probabilities*: for events $A, B \subseteq S$,

$$P(A \cup B) = P(A) + P(B) - P(A \cap B). \tag{1.1}$$

Note that, if $A \cap B = \phi$, then $P(A \cap B) = 0$ and Equation 1.1 simplifies to Axiom A.4.

If we are interested in the probability of occurrence of an event A given that some other event B has definitely occurred, then we compute a *conditional probability*, i.e., the probability of event A given B, denoted $P(A|B)$, is

$$P(A|B) = \frac{P(A \cap B)}{P(B)}, \ P(B) \neq 0; \tag{1.2}$$

and the probability of B given A is

$$P(B|A) = \frac{P(A \cap B)}{P(A)}, \; P(A) \neq 0. \tag{1.3}$$

Then from Equations 1.2 and 1.3, we have the *multiplication rule for probabilities* or

$$P(A \cap B) = P(A|B) \cdot P(B) = P(B|A) \cdot P(A). \tag{1.4}$$

If events $A, B \subseteq S$ are *mutually independent*, then the occurrence of one of them in no way affects the probability of occurrence of the other. In this regard, *A is independent of B* if and only if

$$P(A|B) = P(A), \text{ provided } P(B) > 0; \tag{1.5}$$

and *B is independent of A* if and only if

$$P(B|A) = P(B), \text{ provided } P(A) > 0. \tag{1.6}$$

Given Equations 1.5 and 1.6, it is easily demonstrated that, under independence, the multiplication rule for probabilities become

$$P(A \cap B) = P(A) \cdot P(B). \tag{1.7}$$

It is important to remember that, if $A, B \subseteq S$ are mutually exclusive events, then A and B cannot occur together so that $P(A \cap B) = 0$; but if A and B are independent events, then A and B can occur together ($A \cap B \neq \phi$) so that Equations 1.7 holds.

We mentioned above that $P(A) = n_A/n$ is a theoretical relative frequency. However, n_A/n may also be interpreted as an *empirical relative frequency*, i.e., if A is an event for a random experiment and if that experiment is repeated n times and A is observed in n_A of the n trials, then the empirical relative frequency of event A is n_A/n. Then, via the *frequency limit principle*,

$$P(A) = \lim_{n \to \infty} \frac{n_A}{n}.$$

So if the relative frequencies n_A/n approach some number $P(A)$ (a long-run stable value), then this number is assigned to A as its probability. This method for determining probabilities is called the *frequentist approach.*

Random Variables and Probability Distributions

A random variable X is a real-valued function defined on the sample space S. A random variable X is said to be *discrete* if the number of values it assumes forms a countable set, i.e., the set in question has either a finite number of elements or is countably infinite. Furthermore,

X is said to be a *continuous* random variable if it assumes an infinite or uncountable number of values over some interval.

For X, a discrete random variable defined on S, let the range of X be the set of values $R = \{X_i, i = 1,2,\ldots\}$ and let f be a function that associates with each $X_i \in R$ a number $f(X_i) \in [0,1]$ representing the probability that $X = X_i$, i.e., $P(X = X_i) = f(X_i)$, $i = 1,2,\ldots$. Hence, $f(X_i)$ depicts the *probability mass* at X_i. In general, a function $f(X)$, which assigns a probability mass $f(X_i)$ to each X_i within the range of X, is called a *probability mass function* if

a. $f(X_i) \geq 0$ for all i;
b. $\sum_i f(X_i) = 1$; and
c. $f(X_i) = 0$ for all $X_i \notin R$.

To construct a discrete probability distribution, we need to specify the random experiment and define the random variable X. Given these ingredients, a discrete probability distribution will be fully specified once the sequence of probabilities $f(X_i)$, $i = 1,2,\ldots$, is given. Hence, a discrete probability distribution can be represented by the set of all possible pairs $(X_i, f(X_i))$, $i = 1,2,\ldots$. Once a discrete probability distribution is determined, we may compute, for instance, probabilities such as,

$$P(a \leq X \leq b) = \sum_{a \leq X_i \leq b} f(X_i); \; P(X \geq a) = \sum_{X_i \geq a} f(X_i); \text{ and}$$

$$P(X < a) = 1 - P(X \geq a) = \sum_{X_i < a} f(X_i).$$

Whereas the probability mass function $f(X_i)$ gives $P(X = X_i)$, a related function that yields $P(X \leq X_i)$ is the *cumulative distribution function*:

$$F(X_i) = \sum_{j \leq i} f(X_j), \tag{1.8}$$

with the X_j's placed in increasing order. Its key properties are as follows:

a. $0 \leq F(X_i) \leq 1$ for any X_i;
b. F is a monotone nondecreasing function of X_i;
c. $P(X > X_i) = 1 - F(X_i)$;
d. $P(X = X_i) - F(X_i) - F(X_i - 1)$, X_i an integer;
e. $P(X_s < X \leq X_t) = F(X_t) - F(X_s)$;
f. $F(-\infty) = 0$, $F(+\infty) = 1$; and
g. F is continuous from the right.

Given a discrete random variable X, its *mean* or *expected value* is

$$E(X) = \sum_i X_i f(X_i) = \mu_X; \tag{1.9}$$

and, if $g(X)$ is a single-valued real function of X, then

$$E[g(X)] = \sum_i g(X_i) f(X_i),$$

provided that these sums are finite. Moreover, some important properties of the expectation operator for a discrete random variable X include the following:

a. $E(a) = a$ (a is constant);
b. $E(a \pm bX) = a \pm bE(X)$ (a, b constants); and
c. $E(\sum_{i=1}^{n} X_i) = \sum_{i=1}^{n} E(X_i)$.

The variance of a discrete random variable X is calculated as

$$V(X) = E[(X - E(X))^2] = \sum_i (X_i - E(X))^2 f(X_i) = \sigma_X^2 \tag{1.10}$$

or

$$V(X) = E(X^2) - E(X)^2 = \sum_i X_i^2 f(X_i) - E(X)^2. \tag{1.11}$$

Key properties of the variance of a discrete random variable X are as follows:

a. $V(a) = 0$ (a is constant); and
b. $V(aX) = a^2 V(X)$ (a is constant).

The standard deviation of X is determined as $\sigma_X = +\sqrt{V(X)} \geq 0$.

If X is a discrete random variable whose expectation $E(X)$ exists and whose standard deviation $\sigma_X > 0$, then the *standardized variable* is

$$Z = \frac{X - E(X)}{\sigma_X} \tag{1.12}$$

with

$$Z_i = \frac{X_i - E(X)}{\sigma_X}, \quad i = 1, \ldots.$$

It is readily demonstrated that $E(Z) = \mu_Z = 0$ and $V(Z) = \sigma_Z^2 = 1$.

For X, a continuous random variable defined on the class S of events representable by all open and closed intervals (as well as by all half-open intervals and the rays $(-\infty, a]$, $[b, +\infty)$), let P be a probability measure or set function that associates with each event $A \subseteq S$ a number $P(A)$ representing the *probability* that event A occurs. A function $f(X)$ that defines the probability measure

$$P(A) = P(X \in A) = \int_A f(x)dx, \ A \subseteq S, \tag{1.13}$$

is termed a *probability density function* if

a. $f(x) \geq 0$, $x \in (-\infty, +\infty)$ and $f(x) > 0$, $x \in A$;
b. $\int_A f(x)dx$ exists for every $A \subseteq S$; and
c. $\int_S f(x)dx = \int_{-\infty}^{+\infty} f(x)dx = 1.$

If X is a continuous random variable with probability density function $f(x)$, then for $X = \{x | a \leq x \leq b\}$,

$$P(A) = P(a \leq X \leq b) = \int_a^b f(x)dx. \tag{1.14}$$

Moreover, $P(X = a) = 0$ for any constant a.

A function derived from the probability density function $f(x)$ that yields the probability that continuous X takes on a value less than or equal to some real number t is the cumulative distribution function:

$$F(t) = P(X \leq t) = \int_{-\infty}^{t} f(x)dx. \tag{1.15}$$

Thus, the cumulative distribution function serves as an alternative representation of a continuous probability distribution in that, if $f(x)$ is known, then we can determine its cumulative distribution function $F(t)$; conversely, given $F(t)$, we can determine the probability density $f(t)$ at each of its points of continuity as $dF(t)/dt = f(t)$. Key properties of the cumulative distribution function are as follows:

a. $0 \leq F(t) \leq 1$;
b. F is a nondecreasing function of t;
c. $F(-\infty) = 0$, $F(+\infty) = 1$;
d. F is everywhere continuous from the right at each t;
e. For $a < b$, $P(a \leq X \leq b) = F(b) - F(a)$; and
f. If F has a point of discontinuity at t, then $P(X = t)$ is the size of the jump that occurs at t; and if F is continuous at t, then $P(X = t) = 0$.

If X is a continuous random variable with probability density function $f(x)$, then the *mean* of X or its *mathematical expectation* is

$$E(X) = \int_{-\infty}^{+\infty} f(x)dx; \tag{1.16}$$

and, if $g(x)$ is a single-valued real function of X, then

$$E[g(X)] = \int_{-\infty}^{+\infty} g(x)f(x)dx, \tag{1.17}$$

provided, of course, that these integrals converge to a finite value.

The *variance* of a continuous random variable X is defined as

$$V(X) = \int_{-\infty}^{+\infty} (x - E(X))^2 f(x)dx = \sigma_X^2 \tag{1.18}$$

or

$$V(X) = E(X^2) - E(X)^2 = \int_{-\infty}^{+\infty} x^2 f(x)dx - E(X)^2. \tag{1.19}$$

The Normal Probability Distribution and Derived Distributions

Normal Distribution

The *normal distribution* is a continuous bell-shaped distribution that is symmetrical about its mean and asymptotic to the horizontal axis. Its probability density function is given by

$$f(x;\mu,\sigma) = \frac{1}{\sigma\sqrt{2\pi}} e^{-(x-\mu)^2/2\sigma^2}, -\infty < x < +\infty. \tag{1.20}$$

Its mean is μ and its variance is σ^2. If X is normally distributed, then we may write $X : N(\mu, \sigma)$. If $X : N(\mu,\sigma)$, then $Z = (X - \mu)/\sigma : N(0, 1)$. Here Z is termed a *standard normal* random variable with probability density function

$$f(z;0,1) = \frac{1}{\sqrt{2\pi}} e^{-z^2/2}, -\infty < z < +\infty.$$

Table A.1 of areas under the standard normal distribution is provided in the Appendix found at the end of the text.

We now turn to a collection of probability distributions that are related to or derived from the normal distribution.

Derived Distributions (χ^2, *t*, and *F*)

If $Z_1, Z_2, \ldots, Z_n$ are independent standard normal random variables ($Z_i : N(0,1)$, $i = 1,\ldots,n$), then $Z = \sum_{i=1}^{n} Z_i^2$ is said to have a χ^2 *distribution* with *n* degrees of freedom (d.f.) and written χ_n^2. Hence, the χ^2 distribution is the sum of the squares of *n* independent standard normal random variables. It is important to note that, if $X_1, X_2, \ldots, X_n$ are independent normal random variables with a mean of zero and a standard deviation of σ($X_i : N(0,\sigma)$, $i = 1,\ldots,n$), then $X = \sum_{i=1}^{n} (X_i^2/\sigma^2) : \chi_n^2$. In addition, the χ^2 distribution possesses the *additive property*: if $Z_1 : \chi_n^2$ and $Z_2 : \chi_m^2$ and Z_1 and Z_2 are independently random variables, then $Z = Z_1 + Z_2 : \chi_{n+m}^2$. Table A.3 in the Appendix to the text contains various quantiles of the χ^2 distribution.

Next, suppose $Z : N(0,1)$, $Y : \chi_n^2$, and the random variables Z and Y are independent. Then the quantity $T = X/\sqrt{Y/n}$ has a (Student's) *t distribution* with n degrees of freedom and is written $T : t_n$. Thus, the t distribution is the distribution of a standard normal variable divided by the square root of a χ^2 variable divided by its degrees of freedom, with these variables taken to be independent. Moreover, the t distribution is symmetrical about zero like the normal distribution, but its tails are flatter (and longer) than those of the normal distribution. As the degrees of freedom n approaches $+\infty$, the t distribution approaches the standard normal distribution. Quantiles of the t distribution appear in Table A.2 of the Appendix.

Finally, if $X_1 : \chi_n^2$ and $X_2 : \chi_m^2$ and X_1 and X_2 are independent variables, then $Y = (X_1/n)/(X_2/m)$ has an *F distribution* with degrees of freedom n and m and is written $Y : F_{n,m}$. (Remember that the first and second subscripts refer to the degrees of freedom of the numerator and denominator, respectively.) Thus, the ratio of two independent χ^2 random variables, each divided by its own degrees of freedom, is F distributed. Quantiles of the F distribution are provided in Table A.4 of the Appendix.

Bivariate Random Variables and Probability Distributions

Suppose (X, Y) is a pair of real-valued functions defined on a sample space S. The pair (X, Y) is a *bivariate random variable* if both X and Y map elements in S into real numbers. Thus, pair (X, Y) defines the possible outcomes of some random experiment. Now, if the range of (X, Y) is a discrete ordered set of points $R_{X,Y} = \{(X_i, Y_i), i = 1,\ldots,m\}$ in the Cartesian plane, then (X, Y) is termed a *discrete bivariate random variable.* However, if S consists of a class of events representable by all open and closed rectangles, then X and Y defined on S are termed *continuous bivariate random variables* if they can assume any values within some event $A = \{(X, Y) | a \le X \le b, c \le Y \le d\} \subseteq S$.

For the discrete case, given a random variable $X : X_i, i = 1,\ldots,n$ (with $X_1 < X_2 < \ldots < X_n$) and a second random variable $Y : Y_j, j = 1,\ldots,m$ (with $Y_1 < Y_2 < \ldots < Y_m$), we may denote the *joint probability* that $X = X_i$ and $Y = Y_j$ as $P(X = X_i, Y = Y_j) = f(X_i, Y_j), i = 1,\ldots,n; j = 1,\ldots,m$. Here $f(X_i, Y_j)$ is termed the probability mass at the point (X_i, Y_j). In general, a function $f(X, Y)$ that assigns a probability mass $f(X_i, Y_j)$ within the ranges of the discrete random variables X and Y is called a *bivariate probability mass function* if

a. $f(X_i, Y_j) \ge 0$ for all i, j; and
b. $\sum_i \sum_j f(X_i, Y_j) = 1.$

Then the set of nm events (X_i, Y_j) together with their associated probability masses $f(X_i, Y_j)$, $i = 1,\ldots,n$; $j = 1,\ldots,m$, constitutes the *discrete bivariate probability distribution* of the random variables X and Y. Given a bivariate probability mass function $f(X, Y)$, the joint probability that $X \le X_r$ and $Y \le Y_s$ is provided by the *bivariate cumulative distribution function*

$$F(X_r, Y_s) = P(X \le X_r, Y \le Y_s) = \sum_{i \le r} \sum_{j \le s} f(X_i, Y_j). \tag{1.21}$$

For X and Y discrete random variables with bivariate probability mass function $f(X, Y)$, the univariate *marginal probability mass function of X* is

$$g(X) = \sum_j f(X, Y_j); \tag{1.22}$$

and the univariate *marginal probability mass function of Y* is

$$h(Y) = \sum_j f(X_i, Y). \tag{1.23}$$

Given Equations 1.22 and 1.23, if (X, Y) is any point at which $h(Y) > 0$, then the *conditional probability mass function of X given Y* is

$$g(X|Y) = \frac{f(X,Y)}{h(y)}; \tag{1.24}$$

and, at any point (X, Y) at which $g(X) > 0$, the *conditional probability mass function of Y given X* is

$$h(Y|X) = \frac{f(X,Y)}{g(x)}. \tag{1.25}$$

If we solve for $f(X, Y)$ from both Equations 1.24 and 1.25, then we have the *multiplication theorem for probability mass functions*:

$$f(X, Y) = g(X) \cdot h(Y|X) = g(X|Y) \cdot h(Y). \tag{1.26}$$

Given Equations 1.22 through 1.25, we can state that the *random variable X is independent of the random variable Y* if

$$g(X|Y) = g(X); \tag{1.27}$$

similarly, the *random variable Y is independent of the random variable X* if

$$h(Y|X) = h(Y). \tag{1.28}$$

If Equations 1.27 and 1.28 both hold, then X and Y are said to be mutually *independent random variables* if and only if Equation 1.26 can be written as

$$f(X, Y) = g(X) \cdot h(Y), \tag{1.26.1}$$

i.e., X and Y are independent if and only if the joint probability mass function can be written as the product of their individual marginal probability mass functions.

For the continuous case, let P be a (joint) probability measure or set function that associates with each event $A = \{(X, Y)|a \le X \le b, c \le Y \le d\} \subseteq S$ some number $P(A)$. In general, the function $f(X, Y)$ that defines the probability measure

$$P(A) = P(a \le X \le b,\ c \le Y \le d) = \int_a^b \int_c^d f(x,y)dydx \tag{1.29}$$

is called a *bivariate probability density function* if

a. $f(x,y) \ge 0$ for all real x and y such that $-\infty < x, y < +\infty$ and $f(x, y) > 0$ for $(x, y) \in A$; and
b. $\int_{-\infty}^{+\infty} \int_{-\infty}^{+\infty} f(x,y)dydx = 1.$

So, if the collection of all open and closed rectangles constitutes the sample space S and $P(A)$ is determined by Equation 1.29, then the random variables X and Y follow a *continuous bivariate probability distribution.*

Given a bivariate probability density function $f(x, y)$, the joint probability that $X \le t$ and $Y \le s$ is provided by the *bivariate cumulative distribution function*

$$F(t,s) = P(X \le t,\ Y \le s) = \int_{-\infty}^{t} \int_{-\infty}^{s} f(x,y)dydx. \tag{1.30}$$

$F(t, s)$ is a continuous function of t and s and, at every point of continuity of $f(x, y)$, $\partial^2 F/\partial s \partial t = f(t, s)$. So, if $F(t, s)$ is known, we can determine the probability density function $f(t,s)$ at each of its points of continuity by calculating $\partial^2 F/\partial s \partial t$.

For X and Y continuous random variables with bivariate probability density $f(x, y)$, the *marginal probability density function of X* is

$$g(x) = \int_{-\infty}^{+\infty} f(x,y)dy \tag{1.31}$$

(satisfying $g(x) \ge 0$ and $\int_{-\infty}^{+\infty} g(x)dx = 1$); whereas the *marginal probability density function of Y* is

$$h(y) = \int_{-\infty}^{+\infty} f(x,y)dx \tag{1.32}$$

(with $h(y) \ge 0$ and $\int_{-\infty}^{+\infty} h(y)dy = 1$). In addition, if the marginal probability density functions $g(x)$ and $h(y)$ are known, then we may define, for Y fixed at y, the *conditional probability density function for X given Y* as

$$g(x|y) = \frac{f(x,y)}{h(y)}, h(y) > 0 \tag{1.33}$$

(with $g(x|y) > 0$ and $\int_{-\infty}^{+\infty} g(x|y)dx = 1$); and, for X fixed at x, the *conditional probability density function for Y given X* is

$$h(y|x) = \frac{f(x,y)}{g(x)}, g(x) > 0 \tag{1.34}$$

(satisfying $h(y|x) > 0$ and $\int_{-\infty}^{+\infty} h(y|x)dy = 1$).

If X and Y are continuous random variables with bivariate probability density function $f(x, y)$ and marginal probability densities $g(x)$ and $h(y)$, respectively, then the *random variable X is said to be independent of the random variable Y* if

$$g(x|y) = g(x) \tag{1.35}$$

for all values of X and Y for which both of these functions exist; likewise, the *random variable Y is independent of the random variable X* if

$$h(y|x) = h(y). \tag{1.36}$$

Given, say, Equation 1.33, we may use Equations 1.35 to write

$$f(x, y) = g(x) \cdot h(y), \tag{1.37}$$

i.e., *X and Y are independent random variables* if and only their joint probability density function can be written as the product of their individual marginal probability densities.

Suppose X and Y are discrete random variables with bivariate probability mass function $f(X, Y)$. Then the *expectation of a function of X and Y, $\varphi(X, Y)$,* is

$$E[\varphi(X,Y)] = \sum_i \sum_j \varphi(X_i, Y_j) f(X_i, Y_j). \tag{1.38}$$

Given this expression, the *rth moment of the random variable X about zero* is

$$E(X^r) = \sum_i \sum_j X_i^r f(X_i, Y_j) = \sum_i X_i^r g(X_i); \tag{1.39}$$

the *sth moment of the random variable Y about zero* is

$$E(Y^s) = \sum_i \sum_j Y_j^s f(X_i, Y_j) = \sum_j Y_j^s h(Y_j); \tag{1.40}$$

the *rth and sth product moment of X and Y about the origin* is

$$E(X^r Y^s) = \sum_i \sum_j X_i^r Y_j^s f(X_i, Y_j); \tag{1.41}$$

and, for r and s nonnegative integers, the *rth and sth product moment of X and Y about the mean* is

$$E\left[(X - \mu_X)^r (Y - \mu_Y)^s\right] = \sum_i \sum_j (X_i - \mu_X)^r (Y_j - \mu_Y)^s f(X_i, Y_j), \tag{1.42}$$

where $\mu_X = E(X)$ and $\mu_Y = E(Y)$.

If X and Y are continuous random variables with joint probability density function $f(X, Y)$, then the expectation of a function of X and Y, $\varphi(X, Y)$, is

$$E[\varphi(X,Y)] = \int_{-\infty}^{+\infty}\int_{-\infty}^{+\infty} \varphi(x,y) f(x,y)dydx. \tag{1.43}$$

In this regard, the *rth moment of X about zero* is

$$E(X^r) = \int_{-\infty}^{+\infty}\int_{-\infty}^{+\infty} x^r f(x,y)dydx = \int_{-\infty}^{+\infty} x^r g(x)dx; \tag{1.44}$$

the *sth moment of Y about zero* is

$$E(Y^s) = \int_{-\infty}^{+\infty}\int_{-\infty}^{+\infty} y^s f(x,y)dydx = \int_{-\infty}^{+\infty} y^s h(y)dy; \tag{1.45}$$

the *rth and sth product moment of X and Y about the origin* is

$$E(X^r Y^s) = \int_{-\infty}^{+\infty}\int_{-\infty}^{+\infty} x^r y^s f(x,y)dydx; \tag{1.46}$$

and, for r and s nonnegative integers, the *rth and sth product moment of X and Y about the mean* is

$$E\left[(X-\mu_X)^r (Y-\mu_Y)^s\right] = \int_{-\infty}^{+\infty}\int_{-\infty}^{+\infty} (x-\mu_X)^r (y-\mu_Y)^s f(x,y)dydx. \tag{1.47}$$

We note briefly that, if X_i, $i = 1,\ldots,k$, are discrete or continuous random variables and $\varphi = \sum_{i=1}^{k} a_i X_i$, a_i constant for all i, then

$$E\left(\sum_{i=1}^{k} a_i X_i\right) = \sum_{i=1}^{k} a_i E(X_i). \tag{1.48}$$

Moreover,

1. If $r = s = 1$ in Equations 1.42 or 1.47, then the expression

$$E[(X-\mu_X)(Y-\mu_Y)] = E(XY) - \mu_X \mu_Y = COV(X, Y) = \sigma_{XY} \tag{1.49}$$

is called the *covariance* of X and Y.

2. The *coefficient of correlation* between the random variables X and Y is

$$\rho_{XY} = \frac{\sigma_{XY}}{\sigma_X \sigma_Y}, \tag{1.50}$$

with $-1 \le \rho_{XY} \le 1$ or $|\sigma_{XY}| \le \sigma_X \sigma_Y$.

3. If X and Y are discrete or continuous random variables and $Z_X = (X - \mu_X)/\sigma_X$ and $Z_Y = (Y - \mu_Y)/\sigma_Y$, then

$$\rho_{Z_X Z_Y} = \sigma_{Z_X Z_Y} / \sigma_{Z_X} \sigma_{Z_Y} = \sigma_{Z_X Z_Y}. \tag{1.51}$$

Here the correlation coefficient of two standardized random variables is just their covariance.

4. If X and Y are independent discrete or continuous random variables, then $\sigma_{XY} = 0$ and thus $E(XY) = \mu_X \mu_Y$; but if $\sigma_{XY} = 0$, it does not follow that X and Y are independent random variables unless X and Y are normal.

If φ is a real-valued function of the discrete or continuous random variables X and Y, then the *variance of a function of X and Y*, $\varphi(X, Y)$, is

$$V[\varphi(X,Y)] = E\{[\varphi(X,Y) - E(\varphi(X,Y))]^2\} = \sigma_\varphi^2. \tag{1.52}$$

For a and b constants:

1. If $\varphi(X, Y) = aX + bY$, then

$$V(aX + bY) = a^2\sigma_X^2 + b^2\sigma_Y^2 + 2ab\sigma_{XY}; \tag{1.53}$$

2. If $\varphi(X, Y) = aX - bY$, then

$$V(aX - bY) = a^2\sigma_X^2 + b^2\sigma_Y^2 - 2ab\sigma_{XY}; \text{ and} \tag{1.54}$$

3. If $\varphi(X, Y) = aX \pm bY$ and X and Y are independent random variables, then

$$V(aX \pm bY) = a^2\sigma_X^2 + b^2\sigma_Y^2. \tag{1.55}$$

In general, if X_i, $i = 1,\ldots,k$, are discrete or continuous random variables and $\varphi = \sum_{i=1}^k a_i X_i$, a_i constant for all i, then

$$V\left(\sum_{i=1}^k a_i X_i\right) = \sum_{i=1}^k a_i^2 \sigma_i^2 + 2\sum_{\substack{i=1 \\ i<j}}^k \sum_{j=1}^k a_i a_j \sigma_{ij}; \tag{1.56}$$

and, if X_i are independent random variables, then

$$V\left(\sum_{i=1}^k a_i X_i\right) = \sum_{i=1}^k a_i^2 \sigma_i^2. \tag{1.57}$$

Classical Statistical Inference

In general, *classical statistical inference* is based upon the following notions:

1. Sample data, garnered via random sampling performed under essentially unchanged conditions, constitutes our relevant (and sole) information set; and
2. The determination and assessment of various procedures for making inferences is founded on the long-run behavior of estimators (statistics) computed from sample data.

Classical statistical inference has essentially three components: point estimation, interval estimation, and hypothesis testing.

Point Estimation

We may view a *statistic* T as a characteristic of a sample that is used to estimate a parameter θ. Here T is a function of the sample random variables $X_1, X_2,\ldots,X_n$ and is itself a random variable that does not depend on any unknown parameters. If T renders a single numerical value as the estimate of θ, then T is called a *point estimator* of θ. Because of chance variability in random sampling, T will rarely be exactly "on target" so that sampling error will occur. Here *sampling error* is defined as $SE(T, \theta) = T - \theta$ and is taken to be a random variable because T is random.

What are the desirable properties that a good point estimator T may possess? Such properties are classified as small sample properties or large sample (asymptotic) properties and are framed in terms of the *sampling distribution* of T, which is a probability density function that embodies all relevant statistical information about T; it shows how the sample realizations of T vary under random sampling. Hence, the process of evaluating the "goodness" of an estimator hinges on conceptualizing the entire spectrum of realizations that would be obtained if the estimation rule T were applied repeatedly to many samples taken from a given population; in this context, it is the long-run behavior of an estimator under repeated sampling that matters.

Small Sample Properties of a Point Estimator

A small sample property of a point estimator T of θ is a characteristic of the sampling distribution of T that holds for a fixed sample size n. Small sample properties include the following.

Unbiasedness

T is an *unbiased estimator* for θ if $E(T) = \theta$. Hence, the mean of the sampling distribution of T equals θ, i.e., if T is determined for each of an infinite number of samples, then the average of all these estimates will equal θ. Let us define the *bias* ($B(T, \theta)$) of an estimator as the expected value of its sampling error or $B(T, \theta) = E(T) - \theta$. Hence, an unbiased estimator is one whose bias is zero.

Minimum Variance

Given $V(T) = E(T^2) - E(T)^2$, T is a *minimum variance estimator* for θ if $V(T) \leq V(T')$ where T' is any alternative estimator for θ. A general criterion for assessing the goodness of an estimator T of θ in terms of the relative magnitude of its bias and variance is the *mean squared error* (*MSE*) of T or

$$MSE(T, \theta) = E[(T - \theta)^2] = V(T) + B(T, \theta)^2. \tag{1.58}$$

Hence, an estimator with good *MSE* properties (in terms of bringing us close to θ) is one that has a small combined variance and (squared) bias.

Efficiency

T is termed an *efficient* (or *minimum varianced unbiased* or *best unbiased*) *estimator* of θ if T is unbiased and $V(T) \leq V(T')$ for all values of θ, where T' is any other unbiased estimator of θ.

To set the stage for a discussion of what may be called the *most efficient* estimator of θ, suppose $f(x; \theta)$ depicts the probability density function (or probability mass function) for a population distribution. Then, under random sampling, the outcomes X_i, $i = 1,\ldots,n$, represent a collection of n independent sample random variables each having the same probability density function as the population distribution, so that their *joint probability density function* at $x_1,\ldots, x_n$ is

$$l(x_1,\ldots,x_n;\theta,n) = \prod_{i=1}^{n} f(x_i;\theta), \tag{1.59}$$

where $f(x_i; \theta) = f(x; \theta)$ is the marginal probability density function for the random variables X_i, $i = 1,\ldots,n$. If we rewrite Equation 1.59 as

$$\mathcal{L}(\theta; x_1,\ldots x_n, n) = \prod_{i=1}^{n} f(x_i;\theta), \tag{1.59.1}$$

then this revised expression (simply a reinterpretation of Equation 1.59) is termed the *likelihood function* of the sample. Note that Equation 1.59 depicts a sampling distribution because the X_i's are variable and θ is held constant, but, in Equation 1.59.1, the X_i's are fixed and θ variable. Thus, Equation 1.59.1 yields the likelihood that the random variables $X_1,\ldots X_n$ will assume a certain set of sample realizations, i.e., in terms of θ, it is the a priori probability of obtaining the observed random sample.

For convenience, one typically works with the *log-likelihood function* $\log \mathcal{L}$. In fact, an important application of this function is that it can be used to determine the *minimum variance bound* or *Cramer-Rao lower bond* (denoted $CR(\theta, n)$) for an unbiased estimator of a parameter θ. Specifically, if T is an unbiased estimator of θ, then, under some general regularity conditions, the variance of T must satisfy the *Cramer-Rao inequality*

$$V(T) \geq \frac{1}{-E\left(\dfrac{\partial^2 \log \mathcal{L}}{\partial \theta^2}\right)} = CR(\theta, n), \tag{1.60}$$

which is constant for fixed n. (Note that the form of the population distribution $f(x; \theta)$ must be known.)

On the basis of the preceding discussion, we can now characterize an estimator as *most efficient*.

Most Efficient

If T is an unbiased estimator of θ and Equation 1.60 holds as a strict equality, then T is the *most efficient* or *minimum variance bound estimator* of θ.

Sufficiency

T is a *sufficient estimator* for θ if it uses all the information about θ that appears in a random sample; it does not discard any information about θ. Formally, T is sufficient for θ if the conditional distribution of the sample random variables $X_1,\ldots,X_n$ given T, or $f(X_1,\ldots, X_n|T)$, does not depend on θ.

Large Sample Properties of Point Estimators

A large sample or asymptotic property of a point estimator of θ is actually a property of a sequence of point estimators $\{T_n\}$ indexed by the sample size n; such properties are limiting properties as $n \to \infty$. More specifically, a large sample property of a point estimator is a characteristic of the sampling distribution of the estimator that results when $n \to \infty$. In particular, the *shape* of the sampling distribution of T as $n \to \infty$ is called the *asymptotic* or *limiting distribution* of T, whereas the *asymptotic mean* or *expectation* of $T_n(AE(T_n))$ is the mean of the limiting distribution of T_n and the *asymptotic variance* of $T_n(AV(T_n))$ is the variance of the limiting distribution of T_n.

A desirable asymptotic property of a good point estimator T of θ is that the sampling distribution of T becomes more closely concentrated about θ as the sample size increases without bound. In this context, large sample properties include the following.

Consistency

T_n is a *consistent estimator* of θ if the sequence of estimators $\{T_n\}$ converges in probability to θ as n increases without bound (written $T_n \xrightarrow{p} \theta$), i.e.,

$$\lim_{n\to\infty} P(|T_n - \theta| < \varepsilon) = 1 \tag{1.61}$$

for all admissible θ and all real $\varepsilon > 0$. Specifically, as this equation reveals, the probability that the sampling error $|T_n - \theta|$ is less than any arbitrary small positive constant ε approaches one as n approaches infinity. (A sufficient condition for T to be a consistent estimator for θ is that both the variance of T and its bias should tend to zero as $n \to \infty$; it is not necessary because T can be consistent even if its bias does not approach zero as $n \to \infty$.)

Asymptotic Unbiasedness

T_n is an *asymptotically unbiased estimator* of T if its bias $B(T_n, \theta) = E(T_n) - \theta$ declines to zero as n increases without bound or if $AE(T_n) = 0$. Note that, if an estimator is unbiased, then it is asymptotically unbiased; however, the asymptotic unbiasedness of an estimator does not necessarily imply that the estimator is unbiased (because it may be biased for small samples).

Asymptotic Efficiency

If two estimators of a parameter θ are each asymptotically unbiased, the one with the smaller asymptotic variance is said to be asymptotically more efficient than the other, and the estimator with the smallest asymptotic variance among *all* asymptotically unbiased estimators of θ is said to be *asymptotically efficient*.

Best Asymptotic Normality

Given that $X_1,\ldots,X_n$ constitutes a set of sample random variables drawn from a $N(\mu, \sigma^2)$ population, the sequence of estimators $\{T_n\}$ of a parameter θ is termed *best asymptotically normal* if and only if the following apply:

a. The distribution of $\sqrt{n}(T_n - \theta) \to N(0, V(T_n))$ as $n \to \infty$;
b. $\{T_n\}$ is consistent for θ;
c. For $\{W_n\}$ any other sequence of consistent estimators of θ, the distribution of $\sqrt{n}(W_n - \theta) \to N(0, V(W_n))$ as $n \to \infty$; and
d. $V(T_n) \le V(W_n)$ for all admissible θ.

An important technique for finding "good" point estimators is the *method of maximum likelihood*. Given Equation 1.59.1, the log-likelihood function is

$$\log \mathcal{L}(\theta; x_1, \ldots x_n, n) = \sum\nolimits_{i=1}^{n} \log f(x_i; \theta). \tag{1.62}$$

Then the *principle of maximum likelihood* is as follows: Select as an estimator of θ that value of the parameter, denoted $\hat{\theta}$, that maximizes the probability of observing the given random sample. This can be accomplished by setting $\partial \log \mathcal{L}/\partial\theta = 0$ and solving for the θ value, $\hat{\theta}$, that makes this derivative vanish, provided, of course, that $\partial^2 \log \mathcal{L}/\partial\theta^2 < 0$ at $\hat{\theta}$. Here $\hat{\theta}$ is termed the *maximum likelihood estimator of* θ.

Sampling Distributions of Statistics When Sampling from a Normal Population

The most commonly used sampling distributions involve those determined under sampling from a normal population. Specifically, suppose that $X_1, X_2, \ldots, X_n$ constitute a set of sample random variables drawn from a $N(\mu, \sigma)$ population and consider the statistics $\bar{X} = \sum_{i=1}^{n} X_i/n$ and $S^2 = \sum_{i=1}^{n} (X_i - \bar{X})^2/(n-1)$. Here $\bar{X}$ is the sample mean and S^2 is the sample variance, with $\bar{X}$ and S^2 serving as unbiased estimators of μ and σ^2, respectively.

We know that the sampling distribution of a statistic is a probability distribution that has been generated under random sampling from a given population. Once the sampling distribution of a statistic has been determined, it can be used to make inferences about a parameter from random samples taken from that population. To this end, let us consider the properties of the sampling distributions of $\bar{X}$ and S^2 given above by stating the following two theorems:

1. If $X_1, \ldots, X_n$ are independent normal random variables with $E(X_i) = \mu_i = \mu$ and $V(X_i) = \sigma_i^2 = \sigma^2$ for all $i = 1, \ldots, n$, then the distribution of the sample mean $\bar{X}$ is $N(\mu, \sigma/\sqrt{n})$. Here $\sigma/\sqrt{n}$ is termed the *standard error of the mean* and denoted $\sigma_{\bar{X}}$; and
2. If S^2 is the variance of a random sample of size n taken from a normal population with mean μ and variance σ^2, then the quantity $Y = (n-1)S^2/\sigma^2 : \chi^2_{n-1}$.

What is the relationship between $\bar{X}$ and S^2 when sampling is undertaken from a normal population? Specifically, if $X_1,\ldots, X_n$ constitute a random sample of size n drawn from a $N(\mu, \sigma)$ population, then the statistics $\bar{X}$ and S^2 determined from this sample are independent random variables.

Armed with these considerations, we note that, when sampling from a normal population with σ known, the *standardized sample mean* or the random variable $\bar{Z} = (\bar{X}-\mu)/(\sigma/\sqrt{n}) : N(0,1)$ whereas, from above, $(n-1)S^2/\sigma^2 : \chi^2_{n-1}$. Because $\bar{X}$ and S^2 are independent, the quantity

$$T = \frac{\bar{Z}}{\sqrt{\dfrac{(n-1)S^2/\sigma^2}{n-1}}} = \frac{\bar{X}-\mu}{S/\sqrt{n}} : t_{n-1} \tag{1.63}$$

from the definition of the t distribution (see above, "Derived Distributions (χ^2, t, and F)").

Interval Estimation

Given a set of sample random variables $X_1,\ldots, X_n$, our objective is to specify in advance a *confidence probability* $1 - \alpha$ and then use the sampling distribution of an estimator T to find lower and upper limits $L_1(T)$ and $L_2(T)$, respectively, for a parameter θ, where these *confidence limits* depend on the sample random variables $X_1,\ldots,X_n$ via T and the requirement that $L_1(T) \le L_2(T)$ for all sample points $(X_1,\ldots,X_n)$. Moreover, $(L_1(T), L_2(T))$ is a *confidence interval* in that the confidence limits $L_1(T)$ and $L_2(T)$ are chosen so that the area under each tail of the sampling distribution of T is $\alpha/2$.

Let us consider a confidence interval for μ under random sampling from a normal population with unknown variance. We noted earlier that, when sampling from a normal population with μ and σ^2 unknown, the random variable $T = (\bar{X}-\mu)/(S/\sqrt{n})$ has a t distribution with $n - 1$ degrees of freedom. Then, under centrality,

$$P\left(-t_{\frac{\alpha}{2},n-1} < T < t_{\frac{\alpha}{2},n-1}\right) = P\left(-t_{\frac{\alpha}{2},n-1} < \frac{\bar{X}-\mu}{S/\sqrt{n}} < t_{\frac{\alpha}{2},n-1}\right) = 1-\alpha$$

or, on rearranging terms,

$$P\left(\bar{X} - t_{\frac{\alpha}{2},n-1}\frac{S}{\sqrt{n}} < \mu < \bar{X} + t_{\frac{\alpha}{2},n-1}\frac{S}{\sqrt{n}}\right) = 1-\alpha \tag{1.64}$$

where, before the sample is drawn, Equation 1.64 constitutes a probability statement about a random interval, with $1 - \alpha$ serving as a confidence probability. However, once the sample values are obtained and $\bar{X}$ and S are calculated, Equation 1.64 becomes a *confidence statement* and $1 - \alpha$ is called the *confidence coefficient*. In this regard, a $100(1 - \alpha)$% *confidence interval for* μ is

$$\bar{X} \pm t_{\frac{\alpha}{2},n-1}\frac{S}{\sqrt{n}}, \tag{1.65}$$

which has a long-run relative frequency interpretation : if many samples of size n are taken and the interval of Equation 1.65 is determined for each of them, then, in the long run, $100(1 - \alpha)$% of these intervals would contain μ, and 100α% of them would not.

Next, let us specify a confidence interval for σ^2 under random sampling from a normal population with unknown mean and variance. We also noted earlier that the quantity $Y=(n-1)S^2/\sigma^2 : \chi^2_{n-1}$. Then, again looking to a central confidence interval for σ^2, we start with

$$P\left(-\chi^2_{\frac{\alpha}{2},n-1} < Y < \chi^2_{1-\frac{\alpha}{2},n-1}\right) =$$

$$P\left(-\chi^2_{\frac{\alpha}{2},n-1} < \frac{(n-1)S^2}{\sigma^2} < \chi^2_{1-\frac{\alpha}{2},n-1}\right) = 1-\alpha$$

or

$$P\left(\frac{(n-1)S^2}{\chi^2_{1-\frac{\alpha}{2},n-1}} < \sigma^2 < \frac{(n-1)S^2}{\chi^2_{\frac{\alpha}{2},n-1}}\right) = 1-\alpha. \tag{1.66}$$

Hence, we may be 100(1 – α)% confident that σ^2 is a member of the interval

$$\left(\frac{(n-1)S^2}{\chi^2_{1-\frac{\alpha}{2},n-1}}, \frac{(n-1)S^2}{\chi^2_{\frac{\alpha}{2},n-1}}\right). \tag{1.67}$$

Statistical Hypothesis Testing

Our focus here is to develop the apparatus for testing *parametric statistical hypotheses* or hypotheses that claim that a certain population parameter θ assumes a specific numerical value (or falls within some range of values) given that the form of the population distribution in known.

Statistical hypotheses come in pairs and are complementary. The first hypothesis in the pair is the *null hypothesis* (denoted H_0), which is assumed to be true; it is the hypothesis to be tested and either rejected or not based on sample data. The second hypothesis of the pair is the *alternative hypothesis* (denoted H_1); it states what the population would look like if the null hypothesis were untrue. For instance, the null hypothesis may be $H_0{:}\theta \leq \theta_0$ and thus the alternative hypothesis is $H_1{:}\theta > \theta_0$, where θ_0 is termed the *null value* of θ and H_0 always contains at least an equality.

Because sample information is being used, an element of uncertainty is obviously present; thus it is possible for the test to render a false conclusion. In this regard, we shall identify two types of errors: (1) type I error (TIE), rejecting H_0 when it is actually true; and (2) type II error (TIIE), not rejecting H_0 when it is actually false. Let us specify the risks associated with incorrect decisions as the probabilities of committing type I and type II errors as, respectively:

$$\alpha = P(TIE) = P(\text{reject } H_0 | H_0 \text{ true}),$$
$$\beta = P(TIIE) = P(\text{do not reject } H_0 | H_0 \text{ false})$$

To conduct a test of a statistical hypothesis, we need to specify a *test statistic* (denoted $\hat{\theta}$) : a random variable whose sampling distribution in known under the assumption that $H_0{:}\theta = \theta_0$ is true. Here decision rules concerning the rejection or nonrejection of the null hypothesis can then be established for a given sample outcome by studying the range of $\hat{\theta}$ values. Specifically, this range will be partitioned into two disjoint subsets:

1. R, the *critical region*, which is the region of rejection that contains the sample outcomes least favorable to H_0.
2. A, the *region of nonrejection*, which contains the sample outcomes most favorable to H_0.

Focusing on the critical region R, we need to determine its location and size. To accomplish this, let us use the following *decision rule*: choose α (termed the *level of significance* of the test) small (typically $\alpha = 0.05$ or 0.01) and then, in accordance with H_1, choose R such that the probability of observing a (sample) value of the test statistic in R is α when H_0 is true. Then our test result is said to be *significant at the* α *level*. In summary, H_1 determines the location of R, whereas α specifies its size. Additionally, the value of the test statistic $\hat{\theta}$ that separates A and R is called the *critical value* of $\hat{\theta}$ and denoted $\hat{\theta}_0$. What types of hypothesis tests can be performed? For example, suppose we want to test H_0: $\mu \le \mu_0 = 50$, against $H_1 : \mu > 50$ at the $\alpha = 0.05$ level and we are sampling from a normal population with an unknown standard deviation. Our test statistic is $\bar{X}$, and we know from our previous discussion that $T = (\bar{X} - 50)/(S/\sqrt{n}) : t_{n-1}$. Denoting the critical value of t as $t_{\alpha,n-1} = t_{0.05,n-1}$, Figure 1.1.A displays the critical region R. Then, according to the above decision rule, we will reject H_0 if the value of our test statistic $(\bar{X} - 50)/(S/\sqrt{n}) > t_{0.05,\,n-1}$ because we are now

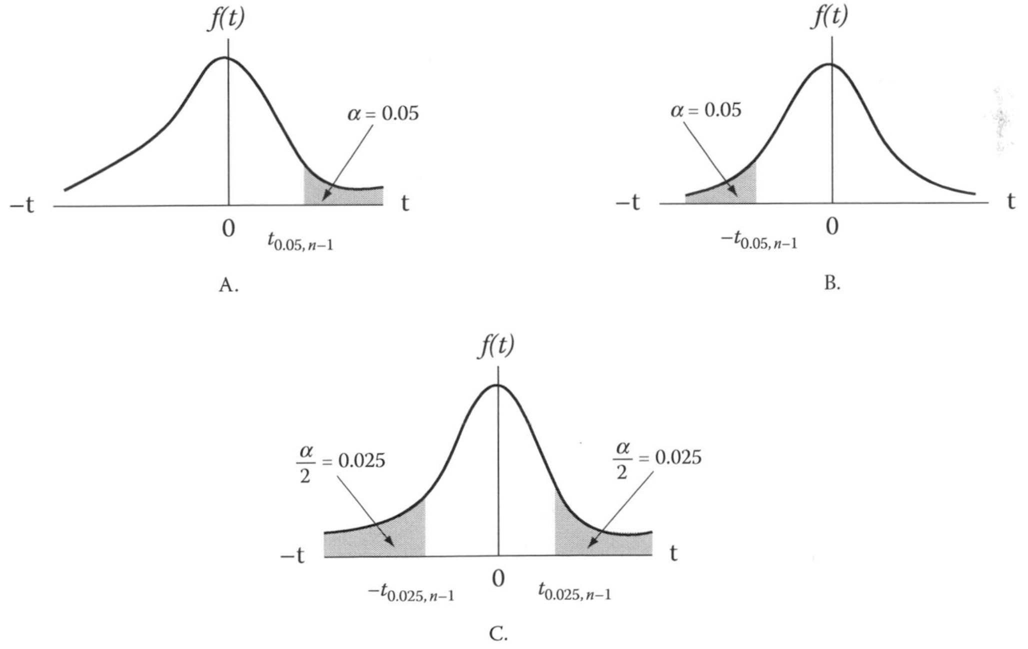

FIGURE 1.1
Critical regions for t tests.

in the critical region $R = \{t | t > t_{0.05,\, n-1}\}$. Here we have a one-tail test involving the upper tail of the t distribution because H_1 is supported by large values of the test statistic. Clearly, R contains the sample outcomes least favorable to H_0. Hence, we may reject H_0 for "large" values of our test statistic. That is, because the preceding inequality can be rewritten as $\bar{X} > 50 + t_{0.05,\, n-1}\frac{S}{\sqrt{n}}$, it follows that we reject H_0 if $\bar{X}$ exceeds 50 by "too much," in which our measure of too much is $t_{0.05,\, n-1}\frac{S}{\sqrt{n}}$ and is termed the *error bound*.

A few additional points regarding types of tests are in order. Specifically, if in the preceding example we had tested $H_0 : \mu \geq \mu_0 = 50$ versus $H_1 : \mu < 50$, then $R = \{t | t < -t_{0.05,n-1}\}$ (Figure 1.1.B). We now have a one-tail test involving the lower tail of the t distribution because H_1 is supported by small values of the test statistic. Thus, H_0 is rejected if $\bar{X} < 50 - t_{0.05,\, n-1}\frac{S}{\sqrt{n}}$, i.e., $\bar{X}$ falls short of 50 by "too much." Now the error bound is $-t_{0.05,\, n-1}\frac{S}{\sqrt{n}}$.

Furthermore, if we had tested $H_0 : \mu = \mu_0 = 50$, against $H_1 : \mu \neq 50$, then $R = \{t \| t | > t_{\frac{\alpha}{2},\, n-1} = t_{0.025,\, n-1}\} = \{t | t < -t_{0.025,\, n-1} \text{ or } t > t_{0.025,\, n-1}\}$ (Figure 1.1.C). Here we have a two-tail test because now H_1 is supported by both large and small values of the test statistic, i.e., H_0 is rejected when $\bar{X} < 50 - t_{0.025,\, n-1}\frac{S}{\sqrt{n}}$ or $\bar{X} > 50 + t_{0.025,\, n-1}\frac{S}{\sqrt{n}}$.

Armed with the essentials of hypothesis testing, we can now look to specifying what is called the *research hypothesis* for a test. The research hypothesis is actually the alternative hypothesis H_1; it is what one wants to demonstrate. The null hypothesis H_0 is a "throw-away" in that we want to be able to reject it in favor of H_1. Also, we can now address the issue of *statistical significance*. Specifically, what does it mean to say that a test outcome is statistically significant? If a test result is deemed statistically significant, then there is only a very small probability that the observed sample result (in this instance, the discrepancy between $\mu_0 = 50$ and $\bar{X}$) arose solely because of chance factors or sampling error: something meaningful or systematic is going on in the population.

How are hypothesis test results typically reported? Remember that α is our *chosen* level of significance. However, the *observed* or *actual* level of significance will be termed the *p-value* of the test. We may view a p-value as the probability of obtaining, say, a calculated $T(= \bar{X} - 50)/(S/\sqrt{n}))$ at least as large as the one observed if the null hypothesis H_0 is true. The role of the p-value is that it lets the data or sample determine the actual level of significance of the test. In this regard, the larger is the p-value, the more likely it is that H_0 is true, and the smaller is the p-value, the more likely it is that H_0 is not true. Hence, "small" p-values (say, < 0.05) lead us to reject H_0.

For example, suppose $H_0 : \mu \leq \mu_0 = 50$, $H_1 : \mu > 50$, $n = 20$, and the realized value of $T = (\bar{X} - 50)/(S/\sqrt{n})$ is $t = 2.47$. Then, looking to the t table (Table A.2 of the Appendix) for $n - 1 = 19$ degrees of freedom, the calculated t falls between 2.093 and 2.539. Hence, $0.001 <$ p-value < 0.025. Because this (one-tail) p-value is less than 0.05, we would reject H_0 in favor of H_1. Note that, if we had conducted a two-tail test (now $H_0 : \mu = \mu_0 = 50$ and $H_1 : \mu \neq 50$), then $0.001 < p\text{-value}/2 < 0.025$ so that $0.02 < p\text{-value} < 0.05$. Again, we would reject H_0.

Thus far, we have focused primarily on α, the level of significance of a test. However, remember that $\beta = P(TIIE)$ (the probability of not rejecting H_0 given that H_0 is false). For a fixed sample size n and α level, the magnitude of β varies for different values of θ subsumed under H_1. Also, the quantity $1 - \beta$ is called the *power* of the test; it is the (a priori) probability of detecting a false null hypothesis (or true alternative hypothesis). Thus, the power of a test cannot be determined unless the alternative hypothesis H_1 is specified (because H_0 false means H_1 is true). Hence, a test that has *high power* has a much greater chance of detecting a false null hypothesis than one of lower power. In this regard, suppose, hypothetically, that $H_0: \theta = \theta_0$ but that the true situation is $H_1: \theta = \theta_1 (\neq \theta_0)$. So for given n and α values, the larger the difference between θ and θ_1 (and thus the smaller the probability of not rejecting H_0 when it is false), the *more powerful* is the test of H_0.

Finally, what is the relationship between confidence interval estimation and hypothesis testing? Suppose we are sampling from a normal population with an unknown variance and we want to determine a 100(1–α)% confidence interval for μ. Suppose also that we are interested in testing $H_0: \mu = \mu_0$, against $H_0: \mu \neq \mu_0$. Then a 100(1–α)% confidence interval for μ contains all those null values that *would not be rejected* at the 100α% level in a two-tailed test involving H_0 and H_1 above. Correspondingly, all null values found outside of a 100(1–α)% confidence interval for μ would be rejected at the 100α% level.

Hypothesis Tests for the Differences of Variances

Suppose we extract random samples of sizes n_X and n_Y from two independent normal populations $N(\mu_X, \sigma_X^2)$ and $N(\mu_Y, \sigma_Y^2)$, respectively, having unknown means and variances. Our goal is to test the null hypothesis $H_0: \sigma_X^2 = \sigma_Y^2$ against any one of the following alternative hypotheses:

Case 1: $H_1: \sigma_X^2 > \sigma_Y^2$;
Case 2: $H_1: \sigma_X^2 < \sigma_Y^2$;
Case 3: $H_1: \sigma_X^2 = \sigma_Y^2$.
Now, it can be demonstrated that the random variable

$$\frac{S_X^2 / \sigma_X^2}{S_Y^2 / \sigma_Y^2} : F_{n_X-1, n_Y-1}, \tag{1.68}$$

where the sample variances S_X^2 and S_Y^2 serve as the best estimators for the population variances σ_X^2 and σ_Y^2, respectively. Under H_0, Equation 1.68 reduces to $f = S_X^2 / S_y^2$. Then we will reject the null hypothesis of equal population variances at the α level if

Case 1: $f = S_X^2 / S_y^2 \geq F_{1-\alpha, n_X-1, n_Y-1}$;
Case 2: $f = S_X^2 / S_y^2 \leq F_{1-\alpha, n_X-1, n_Y-1}$;
Case 3: $f = \dfrac{\text{larger sample variance}}{\text{smaller sample variance}} \geq F_{1-(\alpha/2), v_1, v_2}$,

where v_1 corresponds to the numerator degrees of freedom and v_2 denotes the dominator degrees of freedom.

2

Bivariate Linear Regression and Correlation

The Regression Model

In this and in subsequent chapters, we shall examine statistical techniques that use sample data to investigate the relationship between a group of variables by constructing a *model* (typically linear in form) that can be used to express the value of one variable (the *response* or *dependent* or *explained variable*) in terms of the others (the *independent* or *explanatory variables* or *regressors*). In this regard, the independent variables can then be used to predict the value of the dependent variable. The process of finding the equation that best fits the sample dataset is part and parcel of what is termed *regression analysis* or *regression modeling*.

For instance, suppose our objective is to examine the determinants of the quantity demanded of a given agricultural product. At the conceptual stage, our model might look something like the functional equation

quantity demanded = *f*(product's own price, prices of substitute goods, prices of complementary goods, disposable income, tastes, expectations, seasonal factors, etc.).

Can we reasonably expect to account for the effects of all possible determinants of the quantity demanded of this product? The answer is, obviously not. Given this limitation, let us, for the sake of simplicity, initially focus on the most important explanatory variable, namely the product's own price. But what about the other independent variables? Should they summarily be excluded from the regression equation? Again, the response is, obviously not.

To handle this modification of our original model, let us consolidate all of the variables other than the product's own price into the term ε, i.e., the role of ε is to account for the net effect of all excluded factors on quantity demanded. Can the impact of ε on quantity demanded be accurately predicted? The answer is, certainly not. Hence, ε must be specified as a random variable. Moreover, some of the factors operating in ε may cause quantity demanded to increase, whereas others can cause quantity demanded to decrease. Thus, on balance, the net effect can typically be considered to be very small or we surmise that $E(\varepsilon) = 0$.

In the light of the preceding discussion, our simplified model has quantity demanded (Y) specified as a function of the product's own price (X) and ε or

$$Y = f(X, \varepsilon).$$

Because the product's own price is observable, it makes up what is called the *systematic* or *deterministic portion* of the regression model, and ε constitutes its *random portion*. For the sake of ease of analysis, let us assume that the systematic portion of the model is linear. Hence, the preceding expression can be written as

$$Y = \beta_0 + \beta_1 X + \varepsilon,$$

where X is taken to be fixed. Then $E(Y) = \beta_0 + \beta_1 X$ (because $E(\varepsilon) = 0$) so that our regression model is actually a *probabilistic* or *statistical regression model* and can be expressed as

$$Y = \beta_0 + \beta_1 X + \varepsilon = E(Y) + \varepsilon.$$

Given this equation, the actual value of Y depends on its mean plus a random component ε that accounts for the net effect on Y of all omitted variables. Hence, there is ostensibly a whole distribution of Y's centered at $E(Y)$, with deviations above or below $E(Y)$ dictated by the behavior of ε. So with ε unknown, the best predictor of Y is $E(Y) = \beta_0 + \beta_1 X$. However, β_0 and β_1 are also unknown and must be estimated from sample data. This, in part, is the task of the present chapter. In what follows, we shall consider the rudiments of *simple* or *bivariate regression analysis* (because we have only two variables X, Y). Later in the text, we turn to *multiple regression analysis*, which attempts to disentangle or disaggregate the random error term and explicitly incorporate variables subsumed therein into the systematic portion of the regression equation. Once this is done, the random effect of ε on Y can be significantly tempered.

This said, it is important to note that the impact of ε and Y can mask a multitude of sins, i.e., "good" estimates of β_0 and β_1 obtain only if ε is "well behaved" (it must satisfy certain distributional assumptions). Hence, there are no free lunches in regression analysis: once estimates are obtained, a whole host of diagnostic tests must subsequently be undertaken to determine whether ε, and consequently the estimators for β_0 and β_1, have certain desirable properties. If it appears that such properties do indeed hold, then we can be reasonably confident that our regression modeling exercise has provided us with a useful decision-making tool.

Given the preceding discussion, we can now state that the purpose of regression analysis is twofold: (1) we first study the functional relationship between two variables X and Y, and, if the said relationship is deemed meaningful (in a statistical sense), then (2) we seek to forecast or predict a value of one variable (Y) from a given observation on another variable (X). More specifically, as will be seen below, we can predict either the average value of Y for a given value of X or a particular value of Y corresponding to a given value of X. To obtain said predictions, we need to first determine the association between the X and Y variables.

To this end, let us first assume that X and Y are random variables with joint probability density function $f(x, y)$. Then, for each realization x of X, there is a conditional distribution of Y given $X = x$ or $h(y|x)$. Hence the *regression of* Y *on* $X = x$ is the mean of Y in the conditional density of Y given $X = x$ and appears as

$$j(x) = E(Y|X = x) = \int^{+\infty} yh(y|x)\, dy$$

$$= \int_{-\infty}^{+\infty} y \frac{f(x,y)}{g(x)} dx, \tag{2.1}$$

where $g(x)=\int_{-\infty}^{+\infty} f(x,y)dy$ is the marginal distribution of X. Hence, $y = j(x)$, the locus of means of the conditional distribution $E(Y|X = x)$ when plotted in the x, y plane, represents the *regression curve of y on x* for Y given $X = x$ or the *response function* for Y given $X = x$.

Example 2.1

Suppose

$$f(x, y) = \begin{cases} \frac{4}{3} - \frac{1}{2}xy, & 0 < x < 2,\ 0 < y < 1; \\ 0, & \text{elsewhere.} \end{cases}$$

Then

$$g(x) = \int_0^1 f(x,y)dy = \frac{4}{3} - \frac{1}{4}x$$

and thus

$$h(y|x) = \frac{f(x,y)}{g(x)} = \frac{8-3xy}{8-\frac{3}{2}x}.$$

Hence, the regression curve of y on x is

$$j(x) = E(Y|X = x)$$

$$= \int_0^1 yh(y|x)dy = \left(\frac{4-x}{8-\frac{3}{2}x}\right).$$

Next, if the joint probability density function $f(x, y)$ is unknown, then we can simply assume a particular functional form (typically linear) for the regression curve and then fit the function to a set of observations on the variables X and Y. As will be explained shortly, Y is taken to be a random variable whose values are conditional on a set of *fixed* admissible X values. This second approach to regression analysis (involving random Y, nonrandom X, and a linear regression curve) is the one that will be principally followed throughout this chapter.

The basic problem that we face is to determine the line of best fit through a scatter of n sample points $(X_1, Y_1),\ldots,(X_n, Y_n)$. Moreover, once the equation of the implied line has been empirically determined, we must make an assessment of its *goodness of fit* to the data. If the line is indeed found to fit the data well, in the sense that it closely approximates the true (but unknown) relationship between X and Y, then it can be used for predictive purposes, with the appropriate prediction error reported.

To solve the regression problem, let us begin by obtaining a sample of n observations on a variable Y:$Y_1,\ldots,Y_n$, where

$$Y_i = \textit{systematic component} + \textit{random component}, \qquad i = 1,\ldots,n. \tag{2.2}$$

Here the systematic or deterministic part of Y_i reflects a particular behavioral hypothesis, whereas the random part is unobservable and arises because of some combination of sampling, measurement, and specification error, i.e., it covers a multitude of sins in that it depicts the influence on Y_i of many omitted variables, each presumably exerting an individual small effect. Moreover, as will be seen below, the X_i's, $i = 1,\ldots,n$, are taken to be nonrandom or predetermined.

If we assume a priori a functional relationship between X and Y, and, in particular that Y is linearly related to X, $Y = \beta_0 + \beta_1 X$, where the unknown population parameters β_0 and β_1 are, respectively, the *vertical intercept* and *slope* of the regression line, then Equation 2.2 may be rewritten as

$$Y_i = \beta_0 + \beta_1 X_i + \varepsilon_i, \qquad i = 1,\ldots,n, \tag{2.3}$$

where $\beta_0 + \beta_1 X_i$ is the systematic component and ε_i is a stochastic disturbance or *random error term*. (Strictly speaking, Equation 2.3 is termed a linear regression model because it is linear in the parameters β_0 and β_1.) It is apparent that, because of the random nature of the ε_i values, we must construct, for purposes of estimation and testing, a probabilistic model that incorporates their requisite properties so that inferences from the set of sample data can be made about the unobserved population regression line.

The Strong Classical Linear Regression Model

Given that Y_i, $i = 1,\ldots,n$, is assumed to be statistically (and linearly) related to X and ε according to the specification $Y = \beta_0 + \beta_1 X + \varepsilon$, we will refer to X as the explanatory variable or regressor (remember that X is taken to be nonstochastic, e.g., it is controllable or fully predictable), to ε as a random variable and to Y as the explained variable (Y is also a random variable because ε is stochastic). In this regard, for each fixed X_i, there is a whole probability distribution associated with Y_i, the characteristics of which are determined by X_i and the probability distribution of ε_i, $i = 1,\ldots,n$. For instance, as Figure 2.1 reveals, at each X_i value, we cannot expect Y_i to equal $\beta_0 + \beta_1 X_i$ exactly because of the random behavior of ε_i. Hence, we only require that, on average, Y_i equals $\beta_0 + \beta_1 X_i$ since realized values of Y_i may depart radically from the latter. To illustrate this average relationship, we shall indicate that, on average, the systematic part of Y_i given X_i is $E(Y_i|X_i) = \beta_0 + \beta_1 X_i$, i.e., the conditional expectation of Y_i given X_i. (In this regard, β_1 is the average change in Y per unit change in X.) Hence, the probability distribution of ε_i, $f(\varepsilon_i)$, is concentrated about $E(Y_i|X_i)$. Thus, the regression equation resulting from knowledge of this average relationship is $Y = E(Y|X) + \varepsilon = \beta_0 + \beta_1 X + \varepsilon$, i.e., on average, Y's systematic component is $E(Y|X) = \beta_0 + \beta_1 X$. As indicated in Figure 2.1, under the linearity hypothesis, each of the n means $E(Y_i|X_i)$ lies on this population regression line. Let us now turn to the specific assumptions concerning the probability distribution of ε. Given that Equation 2.3 holds:

A.1. ε_i is normally distributed for all i. (Given that ε_i is composed of a diversity of factors that work in opposite directions, we may expect small values of ε_i to occur more frequently than large ones.)

A.2. ε_i has zero mean or $E(\varepsilon_i) = 0$ for all i. (Positive deviations from $\beta_0 + \beta_1 X$ are just as likely to occur as negative ones so that the random errors ε_i are distributed about a mean of zero.)

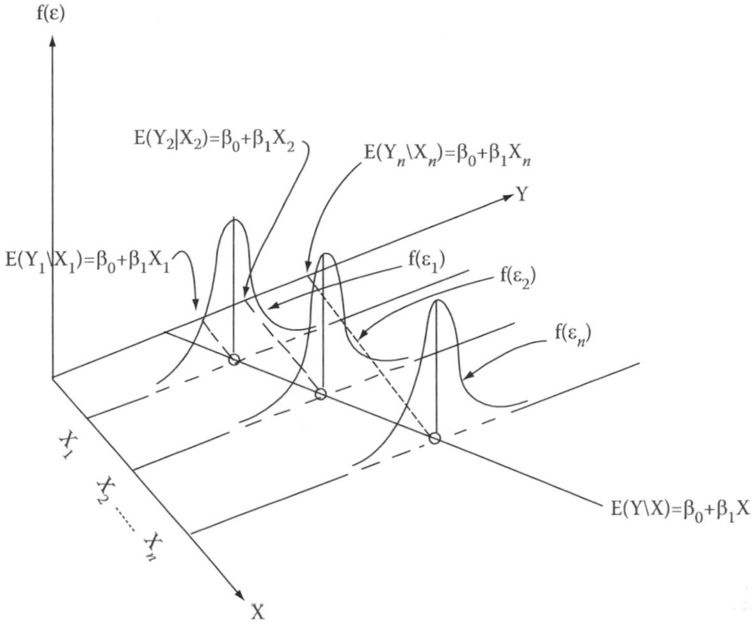

FIGURE 2.1
Conditional expectation of Y given X.

A.3. The errors ε_i are homoscedastic or $V(\varepsilon_i) = E(\varepsilon_i^2) - E(\varepsilon_i)^2 = E(\varepsilon_i^2) = \sigma_\varepsilon^2$ = constant for all i. Each $f(\varepsilon_i)$ distribution has the same constant variance σ_ε^2 whose value is unknown.

A.4. Nonautocorrelation of the ε_i's or $COV(\varepsilon_i, \varepsilon_j) = E[\varepsilon_i - E(\varepsilon_i)][\varepsilon_j - E(\varepsilon_j)] = E(\varepsilon_i\varepsilon_j) = 0, i \neq j$. (Successive ε_i values are uncorrelated.) In addition, X and ε are uncorrelated or $COV(X_i, \varepsilon_j) = 0$ for all i and j.

A.5. X is a nonrandom variable with finite variance. (The X_i values are fixed, i.e., controllable, in repeated sampling from the same population.)[1]

Here Assumptions A.2 and A.4 imply that the ε_i's are uncorrelated, whereas Assumptions A.1, A.2, and A.4 imply that the ε_i's are independent. Moreover, Assumption A.1 allows us to conduct tests of hypotheses and to construct confidence intervals for β_0, β_1, and $E(Y|X) = \beta_0 + \beta_1 X$, as well as determine prediction intervals for either the average value of Y given some X value (call it X_0) or for a particular value of Y corresponding to X_0.

We noted in Assumption A.5 that the X_i, $i = 1,\ldots,n$, values are held fixed. Hence, the sole source of variation in the random variable Y in repeated sampling from the same population is variation in ε so that the probability distribution of Y is identical to that of ε. So, as far as the properties of the probability distribution of Y are concerned:

B.1. $E(Y_i) = E(Y_i/X_i) + E(\varepsilon_i) = \beta_0 + \beta_1 X_i, i = 1,\ldots,n$.

B.2. $V(Y_i) = E[Y_i - E(Y_i)]^2 = E(\beta_0 + \beta_1 X_i + \varepsilon_i - \beta_0 - \beta_1 X_i)^2 = E(\varepsilon_i^2) = \sigma_\varepsilon^2$ = constant for all i, whereas $COV(Y_i, Y_j) = 0, i \neq j$.

[1] If only Assumptions A.2 through A.5 are considered, then the resulting model is simply termed the *weak classical linear regression model*. If Assumption A.1 is also included, then the *strong* case emerges.

B.3. The random variable Y_i is normally distributed because ε_i is normally distributed, $i = 1,\ldots,n$.

B.4. The random variables Y_i are independent because the ε_i are independent, $i = 1,\ldots,n$.

Having discussed the nature of the variables X, ε, and Y, our next set of objectives include the following:

a. Obtain estimates of β_0, β_1 (denoted as $\hat{\beta}_0, \hat{\beta}_1$, respectively).
b. Obtain estimates of the variances of $\hat{\beta}_0, \hat{\beta}_1$, and ε. On the basis of this information, we can next do the following.
c. Test hypotheses concerning the population parameters β_0, β_1.
d. Construct confidence intervals for the population parameters β_0, β_1.
e. Test hypotheses concerning the *entire* population regression line, i.e., perform a significance test of β_0 and β_1 jointly.
f. Determine a prediction interval for the average value of Y given X_0.
g. Construct a confidence band for the entire population regression line proper.
h. Determine a prediction interval for a particular value of Y given X_0.
i. Construct a *partitioned sum-of-squares table.*

Estimating the Slope and Intercept of the Population Regression Line

When the values of β_0 and β_1 are estimated on the basis of sample information, we obtain the *sample regression line* $\hat{Y} = \hat{\beta}_0 + \hat{\beta}_1 X$ that serves as our proxy for the *population regression line* $E(Y/X) = \beta_0 + \beta_1 X$, where $\hat{Y}$ is the fitted or estimated value of Y and $\hat{\beta}_0$ and $\hat{\beta}_1$ represent the estimated population parameters. (Think of $\hat{\beta}_0$ and $\hat{\beta}_1$ as arbitrary estimators of β_0 and β_1, respectively.) Because most (if not all) of the observed Y_i values will not lie exactly on the sample regression line, the values of Y_i and $\hat{Y}_i$ differ. This difference will be denoted as $e_i = Y_i - \hat{Y}_i, i = 1,\ldots,n$, and will be termed the ith *residual* or *deviation* from the sample regression line. Here e_i serves as an estimate of the stochastic or unobserved disturbance ε_i. To directly see the difference between ε_i and e_i, let us examine the hypothetical population and sample regression lines illustrated in Figure 2.2, in which Y_i may be viewed as being determined by either the population or sample regression lines:

$$\text{(population regression line)}\quad Y_i = E(Y_i/X_i) + \varepsilon_i = \beta_0 + \beta_1 X_i + \varepsilon_i,$$

$$\text{(sample regression line)}\quad Y_i = \hat{Y}_i + e_i = \hat{\beta}_0 + \hat{\beta}_1 X_i + e_i.$$

The decision rule or criterion of goodness of fit to be used in estimating β_0 and β_1 is depicted by the *principle of least squares:* to obtain the line of best fit, choose $\hat{\beta}_0$ and $\hat{\beta}_1$ so as to minimize the sum of the squared deviations from the sample regression line, i.e.,

$$\min\left\{\sum_{i=1}^{n} e_i^2 = \sum (Y_i - \hat{Y}_i)^2 = \sum (Y_i - \hat{\beta}_0 - \hat{\beta}_1 X_i)^2 = F(\hat{\beta}_0, \hat{\beta}_1)\right\}.$$

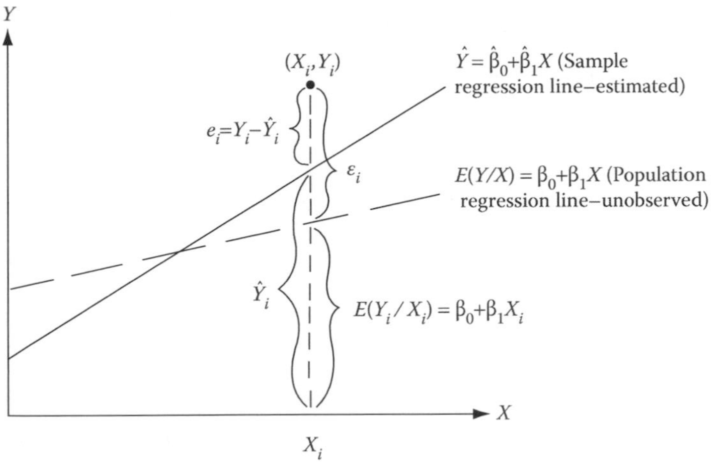

FIGURE 2.2
Population and sample regression lines.

(For convenience, the operator $\sum_{i=1}^{n}$ will at times be simplified to Σ, where it is to be understood that we always sum over all values of the i index, i.e., $i = 1,\ldots,n$.) On setting $\partial F/\partial\hat{\beta}_0 = \partial F/\partial\hat{\beta}_1 = 0$ (it is assumed that the second-order conditions for a minimum are satisfied), the resulting simultaneous linear equation system

$$\begin{aligned}&(a)\ \sum e_i = 0\\ &(b)\ \sum X_i e_i =\end{aligned} \quad \text{or} \quad \begin{cases}(a)\ n\hat{\beta}_0 + \hat{\beta}_1 \sum X_i = \sum Y_i \\ (b)\ \hat{\beta}_0 \sum X_i + \hat{\beta}_1 \sum X_i^2 = \sum X_i Y_i\end{cases} \quad \text{[system of \textit{least squares normal equations}]} \tag{2.4}$$

yields the solution

$$\hat{\beta}_0 = \frac{\left(\sum Y_i\right)\left(\sum X_i^2\right) - \left(\sum X_i Y_i\right)\left(\sum X_i\right)}{n\sum X_i^2 - \left(\sum X_i\right)^2}, \quad \hat{\beta}_1 = \frac{n\sum X_i Y_i - \left(\sum X_i\right)\left(\sum Y_i\right)}{n\sum X_i^2 - \left(\sum X_i\right)^2}.$$

An alternative specification of $\hat{\beta}_0$ and $\hat{\beta}_1$ is in terms of the deviations of X_i and Y_i from their respective means $\bar{X}$ and $\bar{Y}$. That is, if $x_i = X_i - \bar{X}$ (respectively, $y_i = Y_i - \bar{Y}$), $i = 1,\ldots,n$, represents the ith deviation of X_i (respectively, Y_i) from its mean, then

$$\text{(a)}\ \hat{\beta}_1 = \frac{\sum (X_i - \bar{X})(Y_i - \bar{Y})}{\sum (X_i - \bar{X})^2} = \frac{\sum x_i y_i}{\sum x_i^2}, \tag{2.5}$$

$$\text{(b)}\ \hat{\beta}_0 = \bar{Y} - \hat{\beta}_1 \bar{X} \text{ (from Equation 2.4.a).} \tag{2.5}$$

Note that Equation 2.5.b reveals an important property of the ordinary least-squares (OLS) line of best fit $\hat{Y} = \hat{\beta}_0 + \hat{\beta}_1 X$, namely that it passes through the point of means $(\bar{X}, \bar{Y})$ (Figure 2.3).

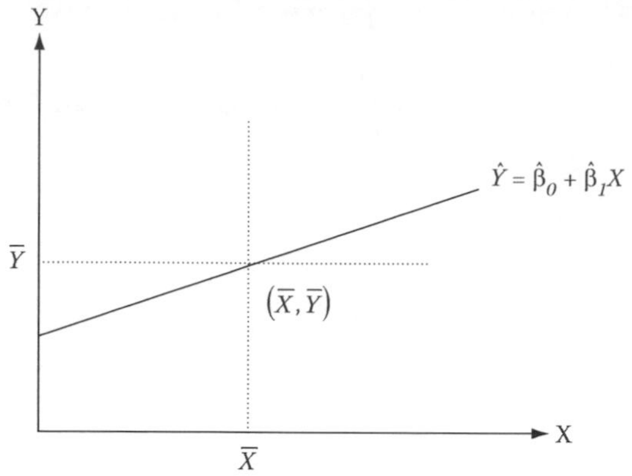

FIGURE 2.3
The OLS regression line passes through the point of means.

What about the statistical properties of the least-squares estimators $\hat{\beta}_0$ and $\hat{\beta}_1$? To answer this question, we shall rely on the *Gauss-Markov Theorem:* if Assumptions A.2 through A.5 above underlying the weak classical linear regression model hold, then, within the class of linear unbiased estimators of β_0 and β_1, the least-squares estimators have minimum variance, i.e., the least-squares estimators are *BLUE* (best linear unbiased estimators). Hence, BLUE requires that the estimator $\hat{\beta}_0$ (respectively, $\hat{\beta}_1$) be expressible as a *linear combination* of the Y_i, $i = 1,\ldots,n$, that it be *unbiased,* and its *variance be smaller* than that of any alternative linear unbiased estimator of β_0 (respectively, β_1). (For a proof of the Gauss-Markov Theorem, see Appendix 2.A.1.) It is important to note that, if we admit Assumption A.1 to our discussion, i.e., if we work with the strong classical linear regression model, then the least-squares estimators $\hat{\beta}_0$ and $\hat{\beta}_1$ are maximum-likelihood estimators of β_0 and β_1, respectively. On all this, see Appendix 2.A.2.

Example 2.2

Let us assume that a certain manufacturing company has compiled a set of data (see columns 1 and 2 of Table 2.1 and Figure 2.4) pertaining to its last 10 years of operation in a certain market. It has available observations on gross sales (in millions of current dollars) and advertising outlay (also expressed in millions of current dollars). Is there a statistically significant linear relationship between gross sales and advertising expenditure? In addition, if the company is considering a new advertising campaign that would involve an outlay of $12 million in 2007, what is the anticipated level of gross sales for that period?

A glance at columns 1 and 2 of Table 2.1 reveals that, with $n = 10$, $\Sigma X_i = 72.10$, and $\Sigma Y_i = 122.95$, it follows that $\bar{X} = 7.210$ and $\bar{Y} = 12.295$. On the basis of these values, we may determine the entries in columns 3-7. And, from the indicated column totals, we obtain

$$\hat{\beta}_1 = \frac{\sum x_i y_i}{\sum x_i^2} = \frac{88.7010}{61.7490} = 1.4365,$$

$$\hat{\beta}_0 = \bar{Y} - \hat{\beta}_1 \bar{X} = 12.295 - 1.4365(7.210) = 1.9379.$$

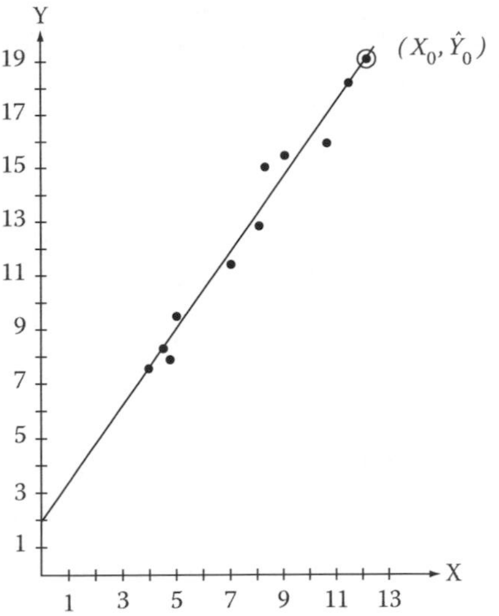

FIGURE 2.4
Regression of gross sales on advertising outlay.

So when X increases by one unit, Y's average increase is 1.4365 units. Moreover, the average value of Y when $X = 0$ is 1.9379. Hence, the estimated regression line is $\hat{Y} = \hat{\beta}_0 + \hat{\beta}_1 X = 1.9379 + 1.4365X$. Furthermore, if advertising outlay increases to $X_0 = 12$, the predicted value of gross sales, $\hat{Y}_0$, is determined from the estimated regression equation to be $\hat{Y}_0 = \hat{\beta}_0 + \hat{\beta}_1 X_0 = 1.9379 + 1.4365(12) = 19.1759$.

How reliable is the predicted value of Y? The answer to this question obviously hinges on how well we have estimated the population regression equation itself, i.e., does the

TABLE 2.1
OLS Calculations

Year	1 Gross Sales (Y)	2 Advertising Outlay (X)	3 $x_i = X_i - \bar{X}$ $= X_i - 7.210$	4 $y_i = Y_i - \bar{Y}$ $= Y_i - 12.295$	5 x_i^2	6 y_i^2	7 $x_i y_i$
1997	7.60	4.00	−3.210	−4.695	10.3041	22.0430	15.0710
1998	8.36	4.50	−2.710	−3.935	7.3441	15.4842	10.6639
1999	8.00	4.60	−2.610	−4.295	6.8121	18.4470	11.2100
2000	9.58	5.00	−2.210	−2.715	4.8841	7.3712	6.0002
2001	11.51	7.00	−0.210	−0.785	0.0441	0.6162	0.1649
2002	13.00	8.00	0.790	0.705	0.6241	0.4970	0.5570
2003	15.10	8.20	0.990	2.805	0.9801	7.8680	2.7770
2004	15.60	9.00	1.790	3.305	3.2041	10.9230	5.9160
2005	16.00	10.50	3.29	3.705	10.8241	13.7270	12.1895
2006	18.20	11.30	4.090	5.905	16.7281	34.8690	24.1515
Totals	$\Sigma Y_i = 122.95$	$\Sigma X_i = 72.10$	$\Sigma x_i = 0$	$\Sigma y_i = 0$	$\sum x_i^2 =$ 61.74.90	$\sum y_i^2 =$ 131.8456	$\Sigma x_i y_i =$ 88.7010

set of sample data offer sufficient evidence to indicate that advertising outlay serves as a reasonably good explanatory variable as far as the behavior of gross sales is concerned or can the observed relationship between these variables be attributed solely to chance? In the next three sections, we shall explicitly address the issue of goodness of fit of the sample regression line.

Mean, Variance, and Sampling Distribution of the Least-Squares Estimators $\hat{\beta}_0$ and $\hat{\beta}_1$

From the Gauss-Markov Theorem, we know that $E(\hat{\beta}_0)=\beta_0$ and $E(\hat{\beta}_1)=\beta_1$. Moreover, it can be shown (see Appendix 2.A.1) that

$$V(\hat{\beta}_0)=\sigma_\varepsilon^2\left(\frac{1}{n}+\frac{\bar{X}^2}{\sum x_i^2}\right),\ V(\hat{\beta}_1)=\frac{\sigma_\varepsilon^2}{\sum x_i^2}, \tag{2.6}$$

where σ_ε^2 is the unknown variance of the random (error) variable ε. (It is evident from these expressions that $V(\hat{\beta}_0)$ and $V(\hat{\beta}_1)$ both vary directly with the spread or dispersion of the unobserved ε_i values about the population regression line, i.e., with σ_ε^2, and inversely with the variation or dispersion of the X_i values about $\bar{X}$, or $\sum x_i^2$). In addition,

$$COV(\hat{\beta}_0,\hat{\beta}_1)=E\left[\hat{\beta}_0-E(\hat{\beta}_0)\right]\left[\hat{\beta}_1-E(\hat{\beta}_1)\right]=E(\hat{\beta}_0-\beta_0)(\hat{\beta}_1-\beta_1)=-\bar{X}\left(\frac{\sigma_\varepsilon^2}{\sum x_i^2}\right),$$

where $\hat{\beta}_0-\beta_0$ and $\hat{\beta}_1-\beta_1$ are the sampling errors of $\hat{\beta}_0$ and $\hat{\beta}_1$, respectively (for the derivation of this result, see Appendix 2.A.3). Clearly, $COV(\hat{\beta}_0,\hat{\beta}_1)<0$ when $\bar{X}>0$, i.e., the said sampling errors are of opposite sign, to wit an understatement of β_0 is accompanied with an overstatement of β_1 and conversely. It should be apparent that neither of the quantities in Equation 2.6 can be directly used as an aid in determining just how precisely the population regression line has been estimated because σ_ε^2 is unknown and must be determined from the sample values. To this end, we shall use

$$S_\varepsilon^2=\frac{\Sigma e_i^2}{n-2}=\frac{\Sigma(Y_i-\hat{Y}_i)^2}{n-2}=\frac{\Sigma y_i^2-\hat{\beta}_1\Sigma x_i y_i}{n-2} \tag{2.7}$$

as an unbiased estimator of σ_ε^2 (see Appendix 2.A.4). If this quantity is inserted into Equation 2.6, it follows that estimators for the variances of $\hat{\beta}_0$ and $\hat{\beta}_1$ are, respectively,

$$S_{\hat{\beta}_0}^2=S_\varepsilon^2\left(\frac{1}{n}+\frac{\bar{X}^2}{\Sigma x_i^2}\right),\quad S_{\hat{\beta}_1}^2=\frac{S_\varepsilon^2}{\Sigma x_i^2}.$$

If we take the positive square root of each of these expressions, then we obtain the estimated standard deviations of $\hat{\beta}_0$ and $\hat{\beta}_1$ or, as they are more commonly called, the estimated *standard errors of the regression coefficients*

$$S_{\hat{\beta}_0} = S_\varepsilon \sqrt{\frac{1}{n} + \frac{\bar{X}^2}{\sum x_i^2}}, \quad S_{\hat{\beta}_1} = \frac{S_\varepsilon}{\sqrt{\sum x_i^2}}. \tag{2.8}$$

In these expressions, S_ε is referred to as the *standard error of estimate* and serves as a measure of the average dispersion of the individual sample points about the estimated regression line. Clearly, the estimated standard errors of $\hat{\beta}_0$ and $\hat{\beta}_1$ vary directly with S_ε, i.e., the greater the dispersion of the sample points about the estimated regression line, the less precise our estimates of β_0 and β_1.

Example 2.3

We may continue with the preceding example by using the sample data presented in Table 2.1 above to find sample realizations of $S_{\hat{\beta}_0}$ and $S_{\hat{\beta}_1}$. Because from Equation 2.7 we have

$$s_\varepsilon = \sqrt{\frac{\sum y_i^2 - \hat{\beta}_1 \sum x_i y_i}{n-2}} = \sqrt{\frac{131.8456 - 1.4365(88.7010)}{8}} = 0.7439,$$

it follows from Equation 2.8 that

$$s_{\hat{\beta}_0} = s_\varepsilon \sqrt{\frac{1}{n} + \frac{\bar{X}^2}{\sum x_i^2}} = 0.7439\sqrt{\frac{1}{10} + \frac{(7.210)^2}{61.7490}} = 0.7219,$$

$$s_{\hat{\beta}_1} = \frac{s_\varepsilon}{\sqrt{\sum x_i^2}} = \frac{0.7439}{\sqrt{61.7490}} = 0.0946.$$

If we now couple Assumption A.1 with the Gauss-Markov results, we see that both $\hat{\beta}_0$ and $\hat{\beta}_1$ are expressible as linear combinations of the independent normal random variables (see Appendix 2.A.5) so that

$$\hat{\beta}_0 : N\left(E(\hat{\beta}_0), \sqrt{V(\hat{\beta}_0)}\right) = N\left(\beta_0, \sigma_\varepsilon \sqrt{\frac{1}{n} + \frac{\bar{X}^2}{\sum x_i^2}}\right),$$

$$\hat{\beta}_0 : N\left(E(\hat{\beta}_1), \sqrt{V(\hat{\beta}_1)}\right) = N\left(\beta_1, \sigma_\varepsilon / \sqrt{\sum x_i^2}\right)$$

and thus

$$\frac{\hat{\beta}_0 - \beta_0}{\sigma_\varepsilon \sqrt{\frac{1}{n} + \frac{\bar{X}^2}{\sum x_i^2}}} : N(0,1) \text{ and } \frac{\hat{\beta}_1 - \beta_1}{\sigma_\varepsilon / \sqrt{\sum x_i^2}} : N(0,1). \tag{2.9}$$

(The notation ":" is read "is distributed as.") If in Equation 2.9 we replace σ_ε by S_ε, then the resulting quantities follow a *t* distribution with $n - 2$ degrees of freedom (d.f.) (because from a sample of size n, we have estimated two population parameters so that the number of independent observations remaining in the sample is $n - 2$), i.e.,

$$\frac{\hat{\beta}_0 - \beta_0}{S_\varepsilon \sqrt{\frac{1}{n} + \frac{\bar{X}^2}{\sum x_i^2}}} = \frac{\hat{\beta}_0 - \beta_0}{S_{\hat{\beta}_0}} : t_{n-2}, \quad \frac{\hat{\beta}_1 - \beta_1}{S_\varepsilon / \sqrt{\sum x_i^2}} = \frac{\hat{\beta}_1 - \beta_1}{S_{\hat{\beta}_1}} : t_{n-2}. \tag{2.10}$$

Precision of the Least-Squares Estimators $\hat{\beta}_0, \hat{\beta}_1$: Confidence Intervals

To determine just how precisely the population parameters have been estimated from the sample data, let us construct the probability statements

$$\text{(a) } P\left(-t_{\alpha/2,n-2} \le \frac{\hat{\beta}_0 - \beta_0}{S_{\hat{\beta}_0}} \le t_{\alpha/2,n-2} \right) = 1 - \alpha, \tag{2.11}$$

$$\text{(b) } P\left(-t_{\alpha/2,n-2} \le \frac{\hat{\beta}_1 - \beta_1}{S_{\hat{\beta}_1}} \le t_{\alpha/2,n-2} \right) = 1 - \alpha, \tag{2.11}$$

where $-t_{\alpha/2,n-2}$, $t_{\alpha/2,n-2}$ are lower and upper percentage points, respectively, of the t distribution (i.e., values that cut off $\alpha/2$ of the total area under the t distribution at each tail end) (Figure 2.5) and $1 - \alpha$ is the confidence probability. After rearranging (Equations 2.11.a and b) and passing to sample realizations, we obtain the following confidence statements:

$$\text{(a) } P\left(\hat{\beta}_0 - t_{\alpha/2,n-2} s_{\hat{\beta}_0} \le \beta_0 \le \hat{\beta}_0 + t_{\alpha/2,n-2} s_{\hat{\beta}_0} \right) = 1 - \alpha, \tag{2.12}$$

$$\text{(b) } P\left(\hat{\beta}_1 - t_{\alpha/2,n-2} s_{\hat{\beta}_1} \le \beta_1 \le \hat{\beta}_1 + t_{\alpha/2,n-2} s_{\hat{\beta}_1} \right) = 1 - \alpha, \tag{2.12}$$

with confidence coefficient $1 - \alpha$ so that, from Equation 2.12.a, a $100(1 - \alpha)\%$ confidence interval for β_0 is

$$\left(\hat{\beta}_0 - t_{\alpha/2,n-2} s_{\hat{\beta}_0}, \hat{\beta}_0 + t_{\alpha/2,n-2} s_{\hat{\beta}_0} \right); \tag{2.13}$$

and from Equation 2.12.b, a $100(1 - \alpha)\%$ confidence interval for β_1 amounts to

$$\left(\hat{\beta}_1 - t_{\alpha/2,n-2} s_{\hat{\beta}_1}, \hat{\beta}_1 + t_{\alpha/2,n-2} s_{\hat{\beta}_1} \right). \tag{2.14}$$

In this regard, Equations 2.13 and 2.14 inform us that we may be $100\ (1 - \alpha)\%$ confident that the true regression intercept β_0 (respectively, slope β_1) lies between $\hat{\beta}_0 \pm t_{\alpha/2,n-2} s_{\hat{\beta}_0}$ (respectively, $\hat{\beta}_1 \pm t_{\alpha/2,n-2} s_{\hat{\beta}_1}$). It should be evident that the narrower the intervals depicted in Equations 2.13 and 2.14, the more precise the estimates of β_0 and β_1, respectively.

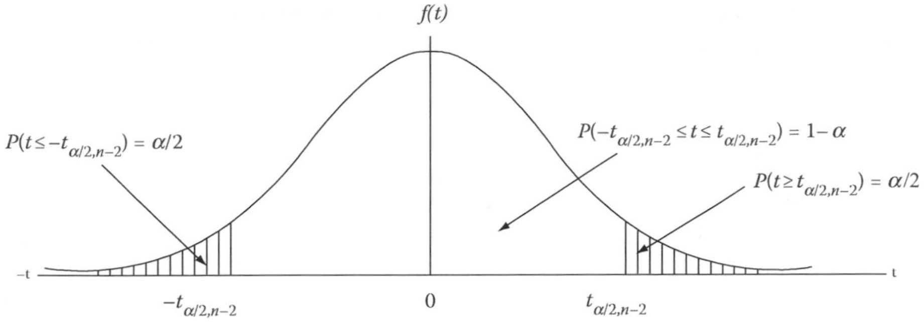

FIGURE 2.5
Percentage points of the t distribution.

Example 2.4

How precisely have we estimated β_0 and β_1 from the above sample of gross sales-advertising outlay data (Table 2.1)? If we desire a 95% confidence interval, then, with $1-\alpha = 0.95$, $\alpha = 0.05$, and thus $\alpha/2 = 0.025$. Hence, $t_{\alpha/2,\, n-2} = t_{0.025,8} = 2.306$. From the above section "Mean, Variance, and Sampling Distribution of the Least-Squares Estimators $\hat{\beta}_0$ and $\hat{\beta}_1$," we found that $s_{\hat{\beta}_0} = 0.7219$, $s_{\hat{\beta}_1} = 0.0946$. Then from Equation 2.13, we have

$$\hat{\beta}_0 - t_{\alpha/2,n-2}s_{\hat{\beta}_0} = 1.9379 - 2.306(0.7219) = 0.2732,$$
$$\hat{\beta}_0 + t_{\alpha/2,n-2}s_{\hat{\beta}_0} = 1.9379 + 2.306(0.7219) = 3.6026.$$

Hence, we may be 95% confident that the true regression intercept β_0 lies between 0.2732 and 3.6026. From Equation 2.14, we get

$$\hat{\beta}_1 - t_{\alpha/2,n-2}s_{\hat{\beta}_1} = 1.4365 - 2.306(0.0946) = 1.4147,$$
$$\hat{\beta}_1 + t_{\alpha/2,n-2}s_{\hat{\beta}_1} = 1.4365 + 2.306(0.0946) = 1.4583.$$

Thus, we may be 95% confident that the true regression slope β_1 lies between 1.4147 and 1.4583.

The confidence statements (Equations 2.12.a and b) for β_0 and β_1, respectively, with $1-\alpha$ representing what we may now call the *statement confidence coefficient*, give rise to what may be termed the *statement confidence intervals* (Equations 2.13 and 2.14). These statement confidence intervals hold separately and, for, say, β_0, the statement confidence coefficient gives the proportion of time, under repeated sampling from the same population, that the realized intervals contain β_0. That is, if $1-\alpha = 0.95$ and many samples of size n were taken from the population at hand and a statement confidence interval was computed for each of them, then, in the long run, 95% of these intervals would bracket β_0 and 5% of them would not.

However, what if we desire to have Equations 2.13 and 2.14 hold simultaneously? (Individual $100(1-\alpha)\%$ confidence intervals for β_0 and β_1 do not provide us with $100(1-\alpha)\%$ confidence that Equations 2.13 and 2.14 hold simultaneously because these estimates of β_0 and β_1 are not independent in that they are made from the same sample.) Given that $\{\hat{\beta}_0, \hat{\beta}_1\}$

can be considered a *family of estimates*, we need to construct joint or family confidence intervals for β_0 and β_1, in which $1-\alpha$ is now to be interpreted as a *family confidence coefficient*.

In this regard, the family confidence coefficient gives the proportion of time, under repeated sampling from the population, that the entire family of realized intervals contains β_0 and β_1, i.e., if again $1-\alpha = 0.95$ and many samples of size n are taken from the given population and statement confidence intervals for both β_0 and β_1 are calculated for each sample, then, in the long run, 95% of the samples would render a family of estimates bracketing both β_0 and β_1, and 5% of them would not.

To guarantee that Equations 2.13 and 2.14 are simultaneously valid with $100(1-\alpha)\%$ confidence, the Bonferroni method of constructing joint or family confidence interval with a given family confidence coefficient of $1-\alpha$ proceeds as follows: adjust each statement confidence interval to a level above $1-\alpha$ so that the family confidence coefficient is at least $1-\alpha$. Specifically, if the individual statement confidence coefficients are each $1-\alpha$, the Bonferroni family confidence coefficient must be at least $1-2\alpha$ for Equations 2.13 and 2.14 to hold simultaneously. That is, if the family confidence coefficient is to be at least $1-\alpha$ for estimating β_0 and β_1 simultaneously, then we must estimate β_0 and β_1 separately with statement confidence coefficients of $1-\frac{\alpha}{2}$ each.

This implies that $\alpha/2$ must be divided between the two tails of the t distribution so that finding $t_{\alpha/4,n-2}$ is required. In this regard, the Bonferroni $100(1-\alpha)\%$ joint or family confidence limits for β_0 and β_1 are, respectively,

$$\hat{\beta}_0 \pm t_{\alpha/4,n-2}s_{\hat{\beta}_0}; \tag{2.13.1}$$

$$\hat{\beta}_1 \pm t_{\alpha/4,n-2}s_{\hat{\beta}_1}. \tag{2.14.1}$$

In general, if r joint confidence interval estimates are desired with a family confidence coefficient of $1-\alpha$, then determining each separate interval estimate with a statement confidence coefficient of $1-\frac{\alpha}{r}$ will do. Hence, α/r must be divided between the two tails of the t distribution, i.e., we must determine $t_{\alpha/2r,\,n-r}$.

Example 2.5

Using the information provided in Example 2.4 above, determine Bonferroni family confidence intervals using a family confidence coefficient of $1-\alpha = 0.95$. Because $t_{\alpha/4,\,n-2} = t_{0.0125,8} \approx 2.896$, Equations 2.13.1 and 2.14.1 become $1.9379 \pm 2.896\ (0.7219)$ and $1.4365 \pm 2.896\ (0.0946)$, respectively. Hence, we may be 95% confident that $-0.1527 \le \beta_0 \le 4.0285$ and $1.1625 \le \beta_1 \le 1.7105$.

Testing Hypotheses Concerning β_0, β_1

One question that naturally arises once β_0 and β_1 have been estimated is whether or not the variables X and Y are truly linearly related. Equivalently, we may inquire as to whether or not the sample dataset exhibits sufficient evidence to indicate that X actually contributes significantly to the prediction of the average value of Y for a given value of X or to the prediction of a particular value of Y corresponding to a given value of X. Conceivably, the observed or fitted linear relationship may simply be the result of chance phenomena. To answer these questions, we shall test a particular null hypothesis concerning β_1 against an appropriate alternative hypothesis, in which the latter is typically dictated on a priori grounds by some supporting scientific or behavioral (e.g., economic) theory relating X and Y. The most common type of hypothesis tested is that there is no linear relationship between X and Y or $E(Y/X) = \beta_0 + \beta_1 X = \beta_0$, i.e., the conditional expectation of Y given X does not depend linearly on X so that the population regression line is horizontal. Hence, the implied null hypothesis of *no linear relationship between X and Y* is $H_0 : \beta_1 = 0$. Consider the following as far as possible alternative hypotheses are concerned:

a. If, say, *a priori* economic theory indicates that X and Y are positively related, we then choose $H_1 : \beta_1 > 0$. In this instance, we have a one-tail alternative involving the right tail of the t distribution (Figure 2.6.A). Here the region of rejection is $\mathcal{R} = \left\{t/t > t_{\alpha,n-2}\right\}$, where $t_{\alpha,n-2}$ is the critical (tabular) value of t. From Equation 2.10, given that H_0 is true, the test statistic is $T = \hat{\beta}_1/S_{\hat{\beta}_1}$, where the sample realization of T will be denoted as $t = \hat{\beta}_1/s_{\hat{\beta}_1}$. Then, if $t > t_{\alpha,n-2}$, we reject H_0 in favor of H_1, i.e., at the 100α% level, we conclude that there exists a statistically significant positive linear relationship between X and Y.
b. However, if a priori economic theory indicates that X and Y are negatively related, we then have $H_1 : \beta_1 < 0$. Relative to this alternative we have a one-tail test involving the left tail of the t distribution (Figure 2.6.B) with $\mathcal{R} = \left\{t/t < -t_{\alpha,n-2}\right\}$. Again the test statistic is $T = \hat{\beta}_1/S_{\hat{\beta}_1}$. In this regard, if $t < -t_{\alpha,n-2}$, we reject H_0 in favor of H_1 and conclude that, at the 100α% level, there exists a statistically significant negative linear relationship between X and Y.
c. If we are uncertain about the relationship between X and Y, then the appropriate alternative hypothesis is $H_1 : \beta_1 \neq 0$. Clearly, this alternative implies a two-tail test with $\mathcal{R} = \left\{t/|t| > t_{\alpha/2,n-2}\right\}$ (Figure 2.6.C). For this test, if $|t| = \left|\hat{\beta}_1/s_{\hat{\beta}_1}\right| > t_{\alpha/2,n-2}$, we reject H_0 in favor of H_1 and conclude that, at the 100α% level, there exists a statistically significant linear relationship between X and Y.

(Note: If we do not reject H_0, it does not mean that X and Y are unrelated but only that there is no significant linear relationship exhibited by the data; the true underlying relationship between X and Y may be highly nonlinear.)

The occasion may arise when we want to test the null hypothesis $H_0 : \beta_1 = \beta_1^0$, where β_1^0 is some specific or anticipated level of β_1 (not necessarily zero). If from Equation 2.10 the sample realization of the test statistic T under H_0 is denoted as $t = \left(\hat{\beta}_1 - \beta_1^0\right)/s_{\hat{\beta}_1}$, then the appropriate set of alternative hypotheses is as follows:

a. $H_1 : \beta_1 > \beta_1^0$. Here $\mathcal{R} = \left\{t/t > t_{\alpha,n-2}\right\}$; or
b. $H_1 : \beta_1 < \beta_1^0$. In this case $\mathcal{R} = \left\{t/t < -t_{\alpha,n-2}\right\}$; or
c. $H_1 : \beta_1 \neq \beta_1^0$. Then $\mathcal{R} = \left\{t/|t| > t_{\alpha/2,n-2}\right\}$.

As far as the true regression intercept is concerned, the relevant null hypotheses to be tested are either $H_0 : \beta_0 = 0$ (i.e., the population regression line passes through the origin) or $H_0 : \beta_0 = \beta_0^0$, where β_0^0 is some specific or anticipated level of β_0 (not necessarily zero). The

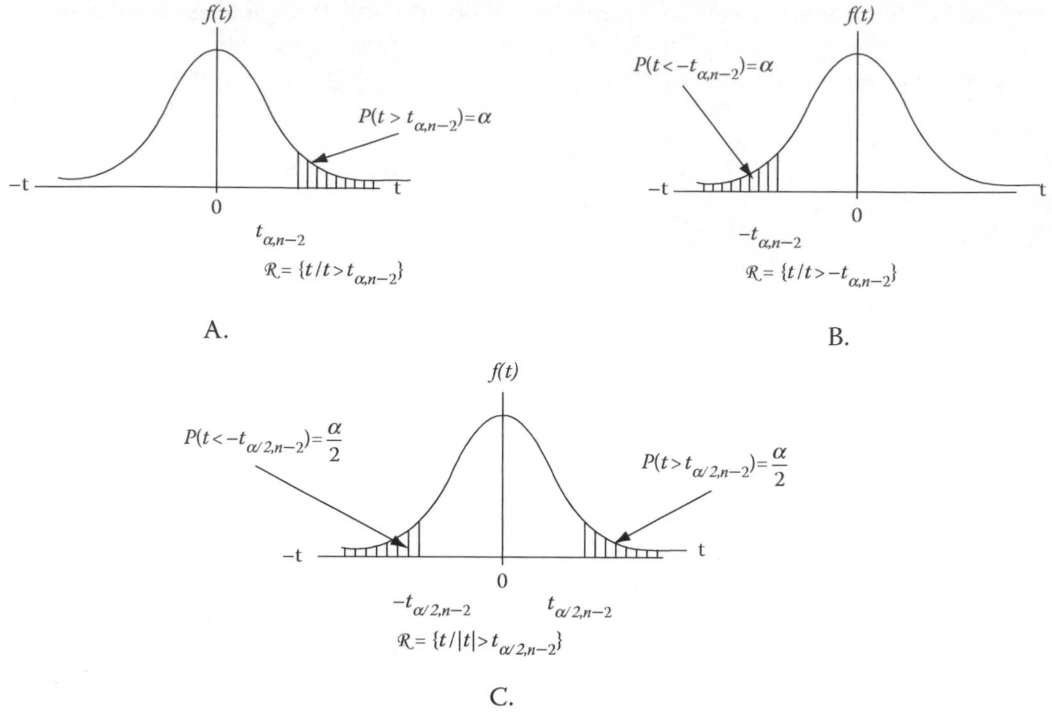

FIGURE 2.6
Critical regions for t tests.

appropriate critical region $\mathcal{R}$ and test statistic T are formed in a manner similar to those specified above when hypotheses about β_1 were tested.

Example 2.6

On the basis of the least-squares estimates $\hat{\beta}_0$ and $\hat{\beta}_1$ obtained from the sample data on advertising outlay (X) and gross sales (Y), can we conclude, at the 5% level, that the true population parameters β_0 and β_1 are significantly different from zero? As far as the population regression intercept is concerned, $H_0 : \beta_0 = 0$ and $H_1 : \beta_0 > 0$ (we anticipate a priori that even if the company did not advertise, the level of gross sales should be positive). With $t_{\alpha,n-2} = t_{0.05,8} = 1.86$, $\mathcal{R} = \{t/t > 1.86\}$. Because $t = \hat{\beta}_0 / s_{\hat{\beta}_0} = 1.9379/0.7219 = 2.6844 \in \mathcal{R}$, we thus reject H_0 in favor of H_1.

Is there a significant positive linear relationship between sales and advertising? To answer this question, we form $H_0 : \beta_1 = 0$, $H_1 : \beta_1 > 0$ (we anticipate a priori that the effect of the company's promotional activities is to increase sales). Because $\mathcal{R} = \{t/t > 1.86\}$ and $t = \hat{\beta}_1 / s_{\hat{\beta}_1} = 1.4365/0.0946 = 15.1849 \in \mathcal{R}$, we reject H_0 in favor of H_1 and conclude that there is a strong (statistically significant) positive linear relationship between X and Y.

At a promotional strategy session of the advertising department, one of the account coordinators stated that, for every \$1 expended on advertising the company would obtain, on average, about a \$2 increase in gross sales. Is this assertion supported by the data at the 5% level? Let us select $H_0 = \beta_1 = \beta_1^0 = 2$ and test it against $H_1 = \beta_1 \neq \beta_1^0 = 2$ (gross sales increases by some figure other than \$2). Because $\mathcal{R} = \left\{t/|t| > t_{\alpha/2,n-2} = t_{0.025,8} = 2.306\right\}$ and $|t| = \left|\left(\hat{\beta}_1 - \beta_1^0\right)/s_{\hat{\beta}_1}\right| = |(1.4365 - 2)/0.0946| = 5.9567 \in \mathcal{R}$, we reject H_0 in favor of H_1, i.e., on average, a \$1 increase in advertising outlay does not precipitate a \$2 increase in gross sales. In fact, the actual increase lies significantly below the \$2 level. (Interestingly enough, we could have reached this same conclusion by simply examining the 95% confidence interval for β_1 computed above in "Precision of the Least-Squares Estimators $\hat{\beta}_0$, $\hat{\beta}_1$: Confidence Intervals." Because 2 is not a member of the said interval, β_1 must depart significantly from 2. Hence, an explicit or formal test at the 100α% level of $H_0 : \beta_1 = \beta_1^0$ against $H_1 : \beta_1 \neq \beta_1^0$ is not necessary if the 100(1 − α)% confidence interval for β_1 is given.)

One final set of test procedures concerning β_0 and β_1 will be briefly mentioned. We have been performing hypothesis tests on β_0 and β_1 separately. However, we can easily test β_0 and β_1 jointly to determine whether or not the entire population regression equation itself is significant, i.e., we can determine whether or not β_0 and β_1 are jointly significantly different from zero. To this end, we form the joint null hypothesis $H_0 : \beta_0 = \beta_1 = 0$ and test it against the joint alternative $H_1 : \beta_0 \neq 0$, $\beta_1 \neq 0$ at the 100α% level. At a particular X_i value, let us compute the difference between the sample and population regression lines as $\hat{Y}_i - E(Y_i/X_i) = \hat{\beta}_0 + \hat{\beta}_1 X_i - (\beta_0 + \beta_1 X_i) = (\hat{\beta}_0 - \beta_0) + (\hat{\beta}_1 - \beta_1)X_i$. Then $\Sigma\left[\hat{Y}_i - E(Y_i/X_i)\right]^2 = \Sigma\left[(\hat{\beta}_0 - \beta_0) + (\hat{\beta}_1 - \beta_1)X_i\right]^2$ serves as a measure of the overall discrepancy between the estimated values of β_0 and β_1 and their true population values for given X_i's. Clearly, this sum varies directly with $|\hat{\beta}_0 - \beta_0|$ and $|\hat{\beta}_1 - \beta_1|$. Furthermore, it can be shown that the quantity

$$F = \frac{\Sigma\left[(\hat{\beta}_0 - \beta_0) + (\hat{\beta}_1 - \beta_1)X_i\right]^2 / 2}{S_\varepsilon^2} : F_{2,n-2}. \tag{2.15}$$

Under H_0, this test statistic becomes

$$F = \frac{\Sigma(\hat{\beta}_0 + \hat{\beta}_1 X_i)^2 / 2}{S_\varepsilon^2} = \frac{(n\bar{Y}^2 + \hat{\beta}_1 \Sigma x_i y_i)/2}{S_\varepsilon^2}, \tag{2.16}$$

with $\mathcal{R} = \{F/F > F_{\alpha,2,n-2}\}$. Here we have a one-tail alternative on the upper tail of the F distribution (Figure 2.7) so that we reject H_0 when the realized F value exceeds $F_{\alpha,2,n-2}$.

It is evident that a more general joint test can be performed by stating the joint null hypothesis as $H_0 : \beta_0 = \beta_0^0, \beta_1 = \beta_1^0$, where β_0^0 and β_1^0 are specific anticipated values (not necessarily zero) of the true regression intercept and slope, respectively. This hypothesis is to be tested against the joint alternative $H_1 : \beta_0 \neq \beta_0^0, \beta_1 \neq \beta_1^0$. Given that H_0 is true, the test statistic may be obtained from Equation 2.15 and written as

$$F = \frac{\left\{n\left[(\hat{\beta}_0 - \beta_0^0) + (\hat{\beta}_1 - \beta_1^0)\bar{X}\right]^2 + (\hat{\beta}_1 - \beta_1^0)^2 \Sigma x_i^2\right\} / 2}{S_\varepsilon^2}.$$

Example 2.7

Do the estimates obtained for β_0 and β_1 from the sample data on gross sales and advertising expenditure warrant the conclusion that the entire population regression line is significant at the 5% level? To answer this, let us set $H_0 : \beta_0 = \beta_1 = 0$ and test it against $H_1 : \beta_0 \neq 0, \beta_1 \neq 0$ (we are testing to determine whether the entire population regression equation is significant). Because $\mathcal{R} = \{F/F > F_{\alpha,2,n-2} = F_{0.05,2,8} = 4.46\}$ and, from Equation 2.16, the realized

$$F = \frac{\left[10(12.295)^2 + 1.4365(88.7010)\right]/2}{0.5533} = 1250.9048 \in \mathcal{R},$$

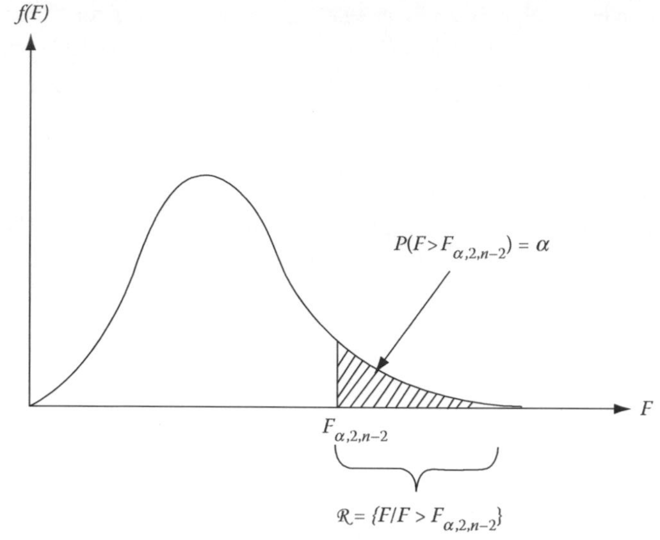

FIGURE 2.7
Critical region for F tests.

we conclude that the entire population regression line is statistically significant, i.e., $|\hat{\beta}_0|$ and $|\hat{\beta}_1|$ are sufficiently large to warrant rejecting H_0 at the 5% level.

The Precision of the Entire Least-Squares Regression Equation: A Confidence Band

We noted previously that the purpose of regression analysis is to predict either the average value of Y for a given value of X or a particular value of Y corresponding to a given value of X. If in this section we restrict our discussion to predicting the average value of Y given $X = X_i$, $E(Y/X_i)$, then we can determine just how precisely we have estimated the population regression line $E(Y/X) = \beta_0 + \beta_1 X$ by constructing a *confidence band* around the sample regression line $\hat{Y} = \hat{\beta}_0 + \hat{\beta}_1 X$. Clearly, the narrower the confidence band, the more precisely the sample regression line estimates the population regression line. Because $\hat{\beta}_0$ and $\hat{\beta}_1$ are unbiased estimators of β_0 and β_1, respectively, it follows that $E(\hat{Y}) = \beta_0 + \beta_1 X$, i.e., an unbiased estimator for the population regression line is the sample regression line. In this regard, because $\hat{Y}_i = \hat{\beta}_0 + \hat{\beta}_1 X_i$ is an estimator for $E(Y_i/X_i) = \beta_0 + \beta_1 X_i$, it can be shown (Appendix 2.B.1) that the variance of $\hat{Y}_i$ may be computed as

$$\sigma^2_{\hat{Y}_i} = \sigma^2_{\varepsilon}\left[\frac{1}{n} + \frac{(X_i - \bar{X})^2}{\Sigma x_i^2}\right]. \tag{2.17}$$

It is evident that Equation 2.17 cannot be directly used to ascertain the precision of our estimate of the population regression line because σ_ε^2 is unknown. As before, let us estimate σ_ε^2 by S_ε^2 (see Equation 2.7). Then from Equation 2.17, the estimated variance of $\hat{Y}_i$ is

$$S_{\hat{Y}_i}^2 = S_\varepsilon^2\left[\frac{1}{n}+\frac{(X_i-\bar{X})^2}{\Sigma x_i^2}\right].$$

If we take the positive square root of this expression, we obtain the estimated standard deviation of $\hat{Y}_i$,

$$S_{\hat{Y}_i} = S_\varepsilon\sqrt{\frac{1}{n}+\frac{(X_i-\bar{X})^2}{\Sigma x_i^2}}. \tag{2.18}$$

By virtue of Assumption A.1 and the Gauss-Markov results, Appendix 2.B.1 also informs us that $\hat{Y}_i = \hat{\beta}_0 + \hat{\beta}_1 X_i$ is a linear combination of normal random variables so that

$$\hat{Y}_i : N\left(E(Y_i/X_i), \sqrt{V(\hat{Y}_i)}\right) = N\left(\beta_0+\beta_1 X_i, \sigma_\varepsilon\sqrt{\frac{1}{n}+\frac{(X_i-\bar{X})^2}{\Sigma x_i^2}}\right)$$

and thus

$$\frac{\hat{Y}_i-(\beta_0+\beta_1 X_i)}{\sigma_\varepsilon\sqrt{\frac{1}{n}+\frac{(X_i-\bar{X})^2}{\Sigma x_i^2}}} : N(0,1). \tag{2.19}$$

If in Equation 2.19 we replace σ_ε by S_ε, the resulting quantity follows a t distribution with $n-2$ degrees of freedom, i.e.,

$$\frac{\hat{Y}_i-(\beta_0+\beta_1 X_i)}{S_\varepsilon\sqrt{\frac{1}{n}+\frac{(X_i-\bar{X})^2}{\Sigma x_i^2}}} = \frac{\hat{Y}_i-(\beta_0+\beta_1 X_i)}{S_{\hat{Y}_i}} : t_{n-2}.$$

To determine just how precisely the population regression line has been estimated from the sample data, let us form the probability statement

$$P\left(-t_{\alpha/2,n-2} \le \frac{\hat{Y}_i-(\beta_0+\beta_1 X_i)}{S_{\hat{Y}_i}} \le t_{\alpha/2,n-2}\right) = 1-\alpha. \tag{2.20}$$

After rearranging Equation 2.20 and using Equation 2.18, we obtain the confidence statement

$$P\left(\hat{Y}_i - t_{\alpha/2,n-2}s_{\hat{Y}_i} \le \beta_0 + \beta_1 X_i \le \hat{Y}_i + t_{\alpha/2,n-2}s_{\hat{Y}_i}\right) = 1-\alpha \tag{2.21}$$

so that, from Equation 2.21, a 100(1 – α)% confidence interval for $E(Y_i/X_i) = \beta_0 + \beta_1 X_i$ is

$$\left(\hat{Y}_i - t_{\alpha/2,n-2}s_{\hat{Y}_i}, \hat{Y}_i + t_{\alpha/2,n-2}s_{\hat{Y}_i}\right). \tag{2.22}$$

In this regard, Equation 2.22 informs us that we may be 100(1 – α)% confident that the true or population average value of Y_i given X_i lies between $\hat{Y}_i - t_{\alpha/2,n-2}s_{\hat{Y}_i}$ and $\hat{Y}_i + t_{\alpha/2,n-2}s_{\hat{Y}_i}$.

A glance at Equation 2.22 reveals that this confidence interval can be calculated for any X_i within the domain of X, and, as we vary X, we can compute a whole set of confidence intervals, with each one corresponding to (i.e., centered around) a point on the sample regression line. The collection of these confidence intervals forms a confidence band about the estimated sample regression line. Note that, because $S_{\hat{Y}_i}$ varies directly with $(X_i - \bar{X})^2$, the farther X_i lies from $\bar{X}$, the greater the estimated standard deviation of Y_i and thus the wider the confidence band at this point. If we use Equation 2.22 to calculate a confidence interval for each X_i, the locus of end points of these intervals would correspond to two branches of a rectangular hyperbola, indicating that we have more confidence in predicting $E(Y_i/X_i)$ near the center of the X_i's than at the extremes of the range of X values (Figure 2.8).

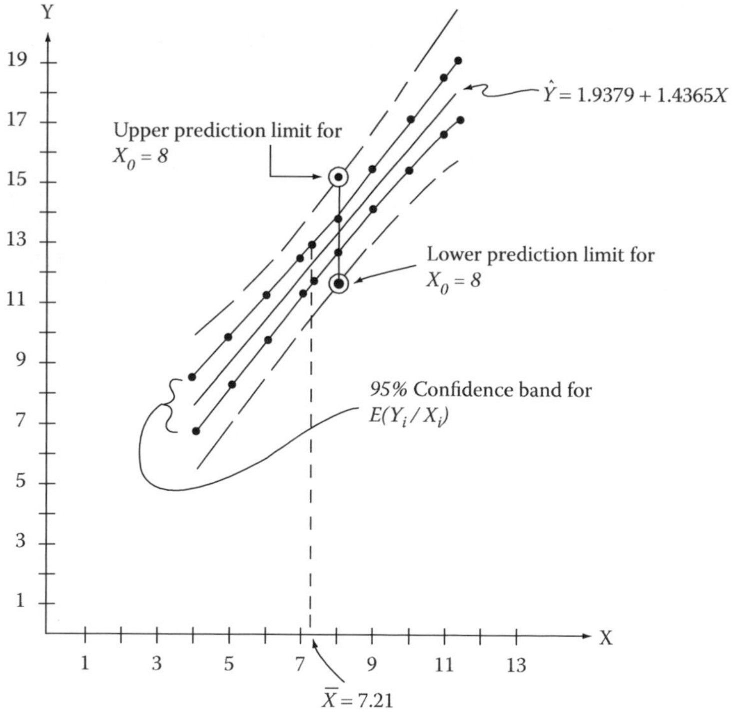

FIGURE 2.8
Confidence band and prediction limits.

We may use the results of the preceding discussion to conduct a test of the hypothesis that $E(Y_i/X_i)$ equals some specific or anticipated value E_i^0, i.e., the null hypothesis is $H_0: E(Y_i/X_i) = E_i^0$. Given that H_0 is true, the test statistic

$$T = \left(\hat{Y}_i - E_i^0\right)/S_{\hat{Y}_i} : t_{n-2}. \tag{2.23}$$

As always, the critical region is determined by the alternative hypothesis offered (see the presentation given above in "Testing Hypotheses Concerning β_0, β_1" concerning the various possible alternative hypotheses usually made).

Example 2.8

Let us construct a 95% confidence band for the population regression equation given that the estimated regression equation determined from the sales/advertising dataset is $\hat{Y} = 1.9379 + 1.4365X$. Using $t_{\alpha/2,\, n-2} = t_{0.025,8} = 2.306$ and $s_{\hat{Y}_i} = 0.7439\sqrt{\frac{1}{10} + \frac{(X_i - 7.210)^2}{61.7490}}$, we may easily complete Table 2.2 below for selected X_i values and thus determine the 95% confidence band. As Figure 2.8 reveals, the confidence band is narrowest at $\bar{X}$, with the width of the individual confidence intervals increasing as we move away from $\bar{X}$ in either direction.

For instance, at a level of, say, advertising outlay of \$10.5 million ($X_i = 10.5$), is there sufficient evidence for concluding, at the 5% level, that, on average, gross sales will exceed \$17 million $\left(E_i^0 = 17.0\right)$. Here $H_0: E_i^0 \le 17.0$, $H_1: E_i^0 > 17.0$, and $t_{\alpha,\, n-2} = t_{0.05,8} = 1.86$. Because $\mathcal{R} = \{t/t > 1.86\}$ and, from Equation 2.23, the realized $T = (17.0212 - 17.0)/0.3903 = 0.0543 \notin \mathcal{R}$, it follows that we cannot reject H_0, i.e., the predicted average value of gross sales does not lie far enough above $E_i^0 = 17$ to enable us to conclude that gross sales at $X_i = 10.5$ will exceed $E_i^0 = 17$.

TABLE 2.2

Calculation of Confidence Limits

			95% Confidence Band		
X_i	$\hat{Y}_i$	$S_{\hat{Y}_i}$	Lower Limit $(\hat{Y}_i - 2.306s_{\hat{Y}_i})$	Upper Limit $(\hat{Y}_i + 2.306s_{\hat{Y}_i})$	Width = Upper Limit Minus Lower Limit
4.00	7.6839	0.3843	6.7977	8.5901	1.7724
5.00	9.1204	0.3148	8.3945	9.8463	1.4518
6.00	10.5569	0.2616	9.9537	11.1602	1.2064
7.00	11.9934	0.2360	11.4492	12.5376	1.0884
7.21 ($=\bar{X}$)	12.2950	0.2352	11.7526	12.8374	1.0848
8.00 ($=X_0$)	13.4299	0.2468	12.8608	13.9990	1.1382
9.00	14.8664	0.2899	14.1979	15.5349	1.3370
10.00	16.3029	0.3537	15.4873	17.1185	1.6312
11.00	17.7394	0.4290	16.7501	18.7289	1.9788
11.30	18.1704	0.4530	17.1258	19.2150	2.0892

The Prediction of a Particular Value of Y Given X

In the preceding section, we considered the problem of predicting the average value of Y_i given X_i. Let us now look to the task of predicting a particular Y_i value from a given X_i. Specifically, for $X = X_0$, let us predict or forecast the value of the random variable $Y_0 = \beta_0 + \beta_1 X_0 + \varepsilon_0$. If the true population regression line were known, the predictor of Y_0 would be $E(Y_0/X_0) = \beta_0 + \beta_1 X_0$, a point on the population regression line corresponding to X_0. Because $E(Y_0/X_0)$ is unknown and must be estimated from the sample data, let us use as an estimator $\hat{Y}_0 = \hat{\beta}_0 + \hat{\beta}_1 X_0$, a point on the sample regression line. In this regard, if $\hat{Y}_0$ is used to estimate Y_0, then the forecast error is the random variable $Y_0 - \hat{Y}_0 = \beta_0 + \beta_1 X_0 + \varepsilon_0 - \left(\hat{\beta}_0 + \hat{\beta}_1 X_0\right)$. Because $\hat{\beta}_0$ and $\hat{\beta}_1$ are unbiased estimators of β_0 and β_1, respectively, and $E(\varepsilon_0) = 0$, the mean of the forecast error is $E\left(Y_0 - \hat{Y}_0\right) = 0$, i.e., $\hat{Y}_0$ is an unbiased estimator of Y_0. [Note that $\hat{Y}_0$ is an unbiased estimator for Y_0 as well as for $E(Y_0/X_0)$.] Moreover, it can be shown (Appendix 2.B.2) that the variance of $Y_0 - \hat{Y}_0$ is $\sigma^2_{(Y_0-\hat{Y}_0)} = \sigma^2_\varepsilon + \sigma^2_{\hat{Y}_0}$, where $\sigma^2_{\hat{Y}_0}$ is determined from Equation 2.17 when $X_i = X_0$. Then

$$\sigma^2_{(Y_0-\hat{Y}_0)} = \sigma^2_\varepsilon \left[1 + \frac{1}{n} + \frac{(X_0 - \bar{X})^2}{\Sigma x_i^2}\right]. \tag{2.24}$$

Because σ^2_ε is unknown, it will be estimated by S^2_ε (Equation 2.7) so that Equation 2.24 becomes

$$S^2_{(Y_0-\hat{Y}_0)} = S^2_\varepsilon \left[1 + \frac{1}{n} + \frac{(X_0 - \bar{X})^2}{\Sigma x_i^2}\right]. \tag{2.25}$$

After taking the positive square root of this expression, we obtain the estimated *standard deviation of the forecast error* $Y_0 - \hat{Y}_0$ or

$$S_{(Y_0-\hat{Y}_0)} = S_\varepsilon \sqrt{1 + \frac{1}{n} + \frac{(X_0 - \bar{X})^2}{\Sigma x_i^2}}. \tag{2.26}$$

If we invoke Assumption A.1 and use the Gauss-Markov Theorem, we see that $Y_0 - \hat{Y}_0$ is a linear combination of the normal random variables Y_0 and $\hat{Y}_0$ so that

$$Y_0 - \hat{Y}_0 : N\left(E\left(Y - \hat{Y}_0\right), \sqrt{V\left(Y - \hat{Y}_0\right)}\right) = N\left(0, \sigma_\varepsilon \sqrt{1 + \frac{1}{n} + \frac{(X_0 - \bar{X})^2}{\Sigma x_i^2}}\right)$$

and thus

$$\frac{Y_0 - \hat{Y}_0}{\sigma_\varepsilon \sqrt{1 + \frac{1}{n} + \frac{(X_0 - \bar{X})^2}{\sum x_i^2}}} : N(0,1). \tag{2.27}$$

If in Equation 2.27 σ_ε is replaced by S_ε, the resulting quantity follows a t distribution with $n - 2$ degrees of freedom, i.e.,

$$\frac{Y_0 - \hat{Y}_0}{S_\varepsilon \sqrt{1 + \frac{1}{n} + \frac{(X_0 - \bar{X})^2}{\sum x_i^2}}} = \frac{Y_0 - \hat{Y}_0}{S_{(Y_0 - \hat{Y}_0)}} : t_{n-2}.$$

To determine just how precisely Y_0 has been estimated from the sample data, let us form the probability statement

$$P\left(-t_{\alpha/2,n-2} \leq \frac{Y_0 - \hat{Y}_0}{S_{(Y_0 - \hat{Y}_0)}} \leq t_{\alpha/2,n-2}\right) = 1 - \alpha. \tag{2.28}$$

After rearranging Equation 2.28 and passing to sample realizations, we obtain

$$P\left(\hat{Y}_0 - t_{\alpha/2,n-2}s_{(Y_0 - \hat{Y}_0)} \leq Y_0 \leq \hat{Y}_0 + t_{\alpha/2,n-2}s_{(Y_0 - \hat{Y}_0)}\right) = 1 - \alpha \tag{2.29}$$

so that, from Equation 2.29, a $100(1 - \alpha)\%$ confidence interval for Y_0 is

$$\left(\hat{Y}_0 - t_{\alpha/2,n-2}s_{(Y_0 - \hat{Y}_0)}, \hat{Y}_0 + t_{\alpha/2,n-2}s_{(Y_0 - \hat{Y}_0)}\right). \tag{2.30}$$

Hence, Equation 2.30 informs us that we may be $100(1 - \alpha)\%$ confident that the true value of Y_0 lies within the interval determined by $\hat{Y}_0 \pm t_{\alpha/2,n-2}s_{(Y_0 - \hat{Y}_0)}$.

When closely examining Equation 2.30, we can easily see that this confidence interval can be calculated for any $X = X_0$ value. In fact, as X_0 is varied, we can generate a whole set of confidence intervals, the collection of which defines a *prediction band* about the estimated sample regression line. Furthermore, because $S_{(Y_0 - \hat{Y}_0)}$ varies directly with $(X_0 - \bar{X})^2$, the farther X_0 lies from $\bar{X}$, the greater is the value of $S_{(Y_0 - \hat{Y}_0)}$ and thus the wider the prediction band at X_0. Hence, we have more confidence in predicting Y_0 near the center of the X_i's than at the extremes of the range of X values. Note also that, because $s_{(Y_0 - \hat{Y}_0)} > s_{\hat{Y}_i}$, the prediction band obtained by using Equation 2.30 is wider than the confidence band obtained from Equation 2.22, i.e., for $X = X_0$, it is intuitively clear that the prediction of an individual value of Y, namely Y_0, should have a greater error associated with it than the error arising from an estimate of the average value of Y given X_0, $E(Y_0/X_0)$.

To test the hypothesis that Y_0 equals some specific or anticipated value Y_0^0, let us form $H_0: Y_0 = Y_0^0$. If H_0 is true, the statistic

$$T = \left(\hat{Y}_0 - Y_0^0\right) / S_{(Y_0 - \hat{Y}_0)} : t_{n-2}. \tag{2.31}$$

Here too, the critical region is determined by the alternative hypothesis made (see above, "Testing Hypotheses Concerning β_0, β_1").

Example 2.9

Let us again look to the gross sales/advertising outlay data presented in Table 2.1. If $X_0 = 8$, determine a 95% prediction interval for Y_0. Because $t_{\alpha/2,\, n-2} = t_{0.025,8} = 2.306$, $\hat{Y}_0 = \hat{\beta}_0 + \hat{\beta}_1 X_0 = 1.9379 + 1.4365(8) = 13.4299$, and

$$s_{(Y_0 - \hat{Y}_0)} = 0.7439\sqrt{1 + \frac{1}{10} + \frac{(8 - 7.210)^2}{61.7490}} = 0.7847,$$

it follows that

$$\hat{Y}_0 - t_{\alpha/2,n-2} s_{(Y_0 - \hat{Y}_0)} = 13.4299 - 2.306(0.7847) = 11.6203,$$

$$\hat{Y}_0 + t_{\alpha/2,n-2} s_{(Y_0 - \hat{Y}_0)} = 13.4299 + 2.306(0.7847) = 15.2394.$$

Thus, we may be 95% confident that the true value of Y_0 lies between 11.6203 and 15.2394 (Figure 2.8). If we vary X_0, then we may construct the entire prediction band (see the dashed lines in Figure 2.8) in a manner similar to that used to derive the confidence band depicted in Table 2.2.

At the 5% level, can we conclude that, for $X_0 = 0$, the predicted value of Y does not exceed 18. Here $H_0: Y_0 \geq 18$ is tested against $H_1: Y_0 < 18$. With $\hat{Y}_0 = 1.9379 + 1.4365(10) = 16.3029$ and $\mathcal{R} = \{t/t < -1.86\}$, we see that, from Equation 2.31, the realized $T = (16.3029 - 18)/0.8237 = -2.0603 \in \mathcal{R}$ so that we may reject H_0 in favor of H_1, i.e., Y_0 lies significantly below $Y_0^0 = 18$ at the 5% level.

Example 2.10

If we now return to the question posed at the beginning of the final paragraph of "Estimating the Slope and Intercept of the Population Regression Line" above, it can be seen that, to answer the same, we must determine the precision of the predicted value of Y. We found previously that, on the basis of the estimated linear relationship between X and Y for 1997–2006, when $X_0 = 12$, it follows that $\hat{Y}_0 = 19.1759$. Then, the 95% lower and upper prediction limits for Y_0 (the true Y value at X_0) are, respectively,

$$\hat{Y}_0 - t_{\alpha/2,n-2} s_{(Y_0 - \hat{Y}_0)} = 19.1759 - 2.306(0.9024) = 17.0949,$$

$$\hat{Y}_0 + t_{\alpha/2,n-2} s_{(Y_0 - \hat{Y}_0)} = 19.1759 + 2.306(0.9024) = 21.2568.$$

Hence, we may be 95% confident that the population value of Y given $X_0 = \$12{,}000{,}000$ lies between \$17,094,900 and \$21,256,800.

Example 2.11

One important application of the statistic $\left(Y_0 - \hat{Y}_0\right)/s_{(Y_0-\hat{Y}_0)}$ is the determination of whether or not a new observed data point comes from the same linear relationship as that obtained from the original set of sample points. For example, let us assume that the aforementioned new advertising campaign for 2007 is actually undertaken and that for this particular year the company finds that, with an advertising outlay of $12 million, the realized or actual level of gross sales is $22 million. Does this increase in gross sales indicate that there has been a statistically significant change in the underlying linear relationship, or did this additional sample point come from the same linear relationship as before? To determine whether or not there has been a change in the basic structure of the gross sales/advertising outlay equation, let $X_0 = 12$, $Y_0 = 22$, and $\hat{Y}_0 = 19.1759$. At the 5% level, is there sufficient evidence to indicate that Y_0 is significantly greater than $\hat{Y}_0$? Let us test $H_0 : Y_0 - \hat{Y}_0 = 0$ against $H_1 : Y_0 - \hat{Y}_0 > 0$. Because $\mathcal{R} = \{t/t > 1.86\}$ and, under H_0, the realized $T = \left(Y_0 - \hat{Y}_0\right)/s_{(Y_0-\hat{Y}_0)} = (22 - 19.1759)/0.9024 = 3.1295 \in \mathcal{R}$, we reject H_0 in favor of H_1 and conclude that the underlying linear relationship that held from 1997 to 2006 did not generate the observed pair (12,22).

Decomposition of the Sample Variation of *Y*

One useful device for summarizing the regression results is provided by the *analysis-of-variance table* depicting what we shall call the partitioned sum of squares. The purpose of this construct is to ultimately determine why the values of Y change between successive X values. If we invoke Assumptions B.1, B.2, and B.4 (see above, "The Strong Classical Linear Regression Model"), then the variation in Y can be attributed partly to changes in the X values and partly to the effect of the random disturbance term ε. In this regard, we may ask the following: how much of the variation in Y can be attributed to the systematic influence of X and how much is attributed to the random effect of the disturbance term?

To answer this question, let us begin by determining the sample variation of Y about its mean (if there is no variation in Y as X increases in value, then all of the Y values lie on the horizontal line $Y = \bar{Y}$). As a concise summary measure of the total variation of the Y_i values about $\bar{Y}$, let us use

$$\Sigma\left(Y_i - \bar{Y}\right)^2 = \Sigma y_i^2. \tag{2.32}$$

For $X = X_i$, Figure 2.9 reveals that $Y_i - \bar{Y} = \left(\hat{Y}_i - \bar{Y}\right) + e_i$. On substituting this expression into Equation 2.32, we obtain

[Total sum of squares (SST)]	[Regression sum of squares (SSR)]	[Error sum of squares (SSE)]

$$\Sigma\left(Y_i - \bar{Y}\right)^2 = \Sigma\left(\hat{Y}_i - \bar{Y}\right)^2 + \Sigma e_i^2 + \underbrace{2\Sigma\left(\hat{Y}_i - \bar{Y}\right)e_i}_{(=0)}. \tag{2.33}$$

This expression reveals that the total sum of squares may be partitioned (Appendix 2.C) into two parts: (1) the regression sum of squares (SSR) (which reflects the variation in Y attributed to the linear influence of X); and (2) the error sum of squares (SSE) (which depicts the variation in Y ascribed to random factors). In this regard, SSR is called the

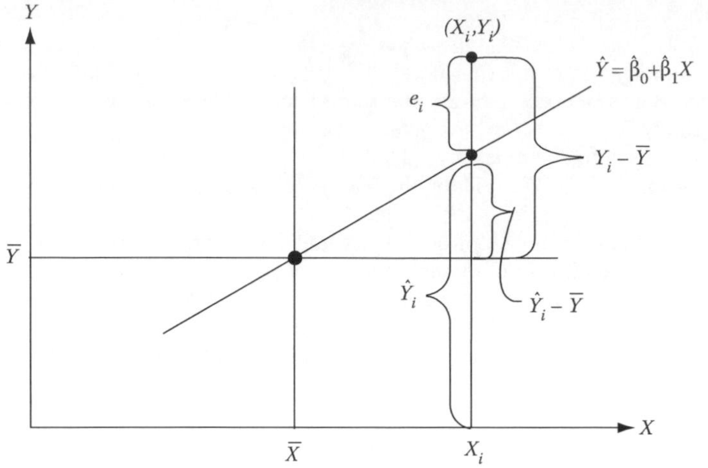

FIGURE 2.9
Partitioning the total sum of squares.

explained sum of squares (explained SS), whereas SSE is referred to as the *unexplained sum of squares* (unexplained SS).

As far as the computation of the various sums of squares is concerned,

$$\begin{aligned} SST &= \Sigma y_i^2, \\ SSR &= \hat{\beta}_1^2 \Sigma x_i^2 \left(or = \hat{\beta}_1 \Sigma x_i y_i \right), \\ SSE &= \Sigma y_i^2 - \hat{\beta}_1 \Sigma x_i y_i. \end{aligned} \tag{2.34}$$

One important application of these sums of squares is the specification of what is called the *coefficient of determination* (see Appendix 2.C):

$$R^2 = \frac{SSR}{SST} = \frac{\text{explained } SS}{\text{total } SS} = \frac{\hat{\beta}_1 \Sigma x_i y_i}{\Sigma y_i^2} \tag{2.35.a}$$

or, because SSR = SST − SSE,

$$R^2 = 1 - \frac{SSE}{SST} = 1 - \frac{\text{unexplained } SS}{\text{total } SS} = 1 - \frac{\Sigma y_i^2 - \hat{\beta}_1 \Sigma x_i y_i}{\Sigma y_i^2}, \tag{2.35.b}$$

where r^2 denotes its realized value. Here R^2 serves as a measure of goodness of fit: it represents the proportion of the total variation in Y, which can be explained by the linear influence of X. (When the number of degrees of freedom is small, R^2 is biased upward. To correct for degrees of freedom, let us compute the *adjusted coefficient of determination* $\bar{R}^2 = 1 - (SSE/SST)[(n-1)/(n-2)]$, which serves as an unbiased estimator of R^2. Clearly, $(n-1)/(n-2) > 1$ so that $\bar{R}^2 < R^2$, and as $n \to \infty$, $\bar{R}^2 - R^2 \to 0$.) From Equation 2.35.b, it is evident that $0 \le R^2 \le 1$, i.e., when $\Sigma e_i^2 = \Sigma y_i^2$ (the sample regression line is $\hat{Y} = \bar{Y}$ and the explained variation in Y or SSR is zero), it follows that $R^2 = 0$ (Figure 2.10.A), whereas for $\Sigma e_i^2 = 0$ (the observed points all lie on the sample regression line and the unexplained variation in Y or SSE is zero), we see that $R^2 = 1$ (Figure 2.10.B). The only time that R^2 is undefined is when $\Sigma y_i^2 = 0$, i.e., there is no variation in Y (Figure 2.10.C).

We noted above that R^2 is a measure of goodness of fit. Hence, $R^2 = 0$ may reflect the fact that a linear function provides a poor fit (here $\Sigma e_i^2 = \Sigma y_i^2$) to an essentially nonlinear scatter of points (Figure 2.10.D).

Based on the sums of squares presented in Equation 2.34, we may construct the partitioned sum-of-squares (or analysis-of-variance) table for our regression results as in Table 2.3. Note that the error mean square presented in Table 2.3 is just S_ε^2. So if the underlying population regression equation is truly linear, the error mean square is an estimate of σ_ε^2. Alternatively, the regression mean square term provides us with an estimate of σ_ε^2 only if $H_0 : \beta_1 = 0$ is true, i.e., only if X is of no use in explaining the variation in Y.

The second important application of the partitioned sum-of-squares notion is to determine whether or not X contributes significantly to the variation in Y. If $H_0 : \beta_1 = 0$ is true (there is no linear relationship between X and Y), then the sole source of variation in Y is the random disturbance term ε because the population regression sum of squares is zero. Now, it can be shown that the statistic

$$\frac{\left(\hat{\beta}_1 - \beta_1\right)^2 \Sigma x_i^2}{\Sigma e_i^2/(n-2)} : F_{1,n-2}.$$

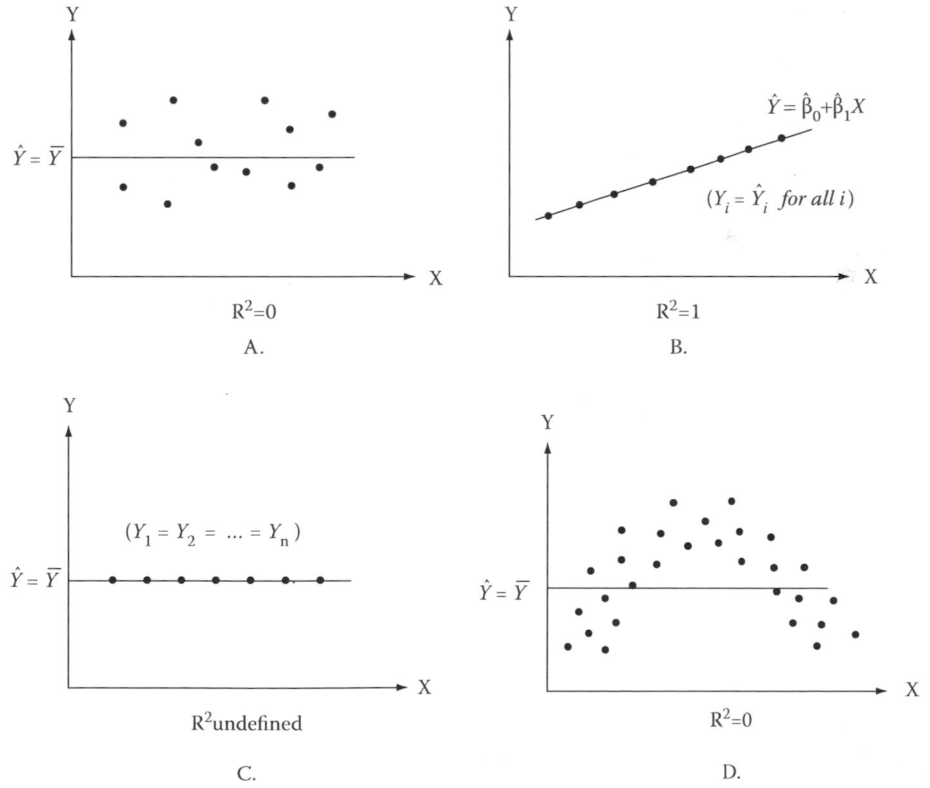

FIGURE 2.10
The behavior of R^2.

TABLE 2.3

Summary Table for the Partitioned Sums of Squares

Source of Variation	Sum of Squares	Degrees of Freedom	MS = SS/d.f.
Regression (explained variability)	(SSR) $\hat{\beta}_1 \Sigma x_i y_i$	1	(Regression MS) $MSR = SSR/1 = \hat{\beta}_1 \Sigma x_i y_i /1 = \hat{\beta}_1 \sum x_i y_i$
Error (unexplained variability)	(SSE) $\Sigma y_i^2 - \hat{\beta}_1 \Sigma x_i y_i$	$n-2$	(Error MS) $MSE = SSE/(n-2) = \left(\Sigma y_i^2 - \hat{\beta}_1 \Sigma x_i y_i\right)/(n-2)$
Total variability in Y	(SST) Σy_i^2	$n-1$	$R^2 = SSR/SST$; $F = \dfrac{MSR}{MSE} = [SSR/1]/[SSE/(n-2)]$

Under H_0: $\beta_1 = 0$, the preceding expression becomes

$$\frac{\hat{\beta}_1 \Sigma x_i y_i}{S_\varepsilon^2} = \frac{\text{regression mean square}}{\text{error mean square}}$$
$$= \frac{SSR/1}{SSE/(n-2)} = \frac{R^2/1}{\left(1-R^2\right)/(n-2)} : F_{1,n-2}.^2 \tag{2.36}$$

Here the appropriate alternative hypothesis is H_1: $\beta_1 \neq 0$ so that $\mathcal{R} = \left\{ F \Big/ \dfrac{SSR/1}{SSE/(n-2)} > F_{\alpha,1,n-2} \right\}$ provides us with a one-tail alternative on the upper tail of the F distribution. So if the realized value of F exceeds the tabular value, we reject H_0 in favor of H_1 and conclude that there exists a statistically significant linear relationship between X and Y, i.e., the observed linear relationship between X and Y did not arise solely because of chance or random factors. It is interesting to note that, if we square the T statistic $T = \hat{\beta}_1 / S_{\hat{\beta}_1}$, we obtain the F statistic presented in Equation 2.36. Thus, under $H_0 : \beta_1 = 0$,

$$T^2 = \left(\frac{\hat{\beta}_1}{S_{\hat{\beta}_1}} \right)^2 = \frac{\hat{\beta}_1^2}{S_\varepsilon^2 / \Sigma x_i^2} = \frac{\hat{\beta}_1 \Sigma x_i y_i}{S_\varepsilon^2} = F.$$

Example 2.12

If $\alpha = 0.05$ and $n-2 = 10$, then the one-tail F value $F_{0.05,1,10} = 4.96$ is equivalent to the two-tail t value squared or $\left(t_{0.025,10}\right)^2 = (2.228)^2 = 4.9639$. Hence, the preceding one-sided F test of $H_0 : \beta_1 = 0$ versus $H_1 : \beta_1 \neq 0$ is equivalent to the two-sided t test of $H_0 : \beta_1 = 0$ versus $H_1 : \beta_1 \neq 0$.

Example 2.13

Using the above set of gross sales/advertising outlay data, construct the analysis-of-variance table for the partitioned sum of squares. What inferences can be made from the information contained within this table? Using Table 2.3 as our guide, we obtain Table 2.3.1. Because r^2=0.9664, we see

---

2 If $H_0 : \beta_1 = 0$ is true then SSR/σ_ε^2 is χ_1^2 and SSE$/\sigma_\varepsilon^2$ is χ_{n-2}^2. Because SSR and SSE are independent, Equation 2.36 follows by definition of the F distribution.

TABLE 2.3.1

Summary Table for the Partitioned Sums of Squares

Source of Variation	Sum of Squares (SS)	Degrees of Freedom (d.f.)	MS = SS/d.f.
Regression (explained variability)	(SSR) $\hat{\beta}_1 \sum x_i y_i =$ $1.4365(88.7010) = 4.4266$	1	(Regression MS) $MSR = SSR/1 =$ 127.4190
Error (unexplained variability)	(SSE) $\Sigma y_i^2 - \hat{\beta}_1 \Sigma x_i y_i =$ $131.8456 - 127.4190 =$ 4.4266	$n - 2 = 8$	(Error MS) $MSE = SSE/(n-2) =$ $4.4266/8 = 0.5533$
Total variability in Y	(SST) $\Sigma y_i^2 = 131.8456$	$n - 1 = 9$	$r^2 = SSR/SST =$ $127.4190/131.8456 =$ 0.9664; $F = \dfrac{MSR}{MSE} =$ $127.4190/0.5533 =$ 230.2892

that approximately 97% of the variation in Y may be explained by the linear influence of X, whereas about 3% is left unexplained. In addition, we may look to the question of whether or not there exists a statistically significant linear relationship between X and Y. That is, for $\alpha = 0.05$, let $H_0: \beta_1 = 0$ and $H_1: \beta_1 \neq 0$ with $\mathcal{R} = \{F/F > F_{\alpha,1,n-2} = F_{0.05,1,8} = 5.32\}$. Because the realized F = regression MS/error $MS = 230.2892 \in \mathcal{R}$, we may conclude that the linear relationship between X and Y is highly significant at the 5% level.

The Correlation Model

The purpose of correlation analysis is as follows: Case A, to determine the degree of covariability between two random variables X and Y (here both X and Y are subject to random errors); or Case B, if only Y is assumed random and is regressed on X, with the values of the latter variable fixed (see above, "The Strong Classical Linear Regression Model"), then correlation serves to measure the goodness of fit of the sample linear regression equation to the scatter of observations on X and Y.

If Case A above is of interest, then we need to determine the direction as well as the strength (i.e., the degree of closeness) of the relationship between the random variables X and Y, where X and Y follow a joint bivariate distribution. This will be accomplished by extracting a sample consisting of the n pairs of values (X_i, Y_i), $i = 1,\ldots,n$, from the said distribution. Moreover, once we compute a *sample correlation coefficient*, we must determine whether or not it serves as a good estimate of the underlying degree of covariation within the population.

For Case B, we need to regress Y on X given the scatter of the n sample points $(X_i\ Y_i)$, $i = 1,\ldots,n$, under the assumptions of the strong classical linear regression model. Once we obtain the line of best fit and construct the analysis-of-variance table, we will be able to directly determine the sample correlation coefficient as well as test how good this measure is as an indicator of goodness of fit of the least-squares regression line to the set of observations on X and Y.

To address the Case A problem, let X and Y be random variables that follow a joint bivariate distribution. If $E(X)$ and $E(Y)$ depict the means of X and Y, respectively, $S(X)$ and $S(Y)$ represent the standard deviations of X and Y, respectively, and $COV(X,Y)$ denotes the covariance between X and Y, then the *population correlation coefficient*, which serves as a measure of the linear association between X and Y, may be depicted as

$$\rho = \frac{COV(X,Y)}{S(X)S(Y)} = \frac{E[X-E(X)][Y-E(Y)]}{S(X)S(Y)}. \tag{2.37}$$

(If we form the standardized variables $\tilde{X} = \frac{X-E(X)}{S(X)}$ and $\tilde{Y} = \frac{Y-E(Y)}{S(Y)}$, then Equation 2.37 can be rewritten as $\rho = COV(\tilde{X},\tilde{Y})$, i.e., we may think of the population correlation coefficient as simply the covariance of the two standardized variables $\tilde{X}$ and $\tilde{Y}$.) As far as the properties of ρ are concerned, (1) it is symmetrical with respect to X and Y (the correlation between X and Y is the same as that between Y and X), (2) it is dimensionless (i.e., a pure number), (3) it is independent of the units or the scale of measurement used, and, (4) by incorporating in its calculation deviations from the means of X and Y, it shifts the origin of the population values to the means of X and Y.

The value of the population correlation coefficient exhibits both the direction and the strength of the linear relationship between the random variables X and Y. That is, if $\rho > 0$, both variables tend to increase or decrease together (we have a *direct relationship* between X and Y), whereas if $\rho < 0$, an increase in one variable is accompanied by a decrease in the other (there exists an *inverse relationship* between X and Y). Clearly, the sign of ρ is determined by the sign of $COV(X,Y)$.

As far as the range of values assumed by ρ is concerned, it can be shown that $|\rho| \leq 1$ or $-1 \leq \rho \leq 1$. In this regard, for $\rho = 1$ or -1, we have perfect positive association or perfect negative association, respectively. If $\rho = 0$, the variables are uncorrelated, thus indicating the absence of any linear relationship between X and Y. (It is important to remember that ρ depicts the strength of the linear relationship between X and Y. If X and Y are independent random variables, then $\rho = 0$ because $COV(X,Y) = 0$. However, the converse is not true, i.e. , we cannot legitimately infer that X and Y are independent if $\rho = 0$ because the true underlying relationship between X and Y may be highly nonlinear.) Furthermore, as ρ increases (respectively decreases) in value from 0 to 1 (respectively, from 0 to –1), the closeness or strength of the linear relationship between X and Y concomitantly increases. It is important to remember that, because two random variables may be highly correlated, the association between them does not allow us to infer anything about cause and effect (both X and Y may be related to a third *unobserved* variable, say, W, which causes movements in both X and Y) or to predict values of one variable from the other. In addition, no functional relationship between X and Y (as was the case for the preceding regression model involving X and Y) is assumed.

Estimating the Population Correlation Coefficient ρ

If a sample of size n consisting of the pairs of observations (X_i, Y_i), $i = 1,\ldots,n$, is extracted from the underlying joint bivariate population of true X and Y values, then we may estimate ρ by using the sample correlation coefficient (called the *Pearson product-moment coefficient of correlation*)

$$R = \frac{\frac{1}{n-1}\Sigma(X_i-\bar{X})(Y_i-\bar{Y})}{\sqrt{\frac{1}{n-1}\Sigma(X_i-\bar{X})^2}\sqrt{\frac{1}{n-1}\Sigma(Y_i-\bar{Y})^2}} = \frac{\Sigma x_i y_i}{\sqrt{\Sigma x_i^2 \Sigma y_i^2}}, \tag{2.38}$$

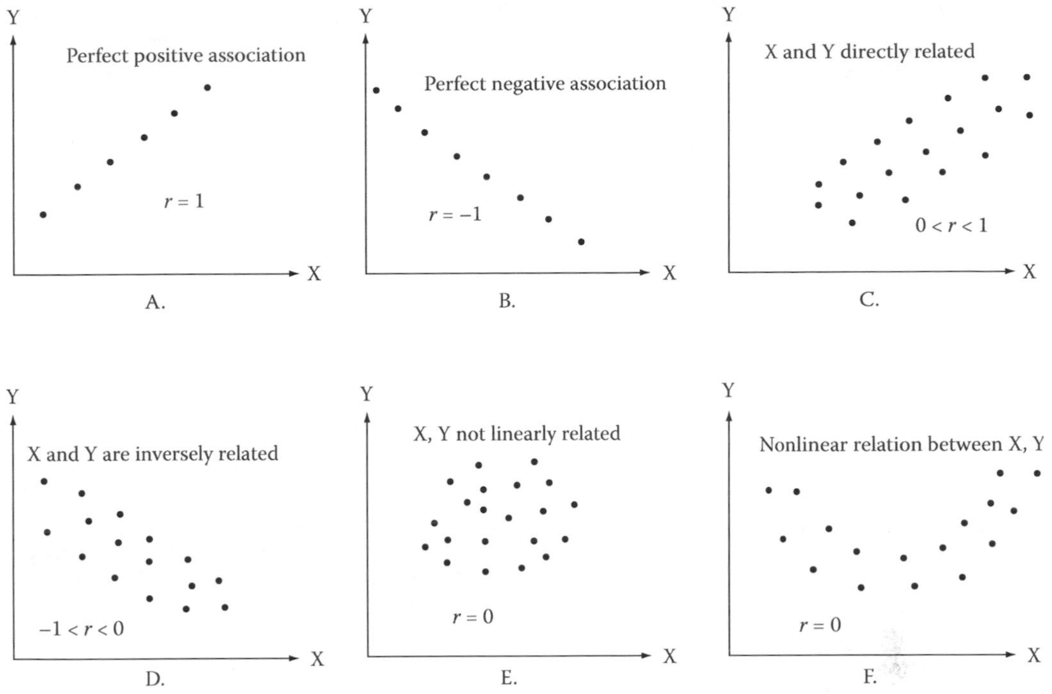

FIGURE 2.11
The behavior of r.

where r denotes its sample realization. Although R is a slightly biased (downward) estimator of ρ for small samples, it is a consistent as well as sufficient estimator. (An unbiased estimator of ρ is $R^* = R\left[1+\frac{1-R^2}{2(n-4)}\right]$.) Our interpretation of R is the same as the one advanced above for ρ. In this regard, as Figures 2.11.A and B reveal, if $r = 1$ (respectively, −1), we have perfect positive (respectively, negative) linear association between X and Y (here all sample points lie on some imaginary positively or negatively sloped line); if $|r| < 1$, the random variables are linearly related but to a lesser degree (Figures 2.11.C and D); and if $r = 0$ (Figure 2.11.E), X and Y are not linearly related at all. Furthermore, as noted above, if $r = 0$, we may not legitimately conclude that X and Y are not related; the true relationship may be nonlinear (Figure 2.11.F).

Example 2.14

Appearing in Table 2.4 is a random sample of observations taken from a joint bivariate distribution for the random variables X and Y. Are these variables linearly related? If so, in what direction?

If the totals of columns 5, 6, and 7 are inserted into Equation 2.38, we have

$$r = \frac{\Sigma x_i y_i}{\sqrt{\Sigma x_i^2}\sqrt{\Sigma y_i^2}} = \frac{-22.9690}{\sqrt{7.3690}\sqrt{12.5511}} = -0.6741.$$

With $r < 0$, it follows that X and Y are negatively related and thus vary inversely.

TABLE 2.4
Calculation of r

(1) X	(2) Y	(3) $x_i = X_i - \bar{X}$ $= X_i - 4.61$	(4) $y_i = Y_i - \bar{Y}$ $= Y_{i-85.27}$	(5) x_i^2	(6) y_i^2	(7) $x_i y_i$
4.5	89.5	−0.11	4.23	0.0121	17.8929	−0.4653
3.8	91.1	−0.81	5.83	0.6561	33.9889	−4.7223
3.8	86.9	−0.81	1.63	0.6561	2.6569	−1.3023
3.6	87.0	−1.01	1.73	1.0201	2.9929	−1.7473
3.5	86.2	−1.11	0.93	1.2321	0.8649	−1.0323
4.9	79.2	0.29	−6.07	0.0841	36.0840	−1.7603
5.9	78.0	1.29	−7.27	1.6641	52.8529	−9.3783
5.6	83.1	0.99	−2.17	0.9801	4.7089	−2.1483
4.9	87.5	0.29	2.23	0.0841	4.9729	0.6467
5.6	84.2	0.99	−1.07	0.9801	1.1449	−1.0593
$\Sigma X_i = 46.1$	$\Sigma Y_i = 852.7$	$\Sigma x_i = 0$	$\Sigma y_i = 0$	$\Sigma x_i^2 =$ 7.3690	$\Sigma y_i^2 =$ 157.5301	$\Sigma x_i y_i =$ −22.9690

Inferences about the Population Correlation Coefficient ρ

The population correlation coefficient ρ may be estimated by R from a set of observations taken from any joint bivariate distribution relating the random variables X and Y. However, if we desire to test hypotheses about ρ or to determine confidence intervals for this parameter, then we must strengthen our assumptions concerning the bivariate population at hand. In particular, we shall assume that X and Y follow a *joint bivariate normal distribution* (whose joint density function is a bell-shaped surface, as in Figure 2.12), the specific properties of which are as follows:

C.1. Both X and Y are random variables with means μ_X and μ_Y, respectively, and possess constant variances σ_X^2 and σ_Y^2, respectively.

C.2. The individual marginal distributions of both X and Y are $N(\mu_X, \sigma_X)$ and $N(\mu_Y, \sigma_Y)$, respectively. (It must be mentioned that, if the individual marginal distributions of two random variables X and Y are normal, it does not necessarily follow that the joint bivariate distribution of X and Y will be normal.)

C.3. The conditional distribution of Y given X is $N\left(\mu_X + \rho\sigma_X\tilde{Y}, \sigma_X\sqrt{1-\rho^2}\right)$, whereas the conditional distribution of X given Y is $N\left(\mu_Y + \rho\sigma_Y\tilde{X}, \sigma_Y\sqrt{1-\rho^2}\right)$, where $\tilde{X} = (X - \mu_X)/\sigma_X$ and $\tilde{Y} = (Y - \mu_Y)/\sigma_Y$.

C.4. The relationship between X and Y is strictly linear and is summarized by the correlation coefficient ρ (we will return to a formal interpretation of C.3 and C.4 when Case B is considered below).

Thus, the joint bivariate normal distribution has five parameters; it is completely specified once we know the means and standard deviations of X and Y as well as ρ.

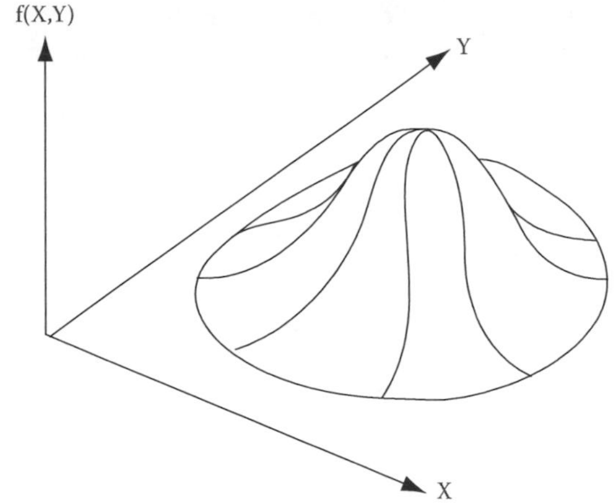

FIGURE 2.12.
Joint birariate normal density function.

We noted above that, if X and Y are independent, then $\rho = 0$. It was also mentioned that, if X and Y follow a joint bivariate distribution and $\rho = 0$, then we cannot generally conclude that X and Y are independent. Interestingly enough, for a joint bivariate normal distribution, independence implies $\rho = 0$ and conversely, i.e., zero covariance is equivalent to independence.

Given that X and Y follow a joint bivariate normal distribution, it follows that, under $H_0 : \rho = 0$,

$$R : N\left(E(R), \sqrt{V(R)}\right) = N\left(0, \sqrt{\frac{1-\rho^2}{n-2}}\right).$$

Then

$$\frac{R - E(R)}{S(R)} = \frac{R\sqrt{n-2}}{\sqrt{1-\rho^2}} : N(0,1). \tag{2.39}$$

If R is used to estimate ρ in Equation 2.39, then the resulting quantity

$$T_R = \frac{R\sqrt{n-2}}{\sqrt{1-R^2}} : t_{n-2}. \tag{2.39.1}$$

We noted above that *lack of linear association* is equivalent to independence if X and Y follow a joint bivariate normal distribution. Hence, testing lack of linear association is equivalent to testing the independence of X and Y. In this regard, we may test the null hypothesis $H_0 : \rho = \rho_0 = 0$ against any of the following alternative hypotheses:

Case I	Case II	Case III
$H_0 : \rho = 0$	$H_0 : \rho = 0$	$H_0 : \rho = 0$
$H_1 : \rho > 0$	$H_1 : \rho < 0$	$H_1 : \rho \neq 0$

The corresponding critical regions are determined in the usual manner, i.e., at the 100α% level, our set of decision rules for rejecting H_0 relative to H_1 is as follows:

a. Case I – reject H_0 if $t_R > t_{\alpha, n-2}$;
b. Case II – reject H_0 if $t_R < -t_{\alpha, n-2}$;
c. Case III – reject H_0 if $|t_R| > t_{\alpha/2, n-2}$,

where t_R is the sample realization of T_R.

The test statistic presented in Equation 2.39.1 is valid only under $H_0 : \rho = \rho_0 = 0$. It cannot be used for testing any other hypothesis concerning ρ. This is because, when $\rho_0 \neq 0$, the sampling distribution of R is highly skewed. To circumvent this problem, let us construct an expression involving R, which may be used to transform R into a test statistic that is similar to Equation 2.39. To this end, we may note that, for moderately large samples ($n \geq 20$) taken from a joint bivariate normal population, the function (known as *Fisher's* ξ *transformation*)

$$\xi = \frac{1}{2}\log_e\left(\frac{1+R}{1-R}\right) \tag{2.40}$$

is approximately

$$N\left(E(\xi), \sqrt{V(\xi)}\right) = N\left(\frac{1}{2}\log_e\left(\frac{1+\rho}{1-\rho}\right), \sqrt{\frac{1}{n-3}}\right) \tag{2.41}$$

for any admissible value of ρ. Hence, the quantity

$$Z = \frac{\xi - E(\xi)}{S(\xi)} \tag{2.41.1}$$

is approximately $N(0, 1)$ and thus, for $H_0 : \rho = \rho_0 \neq 0$ (where ρ_0 is some specific or anticipated nonzero value of ρ),

$$\xi_0 = \frac{1}{2}\log_e\left(\frac{1+\rho_0}{1-\rho_0}\right)$$

so that we can use the test statistic

$$Z_\xi = \frac{\xi - E(\xi)}{S(\xi)} = \frac{\frac{1}{2}\log_e\left(\frac{1+R}{1-R}\right) - \frac{1}{2}\log_e\left(\frac{1+\rho_0}{1-\rho_0}\right)}{1/\sqrt{n-3}}. \tag{2.42}$$

Again we may test the null hypothesis $H_0 : \rho = \rho_0 = 0$ against any of the following alternatives:

Case I	Case II	Case III
$H_0 : \rho = \rho_0$	$H_0 : \rho = \rho_0$	$H_0 : \rho = \rho_0$
$H_1 : \rho > \rho_0$	$H_1 : \rho < \rho_0$	$H_1 : \rho \neq \rho_0$

Then for a test conducted at the 100α% level, our decision rules for rejecting H_0 in favor of H_1 are as follows:

a. Case I – reject H_0 if $z_\xi > z_\alpha$;
b. Case II – reject H_0 if $z_\xi < -z_\alpha$;
c. Case III – reject H_0 if $|z_\xi| \geq z_{\alpha/2}$,

where z_ξ is the sample realization of Z_ξ. Note that, as far as this general test procedure is concerned (it will also be valid when we construct confidence intervals for ρ later on), because ξ is a one-to-one and onto mapping (for each R value, there exists a unique ξ value and conversely), we can convert an R value into a ξ value, make an inference about ξ, and then transform the inference concerning ξ back into an inference about ρ.

Example 2.15

Given the data appearing in Table 2.4, does our estimate for ρ, $r = -0.6741$, lie significantly below zero at the 5% level? In addition, are we safe in concluding that $\rho = -0.7$ at this same level? Under $H_0 : \rho = 0$, let $H_1 : \rho < 0$. Then from Equation 2.39.1, because

$$t_R = \frac{-0.6741\sqrt{8}}{\sqrt{1-0.4544}} = -2.5814 \in \mathcal{R} = \left\{t_R / t_R < -t_{0.05,8} = -1.86\right\},$$

we may reject H_0 in favor of H_1, i.e., at the 5% level the population correlation coefficient lies significantly below zero. Next, let us test $H_0 : \rho = \rho_0 = -0.7$ against $H_1 : \rho \neq \rho_0 = -0.7$. Because

$$\xi = \frac{1}{2}\log_e\left(\frac{1-0.6741}{1+0.6741}\right) = -0.8180$$

and

$$\xi_0 = \frac{1}{2}\log_e\left(\frac{1-0.7}{1+0.7}\right) = -0.8673$$

(see Table A.11 in Appendix), we have, from Equation 2.42,

$$|z_\xi| = \left|\frac{-0.8180 + 0.8673}{1/\sqrt{7}}\right| = 0.1305 \notin \mathcal{R} = \left\{z_\xi / |z_\xi| > 1.96\right\}.$$

Thus, we cannot reject H_0 in favor of H_1; at the 5% level, ρ is not significantly different from −0.7.

To determine just how precisely the population correlation coefficient has been estimated from the sample data, let us construct the probability statement

$$P\left(-z_{\alpha/2} \le \frac{\xi - E(\xi)}{S(\xi)} \le z_{\alpha/2}\right) = 1 - \alpha.$$

After rearranging this expression and passing to sample realizations, we have the confidence statement

$$P\left(\frac{1}{2}\log_e\left(\frac{1+R}{1-R}\right) - \frac{z_{\alpha/2}}{\sqrt{n-3}} \le \frac{1}{2}\log_e\left(\frac{1+\rho}{1-\rho}\right) \le \frac{1}{2}\log_e\left(\frac{1+R}{1-R}\right) + \frac{z_{\alpha/2}}{\sqrt{n-3}}\right) = 1 - \alpha.$$

Hence, a 100(1 − α)% confidence interval for the parameter $\frac{1}{2}\log_e\left(\frac{1+\rho}{1-\rho}\right)$ is

$$\left(\frac{1}{2}\log_e\left(\frac{1+R}{1-R}\right) - \frac{z_{\alpha/2}}{\sqrt{n-3}}, \frac{1}{2}\log_e\left(\frac{1+R}{1-R}\right) + \frac{z_{\alpha/2}}{\sqrt{n-3}}\right). \tag{2.43}$$

If the lower and upper confidence limits indicated in Equation 2.43 are denoted as l_ξ and u_ξ, respectively, then these quantities can be used to determine the lower and upper confidence limits l_ρ and u_ρ, respectively, for the parameter ρ by reversing the ξ transform, i.e., after setting

$$l_\xi = \frac{1}{2}\log_e\left(\frac{1+l_\rho}{1-l_\rho}\right), u_\xi = \frac{1}{2}\log_e\left(\frac{1+u_\rho}{1-u_\rho}\right), \tag{2.44}$$

we may read Table A.11 in reverse to find a 100(1 − α)% confidence interval for ρ. That is, if we find l_ξ (respectively, u_ξ) in the body of this table, we can easily determine the corresponding l_ρ (respectively, u_ρ) value so that the 100(1 − α)% confidence interval for ρ is (l_ρ, u_ρ).

Example 2.16

Using the data presented in Table 2.4, find a 95% confidence interval for the true or population correlation coefficient ρ given that $r = -0.6741$. Using Equation 2.43, it is easily seen that a 95% confidence interval for $\frac{1}{2}\log_e\left(\frac{1+\rho}{1-\rho}\right)$ is, for $z_{\alpha/2} = z_{0.025} = 1.96$,

$$\left[\frac{1}{2}\log_e\left(\frac{1+0.6741}{1-0.6741}\right) - \frac{1.96}{\sqrt{7}}, \frac{1}{2}\log_e\left(\frac{1+0.6741}{1-0.6741}\right) + \frac{1.96}{\sqrt{7}}\right]$$

or, using Table A.11, $l_\xi = -0.8180 - 0.7408 = -1.5588$, $u_\xi = -0.8180 - 0.7408 = -1.0772$. Then

$$-1.5588 = \frac{1}{2}\log_e\left(\frac{1+l_\rho}{1-l_\rho}\right),$$

$$-0.0772 = \frac{1}{2}\log_e\left(\frac{1+u_\rho}{1-u_\rho}\right)$$

and thus a 95% confidence interval for ρ has as its lower and upper limits $l_\rho = -0.9150$ and $u_\rho = -0.0770$, respectively (here we find 1.5588 and 0.0772 in the body of Table A.11 and read it in reverse so as to obtain the appropriate correlation coefficient values). Thus, we may be 95% confident that the population correlation coefficient lies between −0.9150 and −0.0770.

Let us now consider the Case B problem. Our preceding discussion concerning ρ specified that the random variables X and Y followed a joint bivariate distribution (Case A). We then strengthened this assumption by introducing the concept of normality. In this regard, let us, for the moment, retain the requirement that X and Y follow a joint bivariate normal distribution. Because both X and Y are variable, we can regress Y on X and X on Y. In the first instance, the X_i's are held fixed (Y is the dependent variable), whereas in the second, the Y_i's are taken to be fixed (X is now the dependent variable). That is, we want the mean of Y given X as well as the mean of X given Y. Hence, the implied population regression equations are, respectively,

$$\begin{aligned}&\text{(a) } E(Y/X)=\beta_0+\beta_1 X,\\ &\text{(b) } E(X/Y)=\bar{\beta}_0+\bar{\beta}_1 Y.\end{aligned} \tag{2.45}$$

To see this, let us refer to C.3 and C.4 above. As indicated therein, we may form, for instance,

$$\begin{aligned}E(Y/X)&=\mu_Y+\rho\sigma_Y\tilde{X}=\mu_Y+\rho\sigma_Y\left(\frac{X-\mu_X}{\sigma_X}\right)\\ &=\left[\mu_Y-\rho\sigma_Y\frac{\mu_X}{\sigma_X}\right]+\rho\frac{\sigma_Y}{\sigma_X}X\end{aligned} \tag{2.46}$$

(see above, "The Prediction of a Particular Value of Y Given X"). In this regard, if X and Y follow a joint bivariate normal distribution, then ρ^2 can be used as a measure of the goodness of fit of the population regression line to the data points(X_i, Y_i), $i=1,\ldots,n$, in that it indicates the proportion of the variation in Y explained by the linear influence of X; it also serves as a measure of the covariability between the random variables X and Y. This same conclusion is valid if the assumption of normality is dropped and the random variables X and Y simply follow a joint bivariate distribution. Moreover, if the assumptions of the strong classical linear regression model hold so that only Y is a random variable and X is held fixed, then, because $COV(X,Y)$ does not exist, ρ^2 (or ρ) cannot serve as a measure of covariability; it is only indicative of the goodness of fit of the population regression line to the scatter of data points. In this regard, if the strong classical linear regression model is applied and we have obtained R^2 according to $SSR/SST=\hat{\beta}_1\Sigma x_i y_i/\Sigma x_i^2$ (see Table 2.3), then we also have obtained R^2 as an estimate of ρ^2. However, R does not serve as an estimate of ρ in Equation 2.37 because the latter is, strictly speaking, undefined.

Example 2.17

Let us return to the example problem set out above when we discussed the regression model, i.e., we regressed gross sales (Y) on advertising outlay (X). As far as this particular estimation procedure is concerned, X is not a random variable (the X_i values are assumed fixed), and thus the preceding joint bivariate (normal) model is not appropriate. This being the case, R cannot be used as an estimate of ρ. However, let us offer a slightly different data gathering process. That is, we shall not predetermine the X_i values but simply treat the problem as one involving the selection of a random sample of n data points (X_i, Y_i) and record whatever X_i and Y_i sample values emerge. Then, if all of the assumptions underlying the joint bivariate normal model are approximately met, we can calculate R and interpret it as a measure of covariability. In this regard, because $\sqrt{R^2}=\pm R$, the sign actually chosen corresponds to the sign of $\hat{\beta}_1$. Then from Table 2.1 and Equation 2.35.a, $r=0.9831$. Clearly, there exists a high degree of positive association between the variables X and Y.

Are regression and correlation analysis in any way related? The answer is, yes. If we have a sample of n data points (X_i, Y_i), $i=1,\ldots,n$, then, from Equation 2.46, $\hat{\beta}_1 = R\left(S_y/S_X\right)$. Under $H_0: \beta_1 = 0$ we previously used as a test statistic $T = \hat{\beta}_1 / S_{\hat{\beta}_1}$. If the preceding expression for $\hat{\beta}_1$ is substituted into this T value, we obtain

$$T = \frac{\hat{\beta}_1}{S_{\hat{\beta}_1}} = \frac{R\left(S_Y/S_X\right)}{S_\varepsilon / \sqrt{\sum x_i^2}} = \frac{R\left(S_Y/S_X\right)S_X\sqrt{n-1}\sqrt{n-2}}{\sqrt{\sum y_i^2 - \hat{\beta}_1 \sum x_i y_i}}$$

$$= \frac{R\left(S_Y/S_X\right)S_X\sqrt{n-1}\sqrt{n-2}}{\sqrt{(n-1)S_Y^2 - R\left(S_Y/S_X\right)S_X S_Y R(n-1)}} = \frac{R\sqrt{n-2}}{\sqrt{1-R^2}}.$$

But this is the T value obtained above under $H_0: \rho = 0$. Moreover, we previously determined (see above, "The Prediction of a Particular Value of Y Given X") that $F = \hat{\beta}_1 \sum x_i y_i / S_\varepsilon^2$. If $\hat{\beta}_1 = \hat{\rho}\left(S_Y/S_X\right)$ is substituted into this expression, we have

$$F = \frac{\hat{\beta}_1 \sum x_i y_i}{S_\varepsilon^2} = \frac{R\left(S_Y/S_X\right)S_X S_Y R(n-1)(n-2)}{(n-1)S_Y^2 - R\left(S_Y/S_X\right)S_X S_Y R(n-1)} = \frac{R^2(n-2)}{1-R^2},$$

which is the square of the preceding T value. Thus, the regression and correlation t tests and the analysis-of-variance F test are all equivalent ways of testing for a significant linear relationship between X and Y, i.e., testing $H_0: \beta_1 = 0$ is equivalent to testing $H_0: \rho = 0$. In this regard, for the bivariate linear model, if either the t or F test leads us to reject one of the aforementioned null hypotheses, then the other must be rejected also. Hence, we need not perform all three significance tests; only one of them is necessary and any one will suffice.

Regression through the Origin

In this section, we shall consider the special instance when the regression line is known to go through the origin, i.e., when we know a priori that in the population $Y = 0$ when $X = 0$. (Examples of this case arise when we seek to estimate, for instance, a total variable cost function or a total revenue function.) Let us apply the above strong classical linear regression model with $\beta_0 = 0$. Hence, the true population regression equation is of the form $Y_i = E(Y_i/X_i) + \varepsilon_i = \beta_1 X_i + \varepsilon_i$. When the value of β_1 is estimated on the basis of sample information, we obtain the sample regression line $\hat{Y} = \hat{\beta}_1 X$. As before, let the ith residual be $e_i = Y_i - \hat{Y}_i$. If we again apply the technique of least squares, we seek to $\min\left\{\sum e_i^2 = \sum\left(Y_i - \hat{\beta}_1 X_i\right)^2 = F\left(\hat{\beta}_1\right)\right\}$. After setting $\partial F/\partial \hat{\beta}_1 = 0$, we obtain $\sum X_i e_i = 0$, from which we may determine

$$\hat{\beta}_1 = \frac{\sum X_i Y_i}{\sum X_i^2} \tag{2.47}$$

as the best linear unbiased estimator of β_1. Clearly, the variables are not expressed in deviation form as was the case above. (Note that, because the equation $\sum e_i = 0$ is absent, the least-squares regression line under the restriction $\beta_0 = 0$ does not necessarily pass through the point of means $(\bar{X}, \bar{Y})$.)

Let us next look to the precision of the estimate of β_1. Because we have estimated only one parameter in this model, an unbiased estimator of σ_ε^2 is

$$\tilde{S}^2 = \frac{\sum e_i^2}{n-1} = \frac{\sum\left(Y_i - \hat{\beta}_1 X_i\right)^2}{n-1} = \frac{\sum Y_i^2 - \hat{\beta}_1 \sum X_i Y_i}{n-1}.$$

Then the estimated standard deviation of $\hat{\beta}_1$ is

$$\tilde{s}_{\hat{\beta}_1} = \frac{\tilde{s}}{\sqrt{\sum X_i^2}}. \tag{2.48}$$

In view of Equation 2.47, a 100(1 – α)% confidence interval for β_1 is $\hat{\beta}_1 \pm t_{\alpha/2,n-1}\tilde{s}_{\hat{\beta}_1}$.

Next, to test the hypothesis of no linear relationship between X and Y, let us form $H_0: \beta_1 = 0$. Then the appropriate test statistic is $T = \hat{\beta}_1 / \tilde{s}_{\hat{\beta}_1}$. As always, the critical region is determined by the alternative hypothesis made (see above, "Precision of the Least-Squares Estimators $\hat{\beta}_0$, $\hat{\beta}_1$: Confidence Intervals").

If we desire to predict the average value of Y_i given X_i, then we can determine just how precisely we have estimated the population regression line from the sample data by constructing a confidence interval for $E(Y_i/X_i) = \beta_1 X_i$. That is, we can be 100(1 – α)% confident that the true average value of Y_i given X_i lies between $\hat{Y}_i \pm t_{\alpha/2,n-1}\tilde{s}_{\hat{Y}_i}$, where

$$\tilde{s}_{\hat{Y}_i} = \tilde{s}\sqrt{\frac{X_i^2}{\sum X_i^2}} \tag{2.49}$$

(Note that, as X_i varies over the range of X, we may construct a 100(1 – α)% confidence band for the population regression line.) To conduct a test of the hypothesis that $E(Y_i/X_i)$ equals some specific value E_i^0, i.e., $H_0 : E(Y_i/X_i) = E_i^0$, the test statistic $T = \left(\hat{Y}_i - E_i^0\right)/\tilde{S}_{\hat{Y}_i}\ t_{n-1}$.

We next turn to the problem of predicting a particular Y_i value from a given X_i. That is, for $X = X_0$, we seek to determine $Y_0 = \beta_1 X_0 + \varepsilon_0$. An estimate of this quantity obtained from the sample regression line is $\hat{Y}_0 = \hat{\beta}_1 X_0$. To determine just how precisely Y_0 has been estimated, let us construct a confidence interval for Y_0, i.e., we may be 100(1 – α)% confident that the true Y_0 value lies between $\hat{Y}_0 \pm t_{\alpha/2,n-1}\tilde{s}_{(Y_0-\hat{Y}_0)}$, where

$$\tilde{s}_{(Y_0-\hat{Y}_0)} = \tilde{s}\sqrt{1 + \frac{X_0^2}{\sum X_i^2}} \tag{2.50}$$

TABLE 2.5

Partitioned Sums of Squares

Source of Variation	Sum of Squares	Degrees of Freedom	MS = SS/d.f.
Regression (explained variation in Y)	(SSR) $\hat{\beta}_1 \sum X_i Y_i$	1	(Regression MS) MSR = SSR/1 = $\hat{\beta}_1 \sum X_i Y_i / 1$
Error (unexplained variation in Y)	(SSE) $\sum Y_i^2 - \hat{\beta}_1 \sum X_i Y_i$	$n-1$	(Error MS) MSE = SSE/$(n-1)$ = $\left(\sum Y_i^2 - \hat{\beta}_1 \sum X_i Y_i\right) / (n-1)$
Total variation in Y	(SST) ΣY_i^2	n	R^2 = SSR/SST; $F = \frac{\text{MSR}}{\text{MSE}} =$ $[SSR/1]/[SSE/(n-1)]$

To conduct a test of the hypothesis that Y_0 equals some specific value Y_0^0, i.e., $H_0 : Y_0 = Y_0^0$, the appropriate test statistic $T = \left(\hat{Y}_0 - Y_0^0\right) / \tilde{S}_{(Y_0 - \hat{Y}_0)} : t_{n-1}$.

Next, the partitioned sum-of-squares (analysis-of-variance) table for this regression model appears as Table 2.5. An F test for the significance of this regression model can then be conducted in the usual manner. (Finally, for an actual test of $H_0 : \beta_0 = 0$, see above, "Testing Hypotheses Concerning β_0, β_1".)

Reverse Prediction

In some instances, it may be necessary to predict the value X_i of the independent variable X that is expected to occur in the population at a specified level Y_i of the dependent variable Y. Given that the estimated value of Y at X_i is $\hat{Y}_i = \hat{\beta}_0 + \hat{\beta}_1 X_i$, we may rearrange this equation for reverse prediction purposes and obtain

$$\hat{X}_i = \frac{Y_i - \hat{\beta}_0}{\hat{\beta}_1}.$$

If we start with Equation 2.29, i.e., with $\hat{Y}_0 \pm t_{\alpha/2, n-2} s_{(Y_0 - \hat{Y}_0)}$ or

$$\hat{Y}_i = \bar{Y} + \hat{\beta}_1 \left(X_i - \bar{X}\right) \pm t_{\alpha/2, n-2} s_{\varepsilon} \sqrt{1 + \frac{1}{n} + \frac{\left(X_i - \bar{X}\right)^2}{\sum x_i^2}}$$

(here $\hat{\beta}_0 = \bar{Y} - \hat{\beta}_1 \bar{X}$) and rearrange this expression, we ultimately obtain

$$\hat{X}_i = \bar{X} + \frac{\hat{\beta}_1 \left(Y_i - \bar{Y}\right)}{k} \pm \frac{t_{\alpha/2, n-2} s_{\varepsilon}}{k} \sqrt{k\left(1 + \frac{1}{n}\right) + \frac{\left(Y_i - \bar{Y}\right)^2}{\sum x_i^2}}, \tag{2.51}$$

where $k = \hat{\beta}_1^2 - t^2_{\alpha/2,n-2}s^2_{\hat{\beta}_1}$. Note that, although the confidence interval for predicting Y_i for a given level of X_i is symmetrical about $\hat{Y}_i$, the confidence interval for the predicted X_i value, $\hat{X}_i$, at Y_i is symmetrical about $\bar{X} + \left(\hat{\beta}_1\left(Y_i - \bar{Y}\right)/k\right)$, which is close to $\hat{X}_i$.

Example 2.18

From the gross sales/advertising outlay data of Table 2.1, if a gross sales level of 21 (millions of dollars) is the target for 2007, what can we infer about the advertising outlay needed to generate this level of sales? From Equation 2.51, at the 95% level, we have

$$\hat{X}_i = 7.210 + \frac{1.4365(21 - 12.2950)}{2.0162} \pm \frac{(2.306)(0.7439)}{2.0162}\sqrt{2.0162\left(1 + \frac{1}{10}\right) + \frac{(21 - 12.2950)^2}{61.7490}},$$

where $k = (1.4365)^2 - (2.306)^2\,(0.0946)^2 = 2.0162$ and $t_{0.025,8} = 2.306$. After simplifying the preceding expression, we obtain 13.4120 ± 1.5791. Hence, we may be 95% confident that the predicted (and thus required) value of advertising outlay X given sales of Y = \$21 million lies between \$11.8329 million and \$14.9911 million.

Assessing the Appropriateness of the Regression Model

We previously made the assumption that the linear model in Equation 2.3 was appropriate for describing the relationship between the variables X and Y. As we shall soon see, this assumption can actually serve as a hypothesis to be tested against the data. If the hypothesis is rejected, then an alternative functional form can be chosen to represent the relationship between X and Y. As a first step, however, let us determine whether X and Y satisfy any functional relationship. Hence, the initial hypothesis to be tested is that X and Y are *functionally related*. If we accept this hypothesis, then we can next test the follow-up hypothesis that X and Y are *linearly related*.

Testing for a Functional Relationship between *X* and *Y*

One of the requirements that must be met for these test procedures to be performed is that most (but typically all) of the X_i's have replicate Y observations associated with them. (For instance, an X value of 13 may have, say, four replicates, 8, 10, 12, and 14.) In this regard, let us assume that each of the p-values of the variable X has a set of replicates associated with it (see Table 2.6). For $X = X_i$, the replicates on Y are $Y_{i1}, Y_{i2}, \ldots, Y_{in_i}$, where n_i (≥ 2), $i = 1, \ldots, p$, is the number of replicates associated with X_i. Also indicated in Table 2.6 (last column) are the conditional (on X_i) means of the Y_{ij} replicates:

$$\bar{Y}_i = \frac{1}{n_i}\sum_{j=1}^{n_i} Y_{ij},\ i = 1, \ldots, p.$$

Moreover, the grand or overall mean of the Y_{ij}'s is $\bar{Y} = \frac{1}{n}\sum_{i=1}^{p} n_i\bar{Y}_i$, where $n = \sum_{i=1}^{p} n_i$. (If each X_i has the same number m of replicates, then $\bar{Y} = \sum_{i=1}^{p}\bar{Y}_i/p$.)

TABLE 2.6
Y Replicates for Each Value of X

X	Y_{ij}/X_i Replicates, $j = 1, \ldots, n_i$	Conditional (on X_i) Mean of Replicates $\bar{Y}_i$
X_1	$Y_{11}, Y_{12}, \ldots, Y_{1n_1}$	$\bar{Y}_1$
X_2	$Y_{21}, Y_{22}, \ldots, Y_{2n_2}$	$\bar{Y}_2$
$\vdots$	....	$\vdots$
X_p	$Y_{p1}, Y_{p2}, \ldots, Y_{pn_p}$	$\bar{Y}_p$

Let us assume that the dataset in Table 2.6 has been generated by a model of the form $Y_{ij}=\mu_i+\varepsilon_{ij}$, where the population conditional mean $\mu_i=E(Y_i/X_i)=f(X_i)$ is specified as a function f of X_i (assumed fixed) and the random error term ε_{ij} is assumed to satisfy a set of assumptions similar to Assumptions A.1 through A.4 appearing above in "The Strong Classical Linear Regression Model", i.e., ε_{ij} is a normally distributed independent random variable with a zero mean and a constant variance σ^2. Then for any X_i, $\bar{Y}_i = \mu_i + \varepsilon_i$ with $E(\varepsilon_i) = 0$ and thus $E(\bar{Y}_i) = \mu_i$ $\bar{Y}_i$ is an unbiased estimator of μ_i), whereas $V(\varepsilon_i) = V(\bar{Y}_i) = \sigma_i^2/n_i = \sigma_\varepsilon^2/n_i$. In addition, $E(\bar{Y}) = \mu$ so that $\bar{Y}_i$ is an unbiased estimator of the overall population mean μ.

To set the stage for conducting a test of whether X and Y satisfy any functional (regressional) relationship, let us borrow some terminology from the area of (one-way) analysis of variance and express the total deviation of the coordinate of any sample point (X_i, Y_{ij}) from the overall mean $\bar{Y}_i$ as

$$\left(Y_{ij} - \bar{Y}\right) = \left(\bar{Y}_i - \bar{Y}\right) + \left(Y_{ij} - \bar{Y}_i\right). \tag{2.52}$$

[total deviation] [explained deviation] [unexplained deviation]

Given that the sample conditional mean $\bar{Y}_i$ is an unbiased estimator of the population conditional mean μ_i, the first term on the right side of Equation 2.52 can be termed *the explained* (or *between-group*) *deviation* and the second term the *unexplained* (or *within-group*) *deviation* (Figure 2.13).

Because the Y_{ij}'s for a given i have the same constant variance σ_ε^2, the sole source of variation between the p classes is in the means μ_i, $i = 1, \ldots, p$. Moreover, this variation is taken to be completely general in nature and not restricted to any particular functional form.

If we square both sides of Equation 2.52 and then sum over all i and j values, we obtain

$$\sum_{i=1}^{p}\sum_{j=1}^{n_i}\left(Y_{ij} - \bar{Y}\right)^2 = \sum_{i=1}^{p} n_i\left(\bar{Y}_i - \bar{Y}\right)^2 + \sum_{i=1}^{p}\sum_{j=1}^{n_i}\left(Y_{ij} - \bar{Y}_i\right)^2 \tag{2.53}$$

[total variation or total sum of squares (SST)] [explained variation or explained sum of squares (SSR')] [unexplained variation or unexplained sum of squares (SSE')]

Given that the cross-product terms in Equation 2.53 are zero, it is evident that the total sum of squares can be partitioned or decomposed into two parts: (1) an explained (by group

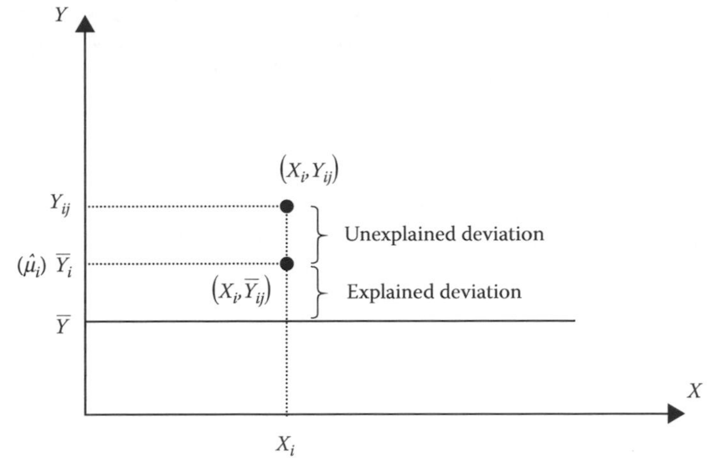

FIGURE 2.13
Total deviation in Y.

means) sum of squares due to between-class variation in group means about the overall mean (explained SS) and (2) an unexplained (by group means) or error sum of squares attributed to the within-class variation in group-values about their means (unexplained SS).

Once the components of Equation 2.53 have been obtained, we may compute the *determination ratio*

$$H^2 = \frac{SSR'}{SST} = \frac{\text{explained (regressional) SS}}{\text{total SS}} \tag{2.54}$$

or, because SSR′ = SST–SSE′,

$$H^2 = 1 - \frac{SSE'}{SST} = 1 - \frac{\text{unexplained SS}}{\text{total SS}}, \tag{2.54.1}$$

with its sample realization denoted as h^2. Here Equation 2.54.1 depicts the estimated *degree of regressional (functional) relationship* between X and Y. It represents the proportion of the variance in Y, which is accounted for by X. (The determination ratio is only similar to the coefficient of determination R^2 and should not be interpreted in the same manner, i.e., SSR′ is based on variations in group means about $\bar{Y}_i$, whereas SSR in R^2 depends on variations in estimated Y values about $\bar{Y}_i$.) If we next take the positive square root of Equation 2.54, then we obtain the *correlation ratio*

$$H = \sqrt{\frac{\text{SSR}'}{\text{SST}}}. \tag{2.55}$$

Unlike the sample correlation coefficient R, which ranges between +1 and –1, the estimated correlation ratio is restricted to the unit interval or $0 \le H \le 1$. That is, if there is no regressional relationship between X and Y (here $\bar{Y}_i = \bar{Y}$ for all i so that explained variation by group means is zero), then $H = 0$, and, if X and Y are strictly functionally related (total variation is explained completely by the between-class variation in group means), then $H = 1$. Moreover, because any general functional relationship connecting X and Y may exist, H does not reveal whether X and Y vary directly or inversely. (This is not true of R itself because it depicts the degree of linear association between X and Y.) Indeed, the regressional relationship may possess an interval of X values over which the variables are positively associated and an interval of X's over which X and Y are negatively associated.

To determine whether X and Y are functionally related, let us test $H_0 : \eta = 0$ (there is no functional relationship between X and Y) against $H_0 : \eta \neq 0$ (there is a functional relationship between X and Y), where η is the population correlation ratio. Note that this null hypothesis is equivalent to $H_0 : \mu_1 = \mu_2 =\ldots= \mu_p = \mu$, i.e., between-class variation in group means is zero. Under H_0, the test statistic

$$F = \frac{\text{regressional mean square}}{\text{error mean square}} = \frac{SSR'/(p-1)}{SSE'/(n-p)} = \frac{H^2/(p-1)}{(1-H^2)/(n-p)} : F_{p-1,n-p}, \tag{2.56}$$

where the critical region $\mathcal{R} = \{F / F > F_{\alpha,p-1,n-p}\}$. Here we are performing a one-tail test using the upper tail of the F distribution so that H_0 is rejected for large realized values of F because, if H_0 is false, the value of SSR′ contains a component of variation in excess of that attributed to between-class variation by group means. This discussion may be summarized by constructing the following partitioned (by group means) sum-of-squares table (Table 2.7).

How may the various sums of squares presented in Table 2.7 be computed? We may first find

$$SST = \sum_{i=1}^{p}\sum_{j=1}^{n_i}\left(Y_{ij} - \bar{Y}\right)^2 = \sum_{i=1}^{p}\sum_{j=1}^{n_i} Y_{ij}^2 - n\bar{Y}^2 \tag{2.57}$$

along with

$$SSR' = \sum_{i=1}^{p} n_i\left(\bar{Y}_i - \bar{Y}\right)^2 = \sum_{i=1}^{p} n_i\bar{Y}_i^2 - n\bar{Y}^2 \tag{2.58}$$

and then determine SSE′ as the residual

$$SSE' = SST - SSR'. \tag{2.59}$$

TABLE 2.7

Summary Table for the Partitioned (by Group Means) Sums of Squares

Source of Variation	Sums of Squares	Degrees of Freedom	MS = SS/d.f.
Regressional (explained by group means)	(SSR′) $\sum n_i\bar{Y}_i^2 - n\bar{Y}^2$	$p-1$	(Regressional MS) SSR′/(p-1)
Error (unexplained variation in Y)	(SSE′) SST-SSR′	$n-p$	(Error MS) (SST - SSR′)/(n-p)
Total variation in Y	(SST)	$n-1$	H^2 = SSR′/SST; $F = \frac{\text{Regressional MS}}{\text{Error MS}}$ $= [SSR'/(p-1)]/[SSE'/(n-p)]$

Example 2.19

Using the data presented in Table 2.8, let us determine whether or not a functional relationship exists between the variables X and Y. If for each of the X_i values we arrange the multiple observations on Y in a manner similar to that appearing in Table 2.6, we obtain column two of Table 2.8. Then from Equations 2.57 through 2.59, respectively,

$$\text{SST} = 11{,}166 - 35\,(16.5142)^2 = 1{,}620.8419,$$
$$\text{SSR}' = 10{,}462.74 - 35\,(16.5142)^2 = 917.5819,$$
$$\text{SSE}' = \text{SST} - \text{SSR}' = 703.26.$$

Based on this information, the partitioned sum-of-squares table appears as Table 2.9.

Because $h^2 = 0.566$, we see that about 57% of the variation in Y is accounted for by X. Additionally, we may answer the above question of whether or not a regressional relationship exists between X and Y by testing, at the $\alpha = 0.05$ level, H_0 (no functional relationship between X and Y) against H_1 (there is a functional relationship between X and Y). In this regard, because the realized F = regression MS/error MS = $5.032 \in \mathcal{R} = \left\{F/F > F_{0.05,7,27} = 2.37\right\}$, we can legitimately conclude that there exists a statistically significant functional expression connecting these variables.

TABLE 2.8

Y Replicates for Each X Value

X	**Y/X Replicates**	$\sum_j Y_{ij}$	$\bar{Y}_i = \sum_j Y_{ij}/n_i$
4	5, 6, 8	19	19/3 = 6.33
8	7, 11	18	18/2 = 9
13	8, 10, 12, 14	44	44/4 = 11
18	12, 13, 16, 18	59	59/4 = 14.75
26	12, 15, 19, 21	67	67/4 = 16.75
31	12, 14, 17, 23, 25	91	91/5 = 18.20
34	13, 15, 17, 20, 23, 26, 29	143	143/7 = 20.42
40	14, 17, 20, 24, 29, 33	137	137/6 = 22.83

TABLE 2.9

Summary Table for Partitioned (by Group Means) Sums of Squares

Source of Variation	**Sum of Squares**	**Degrees of Freedom**	**MS = SS/d.f.**
Regressional (explained by group means)	(SSR') 917.5819	$p - 1 = 8 -\ = 7$	(Regressional MS) $SSR'/(p-1) = 917.5819/7 = 131.0831$
Error (unexplained variation in Y)	(SSE') 703.26	$n - p = 35 - 8 = 27$	(Error MS) $(SST - SSR')/(n - p) = 703.26/27 = 26.0466$
Total variation in Y	(SST) 1,620.8419	$n - 1 = 35 - 1 = 34$	$h^2 = SSR'/SST = 0.566$; F = Regressional MS/Error MS = 5.032

Testing for a Linear Relationship between X and Y

Given that the preceding null hypothesis was rejected, let us now explore the possibility that the implied functional form is linear. In particular, we shall now invoke the linearity assumption and continue our analysis of the relationship between the variables X and Y presented in Table 2.8.

If the linearity assumption is appropriate, then $\mu_i = E(Y_i/X_i) = f(X_i) = \beta_0 + \beta_1 X_i, i=1,\ldots, p$. In this instance, $\hat{Y}_i$ is an estimate of $\bar{Y}_i$ (which in turn is an unbiased estimator of μ_i) and thus the *specification error (lack-of-fit) deviation* $\bar{Y}_i - \hat{Y}_i = 0$ for all i. But what happens if the true population regression equation is nonlinear but we incorrectly specify it as linear? Figure 2.14 provides us with the answer. As can be readily seen therein, $Y_{ij} = \hat{Y}_i + e_i = \hat{Y}_i + (\mu_i - \hat{Y}_i) + \varepsilon_{ij}$.

Clearly, the residual e_i is now forced to do double duty: it must estimate the sum of the *pure* (independent of X_i) *error deviation* ε_{ij} and the *specification error deviation* $\mu_i - \hat{Y}_i$. Hence, there are two possible reasons for a large error sum-of-squares term SSE: (1) there is a considerable amount of pure error variation in the data itself (σ_ε^2 tends to be large), or (2) the variation attributable to specification error or lack of fit is sizeable, thus indicating that the linearity assumption is not appropriate.

The basis of the test of the linearity assumption will be to partition the error sum of squares from the regression of Y on X, SSE, into the sum of a pure error sum of squares (SSE′) and a lack-of-fit sum of squares (SSLF). Our approach will be to first determine SSE′ and then obtain SSLF = SSE − SSE′. Given the set of n_i replicate observations at each X_i, their contribution to the pure error sum of squares is

$$SSE_i' = \sum_{j=1}^{n_i} (Y_{ij} - \bar{Y}_i)^2 = \sum_{j=1}^{n_i} Y_{ij}^2 - n_i \bar{Y}_i^2, i = 1, \ldots, p. \tag{2.60}$$

Then the (pooled) pure error sum of squares is

$$SSE' = \sum_{i=1}^{p} SSE_i'. \tag{2.61}$$

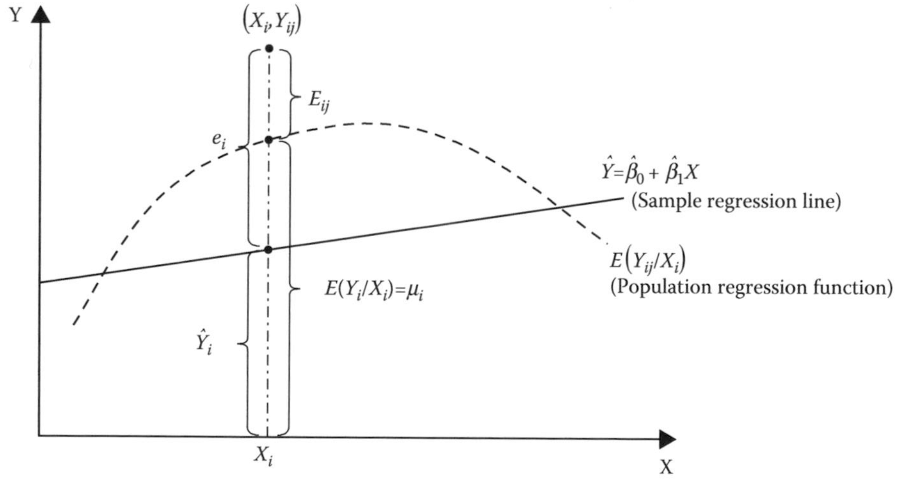

FIGURE 2.14
Specification error deviation.

TABLE 2.10

Summary Table for the Partitioned Error Sum of Squares

Source of Variation	Sum of Squares	Degrees of Freedom	MS = SS/d.f.
Pure error (independent of X_i)	(SSE') $\sum_{i=1}^{p} SSE_i'$	$n-p$	(Pure error MS) $SSE'/(n-p)$
Lack of fit (due to specification error)	(SSLF) SSE − SSE'	$p-2$	(Lack of fit MS) $SSLF'/(p-2)$
Error (total) (unexplained variation in Y)	(SSE) SSE' + SSLF	$n-2$	$F = \frac{\text{lack of fit MS}}{\text{pure error MS}} = \frac{\text{SSLF}/(p-2)}{\text{SSE}'/(n-p)}$

To ascertain whether the linearity assumption is tenable, let us test H_0 (the regression function is linear) against H_1 (the regression function is not linear). Under H_0, the test statistic

$$F = \frac{\text{lack of fit mean square}}{\text{pure error mean square}} = \frac{SSLF/(p-2)}{SSE'/(n-p)} : F_{p-2,n-p}, \tag{2.62}$$

where $n = \sum_{i=1}^{p} n_i$, the lack-of-fit d.f. = error sum-of-squares d.f. − pure error sum-of-squares d.f. = $(n-2)-(n-p) = p-2$, and the critical region $\mathcal{R} = \{F/F > F_{\alpha,p-2,n-p}\}$. Note that the pure error mean square is simply a pooled pure error estimate of σ_ε^2. This one-tail test, which uses the upper tail of the F distribution, rejects H_0 for large realized F values because, if H_0 is false, the value of SSLF is large relative to σ_ε^2, thus indicating that SSLF makes a statistically significant contribution to the error sum-of-squares SSE. This discussion may be summarized by the partitioned error sum-of-squares table (Table 2.10).

Example 2.20

From the data appearing in Table 2.8, we determine whether the linearity assumption underlying the specification of the population regression equation is appropriate. Let us first find SSE'. (This value was determined above as 703.26. However, it is instructive to recompute it in this alternative manner.) Using Equation 2.60, we may determine, say, for $i = 3$ ($X_3 = 13$),

$$SSE'_3 = 504 - 4(11)^2 = 20.$$

The remaining sums of squares are determined in a similar manner (see Table 2.11) so that $SSE' = \sum_{i=1}^{8} SSE_i' = 703.26$. Moreover, if an ordinary least-squares (OLS) regression equation is fit to this dataset, then it can be shown that the resulting line of best fit is

$$\hat{Y} = \hat{\beta}_0 + \hat{\beta}_1 X = 5.4464 + 0.4357X$$

with SSR = 905.388 and SSE = 715.355. Then SSLF = SSE − SSE′ = 12.095. Based on these results, the partitioned error sum-of-squares table becomes Table 2.12.

TABLE 2.11

Calculation of (Pooled) Pure Error Sums of Squares

X	SSE_i'
4	4.7933
8	8.0000
13	20.0000
18	22.7500
26	48.7500
31	126.8000
34	210.1652
40	263.7466
	$SSE' = \sum_{i=1}^{8} SSE_i' = 703.26$

TABLE 2.12

Calculation of Partitioned Error Sums of Squares

Source of Variation	Sum of Squares	Degrees of Freedom	MS = SS/d.f.
Pure error (independent of X_i)	(SSE') 703.26	$n-p=35-8=27$	(Pure error MS) $SSE'/(n-p) = 703.26/27{=}26.046$
Lack of fit (attributable to specification error)	(SSLF) 12.095	$p-2=8-2=6$	(Lack of fit MS) $SSLF/(p-2) = 12.095/6 = 2.0158$
Error (total) (unexplained variation in Y)	(SSE) 715.355	$n-2=33$	$F = \dfrac{\text{Lack of fit MS}}{\text{Pure error MS}} = 0.077$

To determine whether the linearity assumption regarding the population regression function is appropriate, let us test, at the $\alpha = 0.05$ level, H_0 (the regression function is linear) against H_1 (the regression function is not linear). From Table 2.12, because the realized F = lack of fit MS/pure error MS = $0.077 \notin \mathcal{R} = \{F/F > F_{0.05,6,27} = 2.46\}$, we cannot reject the null hypothesis of linearity; we do not have sufficient evidence to warrant abandoning the linearity assumption.

An alternative and extremely illuminating equivalent test of the linearity assumption may be conducted by extending the approach taken above when we tested the more general assumption of a functional relationship between X and Y. As Figure 2.15 reveals, we may rewrite Equation 2.52 by partitioning its explained deviation component into the sum of the linear explained deviation and the specification error or lack-of-fit deviation as

$$\left(Y_{ij}-\bar{Y}\right)=\left(\hat{Y}_i-\bar{Y}\right)+\left(\bar{Y}_i-\hat{Y}_i\right)+\left(Y_{ij}-\bar{Y}_i\right). \tag{2.52.1}$$

[total deviation]	[linear explained deviation]	[specification error deviation]	[unexplained deviation]

We noted above that, if the population regression function is linear, $\hat{Y}_i$ is an estimator for $\bar{Y}_i$, where the latter is an unbiased estimator for μ_i. If a linear specification for the population

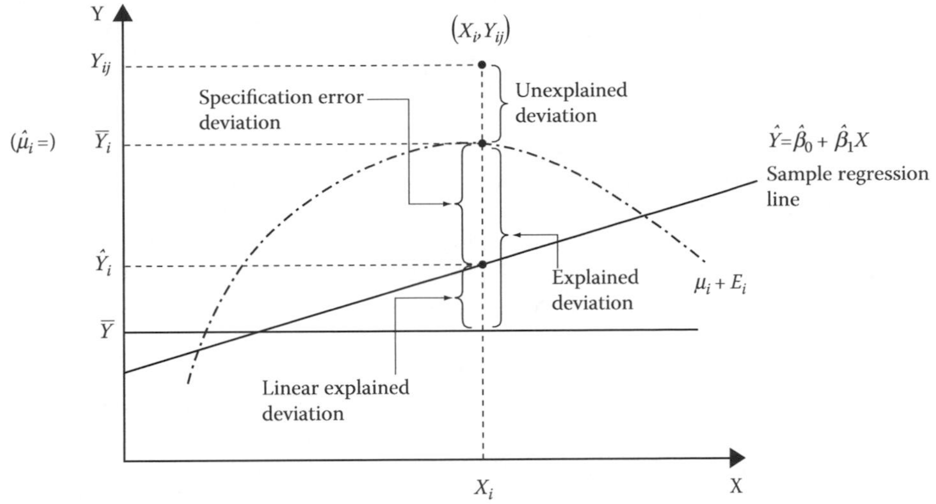

FIGURE 2.15
Specification error deviation.

regression function is erroneous, then the specification error deviation resulting from using $\hat{Y}_i$ to estimate μ_i is $\bar{Y}_i - \hat{Y}_i$. If the said linear specification is correct, linear explained deviation equals explained deviation and $\bar{Y}_i - \hat{Y}_i = 0$.

If we square both sides of Equation 2.52.1 and then sum over all i, j values, we obtain

$$\sum_{i=1}^{p}\sum_{j=1}^{n_i}\left(Y_{ij}-\bar{Y}\right)^2 = \sum_{i=1}^{p} n_i\left(\hat{Y}_i-\bar{Y}\right)^2 + \sum_{i=1}^{p} n_i\left(\bar{Y}_i-\hat{Y}_i\right)^2 + \sum_{i=1}^{p}\sum_{j=1}^{n_i}\left(Y_{ij}-\bar{Y}_i\right)^2. \tag{2.53.1}$$

[total variation or total sum of squares (SST)]	[linear explained variation or linear explained sum of squares (SSR)]	[specification error variation or specification error sum of squares (SSLF)]	[unexplained variation or unexplained sum of squares (SSE')]

Because the cross-product terms in Equation 2.53.1 vanish, it is evident that the total sum of squares can be partitioned into three distinct parts: (1) a linear explained sum of squares attributable to variation in between-group linear regression estimates of Y about the overall mean, (2) a specification error sum of squares attributable to variation of group means about the linear regression line (i.e., a result of differences in group means that are not attributed to linear regression), and (3) an unexplained or error sum of squares attributed to the within-class variation in group-values about their means. In the aggregate, 1 and 2 constitute the between-class variation in group means about the overall mean, i.e., their sum represents the explained SS(SSR′).

From the components of Equation 2.53.1, it is readily seen that the coefficient of determination

$$R^2 = \frac{SSR}{SST} = \frac{\text{linear explained SS}}{\text{total SS}};$$

and the determination ratio

$$H^2 = \frac{\text{SSR} + \text{SSLF}}{\text{SST}} = \frac{\text{linear explained SS + lack of fit SS}}{\text{total SS}}$$

or, in view of the definition of R^2,

$$H^2 = R^2 + \frac{\text{lack of fit SS}}{\text{total SS}} = R^2 + W^2. \tag{2.63}$$

Clearly, $H^2 \geq R^2$ with equality holding only if all group means fall on a linear regression equation, i.e., $W^2 = 0$ so that the population regression equation is linear. Because $W^2 = H^2 - R^2 \geq 0$ is the excess of the proportion of the variation in Y, which is accounted for by X, over the proportion of the variation in Y accounted for by the linear influence of X, it may be thought of as the proportion of the variation in Y explained by the nonlinear influence of X; it is the degree of error resulting from a linear specification of a population regression function that is nonlinear.

The results in this and in the preceding section have indicated that, when a set of replicated Y values are available for each X_i, it is possible to partition SST by using two equivalent specifications, either SST = SSR′ + SSE′ or SST = SSR + SSE, where the former involves an analysis of group means and the latter uses a linear regression approach. The connection between these two representations can readily be seen from the following equation:

Decomposition by group means

SSR′ (over SSR + SSLF)

SST = SSR + SSLF + SSE′

SSE (under SSLF + SSE′)

Decomposition by linear regression

Based on the preceding discussion, we can, as an alternative to Table 2.7, decompose SST by constructing the partitioned (by linear regression) sum-of-squares table (Table 2.13). (Because all of the information required to complete this table has already been provided above in one form or another, the reader can easily fill in the appropriate numerical values.)

Example 2.21

One final set of calculations is warranted. Because r^2 = SSR/SST = 905.388/1,620.743 = 0.5586 and h^2 = SSR′/SST = 917.5819/1,620.8419 = 0.566, it follows that $w^2 = h^2 - r^2 = 0.0074$. Thus, the degree of error resulting from an incorrect (linear) specification of the population regression function is less than 1%.

Comparing Two Linear Regression Equations

Suppose we have two independent datasets with observations (X_{Ai}, Y_{Ai}), $i = 1,\ldots,n_A$, and (X_{Bi}, Y_{Bi}), $i = 1,\ldots,n_B$, and that the sample regression lines

$$\begin{aligned} \hat{Y}_A &= \hat{\beta}_{0A} + \hat{\beta}_{1A} X_A, \\ \hat{Y}_B &= \hat{\beta}_{0B} + \hat{\beta}_{1B} X_B \end{aligned} \tag{2.63}$$

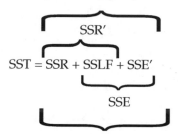

TABLE 2.13

Summary Table for the Partitioned (by Linear Regression) Sum of Squares

Source of Variation	Sum of Squares	Degrees of Freedom	MS = SS/d.f.
Regression (explained variation in Y)	(SSR)	1	(Regression MS) SSR / 1
Error (lack of fit + pure error) (unexplained variation in Y)	(SSE = SSLF + SSE′)	n − 2 = (p − 2) + (n − p)	(Error MS) SSE / (n − 2)
Lack of fit (attributable to specification error)	(SSLF)	p − 2	(Lack of fit MS) SSLF / (p − 2)
Pure error (independent of X)	(SSE′)	n − p	(Pure error MS) SSE′ / (n − p)
Total variation in Y	(SST)	n − 1	$R^2 = SSR/SST$; $H^2 = SSR'/SST = (SSR + SSLF)/SST$ (test for significance of regression slope) $F = \frac{Regression\ MS}{Error\ MS}$ (test for linearity specification) $F = \frac{Lack\ of\ Fit\ MS}{Pure\ Error\ MS} = \frac{W^2/(p-2)}{(1-H^2)/(n-p)}$

are to serve as estimates of the population regression lines

$$E(Y_A/X_A) = \beta_{0A} + \beta_{1A}X_A,$$
$$E(Y_B/X_B) = \beta_{0B} + \beta_{1B}X_B \qquad (2.64)$$

respectively. (It is assumed that the random error terms ε_A and ε_B associated with Y_A and Y_B, respectively, satisfy the assumptions of the strong classical linear regression model.) Once the sample regression values in Equation 2.64 have been obtained (ostensibly by the method of least squares), they may be used to determine whether the population regression equations are similar in terms of their slopes and/or vertical intercepts and correlation coefficients. In what follows, it is assumed that both n_A and n_B are at least 25. Hence, the tests that follow hold for moderately large samples.

Comparing Two Regression Slopes

A comparison of the slopes of these two separate and distinct sample regression equations is essentially a test to determine whether the said lines are parallel. Here we may test the null hypothesis $H_0 : \beta_{1A} = \beta_{1B}$ against any of the alternative hypotheses:

Case I	Case II	Case III
$H_0 : \beta_{1A} = \beta_{1B}$	$H_0 : \beta_{1A} = \beta_{1B}$	$H_0 : \beta_{1A} = \beta_{1B}$
$H_1 : \beta_{1A} > \beta_{1B}$	$H_1 : \beta_{1A} < \beta_{1B}$	$H_1 : \beta_{1A} \neq \beta_{1B}$

Let us denote the estimated variances of the sample regression slopes as

$$S^2_{\hat{\beta}_{1A}} = \frac{S^2_A}{\sum x^2_{iA}},$$
$$S^2_{\hat{\beta}_{1B}} = \frac{S^2_B}{\sum x^2_{iB}} \tag{2.66}$$

respectively, where

$$S^2_A = \frac{\sum e^2_{iA}}{n_A - 2}$$

is the estimated error variance of the sample A regression equation,

$$S^2_B = \frac{\sum e^2_{iB}}{n_B - 2}$$

is the estimated error variance of the sample B regression equation, and the X_A and X_B variables have been expressed in terms of deviations from their respective means or

$$x_{iA} = X_{iA} - \bar{X}_A, \quad i = 1, \ldots, n_A,$$
$$x_{iB} = X_{iB} - \bar{X}_B, \quad i = 1, \ldots, n_A.$$

(Note that the sample A residuals are denoted as $e_{iA} = Y_{iA} - \hat{Y}_{iA}, i = 1, \ldots, n_A$, and the sample B residuals appear as $e_{iB} = Y_{iB} - \hat{Y}_{iB}, i = 1, \ldots, n_B$.)

Given the estimated slopes $\hat{\beta}_{1A}$ and $\hat{\beta}_{1B}$ and Equation 2.66, our large sample test statistic

$$Z = \frac{\hat{\beta}_{1A} - \hat{\beta}_{1B}}{\sqrt{S^2_{\hat{\beta}_{1A}} + S^2_{\hat{\beta}_{1B}}}} : N(0, 1). \tag{2.67}$$

Then for a test conducted at the $\alpha = P(\text{TIE})$ level of significance, the appropriate decision rules for rejecting H_0 in favor of H_1 are as follows:

a. Case I: reject H_0 if $z \geq z_\alpha$;
b. Case II: reject H_0 if $z \leq -z_\alpha$;
c. Case III: reject H_0 if $|z| \geq z_{\alpha/2}$, (2.68)

where z is the realized value of Z in Equation 2.67.

Additionally, we may determine just how precisely the true difference in population slopes has been estimated by constructing the following $100(1 - \alpha)\%$ confidence interval for $\beta_{1A} - \beta_{1B}$ or

$$\left(\hat{\beta}_{1A} - \hat{\beta}_{1B}\right) \pm z_{\alpha/2}\sqrt{s^2_{\hat{\beta}_{1A}} + s^2_{\hat{\beta}_{1B}}}, \tag{2.69}$$

where $S^2_{\hat{\beta}_{1A}}$ and $S^2_{\hat{\beta}_{1B}}$ are the sample realizations of $S^2_{\hat{\beta}_{1A}}$ and, $S^2_{\hat{\beta}_{1B}}$, respectively.

Example 2.22

Table 2.14 contains a set of $n_A = 25$ data points for sample A (columns 1, 2) and $n_B = 25$ observations for sample B (columns 3, 4). The OLS regression results for each of these samples are given below, where the values in parentheses beneath the estimates coefficients are t values.

Sample A	Sample B
$\hat{Y}_A = 8.246 + 0.312X_A,$	$\hat{Y}_B = 14.082 + 0.426X_B,$
(6.847) (8.576)	(14.333) (13.546)
$r_A^2 = 0.76, s_{\hat{\beta}_{0A}} = 1.204, s_{\hat{\beta}_{1A}} = 0.036.$	$r_B^2 = 0.88, s_{\hat{\beta}_{0B}} = 0.983, s_{\hat{\beta}_{1B}} = 0.031.$

For $\alpha = 0.05$, is the regression A slope significantly lower than the regression B slope? Here we want to test $H_0: \beta_{1A} = \beta_{1B}$, against $H_1: \beta_{1A} < \beta_{1B}$ (Case II). From Equation 2.67,

$$z = \frac{0.312 - 0.426}{\sqrt{(0.036)^2 + (0.031)^2}} = -2.40.$$

TABLE 2.14

Data for Testing the Difference between Two Regression Slopes

(1) X_A	(2) Y_A	(3) X_B	(4) Y_B
2	8	2	12
6	10	4	15
8	8	8	15
10	12	12	19
12	11	14	21
16	14	16	24
18	11	18	23
22	16	20	21
26	16	22	22
26	26	24	26
28	17	26	26
30	20	28	30
32	21	30	32
36	18	31	27
36	16	33	27
38	19	34	28
40	20	37	30
41	22	38	32
42	20	39	30
44	19	40	32
45	22	41	29
46	23	45	30
47	22	46	32
47	24	47	34
49	24	49	35

Because this realized Z value is an element of the critical region $\mathcal{R} = \{z/z \le -1.645\}$, we reject H_0 at the 5% level, i.e., the sample A slope is significantly lower than the sample B slope.

Comparing Two Regression Intercepts

Here we desire to test the null hypothesis $H_0: \beta_{0A} = \beta_{0B}$ against the alternative hypotheses

Case I	Case II	Case III
$H_0: \beta_{0A} = \beta_{0B}$	$H_0: \beta_{0A} = \beta_{0B}$	$H_0: \beta_{0A} = \beta_{0B}$
$H_1: \beta_{1A} > \beta_{0B}$	$H_1: \beta_{0A} < \beta_{0B}$	$H_1: \beta_{0A} \ne \beta_{0B}$

Let us express the estimated variances of the sample regression intercepts as

$$S^2_{\hat{\beta}_{0A}} = S^2_A\left(\frac{1}{n_A} + \frac{\bar{X}^2_A}{\sum x^2_{iA}}\right), \quad S^2_{\hat{\beta}_{0B}} = S^2_B\left(\frac{1}{n_B} + \frac{\bar{X}^2_B}{\sum x^2_{iB}}\right) \tag{2.70}$$

respectively. Then given the estimated intercepts $\hat{\beta}_{0A}$ and $\hat{\beta}_{0B}$ and Equation 2.70, the large sample test statistic is

$$Z = \frac{\hat{\beta}_{0A} - \hat{\beta}_{0B}}{\sqrt{S^2_{\hat{\beta}_{0A}} + S^2_{\hat{\beta}_{0B}}}} : N(0,1). \tag{2.71}$$

So for a test conducted at the α% level of significance, the decision rules for rejecting H_0 in favor of H_1 are as follows:

a. Case I: reject H_0 if $z \ge z_\alpha$;
b. Case II: reject H_0 if $z \le -z_\alpha$;
c. Case III: reject H_0 if $|z| \ge z_{\alpha/2}$, (2.72)

where z depicts the sample realization of Z in Equation 2.71.

Moreover, a 100(1 − α)% confidence interval for $\beta_{0A} - \beta_{0B}$ is

$$\left(\hat{\beta}_{0A} - \hat{\beta}_{0B}\right) \pm z_{\alpha/2}\sqrt{s^2_{\hat{\beta}_{0A}} + s^2_{\hat{\beta}_{0B}}}. \tag{2.73}$$

Example 2.23

Given the sample A and sample B regression results provided above (Table 2.14), let us determine whether the regression B intercept lies significantly above the regression A intercept for $\alpha = 0.05$. That is, let us test $H_0: \beta_{0A} = \beta_{0B}$ against the Case II alternative hypothesis. Using Equation 2.71,

$$z = \frac{8.246 - 14.082}{\sqrt{(1.204)^2 + (0.983)^2}} = -3.76.$$

Because this calculated z value is an element of $\mathcal{R}=\{z/z<-1.645\}$, we can reject the null hypothesis of equal regression intercepts at the 5% level; the sample B regression intercept lies significantly above the sample A regression intercept.

We shall return to the problem of comparing two regression equations when the concept of indicator variables is considered in Chapter 15.

Comparing Two Correlation Coefficients

We now look to the issue of whether or not the strength of the linear association between the X and Y variables is the same for the group A and the group B datasets. In what follows, Assumptions A.1 through A.5 (see page 26) underlying the strong classical linear regression model are assumed to hold.

Let us also assume that we actually have two independent random samples, each extracted from a separate bivariate normal distribution, in which the first sample consists of n_A ordered pairs of observations taken from population A and the second sample involves n_B ordered pairs of observations taken from population B. Suppose we are interested in testing the difference between the two population correlation coefficients ρ_A and ρ_B, where the said difference is denoted $\rho_A - \rho_B = \delta_0$. In particular, it is often the case that $\delta_0 = 0$, i.e., we are interested in determining whether the population correlation coefficients ρ_A and ρ_B are equal. In this regard, we may test the null hypotheses $H_0 : \rho_A - \rho_B = 0$ against any of the three alternative hypotheses given in the following:

Case I	Case II	Case III
$H_0 : \rho_A - \rho_B = 0$	$H_0 : \rho_A - \rho_B = 0$	$H_0 : \rho_A - \rho_B = 0$
$H_1 : \rho_A - \rho_B > 0$	$H_1 : \rho_A - \rho_B < 0$	$H_1 : \rho_A - \rho_B \neq 0$

On the basis of our preceding discussion, it is readily seen that testing $H_0 : \rho_A - \rho_B = 0$ is equivalent to testing, via Equation 2.40, $H_0 = \xi_{\rho A} - \xi_{\rho B} = 0$ or H_0: $\frac{1}{2}\ln\left(\frac{1+\rho_A}{1-\rho_A}\right) - \frac{1}{2}\ln\left(\frac{1+\rho_B}{1-\rho_B}\right) = 0.$ Let the sample correlation coefficients be denoted as R_A and R_B, respectively, with r_A and r_A representing their respective sample realizations. Then because $R_A - R_B$ is an estimator for $\rho_A - \rho_B$, Equations 2.41 and 2.41.1 enable us to form, under H_0: $\xi_{\rho A} - \xi_{\rho B} = 0$, the test statistic

$$Z_{\delta_0} = \frac{\frac{1}{2}\ln\left(\frac{1+R_A}{1-R_A}\right) - \frac{1}{2}\ln\left(\frac{1+R_B}{1-R_B}\right)}{\sqrt{\frac{1}{n_A-3}+\frac{1}{n_B-3}}} \quad \text{[sampling from two independent bivariate normal populations]} \tag{2.74}$$

which is N (0,1). Then for a test conducted at the level of significance, the appropriate decision rules for rejection H_0 relative to H_1 are as follows:

a. Case I: reject H_0 if $z_{\delta_0} \geq z_\alpha$;
b. Case II: reject H_0 if $z_{\delta_0} \leq -z_\alpha$;
c. Case III: reject H_0 if $|z_{\delta_0}| \geq z_{\alpha/2}$,

where

$$z_{\delta_0} = \frac{\frac{1}{2}\ln\left(\frac{1+r_A}{1-r_A}\right) - \frac{1}{2}\ln\left(\frac{1+r_B}{1-r_B}\right)}{\sqrt{\frac{1}{n_1-3}+\frac{1}{n_2-3}}}$$

is the sample realization of Z_{δ_0}.

Example 2.16.3

Given the sample A and sample B datasets appearing in Table 2.14, it was previously determined that $r_A = 0.8718$ and $r_B = 0.9381$. If we assume that we actually have independent random samples taken from separate bivariate normal distributions, then, for $\alpha = 0.05$, we can determine whether the sample A correlation coefficient is significantly lower than its sample B counterpart. That is, we seek to test $H_0 : \rho_A - \rho_B = 0$ against $H_1 : \rho_A - \rho_B < 0$ (Case II). From Equation 2.74,

$$z_{\delta_0} = \frac{\frac{1}{2}\ln\left(\frac{1+0.8718}{1-0.8718}\right) - \frac{1}{2}\ln\left(\frac{1+0.9381}{1-0.9381}\right)}{\sqrt{\frac{1}{22}+\frac{1}{22}}} = -1.2652$$

Because this realized Z_{δ_0} value is not an element of $= \left\{z_{\delta_0} \middle| z_{\delta_0} < -1.645\right\}$, we cannot reject H_0 at the 5% level, i.e., the strength of linear association for the sample A dataset is not significantly different from that of the sample B dataset.

Appendix A: Properties of the Least-Squares Estimators

1. Gauss-Markov Theorem

If Assumptions A.2 through A.4 for the weak classical linear regression model hold, then, within the class of linear unbiased estimators of β_0 and β_1, the least-squares estimators $\hat{\beta}_0$ and $\hat{\beta}_1$ have minimum variance, i.e., the least-squares estimators $\hat{\beta}_0$ and $\hat{\beta}_1$ are best linear unbiased estimators for β_0 and β_1, respectively.

To verify this result, we first demonstrate that $\hat{\beta}_0$ and $\hat{\beta}_1$ are linear and unbiased.

$\hat{\beta}_0$ and $\hat{\beta}_1$ Are Linear Functions of the Actual Observations Y_i, $i = 1,\ldots,n$

$$\hat{\beta}_1 = \frac{\sum x_i y_i}{\sum x_i^2} = \frac{\sum x_i Y_i}{\sum x_i^2} - \frac{\bar{Y}\sum x_i}{\sum x_i^2} = \sum w_i Y_i, \tag{2.A.1}$$

where $\sum x_i = 0$ and $w_i = x_i / \sum x_i^2$. Note that, because X_i is fixed, x_i and thus w_i is also fixed in value. In addition, it is easily shown that

$$\sum w_i = 0, \sum w_i x_i = 1, \text{ and } \sum w_i^2 = 1/\sum x_i^2. \tag{2.A.2}$$

Similarly,

$$\hat{\beta}_0 = \bar{Y} - \hat{\beta}_1 \bar{X} = \frac{\sum Y_i}{n} - \bar{X} \sum w_i Y_i = \sum v_i Y_i, \tag{2.A.3}$$

where $v_i = \frac{1}{n} - \bar{X} w_i$.

$\hat{\beta}_0$ and $\hat{\beta}_1$ and Are Unbiased Estimators of $\hat{\beta}_0$ and β_1, Respectively

Because $Y_i = \beta_0 + \beta_1 X_i + \varepsilon_i$ and $\bar{Y} = \beta_0 + \beta_1 \bar{X} + \bar{\varepsilon}$, where $\bar{\varepsilon} = \sum \varepsilon_i / n$, it follows that $Y_i - \bar{Y} = y_i = \beta_1 x_i + \varepsilon_i - \bar{\varepsilon}$ and thus

$$\begin{aligned}\hat{\beta}_1 &= \frac{\sum x_i(\beta_1 x_i + \varepsilon_i - \bar{\varepsilon})}{\sum x_i^2} = \frac{\beta_1 \sum x_i^2 + \sum x_i \varepsilon_i - \bar{\varepsilon} \sum x_i}{\sum x_i^2} \\ &= \beta_1 + \sum w_i \varepsilon_i\end{aligned}$$

because $\sum x_i = 0$. Then

$$E\left(\hat{\beta}_1\right) = \beta_1 + E\left(\sum w_i \varepsilon_i\right) = \beta_1 + \sum w_i E(\varepsilon_i) = \beta_1$$

(because $E(\varepsilon_i) = 0$ for all i) so that $\hat{\beta}_1$ is an unbiased estimator for β_1.
Similarly,

$$\hat{\beta}_0 = \bar{Y} - \hat{\beta}_1 \bar{X} = \beta_0 + \beta_1 \bar{X} + \bar{\varepsilon} - \hat{\beta}_1 \bar{X} = \beta_0 + \left(\beta_1 - \hat{\beta}_1\right)\bar{X} + \bar{\varepsilon} \tag{2.A.5}$$

and thus

$$E(\hat{\beta}_0) = \beta_0 + E[(\beta_1 - \hat{\beta}_1)\bar{X}] + E(\bar{\varepsilon}) = \beta_0$$

because $\bar{X}E(\beta_1 - \hat{\beta}_1) = 0$ and $\frac{1}{n}E\left(\sum \varepsilon_i\right) = \frac{1}{n}\sum E(\varepsilon_i) = 0$. Hence, $\hat{\beta}_0$ is an unbiased estimator for β_0.

Determining the Variances of $\hat{\beta}_0$ and $\hat{\beta}_1$

From Equation 2.A.4,

$$
\begin{aligned}
V(\hat{\beta}_1) &= E(\hat{\beta}_1 - \beta_1)^2 = E\left(\sum w_i \varepsilon_i\right)^2 \\
&= E\left(\sum w_i^2 \varepsilon_i^2 + 2\sum_{\substack{i \\ (i<j)}} \sum_j w_i w_j \varepsilon_i \varepsilon_j\right) = \sum w_i^2 E(\varepsilon_i^2) + 2\sum_{\substack{i \\ (i<j)}} \sum_j w_i w_j E(\varepsilon_i \varepsilon_j) \\
&= \sigma_\varepsilon^2 \sum w_i^2 = \sigma_\varepsilon^2 \Big/ \sum x_i^2, \qquad (2.A.6)
\end{aligned}
$$

because $E(\varepsilon_i^2) = \sigma_\varepsilon^2$ for all i, $E(\varepsilon_i \varepsilon_j) = 0$ for $i \neq j$, and $\sum w_i^2 = 1/\sum x_i^2$. Also from Equation 2.A.5,

$$
\begin{aligned}
V(\hat{\beta}_0) &= E(\hat{\beta}_0 - \beta_0)^2 = E\left[(\beta_1 - \hat{\beta}_1)\bar{X} + \bar{\varepsilon}\right]^2 \\
&= E\left[(\beta_1 - \hat{\beta}_1)^2 \bar{X}^2 + 2\bar{X}(\beta_1 - \hat{\beta}_1)\bar{\varepsilon} + \bar{\varepsilon}^2\right] \\
&= \bar{X}^2 E(\hat{\beta}_1 - \beta_1)^2 - 2\bar{X}E\left[(\hat{\beta}_1 - \beta_1)\bar{\varepsilon}\right] + E(\bar{\varepsilon}^2) \\
&= \bar{X}^2 V(\hat{\beta}_1) + \frac{\sigma_\varepsilon^2}{n} = \sigma_\varepsilon^2\left(\frac{1}{n} + \frac{\bar{X}^2}{\sum x_i^2}\right), \qquad (2.A.7)
\end{aligned}
$$

where

$$
\begin{aligned}
E\left[(\hat{\beta}_1 - \beta_1)\bar{\varepsilon}\right] &= E\left[\left(\sum w_i \varepsilon_i\right)\left(\frac{1}{n}\sum \varepsilon_i\right)\right] = \frac{1}{n}E\left[\sum w_i \varepsilon_i^2 + \sum_{\substack{i \\ (i<j)}} \sum_j (w_i + w_j)\varepsilon_i \varepsilon_j\right] \\
&= \frac{1}{n}\sum w_i E(\varepsilon_i^2) + \sum_{\substack{i \\ (i<j)}} \sum_j (w_i + w_j) E(\varepsilon_i \varepsilon_j) = \frac{\sigma_\varepsilon^2}{n}\sum w_i = 0,
\end{aligned}
$$

(because $\sum w_i = 0$ and $E(\varepsilon_i \varepsilon_j) = 0$, $i \neq j$) and

$$
\begin{aligned}
E(\bar{\varepsilon}^2) &= \frac{1}{n^2}E\left[\left(\sum \varepsilon_i\right)^2\right] = \frac{1}{n^2}E\left(\sum \varepsilon_i^2 + \sum_{\substack{i \\ (i<j)}} \sum_j \varepsilon_i \varepsilon_j\right) \\
&= \frac{1}{n^2}\left(\sum E(\varepsilon_i^2) + \sum_{\substack{i \\ (i<j)}} \sum_j E(\varepsilon_i \varepsilon_j)\right) = \frac{1}{n^2}\sum \sigma_\varepsilon^2 = \sigma_\varepsilon^2 / n.
\end{aligned}
$$

The Least-Squares Estimators $\hat{\beta}_0$ and $\hat{\beta}_1$ of β_0 and β_1, Respectively, Are Best Estimators

Hence, we must demonstrate that, within the class of linear unbiased estimators of the regression parameters, the least-squares estimators have minimum variance. To this end, let $\beta_1' = \sum a_i Y_i$ be any alternative linear estimator of β_1, where the a_i's, $i = 1,\ldots,n$, are fixed values. (Here β_1' is a general linear combination of the Y_i's.) Then

$$
\beta_1' = \sum a_i(\beta_0 + \beta_1 X_i + \varepsilon_i) = \beta_0 \sum a_i + \beta_1 \sum a_i X_i + \sum a_i \varepsilon_i
$$

and

$$E(\beta_1') = \beta_0 \sum a_i + \beta_1 \sum a_i X_i + \sum a_i E(\varepsilon_i)$$
$$= \beta_0 \sum a_i + \beta_1 \sum a_i X_i.$$

For β_1' to be an unbiased estimator of β_1, we must impose the restrictions that

$$\sum a_i = 0 \text{ and } \sum a_i X_i = 1. \tag{2.A.8}$$

(Note that these two conditions simultaneously imply that $\sum a_i x_i = 1$.) Thus, $E(\beta_1') = \beta_1$ under Equation 2.A.8. Given Equation 2.A.8, $\beta_1' = \beta_1 + \sum a_i \varepsilon_i$ and thus

$$V(\beta_1') = E[\beta_1' - E(\beta_1')]^2 = E(\beta_1' - \beta_1)^2 = E\left(\sum a_i \varepsilon_i\right)^2$$
$$= E\left(\sum a_i^2 \varepsilon_i^2 + 2 \sum_{\substack{i \\ (i<j)}} \sum_j a_i a_j \varepsilon_i \varepsilon_j\right)$$
$$= \sum a_i^2 E(\varepsilon_i^2) + 2 \sum_{\substack{i \\ (i<j)}} \sum_j a_i a_j E(\varepsilon_i \varepsilon_j) = \sigma_\varepsilon^2 \sum a_i^2. \tag{2.A.9}$$

To compare β_1' with the least-squares estimator $\hat{\beta}_1$ let us set $a_i = w_i + d_i$ where d_i is constant for all i. (Note that, if $a_i = w_i$, then $\beta_1' = \sum a_i Y_i = \sum w_i Y_i = \sum w_i (y_i + \bar{Y}) = \sum w_i y_i + \bar{Y} \sum w_i$ $= \sum x_i y_i / \sum x_i^2 = \hat{\beta}_1$.) Then from Equation 2.A.9,

$$V(\beta_1') = \sigma_\varepsilon^2 \sum (w_i + d_i)^2 = \sigma_\varepsilon^2 \left(\sum w_i^2 + 2 \sum w_i d_i + \sum d_i^2\right). \tag{2.A.9.1}$$

Because we require that $\sum a_i x_i = \sum (w_i + d_i) x_i = \sum w_i x_i + \sum d_i x_i = 1$ and $\sum w_i x_i = 1$ (see Equation 2.A.2), it follows that $\sum d_i x_i = 0$ so that, from Equations 2.A.9.1 and 2.A.6,

$$V(\beta_1') = \sigma_\varepsilon^2 \left(\sum w_i^2 + 2 \sum \left(\frac{x_i}{\sum x_i^2}\right) d_i + \sum d_i^2\right)$$
$$= V(\hat{\beta}_1) + \sigma_\varepsilon^2 \sum d_i^2. \tag{2.A.9.2}$$

With $\sum d_i^2 \geq 0$ it follows that $V(\beta_1') \geq V(\hat{\beta}_1)$ and $V(\beta_1') = V(\hat{\beta}_1)$ only if $\sum d_i^2 = 0$ or that $d_i = 0$ for all i (provided, of course, that $\sigma_\varepsilon^2 > 0$). However, if $d_i = 0$ then $a_i = w_i$ and thus $\beta_1' = \hat{\beta}_1$ Hence, the linear unbiased estimator of β_1 with the smallest variance is the least-squares estimator $\hat{\beta}_1$.

In a similar manner, it can be demonstrated that the linear unbiased estimator of β_0 with the smallest variance is the least-squares estimator $\hat{\beta}_0$. This completes our proof of the notion that the least-squares estimators $\hat{\beta}_0$ and $\hat{\beta}_1$ are the best linear unbiased estimators of the regression parameters β_0 and β_1, respectively. It is important to note that no distributional assumption such as the normality of ε (Assumption A.1, p. 26) was needed to execute the proof of the Gauss-Markov Theorem. This is because we have restricted our discussion to a relatively small class of estimators, namely those that are linear and unbiased.

2. For Normal ε, the Least-Squares Estimators Are Maximum-Likelihood Estimators

Given that the assumptions of the strong classical linear regression model hold so that the ε_i are independent and normally distributed, then, from Equations 1.59 and 1.20, the joint probability density function of the ε_i's is provided by

$$\prod_{i=1}^{n} f\left(\varepsilon_i;0,\sigma_\varepsilon^2\right)=\left(\sigma_\varepsilon^2 2\pi\right)^{-n/2} e^{-\frac{1}{2\sigma_\varepsilon^2}\sum_{i=1}^{n}\varepsilon_i^2}. \tag{2.A.10}$$

Under Assumptions B.3 and B.4 above and the fact that ε_i transforms linearly to Y_i,[3] the likelihood function for the sample (see Equation 1.59.1) becomes, for $\varepsilon_i = Y_i - \beta_0 - \beta_1 X_i$, $i = 1,\ldots,n$,

$$\mathcal{L}(\beta_0,\beta_1,\sigma_\varepsilon^2;Y_1,\ldots,Y_n,n)=(\sigma_\varepsilon^2 2\pi)^{-n/2} e^{-\frac{1}{2\sigma_\varepsilon^2}\sum(Y_i-\beta_0-\beta_1X_i)^2}$$

or, in log-likelihood form,

$$\log \mathcal{L} = -\frac{n}{2}\log 2\pi - \frac{n}{2}\log\sigma_\varepsilon^2 - \frac{1}{2\sigma_\varepsilon^2}\sum(Y_i-\beta_0-\beta_1X_i)^2. \tag{2.A.12}$$

Differentiating Equation 2.A.12 partially with respect to the parameters β_0, β_1, and σ_ε^2 yields the following:

---

[3] Consider the problem of obtaining the probability distribution of a random variable Y from information about the probability distribution of a random variable X, where $y = g(x)$ is a functional relationship between the values of X and Y. To perform the indicated *change of variable* or transformation we shall use Theorem A.1 as follows.

Theorem A.1

Let the probability density function of the random variable X be given by $f(x)$ and let the function $y = g(x)$ define a one-to-one transformation between X and Y. (A one-to-one transformation implies that g is either an increasing or decreasing function for all admissible x.) In addition, let the unique inverse transformation of g be denoted as $x = w(y)$ and let $dx/dy = w'(y)$ be continuous and not vanish for all admissible y's. Then the probability density function of Y is given by

$$h(y) = |dw(y)/dy|\,f(w(y)),\ dw(y)/dy \neq 0. \tag{2.A.11}$$

Here $|dw(y)/dy|$ denotes the absolute value of $dw(y)/dy$. (For instance, if $f(x)=e^{-x}$, $x \geq 0$, and $y = g(x) = 3x$, then $x = w(y) = \frac{1}{3}y$ and $dw(y)/dy = \frac{1}{3} \neq 0$. Then $h(y) = \frac{1}{3}e^{-\frac{1}{3}y}$ $y \geq 0$.)

On the basis of this theorem, we can determine the form of the probability density function of the random variable Y_i from that of the random variable ε_i whose probability density function $f(\varepsilon_i)$ is $N(0,\sigma_\varepsilon^2)$. Because Y_i and ε_i are linearly related according to $Y_i = \beta_0 + \beta_1X_i + \varepsilon_i = g(\varepsilon_i)$ and $\varepsilon_i = Y_i - \beta_0 - \beta_1X_i = w(Y_i)$ with $d\varepsilon_i/dY_i = 1$, Equation 2.A.11 renders $h(y_i) = |d\varepsilon_i/dy_i| \cdot f(\varepsilon_i) = f(\varepsilon_i)$ for all i. So given that ε_i transforms linearly to Y_i ($|d\varepsilon_i/dY_i| = 1$), it follows that the random variables Y_i must be mutually independent and normally distributed because $h(y_i) = f(\varepsilon_i)$, $i = 1,\ldots,n$.

$$\begin{aligned}
&\text{(a)}\quad \partial \log \mathcal{L} \big/ \partial \beta_0 = \frac{1}{\sigma_\varepsilon^2}\sum (Y_i - \beta_0 - \beta_1 X_i) \\
&\text{(b)}\quad \partial \log \mathcal{L} \big/ \partial \beta_1 = \frac{1}{\sigma_\varepsilon^2}\sum X_i(Y_i - \beta_0 - \beta_1 X_i) \\
&\text{(c)}\quad \partial \log \mathcal{L} \big/ \partial \sigma_\varepsilon^2 = -\frac{n}{2\sigma_\varepsilon^2} + \frac{1}{2\sigma_\varepsilon^4}\sum (Y_i - \beta_0 - \beta_1 X_i)^2.
\end{aligned} \tag{2.A.13}$$

After equating each of these equations to zero (it is assumed that the second-order conditions for a maximum of log $\mathcal{L}$ hold) and placing a tilde over each unknown parameters so as to indicate a maximum-likelihood estimator of that parameter, we see that Equations 2.A.13.a and b reduce to the system of least-squares normal Equations 2.4 or

$$\begin{aligned}
n\tilde{\beta}_0 + \tilde{\beta}_1 \sum X_i &= \sum Y_i \\
\tilde{\beta}_0 \sum X_i + \tilde{\beta}_1 \sum X_i^2 &= \sum X_i Y_i
\end{aligned} \tag{2.A.14}$$

Hence, the maximum-likelihood estimators $\tilde{\beta}_0$ and $\tilde{\beta}_1$ of β_0 and β_1, respectively, are the same as the least-squares estimators of these parameters. From Equation 2.A.13.c, we obtain the maximum-likelihood estimator of σ_ε^2 or

$$\tilde{\sigma}_\varepsilon^2 = \frac{1}{n}\sum (Y_i - \tilde{\beta}_0 - \tilde{\beta}_1 X_i)^2.$$

Because $\tilde{\beta}_0$ and $\tilde{\beta}_1$ coincide with the least-squares estimators of the regression parameters, it follows that

$$\tilde{\sigma}_\varepsilon^2 = \frac{1}{n}\sum e_i^2, \tag{2.A.15}$$

i.e., the maximum-likelihood estimator of σ_ε^2 is the sample variance of the least-squares residuals e_i.

In summary, for the strong classical linear regression model, the least-squares estimators are equivalent to the maximum-likelihood estimators of the regression intercept and slope. Moreover, although the least-squares and maximum-likelihood methods both provide us with formulas for the estimators for β_0 and β_1, the maximum-likelihood routine goes a step further: it provides us with a formula for an estimator of σ_ε^2, namely Equation 2.A.15.

It was mentioned above in "Gauss-Markov Theorem" that, by virtue of the Gauss-Markov theorem, the least-squares estimators for β_0 and β_1 possess key finite-sample properties in that they are best linear unbiased estimators, i.e., of the class of unbiased linear estimators for the regression parameters, the least-squares estimators have minimum variance. Moreover, it can be shown that the least-squares estimators are the most efficient estimators for β_0 and β_1 in that their variances satisfy the Cramer-Rao lower bounds for unbiased estimators. Furthermore, because the least-squares estimators are minimum variance bound estimators, they are also sufficient estimators of the regression parameters.

Finally, the least-squares estimators, being the same as the maximum-likelihood estimators for β_0 and β_1, possess certain asymptotic or large-sample properties of a good estimator because, under certain regularity conditions imposed on the likelihood function, the maximum likelihood (and thus least-squares) estimators are asymptotically unbiased, consistent, asymptotically efficient, and asymptotically normal.

3. Determining $COV(\hat{\beta}_0, \hat{\beta}_1)$

From Equation 2.A.5, we have $\hat{\beta}_0 - \beta_0 = -(\hat{\beta}_1 - \beta_1)\bar{X} + \bar{\varepsilon}$. Then

$$(\hat{\beta}_0 - \beta_0)(\hat{\beta}_1 - \beta_1) = -(\hat{\beta}_1 - \beta_1)^2 \bar{X} + (\hat{\beta}_1 - \beta_1)\bar{\varepsilon}$$

and thus

$$COV(\hat{\beta}_0, \hat{\beta}_1) = E(\hat{\beta}_0 - \beta_0)(\hat{\beta}_1 - \beta_1) = -\bar{X}E(\hat{\beta}_1 - \beta_1)^2 + E(\hat{\beta}_1 - \beta_1)\bar{\varepsilon} = \frac{-\bar{X}\sigma_\varepsilon^2}{\sum x_i^2} \tag{2.A.16}$$

given Equation 2.A.6 and $E(\hat{\beta}_1 - \beta_1)\bar{\varepsilon} = 0$ (that this latter expression is zero follows from the derivation immediately following Equation 2.A.7.

4. Finding an Estimator for σ_ε^2

As an estimator of the variance of the random disturbances ε_i, let us use the variance of the estimated residuals e_i or

$$d^2 = \frac{1}{n}\sum (e_i - \bar{e})^2 = \frac{1}{n}\sum e_i^2, \tag{2.A.17}$$

where $\bar{e} = \frac{1}{n}\sum e_i = \frac{1}{n}\sum (Y_i - \hat{\beta}_0 - \hat{\beta}_1 X_i) = \bar{Y} - \hat{\beta}_0 - \hat{\beta}_1\bar{X} = \bar{Y} - (\bar{Y} - \hat{B}_1\bar{X}) - \hat{\beta}_1\bar{X} = 0.$ (Note that Equation 2.A.17 corresponds to the maximum-likelihood estimator $\tilde{\sigma}_\varepsilon^2$ provided by Equation 2.A.15.) Because

$$\begin{aligned} e_i &= Y_i - \hat{\beta}_0 - \hat{\beta}_1 X_1 = Y_i - (\bar{Y} - \hat{\beta}_1\bar{X}) - \hat{\beta}_1 X_i \\ &= y_i - \hat{\beta}_1 x_i = \beta_1 x_i + \varepsilon_i - \bar{\varepsilon} - \hat{\beta}_1 x_i \\ &= -(\hat{\beta}_1 - \beta_1)x_i + \varepsilon_i - \bar{\varepsilon}, \end{aligned} \tag{2.A.18}$$

it follows from Equation 2.A.17 that

$$\begin{aligned} d^2 &= \frac{1}{n}\sum \left[-(\hat{\beta}_1 - \beta_1)x_i + \varepsilon_i - \bar{\varepsilon}\right]^2 \\ &= \frac{1}{n}\left[(\hat{\beta}_1 - \beta_1)^2 \sum x_i^2 + \sum \varepsilon_i^2 + n\bar{\varepsilon}^2 - 2(\hat{\beta}_1 - \beta_1)\sum x_i\varepsilon_i + 2\bar{\varepsilon}(\hat{\beta}_1 - \beta_1)\sum x_i - 2\bar{\varepsilon}\sum \varepsilon_i\right] \\ &= \frac{1}{n}\left[-(\hat{\beta}_1 - \beta_1)^2 \sum x_i^2 + \sum \varepsilon_i^2 - n\bar{\varepsilon}^2\right] \end{aligned} \tag{2.A.17.1}$$

because $\sum x_i\varepsilon_i = (\hat{\beta}_1 - \beta_1)\sum x_i^2$, $\sum x_i = 0$, and $\sum \varepsilon_i = n\bar{\varepsilon}$. Then

$$\begin{aligned} E(d^2) &= \frac{1}{n}\left[-\sum x_i^2 E(\hat{\beta}_1 - \beta_1)^2 + \sum E(\varepsilon_i^2) - nE(\bar{\varepsilon}^2)\right] \\ &= \left(\frac{n-2}{n}\right)\sigma_\varepsilon^2 \end{aligned} \tag{2.A.19}$$

because $E(\hat{\beta}_1 - \beta_1)^2 = \sigma_\varepsilon^2 / \sum x_i^2$, $\sum E(\varepsilon_i^2) = n\sigma_\varepsilon^2$, and $E(\bar{\varepsilon}^2) = \sigma_\varepsilon^2 / n$.

As Equation 2.A.19 reveals, d^2 is a biased estimator of σ_ε^2, i.e., on average, d^2 underestimates σ_ε^2. However, for large n, the bias will be small. Moreover, $(n-2)/n = 1 - \frac{2}{n} \to 1$ as $n \to \infty$ so that d^2 is an asymptotically unbiased estimator for σ_ε^2. To correct for the bias, let us consider the estimator $S_\varepsilon^2 = (\frac{n}{n-2})d^2$. Clearly, $E(S_\varepsilon^2) = (\frac{n}{n-2})E(S^2) = \sigma_\varepsilon^2$. Hence,

$$S_\varepsilon^2 = \frac{\sum e_i^2}{n-2} \tag{2.A.20}$$

is an unbiased estimator of σ_ε^2.

5. The Sampling Distribution of the Least-Squares Estimators $\hat{\beta}_0, \hat{\beta}_1$

From Equation 2.A.4, $\hat{\beta}_1 = \beta_1 + \sum w_i \varepsilon_i$. Suppose we invoke the assumptions of the strong classical linear regression model. Then ε_i is $N(0, \sigma_\varepsilon^2)$ and thus, by virtue of Assumptions A.1 and A.4, $\hat{\beta}_1$ is a linear combination of independent normally distributed random variables. Hence, $\hat{\beta}_1$ is $N(\beta_1, \sqrt{V(\hat{\beta}_1)})$. Similarly, from Equation 2.A.5, $\hat{\beta}_0 = \beta_0 + (\beta_1 - \hat{\beta}_1)\bar{X} + \bar{\varepsilon}$. Because each ε_i is normal, then so is $\bar{\varepsilon} = \sum \varepsilon_i / n$. Moreover, because $\hat{\beta}_1$ and $\bar{\varepsilon}$ are each normal and $COV\left[(\hat{\beta}_1 - \beta_1), \bar{\varepsilon}\right] = E\left[(\hat{\beta}_1 - \beta_1)\bar{\varepsilon}\right] = 0$, it follows that $\hat{\beta}_1 - \beta_1$ and $\bar{\varepsilon}$ are independent. Thus, $\hat{\beta}_0$ is also a linear combination of independent normally distributed random variables so that $\hat{\beta}_0$ is $N(\beta_0, \sqrt{V(\hat{\beta}_0)})$. Given this set of results, it then follows that the quantities $(\hat{\beta}_1 - \beta_1)/\sqrt{V(\hat{\beta}_1)}$ and $(\hat{\beta}_0 - \beta_0)/\sqrt{V(\hat{\beta}_0)}$ are each N (0, 1).

Let us examine the sampling distribution of S_ε^2. From Equation 2.A.17.1,

$$(n-2)S_\varepsilon^2 = \sum e_i^2 = nd^2 = -(\hat{\beta}_1 - \beta_1)\sum x_i^2 + \sum \varepsilon_i^2 - n\bar{\varepsilon}^2$$

or

$$(n-2)S_\varepsilon^2 + (\hat{\beta}_1 - \beta_1)^2 \sum x_i^2 + n\bar{\varepsilon}^2 = \sum \varepsilon_i^2$$

and thus

$$(n-2)\frac{S_\varepsilon^2}{\sigma_\varepsilon^2} + \left(\frac{\hat{\beta}_1 - \beta_1}{\sigma_\varepsilon / \sqrt{\sum x_i^2}}\right)^2 + \left(\frac{\bar{\varepsilon}}{\sigma_\varepsilon / \sqrt{n}}\right)^2 = \sum \left(\frac{\varepsilon_i}{\sigma_\varepsilon}\right)^2. \tag{2.A.21}$$

We must now scrutinize the individual terms on the left side of Equation 2.A.21. It was demonstrated previously that $COV\left[(\hat{\beta}_1 - \beta_1), \bar{\varepsilon}\right] = E\left[(\hat{\beta}_1 - \beta_1)\bar{\varepsilon}\right] = 0$. Because $\hat{\beta}_1 - \beta_1$ and $\bar{\varepsilon}$ are both normally distributed, it follows that they are also independent by virtue of Theorem 13.A.1. Next, because $e_i = \left(\beta_1 - \hat{\beta}_1\right)x_i + \varepsilon_i - \bar{\varepsilon}$, it follows that $COV\left(e_i, \bar{\varepsilon}\right) = E\left(e_i\bar{\varepsilon}\right) = E\left[x_i(\beta_1 - \hat{\beta}_1), \bar{\varepsilon}\right] = E(\varepsilon_i\bar{\varepsilon}) - E(\bar{\varepsilon}^2) = 0$ (because $E\left(\varepsilon_i\bar{\varepsilon}\right) = E\left(\bar{\varepsilon}^2\right) = \sigma_\varepsilon^2/n$). Given that e_i and $\bar{\varepsilon}$ are normally distributed, a second application of Theorem 13.A.1 enables us to conclude that e_i and $\bar{\varepsilon}$ are also independent. Hence, e_i^2 and $\bar{\varepsilon}^2$ must be independent so that we thus conclude that s_ε^2 and $\bar{\varepsilon}^2$ are independent. Finally, given e_i,

$COV\left[(\hat{\beta}_1-\beta_1),e_i\right]=E\left[(\hat{\beta}_1-\beta_1)e_i\right]=x_iV(\hat{\beta}_1)+E(\beta_1-\hat{\beta}_1)(\varepsilon_i-\bar{\varepsilon})=0$. With $\beta_1-\hat{\beta}_1$ and e_i normal, it again follows from Theorem 13.A.1 that $\beta_1-\hat{\beta}_1$ and e_i are independent and thus $(\hat{\beta}_1-\beta_1)^2$ and s_ε^2 are independent. In the light of this discussion, we thus conclude that the three terms on the left side of Equation 2.A.21 are mutually independent.

Looking to the right side of Equation 2.A.21, we have the sum of the squares of n independent standard normal random variables (under Assumptions A.1 through A.4 of the strong classical linear regression model). Hence, from Appendix 12.E, the right side of Equation 2.A.21 is χ^2 distributed with n degrees of freedom. In addition, a moment of reflection reveals that both the second and third terms on the left side of Equation 2.A.21 are squares of standard normal random variables and thus, by virtue of Appendix 12.E, each is χ^2 distributed with one degree of freedom. Because all three terms on the left side of Equation 2.A.21 are independent, the first term must be χ^2 distributed with $n-2$ degrees of freedom.

Based on the preceding results, we immediately see that, because $\left(\hat{\beta}_1-\beta_1\right)/\left(\sigma_\varepsilon/\sqrt{x_i^2}\right):N(0,1)$ and the quantity $(n-2)S_\varepsilon^2/\sigma_\varepsilon^2:\chi_{n-2}^2$, it also follows from Appendix 12.E that with S_ε^2 and $\hat{\beta}_1-\beta_1$ independent,

$$\frac{\left(\hat{\beta}_1-\beta_1\right)/\left(\sigma_\varepsilon/\sqrt{\sum x_i^2}\right)}{\sqrt{\dfrac{(n-2)S_\varepsilon^2/\sigma_\varepsilon^2}{n-2}}}=\frac{\hat{\beta}_1-\beta_1}{S_\varepsilon/\sqrt{\sum x_i^2}}:t_{n-2}.$$

By a similar argument, we can also conclude that

$$\frac{\hat{\beta}_0-\beta_0}{S_\varepsilon\sqrt{\dfrac{1}{n}+\dfrac{\bar{X}^2}{\sum x_i^2}}}:t_{n-2}.$$

Appendix B: Sampling Distribution of $\hat{Y}_i$ and $Y_0-\hat{Y}_0$

1. Sampling Distribution of $\hat{Y}_i$

Given $X=X_i$, we may estimate $E(Y_i/X_i)=\beta_0+\beta_1X_i$ by using $\hat{Y}_i=\hat{\beta}_0+\hat{\beta}_1X_i$. (Clearly, $\hat{Y}_i$ is an unbiased estimator for $E(Y_i/X_i)$ because $E\left(\hat{Y}_i\right)=E\left(\hat{\beta}_0\right)+E\left(\hat{\beta}_1\right)X_i=\beta_0+\beta_1X_i=E\left(Y_i/X_i\right)$.) For X_i fixed,

$$\begin{aligned}V\left(\hat{Y}_i\right)&=E\left[\hat{Y}_i-E\left(\hat{Y}_i\right)\right]^2=E\left[\left(\hat{\beta}_0-\beta_0\right)+\left(\hat{\beta}_1-\beta_1\right)X_i\right]^2\\&=E\left[\left(\hat{\beta}_0-\beta_0\right)^2+X_i^2\left(\hat{\beta}_1-\beta_1\right)^2+2X_i\left(\hat{\beta}_0-\beta_0\right)\left(\hat{\beta}_1-\beta_1\right)\right]\\&=V\left(\hat{\beta}_0\right)+X_i^2V\left(\hat{\beta}_1\right)+2X_iCOV\left(\hat{\beta}_0,\hat{\beta}_1\right).\end{aligned}\tag{2.B.1}$$

From Appendix A.3, $COV\left(\hat{\beta}_0,\hat{\beta}_1\right)=-\bar{X}\sigma_\varepsilon^2/\sum x_i^2$. After combining this result with Equations 2.A.6 and 2.A.7, we obtain

$$V\left(\hat{Y}_i\right)=\sigma_\varepsilon^2\left(\frac{1}{n}+\frac{\bar{X}^2}{\sum x_i^2}\right)+\frac{X_i^2\sigma_\varepsilon^2}{\sum x_i^2}-2\frac{X_i\bar{X}\sigma_\varepsilon^2}{\sum x_i^2}$$
$$=\sigma_\varepsilon^2\left(\frac{1}{n}+\frac{\left(X_i-\bar{X}\right)^2}{\sum x_i^2}\right). \qquad (2.B.1.1)$$

Because $\hat{Y}_i$ is a linear combination of the normally distributed random variables $\hat{\beta}_0$ and $\hat{\beta}_1$, it follows that $\hat{Y}_i : N\left(\beta_0+\beta_1X_i,\sqrt{V\left(\hat{Y}_i\right)}\right)$ and thus

$$\frac{\hat{Y}_i-\beta_0-\beta_1X_i}{\sqrt{V\left(\hat{Y}_i\right)}}:N(0,1).$$

Then from Equation 2.7 and Appendix A.5,

$$\frac{\hat{Y}_i-\beta_0-\beta_1X_i}{S_\varepsilon\sqrt{\frac{1}{n}+\frac{\left(X_i-\bar{X}\right)^2}{\sum x_i^2}}}:t_{n-2}. \qquad (2.B.2)$$

2. Sampling Distribution of $Y_0-\hat{Y}_0$

In the preceding section, we concentrated on predicting the average value of Y for a given $X=X_i$ or we used $\hat{Y}_i(=\hat{\beta}_0+\hat{\beta}_1X_i)$ as a predictor of $E(Y_i/X_i)=\beta_0+\beta_1X_i$. Now, let us predict a specific Y corresponding to a given X value. That is, our goal is to predict $Y_0=\beta_0+\beta_1X_0+\varepsilon_0$. ($Y_0$ is thus a random variable) using the estimator $\hat{Y}_0=\hat{\beta}_0+\hat{\beta}_1X_0$. Consider the random variable $Y_0-\hat{Y}_0=\beta_0+\beta_1X_0+\varepsilon_0-\hat{\beta}_0-\hat{\beta}_1X_0$. Because $\hat{\beta}_0$ and $\hat{\beta}_1$ are unbiased estimators for β_0 and β_1, respectively, and $E(\varepsilon_0)=0$, it follows that $E\left(Y_0-\hat{Y}_0\right)=0$ so that $\hat{Y}_0$ is an unbiased estimator for Y_0. Hence,

$$V\left(Y_0-\hat{Y}_0\right)=E\left(Y_0-\hat{Y}_0\right)^2=E\left[\left(\hat{\beta}_0-\beta_0\right)+\left(\hat{\beta}_1-\beta_1\right)X_0-\varepsilon_0\right]^2$$
$$=E\left[\left(\hat{\beta}_0-\beta_0\right)^2+X_0^2\left(\hat{\beta}_1-\beta_1\right)^2+\varepsilon_0^2+2X_0\left(\hat{\beta}_0-\beta_0\right)\left(\hat{\beta}_1-\beta_1\right)\right.$$
$$\left.-2\left(\hat{\beta}_0-\beta_0\right)\varepsilon_0-2X_0\left(\hat{\beta}_1-\beta_1\right)\varepsilon_0\right]$$
$$=V\left(\hat{\beta}_0\right)+X_0^2V\left(\hat{\beta}_1\right)+V\left(\varepsilon_0\right)+2X_0COV\left(\hat{\beta}_0,\hat{\beta}_1\right) \qquad (2.B.3)$$

because $E\left(\hat{\beta}_0-\beta_0\right)\varepsilon_0=E\left(\hat{\beta}_1-\beta_1\right)\varepsilon_0=0$. Then, as argued in the preceding section,

$$V\left(Y_0-\hat{Y}_0\right)=\sigma_\varepsilon^2\left(\frac{1}{n}+\frac{\bar{X}^2}{\sum x_i^2}\right)+X_0^2\frac{\sigma_\varepsilon^2}{\sum x_i^2}+\sigma_\varepsilon^2-2\frac{\bar{X}X_0\sigma_\varepsilon^2}{\sum x_i^2}$$
$$=\sigma_\varepsilon^2\left(1+\frac{1}{n}+\frac{\left(X_0-\bar{X}\right)^2}{\sum x_i^2}\right)=\sigma_\varepsilon^2+\sigma_{\hat{Y}_0}^2, \qquad (2.B.3.1)$$

where $\sigma_{\hat{Y}_0}^2$ is obtained from Equation 2.B.1.1 for $X_i=X_0$.

Given that $\hat{Y}_0$ and Y_0 are both normal variates, it follows that $Y_0 - \hat{Y}_0 : N\left(0, \sqrt{V\left(Y_0 - \hat{Y}_0\right)}\right)$ and thus

$$\frac{Y_0 - \hat{Y}_0}{\sqrt{V\left(Y_0 - \hat{Y}_0\right)}} : N(0,1)).$$

Then via Equation 2.7 and Appendix A.5,

$$\frac{Y_0 - \hat{Y}_0}{S_\varepsilon \sqrt{1 + \frac{1}{n} + \frac{\left(X_0 - \bar{X}\right)^2}{\sum x_i^2}}} : t_{n-2}. \tag{2.B.4}$$

Appendix C: Partitioning the Sample Variation of Y

For $Y_i = \hat{\beta}_0 + \hat{\beta}_1 X_i + e_i$ and $\bar{Y} = \hat{\beta}_0 + \hat{\beta}_1 \bar{X}$, it readily follows that $Y_i - \bar{Y} = \hat{\beta}_1\left(X_i - \bar{X}\right) + e_i$ or

$$y_i = \hat{\beta}_1 x_i + e_i. \tag{2.C.1}$$

After squaring this expression and summing over all i values, we obtain

$$\begin{aligned}\sum y_i^2 &= \hat{\beta}_1^2 \sum x_i^2 + \sum e_i^2 + 2\hat{\beta}_1 \sum x_i e_i \\ &= \hat{\beta}_1 \sum x_i y_i + \sum e_i^2 + 2\hat{\beta}_1 \sum x_i e_i.\end{aligned} \tag{2.C.2}$$

Because $\sum x_i e_i = \sum x_i\left(y_i - \hat{\beta}_1 x_i\right) = \sum x_i y_i - \hat{\beta}_1 \sum x_i^2 = 0$, it follows that Equation 2.C.2 reduces to

$$\sum y_i^2 = \hat{\beta}_1 \sum x_i y_i + \sum e_i^2. \tag{2.C.2.1}$$

[Total sum of squares (SST)]	[Regression sum of squares (SSR)]	[Error sum of squares (SSE)]

Next, because $\sum y_i^2 - \hat{\beta}_1^2 \sum x_i^2 = \sum e_i^2 \geq 0$ (where $\sum y_i^2 = \sum e_i^2$ holds only if $\hat{\beta}_1 = 0$ or if the X_i's, $i = 1, \ldots, n$ are all equal in value), it follows that $\sum y_i^2 \geq \sum e_i^2 \geq 0$ or $1 \geq \sum e_i^2 / \sum y_i^2 \geq 0$. Now, it should be obvious that the smaller the magnitude of $\sum e_i^2$ the better the fit of the sample regression line to the dataset. Hence, our measure of goodness of fit will be constructed as

$$R^2 = 1 - \frac{\sum e_i^2}{\sum y_i^2} = 1 - \frac{SSE}{SST}, 0 \leq R^2 \leq 1, \tag{2.C.3}$$

and termed the coefficient of determination. (Note that $R^2 = 1$ when $\sum e_i^2 = 0$; $R^2 = 0$ when $\sum e_i^2 = \sum y_i^2$.) Alternatively, Equation 2.C.3 can be rewritten as

$$R^2 = \frac{SST - SSE}{SST} = \frac{SSR}{SST} \tag{2.C.3.1}$$

or

$$R^2 = \frac{\hat{\beta}_1 \sum x_i y_i}{\sum y_i^2} \left(= \frac{\hat{\beta}_1^2 \sum x_i^2}{\sum y_i^2} \right). \tag{2.C.3.2}$$

Appendix D: Linear Regression Using the SAS System

Example 2.D.1

Suppose we have $n = 20$ data points (X_i, Y_i) $i = 1,\ldots,20$, on the variables X and Y and that we want to perform a basic regression study using the SAS System. The program (written in SAS code) appears in Exhibit 2.D.1. Here we invoke the PRINT, PLOT, and REG procedures (procedure is abbreviated as *proc* in SAS code). The output generated by this program appears as Table 2.D.1, Table 2.D.2, and Figure 2.D.1.

Exhibit 2.D.1 SAS Code for PROC REG

```
data reg1; ①
   input x y; ②
datalines;
 1    4
 4    7
 5    8
 5    9
 5   11
 6    9
 6   10
 6   12
 6   18
 7    8 ③
 7   10
 7   12
 7   13
 8    9
```

① We name the dataset reg1.

② We name the variables *X* and *Y*.

③ Each variable gets its own column, and the columns must correspond to the order in which the variables appear in the input statement (②).

④ A semicolon indicates that the data stream has ended.

⑤ We request a printout of the reg1 dataset (see Table 2.D.1).

⑥ We request a plot of the reg1 dataset.

⑦ The variable listed first (second) is plotted on the vertical (horizontal) axis. We also request that each data point be indicated by a * (see Figure 2.D.1).

⑧ We request the fitting of a regression equation using the reg1 dataset.

```
 8     12
 8     13
 9     13
10     13
11      6
17     20
; ④
proc print data = reg1; ⑤
proc plot data = reg1; ⑥
   plot y*x = '*'; ⑦
proc reg data = reg1; ⑧
   model y = x/clb; ⑨
run;
```

⑨ The model specifies Y as the dependent variable and X as the independent variable. Confidence limits for β_0 and β_1 are also requested using the option *clb*.

A glance at Table 2.D.2 reveals the following features of the SAS output:

① Degrees of freedom for the error sum of squares(SSE) term is $n-2$ = 18, while degrees of freedom for the total sum of squares (SST) is $n-1$ = 19.

② Regression sum of squares (SSR) = 102.477.

③ SSE = 168.102.

④ SST = 270.550.

⑤ Regression mean square MSR = SSR/1 = 102.477.

⑥ Error mean square MSE = SSE/$(n-2)$ = 9.339.

⑦ Under $H_0:\beta_1 = 0$, the sample F value is F = MSR/MSE = 10.97.

TABLE 2.D.1

Output for Example 2.D.1 (PROC PRINT)

The SAS System

Obs	x	y
1	1	4
2	4	7
3	5	8
4	5	9
5	5	11
6	6	9
7	6	10
8	6	12
9	6	18
10	7	8
11	7	10
12	7	12
13	7	13
14	8	9
15	8	12
16	8	13
17	9	13
18	10	13
19	11	6
20	17	20

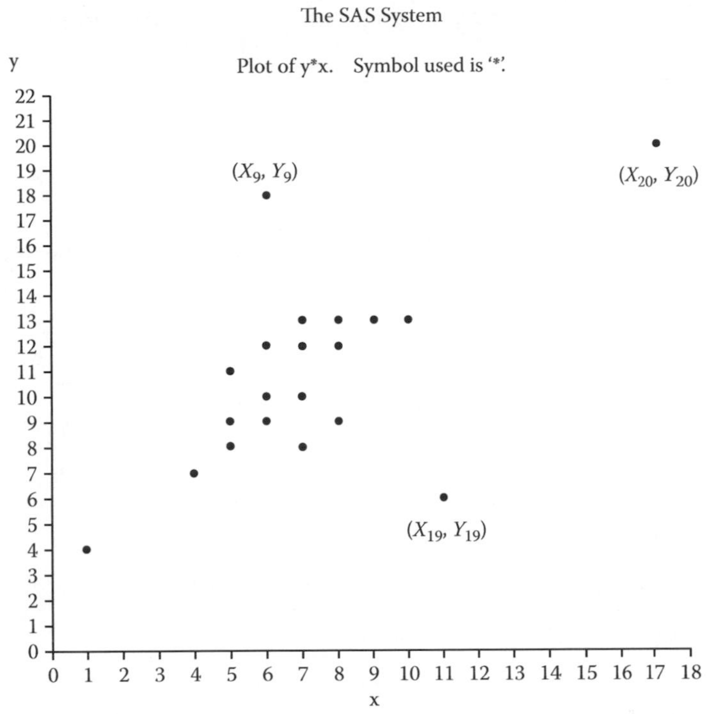

FIGURE 2.D.1
Plot of Y versus X.

TABLE 2.D.2

Output for Example 2.D.1 (PROC REG)

The SAS System
The REG Procedure
Model: MODEL1
Dependent Variable: *y*

Analysis of Variance

Source	**DF**	**Sum of Squares**	**Mean Square**	***F* Value**	**Pr > *F***
Model	1	102.44717 ②	102.44717 ⑤	10.97 ⑦	0.0039 ⑧
Error	18 ①	168.10283 ③	9.33905 ⑥		
Corrected total	19	270.55000 ④			
	Root MSE	3.05599 ⑨	R^2	0.3787 ⑩	
	Dependent mean	10.85000	Adj R^2	0.3441	
	Coeff Var	28.16576			

Parameter Estimates

Variable	**DF**	**Parameter Estimate**	**Standard Error**	**t Value**	**Pr > |t|**	**95% Confidence Limits**	
Intercept	1	5.63464 ⑪	1.71653 ⑬	3.28 ⑮	0.0041 ⑰	2.02833 ⑲	9.24095
x	1	0.72942 ⑫	0.22023 ⑭	3.31 ⑯	0.0039 ⑱	0.26673	1.19211

⑧ The p-value accompanying the calculated F value in ⑦ is very small, indicating a highly significant linear relationship between X and Y (there is only about a 0.4% chance of finding a calculated F value at least as large as 10.97 if $H_0: \beta_1 = 0$ is true).

⑨ $\sqrt{MSE} = s_\varepsilon = \sqrt{9.33905} = 3.056$.

⑩ $R^2 = 0.3787$; about 38% of the variation in Y is explained by the linear influence of X, whereas 62% is attributed to random factors.

⑪, ⑫ The estimated regression equation has the form $\hat{Y} = \hat{\beta}_0 + \hat{\beta}_1 X = 5.634 + 0.729X$. Here the average value of Y when $X = 0$ is 5.634, whereas when X increases by one unit, Y increases by 0.73 units on average.

⑬, ⑭ The estimated standard errors of the regression coefficients are $s_{\hat{\beta}_0} = 1.717$ and $s_{\hat{\beta}_1} = 0.220$.

⑮ Under $H_0: \beta_1 = 0$, the calculated t value for the estimated intercept is $\hat{\beta}_0 / s_{\hat{\beta}_0} = 3.28$.

⑯ Under $H_0: \beta_1 = 0$, the calculated t value for the estimated slope is $\hat{\beta}_1 / s_{\hat{\beta}_1} = 3.31$.

⑰ The very small p-value accompanying the calculated t value for the estimated intercept indicates that the intercept is significantly different from zero (there is only about a 0.41% chance of finding a t value at least as large as 3.28 if $H_0: \beta_0 = 0$ is true).

⑱ The very small p-value supporting the calculated t value for the estimated slope reveals that the slope is significantly different from zero (there is only about a 0.40% chance of finding a t value at least as large as 3.31 if $H_0: \beta_1 = 0$ is true).

⑲ We may be 95% confident that $2.028 \le \beta_0 \le 9.241$, and we are likewise 95% confident that $0.267 \le \beta_1 \le 1.192$.

2. Additional SAS System Regression Program Elements

Standardizing the Regression Slope

If we desire to express the regression slope in standard deviation units, then we can use the *stb* option. To accomplish this, we would adjust the model statement in Exhibit 2.D.1 as follows: model *y = x/clb stb*; In addition to the normal regression output, we would also obtain the value of the standardized regression coefficient $\hat{\beta}_1^*$, i.e., when X increases by one standard deviation, Y changes by $\hat{\beta}_1^*$ standard deviations on average (see Exercise 2.14).

Suppressing the Regression Intercept

If we desire to force the estimated regression equation to pass through the origin, then we need only use the *noint* option in the model statement of Exhibit 2.D.1, i.e., model *y = x/clb noint*;

Predicted Values and Residuals

For each data point (X_i, Y_i), i = 1,…,n, we can request a print of the observed value of the dependent variable Y_i, the predicted value $\hat{Y}_i$, and the residual $e_i = Y_i - \hat{Y}_i$. This is accomplished by using the *p* option in the model statement of Exhibit 2.D.1 or model $y = x$ / *clb p*.

95% Confidence Interval for the Mean of Y Given X

A 95% confidence interval for $E(Y_i | X_i) = \beta_0 + \beta_1 X_i$ (see Equation 2.22) can be determined by using the *clm* option in the model statement of Exhibit 2.D.1, i.e.,

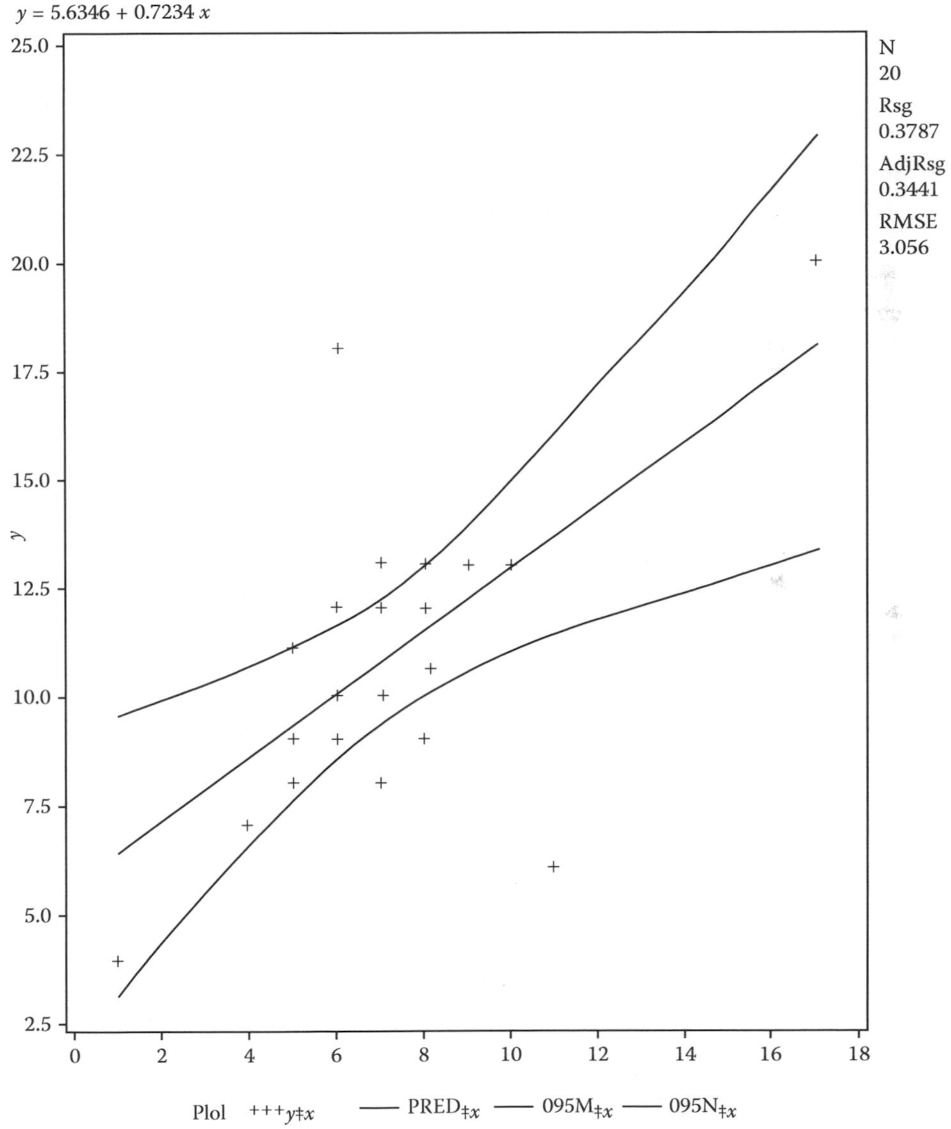

FIGURE 2.D.2
95% Confidence band for the mean of *Y* given *X*.

model $y = x$/*clb clm*;
plot y^*x/conf;

If a plot of the resulting confidence band is desired, then the indicated plot statement follows the model statement.

95% Prediction Interval for a Particular Value of Y Given X

A 95% prediction interval for Y_0 given X_0 (see Equation 2.30) can be formed by using the *cli* option in the model statement of Exhibit 2.D.1. Here,

model $y = x$/*clb cli*;
plot y^*x/pred;

If a plot of the resulting prediction band is desired, then the indicated plot statement follows the model statement.

TABLE 2.D.3

Output for Example 2.D.2

The SAS System
The REG Procedure
Model: MODEL1
Dependent Variable: *y*
Output Statistics

Obs	Dep Var y	(20) Predicted Value	(21) Std Error Mean Predict	(22) 95% CL Mean		(23) 95% CL Predict		(24) Residual
1	4.0000	6.3641	1.5170	3.1769	9.5512	−0.8039	13.5320	−2.3641
2	7.0000	8.5523	0.9738	6.5065	10.5981	1.8139	15.2908	−1.5523
3	8.0000	9.2817	0.8314	7.5351	11.0284	2.6280	15.9355	−1.2817
4	9.0000	9.2817	0.8314	7.5351	11.0284	2.6280	15.9355	−0.2817
5	11.0000	9.2817	0.8314	7.5351	11.0284	2.6280	15.9355	1.7183
6	9.0000	10.0112	0.7288	8.4801	11.5422	3.4107	16.6116	−1.0112
7	10.0000	10.0112	0.7288	8.4801	11.5422	3.4107	16.6116	−0.0112
8	12.0000	10.0112	0.7288	8.4801	11.5422	3.4107	16.6116	1.9888
9	18.0000	10.0112	0.7288	8.4801	11.5422	3.4107	16.6116	7.9888
10	8.0000	10.7406	0.6841	9.3033	12.1779	4.1613	17.3199	−2.7406
11	10.0000	10.7406	0.6841	9.3033	12.1779	4.1613	17.3199	−0.7406
12	12.0000	10.7406	0.6841	9.3033	12.1779	4.1613	17.3199	1.2594
13	13.0000	10.7406	0.6841	9.3033	12.1779	4.1613	17.3199	2.2594
14	9.0000	11.4700	0.7085	9.9815	12.9585	4.8793	18.0607	−2.4700
15	12.0000	11.4700	0.7085	9.9815	12.9585	4.8793	18.0607	0.5300
16	13.0000	11.4700	0.7085	9.9815	12.9585	4.8793	18.0607	1.5300
17	13.0000	12.1994	0.7956	10.5280	13.8709	5.5650	18.8338	0.8006
18	13.0000	12.9288	0.9279	10.9795	14.8782	6.2191	19.6386	0.0712
19	6.0000	13.6583	1.0890	11.3704	15.9461	6.8424	20.4741	−7.6583
20	20.0000	18.0348	2.2744	13.2565	22.8131	10.0315	26.0381	1.9652

Sum of residuals	0
Sum of squared residuals	168.10283
Predicted residual SS (PRESS)	218.81430

Example 2.D.2

For the dataset involving the $n = 20$ data points provided in Example 2.D.1, let us modify the model statement of Exhibit 2.D.1 as follows:

```
model y = x / clb p clm  cli;
plot y*x/conf;
plot y*x/pred;
```

The output for this revised program appears as Table 2.D.3. (Note that only the new output for PROC REG is given.) Looking to the features of Table 2.D.3, we have the following:

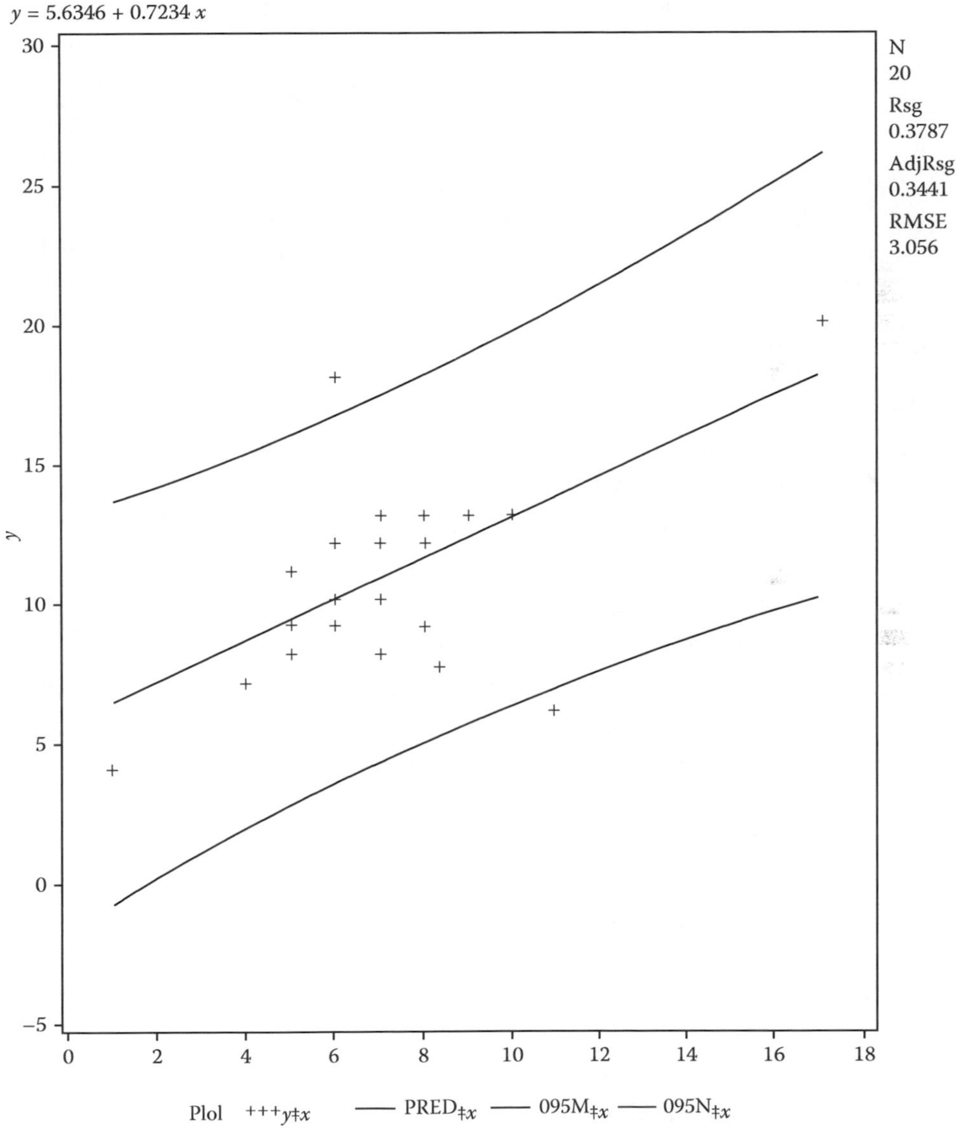

FIGURE 2.D.3
95% Prediction band for Y_0 given X_0.

(20) Given that the coordinates of the data points $(X_i, Y_i), i = 1, \ldots, 20$ are provided in Table 2.D.1, it follows, for observation 1, that $\hat{Y}_i = \hat{\beta}_0 + \hat{\beta}_1 X_1 = 5.634 + 0.729(1) = 6.3641$. (The other predicted values of Y are determined in a similar manner.)

(21) Because $\hat{Y}_i = \hat{\beta}_0 + \hat{\beta}_1 X_i$ is an estimator for $E(Y_i | X_i) = \hat{\beta}_0 + \hat{\beta}_1 X_i$, the estimated standard error of $\hat{Y}_i$ is provided by Equation 2.18. For observation 2, $X_2 = 4$, $\hat{Y}_2 = 8.5523$, and $S_{\hat{Y}_2} = 0.9738$.

(22) 95% confidence limits for $E(Y_i | X_i) = \beta_0 + \beta_1 X_i$ are provided by Equation 2.22. For observation 2, we have $\hat{Y}_2 \pm t_{\alpha/2, n-2} S_{\hat{Y}_2}$ or $8.5523 \pm 2.306(0.9738)$ or $(6.5065, 10.5981)$. That is, we may be 95% confident that the population average value of Y given $X = 4$ lies between 6.5065 and 10.5981 (Figure 2.D.2).

(23) A 95% prediction interval for Y_0 given X_0 is provided by Equation 2.30. For observation no.2, $X_0 = 4$, $\hat{Y}_0 = 8.5523$, and the prediction limits are $(1.8139, 15.2908)$, i.e., we may be 95% confident that Y_0 given $X_0 = 4$ lies between 1.8139 and 15.2908 (Figure 2.D.3).

(24) For observation 1, the observed Y value is $Y_1 = 4.0000$, the predicted Y is $\hat{Y}_1 = 6.3641$, and thus the residual $e_1 = Y_1 - \hat{Y}_1 = -2.3641$. (The remaining residuals are determined in a similar manner.)

Exercises

2-1. Given the following production dataset pertaining to "batch size" (X) and "total defects per batch" (Y) (Table E.2-1):

a. Determine $\sum x_i^2, \sum y_i^2$, and $\sum x_i y_i$ given $\overline{X}=50$, $\overline{Y}=11.1$, $\sum X_i^2=28400$, $\sum Y_i^2=1373$, and $\sum X_i Y_i=6240$.

b. Establish the OLS regression equation that best fits the dataset.

c. Interpret the estimated intercept and slope.

d. Verify that the estimated regression equation passes through the "point of means."

e. Find the residuals e_i. Verify that $\sum e_i = 0$ (approximately).

2-2. Using the Exercise 2-1 dataset:

a. Establish S_ε^2, SSE, MSE, and $COV\left(\hat{\beta}_0, \hat{\beta}_1\right)$.

b. Suppose it is assumed that ε_i are independent $N(0, \sigma_\varepsilon^2)$. What is known about the form of the sampling distributions of $\hat{\beta}_1$ and $\hat{\beta}_0$?

TABLE E.2-1

Defects per Batch (Y) and Batch Size (X)

X	30	20	60	80	40	50	60	30	70	60
Y	7	5	13	17	9	11	14	7	15	13

c. Determine $S_{\hat{\beta}_1}$ and $S_{\hat{\beta}_0}$.

d. What is known about the form of the sampling distributions of $(\hat{\beta}_1 - \beta_1)/S_{\hat{\beta}_1}$ and $(\hat{\beta}_0 - \beta_0)/S_{\hat{\beta}_0}$?

e. Find 95% confidence limits for β_1 and β_0. How precisely have we estimated these parameters? Also, determine Bonferroni 95% joint confidence limits for β_1 and β_0.

f. Test H_0: $\beta_1 = 0$ against H_0: $\beta_1 > 0$ for α=0.01. What can be said about the p-value of test?

g. Test H_0: $\beta_0 = 0$ against H_1: $\beta_1 > 0$ for α=0.01. What can be said about this test's p-value?

h. Is the entire population regression equation significant? Answer this question by testing β_1 and β_0 jointly, i.e., test H_0: $\beta_0 = \beta_1 = 0$ against H_0: $\beta_0 \neq 0,\ \beta_1 \neq 0$.

2-3. Using the Exercise 2-1 dataset, determine how precisely the population regression equation has been estimated from the sample data, i.e., using Equation 2.22, find a 95% confidence band for $E(Y_i|X_i = \beta_0 + \beta_1 X_i$. Also, given $X_0 = 40$, determine how precisely Y_0 has been estimated from the sample data, i.e., using Equation 2.30, find a 95% confidence interval for Y_0. Finally, test H_0: $Y_0 \leq 10$ against H_1: $Y_0 > 10$ using α=0.05.

2-4. Using the Exercise 2-1 dataset:

a. Find SST, SSR, SSE,$R^2, \bar{R}^2$.

b. Use this information to construct a partitioned sums-of-squares table.

c. Determine whether or not X contributes significantly to the variation in Y, i.e., test $H_0: \beta_1 = 0$ versus $H_0: \beta_1 \neq 0$ using Equation 2.36. (Can one also make use of the fact that, under $H_0: \beta_1 = 0$, t^2 = F?)

2-5. Using the Exercises 2-1 dataset:

a. Determine the value of the Pearson product-moment correlation coefficient.

b. Test $H_0: \rho_1 = \rho_0 = 0$ against $H_1: \rho >= 0$ for $\alpha = 0.05$.

c. Test $H_0: \rho = \rho_0 = 0.8$ against $H_0: \rho \neq 0.8$ for $\alpha = 0.05$.

d. Find a 95% confidence interval for ρ or $((l_\rho, u_\rho)$.

2-6. Given the following dataset involving "gross sales revenue" (Y) and "units sold" (X) (Table E.2-6) and the restriction that $\beta_l = 0$:

a. A 95% confidence interval for $\beta_l = 0$ (given that $\sum X_i^2 = 9625$, $\sum Y_i^2 = 96550421$, and $\sum X_i Y_i = 963990$).

b. For $\alpha = 0.01$, test $H_0: \beta_1 = 0$ versus $H_1: \beta_1 > 0$.

c. Find a 95% confidence band for the population regression line.

d. Test $H_0: E(Y_3|X_3 = 15) = 1525$ against $H_1: E(Y_3|X_3 = 15) \neq 1525$ using $\alpha = 0.05$

e. Suppose $X = X_0 = 40$. Find $\hat{Y}_0$ along with a 95% confidence interval for the true Y_0 value.

f. Construct the partitioned sums-of-squares table and find r^2. Determine via an F test if X contributes significantly to the variation in Y.

TABLE E.2-6

Gross Sales Revenue (Y) and Units Sold (X)

X	5	10	15	20	25	30	35	40	45	50
Y	525	990	1510	1985	2520	2999	3520	3989	4521	5000

2-7. Given the dataset presented in Exercise 2-1, if 10 defects per batch is to become the standard, make an inference concerning the batch size that will generate this number of defects. Use $1-\alpha = 0.95$.

2-8. Given the accompanying set of observations on X and Y (Table E.2-8):

a. Determine whether a regressional relationship exists between these variables. (Hint: for $\alpha = 0.05$, test H_0: no functional relationship between X and Y versus H_1: not H_0.) What proportion of the variation in Y is accounted for by X?

b. If the null hypothesis in part a is rejected, determine whether the implied functional form is linear. (Hint: for $\alpha = 0.05$, test H_0: the regression function is linear versus H_1: not H_0. Use Equation 2.62 and the partitioned error sums-of-squares table (Table 2.10).)

c. Construct the partitioned (by linear regression) sums-of-squares table (Table 2.13) and, using $\alpha = 0.05$, test for the significance of the slope and for linearity. Find R^2, H^2, and W^2. Interpret your results.

2-9. Given the following datasets involving independent samples extracted from populations A and B (Table E.2-9):

a. Determine whether the regression slopes are equal.

b. Determine whether the regression intercepts are equal.

c. Is the strength of the linear association between the X and Y variables the same for each population? Use $\alpha = 0.05$. (What assumptions about the A and B populations must be made?)

2-10. Resolve Exercises 1-4 using SAS PROC REG. Fully explain your results.

TABLE E.2-8

Observations on X and Y

X	Y|X Replicates				
1	21,	23,	27		
2	21,	24,	25		
4	16,	18,	19		
7	9,	10,	12,	14	
10	3,	5,	6,	8,	9
12	3,	6,			
13	1,	3,	4		

TABLE E.2-9

Observations on X and Y

Population A			Population B		
X_A	Y_A		X_B	Y_B	
2	4		3	8	
4	5	$n_A=20$	5	11	$n_B=20$
6	3	$\bar{X}_A = 21$	7	13	$\bar{X}_B = 22$
8	4	$\bar{Y}_A = 9.1$	9	12	$\bar{Y}_B = 23.8$
10	7	$\sum X_{Ai}^2 = 2660$	11	17	$\sum X_{Bi}^2 = 2660$
12	8	$\sum Y_{Ai}^2 = 261.8$	13	16	$\sum Y_{Bi}^2 = 1739.2$
14	6	$\sum X_{Ai} Y_{Ai} = 776$	15	21	$\sum X_{Bi} Y_{Bi} = 2112$
16	9		17	18	
18	7		19	19	
20	9		21	24	
22	10		23	22	
24	9		25	26	
26	10		27	30	
28	11		29	28	
30	13		31	34	
32	10		33	30	
34	16		35	35	
36	12		37	34	
38	15		39	39	
40	14		41	39	

2-11. Resolve Exercise 6 using SAS PROC REG. Fully explain your results.

2-12. Given the sample dataset in Table E.2-12, find:

a. Least-squares estimates of β_0 and β_1.

b. An estimate of σ_ε^2.

c. Standard error of estimate.

d. Estimates of the standard errors of the regression coefficients.

TABLE E.2-12

Observations on X and Y

Y	4	4	6	6	10	8	10	14	10	16
X	2	4	6	8	9	10	11	12	14	16

2-13. Using the data presented in Exercise 2-12:

a. Find 95% confidence intervals for β_0 and β_1.

b. For $\alpha = 0.01$, test $H_0: \beta_1 = 0$, against $H_1: \beta_1 > 0$.

c. Construct a hypothesis test to determine whether the population regression line passes through the origin.

d. For $\alpha = 0.05$, test $H_0: \beta_1 = \beta_1^0 = 0.65$, against $H_1: \beta_1 < 0.65$.

e. Determine whether β_0 and β_1 are jointly significantly different from zero.

2-14. The *point elasticity* of Y with respect to X is defined as $\eta = (dY/dX)(X/Y)$. For the population regression equation estimated in Exercise 2-12, find an estimate of the point elasticity coefficient η at the point of means of the variables; that is, find $\hat{\eta} = (d\hat{Y} / dX)(\bar{X} / \bar{Y})$. Interpret your result. (Hint: η serves as an index of the responsiveness of Y to a 1% change in X. So if, say, $\hat{\eta} = 1.35$, then a 1% increase in X leads to a 1.35% increase in Y (the *elastic case*, because $\hat{\eta} > 1$). If $\hat{\eta} = 0.80$, a 1% increase in X precipitates only a 0.80% increase in Y (the *inelastic case*, because $\hat{\eta} < 1$)).

2-15. Using the data presented in Exercise 2-12:

a. Find a 95% confidence band for the population regression equation using the following X values: 2, 4, 6, 8, 10, 12, 14, and 16.

b. For $X = 9$, can we conclude at the $\alpha = 0.05$ level that, on average, Y will exceed 10.5?

c. If $X_0 = 17$, determine a 95% prediction interval for Y_0.

d. For $\alpha = 0.05$, can we conclude that, for $X_0 = 17$, the predicted value of Y does not exceed 18?

e. For $\alpha = 0.05$, determine whether the underlying regression relationship generated the point $(X_0, Y_0) = (18, 18)$.

2-16. Using the data presented in Exercise 2-12:

a. Construct the analysis-of-variance table for the partitioned sum of squares.

b. What is the value of the coefficient of determination?

c. Use the F test with $\alpha = 0.05$ to determine whether there exists a significant linear relationship between X and Y.

2-17. For the Exercise 2-12 dataset:

a. Find $\hat{\rho}$ using Equation 16.37.

b. For $\alpha = 0.01$, test $H_0: \rho = 0$, against $H_1: \rho > 0$.

c. Using $\alpha = 0.05$, test $H_0: \rho = \rho_0 = 0.85$, against $H_1: \rho < 0.85$.

d. Find a 95% confidence interval for ρ.

2-18. Given the sample dataset in Table E.2-18:

a. Perform a regression and correlation study. Use $\alpha = 0.05$.

b. For $X_0 = 23$, find 95% prediction interval for Y_0.

TABLE E.2-18

Observations on X and Y

Y	4	2	4	6	10	6	8	12	16	14
X	2	4	6	8	10	12	14	16	18	20

2-19. Our operating assumption in this chapter is that Y is a linear function of X (plus an additive error term). What is important is not that the regression model is linear in the variables but that it must be *linear in the parameters* to be estimated. For example,

a. $Y = \beta_0 + \beta_1 X^{-1}$ is nonlinear in X but linear in the parameters; and

b. $Y = \beta_0 X^{\beta_1}$ is nonlinear in both X and in the parameters yet $\ln Y = \ln \beta_0 + \beta_1 \ln X$ is linear in the parameters.

For case a:

1. Show that the slope is everywhere negative and decreases in absolute value as X increases.
2. Graph the function for $X > 0$. Verify that as $X \to 0, Y \to \infty$, and, as $X \to \infty, Y \to \beta_0$.
3. What is the interpretation of β_0?
4. Convert Y into a linear estimating equation by introducing a new variable $Z = X^{-1}$. Estimate β_0 and β_1 via least squares using the data in Table E.2-19.A.

For case b:

1. Show that, if $\beta_1 > 0$, the slope is always positive and $Y \to \infty$ as $X \to \infty$. If $\beta_1 > 1$, the slope is monotonically increasing as X increases; if $0 < \beta_1 < 1$, the slope is monotonically decreasing as X increases. If $\beta_1 < 0$, the slope is always negative as X increases.
2. Graph the function for $\beta_1 = -1$ and $X > 0$. What is this type of expression called?
3. Verify that this function has a constant elasticity equal to β_1.
4. Using the data in Table E.2-19.B, estimate this expression via least squares by defining new variables $W = \ln Y$ and $Z = \ln X$. How is β_0 obtained?

2-20. Use the methodology of the preceding exercise and the accompanying dataset given here (Table E.2-20) to estimate the parameters β_0 and β_1 of the expression $Y = e^{\beta_0 + \beta_1 X}, Y > 0$, where the vertical intercept is e^{β_0}. The slope of this function is positive (respectively, negative) if $\beta_1 > 0$ (respectively, < 0). Because $\beta_1 = (1/Y)(dY/dX)$, this function is alternatively referred to as the *constant growth curve*; that is, the proportionate rate of change in Y per unit change in X is the constant β_1.

TABLE E.2-19

Observations on X and Y

A		B	
X	Y	X	Y
1	14	1	2
2	10	2	5
3	8	3	8
4	6	4	12
5	5.8	5	17
6	5.5	6	25
7	5.3	7	37
8	4.9	8	52

TABLE E.2-20

Observations on X and Y

Y	8	11	14	18	23	31	43	58
X	1	2	3	4	5	6	7	8

2-21. What if, instead of regressing a dependent variable Y on an explanatory variable X, we regress the Z-scores of Y on the Z-scores of X? From $Y_i = \beta_0 + \beta_1 X_i + \varepsilon_i$ and $\bar{Y} = \beta_0 + \beta_1 \bar{X} + \bar{\varepsilon}$,

$$\frac{Y_i - \bar{Y}}{\sigma_Y} = \beta_1 \frac{X_i - \bar{X}}{\sigma_X}\frac{\sigma_X}{\sigma_Y} + \frac{\varepsilon_i - \bar{\varepsilon}}{\sigma_Y}$$

or

$$Z_{iY} = \beta_1 \left(\frac{\sigma_X}{\sigma_Y}\right) Z_{iX} + u_i = \beta_1^* Z_{iX} + u_i. \tag{2.E.1}$$

Applying ordinary least squares to 2.E.1 renders

$$\hat{Z}_{iY} = \hat{\beta}_1^* Z_{iX}.$$

(What assumptions must u_i satisfy?)

The quantity β_1^* is termed a *standardized regression coefficient* or simply a *beta coefficient*. So although the usual regression slope $\hat{\beta}_1$ represents the average rate of change in Y per unit change in X, we see that the beta coefficient $\hat{\beta}_1^*$ measures the said changes in *standard deviation units*; that is, when X increases by one standard deviation, Y changes by $\hat{\beta}_1^*$ standard deviations on average. Hence, the beta coefficient is independent of units. Note that we can easily recover the usual regression slope as $\hat{\beta}_1 = \hat{\beta}_1^* (\sigma_Y / \sigma_X)$. A glance at the structure of $\hat{\beta}_1^*$ reveals that it is simply the estimated coefficient of correlation between the variables X and Y.

For the following dataset (Table E.2-21):

a. Transform the X and Y variables to Z scores.

b. Use least squares to regress the Y Z-score on the X Z-score with the intercept suppressed.

c. From $\hat{\beta}_1^*$, determine $\hat{\beta}_1$.

d. Calculate the coefficient of correlation between X and Y directly and verify that it equals $\hat{\beta}_1^*$.

TABLE E.2-21

Observations on X and Y

X	1	2	3	4	5	6	7	8
Y	2	4	3	5	6	5	4	7

2-22. Testing for a *structural break*, or testing for the equality of two regression equations, can be accomplished by performing the so-called *Chow test*. For instance, suppose we have a sample of n_1 observations on the variables X and Y for one time period and a second sample of n_2 data points on these variables for another time period (e.g., we can model some measurable characteristic before 9/11/01 and after 9/11/01). The relevant question is then, "Is there a change (in intercepts, or slopes, or both) in the response of Y to X between the two periods?" That is, "Can each regression equation be viewed as coming from the same population?" Looked at a third way, "Is the regression relationship between X and Y structurally stable over the two time periods?"
To perform the Chow test for equality of two regression equations

$$Y_{i1} = \beta_{01} + \beta_{11}X_{i1} + \varepsilon_{i1}, i = 1,\ldots,n_1; \text{ [Period 1]} \quad (2.E.2)$$
$$Y_{j2} = \beta_{02} + \beta_{12}X_{j2} + \varepsilon_{j2}, j = 1,\ldots,n_2 \text{ [Period 2]} \quad (2.E.3)$$

(here the second subscript refers to the subsample or period):

1. STEP 1. Pool all $n_1 + n_2 = n$ observations and estimate a single regression equation $Y = \beta_0 + \beta_1 X + \varepsilon$ over the combined sample period. From this regression, obtain the residual sum of squares SSE_C with $n_1 + n_2 - 2$ degrees of freedom.
2. STEP 2. Estimate the two subperiod regressions 2.E.2 and 2.E.3 separately and obtain the residual sum of squares for each, denoted SSE_1 with degrees of freedom $n_1 - 2$ and SSE_2 with degrees of freedom $n_1 - 2$, respectively. Form $SSE_+ = SSE_1 + SSE_2$ with $n_1 + n_2 - 4$ degrees of freedom.
3. STEP 3. To test whether $H_0 : \beta_{01} = \beta_{02}$ and $\beta_{11} = \beta_{12}$ against $H_1 : H_0$ is not true, we use the test statistic

$$F = \frac{(SSE_C - SSE_+)/2}{SSE_+/(n_1 + n_2 - 4)} : F_{2,n_1+n_2-4,} \quad (2.E.4)$$

where

$$R = \left\{F \middle| F > F_{1-\alpha,2,n_1+n_2-4}\right\}.$$

For the dataset given here (Table E.2-22), use the Chow test to determine whether, for $\alpha = 0.05$, the two regression structures are the same. That is, at this level of significance, can we conclude that the two subsamples have been drawn from the same population regression structure?

2-23. *Binary*, *dummy*, or *count* variables are used to depict the situation in which some attribute is either present or absent. For instance, in taking the medical history of a patient in a hospital, we may ask whether he or she is a smoker. If the individual answers yes, we can record a 1; if the answer is no, we record a 0. In general, a 1 indicates the presence of some attribute and 0 indicates its absence. More formally, if X_i denotes a binary variable, then

$$X_i = \begin{cases} 1 & \text{if an attribute is present;} \\ 0 & \text{if an attribute is not present.} \end{cases}$$

TABLE E.2-22

Observations on X and Y

Period 1		Period 2	
X	Y	X	Y
88	36	166	90
94	21	176	95
100	18	186	82
105	20	196	104
109	10	211	153
118	12	227	194
127	41	239	175
134	50	250	199
141	43		
155	59		

Other instances include the attributes of sex ($X_i = 1$) if the person is male; $X_i = 0$ if the person is female), automobile ownership ($X_i = 1$ if an individual owns an automobile; $X_i = 0$ if not), and so on. Suppose we are interested in comparing the average salaries of males versus females at the XYZ Corporation, where μ_M (respectively, μ_F) denotes the true or population mean salary level for males (respectively, females). If Y represents employee salaries (assume Y is $N(\mu,\sigma)$, $\sigma =$ constant, with overall mean μ), then a salary comparison between males and females can be undertaken by formulating the following linear regression model:

$$Y_i = \beta_0 + \beta_1 X_i + \varepsilon_i, i = 1,\ldots,n, \tag{2. E.5}$$

where Y_i is the salary of the ith employee in a sample of size n, X_i is a dummy variable such that

$$X_i = \begin{cases} 1 & \text{if the employee is male (M);} \\ 0 & \text{if the employee is female (F),} \end{cases}$$

and ε_i is a random error term satisfying all of the assumptions of the (strong) classical linear regression model. In this regard, because

$$E(Y_i | X_i = 0) = \mu_F = \beta_0,$$
$$E(Y_i | X_i = 1) = \mu_M = \beta_0 + \beta_1,$$

it follows that

$$\beta_1 = \mu_M - \mu_F;$$

that is, the intercept of the population regression equation measures the mean salary of females, and the slope of the same measures the difference between the mean salary of males and the mean salary of females. So to compare the average salaries of males and females, let us test $H_0 : \beta_1 = 0$ (there is no difference between the average salaries of males and females), against any of the following alternative hypotheses:

TABLE E.2-23

Absenteeism (Y) versus Smoking (X)

X	Y (Days Absent)
0	3
0	2
1	5
0	2
1	7
1	9
1	11
0	4
0	3
0	2
1	6
0	0
0	1
0	3
1	5

Case I	Case II	Case III
$H_0 : \beta_1 = 0$	$H_0 : \beta_1 = 0$	$H_0 : \beta_1 = 0$
$H_1 : \beta_1 > 0$	$H_1 : \beta_1 < 0$	$H_1 : \beta_1 \neq 0$
(or $\mu_M > \mu_F$)	(or $\mu_M < \mu_F$)	(or $\mu_M \neq \mu_F$)

The coefficient of Equation 2.E.5 can be estimated via least squares and all of the standard tests and interval estimates apply. (Note that, because it can be readily demonstrated that $\hat{\beta}_1 = \bar{Y}_M - \bar{Y}_F$ and $\hat{\beta}_0 = \bar{Y}_F$, where $\bar{Y}_M$ is the sample mean salary of males and $\bar{Y}_F$ is the sample mean salary of females, it follows that testing $H_0 : \beta_1 = 0$ is equivalent to the two-sample t test of $H_0 : \mu_M - \mu_F = 0$.)

For the following dataset pertaining to employee absenteeism (Y) for a one-month period at ABC Inc. and smoking (X) (Table E.2-23), determine whether smokers have a significantly higher rate of absenteeism relative to nonsmokers over this time period using $\alpha = 0.05$. Let

$$X_i = \begin{cases} 1 & \text{if the employee is a smoker;} \\ 0 & \text{if the employee is a nonsmoker.} \end{cases}$$

2-24. Demonstrate that, if R is the coefficient of correlation between the n pairs (X_i, Y_i), $i = 1, \ldots, n$, then for constants a, b, c, and d, the correlation coefficient between the n pairs $(a + bX_i, c + dY_i)$, $i = 1, \ldots, n$, is also R.

3

Misspecified Disturbance Terms

The ε_i Are Not Normally Distributed

In the preceding chapter, we specified the assumptions underlying the strong classical linear regression model as follows:

A.1. ε_i is normally distributed for all i;

A.2. $E(\varepsilon_i) = 0$ for all i;

A.3. $V(\varepsilon_i) = \sigma_\varepsilon^2 =$ constant for all i;

A.4. $COV(\varepsilon_i, \varepsilon_j) = 0, i \neq j$, and $COV(X_i, \varepsilon_i) = 0$ for all i and j; and

A.5. X is nonrandom.

In this chapter, we seek to determine the implications of the violations of these assumptions on the properties of the least-squares estimators of the population regression parameters β_0 and β_1. (In what follows, these estimators will be termed OLS estimators.)

If Assumptions A.1 through A.5 are valid, then the OLS estimators for β_0 and β_1 are deemed good estimators. In the absence of any compelling evidence to the contrary, these assumptions are assumed to hold. But what if one or more of these assumptions are thought to be invalid?

Assumptions A.1 through A.5 may be challenged on a priori grounds (the process that generates the ε_i's and/or X may be called into question) or, as is usually the case, on a posteriori grounds (from an examination of the behavior of the OLS residuals, e_i, $i = 1,\ldots,n$). If a test of the regression residuals leads us to reject one or more of Assumptions A.1 through A.5, then some of the desirable properties of the OLS estimators may be compromised and we will have to make appropriate adjustments to our data or use some alternative estimation technique.

If Assumption A.1 is dropped (the weak classical linear regression model obtains), then $\hat{\beta}_0$ and $\hat{\beta}_1$ are still BLUE (in the sense of having the smallest variance among all linear unbiased estimators of β_0 and β_1), and S_ε^2 is still an unbiased estimators of σ_ε^2. Moreover, these estimators are all consistent. If the sample size n is small, then the use of the t and F statistics are not fully justified in testing hypotheses concerning the regression parameters. However, if n is large, then the sampling distributions of $\hat{\beta}_0$ and $\hat{\beta}_1$ are approximately normal and thus the use of the t and F tests (although not strictly correct) are justified. In fact, the t test is quite robust in that it is not appreciably affected by departures from normality.

If Assumptions A.2 through A.5 are valid, then $\hat{\beta}_0$ and $\hat{\beta}_1$ are best unbiased estimators of β_0 and β_1, respectively. However, if Assumptions A.3 and A.4 are not also valid, then, as will be seen below, these estimators are not best, and the usual formulas for their standard errors are not applicable.

E (ε_i) ≠ 0 for All *i*

Unfortunately, $E(\varepsilon_i) = 0$ cannot be tested directly. In fact, $E(\varepsilon_i) = 0$ cannot even be verified by an analysis of the residuals because $\sum_{i=1}^{n} e_i = 0$ always holds under the method of least squares (see the system of least-squares normal Equation 2.4). What is the consequence of Assumption A.2 being invalid?

Let $\varepsilon_i = h + u_i$, where h is a nonzero constant and u_i is a random error term that obeys all of the assumptions underlying the weak classical linear regression model. Clearly, $E(\varepsilon_i) = h \neq 0$. From Equation 2.3,

$$\begin{aligned} Y_i &= \beta_0 + \beta_1 X_i + \varepsilon_i \\ &= (\beta_0 + h) + \beta_1 X_i + u_i. \end{aligned}$$

Then using OLS to estimate the constant term in this expression results in a biased estimate of β_0. Hence, the effect of $E(\varepsilon_i) = h \neq 0$ is to shift the sample regression line upward (respectively, downward) by $h > 0$ (respectively, $h < 0$) units. So, whereas the least-square estimator of the constant term is biased upward ($h > 0$) or downward ($h < 0$), the least-square estimate of $\hat{\beta}_1$ is unaffected, i.e., $\hat{\beta}_1$ is still BLUE.

Heteroscedasticity

By Assumptions A.2 and A.3 of the weak classical linear regression model, we have $V(\varepsilon_i) = E(\varepsilon_i^2) = \sigma_\varepsilon^2$ = constant for all $i = 1,\ldots,n$. This constant error variance assumption is known as *homoscedasticity*, But, if the variance of the regression disturbance term ε is not constant for all i, then we have $V(\varepsilon_i) = \sigma_i^2$, the case of heteroscedasticity. Here the error variance varies from one data point to another. How does a nonconstant error variance affect the properties of the least-squares estimators of the population regression coefficients β_0 and β_1?

It was determined in Appendix 2.A that $\hat{\beta}_1 = \beta_1 + \sum w_i \varepsilon_i$ (Equation 2.A.4) and $\hat{\beta}_0 = \beta_0 + \left(\beta_1 - \hat{\beta}_1\right)\bar{X} + \bar{\varepsilon}$ (Equation 2.A.5). It was also verified therein that $E\left(\hat{\beta}_0\right) = \beta_0$ and $E\left(\hat{\beta}_1\right) = \beta_1$. Hence, the OLS estimators of β_0 and β_1 are unbiased even if the random error term ε is heteroscedastic.

Next, let us determine whether, in the presence of heteroscedasticity, the OLS estimators of the regression parameters are still best linear unbiased estimators or BLUE. To accomplish this task, let us start with a regression equation exhibiting a heteroscedastic disturbance term (it is assumed that the remaining assumptions of the weak classical linear regression model still hold) or

$$Y_i = \beta_0 + \beta_1 X_i + \varepsilon_i, V(\varepsilon_i) = \sigma_i^2, \quad i = 1,\ldots,n, \tag{3.1}$$

where the values of the σ_i's are known. Then, under a suitable weighting process, we may transform this expression into an equivalent regression equation having a homoscedastic disturbance term

$$\frac{Y_i}{\sigma_i} = \beta_0\left(\frac{1}{\sigma_i}\right) + \beta_1\left(\frac{X_i}{\sigma_i}\right) + \frac{\varepsilon_i}{\sigma_i}, \quad \sigma_i \neq 0, \tag{3.2}$$

or

$$Y_i^* = \beta_0 V_i^* + \beta_1 X_i^* + \varepsilon_i^*, \tag{3.2.1}$$

where $Y_i^* = Y_i/\sigma_i$, $V_i^* = \sigma_i^{-1}$, $X_i^* = X_i/\sigma_i$ and $\varepsilon_i^* = \varepsilon_i/\sigma_i$. Note that, because the σ_i's vary or are generated from different conditional distributions having different standard deviations, this procedure assigns more weight to the observations having less dispersion relative to those data points exhibiting more dispersion.

If the values of the explanatory variables V_i^* and X_i^* are held fixed for all i, then clearly,

$$E(\varepsilon_i^*) = \frac{E(\varepsilon_i)}{\sigma_i} = 0,$$

$$V(\varepsilon_i^*) = \frac{V(\varepsilon_i)}{\sigma_i^2} = 1, \text{ and}$$

$$COV(\varepsilon_i^*, \varepsilon_j^*) = \frac{E(\varepsilon_i \varepsilon_j)}{\sigma_i \sigma_j} = 0,$$

i.e., all of the assumptions underlying the weak classical linear regression model hold, and thus, via the Gauss-Markov Theorem, we may apply the principle of least squares to Equation 3.2 so as to derive *weighted least-squares (WLS) estimators* for β_0 and β_1.

What is the rationale for working with a regression equation such as Equation 3.2? As we shall now see, a specification such as Equation 3.2 arises quite naturally if we assume that the ε_i are $N(0,\sigma_i)$ for all i so that the method of maximum-likelihood estimation can be applied. Suppose that the assumptions underlying the strong classical linear regression model hold save for homoscedasticity. Then, given $Y_i = \beta_0 + \beta_1 X_i + \varepsilon_i$ and $\varepsilon_i \sim N(0,\sigma_i)$, $i = 1,\ldots,n$, with the σ_i known, we may write the joint probability density function of the ε_i's as

$$\prod_{i=1}^{n} f(\varepsilon_i; 0, \sigma_1^2, \ldots, \sigma_n^2) = (2\pi)^{-n/2} \prod_{i=1}^{n} (\sigma_i^2)^{-\frac{1}{2}} e^{-\frac{1}{2}\sum \varepsilon_i^2/\sigma_i^2}.$$

Then the likelihood function for the sample becomes, for $\varepsilon_i = Y_i - \beta_0 - \beta_1 X_i$, $i = 1,\ldots,n$,

$$\mathcal{L}(\beta_0, \beta_1, \sigma_1^2, \ldots, \sigma_n^2; Y_1, \ldots, Y_n, n) = (2\pi)^{-n/2} \prod_{i=1}^{n} (\sigma_i^2)^{-\frac{1}{2}} e^{-\frac{1}{2}\sum (Y_i - \beta_0 - \beta_1 X_i)^2/\sigma_i^2}$$

or, in log-likelihood form,

$$\log \mathcal{L} = -\frac{n}{2}\log 2\pi - \frac{1}{2}\sum_{i=1}^{n}\log \sigma_i^2 - \frac{1}{2}\sum \frac{(Y_i - \beta_0 - \beta_1 X_i)^2}{\sigma_i^2}.$$

Because only the last term on the right side of this expression involves β_0 and β_1, to maximize $\log \mathcal{L}$, we may choose β_0 and β_1 so as to minimize the *weighted sum of squares*

$$\begin{aligned}\sum \frac{(Y_i - \beta_0 - \beta_1 X_i)^2}{\sigma_i^2} &= \sum \left[\left(\frac{Y_i}{\sigma_i}\right) - \beta_0\left(\frac{1}{\sigma_i}\right) - \beta_1\left(\frac{X_i}{\sigma_i}\right)\right]^2 \\ &= \sum \left(Y_i^* - \beta_0 V_i^* - \beta_1 X_i^*\right)^2 \\ &= \sum e_i^{*2} = F^*(\beta_0, \beta_1),\end{aligned} \tag{3.3}$$

where e_i^* is the ith residual or deviation from the weighted sample regression plane.

To this end, after setting $\partial F^*/\partial\beta_0 = \partial F^*/\partial\beta_1 = 0$, we obtain the system of *WLS normal equations*:

$$\left.\begin{aligned}\sum V_i^* e_i^* = 0 \\ \sum X_i^* e_i^* = 0\end{aligned}\right\} \text{ or } \left\{\begin{aligned}\tilde{\beta}_0 \sum V_i^{*2} + \tilde{\beta}_1 \sum V_i^* X_i^* = \sum V_i^* Y_i^* \\ \tilde{\beta}_0 \sum V_i^* X_i^* + \tilde{\beta}_1 \sum X_i^{*2} = \sum X_i^* Y_i^*\end{aligned}\right. . \tag{3.4}$$

If we set $v_i = 1/\sigma_i^2$, then the preceding system becomes

$$\begin{aligned}\tilde{\beta}_0 \sum v_i + \tilde{\beta}_1 \sum v_i X_i &= \sum v_i Y_i \\ \tilde{\beta}_0 \sum v_i X_i + \tilde{\beta}_1 \sum v_i X_i^2 &= \sum v_i X_i Y_i\end{aligned}, \tag{3.4.1}$$

and thus it can be shown (by solving Equations 3.4.1 simultaneously) that the WLS estimators for the population regression parameters β_1 and β_0 are, respectively,

$$\tilde{\beta}_1 = \frac{\sum v_i (X_i - \tilde{X})(Y_i - \tilde{Y})}{\sum v_i (X_i - \tilde{X})^2} \tag{3.5}$$

and

$$\tilde{\beta}_0 = \tilde{Y} - \tilde{\beta}_1 \tilde{X}, \tag{3.6}$$

where $\tilde{X} = \sum v_i X_i / \sum v_i$ and $\tilde{Y} = \sum v_i Y_i / \sum v_i$. (Note that $\tilde{X}$ is the weighted mean of the X_i's, and $\tilde{Y}$ is the weighted mean of the Y_i's, where the weights are the v_i's. Moreover, if $\sigma_i^2 = \sigma_\varepsilon^2$ = constant for all i, then $v_i = v$ = constant for all i, and thus $\tilde{X} = \bar{X}$ with $\tilde{Y} = \bar{Y}$. Hence, the WLS estimators $\tilde{\beta}_0$ and $\tilde{\beta}_0$ reduce to the OLS estimators $\hat{\beta}_0$ and $\hat{\beta}_1$, respectively.)

Because the WLS estimators of β_0 and β_1 are BLUE (according to the Gauss-Markov Theorem) and obviously differ from the formulas for the OLS estimators of these parameters (see Equations 2.5.a and b), it follows that, in the presence of heteroscedasticity, the OLS estimators $\hat{\beta}_0$ and $\hat{\beta}_1$ are not *BLUE*. Furthermore, because these OLS estimators do not have minimum variance among all unbiased linear estimators of the population regression parameters, they are not efficient. However, it can be demonstrated that, in the presence of heteroscedasticity, the OLS estimators are consistent, although not asymptotically efficient. In summary, if the random error term ε is heteroscedastic, the OLS estimators of the regression parameters are unbiased and consistent; they are not BLUE (and thus not efficient) or even asymptotically efficient.

Variances of the WLS and OLS Estimators under Heteroscedasticity

Next, it can be shown (via Appendix 3.A.1) that the variances of the WLS estimators $\hat{\beta}_1$ and $\tilde{\beta}_0$ are, respectively,

$$V(\tilde{\beta}_1) = \frac{1}{\sum v_i (X_i - \tilde{X})^2}, \tag{3.7}$$

$$V(\tilde{\beta}_0) = \frac{1}{\sum v_i} + \frac{\tilde{X}^2}{\sum v_i (X_i - \tilde{X})^2}. \tag{3.8}$$

(Note that if, $v_i = v = 1/\sigma_\varepsilon^2 = \text{constant}$ for all i, then Equations 3.7 and 3.8 reduce to 2.A.6 and 2.A.7, respectively, the variances of the OLS estimators $\hat{\beta}_1$ and $\tilde{\beta}_0$.)

We now turn to an examination of the variances of the OLS estimators of β_1 and β_0 when the error variances of the ε_i are heteroscedastic. From the first line of Equation 2.A.6, we have

$$V(\hat{\beta}_1) = E\left(\sum w_i \varepsilon_i\right)^2.$$

If $E(\varepsilon_i^2) = \sigma_i^2$ and $E(\varepsilon_i \varepsilon_j) = 0$, then the proceeding expression becomes

$$V(\hat{\beta}_1)_{het} = \sum w_i^2 \sigma_i^2, \tag{3.9}$$

where $w_i = x_i / \sum x_i^2$ and $x_i = X_i - \bar{X}$. From the next to the last line of the derivation of Equation 2.A.7,

$$\begin{aligned} V(\hat{\beta}_0)_{het} &= \bar{X}^2 V(\hat{\beta}_1) - 2\bar{X} E\left[(\hat{\beta}_1 - \beta_1)\bar{\varepsilon}\right] + E(\bar{\varepsilon}^2) \\ &= \bar{X}^2 V(\hat{\beta}_1) - 2\bar{X}\left(\frac{1}{n}\right)\sum w_i \sigma_i^2 + \frac{1}{n^2}\sum \sigma_i^2 \\ &= \bar{X}^2 \sum w_i^2 \sigma_i^2 - 2\bar{X}\left(\frac{1}{n}\right)\sum w_i \sigma_i^2 + \frac{1}{n^2}\sum \sigma_i^2 \\ &= \sum \sigma_i^2 \left(\frac{1}{n} - \bar{X} w_i\right)^2. \end{aligned} \tag{3.10}$$

(Note that, if $\sigma_i^2 = \sigma_\varepsilon^2 =$ constant for all i, Equations 3.9 and 3.10 simplify to 2.A.6 and 2.A.7, respectively.)

A comparison of Equations 3.7 and 3.9 reveals that WLS estimator $\tilde{\beta}_1$ is more efficient (it possesses a smaller variance) than the OLS estimator $\hat{\beta}_1$ if the random error term ε is heteroscedastic (see Appendix 3.A.2). A similar comparison can be made between Equations 3.8 and 3.10.

Properties of the Estimated Variances of the OLS Estimators under Heteroscedasticity

It was determined above that, in the presence of heteroscedasticity of the random error term ε, the OLS estimators $\hat{\beta}_1$ and $\hat{\beta}_0$ of the regression parameters β_1 and β_0, respectively, are unbiased and consistent but not efficient or even asymptotically efficient. Because we ultimately want to test hypotheses about the regression parameters as well as obtain confidence interval estimates of the same, we need to know whether the variances of $\hat{\beta}_1$ and $\hat{\beta}_0$ are biased or unbiased under heteroscedasticity. If the said variances are biased, then obviously any hypothesis tests and confidence intervals determined from them are also biased and thus invalid.

For instance, to determine whether $S_{\hat{\beta}_1}^2 = S_\varepsilon^2 / \sum x_i^2$ (an unbiased estimator of $V(\hat{\beta}_1)$ under the homoscedasticity of ε) remains unbiased for $V(\hat{\beta}_1)$ in the presence of heteroscedasticity, let us consider $E(S_\varepsilon^2)$ given that ε is heteroscedastic. From Equations 2.A.18 and 2.A.20,

$$E(S_\varepsilon^2) = \frac{1}{n-2} E\left[-(\hat{\beta}_1 - \beta_1)x_i + \varepsilon_i - \bar{\varepsilon}\right]^2. \tag{3.11}$$

Then, if $E(\varepsilon_i^2) = \sigma_i^2$ for all i, it follows from Equation 2.A.17.1 that

$$E(S_\varepsilon^2) = \frac{1}{n-2}\left[-\sum x_i^2 - V(\hat{\beta}_1)_{het} + \sum \sigma_i^2 - \frac{\sum \sigma_i^2}{n}\right]$$

$$= \frac{1}{n-2}\left[-\frac{\sum x_i^2 \sigma_i^2}{\sum x_i^2} + \left(\frac{n-1}{n}\right)\sum \sigma_i^2\right]. \tag{3.11.1}$$

Hence, by virtue of this expression and Equation 2.8,

$$E(S_{\hat{\beta}_1}^2) = E\left(\frac{S_\varepsilon^2}{\sum x_i^2}\right) = \frac{1}{n-2}\left[-\frac{\sum x_i^2 \sigma_i^2}{(\sum x_i^2)^2} + \left(\frac{n-1}{n}\right)\frac{\sum \sigma_i^2}{\sum x_i^2}\right]$$

$$= \frac{1}{n-2}\left[-V(\hat{\beta}_1)_{het} + \left(\frac{n-1}{n}\right)\frac{\sum \sigma_i^2}{\sum x_i^2}\right]. \tag{3.12}$$

Clearly, $E(S_{\hat{\beta}_1}^2) \neq \sigma_\varepsilon^2 / \sum x_i^2$, where $\sigma_\varepsilon^2 / \sum x_i^2$ is the variance of $\hat{\beta}_1$ under homoscedasticity. In fact, it is easily demonstrated that $E(S_{\hat{\beta}_1}^2) = \sigma_\varepsilon^2 / \sum x_i^2$ only if $\sigma_i^2 = \sigma_\varepsilon^2 =$ constant for all i. We thus conclude that, if the random disturbance term ε is heteroscedastic, the OLS estimator of $V(\hat{\beta}_1)$ is biased. A similar conclusion holds for $V(\hat{\beta}_0)$. Given the presence of this bias, any confidence interval estimates or hypothesis tests concerning β_0 and β_1 will be distorted and thus unreliable.

What is the direction of the bias implied by Equation 3.12? Using Equation 3.9, let us express the *bias of the variance of* $\hat{\beta}_1$ as

$$E(S^2_{\hat{\beta}_1}) - V(\hat{\beta}_1)_{het} = \frac{(n-1)\left[(\sum x_i^2)(\sum \sigma_i^2) - n\sum x_i^2\sigma_i^2\right]}{n(n-2)(\sum x_i^2)^2}. \tag{3.13}$$

Then, for $n > 2$, the direction (or sign) of this bias depends on the sign of the square-bracketed term in the numerator of Equation 3.13. In this regard, it can be demonstrated that, if $COV(x_i^2, \sigma_i^2) > 0$ (x_i^2 and σ_i^2 are positively associated), then the bias is negative, i.e., using OLS in the presence of heteroscedasticity artificially reduces the magnitude of the standard error of $\hat{\beta}_1$, thus favoring the terms on which the null hypothesis of no linear relationship between X and Y for a given α is rejected (because the value of the relevant t statistic is inflated). In this instance, the probability of rejecting H_0: $\beta_1 = 0$ will be higher than indicated by the chosen α or the p-value will be smaller than it should be. In addition, because the standard error of $\hat{\beta}_1$ is artificially lower than it should be, the width of a 100 $(1-\alpha)$% confidence interval for β_1 is shorter than it should be: the precision of our estimate of β_1 is artificially enhanced. Alternatively, if x_i^2 and σ_i^2 are negatively associated, then the bias provided by Equation 3.13 is positive; the magnitude of the standard error of $\hat{\beta}_1$ is artificially inflated and thus β_1 is estimated in a less precise manner (a confidence interval for β_1 is wider than it should be), and the likelihood that the null hypothesis of no linear relationship between X and Y will be rejected is decreased: the probability of rejecting the said null hypothesis will be lower than that indicated by the specified α or the p-value will be larger than warranted.

Given that OLS estimators of the regression parameters β_0 and β_1 in the presence of heteroscedasticity are only unbiased and consistent and possess variances that are biased, we need to develop an alternative estimation technique that would yield valid or untainted hypothesis tests and confidence interval estimates of β_0 and β_1. Fortunately, our discussion on weighted least squares presented above (see "Heteroscedasticity") easily forms the basis for an estimation method that can be used to correct for the effects of heteroscedastic disturbances. Because the weights were specified as $v_i = 1/\sigma_i^2$, it is evident that, to obtain a viable estimation scheme, we need to make some assumption(s) about the behavior of σ_i^2.

Assumptions about the Nature of σ_i^2

Suppose the random error variance is associated with some variable Z, where the Z_i values are known. Specifically, let

$$\sigma_i^2 = \sigma^2 Z_i^{\delta}, \qquad \text{[multiplicative heteroscedasticity]} \tag{3.14}$$

where σ^2 and δ are (unknown) parameters. (Note that, if $\delta = 0$, the disturbances are homoscedastic.) Hence, $\delta \neq 0$ serves as a measure of the strength of heteroscedasticity, i.e., the larger the value of $|\delta|$, the more pronounced are the differences between the error variances.

Moreover, if $\delta > 0$ (respectively, < 0), then the σ_i^2 are increasing (respectively, decreasing) with increasing values of Z_i.) An alternative specification of multiplicative heteroscedasticity is

$$\ln \sigma_i^2 = \ln \sigma^2 + \delta \ln Z_i, \quad Z_i > 0. \tag{3.14.1}$$

Key elements in the application of Equation 3.14 (or 3.14.1) are the choice of the variable Z and the specification of the level of the parameter δ. One common selection scheme is $Z_i = X_i$ and $\delta = 2$ so that Equation 3.14 appears as

$$\sigma_i^2 = \sigma^2 X_i^2, \tag{3.14.2}$$

i.e., σ_i^2 is known up to a multiplicative constant. Then, from Equations 3.2 and 3.14.2, a regression equation exhibiting homoscedastic disturbances is

$$\frac{Y_i}{X_i} = \beta_0\left(\frac{1}{X_i}\right) + \beta_1 + \frac{\varepsilon_i}{X_i} \quad X_i \neq 0, \tag{3.15}$$

or

$$Y_i' = \beta_1 + \beta_0 X_i' + \varepsilon_i', \tag{3.15.1}$$

where $Y_i' = Y_i/X_i$, $X_i' = X_i^{-1}$, and the new random error term $\varepsilon_i' = \varepsilon_i/X_i$ satisfies all of the assumptions of the weak classical linear regression model. Note that, in this formulation of the regression equation, the values of β_0 and β_1 are reversed, i.e., the vertical intercept is β_1 and the regression slope is now β_0. Here Equation 3.15.1 may be estimated by OLS, and, if the specification of the nature of heteroscedasticity using Equation 3.14.2 is correct, then the OLS estimators $\hat{\beta}_0$ and $\hat{\beta}_1$ of β_0 and β_1, respectively, possess all of the desirable properties of a good estimator (in particular, they are BLUE) and, because $\hat{\beta}_0$ and $\hat{\beta}_1$ are efficient estimators of the regression parameters, valid or unbiased hypothesis tests and confidence interval estimates can safely be made.

A Test for Homoscedasticity

How may we determine whether the random error term ε in the linear regression model is homoscedastic (has a constant variance over all data points)? The test for homoscedasticity used herein is the *Goldfeld-Quandt (G-Q) test* [Goldfeld and Quandt 1965] of the null hypothesis $H_0: \sigma_1^2 = \sigma_2^2 = \ldots = \sigma_n^2$, against H_1: not all of the σ_i^2's are equal. The G-Q test extracts two subsamples from the original n data points: one corresponding to large X_i values and the other corresponding to small X_i values. The motivation for the test is that, if the sample variances are homoscedastic, then the variance of the disturbances for one subsample must be the same as the variance of the disturbances for the other subsample or $H_0: \sigma_1^2 = \sigma_2^2$. So if H_0 is true, then subsample variances will differ only because of chance factors.

In the light of the preceding remarks, it should be clear that the G-Q test for homoscedasticity ultimately reduces to a test for the equality of two independent sample variances or, equivalently, a test based on the ratio of two independent sample variances as presented in Chapter 1 (see "Hypothesis Tests for the Differences of Variances"). As noted therein, if ε is normally distributed, then, under H_0, the ratio of two independent χ^2 random variables, each divided by its respective degrees of freedom, follows an F distribution with degrees of freedom determined by the degrees of freedom specified in the numerator and in the denominator. The two subsample variances will indeed be independent if, as the G-Q test requires, we actually estimate, via OLS, two separate linear regression equations, one for each of the subsamples selected.

An outline of the steps involved in performing the G-Q test now follows. Specifically, if Equation 3.14.2 applies, then

a. Order the n observations on the variable X from smallest to largest.
b. Omit p central observations that are not to be included in either subsample. Hence, the first subsample (containing the smallest X_i's) is indexed as $i = 1,2, \ldots, n_1$, whereas the second subsample (containing the largest X_i's) is indexed as $i = n_1 + p + 1, n_1 + p + 2, \ldots, n_1 + p + n_2$.
c. Fit separate OLS regressions to the n_1 observations within the first subsample and to the n_2 observations constituting the second subsample.
d. Let $S_1^2 = SSE_1/(n_1 - 2) = MSE_1$ and $S_2^2 = SSE_2/(n_2 - 2) = MSE_2$ denote the error mean squares from the two regressions. (1 is the subscript from the first subsample or the set of smaller X_i's, and 2 is the subscript from the second subsample or the set of larger X_i's.)
e. Form the G-Q test statistic

$$g = MSE_2/MSE_1. \tag{3.16}$$

(We expect MSE_2 to be larger than MSE_1.) Suppose the random error term ε is normally distributed. Then, under H_0, g is F distributed with $n_2 - 2$ and $n_1 - 2$ degrees of freedom. For a given $\alpha = P(TIE)$, the critical region is $\mathcal{R} = \{g \,/\, g > F_{1-\alpha, n_2-2, n_1-2}\}$.

A few comments pertaining to the G-Q test are in order. First, this test can be used when the sample size n is small. Next, experience dictates that the p middle data points not appearing in either subsample should be between 20 and 30% of the total sample size n. Because the choice of p is arbitrary, a weakness of the G-Q test is that the results can be sensitive to the choice of the observations deleted from the sample. Third, the power of this test (the probability of rejecting H_0 when it is false) is diminished somewhat if the number of central observations excluded is sizeable (in this instance, MSE_1 and MSE_2 have too few degrees of freedom) or if the number of central observations is too small (so that the contrast between MSE_1 and MSE_2 is too modest). Finally, one can apply the G-Q test using the residuals from the transformed Equation 3.15.1 to determine whether this transformation of the variables has helped reduce the amount of heteroscedasticity. This time, the observations are arranged in order of magnitude of $1/X_i$.

Example 3.1

To see exactly how the G-Q test procedure works, let us first fit, via OLS, a linear regression equation to the dataset presented in the first two columns of Table 3.1. The regression results are

$$\hat{Y} = 1.4650 + 0.1165X,$$
$$(1.947) \quad (2.303)$$

$$r^2 = 0.19, \; MSE = 3.33,$$

where the numbers in parentheses beneath the estimated coefficients are associated t values.

TABLE 3.1
Data for G-Q Test

(1) X_i	(2) Y_i	(3) Ordered X_i's	(4) Y_i	
5	1.5	1	2	Subsample 1 ($n_1 = 10$)
7	4	2	1	
1	2	3	1.5	
4	2	4	2	
13	4	5	1.5	
15	4	6	3	
10	1.5	7	4	
2	1	8	1.5	
6	3	9	3	
3	1.5	10	1.5	
16	3	11	4	$p = 5$ observations deleted
8	1.5	12	1.5	
12	3	13	4	
11	4	14	1.5	
9	3	15	4	
19	6	16	3	Subsample 2 ($n_2 = 10$)
17	1	17	1	
14	1.5	18	3	
22	8	19	6	
25	8	20	1	
21	4	21	4	
18	3	22	8	
20	1	23	2	
24	1	24	1	
23	2	25	8	

To conduct a test of homoscedasticity of the random error term ε, let us order the data points according to the magnitude of the X values (see Table 3.1, columns 3 and 4) and, after deleting the middle $p = 5$ observations, form the indicated subsamples, where $n_1 = n_2 = 10$. From the subsample 1 dataset, we obtain the regression results

$$\hat{Y} = 1.5333 + 0.1030X, \quad MSE_1 = 0.878,$$

and from the subsample 2 data values, the following regression output emerges

$$\hat{Y} = -2.3879 + 0.2970X, \quad MSE_2 = 7.601.$$

Then, for $H_0: \sigma_1^2 = \sigma_2^2$ and $H_1: \sigma_1^2 \neq \sigma_2^2$, Equation 3.16 yields

$$g = \frac{MSE_2}{MSE_1} = \frac{7.601}{0.878} = 8.657$$

and, for $\alpha = 0.05$, $\mathcal{R} = \left\{g/g \geq F_{0.95,8,8} = 3.44\right\}$. Because our sample g is a member of the critical region, we reject the null hypothesis of homoscedasticity in favor of the alternative hypothesis of unequal subsample variances.

Given that the error variances of the subsamples are unequal, let us use transformation Equation 3.15.1 so as to offset the impact of heteroscedasticity on our regression results. To this end, our transformed dataset occupies columns 1 and 2 of Table 3.2. Let us now determine whether the transformed dataset exhibits heteroscedasticity by performing the G-Q test using the two subsamples appearing in columns 3 and 4 of Table 3.2. For subsample 1, the fitted equation has the form

$$\hat{Y}' = 0.2836 - 2.1328X', \quad MSE_1 = 0.016,$$

and for subsample 2,

$$\hat{Y}' = 0.0381 + 1.7619X', \quad MSE_2 = 0.046.$$

A second application of Equation 3.16 under the above null hypothesis yields

$$g = \frac{MSE_2}{MSE_1} = \frac{0.046}{0.016} = 2.875,$$

TABLE 3.2

Transformed Data for G-Q Test

(1) $1/X_i$	(2) Y_i/X_i	(3) Ordered X_i^{-1}'s	(4) Y_i/X_i	
0.200	0.300	0.040	0.320	Subsample 1 ($n_1 = 10$)
0.143	0.571	0.042	0.042	
1.000	2.000	0.044	0.087	
0.250	0.500	0.046	0.363	
0.077	0.308	0.047	0.191	
0.067	0.267	0.050	0.050	
0.100	0.150	0.052	0.315	
0.500	0.500	0.056	0.167	
0.167	0.500	0.058	0.058	
0.333	0.500	0.063	0.188	
0.063	0.188	0.067	0.267	$p = 5$ observations deleted
0.125	0.188	0.071	0.107	
0.083	0.250	0.077	0.308	
0.091	0.363	0.083	0.250	
0.111	0.333	0.091	0.363	
0.052	0.315	0.100	0.150	Subsample 2 ($n_2 = 10$)
0.058	0.058	0.111	0.333	
0.071	0.107	0.125	0.188	
0.046	0.363	0.143	0.571	
0.040	0.320	0.167	0.500	
0.047	0.191	0.200	0.300	
0.056	0.167	0.250	0.500	
0.050	0.050	0.333	0.500	
0.042	0.042	0.500	0.500	
0.044	0.087	1.000	2.000	

and, for $\alpha = 0.05$, our calculated g value is not an element of the critical region $\mathcal{R}$. Hence, we cannot reject the null hypothesis of homoscedasticity of the random error term. Given that our data transformation Equation 3.15.1 has apparently worked (in the sense of making the estimated variances of the ε_i's much more homogeneous), let us now reestimate our regression equation using all 25 transformed data points by regressing Y_i/X_i on X_i^{-1} (Table 3.2, columns 1 and 2). We subsequently obtain

$$\hat{Y}' = \underset{(3.15)}{0.1118} + \underset{(4.65)}{1.5258}X',$$

$$r^2 = 0.301, \quad \text{MSE} = 0.4512.$$

Note that Equation 3.15.1 renders $\hat{\beta}_0 = 1.5258$ and $\hat{\beta}_1 = 0.1118$.

Autocorrelation

Given the assumptions of the strong classical linear regression model (Chapter 2, "The Strong Classical Linear Regression Model"), $COV(\varepsilon_i, \varepsilon_j) = E(\varepsilon_i\varepsilon_j) = 0$ for all $i \neq j$ (because $E(\varepsilon_i) = E(\varepsilon_j) = 0$). Moreover, because the ε_i's, $i = 1,\ldots,n$, are taken to be normally distributed, it follows that the zero covariance of ε_i and ε_j implies that these random error terms are also independent (see Chapter 2, "Inferences about the Population Correlation Coefficient ρ", and Theorem 13.A.1), i.e., the disturbance occurring at one data point is not associated with the disturbance occurring at any other such point. This independence property of the random error term is referred to as *nonautocorrelation*.

Under what circumstances can we expect the assumption of nonautocorrelation to be violated? It is typically the case that violations occur when regression estimates are made using *time-series data* (the data points are ordered in time). This being the case, let us replace the general i index used above by t (obviously chosen to represent time) to reflect the occurrence of this phenomenon. In this regard, if the disturbance ε_t is associated with the disturbance ε_{t-s}, then the disturbances are said to be *autocorrelated* and

$$COV(\varepsilon_t, \varepsilon_{t-s}) = E(\varepsilon_t\varepsilon_{t-s}) \neq 0, \quad t > s, \tag{3.17}$$

i.e., the effect of a myriad of random factors influencing the behavior of ε in period $t - s$ carry forward to affect the behavior of ε in period t.

To determine the impact of autocorrelation on our OLS regression results, let us be a bit more specific about the nature of Equation 3.17. In particular, what is our choice for s? It is usually the case that a *first-order (Markov) autoregressive scheme* (denoted as AR(1)) is posited, or

$$AR(1): \varepsilon_t = \rho\varepsilon_{t-1} + u_t \tag{3.18}$$

for all t, where the parameter ρ is such that $|\rho| < 1$ and the new disturbance or random error term u_t (called an *innovation*) is taken to be *white noise* (it obeys the strong classical assumptions), i.e., u_t is normally and independently distributed with a zero mean, a constant variance σ_u^2, and is independent of ε_{t-1}, or

$$\begin{aligned}&\text{(a) } u_t \text{ is } N(0, \sigma_u) \text{ for all } t\\&\text{(b) } COV(u_t, u_s) = E(u_t u_s) = \begin{cases} V(u_t) = \sigma_u^2, & t = s;\\ 0, & t \neq s\end{cases}\\&\text{(c) } COV(u_t, \varepsilon_{t-1}) = 0 \text{ for all } t.\end{aligned} \tag{3.19}$$

(Note that, if $\rho = 0$, $\varepsilon_t = u_t$ and $V(\varepsilon_t) = V(u_t) = \sigma_\varepsilon^2$ = constant for all t so that all of the basic assumptions of the strong classical linear regression model hold.) Here ρ is termed the *autoregression coefficient*; it represents the change in ε_t per unit change in ε_{t-1}. Additionally, Equation 3.18 implies that ε in any period t is equal to only a fraction ρ of the value of ε in the preceding period $t-1$ plus the contemporaneous effect of the random error term u_t.

Given that ε_t is AR(1) and Equation 3.19 holds, it can be demonstrated (Appendix 3.B.1) that

$$\begin{aligned}&\text{(a) } V(\varepsilon_t) = \sigma_\varepsilon^2 = \frac{\sigma_u^2}{1-\rho^2}; \text{ and}\\&\text{(b) } COV(\varepsilon_t, \varepsilon_{t-s}) = \rho^s \sigma_\varepsilon^2.\end{aligned} \tag{3.20}$$

As Equation 3.20.b reveals,

$$\rho^s = \frac{COV(\varepsilon_t, \varepsilon_{t-s})}{\sqrt{V(\varepsilon_t)}\sqrt{V(\varepsilon_{t-s})}}, \quad |\rho| < 1, \tag{3.21}$$

represents the coefficient of correlation between ε_t and ε_{t-s} because $\sigma_\varepsilon^2 = V(\varepsilon_t) = V(\varepsilon_{t-s})$. In particular, for $s = 1$, $COV(\varepsilon_t, \varepsilon_{t-1}) = \rho\sigma_\varepsilon^2$ and thus Equation 3.21 becomes

$$\rho = \frac{COV(\varepsilon_t, \varepsilon_{t-1})}{\sigma_\varepsilon^2} = \frac{COV(\varepsilon_t, \varepsilon_{t-1})}{\sqrt{V(\varepsilon_t)}\sqrt{V(\varepsilon_{t-1})}}, \quad |\rho| < 1, \tag{3.21.1}$$

the coefficient of correlation between successive values of the disturbance term or between ε_t and ε_{t-1}.

It is important to note that, within the AR(1) process specified above, there is no theoretical lower limit to the value of t, i.e., $t = -\infty$ is consistent with the beginning of this first-order autoregressive scheme. However, the first observation in our *finite* sample must be indexed at $t = 1$. To take account of the effect of previous disturbances on ε_t, let us specify the first sample disturbance term as

$$\varepsilon_1 = \frac{u_1}{\sqrt{1-\rho^2}}, \tag{3.22}$$

with $\varepsilon_t = \rho\varepsilon_{t-1} + u_t$ for $t = 2,3,\ldots$. By virtue of this specification of ε_1, all of the requisite features of ε_t, as specified by Equations 3.18 and 3.20) above, still hold, i.e., $E(\varepsilon_1) = 0$, $V(\varepsilon_1) = \sigma_\varepsilon^2 = \sigma_u^2/(1-\rho^2)$, and $COV(\varepsilon_1, \varepsilon_t) = \rho^{t-1}\sigma_\varepsilon^2$, $t = 2,3,\ldots$. So given Equation 3.22, a sample of n data points is associated with exactly n random disturbance terms.

Properties of the OLS Estimators When ε Is AR(1)

Let us now examine the properties of the OLS estimators $\hat{\beta}_1$ and $\hat{\beta}_0$ of β_1 and β_0, respectively, when the random error terms ε_t are autocorrelated. Suppose

$$Y_t = \beta_0 + \beta_1 X_t + \varepsilon_t \tag{3.23}$$

and that ε_t follows the first-order autoregressive scheme provided by Equation 3.18. Then by an argument similar to that used above (see "Heteroscedasticity"), it is easily demonstrated that the OLS estimators are unbiased when the disturbances are AR(1).

To determine whether the OLS estimators $\hat{\beta}_1$ and $\hat{\beta}_0$ are still BLUE in the presence of an autoregressive structure for the random error term ε, we need to transform Equation 3.23 to an alternative regression equation that has a disturbance term that is not autocorrelated. Then the transformed equation displaying independent disturbances can be used to obtain OLS estimators for β_1 and β_0. To this end, suppose we lag Equation 3.23 by one period, multiply the lagged equation by $\rho(\neq 0)$, and then subtract the result from Equation 3.23 itself so as to obtain the *quasi-first difference*

$$\begin{aligned} Y_t - \rho Y_{t-1} &= \beta_0(1-\rho) + \beta_1(X_t - \rho X_{t-1}) + \varepsilon_t - \rho\varepsilon_{t-1} \\ &= \beta_0(1-\rho) + \beta_1(X_t - \rho X_{t-1}) + u_t \end{aligned}$$

or

$$Y_t^* = \beta_0(1-\rho) + \beta_1 X_t^* + u_t, t = 2,3,..., \tag{3.24}$$

where $Y_t^* = Y_t - \rho Y_{t-1}$ and $X_t^* = X_t - \rho X_{t-1}$. Note that, under this transformation, one observation pertaining to $t = 1$ has been lost. To recover the first observation (X_1^*, Y_1^*) as well as u_1, let us use Equation 3.22 so as to obtain

$$Y_1 = \beta_0 + \beta_1 X_1 + \varepsilon_1 = \beta_0 + \beta_1 X_1 + \frac{u_1}{\sqrt{1-\rho^2}}.$$

Then

$$Y_1\sqrt{1-\rho^2} = \beta_0\sqrt{1-\rho^2} + \beta_1\sqrt{1-\rho^2}X_1 + u_1 \tag{3.25}$$

so that $Y_1^* = Y_1\sqrt{1-\rho^2}$ and $X_1^* = X_1\sqrt{1-\rho^2}$. Hence, from Equations 3.24 and 3.25, the complete transformation of Equation 3.23 to a regression equation in which the disturbance term u_t satisfies the strong classical assumptions appears as

$$Y_t^* = \beta_0 W_t^* + \beta_1 X_t^* + u_t, t = 1,2,...,n, \tag{3.26}$$

where, for $t = 1$,

$$Y_1^* = Y_1\sqrt{1-\rho^2}, W_1^* = \sqrt{1-\rho^2}, \text{ and } X_1^* = X_1\sqrt{1-\rho^2}; \tag{3.27}$$

and for $t = 2,3,\ldots,$

$$Y_t^* = Y_t - \rho Y_{t-1},\; W_1^* = 1-\rho^2, \text{ and } X_t^* = X_t - \rho X_{t-1}. \tag{3.28}$$

Using Equation 3.26, let us express the tth residual or deviation from the transformed regression plane as $u_t^* = Y_t^* - \beta_0 W_t^* - \beta_1 X_t^*$. Then, to apply the principle of least squares, let us minimize $F^*(\beta_0, \beta_1) = \sum u_t^{*2}$ with respect to the arguments β_0 and β_1. To this end, after setting $\partial F^*/\partial\beta_0 = \partial F^*/\partial\beta_1 = 0$, we obtain the system of least-squares normal equations

$$\left.\begin{aligned}\sum W_t^* u_t^* &= 0\\ \sum X_t^* u_t^* &= 0\end{aligned}\right\} \text{ or } \left\{\begin{aligned}\tilde{\beta}_0 \sum W_t^{*2} + \tilde{\beta}_1 \sum W_t^* X_t^* &= \sum W_t^* Y_t^*\\ \tilde{\beta}_0 \sum W_t^* X_t^* + \tilde{\beta}_1 \sum X_t^{*2} &= \sum X_t^* Y_t^*\end{aligned}\right. \tag{3.29}$$

where, for $t = 1$, Equation 3.27 holds, and, for $t = 2,3,\ldots$, Equation 3.28 holds. In particular, we may explicitly illustrate the specific impact of the parameter ρ within the AR(1) structure on these least-squares normal equations by rewriting, via Equations 3.27 and 3.28, Equation 3.29 as

$$\begin{aligned}\tilde{\beta}_0\left[(1-\rho^2)+(n-1)(1-\rho)^2\right]+\tilde{\beta}_1\left[(1-\rho^2)X_1+(1-\rho)\sum_{t=2}^{n} X_t^*\right] &= (1-\rho^2)Y_1+(1-\rho)\sum_{t=2}^{n} Y_t^*\\ \tilde{\beta}_0\left[(1-\rho^2)X_1+(1-\rho)\sum_{t=2}^{n} X_t^*\right]+\tilde{\beta}_1\left[(1-\rho^2)X_1^2+\sum_{t=2}^{n} X_t^{*2}\right] &= (1-\rho^2)X_1Y_1+\sum_{t=2}^{n} X_t^* Y_t^*\end{aligned} \tag{3.30}$$

(Note that, if $\rho = 0$, system Equation 3.30 reduces to system Equation 2.4, the solution of which yields the OLS estimators $\hat{\beta}_1$ and $\hat{\beta}_0$.)

A few comments pertaining to Equation 3.30 are in order. First, these equations cannot be solved unless ρ is specified. Second, a comparison of Equation 3.30 with Equation 3.4 reveals that the former system is also a WLS equation system, in which the weights for the first data point are provided by Equation 3.27 and the weights for all subsequent observations are given by Equation 3.28. Because the solution of Equation 3.30 for the (WLS) estimators $\tilde{\beta}_0$ and $\tilde{\beta}_1$ obviously involves the parameter ρ and because these estimators are BLUE, by virtue of the Gauss-Markov Theorem, it follows that the OLS estimators $\hat{\beta}_0$ and $\hat{\beta}_1$ are not BLUE because ρ is not one of their arguments. So when the disturbances ε_t are autocorrelated, the OLS estimators, although unbiased, are not best unbiased among all linear estimators of β_0 and β_1 and thus are inefficient. Moreover, it can be demonstrated that the OLS estimators in the presence of autocorrelated errors are, for large n, consistent although not asymptotically efficient.

Variances of the OLS Estimators When ε Is Autoregressive

Under the autoregressive structure (Equation 3.18), it can be demonstrated (Appendix 3.B.2) that the variances of the usual OLS estimators $\hat{\beta}_1$ and $\hat{\beta}_0$ are, respectively,

$$V\left(\hat{\beta}_1\right)_{auto} = \frac{\sigma_\varepsilon^2}{\sum x_i^2} + \Delta_1 \tag{3.31}$$

where

$$\Delta_1 = 2 \sum_{\substack{i \\ (i<j)}} \sum_j w_i w_j E(\varepsilon_i \varepsilon_j); \tag{3.32}$$

and

$$V(\hat{\beta}_0)_{auto} = \sigma_\varepsilon^2 \left(\frac{1}{n} + \frac{\bar{X}^2}{\sum x_i^2} \right) + \Delta_2, \tag{3.33}$$

where

$$\Delta_2 = \sum_{\substack{i \\ (i<j)}} \sum_j (w_i + w_j) E(\varepsilon_i \varepsilon_j) + \frac{1}{n^2} \sum_{\substack{i \\ (i<j)}} \sum_j E(\varepsilon_i \varepsilon_j). \tag{3.34}$$

(For a fuller specification of these variances when ε is AR(1), see Equations 3.B.7 and 3.B.11 of Appendix 3.B.)

What is our expectation of the signs of Δ_1 and Δ_2? If ρ is positive and the observations on X are positively correlated (i.e., X_i is positively associated with $X_{i+1}, X_{i+2},\ldots$), then Δ_1 will have a positive sign. (Actually, for $0 < \rho < 1$, the terms $w_i w_j$ will be weighted more heavily by $E(\varepsilon_i \varepsilon_j)$ for i near j than for i far away from j. Hence, if we expect $w_i w_j$ to be positive for i near j, then we should expect that, on balance, $\Delta_1 > 0$.) So as Equation 3.31 reveals, for $\Delta_1 > 0$, $V(\hat{\beta}_1)_{auto} > V(\hat{\beta}_1)$ i.e., the ordinary OLS variance understates the true variance of $\hat{\beta}_1$ when the random errors are autoregressive.

By a similar argument, we should expect that $\Delta_2 > 0$. Hence, $V(\hat{\beta}_0)_{auto} > V(\hat{\beta}_0)$ and the usual OLS variance understates the true variance of $\hat{\beta}_0$ when ε is AR(1).

Properties of the Estimated Variances of the OLS Estimators When ε Is Autocorrelated

We noted above that, when the random disturbances were heteroscedastic, the variances of the OLS estimators were biased and thus were not appropriate for testing hypotheses about the regression parameters or computing confidence intervals for the same. As one might have anticipated, a similar conclusion emerges when the random error terms are autocorrelated.

Given $S_\varepsilon^2 = \sum e_i^2/(n-2)$ (the usual OLS estimator of σ_ε^2), it can be verified (Appendix 3.B.3) that, for ε AR(1),

$$E(S_\varepsilon^2) = \sigma_\varepsilon^2 - \left[\frac{\Delta_1 \sum x_i^2}{n-2} + \frac{2\rho\sigma_\varepsilon^2}{n(n-2)(1-\rho)} \right]. \tag{3.35}$$

For both ρ and Δ_1 positive, Equation 3.35 reveals that $E(S_\varepsilon^2) < \sigma_\varepsilon^2$, i.e., the traditional OLS residual variance formula will understate the true error variance σ_ε^2. Clearly, S_ε^2 is biased downward in the presence of autocorrelated disturbances. Moreover, this bias is transmitted to $V(\hat{\beta}_1)$. That is, from Equations 2.8 and 3.35,

$$E\left(S_{\hat{\beta}_1}^2\right) = E\left(\frac{S_\varepsilon^2}{\sum x_i^2}\right)$$
$$= \frac{\sigma_\varepsilon^2}{\sum x_i^2} - \left[\frac{\Delta_1}{n-2} + \frac{2\rho}{n(n-2)(1-\rho)}\left(\frac{\sigma_\varepsilon^2}{\sum x_i^2}\right)\right]. \tag{3.36}$$

Because from this expression we have, for both ρ and Δ_1 positive, $E\left(S_{\hat{\beta}_1}^2\right) < V\left(\hat{\beta}_1\right) = \sigma_\varepsilon^2/\sum x_i^2$ (the variance of $\hat{\beta}_1$ in the absence of autocorrelated error terms), we thus conclude that, if the random error term ε is autoregressive, the usual OLS estimator of $V\left(\hat{\beta}_1\right)$ is negatively biased, i.e., the traditional OLS variance formula for $\hat{\beta}_1$ understates the true error variance $\sigma_\varepsilon^2/\sum x_i^2$. A similar conclusion holds for $V\left(\hat{\beta}_0\right)$.

Given the presence of this negative bias, any confidence interval estimates or hypothesis tests concerning β_0 and β_1 will be unreliable. In this regard, because the OLS standard errors of the estimated regression coefficients can be seriously understated for ε AR(1), any hypothesis tests concerning β_0 and β_1 will be biased (the calculated *t* statistics are inflated, thus overstating the statistical significance of the estimators of the regression parameters) and the confidence intervals for β_0 and β_1 will be narrower than they should be, thus inflating the precision of our estimates of these regression parameters.

Testing for the Absence of Autocorrelation: The Durbin-Watson Test

Having discussed the consequences of autocorrelated disturbances for OLS estimation of the regression parameters when the random error term ε is specified as AR(1), let us now look to our sample information to detect the presence of first-order autocorrelation. In particular, we shall work with the OLS residuals e_t so as to test the statistical significance of the AR(1) structure. To this end, we may test the null hypothesis that ε is not AR(1) or $H_0: \rho = 0$, against any of the following alternative hypotheses:

Case I	Case II	Case III
$H_0: \rho = 0$	$H_0: \rho = 0$	$H_0: \rho = 0$
$H_1: \rho > 0$ (positive autocorrelation)	$H_1: \rho < 0$ (negative autocorrelation)	$H_1: \rho \neq 0$ (ε is autoregressive)

Our test statistic will be the *Durbin-Watson* (DW) *statistic*

$$DW = \frac{\sum_{t=2}^{n}(e_t - e_{t-1})^2}{\sum_{t=1}^{n} e_t^2}, \quad 0 \le d \le 4, \tag{3.37}$$

where e_t is the usual OLS residual. After expanding the numerator of *DW*, we obtain

$$DW = \frac{\sum_{t=2}^{n} e_t^2 + \sum_{t=2}^{n} e_{t-1}^2 - 2\sum_{t=2}^{n} e_t e_{t-1}}{\sum_{t=1}^{n} e_t^2}.$$

If we let

$$\hat{\rho} = \sum_{t=2}^{n} e_t e_{t-1} \Big/ \sum_{t=1}^{n} e_t^2 \tag{3.21.2}$$

depict the sample correlation coefficient between the OLS residuals e_t and e_{t-1} (it serves as a consistent estimator of ρ in the AR(1) scheme), then, for large n, $\sum_{t=2}^{n} e_t^2 \approx \sum_{t=2}^{n} e_{t-1}^2 \approx \sum_{t=1}^{n} e_t^2$ so that this expression simplifies to

$$DW \approx 2 - 2\hat{\rho}. \tag{3.38}$$

(Note that, if $H_0: \rho = 0$ is true and $n \to \infty$, DW assumes the value 2 with unitary probability.) So if e_t and e_{t-1} are not related, their sample covariance will be near zero so that $\hat{\rho} \approx 0$ and thus $DW \approx 2$. Furthermore, if e_t and e_{t-1} exhibit high negative association, then $\hat{\rho} \approx -1$ and thus $DW \approx 4$.

It is instructive to consider the graphical behavior of the e_t's under positive or negative first-order autocorrelation. Given that the average of the e_t's is zero, these OLS residuals will be scattered about the horizontal (time) axis. If there exists first-order positive autocorrelation, successive e_t's will tend to be close to each other, thus exhibiting runs above and below the horizontal axis (Figure 3.1.A). In the presence of such runs, $e_t - e_{t-1}$ will tend to be smaller in magnitude than e_t itself, thus leading (over all t) to a relatively small value of DW. If negative first-order autocorrelation occurs, successive e_t's tend to be on opposite sides of the horizontal axis (Figure 3.1.B). Here, $e_t - e_{t-1}$ will tend to be numerically larger than e_t, thus leading (over all t) to a relatively large DW value. If neither positive nor negative first-order autocorrelation occurs, then the e_t's are randomly distributed about the horizontal axis and thus DW assumes an intermediate value (Figure 3.1.C).

Under $H_0: \rho = 0$, the exact sampling distribution of DW depends on the values of the explanatory variable X. Because the critical values of DW are sensitive to the choice of the

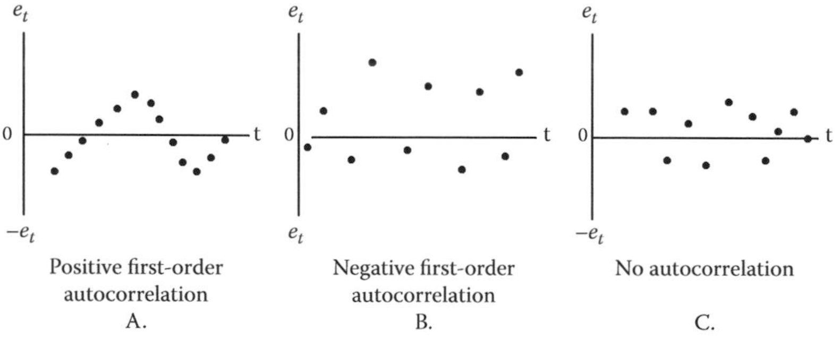

FIGURE 3.1
Behavior of the OLS residuals.

X_t's, obviously a single table covering all possible choices is not available. However, Durbin and Watson [1951] have shown that the actual sampling distribution of DW is bracketed by (it lies between) two limiting distributions termed simply *lower* and *upper* distributions. For these lower and upper distributions, Durbin and Watson establish lower (d_l) and upper (d_u) bounds for the critical values of DW that depend only on the sample size (n) and the number of explanatory variables (k'). Then, for a DW test of $H_0: \rho = 0$ conducted at the $\alpha =$ P(TIE) level of significance, a set of decision rules for rejecting H_0 relative to H_1 consists of:

Case I:
a. If $DW < d_l$, reject H_0 (ε is not AR(1)) in favor of H_1 (ε exhibits positive first-order autocorrelation);
b. If $DW > d_u$, do not reject H_0; and
c. If $d_l < DW < d_u$, the test is inconclusive.

Case II:
a. If $DW > 4 - d_l$, reject H_0 (ε is not AR(1)) in favor of H_1 (ε exhibits negative first-order autocorrelation);
b. If $DW > 4 - d_u$, do not reject H_0; and
c. If $4 - d_u < DW < 4 - d_l$, the test is inconclusive.

Case III.
a. If $DW < d_l$, or if $DW > 4 - d_l$, reject H_0 (ε is not AR(1)) in favor of H_1 (ε is autoregressive);
b. If $d_u < DW < 4 - d_u$, do not reject H_0; and
c. If $d_l \le DW \le d_u$, or if $4 - d_u \le DW \le 4 - d_l$, the test is inconclusive.

A description of the entire range of DW values (incorporating both regions of rejection and nonrejection of $H_0: \rho = 0$) is illustrated in Figure 3.2. DW tables exhibiting significance points of d_l and d_u for 5 and 1% levels of significance appear collectively as Table A.5 of the Appendix.

We now turn to a few caveats pertaining to the conventional DW test:

1. A constant term must be included in the regression equation.
2. The explanatory variable X must be nonstochastic.

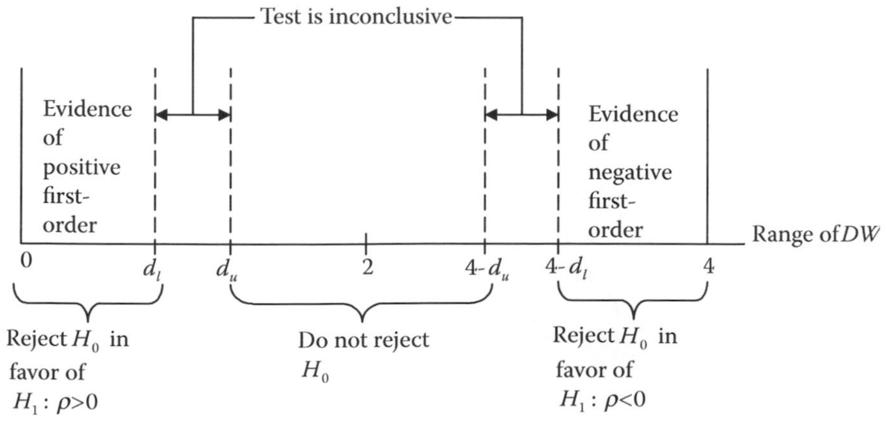

FIGURE 3.2
Range of Durbin-Watson (DW) values.

3. The size of the inconclusive region becomes fairly large for low degrees of freedom.
4. The test can be performed only for a limited number of significance levels (typically 1, 2.5, or 5%).

Example 3.2

Using the $n = 22$ data points presented in columns 1 and 2 of Table 3.3, use the OLS residuals e_t from the estimated regression equation to determine whether the random error term ε is AR(1). Given these observations on X and Y, it can be demonstrated that the estimated line of best fit is

$$\hat{Y} = -12.1429 + 32.4608X, \quad r^2 = 0.9797,$$
$$(-4.283) \quad (31.104)$$

where the numbers in parentheses beneath the parameter estimates are the associated t values. Next, column 3 displays the estimated Y values or $\hat{Y}_t = -12.1429 + 32.4608X_t$ (e.g., $\hat{Y}_1 = -12.1429 + 32.4608X_1 = -12.1429 + 32.4608(1.33) = 31.03$), and column 4 contains the OLS residuals $e_t = Y_t - \hat{Y}_t$ (e.g., $e_1 = Y_1 - \hat{Y}_1 = 19.3 - 31.03 = -11.73$). Column 5 contains the lagged residuals or the set of e_{t-1} values. From columns 6–8 of Table 3.3 and Equation 3.37, we have $DW = 304.9578/876.5299 = 0.348$. Let us test $H_0: \rho = 0$, against $H_1: \rho > 0$ for $\alpha = 0.05$. Then, from Table

TABLE 3.3

Data for Durbin-Watson Test

(1) X_t	(2) Y_t	(3) $\hat{Y}t$	(4) e_t	(5) e_{t-1}	(6) e_t-e_{t-1}	(7) $(e_t-e_{t-1})^2$	(8) e_t^2	(9) $e_t e_{t-1}$
1.33	19.3	31.03	−11.73				137.59	
1.29	20.4	29.73	−9.33	−11.73	2.40	5.76	87.04	109.44
1.25	20.9	28.43	−7.53	−9.33	1.80	3.24	56.70	70.25
1.21	21.9	27.13	−5.23	−7.53	2.30	5.29	27.35	39.36
1.19	23.4	26.49	−3.09	−5.23	2.14	4.58	9.54	16.16
1.19	24.5	26.49	−1.99	−3.09	1.10	1.21	3.96	6.15
1.21	25.8	27.13	−1.33	−1.97	0.66	0.44	1.76	2.64
1.29	30.5	29.73	0.77	−1.33	2.10	4.41	0.59	−1.02
1.33	33.3	31.03	2.27	0.77	1.50	2.25	5.15	1.75
1.42	37.2	33.95	3.24	2.27	0.97	0.94	10.50	7.35
1.52	42.5	37.20	5.30	3.24	2.06	4.24	28.09	17.17
1.59	48.8	39.47	9.83	5.30	4.03	16.24	87.04	49.45
1.84	55.4	47.59	7.82	9.33	−1.51	2.28	61.15	72.96
2.17	64.3	58.30	6.00	7.82	−1.82	3.31	36.00	46.92
2.55	78.9	70.63	8.27	6.00	2.27	5.15	68.39	49.62
2.97	86.5	84.27	2.23	8.27	−6.04	36.48	4.97	18.44
3.70	114.6	108.0	6.64	2.23	4.41	19.44	44.09	14.81
4.10	129.7	120.9	8.75	6.64	2.11	4.45	76.56	58.10
4.34	126.1	128.7	−2.64	8.75	−11.39	129.73	6.97	−23.10
4.71	132.0	140.7	−8.75	−2.64	−6.11	37.33	76.56	23.10
4.82	138.1	144.3	−6.22	−8.75	2.53	6.40	38.68	54.43
4.81	141.2	144.0	−2.79	−6.22	3.43	11.76	7.78	17.35
					Totals	304.96	876.53	651.36

A.5, we obtain, for $n = 22$ and $k' = 1$, $d_l = 1.24$ and $d_u = 1.43$. Because $DW < d_l$, the Case I decision rule (see also Figure 3.2) tells us to reject H_0 in favor of H_1; the errors ε_t exhibit positive first-order autocorrelation. As indicated above, in this circumstance, the usual confidence interval estimates and hypothesis tests are unreliable.

Estimation with Autocorrelated Disturbances

Once the presence of first-order autocorrelation has been detected (via the DW test), remedial steps must be undertaken so as to generate estimates that are efficient. One estimation method designed to accomplish this task is the *Cochrane-Orcutt (C-O)* [Cochrane and Orcutt 1949] *two-step iterative routine*, in which the first step involves getting an estimate of ρ and the second step involves obtaining estimates of the regression parameters. In what follows, we shall use the quasi-first difference (Equation 3.24) as well as a simple transformation of Equation 3.38.

To set the stage for executing the C-O procedure when ε is AR(1), let us initially perform the usual OLS regression of Y_t on X_t, $t = 1,\ldots,n$. We then undertake the following:

STEP 1. Find

$$\hat{\rho} \approx 1 - \frac{1}{2} DW \tag{3.39}$$

and transform the variables Y_t and X_t to

$$\begin{aligned} Y_t^* &= Y_t - \hat{\rho} Y_{t-1}, \\ X_t^* &= X_t - \hat{\rho} X_{t-1} \end{aligned} \tag{3.40}$$

respectively.

STEP 2. Perform OLS regression using the transformed variables (Equation 3.40), i.e., regress Y_t^* on X_t^*, $t = 2,3,\ldots$. Conduct a DW test using the new DW statistic DW*. If $H_0: \rho = 0$ is not rejected, stop. If $H_0: \rho = 0$ is rejected, repeat step 1 and find a new $\hat{\rho}$ using Equation 3.39.

(That is, repeating step 1 yields $\hat{\rho}^* \approx 1 - \frac{1}{2} DW^*$ with $Y_t^{**} = Y_t^* - \hat{\rho}^* \left(Y_{t-1}^*\right)$ and $X_t^{**} = X_t^* - \hat{\rho}^* \left(X_{t-1}^*\right)$. Then a repeat of step 2 has us regress Y_t^{**} on X_t^{**}.) This two-step process is continued until we can no longer reject the null hypothesis of no autocorrelation of the random errors ε_t or successive values of $\hat{\rho}$ are approximately the same. (Note that the C-O procedure is characterized as being *iterative* in that, once we conduct an OLS regression of Y_t^* on X_t^*, we recompute $\hat{\rho}$, retransform the variables, and then recompute the OLS estimates of the regression parameters.)

The appeal of the C-O routine stems from the fact that the OLS estimates obtained from this two-step process are efficient, i.e., for large n, the variances of the OLS estimators obtained from the C-O scheme are less than or equal to the variances obtained from any alternative unbiased estimators of the regression parameters. So, under this estimation technique, we can, in principle, conduct legitimate hypothesis tests about and construct valid confidence intervals for the regression parameters.

It is important to remember that the OLS intercept obtained from the C-O procedure is $\hat{\beta}_0^* = \hat{\beta}_0(1-\hat{\rho})$. Hence, our estimate of the original intercept parameter β_0 is $\hat{\beta}_0 = \hat{\beta}_0^*/(1-\hat{\rho})$. Additionally, because $\hat{\beta}_0$ is linearly related to $\hat{\beta}_0^*$, it follows that

$$V(\hat{\beta}_0) = \left(\frac{1}{1-\hat{\rho}}\right)^2 V(\hat{\beta}_0^*). \tag{3.41}$$

A couple of important modifications of the preceding C-O routine are as follows. (1) To avoid the loss of the first observation and to improve the efficiency of the OLS estimates, we may use Equation 3.27 so as to transform Y_1 and X_1 to

$$\begin{aligned} Y_1^* &= Y_1\sqrt{1-\hat{\rho}^2}, \\ X_1^* &= X_1\sqrt{1-\hat{\rho}^2} \end{aligned} \tag{3.42}$$

respectively. This adjustment of the first observation is known as the *Prais-Winsten modification* [Prais and Winsten 1954]. (2) Because Equation 3.40 holds only approximately (and thus may be problematic for small samples), Theil and Nagar [1961] have suggested using

$$\hat{\rho} = \frac{n^2\left(1-\dfrac{DW}{2}\right)+k^2}{n^2-k^2}, \tag{3.39.1}$$

where k is the number of regression parameters estimated (including the intercept). For our purposes,

$$\hat{\rho} = \frac{n^2\left(1-\dfrac{DW}{2}\right)+4}{n^2-4} \tag{3.39.2}$$

because only two parameters are estimated.

It is important to note that, under the C-O iteration scheme, conventional finite sample test procedures applied to the final OLS regression are no longer exact but are only legitimate asymptotically. Hence, the usual OLS variance formulas will generally yield consistent estimators of the asymptotic variances. In fact, the two-step C-O estimators generally possess all requisite desirable asymptotic properties.

Example 3.3

We found above (see "Variances of the OLS Estimators When ε Is AR (1)") that the OLS regression results generated from the X_t and Y_t observations provided in Table 3.3 were tainted because of the presence of (positively) autocorrelated disturbances. Using the C-O routine, let us use the first-order autocorrelation coefficient (Equation 3.21.2) or $\hat{\rho} = \sum_{t=2}^{n} e_t e_{t-1} \Big/ \sum_{t=1}^{n} e_t^2 = 541.9216/876.5299 = 0.743$ to transform, via Equations 3.40 and 3.42, the said X_t and Y_t values and reestimate the regression equation. (Note that Equation 3.39 yields $\hat{\rho} \approx 1 - \frac{1}{2}(0.348) = 0.826$.) The transformed X_t and Y_t values appear as columns 1 and 2 of Table 3.4. (For example, $X_1^* = 1.33\sqrt{1-0.552} = 0.89$ and $Y_1^* = 19.3\sqrt{1-0.552} = 12.91$, whereas $X_2^* = X_2 - 0.743X_1 = 1.29 - 0.743(1.33) = 0.30$ and $Y_2^* = Y_2 - 0.743Y_1 = 20.4 - 0.743(19.3) = 6.06$, etc.)

TABLE 3.4

Cochrane-Orcutt Data Transformations

(1)[a] $X_t^* = X_t - \hat{\rho}X_{t-1}$ $= X_t - 0.743X_{t-1}$	**(2)**[a] $Y_t^* = Y_t - \hat{\rho}Y_{t-1}$ $= Y_t - 0.743Y_{t-1}$
0.89	12.91
0.30	6.60
0.29	5.74
0.28	6.37
0.29	7.12
0.31	7.11
0.33	7.59
0.39	11.33
0.37	10.63
0.43	12.45
0.46	14.86
0.46	17.22
0.65	19.14
0.80	23.14
0.94	31.13
1.08	27.88
1.35	50.33
1.29	44.55
1.49	29.73
1.32	38.31
1.22	40.02
1.23	38.60

[a] $X_1^* = X_1\sqrt{1-\hat{\rho}^2}$, $Y_1^* = Y_1\sqrt{1-\hat{\rho}^2}$.

If we now regress Y_t^* on X_t^*, we obtain

$$\hat{Y}^* = -13.4753 + 32.0561X^*, \quad r^2 = 0.92.$$
$$(-1.859) \qquad (14.352)$$

Notice that, as expected, the t values of the regression coefficients have decreased because, as noted above, the effect of autocorrelated disturbances is to inflate somewhat the estimated standard errors of the OLS regression coefficients. Moreover, it can be shown that the D-W statistic determined from the residuals of this new regression equation turns out to be DW = 1.57. Because $d_u = 1.43 < DW < 4 - d_u = 2.57$ (see Figure 3.2), the Case I decision rule inform us that the null hypothesis that ε is not AR(1) should not be rejected.

Prediction with Autocorrelated Disturbances

Suppose we are interested in making a prediction using the linear regression model $Y_t = \beta_0 + \beta_1 X_t + \varepsilon_t$, $t = 1,\ldots,n$, and ε is AR(1) or $\varepsilon_t = \rho\varepsilon_{t-1} + u_t$, where u_t satisfies Equation 3.19.

More specifically, we desire to predict Y_{n+1} given X_{n+1}. Then under the AR(1) autoregressive scheme,

$$
\begin{aligned}
E(Y_{n+1}/\varepsilon_1,\ldots,\varepsilon_n) &= \beta_0 + \beta_1 X_{n+1} + E(\varepsilon_{n+1}/\varepsilon_1,\ldots,\varepsilon_n) \\
&= \beta_0 + \beta_1 X_{n+1} + \rho\varepsilon_n.
\end{aligned} \tag{3.43}
$$

Substituting $\varepsilon_n = Y_n - \beta_0 - \beta_1 X_n$ into the preceding expression yields

$$
\begin{aligned}
E(Y_{n+1}/\varepsilon_1,\ldots,\varepsilon_n) &= \beta_0 + \beta_1 X_{n+1} + \rho(Y_n - \beta_0 - \beta_1 X_n) \\
&= \beta_0(1-\rho) + \beta_1(X_{n+1} - \rho X_n) + \rho Y_n
\end{aligned}
$$

or

$$
E(Y_{n+1} - \rho Y_n/\varepsilon_1,\ldots,\varepsilon_n) = \beta_0(1-\rho) + \beta_1(X_{n+1} - \rho X_n). \tag{3.43.1}
$$

Hence, given known ρ, we may rewrite Equation 3.43.1 as the quasi-first difference (Equation 3.24) or

$$
Y_t - \rho Y_{t-1} = \beta_0(1-\rho) + \beta_1(X_t - \rho X_{t-1}) + u_t. \tag{3.44}
$$

Here, Equation 3.44 has independent disturbances and thus can be used to obtain OLS estimates for the regression parameters.

In view of Equation 3.44, the best predictor of $Y_{n+1} - \rho Y_n$ is $\hat{\beta}_0(1-\rho) + \hat{\beta}_1(X_{n+1} - \rho X_n)$, where $\hat{\beta}_0$ and $\hat{\beta}_1$ are the OLS estimators of β_0 and β_1, respectively. Equivalently, the best linear unbiased predictor of Y_{n+1} (denoted $\hat{Y}_{n+1}$) must be

$$
\hat{Y}_{n+1} = \hat{\beta}_0(1-\rho) + \hat{\beta}_1(X_{n+1} - \rho X_n) + \rho Y_n \tag{3.45}
$$

or

$$
\begin{aligned}
\hat{Y}_{n+1} &= \hat{\beta}_0 + \hat{\beta}_1 X_{n+1} + \rho\left(Y_n - \hat{\beta}_0 - \hat{\beta}_1 X_n\right) \\
&= \hat{\beta}_0 + \hat{\beta}_1 X_{n+1} + \rho e_n,
\end{aligned} \tag{3.45.1}
$$

where $e_n = Y_n - \hat{\beta}_0 - \hat{\beta}_1 X_n$ is the estimated disturbance or residual for period n.

If ρ is unknown, then the recommended predictor of Y_{n+1} is

$$
\hat{Y}_{n+1} = \hat{\beta}_0 + \hat{\beta}_1 X_{n+1} + \hat{\rho} e_n, \tag{3.45.2}
$$

where $\hat{\rho} = \sum_{t=2}^{n} e_t e_{t-1} \Big/ \sum_{t=1}^{n} e_t^2$. In this instance, all of the desirable properties of this predictor hold only asymptotically. In general, the prediction of the value of Y for s periods ahead is

$$
\hat{Y}_{n+s} = \hat{\beta}_0 + \hat{\beta}_1 X_{n+s} + \hat{\rho}^s e_n, \quad s \geq 1. \tag{3.46}
$$

As the last term on the right side of this expression reveals, the further ahead our prediction, the smaller the impact of the estimated AR(1) autoregression coefficient $\hat{\rho}$ because $|\hat{\rho}| < 1$.

Dependence between X_i and ε_i

Suppose the X_i's are random variables that are correlated with the ε_i's, $i = 1,\ldots,n$, so that Assumption A.4 above does not hold or $COV(X_i, \varepsilon_i) \neq 0$. (This dependency between X_i and ε_i cannot be easily detected by sample information because, according to the system of least-squares normal equations (Equation 2.4), X_i is required to be uncorrelated with the residual e_i given that $\sum_{i=1}^{n} X_i e_i = 0$.)

Given Equation 2.3, suppose the ε_i's are independent ($COV(\varepsilon_i, \varepsilon_j) \neq 0, i \neq j$) and that ε_i is distributed as $N(0, \sigma_\varepsilon)$ with X_i and ε_i not independent. As we shall now demonstrate, under this set of circumstances, the OLS estimators $\hat{\beta}_1$ and $\hat{\beta}_0$ of $\beta_1$and β_0, respectively, are biased and inconsistent.

To see this, let us use Equation 2.A.4 or

$$\hat{\beta}_1 = \beta_1 + \sum_{i=1}^{n} w_i \varepsilon_i, \tag{2.A.4}$$

where $w_i = x_i / \sum_{i=1}^{n} x_i^2$. If X_i and ε_i are independent random variables, then so are w_i and ε_i. Hence, $E(w_i \varepsilon_i) = E(w_i) \cdot E(\varepsilon_i) = 0$ via Assumption A.2 so that $E(\hat{\beta}_1) = \beta_1$ ($\hat{\beta}_1$ is an unbiased estimators of β_1). However, if x_i and ε_i are correlated, then $E(w_i \varepsilon_i) \neq E(w_i) \cdot E(\varepsilon_i)$ and thus $E(\hat{\beta}_1) \neq \beta_1$ ($\hat{\beta}_1$ is a biased estimators of β_1).

More specifically, let us rewrite the preceding expression for $\hat{\beta}_1$ as

$$\hat{\beta}_1 = \beta_1 + \frac{\sum_{i=1}^{n} x_i \varepsilon_i / (n-1)}{\sum_{i=1}^{n} x_i^2 / (n-1)} = \beta_1 + \frac{S(X,\varepsilon)}{S^2(X)}, \tag{3.47}$$

where $S(X, \varepsilon)$ denotes of the sample covariance between X and ε, and $S^2(X)$ denotes the sample variance of X. Then, because $S^2(X) > 0$, $\hat{\beta}_1$ overstates (respectively, understates) β_1 if $S(X, \varepsilon) > 0$ (respectively, $S(X, \varepsilon) < 0$), i.e., $\hat{\beta}_1$ is a positively biased (respectively, negatively biased) estimator of β_1 if X and ε are positively correlated (respectively, negatively correlated).

Next, suppose $n \to \infty$ in Equation 3.47. Then $S^2(X) \to \sigma_X^2$, whereas $S(X, \varepsilon) \to \sigma_{X\varepsilon} \neq 0$ (because X and ε are correlated). So even for large samples, $\hat{\beta}_1$ will either overstate or understate β_1 and thus, when Equation A.4 is violated in the sense that $COV(X_i, \varepsilon_i) \neq 0$, $\hat{\beta}_1$ is an inconsistent estimator for β_1. What about $\hat{\beta}_0$ as an estimator for β_0? Suffice it to say that, when X_i and ε_i are positively correlated, $\hat{\beta}_1$ overstates β_1 and thus $\hat{\beta}_0$ must understate β_0, and, if X_i and ε_i are negatively correlated, then $\hat{\beta}_1$ understates β_1 and consequently $\hat{\beta}_0$ must overstate β_0.

If Assumption A.4 is not valid, we generally cannot find unbiased estimators for the population parameters β_0 and β_1. However, consistent estimators for the population parameters β_0 and β_1 can be determined. Specifically, we can use what is called the *instrumental variable technique*. That is, consider a new variable Z that is either random or nonrandom. Then Z will be termed an instrumental variable (or *instrument* for short) if it is highly correlated with X and uncorrelated with ε. In what follows, the deviation of Z_i from its mean will be denoted as $z_i = Z_i - \bar{Z}, i = i,\ldots,n$.

Mirroring Equation 2.4, the least-square normal equations for instrumental variable estimation of β_0 and β_1 are

$$\sum_{i=1}^{n} e_i^* = 0 \qquad \sum_{i=1}^{n} Z_i e_i^* = 0 \qquad \text{[system of least squares normal equations } (Z_i \text{ an instrument)]} \tag{3.48}$$

where $e_i^* = Y_i - Y_i^* = Y_i - \beta_0^* - \beta_1^* X_i$ and Y_i^* is the estimated value of Y_i using instrument Z. Then the instrumental variable estimator for β_1 is

$$\beta_1^* = \frac{\sum_{i=1}^{n} z_i y_i}{\sum_{i=1}^{n} z_i x_i}; \tag{3.49}$$

and the instrumental variable estimator for β_0 is

$$\beta_0^* = \bar{Y} - \beta_1^* \bar{X}. \tag{3.50}$$

Next, adjusting Equation 3.47 to reflect the use of the instrument Z enables us to write

$$\beta_1^* = \beta_1 \frac{\sum_{i=1}^{n} z_i \varepsilon_i / (n-1)}{\sum_{i=}^{n} z_i x_i / (n-1)} = \beta_1 + \frac{S(Z,\, \varepsilon)}{S(Z,\, X)}. \tag{3.51}$$

Then

$$E(\beta_1^*) = \beta_1 + E\left[\frac{S(Z,\, \varepsilon)}{S(Z,\, X)}\right].$$

Because X and ε are not independent, it follows that $S(Z, \varepsilon)$ and $S(Z, X)$ are not independent. So, although $E[S(Z, \varepsilon)] = 0$ (remember that Z and ε are uncorrelated), it does not follow that $E\left[\frac{S(Z,\, \varepsilon)}{S(Z,\, X)}\right] = 0$. Hence, the instrumental variable estimator β_1^* of β_1 is biased.

Suppose we let $n \to \infty$ in Equation 3.51. Then $S(Z, \varepsilon) \to \sigma_{Z\varepsilon} = 0$ whereas $S(Z, X) \to \sigma_{zx}$ and thus $S(Z, \varepsilon)/S(Z, X) \to 0$. Hence, the sampling distribution of β_1^* collapses on β_1, and thus the instrumental variable estimator β_1^* of β_1 is consistent. In summary, when X is correlated with ε, OLS estimators are biased and inconsistent . By using an instrumental variable, a consistent albeit biased estimator for β_1 can be found.

Under what circumstances can we expect X and ε to be correlated? Suppose X is measured with error in that, instead of observing X_i, we observe $X_i^* = X_i + u_i$, where it is assumed that the measurement error u_i is $N(0, \sigma_u)$ with σ_u^2 = constant, $E(u_i u_j) = 0, i \neq j$, and $E(u_j \varepsilon_j) = 0$. Moreover, it is also assumed that the measurement error is independent of the values of the X and Y variables. Then substituting for X_i in Equation 2.3 yields.

$$Y_i = \beta_0 + \beta_1 X_i^* + \xi_i,$$

where $\xi_i = \varepsilon_i - \beta_1 u_i$. Because X_i^* is contemporaneously correlated with ξ_i (it is easily demonstrated that $COV(X_i^*, \xi_i) = -\beta_1 \sigma_u^2 \neq 0$), it follows that the OLS estimators for the regression parameters β_0 and β_1 are inconsistent. In this instance, estimation by the instrumental variable method is warranted.

If the dependent variable Y is measured with error in that we observe $Y_i^* = Y_i + v_i$ instead of Y_i, where v_i is $N(0, \sigma_v)$ with σ_v^2 = constant, $E(v_i v_j) = 0, i \neq j$, and $E(v_i \varepsilon_j) = 0$, then the OLS estimators of β_0 and β_1 have the usual desirable properties because this case is formally equivalent to Equation 2.3 itself. That is, if Y_i is substituted into Equation 2.3, we obtain

$$Y_i^* = \beta_0 + \beta_1 X_i + \eta_i,$$

where $\eta_i = \varepsilon_i + v_i$ is distributed as $N(0, \sigma_\eta)$ with $\sigma_n^2 = \sigma_\varepsilon^2 + \sigma_u^2$. Hence, η_i behaves exactly like ε_i.

Stochastic Explanatory Variable *X*

Assumption A.5, given in Chapter 2 (see "The Strong Classical Linear Regression Model) and restated at the outset of this chapter, specified X as a nonrandom variable, i.e., the X_i's, $i = 1,\ldots,n$, were taken to be fixed in repeated sampling from the same population so that the sampling distributions of the parameter estimators could be specified and conditional (on X) expectations could be determined. If Assumption A.5 is violated and X is random or stochastic (now unconditional expectations are calculated, with terms involving X replaced by their expectations), then we need to take account of X's relationship with the random disturbance term ε (in which we are still assuming that $E(\varepsilon_i) = 0$, $V(\varepsilon_i) = \sigma_\varepsilon^2$, and $E(\varepsilon_i \varepsilon_j) = 0$, $i \neq j$). In this regard, suppose X and ε are taken to be independent random variables.

What about the finite sample properties of the OLS estimators $\hat{\beta}_0$ and $\hat{\beta}_1$ if X is stochastic? We note first that Equations 2.A.1 and 2.A.3 no longer hold, i.e., with X random, $\hat{\beta}_0$ and $\hat{\beta}_1$ are no longer functions of the Y_i's, $i = 1,\ldots,n$, because now the w_i's are not constants. Second, given that X and ε are mutually independent, it is easily demonstrated that the OLS estimators $\hat{\beta}_0$ and $\hat{\beta}_1$ are unconditionally unbiased (simply take the expectations of Equations 2.A.4 and 2.A.5 with $E(\varepsilon_i) = 0$ for all i). Additionally, it can be demonstrated that the variances of $\hat{\beta}_0$ and $\hat{\beta}_1$ are unconditionally efficient. Next, for stochastic X independent of ε, the maximum-likelihood estimators of β_0 and β_1 coincide with the OLS estimators of the same. Finally, all of the confidence intervals and hypothesis tests for fixed X are still valid for X stochastic (the same confidence coefficients and significance levels hold for each case).

Looking to the asymptotic properties of the OLS estimators $\hat{\beta}_0$ and $\hat{\beta}_1$, the maximum-likelihood estimators (and thus OLS estimators) still exhibit all of their desirable asymptotic properties when X is stochastic and independent of ε. As long as the probability distribution of X does not involve any of the parameters β_0, β_1, and σ_ε^2, the OLS estimators are consistent, asymptotically efficient, and asymptotically normal.

Detection of Outliers

It is not unusual for a regression study to be performed using a dataset containing outlier observations, i.e., data points that are extreme in the sense of being remote or distinctly separated from the main scatter of observations. Clearly, outlier points can have a significant effect on the regression results: they can exhibit inordinately large residuals that, in turn, can impact the location of the estimated regression line.

How can outliers be identified? Moreover, if a sample point is deemed an outlier, should it be retained or eliminated? A data point can be an outlier with respect to its X coordinate, its Y coordinate, or both (Figure 3.3). Point (X_1, Y_1) in panel A of Figure 3.3 is an outlier with respect to both its X and Y coordinates because the said coordinates are well outside of the main scatter of points, point (X_2, Y_2) in panel B is an outlier with respect to its Y coordinate (its X coordinate is near the middle of the range of X values), and point (X_3, Y_3) in panel C is an outlier with respect to its X coordinate (the Y coordinate is near the middle of the range of Y values).

Outlying points may not be equally influential on the regression results. That is, although (X_1, Y_1) is consistent with the regression equation passing through the non-outlying points, this will not be the case for (X_4, Y_4) (see Figure 3.3.D).

It may be the case that an outlier is simply a *bad value* that results because of a faulty or imprecise measuring device or procedure. In this instance, the source of the outlier is

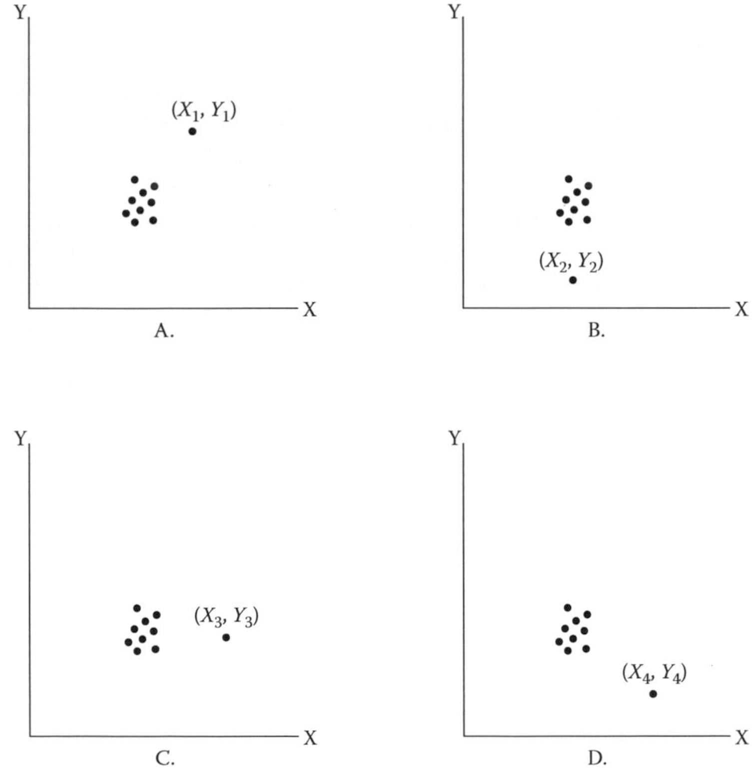

FIGURE 3.3
Detection of influential outliers.

explainable on nonstatistical grounds and thus the errant data point should be adjusted or discarded.

However, an outlier may be a perfectly legitimate observation. In this regard, the presence of one or more outliers may be attributed to, say, a severe degree of skewness of the probability distribution of the random error term ε or it may be that the regression model itself is suspect, i.e., the model may be overly sensitive to a small subset of X values or it may not adequately predict the Y variable over a specific range of the X variable. Points that are plausible (albeit unusual) should not be summarily deleted without additional scrutiny. (In fact, discarding such points could lead to the undesirable consequence of increased standard errors of the estimated regression coefficients.)

Our search for outliers will involve examining the set of OLS residuals e_i, $i = 1,\ldots,n$, as well as any unusual extreme values of the X and Y variables. Our goal is to identify data points that might significantly affect the regression results. Moreover, if an outlier is detected, exactly what is the extent of its impact on goodness of fit, the regression coefficients, and attendant test statistics?

When fitting a regression model to a sample dataset, it may be the case that some data points have an undue influence or exert a disproportionate weight in determining the regression outcome. For instance, although an outlying point such as (X_1, Y_1) in panel a of Figure 3.3 is said to be a *high leverage point* in that it displays a remote X value, it does not affect the estimated intercept or slope (but it does affect goodness of fit and the estimated standard errors of the regression coefficients). However, (X_4, Y_4) in panel d of Figure 3.3 is termed a *high influence point* in that it significantly impacts the estimated regression coefficients by tilting or pulling the regression equation in its direction. In this regard, an outlier is deemed *influential* if its exclusion precipitates major changes in the fitted regression equation. Clearly, not all outliers will be found to be influential (e.g., although (X_1, Y_1) in panel a of Figure 3.3 has high leverage, it is not influential).

Based on the preceding discussion, it should be evident that our study of the impact of an outlier must involve the assessment of its leverage and influence. In what follows then, we shall offer a meaningful way to measure the degree of leverage associated with an outlier point along with a host of so-called *deletion diagnostics* designed to assess the influence of the outlier on the regression results, specifically

1. To assess the impact of an outlying point among the X values, measure its *leverage*.
2. To assess the impact of an outlying point among the Y values, determine its *studentized deleted residual*.
3. To assess the regression model's ability to predict new data or its overall quality, calculate the *PRESS statistic*.
4. To determine how much a regression coefficient changes if the ith observation is deleted, find *DFBETA*.
5. To determine the impact of deleting the ith observation on predicting the estimated Y value, find *DFFITS*.
6. To determine an overall measure of the combined effect of deleting the ith observation on all of the estimated regression coefficients, use *Cook's distance measure*.
7. To assess the impact of the ith data point on the precision of estimation, determine *COVRATIO*.

Identifying Outlying X Values

It can be shown (see Appendix 3.C) that the variance of the residual e_i can be expressed as

$$\sigma_{e_i}^2 = \sigma_\varepsilon^2(1-h_{ii}), \tag{3.52}$$

where

$$h_{ii} = \frac{1}{n} + \frac{x_i^2}{\sum_{i=1}^{n} x_i^2}, \quad i = 1,\ldots,n, \tag{3.53}$$

with $0 \le h_{ii} \le 1$ and $\sum_{i=1}^{n} h_{ii}$ = number of parameters estimated = 2. The quantity h_{ii}, called the leverage of X_i, is useful in indicating whether or not X_i is outlying because it essentially measures the distance of X_i from the center of all X values. Hence, a large h_{ii} value indicates that X_i is far removed from $\bar{X}$.

We may think of h_{ii} as the amount of leverage exerted by Y_i on the fitted value $\hat{Y}_i$ because, from Equation 3.C.7,

$$\hat{Y}_i = h_{ii}Y_i + \sum_{j \ne i} h_{ij}Y_j,$$

i.e., because h_{ii} is a function of the set of fixed X values, it measures the role of the X_i's in determining the importance of Y_i in the calculation of $\hat{Y}_i$. So the larger is h_{ii}, the greater the weight (with respect to the X_i's) exerted by Y_i in determining $\hat{Y}_i$. Hence, large h_{ii}'s serve to reveal high leverage X_i's that are potentially influential because they are remote from $\bar{X}$. In addition, the larger is h_{ii}, the smaller is $\sigma_{e_i}^2$ and thus the closer $\hat{Y}_i$ is to Y_i. In the extreme, when $h_{ii} = 1, \sigma_{e_i}^2 = 0$ and thus $Y_i - \hat{Y}_i = e_i = 0$. Because h_{ii} and e_i vary inversely, the detection of outliers cannot be left to simply an examination of the OLS residuals e_i.

A leverage value h_{ii} is deemed large if it is more than twice as large as the mean leverage value $\bar{h} = \sum_{i=1}^{n} {h_{ii}}/{n} = {2}/{n}$. That is, X_i is considered to be a high leverage point if $h_{ii} > {4}/{n}$; X_i is thus remote enough from the remaining X values to be considered an outlier. Alternatively, it is sometimes suggested that X_i exhibits high leverage if $h_{ii} > 0.5$, whereas X_i exhibits moderate leverage if $0.2 \le h_{ii} \le 0.5$.

It was mentioned previously that h_{ii}'s examine the position of observations among the set of X values. We next turn to the search of remote points among the set of Y values.

Identifying Outlying Y Values

We noted above that the variance of e_i could be expressed as $\sigma_{e_i}^2 = \sigma_\varepsilon^2(1-h_{ii})$, where the magnitude of the leverage of X_i, h_{ii}, reflects the location of X_i in X space. In this regard, high leverage points will have a smaller $\sigma_{e_i}^2$ value than points close to $\bar{X}$. To avoid this difficulty, let us convert e_i to a form that has a constant variance regardless of where X_i lies in X space. We can do so by scaling e_i by $S_{e_i} = \sqrt{MSE(1-h_{ii})}$ so as to form the *studentized residual*

$$r_i = \frac{e_i}{\sqrt{MSE(1-h_{ii})}}, \quad i = 1,\ldots,n, \tag{3.54}$$

where now $\sigma_{r_i}^2 = 1$ at each X_i in X space, provided, of course , that the regression model is correctly specified (i.e., the linearity assumption is appropriate). Clearly, any point with a large e_i and/or a large h_{ii} value is potentially influential in determining the OLS fit. Hence, r_i is useful for identifying a remote Y value.

Let us now consider two refinements to Equation 3.54. First, we need to replace e_i by its deleted variant. Specifically, a *deleted residual* (sometimes called a PRESS residual) is denoted as

$$e_{i(-i)} = Y_i - \hat{Y}_{i(-i)}, \quad i = 1,\ldots,n, \tag{3.55}$$

where $\hat{Y}_{i(-i)}$ is the estimated Y value obtained by deleting (X_i, Y_i) from the sample dataset and fitting a regression equation to the remaining $n-1$ observations. (Note that, once the ith data point is deleted and the adjusted regression equation is obtained, $\hat{Y}_{i(-i)}$ is the point estimate that results when the X coordinate of (X_i, Y_i) is inserted into the adjusted regression equation.)

Deleted residuals are important in that, if Y_i is an outlying value of Y, then the estimated regression equation based on all n data points will be pulled close to Y_i, thus yielding a small residual e_i and consequently masking the fact that Y_i is an outlier. However, if the ith data point is deleted and the regression equation is fitted to the remaining $n-1$ observations, the estimated Y value, $\hat{Y}_{i(-i)}$, is not influenced by the remoteness of Y_i and thus $e_{i(-i)}$ will tend to be larger then e_i and consequently more likely to reveal the presence of an outlying Y_i observation. In this context, $e_{i(-i)}$ is termed the *prediction error.*

Interestingly enough, an equivalent representation of Equation 3.55 that does not require the fitting of the adjusted regression equation is

$$e_{i(-i)} = \frac{e_i}{1-h_{ii}}, \quad i = 1,\ldots,n. \tag{3.55.1}$$

Clearly, the larger is h_{ii} (thus, X_i is a high leverage value of X), the larger is $e_{i(-i)}$ relative to e_i and thus the more likely it is that Y_i is an outlying value of Y and that (X_i, Y_i) is a high influence point.

Second, the scaling used to obtain Equation 3.54 is called a *internal scaling* of e_i because MSE has been internally estimated by fitting the regression equation to all n sample points. As an alternative approach to scaling, let us obtain an estimate of σ_ε^2 by deleting the ith sample point and then fitting the regression equation to the remaining $n-1$ data points, the rationale being that, if the ith data point is indeed influential, then the values of σ_ε^2 estimated with and without the ith observation should differ substantially.

Let us denote the estimate of σ_ε^2 obtained without the ith data points as

$$MSE_{i(-i)} = \frac{(n-p)MSE - e_i^2/(1-h_{ii})}{n-p-1}, \tag{3.56}$$

where, in general, p is the number of parameters estimated (here $p = 2$). Hence, the contribution of the ith data point to MSE is eliminated and thus the effect of X_i on MSE is exposed. Then the *externally scaled* version of e_i, called the ith studentized deleted residual (or ith *R student statistic* or ith *jackknife residual*), appears as

$$\begin{aligned} r_{i(-i)} &= \frac{e_i}{\sqrt{MSE_{i(-i)}(1-h_{ii})}} \\ &= e_i\left[\frac{n-p-1}{SSE(1-h_{ii})-e_i^2}\right]^{\frac{1}{2}}, \quad i = 1,\ldots,n, \end{aligned} \tag{3.57}$$

because MSE = SSE/$(n-p)$. If the ith observation is far removed from the other data points, then $MSE_{i(-i)}$ will tend to be much smaller than MSE, which in turn will make $r_{i(-i)}$ larger compared with r_i. Also, larger leverage values h_{ii} associated with high leverage data points tend to admit correspondingly larger $r_{i(-i)}$ values relative to r_i itself. This obviously serves to help identify an outlier in Y space.

Under the usual regression assumptions (see Assumptions A.1 through A.5 underlying to strong classical linear regression model presented in Chapter 2), $r_{i(-i)}$ follows a t distribution with $n-p-1$ degrees of freedom. Then $|r_{i(-i)}|$ can be compared with, say, the upper 5% quantile of the t distribution, i.e., a Y_i value is considered to be an outlier if $|r_{i(-i)}| > t_{0.05,\, n-p-1}$.

The PRESS Statistic

We noted in the preceding section that a deleted or PRESS residual has the form $e_{i(-i)} = {e_i}/{(1-h_{ii})}$, $i = 1,\ldots,n$. When h_{ii} is large, $e_{i(-i)}$ will also tend to be large so that (X_i, Y_i) will be a high influence point, i.e., it is a point at which the regression model does not fit the data well, and thus, in the presence of the said point, the model predicts poorly.

In the light of this discussion, an overall measure of regression model *quality* (stated in terms of how well it will predict new observations) is the *prediction error sum of squares* (calculated as the sum of the squared PRESS residuals) or

$$PRESS = \sum_{i=1}^{n} e^2_{i(-i)} = \sum_{i=1}^{n} \left(\frac{e_i}{1-h_{ii}} \right)^2. \tag{3.58}$$

It should be intuitively clear that a regression model with a small realized value of the PRESS statistic is preferable to one having a large PRESS statistic realization.

The PRESS statistic can be used to obtain an overall measure of the ability of the regression model to predict new observations by forming

$$R^2_{PRESS} = 1 - \frac{PRESS}{SST}, \quad 0 \le R^2_{PRESS} \le 1, \tag{3.59}$$

with sample realization r^2_{PRESS}. Hence, PRESS replaces SSE in Equation 2.35.b so as to yield a statistic similar to R^2 (a measure of goodness of fit). In this regard, one could expect the regression model to explain about 100 r^2_{PRESS} % of the variability in predicting new observations.

DFFITS$_{i(-i)}$

To assess the influence that the ith data point has on the estimated value of $Y, \hat{Y}_i$, let us first determine $\hat{Y}_{i(-i)}$, the estimated value of Y_i obtained by deleting the ith observation (X_i, Y_i) and fitting the regression equation to the remaining $n-1$ data points. Given $\hat{Y}_{i(-i)}$, let us standardize $\hat{Y}_i - \hat{Y}_{i(-i)}$, by $\sqrt{V(\hat{Y}_i)} = \sqrt{\sigma^2_\varepsilon h_{ii}}$ so as to obtain

$$DFFITS_{i(-i)} = \frac{\hat{Y}_i - \hat{Y}_{i(-i)}}{\sqrt{MSE_{i(-i)} h_{ii}}}, \quad i = 1,\ldots,n, \tag{3.60}$$

where, as in Equation 3.57, σ_ε^2 is estimated by $MSE_{i(-i)}$. Here $DFFITS_{i(-i)}$ indicates how much the estimated Y value $\hat{Y}_i$ changes, in standard deviation units, if the ith observation is deleted. For computational purposes, we may rewrite Equation 3.60 as

$$DFFITS_{i(-i)} = r_{i(-i)} \left(\frac{h_{ii}}{1 - h_{ii}} \right)^{\frac{1}{2}}, \quad i = 1, \ldots, n, \tag{3.60.1}$$

where $r_{i(-i)}$ is the ith studentized deleted residual.

A glance at the structure of Equation 3.60.1 reveals that the magnitude of this deletion diagnostic is affected by both leverage and prediction error, i.e., if the ith data point (X_i, Y_i) is an outlier, then $r_{i(-i)}$ tends to be large, and if this point is a high leverage point, then h_{ii} will be large (and thus near unity). In either case, $DFFITS_{i(-i)}$ will correspondingly tend to be large in value, thus prompting us to subject the ith data point to additional scrutiny. In fact, an observation (X_i, Y_i) is considered influential if $|DFFITS_{i(-i)}| > 1$ for small- to medium-sized datasets, whereas for larger datasets, any data point for which $|DFFITS_{i(-i)}| > 2\sqrt{p/n}$ (p is the number of parameters estimated) is considered influential [for additional computational details, see Belsley, Kuhn, and Welsh 1980].

$DFBETAS_{j(-i)}$

To determine the influence of the ith data point on the estimate of the jth regression parameter $\beta_j, \hat{\beta}_j$, let us first find $\hat{\beta}_{j(-i)}$, the estimated value of β_j obtained by deleting the ith observation (X_i, Y_i). Given $\hat{\beta}_{j(-i)}$, once $\hat{\beta}_j - \hat{\beta}_{j(-i)}$ is standardized, we obtain

$$DFBETAS_{j(-i)} = \frac{\hat{\beta}_j - \hat{\beta}_{j(-i)}}{\sqrt{MSE_{(-i)} C_{jj}}}, \quad j = 0, 1, \ldots, \tag{3.61}$$

where C_{jj} is the jth diagonal element within the matrix $(\mathbf{X'X})^{-1}$ (see Appendix 3.C). (Note that the denominator in Equation 3.61 is an estimate of the standard error of $\hat{\beta}_j$ based on only $n - 1$ data points.)

A large value of $DFBETAS_{j(-i)}$ reveals that the ith data point exerts considerable influence on the estimate of the jth regression parameter β_j, $j = 0, 1, \ldots, p$. So if $|DFBETAS_{j(-i)}| > 1$ for small- to medium-sized datasets, then the ith observation is considered influential. If $|DFBETAS_{j(-i)}| > 2/\sqrt{n}$ for large datasets, then again the ith observation is deemed influential. In either instance, the ith data point merits additional consideration [see Belsley, Kuhn, and Welsh 1980].

Cook's Distance Measure

For the ith data point (X_i, Y_i), Cook's measure of influence considers both the leverage of X_i (i.e., the location of X_i in X space) as well as the remoteness of Y_i in Y space. It does so by using a measure of the squared distance between the OLS parameter estimates based on all n sample data points $(\hat{\beta}_0, \hat{\beta}_1)$and the estimates obtained by deleting the ith observation $(\hat{\beta}_{0(-i)}, \hat{\beta}_{1(-i)})$. Hence, Cook's deletion diagnostic, which focuses on both the differences $\hat{\beta}_0 - \hat{\beta}_{0(-i)}$ and $\hat{\beta}_1 - \hat{\beta}_{1(-i)}$ at the same time, measures the influence of an observation by considering the extent to which the regression coefficients simultaneously change when that observation is deleted from the dataset; it thus serves as a global measure of the combined impact of the ith observation on all of the estimated regression coefficients.

For the ith data point and p parameters estimated, we may express *Cook's distance* as

$$D_i = \frac{r_i^2}{p}\left(\frac{h_{ii}}{1-h_{ii}}\right)$$

$$= \frac{e_i^2}{pMSE}\left[\frac{h_{ii}}{(1-h_{ii})^2}\right], \quad i = 1, 2, \ldots, n. \tag{3.62}$$

Note that, structurally, D_i consists of a component that indicates how well the regression model fits the Y_i values (the ith studentized residual) and a component that reflects the location of X_i in X space (the leverage of X_i). Hence, D_i will be large if either $r_{i(-i)}$ or h_{ii} (or both) is large.

A data point with a large D_i value is said to exert considerable influence on the OLS estimates of β_0 *and* β_1 taken together. In fact, any observation (X_i, Y_i) for which $D_i > 1$ is considered to be influential in that the set of regression coefficients determined by using all n data points differs considerably from the set of coefficient estimates with the ith observation deleted [for additional details, see Cook 1977, 1979].

COVRATIO$_i$

A measure of regression model performance, in terms of the effect of the ith data point on the overall precision of estimation, is

$$COVRATIO_i = \left(\frac{MSE_{i(-i)}}{MSE}\right)^2 \left(\frac{1}{1-h_{ii}}\right). \tag{3.63}$$

If $COVRATIO_i > 1$ (respectively, <1), then the inclusion of the ith data point improves (adversely affects) the precision of estimation. Note that $COVRATIO_i$ varies directly with h_{ii} so that a high leverage observation improves the precision of estimation (unless the point is an outlier in Y space). Note also that, if the ith data point is an outlier, $MSE_{i(-i)}/MSE < 1$. For large samples, the ith data point is considered to be influential if

$$COVRATIO_i > 1 + \frac{3p}{n} \quad \text{or} \quad COVRATIO_i < 1 - \frac{3p}{n},$$

where p is the number of parameters estimated [Belsley, Kuhn, and Welsh 1980].

Appendix A: Properties of the WLS Estimators

The Variances of the WLS Estimators $\tilde{\beta}_1$ and $\tilde{\beta}_0$

Given $Y_i = \beta_0 + \beta_1 X_i + \varepsilon_i$ and $\tilde{Y} = \beta_0 + \beta_1 \tilde{X} + \tilde{\varepsilon}$, where $\tilde{\varepsilon} = \sum v_i \varepsilon_i / \sum v_i$, we have (from Equation 3.5)

$$\tilde{\beta}_1 = \frac{\sum v_i (X_i - \tilde{X})\left[\beta_1 (X_i - \tilde{X}) + \varepsilon_i - \tilde{\varepsilon}\right]}{\sum v_i (X_i - \tilde{X})^2} = \beta_1 + \frac{\sum v_i (X_i - \tilde{X})(\varepsilon_i - \tilde{\varepsilon})}{\sum v_i (X_i - \tilde{X})^2}$$

$$= \beta_1 + \frac{\sum v_i (X_i - \tilde{X})\varepsilon_i}{\sum v_i (X_i - \tilde{X})^2} \tag{3.A.1}$$

because

$$\sum v_i\left(X_i-\tilde{X}\right)\tilde{\varepsilon}=\sum v_i\left(X_i-\tilde{X}\right)\left(\tilde{Y}-\beta_0-\beta_1\tilde{X}\right)$$
$$=\sum v_i\left(X_i-\tilde{X}\right)\left[\frac{\sum v_iY_i}{\sum v_i}-\beta_0-\beta_1\frac{\sum v_iX_i}{\sum v_i}\right]$$
$$=\sum v_i\left(X_i-\tilde{X}\right)\left[\frac{\sum v_iY_i-\beta_0\sum v_i-\beta_1\sum v_iX_i}{\sum v_i}\right]=0$$

by virtue of the first WLS normal equation within Equation 3.4.1.

Then, from Equation 3.A.1 (remember that $E\left(\varepsilon_i^2\right)=\sigma_i^2$, $E(\varepsilon_i\varepsilon_j)=0$, and $v_i=1/\sigma_i^2$),

$$V\left(\tilde{\beta}_1\right)=E\left(\tilde{\beta}_1-\beta_1\right)^2=E\left[\frac{\sum v_i\left(X_i-\tilde{X}\right)\varepsilon_i}{\sum v_i\left(X_i-\tilde{X}\right)^2}\right]^2$$
$$=\frac{\sum v_i^2\left(X_i-\tilde{X}\right)^2\sigma_i^2}{\left[\sum v_i\left(X_i-\tilde{X}\right)^2\right]^2}=\frac{\sum v_i\left(X_i-\tilde{X}\right)^2}{\left[\sum v_i\left(X_i-\tilde{X}\right)^2\right]^2}$$
$$=\frac{1}{\sum v_i\left(X_i-\tilde{X}\right)^2}. \tag{3.A.2}$$

Next, from Equation 3.6,

$$\tilde{\beta}_0=\tilde{Y}-\tilde{\beta}_1\tilde{X}=\beta_0+\beta_1\tilde{X}+\tilde{\varepsilon}-\tilde{\beta}_1\tilde{X}$$
$$=\beta_0-\left(\tilde{\beta}_1-\beta_1\right)\tilde{X}+\tilde{\varepsilon}$$

and thus

$$V\left(\tilde{\beta}_0\right)=E\left(\tilde{\beta}_0-\beta_0\right)^2=E\left[-\left(\tilde{\beta}_1-\beta_1\right)\tilde{X}+\tilde{\varepsilon}\right]^2$$
$$=E\left[\left(\tilde{\beta}_1-\beta_1\right)^2\tilde{X}^2-2\left(\tilde{\beta}_1-\beta_1\right)\tilde{X}\tilde{\varepsilon}+\tilde{\varepsilon}^2\right]$$
$$=\tilde{X}^2V\left(\tilde{\beta}_1\right)-2\tilde{X}E\left[\left(\tilde{\beta}_1-\beta_1\right)\tilde{\varepsilon}\right]+E\left(\tilde{\varepsilon}^2\right)$$
$$=\frac{1}{\sum v_i}+\frac{\tilde{X}^2}{\sum v_i\left(X_i-\tilde{X}\right)^2}, \tag{3.A.3}$$

where

$$E\left[\left(\tilde{\beta}_1-\beta_1\right)\tilde{\varepsilon}\right]=E\left[\left(\frac{\sum v_i\left(X_i-\tilde{X}\right)\varepsilon_i}{\sum v_i\left(X_i-\tilde{X}\right)^2}\right)\left(\frac{\sum v_i\varepsilon_i}{\sum v_i}\right)\right]$$
$$=\frac{\sum v_i^2\left(X_i-\tilde{X}\right)\sigma_i^2}{\left[\sum v_i\left(X_i-\tilde{X}\right)^2\right]\sum v_i}=\frac{\sum v_i\left(X_i-\tilde{X}\right)}{\left[\sum v_i\left(X_i-\tilde{X}\right)^2\right]\sum v_i}=0$$

(because $E(\varepsilon_i\varepsilon_j) = 0$, $v_i = 1/\sigma_i^2$, and $\sum v_i(X_i - \tilde{X}) = 0$) and

$$E(\tilde{\varepsilon}^2) = E\left(\frac{\sum v_i\varepsilon_i}{\sum v_i}\right)^2 = \frac{\sum v_i^2\sigma_i^2}{\left(\sum v_i\right)^2} = \frac{\sum v_i}{\left(\sum v_i\right)^2} = \frac{1}{\sum v_i}.$$

The Relative Efficiency of the WLS Estimator $\tilde{\beta}_1$ and the OLS Estimator $\hat{\beta}_1$

Given Equations 3.7 and 3.10, we can now demonstrate that the WLS estimator $\tilde{\beta}_1$ is more efficient (has a smaller variance) relative to its OLS counterpart in the presence of a heteroscedastic disturbance term. To this end, let us form

$$\frac{V(\tilde{\beta}_1)}{V(\hat{\beta}_1)_{het}} = \frac{\left(\dfrac{1}{\sum v_i(X_i - \tilde{X})^2}\right)}{\sum w_i^2\sigma_i^2} = \frac{\left(\sum x_i^2\right)^2}{\left[\sum v_i(X_i - \tilde{X})^2\right]\sum x_i^2\sigma_i^2}, \tag{3.A.4}$$

where $v_i = 1/\sigma_i^2$. Now, suppose that σ_i^2 is associated with some variable Z having known values Z_i and that σ_i^2 is proportional to Z_i^2 or

$$\sigma_i^2 = \sigma^2 Z_i^2, \tag{3.A.5}$$

i.e., σ_i^2 is known up to a multiplicative constant (the case of *multiplicative heteroscedasticity*) and increases with increasing Z_i values. Then Equation 3.A.4 becomes

$$\frac{V(\tilde{\beta}_1)}{V(\hat{\beta}_1)_{het}} = \frac{\left(\sum x_i^2\right)^2}{\left[\sum \dfrac{(X_i - \tilde{X})^2}{Z_i^2}\right]\sum x_i^2 Z_i^2}. \tag{3.A.4.1}$$

In this instance, $\tilde{X} = \sum\left(X_i/Z_i^2\right)/\sum\left(1/Z_i^2\right) < \bar{X}$ because the weights (the $1/Z_i^2$ values) decrease with increasing Z_i. Moreover, because $\sum x_i^2 < \sum x_i^2 Z_i^2$ and $X_i - \bar{X} < X_i - \tilde{X}$, it follows that $\sum x_i^2 < \sum[(X_i - X)^2 / Z_i^2]$ so that the denominator of Equation 3.A.4.1 is larger than the numerator or $V(\tilde{\beta}_1)/V(\hat{\beta}_1)_{het} < 1$. That is, the OLS estimator of β_1 is less efficient that the WLS estimator of the same. In fact, if $\sigma_i^2 = \sigma_\varepsilon^2 =$ constant for all i (the case of homoscedastic disturbances), then $V(\tilde{\beta}_1)/V(\hat{\beta}_1)_{het} = 1$.

Appendix B: Variances and Covariances of ε_t

Variances and Covariances of the ε_t's

Given the AR(1) scheme $\varepsilon_t = \rho\varepsilon_{t-1} + u_t$ for all t, it follows by successive substitution that

$$\begin{aligned}\varepsilon_t &= \rho(\rho\varepsilon_{t-2} + u_{t-1}) + u_t \\ &= \rho^2\varepsilon_{t-2} + \rho u_{t-1} + u_t \\ &= \rho^2(\rho\varepsilon_{t-3} + u_{t-2}) + \rho u_{t-1} + u_t \\ &= \rho^3\varepsilon_{t-3} + \rho^2 u_{t-2} + \rho u_{t-1} + u_t \\ &\vdots \\ &= \rho^r\varepsilon_{t-r} + \rho^{r-1}u_{t-r+1} + \rho^{r-2}u_{t-r+2} + \cdots + \rho u_{t-1} + u_t.\end{aligned}$$

Note that, because $|\rho| < 1$, $\rho^r \to 0$ as $r \to \infty$ so that the preceding expression becomes

$$\varepsilon_t = \sum_{r=0}^{\infty}\rho^r u_{t-r}. \tag{3.B.1}$$

As this representation of ε_t indicates, the effect of lagged u's in the AR(1) structure diminishes gradually because $|\rho^r|$ decreases monotonically.

From Equation 3.B.1, we may determine

$$\begin{aligned}V(\varepsilon_t) &= \sum_{r=0}^{\infty}\rho^{2r}V(u_{t-r}) \\ &= V(u_t) + \rho^2 V(u_{t-1}) + \rho^4 V(u_{t-2}) + \cdots \\ &= \sigma_u^2\left(1 + \rho^2 + \rho^4 + \cdots\right)\end{aligned} \tag{3.B.2}$$

(because the u's are independent random variables) or

$$\sigma_\varepsilon^2 = \frac{\sigma_u^2}{1-\rho^2}.[1] \tag{3.B.2.1}$$

Next, Equation 3.B.1 also enables us to determine, via Equation 3.19.b,

$$COV(\varepsilon_t, \varepsilon_{t-s}) = E\left[\left(\sum_{r=0}^{s-1}\rho^r u_{t-r} + \rho^S u_{t-S} + \rho^{s+1}u_{t-s-1} + \rho^{s+2}u_{t-s-2} + \cdots\right)\times \left(u_{t-s} + \rho u_{t-s-1} + \rho^2 u_{t-s-2} + \cdots\right)\right]$$

[1] Because we are summing a geometric series with common ratio ρ^2 (the quotient of any term except the first and the next preceding term is the same) and $|\rho| < 1$, the series converges to $\sigma_u^2 / = (1 - \rho^2)$. (If $\rho \geq 1$, the series does not converge.) Hence, $|\rho| < 1$ is termed the *stationarity condition*: it ensures that $V(\varepsilon_t)$ tends to a limiting value, σ_ε^2, as $t \to \infty$.

$$= \rho^s\sigma_u^s + \rho^{s+2}\sigma_u^2 + \rho^{s+4}\sigma_u^2 + \cdots$$
$$= \rho^s\sigma_u^s\left(1+\rho^2+\rho^4+\cdots\right)$$
$$= \rho^s\left(\frac{\sigma_u^2}{1-\rho^2}\right) = \rho^s\sigma_\varepsilon^2 \tag{3.B.3}$$

by virtue of Equation 3.B.2.1.

Variances of the OLS Estimators When ε Is Autoregressive

We previously determined from Appendix 2.A (Equation 2.A.6) that the variance of the OLS estimator $\hat{\beta}_1$ is

$$V\left(\hat{\beta}_1\right) = \sum w_i^2 E\left(\varepsilon_i^2\right) + 2\sum_{\substack{i\\(i<j)}}\sum_j w_i w_j E\left(\varepsilon_i\varepsilon_j\right). \tag{3.B.4}$$

If $E\left(\varepsilon_i^2\right) = \sigma_\varepsilon^2$ for all i and $COV(\varepsilon_i, \varepsilon_j) = E(\varepsilon_i\,\varepsilon_j) = 0$, $i \neq j$, then the preceding expression reduces to $V\left(\hat{\beta}_1\right) = \sigma_\varepsilon^2/\sum x_i^2$ for $w_i = x_i/\sum x_i^2$. However, if the ε_i are autocorrelated, then $E(\varepsilon_i\,\varepsilon_j) \neq 0$, $i \neq j$ and thus Equation 3.B.4 becomes

$$V\left(\hat{\beta}_1\right)_{\text{auto}} = \frac{\sigma_\varepsilon^2}{\sum x_i^2} + \Delta_1, \tag{3.B.5}$$

where

$$\Delta_1 = 2\sum_{\substack{i\\(i<j)}}\sum_j w_i w_j E\left(\varepsilon_i\varepsilon_j\right). \tag{3.B.6}$$

Given $w_i = x_i/\sum x_i^2$ and Equation 3.B.3, we may rewrite Equation 3.B.6 as

$$\Delta_1 = 2\left[\sum_{i<j} w_i w_2 E\left(\varepsilon_i\varepsilon_2\right) + \sum_{i<j} w_i w_3 E\left(\varepsilon_i\varepsilon_3\right) + \cdots + \sum_{i<j} w_i w_n E\left(\varepsilon_i\varepsilon_n\right)\right]$$
$$= 2\left[\sum_{i=1}^{n-1} w_i w_{i+1} E\left(\varepsilon_i\varepsilon_{i+1}\right) + \sum_{i=1}^{n-2} w_i w_{i+2} E\left(\varepsilon_i\varepsilon_{i+2}\right) + \cdots + w_1 w_n E\left(\varepsilon_1\varepsilon_n\right)\right]$$
$$= 2\left[\frac{\sum_{i=1}^{n-1} x_i x_{i+1}}{\left(\sum x_i^2\right)^2}\left(\rho\sigma_\varepsilon^2\right) + \frac{\sum_{i=1}^{n-2} x_i x_{i+2}}{\left(\sum x_i^2\right)^2}\left(\rho^2\sigma_\varepsilon^2\right) + \cdots + \frac{x_1 x_n}{\left(\sum x_i^2\right)^2}\left(\rho^{n-1}\sigma_\varepsilon^2\right)\right]. \tag{3.B.6.1}$$

Then a substitution of this expression into Equation 3.B.5 ultimately yields

$$V\left(\hat{\beta}_1\right)_{auto} = \frac{\sigma_\varepsilon^2}{\sum x_i^2}\left\{1 + 2\left[\rho\frac{\sum_{i=1}^{n-1} x_i x_{i+1}}{\sum x_i^2} + \rho^2\frac{\sum_{i=1}^{n-2} x_i x_{i+2}}{\sum x_i^2} + \cdots + \rho^{n-1}\frac{x_1 x_n}{\sum x_i^2}\right]\right\}. \tag{3.B.7}$$

In a similar manner, we can demonstrate, using Equation 2.A.7, that

$$V(\hat{\beta}_0)_{auto} = \sigma_\varepsilon^2\left(\frac{1}{n}+\frac{\bar{X}^2}{\sum x_i^2}\right)+\Delta_2, \tag{3.B.8}$$

where

$$\Delta_2 = \underset{(i<j)}{\sum_i\sum_j}(w_i+w_j)E(\varepsilon_i\varepsilon_j)+\frac{1}{n^2}\underset{(i<j)}{\sum_i\sum_j}E(\varepsilon_i\varepsilon_j). \tag{3.B.9}$$

If we again set $w_i = x_i/\sum x_i^2$, then an additional application of Equation 3.B.3 allows us to write Equation 3.B.9 as

$$\begin{aligned}
\Delta_2 &= \left\{\sum_{i<j}(w_i+w_2)E(\varepsilon_i\varepsilon_2)+\sum_{i<j}(w_i+w_3)E(\varepsilon_i\varepsilon_3)+\cdots+\sum_{i<j}(w_i+w_n)E(\varepsilon_i\varepsilon_n)\right.\\
&\quad\left.+\frac{1}{n^2}\left[\sum_{i<j}E(\varepsilon_i\varepsilon_2)+\sum_{i<j}E(\varepsilon_i\varepsilon_3)+\cdots+\sum_{i<j}E(\varepsilon_i\varepsilon_n)\right]\right\}\\
&= \left\{\sum_{1=1}^{n-1}(w_i+w_{i+1})E(\varepsilon_i\varepsilon_{i+1})+\sum_{1=1}^{n-2}(w_i+w_{i+2})E(\varepsilon_i\varepsilon_{i+2})+\cdots+(w_i+w_n)E(\varepsilon_i\varepsilon_n)\right.\\
&\quad\left.+\frac{1}{n^2}\left(\rho\sigma_\varepsilon^2+\rho^2\sigma_\varepsilon^2+\cdots+\rho^{n-1}\sigma_\varepsilon^2\right)\right\}\\
&= \left\{\frac{\sum_{i=1}^{n-1}(x_i+x_{i+1})}{\sum x_i^2}(\rho\sigma_\varepsilon^2)+\frac{\sum_{i=1}^{n-2}(x_i+x_{i+2})}{\sum x_i^2}(\rho^2\sigma_\varepsilon^2)+\cdots+\frac{(x_i+x_n)}{\sum x_i^2}(\rho^{n-1}\sigma_\varepsilon^2)+\frac{\sigma_\varepsilon^2}{n^2}\left(\frac{\rho}{1-\rho}\right)\right\},
\end{aligned} \tag{3.B.10}$$

where the last term within the braces follows from Equation 3.B.2.1 and footnote 1. After substituting this version of Δ_2 into Equation 3.B.8, we obtain

$$\begin{aligned}
V(\hat{\beta}_0)_{auto} &= \sigma_\varepsilon^2\left\{\frac{1}{n}\left[1+\frac{\rho}{n(1-\rho)}\right]+\frac{\bar{X}^2}{\sum x_i^2}+\rho\frac{\sum_{i=1}^{n-1}(x_i+x_{i+1})}{\sum x_i^2}+\rho^2\frac{\sum_{i=1}^{n-2}(x_i+x_{i+2})}{\sum x_i^2}\right.\\
&\quad\left.+\cdots+\rho^{n-1}\frac{(x_i+x_n)}{\sum x_i^2}\right\}.
\end{aligned} \tag{3.B.11}$$

Appendix C: Calculation of Leverage Values

Let us denote the matrix of X values (including a column of 1's to reflect the inclusion of an intercept term in the regression equation) and the vector of Y values as

$$\boldsymbol{X} = \begin{bmatrix} 1 & X_1 \\ 1 & X_2 \\ \cdot & \cdot \\ \cdot & \cdot \\ \cdot & \cdot \\ 1 & X_n \end{bmatrix} \quad \text{and} \quad \boldsymbol{Y} = \begin{bmatrix} Y_1 \\ Y_2 \\ \cdot \\ \cdot \\ \cdot \\ Y_n \end{bmatrix}$$

respectively. Then

$$\boldsymbol{X'X} = \begin{bmatrix} n & \sum X_i \\ \sum X_i & \sum X_i^2 \end{bmatrix} \quad \text{and} \quad \boldsymbol{X'Y} = \begin{bmatrix} \sum Y_i \\ \sum X_i Y_i \end{bmatrix},$$

where a prime denotes transposition, and, for notational convenience, the i index has been suppressed on the summation operator.

Next, it can be shown that the inverse of the matrix $\boldsymbol{X'X}$ is

$$(\boldsymbol{X'X})^{-1} = \frac{1}{n\sum X_i^2 - (\sum X_i)^2} \begin{bmatrix} \sum X_i^2 & -\sum X_i \\ -\sum X_i & n \end{bmatrix}; \tag{3.C.1}$$

the vector of OLS estimates of the regression parameters is

$$\hat{\boldsymbol{\beta}} = \begin{bmatrix} \hat{\beta}_0 \\ \hat{\beta}_1 \end{bmatrix} = (\boldsymbol{X'X})^{-1}\boldsymbol{X'Y} \tag{3.C.2}$$

(it can be demonstrated that Equation 3.C.2 simplifies to Equation 2.4.1 above); the vector of calculated or predicted values of Y is

$$\hat{Y} = \boldsymbol{X}\hat{\boldsymbol{\beta}} = \boldsymbol{X}(\boldsymbol{X'X})^{-1}\boldsymbol{X'Y} = \boldsymbol{HY}; \tag{3.C.3}$$

and the vector of OLS residuals is

$$e = \boldsymbol{Y} - \hat{Y} = \boldsymbol{Y} - \boldsymbol{X}\hat{\boldsymbol{\beta}} = (\boldsymbol{I} - \boldsymbol{H})\boldsymbol{Y}, \tag{3.C.4}$$

where $\boldsymbol{I}$ is an identity matrix and

$$\boldsymbol{H} = \boldsymbol{X}(\boldsymbol{X}'\boldsymbol{X})^{-1}\boldsymbol{X}' \tag{3.C.5}$$

is termed the *hat matrix* because it expresses $\hat{\boldsymbol{Y}}$ as a linear combination of the components of the observed $\boldsymbol{Y}$ vector.

To determine the elements of $\boldsymbol{H}$, let us form

$$\boldsymbol{H} = \frac{1}{n\left(\sum X_i^2 - \bar{X}\sum X_i\right)} \begin{bmatrix} 1 & X_1 \\ 1 & X_2 \\ \cdot & \cdot \\ \cdot & \cdot \\ \cdot & \cdot \\ 1 & X_n \end{bmatrix} \begin{bmatrix} \sum X_i^2 & -\sum X_i \\ -\sum X_i & n \end{bmatrix} \begin{bmatrix} 1 & 1 & \ldots & 1 \\ X_1 & X_2 & \ldots & X_n \end{bmatrix}$$

$$= \frac{1}{n\left(\sum X_i^2 - \bar{X}\sum X_i\right)} \begin{bmatrix} \left(\sum X_i^2 - 2n\bar{X}X_1 + nX_1^2\right) \ldots \left(\sum X_i^2 - n\bar{X}(X_1 + X_n) + nX_1X_2\right) \\ \left(\sum X_i^2 - n\bar{X}(X_1 + X_n) + nX_1X_2\right) \ldots \left(\sum X_i^2 - 2n\bar{X}X_n + nX_n^2\right) \end{bmatrix}$$

$$= \begin{bmatrix} h_{11} & . & . & . & h_{1n} \\ \cdot & & & & \cdot \\ \cdot & & & & \cdot \\ \cdot & & & & \cdot \\ h_{n1} & . & . & . & h_{nn} \end{bmatrix},$$

where the i the diagonal element of $\boldsymbol{H}$ is

$$h_{ii} = \frac{\sum X_i^2 - 2n\bar{X}X_i + nX_i^2}{n\left(\sum X_i^2 - \bar{X}\sum X_i\right)}$$

$$= \frac{1}{n} + \frac{x_i^2}{\sum x_i^2},\ i = 1,\ldots,n, \tag{3.C.6}$$

and $x_i = X_i - \bar{X}, i = 1,\ldots,n$.

Given Equations 3.C.4 and 3.C.5, it follows that

$$E(e_i) = 0 \text{ and } \sigma_{e_i}^2 = \sigma_\varepsilon^2(1 - h_{ii})$$

whereas, from Equation 3.C.3,

$$\hat{Y}_i = \sum_{i=1}^{n} h_{ii}Y_i. \tag{3.C.7}$$

Appendix D: Analysis of Residuals Using the SAS System

Outliers and Influence

Example 3.D.1

Let us return to the $n = 20$ sample data points specified in Example 2.D.1 above when we first examined the use of the SAS System for handling regression problems. For convenience, the entire dataset along with all previous SAS code lines is repeated here as Exhibit 3.D.1. However, our analysis of this exhibit will be restricted to only the new lines of SAS code as well as to the output generated by the same.

Exhibit 3.D.1 SAS Code for Analysis of Residuals

```
data reg1;
   input x y;
datalines;
 1   4
 4   7
 5   8
 5   9
 5   11
 6   9
 6   10
 6   12
 6   18
 7   8
 7   10
 7   12
 7   13
 8   9
 8   12
 8   13
 9   13
10   13
11   6
17   20
;
proc print data=reg1;
 proc plot data=reg1;
  plot y*x= '*';
proc reg data=reg1;
     model y=x/ clb p
       clm cli r ①
         influence; ②
     output out=new
       p=py r=ry rstudent= rstudent; ③
```

① The *r* option provides everything that the *p* option prints plus the standard errors of the predicted values, the standard errors of the residuals e_i, the studentized residuals $r_i (= e_i / S_{e_i})$, and the values of Cook's D_i statistic. Also printed is a plot (headed "-2 -1 0 1 2") of the studentized residuals versus the observation number, in which asterisks are used to form a bar the length of which approximates the value of r_i. (Here one asterisk is used for each 0.5 portion of the studentized residual, e.g., if $r_i = 2.00$, then four asterisks are plotted to the right of the vertical zero line.)

② The *influence* option prints the residual along with the influence statistics: *R*-student; the diagonal elements h_{ii} of the hat matrix; $COVRATIO_i$; $DFFITS_{i(-i)}$; and $DFBETAS_{j(-i)}$.

③ The *output* statement output out = new creates a new dataset named *new* containing the original X_i's and Y_i's along with predicted values for the dependent variable Y (requested by *p=py*), residuals e_i (requested by *r=ry*), and studentized deleted residuals (requested by *rstudent=rstudent*).

④, ⑤ The *univariate* procedure is applied to the *new* dataset to determine whether the studentized deleted residuals (the values of the variable *rstudent*) follow a normal distribution. The *normal* option provides a variety of test statistics and their accompanying *p*-values, whereas the *plot* option produces a normal probability plot of the studentized deleted residuals. (The *proc univariate* output appears in Example 3.D.2 of Appendix 3.D.)

```
     plot y*x/conf;
     plot y*x/pred;
proc univariate data=new normal plot; ④
     var rstudent; ⑤
run;
```

Looking at Table 3.D.1, our interpretation of the SAS output now follows:

① The standard error of the ith residual e_i is $S_{e_i} = \sqrt{MSE(1-h_{ii})}$, $i = 1,\ldots,n$.

②,③ The studentized residuals $r_i = e_i / S_{e_i}$, $i = 1,\ldots,n$, follow student's t distribution and, for degrees of freedom greater than 10, an observation with a studentized residual having at least five asterisks associated with it (this correspond to a t value exceeding 2.50) is potentially an outlier/influential observation. This requires additional scrutiny. Clearly, observations 9 and 19 warrant additional investigation (see Figure 2.D.1).

④ Because Cook's $D_i < 1$ for all values of i, it appears that no observation exerts a marked impact on the OLS estimates of β_0 and β_1 simultaneously when that observation is deleted from the dataset.

⑤ The R-student statistics $r_{i(-i)}$ must be compared with the upper 5% quantile of the t statistic with $n - 3$ degrees of freedom to determine whether the point (X_i, Y_i) is an outlier in Y space. In this regard, if $|r_{i(-i)}| > t_{0.05,n-3} = t_{0.05,17} = 2.11$, then Y_i is considered to be an outlying value of Y. Because the R-student (absolute) values for both observations 9 and 19 exceed 2.11, it follows that Y_9 and Y_{19} are outlying in Y space (Figure 2.D.1).

⑥ Because $h_{11} = 0.2464 > 0.20$, we see that X_1 exhibits moderate leverage among the X Values, and, with $h_{20,20} = 0.5539 > 0.50$, we can conclude that X_{20} is a high leverage point in X space (Figure 2.D.1).

⑦ With $COVRATIO_9$ and $COVRATIO_{19}$ each less than unity, it appears that the inclusion of the data points 9 and 19 adversely affects the overall precision of the regression equation.

⑧ Because $|DFFITS_{19(-19)}|$ and $|DFFITS_{20(-20)}|$ each exceeds unity, observations 19 and 20 are considered influential. That is, if (X_{19}, Y_{19}) is deleted, then $\hat{Y}_{19}$ decreases by 1.2829 standard deviations, and, if (X_{20}, Y_{20}) is deleted, $\hat{Y}_{20}$ increases by 1.0705 standard deviations.

⑨ With $|DFBETAS_{1(-20)}| > 1$, it appears that data point 20 is influential, i.e., if (X_{20}, Y_{20}) is deleted, $\hat{\beta}_1$ increases by 1.0210 standard deviations.

⑩ Although the PRESS statistic is typically used to compare the overall quality (in predicting new observations) of two (or more) regression models, for our purposes, let us compute, via Equation 3.59,

$$r^2_{PRESS} = 1 - \frac{218.8143}{270.5500} = 0.1912,$$

i.e., the regression model explains about 19.12% of the variation in predicting new data points.

To summarize, we have determined that observations 9, 19, and 20 deserve additional scrutiny. Specifically, observations 9 and 19 had large studentized residuals as well as large R-student values (these two observations are outliers in Y space), observation 20 was found to be a high leverage point in X-space, observations 19 and 20 were considered influential on the estimated values of Y, whereas observation 20 was deemed influential in the estimation of β_1.

What is the effect on the estimated regression coefficients and goodness of fit if one or more of these three data points is actually deleted? As Table 3.D.2 reveals (p-values appear in parentheses next to the estimated regression parameters), when compared with the full-sample regression results, excluding observations 9 and 19 separately as well as in combination improves the fit, as does excluding data points 9, 19, and 20 all at the same time. The deletion of observation 20 and observations 9 and 20 taken together has an adverse effect on the regression results.

Tests for Normality

Example 3.D.2

In what follows, only an abbreviated version of the *proc univariate* printout is presented. Only those portions that display the output of the *normal* option and *plot* option will be offered. To this end, Table 3.D.3 displays the output of the *tests for normality* section. Here we have an assortment of test statistics and their associated p-values. In particular, we shall focus on the result of

TABLE 3.D.1

Output for Example 3.D.1

The SAS System
The REG Procedure
Dependent Variable: y
Output Statistics

<table>
<tr><th></th><th></th><th>①
Std Error</th><th>②
Student</th><th>③</th></tr>
<tr><th>Obs</th><th>Residual</th><th>Residual</th><th>Residual</th><th>−2 −1 0 1 2</th></tr>
<tr><td>1</td><td>−2.3641</td><td>2.653</td><td>−0.891</td><td>| *| |</td></tr>
<tr><td>2</td><td>−1.5523</td><td>2.897</td><td>−0.536</td><td>| *| |</td></tr>
<tr><td>3</td><td>−1.2817</td><td>2.941</td><td>−0.436</td><td>| | |</td></tr>
<tr><td>4</td><td>−0.2817</td><td>2.941</td><td>−0.0958</td><td>| | |</td></tr>
<tr><td>5</td><td>1.7183</td><td>2.941</td><td>0.584</td><td>| |* |</td></tr>
<tr><td>6</td><td>−1.0112</td><td>2.968</td><td>−0.341</td><td>| | |</td></tr>
<tr><td>7</td><td>−0.0112</td><td>2.968</td><td>−0.0038</td><td>| | |</td></tr>
<tr><td>8</td><td>1.9888</td><td>2.968</td><td>0.670</td><td>| |* |</td></tr>
<tr><td>9</td><td>7.9888</td><td>2.968</td><td>2.692</td><td>| |***** |</td></tr>
<tr><td>10</td><td>−2.7406</td><td>2.978</td><td>−0.920</td><td>| *| |</td></tr>
<tr><td>11</td><td>−0.7406</td><td>2.978</td><td>−0.249</td><td>| | |</td></tr>
<tr><td>12</td><td>1.2594</td><td>2.978</td><td>0.423</td><td>| | |</td></tr>
<tr><td>13</td><td>2.2594</td><td>2.978</td><td>0.759</td><td>| |* |</td></tr>
<tr><td>14</td><td>−2.4700</td><td>2.973</td><td>−0.831</td><td>| *| |</td></tr>
<tr><td>15</td><td>0.5300</td><td>2.973</td><td>0.178</td><td>| | |</td></tr>
<tr><td>16</td><td>1.5300</td><td>2.973</td><td>0.515</td><td>| |* |</td></tr>
<tr><td>17</td><td>0.8006</td><td>2.951</td><td>0.271</td><td>| | |</td></tr>
<tr><td>18</td><td>0.0712</td><td>2.912</td><td>0.0244</td><td>| | |</td></tr>
<tr><td>19</td><td>−7.6583</td><td>2.855</td><td>−2.682</td><td>| *****| |</td></tr>
<tr><td>20</td><td>1.9652</td><td>2.041</td><td>0.963</td><td>| |* |</td></tr>
</table>

TABLE 3.D.1 *(continued)*

The SAS System
The REG Procedure
Dependent Variable: y
Output Statistics

Obs	④ Cook's *D*	⑤ *R* Student	⑥ Hat Diag *H*	⑦ Cov Ratio	⑧ DFFITS	⑨ DFBETAS Intercept	x
1	0.130	−0.8858	0.2464	1.3594	−0.5065	−0.5057	0.4522
2	0.016	−0.5250	0.1015	1.2083	−0.1765	−0.1646	0.1257
3	0.008	−0.4258	0.0740	1.1853	−0.1204	−0.1023	0.0686
4	0.000	−0.0931	0.0740	1.2095	−0.0263	−0.0224	0.0150
5	0.014	0.5733	0.0740	1.1652	0.1621	0.1377	−0.0923
6	0.003	−0.3322	0.0569	1.1734	−0.0816	−0.0565	0.0283
7	0.000	−0.8858	0.0569	1.1887	−0.0009	−0.0006	0.0003
8	0.014	0.6595	0.0569	1.1301	0.1620	0.1121	−0.0563
9	0.218	3.3844	0.0569	0.4243	0.8311	0.5057	−0.2888
10	0.022	−0.9160	0.0501	1.0718	−0.2104	−0.0930	0.0102
11	0.002	−0.2421	0.0501	1.1722	−0.0556	−0.0246	0.0027
12	0.005	0.4130	0.0501	1.1569	0.0949	−0.0419	−0.0046
13	0.015	0.7493	0.0501	1.1060	0.1721	−0.0761	−0.0083
14	0.020	−0.8234	0.0538	1.0957	−0.1963	−0.0278	−0.0519
15	0.001	0.1734	0.0538	1.1806	0.0413	0.0059	0.0109
16	0.008	0.5039	0.0538	1.1502	0.1201	0.0170	0.0317
17	0.003	0.2642	0.0678	1.1928	0.0712	−0.0091	0.0365
18	0.000	0.0237	0.0922	1.2349	0.0076	−0.0025	0.0051
19	0.523	−3.3639	0.1270	1.4629	−1.2829	0.5959	−0.9989
20	0.575	0.9607	0.5539	2.2609	0.0705	−0.8086	1.0210

Sum of residuals	0
Sum of squared residuals	168.10283
Predicted residuals SS (PRESS)	218.81430

TABLE 3.D.2

Deletion of Outliers/Influential Data Points

Estimation With:	$\hat{\beta}_0$		$\hat{\beta}_1$		$\sqrt{MSE}$	R^2
Full sample	5.6346	(0.0041)	0.7294	(0.0039)	3.056	0.3787
No. 9 out	4.8494	(0.0027)	0.7800	(0.0004)	2.430	0.5366
No. 19 out	4.8191	(0.0029)	0.9048	(0.001)	2.437	0.5894
No. 20 out	7.0256	(0.0062)	0.5041	(0.1360)	3.063	0.1259
Nos. 9, 19 out	4.0801	(0.003)	0.9488	(0.0001)	1.524	0.8101
Nos. 9, 20 out	6.1913	(0.0031)	0.5630	(0.0410)	2.4034	0.2358
Nos. 19, 20 out	4.6363	(0.0342)	0.9351	(0.0066)	2.5101	0.3877
Nos. 9,19, 20 out	3.884	(0.0076)	0.9807	(0.0001)	1.5712	0.6455

the Shapiro-Wilk (S-W) test for normality (of the least-squares studentized deleted residuals). The null hypothesis is that the residuals constitute a random sample drawn from a normal distribution function with unknown mean and variance, whereas the alternative hypothesis is that the said residuals are not a random sample taken from a normal distribution function.

TABLE 3.D.3

Output from PROC UNIVARIATE (Test for Normality)

The SAS System

The UNIVARIATE Procedure

Variable : R Student (Studentized Residual Without Current obs)

Test for Normality

Test	Statistic		p-Value	
① Shapiro-Wilk	W	0.863153	Pr < W	0.0089
Kolmogorov-Smirnov	D	0.176006	Pr > D	0.1002
Cramer-von Mises	W^2	0.147944	Pr > W^2	0.0232
Anderson-Darling	A^2	1.009546	Pr > A^2	0.0093

Accompanying the value of the S-W statistic W is its p-value (the probability of obtaining the realized value of W if the null hypothesis were true). In general, the smaller this p-value, the less likely it is that the dataset was drawn from a normal population. If we are comfortable with selecting, say, $\alpha = 0.05$, then a p-value < 0.05 leads us to reject the null hypothesis of normality in favor of the alternative hypothesis of nonnormality. So from Table 3.D.3 we get the following:

① Because the p-value of 0.0089 for the realized S-W test statistic is very low, we see that there is some evidence that the null hypothesis of normality of the least-squares studentized deleted residuals should be rejected; there is only about a 0.9% chance of obtaining a value of W equal to 0.8632 if the sample of studentized deleted residuals were drawn from a normally distributed population.

Next, the normal probability plot of the studentized deleted residuals appears in Figure 3.D.1. In general, sample data points are represented by an asterisk, and + signs are used to denote a diagonal reference line that depicts perfect normality. Hence, the studentized deleted residuals ranked in increasing order are plotted against cumulative normal probability, and the resulting points (asterisks), under normality of these residuals, should graph as a straight line. Usually $n \geq 20$ is required to get a reasonable plot. If the majority of the asterisks do not cover or fall on the + signs, then we may view the residuals as being drawn from a non-normal population. Looking to Figure 3.D.1, we see the following:

② Because we have a large number of visible + signs, we may conclude that the studentized deleted residuals probably do not follow a normal distribution.

Appendix E: The Heteroscedasticity Issue Revisited

In text, we addressed an assortment of issues surrounding the OLS estimation of regression parameters in the presence of heteroscedasticity of the random error terms $\varepsilon_i, i = 1,\ldots,n$. What we now want to do is broaden our discussion of this topic somewhat as well as touch on the possibility of the presence of certain other model inadequacies. As noted previously, the G-Q test was used to detect the presence of heteroscedasticity. Let us now consider some other approaches to determining whether the ε_i are heteroscedastic.

Specifically, an extremely successful methodology is to first examine a plot of the studentized deleted residuals or R-student values $r_{i(-i)}$ against the predicted or estimated values

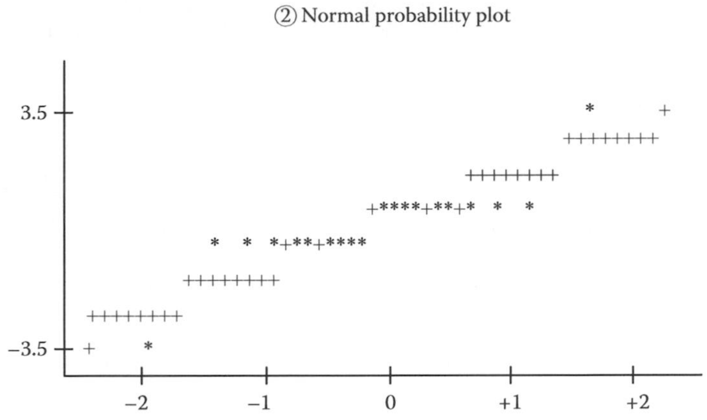

FIGURE 3.D.1
Normal probability plot.

$\hat{Y}_i$, $i = 1,\ldots,n$. Ideally, the implied graph should look like Figure 3.E.1.a. Here the studentized deleted residuals are randomly dispersed about a mean of zero and are contained within a horizontal band. In this instance, heteroscedasticity is not a problem.

However, specific instances in which the homogeneity of error variances assumption is violated are illustrated in Figures 3.E.1.b through d. In panel b of Figure 3.E.1, the outward-pointing funnel pattern indicates that σ_ε^2 is not constant but varies directly with the dependent variable Y. Panel c of Figure 3.E.1 indicates just the opposite phenomenon: the inward-constricting funnel pattern reveals that again σ_ε^2 is not constant but varies inversely with Y. The double-bowed pattern depicted in panel d of Figure 3.E.1 indicates that perhaps the Y values are drawn from a binominal population because, in this instance, σ_ε^2 near 0.5 is greater than σ_ε^2 near 0 or 1.

Finally, *model correctness* comes into question if the patterns depicted in Figures 3.E.1.e and f are obtained. Here one is attempting to fit a linear equation to a nonlinear scatter of data points. Clearly, the linearity assumption is inappropriate.

Example 3.E.1

Let us now use the SAS System to execute a plot of the studentized deleted residuals $r_{i(-i)}$ against the predicted values $\hat{Y}_i$, $i = 1,\ldots,n$. The dataset used is that supporting Example 3.1. Exhibit 3.E.1 provides the details of the requisite SAS code, and Figure 3.E.2 displays the requested plot. Obviously, the case depicted in Figure 3.E.1.b emerges, thus enabling us to conclude that the random error terms ε_i are heteroscedastic.

Exhibit 3.E.1 SAS Code for Studentized Deleted Residuals

```
data regh;
   input x y;
datalines;
```

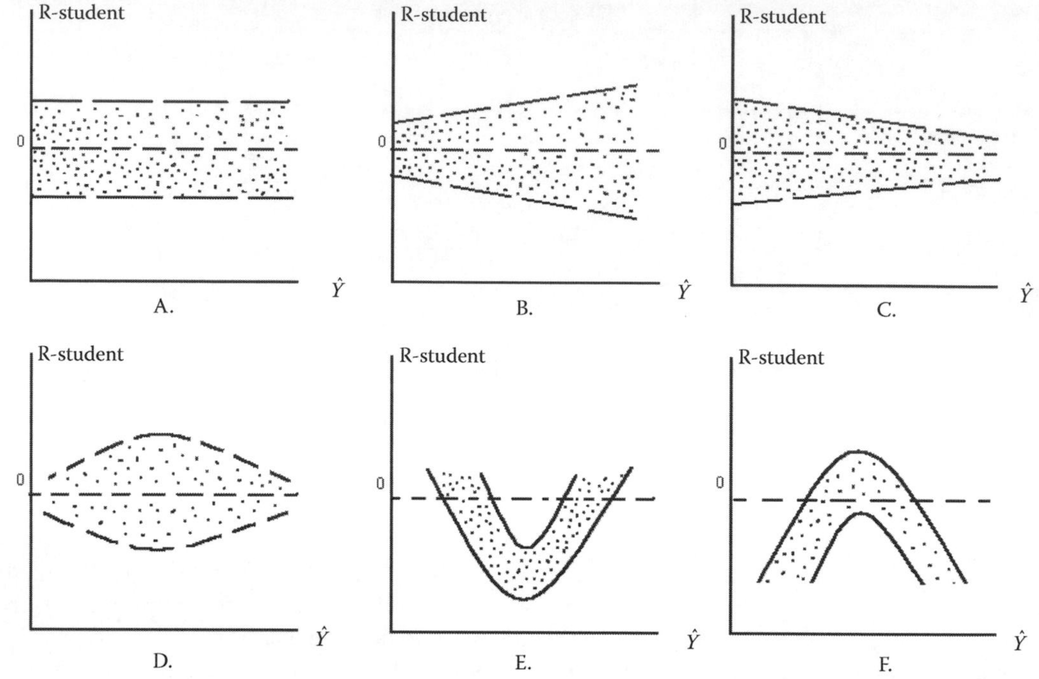

FIGURE 3.E.1
Behavior of studentized deleted residuals.

```
 5   1.5
 7   4
 1   2
 4   2
13   4
15   4
10   1.5
 2   1
 6   3
 3   1.5
16   3
 8   1.5
12   3
11   4
 9   3
19   6
17   1
14   1.5
22   8
25   8
21   4
18   3
20   1
24   1
```

① The *SPEC* option performs White's model specification test. (The output of the *spec* option appears in Example 3.E.2 of Appendix 3.E.)

② A plot of *R*-student versus predicted values is requested.

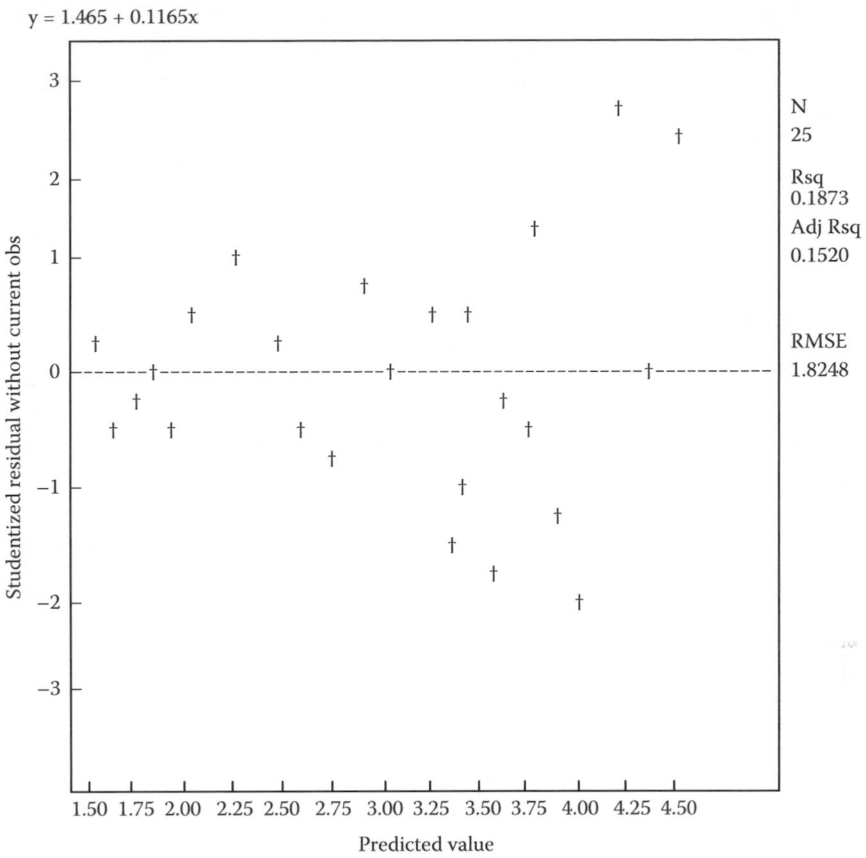

FIGURE 3.E.2
Studentized deleted residuals versus predicted *Y* values.

```
23   2
;
proc reg data=regh;
    model y=x/spec; ①
      plot rstudent.*predicted.; ②
run;
```

Detection of Heteroscedasticity

Although a plot such as the one depicted in Figure 3.E.2 can be quite revealing, let us now look at a more formal procedure for detecting the presence of heteroscedasticity. Specifically, we shall examine *White's test* [White 1980] for heteroscedasticity (alternatively called a *test of specification*). This test is general in that it does not make any formal assumptions about the nature of the heteroscedasticity, i.e., about the specific behavior of σ_i^2, $i = 1,..., n$. Here the null hypothesis is $H_0 : \sigma_i^2 = \sigma_\varepsilon^2$ and ε_i is independent of X, $i = 1,\ldots,n$; the alternative hypothesis is H_1: not H_0. When the regression model is correctly specified and the errors ε_i are independent of X, then rejection of H_0 is evidence of heteroscedasticity. For a single explanatory variable X, White's test statistic follows a χ^2 distribution with 1 degree of freedom.

Example 3.E.2

As indicated in Exhibit 3.E.1, White's specification test is requested by the *SPEC* option in the model statement. Table 3.E.1 reveals the following:

① The p-value for White's specification test is 0.0311, indicating that we should reject the null hypothesis (that the errors are homoscedestic and independent of X) in favor of the alternative hypothesis of heteroscedasticity. Note that, even if H_0 is rejected in White's test and it is concluded that the errors are heteroscedastic, we still need to decide what remedial measures best suit our data transformation requirements. This is the topic of the next section.

Variance Stabilizing Transformations

White's test for heteroscedasticity is very general in that, if the null hypothesis (of homoscedasticity and ε_i independent of X) is rejected, we are not given any information/guidance concerning the nature of the problem. Suppose then that heteroscedasticity is detected by either a plot of $r_{i(-i)}$ versus $\hat{Y}_i$ or using White's test result.

To correct for the effects of the heteroscedasticity, we must first determine its specification. Following Glejser [1969], we may use a *two-step weighted least-squares (TSWLS)* procedure :

TABLE 3.E.1

Output for White's Specification Test

The SAS System
The REG Procedure
Model: MODEL 1
Dependent Variable: y
Analysis of Variance

Source	DF	Sum of Squares	Mean Squares	F Value	Pr > F
Model	1	17.65558	17.65558	5.30	0.0307
Error	23	76.58442	3.32976		
Model	24	94.24000			

Root MSE	1.82476	R-Square	0.1873
Dependent mean	2.98000	Adj R-Sq	0.1520
Coeff Var	61.23363		

Parameter Estimates

Variable	DF	Parameter Estimate	Standard Error	t Value	Pr > |t|
Intercept	1	1.46500	0.75237	1.95	0.0638
x	1	0.11654	0.05061	2.30	0.0307

Test of First and Second Moment Specification

DF	Chi-Square	Pr > ChiSq
2	6.94	0.0311

STEP 1:

1. Fit $Y = \beta_0 + \beta_1 X + \varepsilon$ to obtain the OLS studentized deleted residuals (*R*-studentized values) $r_{i(-i)}$.
2. Suppose σ_i^2 is known up to a multiplicative constant. Then we may perform the following regression :

   a. If $\sigma_i^2 = \sigma^2 X_i^2$, fit $|r_{i(-i)}| = \alpha_0 + \alpha_1 X$.

   b. If $\sigma_i^2 = \sigma^2 / X_i^2$, fit $|r_{i(-i)}| = \alpha_0 + \alpha_1 X^{-1}$.

   c. If $\sigma_i^2 = \sigma^2 X_i$, fit $|r_{i(-i)}| = \alpha_0 + \alpha_1 X^{1/2}$.

   d. If $\sigma_i^2 = \sigma^2 / X_i$, fit $|r_{i(-i)}| = \alpha_0 + \alpha_1 X^{-1/2}$.

3. Choose the regression in 2.a through 2.d with the most significant t value for $\hat{\alpha}_1$ to determine the weights w_i.

STEP 2 :

1. Suppose regression 2.a is selected. Then set $w_i = 1/X_i$. If 2.b is chosen, set $w_i = X_i$. If 2.c fits best, set $w_i = 1/\sqrt{X_i}$, and if 2.d is relevant, set $w_i = \sqrt{X_i}$.

2. Replace $Y_i = \beta_0 + \beta_1 X_i + \varepsilon_i$ by

$$w_i Y_i = \beta_0 w_i + \beta_1 w_i X_i + w_i \varepsilon_i, \tag{3.E.1}$$

   i.e., use OLS to regress $w_i Y_i$ on w_i and $w_i X_i$, where the transformed random error term is $w_i \varepsilon_i$ (and ostensibly homosecdastic). Note that this process is equivalent to minimizing the weighted sum of squares $\sum_{i=1}^{n} w_i (Y_i - \hat{\beta}_0 - \hat{\beta}_1 X_i)^2$ with respect to $\hat{\beta}_0$ and $\hat{\beta}_1$ so as to obtain weighted least-squares estimates of β_0 and β_1. For instance, if $\sigma_i^2 = \sigma^2 X_i^2$, then, via Equation 3.E.1, we have Equation 3.15 again or

$$\frac{Y_i}{X_i} = \beta_0 \left(\frac{1}{X_i} \right) + \beta_1 + \frac{\varepsilon_i}{X_i}, \quad X_i \neq 0;$$

   and if $\sigma_i^2 = \sigma^2 X_i$, then Equation 3.E.1 reduces to

$$\frac{Y_i}{\sqrt{X_i}} = \beta_0 \left(\frac{1}{\sqrt{X_i}} \right) + \beta_1 \frac{X_i}{\sqrt{X_i}} + \frac{\varepsilon_i}{\sqrt{X_i}}, \quad X_i \neq 0.$$

(Note that, in this expression, there is no constant term and there are two explanatory variables, namely $1/\sqrt{X_i}$ and $X_i/\sqrt{X_i}$. Fitting this expression by OLS will require multiple regression techniques offered in Chapter 12.)

An alternative two-step WLS routine has us first take the following:

STEP 1:

1. Fit $Y = \beta_0 + \beta_1 X + \varepsilon$ to obtain the OLS studentized deleted residuals $r_{i(-i)}$. We then perform the following regressions :

   a. $|r_{i(-i)}| = \gamma_0 + \gamma_1 X_i$.
   b. $|r_{i(-i)}| = \gamma_0 + \gamma_1 X_i + \gamma_2 X_i^2$.
   c. $|r_{i(-i)}| = \gamma_0 + \gamma_1 X_i + \gamma_2 X_i^2 + \gamma_3 X_i^3$.

2. Choose one of these regression equations on the basis of t and R^2 considerations.

STEP 2 :

1. Suppose regression 1.b is selected. Set

$$w_i = \hat{\gamma}_0 + \hat{\gamma}_1 X_i + \hat{\gamma}_2 X_i^2.$$

2. Estimate the regression equation

$$\frac{Y_i}{w_i} = \beta_0 \left(\frac{1}{w_i} \right) + \beta_1 \frac{X_i}{w_i} + \frac{\varepsilon_i}{w_i}$$

   using OLS.

Experience dictates that, at times, the homoscedasticity problem may be solved by a transformation of one or both of the regression variables to logarithms. That is to say, instead of estimating $Y_i = \beta_0 + \beta_1 X_i + \varepsilon_i$, we can fit either

$$\log Y_i = \beta_0 + \beta_1 \log X_i + \varepsilon_i$$

or

$$\log Y_i = \beta_0 + \beta_1 X_i + \varepsilon_i.$$

Clearly, great care has to be taken in rationalizing the use of these transformed regression as well as interpreting the regression parameters and the random error terms.

Example 3.E.3

The preceding two examples have enabled us to conclude that heteroscedasticity is a problem that must be corrected if we are to efficiently estimate β_0 and β_1 from the *regh* dataset. As a first step, we need to determine the nature of the heteroscedasticity problem. To this end, let us examine Exhibit 3.E.2. Here we are going to perform STEP 1 of the TSWLS procedure. The output of item ⑦ of this step is summarized in Table 3.E.2. Clearly, the model involving $|r_{i(-i)}| = \alpha_0 + \alpha_1 X$ with $\sigma_i^2 = \sigma^2 X_i^2$ fits best.

Exhibit 3.E.2 SAS Code for STEP1 of TSWLS Procedure

```
data regh;
   input x y;
datalines;
 5   1.5
 7   4
 1   2
     .
     .
     .
20   1
24   1
23   2
;
proc reg
  data=regh;
   model y=x/
influence; ①
output out=new
rstudent=rstudent; ②
data regh2; set new; ③
absr=abs (rstudent); ④
x1=1/x;
x2=sqrt (x); ⑤
x3=1/x2;
proc reg data=regh2; ⑥
       model absr=x;
       model absr=x1;
       model absr=x2; ⑦
       model absr=x3;
run;
```

① The *influence* option is requested in the MODEL statement (among other statistics, the studentized deleted residuals or *R*-student values are generated).
② A dataset named *new* is formed that contains the variable *rstudent* as well as all variables in *regh*.
③ For regression purposes, the dataset *regh2* is formed that contains all variables found in dataset *new*.
④ A variable called *absr* is formed as the absolute value of the studentized deleted residuals.
⑤ Transformations of the variable *X* are requested.
⑥ The regression procedure is invoked and applied to the *regh2* dataset.
⑦ Various model specifications concerning the behavior of the variable *absr* as a function of *X* are estimated.

We now turn to STEP 2 of the TSWLS routine. On the basis of the STEP 1 results, we define weights $w_i = 1/X_i$. The requisite SAS code is provided by Exhibit 3.E.3, and the output of this program is presented as Table 3.E.3. As item ① reveals, the *p*-value for White's test of heteroscedasticity is 0.1124, thus prompting us to not reject the null hypothesis of homoscedasticity and ε_i independent of *X*. (This conclusion is also confirmed by the plot depicted by Figure 3.E.3). Hence, the estimates for β_0 and β_1 given in Table 3.E.3 (items ② and ③) are now considered BLUE.

TABLE 3.E.2

Output of TSWLS Procedure (STEP1)

Model	*t* Value for $\hat{\alpha}_1$	*p*-Value
$absr = x$	4.47	0.0002
$absr = x_1$	−1.77	0.0894
$absr = x_2$	3.88	0.0008
$absr = x_3$	−2.40	0.0251

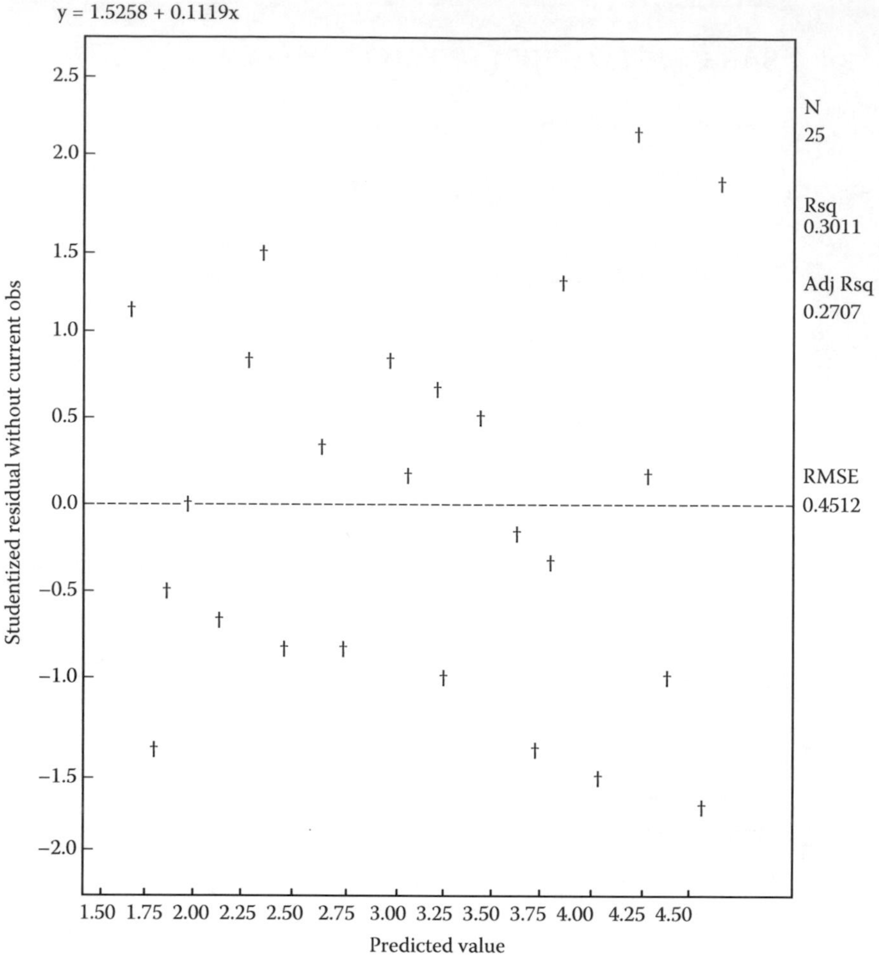

FIGURE 3.E.3
Studentized deleted residuals versus predicted *Y* values.

Exhibit 3.E.3 SAS Code for STEP2 of TSWLS Procedure

```
data regh;
   input x y;
w=1/x; ①
datalines;
 5   1.5
 7   4
 1   2
     .
     .
     .
20   1
```

① A weight variable *W* with values $w_i = 1/X_i$, $w_i \geq 0$, is defined.

② The *weight* statement invokes the WLS regression procedure with weight variable *W*.

③ WLS is used to estimate the stated regression model, and White's test of specification is requested in the model statement.

TABLE 3.E.3

Output of TSWLS Procedure (STEP2)

The SAS System
The REG Procedure
Model: MODEL 1
Dependent Variable: *y*
Analysis of Variance

Source	**Sum of DF**	**Mean Squares**	**Squares**	***F* Value**	**Pr > *F***
Model	1	2.01729	2.01729	9.91	0.0045
Error	23	4.68289	0.20360		
Corrected total	24	6.70018			

Root MSE	0.45122	*R*-Square	0.3011
Dependent mean	2.25865	Adj *R*-Sq	0.2707
Coeff Var	19.97763		

Parameter Estimates

Variable	**DF**	**Parameter Estimate**	**Standard Error**	***t* Value**	**Pr > |*t*|**
Intercept	1	② 1.52579	0.32797	4.65	0.0001
x	1	③ 0.11186	0.03554	3.15	0.0045

Test of First and Second Moment Specification

① DF	Chi-Square	Pr > ChiSq
2	4.37	0.1124

```
24   1          ④ A plot of the studentized deleted residuals versus
23   2             the predicted values is requested.
;
proc reg data=regh; weight w; ②
   model y=x/spec; ③
   plot rstudent.*predicted.; ④
run;
```

Appendix F: Detecting Autocorrelated Errors

In Example 3.2 above, we concluded that the random errors ε_t exhibited positive first-order autocorrelation. Let us see how the SAS System can be used to establish this same result. Exhibit 3.F.1 displays the requisite SAS code. (For convenience, the observations on the *X* and *Y* variables are included.)

The output of this program is given in Table 3.F.1 and Figure 3.F.1. As Table 3.F.1 reveals, from items ① and ②, DW = 0.384 and $\hat{\rho} = 0.743$, respectively. At this point, either a direct DW test of $H_0: \rho = 0$ against $H_1:$ the errors ε_t exhibit positive first-order autocorrelation (see above, "Testing for the Absence of Autocorrelation: The Durbin-Watson Test") can be performed or we may simply note that the value of $\hat{\rho}$ is large enough for us to reasonably expect that positive first-order autocorrelation exists (i.e., e_t tends to have the same sign as e_{t-1}). Hence, the estimated standard errors of $\hat{\beta}_0$ and $\hat{\beta}_1$ are incorrect, and thus the OLS estimates of β_0 and β_1 are inefficient.

Exhibit 3.F.1 SAS Code for Durbin-Watson Test

```
data auto;
   input x y;
datalines;
1.33      19.3
1.29      20.4
1.25      20.9
1.21      21.9
1.19      23.4
1.20      24.5
1.21      25.8
1.29      30.5
1.33      33.3
1.42      37.2
1.52      42.5
1.59      48.8
1.84      55.4
2.17      64.3
2.55      78.9
2.97      86.5
3.70      114.6
4.10      129.7
4.34      126.1
4.71      132.0
4.82      138.1
4.81      141.2
;
proc reg data= auto;
   model y= x/dw; ①
     output out= new r= ry; ②
     data auto1; set new; ③
     n= _n_; ④
proc plot data= auto1; ⑤
   plot ry*n= '*'; ⑥
run;
```

① The Durbin-Watson statistic *DW* is requested as an option in the MODEL statement.
② A dataset *new* is created that contains the variable *ry* as well as all variables in dataset *auto*.
③ For plotting purposes, the dataset *auto1* is formed that contains all variables found in dataset *new*.
④ After the *data auto1* step, we use the automatic observation variable *_n_* to create a sequential period indicator *n*.
⑤,⑥ We invoke the *plot* procedure using the dataset *auto1*, and we plot the residuals *ry* against the sequential period indicator *n*.

Estimation with Autocorrelated Errors

To correct for the impact of autocorrelated disturbances on the efficiency of the OLS estimates of β_0 and β_1 given in Table 3.F.1, let us again appeal to the SAS System. Here the *AUTOREG* procedure will be called on. In general, this routine is used to estimate linear

TABLE 3.F.1

Output for Durbin-Watson Test

The SAS System
The REG Procedure
Model: MODEL 1
Dependent Variable: y
Analysis of Variance

Source	DF	Sum of Squares	Mean Squares	*F* Value	Pr > *F*
Model	1	42404	42404	967.45	<0.0001
Error	20	876.61778	43.83089		
Corrected total	21	43281			

Root MSE	6.62049	*R*-Square	0.9797
Dependent mean	64.33182	Adj *R*-Sq	0.9787
Coeff Var	10.29116		

Parameter Estimates

Variable	DF	Parameter Estimate	Standard Error	*t* Value	Pr > |*t*|
Intercept	1	−12.14291	2.83504	−4.28	0.0004
x	1	32.46082	1.04362	31.10	<0.0001

Durbin-Watson D	0.348 ①
Number of observations	22
1st order autocorrelation	0.743 ②

regression models for time series data when the random errors are autocorrelated. It uses an autoregressive model to correct for the effects of autocorrelation. The procedure first tests for the presence of autocorrelation by calculating DW statistics and their associated *p*-values. By simultaneously estimating the regression parameters and the autoregressive model parameters, *AUOTREG* corrects the regression parameter estimates for the effects of autocorrelation.

The *AUTOREG* procedure produces the following sequence of general output features:

A. Statistics for the OLS residuals of the regression model ($Y = \beta_0 + \beta_1 X + \varepsilon$).

B. OLS estimates of the regression parameters along with their standard errors and *t* values.

C. Estimates of autocorrelations determined from the OLS residuals;

D. Estimates of autoregression parameters $\rho_1, \rho_2, \ldots, \rho_s$ from the AR(s) specification $\varepsilon_t = \rho_1\varepsilon_{t-1} + \rho_2\varepsilon_{t-2} + \cdots + \rho_s\varepsilon_{t-s} + u_t$, where u_t is taken to be white noise. (If the *backstep* option is used, only significant autoregressive parameters are shown.)

E. Maximum-likelihood estimates of statistics for residuals (assuming that the maximum-likelihood option is chosen).

F. Final estimates of the regression parameters and their standard errors recomputed under the assumption that the autoregression parameter estimates equal their population counterparts.

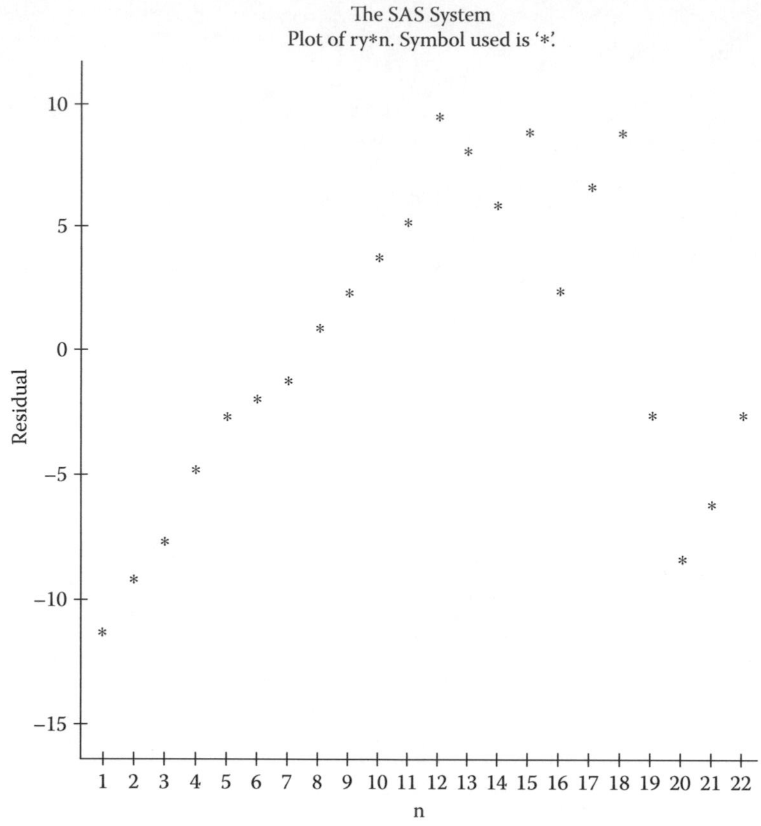

FIGURE 3.F.1
Plot of *Y* residuals over time.

Exhibit 3.F.2 specifies the SAS code supporting the implementation of *PROC AUTOREG.*

Exhibit 3.F.2. SAS Code for PROC AUTOREG

```
data auto;
   input x y;
datalines;
1.33    19.3
1.29    20.4
1.25    20.9
1.21    21.9
1.19    23.4
1.19    24.5
1.19    25.8
1.29    30.8
1.33    33.3
1.42    37.2
1.52    42.5
1.59    48.8
1.84    55.4
2.17    64.3
2.55    78.9
```

① The *AUTOREG* procedure is invoked.

② The MODEL statement option *nlag= 4* posits an AR(4) autoregressive model $\varepsilon_t = \rho_1\varepsilon_{t-1} + \rho_2\varepsilon_{t-2} + \rho\varepsilon_{t-3} + \rho_4\varepsilon_{t-4} + u_t$.

③ The MODEL statement option *backstep* requests that a stepwise autoregression be performed, i.e., in general, a high-order autoregression model having many lags is initially estimated, and then, sequentially, autoregression parameters are removed until only those autoregressive parameters having significant *t* values ($\alpha = 0.05$) are retained.

④ The maximum-likelihood option *method= ml* is requested in the MODEL statement.

⑤,⑥ The *dw= 4* and *dwprob* options in the model statement request D-W statistics for lags 1-4 and their *p*-values, respectively.

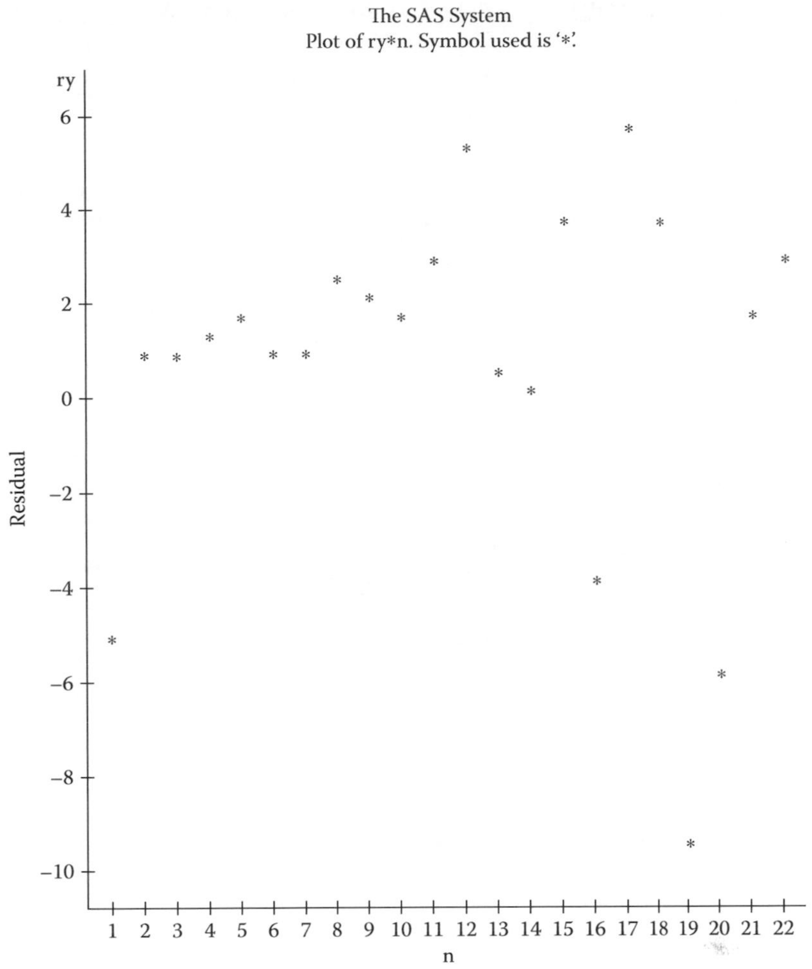

FIGURE 3.F.2
Plot of Y residuals over time.

```
2.97    86.5
3.70    114.6
4.10    129.7
4.34    126.1
4.71    132.0
4.82    138.1
4.81    141.2
;       ①
proc autoreg
data= auto;
                    ②           ③           ④          ⑤        ⑥
    model y= x/nlag= 4  backstep  method= ml  dw= 4  dwprob;
    output out= new r= ry; ⑦
    data auto1; set new; ⑧
    n= _n_; ⑨
proc plot data= auto1; ⑩
    plot ry*n= '*'; ⑪
run;
```

⑦ A dataset *new* is created that contains the residuals variable *ry* as well as all variables found in dataset *auto*.

⑧ For plotting purposes, the dataset *auto1* is formed that contains all variables found in dataset *new*.

⑨ After the *data auto1* step, we use the automatic observation variable *_n_* to create a sequential period indicator *n*.

⑩, ⑪ We invoke the *plot* procedure using the dataset *auto1* and we plot the residuals *ry* against the sequential period indicator *n*.

The output of this *AUTOREG* procedure is presented in Table 3.F.2 and Figure 3.F.2. Specifically,

Ⓐ ① Regress *R*-Square depicts the R^2 statistic for the transformed regression equation, i.e., it is a measure of goodness of fit of the systematic part of the regression model after the variables are adjusted for the effects of the estimated autocorrelation.

② Total *R*-Square is a measure of the contribution of both the systematic part of the regression model and past residual values in predicting the next successive value of the dependent variable *Y*. (Note that Regress *R*-Square and Total *R*-Square must be equal when OLS estimation is used because no adjustment for autocorrelation has been undertaken.)

③ A glance at the set of DW test *p*-values for detecting positive autocorrelation indicates that a third-order autoregressive model

TABLE 3.F.2
Output for AUTOREG Procedure

The SAS System
The AUTOREG Procedure
Dependent Variable: y

Ⓐ

Ordinary Least-Squares Estimates

SSE	876.617782	DFE	20
MSE	43.83089	Root MSE	6.62049
SBC	149.68601	AIC	147.503925
① Regress R-Square	0.9797	② Total R-Square	0.9797

③ Durbin-Watson Statistics

Order	**DW**	**Pr < DW**	**Pr > DW**
1	0.3477	<0.0001	1.0000
2	0.7509	0.0005	0.9995
3	0.9017	0.0058	0.9942
4	1.1649	0.0659	0.9341

Note: Pr < DW is the *p*-value for testing positive autocorrelation, and Pr > DW is the *p*-value for testing negative autocorrelation.

Ⓑ

Variable	**DF**	**Estimate**	**Standard Error**	***t* Value**	**Aprox Pr > |*t*|**
Intercept	1	−12.1429	2.8350	−4.28	0.0004
x	1	32.4608	1.0436	31.10	<0.0001

Ⓒ

Estimates of Autocorrelations

<table>
<tr><th>Lag</th><th>Covariance</th><th>Correlation</th><th>−1 9 8 7 6 5 4 3 2 1 0 1 2 3 4 5 6 7 8 9 1</th></tr>
<tr><td>0</td><td>39.8463</td><td>1.000000</td><td>| |********************|</td></tr>
<tr><td>1</td><td>29.6148</td><td>0.743226</td><td>| |***************|</td></tr>
<tr><td>2</td><td>18.7238</td><td>0.469900</td><td>| |*********|</td></tr>
<tr><td>3</td><td>12.6907</td><td>0.318493</td><td>| |******|</td></tr>
<tr><td>4</td><td>6.6665</td><td>0.167306</td><td>| |***|</td></tr>
</table>

TABLE 3.F.2 *(continued)*

The SAS System
The AUTOREG Procedure
Backward Elimination of Autoregressive Terms

Lag	Estimate	*t* Value	Pr > |*t*|
4	0.149451	0.60	0.5539
3	−0.096787	−0.40	0.6935
2	0.184275	0.80	0.4367
Preliminary MSE	17.8358		

Ⓓ

Estimates of Autoregressive Parameters

Lag	Coefficient	Standard Error	*t* Values
1	−0.743226	0.153488	−4.84

Algorithm converged

Ⓔ

Maximum-Likelihood Estimates

SSE	292.705084	DFE	19
MSE	15.40553	Root MSE	3.92499
SBC	129.885323	AIC	126.612195
① Regress R-Square	0.9156	② Total R-Square	0.9932

③ Durbin-Watson Statistics

Order	DW	Pr < DW	Pr > DW
1	0.5705	0.1124	0.8876
2	2.1449	0.6653	0.3347
3	1.4391	0.1545	0.8455
4	1.3073	0.1395	0.8605

Note: Pr < DW is the *p*-value for testing positive autocorrelation, and Pr > DW is the *p*-value for testing negative autocorrelation.

Variable	DF	Estimate	Standard Error	*t* Value	Approx Pr > |*t*|
Intercept	1	−13.4753	8.4834	−1.59	0.1287
x	1	32.0561	2.7876	11.50	<0.0001
AR1	1	−0.8430	0.1396	−6.04	<0.0001

Ⓕ

Autoregressive Parameters Assumed Given

Variable	DF	Estimate	Standard Error	*t* Value	Approx Pr > |*t*|
Intercept	1	−13.4753	7.2500	−1.86	0.0786
x	1	32.0561	2.2336	14.35	<0.0001

is adequate because the *p*-value for the fourth-order term is $0.066 > \alpha = 0.05$.

Ⓑ OLS parameter estimates are printed along with their *t* statistics and *p*-values.

Ⓒ Here the estimated autocorrelation coefficients are $\hat{\rho}_1 = 0.743$, $\hat{\rho}_2 = 0.473$, $\hat{\rho}_3 = 0.318$, and $\hat{\rho}_4 = 0.167$. As expected, the value of $\hat{\rho}_4$ is rather low.

Ⓓ The *backstep* option has eliminated the second-, third-, and fourth-order terms in the specified AR(4) autoregression model. Hence, the estimate of $-\hat{\rho}_1 = 0.743$. (Note that SAS writes the coefficient of ε_{t-r} in an autoregressive process as $-\rho_r$. This accounts for the minus sign on the estimated coefficient of ε_{t-1} so that $-\hat{\rho}_1 = 0.743$.)

Ⓔ ① The maximum-likelihood estimate of Regress *R*-Square is lower than its OLS counterpart determined in Ⓐ above. This reflects the adjustment in the regression variables for the estimated autocorrelation.

② The maximum-likelihood estimate of Total *R*-Square is higher than that obtained by OLS in Ⓐ above. This reflects the improved fit obtained when past residuals are used to help predict the next *Y* value.

③ Under the adjustment for autocorrelation, none of the *p*-values for the new set of DW statistics used in testing for positive autocorrelation are below $\alpha = 0.05$. Hence, autocorrelation is no longer a problem.

Ⓕ Efficient estimates for both β_0 and β_1 and their corresponding *t* statistics and *p*-values occupy the final portion of the *AUTOREG* output.

The plot of *ry* versus *n* requested in Exhibit 3.F.2 appears as Figure 3.F.2. This plot depicts a much more random pattern to the ordered residuals than that found previously in Figure 3.F.1.

Appendix G: Yet Another Look at the Heteroscedasticity Issue

Autoregressive Conditional Heteroscedastic Processes [Engle 1982]

Above (see "Assumptions about the Nature of σ_i^2), we dealt with the case of heteroscedastic disturbances of the multiplicative variety in that the error variance was not constant but varied proportionately with a function of the explanatory variable *X* (e.g., we posited that $\sigma_i^2 = \sigma^2 X_i^2$, $i = 1,\ldots,n$, σ^2 a constant). However, let us now examine an alternative form of heteroscedastic error architecture. Specifically, suppose that the error variance does not vary with *X* but, instead, is *time dependent*. That is , the error variance changes over time in a manner that depends on the volatility of past error terms, the so-called *ARCH effect* (for autoregressive conditional heteroscedastic).

This specialized type of heteroscedastic process is one in which $V(\varepsilon_t) = \sigma_t^2$ (note that we have traded the *i* subscript for a *t* subscript) depends on the magnitudes of the errors experienced in the recent past and thus can be modeled as, say,

$$\sigma_t^2 = \alpha_0 + \alpha_1 \varepsilon_{t-1}^2 + \alpha_2 \varepsilon_{t-2}^2 + \cdots + \alpha_q \varepsilon_{t-q}^2. \qquad (3.G.1)$$

This expression is termed a *qth-order ARCH process* and is denoted ARCH(q). Clearly, σ_t^2 changes with each observation so that the error variance is not constant.

The motivation for this heteroscedastic structure comes from forecasters' observations that, for some time series datasets (e.g., for a great many financial series), there often occurs a bunching together of large and small error terms over certain subperiods within the entire time span studied (Figure 3.G.1). Hence, an ARCH process should be modeled when the ability to predict a variable Y at a given point in time varies from one subperiod to another because $V(\varepsilon_t)$ changes over time in response to the *volatilities* $\varepsilon_{t-1}^2, \varepsilon_{t-2}^2, \ldots$, displayed by past error terms.

Let us now examine the structure of the ARCH regression model. To keep our analysis manageable, we shall focus on the ARCH(1) case. Let the data-generating process be given as $Y_t = \beta_0 + \beta_1 X_t + \varepsilon_t$, where the X_t's are fixed, $E(X_t\, \varepsilon_s) = 0$ for all t and s ($t \neq s$), and ε_t is ARCH(1) or $\varepsilon_t = v_t(\alpha_0 + \alpha_1 \varepsilon_{t-1}^2)^{1/2}$ with $\alpha_0 > 0$, $0 \leq \alpha_1 < 1$. Furthermore, v_t and ε_{t-1} are independent of each other, and the v_t's are independent identically distributed $N(0, 1)$ random variables.

Given the ARCH(1) specification for ε_t, the conditional mean and variance of ε_t are, respectively,

$$\begin{aligned} &\text{a. } E(\varepsilon_t | \varepsilon_{t-1}) = E(v_t(\alpha_0 + \alpha_1 \varepsilon_{t-1}^2)) = 0. \\ &\text{b. } V(\varepsilon_t | \varepsilon_{t-1}) = E(\varepsilon_t^2 | \varepsilon_{t-1}) = E(v_t^2(\alpha_0 + \alpha_1 \varepsilon_{t-1}^2)) \\ &\qquad = \alpha_0 + \alpha_1 \varepsilon_{t-1}^2 . \end{aligned} \tag{3.G.2}$$

Here Equation 3.G.2.b reveals that ε_t conditional on ε_{t-1} is clearly heteroscedastic.

Next, the conditional mean and variance of Y_t are, respectively,

$$\begin{aligned} &\text{a. } E(Y_t | Y_{t-1}) = \beta_0 + \beta_1 X_t. \\ &\text{b. } V(Y_t | Y_{t-1}) = \alpha_0 + \alpha_1 \varepsilon_{t-1}^2 . \end{aligned} \tag{3.G.3}$$

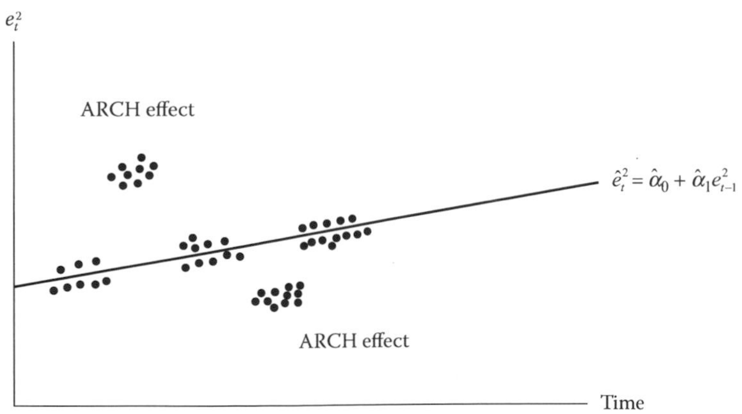

FIGURE 3.G.1
ARCH effects over time.

Clearly, the conditional forecast variance $V(Y_t|Y_{t-1})$ is not constant when ε_t is ARCH(1). Note that the unconditional means of ε_t and Y_t are, respectively, $E(\varepsilon_t)=0$ and $E(Y_t)=\beta_0+\beta_1X_t$, whereas the unconditional variances of ε_t and Y_t are equal, i.e., $V(\varepsilon_t)=\sigma_\varepsilon^2=V(Y_t)=\alpha_0/(1-\alpha_1), \alpha_1<1$. To see this, set

$$V(\varepsilon_t)=V(E(\varepsilon_t|\varepsilon_{t-1}))+E(V(\varepsilon_t|\varepsilon_{t-1}))$$

$$=\alpha_0+\alpha_1E(\varepsilon_{t-1}^2)=\alpha_0+\alpha_1V(\varepsilon_{t-1}).$$

If the process-generating ε_t is *variance stationary* (i.e., $V(Y_t)$ is constant and independent of t), then

$$V(\varepsilon_t)=V(\varepsilon_{t-1})=\alpha_0+\alpha_1V(\varepsilon_t)$$

or

$$V(\varepsilon_t)=\alpha_0/(1-\alpha_1), \quad \alpha_1<1.$$

Because $E(\varepsilon_t)=0$ and $E(\varepsilon_t^2)=\sigma_\varepsilon^2=\alpha_0/(1-\alpha_1)$, $\alpha_1<1$, the ε_t's are not autocorrelated.

Given these developments, we may summarize the *ARCH(1) regression model* as

$$\begin{aligned} &Y_t|\psi_{t-1}: N(\beta_0+\beta_1X_t,\sigma_t) \\ &\sigma_t^2=\alpha_0+\alpha_1\varepsilon_{t-1}^2 \\ &\varepsilon_t=Y_t-\beta_0-\beta_1X_t, \end{aligned} \tag{3.G.4}$$

where ψ_{t-1} depicts the *information set* containing all available information at time $t-1$. Hence, the ARCH(1) model allows us to simultaneously estimate its mean and variance given:

a. $Y_t=\beta_0+\beta_1X_t+\varepsilon_t$ (the model of the mean) and $\varepsilon_t=v_t(\alpha_0+\alpha_1\varepsilon_{t-1}^2)^{1/2}$, $v_t: N$ (0, 1) (the model of the variance).
b. The conditional and unconditional means of ε_t are both zero.
c. The ε_t's are nonautocorrelated.
d. The ε_t's are not independent but are conditionally heteroscedastic because they are related via the conditional variance equation

$$E(\varepsilon_t^2|\varepsilon_{t-1})=\alpha_0+\alpha_1\varepsilon_{t-1}^2.$$

e. The conditional and unconditional means of Y_t are both $\beta_0+\beta_1X_t$.
f. The conditional variance of Y_t is $\alpha_0+\alpha_1\varepsilon_{t-1}^2$, whereas the unconditional variance of both ε_t and Y_t is $\alpha_0/(1-\alpha_1)$, $\alpha_1<1$.

Looking to the estimation of β_0 and β_1 under the ARCH(1) scheme depicted by Equation 3.G.4, we may note briefly that, if $|\alpha_1| < 1$, then, based on the preceding results, unconditionally $\varepsilon_t : N\left(0, \frac{\alpha_0}{1-\alpha_1}\right)$. Hence, the model obeys the assumptions of the strong classical linear regression model so that OLS estimates for β_0 and β_1 are BLUE, i.e., OLS yields the most efficient linear unbiased estimates of the regression parameters. However, it can be shown that maximum-likelihood estimates of β_0 and β_1 are asymptotically more efficient than OLS.

Engle suggests performing what is called a *Lagrange multiplier (LM) test* to determine whether the disturbances ε_t follow an ARCH(1) process. Specifically, let us first regress Y on X and calculate the resulting OLS residuals $e_t = \hat{Y}_t - \hat{\beta}_0 - \hat{\beta}_1 X_t$. Next, we estimate via OLS the *ARCH(1) residual diagnostic equation* $\varepsilon_t^2 = \alpha_0 + \alpha_1 e_{t-1}^2 + u_t$ (u_t is taken to be white noise) and reserve its R^2 value. The test statistic is then nR^2. This statistic is asymptotically distributed as χ_1^2 under $H_0 : \varepsilon_t$ is conditionally homoscedastic versus $H_1 : \varepsilon_t$ is conditionally heteroscedastic (or under $H_0 : \alpha_1 = 0$, versus $H_1 : \alpha_1 > 0$).

We may extend this model to q lags by assuming that

$$\sigma_t^2 = V(\varepsilon_t | \varepsilon_{t-1}, \varepsilon_{t-2}, \ldots, \varepsilon_{t-q})$$

$$= \alpha_0 + \alpha_1 \varepsilon_{t-1}^2 + \alpha_2 \varepsilon_{t-2}^2 + \ldots + \alpha_q \varepsilon_{t-q}^2.$$

In this instance, the test statistic nR^2 is asymptotically distributed as χ_q^2, where R^2 is determined via a regression of e_t^2 on a constant and $e_{t-1}^2, e_{t-2}^2, \ldots, e_{t-q}^2$. Here we are testing $H_0 : \alpha_i = 0$ for all i against H_1 : at least one $\alpha_i \neq 0$.

Example 3.G.1

A convenient feature of the AUTOREG procedure is that we can easily test for ARCH effects by using the *archtest* option in the MODEL statement. For instance, if the MODEL statement appearing in Exhibit 3.F.2 is replaced by

model y= x/nlag= 4 backstep method= ml archtest dw= 4 dwprob;

then the output of the *archtest* option appears in Table 3.G.1.

Here the LM statistics are computed from the OLS residuals obtained from estimates of first-order through twelfth-order ARCH processes and denoted as $LM(q) = nR(q)^2$, $q = 1,\ldots,12$, where $R(q)^2$ is the R^2 value resulting from the estimation of the ARCH(q) process. That is,

① LM(1)=0.4066 is obtained by fitting the ARCH(1) process $e_t^2 = \alpha_0 + \alpha_1 e_{t-1}^2 + u_t$; LM(2)=0.6468 results from estimating the ARCH(2) process $e_t^2 = \alpha_0 + \alpha_1 e_{t-1}^2 + \alpha_2 e_{t-2}^2 + u_t$, etc.

② Clearly, none of the p-values associated with any of the LM(q), $q = 1,\ldots,12$, values warrants rejecting the null hypothesis that the ε_t's are conditionally homosedatsic. Hence, ARCH effects (at least for orders 1–12) are not present.

Generalized Autoregressive Conditional Heteroscedastic Processes [Bollerslev 1986]

The preceding section has addressed the issue of effectively modeling the clustering together of large and small error terms by specifying an ARCH process, which allows the

TABLE 3.G.1

Output for *Auchtest* Option

The SAS System

The AUTOREG Procedure

Q and LM Test for ARCH Disturbances

Order	Q	Pr > Q	① LM	② Pr > LM
1	2.4498	0.1175	0.6888	0.4066
2	3.0177	0.2212	0.8713	0.6468
3	3.3048	0.3470	1.1059	0.7757
4	6.8720	0.1428	6.7430	0.1501
5	7.8203	0.1664	6.8438	0.2325
6	9.1387	0.1659	8.2243	0.2221
7	13.2935	0.0653	10.4583	0.1641
8	14.3121	0.0740	11.0080	0.2012
9	15.1180	0.0877	11.5350	0.2408
10	15.1875	0.1254	11.5721	0.3147
11	16.6736	0.1179	11.7332	0.3840
12	16.9616	0.1510	12.2535	0.4255

conditional error variance to change over time with past volatilities, leaving the unconditional variance constant. The basic ARCH model can typically be viewed as a *short memory process* in that it incorporates only the most recent volatilities (squared residuals) in the estimation of the changing conditional error variance.

Let us now extend the ARCH model to the case where a *long memory process* involving all past volatilities is used to estimate the current conditional error variance. That is, we shall now consider what is called the generalized autoregressive conditional heteroscedastic (GARCH) process that includes within the current conditional error variance function not only past volatilities but also past conditional error variances. Hence, the GARCH process is structured so that changes in conditional error variance depend on the realizations of past volatilities, with the said changes constituting temporary albeit random departures from a constant unconditional error variance.

In this regard, let us now express

$$V(\varepsilon_t) = \sigma_t^2 = \alpha_0 + \underbrace{\sum_{i=1}^{q} \alpha_i \varepsilon_{t-i}^2}_{ARCH} + \underbrace{\sum_{j=1}^{p} \gamma_j \sigma_{t-j}^2}_{GARCH}, \tag{3.G.5}$$

$$\alpha_0 > 0,\ p \geq 0,\ q > 0,\ \alpha_i \geq 0 \text{ and } \gamma_j \geq 0.$$

This equation is characterized as a GARCH pth-, qth-order process and denoted as GARCH(p, q). Here the error variance is conditional on past volatilities (the ARCH term) as well as on past lagged conditional error variances (the GARCH term). Suppose $\varepsilon_t = v_t(\sigma_t^2)^{1/2}$, where σ_t^2 is determined from Equation 3.G.5 and the v_t are independent identically distributed $N(0, 1)$ random variables. The full GARCH(p, q) regression model can now be expressed as

$$Y_t|\psi_{t-1} \sim N(\beta_0 + \beta_1 X_t, \sigma_t)$$

$$\sigma_t^2 = \alpha_0 + \sum_{i=1}^{q} \alpha_i \varepsilon_{t-i}^2 + \sum_{j=1}^{p} \gamma_j \sigma_{t-j}^2 \tag{3.G.6}$$

$$\varepsilon_t = Y_t - \beta_0 - \beta_1 X_t,$$

where again ψ_{t-1} represents the information set containing all available information at time $t - 1$. For $p = 0$, Equation 3.G.6 reduces to the ARCH(q) case, whereas for $p = q = 0$, ε_t is taken to be white noise.

An important special case of Equation 3.G.6 is the GARCH(1,1) regression model:

$$Y_t|\psi_{t-1} : N(\beta_0 + \beta_1 X_t, \sigma_t)$$

$$\sigma_t^2 = \alpha_0 + \alpha_1 \varepsilon_{t-1}^2 + \gamma_1 \sigma_{t-1}^2 \tag{3.G.6.1}$$

$$\varepsilon_t = Y_t - \beta_0 - \beta_1 X_t.$$

For this specification, we again simultaneously estimate the mean and variance given the following:

a. $Y_t = \beta_0 + \beta_1 X_t$ (the model of the mean) and $\varepsilon_t = v_t(\alpha_0 + \alpha_1 \varepsilon_{t-1}^2 + \gamma_1 \sigma_{t-1}^2)^{1/2}$, $v_t : N(0, 1)$ (the model of the variance).
b. The conditional and unconditional means of ε_t are both zero.
c. The ε_t's are nonautocorrelated.
d. The conditional variance of ε_t is not constant but expressible as $\sigma_t^2 = \alpha_0 + \alpha_1 \varepsilon_{t-1}^2 + \gamma_1 \sigma_{t-1}^2$, whereas the unconditional variance of ε_t is $\alpha_0/(1 - \alpha_1 - \gamma_1)$, $\alpha_1 + \gamma_1 < 1$.
e. The conditional and unconditional means of Y_t are both $\beta_0 + \beta_1 X_t$. The conditional variance of Y_t is $\sigma_t^2 = \alpha_0 + \alpha_1 \varepsilon_{t-1}^2 + \gamma_1 \sigma_{t-1}^2$, whereas the unconditional variance of Y_t is $\alpha_0/(1 - \alpha_1 - \gamma_1)$, $\alpha_1 + \gamma_1 < 1$.

As a final point, if we combine an mth-order autoregressive model with a GARCH(p, q) model, Equation 3.G.6 becomes

$$Y_t|\psi_{t-1} : N(\beta_0 + \beta_1 X_t, \sigma_t)$$

$$\varepsilon_t = u_t + \rho_1 \varepsilon_{t-1} + \rho_2 \varepsilon_{t-2} + \cdots + \rho_m \varepsilon_{t-m} \tag{3.G.7}$$

$$u_t = v_t(\sigma_t^2)^{1/2}, \ v_t : N(0,1)$$

Example 3.G.2

To see exactly how the GARCH(p, q) regression model is used, let us consider the SAS code provided in Exhibit 3.G.1. The output generated by this program is given as Table 3.G.2:

① The *AIC*(=151.918) is the *Akaike Information Criterion.* It is calculated as $-\ln \mathcal{L} + 2k$, where $\mathcal{L}$ (=-70.959) is the value of the likelihood function at the parameter estimates and k (=5) is the number of parameters estimated.

The *SBC*(=158.013) is the *Schwartz Bayesian Criterion.* It is calculated as $-2\ln \mathcal{L} + k\ln(n)$.

Both the AIC and SBC are used as inputs into the *model selection process,* i.e., both are used to compare alternative or competing models that are estimated from the same dataset. In this regard, the model with the smaller AIC (or SBC) is said to provide a better fit. As model selection criteria, both the AIC and SBC trade off a reduction in SSE against a reduction in degrees of freedom as additional lags are added to the model. So if the order of the lags is increased but these additional regressors display little or no explanatory power, the AIC (SBC) will increase. Hence, it is important to ensure that the t statistic for each additional lag is statistically significant. Experience dictates that the AIC, although typically selecting an overparamaterized model, works well with small samples; the SBC has far better large-sample properties.

Total R-Square (=0.2939) is a measure of how well the next Y value can be predicted by using the structural portion of the regression model as well as past values of the residuals. If no correction for autocorrelation is made, the Total R-Square and ordinary regression R-square (R^2) values will be equal (as indicated in the OLS Estimates section displayed at the beginning of Table 3.G.2).

The *normality test* is a specification test of H_0: the standardized residuals from the GARCH(p, q) model ($\hat{v}_t = e_t / \sqrt{\hat{\sigma}_t^2}$) are normally distributed, against H_1 : not H_0. If the p-value exceeds, say, 0.05, then H_0 is not rejected.

② The p-values for both ARCH terms as well as the GARCH term are insignificant. Hence, the GARCH(1,1) model does not fit the data well.

③ The estimated conditional error standard deviation values appear as the final part of this SAS output (see also Figure 3.G.2).

Exhibit 3.G.1 SAS Code for GARCH Effects

```
data garch1;
   input x y;
datalines;
1.23      19.3
1.24      20.4
1.25      20.9
1.26      19.9
1.27      23.4
1.29      23.4
1.30      24.5
1.31      20.0
1.32      18.0
1.33      18.7
1.42      27.2
1.52      30.5
1.59      31.0
1.84      25.5
```

① In the MODEL statement, we specify a GARCH(1,1) conditional error variance regression model.

② We specify a maximum number of iterations to be performed to achieve convergence to stable parameter values. The default is maxiter = 50. (Note: *METHOD=ml* does not need to be specified because the default is the maximum-likelihood routine.)

③ *cev=ecev* writes to the output dataset *new* the values of the variable *ecev* that represent the estimated conditional error variance at each time period.

④ We transform the estimated conditional error variances *ecev* to standard deviations by forming a new variable *estd* with values *sqrt(ecev)*.

⑤ We request a plot of the *estd* series against the sequential period indicator *n*.

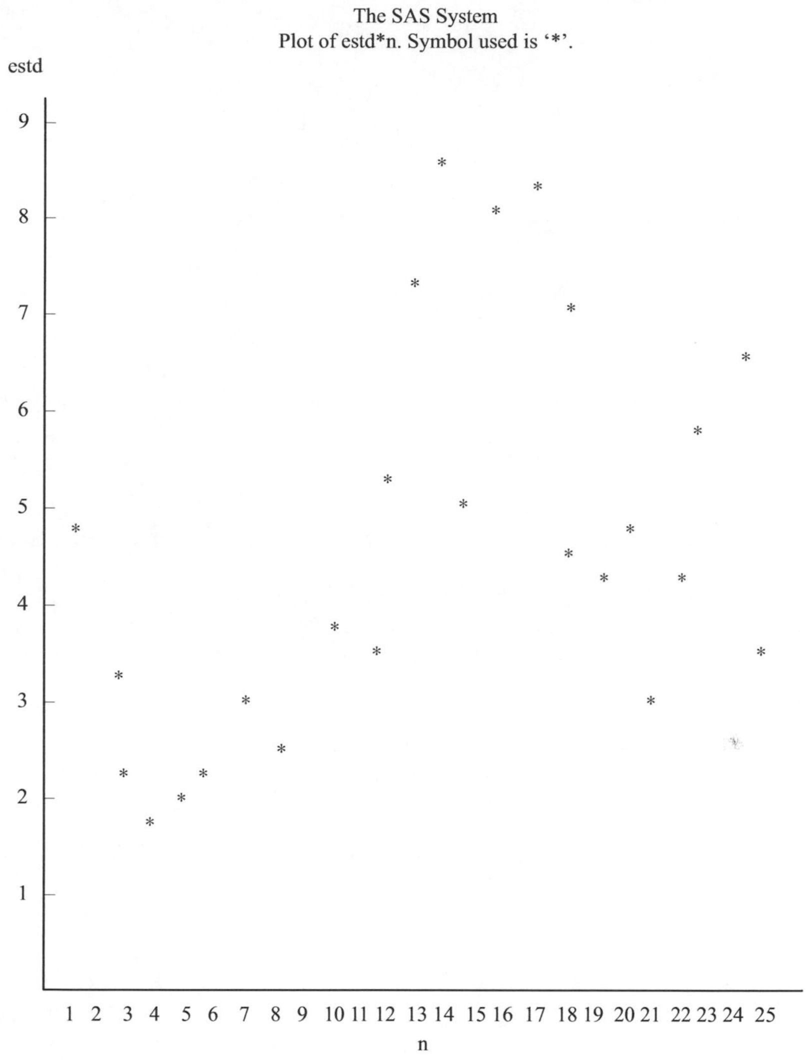

FIGURE 3.G.2
Plot of estimated standard deviation values over time.

2.17	32.5
2.55	32.9
2.97	31.7
3.70	30.0
4.10	32.1
4.34	33.1
4.71	31.9
4.82	26.1
4.88	24.7
5.32	25.0

TABLE 3.G.2

Output for Example 3.G.2

The SAS System
The AUTOREG Procedure
Dependent Variable: *y*
Ordinary Least-Squares Estimates

SSE	466.178634	DFE	23
MSE	20.26864	Root MSE	4.50207
SBC	150.527005	AIC	148.089253
Regress *R*-Square	0.3830	Total *R*-Square	0.3830

Variable	DF	Estimate	Standard Error	*t* Value	Approx Pr > |*t*|
Intercept	1	21.2652	1.7105	12.43	<0.0001
x	1	1.9591	0.5185	3.78	0.0010

Algorithm converged

①
GARCH Estimates

SSE	533.436936	Observations	25
MSE	21.33748	Uncond var	146.71695
Log likelihood	–70.959435	Total *R*-Square	0.2939
SBC	158.013249	AIC	151.91887
Normality test	2.1156	Pr > ChiSq	0.3472

Variable	DF	Estimate	Standard Error	Approx *t* Value	Pr > |*t*|
Intercept	1	18.6242	1.2982	14.35	<0.0001
x	1	2.3759	0.4705	5.05	<0.0001
ARCH0	1	1.9223	6.4617	0.30	0.7661
② ARCH1	1	0.7536	0.7930	0.95	0.3420
GARCH1	1	0.2333	0.3592	0.65	0.5160

```
6.10     34.1
6.55     35.6
;
proc autoreg data=garch1; ①         ②
   model y=x/garch=(p=1,q=1)  maxiter=100;
     output out=new  cev=ecev; ③
     data garch2;  set  new;
     estd=sqrt (ecev); ④
     n=_n_;
proc print data=garch2;
proc plot data=garch2;
     plot estd*n='*'; ⑤
run;
```

TABLE 3.G.2 (*continued*)

The SAS System

Obs	ecev	x	y	③ estd	n
1	22.4295	1.23	19.3	4.73598	1
2	10.9584	1.24	20.4	3.31035	2
3	5.5110	1.25	20.9	2.34755	3
4	3.5710	1.27	19.9	1.88971	4
5	5.0412	1.29	23.4	2.24526	5
6	5.3042	1.30	24.5	2.30309	6
7	9.0137	1.31	20.0	3.00229	7
8	6.2979	1.32	18.0	2.50957	8
9	14.0480	1.33	18.7	3.74807	9
10	12.3679	1.42	27.2	3.51680	10
11	25.2009	1.52	30.5	5.02005	11
12	59.2732	1.59	31.0	7.69891	12
13	71.4624	1.84	25.5	8.45355	13
14	23.3196	2.17	32.5	4.82903	14
15	64.6668	2.55	32.9	8.04156	15
16	67.8945	2.97	31.7	8.23981	16
17	45.0669	3.70	30.0	6.71319	17
18	17.4717	4.10	32.1	4.17992	18
19	16.5092	4.34	33.1	4.06315	19
20	18.8430	4.71	31.9	4.34086	20
21	9.5954	4.82	26.1	3.09764	21
22	16.0743	4.88	24.7	4.00928	22
23	28.6230	5.32	25.0	5.35005	23
24	38.1692	6.10	34.1	6.17812	24
25	11.5548	6.55	35.6	3.39924	25

Example 3.G.3

If we want to estimate the parameters of an ARCH(q) regression model, then we need only use the GARCH=(q=number) option in the MODEL statement. For $q = 3$, this is provided by the SAS program given in Exhibit 3.G.2.

Exhibit 3.G.2 SAS Code for Estimating an ARCH(3) Process

```
data arch1;
   input x y;
    datalines;
1.33    19.3        ① In the MODEL statement, option GARCH =(q =3) is
```

```
1.29   20.4          used to request the estimation of an
1.25   20.9          ARCH(3) conditional error variance regression model.
1.21   21.9       ② The archtest option requested in the MODEL statement
1.19   23.4          enables us to actually test for first-order through
1.19   24.5          twelfth-order ARCH effects. Any statistically
1.20   25.8          significant ARCH effects beyond q = 3 can be
1.29   30.5          estimated by respecifying the value of q in the
1.33   33.3          GARCH = (q = number) option.
1.42   37.2     ③,④ These items have been explained above in Exhibit
1.52   42.5          3.G.1.
1.59   48.8
1.84   55.4
2.17   64.3
2.55   78.9
2.97   86.5
3.70  114.6
4.10  129.7
4.34  126.1
4.71  132.0
4.82  138.1
4.81  141.2
;
proc autoreg data=arch1;
        ①                ②
   model y=x/garch=(q=3)  archtest maxiter=100;
     output out=new  cev=ecev; ③
     data arch2;  set  new;
     estd=sqrt (ecev); ④
proc print data=arch2;
run;
```

The resulting output occupies Table 3.G.3:

① The *p*-values associated with the LM statistics all exceed 0.05. Hence, no significant ARCH effects are indicated.

② Because the *p*-value for the normality test exceeds 0.05, we do not reject the null hypothesis that the standardized residuals $\hat{v}_t = e_t / \sqrt{\hat{\sigma}_t^2}$ are normally distributed.

③ The *p*-values associated with the *t* statistics for the ARCH0,…,ARCH3 estimated coefficients all exceed 0.05, thus confirming the absence of any significant ARCH effects as determined in item ① above. Additionally, in the presence of significant ARCH effects, the AIC can be used to determine the appropriate lag order, i.e., select *q* to minimize AIC(*q*).

④ The estimated conditional error standard deviation values display limited volatility.

TABLE 3.G.3

Output for Example 3.G.3

The SAS System
The AUTOREG Procedure
Dependent Variable: *y*
Ordinary Least-Squares Estimates

SSE	875.848621	DFE	20
MSE	43.79243	Root MSE	6.61758
SBC	149.666698	AIC	147.484613
Regress R-Square	0.9798	Total R-Square	0.9798
Durbin-Watson	0.3471		

①

***Q* and LM Tests for ARCH Disturbances**

Order	Q	Pr > Q	LM	Pr > LM
1	2.4875	0.1148	0.7078	0.4002
2	3.0694	0.2155	0.8880	0.6415
3	3.3576	0.3397	1.1261	0.7708
4	6.8873	0.1420	6.7389	0.1503
5	7.8568	0.1643	6.8496	0.2321
6	9.2089	0.1622	8.2583	0.2198
7	13.3255	0.0646	10.4793	0.1630
8	14.3517	0.0730	11.0660	0.1980
9	15.1430	0.0871	11.6210	0.2355
10	15.2121	0.1245	11.6740	0.3075
11	16.6654	0.1182	11.8108	0.3780
12	16.9526	0.1514	12.3544	0.4177

Variable	DF	Estimate	Standard Error	*t* Value	Approx Pr > |*t*|
Intercept	1	−12.1070	2.8328	−4.27	0.0004
x	1	32.4581	1.0429	31.12	<0.0001

Algorithm converged

②

GARCH Estimates

SSE	1019.81478	Observations	22
MSE	46.35522	Uncond Var	.
Log likelihood	−69.076542	Total R-Square	0.9764
SBC	153.608297	AIC	153.153085
Normality test	3.6842	Pr > ChiSq	0.1585

(continued)

TABLE 3.G.3 *(continued)*

The SAS System
The AUTOREG Procedure

Variable	DF	Estimate	Standard Error	*t* Value	Approx Pr > |*t*|
Intercept	1	−11.9695	2.2764	−5.26	<0.0001
x	1	31.4665	0.8529	36.89	<0.0001
ARCH0	1	2.1999	8.2540	0.27	0.7898
ARCH1	1	1.0407	0.9812	1.06	0.2889
③ ARCH2	1	0.0751	0.4320	0.17	0.8619
ARCH3	1	1.138E-18	1.689E-15	0.00	0.9995

Obs	ecev	x	y	④ estd
1	51.063	1.33	19.3	7.1459
2	121.999	1.29	20.4	11.0453
3	80.967	1.25	20.9	8.9982
4	50.757	1.21	21.9	7.1244
5	23.740	1.19	23.4	4.8724
6	8.012	1.19	24.5	2.8306
7	3.514	1.20	25.8	1.8746
8	2.272	1.29	30.5	1.5072
9	5.869	1.33	33.3	2.4226
10	14.630	1.42	37.2	3.8249
11	24.030	1.52	42.5	4.9021
12	49.600	1.59	48.8	7.0427
13	125.500	1.84	55.4	11.2027
14	104.212	2.17	64.3	10.2084
15	75.328	2.55	78.9	8.6792
16	124.581	2.97	86.5	11.1616
17	36.852	3.70	114.6	6.0706
18	111.160	4.10	129.7	10.5432
19	176.637	4.34	126.1	13.2905
20	16.593	4.71	132.0	4.0735
21	21.059	4.82	138.1	4.5891
22	6.211	4.81	141.2	2.4921

Exercises

3-1. Given the sample dataset in Table E.3-1:

a. Estimate the regression equation for the entire sample of n = 41 data points via OLS. In particular, find $\hat{\beta}_0$ and $\hat{\beta}_1$ and their associated *t* values (under $H_0 : \beta_i = 0$, $i = 0, 1$) along with r^2 and MSE.

b. Conduct a G-Q test of homoscedasticity of the random error term ε. (Hint: place the X_i's in increasing order and eliminate the p=10 observations associated with X = 50 and X = 60. Hence $n_1 = 14$ and $n_2 = 17$ with $n_1 + p + n_2 = 41$.)

c. If the null hypothesis of homoscedasticity is rejected, use transformation (3.15.1) to reestimate the subsample 1 and subsample 2 regression equation. (Hint: form $X_i' = 1/X_i$ and $Y_i' = Y_i/X_i$ and place the X_i''s in increasing order. With p=10, determine MSE_1, and MSE_2 from the subsample 1 and subsample 2 regressions.)
d. Given the results of part c, perform a second G-Q test to see if the data transformation has helped to diminish the degree of heterogeneity.
e. If the null hypothesis of homogeneity is not rejected in part d, reestimate the regression equation using all n=41 data points by regressing Y_i' on X_i'. Compare these results to those obtained in part a above.

TABLE E.3-1

Observations on X and Y

X	Y
10	7
20	15
10	11
10	8
20	10
30	11
20	11
50	21
30	15
30	18
40	12
50	26
50	31
60	11
30	10
50	12
60	25
70	15
60	18
70	40
40	10
40	24
50	17
40	16
60	30
80	12
90	32
80	19
70	19
70	31
60	35
80	26

(continued)

TABLE E.3-1 *(continued)*

X	*Y*
90	46
80	43
80	39
90	11
90	20
100	10
90	27
100	50
100	27

3-2. Use the G-Q test to determine if the dataset in Table E.3-2 displays homoscedasticity. If not, make the appropriate transformation of variables to decrease the degree of heteroscedasticity. (Hint: follow the procedure outlined in the preceding exercise by dropping the $p = 5$ data points for $X = 15$. Thus $n_1 = 13$ and $n_2 = 11$.)

TABLE E.3-2

Observations on X and Y

X	Y
5	20
5	15
10	31
10	23
10	27
5	8
10	6
15	12
15	16
10	11
5	13
5	25
20	4
20	13
20	16
25	14
5	40
5	35
10	16
15	23
20	10
15	9
15	5
20	3
25	10
25	4
25	6
25	7
25	12

3-3. Given the time series data set appearing in Table E.3-3 (here X_t is the bank prime rate and Y_t is household debt outstanding as a percent of disposable income):

a. Find the OLS line of best fit for $Y_t = \beta_0 + \beta_1 X_t + \varepsilon_t$.

b. Determine the estimated values of Y ($\hat{Y}_t$ for all t).

c. Determine the set of OLS residuals e_t and the D-W statistic.

d. Test for first-order autocorrelation ($H_0: \rho = 0$ vs. $H_1: \rho > 0$) using $\alpha = 0.05$.

e. If H_0 is rejected in part d, reestimate this model via the C-O two-step routine. Now what is your conclusion as to the presence of first-order autocorrelation? (Hint: the reader should be able to demonstrate that the initial OLS line of best fit is

$$\hat{Y}_t = 134.8016 - 5.1460X_t, r^2 = 0.4140,$$

with DW = 0.520 and $\hat{\rho} = 0.526$. Then $\hat{\rho}$ can be used to determine the transformed variables X_t^* and Y_t^*.)

3-4. Given the set of time-series observations appearing in Table E.3-4 (with X_t depicting the % of the civilian labor force unemployed and Y_t representing personal saving in billions of current dollars):

a. Find the OLS line of best fit along with DW and $\hat{\rho}$.

b. Test for first-order autocorrelation.

c. If warranted, reestimate the model using the C-O iterative scheme and again test for first-order autocorrelation.

TABLE E.3-3

Observations on X and Y

Year	X_t	Y_t
1990	10.01	83.9
1991	8.46	84.8
1992	6.25	83.8
1993	6.00	85.9
1994	7.15	88.0
1995	8.83	89.8
1996	8.27	91.2
1997	8.44	91.7
1998	8.35	92.5
1999	8.00	95.8
2000	9.23	97.4
2001	6.91	102.3
2002	4.67	108.2
2003	4.12	116.0
2004	4.34	121.8
2005	6.19	130.8

TABLE E.3-4

Observations on X and Y

Year	X_t	Y_t
1990	5.6	299.4
1991	6.8	324.2
1992	7.5	366.0
1993	6.9	284.0
1994	6.1	249.5
1995	5.6	250.9
1996	5.4	228.4
1997	4.9	218.3
1998	4.5	276.8
1999	4.2	158.6
2000	4.0	168.5
2001	4.7	132.3
2002	5.8	184.7
2003	6.0	174.9
2004	5.5	174.3
2005	5.1	-34.8

3-5. Given the data set appearing in Table E.2-1, use (3.C.6) to determine the set of leverage values $h_{ii}, i = 1,\ldots,10$. Also, demonstrate that for $i = 2, \hat{Y}_2 = \sum_{i=1}^{10} h_{ii} Y_i$.

3-6. For the data values housed in Table E.3-6, use the SAS system to conduct an "outlier and influence study." (Hint: let Exhibit 3.D.1 be your guide.) Fully interpret your results.

TABLE E.3-6

Observations on X and Y

X	Y
1.6	18.1
1.5	18.0
2.0	16.0
2.2	18.0
2.6	30.0
3.0	16.1
4.0	14.0
2.5	6.0
5.0	14.0
5.5	11.8
6.8	12.0
7.4	10.0
8.3	10.0
8.2	7.3
13.0	2.0

3-7. Using the SAS system, conduct a test to see if the OLS regression residuals determined from the preceding exercise are normally distributed. What are the appropriate null and alternative hypotheses?

3-8. For Exercises 3-1 and 3-2, use the SAS system to perform White's model specification test for heteroscedasticity. What are the null and alternative hypotheses? Interpret your results.

3-9. Given the dataset in Table E.3-1, correct for the effects of heteroscedasticity by employing the TSWLS procedure. (Hint: let Example 3.E.3 be your guide.)

3-10. Given the dataset in Table E.3-2, correct for the effects of heteroscedasticity by employing the TSWLS routine. (Hint: let Example 3.E-3 serve as your guide.)

3-11. Given the dataset in Table E.3-3, correct for the impact of autocorrelation by employing the AUTOREG procedure in the SAS system.

3-12. For the dataset in Table E.3-4, correct for the impact of autocorrelation by applying PROC AUTOREG in the SAS system.

3-13. Durbin's *h* Statistic [1970]
Suppose we have a regression model with a lagged dependent variable serving as a regressor or

$$\begin{aligned} &\text{a. } Y_t = \alpha Y_{t-1} + \beta X_t + \varepsilon_t, |\alpha| < 1, \\ &\text{b. } \varepsilon_t = \rho \varepsilon_{t-1} + \eta_t, |\rho| < 1, \end{aligned} \tag{3.E.1}$$

where the η_t's are independent $N(0,\sigma^2)$ random variables and ρ is a first-order autocorrelation coefficient. Under this specification, since ε_t depends on ε_{t-1} and Y_{t-1} depends on ε_{t-1}, it follows that Y_{t-1} and ε_t will be correlated and thus the OLS estimator $\hat{\alpha}$ will be inconsistent.

Furthermore, it can be demonstrated that if, say, $\rho > 0$, then the OLS estimate of α is biased upwards and the OLS estimate of ρ is biased downwards so that $DW \approx 2(1-\hat{\rho})$ is biased towards 2, thus prompting us to not reject the null hypothesis of no autocorrelation even though the ε_t's may actually be highly correlated. Hence the standard DW test is not applicable under the Equation 3.E.1 structure.

To overcome this difficulty, let us employ *Durbin's h Statistic*

$$h = \hat{\rho}\sqrt{\frac{n}{1-nS_{\hat{\alpha}}^2}} : N(0,1), nS_{\hat{\alpha}}^2 < 1, \tag{3.E.2}$$

where $S_{\hat{\alpha}}^2$ is the OLS estimate of the variance of $\hat{\alpha}, \hat{\rho}$ is determined from the OLS fit of $e_t = \rho e_{t-1} + \eta_t, |\rho| < 1$, and the e_t's represent the OLS residuals obtained from estimating (3.E.1.a). For $H_0: \rho = 0$ vs. $H_1: \rho > 0$ (positive first-order autocorrelation) and $\alpha = 0.05$, we will reject H_0 if $h > Z_{0.05} = 1.645$. (If $h < 0$, then we can test $H_0: \rho = 0$, against $H_1: \rho < 0$ (negative first-order autocorrelation) at the $\alpha = 0.05$ level and consequently reject H_0 if $h < -1.645$.)

Given the data set presented in Table E.3-13, estimate equation 3.E.1.a and test for the absence of first-order autocorrelation using Durbin's h statistic. Two separate sets of SAS code are to be employed. First, calculate h directly from the output determined from Exhibit 3.E-1:

TABLE E.3-13

Observations on X and Y

X	Y
1.33	19.3
1.29	20.1
1.25	20.4
1.21	21.0
1.19	20.9
1.19	23.5
1.20	22.8
1.29	24.5
1.33	26.3
1.42	27.2
1.52	32.5
1.59	31.8
1.84	36.4
2.17	35.3
2.55	38.9
2.97	40.5
3.70	44.6
4.10	42.7
4.34	46.1
4.71	50.0
4.82	53.1
4.81	55.2

Exhibit 3.E-1

```
data autoh;
     input x   y;
ly=lag (y); j
datalines;
1.33  19.3
1.29  20.1
   .     .
   .     .
   .     .
4.81  55.2
```

① We create the lagged dependent variable Y_{t-1} by using the lag operator lag (Y).

② We invoke the *regression procedure*.

③ We estimate the model $Y_t = \alpha Y_{t-1} + \beta X_t + \varepsilon_t$ or (3.E.1.a).

④ We house the residual variable *rsd* in the output data set *a*.

⑤ The output data set *a* becomes the new SAS data set *outa*.

```
;
proc reg data=autoh; ②
     model y=ly x; ③
     output out=a r=rsd;④
data outa; set a; ⑤
lrsd=lag (rsd); ⑥
proc reg data=outa; ⑦
     model rsd=lrsd; ⑧
run;
```

⑥ We create the lagged residual variable e_{t-1} by using the lag operator lag (rsd).
⑦ We apply the regression procedure to the new SAS data set *outa*.
⑧ We estimate the model $e_t = \rho e_{t-1} + \eta_t$ or (3.E.1.b).

Once $\hat{\alpha}$ and $\hat{\rho}$ are obtained, Durbin's h can be readily calculated via Equation 3.E.2. Use $\alpha = 0.05$ to test H_0 vs. H_1 as stated above.

Second, let us consider a simplification of the preceding lines of SAS code:

Exhibit 3.E-2

```
data autoh;
     input  x  y;
ly=lag (y);
datalines;
1.33    19.3
1.29    20.1
  .       .
  .       .
  .       .
 4.81   55.2
;
proc autoreg data= autoh; ①
      model y=ly  x/lagdep=ly; ②
run;
```

① We invoke *proc autoreg* and use only the original SAS data set *autoh*.
② For Durbin's h test, we need only specify the name of the lagged dependent variable in *LAGDEP=* option.

Now Durbin's h statistic along with its associated p-value is automatically calculated. What is your conclusion regarding H_0 vs. H_1?

4

Nonparametric Regression

The Nonparametric Regression Model [Theil 1950a,b,c; Jaeckel 1972; Hettmansperger and McKean 1977; Hettmansperger 1984; Puri and Sen 1985]

We noted in previous chapters that the method of least squares provides us with good estimates of the regression intercept and slope (β_0 and β_1, respectively) if the random error term ε is assumed to follow a normal distribution having zero mean and constant variance. But what if we are unsure as to the form of the probability density function of ε? Obviously we need a regression procedure that will yield reasonable estimates of β_0 and β_1 for a whole host of possible (unknown) random error distributions. Such regression procedures are said to be *distribution free* or *nonparametric* in nature and, as we shall now see, are, for the most part, based on the ranks of certain values determined from the sample observations. Two *rank-oriented regression methods* will be presented shortly; one uses the ordinary median of a certain set of slope estimates and the other uses the *weighted median* of these slope estimates.

Given a set of n fixed X_i values on the explanatory variable X, with $X_1 < X_2 < \ldots < X_n$, let Y_i, $i = 1,\ldots, n$, represent the value of the explained variable Y when $X = X_i$, $i = 1,\ldots, n$. Suppose, as assumed previously, that Y_i has a systematic component $\beta_0 + \beta_1 X_i$ as well as a random component ε_i, $i = 1,\ldots, n$, where the ε_i's depict a set of sample random variables drawn from a continuous population having a zero median. If X and Y are linearly related, then our additive regression model assumes the form

$$Y_i = \beta_0 + \beta_1 X_i + \varepsilon_i, \quad i = 1,\ldots, n. \tag{4.1}$$

Estimating the Slope (β_1) and Intercept (β_0) Parameters via the Ordinary Median

Suppose the X_i values are distinct, $i = 1,\ldots, n$. To estimate β_1 in Equation 4.1, let us compute the $C = {}_nC_2 = n(n-1)/2$ pairwise sample slopes

$$S_{ij} = \frac{Y_j - Y_i}{X_j - X_i}, \quad 1 \le i < j \le n. \tag{4.2}$$

For instance, given $n = 5$, there are $C = 10$ pairs of subscripts (i, j) with $1 \le i < j \le 5$, i.e.,

$i = 1$, $j = 2, 3, 4, 5$;
$i = 2$, $j = 3, 4, 5$;

$i = 3,\ j = 4, 5;$
$i = 4,\ j = 5.$

Hence, the corresponding (i, j) pairs are as follows:

(1, 2), (1, 3), (1, 4), (1, 5);
(2, 3), (2, 4), (2, 5);
(3, 4), (3, 5);
(4, 5).

Then for each such pair, the slope S_{ij} is computed according to Equation 4.2.

Given the S_{ij}'s, the estimate of β_1 is taken to be

$$\hat{\beta}_1 = median\{s_{ij}, 1 \le i < j \le n\}, \tag{4.3}$$

where s_{ij} is the realized value of S_{ij}. Let us write the ordered values of the s_{ij} sample slope realizations as $s_{(1)}, s_{(2)}, \ldots, s_{(C)}$. Then as dictated by Equation 4.3,

$$\hat{\beta}_1 = \begin{cases} s_{((C+1)/2)}, \text{if } C \text{ is odd;} \\ \\ \left[s_{(C/2)} + s_{((C/2)+1)}\right]/2, \text{ if } C \text{ is even.} \end{cases} \tag{4.3.1}$$

So if C is odd, $\hat{\beta}_1$ corresponds to the s_{ij} value that occupies position $(C + 1)/2$ in the set of ordered sample s_{ij}'s, and, if C is even, $\hat{\beta}_1$ is the average of the two s_{ij} values that occupy positions $C/2$ and $(C/2) + 1$ in the set of ordered sample s_{ij}'s.

Next, let us define

$$A_i = Y_i - \hat{\beta}_1 X_i,\ i = 1, \ldots, n. \tag{4.4}$$

As an *estimate of* β_0, let us use

$$\hat{\beta}_0 = median\{a_i, i = 1, \ldots, n\}, \tag{4.5}$$

where a_i is the realized value of A_i. Let us write the ordered values of the a_i's as $a_{(1)}, a_{(2)}, \ldots, a_{(n)}$. Then as required by Equation 4.5,

$$\hat{\beta}_0 = \begin{cases} a_{((n+1)/2)}, \text{if } n \text{ is odd;} \\ \left[a_{(n/2)} + a_{((n/2)+1)}\right]/2, \text{if } n \text{ is even.} \end{cases} \tag{4.5.1}$$

So for n odd, $\hat{\beta}_0$ corresponds to the a_i value that holds position $(n + 1)/2$ in the set of ordered a_i's, and, if n is even, $\hat{\beta}_0$ is the average of the two a_i values that hold positions $n/2$ and $(n/2) + 1$ in the set of ordered a_i's.

From Equations 4.3.1 and 4.5.1, the estimated line of best fit for the collection of sample points (X_i, Y_i), $i = 1,\ldots, n$, thus appears as

$$median\,(Y|X) = \hat{\beta}_0 + \hat{\beta}_1 X. \tag{4.6}$$

Precision of the Slope Estimate

To assess the precision of our nonparametric slope estimate $\hat{\beta}_1$, let us form a symmetric confidence interval for β_1 with confidence coefficient $1-\alpha$. As a first step, let us determine, from Table A.10, the upper $\alpha/2$ percentage point $d_{\alpha/2}$ and form the quantity $D_\alpha = d_{\alpha/2} - 2$. If we now set

$$l = \frac{C - D_\alpha}{2} \tag{4.7}$$

and

$$u = \frac{C + D_\alpha}{2} = l + D_\alpha, \tag{4.8}$$

then a 100 $(1-\alpha)$% confidence interval for the slope β_1 is

$$(\beta_{1l}, \beta_{1u}) = (s_{(l)}, s_{(u+1)}), \tag{4.9}$$

where, as defined above, $s_{(1)}, s_{(2)},\ldots, s_{(C)}$ are the ordered values of the s_{ij} sample slope realizations. Thus, the lower confidence limit β_{1l} is the sample slope realization s_{ij} that holds position l (Equation 4.7) in the set of C ordered sample slopes, and β_{1u} is the realized sample slope value s_{ij} that assumes position u (Equation 4.8) in the set of ordered sample slopes.

Testing the Significance of the Slope

Let us test the null hypothesis $H_0 : \beta_1 = \beta_1^0$ against any of the following alternative hypotheses:

Case I	Case II	Case III
$H_0 : \beta_1 = \beta_1^0$	$H_0 : \beta_1 = \beta_1^0$	$H_0 : \beta_1 = \beta_1^0$
$H_1 : \beta_1 > \beta_1^0$	$H_1 : \beta_1 < \beta_1^0$	$H_1 : \beta_1 \neq \beta_1^0$

To specify *Theil's test statistic D*, let us first determine the n differences

$$D_i = Y_i - \beta_1^0 X_i,\ i = 1,\ldots, n. \tag{4.10}$$

Then we may form

$$D = \sum_{i=1}^{n-1} \sum_{j=i+1}^{n} \delta(D_j - D_i), \tag{4.11}$$

where, for each pair of subscripts (i, j), $1 \le i < j \le n$,

$$\delta(D_j - D_i) = \begin{cases} -1, D_j - D_i < 0; \\ 0, D_j - D_i = 0; \\ 1, D_j - D_i > 0. \end{cases} \tag{4.12}$$

Clearly, D will be large when $D_j - D_i > 0$ for many (i, j) pairs of subscripts. In addition, because under $H_0 : \beta_1 = \beta_1^0$ we have

$$D_j - D_i = Y_j - Y_i + \beta_1^0 (X_i - X_j),$$

it follows that, if the null value $\beta_1^0 = 0$, then $D_j - D_i = Y_j - Y_i$. Hence, testing $H_0 : \beta_1 = \beta_1^0 = 0$ is tantamount to testing for linear association between X and Y.

Relative to Cases I–III above, let us now look to the specification of the corresponding critical regions, i.e., at the $100\alpha\%$ level, our set of decision rules for rejecting H_0 relative to H_1 is as follows:

a. Case I: reject H_0 if $d > d_\alpha$.
b. Case II: reject H_0 if $d < -d_\alpha$.
c. Case III: reject H_0 if $|d| > d_{\alpha/2}$.

where d is the sample realization of D with d_α and $d_{\alpha/2}$ determined from Table A.10.

Large-Sample Approximations in Assessing Significance and Precision

Given $H_0 : \beta_1 = \beta_1^0 = 0$, it can be shown that the *null distribution* of D has

$$E(D) = 0 \text{ and } V(D) = \frac{n(n-1)(2n+5)}{18}. \tag{4.13}$$

Then the *standardized D statistic* is asymptotically standard normal or, as $n \to \infty$,

$$D^* = \frac{D - E(D)}{\sqrt{V(D)}} = \frac{D}{[n(n-1)(2n+5)/18]^{1/2}} : N(0, 1), \tag{4.14}$$

where d^* denotes the sample realization of D^*. In this instance, the Case I–III critical regions assume the following form:

a. Case I: reject H_0 if $d^* > z_\alpha$.
b. Case II: reject H_0 if $d^* < -z_\alpha$.
c. Case III: reject H_0 if $|d^*| > z_{\alpha/2}$.

In addition, for large n, the quantity D_α may be approximated as

$$D_\alpha \approx z_{\alpha/2} [n(n-1)(2n+5)/18]^{1/2} = D'. \tag{4.15}$$

Actually, let us set $D_\alpha = [D']$, where $[D']$ is the algebraically largest integer that does not exceed D' (e.g., [2] = 2, [2.5] = 2). Then Equations 4.7 and 4.8 become

$$l' = \frac{C - [D']}{2} \tag{4.7.1}$$

and

$$u' = \frac{C + [D']}{2} = l' + [D'], \tag{4.8.1}$$

respectively. Then a large-sample 100 (1–α)% confidence interval for the slope β_1 is

$$(\beta_{1l'}, \beta_{1u'}) = (s_{(l')}, s_{(u'+1)}). \tag{4.16}$$

Example 4.1

Table 4.1 contains the coordinates of $n = 6$ sample points (X_i, Y_i), $i = 1,\ldots, 6$. To determine the distribution-free equation of the line that best fits this set of points, let us first determine $C = 6(5)/2 = 15$ pairs of subscripts (i, j), $1 \le i < j \le 6$ (column 1 of Table 4.2). For each such pair, column 2 of Table 4.2 lists the realized samples slopes S_{ij}, and column 3 displays the ordered sample slope realizations. Because C is odd, Equation 4.3 yields $\hat{\beta}_1 = s_{((15+1)/2)} = 0.900$.

TABLE 4.1

Observations on Variables X and Y

X_i	Y_i
1	5.2
2	6.0
3	10.1
4	8.8
5	8.7
6	10.5

TABLE 4.2

Calculations for Line of Best Fit and Theil's D

(1) (i, j)	(2) S_{ij}	(3) Ordered S_{ij}'s	(4) $D_j - D_i$	(5) $\delta(D_j - D_i)$
(1, 2)	0.800	−1.300	0.80	1
(1, 3)	2.450	−0.700	4.90	1
(1, 4)	1.200	−0.100	3.60	1
(1, 5)	0.875	0.133	3.50	1
(1, 6)	1.060	0.596	5.30	1
(2, 3)	4.100	0.800	4.10	1
(2, 4)	1.400	0.875	2.80	1
(2, 5)	0.900	0.900	2.70	1
(2, 6)	1.125	1.060	4.50	1
(3, 4)	−1.300	1.125	−1.30	−1
(3, 5)	−0.700	1.200	−1.40	−1
(3, 6)	0.133	1.400	0.40	1
(4, 5)	−0.100	1.800	−0.10	−1
(4, 6)	0.596	2.450	1.70	1
(5, 6)	1.800	4.100	1.80	1

Next, from Equation 4.4, the six realized A_i values are 4.3, 4.2, 7.4, 5.2, 4.2, and 5.1, whereas the ordered a_i's, $i = 1,\ldots, 6$, are 4.2, 4.2, 4.3, 5.1, 5.2, and 7.4. Then from Equation 4.5.1, because n is even, $\hat{\beta}_0 = \left[a_{(6/2)} + a_{((6/2)+1)}\right]/2 = 4.7$. Then from Equation 4.6, the estimated line of best fit appears as

$$median\,(Y|X) = \hat{\beta}_0 + \hat{\beta}_1 X = 4.7 + 0.90X.$$

To assess the precision of our estimate of β_1, let us use Table A.10 with $n = 6$ and $\alpha/2 = 0.028$ (hence, $\alpha = 0.056$ and $1 - \alpha = 0.944$). Then from Equations 4.7 and 4.8, given that the upper 0.028 percentage point is $d_{0.028} = 11$, so that $D_{0.056} = d_{0.028} - 2 = 11 - 2 = 9$, we find that

$$l = \frac{15-9}{2} = 3 \text{ and } u = \frac{15+9}{2} = 12.$$

Then a 94.4% confidence interval for $\hat{\beta}_1$ is $(\beta_{1l}, \beta_{1u}) = (s_{(3)}, s_{(13)}) = (-0.100, 1.800)$.

Finally, let us test the significance of the slope given Case III above with $\beta_1^0 = 0$. Then via Equation 4.10, $D_i = Y_i$, $i = 1,\ldots, 6$, and, for each pair of subscripts (i, j), $1 \le i < j \le 6$, the $D_j - D_i$ values as well as the $\delta(D_j - D_i)$'s appear in columns 4 and 5, respectively, of Table 4.2. So from Equation 4.11, $D = 9$. Again looking to Table A.10 for $n = 6$ and $D = 9$, the (p-value)/2 amounts to 0.068 and thus the p-value itself is 0.136. Hence, under our two-tail test, the probability of obtaining a D value at least as large as 9 is in excess of 13%. Clearly, we do not have a statistically significant linear relationship between X and Y.

Estimating the Slope (β_1) and Intercept (β_0) Parameters via the Weighted Median

When we estimated the population intercept and slope β_0 and β_1, respectively, using the method of least squares, we chose estimates $\hat{\beta}_0$ and $\hat{\beta}_1$ so as to minimize $\sum_{i=1}^{n} e_i^2$, where the ith residual $e_i = Y_i - \hat{Y}_i = Y_i - \left(\hat{\beta}_0 + \hat{\beta}_1 X_i\right)$, $i = 1,\ldots, n$. An alternative decision criterion is to, say, minimize a weighted sum of the absolute values of the residuals or to find $\min\left\{W = \sum_{i=1}^{n} w_i |e_i|\right\}$, where the ith weight $w_i \geq 0$. (If $w_i = |e_i|$, then the preceding objective function is $\sum_{i=1}^{n} e_i^2$, and the principle of least squares again applies.)

How should the weights w_i, $i = 1,\ldots, n$, be chosen? If we weight the residuals e_i by the ranks of their absolute values, i.e., $w_i = \text{rank}\,(|e_i|)$, $i = 1,\ldots, n$, then, because $\text{rank}\,(|e_i|) \leq n$ for all i, this restriction tends to dampen the effect of large residuals so that they cannot exert undue influence on

$$W = \sum_{i=1}^{n} w_i |e_i| = \sum_{i=1}^{n} \text{rank}\left(|e_i|\right) \cdot |e_i|. \tag{4.17}$$

(Note that large e_i's in $\sum_{i=1}^{n} e_i^2$ tend to dominate this sum when the principle of least squares is invoked.)

Let us transform Equation 4.17 into an equivalent expression that is a bit easier to apply. If we assume that the probability density function of ε_i, $i = 1,\ldots, n$, is approximately symmetric so that the median of the e_i's should be approximately zero, then $\frac{n+1}{2}$ is the rank of the median residual.[1] Clearly, negative residuals have rank less than $\frac{n+1}{2}$; positive residuals have rank greater than $\frac{n+1}{2}$. In this regard, we may write

$$w_i = \text{rank}\left(|e_i|\right) \approx 2\left|\text{rank}\left(e_i\right) - \frac{n+1}{2}\right|.$$

Because $\text{rank}\left(e_i\right) - \frac{n+1}{2}$ has the same sign as e_i itself, the objective function in Equation 4.17 can be rewritten as

$$W = \sum_{i=1}^{n} \text{rank}\left(|e_i|\right) \cdot |e_i| \approx 2\sum_{i=1}^{n} \left|\text{rank}\left(e_i\right) - \frac{n+1}{2}\right| \cdot |e_i| =$$

$$2\sum_{i=1}^{n} \left[\text{rank}\left(e_i\right) - \frac{n+1}{2}\right] \cdot e_i. \tag{4.18}$$

[1] Under symmetry, the median of the ranks of the e_i's equals its mean. Given that the weights $w_i = \text{rank}\,(|e_i|)$ represent a permutation of the integers $1,\ldots, n$, the average weight or rank is the average of the integers $1,\ldots, n$, i.e., $\bar{w} = \sum_{i=1}^{n} i/n = \frac{1}{n}\sum_{i=1}^{n} i = \frac{1}{n}\left(\frac{n(n+1)}{2}\right) = \frac{n+1}{2}$.

Given that $e_i = Y_i - \left(\hat{\beta}_0 + \hat{\beta}_1 X_i\right)$, a moment of reflection reveals that the value of W in Equation 4.18 is unaffected by the value of $\hat{\beta}_0$. (Adjusting all residuals e_i by the same amount $\hat{\beta}_0$ does not affect their ranks so that rank $(e_i) = \text{rank}\left(Y_i - \hat{\beta}_1 X_i\right)$.) In addition, because under this adjustment W is a function of $\hat{\beta}_1$ alone, the $\hat{\beta}_1$ value that minimizes W will be the same as the one that minimizes $W' = W/2$. Hence, under these two modifications, we finally obtain an expression equivalent to Equation 4.18 or

$$W' = \sum_{i=1}^{n}\left[\text{rank}\left(Y_i - \hat{\beta}_1 X_i\right) - \frac{n+1}{2}\right]\left(Y_i - \hat{\beta}_1 X_i\right). \tag{4.18.1}$$

To determine the value of the $\hat{\beta}_1$ that minimizes W', let us choose $\hat{\beta}_1$ to be the weighted median of the pairwise sample slopes $S_{ij} = (Y_j - Y_i)/(X_j - X_i)$, $1 \le i \le j \le n$ (see Equation 4.2), with the weights proportional to $|X_j - X_i|$. Hence, $\hat{\beta}_1$ chosen in this manner is termed the *rank-estimate of* β_1. Because W' (with argument $\hat{\beta}_1$) is a piecewise continuous function consisting of a series of line segments connected at the points $\hat{\beta}_1 = S_{ij}$, the S_{ij}'s represent the points at which the slope of Equation 4.18.1 exhibits an abrupt change or finite discontinuity in its value.

How should the weights w_{ij} that we attached to the S_{ij}'s be selected? Because we require that each w_{ij} be proportional to $|X_j - X_i|$, let us set

$$w_{ij} = \frac{|X_j - X_i|}{\sum |X_j - X_i|}, \quad 1 \le i < j \le n. \tag{4.19}$$

Given the set of w_{ij}'s provided by Equation 4.19, how can they be used to determine the weighted median of the S_{ij}'s? To answer this question, we must first define the concept of a weighted median. Specifically, for a collection of values X_i with associated weights w_i such that $w_i \ge 0$ and $\sum_{i=1}^{n} w_i = 1$, let us first arrange the X_i's in an increasing sequence $X_{(1)} \le X_{(2)} \le \ldots \le X_{(n)}$. If there exists an index k such that

$$w_{(1)} + w_{(2)} + \ldots + w_{(k-1)} < 0.5 \text{ and}$$

$$w_{(1)} + w_{(2)} + \ldots + w_{(k-1)} + w_{(k)} > 0.5, \tag{4.20}$$

then $X_{(k)}$ is the weighted median of the X_i's, where $w_{(1)}$ is the weight attached to $X_{(1)}$, $w_{(2)}$ is the weight attached to $X_{(2)}$, and so on. If there exists an index k such that $\sum_{i=1}^{k-1} w_{(i)} = 0.5$, then the weighted median is taken to be $(X_{(k-1)} + X_{(k)})/2$. (If $w_i = 1/n$ for all i, then the weighted median is simply the ordinary median of the X_i's.)

Example 4.2

Table 4.3 lists $n = 7$ values X_i of a variable X along with their associated weights w_i, $i = 1,\ldots, 7$. Because

$$X_{(1)} = 1,\ X_{(2)} = 2,\ X_{(3)} = 7,\ X_{(4)} = 9,\ X_{(5)} = 14,\ X_{(6)} = 17,\ X_{(7)} = 26,$$

it follows that

TABLE 4.3

Calculations for Weighted Median

X	w_i
$X_1 = 26$	$w_1 = 0.25$
$X_2 = 9$	$w_2 = 0.17$
$X_3 = 14$	$w_3 = 0.08$
$X_4 = 1$	$w_4 = 0.10$
$X_5 = 7$	$w_4 = 0.05$
$X_6 = 2$	$w_6 = 0.15$
$X_7 = 17$	$w_7 = 0.20$
	1

$$w_{(1)} = 0.10,\ w_{(2)} = 0.15,\ w_{(3)} = 0.05,\ w_{(4)} = 0.17,\ w_{(5)} = 0.08,\ w_{(6)} = 0.20,\ w_{(7)} = 0.25.$$

Then, from Equation 4.20,

$$0.10 + 0.15 + 0.05 + 0.17 = 0.47 < 0.5 \text{ and}$$

$$0.10 + 0.15 + 0.05 + 0.17 + 0.08 = 0.55 > 0.5.$$

(Here $\sum_{i=1}^{4} w_{(i)} < 0.5$ and $\sum_{i=1}^{5} w_{(i)} > 0.5$.) Hence, $X_{(k)} = X_{(5)} = 14$ is the weighted median of this data set.

Once $\hat{\beta}_1$ is taken to be the weighted median of the $\hat{S}_{ij}$'s, we can then calculate the estimated intercept $\hat{\beta}_0$ as the ordinary median of the $Y_i - \hat{\beta}_1 X_i$ differences (because these differences are centered at β_0), i.e.,

$$\hat{\beta}_0 = \text{median}\left\{Y_i - \hat{\beta}_1 X_i, i = 1, \ldots, n\right\}. \tag{4.21}$$

Then from Equations 4.2, 4.19, and 4.21, the estimated line of best fit appears as

$$\text{weighted median}\left(Y \middle| X\right) = \hat{\beta}_0 + \hat{\beta}_1 X.$$

On all this, see Example 4.3.

Testing the Significance of the Slope under Weighting

Let us test $H_0 : \beta_1 = 0$, against $H_1 : \beta_1 \neq 0$ by using the *rank test statistic*

$$U = \sum_{i=1}^{n} \left(X_i - \overline{X}\right) \text{rank}\left(Y_i\right) = \sum_{i=1}^{n} \left[\text{rank}\left(Y_i\right) - \frac{n+1}{2}\right] X_i. \tag{4.22}$$

(If ties occur among the Y values, we can simply assign the tied Y_i's the appropriate midrank value, e.g., for Y: 1, 3, 6, 6, 9, the associated ranks are 1, 2, 3.5, 3.5, 5 because the 6's occupy the third and fourth positions in this increasing sequence.)

Then, under $H_0 : \beta_1 = 0$, it can be demonstrated that $E(U) = 0$ and

$$V(U) = \frac{n(n+1)}{12}\sum_{i=1}^{n}\left(X_i - \overline{X}\right)^2.$$

Hence, our test statistic can be written as

$$|R| = \left|\frac{U}{S(U)}\right| = \frac{|U|}{\sqrt{\frac{n(n+1)}{12}\sum_{i=1}^{n}\left(X_i - \overline{X}\right)^2}}, \tag{4.23}$$

where $S(U)$ denotes the standard deviation of U. For large n, R is asymptotically $N(0, 1)$ under $H_0 : \beta_1 = 0$, and thus the p-value associated with the sample realization $|r|$ of $|R|$ is $P(|Z| \geq |r|)$. For n small, we could use the t distribution with $n - 2$ degrees of freedom and determine the p-value as $P(|t| \geq |r|)$.

Example 4.3

Given the sample of $n = 8$ data points (X_i, Y_i), $i = 1, \ldots, 8$, exhibited in Table 4.4, our objective is to determine the distribution-free line of best fit for this set of points via the weighted median method. There are $C = \frac{n(n-1)}{2} = 28$ pairs of subscripts (i, j) with $1 \leq i < j \leq 8$, i.e.,

$i = 1$, $j = 2, 3, 4, 5, 6, 7, 8$;
$i = 2$, $j = 3, 4, 5, 6, 7, 8$;
$i = 3$, $j = 4, 5, 6, 7, 8$;
$i = 4$, $j = 5, 6, 7, 8$;
$i = 5$, $j = 6, 7, 8$;
$i = 6$, $j = 7, 8$;
$i = 7$, $j = 8$;

Then the corresponding (i, j) pairs are

(1, 2), (1, 3),…, (1, 8);
(2, 3), (2, 4),…, (2, 8);
(3, 4), (3, 5),…, (3, 8);
(4, 5), (4, 6),…, (4, 8);
(5, 6), (5, 7), (5, 8);
(6, 7), (6, 8);
(7, 8).

Column 1 of Table 4.5 lists the 28 pairs of subscripts (i,j), $1 \leq i < j \leq 8$. Columns 2 and 3 depict the realized differences $X_j - X_i$ and $Y_j - Y_i$, respectively, and column 4 houses the realized pairwise sample slopes S_{ij}. Column 5 is used to determine the weights w_{ij} (note that $\Sigma|X_i - X_j| = 115$) appearing in column 6.

Next, the first column of Table 4.6 lists the ordered S_{ij}'s, and the second column displays the weights attached to the ordered S_{ij}'s. If we now start cumulating the indicated weights in the second column of Table 4.6, we have, according to Equation 4.20,

$$0.0088 + 0.0177 + \ldots + 0.0619 = 0.4510 < 0.5$$

$$0.0088 + 0.0177 + \ldots + 0.0619 + 0.0796 = 0.5306 > 0.5.$$

TABLE 4.4

Observations on Variables X and Y

Point number	X	Y
1	1	1
2	3	5
3	5	5
4	6	6
5	7	4
6	8	6
7	10	8
8	11	12

TABLE 4.5

Calculation of Weights of Pairwise Sample Slopes

(1) (i,j)	(2) $X_j - X_i$	(3) $Y_j - Y_i$	(4) S_{ij}	(5) $\|X_i - X_j\|$	(6) $w_{ij} = \|X_i - X_j\|/115$
(1, 2)	2	4	2.0000	2	0.0174
(1, 3)	4	4	1.0000	4	0.0348
(1, 4)	5	5	1.0000	5	0.0435
(1, 5)	6	3	0.5000	6	0.0522
(1, 6)	7	5	0.7143	7	0.0609
(1, 7)	9	7	0.7778	9	0.0783
(1, 8)	10	11	1.1000	10	0.0870
(2, 3)	2	0	0.0000	2	0.0174
(2, 4)	3	1	0.3333	3	0.0261
(2, 5)	4	−1	−0.2500	4	0.0348
(2, 6)	5	1	0.2000	5	0.0435
(2, 7)	7	3	0.4286	7	0.0609
(2, 8)	8	7	0.8750	8	0.0696
(3, 4)	1	1	1.0000	1	0.0087
(3, 5)	2	−1	−0.5000	2	0.0174
(3, 6)	3	1	0.3333	3	0.0261
(3, 7)	5	3	0.6000	5	0.0435
(3, 8)	6	7	1.1667	6	0.0522
(4, 5)	1	−2	−2.0000	1	0.0087
(4, 6)	2	0	0.0000	2	0.0174
(4, 7)	4	2	0.5000	4	0.0348
(4, 8)	5	6	1.2000	5	0.0435
(5, 6)	1	2	2.0000	1	0.0087
(5, 7)	3	4	1.3333	3	0.0261
(5, 8)	4	8	2.0000	4	0.0348
(6, 7)	2	2	1.0000	2	0.0174
(6, 8)	3	6	2.0000	3	0.0261
(7, 8)	1	4	4.0000	1	0.0087
Total				115	1.0000

Hence, the weighted median of the collection of pairwise slopes S_{ij} is $0.7728 = \hat{\beta}_1$, our estimate of the population slope β_1.

To estimate the population intercept β_0, let us determine the set of differences $Y_i - \hat{\beta}_1 X_i$ for each of the $n = 8$ data points (column 1 of Table 4.7). Then column 2 of this table displays the ordered $Y_i - \hat{\beta}_1 X_i$ values. Clearly, the median of these ordered differences is the average of the two middle terms or

$$\hat{\beta}_0 = \frac{0.2720 + 1.1360}{2} = 0.7040.$$

Hence our distribution-free or nonparametric line of best fit is of the form

$$\text{weighted median } (Y/X) = 0.7040 + 0.7728X.$$

TABLE 4.6

Calculation of Weighted Median of Pairwise Sample Slopes

	Ordered S_{ij}'s	Weights of the Ordered S_{ij}'s
	−2.0000	0.0087
	−0.5000	0.0174
	−0.2500	0.0348
	0.0000	0.0174
	0.0000	0.0174
	0.2000	0.0435
	0.3333	0.0261
	0.3333	0.0261
	0.4286	0.0609
	0.5000	0.0348
	0.5000	0.0522
	0.6000	0.0435
	0.7143	0.0609 (Σ ordered w_{ij}'s $= 0.4437 < 0.5$)
weighted median of the S_{ij}'s	0.7728	0.0783 ($0.4437 + 0.0783 = 0.5220 > 0.5$)
	0.8750	0.0696
	1.0000	0.0174
	1.0000	0.0087
	1.0000	0.0348
	1.0000	0.0435
	1.1000	0.0870
	1.6670	0.0522
	1.2000	0.0435
	1.3333	0.0261
	2.0000	0.0174
	2.0000	0.0087
	2.0000	0.0348
	2.0000	0.0261
	4.0000	0.0088

TABLE 4.7
Calculation of Regression Intercept

$Y_i - \hat{\beta}_1 X_i =$ $Y_i - 0.7728X_i$	Ordered $Y_i - 0.7728\ X_i$ values
0.2272	−1.4096
2.6816	−0.1824
1.1360	0.2272
1.3632	0.2720 }average is 0.7040
−1.4096	1.1360
−0.1824	1.3632
0.2720	2.6816
3.4992	3.4492
0.2272	−1.4096

TABLE 4.8
Determining Ranks of Ordered *Y* Values

Ordered Y_i's	Rank (Y_i)
1	1
4	2
5	3.5
5	3.5
6	5.5
6	5.5
8	7
12	8

Does this estimated regression equation represent a statistically significant linear relationship between the *X* and *Y* variables? From Equation 4.22, we obtain the realization $u = 56.50$ (the ordered *Y* values along with their ranks appear in Table 4.8). In addition, other requisite sample realizations are $\sum_{i=1}^{n}(x_i - \bar{x})^2 = 79.8748$, $s(u) = 21.8918$, and thus, from Equation 4.23, $|r| = 56.50/21.8918 = 2.5809$. From the *t* table with $n - 2 = 6$ degrees of freedom, we find that $0.02 < p\text{-value} < 0.05$. Hence, *X* and *Y* are significantly related.

Exercises

4-1. Given the sample dataset in Table E.4-1:

TABLE E.4-1
Sample Dataset

X	2	3	4	5	6	7
Y	4	6	5.5	6.2	6.5	8

Estimate the line of best fit using the ordinary median approach.

a. How precisely has the regression slope β_1 been estimated?
b. Using the D statistic, test the significance of the slope.
c. Using the D statistic, test $H_0 : \beta_1 = 3$ and $H_1 : \beta_1 < 3$.

4-2. Given the sample dataset in Table E.4-2:

a. Estimate the line of best fit using the weighted median procedure.
b. Using Equation 4.23, test the significance of the slope β_1 via the t test. (Hint: determine the associated p-value.)

TABLE E.4-2

Sample Dataset

X	1	7	3	16	11	25	18
Y	3	6	5	8	9	15	12

5

Logistic Regression

The Logistic Regression Setting

The preceding set of regression methods involved a (linear) relationship between a response variable Y and a regressor X, where Y was assumed to be continuous. However, when dealing with logistic regression, the response variable is taken to be dichotomous or binary (it takes on only two possible values), i.e., $Y_i = 0$ or 1 for all $i = 1, \ldots, n$. For instance, we can have a situation in which the outcome of some process of observation is either a success (we record a 1) or failure (we record a 0), or we observe the presence (1) or absence (0) of some characteristic or phenomenon. In addition, dichotomous variables are useful for making predictions, e.g. we may ask the following:

1. Will an individual make a major purchase of a particular item in the near future? Here

$$Y_i = \begin{cases} 1, \text{yes;} \\ 0, \text{no.} \end{cases}$$

2. Will a person develop heart disease given that he/she is a smoker? Now

$$Y_i = \begin{cases} 1, \text{yes;} \\ 0, \text{no.} \end{cases}$$

3. Will an individual complete a certain task in a given period of time? Again our code is

$$Y_i = \begin{cases} 1, \text{yes;} \\ 0, \text{no.} \end{cases}$$

Additional examples abound. Given that the response variable Y is binary, standard (OLS) regression techniques are not directly applicable. To see why this is the case, let us turn to one of the main differences between ordinary linear regression and logistic regression. Other differences will be explored in the next section.

When we applied OLS to estimate the parameters of the model $Y = \beta_0 + \beta_1 X + \varepsilon$, we assumed that the population regression equation had the form $E(Y|X) = \beta_0 + \beta_1 X$, where the value of the response function $E(Y|X)$ was unrestricted for $-\infty < X < +\infty$. However, if $Y_i = 0$ or 1 for all i, then we must have $0 \leq E(Y|X) \leq 1$. To see this, let us consider the dataset appearing in Table 5.1 wherein X represents annual income in thousands of dollars and Y depicts the response ($Y_i = 1$ if "yes;" $Y_i = 0$ if "no") given by recent college graduates to the question "Do you expect to purchase a new vehicle about every three years?"

A plot of the data points appearing in Table 5.1 (Figure 5.1) immediately reveals the dichotomous nature of the response variable Y. However, the pattern that emerges does not provide us with any hint as to the structure of the functional relationship between these variables. To gain some additional insight into the relationship between X and Y, let us group the observations on X into a set of income classes (Table 5.2) and calculate, for each such class, the average number of successes or "yes" responses. (Note that this is just the proportion of successes for the class.) We can then plot each of these means (our estimate of $E(Y|X)$) against the midpoint of the associated class of X (Figure 5.2) to get a feel for the association between the proportion of "yes" responses and income level.

Note that, as X increases in value, the path traced out by combinations of $E(Y|X)$ and X levels is "tilted S-shaped," a form that can be nicely approximately by the logistic regression function. It is this function (presented below) that provides us with a vehicle for studying the relationship between a dichotomous response variable Y and a regressor X.

TABLE 5.1

Prospective Vehicle Purchases versus Income

X (Income)	Y (Response)	X (Income)	Y (Response)
30.0	0	71.2	0
31.1	0	73.1	0
31.8	0	73.7	1
34.5	0	74.6	0
36.6	1	75.8	1
38.0	0	76.1	1
40.6	0	78.0	1
45.8	1	79.0	1
47.1	1	79.1	1
48.3	0	83.0	1
49.6	0	84.1	0
52.5	1	85.1	0
53.0	1	86.7	1
54.1	0	87.8	1
55.0	0	88.8	1
57.8	0	89.1	1
58.0	1	91.6	1
58.5	0	93.8	0
61.7	0	95.4	1
63.0	1	95.9	1
63.3	0	100.6	1
63.7	1	105.1	1
63.8	0	108.3	1
67.8	1	110.2	1
69.1	1	117.0	1

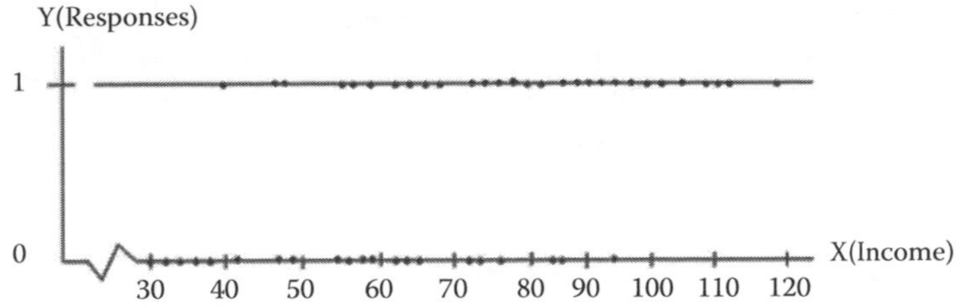

FIGURE 5.1
Responses versus income.

The Logistic Regression Model

As stated at the outset of Chapter 2, the assumptions underlying the strong classical linear regression model are as follows. For a sample of n data points (X_i, Y_i):

1. $Y_i = \beta_0 + \beta_1 X_i + \varepsilon_i$, X_i fixed, $i = 1,\ldots, n$.
2. $E(\varepsilon_i) = 0$, $i = 1,\ldots, n$.
3. $V(\varepsilon_i) = \sigma_\varepsilon^2$ = constant, $i = 1,\ldots, n$.
4. $COV(\varepsilon_i, \varepsilon_j) = 0$, $i \neq j$, with $i, j = 1,\ldots, n$.
5. ε_i is normally distributed, $i = 1,\ldots, n$.

Under these assumptions, it was determined that the OLS estimators of β_0 and β_1 were BLUE. For this model, both the explanatory variable X and the explained variable Y were treated as continuous. However, Exercise 23 of Chapter 2 involved the OLS estimation of the regression parameters β_0 and β_1 when X was assumed binary or dichotomous, i.e., $X = 1$ (some attribute is present) or $X = 0$ (the attribute is absent).

TABLE 5.2
Income Classes versus Mean Response

Income Class	Frequency	Response Yes	Response No	Mean Response
30–39.5	6	1	5	0.166
40–49.5	5	1	4	0.200
50–59.5	7	2	5	0.286
60–69.5	7	4	3	0.571
70–79.5	9	6	3	0.666
80–89.5	7	5	2	0.714
90–99.5	4	3	1	0.750
100–109.5	3	3	0	1.000
110–119.5	2	2	0	1.000
	50	27	23	

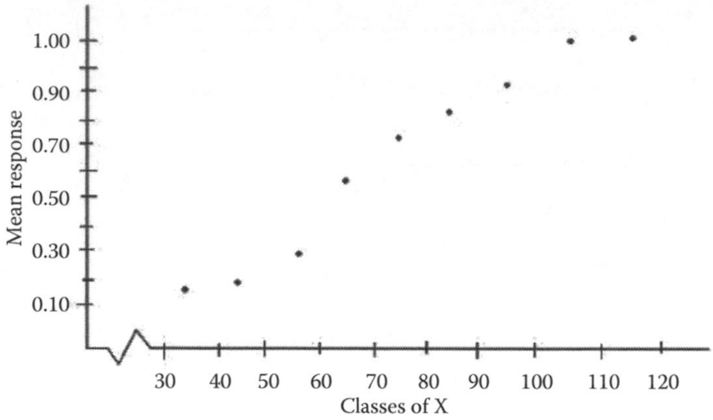

FIGURE 5.2
Mean response versus classes of X.

In this section, we shall consider the case in which Y is dichotomous and X is either a dichotomous or continuous variable. To this end, suppose that Assumptions 1 and 2 above hold so that $E(Y_i|X_i) = \beta_0 + \beta_1 X_i$, $i = 1,\ldots, n$.

Let Y_i be specified as a dichotomous or *Bernoulli random variable* with probability mass function appearing in Table 5.3, where $f(Y_i)$ is the probability mass at Y_i. For the Bernoulli case, we know that $E(Y_i) = p_i$ and $V(Y_i) = p_i\,(1 - p_i)$ for all i. Hence, it must be the case that $p_i = E(Y_i|X_i) = \beta_0 + \beta_1 X_i$ or the conditional mean of Y_i given X_i is p_i, the probability that $Y_i = 1$ when $X = X_i$. Clearly p_i is a linear function of X_i when $Y_i = 1$. Moreover, $V(\varepsilon_i) = V(Y_i) = p_i\,(1 - p_i) = (\beta_0 + \beta_1 X_i)(1 - \beta_0 - \beta_1 X_i)$, , $i = 1,\ldots, n$, i.e., $V(\varepsilon_i)$ is not a constant but varies with X_i so that Assumption 3 above is violated. Hence, OLS estimators for β_0 and β_1 are not efficient when Y_i is dichotomous.

Although Assumption 4 above is still tenable when Y_i is dichotomous, Assumption 5 no longer holds. Given $\varepsilon_i = Y_i - \beta_0 - \beta_1 X_i$, it follows that

if $Y_i = 1$, $\varepsilon_i = 1 - \beta_0 - \beta_1 X_i$; and
if $Y_i = 0$, $\varepsilon_i = -\beta_0 - \beta_1 X_i$.

Because ε_i can assume only these two possible values with probabilities p_i and $1 - p_i$, respectively, it is evident that ε_i is not normally distributed when Y_i is dichotomous and thus Assumption 5 above is untenable. Keep in mind, however, that this poses a problem only if n is small.

However, this is not the whole story. If, as indicated above, $p_i = \beta_0 + \beta_1 X_i$ and X_i is a continuous variable such that $-\infty < X_i < +\infty$, then, for virtually any value of β_1, there may be values of X_i for which p_i exceeds 1 or falls below 0. Obviously, this is not allowed because

TABLE 5.3
Bernoulli Probability Mass Function

Y_i	$f(Y_i)$
1	$f(1) = p_i$
0	$f(0) = \underline{1 - p_i}$
	1

we must have $0 \le p_i \le 1$ for all i. So for Y_i dichotomous, an X_i can predict a p_i outside of the unit interval [0, 1].

The Logistic Regression Function

To avoid the aforementioned difficulty that p_i might lie outside the unit interval for dichotomous Y_i, we actually need to use a *nonlinear model* that constraints p_i or the conditional mean of Y_i given X_i to lie between 0 and 1. Specifically, we shall use a *simple logistic regression function* of the form

$$p_i = E(Y_i|X_i) = \frac{\exp(\beta_0 + \beta_1 X_i)}{1+\exp(\beta_0 + \beta_1 X_i)}, \tag{5.1}$$

or, equivalently,

$$p_i = E(Y_i|X_i) = \frac{1}{1+\exp(-\beta_0 - \beta_1 X_i)}. \tag{5.1.1}$$

Here p_i or the conditional mean of Y_i is the probability that $Y_i = 1$.

This logistic function has the following properties:

a. It is monotonically increasing in X_i if $\beta_1 > 0$ and monotonically decreasing in X_i if $\beta_1 < 0$.
b. It is approximately linear when $0.2 < p_i < 0.8$ and has 0 and 1 as horizontal asymptotes.
c. For any β_0 and β_1 values, we have $0 < p_i < 1$ so that the constraint $0 \le p_i \le 1$ is satisfied.
d. The impact on p_i of a per unit change in X_i can be shown to equal $\partial p_i / \partial X_i = \beta_1 p_i (1 - p_i)$. Clearly, the magnitude of this slope expression depends on the value of p_i given a fixed β_1 value, i.e., when p_i is near 0.5, the effect is large, and when p_i is near 0 or 1, the effect is small.

For additional details, see Figure 5.3 for $\beta_1 > 0$.

To linearize Equation 5.1, set

$$\frac{1}{p_i} = \frac{1}{\exp(\beta_0 + \beta_1 X_i)} + 1$$

so that

$$\frac{1}{\exp(\beta_0 + \beta_1 X_i)} = \frac{1 - p_i}{p_i}$$

or

$$\frac{p_i}{1 - p_i} = \exp(\beta_0 + \beta_1 X_i). \tag{5.2}$$

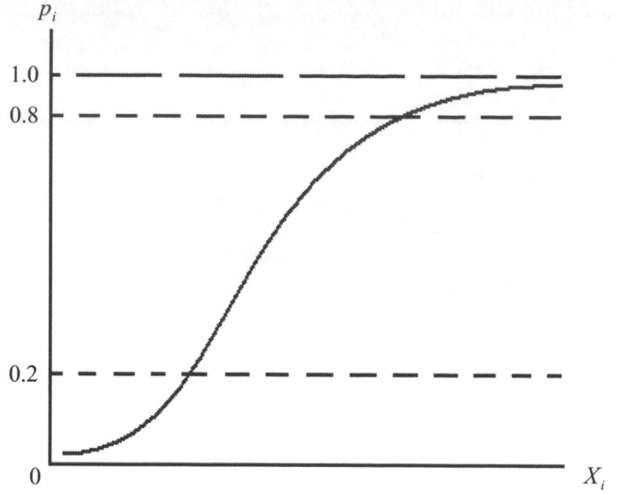

FIGURE 5.3
Logistic function.

Here p_i / $(1 - p_i)$ is called the *odds ratio*, whereas the *log odds function* or *logit transformation of the probability* p_i (or *logit function* for short) is

$$p_i' = \ln\left(\frac{p_i}{1-p_i}\right) = \beta_0 + \beta_1 X_i. \tag{5.3}$$

Note that $-\infty < p_i' < +\infty$ as $-\infty < X_i < +\infty$.

The Odds Ratio

What have we gained by working with the odds ratio p_i / $(1 - p_i)$ rather than with p_i itself? As mentioned previously, we need to avoid the potential difficulty that, for certain X_i's and a given β_1 value, p_i may exceed 1 or fall below 0. (Remember that p_i is the probability that $Y_i = 1$ so that we require $0 \le p_i \le 1$.)

An alternative device for assessing the probability that $Y_i = 1$ is the odds of this event occurring. That is, the odds of event A is the expected number of times A will occur divided by the expected number of times A will not occur or $P(A)/P(\bar{A}) = P(A)/(1-P(A))$. Hence, odds compare the chances in favor of an event to the chances against it. For instance, if the odds of an event are 3, then we expect three times as many occurrences as nonoccurrences of the event, and if the odds of some event are ½, we expect only one-third as many occurrences as nonoccurrences.

The connection between probability and odds can be expressed as follows. In general, if the odds are $x : y$ in favor of event A, then $P(A) = x/x + y$, and, if the probability of an event is $P(A)$, then the odds in favor of the occurrence of A are $P(A)$:$(1 - P(A))$.

So for Y, a dichotomous variable, the odds that $Y_i = 1$ are $O_i = P(Y_i = 1)/ P(Y_i = 0) = p_i/(1 - p_i)$, and the probability that $Y_i = 1$ given O_i is $p_i = O_i /(1 + O_i)$. In summary, probability and odds for dichotomous Y are related as

$$O_i = \frac{p_i}{1-p_i}, p_i = \frac{O_i}{1+O_i}, \tag{5.4}$$

where $P(Y_i = 1) = p_i$.

It can readily be demonstrated that odds less than one correspond to probabilities below 0.5, and odds greater than one correspond to probabilities above 0.5. Whereas odds, like probabilities, have a lower bound of zero, odds values, unlike probabilities, exhibit no upper bound. For example , if $p = 0.6$, $O_i = 1.5$, and as $p_i \rightarrow 1$ (from below), $O_i \rightarrow +\infty$. So when working on the *probability scale*, $0 \leq p_i \leq 1$, but when working on the *odds scale*, $0 \leq O_i \leq +\infty$. In addition, if we transform O_i to log odds or $\ln O_i = \ln (p_i / (1 - p_i))$, then we also eliminate the lower bound on O_i so that, on the *log odds scale*, $-\infty < \ln O_i < +\infty$. These observations thus offer a rationale for working with the logit function Equation 5.3.

Working with odds (and ultimately odds ratios) enables us to assess the relationship between a pair of events when a simple comparison of their probabilities leads to unrealistic results. For example, suppose that, for events A and B, we know that event A is twice as likely to occur as event B. So if $P(B) = 0.10$, then obviously $P(A) = 0.20$. However, if $P(B) = 0.75$, we cannot legitimately set $P(A) = 1.50$ because we must have $0 \leq P(A) \leq 1$. To avoid this difficulty, let us switch to the calculation of odds. That is, a probability of 0.75 for event B corresponds to odds of $O_B = 0.75 / 0.25 = 3$. Doubling this value yields $O_A = 6$. Converting back probabilities renders, via Equation 5.4, $6/7 = 0.857 = P(A)$.

Interpreting the Logistic Regression Parameters

In the basic regression model, the slope coefficient β_1 represents the average rate of change in Y per unit increase in X. Here β_1 is constant for every admissible value of X. However, as we shall now see, for logistic regression, the usual slope interpretation does not apply. Because the logistic function is nonlinear, the slope of $p(X)$ or the change in $p(X)$ per unit change in X obviously depends on where one starts. As will now be demonstrated, β_1 represents the change in the logit precipitated by a one unit change in X. This result consequently enables us to offer an interpretation of β_1 in terms of odds.

Let us express the log odds or logit function as

$$p'(X) = \ln\left(\frac{p(X)}{1-p(X)}\right) = \ln O(X) = \beta_0 + \beta_1 X. \tag{5.3.1}$$

Evaluating this expression at $X = X_j$ and at $X = X_h$ yields $p'(X_j) = \beta_0 + \beta_1 X_j$ and $p'(X_h) = \beta_0 + \beta_1 X_h$, respectively. Then the *logit difference* is

$$p'(X_j) - p'(X_h) = \beta_1(X_j - X_h). \tag{5.5}$$

But

$$p'(X_j) - p'(X_h) = \ln O(X_j) - \ln O(X_h)$$
$$= \ln(O(X_j)/\ O(X_h)) = \beta_1(X_j - X_h). \tag{5.5.1}$$

Let us define the odds ratio or

$$\Psi(X_j, X_h) = O(X_j)/O(X_h) \tag{5.6}$$

as the ratio of the odds for X_j to the odds for X_h, i.e., $\Psi(X_j, X_h)$ indicates how much more likely it is for some attribute to be present when $X = X_j$ than when $X = X_h$. Then the log odds ratio can be represented as

$$\ln \Psi(X_j, X_h) = \ln(O(X_j)/O(X_h)) \tag{5.7}$$

so that, from Equation 5.5.1,

$$\ln \Psi(X_j, X_h) = \beta_1(X_j - X_h) \tag{5.8}$$

or

$$\Psi(X_j, X_h) = \exp(\beta_1(X_j - X_h)). \tag{5.8.1}$$

Thus, the logit difference Equation 5.5 or change in log odds is the log odds ratio Equation 5.8, where the right side is obtained from the logit function via Equation 5.5.1.

Now, suppose that $X_j = X_h + 1$. Then Equation 5.8 becomes

$$\ln \Psi(X_h + 1, X_h) = \beta_1 \tag{5.9}$$

or

$$\Psi(X_h + 1, X_h) = \exp(\beta_1) \tag{5.9.1}$$

i.e., the logit function slope β_1 is the log odds ratio for a one unit increase in X from X_h to $X_h + 1$. Alternatively stated, β_1 represents the change in the log odds when X increases by one unit.

Note also that, if X is a dichotomous variable, then

$$\ln \Psi(1, 0) = \beta_1 \tag{5.10}$$

or

$$\Psi(1, 0) = \exp(\beta_1). \tag{5.10.1}$$

In this instance, $\Psi(1, 0)$ indicates how much more likely it is for some attribute to be present when $X = 1$ than when $X = 0$.

In summary, let the odds that $Y = 1$ be given by $O = P(Y = 1)/P(Y = 0)$. Then,

1. For X continuous,
   a. $\exp(\beta_1)$ is the increase in the odds that $Y = 1$ with each additional unit increase in X.
   b. $\exp(\beta_1(X_j - X_h))$ is the increase in the odds ratio Ψ when X changes by $X_j - X_h$ units.
2. For X dichotomous, $\exp(\beta_1)$ is the increase in the odds that $Y = 1$ when $X = 1$ relative to when $X = 0$.

Example 5.1

Suppose that X is a continuous variable and that the estimated logit function has the form

$$\hat{p}'(X) = \hat{\beta}_0 + \hat{\beta}_1 X = -3.1471 + 0.1124X.$$

Then from Equation 5.9.1, $\hat{\psi} = \exp(\hat{\beta}_1) = e^{0.1124} = 1.1190$. That is, the odds that $Y = 1$ are estimated to increase by 11.9% with each unit increase in X. More formally, to attain this result, we subtract 1 from the odds ratio and multiply the difference by 100 to obtain $100(e^{\hat{\beta}_1} - 1)\% = 100(1.1190 - 1)\% = 11.9\%$. This gives us the percentage change in odds per unit increase in X. Additionally, from Equation 5.8.1, if $X_j - X_h = 5$, then $\hat{\psi} = \exp(\hat{\beta}_1(X_j - X_h)) = \exp(0.1124(5)) = e^{0.1124(5)} = 1.7542$. So for every increase of five units in X, the odds that $Y = 1$ are estimated to increase by 75.42%. (In general, the said increase is determined as $100(e^{\hat{\beta}_1(X_j - X_h)} - 1)\%$.)

From Equation 5.1, the estimated logistic regression function, expressed in the original units of X, is

$$\hat{p}(X) = \frac{\exp(-3.1471 + 0.1124X)}{1 + \exp(-3.1471 + 0.1124X)}.$$

(Alternatively, once the estimated logit function $\hat{p}'(X) = \hat{\beta}_0 + \hat{\beta}_1 X$ has been obtained, we may transform back to the original units using $\hat{p}'(X) = \ln[\hat{p}(X)/(1 - \hat{p}(X))]$, i.e., the estimated logistic regression function is obtained by solving for $\hat{p}(X)$ from $\exp(\hat{p}'(X)) = \hat{p}(X)/(1 - \hat{p}(X))$ or $\hat{p}'(X) = \exp(\hat{p}'(X))/(1 + \exp(\hat{p}'(X))$.) For instance, suppose that $X = 10$. Then the probability that $Y = 1$ at $X = 10$ is $\hat{p}(10) = 0.1168$, i.e., there is about an 11.68% chance that $Y = 1$ when $X = 10$.

Finally, for X dichotomous, suppose that

$$\hat{p}'(X) = 2.5828 + 0.5352X.$$

Then $\hat{\psi} = \exp(\hat{\beta}_1) = e^{0.5352} = 1.7077$. Hence, the odds that $Y = 1$ when $X = 1$ are estimated to be about 1.71 times higher than when $X = 0$, i.e., the odds that $Y = 1$ when $X = 1$ are estimated to be about 71% higher than when $X = 0$.

Inferences about the Logistic Regression Slope

In what follows, we shall assume that the parameters of the logistic regression function have been estimated by the maximum-likelihood technique. (Details pertaining to the derivation of the likelihood function for the observed sample dataset are presented in

Appendix 5.A.) As noted in Chapter 2, for large samples, maximum-likelihood estimators are consistent, asymptotically efficient, and asymptotically normal. Moreover, even for small samples, the method of maximum likelihood can be expected to provide us with good parameter estimates.

To assess the significance of the estimated slope of the logistic regression function, let us form the test statistic

$$Z^* = \frac{\hat{\beta}_1 - \beta_1}{S_{\hat{\beta}_1}}. \tag{5.11}$$

For large n, it is known that $Z^*:N(0, 1)$. As usual, we may test $H_0 : \beta_1 = 0$ against any one of the following alternative hypotheses:

Case I	Case II	Case III
$H_0 : \beta_1 = 0$	$H_0 : \beta_1 = 0$	$H_0 : \beta_1 = 0$
$H_0 : \beta_1 > 0$	$H_1 : \beta_1 < 0$	$H_1 : \beta_1 \neq 0$

Then the decision rules for rejecting H_0 relative to the above-stated alternatives at the $100\alpha\%$ level are as follows:

a. Case I: reject H_0 if $z^* > z_\alpha$;
b. Case II: reject H_0 if $z^* < -z_\alpha$;
c. Case III: reject H_0 if $|z^*| > z_{\alpha/2}$; (5.12)

where z^* depicts sample realization of Z^*.

To determine how precisely β_1 has been estimated from the sample, let us form a $100(1 - \alpha)\%$ confidence interval for β_1 as

$$\hat{\beta}_1 \pm Z_{\alpha/2} S_{\hat{\beta}_1}. \tag{5.13}$$

Corresponding to this interval estimate of β_1 is a $100(1 - \alpha)\%$ confidence interval for the associated odds ratio $\exp(\beta_1)$ or

$$\exp(\hat{\beta}_1 \pm Z_{\alpha/2} S_{\hat{\beta}_1}). \tag{5.14}$$

That is, if we are $100(1 - \alpha)\%$ confident that $l_1 < \beta_1 < l_2$ (where $l_1 = \hat{\beta}_1 - Z_{\alpha/2} S_{\hat{\beta}_1}$ and $l_2 = \hat{\beta}_1 + Z_{\alpha/2} S_{\hat{\beta}_1}$), then we are also $100(1 - \alpha)\%$ confident that

$$\exp(l_1) < \Psi < \exp(l_2). \tag{5.15}$$

In this regard, we are $100(1 - \alpha)\%$ confident that the odds that $Y = 1$ increase by between $\exp(l_1)$ and $\exp(l_2)$ with each additional unit increase in X.

When X changes by an amount $X_j - X_h = \delta$ units, then Equation 5.14 becomes

$$\exp(\hat{\beta}_1 \delta \pm Z_{\alpha/2} |\delta| S_{\hat{\beta}_1}) \tag{5.14.1}$$

or Equation 5.15 becomes

$$\exp(\delta l_1) < \Psi < \exp(\delta l_2). \tag{5.15.1}$$

In this instance, we are 100(1 – α)% confident that the odds that $Y = 1$ increase by between $\exp(\delta l_1)$ and $\exp(\delta l_2)$ when X changes by δ units.

Example 5.2

Using the SAS System's PROC LOGISTIC: As indicated above, PROC LOGISTIC uses the method of maximum likelihood to obtain estimates of β_0 and β_1. It also provides an array of useful diagnostics for assessing the efficacy of the logistic regression model. The appropriate SAS code for invoking PROC LOGISTIC appears in Exhibit 5.1. (The output generated by this program appears as Table 5.4.). The dataset first constitutes a random sample of n = 17 mothers chosen from the local PTA roster for which the dependent variable *grad* pertains to whether or not a woman is a college graduate (grad = 1 if yes; grad = 0 if no), whereas the explanatory variable *age* indicates the age at which she had her first child.

Exhibit 5.1 SAS Code for Logistic Regression

```
data first; input grad
age;
datalines;
0  22
1  24
1  27
0  25
1  27
0  25
0  19
1  30
1  27
0  24
0  22
0  21
1  29
0  23
1  35
1  23
1  24
;
             ①                 ②
proc logistic data=first des;
       ③          ④          ⑤             ⑥       ⑦     ⑧          ⑨
model grad=age/clparm=wald clodds=wald lackfit rsq ctable pprob=(0.05 to
                       ⑩          ⑪
1.00 by 0.05) influence iplots;
```

① We invoke the LOGISTIC procedure so as to handle the dichotomous response variable Y(= 0 or 1). SAS characterizes Y = 1 as an *event* and Y = 0 as a *nonevent*.

② PROC LOGISTIC, by default, models the probability of a *nonevent* (the lower ordered value of Y).
Requesting DES (for DESCENDING) reverses the order in which the levels of the response variable Y are sorted; DES models the probability of an *event* or the probability that Y = 1 given X.

③ We model *grad* (Y) as a function of *age* (X).

④ CLPARM = WALD requests 95% conventional (symmetrical) confidence intervals for the logistic regression parameters.

⑤ CLODDS = WALD requests 95% conventional (symmetrical) confidence intervals for the odds ratio.

```
          ⑫           ⑬                    ⑭                 ⑮
output out=pred p=phat predprobs=(individual) xbeta=logit;
proc print data=pred; ⑯
proc plot data=pred; ⑰
plot logit*age= '*' ; ⑱
plot phat*age= '*' ; ⑲
run;
```

⑥ LACKFIT requests the Hosmer-Lemeshow (H-L) goodness-of-fit test for a binary or dichotomous response model. The H-L statistic is approximately χ^2 distributed under H_0: the model is correct. Hence, a small *p*-value leads us to reject H_0 and thus conclude that the fitted model is inadequate.

⑦ RSQ yields a generalization of the conventional R^2 statistic for the estimated model. Here we are concerned with how well we can predict the *Y* response or event given values of *X*, i.e., we seek to determine how well the model fits the data.

⑧ The CTABLE option creates a (bias-adjusted) *classification table* that, for binary response data, (a) uses the estimated logistic model to classify observations as events or nonevents and (b) gauges the predictive accuracy of the estimated model.

⑨ The PROB = option specifies probability cut points $c_j \in [0, 1]$, $j = 1,\ldots, J$, for the CTABLE option, with each probability cut point producing a separate classification table (there will be *J* such tables). For each c_j, if the predicted or estimated event probability $\hat{p}_i$ (obtained from Equation 5.1.1) $\geq c_j$, then the *i*th observation is classified as an event; if $\hat{p}_i < c_j$, then the *i*th observation is classified as a nonevent, $i = 1,\ldots, n$. For each c_j a 2×2 absolute frequency (classification) table is constructed by cross-classifying the observed ($Y = 0, 1$) and predicted responses (if $\hat{p}_i \geq c_j$, the model predicts 1; if $\hat{p}_i < c_j$, the model predicts 0).

⑩ The INFLUENCE option provides diagnostic measures that depict how influential each observation is to the fit of a binary response regression model, i.e., these measures essentially indicate the impact on the estimated model of deleting a particular observation.

⑪ IPLOTS produces an index plot for each regression diagnostic provided by the INFLUENCE option, i.e., it renders a set of scatter diagrams wherein a regression diagnostic is plotted on the vertical axis and the case or observation number appears on the horizontal axis.

⑫,⑬ We use the OUTPUT statement with the P (for *predicted*) = PHAT option to create an output dataset that contains the values of the original *X* and *Y* variables along with two new variables, _LEVEL_ and PHAT, where the former indicates the value of the response variable being modeled (the response level $Y = 1$) and the latter yields the probability that each observation has the response level indicated by _LEVEL_, i.e., we obtain an estimate of the probability that each observation is an event. Hence, the PHAT values $\hat{p}_i$ (see Equation 5.1.1) render the probability that $Y = 1$ at each X_i, $i = 1,\ldots, n$.

⑭ The PREDPROBS = (INDIVIDUAL) option requests individual predicted probabilities for each response level of *Y*.

⑮ The XBETA statistic determines the predicted log odds or logit value for each case or observation corresponding to the ordered value of the variable _LEVEL_ (here the response level is $Y = 1$).

⑯ We request a PRINT of the output dataset.

⑰ We request a PLOT of variables appearing in the output dataset.

⑱ We request a PLOT of the variable LOGIT on the vertical axis against *age* (*X*) on the horizontal axis.

⑲ We request a PLOT of the variable PHAT on the vertical-axis against *age* (*X*) on the horizontal axis.

The output generated by the preceding program appears as Table 5.4.

TABLE 5.4

Output for Example 5.2

The SAS System
The LOGISTIC Procedure

①
Model Information

Dataset	WORK. FIRST
Response variable	grad
Number of response levels	2
Number of observations	17
Model	binary logit
Optimization technique	Fisher's scoring

②
Response Profile

Ordered Value	Grad	Total Frequency
1	1	9
2	0	8

Probability modeled is grad = 1

Model convergence status

Convergence criterion (GCONV = 1E-8) satisfied

③
Model Fit Statistics

Criterion	Intercept Only	Intercept and Covariates AIC
AIC	25.508	17.244
SC	26.341	18.911
−2 Log L	23.508	13.244

④

***R*-Square**	**0.4533**	**Max-rescaled *R*-Square**	**0.6050**

⑤
Testing Global Null Hypothesis: BETA = 0

Test	Chi-Square	DF	Pr > ChiSq
Likelihood ratio	10.2641	1	0.0014
Score	6.8279	1	0.0090
Wald	3.9248	1	0.0476

⑥
Analysis of Maximum Likelihood Estimates

Parameter	DF	Estimate	Standard Error	Wald Chi-Square	Pr > ChiSq
Intercept	1	−20.3209	10.2086	3.9624	0.0465
Age	1	0.8340	0.4210	3.9248	0.0476

⑦
Odds Ratio Estimates

Effect	Points Estimate	95% Wald Confidence Limits	
Age	2.303	1.009	5.255

(continued)

TABLE 5.4 *(continued)*

The SAS System
The LOGISTIC Procedure

⑧
Association of Predicted Probabilities and Observed Responses

Percent concordant	86.1	Somers' *D*	0.764
Percent discordant	9.7	Gamma	0.797
Percent tied	4.2	Tau-a	0.404
Pairs	72	*c*	0.882

⑨
Wald Confidence Interval for Parameters

Parameter	Estimate	95% Confidence Limits	
Intercept	–20.3209	–40.3295	–0.3124
Age	0.8340	0.00891	1.6591

⑩
Wald Confidence Interval for Adjusted Odds Ratios

Effect	Unit	Estimate	95% Confidence Limits	
Age	**1.0000**	**2.303**	**1.009**	**5.255**

⑪
Partition for the Hosmer and Lemeshow Test

		grad = 1		grad = 0	
Group	**Total**	**Observed**	**Expected**	**Observed**	**Expected**
1	2	0	0.07	2	1.93
2	2	0	0.24	2	1.76
3	2	1	0.49	1	1.51
4	3	2	1.27	1	1.73
5	2	0	1.26	2	0.74
6	3	3	2.70	0	0.30
7	2	2	1.97	0	0.03
8	1	1	1.00	0	0.00

Hosmer and Lemeshow Goodness-of-Fit Test

Chi-Square	DF	Pr > ChiSq
5.5500	6	0.4754

⑫
Classification Table

(A)	(B)		(C)					(D)	
	Correct		Incorrect					Percentages	
Prob Level	Event	Nonevent	Event	Nonevent	Correct	Sensitivity	Specificity	False POS	False NEG
0.050	9	1	7	0	58.8	100.0	12.5	43.8	0.0
0.100	9	2	6	0	64.7	100.0	25.0	40.0	0.0
0.150	8	4	4	1	70.6	88.9	50.0	33.3	20.0

TABLE 5.4 *(continued)*

⑬

Regression Diagnostics

0.200	8	4	4	1	70.6	88.9	50.0	33.3	20.0
0.250	8	4	4	1	70.6	88.9	50.0	33.3	20.0
0.300	8	5	3	1	76.5	88.9	62.5	27.3	16.7
0.350	8	5	3	1	76.5	88.9	62.5	27.3	16.7
0.400	6	5	3	3	64.7	66.7	62.5	33.3	37.5
0.450	6	5	3	3	64.7	66.7	62.5	33.3	37.5
0.500	6	6	2	3	70.6	66.7	75.0	25.0	33.3
0.550	6	6	2	3	70.6	66.7	75.0	25.0	33.3
0.600	6	6	2	3	70.6	66.7	75.0	25.0	33.3
0.650	6	6	2	3	70.6	66.7	75.0	25.0	33.3
0.700	6	6	2	3	70.6	66.7	75.0	25.0	33.3
0.750	6	8	0	3	82.4	66.7	100.0	0.0	27.3
0.800	6	8	0	3	82.4	66.7	100.0	0.0	27.3
0.850	6	8	0	3	82.4	66.7	100.0	0.0	27.3
0.900	3	8	0	6	64.7	33.3	100.0	0.0	42.9
0.950	3	8	0	6	64.7	33.3	100.0	0.0	42.9
1.000	0	8	0	9	47.1	0.0	100.0	.	52.9

The SAS System
The LOGISTIC Procedure

	Covariates	(a) Pearson Residual		(b) Deviance Residual	
			(1 Unit = 0.22)		(1 Unit = 0.21)
Case Number	Age	Value	−8 −4 0 2 4 6 8	Value	−8 −4 0 2 4 6 8
1	22.0000	−0.3729	| *| |	−0.5104	| *| |
2	24.0000	1.1645	| | * |	1.3092	| | * |
3	27.0000	0.3333	| | * |	0.4590	| | * |
4	25.0000	−1.3031	| * | |	−1.4089	| * | |
5	27.0000	0.3333	| | * |	0.4590	| | * |
6	25.0000	−1.3031	| * | |	−1.4089	| * | |
7	19.0000	−0.1067	| * |	−0.1505	| *| |
8	30.0000	0.0954	| * |	0.1346	| | * |
9	27.0000	0.3333	| |* |	0.4590	| | * |
10	24.0000	−0.8587	| * | |	−1.0511	| * | |
11	22.0000	−0.3729	| * | |	−0.5104	| *| |
12	21.0000	−0.2458	| * | |	−0.3425	| * | |
13	29.0000	0.1448	| |* |	0.2036	| |* |
14	23.0000	−0.5659	| * | |	−0.7454	| * | |
15	35.0000	0.0119	| * |	0.0168	| * |
16	23.0000	1.7670	| | *|	1.6831	| | *|
17	24.0000	1.1645	| | * |	1.3092	| | * |

(continued)

TABLE 5.4 *(continued)*

The SAS System
The LOGISTIC Procedure
Regression Diagnostics

Case Number	(c) Hat Matrix Diagonal Value	(1 Unit = 0.01) 0 2 4 6 8 12 16	(d) Intercept DfBeta Value	(1 Unit = 0.06) −8 −4 0 2 4 6 8
1	0.1415	*	−0.1377	*
2	0.1155	*	0.0831	*
3	0.1674	*	−0.1360	*
4	0.1355	*	0.2128	*
5	0.1674	*	−0.1360	*
6	0.1355	*	0.2128	*
7	0.0585	*	−0.0267	*
8	0.0575	*	−0.0229	*
9	0.1674	*	−0.1360	*
10	0.1155	*	−0.0613	*
11	0.1415	*	−0.1377	*
12	0.1222	*	−0.0901	*
13	0.0916	*	−0.0446	*
14	0.1322	*	−0.1531	*
15	0.00297	*	−0.00063	*
16	0.1322	*	0.4780	*
17	0.1155	*	0.0831	*

Case Number	(e) Age DfBeta Value	(1 Unit = 0.05) −8 4 0 2 4 6 8	(f) Confidence Interval Displacement C Value	(1 Unit = 0.03) 0 2 4 6 8 12 16
1	0.1315	*	0.0267	*
2	−0.0535	*	0.2003	*
3	0.1418	*	0.0268	*
4	−0.2466	*	0.3079	*
5	0.1418	*	0.0268	*
6	−0.2466	*	0.3079	*
7	0.0262	*	0.000752	*
8	0.0234	*	0.000589	*
9	0.1418	*	0.0268	*
10	0.0395	*	0.1089	*
11	0.1315	*	0.0267	*
12	0.0873	*	0.00958	*
13	0.0457	*	0.00233	*
14	0.1407	*	0.0562	*
15	0.000641	*	4.206E-7	*
16	0.4392	*	0.5480	*
17	0.0535	*	0.2003	*

TABLE 5.4 *(continued)*

The SAS System
The LOGISTIC Procedure
Regression Diagnostics

Case Number	(g) Confidence Interval Displacement CBar (1 Unit = 0.03) Value	0	2	4	6	8	12	16	(h) Delta Deviance (1 Unit = 0.21) Value	0	2	4	6	8	12	16
1	0.0229	*							0.2834		*					
2	0.1771				*				1.8911					*		
3	0.0223	*							0.2330		*					
4	0.2662					*			2.2511						*	
5	0.0223		*						0.2330		*					
6	0.2662					*			2.2511						*	
7	0.000708	*							0.0234	*						
8	0.000555	*							0.0187	*						
9	0.0223		*						0.2330		*					
10	0.0963		*						1.2011				*			
11	0.0229		*						0.2834		*					
12	0.00841	*							0.1257		*					
13	0.00211	*							0.0436	*						
14	0.0488		*						0.6044		*					
15	4.206E-7	*							0.000282	*						
16	0.4755							*	3.3084							*
17	0.1771				*				1.8911					*		

(i) Delta Chi-Square (1 Unit = 0.22)

Case Number	Value	0	2	4	6	8	12	16
1	0.1620		*					
2	1.5332					*		
3	0.1334		*					
4	1.9641						*	
5	0.1334		*					
6	1.9641						*	
7	0.0121	*						
8	0.00966	*						
9	0.1334		*					
10	0.8337			*				
11	0.1620		*					
12	0.0688	*						
13	0.0231	*						
14	0.3690		*					
15	0.000141	*						
16	3.5980							*
17	1.5332					*		

(continued)

TABLE 5.4 *(continued)*

The SAS System
The LOGISTIC Procedure
(a)

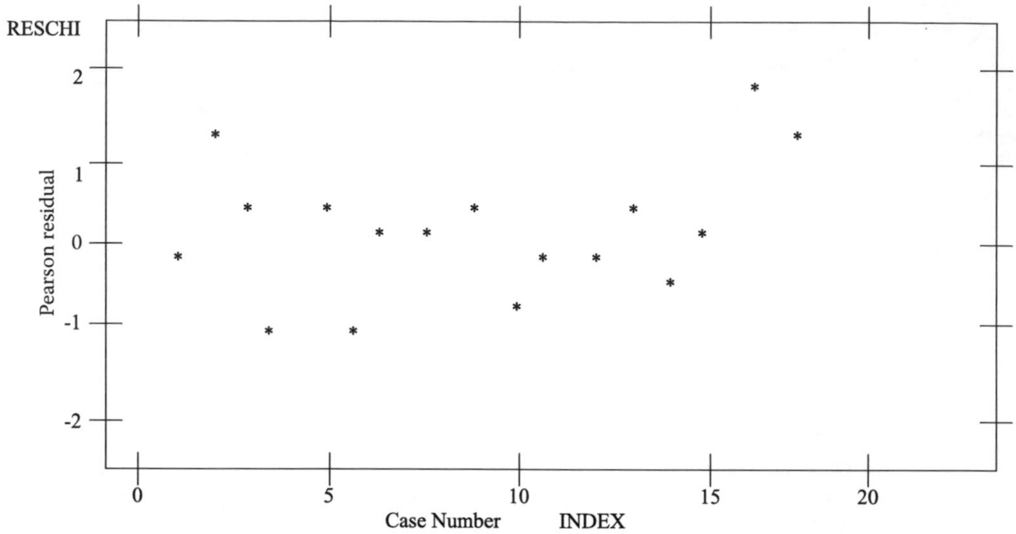

(b)

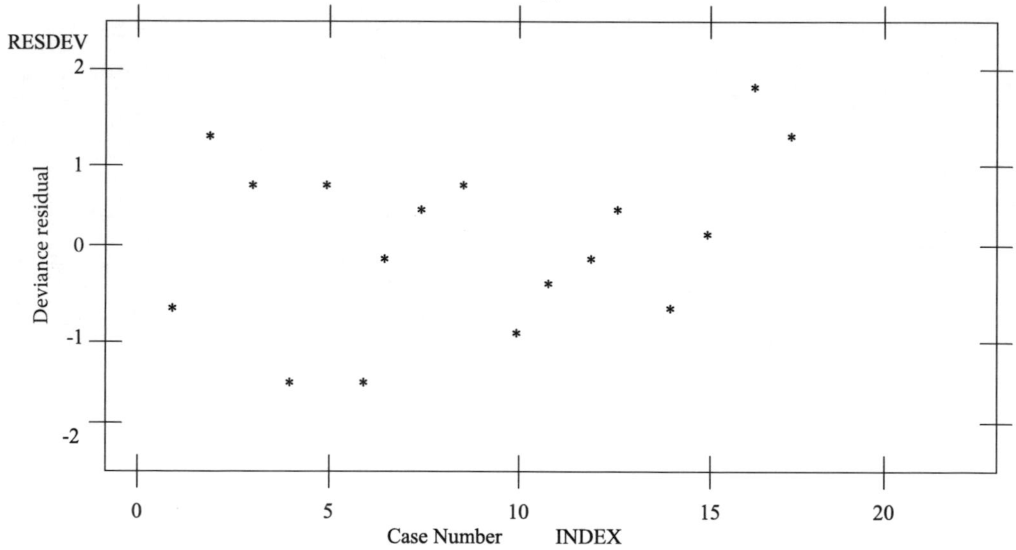

TABLE 5.4 *(continued)*

The SAS System
The LOGISTIC Procedure

(c)

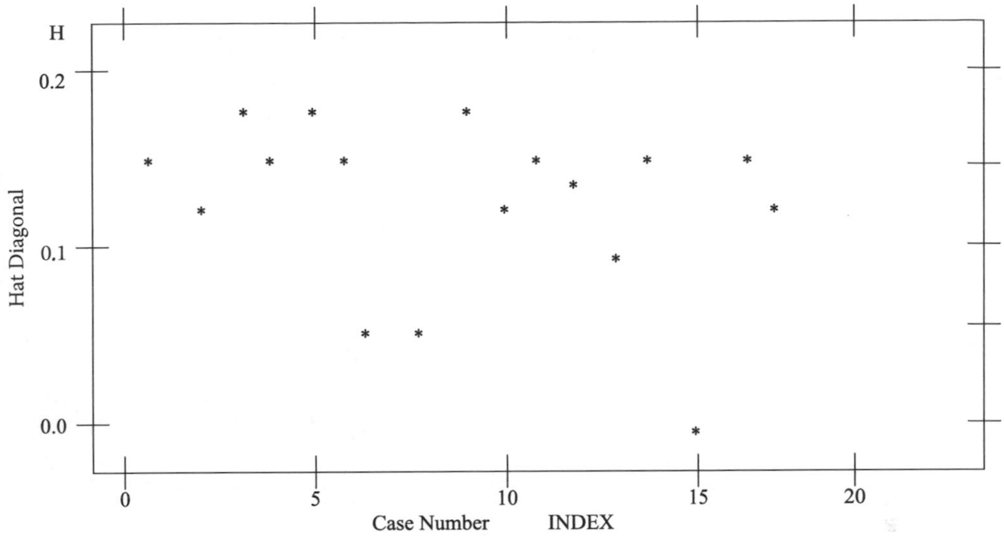

(d)

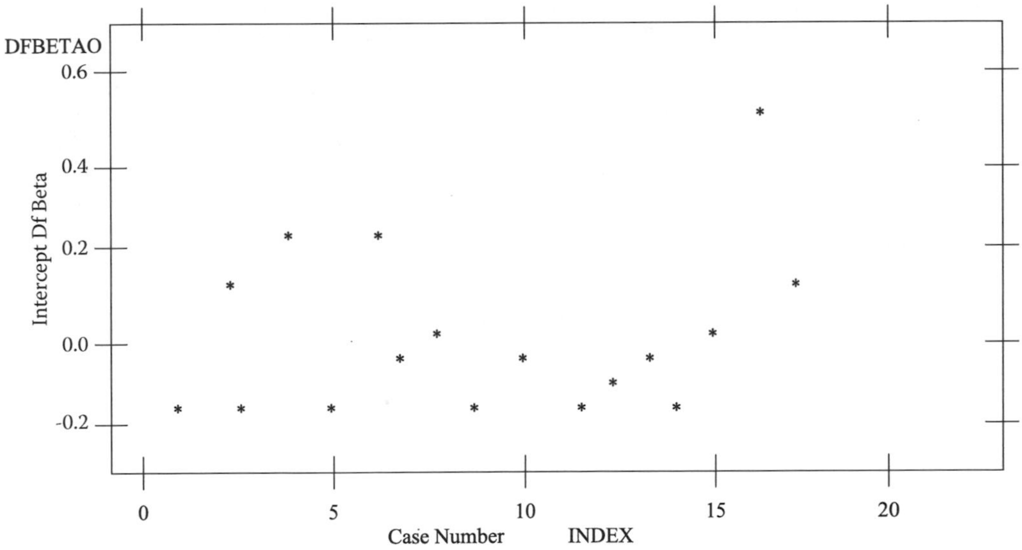

(continued)

TABLE 5.4 *(continued)*

The SAS System
The LOGISTIC Procedure
(e)

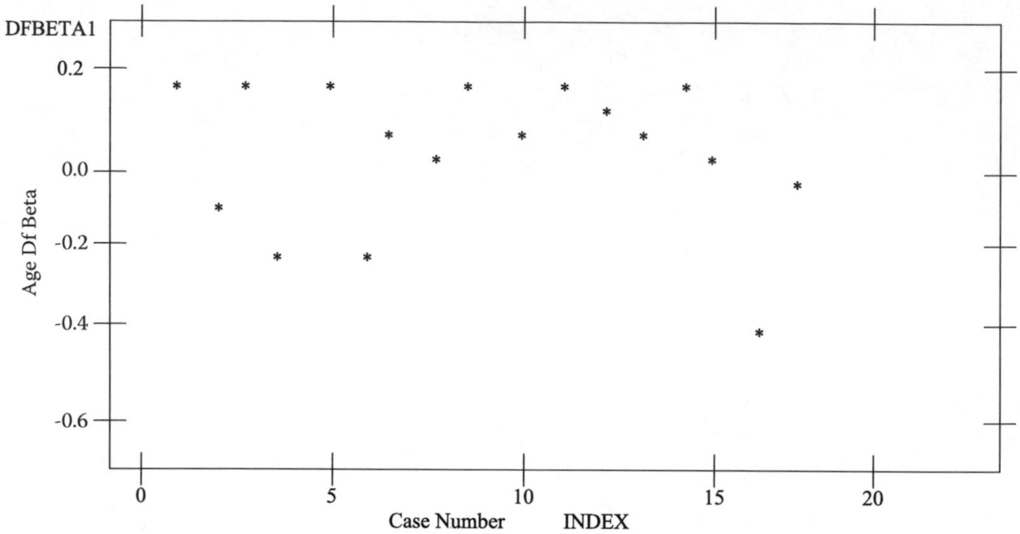

(f)

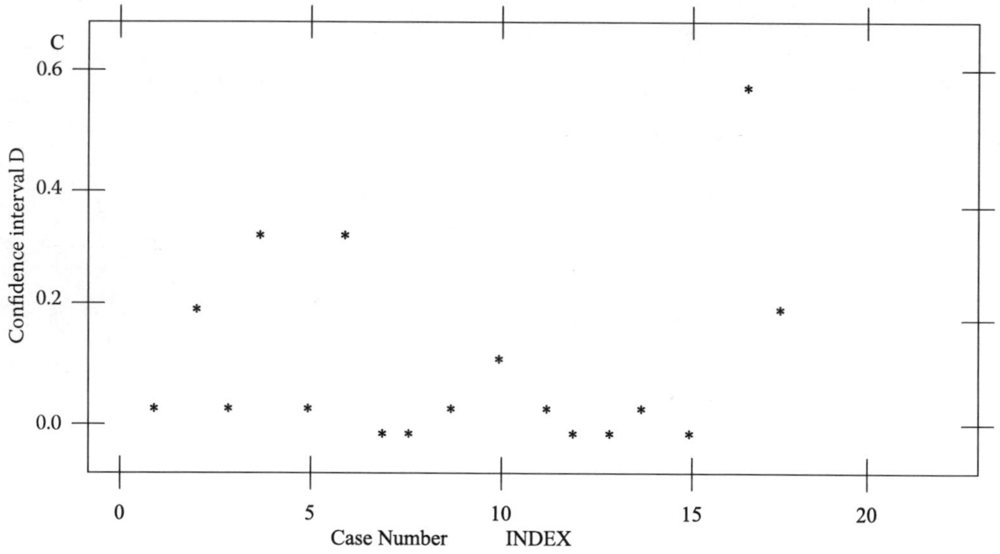

TABLE 5.4 *(continued)*

The SAS System
The LOGISTIC Procedure
(g)

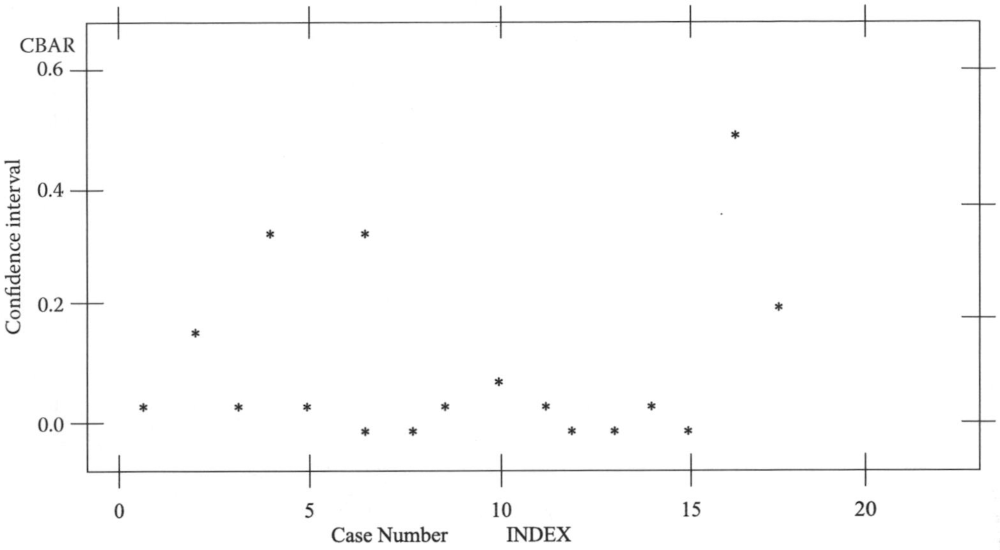

(h)

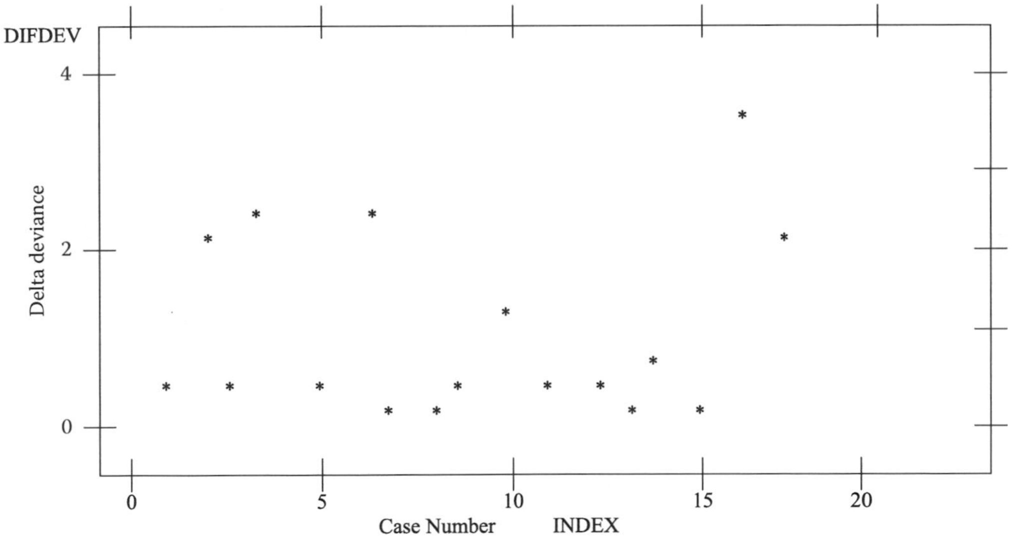

(continued)

TABLE 5.4 *(continued)*

The SAS System
The LOGISTIC Procedure
(i)

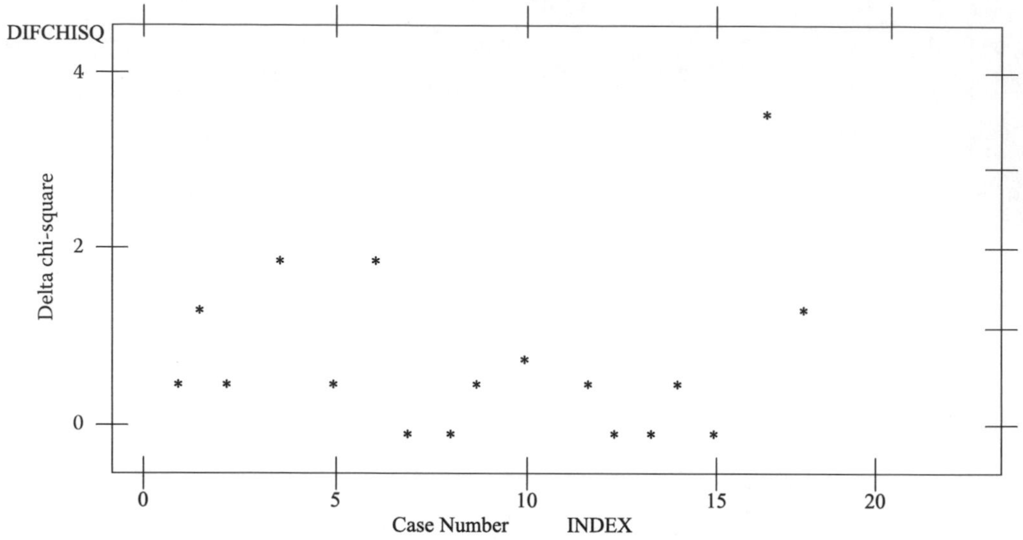

⑮
The SAS System

Obs	grad	age	_FROM_	_INTO_	IP_1	IP_0	_LEVEL_	logit	phat
1	0	22	0	0	0.12211	0.87889	1	−1.97262	0.12211
2	1	24	1	0	0.42443	0.57557	1	−0.30459	0.42443
3	1	27	1	1	0.90002	0.09998	1	2.19745	0.90002
4	0	25	0	1	0.62935	0.37065	1	0.52942	0.62935
5	1	27	1	1	0.90002	0.09998	1	2.19745	0.90002
6	0	25	0	1	0.62935	0.37065	1	0.52942	0.62935
7	0	19	0	0	0.01127	0.98873	1	−4.47467	0.01127
8	1	30	1	1	0.99098	0.00902	1	4.69949	0.99098
9	1	27	1	1	0.90002	0.09998	1	2.19745	0.90002
10	0	24	0	0	0.42443	0.57557	1	−0.30459	0.42443
11	0	22	0	0	0.12211	0.87789	1	−1.97262	0.12211
12	0	21	0	0	0.05697	0.94303	1	−2.80664	0.05697
13	1	29	1	1	0.97948	0.02052	1	3.86548	0.97948
14	0	23	0	0	0.24258	0.75742	1	−1.13861	0.24258
15	1	35	1	1	0.99986	0.00014	1	8.86957	0.99986
16	1	23	1	0	0.24258	0.75742	1	−1.13861	0.24258
17	1	24	1	0	0.42443	0.57557	1	−0.30459	0.42443

① We check the preliminary model information to ensure that we are fitting the correct model.

② The response profile table indicates whether PROC LOGISTIC is ordering the dependent variable *grad* in the desired manner. We requested a logit model that predicts the probability that *grad* = 1 or logit(p) = $\beta_0 + \beta_1 X$. (Note that, because logit(p) = −logit(1 − p), the effect of modeling the response grad = 0 instead of grad = 1 reverses the signs of β_0 and β_1 because logit(1 − p) = $-\beta_0 - \beta_1 X$.)

③ The Akaike information criterion (AIC) and Schwartz criterion (SC) are two goodness-of-fit statistics that are useful for comparing the *relative fit* of different logit models in that the lower the values of AIC and SC, the more desirable the model. For a fixed n, each provides an adjustment (a penalty) to $-2\log\mathcal{L}$ based on the number of estimated parameters (k) in the model because

$$AIC = -2\log\mathcal{L} + 2k,$$

$$SC = -2\log\mathcal{L} + k\log n.$$

The statistic $-2\log\mathcal{L}$ has a χ^2 distribution under $H_0 : \beta_1 = 0$. For the *intercept only* or *reduced* model, $-2\log\mathcal{L}\left(\tilde{\beta}_0, 0\right) = 23.508$; for the *intercept and covariates* or *full* model, $-2\log\mathcal{L}\left(\tilde{\beta}_0, \tilde{\beta}_1\right) = 13.244$.

④ For discrete models, the *generalized coefficient of determination* R^2 = 0.4503 = $1-\left[\mathcal{L}\left(\tilde{\beta}_0,0\right)/\mathcal{L}\left(\tilde{\beta}_0,\tilde{\beta}_1\right)\right]^{2/n} < 1$. Because the maximum value of R^2 is $R^2_{max} = 1-\left[\mathcal{L}\left(\tilde{\beta}_0,0\right)\right]^{2/n} < 1$, an adjusted R^2 coefficient that has an upper bound of 1 is provided by $\tilde{R}^2$ = max− rescaled R − square = R^2/R^2_{max}=0.6050. (Note that $\tilde{R}^2$ does not represent the proportion of the variation in Y explained by the linear influence of X as under OLS estimation.)

⑤ PROC LOGISTIC provides two global χ^2 statistics that each test $H_0 : \beta_1 = 0$, against $H_1 : \beta_1 \neq 0$. The first is the *likelihood ratio* χ^2 *statistic* $X^2 = -2\left[\log\ell\left(\tilde{\beta}_0,0\right) - \log\ell\left(\tilde{\beta}_0,\tilde{\beta}_1\right)\right] = 23.508 - 13.244 = 10.2641$ with degrees of freedom $(n-1)-(n-2)=1$. For large n, X^2 is χ^2_1 under H_0. Note that, if $\mathcal{L}\left(\tilde{\beta}_0,0\right)/\mathcal{L}\left(\tilde{\beta}_0,\tilde{\beta}_1\right)$ is small, then X^2 is large and thus H_1 is supported. Equivalently, a small p–value leads us to reject H_0 in favor of H_1 and thus conclude that the model is acceptable. This is clearly the case here because the p-value is 0.0041. The second global χ^2 statistic is the *score statistic*, which performs the same function as X^2. Here too the p-value (=0.0090) is small, thus indicating that the model performs adequately.

⑥ The estimated intercept and slope of the logistic regression equation are $\hat{\beta}_0 = -20.3209$ and $\hat{\beta}_1 = 0.834$, respectively. Moreover, their estimated standard errors are $s_{\hat{\beta}_0} = 10.2086$ and $s_{\hat{\beta}_1} = 0.4210$, respectively. Then the test statistics (called *Wald* χ^2) $\left(\hat{\beta}_0/s_{\hat{\beta}_0}\right)^2, \left(\hat{\beta}_1/s_{\hat{\beta}_1}\right)^2$ are approximately χ^2 distributed with one degree of freedom for large n under H_0: the population regression coefficient is zero. Their associated p-values are 0.0465 and 0.0476, respectively. Because each p-value is less than 0.05, we shall reject H_0 and conclude that both the intercept and slope are statistically significant.

⑦ Given $\hat{\beta}_1 = 0.834$, we see that the estimated odds ratio is $\hat{\psi} = e^{\hat{\beta}_1} = 2.303$, i.e., the odds of a woman being a college graduate increased by a factor of 2.303 for each unit (per year) increase in the age at which she had her first child.

⑧ This section considers the association between the set of predicted probabilities $\hat{p}_i$ (see Equation 5.1.1), $i = 1,\ldots, n$, and the observed responses ($Y = 0$ or 1). In particular, SAS computes four separate rank correlation measures to assess the quality, in terms of predictive power, of the estimated logistic regression model. These ordinal measures (given below) are computed from the numbers of concordant and discordant pairs of observations, where, for all pairs of observations with different Y values or responses,

a. A pair of observations is termed *concordant* if the observation exhibiting $Y = 1$ has a higher predicted event probability than does the observation with $Y = 0$;
b. A pair of observations is said to be *discordant* if the observation exhibiting $Y = 1$ has a lower predicted event probability than does the observation with $Y = 0$; and
c. A *tie* is recorded if neither pair of observations is concordant or discordant, i.e., each observation has the same predicted event probability.

If c is the number of concordant pairs, d is the number of discordant pairs, and t is the number of tied pairs of observations, then:

Somer's D $= (c - d)/ (c + d + t) = 0.764$;

Goodman-Kruskall Gamma $= (c - d)/(c + d) = 0.797$,

Kendall's tau-a $= 2(c - d)/n(n - 1) = 0.404$; and

$$c = \frac{1}{2}(1 + \text{Somer's D}) = 0.882.$$

Each of these indices varies between zero and one and the closer its value is to unity, the stronger the degree of association between predicted probabilities and observed responses. Except for Kendall's tau-a, these measures indicate a reasonably high level of association between predicted probabilities and observed responses.

⑨ A 95% conventional confidence interval for β_1 (called a *Wald confidence interval* for β_1) is based on the asymptotic normality of the maximum-likelihood estimator for β_1 and appears (see Equation 5.13) as

$$0.8340 \pm 1.93(0.4210) \rightarrow (0.00891, 1.6591).$$

⑩ Corresponding to our preceding 95% confidence interval estimate for β_1, we have a 95% conventional confidence interval for the associated odds ratio $\psi = e^{\beta_1}$ (termed a Wald confidence interval for ψ) (see Equation 5.14) or

$$\exp\,(0.8340 \pm 1.96(0.4210)) \rightarrow (1.009, 5.255).$$

So if we are 95% confident that $0.00891 < \beta_1 < 1.6591$, then we are also 95% confident that $1.009 < \psi < 5.225$. (Note that the *unit column* indicates that the amount X is incremented to yield the estimated odds ratio (the default is one unit).)

⑪ The Hosmer-Lemeshow goodness-of-fit test (applicable when modeling a binary response variable) partitions the observations into approximately 10 classes of about the same size based in the percentiles of the ordered estimated event probabilities $\hat{p}_i, i = 1,\ldots,n$. Within each interval, the expected frequencies are determined as the product of the number of observations in the class and the average estimated event probability for that class. A comparison between these expected frequencies (taken to be the number of event outcomes for the class) is made using the conventional Pearson χ^2 statistic. This statistic is approximately

χ^2_v (v = number of classes − 2) under H_0: the model is correct. A small p-value leads us to reject H_0 and consequently conclude that the estimated logistic regression model is inadequate or does not provide a good fit to the data. Because the reported p-value = 0.4754, we may conclude that the estimated model fits the data well.

⑫ We generate a set of classification tables that enable us to determine the *predictive accuracy* of the binary logistic regression model in terms of its ability to classify observations as events or nonevents at each selected probability cut point. In this regard, the ith observation is classified as an event if its estimated probability $\hat{p}_i \geq$ a given probability cut point $c_j \in [0,1]$, and the ith observation is classified as a nonevent if $\hat{p}_i < c_j$, $i=1,\ldots,n$; $j=1,\ldots,J$. As we shall now see, these classification tables indicate how well the predicted classifications (events or nonevents) correspond to the actual event or nonevent status of each observation.

For instance, in Table 5.4, section 12.A ("Prob Level") lists the chosen probability cut points c_j, $j=1,\ldots,20$. Sections 12.B and C ("Correct" and "Incorrect") display the frequencies with which the observations are correctly or incorrectly classified as event or nonevents for each probability cut point. In section 12.D (percentages for predictive accuracy), the "Correct" column gives the probability that the logistic regression model correctly classifies the observations for each probability cut point, the "Sensitivity" column gives the probability of predicting an event correctly, the "Specificity" column yields the probability of predicting a nonevent correctly, the "False Positive" column gives the probability of incorrectly classifying observed nonevents as predicted events, and the "False Negative" column gives the probability of incorrectly classifying observed events as predicted nonevents. For example, at the probability cut point of, say, 0.30, the percentage correct is

$$\frac{\text{number of correctly classified observations}}{\text{total number of observations}} \times 100 = \frac{8+5}{17} \times 100 = 76.5\%;$$

the sensitivity rate is

$$\frac{\text{number of correctly classified events}}{\text{total number of events}} \times 100 = \frac{8}{8+1} \times 100 = 88.9\%;$$

the specificity rate is

$$\frac{\text{number of correctly classified nonevents}}{\text{total number of nonevents}} \times 100 = \frac{5}{5+3} \times 100 = 62.5\%;$$

the false positive rate is

$$\frac{\text{number of nonevents incorrectly classified as events}}{\text{number of observations classified as events}} \times 100 = \frac{3}{8+3} \times 100 = 27.3\%;$$

and the false negative rate is

$$\frac{\text{number of events incorrectly classified as nonevents}}{\text{number of observations classified as nonevents}} \times 100 = \frac{1}{5+1} \times 100 = 16.7\%.$$

Note that, as the probability cut points increase in value, (a) the rate of sensitivity decreases, i.e., the less likely it is that an observation is correctly classified as an event, and (b) the rate of specificity increases, i.e., the more likely it is that an observation is correctly classified as a nonevent.

⑬ PROC LOGISTIC offers an assortment of regression diagnostics that indicate how influential each observation is in terms of its impact on the estimated binary logistic regression model, i.e., these influence statistics provide us with an approximation as to how some feature of the model changes when a particular data point is deleted from the sample and the fit is made using only the remaining data points. Although the discussion on influential observations that now follows [see also Collett 1993, 120–146; Pregibon 1981, 705–724] closely parallels that provided in Chapter 3, we note briefly that the list of regression diagnostics includes the following:

a., b. Pearson residual (RESCHI) and deviance[1] residual (RESDEV): both are useful for identifying poorly fitting observations, i.e., observations that are not adequately explained by the model.

c. Hat matrix diagonal (H): these elements are useful in detecting extreme points in X space.

To detect observations that constitute a source of instability in the estimated regression parameters, we look to d and e:

d. Intercept DfBeta (DFBETA0): reports the standardized difference in the estimated intercept $\hat{\beta}_0$ attributable to deleting an individual observation.

e. Slope DfBeta (DFBETA1): reports the standardized difference in the estimated slope $\hat{\beta}_1$ attributable to deleting an individual observation.

f. Confidence interval displacement (C): measures the influence of an individual observation on the parameter estimates $\hat{\beta}_0$ and $\hat{\beta}_1$ simultaneously.

g. Confidence interval displacement (CBAR): measures the influence of an individual observation on the overall fit of the model.

To detect observations that are associated with a marked disparity between the sample data and the predicted values determined from the estimated model, we examine h and i:

h. Delta deviance (DIFDEV): specifies the change in the deviance attributable to deleting an individual observation.

i. Delta chi-square (DIFCHISQ): specifies the change in the Pearson χ^2 statistic attributable to deleting an individual observation.

A glance at section 13.c in Table 5.4 reveals that the observations (cases) numbered 3, 5, and 9 can be viewed as extreme in age space. Moreover, the remaining panels of Section 13 indicate that the observations numbered 4, 6, and 16 appear to be influential data points and thus warrant further scrutiny.

⑭ IPLOTS generates an *index plot* for each of the regression diagnostics found under the INFLUENCE option, i.e., realizations of the various influence statistics appear on the vertical axis and the corresponding observation numbers (the index values) appear on the horizontal axis.

⑮ PREDPROBS = INDIVIDUAL automatically generates two new variables:

---

[1] In general, a *deviance statistic* compares the observed Y_i's with their corresponding estimated values $\hat{Y}_i$ obtained from the fitted model. It does so by comparing the current estimated model with a reference or baseline model (called the *full* or *saturated model*) that fits the data perfectly because the estimated values are used in place of the actual observations. If the value of the likelihood function for the current estimated model is denoted as $\mathcal{L}_c$ and the value of the likelihood function for the saturated model is devoted as $\mathcal{L}_s$, then the deviance can be expressed as

$$D = -2\log(\mathcal{L}_c/\mathcal{L}_s).$$

Thus, D measures the extent to which the current model *deviates* from the saturated model, e.g., if $\mathcal{L}_c$ is small relative to $\mathcal{L}_s$, then D is large, thus indicating a poor fit to the sample data. If $\mathcal{L}_c \approx \mathcal{L}_s$, then D is small and thus indicative of a good fit to the data. For a binary logistic regression model, it can be demonstrated that

$$D = -2\sum_{i=1}^{n}\left\{\hat{p}_i \operatorname{logit}\left(\hat{p}_i\right) + \log\left(1-\hat{p}_i\right)\right\}.$$

1. _FROM_: has values that correspond to the formatted values of the response variable Y (Y has the response levels 0 or 1);
2. _INTO_: has values that correspond to the formatted value of the response level with the largest individual predicted probability,

where the individual predicted probabilities appear under the variable headings:

$$IP\_1 = P\,(Y{=}1) \text{ (see Equation 5.1.1);}$$

$$IP\_0 = P(Y = 0) = 1 - IP\_1.$$

For instance, for observation 2, _FROM_= 1. However, because $IP\_0 = 0.57557 > IP\_1 = 0.42443$, it follows that _INTO_=0. For observation 3, _FROM_= 1, but with $IP\_1 = 0.90002 > IP\_0 = 0.9998$, it follows that _INTO_=1.

Next, the variable _LEVEL_ indicates the level of the response variable being modeled, i.e., we are modeling logit(p)= $\beta_0 + \beta_1 X$ or predicting the probability that $Y = 1$.

Given _LEVEL_= 1 for all i, the variable XBETA = LOGIT provides us with a set of predicted logit values determined from the estimated logistic regression equation

$$\begin{aligned}\text{logit}\,(\hat{p}) &= \hat{\beta}_0 + \hat{\beta}_0 X \\ &= -20.3209 + 0.8340X.\end{aligned}$$

For instance, for observation 1, the estimated value for $X = 22$ is $-20.3209+0.8340(22) = -1.9726$.

Finally, the variable PHAT has as its set of realizations the predicted probabilities $\hat{p}_i$, $i = 1,\ldots, n$ (see Equation 5.1.1), determined for the response level specified by the variable _LEVEL_ (as stated previously, we are predicting the probability that $Y = 1$). Note that, under this restriction, the values of the variables IP_1 and PHAT must coincide.

⑯ A plot of the variable LOGIT versus age (X) (Figure 5.4) reveals that logit ($\hat{p}$) is a linear function of X.

⑰ A plot of the variable PHAT versus age (X) (Figure 5.5) indicates that the simple logistic regression has the expected shape. In addition, as anticipated above, it is approximately linear for $0.2 < \hat{p}_i < 0.8$ and has 0 and 1 as horizontal asymptotes.

Additional PROC LOGISTIC Program Elements: Although the SAS program depicted in Exhibit 5.1 has exposed the reader to the basics of binary logistic regression analysis, various modifications of this program can easily be made. For instance,

1. a. Instead of requesting 95% Wald confidence intervals for the logistic regression parameters, we may, instead, opt for 95% profile likelihood confidence intervals for the same by using CLPARM = PL in the model statement. CDPRM = BOTH requests 95% confidence intervals for each type.
   b. Instead of requesting a 95% Wald confidence interval for the odds ratio, we may, as an alternative, use CLODDS = PL to request a 95% profile likelihood confidence interval for the odds ratio. CLODDS = BOTH requests a 95% confidence interval of each type.
2. Appearing in section 10 of Table 5.4 is a column headed "Unit" with the number 1 (the default) contained therein. This unity element indicates how much X is incremented to produce the estimated odds ratio. However, if $\psi = e^{\beta_1}$ is the odds ratio for a one-unit increment in X, then $\psi^{\delta} = e^{(\delta\beta_1)}$ is the odds ratio for a δ-unit increment in X. In this regard, for $\delta > 0$, the associated 95% (Wald) confidence interval for ψ is $(e^{\delta l_1}, e^{\delta l_2})$ (see Equation 5.15.1). For $\delta < 0$, the 95% (Wald) confidence interval for ψ is $(e^{\delta l_2}, e^{\delta l_1})$. So to obtain, for instance, the odds ratio as well as 95% Wald confidence limits for the same for a two-unit increase (as well as a two-unit decrease) in age, we would insert the statement

```
UNITS    age = 2   -2;
```

into the program immediately after the MODEL statement.

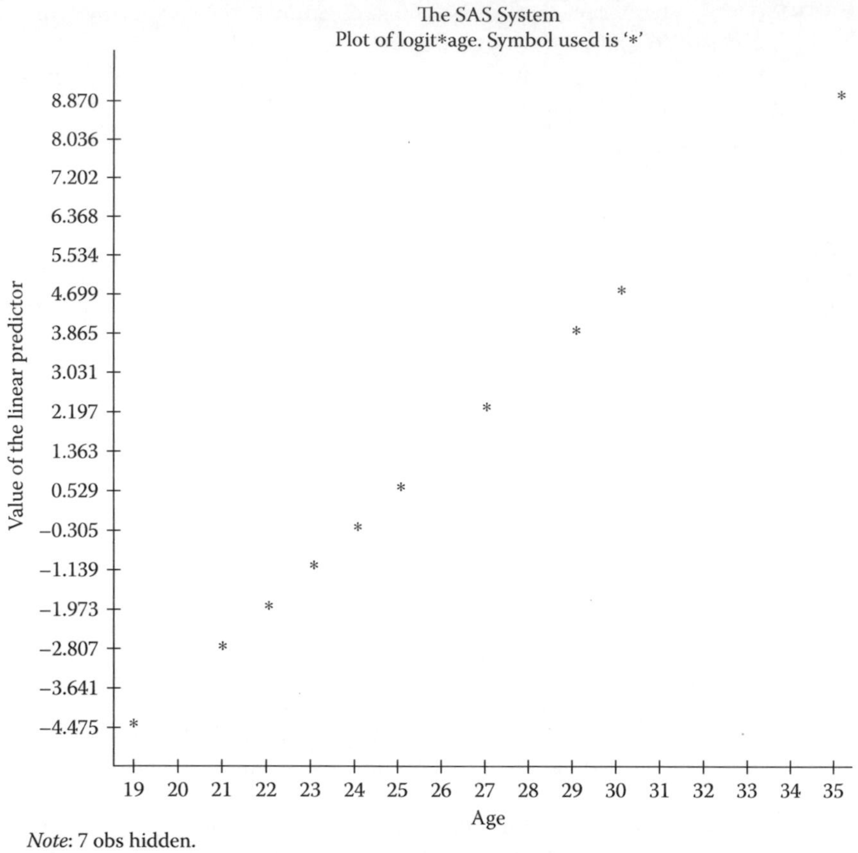

FIGURE 5.4
Plot of logit versus age.

3. In the CTABLE option within the MODEL statement, we used PPROB= option to specify probability cut points for the classification table. We could also specify *prior probabilities* for some event of interest by inserting the PEVENT = option immediately after PPROB = option. For example,

   PEVENT = 0.25 0.50 0.75

   could have been used. The SAS output produces a separate classification table for each of these previous probabilities. If PEVENT = option is not included (as is the case for the above SAS program), PROC LOGISTIC uses the observed sample proportion of event responses as the previous probability.

4. The *receiver operating characteristic (ROC)* curve graphically displays the predictive accuracy of the logistic regression model. In section 12 in Table 5.4 of the SAS output, we encountered the terms *sensitivity* and *specificity* (the reader should take a moment to review these notions). As observed previously, the sensitivity and specificity rates vary according to the probability cut point used for ascertaining whether an observation is an event or nonevent, i.e., as the value of the probability cut point increases from 0 to 1, the sensitivity rate decreases and the specificity rate increases. Obviously, one would like to have high rates of both sensitivity and specificity so that the logistic regression model can accurately predict both events

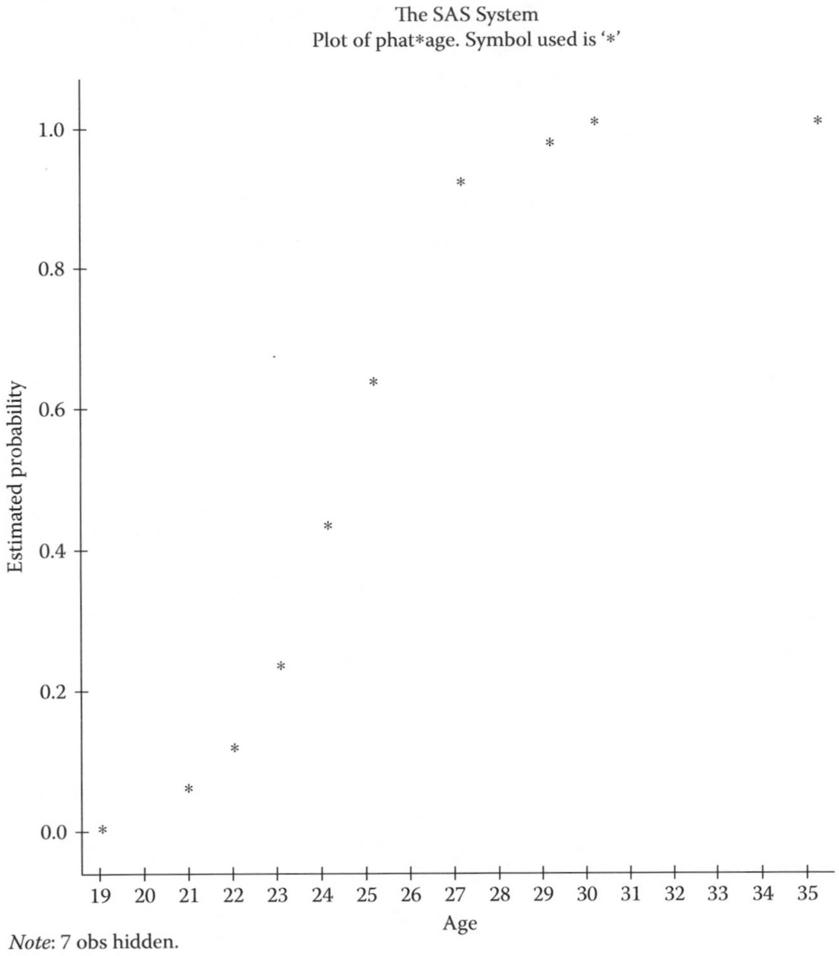

FIGURE 5.5
Plot of $\hat{p}$ versus age.

and nonevents. An ROC curve enables us to track the accuracy of the model. Once an output dataset has been created using the OUTROC= option, the ROC curve graphs sensitivity against 1–specificity (the proportion of nonevent observations that were incorrectly predicted as events), with both of these rates increasing as the estimated probability cut points decrease. If the logistic regression model has high (low) predictive accuracy, than the ROC curve rises quickly (slowly), e.g., for a model with high predictive accuracy, both sensitivity and specificity are high (thus 1–specificity is low) for higher estimated probability cut points so that the ROC curve tends to rise quickly.

To create the output dataset necessary to generate an ROC curve, let us insert

OUTROC=ROCDATA

into the MODEL statement. A print of the ROCDATA dataset is then obtained by inserting the line

PROC PRINT DATA=ROCDATA;

into the program after the MODEL statement. The ROCDATA dataset contains the following variables:

PROB: the estimated probability cut points.

POS: the number of correctly predicted events.

NEG: the number of correctly predicted nonevents.

FALPOS: the number of incorrectly predicted events.

FALNEG: the number of incorrectly predicted nonevents.

SENSIT: sensitivity

1MSPEC: 1–specificity.

A plot of the ROC curve can now be obtained by inserting

```
PROC PLOT DATA = ROCDATA;
PLOT _SENSIT_*_1MSPEC_='*';
```

into the SAS program.

Logistic Regression: The Proportional Odds Model

In "The Logistic Regression Setting" (see page 203), we considered a logistic regression model in which the response variable Y was binary in nature, i.e., $Y = 0$ or 1. Let us now consider a model in which logistic regression is applied to a situation in which Y displays multilevel responses that are ordinal in nature. This new model, which displays cumulative logits, is referred to as the *proportional odds model.*

Let us assume that the response variable Y can take on k possible values that are measured on an ordinal scale. For instance, suppose a group of customers is asked a question concerning the quality of the service in a particular restaurant and one of $k = 4$ possible responses or outcomes is given by each individual (see Table 5.5). Note that these categories are ordered in terms of increasing service quality. Then we can assign the ordered Y values as indicated (e.g., if a person responds "poor", then $Y = 1$, etc.). Next come the category probabilities $p_j = P(Y = j|X)$, $j = 1, 2, 3, 4$ $(=k)$.

Because the categories are ordered in the sequence $j = 1,\ldots, k$, let us express the cumulative probability or the probability that an observation or response falls in the jth or lower category as $F_j = P(Y \le j) = \sum_{r=1}^{j} p_r, j = 1,\ldots,k$. Clearly, each F_j corresponds to a specific dichotomization of Y. In terms of the entries in Table 5.5,

TABLE 5.5

Multilevel Ordinal Responses

Response Category	Ordered Value	Category Probability
Poor	$Y=1$	$p_1 = P(Y=1 \mid X)$
Fair	$Y=2$	$p_2 = P(Y=2 \mid X)$
Good	$Y=3$	$p_3 = P(Y=3 \mid X)$
Very good	$Y=4$	$p_4 = P(Y=4 \mid X)$

$F_1 = P(Y \le 1) = p_1$ (the probability of responding poor);
$F_2 = P(Y \le 2) = p_1 + p_2$ (the probability of responding poor or fair);
$F_3 = P(Y \le 3) = p_1 + p_2 + p_3$ (the probability of responding poor or fair or good);
$F_4 = P(Y \le 4) = 1.$

Note that, if there are k response levels or categories, then we will need only $k - 1$ *cumulative logits* or *log cumulative odds* because $\sum_{j=1}^{k} p_j = 1$. For the preceding set of cumulative probabilities, we need $k - 1 = 3$ cumulative logits:

$$\text{logit } F_1 = \log\left(\frac{F_1}{1-F_1}\right) = \log\left(\frac{p_1}{1-p_1}\right) = \beta_{01} + \beta_{11}X;$$

(log odds of responding poor to fair or good or very good)

$$\text{logit } F_2 = \log\left(\frac{F_2}{1-F_2}\right) = \log\left(\frac{p_1+p_2}{1-p_1-p_2}\right) = \beta_{02} + \beta_{12}X;$$

(log odds of responding poor or fair to good or very good)

$$\text{logit } F_3 = \log\left(\frac{F_3}{1-F_3}\right) = \log\left(\frac{p_1+p_2+p_3}{1-p_1-p_2-p_3}\right) = \beta_{03} + \beta_{13}X.$$

(log odds of responding poor or fair or good to very good).
(Here all log cumulative odds equations are structured in terms of more favorable to less favorable responses.) In general,

$$\text{logit } F_l = \log\left(\frac{F_l}{1-F_l}\right) = \log\left(\frac{\sum_{r=1}^{l} p_r}{1-\sum_{r=1}^{l} p_r}\right) = \beta_{0l} + \beta_{1l}X, l = 1,\ldots,k-1.$$

Let us now make what is called the *proportional odds assumption*: $\beta_{1l} = \beta_1$, $l = 1,\ldots, k-1$, where β_1 depicts the *common slope parameter*. Hence, under this assumption, the *proportional odds model* appears as

$$\text{logit}(F_l) = \beta_{0l} + \beta_1 X, l = 1,\ldots, k-1, \tag{5.16}$$

i.e., Equation 5.16 represents a parallel lines logistic regression model (all regression slopes are hypothesized as being equal) based on the cumulative distribution probabilities of the response levels. Hence, only the logit intercepts can change.

As this proportional odds model implies, the ratio of the log cumulative odds of the event $Y \le l$ or $\log\left(\frac{F_l}{1-F_l}\right)$ is independent of the level j, i.e., the log cumulative odds ratio is constant for all response levels. In fact, to illustrate the proportional odds property implied by Equation 5.16, let us take the difference in logits in Equation 5.16 for different values of X, i.e., at X_s and X_t, $\text{logit}(F_{ls}) = \beta_{0l} + \beta_1 X_s$ and $\text{logit}(F_{lt}) = \beta_{0l} + \beta_1 X_t$, $l = 1,\ldots, k-1$. Then

$$\text{logit}(F_{ls}) - \text{logit}(F_{lt}) = \log\left[\frac{F_{ls}/(1-F_{ls})}{F_{lt}/(1-F_{lt})}\right] = \beta_1(X_s - X_t), l = 1,\ldots,k-1.$$

TABLE 5.6

Categories and Ordered Values for Y

Response Category	Ordered Value
Less hopeful	1
No different	2
More hopeful	3

So for each value of l, the log cumulative odds are proportional to the difference between the explanatory variable values at different data points, where the common factor of proportionality is β_1. Moreover, this proportionality holds for "each value of l," i.e., the influence of X is independent of the response levels at which successive logits are defined.

Example 5.3

Suppose we are interested in assessing the attitude of a local community concerning how hopeful individuals are about their economic prospects for the coming year (the dependent variable Y is *hope*) in relation to their most recent percentage increase in salary (the explanatory variable X is *pct*). Table 5.6 indicates three response categories along with their accompanying ordered values for the response variable Y. Clearly, the categories are ordered in terms of increasing levels of optimism about economic prospects for the coming year. The complete dataset *year* (Exhibit 5.2) follows from a random sample of $n = 30$ respondents. The SAS code for the application of PROC LOGISTIC to the solution of this problem appears in Exhibit 5.2. (It is important to note that, whenever PROC LOGISTIC encounters more than two categories for the response variable Y, it estimates, by default, a cumulative logit or proportional odds model.)

Exhibit 5.2 SAS Code for Proportional Odds Model

```
data year;
  input hope pct;
datalines;
1   2
3   4
3   3.5
1   1
2   3
2   3.33
2   3.5
1   0.5
2   2
1   0.8
3   4.25
1   0
3   4.5
3   3.75
3   5
1   3
1   2.5
```

① The DES option is used to model the probability of *increasing hopefulness* by reversing the order in which the ordered response levels for Y are sorted.

② PROC LOGISTIC automatically fits $k - 1$ cumulative logits whenever it encounters $k > 2$ response categories. Under the DES option, PROC LOGISTIC models (for $k = 3$)
logit $(p_3) = \beta_{01} + \beta_1 X$,
logit $(p_2 + p_3) = \beta_{02} + \beta_1 X$,
i.e., we require the logistic model to predict the probability of being in a *higher hope* category.

```
1    1.5
2    2.25
2    2.4
2    2.5
2    3.7
1    1.75
1    2.4
3    5.1
3    4.7
2    2.9
2    3.2
3    3.7
3    4.1
;
proc logistic data=year des; ①
   model hope=pct; ②
   output out=pred p=prob xbeta=logit; ③
proc print data=pred;
proc plot data=pred;
   plot logit *pct=_level_; ④
   plot prob *pct=_level_; ⑤
data more (rename=(prob=F3)) nodiff (rename=(prob=(F2)); ⑥
   set pred;
   if mod (_n_,2)=1 then output more; ⑦
   else output nodiff;
data pred2; ⑧
  merge more nodiff;
   keep pct prob_3 prob_2 prob_1;
   prob_3=F3; ⑨
   prob_2=F2-F3;
   prob_1=1-F2;
proc plot data=pred2;
  plot prob_3*pct="3"
   prob_2*pct="2" ⑩
   prob_1*pct="1"/overlay;
run;
```

③ The output statement with the P=PROB option is used to create the output dataset *pred* that contains the values of the original X and Y variables along with two new variables _LEVEL_ and PROB, in which the former indicates the value of the response variable being modeled (under the DES option the response level for the first logit given in ② is $Y = 3$ and the response level for the second logit is $Y = 2$ or 3) and the latter yields the probability that each observation has the response level indicated by _LEVEL_, i.e., we obtain an estimate of the probability that $Y = 2$ or 3 at each X_i, $i = 1,\ldots, n$.

The XBETA statistic yields the predicted log odds or logit value for each case or observation corresponding to the ordered value of the variable _LEVEL_ (for $Y = 3$, we get an estimate of the logit value logit (p_3), and for $Y = 2$ or 3, we get an estimate of the logit value logit $(p_2 + p_3)$).

④ We request a plot of the variable LOGIT on the vertical axis against *pct* (X) on the horizontal axis. Under the DES option, we get plots of each of the estimated cumulative logit response functions

$$\text{logit}\left(\hat{p}_3\right) = \hat{\beta}_{01} + \hat{\beta}_1 X,$$

$$\text{logit}\left(\hat{p}_2 + \hat{p}_3\right) = \hat{\beta}_{02} + \hat{\beta}_1 X.$$

Note that the proportional odds assumption forces these lines to be parallel.

⑤ We request a plot of the variable PROB on the vertical axis against *pct* (X) on the horizontal axis. With the DES option, we actually get plots of each of the estimate probabilities $\hat{p}_3$ and $\hat{p}_2+\hat{p}_3$ obtained from the cumulative logit response functions

$$p_3 = \frac{e^{\beta_{01}+\beta_1 X}}{1+e^{\beta_{01}+\beta_1 X}},$$

$$p_2 + p_3 = \frac{e^{\beta_{02}+\beta_1 X}}{1+e^{\beta_{02}+\beta_1 X}},$$

respectively.

⑥, ⑦, ⑧, The output dataset PRED contains the predicted probabilities associated with the two values of the variable _LEVEL_ : for _LEVEL_ = 3, $\hat{p}_3 = P(Y=3|X)$, and for _LEVEL_ = 2,$\hat{p}_2+\hat{p}_3 = P(Y = 2 \text{ or } 3|X)$. The *modulo function*[2] MOD(_n_, 2) = 1 is used to create separate datasets for $\hat{p}_2$ and $\hat{p}_3$. These datasets are renamed and then merged to form the dataset PRED2.

⑨ We compute the individual estimated probabilities prob_1 = $\hat{p}_1$, prob_2 = $\hat{p}_2$, and prob_3 = $\hat{p}_3$ associated with the response levels Y =1, 2, and 3, respectively. These are obtained under the DES option from the cumulative probabilities:

$p_3 = F_3$;
$p_2 + p_3 = F_2$ or $p_2 = F_2 - F_3$; and because
$p_1 + p_2 + p_3 = 1$, $p_1 = 1 - (p_2 + p_3) = 1 - F_2$.

(if the DES option is not used, the appropriate cumulative probabilities are

$p_1 = F_1$;
$p_2 + p_3 = F_2$ or $p_2 = F_2 - F_1$; and
$p_3 = 1 - F_2$).

⑩ We request a plot of the individual estimated probabilities on the vertical axis against *pct* (X) on the horizontal axis.

Let us now examine the output product by the preceding program. All this appears in Table 5.7.

TABLE 5.7

Output for Example 5.3

The SAS System
The LOGISTIC Procedure
Model Information

Dataset	WORK.YEAR
Response variable	hope
Number of response levels	3
Number of observations	30
Model	cumulative logit
Optimization technique	Fisher's scoring

[2] Given that _n_ is a count variable (in this instance, _n_ goes from 1 to 60 because _LEVEL_ renders two values for each data point), the MOD function computes the remainder of the division of _n_ by 2. For instance, if _n_= 1, the remainder is 1 because 1 = 0 · 2 + 1. When _n_= 2, the remainder is 0 because 2 = 2 · 1 + 0. For _n_= 3, the remainder is 1 because 3 = 2 · 1 + 1, etc.

TABLE 5.7 *(continued)*

The SAS System
The LOGISTIC Procedure
①
Response Profile

Ordered Value	Hope	Total Frequency
1	3	10
2	2	10
3	**1**	**10**

Probabilities modeled are cumulated over the lower ordered values.

Model convergence status

Convergence criterion (GCONV = 1E−8) satisfied.

②

Score Test for the Proportional Odds Assumption

Chi-Square	DF	Pr > ChiSq
2.0592	1	0.1513

Model Fit Statistics

Criterion	Intercept Only	Intercept and Covariates
AIC	69.917	32.180
SC	72.719	36.383
−2 Log L	65.917	26.180

③

Testing Global Null Hypothesis: BETA = 0

Test	Chi-Square	DF	Pr > ChiSq
Likelihood ratio	39.7371	1	<0.0001
Score	21.7338	1	<0.0001
Wald	10.3422	1	0.0013

④

Analysis of Maximum-Likelihood Estimates

Parameter	DF	Estimate	Standard Error	Wald x^2	Pr > x^2
Intercept 3	1	−13.8082	4.2834	10.3919	0.0013
Intercept 2	1	−8.9963	2.9812	9.1063	0.0025
pct	1	3.8471	1.1963	10.3422	0.0013

⑤

Odds Ratio Estimates

Effect	Point Estimate	95% Wald Confidence Limits	
pct	46.859	4.493	488.722

(continued)

TABLE 5.7 *(continued)*

The SAS System
The LOGISTIC Procedure

Association of Predicted Probabilities and Observed Responses

Percent concordant	94.3	Somers' *D*	0.907
Percent discordant	3.7	Gamma	0.925
Percent tied	2.0	Tau-a	0.625
Pairs	300	*c*	0.953

⑥

Obs	Hope	pct	_LEVEL_	logit	prob
1 } respondent 1	1	2.00	3	−6.1140	0.00221
2	1	2.00	2	−1.3021	0.21382
3 } respondent 2	3	4.00	3	1.5803	0.82925
4	3	4.00	2	6.3922	0.99833
5	3	3.50	3	−0.3433	0.41502
6	3	3.50	2	4.4686	0.98867
7	1	1.00	3	−9.9611	0.00005
8	1	1.00	2	−5.1492	0.00577
9	2	3.00	3	−2.2668	0.09391
10	2	3.00	2	2.5451	0.92724
11	2	3.33	3	−0.9973	0.26948
12	2	3.33	2	3.8146	0.97843
13	2	3.50	3	−0.3433	0.41502
14	2	3.50	2	4.4686	0.98867
15	1	0.50	3	−11.8847	0.00001
16	1	0.50	2	−7.0728	0.00085
17	2	2.00	3	−6.1140	0.00221
18	2	2.00	2	−1.3021	0.21382
19	1	0.80	3	−10.7305	0.00002
20	1	0.80	2	−5.9186	0.00268
21	3	4.25	3	2.5421	0.92704
22	3	4.25	2	7.3540	0.99936
23	1	0.00	3	−13.8082	0.00000
24	1	0.00	2	−8.9963	0.00012
25	3	4.50	3	3.5039	0.97080
26	3	4.50	2	8.3158	0.99976
27	3	3.75	3	0.6185	0.64988
28	3	3.75	2	5.4304	0.99564
29	3	5.00	3	5.4274	0.99562
30	3	5.00	2	10.2393	0.99996
31	1	3.00	3	−2.2668	0.09391
32	1	3.00	2	2.5451	0.92724
33	1	2.50	3	−4.1904	0.01491
34	1	2.50	2	0.6215	0.65056
35	1	1.50	3	−8.0375	0.00032

TABLE 5.7 *(continued)*

Obs	Hope	pct	_LEVEL_	logit	prob
36	1	1.50	2	−3.2256	0.03812
37	2	2.25	3	−5.1522	0.00575
38	2	2.25	2	−0.3403	0.41574
39	2	2.40	3	−4.5751	0.01020
40	2	2.40	2	0.2368	0.55892
41	2	2.50	3	−4.1904	0.01491
42	2	2.50	2	0.6215	0.65056
43	2	3.70	3	0.4262	0.60496
44	2	3.70	2	5.2381	0.99472
45	1	1.75	3	−7.0757	0.00084
46	1	1.75	2	−2.2638	0.09416
47	1	2.40	3	−4.5751	0.01020
48	1	2.40	2	0.2368	0.55892
49	3	5.10	3	5.8122	0.99702
50	3	5.10	2	10.6241	0.99998
51	3	4.70	3	4.2733	0.98626
52	3	4.70	2	9.08520	0.99989
53	2	2.90	3	−2.65154	0.06589
54	2	2.90	2	2.16036	0.89663
55	2	3.20	3	−1.49740	0.18281
56	2	3.20	2	3.31450	0.96492
57	3	3.70	3	0.42617	0.60496
58	3	3.70	2	5.23807	0.99472
59	3	4.10	3	1.96502	0.87708
60	3	4.10	2	6.77692	0.99886

① We requested a model that sorts the response levels of the dependent variable *hope* in descending order, i.e., we desired to predict the probability of increasing hopefulness. The *response profile* indicates that the ordered values are as required. (It is important to note that, whenever PROC LOGISTIC encounters more than two categories for the dependent variable *Y*, it estimates, by default, a cumulative logit or proportional odds model.)

② The *score test* is a χ^2 test of the proportional odds assumption, i.e., it tests whether or not the ordinal restrictions (Equation 5.16) are valid. In this regard, we test $H_0 : \beta_{1l} = \beta_1$ for all $l = 1, \ldots, k-1$ cumulative logits versus $H_1 : \beta_{1l} \neq \beta_1$ for at least one l. Thus, we hypothesize that there is a common slope parameter β_1 for each of the cumulative logit regression equations instead of $k-1$ distinct slopes. Here, rejecting H_0 rejects the proportional odds assumption, and H_0 will be rejected for a small p-value. Because the p-value = 0.1513, we do not reject H_0; the proportional odds assumption is valid.

③ A global χ^2 statistic that tests $H_0 : \beta_1 = 0$, against $H_0 : \beta_1 \neq 0$ is the likelihood ratio χ^2 X^2 = 39.7371. Because the associated p-value is less than 0.001, we can safely reject H_0 in favor of H_1. This leads us to conclude that the explanatory variable *pct* is important in predicting the response level or category membership.

④ The estimated cumulative logit response functions are

$$\text{logit}\left(\hat{p}_3\right) = -13.8082 + 3.8471X;$$

$$\text{logit}\left(\hat{p}_2 + \hat{p}_3\right) = -8.9963 + 3.8471X.$$

Because $\hat{\beta}_1 = 3.8471$ is positive and statistically significant (the *p*-value associated with the realized Wald χ^2 statistic is 0.0013), it follows that higher levels of *pct* (*X*) are associated with a higher probability of being hopeful.

How should we interpret the two estimated intercepts? The first intercept $\hat{\beta}_{01} = -13.8082$ is the predicted log odds of being in category 3 rather than in either category 1 or category 2 when $X = 0$. Then $e^{\hat{\beta}_{01}} = 0.00000101$ so that the predicted odds of being in category 3 versus being in categories 1 or 2 is 0.00000101 when $X = 0$. In addition, the second intercept $\hat{\beta}_{02} = -8.9963$ is the predicted log odds of being in either category 2 or category 3 rather than category 1 when $X = 0$. Hence, $e^{\hat{\beta}_{02}} = 0.0001$ so that the predicted odds of being in category 2 or 3 relative to category 1 is 0.0001 when $X = 0$.

⑤ For the proportional odds model with the DES option, the odds ratio represents the effect of the explanatory variable *X* on the odds of being in a higher rather than in a lower category. Moreover, the said odds is independent of how the dependent variable is dichotomized. So given $\hat{\beta}_1 = 3.8471$, the estimated odds ratio is $\hat{\Psi} = e^{\hat{\beta}_1} = 46.859$ That is, for each 1% increase in *pct*, the odds of being in a higher category are about 47 times greater than the odds of being in a lower category.

⑥ The output dataset *pred* has $2n = 60$ observations because, with two cumulative logit functions estimated, there are two observations determined for each of the 30 respondents. In this regard, the observation with _LEVEL_= 3 (more hopeful) gives the estimated probability that a respondents economic prospects are *more hopeful*, and the observation with _LEVEL_=2 (no different) yields the estimated probability that a respondents economic prospects are either *more hopeful* or *no different*. For instance, respondent 1 (who has indicated a 2% salary increase) has an estimated probability of 0.00221 of being more hopeful about economic prospects, and this respondent has an estimated probability of 0.21382 of feeling either more hopeful or no different about economic prospects. Note also that respondent 2 (who has a salary increase of 4% or twice that of respondent 1) has an estimated probability of 0.82925 of being more hopeful about economic prospects, and respondent 2 has an estimated probability of 0.99833 of feeling either more hopeful or no different concerning economic prospects.

⑦ Figure 5.6 displays a plot of the two estimated cumulative logit response functions presented in item 4 above. For _LEVEL_= 3,

$$\text{logit}(\hat{p}_3) = -13.8082 + 3.8471X;$$

and for _LEVEL_= 2,

$$\text{logit}(\hat{p}_2 + \hat{p}_3) = -8.9963 + 3.8471X.$$

Note that, under the proportional odds assumption, these equations run parallel to each other because they have the same slope.

⑧ Figure 5.7 displays a plot of the two predicted probability functions (see Equation 5.1) obtained from the estimated cumulative logit response functions given in item 4. That is, for _LEVEL_= 3,

$$\hat{p}_3 = \frac{\exp\{-13.8082 + 3.8471X\}}{1 + \exp\{-13.8082 + 3.8471X\}};$$

and for _LEVEL_= 2,

$$\hat{p}_2 + \hat{p}_3 = \frac{\exp\{-8.9963 + 3.8471X\}}{1 + \exp\{-8.9963 + 3.8471X\}}.$$

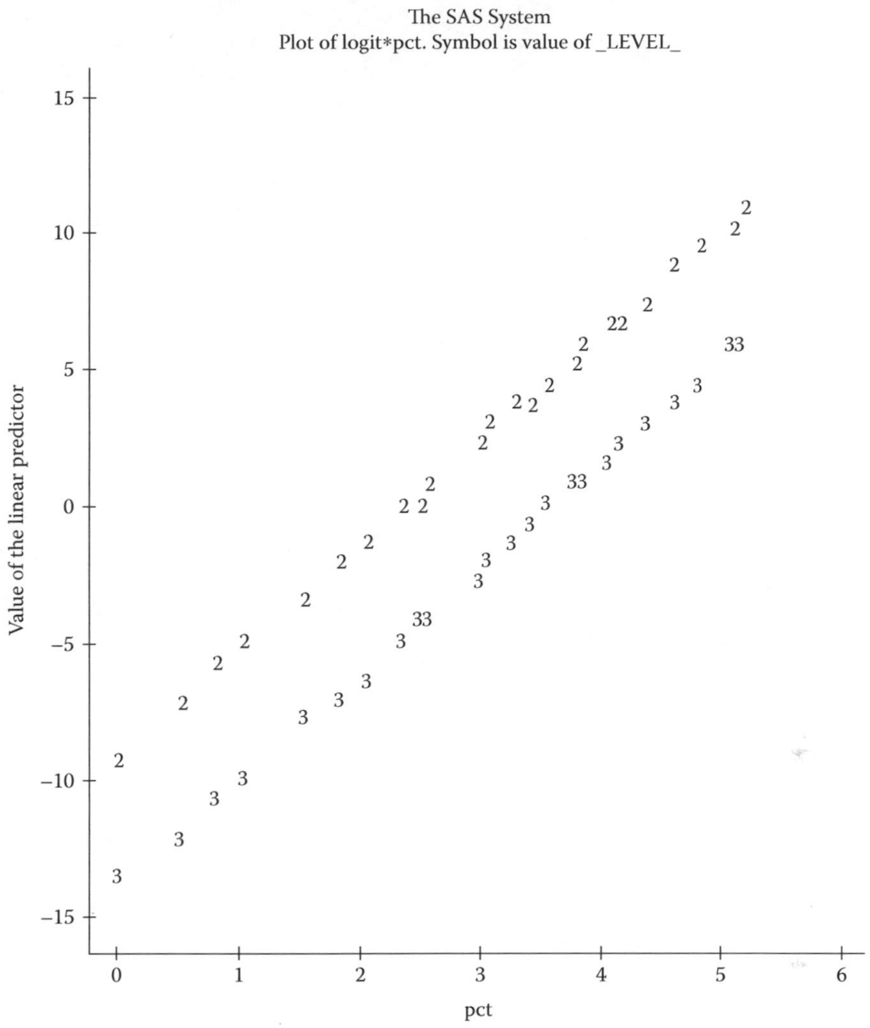

FIGURE 5.6
Plot of estimated logit values versus *X*.

⑨ Figure 5.8 displays a plot of the individual probability functions for the three levels of *Y*:

$$\text{prob\_3} = \hat{p}_3 = \frac{\exp\{-13.8082 + 3.8471X\}}{1 + \exp\{-13.8082 + 3.8471X\}};$$

$$\text{prob\_2} = \hat{p}_2 = \frac{\exp\{-8.9963 + 3.8471X\}}{1 + \exp\{-8.9963 + 3.8471X\}} - \hat{p}_3;$$

$$\text{prob\_1} = \hat{p}_1 = 1 - \hat{p}_2 - \hat{p}_3.$$

It is interesting to note that the set of *X* cut points or thresholds that separate the response categories of less hopeful, no different, and more hopeful can be determined by finding the *X* coordinate

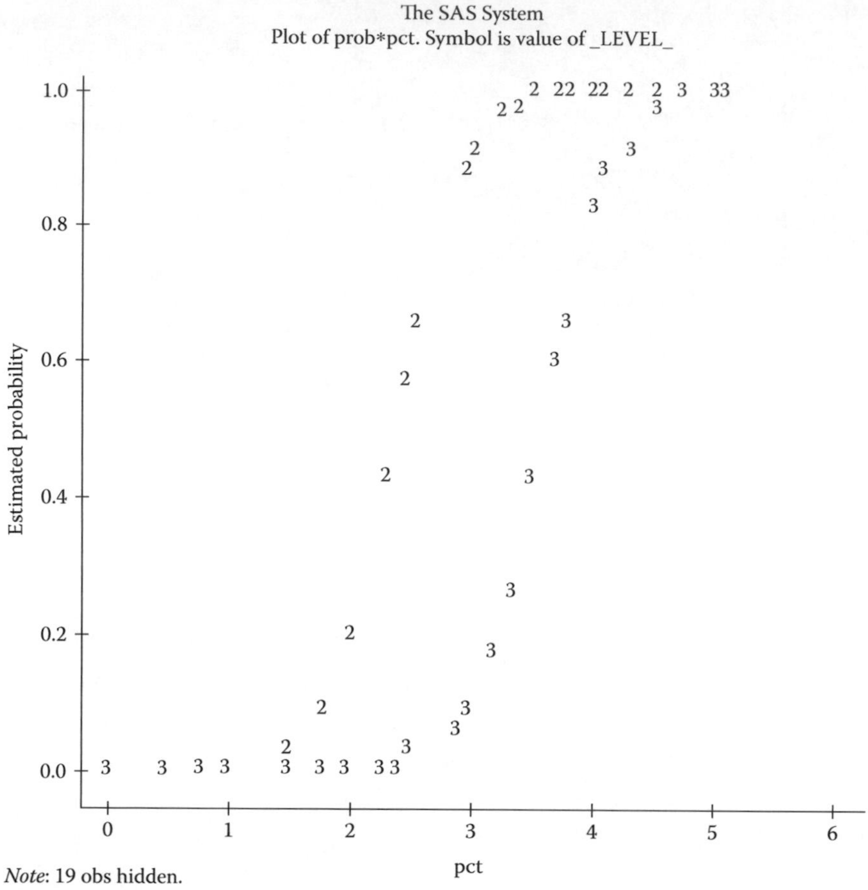

FIGURE 5.7
Plot of estimated probabilities versus X.

of the points A and B in which the probability functions appearing in Figure 5.8 intersect. To determine the X coordinate of point A, we simply set $\hat{p}_1 = \hat{p}_2$ to obtain

$$X_A = \log\left[\frac{1}{e^{\hat{\beta}_{02}} - 2e^{\hat{\beta}_{01}}}\right] \Big/ \hat{\beta}_1.$$

To obtain the X coordinate of point B, we set $\hat{p}_2 = \hat{p}_3$ to find

$$X_B = \log\left[\frac{e^{\hat{\beta}_{02}} - 2e^{\hat{\beta}_{01}}}{e^{\hat{\beta}_{01}+\hat{\beta}_{02}}}\right] \Big/ \hat{\beta}_1.$$

Given the above set of parameter estimates, we readily obtain $X_A = 2.3994$ and $X_B = 3.5283$. So if $pct < 2.40$, the model predicts that the respondent is *less hopeful*. If $2.40 < pct < 3.53$, the model predicts that the respondent's view is *no different*, and if $pct > 3.53$, the model predicts that the respondent is *more hopeful*.

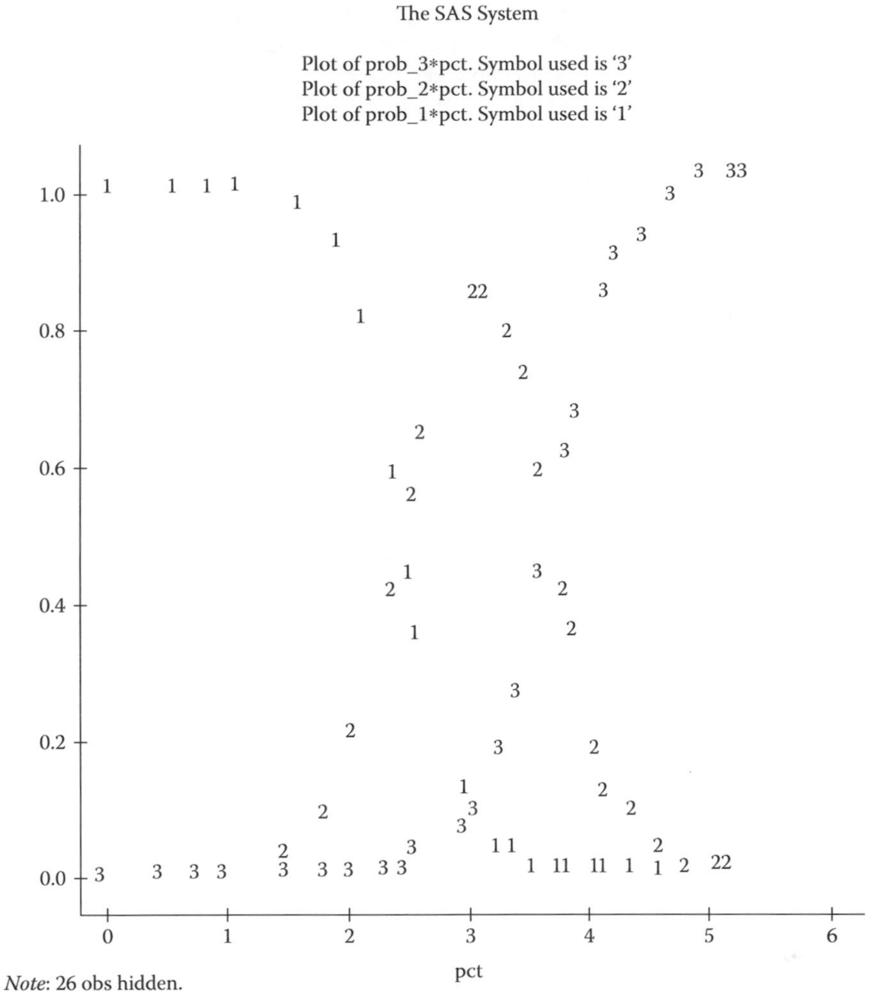

FIGURE 5.8
X cut points A and B.

Logistic Regression: Generalized Logits

If the proportional odds assumption is untenable (the odds ratio changes from category to category), then we can fit a set of logistic regression equations exhibiting individual slope parameters as well as individual intercept parameters. This alternative model reflects the fact that various categories for Y are treated as *unordered* rather than as ordered, i.e., the Y responses are *nominal* in nature.

In this regard, suppose we have k unordered response categories with associated *category probabilities* $p_j = P(Y = j|X)$, $j = 1,\ldots, k$. Then the set of $k - 1$ *generalized* or *multinomial logits* appears as

$$\log\left(\frac{p_j}{p_k}\right) = \beta_{0j} + \beta_{1j}X,\ j = 1,\ldots k-1, \tag{5.17}$$

where each successive category from 1 to $k - 1$ is compared with the kth or last category. Clearly, this system admits $k - 1$ distinct intercepts and slopes. Once individual intercept and slope estimates are obtained, Equation 5.17 can be used to compare any two categories (say categories s and t) by forming the logistic difference $\log(p_s/p_k) - \log(p_t/p_k)$ or

$$\log\left(\frac{p_s}{p_t}\right) = (\beta_{0s} - \beta_{0t}) + (\beta_{1s} - \beta_{1t})X. \tag{5.18}$$

What is important to remember here is that the slope coefficients in Equations 5.17 and 5.18 represent the effect of X on contrasts between *pairs of categories.*

Additionally, specific category probabilities can be determined in the following manner. To keep this analysis manageable, suppose $k = 3$. Then from Equation 5.17,

$$\frac{p_1}{p_3} = \exp\{\beta_{01} + \beta_{11}X\},$$
$$\frac{p_2}{p_3} = \exp\{\beta_{02} + \beta_{12}X\}.$$

Let us form the sum

$$\frac{p_1}{p_3} + \frac{p_2}{p_3} = \exp\{\beta_{01} + \beta_{11}X\} + \exp\{\beta_{02} + \beta_{12}X\},$$

or

$$\begin{aligned} p_1 + p_2 &= p_3\,[\exp\{\beta_{01} + \beta_{11}X\} + \exp\{\beta_{02} + \beta_{12}X\}] \\ &= p_3\,[.]. \end{aligned}$$

Because $\sum_{j=1}^{3} p_j = 1$, set $p_1 + p_2 = 1 - p_3$. Then

$$p_3 = \frac{1}{1 + [\cdot]}.$$

With $p_1 = p_3 \exp\{\beta_{01} + \beta_{11}X\}$, it follows that

$$p_1 = \frac{\exp\{\beta_{01} + \beta_{11}X\}}{1 + [\cdot]};$$

and with $p_2 = p_3 \exp\{\beta_{02} + \beta_{12}X\}$,

$$p_2 = \frac{\exp\{\beta_{02} + \beta_{12}X\}}{1 + [\cdot]}.$$

In general, given that $\sum_{j=1}^{k} p_j = 1$,

$$p_k = \frac{1}{1 + \sum_{j=1}^{k-1} \exp\{\beta_{0j} + \beta_{1j}X\}} \tag{5.19}$$

and

$$\begin{aligned} p_j &= p_k \exp\{\beta_{0j} + \beta_{1j}X\} \\ &= \frac{\exp\{\beta_{0j} + \beta_{1j}X\}}{1 + \sum_{j=1}^{k-1} \exp\{\beta_{0j} + \beta_{1j}X\}},\ j = 1, \ldots, k-1. \end{aligned} \tag{5.20}$$

Example 5.4

Suppose a random sample of $n = 30$ adults were asked to indicate their marital status (Y) and the number of alcoholic beverages they consume per week (X). The nominal response categories for Y are as follows: 1, married; 2, single; and 3, divorced.

To determine the relationship between the Y responses and X, let us use the following SAS code (Exhibit 5.3). Here *marstat* (Y) is the dependent variable and *drinks* (X) is the explanatory variable.

Exhibit 5.3. SAS Code for Generalized Logits Model

```
data marstat;input marstat
drinks; datalines;
1 7
2 21
1 5
1 10
3 14
3 20
3 10
3 12
1 10
2 15
2 25
2 21
3 8
2 14
1 3
2 5
3 8
3 20
1 2
3 5
2 10
3 9
2 25
2 30
1 7
3 9
1 7
3 12
2 15
;
        ①
proc catmod data=marstat;
        direct drinks;  ②
      ③ response logits / out=mar_prob;④
        model marstat=drinks;
proc print data mar_prob;
run;
```

① CATMOD is a general purpose procedure for categorical data analysis. Maximum-likelihood estimation of the multinational logit or generalized logits is the default.

② By default, PROC CATMOD treats all variables as categorical. To indicate that X is a quantitative variable and not categorical, a DIRECT statement must preceed the MODEL statement.

③ The RESPONSE statement specifies the form of the response functions (written in terms of response probabilities) to be modeled. In this instance, the response functions are generalized logits given by Equation 5.17.

④ The SAS output dataset OUT = *mar_prob* contains the values of the X and Y variables, the observed and predicted values of the response functions, their standard errors, and residuals. When generalized logits are selected as the response functions, this dataset also contains observed and predicted values for the cell probabilities, together with their standard errors.

The output generated by this SAS program appears in Table 5.8.

TABLE 5.8

Output for Example 5.4

The SAS System
The CATMOD Procedure

Data Summary

Response	Marstat	Response levels	3
Weight variable	None	Populations	14
Dataset	MARSTAT	Total frequency	30
Frequency missing	0	Observations	30

①

Population Profiles

Sample	Drinks	Sample size
1	2	1
2	3	1
3	5	3
4	7	4
5	8	2
6	9	2
7	10	4
8	12	2
9	14	2
10	15	2
11	20	2
12	21	2
13	25	2
14	30	1

Response Profiles

Response	Marstat
1	1
2	2
3	3

Maximum-Likelihood Analysis

				Parameter Estimates			
Iteration	Sub Iteration	−2 Log Likelihood	Convergence Criterion	1	2	3	4
0	0	65.916737	1.0000	0	0	0	0
1	0	51.431495	0.2198	0.9098	−1.7473	−0.0910	0.1350
2	0	47.742145	0.0717	1.9083	−2.2711	−0.2284	0.1491
3	0	46.728675	0.0212	2.7402	−2.5330	−0.3395	0.1679
4	0	46.623178	0.002258	3.1222	−2.5902	−0.3914	0.1720
5	0	46.621783	0.0000299	3.1733	−2.5935	−0.3983	0.1722
6	0	46.621783	6.1144E-9	3.1741	−2.5935	−0.3985	0.1722

Maximum likelihood computations converged.

TABLE 5.8 *(continued)*

The SAS System
The CATMOD Procedure

②

Maximum-Likelihood Analysis of Variance

Source	DF	χ^2	Pr > χ^2
Intercept	2	8.37	0.0152
Drinks	2	8.46	0.0146
Likelihood ratio	24	28.94	0.2224

③

Analysis of Maximum-Likelihood Estimates

Parameter	Function Number	Estimate	Standard Error	χ^2	Pr > χ^2
Intercept	1	3.1741	1.6528	3.69	0.0548
	2	–2.5935	1.3196	3.86	0.0494
Drinks	1	–0.3985	0.1966	4.11	0.0427
	2	0.1722	0.0864	3.97	0.0462

④

obs	drinks	marstat	_SAMPLE_	_TYPE_	_NUMBER_	_OBS_	_SEOBS_	_PRED_	_SEPRED_	_RESID_
1	2	.	1	FUNCTION	1	.	.	2.37718	1.28743	.
2	2	.	1	FUNCTION	2	.	.	2.24908	1.16148	.
3	2	1	1	PROB	1	1.00000	0.00000	0.90694	0.10792	0.09306
4	2	2	1	PROB	2	0.00000	0.00000	0.00888	0.01365	–0.00888
5	2	3	1	PROB	3	0.00000	0.00000	0.08418	0.09839	–0.08418
6	3	.	2	FUNCTION	1	.	.	1.97873	1.11185	.
7	3	.	2	FUNCTION	2	.	.	–2.07685	1.08411	.
8	3	1	2	PROB	1	1.00000	0.00000	0.86537	0.12835	0.13463
9	3	2	2	PROB	2	0.00000	0.00000	0.14990	0.02002	–0.01499
10	3	3	2	PROB	3	0.00000	0.00000	0.11963	0.11539	–0.11963
11	5	.	3	FUNCTION	1	0.00000	1.41421	1.18183	0.79093	–1.18183
12	5	.	3	FUNCTION	2	0.00000	1.41421	–1.73239	0.93412	1.73239
13	5	1	3	PROB	1	0.33333	0.27217	0.73477	0.15129	–0.40144
14	5	2	3	PROB	2	0.33333	0.27217	0.39860	0.03869	0.29347
15	5	3	3	PROB	3	0.33333	0.27217	0.22537	0.13284	0.10797
16	7	.	4	FUNCTION	1	.	.	0.38493	0.56932	.
17	7	.	4	FUNCTION	2	.	.	–1.38793	0.79346	.
18	7	1	4	PROB	1	1.00000	0.00000	0.54044	0.13604	0.45956
19	7	2	4	PROB	2	0.00000	0.00000	0.09179	0.06445	–0.09179
20	7	3	4	PROB	3	0.00000	0.00000	0.36777	0.12319	–0.36777
21	8	.	5	FUNCTION	1	.	.	–0.01352	0.53791	.

(continued)

TABLE 5.8 *(continued)*

obs	drinks	marstat	_SAMPLE_	_TYPE_	_NUMBER_	_OBS_	_SEOBS_	_PRED_	_SEPRED_	_RESID_
22	8	.	5	FUNCTION	2	.	.	−1.21570	0.72834	.
23	8	1	5	PROB	1	0.00000	0.00000	0.43212	0.12607	−0.43212
24	8	2	5	PROB	2	0.00000	0.00000	0.12987	0.07881	−0.12987
25	8	3	5	PROB	3	1.00000	0.00000	0.43801	0.12077	0.56199
26	9	.	6	FUNCTION	1	.	.	−0.41197	0.57609	.
27	9	.	6	FUNCTION	2	.	.	−1.04347	0.66805	.
28	9	1	6	PROB	1	0.00000	0.00000	0.32878	0.12176	−0.32878
29	9	2	6	PROB	2	0.00000	0.00000	0.17484	0.09246	−0.17484
30	9	3	6	PROB	3	1.00000	0.00000	0.49638	0.12385	0.50362
31	10	.	7	FUNCTION	1	0.69315	1.22474	−0.81042	0.67208	1.50357
32	10	.	7	FUNCTION	2	0.00000	1.41421	−0.87124	0.61403	0.87124
33	10	1	7	PROB	1	0.50000	0.25000	0.23867	0.11797	0.26133
34	10	2	7	PROB	2	0.25000	0.21651	0.22459	0.10362	0.02541
35	10	3	7	PROB	3	0.25000	0.21651	0.53674	0.12724	−0.28674
36	12	.	8	FUNCTION	1	.	.	−1.60733	0.96085	.
37	12	.	8	FUNCTION	2	.	.	−0.52678	0.53227	.
38	12	1	8	PROB	1	0.00000	0.00000	0.11191	0.09352	−0.11191
39	12	2	8	PROB	2	0.00000	0.00000	0.32972	0.11579	−0.32972
40	12	3	8	PROB	3	1.00000	0.00000	0.55837	0.12485	0.44163
41	14	.	9	FUNCTION	1	.	.	−2.40423	1.30536	.
42	14	.	9	FUNCTION	2	−0.00000	1.41421	−0.18233	0.49929	0.18233
43	14	1	9	PROB	1	0.00000	0.00000	0.04696	0.05764	−0.04696
44	14	2	9	PROB	2	0.50000	0.35355	0.43320	0.12114	0.06680
45	14	3	9	PROB	3	0.50000	0.35355	0.51984	0.12176	−0.01984
46	15	.	10	FUNCTION	1	.	.	−2.80268	1.48670	.
47	15	.	10	FUNCTION	2	.	.	−0.01010	0.50466	.
48	15	1	10	PROB	1	0.00000	0.00000	0.02958	0.04222	−0.02958
49	15	2	10	PROB	2	1.00000	0.00000	0.48276	0.12469	0.51724
50	15	3	10	PROB	3	0.00000	0.00000	0.48766	0.12378	−0.48766
51	20	.	11	FUNCTION	1	.	.	−4.79493	2.42926	.
52	20	.	11	FUNCTION	2	.	.	0.85105	0.71099	.
53	20	1	11	PROB	1	0.00000	0.00000	0.00247	0.00600	−0.00247
54	20	2	11	PROB	2	0.00000	0.00000	0.69906	0.14927	−0.69906
55	20	3	11	PROB	3	1.00000	0.00000	0.29847	0.14851	0.70153
56	21	.	12	FUNCTION	1	.	.	−5.19338	2.62133	.
57	21	.	12	FUNCTION	2	.	.	1.02328	0.77489	.
58	21	1	12	PROB	1	0.00000	0.00000	0.00147	0.00386	−0.00147
59	21	2	12	PROB	2	1.00000	0.00000	0.73453	0.15090	0.26547
60	21	3	12	PROB	3	0.00000	0.00000	0.26400	0.15035	−0.26400
61	25	.	13	FUNCTION	1	.	.	−6.78718	3.39481	.
62	25	.	13	FUNCTION	2	.	.	1.71220	1.06279	.
63	25	1	13	PROB	1	0.00000	0.00000	0.00017	0.00060	−0.00017
64	25	2	13	PROB	2	1.00000	0.00000	0.84698	0.13772	0.15302
65	25	3	13	PROB	3	0.00000	0.00000	0.15285	0.13760	−0.15285
66	30	.	14	FUNCTION	1	.	.	−8.77944	4.36814	.

TABLE 5.8 *(continued)*

obs	drinks	marstat	_SAMPLE_	_TYPE_	_NUMBER_	_OBS_	_SEOBS_	_PRED_	_SEPRED_	_RESID_
67	30	.	14	FUNCTION	2	.	.	2.57334	1.45832	.
68	30	1	14	PROB	1	0.00000	0.00000	0.00001	0.00005	–0.00001
69	30	2	14	PROB	2	1.00000	0.00000	0.92912	0.09604	0.07088
70	30	3	14	PROB	3	0.00000	0.00000	0.07087	0.09603	–0.07087

① Before maximum-likelihood estimation, PROC CATMOD groups the observations into a contingency table, i.e., in this instance, the 30 observations are grouped into *population profiles*, with each profile corresponding to one of the values of the explanatory variable *X* (*drinks*). The "Sample Size" column provides a frequency count of the number of different *X* values.

② The maximum-likelihood analysis-of-variance table provides us with a global test of the impact of *drinks* on the response variable *marstat* (*Y*). Here $H_0 : \beta_1 = 0$ is tested against $H_1 : \beta_1 \neq 0$. Because the Wald χ^2 statistic is highly significant (*p*-value = 0.0146), we reject H_0 and conclude that *drinks* has a statistically significant effect on the *Y* response.

An alternative test of the *model* is provided by the likelihood ratio χ^2 statistic under H_0: the model is correct. Because the *p*-value = 0.2224 > 0.05, we cannot reject H_0; the model provides a good fit to the *X* and *Y* values.

③ The parameter estimates for the two generalized logits are

$$\log\left(\frac{\hat{p}_1}{\hat{p}_3}\right) = \hat{\beta}_{01} + \hat{\beta}_{11}X = 3.1741 - 0.3985X;$$

$$\log\left(\frac{\hat{p}_2}{\hat{p}_3}\right) = \hat{\beta}_{02} + \hat{\beta}_{12}X = -2.5935 + 0.1722X.$$

Because the *p*-values on both $\hat{\beta}_{11}$ and $\hat{\beta}_{12}$ are each less than 0.05, these slope coefficients are statistically significant, i.e., *X* has a significant impact on the contrasts between the category 1 and (the reference) category 3 and between category 2 and category 3. More specifically, $e^{\hat{\beta}_{11}} = e^{-0.3985} = 0.6713$ so that each one-unit increase in *X* multiplies the odds of being married versus divorced by 0.6713. Because 100(0.6713 – 1)% = –33%, the odds that a respondent is married relative to being divorced are expected to decrease by about 33% with each unit increase in *X*.

Similarly, $e^{\hat{\beta}_{12}} = e^{0.1722} = 1.1879$. Thus, each one-unit increase in *X* multiplies the odds of being single versus divorced by 1.1879. With 100(1.1879 – 1)% = 19%, the odds that a respondent is single relative to being divorced are expected to increase by about 19% with each unit increase in *X*.

After subtracting the preceding two generalized logits we obtain

$$\log\left(\frac{\hat{p}_1}{\hat{p}_2}\right) = 5.7676 - 0.5707X$$

with $e^{-0.5707} = 0.5651$. Thus, each one-unit increase in *X* multiplies the odds of being married versus single by 0.5651. Given that 100(0.5651 – 1)% = –43%, the odds that a respondent is married relative to being single are expected to decrease by about 43% with each unit increase in *X*.

④ For each value of *X* (*drinks*), there are five predicted values (indicated in the _PRED_ column) along with their associated standard errors (appearing in the _SEPRED_ column). The source of each of these five predicted values appears under the _TYPE_ and _NUMBER_ column headings. That is, "Function 1" refers to the first generalized logit and "Function 2" refers to the second generalized logit. In addition, "PROB 1" represents p_1, "PROB 2" depicts p_2, and "PROB 3" refers to p_3.

For instance, at $X = 2$ (see Equations 5.17, 5.19, and 5.20),

TYPE	_NUMBER_	_PRED_
FUNCTION	1	$\log\left(\frac{\hat{p}_1}{\hat{p}_3}\right) = 3.1741 - 0.3985(2) = 2.37718$
FUNCTION	2	$\log\left(\frac{\hat{p}_2}{\hat{p}_3}\right) = -2.5935 + 0.1722(2) = -2.24908$
PROB	1	$\hat{p}_1 = \frac{\exp\{3.1744 - 0.3985(2)\}}{1 + [\exp\{3.1749 - 0.3985(2)\} + \exp\{-2.5935 + 0.1722(2)\}]} = 0.90694$
PROB	2	$\hat{p}_2 = \frac{\exp\{-2.5935 + 0.1722(2)\}}{1 + [\exp\{3.1741 - 0.3985(2)\} + \exp\{-2.5935 + 0.1722(2)\}]} = 0.00888$
PROB	3	$\hat{p}_3 = \frac{1}{1 + [\exp\{3.1749 - 0.3985(2)\} + \exp\{-2.5935 + 0.1722(2)\}]} = 0.08418$

Poisson Regression

Suppose we are faced with a situation in which the dependent or response variable Y is restricted to a set of nonnegative integer values 0,1,2,..., i.e., Y is a count variable of some sort (such as the number of arrivals at a checkout counter in a supermarket during a one-hour period, the number of defects per 10-mile strip of new highway after a one-month period, the number of telephone calls arriving at a hotel reservation desk in a five-minute period, etc.). For this type of response variable, a Possion regression model is appropriate.

If Y follows a Poisson distribution, then its probably mass function has the form

$$f(Y;\mu) = \frac{\mu^Y e^{-\mu}}{Y!}, \quad Y = 0,1,2,\ldots; \ \mu > 0, \tag{5.21}$$

with $E(Y) = V(Y) = \mu$. Moreover, if Y has a Poisson distribution conditional on some explanatory variable X, then we must determine the relationship of the mean of the Y count to the behavior of X. As we shall now see, this can easily be accomplished using the SAS System's GENMOD procedure.

The GENMOD procedure fits what are called generalized linear models (which include classical linear models, logistic and probit models involving binary data, log-linear models for multinomial data, and Poisson models for count or rate of incidence data). The principal components of a generalized linear model include the following:

a. A set of response or dependent variables Y_i that are independent and that follow a probability distribution taken from an exponential family of distributions[3]
b. A *linear predictor* $\eta_i = \beta_0 + \beta_1 X_i$
c. A monotonic differentiable *link function* that relates the expected value of Y_i, $E(Y_i) = \mu_i$, to the linear predictor X_i or $g(\mu_i) = \beta_0 + \beta_1 X_i$, $i = 1,\ldots, n$

For instance, for the strong classical linear regression model, Y_i's probability distribution is normal and the link function is the identity $g(\mu_i) = \mu_i$, and, for logistic regression, the probability distribution is binominal and the link function is the logit $g(\mu_i) = \log\left(\frac{\mu_i}{1-\mu_i}\right)$. For Poisson regression, the probability distribution is Poisson, and the link function is logarithmic or $g(\mu_i) = \log(\mu_i)$. Using this log link obviously guarantees that μ_i cannot be negative.

More specifically, let us formulate the *Poisson regression model* as

$$Y_i = E(Y_i) + \varepsilon_i,\ i = 1,\ldots, n,$$

3 The response variable Y_i is a member of the exponential family of probability distributions (which includes, among others, the normal, binomial, and Poisson distributions) if the density function (for Y_i continuous) or the probability mass function (for Y_i discrete) has the general form

$$f(Y_i;\theta_i,\phi) = \exp\left\{\frac{Y_i\theta_i - b(\theta_i)}{a(\phi)} + c(Y_i,\phi)\right\}, \tag{5.23}$$

where *a*, *b*, and *c* are known (stylized) functions chosen to determine a particular probability distribution and

$$\mu_i = E(Y_i) = \frac{db(\theta_i)}{d\theta_i}, V(Y_i) = \frac{d^2b(\theta_i)}{d\theta_i^2}a(\phi) = \frac{d\mu_i}{d\theta_i}a(\phi).$$

If we set $d\mu_i/d\theta_i = V(\mu_i)$, then $V(Y_i) = V(\mu_i)a(\varnothing)$. So to obtain the Poisson specialization of Equation 5.23, let us rewrite Equation 5.21 as

$$f(Y;\mu) = \frac{\mu^Y e^{-\mu}}{Y!} = \exp[Y\ln(\mu) - \mu - \ln(Y!)]. \tag{5.21.1}$$

Set $\theta_i = \ln(\mu)$ so that $\mu = e^{\theta_i}$; $b(\theta_i) = \mu$; $a(\varnothing) = 1$ and $c\,(Y_i, \varnothing) = -\ln(Y!)$. Then, as defined above,

$$E(Y) = \frac{db(\theta_i)}{d\theta_i} = \frac{db(\theta_i)}{d\mu}\frac{d\mu}{d\theta_i} = (1).e^{\theta_i} = (1).e^{\ln(\mu)} = \mu;$$

and

$$V(Y) = \frac{d\mu}{d\theta_i}a(\phi) = e^{\ln(\mu)}.(1) = \mu.$$

Note that, for a fixed $\varnothing$, Equation 5.23, and thus Equation 5.21.1, is a one-parameter family of distributions, in which the single parameter is the mean μ.

where ε_i is a random error term. With $E(Y_i) = \mu_i$ and g a link function, we have $g(\mu_i) = \eta_i = \beta_0 + \beta_1 X_i$. Clearly, the relationship between the mean of Y_i, μ_i, and the linear predictor X_i is determined as

$$\mu_i = g^{-1}(\eta_i) = g^{-1}(\beta_0 + \beta_1 X_i) \tag{5.24}$$

(g^{-1} exists because g is monotonic). So if $g(\mu_i) = \log(\mu_i) = \eta_i = \beta_0 + \beta_1 X_i$, then μ_i and η_i are related as

$$\mu_i = g^{-1}(\eta_i) = e^{\beta_0 + \beta_1 X_i}. \tag{5.24.1}$$

To estimate the parameters β_0 and β_1 of the linear response η_i, PROC GENMOD uses the method of maximum likelihood. For (X_i, Y_i), $i = 1,\ldots, n$, a random sample drawn from the variables X and Y, the likelihood function is

$$\mathcal{L}(\beta_0, \beta_2; Y_1, \ldots, Y_n, n) = \prod_{i=1}^{n} f(Y_i; \mu_i) = \prod_{i=1}^{n} \frac{\mu_i^{Y_i} e^{-\mu_i}}{Y!} = \frac{\prod_{i=1}^{n} \mu_i^{Y_i} e^{-\sum_{i=1}^{n} \mu_i}}{\prod_{i=1}^{n} Y_i!}$$

or, in log-likelihood form,

$$\log \mathcal{L} = \sum_{i=1}^{n} Y_i \log(\mu_i) - \sum_{i=1}^{n} \mu_i - \sum_{i=1}^{n} \log(Y_i!),$$

where, from Equation 5.24, $\mu_i = g^{-1}(\eta_i) = g^{-1}(\beta_0 + \beta_1 X_i)$. For $g(\mu_i)$ the log-link function, $\log(\mu_i) = \eta_i = \beta_0 + \beta_1 X_i$, and thus the log-likelihood function becomes

$$\log \mathcal{L} = \sum_{i=1}^{n} Y_i \log(\beta_0 + \beta_1 X_i) - \sum_{i=1}^{n} e^{\beta_0 + \beta_1 X_i} - \sum_{i=1}^{n} \log(Y_i!).$$

To maximize the log-likelihood function, we need to solve for the values of β_0 and β_1 which satisfy the necessary conditions $\partial \log \mathcal{L} / \partial \beta_0 = 0$, $\partial \log \mathcal{L} / \partial \beta_1 = 0$. However, because this simultaneous equation system is highly nonlinear and does not admit a closed form solution, PROC GENMOD uses an iterative technique to obtain the parameter estimates $\hat{\beta}_0$ and $\hat{\beta}_1$. Then $\hat{Y}_i = g^{-1}(\hat{\beta}_0 + \hat{\beta}_1 X_i) = e^{\hat{\beta}_0 + \hat{\beta}_1 X_i}$.

One important consideration regarding any set of Poisson regression results is the potential effect of what is termed *overdispersion* on the estimated standard errors, χ^2 values, and p-values associated with $\hat{\beta}_0$ and $\hat{\beta}_1$. This phenomenon, which is reflective of a poor fit to the data, can occur because the basic model is incorrectly specified or the observations lack independence. For instance, it may be the case that regressing $\log(\mu)$ on X alone and omitting certain key explanatory variables can result in a residual correlation among the data points which, in turn, would cause $V(Y)$ to exceed $\mu = E(Y)$. Such overdispersion leads to the underestimation of the standard errors of $\hat{\beta}_0$ and $\hat{\beta}_1$ and thus to an overestimation of (Wald) χ^2 statistics. In this regard, although the maximum-likelihood estimates of β_0 and β_1 are unbiased, they are certainly inefficient.

The detection of any potential overdispersion is straightforward: if the *estimate of model dispersion*, as determined by Pearson's χ^2 divided by degrees of freedom, greatly exceeds unity (respectively, falls short of unity), then the data are overdispersed (respectively, underdispersed).

To address the issue of overdispersion, let us respecify the variance of the Poisson random variable Y as $V(Y) = \o\mu$, where $\o$, a *multiplicative dispersion factor (parameter)*, is directly estimated as $\hat{\phi}$ = Pearson goodness-of-fit χ^2 divided by its degrees of freedom $(n - 2)$. Then all of the individual regression coefficient χ^2 are scaled or divided by $\hat{\phi}$ (or, equivalently, the individual regression coefficient standard errors are multiplied by $\sqrt{\hat{\phi}}$).

We now turn to the following example problem to see exactly how PROC GENMOD is applied to a set of data involving a Poisson response (count) variable Y and an explanatory variable X.

Example 5.5

Suppose a count variable Y represents the number of typographic errors per 100 pages of printed text and the explanatory variable X depicts months of experience at keyboard typesetting for a random sample of 19 typesetters. Exhibit 5.4 describes the SAS code required to perform a Poisson regression of Y on X.

Exhibit 5.4 SAS Code for Poisson Regression

```
data=typo;
   input errors months;
datalines;
39  11
44  9
37  13
36  14
29  19
24  24
49  4
46  6
50  4
46  7
53  3
40  10
30  18
21  28
18  30
29  18
35  15
45  9
26  23
;
          ①
proc genmod data=typo;
                 ②                    ③      ④         ⑤
      model errors=months/dist=poisson link=log pscale;
                        ⑥
      output out=resid pred=pred reschi=reschi stdreschi=stdreschi
xbeta=xbeta stdxbeta=stdxbeta;
data typo1; set resid;⑦
proc print data=typo1;
```

① We invoke PROC GENMOD to perform a generalized linear regression.

② The model is specified as regressing errors (Y) on months (X).

③ Y is characterized as following a Poisson probability distribution with mean $E(Y) = \mu$.

④ The link function is specified as logarithmic. Hence, the Poisson mean μ is related to the linear predictor by $\log(\mu) = \beta_0 + \beta_1 X$.

```
proc plot data=typo1;
       plot pred*months='*';⑧

       plot stdreschi*months='*';
run;
```

⑤ The PSCALE option (involving "P" for Pearson) corrects the standard errors and (Wald) χ^2 for overdispersion, in which the scale factor (parameter) ø is *indirectly* estimated as $\hat{\phi} =$ the square root of the Pearson χ^2 divided by degrees of freedom. More specifically, in the estimation process, the scale parameter ø is set at unity. Once the intercept and slope parameter estimates are obtained, $\hat{\phi}$ is then used to adjust the standard errors and (Wald) χ^2 in the manner described previously.

⑥ We create an OUTPUT dataset named RESID that contains the following statistics:

PRED:	the predicted value of the mean;
RESCHI:	Pearson residual (the square root of the *i*th contribution to the Pearson χ^2 statistic);
STDRESCHI:	standardized Pearson residual (these residuals have unit asymptotic variance);
XBETA:	the value of the linear predictor $\hat{\beta}_0 + \hat{\beta}_1 X$;
STDXBETA:	standard error of the linear predictor.

⑦ We create a new dataset named *TYPO1* that has the same statistics as the OUTPUT dataset *RESID*.

⑧ We request a plot of the predicted mean and the standardized Pearson residual against the explanatory variable *months*.

The output generated by this program is presented in Table 5.9.

TABLE 5.9

Output for Example 5.5

The SAS System
The GENMOD Procedure

Model Information

Dataset	WORK.TYPO
Distribution	Poisson
Link function	Log
Dependent variable	Errors
Observations used	19

①

Criteria for Assessing Goodness of Fit

Criterion	DF	Value	Value/DF
Deviance	17	0.6428	0.0378
Scaled deviance	17	16.9764	0.9986
Pearson χ^2	17	0.6437	0.0379
Scaled Pearson χ^2	17	17.0000	1.0000
Log likelihood		48626.0880	

Algorithm converged.

TABLE 5.9 *(continued)*

②

Analysis of Parameter Estimates

Parameter	DF	Estimate	Standard error	Wald 95% confidence limits		χ^2	Pr > χ^2
Intercept	1	4.0809	0.0140	4.0533	4.1084	84514.2	<.0001
Months	1	–0.0373	0.0010	–0.0393	–0.0353	1351.73	<.0001
Scale	0	0.1946	0.0000	0.1946	0.1946		

Note: The scale parameter was estimated by the square root of Pearson's χ^2/DF.

③

The SAS System

Obs	Errors	Months	($e^{\hat{\eta}_i}$) pred	reschi	stdreschi	($\hat{\eta}_i$) xbeta	($S_{\hat{\eta}_i}$) stdxbeta
1	39	11	39.2743	–0.04378	–0.23167	3.67057	0.007413
2	44	9	42.3161	0.25886	1.37913	3.74517	0.007890
3	37	13	36.4512	0.09089	0.48020	3.59598	0.007474
4	36	14	35.1167	0.14905	0.78802	3.55868	0.007708
5	29	19	29.1422	–0.02634	–0.14135	3.37219	0.010394
6	24	24	24.1841	–0.03744	–0.20662	3.18570	0.014426
7	49	4	50.9915	–0.27889	–1.56094	3.93166	0.010796
8	46	6	47.3261	–0.19277	–1.05055	3.85706	0.009416
9	50	4	50.9915	–0.13885	–0.77713	3.93166	0.010796
10	46	7	45.5935	0.06021	0.32502	3.81976	0.008821
11	53	3	52.9293	0.00972	0.05540	3.96896	0.011558
12	40	10	40.7669	–0.12011	–0.63730	3.70787	0.007588
13	30	18	30.2497	–0.04539	–0.24257	3.40948	0.009705
14	21	28	20.8323	0.03675	0.20841	3.03650	0.018034
15	18	30	19.3348	–0.30356	–1.74664	2.96191	0.019903
16	29	18	30.2497	–0.22721	–1.21423	3.40948	0.009705
17	35	15	33.8311	0.20097	1.06418	3.52138	0.008064
18	45	9	42.3161	0.41258	2.19814	3.74517	0.007890
19	26	23	25.1032	0.17900	0.98169	3.22299	0.013564

① Our assessment of goodness-of-fit can be made by comparing the scaled Pearson χ^2 value with the appropriate fractile of the limiting χ^2 distribution with $n - 2 = 17$ degrees of freedom. Here we test H_0 : the model is correct. Because the p-value (the probability that the scaled Pearson χ^2 value of 17 is larger than some asymptotic χ^2 fractile with 17 degrees of freedom) is in excess of 0.45, we cannot reject H_0 ; we thus conclude that the estimated model fits the data well.

② The direct estimate of the overdispersion parameter ϕ is
$\hat{\phi} = \left[(\text{Pearson chi-square})/(\text{degrees of freedom})\right]^{1/2} = (0.0379)^{1/2} = 0.1946.$

The estimated equation of best fit is $\log(\hat{\mu}) = \hat{\beta}_0 + \hat{\beta}_1 X = 4.0809 - 0.0373X$. Clearly, both the estimated intercept and slope are highly significant because the p-value associated with each (scaled) Wald χ^2 value is below 0.0001. Given $\hat{\beta}_1 = -0.0373$, let us determine $100\left(e^{\hat{\beta}_1} - 1\right)\% = 100\left(e^{-0.0373} - 1\right)\% = -3.66\%$. Hence, the expected number of typographic errors per 100 pages of printed text decreases by 3.66% with each additional month of experience.

Moreover, the (scaled) 95% Wald confidence limits for β_1 are provided by $\hat{\beta}_1 \pm z_{\alpha/2} S_{\hat{\beta}_1}$ ($S_{\hat{\beta}}$ is the scaled standard error of $\hat{\beta}_1$) or −0.0373±1.96 (0.001) or (−0.0393, −0.0353).

③ PRED represents the predicted values of the mean μ_i, and XBETA depicts the value of the linear predictor $\eta_i = \beta_0 + \beta_1 X_i$ at each X_i value. That is, for $X = X_1 = 11$, $\log(\hat{\mu}_1) = 4.0809 - 0.0373(11) = 3.6706 = \hat{\eta}_1$ and thus $\hat{\mu}_1 = e^{\hat{\eta}_1} = e^{3.6706} = 39.2743$. A graph of the entire set of predicted means versus months (X) appears in Figure 5.9.

Additionally, an approximate 100(1−α)% confidence interval for μ_i at X_i is provided by $\exp\left\{\hat{\eta}_i \pm z_{\frac{\alpha}{2}} S_{\hat{\eta}_i}\right\}$, where $S_{\hat{\eta}_i}$ (see STDXBETA) is the estimated standard error of the linear predictor $\eta_i = \hat{\beta}_0 + \hat{\beta}_1 X$. So for $X = X_1 = 11$ and $1-\alpha = 0.95$, exp{3.6706±1.96(0.0074)}→ ($e^{3.6567}$, $e^{3.6851}$) or (38.7101, 39.8491).

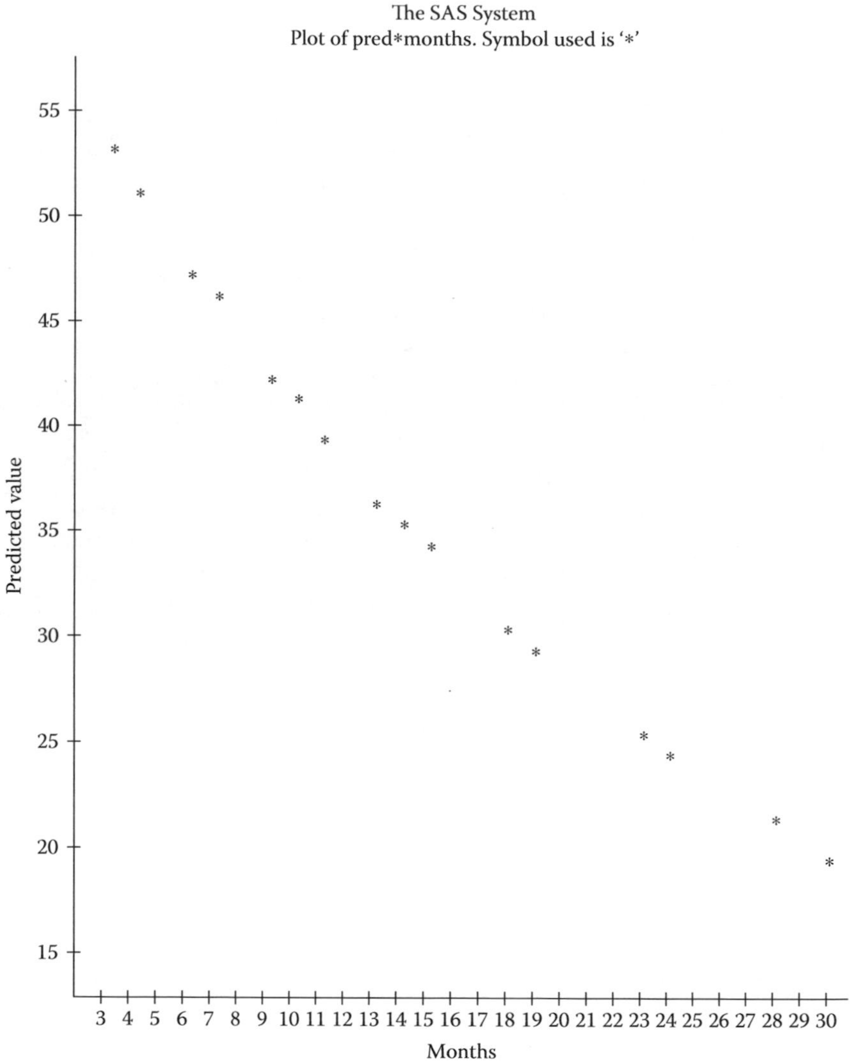

FIGURE 5.9
Plot of predicted means versus X.

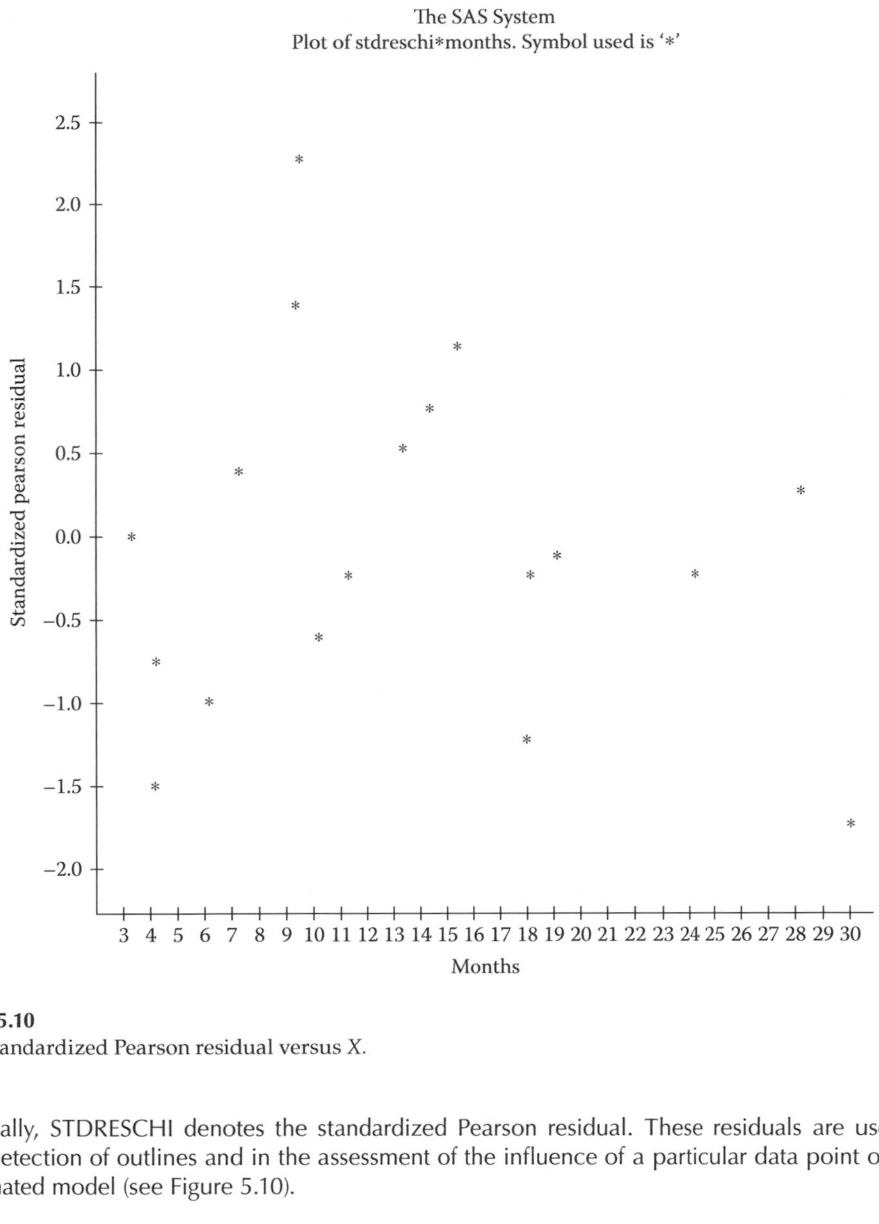

FIGURE 5.10
Plot of standardized Pearson residual versus X.

Finally, STDRESCHI denotes the standardized Pearson residual. These residuals are used in the detection of outlines and in the assessment of the influence of a particular data point on the estimated model (see Figure 5.10).

Appendix A: Maximum-Likelihood Estimation of the Parameters of a Logistic Regression Function

When the dependent variable Y_i is dichotomous with probability mass function appearing in Table 5.3, then obviously Y_i is a Bernoulli random variable with mean given by Equation 5.1 or

$$p_i = E(Y_i|X_i) = \frac{\exp(\beta_0 + \beta_1 X_i)}{1 + \exp(\beta_0 + \beta_1 X_i)}. \tag{5.A.1}$$

Because each sample observation Y_i follows a Bernoulli probability distribution, its associated probability mass function has the form

$$f^i(Y_i) = p_i^{Y_i}(1-p_i)^{1-Y_i},\ i = 1,\ldots,n, \tag{5.A.2}$$

where $Y_i = 0$ or 1 for all i, and because Y_i are independent, their joint probability mass function can be written as

$$g(Y_1,\ldots,Y_n) = \prod_{i=1}^{n} f^i(Y_i) = \prod_{i=1}^{n} p_i^{Y_i}(1-p_i)^{1-Y_i} \tag{5.A.3}$$

or

$$\ln g(Y_1,\ldots,Y_n) = \sum_{i=1}^{n} Y_i \ln\left(\frac{p_i}{1-p_i}\right) + \sum_{i=1}^{n} \ln(1-p_i). \tag{5.A.3.1}$$

Because the logit function in Equation 5.3 has the form

$$p_i' = \ln\left(\frac{p_i}{1-p_i}\right) = \beta_0 + \beta_1 X_i$$

and, from Equation 5.A.1, $1 - p_i = (1 + \exp(\beta_0 + \beta_1 X_i))^{-1}$, it follows that Equation 5.A.3.1 can be rewritten as

$$\ln g(Y_1,\ldots,Y_n) = \sum_{i=1}^{n} Y_i(\beta_0 + \beta_1 X_i) - \sum_{i=1}^{n} \ln(1+\exp(\beta_0 + \beta_1 X_i)). \tag{5.A.5}$$

Given the n sample data points (X_i, Y_i), $i = 1,\ldots, n$, Equation 5.A.5 can be viewed as a function of the parameters β_0 and β_1, i.e., it can be taken to represent the log-likelihood function for the given sample and consequently can be expressed as

$$\begin{aligned}\ln \mathcal{L}(\beta_0,\beta_1;Y_1,\ldots,Y_n,X_1,\ldots,X_n) &= \sum_{i=1}^{n} Y_i(\beta_0 + \beta_1 X_i) \\ &-\sum_{i=1}^{n} \ln(1+\exp(\beta_0 + \beta_1 X_i)).\end{aligned} \tag{5.A.5.1}$$

Maximum-likelihood estimators of β_0 and β_1 are those parameters values that maximize the log-likelihood function or the β_0 and β_1 values (denoted $\tilde{\beta}_0$ and $\tilde{\beta}_1$, respectively) for which

$$\begin{aligned}&\text{(a)}\quad \frac{\partial \ln \mathcal{L}(\tilde{\beta}_0,\tilde{\beta}_1)}{\partial \beta_0} = \sum_{i=1}^{n} Y_i - \sum_{i=1}^{n} \frac{\exp(\tilde{\beta}_0 + \tilde{\beta}_1 X_i)}{1+\exp(\tilde{\beta}_0 + \tilde{\beta}_1 X_i)} = 0 \\ &\text{(b)}\quad \frac{\partial \ln \mathcal{L}(\tilde{\beta}_0,\tilde{\beta}_1)}{\partial \beta_1} = \sum_{i=1}^{n} X_i Y_i - \sum_{i=1}^{n} \frac{X_i \exp(\tilde{\beta}_0 + \tilde{\beta}_1 X_i)}{1+\exp(\tilde{\beta}_0 + \tilde{\beta}_1 X_i)} = 0.\end{aligned} \tag{5.A.6}$$

Because Equation 5.A.6 is highly nonlinear in the unknowns $\tilde{\beta}_0$ and $\tilde{\beta}_1$, special numerical methods need to be applied to obtain a solution to this system. This is the role of the SAS System procedure *LOGISTIC*. This procedure involves the application of iterative numerical routines that are designed to home in on the parameters estimates via a set of successive approximations using an initial starting point.

Convergence to a stable solution point, provided one exists, occurs when the distance between successive sets of coefficient estimates is found to be less than some prespecified tolerance level.

Exercises

5-1. If we expect five times as many occurrences as nonoccurrences of an event *A*, then what are the odds of this event? If we expect one-fifth as many occurrences as non-occurrences of event *A*, then what are the odds of *A*?

5-2. If the odds are 2:1 in favor of event *A*, determine *P* (*A*). If *P* (*A*)=2/3, what are the odds in favor of the occurrence of *A*?

5-3. If *Y* is a dichotomous random variable and $P(Y_i = 1) = \frac{1}{6}$, find the odds O_i that $Y_i = 1$. If$O_i = \frac{1}{3}$, find $p_i = P(Y_i = 1)$.

5-4. If the probability of event *A* is 0.60, find the odds of *A* or O_A.

5-5. Suppose event *A* is four times as likely to occur as event *B* when *P(B)*=0.80. Find *P(A)*.

5-6. Suppose the estimated logit function is $\hat{p}'(X) = \hat{\beta}_0 + \hat{\beta}_1 X = -2.6103 + 0.5122X$ for continuous X. Find (a) the percentage change in odds per unit increase in X, (b) the increase in the odds that $Y = 1$ when X increases by three units, (c) the estimated logistic regression function, and (d) the probability that $Y = 1$ at $X = 15$.

5-7. Let the estimated logit function be $\hat{p}'(X) = 1.9386 + 0.2131X$ for dichotomous X. What are the odds that $Y = 1$ when $X = 1$ relative to when $X = 0$?

5-8. Given the following accident dataset involving single vehicles (Table E.5-8), use PROC LOGISTIC to explore the relationship between the variables *X* (estimated speed at impact in miles per hour) and *Y* ($Y = 1$ if at least one fatality; $Y = 0$ if no fatalities). Fully interpret your results. (Hint: let Exhibit 5.1 serve as your guide.)

5-9. Table E.5-9 provides data on home ownership *Y* ($Y = 1$ if yes; $Y = 0$ if no) and household gross income *X* (measured in thousands of current dollars). Use PROC LOGISTIC to perform a logistic regression study and fully interpret the results. (Hint: use Exhibit 5.1 as your guide.)

5-10. Given the dichotomization of the variable *Y* appearing in Table E.5-10, find $F_j = P(Y \le j)$ $j = 1, \ldots, 6$. Interpret your results. How many cumulative logits will be defined? Find a general expression for logit F_4.

5-11. Given Tables E.5-11.a and E.5-11.b involving the variables *Y* (ordered levels of earning power relative to one's parents) and *X* (years of college), fit a proportional odds model and fully interpret your results. (Hint: let Exhibit 5.2 guide your efforts.)

TABLE E.5-8

Fatalities (Y) and Speed (X)

Y	X
1	60
0	40
0	61
1	65
0	39
1	71
0	52
0	60
0	55
1	75
1	70
1	90
1	80
0	49
1	60
0	51
0	57
1	69
1	64
0	46
0	30
0	58
1	67
0	63
1	82

TABLE E.5-9

Home Ownership (Y) and Income (X)

Y	X
0	48.0
1	61.3
0	49.4
1	53.4
0	57.9
0	63.0
1	51.8
0	50.8
1	56.0
1	62.4
1	48.9
0	50.1
1	59.5
0	48.0
1	53.0
1	65.0
1	61.7
0	49.2
0	50.1
1	63.8

TABLE E.5-10

Dichotomization of Y

Response Category	Ordered Value	Probability
Terrible	$Y = 1$	0.10
Poor	$Y = 2$	0.15
Fair	$Y = 3$	0.20
Good	$Y = 4$	0.25
Very good	$Y = 5$	0.20
Excellent	$Y = 6$	0.10

TABLE E.5-11.A

Earning Power Relative to One's Parents (Y)

Response Category	Ordered Value
Less earning power	1
About the same	2
Greater earning power	3

TABLE E.5-11.B

Earning Power (Y) and Years of College (X)

Y	X
1	0.0
2	3.0
2	4.0
3	5.0
1	1.5
3	4.0
3	4.0
3	5.0
3	5.5
1	2.0
2	3.0
1	0.0
2	4.0
2	4.0
2	3.5
1	3.0
1	0.0
2	2.0
1	2.0
1	2.5
1	2.0
1	2.5
2	4.0
3	6.0
2	4.0
3	7.0
3	4.0
2	3.5
3	4.0
3	4.0

5-12. Given the following parameter estimates for the two generalized logits appearing below (in which the nominal response categories of Y are as follows: 1, SUV; 2, pickup truck; and 3, sedan); and X is the number of speeding tickets issued to the principal driver over a recent three-year period):

a. Interpret $e^{\hat{\beta}_{11}}$ and $e^{\hat{\beta}_{12}}$.

b. Find $\log(\hat{p}_1 / \hat{p}_2)$. Interpret $e^{\hat{\beta}_{11}-\hat{\beta}_{12}}$.

$$\log(\hat{p}_1 / \hat{p}_3) = \hat{\beta}_{01} + \hat{\beta}_{11}X = -4.8183 + 0.5611X;$$
$$\log(\hat{p}_2 / \hat{p}_3) = \hat{\beta}_{02} + \hat{\beta}_{12}X = 1.1866 - 0.2653X.$$

5-13. Given the set of observations involving professional football injuries appearing in Table E.5-12 (here Y represents the nominal response categories: 1, interior lineman; 2, wide receiver; and 3, quarterback/running back; and X depicts the number of distinct injuries affecting quality of life at least 10 years after retirement), fit a set of generalized logits and explain fully your results. (Hint: follow the SAS code presented in Exhibit 5.3.)

TABLE E.5-12

Football Player Position (Y) and Number of Injuries (X)

Y	X
1	4
1	3
3	1
2	2
1	5
2	2
2	1
1	6
3	2
3	2
3	1
1	2
1	2
1	4
2	3
2	2
2	1
3	2
3	1
2	3
2	1
2	2
3	3
3	2
1	3
3	1
1	4
1	5
2	1
3	0

5-14. Given the Table E.5-14 dataset pertaining to the number of failures (Y) of a certain type of magnetic switch in a piece of electronic equipment and the length of time in months (X) since the switch was installed, perform a Poisson regression of Y on X and explain fully your findings. (Hint: let Exhibit 5.4 be your guide.)

TABLE E.5-14

Switch Failures (Y) and Months of Use (X)

Y	X
1	8
0	5
0	10
5	19
3	15
0	11
1	14
4	22
0	7
0	12
0	3
0	9
7	28
0	2
1	7
0	2
1	7
2	12
0	6
3	14

6

Bayesian Regression

Classical versus Bayesian Methodology

In the preceding chapters, the approach typically taken to solve the regression problem was characterized as *classical* or *frequentist* in nature, e.g., under the assumptions of the strong classical linear regression model, the method of maximum likelihood enabled us to determine point estimators for β_0 and β_1 (denoted $\tilde{\beta}_0$ and $\tilde{\beta}_1$, respectively), which coincided with OLS estimators for these parameters. Moreover, by virtue of these assumptions, the (unknown) sampling distributions of these estimators are normally distributed. (Remember that these sampling distributions describe the behavior of the relative frequencies of these estimators under hypothetical repeated sampling from the same population.) In this regard, any particular estimate of, say, β_1, is treated as having been generated via a random drawing from $\tilde{\beta}_1$'s sampling distribution, and the use of $\tilde{\beta}_1$ as a point estimator of β_1 is legitimized by the fact that $\tilde{\beta}_1$ has certain desirable properties, e.g., $\tilde{\beta}_1$ is BLUE for β_1, among others.

The *Bayesian approach* to solving the regression problem stands in marked contrast to the aforementioned classical methodology. The *Bayesian modus operandi* starts with prior information on the parameter vector $\beta' = (\beta_0, \beta_1)$, where, for convenience, it is assumed that σ_ε^2 is known. It incorporates this prior information on β in the form of a *prior probability density function* $f(\beta)$. When sample information $Y = (Y_1, Y_2, \ldots, Y_n)$ is subsequently obtained, it is introduced into the Bayesian calculus in the form of the likelihood function $\mathcal{L}(\beta;Y)$. Then, as will be demonstrated below, this likelihood function is used, via Bayes' rule, to modify or revise our prior knowledge about β, thus resulting in the *posterior probability density function* $f(\beta|Y) \propto \mathcal{L}(\beta;Y) \cdot f(\beta)$, i.e., the posterior probability density function is taken to be proportional to the product of the likelihood function and the prior probability density function, where the symbol $\propto$ is read "is proportional to." This posterior probability density function thus encompasses all of our current knowledge (involving both prior and sample information) about β and thus obtains once the incorporation of prior information is effected and the sampling experiment has been terminated.

Given the posterior density function, a Bayesian estimator for β is a single value that is drawn from this density under some particular selection rule. So instead of generating a point estimator for β as in the classical case, the Bayesian approach produces an entire posterior probability density function. Furthermore, this posterior density is not a sampling distribution for the Bayesian estimator of β.

How is the prior probability density function selected? It is at this point in our analysis of Bayesian methodology that the notion of *subjectivist probability* comes to the fore. This type of probability depicts what one can call a "quantified degree of belief" about β; it simply represents the subjective odds one would offer when betting on what the true β might be before the sample is drawn. Clearly, this "best guess" probability is quite different from the classical concept of long-run relative frequency. Given that the prior density of β, when weighted by the likelihood function, results in the posterior probability density function, it is evident that the Bayesian approach mirrors how sampling results can modify one's prior beliefs about the parameter values.

Bayes' Rule for Random Variables [Bolstad 2004; Box and Tiao 1973; Lancaster 2004; Lee 2004; Zellner 1971, 1975]

In our discussion of probability theory in Chapter 1, Bayes' Theorem for events was introduced in the context of an application of the definition of conditional probability. Specifically, for events A and B within the sample space S,

$$P(A/B) = \frac{P(A \cap B)}{P(B)}, P(B) \neq 0.$$

However, because $P(A \cap B) = P(B/A) \cdot P(A)$, the preceding expression becomes *Bayes' Rule*:

$$P(A/B) = \frac{P(B/A) \cdot P(A)}{P(B)}, P(B) \neq 0. \tag{6.1}$$

How can Bayes' Rule be used for purposes of parameter estimation? Quite generally, suppose event A is replaced by a parameter m-tuple $\Theta = (\Theta_1,\ldots,\Theta_m) \in \Phi$, where Φ is an admissible parameter set, and event B depicts sample information incorporated in the n-tuple $Y = (Y_1,\ldots,Y_n)$, where the Y_i, $i = 1,\ldots,n$, constitute a set of sample random variables. If Θ and Y are now viewed as continuous random variables defined on a sample space S', then their joint probability density function can be expressed as $f(\theta, y)$. (It is important to note that, at this point in our discussion, the sample values are yet to be realized.) Given $f(\theta, y)$, let the marginal probability density functions be depicted as $f(\theta)$ and $f(y)$ with the conditional probability density functions expressed as $f(\theta/y)$ and $f(y/\theta)$.

Then mirroring the derivation of Equation 6.1 because, $f(\theta, y) = f(\theta/y) \cdot f(y) = f(y/\theta) \cdot f(\theta)$, we obtain *Bayes' Rule for continuous random variables*

$$f(\theta/y) = \frac{f(y/\theta) \cdot f(\theta)}{f(y)}, f(y) \neq 0.^{1} \tag{6.2}$$

In this version of Bayes' Rule, $f(\theta)$ is called the prior probability density function for Θ; $f(y/\theta)$ is the *conditional distribution* of Y given Θ, i.e., it represents a probability density function for the sample random variables $Y_1,\ldots,Y_n$ for a fixed Θ, and $f(\theta/y)$ is the *posterior probability density function* for Θ, given sample information Y. How shall we interpret these three expressions?

The prior density $f(\theta)$ over Φ represents one's beliefs about the possible values of Θ "prior to" actually observing the sample data. These beliefs may be based on prior information

[1] Here $f(y) = \int_\Phi f(\theta) \cdot f(y/\theta) d\theta$ is the reciprocal of the normalizing constant k, which makes the probability density function $f(y/\theta) = k f(\theta) \cdot f(y/\theta)$ integrate to unity.

obtained from previously extracted random samples, theory, or general observations or intuition. (It is assumed that $f(\theta)$ contains no unknown parameters.)

Next, because the random variable Y admits a specific probability distribution for each given value of $\Theta \in \Phi$, the conditional density $f(y/\theta)$ represents the joint probability distribution of the sample random variables Y_i, $i = 1,\ldots,n$, that one "expects" to observe, i.e., it depicts one's beliefs about the values of Y conditional on Θ. So given Θ, $f(y/\theta)$ reflects our convictions about what the sample realizations should look like "before" the random sample is drawn.

Because $f(y/\theta)$ depicts the conditional density of Y given Θ, it is actually a member of a family of sampling distributions indexed by each $\Theta \in \Phi$. With $f(y/\theta)$ the joint probability density of the sample random variables $Y_1,\ldots,Y_n$ given Θ, we may express the same as $f(y/\theta) = \prod_{i=1}^{n} f(y_i/\theta)$.

Suppose we now extract a random sample of size n from $f(y/\theta)$, the sampling distribution of Y conditional on Θ. Once we obtain the realizations of the sample random variables Y_i, $i = 1,\ldots,n$, the expression $f(y/\theta)$ becomes the likelihood function and is rewritten as $\mathcal{L}(\theta; y_1,\ldots,y_n) = \prod_{i=1}^{n} f(y_i/\theta)$, i.e., $\mathcal{L}$ is a function of θ with the y_i's, $i = 1,\ldots,n$, serving as parameters. In this regard, the likelihood function is not, strictly speaking, a probability density function. Here $\mathcal{L}$ indicates how likely the realized sample result is given our prior beliefs concerning Θ.

Finally, the posterior distribution of Θ is the conditional density of Θ given $Y = y_i$, $i = 1,\ldots,n$. It combines prior information about Θ, $f(\theta)$, with the sample information contained in the likelihood function, in which the latter includes the sample realizations as well as the form of the sampling distribution chosen for Y. Clearly, the posterior distribution summarizes all that we know about Θ after a random sample has been obtained, i.e., it encompasses subjective information as well as objective sample information; it enables us to revise our prior beliefs concerning Θ in the light of sample evidence. Operationally, one typically engages in the (sequential), *Bayesian Inferential Process*: prior beliefs about $\Theta \to$ sample data observed $\to$ posterior beliefs about Θ (which are modified or updated prior beliefs about Θ).

To make the chain of reasoning underlying Bayes' Rule a bit more explicit, let us rewrite Equation 6.2 as

$$f(\theta/y_1,\ldots,y_n) = \frac{\mathcal{L}(\theta; y_1,\ldots,y_n) \cdot f(\theta)}{f(y_1,\ldots,y_n)}, \tag{6.2.1}$$

where $f(y_1,\ldots,y_n) \neq 0$. See also Figure 6.1. (Bayes' Rule combines the prior distribution of Θ and the likelihood function to obtain the posterior or conditional distribution of Θ given sample data)

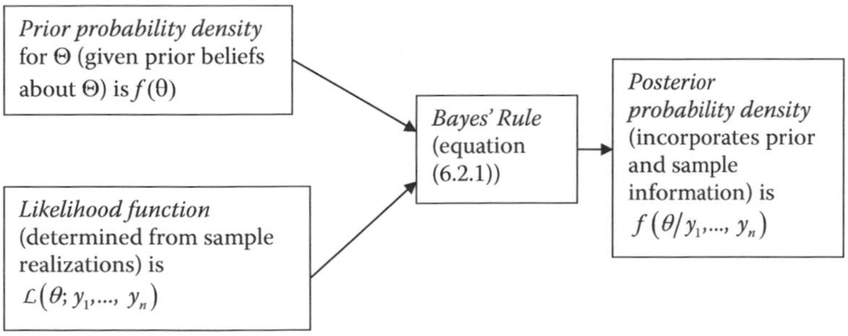

FIGURE 6.1
Bayes' Rule.

Because the marginal density function $f(y)$ does not involve the parameter θ, it can be absorbed into any multiplicative constant components appearing on the right side of Equation 6.2.1. Hence, for estimation purposes, Bayes' Rule can be expressed in its *proportional form* as

$$f(\theta/y_1,\ldots,y_n) \propto \mathcal{L}(\theta;y_1,\ldots,y_n)\cdot f(\theta) \tag{6.2.2}$$

or

$$\text{posterior} \propto \text{likelihood} \times \text{prior},$$

where again the symbol $\propto$ is read as "is proportional to." So if we ignore any multiplicative terms on the right side of Equation 6.2.1 that do not involve θ, then the expression on the right side of Equation 6.2.2 is termed the *kernal* of the posterior distribution.

Bayes' Rule and the Strong Classical Linear Regression Model

For the strong classical linear regression model, let $Y_i = \beta_0 + \beta_1 X_i + \varepsilon_i$, $i = 1,\ldots,n$, where the random error term ε_i is $N(0, \sigma_\varepsilon)$ for all i, σ_ε^2 is the constant variance of each ε_i, and successive ε_i's are independent of each other. For convenience, let us reparameterize the preceding regression equation by measuring the X and Y variables in terms of derivations from their respective means $\overline{X}$ and $\overline{Y}$ as

$$\begin{aligned} Y_i &= \beta_0 + \beta_1 X_i + \varepsilon_i = \beta_0 + \beta_1\overline{X} - \beta_1\overline{X} + \beta_1 X_i + \varepsilon_i \\ &= \left(\beta_0 + \beta_1\overline{X}\right) + \beta_1\left(X_i - \overline{X}\right) + \varepsilon_i \\ &= \beta_* + \beta_1 x_i + \varepsilon_i, i = 1,\ldots,n, \end{aligned} \tag{6.3}$$

where $\beta_* = \beta_0 + \beta_1\overline{X}$ and $x_i = X_i - \overline{X}$. (The reason for this reparameterization will be explained below.)

Then for X_i fixed for all i, the population regression equation is $E(Y|X) = \beta_* + \beta_1 x$, and OLS estimates of β_* and β_1 are, respectively, $\hat{\beta}_* = \overline{Y}$ and $\hat{\beta}_1 = \sum_{i=1}^{n} x_i y_i \Big/ \sum_{i=1}^{n} x_i^2$, where $y_i = Y_i - \overline{Y}$. Thus, the OLS line of best fit $\hat{Y} = \hat{\beta}_* + \hat{\beta}_1 x$ passes through the point of means $\left(\overline{X}, \overline{Y}\right)$ with $\hat{\beta}_*$ representing the estimated intercept of the vertical line $X = \overline{X}$.

The Joint Likelihood Function

Let us now apply the proportionate form of Bayes' Rule (Equation 6.2.2) to the reparameterized model (Equation 6.3). We shall first specify the likelihood portion of this expression and then choose a prior distribution.

Given the preceding set of assumptions on ε_i, $i = 1,\ldots,n$, the joint likelihood function for the sample is

$$\mathcal{L}(\beta_*,\beta_1;Y_1,\ldots,Y_n,n) = \left(\frac{1}{\sqrt{2\pi}\sigma_\varepsilon}\right)^n e^{-\frac{1}{2\sigma_\varepsilon^2}\sum_{i=1}^{n}\varepsilon_i^2} = \left(\frac{1}{\sqrt{2\pi}\sigma_\varepsilon}\right)^n e^{-\frac{1}{2\sigma_\varepsilon^2}\sum_{i=1}^{n}(Y_i-\beta_*-\beta_1 x_i)^2}. \tag{6.4}$$

Let us rewrite the sum-of-squares expression involving β_* and β_1 as

$$\sum_{i=1}^{n}\left(Y_i+\overline{Y}-\overline{Y}-\beta_*-\beta_1 x_i\right)^2=\sum_{i=1}^{n}\left\{y_i+\left[\overline{Y}-(\beta_*+\beta_1 x_i)\right]\right\}^2$$

$$=\sum_{i=1}^{n} y_i^2+2\sum_{i=1}^{n} y_i\left[\overline{Y}-(\beta_*+\beta_1 x_i)\right]+\sum_{i=1}^{n}\left[\overline{Y}-(\beta_*+\beta_1 x_i)\right]^2$$

$$=\sum_{i=1}^{n} y_i^2-2\beta_1\sum_{i=1}^{n} x_i y_i+\beta_1^2\sum x_i^2+n\left(\beta_*-\overline{Y}\right)^2.$$

If we set $\sum_i^n y_i^2=SS_Y$, $\sum_{i=1}^n x_i y_i=SS_{XY}$, and $\sum_{i=1}^n x_i^2=SS_X$, then Equation 6.4 becomes

$$\mathcal{L}\left(\beta_*,\beta_1;Y_1,\ldots,Y_n,n\right)=\left(\frac{1}{\sqrt{2\pi}\sigma_\varepsilon}\right)^n e^{-\frac{1}{2\sigma_\varepsilon^2}\left[SS_Y-2\beta_1 SS_{XY}+\beta_1^2 SS_X+n\left(\beta_*-\overline{Y}\right)^2\right]}$$

$$=\left(\frac{1}{\sqrt{2\pi}\sigma_\varepsilon}\right)^n e^{-\frac{1}{2\sigma_\varepsilon^2}\sum_{i=1}^{n}\varepsilon_i^2\left(SS_Y-2\beta_1 SS_{XY}+\beta_1^2 SS_X\right)} e^{-\frac{1}{2\sigma_\varepsilon^2}\left[n\left(\beta_*-\overline{Y}\right)^2\right]}. \tag{6.4.1}$$

Furthermore, let us rewrite the parenthetical expression involving β_1 on the right side of Equation 6.4.1 as

$$SS_X\beta_1^2-2SS_{XY}\beta_1+SS_Y=SS_X\left[\beta_1^2-2\frac{SS_{XY}}{SS_X}\beta_1+\frac{SS_Y}{SS_X}\right]=$$

$$SS_X\left[\left(\beta_1-\frac{SS_{XY}}{SS_X}\right)^2-\left(\frac{SS_{XY}}{SS_X}\right)^2+\frac{SS_Y}{SS_X}\right],\ SS_X\neq 0.^2 \tag{6.5}$$

If Equation 6.5 is substituted into Equation 6.4.1 and all terms not involving β_* and β_1 are absorbed into a proportionality constant, then the kernal of Equation 6.4.1 becomes

$$\mathcal{L}\left(\beta_*,\beta_1;Y_1,\ldots,Y_n,n\right)\propto e^{-\frac{SS_X}{2\sigma_\varepsilon^2}\left(\beta_1-\frac{SS_{XY}}{SS_X}\right)^2} e^{-\frac{n}{2\sigma_\varepsilon^2}\left(\beta_*-\overline{Y}\right)^2} \tag{6.4.2}$$

or

$$\mathcal{L}\left(\beta_*,\beta_1;Y_1,\ldots,Y_n,n\right)\propto\mathcal{L}\left(\beta_1;Y_1,\ldots,Y_n,n\right)\times\mathcal{L}\left(\beta_*;Y_1,\ldots,Y_n,n\right), \tag{6.4.3}$$

i.e., the joint likelihood function is the product of the individual (independent) likelihood functions

$$\mathcal{L}\left(\beta_1;Y_1,\ldots,Y_n,n\right)\propto e^{-\frac{SS_X}{2\sigma_\varepsilon^2}\left(\beta_1-\frac{SS_{XY}}{SS_X}\right)^2}$$

---

[2] Equation 6.5 has been obtained by a process known as *completing the square*. For instance, suppose we have the quadratic equation $ax^2+bx+c=0$, $a\neq 0$. Then $x^2+\frac{b}{a}x=-\frac{c}{a}$ or $x^2+\frac{b}{a}x+\left(\frac{b}{2a}\right)^2-\left(\frac{b}{2a}\right)^2=-\frac{c}{a}$ or $\left(x+\frac{b}{2a}\right)^2=\left(\frac{b}{2a}\right)^2=-\frac{c}{a}$. The left side of this expression thus corresponds to $\left(\beta_1-\frac{SS_{XY}}{SS_X}\right)^2$. Clearly, $\frac{b}{2a}=-2SS_{XY}/2SS_X=-SS_{XY}/SS_X$.

and

$$\mathcal{L}\left(\beta_*;Y_1,\ldots,Y_n,n\right)\propto e^{-\frac{n}{2\sigma_\varepsilon^2}\left(\beta_*-\bar{Y}\right)^2},$$

where obviously these individual sample likelihoods are $N\left(\hat{\beta}_1,\sigma_\varepsilon/\sqrt{SS_X}\right)$ and $N\left(\hat{\beta}_*,\sigma_\varepsilon/\sqrt{n}\right)$, respectively. Here $\hat{\beta}_1=SS_{XY}/SS_X$ is the OLS estimate of the slope and $\hat{\beta}_*=\bar{Y}$ depicts the OLS estimate of the intercept of the reparameterized model.

The Joint Prior Distribution for β_*, β_1

Let us adopt the convention of choosing a joint prior distribution for β_* and β_1 that readily combines with the likelihood function for the sample. To accomplish this, one need only remember that Bayes' Rule in its proportional form appears as posterior $\propto$ likelihood $\times$ prior. Hence, we can select a joint prior probability density function from a family of density functions that, after forming the product likelihood $\times$ prior, renders a posterior probability density function that is a member of what is called a family of *natural conjugate priors*. (For an alternative choice of a prior distribution, see Appendix 6.A.) Hence, the priors within a natural conjugate family of distributions can be combined with a given likelihood function that exhibits the same general form as the prior distribution. It is important to remember that a conjugate family of prior distributions is *determined relative to the likelihood function*, i.e., a prior distribution is conjugate only with respect to a unique likelihood function that represents some chosen statistical (regression) model.

In this regard, because the right side of Equation 6.4.3 involves the product of two normal sample likelihood functions, it seems reasonable to choose the joint prior distribution for β_* and β_1 as the product of two individual (independent) normal marginal density priors or

$$f\left(\beta_*,\beta_1\right)=f\left(\beta_*\right)\cdot f\left(\beta_1\right), \tag{6.6}$$

where

$$f\left(\beta_*\right)=\frac{1}{\sqrt{2\pi}\sigma_{\beta_*}}e^{-\frac{1}{2\sigma_{\beta_*}^2}\left(\beta_*-\mu_{\beta_*}\right)^2} \tag{6.7.a}$$

is $N\left(\mu_{\beta_*},\sigma_{\beta_*}\right)$, and

$$f\left(\beta_1\right)=\frac{1}{\sqrt{2\pi}\sigma_{\beta_1}}e^{-\frac{1}{2\sigma_{\beta_1}^2}\left(\beta_1-\mu_{\beta_1}\right)^2} \tag{6.7.b}$$

is $N\left(\mu_{\beta_1},\sigma_{\beta_1}\right)$. (See Appendix 6.B for the connection between the above reparameterization of the regression model and the choice of the joint prior distribution.) Substituting Equations 6.7.a and 6.7.b into Equation 6.6 and absorbing all terms not involving β_* and β_1 into a proportionality constant allows us to rewrite the kernal of Equation 6.6 as

$$f\left(\beta_*,\beta_1\right)\propto e^{-\frac{1}{2\sigma_{\beta_*}^2}\left(\beta_*-\mu_{\beta_*}\right)^2}\,e^{-\frac{1}{2\sigma_{\beta_1}^2}\left(\beta_1-\mu_{\beta_1}\right)^2}. \tag{6.6.1}$$

The Joint Posterior Distribution for β_*, β_1

Because Equation 6.2.2 informs us that the joint posterior distribution is proportional to the joint likelihood function times the joint prior distribution, let us form, from Equations 6.4.2, 6.4.3, and 6.6.1, the kernal of the posterior density function:

$$f(\beta_*,\beta_1|Y_1,\ldots,Y_n) \propto \mathcal{L}(\beta_*,\beta_1;Y_1,\ldots,Y_n,n)\times f(\beta_*,\beta_1)=$$
$$\left[\mathcal{L}(\beta_*;Y_1,\ldots,Y_n,n)\cdot f(\beta_*)\right]\times\left[\mathcal{L}(\beta_1;Y_1,\ldots,Y_n,n)\cdot f(\beta_1)\right]=$$
$$e^{-\frac{1}{2}\left[\frac{n}{\sigma_\varepsilon^2}(\beta_*-\bar{Y})^2+\frac{1}{\sigma_{\beta_*}^2}(\beta_*-\mu_{\beta_*})^2\right]}\times e^{-\frac{1}{2}\left[\frac{SS_X}{\sigma_\varepsilon^2}\left(\beta_1-\hat{\beta}_1\right)^2+\frac{1}{\sigma_{\beta_1}^2}(\beta_1-\mu_{\beta_1})^2\right]}. \tag{6.8}$$

Additional simplification of Equation 6.8 (the details of which appear in Appendix 6.C) ultimately results in the expression

$$f(\beta_*,\beta_1|Y_1,\ldots,Y_n) \propto e^{-\frac{1}{2(\sigma'_{\beta_*})^2}(\beta_*-\mu'_{\beta_*})^2}\times e^{-\frac{1}{2(\sigma'_{\beta_1})^2}(\beta_1-\mu'_{\beta_1})^2}. \tag{6.8.1}$$

Clearly, this joint posterior distribution is the product of two independent marginal posterior distributions, in which the first marginal posterior density function in the product on the right side of Equation 6.8.1 is $N\left(\mu'_{\beta_*},\sigma'_{\beta_*}\right)$ with

$$\mu'_{\beta_*}=\frac{n\sigma_{\beta_*}^2\bar{Y}+\sigma_\varepsilon^2\mu_{\beta_*}}{n\sigma_{\beta_*}^2+\sigma_\varepsilon^2},\ \left(\sigma'_{\beta_*}\right)^2=\frac{\sigma_\varepsilon^2\sigma_{\beta_*}^2}{n\sigma_{\beta_*}^2+\sigma_\varepsilon^2}; \tag{6.9}$$

and the second marginal posterior density function is $N\left(\mu'_{\beta_1},\sigma'_{\beta_1}\right)$ with

$$\mu'_{\beta_1}=\frac{SS_X\sigma_{\beta_1}^2\hat{\beta}_1+\sigma_\varepsilon^2\mu_{\beta_1}}{SS_X\sigma_{\beta_1}^2+\sigma_\varepsilon^2},\ \left(\sigma'_{\beta_1}\right)^2=\frac{\sigma_\varepsilon^2\sigma_{\beta_1}^2}{SS_X\sigma_{\beta_1}^2+\sigma_\varepsilon^2}. \tag{6.10}$$

Notice again how we have developed Equation 6.8.1. The joint posterior distribution is taken as proportional to the joint likelihood function times the joint prior distribution or (using abbreviated notation)

$$f(\beta_*,\beta_1|Y_1,\ldots,Y_n) \propto L(\beta_*,\beta_1)\times f(\beta_*,\beta_1).$$

The joint likelihood function of the sample is the product of independent normal likelihood functions, whereas the joint prior (taken from the normal family of natural conjugate priors) is the product of independent normal prior densities, in which each parameter gets its own separate normal prior density function. Then the preceding expression becomes

$$f(\beta_*,\beta_1|Y_1,\ldots,Y_n) \propto \mathcal{L}(\beta_*)\cdot\mathcal{L}(\beta_1)\times f(\beta_*)\cdot f(\beta_1)$$

or

$$f(\beta_*,\beta_1|Y_1,\ldots,Y_n) \propto \left[\mathcal{L}(\beta_*)\cdot f(\beta_*)\right]\times\left[\mathcal{L}(\beta_1)\cdot f(\beta_1)\right]=$$
$$f(\beta_*|Y_1,\ldots,Y_n)\cdot f(\beta_1|Y_1,\ldots,Y_n).$$

Clearly, this joint posterior distribution has been written as the product of two independent marginal posterior densities, one for each parameter to be estimated. In summary, using Bayes' Rule, we started with $N(\mu_{\beta_*}, \sigma_{\beta_*})$ and $N(\mu_{\beta_1}, \sigma_{\beta_1})$ independent marginal conjugate priors and ended up with $N(\mu'_{\beta_*}, \sigma'_{\beta_*})$ and $N(\mu'_{\beta_1}, \sigma'_{\beta_1})$ independent marginal posterior densities. So if we rely on the (normal) family of natural conjugate priors, we simply move from one member of that family to another via the so-called *updating values* provided by Equations 6.9 and 6.10; only the parameters of the prior densities change with the introduction of sample data (provided by the likelihood function), not their mathematical structure.

It is the posterior distribution that now houses all relevant information about β_* and β_1; it summarizes the entire body of evidence, consisting of our prior beliefs (the prior distribution) and the sample dataset (incorporated in the likelihood function) about the regression parameters. So although the prior distribution is selected before any sample data are examined, it is the sample information that is used to update, via Bayes' Rule, the prior distribution. Hence, the posterior distribution is nothing but an updated prior distribution; the likelihood function enables us to modify or revise our prior beliefs about β_* and β_1. Because the posterior distribution summarizes all that we know about β_* and β_1 after the sample is drawn, it can be viewed as replacing the likelihood function as the construct that incorporates all relevant information about the regression parameters.

Rules for Updating Our Prior Beliefs about β_* and β_1

Let us examine the structure of the updating Equations 6.9 and 6.10 developed above. To aid us in this undertaking, we shall define the notion of *precision* as the reciprocal of the variance. In this regard, the usual convention is to rewrite Equations 6.9 and 6.10, respectively, in terms of precisions, as

$$\mu'_{\beta_*} = \left[\frac{n/\sigma_\varepsilon^2}{1/\left(\sigma'_{\beta_*}\right)^2}\right]\overline{Y} + \left[\frac{1/\sigma_{\beta_*}^2}{1/\left(\sigma'_{\beta_*}\right)^2}\right]\mu_{\beta_*}, \quad \frac{1}{\left(\sigma'_{\beta_*}\right)^2} = \frac{n}{\sigma_\varepsilon^2} + \frac{1}{\sigma_{\beta_*}^2}; \tag{6.9.1}$$

$$\mu'_{\beta_1} = \left[\frac{SS_X/\sigma_\varepsilon^2}{1/\left(\sigma'_{\beta_1}\right)^2}\right]\hat{\beta}_1 + \left[\frac{1/\sigma_{\beta_1}^2}{1/\left(\sigma'_{\beta_1}\right)^2}\right]\mu_{\beta_1}, \quad \frac{1}{\left(\sigma'_{\beta_1}\right)^2} = \frac{SS_X}{\sigma_\varepsilon^2} + \frac{1}{\sigma_{\beta_1}^2}. \tag{6.10.1}$$

Here Equations 6.9.1 and 6.10.1 will be termed *Bayes' estimators* for β_* and β_1, respectively.

Then from Equation 6.9.1, the posterior precision of the intercept β_* (which is $\overline{Y}$ under the reparameterization of the regression model) or $1/\left(\sigma'_{\beta_*}\right)^2$ is the precision of the sample $\left(n/\sigma_\varepsilon^2\right)$ plus the precision of the posterior distribution $1/\sigma_{\beta_*}^2$. Moreover, the posterior mean μ'_{β_*} is a weighted average of the sample mean $\left(\overline{Y}\right)$ and the prior mean $\left(\mu_{\beta_*}\right)$, in which the weights are the proportions of the sample and prior precisions in posterior precision.

Similarly, from Equation 6.10.1, the posterior precision of the slope or $1/\left(\sigma'_{\beta_1}\right)^2$ is the precision of the sample $\left(SS_X/\sigma_\varepsilon^2\right)$ plus the precision of the prior distribution $\left(1/\sigma_{\beta_1}^2\right)$. In addition, the posterior mean μ'_{β_1} is a weighted average of the estimated or OLS slope $\left(\hat{\beta}_1\right)$ and the prior mean $\left(\mu_{\beta_1}\right)$, in which the weights are the proportions of the sample and prior precisions in posterior precision.

Let us consider a few of the salient features of Equations 6.9.1 and 6.10.1:

1. The Bayes' estimator is always a value between the OLS estimator and our prior belief about the value of the regression parameter.
2. Posterior precision is larger than the precision of the sample taken by itself.
3. As the prior variance tends to $+\infty$ (meaning that our prior information becomes more vague or diffuse and thus more imprecise), posterior precision tends to sample precision.
4. As the prior variance tends to $+\infty$, the Bayes' estimator tends to the sample or OLS estimator, i.e., as our prior information becomes less reliable or diffuse, the Bayes' estimator assigns more weight to the sample estimator. However, if our prior information is very precise, so that the sample variance exceeds the prior variance, then more weight is assigned to the prior value of the regression parameter.
5. The Bayes' estimator gets closer to the OLS solution as n increases (SS_X concomitantly increases), or when prior information becomes very imprecise.
6. If the information expressed by the prior distribution is accurate, then inferences about the regression parameters based on the posterior or Bayes' estimator will be more precise than those based on sample information alone. However, as just stated above, the more diffuse or vague the prior distribution, the less influence the prior has on the parameter estimates.

Because σ_ε^2 is typically unknown, it will be estimated by

$$S_\varepsilon^2 = \sum_{i=1}^{n}\left[Y_i - \left(\bar{Y} + \hat{\beta}_1 x_i\right)\right]^2 \Big/ (n-2) = \sum_{i-1}^{n}\left(Y_i - \hat{Y}_i\right)^2 \Big/ (n-2).$$

Then from Equation 6.9.1, our estimate of the posterior precision of β_* is

$$\frac{1}{\left(S'_{\beta_*}\right)^2} = \frac{n}{S_\varepsilon^2} + \frac{1}{\sigma_{\beta_*}^2} \tag{6.11}$$

and the Bayes' estimate of β_* or the realized posterior mean of β_* is

$$m'_{\beta_*} = \left[\frac{n/S_\varepsilon^2}{1/\left(S'_{\beta_*}\right)^2}\right]\bar{Y} + \left[\frac{1/\sigma_{\beta_*}^2}{1/\left(S'_{\beta_*}\right)^2}\right]\mu_{\beta_*}. \tag{6.12}$$

Similarly, from Equation 6.10.1, the estimate of the posterior precision of β_1 is

$$\frac{1}{\left(S'_{\beta_*}\right)^2} = \frac{SS_X}{S_\varepsilon^2} + \frac{1}{\sigma_{\beta_1}^2} \tag{6.13}$$

with the Bayes' estimate of β_1 or the realized posterior mean of β_1 appearing as

$$m'_{\beta_1} = \left[\frac{SS_X/S_\varepsilon^2}{1/\left(S'_{\beta_1}\right)^2}\right]\hat{\beta}_1 + \left[\frac{1/\sigma_{\beta_1}^2}{1/\left(S'_{\beta_1}\right)^2}\right]\mu_{\beta_1}. \tag{6.14}$$

TABLE 6.1

Observations on the Variables X and Y

X	Y	x_i	y_i	x_i^2	x_iy_i	$\hat{\beta}_1 x_i$	$\hat{Y} = \overline{Y} + \hat{\beta}_1 x_i$	$Y_i - \hat{Y}_i$	$(Y_i - \hat{Y}_i)^2$
20	35	−2	−10.2	4	20.4	−2.6211	42.5789	7.5789	57.4397
40	65	18	19.8	324	356.4	23.5895	68.7895	3.7895	14.3603
20	40	−2	−5.2	4	10.4	−2.6211	42.5789	2.5789	6.6507
30	60	8	14.8	64	118.4	10.4842	55.6842	4.3158	18.6261
10	30	−12	−15.2	144	182.4	−15.7263	29.4737	0.5263	0.2770
10	35	−12	−10.2	144	122.4	−15.7263	29.4737	5.5263	30.5400
20	40	−2	−5.2	4	10.4	−2.6211	42.5789	2.5789	6.6507
20	45	−2	−0.2	4	0.4	−2.6211	42.5789	2.4211	5.8617
20	37	−2	−8.2	4	16.4	−2.6211	42.5789	5.5789	31.1241
30	65	8	19.8	64	158.4	10.4842	55.6842	9.3158	86.7841
220	452	0	0	760	996				258.3144

Example 6.1

Suppose we have observations on a group of $n = 10$ salesmen for the ABC Corp. The dataset (see Table 6.1) pertains to months of experience (X) and number of units of a product sold (Y) for a one-month sample period.

For this set of observations: $\overline{X} = 22$, $\overline{Y} = 45.2$, $SS_X = \sum_{i=1}^{10} x_i^2 = 760$, and $SS_{XY} = \sum_{i=1}^{10} x_i y_i = 996$. Then the OLS estimate of the regression slope is $\hat{\beta}_1 = 996/760 = 1.3105$ and thus the OLS line of best fit is

$$\hat{Y} = \hat{\beta}_* + \hat{\beta}_1 (X - \overline{X}) = 45.2 + 1.3105(X - 22).$$

In addition, $S_\varepsilon^2 = \sum_{i-1}^{10} (Y_i - \hat{Y}_i)^2 / 8 = 258.3144/8 = 32.2893$. Let us assume that our prior beliefs dictate that the (independent) prior distributions for β_* and β_1 are $N(40, 6)$ and $N(1, 0.4)$, respectively. Then from Equation 6.11, our estimate of the posterior precision of β_* is

$$\frac{1}{(S'_{\beta_*})^2} = \frac{10}{35.2893} + \frac{1}{36} = 0.3375;$$

and from Equation 6.12, Bayes' estimate of β_* or the realized posterior mean of β_* is

$$m'_{\beta_*} = \left[\frac{10/32.2893}{0.3375}\right](45.2) + \left[\frac{1/36}{0.3375}\right](40) = 44.7691.$$

Similarly, from Equation 6.13, our estimate of the posterior precision of β_1 is

$$\frac{1}{(S'_{\beta_1})^2} = \frac{760}{32.2893} + \frac{1}{0.1600} = 29.7872;$$

whereas from Equation 6.14, the Bayes' estimate of β_1 or the realized posterior mean of β_1 is

$$m'_{\beta_1} = \left[\frac{760/32.2893}{29.7872}\right](1.3105) + \left[\frac{1/0.1600}{29.7872}\right](1) = 1.2453.$$

Bayes' Credible Intervals

We found that the Bayes' estimates in Equations 6.12 and 6.14 determined from independent marginal posterior distributions are essentially the *full* or *complete estimates* made for β_* and β_1 given the sample data, i.e., the pairs of values $\left(m'_{\beta_*}, \left(S'_{\beta_*}\right)^2\right)$ and $\left(m'_{\beta_1}, \left(S'_{\beta_1}\right)^2\right)$ summarize the totality of our knowledge about β_* and β_1 once our prior beliefs are augmented by sample information. Let us now translate our posterior point estimators of the regression parameters into a plausible range of values that appear likely, in terms of probability, to contain these parameters given the sample evidence. This type of interval is called a *Bayes' credible interval;* it indicates a range of values for a parameter that are conceivable, in light of prior beliefs and current sample data, at some chosen probability level.

Let us determine a Bayes' credible interval for β_1. Because all information pertaining to β_1 is contained in the marginal posterior probability density function $f(\beta_1 | Y_1,\ldots,Y_n)$, if a credible interval estimate of β_1 is desired, we need to find two functions of $Y_1,\ldots,Y_n$, and call them $l = l(Y_1,\ldots,Y_n)$ and $u = u(Y_1,\ldots,Y_n)$, such that $\int_l^u f(\beta_1 | Y_1,\ldots,Y_n) d\beta_1 = 1-\alpha$. Thus, $[l, n]$ is an interval estimate of β_1 with $1-\alpha$ representing the posterior probability that $\beta_1 \in [l, n]$, i.e., $[l, n]$ contains $100(1-\alpha)\%$ of the posterior probability and thus represents an interval that we believe will contain the true value of β_1.

Based on the results from "Rules for Updating Our Prior Belief about β_* and β_1," a $100(1-\alpha)\%$ Bayes' credible interval for the slope parameter β_1 is

$$m'_{\beta_1} \pm t_{\alpha/2, n-2} S'_{\beta_1}, \tag{6.15}$$

where (repeated for convenience)

$$m'_{\beta_1} = \left[\frac{SS_X / S_\varepsilon^2}{1/\left(S'_{\beta_1}\right)^2}\right]\hat{\beta}_1 + \left[\frac{1/\sigma_{\beta_1}^2}{1/\left(S'_{\beta_1}\right)^2}\right]\mu_{\beta_1}$$

and

$$\frac{1}{\left(S'_{\beta_1}\right)^2} = \frac{SS_X}{\sigma_\varepsilon^2} + \frac{1}{\sigma_{\beta_1}^2}, \quad S_\varepsilon^2 = \frac{\sum_{i=1}^{n}\left[Y_i - \left(\bar{Y} + \hat{\beta}_1 x_i\right)\right]^2}{n-2}.$$

Similarly, a $100(1-\alpha)\%$ Bayes' credible interval for β_* is

$$m'_{\beta_*} \pm t_{\alpha/2, n-2} S'_{\beta_*}, \tag{6.16}$$

with

$$m'_{\beta_*} = \left[\frac{n/S_\varepsilon^2}{1/\left(S'_{\beta_*}\right)^2}\right]\bar{Y} + \left[\frac{1/\sigma_{\beta_*}^2}{1/\left(S'_{\beta_*}\right)^2}\right]\mu_{\beta_*}$$

and

$$\frac{1}{\left(S'_{\beta_*}\right)^2} = \frac{n}{S_\varepsilon^2} + \frac{1}{\sigma_{\beta_*}^2}.$$

How does a 100(1 – α)% Bayes' credible interval for, say, β_1, differ from the classical or frequentist 100(1 – α)% confidence interval for β_1? In the frequentist approach, β_1 is taken to be a fixed and unknown constant and 1 – α is the probability that we will obtain a random sample chosen from the sampling distribution of β_1 such that the interval, once calculated, contains β_1. Before the sample is drawn,

$$P\left(\hat{\beta}_1 - t_{\alpha/2,n-2}S_{\hat{\beta}_1} < \beta_1 < \hat{\beta}_1 + t_{\alpha/2,n-2}S_{\hat{\beta}_1}\right) = 1-\alpha$$

is a *random interval* and 1 – α is a probability because $\hat{\beta}_1$ is a random variable. However, once the sample realizations are obtained and the interval endpoints are determined, 1 – α is simply called a *confidence coefficient*; it is not a probability. It is, however, given a long-run relative frequency interpretation, i.e., if, say, 1 – α = 0.95 and many random samples are taken and an interval for each is calculated, then, in the long-run, 95% of these intervals would contain β_1 and 5% of them would not.

In the Bayes' counterpart to classical interval estimation, an actual probability is calculated using the posterior distribution of β_1 conditional on the realized sample values. It is a legitimate (albeit subjective) probability because β_1 is a random variable. Hence,

$$P(l < \beta_1 < u) = \int_l^u f(\beta_1 | Y_1,\ldots,Y_n)\, d\beta_1 = 1-\alpha$$

gives us the probability that any value contained within (l,u) has credibility as a possible value for β_1. Note that nothing is said about the behavior of a credible interval under repeated sampling from some sampling distribution. In addition, as the sample size $n \to \infty$, the posterior distribution becomes more dependent on the sample information or likelihood function (or less dependent on the chosen prior density function), and thus a credible interval will converge to the same range of values as exhibited by a classical confidence interval.

If the prior distribution is diffuse or uninformative (as defined in Appendix 6.A), we will obtain a Bayes' credible interval that is identical to the classical or frequentist confidence interval. However, the interpretations still differ: the frequentist concludes that, in the long run, 100(1 – α)% of all such intervals will contain the true value of, say, β_1, and the Bayesian concludes that the probability is 100(1 – α)% that β_1 lies within (l,u).

Example 6.2

For the dataset presented in Example 6.1, let us find 95% Bayes' credible intervals for β_1 and β_*. Because $1/\left(S'_{\beta_1}\right)^2 = 29.7872$, $S'_{\beta_1} = 0.1832$ and thus, for $t_{0.025,8} = 2.306$, Equation 6.15 becomes 1.2453 ± 2.306(0.1832) or (0.8228,1.6678). Thus, the probability is 0.95 that β_1 lies between 0.8228 and 1.6678.

Next, because $1/\left(S'_{\beta_*}\right)^2 = 0.3375$, $S'_{\beta_*} = 1.7213$ and thus, from Equation 6.16, 44.7691 ± 2.306(1.7213) or (40.7997,48.7385). That is, the probability is 0.95 that β_* lies between 40.7997 and 48.7385.

Note that, because $S_{\hat{\beta}_1} = S_\varepsilon/\sqrt{\sum x_i^2} = 5.6824/\sqrt{760} = 0.2061$, a 95% classical confidence interval for β_1 is $\hat{\beta}_1 \pm t_{\frac{\alpha}{2},n-2} S_{\hat{\beta}_1}$ or $1.3105 \pm 2.306(0.2061)$ or (0.8352,1.7858). Hence, we may be 95% confident that β_1 lies between 0.8352 and 1.7858. Obviously, the 95% Bayes' credible interval for β_1 is a bit narrower than the 95% classical confidence interval. This is because it incorporates the additional prior information into its calculation and thus renders a somewhat more precise estimate of β_1. If no prior beliefs were used, the frequentist would need a larger sample to attain the same level of precision as that exhibited by the Bayes' credible interval.

Bayes' Hypothesis Tests

Let us first consider testing a one-sided hypothesis about the slope parameter β_1. Specifically, suppose we test $H_0: \beta_1 \leq \beta_1^0$, against $H_1: \beta_1 > \beta_1^0$ (β_1^0 is the null value) at the α level of significance by computing the posterior probability:

$$P(H_0) = P(\beta_1 \leq \beta_1^0 | Y_1, \dots, Y_n)$$
$$= \int_{-\infty}^{\beta_1^0} f(\beta_1 | Y_1, \dots, Y_n) d\beta_1 = P\left(T \leq \frac{\beta_1^0 - m'_{\beta_1}}{S'_{\beta_1}}\right), \tag{6.17}$$

with $P(H_1) = 1 - P(H_0)$. Hence, this probability serves as a measure of our degree of belief that H_0 is true given our chosen prior and sample information. Then if $P(H_0)$ exceeds our chosen α, we will reject H_0 in favor of H_1. (Note that, if we test $H_0: \beta_1 \geq \beta_1^0$ versus $H_1: \beta_1 < \beta_1^0$ at the α level, then we must reverse the sense of the inequalities in Equation 6.17 and the limits of integration go from β_1^0 to $+\infty$.)

Next, we can easily test the two-sided hypothesis $H_0: \beta_1 = \beta_1^0$, against $H_1: \beta_1 \neq \beta_1^0$ at the α level by determining if the null value β_1^0 lies outside of a $100(1 - \alpha)\%$ Bayes' credible interval for β_1. If it does, then we will reject H_0 in favor of H_1. If it does not, then we may conclude that β_1^0 is a credible value for β_1.

Suppose our OLS estimate of β_1 is $\hat{\beta}_1$ and we again choose to test $H_0: \beta_1 \leq \beta_1^0$ versus $H_1: \beta_1 > \beta_1^0$ at the α level of significance. From the classical or frequentist point of view, which uses sample information exclusively, the

$$p - value = P\left(T \geq \frac{\hat{\beta}_1 - \beta_1^0}{S_{\hat{\beta}_1}} \middle| H_0 \text{ true}\right), \tag{6.18}$$

i.e., we are interested in the probability of obtaining a calculate t value at least as large as the one observed if H_0 is true. This p-value, which is determined from the sampling distribution of $\hat{\beta}_1$, indicates how extreme the sample result is if H_0 is true. Moreover, this p-value bears no relationship to $P(H_0)$ given in Equation 6.17 if the chosen prior distribution is not vague or diffuse. So whereas the frequentist determines a test statistic and its p-value, with the letter expressed in terms of $\hat{\beta}_1$, the Bayesian calculates a posterior probability expressed in terms of β_1.

If a diffuse or noninformative prior probability density function is used to form a posterior distribution, then, numerically, Equations 6.17 and 6.18 yield identical results with $P(H_0) = p\text{-}value$. In this circumstance, however, as was the case with frequentist confidence intervals versus Bayes' credible intervals, the interpretations differ. As mentioned above, the p-value is obtained using the sampling distribution of $\hat{\beta}_1$. Specifically, the sample realization of $\hat{\beta}_1$ is compared, in probability terms, with other $\hat{\beta}_1$'s that might have been observed to determine how extreme the sample $\hat{\beta}_1$ is relative to those $\hat{\beta}_1$'s that did not occur. The Bayesian focuses only on the $\hat{\beta}_1$ value actually obtained from the given sample: no comparison is made with other $\hat{\beta}_1$'s that might have been observed.

Example 6.3

To conduct a significance test for the population regression slope β_1, let us consider H_0: $\beta_1 = 0$, versus H_1: $\beta_1 \neq 0$ at the $\alpha = 0.05$ level. Because we determined in Example 6.2 that $P(0.8228 < \beta_1 < 1.6678) = 0.95$, we may reject H_0 in favor of H_1 at the 5% level of significance. That is, for $\alpha = 0.05$, we have a statistically significant linear relationship between X and Y.

Next, let us test H_0: $\beta_1 = 1$, against H_1: $\beta_1 > 1$ at the 5% level. From the information generated in Examples 6.1 and 6.2 and from Equation 6.17, let us determine, for $n - 1 = 8$,

$$P(H_0) = P\left(T \leq \frac{1 - 1.2453}{0.1823}\right) = P(T \leq -1.3455).$$

It is easily verified that $P(H_0) > 0.10$. Because this probability exceeds 0.05, we will reject H_0 in favor of H_1.

Appendix A: Diffuse Prior Information

When the random error term ε_i is drawn from a population that is $N(0, \sigma_\varepsilon)$ for all $i = 1,\ldots,n$, the natural conjugate family of priors under the reparameterization of the regression model (Equation 6.3) was normal, i.e., we were able to express the joint prior distribution of β_* and β_1 as the product of two independent normal marginal prior density functions or $f(\beta_*, \beta_1) = f(\beta_*) \cdot f(\beta_1)$, where $f(\beta_*)$ is $N\left(\mu_{\beta_*}, \sigma_{\beta_*}\right)$ and $f(\beta_1)$ is $N\left(\mu_{\beta_1}, \sigma_{\beta_1}\right)$. This was most convenient because the posterior distribution for each independent parameter is a member of the same family of distributions. However, what happens when we do not have such specific beliefs about how the parameters are distributed?

It may be the case that a researcher has little or no information at hand to estimate a set of regression parameters, i.e., the prior information is *vague* or *diffuse*. In this instance, we shall use what is termed a *diffuse prior probability density function*. To depict *knowing little* about, say, the slope parameter β_1, we shall represent the prior probability density function as

$$f(\beta_1) \propto \text{constant}, -\infty < \beta_1 < +\infty. \tag{6.A.1}$$

Here β_1 is said to have a *noninformative prior*. Note that this specification implies that β_1 follows a uniform prior distribution on the real line. Although Equation 6.A.1 is not a legitimate probability distribution (this prior is improper in that the integral $\int_{-\infty}^{+\infty} f(\beta_1)d\beta_1$ is not finite or does not converge), when combined with the likelihood function $\mathcal{L}(\beta_1)$, the product $f(\beta_1) \cdot \mathcal{L}(\beta_1)$ is proportional to a valid probability density function $f(\beta_1 | Y_1,\ldots,Y_n)$, thus yielding a valid marginal posterior distribution.

If the prior is diffuse relative to the sample information, then the posterior distribution is approximately proportional to the likelihood function. In this instance, because the posterior density depends almost exclusively on sample information, any Bayesian inferences based on the posterior will essentially correspond, numerically, to frequentist inferences based on sample data alone.

A glance back at, say, Equation 6.10.1, reveals that the posterior precision for β_1 is larger than the precision of $\hat{\beta}_1$ alone (or the posterior variance of β_1 is smaller than the variance of $\hat{\beta}_1$ alone). So if our beliefs about the prior marginal distribution of β_1 are "correct," inferences about β_1 based on the posterior distribution of β_1 will be more precise than inferences based exclusively on sample data; the posterior distribution combines estimates from both the sample and one's judgment and thus incorporates more information than from either source taken individually. However, the more noninformative or diffuse the prior distribution (i.e., the larger is $\sigma^2_{\beta_1}$), the less influence the prior has in the determination of the revised mean μ'_{β_1}.

Appendix B: Reparameterizing the Regression Model

In the parameterization of the linear regression equation, we replaced β_0 by $\beta_* = \beta_0 + \beta_1\overline{X}$ (see Equation 6.3). Hence, our prior information about β_0 and β_1 was transformed into a prior distribution concerning β_* and β_1. This was done because prior information about β_* is independent of prior information about β_1, i.e., in the strong classical bivariate linear regression model, the OLS estimates of $\beta_* = \overline{Y}$ and β_1 are independent. (This is not the case for β_0 and β_1 because $COV(\beta_0, \beta_1) = -\overline{X}^2\sigma_\varepsilon^2 \big/ \sum_{i=1}^n x_i^2 \neq 0$.) Hence, under this independence, the joint likelihood function of β_* and β_1 is the product of their individual independent likelihood functions. This being the case, it is also true that the joint prior density of β_* and β_1 can be written as the product of their individual independent prior densities.

Specifically, from Equation 2.A.1, we wrote the OLS estimator of β_1 as a linear combination of the Y_i's or $\hat{\beta}_1 = \sum_{i=1}^n w_iY_i$, where $w_i = x_i \big/ \sum_{i=1}^n x_i^2$ and $\sum_{i=1}^n w_i = 0$. Moreover, $\overline{Y}$ is also a linear combination of the Y_i's because $\overline{Y} = \sum_{i=1}^n \left(\frac{1}{n}\right) Y_i$. In this regard, let us now apply the following well-known result concerning the covariance of two linear functions of a set of random variables. In Equation 6.B.1, if the Y_i, $i = 1,\ldots,n$, are random variables and both ϕ_1 and ϕ_2 are linear combinations of the Y_i or $\varphi_1 = \sum_{i=1}^n a_iY_i$ and $\varphi_2 = \sum_{i=1}^n b_iY_i$, with a_i and b_i constant for all i, then

$$COV(\varphi_1, \varphi_2) = \sum_{i=1}^n a_ib_i\sigma_i^2 + \sum_{i=1}^n \sum_{\substack{j=1 \\ i<j}}^n (a_ib_j + a_jb_i)\sigma_{ij}. \tag{6.B.1}$$

Then, for $a_i = \frac{1}{n}$ for all i and $b_i = w_i$, Equation 6.B.1 yields

$$\begin{aligned} COV(\overline{Y}, \hat{\beta}_1) &= COV\left(\frac{1}{n}\sum_{i=1}^n Y_i, \sum_{i=1}^n w_iY_i\right) \\ &= \frac{1}{n}\sigma^2 \sum_{i=1}^n w_i + \sum_{i=1}^n \sum_{\substack{j=1 \\ i<j}}^n \left(\frac{w_j}{n} + \frac{w_i}{n}\right)\sigma_{ij} = 0 \end{aligned}$$

because $\sum_{i=1}^n w_i = 0$ and $\sigma_{ij} = 0$, $i \neq j$. So with the Y_i's normally distributed, the linear combinations $\overline{Y}$ and $\hat{\beta}_1$ are independent if and only if $COV(\overline{Y}, \hat{\beta}_1) = 0$.

Appendix C: The Joint Posterior Distribution

To obtain Equation 6.8.1, let us work with the individual terms on the right side of Equation 6.8. Specifically, let us first consider the second term in this product or the one involving β_1. Its expanded form appears as

$$e^{-\frac{1}{2}\left[\frac{SS_X\sigma_{\beta_1}^2\left(\beta_1^2-2\beta_1\hat{\beta}_1+\hat{\beta}_1^2\right)+\sigma_\varepsilon^2(\beta_1^2-2\beta_1\mu_{\beta_1}+\mu_{\beta_1}^2)}{\sigma_\varepsilon^2\sigma_{\beta_1}^2}\right]}=$$

$$e^{-\frac{1}{2}\left[\frac{(SS_X\sigma_{\beta_1}^2+\sigma_\varepsilon^2)\beta_1^2-2\left(SS_X\sigma_{\beta_1}^2\hat{\beta}_1+\sigma_\varepsilon^2\mu_{\beta_1}\right)\beta_1+\left(SS_X\sigma_{\beta_1}^2\hat{\beta}_1^2+\sigma_\varepsilon^2\mu_{\beta_1}^2\right)}{\sigma_\varepsilon^2\sigma_{\beta_1}^2}\right]}.$$

If we now factor out $\left(SS_X\sigma_{\beta_1}^2+\sigma_\varepsilon^2\right)/\sigma_\varepsilon^2\sigma_{\beta_1}^2$, complete the square, and absorb any term that does not involve β_1 into a proportionality constant, then the preceding expression becomes

$$e^{-\frac{SS_X\sigma_{\beta_1}^2+\sigma_\varepsilon^2}{2\sigma_\varepsilon^2\sigma_{\beta_1}^2}\left[\beta_1^2-2\frac{\left(SS_X\sigma_{\beta_1}^2\hat{\beta}_1+\sigma_\varepsilon^2\mu_{\beta_1}\right)}{SS_X\sigma_{\beta_1}^2+\sigma_\varepsilon^2}\beta_1+\left(\frac{SS_X\sigma_{\beta_1}^2\hat{\beta}_1+\sigma_\varepsilon^2\mu_{\beta_1}}{SS_X\sigma_{\beta_1}^2+\sigma_\varepsilon^2}\right)^2\right]}=$$

$$e^{-\frac{1}{2\sigma_\varepsilon^2\sigma_{\beta_1}^2/(SS_X\sigma_{\beta_1}^2+\sigma_\varepsilon^2)}\left[\beta_1-\frac{SS_X\sigma_{\beta_1}^2\hat{\beta}_1+\sigma_\varepsilon^2\mu_{\beta_1}}{SS_X\sigma_{\beta_1}^2+\sigma_\varepsilon^2}\right]^2}. \tag{6.C.1}$$

Clearly, the form of this marginal posterior density function is $N\left(\mu'_{\beta_1},\sigma'_{\beta_1}\right)$, where the posterior mean is

$$\mu'_{\beta_1}=\frac{SS_X\sigma_{\beta_1}^2\hat{\beta}_1+\sigma_\varepsilon^2\mu_{\beta_1}}{SS_X\sigma_{\beta_1}^2+\sigma_\varepsilon^2};$$

and the posterior variance is

$$\left(\sigma'_{\beta_1}\right)^2=\frac{\sigma_\varepsilon^2\sigma_{\beta_1}^2}{SS_X\sigma_{\beta_1}^2+\sigma_\varepsilon^2}.$$

A similar set of calculations applied to the first term in the product on the right side of Equation 6.8 enables us to transform the said term to

$$e^{-\frac{1}{2\sigma_\varepsilon^2\sigma_{\beta_*}^2/(n\sigma_{\beta_*}^2+\sigma_\varepsilon^2)}\left[\beta_*-\frac{n\sigma_{\beta_*}^2\bar{Y}+\sigma_\varepsilon^2\mu_{\beta_*}}{n\sigma_{\beta_*}^2+\sigma_\varepsilon^2}\right]^2}. \tag{6.C.2}$$

The resulting marginal posterior density function is obviously $N\left(\mu'_{\beta_*},\sigma'_{\beta_*}\right)$, with posterior mean

$$\mu'_{\beta_*} = \frac{n\sigma^2_{\beta_*}\bar{Y} + \sigma^2_{\varepsilon}\mu_{\beta_*}}{n\sigma^2_{\beta_*} + \sigma^2_{\varepsilon}};$$

and posterior variance

$$\left(\sigma'_{\beta_*}\right)^2 = \frac{\sigma^2_{\varepsilon}\sigma^2_{\beta_*}}{n\sigma^2_{\beta_*} + \sigma^2_{\varepsilon}}.$$

Exercises

6-1. An agronomist is studying the relationship between yield of a certain variety of agricultural crop (Y) and level(s) of fertilizer applied (X). Ten plots of land of equal size have different levels of fertilizer mixed into the soil. The experimental results appear in Table E.6-1.

$$\bar{X} = 28.9, \bar{Y} = 31.1, SS_X = 774.9, SS_{XY} = 293.1, \text{ and } s^2_{\varepsilon} = 0.7546.$$

a. Determine the OLS line of best fit.

b. Let the prior distributions for β_* and β_1 be $N(18.9, 1.3)$ and $N(0.5, 0.05)$, respectively. Estimate the posterior precision of β_* and the Bayes' estimate of β_*. Also, estimate the posterior precision of β_1 as well as the Bayes' estimate of β_1.

c. What are the 95% Bayes' credible intervals for β_1 and β_*?

d. Conduct a Bayesian test of H_0: $\beta_1 = 0$ versus H_1: $\beta_1 \neq 0$ at the 0.05 level. Also, test H_0: $\beta_1 \leq 1$ versus H_1: $\beta_1 > 1$ for $\alpha = 0.05$.

6-2. The DOT is interested in the relationship between distance traveled in miles (Y) and speed in miles per hour (X) for a particular make of light truck. Ten runs on a closed track are made, each at a different speed and with one gallon of regular gasoline. The trial results appear in Table E.6-2.

$$\bar{X} = 72.5, \bar{Y} = 19.75, SS_X = 2062.5, SS_{XY} = -257.25, s^2_{\varepsilon} = 0.5298.$$

a. Estimate the OLS line of best fit.

b. Suppose the prior distributions for β_* and β_1 are $N(30, 1.21)$ and $N(-0.07, 0.03)$, respectively. Estimate the posterior precision of β_* and β_1 as well as the Bayes' estimates of β_* and β_1.

c. Find 90% Bayes' credible intervals for β_1 and β_*. Compare them with the 90% classical confidence intervals for these parameters.

d. Conduct a Bayesian test of H_0: $\beta_1 = 0$, against H_1: $\beta_1 \neq 0$ at the $\alpha = 0.01$ level. In addition, test H_0: $\beta_1 \leq 0.05$, against H_0: $\beta_1 > -0.05$, for $\alpha = 0.01$.

TABLE E.6-1

Crop Yield (Y) and Level of Fertilizer (X)

X	10	27	20	28	25	40	39	30	38	32
Y	25	30	27	31	28	36	35	32	34	33

TABLE E.6-2

Distance Traveled in Miles (Y) and Speed in Miles per Hour (X)

X	50	55	60	65	70	75	80	85	90	95
Y	22.7	22.3	21.9	20.2	19.9	19.1	18.3	17.6	17.0	18.5

6-3. Predictive distribution for an additional observation:

Suppose we have established the existence of a statistically significant linear relationship between X and Y and we want to use this expression to forecast or predict a new value of Y (denoted Y_{n+1}) given an additional value of the explanatory variable X (call it X_{n+1}). As expected, the best prediction of Y_{n+1} is

$$Y_{n+1} = \hat{\beta}_* + \hat{\beta}_1(X_{n+1} - \bar{X}). \tag{E.6.1}$$

To assess how good this prediction is, we need to examine the *predictive distribution* of Y_{n+1} given X_{n+1} or $f(Y_{n+1}|X_{n+1}, data)$, where data refers to the initial set of sample observations (X_i, Y_i), $i = 1,\ldots,n$. To this end, let us write the joint posterior distribution of β_*, β_1, and Y_{n+1} given X_{n+1} and the data as

$$f(Y_{n+1}, \beta_*, \beta_1|X_{n+1}, data) = f(Y_{n+1}|\beta_*, \beta_1, X_{n+1}, data)$$

$$\times f(\beta_*, \beta_1|X_{n+1}, data). \tag{E.6.2}$$

Given β_* and β_1, Y_{n+1} is independent of the "data." Also, the posterior density of β_* and β_1 given the data and X_{n+1} is independent of X_{n+1} Hence, Equation E.6.2 simplifies to

$$f(Y_{n+1}, \beta_*, \beta_1|X_{n+1}, data) = f(Y_{n+1}|\beta_*, \beta_1, X_{n+1})$$

$$\times f(\beta_*, \beta_1|data). \tag{E.6.2.1}$$

Thus, the joint posterior distribution of β_*, β_1 and Y_{n+1} is expressible as the product of the distribution of Y_{n+1} given β_*, β_1, and X_{n+1} and the posterior density of β_* and β_1 given our initial sample. Under normality, the random observation $Y_{n+1}|\beta_*, \beta_1, X_{n+1}$ drawn from the population regression model has mean $\mu_{n+1} = \beta_* + \beta_1(X_{n+1} - \bar{X})$ and variance σ_ε^2, *i.e.*, $Y_{n+1}|\beta_*, \beta_1, X_{n+1} : N(\mu_{n+1}, \sigma_\varepsilon^2)$.

We noted above (see "Bayes' Rule and the Strong Classical Linear Regression Model") that the posterior distributions of β_* and β_1 are independent $N(\mu'_{\beta_*}, (\sigma'_{\beta_*})^2)$ and $N(\mu'_{\beta_1}, (\sigma'_{\beta_1})^2)$, respectively. Because the right side of $\mu_{n+1} = \beta_* + \beta_1(X_{n+1} - \bar{X})$ consists of the sum of two independent terms, it follows that the posterior distribution of μ_{n+1} is $N(\mu'_{n+1}, (\sigma'_{n+1})^2) = N(\mu'_{\beta_*} + \mu'_{\beta_1}(X_{n+1} - \bar{X}), (\sigma'_{\beta_*})^2 + (\sigma'_{\beta_1})^2(X_{n+1} - \bar{X})^2)$. Then it can be shown that the predictive distribution of Y_{n+1} given X_{n+1} is

$$f(Y_{n+1}|X_{n+1}, data) \propto e^{-\frac{1}{2(\sigma_\varepsilon^2 + (\sigma'_{n+1})^2)}(Y_{n+1} - \mu'_{n+1})^2}, \tag{E.6.3}$$

i.e., $N(\mu'_{n+1}, \sigma_\varepsilon^2 + (\sigma'_{n+1})^2)$. Thus, the predictive distribution of Y_{n+1} at X_{n+1} is normal with mean equal to the posterior mean of μ_{n+1} and variance equal to σ_ε^2 plus the posterior variance of μ_{n+1}.

To find a 100(1 – α)% Bayes' credible predictive interval for Y_{n+1} given X_{n+1}, let us use Equation E.6.3 to find

$$m'_{n+1} \pm t_{\alpha/2,n-2} S'_{n+1}, \tag{E.6.4}$$

where

$$m'_{n+1} = m'_{\beta_*} + m'_{\beta_1}(X_{n+1} - \bar{X}),$$

$$(S'_{n+1}) = (S'_{\beta_*})^2 + (S'_{\beta_1})^2 (X_{n+1} - \bar{X}).$$

Given Equation E.6.4,

a. Use the Table E.6-1 dataset to find a 95% Bayes' credible predictive interval for Y_{n+1} given $X_{n+1} = 37$. Interpret your result.
b. Use the Table E.6-2 dataset to find a 99% Bayes' credible predictive interval for Y_{n+1} given $X_{n+1} = 78$. Interpret your result.

7

Robust Regression

Robust Regression [Holland and Welsch 1977; Rousseeuw and Leroy 1987; Huber 1996, 2004]

We noted previously (Chapter 2) that, under the assumptions of the strong classical linear regression model, the OLS estimates of the regression intercept (β_0) and slope (β_1) have certain desirable or optimal properties. For this model, a key assumption concerning the random error term ε was that it followed a normal distribution with zero mean and constant variance. (Under normality, the OLS estimates are also maximum-likelihood estimates.) We also noted (Chapter 4) that, if we were unsure of the form of ε's probability density function, then a nonparametric (e.g., rank-based) regression method was called for. However, it may be the case that the probability density function of ε departs only slightly from the aforementioned normal form. In this situation, a regression method intermediate to OLS and nonparametric regression is warranted. Such a method is referred to as *robust regression.*

Following Huber [1996, 2004], we shall employ a narrow definition of *robustness* as the insensitivity of the regression estimates to "small" departures from the usual OLS (strong) assumptions. In particular, we shall be primarily concerned with *distributional robustness,* i.e., we seek to assess the impact on the regression estimates when the form or shape of the random error distribution derivates slightly from the assumed normal form.

A particularly thorny problem that often emerges in regression analysis is that the random error distribution has "heavy tails." That is, the tails of the probability density function of ε are longer than those of the normal distribution, and thus, if both are taken over the same fixed (finite) interval, the heavy-tailed density encompasses much more area or probability than the normal density. (For instance, lengthening the tails of the error distribution beyond those of the normal distribution causes a sizeable increase in the variability of the estimate of σ_ε, which obviously decreases the precision of the OLS estimates of β_0 and β_1.)

The impact of heavy tails is a tendency for the random process that determines ε to produce certain errors ε_k that are larger than those that would be produced by a normal error distribution, i.e., heavy-tailed error distributions have a tendency to generate outliers. As we found previously (Chapter 3), influential outliers adversely affect the OLS intercept and slope estimates by pulling the regression line in their direction. Moreover, we cannot simply eliminate these offending sample points and estimate, via OLS, a regression equation based on the remaining observations because no satisfactory benchmark (e.g., robust) parameter estimates are typically available. In this regard, a distributionally robust regression procedure is one that dampens the effect of influential observations on the OLS regression estimates.

What are the trappings or characteristics of a useful robust regression procedure? Any such method should

1. Be expected to perform reasonably well if ε is $N(0, \sigma_\varepsilon)$, i.e., it should yield estimates that are approximately the same as those obtained by OLS when ε is normal and no outliers are present.
2. Be fairly insensitive to "small" departures from error term normality, whereas slightly larger departures from the same should not render the regression estimates meaningless.
3. Have at least near-optimal efficiency for the specific model.
4. Exhibit good asymptotic performance.
5. Perform reasonably well if ε is actually $N(0, \sigma_\varepsilon)$ but becomes tainted by large ε_k's when the random process that generates ε at times lacks control and subsequently causes ε to follow some arbitrary and unknown density function.

In what follows, we shall use a robust regression technique called *M-regression* to obtain so-called *M-estimates* of the regression parameters β_0 and β_1. As will be explained in the next few sections, our robust regression method renders *M*-estimates $\hat{\beta}_0$ and $\hat{\beta}_1$ of β_0 and β_1, respectively, by using a weighting scheme that weights the data points unequally (OLS weights each observation equally), in which observations that produce large residuals are appropriately down weighted, with the result that such sample points cannot unduly influence the regression estimates by tilting the regression line in their direction.

M-Estimators [Huber 1996, 2004]

Any estimator T_n (dependent on the sample size n) defined by the minimization problem

$$\min \sum_{i=1}^{n} \rho(X_i; T_n), \tag{7.1}$$

in which ρ is an arbitrary function, is termed an *M*-estimator, where *M* signifies that T_n is a "maximum-likelihood-type estimator." That is, if θ is a fixed parameter in the probability density function $f(x; \theta)$ and a sample of size n is drawn from this distribution, then the log-likelihood function of the sample is of the form

$$\log \mathcal{L}(\theta; x_1, \ldots, x_n, n) = \sum_{i=1}^{n} \log f(x_i; \theta)$$

and thus using $\rho(x_i; \theta) = -\log f(x_i; \theta)$ enables us to find the ordinary maximum-likelihood estimate $\hat{\theta}$ of θ (because the minimum of f occurs at the same value of θ as the maximum of f). Alternatively, for $\psi(X_i; T_n) = \partial\rho(X_i; T_n)/\partial T_n$, an *M*-estimator T_n can be defined as one that provides the solution to the implicit equation

$$\sum_{i=1}^{n} \psi(X_i; T_n) = 0. \tag{7.2}$$

For instance, for T_n, a *location M-estimator*, Equations 7.1 and 7.2 become, respectively,

$$\min \sum_{i=1}^{n} \rho(X_i - T_n) \tag{7.1.1}$$

and

$$\sum_{i=1}^{n} \psi(X_i - T_n) = 0. \tag{7.2.1}$$

If Equation 7.2.1 is rewritten as

$$\sum_{i=1}^{n} \frac{\psi(X_i - T_n)}{X_i - T_n}(X_i - T_n) = \sum_{i=1}^{n} w_i (X_i - T_n) = 0, \tag{7.2.2}$$

with weights (dependent on the sample data)

$$w_i = \frac{\psi(X_i - T_n)}{X_i - T_n}, \quad i = 1, \ldots, n, \tag{7.3}$$

then Equation 7.2.2 enables us to represent T_n as a weighted mean $T_n = \sum_{i=1}^{n} w_i X_i \Big/ \sum_{i=1}^{n} w_i$.

Although the function ρ is arbitrary, it essentially serves as an indicator or measure of "size." Huber recommends using

$$\rho(x) = \begin{cases} x^2/2, & |x| \le c; \\ c|x| - c^2/2, & |x| > c \end{cases} \tag{7.4}$$

(when x is "large" or $|x| > c$, $\rho(x)$ does not increase as fast as when x is small or $|x| \le c$), where $|x| \le c$ means $-c \le x \le c$ and $|x| > c$ implies $x < -c$ or $x > c$, c a constant. Equivalently,

$$\psi(x) = \begin{cases} -c, & x < -c; \\ x, & -c \le x \le c; \\ c, & x > c. \end{cases} \tag{7.5}$$

Then the weights associated with these specifications of ρ and ψ are

$$w_i = \begin{cases} 1, & |X_i - T_n| \le c; \\ \dfrac{c}{|X_i - T_n|}, & |X_i - T_n| > c. \end{cases} \tag{7.3.1}$$

For example, when $|X_i - T_n| \le c$, $\rho(X_i - T_n) = (X_i - T_n)^2/2$ and thus Equation 7.1.1 becomes

$$\min\left\{\frac{1}{2}\sum_{i=1}^{n}(X_i - T_n)^2\right\}$$

(we seek the least-squares estimator T_n) or Equation 7.2.2 appears as

$$\sum_{i=1}^{n}(X_i - T_n) = 0$$

via Equation 7.3.1, i.e., $w_i = 1$ for all i. Then the implied location M-estimator is $T_n = \overline{X}$, the ordinary sample mean.

We note briefly that M-estimates of location are not scale invariant (e.g., if the X_i's, $i = 1,\ldots,n$, are multiplied by some constant, the new estimator T_n' might not be the same as T_n itself). To make them so, we may combine them with a measure of scale S_n so as to obtain either

$$\min \sum_{i=1}^{n}\rho\left(\frac{X_i - T_n}{S_n}\right)$$

or

$$\sum_{i=1}^{n}\psi\left(\frac{X_i - T_n}{S_n}\right) = 0,$$

where Equations 7.4 and 7.5 are adjusted accordingly. Huber suggested as a robust estimator of scale the *median absolute deviation* (MAD):

$$S_n = MAD_n = median\left\{\left|X_i - median\ (X_i)\right|\right\}.$$

Given this brief introduction to *M*-estimation, let us now apply some of its essential features to the solution of the linear regression problem, i.e., we next look to the *M*-regression method. As we shall now see, the resulting estimates are indeed maximum-likelihood-type estimates.

M-Regression

We found in Chapter 2 that, if the random error terms $\varepsilon_i = Y_i - (\beta_0 + \beta_1 X_i)$, $i = 1,\ldots,n$, are independent and identically distributed with probability density function $f(\varepsilon)$, then the maximum-likelihood estimates of the regression parameters β_0 and β_1 are those values $\hat{\beta}_0$ and $\hat{\beta}_1$, respectively, which maximize the log-likelihood function

$$\begin{aligned}\log \mathcal{L}(\beta_0,\beta_1;X_1,\ldots,X_n,n) &= \sum_{i=1}^{n}\log f(e_i)\\ &= \sum_{i=1}^{n}\log f\left(Y_i-(\beta_0+\beta_1X_i)\right),\end{aligned} \tag{7.6}$$

where e_i is the *i*th residual or deviation from the sample regression line. If ε is $N(0,\sigma_\varepsilon)$, then Equation 7.6 takes on the specific functional form

$$\begin{aligned}\log \mathcal{L}(\beta_0,\beta_1,\sigma_\varepsilon;\,X_1,\ldots,X_n,n) = -\left(\frac{n}{2}\right)\ln 2\pi-\left(\frac{n}{2}\right)\ln \sigma^2\\ -\left(\frac{1}{2\sigma^2}\right)\sum_{i=1}^{n}(Y_i-(\beta_0+\beta_1X_i))^2.\end{aligned} \tag{7.6.1}$$

Then it is readily shown that the maximum-likelihood estimates $\hat{\beta}_0$ and $\hat{\beta}_1$ are coincident with the OLS estimates of the regression parameters.

Let us generalize this approach to the robust regression case by simply replacing $\log f(e_i)$ in Equation 7.6 by the arbitrary or unspecified *weighting function* $\rho(e_i/S_\varepsilon)$, where S_ε is a robust estimate of scale. All we require of ρ is that it is smooth, symmetric ($\rho(-e_i/S_\varepsilon) = \rho(e_i/S_\varepsilon)$) for all e_i/S_ε, and, for large derivations e_i, increases less rapidly than $(e_i/S_\varepsilon)^2$, $i = 1,\ldots,n$. Then $\hat{\beta}_0$ and $\hat{\beta}_1$ are robust or *M*-estimates of β_0 and β_1, respectively, if they solve the problem

$$\min_{\beta_0,\beta_1}\sum_{i=1}^{n}\rho\left(e_i/S_\varepsilon\right) = \min_{\beta_0,\beta_1}\sum_{i=1}^{n}\rho\left(\frac{Y_i-(\beta_0+\beta_1X_i)}{S_\varepsilon}\right). \tag{7.7}$$

It is important to note that the underlying distribution of the random variable ε and the functional form of ρ are intimately related. For instance, we noted above that, if $\varepsilon{:}N(0,\sigma_\varepsilon)$, then the weighting function $\rho = \frac{1}{2}\left(e_i/S_\varepsilon\right)^2$. In general, knowing $f(\varepsilon)$ immediately specifies $\rho = (e_i/S_\varepsilon)$. However, if no compelling distributional assumption concerning ε presents itself or seems appropriate, then we must, as a practical matter, use those ρ's that have exhibited reasonable success in solving robust regression problems. Two such weighting schemes are provided in Table 7.1 , where $\rho(u)$ has as its argument the scaled residual $u = e_i/S_\varepsilon$ with $\psi(u) = \partial\rho/\partial u$.

TABLE 7.1

Weighting Functions for Robust Regression

Weighting Function $(\rho(u))$	Weight $(w(u)=\psi(u)/u)$	Tuning Constant (a)
	[Huber 1964, 1996, 2004]	
$\rho(u)=\begin{cases}\frac{1}{2}u^2, & \lvert u\rvert \le a;\\ a\lvert u\rvert-\frac{1}{2}a^2, & u<-a \text{ or } u>a.\end{cases}$	$w(u)=\begin{cases}1, & \lvert u\rvert \le a;\\ a/\lvert u\rvert, & u<-a \text{ or } u>a\end{cases}$	$a=2$
	[Beaton and Tukey 1974]	
$\rho(u)=\begin{cases}\frac{1}{2}u^2-\frac{1}{2}\left(\frac{u^2}{a}\right)^2+\frac{1}{6}\left(\frac{u^3}{a^2}\right)^2, & \lvert u\rvert \le a;\\ (a/2)^2, & u<-a \text{ or } u>a.\end{cases}$	$w(u)=\begin{cases}\left(1-u^2/a^2\right)^2, & \lvert u\rvert \le a;\\ 0, & u<-a \text{ or } u>a.\end{cases}$	$5\le a\le 6$

We noted previously that, under OLS estimation, the appropriate weighting function appears as $\rho(u)=\frac{1}{2}u^2$. In this instance, $w(u) = \psi(u)/u = 1$ for each scaled residual $u = e_i/S_\varepsilon$. Because OLS assigns each scaled residual the same weight, large scaled residuals tend to have a rather sizeable impact on the regression coefficients relative to small ones. However, the $\rho(u)$'s in Table 7.1 are structured so that the resulting $w(u)$'s weight the scaled residuals unequally, i.e., they tend to downweight or dampen the effect of large scaled residuals that are associated with influential observations.

The *M*-Estimation Procedure

Let us now see exactly how the weighting function is used in the determination of robust regression estimates. Instead of working with $\rho(u)$ directly, we shall now demonstrate that M-estimates of β_0 and β_1 can be expressed in terms of the derivative of $\rho(u)$ or $\psi(u) = \partial\rho/\partial u$. That is, instead of solving the minimization problem in Equation 7.6, we can, alternatively, set the derivatives of $\sum_{i=1}^{n}\rho\left(e_i/S_\varepsilon\right)$ with respect to the arguments $\hat{\beta}_0$ and $\hat{\beta}_1$ equal to zero and then solve the resulting set of simultaneous equations for $\hat{\beta}_0$ and $\hat{\beta}_1$. To this end, given that $e_i = Y_i - \left(\hat{\beta}_0 + \hat{\beta}_1 X_i\right)$, a necessary or first-order condition for a minimum is

$$\frac{\partial}{\partial\hat{\beta}_0}\left(\sum_{i=1}^{n}\rho\left(e_i/S_\varepsilon\right)\right)=\sum_{i=1}^{n}\frac{\partial\rho}{\partial u}\frac{\partial u}{\partial\hat{\beta}_0}=\sum_{i=1}^{n}\left(-\frac{1}{S_\varepsilon}\right)\frac{\partial\rho}{\partial u}=0$$

$$\frac{\partial}{\partial\hat{\beta}_1}\left(\sum_{i=1}^{n}\rho\left(e_i/S_\varepsilon\right)\right)=\sum_{i=1}^{n}\frac{\partial\rho}{\partial u}\frac{\partial u}{\partial\hat{\beta}_1}=\sum_{i=1}^{n}\left(-\frac{X_i}{S_\varepsilon}\right)\frac{\partial\rho}{\partial u}=0$$

or

$$\begin{aligned}&\sum_{i=1}^{n}\psi\left(e_i/S_\varepsilon\right)\\&\sum_{i=1}^{n}X_i\psi\left(e_i/S_\varepsilon\right)=0.\end{aligned} \tag{7.8}$$

Let us rewrite the system in Equation 7.8 as

$$\begin{aligned}&\sum_{i=1}^{n}\frac{\left\{\psi\left(e_i/S_\varepsilon\right)/\left(e_i/S_\varepsilon\right)\right\}e_i}{S_\varepsilon}=0\\&\sum_{i=1}^{n}\frac{X_i\left\{\psi\left(e_i/S_\varepsilon\right)/\left(e_i/S_\varepsilon\right)\right\}e_i}{S_\varepsilon}=0.\end{aligned} \tag{7.8.1}$$

If we define weights

$$w_i=\frac{\psi\left(e_i/S_\varepsilon\right)}{e_i/S_\varepsilon},\quad i=1,\ldots,n, \tag{7.9}$$

with $w_i = 1$ if $e_i = 0$, then the system in Equation 7.8.1 simplifies to

$$\begin{aligned}&\sum_{i=1}^{n}w_ie_i=0\\&\sum_{i=1}^{n}X_iw_ie_i=0.\end{aligned} \tag{7.8.2}$$

Here Equation 7.8.2 can be characterized as a system of *weighted least-squares normal equations*; it typically is a highly nonlinear system of simultaneous equations that does not readily admit an explicit solution. Hence, we must solve Equation 7.8.2 via an iterative numerical technique to obtain *M*-estimates of β_0 and β_1. In particular, we shall use a method called *iteratively reweighted least squares* (IRLS).

The Iteratively Reweighted Least-Squares Algorithm

To see exactly how an IRLS routine works, let us consider the IRLS algorithm. The algorithm generates a sequence of parameter estimates from the following steps:

STEP 0. Initialize.

Choose the OLS estimates $\hat{\beta}_0^0$, $\hat{\beta}_1^0$ as the initial estimates of β_0, β_1, respectively, and set $\hat{\beta}_0^k = \hat{\beta}_0^0$, $\hat{\beta}_1^k = \hat{\beta}_1^0$. Fix the estimate S_ε of σ_ε at the initial value S_ε^0.[1]

STEP 1. Given $\hat{\beta}_0^k$ and $\hat{\beta}_1^k$, calculate weights

$$w_i^k = \frac{\psi\left(e_i^k / S_\varepsilon^0\right)}{e_i^k / S_\varepsilon^0}, \quad i = 1, \ldots, n,$$

where

$$e_i^k = Y_i - \left(\hat{\beta}_0^k + \hat{\beta}_1^k X_i\right).$$

STEP 2. Solve the weighted least-squares normal equations (Equation 7.8.2)

$$\sum_{i=1}^{n} w_i^k e_i = 0$$
$$\sum_{i=1}^{n} X_i w_i^k e_i = 0$$

for $\hat{\beta}_0^{k+1}$ and $\hat{\beta}_1^{k+1}$.

STEP 3. The iterations stop when, say, the relative difference for the jth parameter or

$$\frac{\left|\hat{\beta}_j^k - \hat{\beta}_j^{k-1}\right|}{\left|\hat{\beta}_j^{k-1} + 1E-6\right|} \le \tau, \quad j = 0,1,$$

where $\tau > 0$ is a small prespecified tolerance level. Otherwise, proceed to the next step.

STEP 4. Advance k to k+1.

Return to step 1.

An examination of the IRLS algorithm indicates that, when convergence to some stable set of parameter values occurs, those values actually obtained depend on which weighting scheme was selected and how S_ε as well as the tuning constant (that essentially dictates step length at each iteration) were chosen. Once a weighting function $\rho(u)$ with argument $u = e_i/S_\varepsilon$ is chosen, the associated weights w_i (also functions of the scaled residuals e_i/S_ε) are uniquely determined.

Table 7.1 presented two common types of weighting functions and weight patterns. First, for the Huber case, we can set

---

[1] A suggested robust estimator for the scale factor S_ε is the MAD (see above, "M-Estimators"). Also, instead of fixing the value of S_ε at S_ε^0 for each iteration, Huber (1964, p. 96) proposes a way of updating S_ε for each round of calculations. See his "proposal 2."

$$w_i = \begin{cases} 1, & \dfrac{|e_i|}{S_\varepsilon} \le 2; \\ \dfrac{2S_\varepsilon}{|e_i|}, & \dfrac{|e_i|}{S_\varepsilon} > 2. \end{cases} \tag{7.10}$$

Next, looking to the Beaton and Tukey version,

$$w_i = \begin{cases} \left(1 - \dfrac{(e_i/S_\varepsilon)^2}{25}\right)^2, & \dfrac{|e_i|}{S_\varepsilon} \le 5; \\ 0, & \dfrac{|e_i|}{S_\varepsilon} > 5. \end{cases} \tag{7.11}$$

Having developed the essentials of robust regression analysis, our task is now to see exactly how IRLS is used to generate *M*-estimates of β_0 and β_1. To this end, we look to the following example problem.

Example 7.1

Suppose we have a set of n = 15 observations on the variables X and Y (see Table 7.2 and Figure 7.1) and we want to estimate the parameters of the linear regression model $Y = \beta_0 + \beta_1 X + \varepsilon$, where we cannot blithely assume that the random error term $\varepsilon : N(0,\sigma_\varepsilon)$ The OLS fit to this dataset is indicated by the dashed line presented in Figure 7.1. The SAS code for this exercise appears as Exhibit 7.1.

As we shall now see, PROC REG will be run first to generate initial OLS estimates $\hat{\beta}_0$ and $\hat{\beta}_1$ of β_0 and β_1, respectively. A macro (provided by Kathleen Kiernan of the SAS Institute, Cary, NC) subsequently feeds these initial values into PROC NLIN. This latter procedure uses IRLS to determine the final set of parameter estimates.

Exhibit 7.1 SAS Code for IRLS Regression

① The SAS dataset is named *rewttuk*, and the input variables are *Y* and *X*.

② We invoke PROC REG to get initial OLS estimates of the regression parameters β_0 and β_1.

③ The results of the PROC REG estimation are written to the temporary file *ests* by the *outest* option.

④ We request a print of the temporary file *ests*.

⑤, ⑥ We specify SAS dataset *ests*, the contents of which are the entries in the temporary file *ests*.

⑦ We use a macro to identify the initial OLS estimates of β_0 and β_1 from PROG REG.

```
data rewttuk;
   input x y;   ①
datalines;
 1   2
```

TABLE 7.2
Observations on the Variables X and Y

X	Y
1	2
2	6
3	6
4	10
5	3
5	12
6	12
7	25
8	16
8	20
10	22
10	24
11	13
12	22
14	26

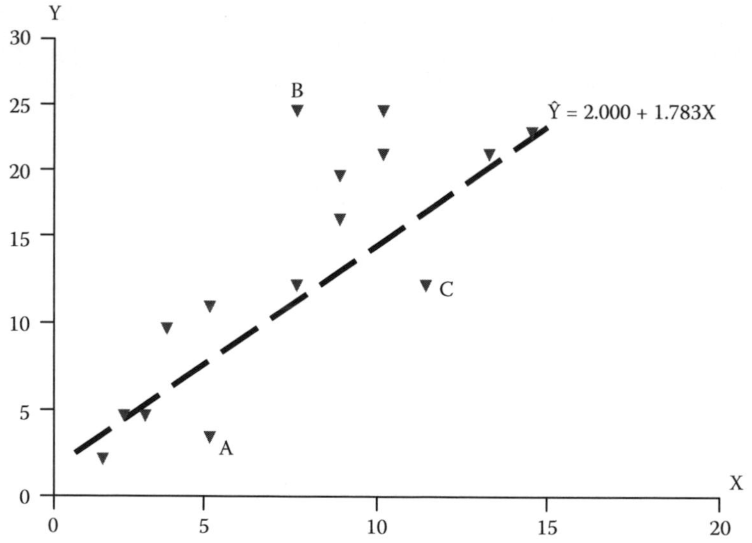

FIGURE 7.1
OLS regression line.

```
2   6
.   .
.   .
.   .
14  26
;
②   ③
proc reg
   data=rewttuk
   outest=ests;
     model y=x;
       run;
proc print data=ests;④
run;
data ests;⑤
set ests; ⑥
     call symput ('init_int ', intercept);  }
                                            } ⑦
     call symput ('init_x', trim(left(x))); }
run;
        ⑧     ⑨     ⑩
proc nlin data = rewttuk hougaard;
     parms intercept=&init_int beta1=&init_x; ⑪
     model y=intercept +bet1*x ; ⑫
     resid=y-model.y; ⑬
     sigma=2; ⑭
     a=5; ⑮
     r=abs(resid/sigma); ⑯
     if r<=a then _weight_=(1- (r/a)**2)**2; }
                                              } ⑰
     else _weight_=0;                         }
     output out=c r = rai;    ⑱
     run;
data c; ⑲
set c; ⑳
     sigma=2;                                 }
     a=5;                                     }
     r=abs(rai/sigma);                        } ㉑
     if r<=a then _weight_=(1-(r/a)**2)**2;   }
     else _weight_=0;                         }
proc print; ㉒
run;
```

⑧ We invoke PROC NLIN to run the IRLS routine.

⑨ The SAS dataset to be used in PROC NLIN is *rewttuk*.

⑩ The *hougaard* option requests a measure of skewness (for each parameter) to determine if the parameter is close to linear or whether it exhibits considerable nonlinearity.[2]

⑪ *Parms* receives, via the macro in ⑦, the initial OLS estimates appearing in the temporary file *ests*.

⑫ We explicitly specify the actual *model* to be estimated in terms of the parameters named in *parms*.

⑬ *Resid* requests the residuals e_i, where *model.y* is a compound variable name that holds the predicted values Y_i for each X_i, $i = 1, \ldots, n$.

⑭ *Sigma* sets the value of σ_ε at 2.

[2] It is important to note that the regression equation to be estimated is *linear in the parameters,* e.g., both $Y = \beta_0 + \beta_1 X$ and $Y = \beta_0 + \beta_1 X^{-1}$ are linear in the parameters, whereas $Y = \beta_0 + \beta_1^{-2} X$ is nonlinear in β_1.

⑮ The constant a specifies the value of the tuning constant a at 5.
⑯ r denotes the absolute value of the scaled residual $u = |e_i/\sigma_\varepsilon| = |e_i|/\sigma_\varepsilon$.
⑰ The Beaton and Tukey weighting scheme (Equation 7.11) is specfied.
⑱ The output dataset c containing the scaled residuals r = rai is named.
⑲, ⑳ We define SAS dataset c, the contents of which are the entries in the output dataet c.
㉑ We iterate until a variety of convergence criteria are satisfied (or lack of convergence is indicatd).
㉒ We request a print of the IRLS results.

Let us now turn to the output generated by this SAS program. The IRLS results are presented in Table 7.3.

① The overall model fit via OLS is adequate because (1)$R^2 = 0.6890$, i.e., about 69% of the variation in *Y* is explained by the linear influence of X, and (2) the *p-value* = 0.0001 associated with the calculated *F* value is well below our usual chosen level of significance of $\alpha = 0.05$.

② The equation of the OLS line of best fit is $\hat{Y} = 2.000 + 1.783X$. Here the *p*-value corresponding to the calculated *t* value for the intercept is 0.4638 (thus, the intercept is not statistically

TABLE 7.3

Output for Example 7.1

The SAS System
The REG Procedure
Model: MODEL 1
Dependent Variable: y

①

Analysis of Variance

Source	DF	Sum of Squares	Mean Square	*F* Value	Pr > *F*
Model	1	651.51509	651.51509	28.80	0.0001
Error	13	294.08491	22.62192		
Corrected total	14	945.60000			

Root MSE	4.75625	*R*-Square	0.6890
Dependent mean	14.60000	Adj *R*-Sq	0.6651
Coeff Var	32.57706		

②

Parameter Estimates

Variable	DF	Parameter Estimates	Standard Error	*t* Value	Pr > | *t* |
Intercept	1	2.00000	2.64964	0.75	0.4638
X	1	1.78302	0.33224	5.37	0.0001

TABLE 7.3 *(continued)*

③

Obs	_MODEL_	_TYPE_	_DEPVAR_	_RMSE_	Intercept	*X*	*y*
1	MODEL1	PARMS	*y*	4.75625	2	1.78302	–1

④

The NLIN Procedure
Dependent Variable y
Method: Gauss-Newton
Iterative Phase

Iter	Intercept	beta1	Weighted SS
0	2.0000	1.7830	50.3068
1	1.6801	1.8564	46.6393
2	1.6278	1.8754	45.6955
3	1.6173	1.8809	45.4167
4	1.6147	1.8826	45.3304
5	1.6140	1.8831	45.3033
6	1.6138	1.8833	45.2948
7	1.6138	1.8833	45.2921
8	1.6138	1.8833	45.2913
9	1.6137	1.8833	45.2910

Note: Convergence criterion met.

Estimation Summary	
Method	Gauss-Newton
Iterations	9
R	5.365E-6
PPC(intercept)	1.225E-6
RPC(intercept)	3.894E-6
Object	5.911E-6
Objective	45.291
Observations read	15
Observations used	14
Observations missing	1

(continued)

TABLE 7.3. *(continued)*

The NLIN Procedure
Dependent Variable y
Method: Gauss-Newton
Iterative Phase

⑤

Source	DF	Sum of Squares	Mean Square	*F* value	Approx Pr > *F*
Regression	2	2947.3	1473.7	163.20	<0.0001
Residual	12	45.2910	3.7743		
Uncorrected total	14	2992.6			
Corrected total	13	661.3			

Parameter	Estimate	Approx Std Error	Approximate 95% Confidence Limits		Skewness
Intercept	1.6137	1.1513	−0.8948	4.1223	0
beta1	1.8833	0.1474	1.5621	2.2045	0

Approximate Correlation Matrix

	Intercept	beta1
Intercept	1.0000000	−0.8653378
beta1	−0.8653378	1.0000000

⑥

Obs	*x*	*y*	RAI	Sigma	*a*	*r*	_weight_
1	1	2	−1.4971	2	5	0.74854	0.95568
2	2	6	0.6196	2	5	0.30979	0.99234
3	3	6	−1.2638	2	5	0.63188	0.96831
4	4	10	0.8529	2	5	0.42646	0.98550
5	5	3	−8.0304	2	5	4.01521	0.12611
6	5	12	0.9696	2	5	0.48479	0.98129
7	6	12	−0.9138	2	5	0.45688	0.98337
8	7	25	10.2029	2	5	5.10145	0.00000
9	8	16	−0.6804	2	5	0.34022	0.99076
10	8	20	3.3196	2	5	1.65978	0.79175
11	10	22	1.5529	2	5	0.77645	0.95235
12	10	24	3.5529	2	5	1.77645	0.76347
13	11	13	−9.3304	2	5	4.66522	0.01675
14	12	22	−2.2138	2	5	1.10689	0.90439
15	14	26	−1.9804	2	5	0.99022	0.92309

significant), whereas the *p*-value for the slope *t* is 0.0001, thus indicating that the slope is statistically significant.

③ We get a print of the contents of the temporary file *ests* or the single observation in the SAS dataset *ests*.

④ PROC NLIN uses the (default) Gauss-Newton method to perform the (nine) iterations. Four separate convergence criteria (R, PPC, RPC, OBJECT) are displayed e.g., RPC (intercept) = 0.0000039 indicates that $\hat{\beta}_0$ changed by 0.0000039 relative to its value at the last iteration. Note that RPC (as well as PPC) is reported only for the parameter estimate with the *largest* PPC value.

⑤ The estimated *M*-regression equation is of the form $\hat{Y} = 1.6137 + 1.8833X$. Clearly, the *M*-estimate of the intercept decreases whereas the *M*-estimate of the slope increases relative to those estimates reported by the OLS procedure. (A glance at the scatter of sample points in Figure 7.1 suggests that this is a reasonable improvement over the OLS estimates.)

In addition, the overall *M*-regression fit is quite good because $R^2 = SSR/SST = 2947.3/2992.6 = 0.9849$ and the $p - value$ associated with the calculated *F* value is less than $0.001 < \alpha = 0.05$; and $H_0 : \beta_1 = 0$ is rejected relative to $H_1 : \beta_1 \neq 0$ for $\alpha = 0.05$ because the Wald approximate 95% confidence interval does not contain the null value of zero. In fact, based on our IRLS estimate of the slope parameter, we may be 95% confident that β_1 lies between 1.5621 and 2.2045.

Finally, the *Hougaard measures of skewness*[3] for both the intercept and slope are zero, thus indicating that each parameter is linear. Hence, the usual hypothesis tests and confidence interval estimates are valid.

⑥ This section displays, for each data point, the final set of scaled residuals $r = |RAI/2|$ as well as the final set of weighting scheme values _WEIGHT_. Note that, for observations 5, 8, and 13 (depicted in Figure 7.1 by points A, B, and C, respectively), the weight values are either very small or zero. This is consistent with the assertion made previously that a robust regression procedure is designed to dampen the impact of influential observations on the OLS estimates: it down weights observations displaying large (scaled) residuals.

Least Median of Squares Estimator [Rousseeuw 1984; Rousseeuw and Yohai 1984; Rousseeuw and Leroy 2003]

We noted previously that a robust estimator of the regression parameters renders estimates that are highly resistant to contamination by outliers. It was also mentioned that even a single outlier point can compromise the OLS estimates of the regression slope and intercept. Indeed, OLS regression estimates cannot readily identify troublesome outlying points.

A robust technique called *M*-estimation for overcoming this shortcoming of OLS was shown above (see "M-Regression," "The M-Estimator Procedure," and "The

[3] If a nonlinear regression model is deemed *close to linear,* then the estimates obtained by the said model have properties similar to those generated using OLS (i.e., they are *close to BLUE*). Hougaard's measure of skewness (computed for each parameter estimated and denoted $H_j, j = 0,1$) is used to determine whether a parameter is close to linear or whether it displays a marked nonlinearity. In this regard, if $|H_j| < 0.1$, the estimator $\hat{\beta}_j$ of β_j is very close to linear in form; if $0.1 < |H_j| < 0.25$, the estimator is reasonably close to linear; if $|H_j| > 0.25$, the estimate is moderately skewed; and if $|H_j| > 1$ the estimator is highly nonlinear. On all this, see Ratakowsky (1990).

Interatively Reweighted Least-Squares Algorithm"). Another way of addressing this issue is to consider a robust estimator that can handle datasets containing a reasonably high percentage of outliers. Specifically, we shall now introduce an estimator that exhibits what is termed a *high breakdown point*. That is, if $Z_n = \{(X_i, Y_i), i = 1,\ldots,n\}$ represents a sample of size n and T is a regression estimator applied to Z_n, then we obtain the estimate $t(Z_n) = (\hat{\beta}_0, \hat{\beta}_1)$, where $t(Z_n)$ is the sample realization of T at Z_n. If we now consider all possible corrupted samples that are generated by replacing any m of the original data points by severely outlying points, then the maximum bias resulting from this replacement can be denoted as *bias* $(m{:}T, Z_n)$. If this bias is infinite, then the m outliers obviously have a significant effect on T, thus causing T to "break down." Hence, the breakdown point of the estimator T at sample Z_n is $\gamma(T, Z_n) = \min\{m/n,$ $bias(m;T, Z_n)$ is infinite$\}$; it is the smallest fraction of contamination that induces T to assume values arbitrarily far from $t(Z_n)$.

Interestingly enough, it can be demonstrated that the highest possible breakdown point for a regression estimator T is 50% (the OLS breakdown point is 0%, thus reflecting the inordinately high sensitivity of OLS to outliers). This said, we now turn to the application of a high breakdown regression technique known as *least median of squares* (LMS). The LMS estimator is given by

$$\min_{\hat{\beta}_o, \hat{\beta}_1} \quad \underset{i}{median}\ e_i^2, \tag{7.12}$$

where $e_i = Y_i - \hat{\beta}_o - \hat{\beta}_1 X_i$, $i = 1,\ldots,n$. Geometrically, finding Equation 7.12 amounts to determining the narrowest band or strip (whose thickness or width is measured vertically) covering about half[4] of the observations with the estimated LMS line located exactly at the center of this strip.

Looking to the properties of the LMS estimator in Equation 7.12, we note briefly the following:

1. Equation 7.12 always has a solution.
2. The regression fit determined by Equation 7.12 is independent of the choice of units in which Y is measured.
3. A linear transformation of the X_i's transforms the LMS estimator in the same manner.
4. The LMS estimator is robust with respect to outliers in both the X and Y directions.
5. The LMS estimator has a breakdown point of 50%.
6. The LMS estimator satisfies the *exact fit property*, i.e., if the majority of the data points satisfy a linear equation exactly, then a robust regression fit should yield this equation.

[4] The fraction of the total number of observations covered by the said band is $[n/2]+1$, where $[n/2]$ is the integer portion of $n/2$. For instance, if $n = 35$ than $[35/2]+1 = [17.5]+1 = 17+1 = 18$.

7. Whereas the *M*-estimator operates to reduce the influence of outliers, the LMS estimator is able to detect the presence of influential outliers and exclude them from the estimation of the model; it performs a least-squares fit on the surviving or remaining observations.

To obtain robust regression estimates via LMS, let us use the SAS/IML subroutine based on the PROGRESS program originally developed by Rousseeuw and Leroy [1987] and modified by Rousseeuw and Hubert [1996]. Here the objective is to minimize the *h*th ordered squared residual, where

$$\frac{n}{2}+1 \le h \le \frac{3n}{4}+\frac{p+1}{4}$$

and *p* is the number of parameters to be estimated. The value of *h* specifies the breakdown point, with its default being $h=\left[\frac{n+p+1}{2}\right]$. Hence, $e^2_{h:n}$ is the *h*th element among the set of ordered squared residuals $e^2_{1:n} \le \ldots \le e^2_{n:n}$. Then, as noted above, the LMS objective is to minimize the *h*th ordered squared residuals $e^2_{h:n}$.

To execute the LMS (PROGRESS) algorithm, either ${}_nC_2 = n!/2!(n-2)!$ subsets containing two observations each (because there are two parameters) are selected from the *n* observations, or, for very large datasets, only a random collection of subsets of two observations each is taken. For each subset or subsample of two different observations each, both $\hat{\beta}_o$ and $\hat{\beta}_1$ are determined, with $\beta=\left(\hat{\beta}_o, \hat{\beta}_1\right)$ termed a *trial solution*. For each such solution, we also determine $e^2_{h:n}$ and the trial estimate for which min $e^2_{h:n}$ is retained.

The LMS routine then determines a robust estimate of σ_ε using the minimal median and a finite sample correction factor of the form 1 + 5/(n – 2), i.e., the preliminary scale estimate is given by

$$S^o_\varepsilon = 1.4826\left(1+\frac{5}{n-2}\right)\sqrt{\underset{i}{median}\ e^2_i} \tag{7.13}$$

assuming normal errors. Here $1.4826 = 1/\Phi^{-1}(0.75)$ (determined from the cumulative standard normal distribution) is used because $\underset{i}{median}|z_i| / \Phi^{-1}(0.75)$ is a consistent estimator for σ_ε when the z_i are $N(0,\sigma_\varepsilon)$. Then S^o_ε is used to determine the weights w_i for the *i*th data point according to

$$w_i = \begin{cases} 1, & \left| e_i / S^0_\varepsilon \right| \le 2.5; \\ 0 & \text{otherwise.} \end{cases} \tag{7.14}$$

Hence, the final scale estimate (devoid of influential outlines) for LMS estimation is

$$\sigma_\varepsilon^* = \sqrt{\sum_i w_i e_i^2 \Big/ \left(\sum_i w_i - 2\right)} \tag{7.15}$$

Next comes a set of reweighted least-squares estimates of $\hat{\beta}_0$ and $\hat{\beta}_1$ based on the preceding LMS results, i.e., here the objective is to

$$\min_{\beta_o, \beta_1} \sum_i w_i^* e_i^2, \tag{7.16}$$

where the weights w_i^* are determined as in Equation 7.14 but with σ_ε^* replacing S_ε^o. Clearly, the effect of this weighting scheme ($w_i = 0$ or 1) is the same as deleting influential outliers and performing OLS regression on the remaining observations; it is a procedure that eliminates the impact of troublesome outliers on our parameter estimates.

Having discussed the salient features of the LMS algorithm, let us now see exactly how this routine is used to determine robust estimates of β_0 and β_1. As the following example problem reveals, the LMS procedure first determines OLS regression estimates, then applies the LMS criterion, and, finally, uses reweighted least squares to generate the final fit to the data points.

Example 7.2

Let us use the $n = 15$ observations on the variables X and Y presented in Table 7.2 and illustrated in Figure 7.1 to obtain the output of the LMS (PROMISE) program. The requisite SAS code for this exercise is provided by Exhibit 7.2.

Exhibit 7.2 SAS Code for LMS Regression

```
proc iml; ①
ab = {1 1 2, 2 2 6, 3 3 6, 4 4 10, 5 5 3, 6 5 12,7 6 12, 8 7 25, 9 8
     16, 10 8 20, 11 10 22, 12 10 24, 13 11 13, 14 12 22, 15 14 26};②
a = ab[,2]; ③
b = ab[,3]; ④
print "*** X-Y: do lms ***";⑤
optn = j(8, 1, .);⑥
optn[2]=3;⑦
optn[3]=3;⑧
optn[8]=3;⑨
           ⑩     ⑪      ⑫
call lms(sc, coef, wgt, optn, b, a); ⑬
run;
```

① We invoke the SAS/IML System.
② *ab* is an array of data points, separated by commas, in which each point is a triple consisting of observation number, *X* coordinate, and *Y* coordinate.
③ *a* contains the second component or the *X* coordinate of the *ab* data array.
④ *b* contains the third component or the *Y* coordinate of the *ab* data array.
⑤ The *X* and *Y* data values are to serve as inputs into the lms subroutine.
⑥ Options 1 through 8, taken sequentially and incremented by one, are to be executed. Unless otherwise explicitly stated, the default for each option applies. Specifically,

optn refers to the *options menu* listing the following elements[5]:

opt[1]=0 is the default: it specifies that an intercept be included in the model.

opt[2]=3 prints arrays such as weights, residuals, and diagnostics; it also prints case numbers of the observations in the best subset and some basic history of the optimization process as well as subsets that result in singular linear systems. (Default is *opt*[2]=0 – no output except error messages.)

opt[3]=3 computes LMS regression along with both unweighted and weighted least-squares regression.

opt[4] specifies the quantile *h* to be minimized. The default is $\left[\frac{n}{2}+1\right]$ or the median (the highest breakdown value).

opt[5]=0 calls for the default number of cases or subgroups to be generated. That is, for two parameters to be estimated, if $n < 50$ all possible or ${}_nC_2$ subgroups are used; otherwise, 1000 subgroups are drawn randomly. An exhaustive search is performed when *opt*[6] = −1.

opt[7] specifies an array containing the parameters β_o, β_1 (rather than the specific subset) for which the objective function is to be evaluated.

opt[7]=3 computes and prints both the covariance matrix and the asymptotic standard errors. (Default is *opt*[8]=0; it does not compute the covariance matrix and asymptotic standard errors.)

⑦ See above.
⑧ See above.
⑨ see above.
⑩ *sc* is an array containing the following:

1. For LMS regression: the quantile *h* used in the objective function; the number of subgroups generated ; the number of subgroups with singular linear systems; the number of nonzero weights w_i; the minimum of the objective function; preliminary scale estimate S_ε^o; final scale estimate σ_ε^*; robust R^2; and an asymptotic consistency factor.

2. If *opt*[3]=3, we obtain, for the weighted least-squares output: the sum of squared residuals; the scale estimate; R^2; and the *F* value.

⑪ *coef* is an array providing the following:

1. the LMS parameter estimates; and the indices of the observations in the best subgroup.

2. if *opt*[3]=3, we obtain, for the weighted least squares output: weighted least squares parameter estimates; approximate standard errors; t-and p-values; and Wald (95%) approximate confidence intervals.

⑫ *wgt* is an array providing the following: weights (with $w_i = 1$ for small residuals and $w_i = 0$ for large residuals); residuals e_i; and a resistant or robust diagnostc RD_i.
⑬ The input/output specifications of the LMS subroutine are completely specified.

Let us now examine the output generated by this SAS/IML program. The results are displayed in Table 7.4.

[5] Although a variety of options exist, the ones to follow give a considerable amount of output and are usually selected. For additional options, see SAS/IML Users Guide Version 8, pp. 600–603.

TABLE 7.4
Output for Example 7.2

The SAS System

①

***** X-Y: do lms *****

LMS: The 9th ordered squared residual will be minimized.

②

Median and Mean

	Median	Mean
VAR1	7	7.0666666667
Intercep	1	1
Response	13	14.6

Dispersion and Standard Deviation

	Dispersion	StdDev
VAR1	4.4478066555	3.8259763771
Intercep	0	0
Response	10.37821553	8.2184461514

Unweighted Least-Squares Estimation

③

LS Parameter Estimates

Variable	Estimate	Approx Std Err	*t* Value	Pr > | *t* |	Lower WCI	Upper WCI
VAR1	1.78301887	0.33224495	5.37	0.0001	1.13183073	2.434207
Intercept	2	2.64964046	0.75	0.4638	–3.1931999	7.19319987

The SAS System

Sum of Squares = 294.08490566

Degrees of Freedom = 13

LS Scale Estimate = 4.7562501848

Cov Matrix of Parameter Estimates

	VAR1	Intercept
VAR1	0.110386707	–0.780066063
Intercept	–0.780066063	7.0205945648

R-Squared = 0.6889965042

F(1,13) statistic = 28.800173227

Probability = 0.0001282868

TABLE 7.4. *(continued)*

The SAS System

LS Residuals

N	Observed	Estimated	Residual	Res/S
1	2.000000	3.783019	−1.783019	−0.374879
2	6.000000	5.566038	0.433962	0.091240
3	6.000000	7.349057	−1.349057	−0.283639
4	10.000000	9.132075	0.867925	0.182481
5	3.000000	10.915094	−7.915094	−1.664146
6	12.000000	10.915094	1.084906	0.228101
7	12.000000	12.698113	−0.698113	−0.146778
8	25.000000	14.481132	10.518868	2.211588
9	16.000000	16.264151	−0.264151	−0.055538
10	20.000000	16.264151	3.735849	0.785461
11	22.000000	19.830189	2.169811	0.456202
12	24.000000	19.830189	4.169811	0.876701
13	13.000000	21.613208	−8.613208	−1.810924
14	22.000000	23.396226	−1.396226	−0.293556
15	26.000000	36.962264	−0.962264	−0.202316

Distribution of Residuals

MinRes	1st Qu.	Median	Mean	3rd Qu.	MaxRes
−8.613207547	−1.396226415	−0.264150943	−1.27676E-15	1.0849056604	10.518867925

The SAS System

④

There are 105 subsets of 2 cases out of 15 cases.

All 105 subsets will be considered.

Complete Enumeration for LMS

Subset	Singular	Best Criterion	Percent
27	0	0.131394	25
53	1	0.131394	50
79	1	0.131394	75
105	3	0.131394	100

(continued)

TABLE 7.4 *(continued)*

Minimum criterion = 0.1313941072
LMS method
Minimizing 9th ordered squared residual
Highest possible breakdown value = 46.67 %
Selection of all 105 subsets of 2 cases out of 15
Among 105 subsets 3 are singular

Observations of Best Subset	
1	14

The SAS System
Estimated Coefficients

VAR1	Intercept
1.8181818182	1.5454545455

LMS objective function = 1.3636363636
Preliminary LMS scale = 2.1187877778
Robust R^2 = 0.9559880679
Final LMS scale = 2.2036783674

LMS Residuals

N	Observed	Estimated	Residual	⑤ Res/S
1	2.000000	3.363636	−1.363636	−0.618800
2	6.000000	5.181818	0.818182	0.371280
3	6.000000	7.000000	−1.000000	−0.453787
4	10.000000	8.818182	1.181818	0.536293
5	3.000000	10.636364	−7.636364	−3.465280
6	12.000000	10.636364	1.363636	0.618800
7	12.000000	12.454545	−0.454545	−0.206267
8	25.000000	14.272727	10.727273	4.867894
9	16.000000	16.090909	−0.090909	−0.041253
10	20.000000	16.090909	3.909091	1.773894
11	22.000000	19.727273	2.272727	1.031333
12	24.000000	19.727273	4.272727	1.938907
13	13.000000	21.545455	−8.545455	−3.877814
14	22.000000	23.363636	−1.363636	−0.618800
15	26.000000	27.000000	−1.000000	−0.453787

Distribution of Residuals

MinRes	1st Qu.	Median	Mean	3rd Qu.	MaxRes
−8.545454545	−1.363636364	−0.090909091	0.2060606061	1.3636363636	10.727272727

TABLE 7.4 *(continued)*

The SAS System
Resistant Diagnostic

N	*U*	⑥ Resistant Diagnostic
1	2.658537	1.000000
2	2.058824	0.774420
3	1.975610	0.743119
4	1.352941	0.508904
5	5.600000	2.106422
6	1.000000	0.376147
7	1.022472	0.384600
8	7.866667	2.959021
9	1.500000	0.564220
10	2.866667	1.078287
11	2.176471	0.818672
12	3.235294	1.216945
13	7.400000	2.783486
14	3.500000	1.316514
15	4.000000	1.504587

Median (*U*) = 2.6585365854

⑦
Weighted Least-Squared Estimation
RLS Parameter Estimates Based on LMS

Variable	Estimate	Approx Std Err	*t* Value	Pr > | *t* |	Lower WCI	Upper WCI
VAR1	1.88643533	0.15001603	12.57	<0.0001	1.59240932	2.18046134
Intercept	1.78548896	1.1930755	1.50	0.1654	–0.552896	4.12387397

Weighted Sum of Squares = 41.615141956
Degrees of Freedom = 10
RLS Scale Estimate = 2.0399789694

Cov Matrix of Parameter Estimates

	VAR1	Intercept
VAR1	0.0225048086	–0.15565826
Intercept	–0.15565826	1.4234291458

(continued)

TABLE 7.4 *(continued)*

Weighted R^2 = 0.9405214741
F(1,10) Statistic = 158.1279058
Probability = 1.8791135E-7
There are 12 points with nonzero weight.
Average weight = 0.8

Weighted LS Residuals

N	Observed	Estimated	Residual	Res/S	Weight
1	2.000000	3.671924	−1.671924	−0.819579	1.000000
2	6.000000	5.558360	0.441640	0.216493	1.000000
3	6.000000	7.444795	−1.444795	−0.708240	1.000000
4	10.000000	9.331230	0.668770	0.327832	1.000000
5	3.000000	11.217666	−8.217666	−4.028309	0
6	12.000000	11.217666	0.782334	0.383501	1.000000
7	12.000000	13.104101	−1.104101	−0.541232	1.000000
8	25.000000	14.990536	10.009464	4.906650	0
9	16.000000	16.876972	−0.876972	−0.429892	1.000000
10	20.000000	16.876972	3.123028	1.530912	1.000000
11	22.000000	20.649842	1.350158	0.661849	1.000000
12	24.000000	20.649842	3.350158	1.642251	1.000000
13	13.000000	22.536278	−9.536278	−4.674694	0
14	22.000000	24.422713	−2.422713	−1.187617	1.000000
15	26.000000	28.195584	−2.195584	−1.076278	1.000000

Distribution of Residuals

MinRes	1st Qu.	Median	Mean	3rd Qu.	MaxRes
−9.536277603	−2.195583596	−0.876971609	−0.516298633	0.7823343849	10.009463722

The run has been executed successfully.

① The ninth ordered squared residual $e^2_{9:15}$ will be minimized because $h = \left[\dfrac{n+p+1}{2}\right] = \left[{}^{18}\!/_{2}\right] = 9$.

② Descriptive statistics such as the median and standard deviation of X and Y are determined.

③ The OLS line of best fit is $\left(\hat{Y} = 2.000 + 1.783X\right)$ with the p-values for $\hat{\beta}_o$ and $\hat{\beta}_1$ equal to 0.4638 and 0.0001, respectively. Clearly, only the estimated slope displays statistical significance. In addition, R^2 = 0.689.

④ Because $n < 50$, all ${}_{15}C_2 = 105$ subsets of two observations each will be completely enumerated and used for LMS estimation. The minimum of the ninth ordered squared residual is 0.1314, the breakdown value is 46.67%, and the two data points within the best subset are 1 (with coordinates (1,2)) and 14 (with coordinates (12,22)). For this subset, the LMS estimates of β_0 and β_1 are 1.545 and 1.818, respectively. The preliminary scale estimate (see Equation 7.13) is $s_\varepsilon^o = 2.1187$. Weights (see Equation 7.14) are determined from the set of e_i/s_ε^o values, and the final scale estimate (see Equation 7.15) is 2.2037. Additionally, a robust R^2 estimate of 0.956 is obtained using

$$R^2 = 1 - \left[\frac{median\left|e_i\right|^2}{MAD(Y_i)}\right]^2,$$

where $MAD(Y_i)$ denotes the *median absolute deviation* of the Y_i's and is calculated as

$$MAD(Y_i) = \underset{i}{median}\{|Y_i - median(Y_i)|\}.$$

⑤ The final scale estimate σ_ε^* is used to find e_i/σ_ε^* (e.g., for observation no. 2, $0.371280 = 0.818182/2.203678$) and thus weights

$$w_i^* = \begin{cases} 1, & \left|e_i/\sigma_\varepsilon^*\right| \le 2.5; \\ 0 & \text{otherwise} \end{cases}$$

will be used in the weighted least-squares estimates generated in ⑦ below.

⑥ The *resistant diagnostic* RD_i is used to identify points that are either outliers in the Y direction or leverage points (whether "harmless" or "bad"). It is defined as

$$RD_i = \frac{u_i}{\underset{j=1,\ldots,n}{median}\, u_j}, \quad i = 1,\ldots,n,$$

where u_i is termed the *outlyingness* of observation i [see Rousseeuw and Leroy 2003, 238–239] and $\underset{j=1,\ldots,15}{median}\, u_j = 2.6585$. The value of this statistic is considered to be large if it exceeds 2.5. It is important to note that, if RD_i is large, the ith observation is not necessarily outlying in that it has a large residual; RD_i will also be large for potentially influential leverage points. Hence, an observation that has a large standardized residual also has a large RD_i, but the converse is not necessarily true, i.e., if $RD_i > 2.5$, the ith observations standardized residual may be small so that this observation is a harmless leverage point that does not influence the overall regression fit. Note that observations 5, 8, and 13 all have fairly large resistant diagnostics as well as large standardized residuals (as indicated by ⑤).

⑦ Weighted least-squares estimates of β_0 and β_1 are obtained by applying the criterion in Equation 7.16, in which observations 5, 8, and 13 each receive a weight of zero. Hence, the weighted least-squares regression equation appears as $\hat{Y} = 1.785 + 1.8864X$. Note that the estimated slope is highly significant (its p-value is <0.0001), whereas the estimated intercept has a p-value far in excess of 0.05 and thus is not statistically significant.

Moreover, a 95% Wald approximate confidence interval for β_1 is (1.5924, 2.1804), the realized weighted R^2 value is 0.941, and the weighted least-squares scale estimate is $s_\varepsilon^w = 2.04$, where

$$S_\varepsilon^w = \sqrt{\frac{\sum_i w_i e_i^2}{\sum_i w_i - 2}}$$

(the OLS scale estimate has $w_i = 1$ for all i).

It should be evident that the estimation results using the LMS technique are very similar to those obtained above when we applied the M-estimation method (Example 7.1).

Exercises

7-1. Given the dataset appearing in Table E.7-1, determine the estimated M-regression equation given that we cannot simply assume that the random error term is normally distributed with a zero mean and a constant variance. Can you detect any "influential" observations? Do the regression parameters exhibit any nonlinearity? (Hint: let Exhibit 7.1 serve as your guide.)

7-2. Determine robust estimates of β_0 and β_1 via the LMS procedure using the Table E.7-1 dataset. In particular, identify the following:

a. The OLS line of best fit and the associated R^2 value.

b. The value of the hth ordered squared residual. What data points are within the "best subset?" What are the LMS estimates of β_0 and β_1?

c. The observations, if any, that have large resistant diagnostics?

d. The WLS regression equation and the associated value of R^2. What is the 95% Wald approximate confidence interval for the regression slope? (Hint: let Exhibit 7.2 guide you.)

TABLE E.7-1

Observations on Variables X and Y

X	2	6	6	7	10	10	15	16	18	20	22	24
Y	22	17	6	19	2	16	15	12	14	10	10	8

8

Fuzzy Regression

Introduction

We noted at the outset of our specification of the (strong) classical linear regression model that the role of the random error term ε_i within $Y_i = \beta_0 + \beta_i X_i + \varepsilon_i$ was to account for the net effect of all excluded variables. Moreover, ε_i is typically taken to be a (normally distributed) random variable with zero mean and a constant variance.

Hence, the operation of $\varepsilon_i = Y_i - E(Y_i/X_i)$ allows for the inexact linear relationship between X_i (call it the *input variable*) and Y_i (termed the *output variable*) attributable to, say, incomplete theory, model misspecification, observation error(s), etc. As noted previously, these modeling/data issues manifest themselves in the behavior of the residuals $e_i = Y_i - \widehat{Y}_i$, where e_i is the difference between the observed and estimated values of Y at the ith data point.

Instead of using $Y_i = \beta_0 + \beta_i X_i + \varepsilon_i$ to model an inexact linear relationship between X and Y, it may be the case that the ε_i's reflect ambiguities inherent in the model or system structure itself, ambiguities that can be traced to the "fuzziness" of the model parameters. This vagueness, resulting from the fuzziness of the system coefficients, leads to the specification of a fuzzy linear regression model, which will be written as $Y = A_0 + A_1 x$, with A_0 and A_1 represented by so-called symmetric triangular fuzzy numbers. So if some system or phenomenon under study has fuzzy or vague properties (in addition to stochastic variability), then the specification of a fuzzy functional relationship between the X and Y variables is warranted. As will be indicated below, the input data (X) is assumed to be crisp (nonfuzzy), whereas the output data (Y) can be either crisp or fuzzy.

When modeling complex systems, exact specifications of such systems are difficult if not impossible to obtain, especially when the modeling involves subjective judgment and intuition. In this circumstance, considerable ambiguity or vagueness may be the norm when it comes to formulating system structure and system parameters. Hence, a vague or imprecise phenomenon is modeled by a *fuzzy function* having fuzzy parameters, with the latter being identified as fuzzy sets.

So when should a fuzzy regression analysis and not, say, OLS, be used? The general consensus is that:

a. Fuzzy linear regression is relevant when we have a vague relationship between X and Y and/or poor model specification. In fact, these variables may be interacting in an uncertain and possibly qualitative manner.

b. Fuzzy linear regression is useful when the model lacks specificity, i.e., there is no sharp transition from nonmembership to membership in, say, one or more groups or categories of, for instance, the output variable Y.

c. Fuzzy linear regression is preferred when the size of the dataset is scant or insufficient to support the (strong) classical linear regression assumptions involving ε. In fact, ε may actually be nonrandom. In addition, one may be forced to work with only a small set of imprecise observation resulting from an ambiguous data generating process or from (questionable) human judgment.

This said, it is generally agreed that classical regression techniques are superior to fuzzy ones in terms of their predictive ability. Classical methods are preferred when the underlying (strong) statistical assumptions can be taken to hold, when the relationship between the model variables is sharply defined, and when we have a significant amount of data. Moreover, fuzzy regression techniques should not be used in the presence of outliers or when there is considerable variability in the data.

In the materials to follow, we first discus the notion of fuzzy sets and their operations, fuzzy functions, and fuzzy arithmetic involving fuzzy numbers. An application of these concepts to the area of fuzzy regression is then offered in subsequent sections.

The Notion of Fuzziness [Ross, Booker, and Parkinson 2002; Zadeh 1964]

Let X represent some universe of elements x with A a nonempty subset of X. In classical set theory, A is taken to be a *crisp set*. That is, an element $x \in X$ is either a member of A or it is not a member of A. Here set membership is on an all-or-nothing basis: there is a precise requirement for membership in A because A, being crisp, has well-defined boundaries. This *binary* notion of set membership is depicted, for $x \in X$, by the *indicator* or *characteristic function*.

$$\chi_A(x) = \begin{cases} 1, & x \in A; \\ 0, & x \notin A. \end{cases} \tag{8.1}$$

This indicator function can also be expressed by the mapping

$$\chi_A : X \to \{0, 1\}.$$

Example 8.1

Suppose $X = \{x|\ 2 \le x \le 6\}$ and $A = \{x|\ 3 \le x \le 5\}$ (the indicator function for membership in A is illustrated in Figure 8.1). So if $x = 4$, $\chi_A(4) = 1$; and if $x = 7$, $\chi_A(7) = 0$.

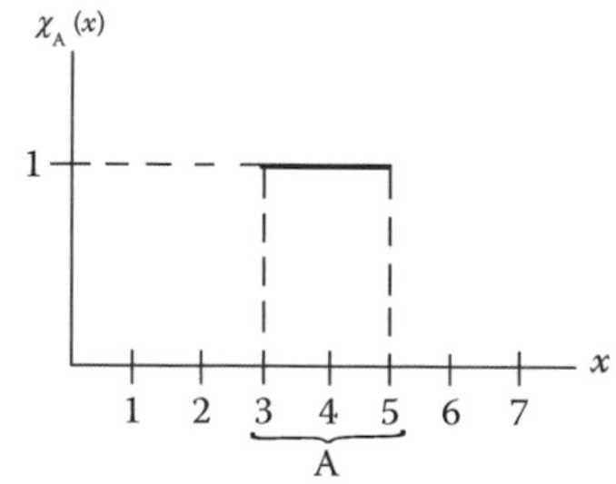

FIGURE 8.1
Indicator function for crisp set A.

However, what if the boundaries of A are not sharply defined or the condition for membership in A is vague. In this circumstance, A is termed a *fuzzy set* (Zadeh 1964). Such sets display imprecise criteria for membership so that set membership can actually be approximate. For example, whereas the set of values "from 3 to 5" (set A of Figure 8.1) is crisp, the set of values "near 4" (call it set B) is less so. In this regard, the degree to which some number x is "near 4" is relative and thus said to be *fuzzy*. In fact, if we again set $A = \{x|\ 3 \le x \le 5\}$, then the complement of A, denoted $\bar{A}$, is defined as all elements within X lying outside of A. Clearly, $A \cap \bar{A} = \varnothing$. However, the set of values $B = \{x|x$ is "near 4"$\}$ and its complement $\bar{B}$ can overlap so that, for fuzzy sets B and $\bar{B}$, $B \cap \bar{B} \neq \varnothing$.

If a fuzzy set has no well-defined boundaries, then the transition from nonmembership to membership must be gradual and not abrupt as in the case of a crisp set.

In this regard, let us modify the notion of binary membership by admitting varying degrees of membership, i.e., let us replace {0, 1} by [0, 1]. Then fuzzy set A in X is represented by the *membership function*:

$$\mu_A : X \to [0, 1], \tag{8.2}$$

where the image of $x \in X$, $\mu_A(x)$, is called the *membership value* of x, which depicts the degree or intensity of membership of x in A. (So if A denotes the set of "old" people, then μ_A gives "the degree of oldness" associated with each element of A.) When $\mu_A(x) = 0$, x is not a member of A; when $\mu_A(x) = 1$, x has complete or full membership in A; and for $0 < \mu_A(x) < 1$, the membership value gives the varying intensity of membership of x in A, i.e., the closer the membership value $\mu_A(x)$ is to 1, the greater the degree of membership of x in A.

We noted above that a fuzzy set A in the universe X is a function $\mu_A : X \to [0, 1]$ that maps elements $x \in X$ onto the unit interval [0, 1], i.e., fuzzy A is the set of ordered pairs $\{(x, \mu_A(x)), x \in X\}$, where $\mu_A(x)$ depicts the grade of membership of x in A. What basic properties should the membership function μ_A possess? If A is the set of values "near a," then μ_A

1. Is not unique: it is subjective and context dependent
2. Displays *normality*, i.e., $\mu_A(a) = 1$
3. Displays *monotonicity* in that the closer $x \in X$ is to a, the closer $\mu_A(x)$ is to 1 and
4. Displays *symmetry* in that $\mu_A(x) = \mu_A(-x)$, $x \in X$

Properties 2 through 4 are summarized in Figure 8.2.

How shall we represent a fuzzy set? Following Zadeh [1964], a useful *function-theoretic notation* for a fuzzy set A on X when X is a finite set $\{x_1, x_2, \ldots, x_n\}$ is

$$\begin{aligned} A &= \sum_{i=1}^{n} \left(\mu_A(x_i) \big/ x_i \right) \\ &= \mu_A(x_1) \big/ x_1 + \mu_A(x_2) \big/ x_2 + \ldots + \mu_A(x_n) \big/ x_n , \end{aligned} \tag{8.3}$$

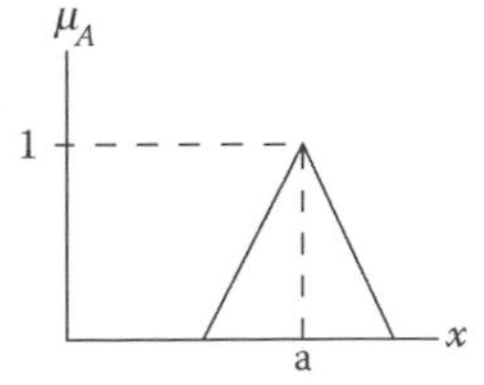

FIGURE 8.2
Typical membership function for fuzzy set A.

where the singleton $\mu_A(x_i)/x_i$ is actually a pair consisting of "membership value $\mu_A(x_i)$ and element x_i"; the slash does not mean quotient but, rather, is a delimiter, the summation sign is used to denote a collection of elements and not algebraic summation, and "+" is meant only in a set-theoretic sense (it is a function-theoretic union and not meant to convey algebraic addition). By convention, singletons with $\mu_A(\cdot) = 0$ are omitted.

Example 8.2

Suppose $X = \{1, 2, \ldots, 10\}$. If A is the set of "small numbers," then clearly A is a fuzzy set and can be written, subjectively, via Equation 8.3, as

$$A = 1/1 + 1/2 + 0.8/3 + 0.5/4 + 0.1/5 + 0.02/6.$$

Properties of and Operations on Fuzzy Sets [Li and Yen 1995; Nguyen and Walker 2000; Mares 1994]

In classical set theory, we are able to perform set-theoretic operations such as the union and intersection of sets as well as finding the complement of a set. However, if, say, A and B are fuzzy sets within some universe X, then the comparable function-theoretic operations on X are stated in terms of membership functions. That is, for fuzzy sets A and B and all $x \in X$, the *union* of A and B, $A \cup B$, is also a fuzzy set with

$$\mu_{A\cup B}(x) = \mu_A(x) \vee \mu_B(x) = \max\{\mu_A(x), \mu_B(x)\}; \tag{8.4}$$

the *intersection* of A and B, $A \cap B$, is also a fuzzy set with

$$\mu_{A\cap B}(x) = \mu_A(x) \wedge \mu_B(x) = \min\{\mu_A(x), \mu_B(x)\};^{1} \tag{8.5}$$

and the *complement* of A, $\bar{A}$, is fuzzy with

$$\mu_{\bar{A}}(x) = 1 - \mu_A(x). \tag{8.6}$$

These membership functions are illustrated in Figure 8.3, A, B, and C, respectively. In addition, for fuzzy set $A \subset X$ and scalar $\alpha \in [0, 1]$, let us define the *scalar product* of A as a new fuzzy set αA with membership function

$$\begin{aligned}\mu_{\alpha A}(x) &= \alpha \wedge \mu_A(x)\\ &= \min\{\alpha, \mu_A(x)\}, x \in X.\end{aligned} \tag{8.7}$$

[1] Here the operations $\vee$ and $\wedge$ are called the *join* and *meet*, respectively. In general, for elements $x, y \in S$, $x \vee y$ is the sup $(x, y) = \max(x, y)$ and $x \wedge y$ is the inf $(x, y) = \min(x, y)$, where "sup" is short for *supremum* (the least upper bound) and "inf" is short for *infimum* (the greatest lower bound).

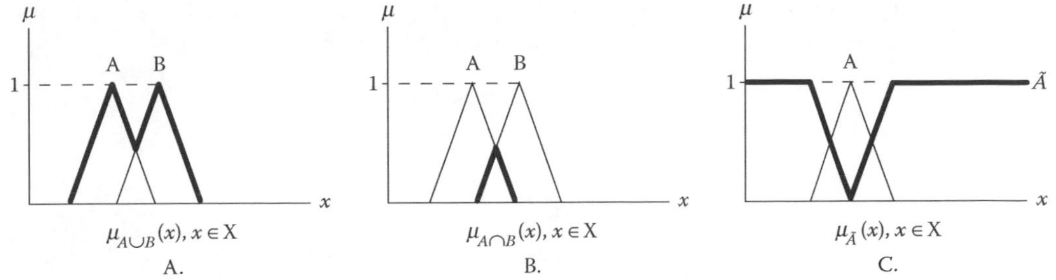

FIGURE 8.3
Membership functions for the union intersection and complement.

So if $\alpha_1 \leq \alpha_2$, then $\alpha_1 A \subset \alpha_2 A$; and for fuzzy $B \subset X$, if $A \subset B$, then $\alpha A \subset \alpha B$.

Next, for fuzzy sets A and B in X, A is a subset of B, or $A \subset B$, if and only if for all $x \in X$, $\mu_A(x) \leq \mu_B(x)$. In addition, A and B are equal, or $A = B$, if and only if $A \subset B$ and $B \subset A$ or, for all $x \in X$, $\mu_A(x) = \mu_B(x)$.

Because a fuzzy set A defined on X is a subset of X, it follows that $A \subset X$ implies $\mu_A(x) \leq \mu_X(x)$. So for all $x \in X$, $\varnothing \subset X$ with $\mu_{\varnothing}(x) = 0$; and $X \subset X$ with $\mu_X(x) = 1$. Moreover, because all fuzzy sets can overlap (thus a set and its complement can overlap), the *complementary laws* of classical set theory do not apply, i.e., $A \cup \bar{A} \neq X$ and $A \cap \bar{A} \neq \varnothing$. That is, given $x_0 \in X$ with $0 < \mu_A(x_0) < 1$,

$$\mu_{\bar{A}}(x_0) \vee (1 - \mu_{\bar{A}}(x_0)) \leq 1 \quad \text{and}$$

$$\mu_{\bar{A}}(x_0) \wedge (1 - \mu_{\bar{A}}(x_0)) > 0$$

so that

$$\mu_{A \cup \bar{A}}(x_0) \neq 1 \text{ and } \mu_{A \cap \bar{A}}(x_0) \neq 0.$$

The membership functions for $A \cup \bar{A}$ and $A \cap \bar{A}$ are illustrated in Figure 8.4.A and 8.4.B, respectively.

An *α-cut* (or *α-level set*) of a fuzzy set A in X is the ordinary or crisp set $A_\alpha \subset X$ such that, for $x \in X$ and $\alpha \in [0, 1]$,

$$A_\alpha = \{x \mid \mu_A(x) \geq \alpha\}. \tag{8.8}$$

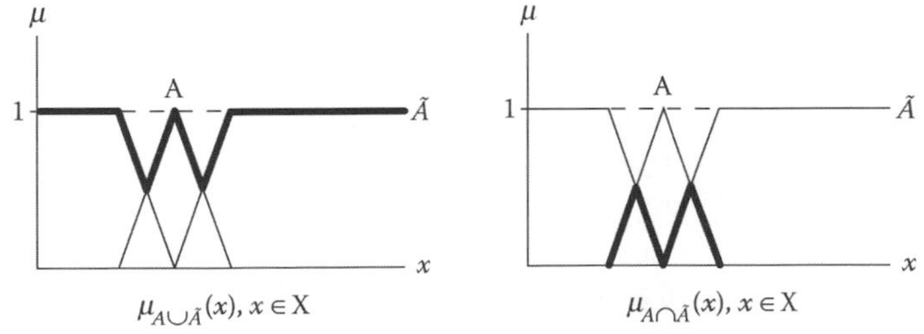

FIGURE 8.4
Membership functions for operations with A and Ā.

A *strong* α*-cut* is obtained when strict inequality holds in Equation 8.8. The *support* of a fuzzy set A in X is the ordinary or crisp set supp(A) $\subset X$ such that, for $x \in X$,

$$\text{supp}(A) = \{x \mid \mu_A(x) \geq 0\}. \tag{8.9}$$

Example 8.3

Suppose $X = \{1, 2, \ldots, 10\}$ with A the set of "small numbers" and B the set of "large numbers." Then A and B can possibly (subjectively) be depicted as

$$A = 1/1 + 1/2 + 0.8/3 + 0.5/4 + 0.2/5 + 0.07/6,$$
$$B = 0.06/5 + 0.3/6 + 0.6/7 + 0.8/8 + 1/9 + 1/10.$$

Then

$$A \cup B = 1/1 + 1/2 + 0.8/3 + 0.5/4 + 0.2/5 + 0.3/6 + 0.6/7 + 0.8/8 + 1/9 + 1/10;$$

$$\begin{aligned}
A \cap B &= 0.06/5 + 0.07/6; \\
\bar{A} &= 1-1/1 + 1-1/2 + 1-0.8/3 + 1-0.5/4 + 1-0.2/5 + 1-0.07/6 + 1-0/7 + 1-0/8 \\
&\quad + 1-0/9 + 1-0/10 \\
&= 0.2/3 + 0.5/4 + 0.8/5 + 0.93/6 + 1/7 + 1/8 + 1/9 + 1/10; \\
\bar{B} &= 1-0/1 + 1-0/2 + 1-0/3 + 1-0/4 + 1-0.06/5 + 1-0.3/6 + 1-0.6/7 + 1-0.8/8 \\
&\quad + 1-1/9 + 1-1/10 \\
&= 1/1 + 1/2 + 1/3 + 1/4 + 0.94/5 + 0.7/6 + 0.4/7 + 0.2/8.
\end{aligned}$$

Also, for A above,

$$A_{0.2} = \{1, 2, 3, 4, 5\},\ A_{0.6} = \{1, 2, 3\}, \text{ and } A_{0.9} = \{1, 2\};$$

$$\text{supp(A)} = \{1, 2, 3, 4, 5, 6\}.$$

For a fuzzy set $A \subset X$ and a scalar $\alpha \in [0, 1]$, we defined an α-cut as the crisp or ordinary set A_α defined by Equation 8.8. We now stipulate via the following *resolution principle* that a fuzzy set $A \subset X$ is decomposable into α cuts, i.e.,

$$\text{if} \quad A \subset X, \text{ then } A = \bigcup_{\alpha \in [0,1]} \alpha A_\alpha. \tag{8.10}$$

[Resolution Principle]

Here Equation 8.10 informs us that a fuzzy set can be represented by a family of crisp sets $\{A_\alpha, \alpha \in [0, 1]\}$. These sets are said to be resolutions of A for various α's, i.e., as α increases (decreases) the A_α's contract (expand). Additionally, because A_α is crisp, the membership function of A in Equation 8.10 can be expressed in terms of the characteristic functions of its α-cuts or

$$\mu_A(x) = \sup \min\{\alpha, \chi_{A_\alpha}(x)\}, \tag{8.11}$$

where

$$\chi_{A_\alpha} = \begin{cases} 1, & x \in A_\alpha; \\ 0, & x \notin A_\alpha. \end{cases}$$

Equivalently, Equation 8.11 can be written as

$$\begin{aligned} \mu_A(x) &= \mu_{\cup \alpha A_\alpha}(x) = \vee\, \mu_{\alpha A_\alpha}(x) \\ &= \sup\{\mu_{A_\alpha}(x)\}, \end{aligned} \tag{8.11.1}$$

where

$$\mu_{\alpha A_\alpha} = \begin{cases} \alpha, & x \in A_\alpha; \\ 0, & x \notin A_\alpha. \end{cases}$$

Fuzzy Relations and Functions [Dubois and Prade 1980]

We first define the direct or *Cartesian product* between two universes and then describe a *fuzzy relation* as a mapping of elements (pairs) in the Cartesian product to the unit interval. In this regard, a fuzzy relation is itself a fuzzy set in the sense defined above. We also specify the Cartesian product between two fuzzy sets within the universes. A variety of definitions of fuzzy functions then follows.

To this end, let X and Y be two distinct universes with elements x and y, respectively. The Cartesian product between X and Y, $X \times Y$, is the set of all ordered pairs (x,y) or $X \times Y = \{(x, y) | x \in X, y \in Y\}$.

Next, let X and Y be two distinct universes with $(x, y) \in X \times Y$. A fuzzy relation on $X \times Y$ is a fuzzy set S with membership function $\mu_S : X \times Y \to [0, 1]$. For each ordered pair (x,y) in $X \times Y$, $\mu_S(x, y)$ gives the strength of the relation between the ordered pair (x, y) from the two universes.

For ordinary or crisp sets, elements x and y are either related or not related. However, for fuzzy S, there is an infinite number of degrees of relationship between x and y because μ_S is a mapping onto [0, 1]. Moreover, the strength of the relationship between x and y is determined by the membership value on [0, 1].

We now define the Cartesian product between two fuzzy sets A and B. Specifically, let X and Y be two universes with $(x, y) \in X \times Y$ and let A and B be fuzzy sets with $A \subset X$ and $B \subset Y$. The Cartesian product between A and B is a fuzzy relation $S = A \times B \subset X \times Y$ with membership function

$$\begin{aligned} \mu_S(x, y) &= \mu_{A \times B}(x, y) = \mu_A(x) \wedge \mu_B(y) \\ &= \min\{\mu_A(x), \mu_B(y)\}. \end{aligned}$$

Fuzzy functions can be defined in a variety of ways depending on where the fuzziness occurs. Following Dubois and Prade [1980], we first consider an ordinary function having "fuzzy properties." Specifically, let X and Y be two universes with f an ordinary function from X to Y or f: $X \to Y$ (i.e., f is a rule that associates with each $x \in X$ a unique point $y = f(x) \in Y$). Let A and B be fuzzy sets with $A \subset X$ and $B \subset Y$. Then f has a fuzzy domain A and a fuzzy range B if and only if for all $x \in X$,

$$\mu_B(f(x)) \geq \mu_A(x). \tag{8.12}$$

Here Equation 8.12 is a constraint on the determination of $f(x)$ for each value of x. For example, suppose we have the proposition that "fast aircraft need long runways." Let X be the set of aircraft, Y the set of runway lengths, and f a rule for assigning runway length $f(x)$ to each aircraft x. Then A is the fuzzy set of fast aircraft, B is the fuzzy set of long runways, and the restriction implied by Equation 8.12 ensures that the faster the aircraft, the longer its runway.

We next consider the case in which a nonfuzzy function imports the fuzziness of its fuzzy arguments without generating any additional fuzziness via the mapping process [Dubois and Prade 1980]. That is, we specify the fuzzy extension of a nonfuzzy function so that fuzziness is unaltered under f.[2] In this regard, let X and Y be two distinct universes and let $F(X)$ (respectively, $F(Y)$) denote the set of fuzzy sets in X (respectively, in Y). Additionally, let f be a nonfuzzy mapping from X to Y, where the image of a fuzzy set $\tilde{x} \in F(\mathcal{X})$ is defined using the extension principle, i.e., $f(\tilde{x}) \in F(\mathcal{Y})$ is defined as

$$\mu_{f(\tilde{x})}(y) = \begin{cases} \sup\limits_{x \in f^{-1}(y)} \mu_{\tilde{x}}(x); \\ 0 \text{ if } f^{-1}(y) = \varnothing, \end{cases} \tag{8.13}$$

where $f^{-1}(y)$ is the set of $\tilde{x}$'s that were mapped into y's. Thus, a function of a fuzzy variable from $F(X)$ to $F(Y)$ is constructed.

Finally, we consider the notion of a fuzzy function of a nonfuzzy variable. Here the image $f(x)$ of nonfuzzy $x \in X$ is "blurred" under the mapping process so that $f(x) \subset F(Y)$ itself is a fuzzy set on Y. In this regard, a *fuzzifying function* f from X to Y is an ordinary function from X to $F(Y)$ or $f : x \to f(x) \subset F(Y)$. Note that the image of $x \in X$ is a fuzzy set $f(x)$ on Y. In addition, the image of a fuzzifying function is equivalent to that of a fuzzy relation, i.e., for all $(x, y) \in X \times Y$, $\mu_{f(x)}(y) = \mu_f(x, y)$, where the membership value depicts the degree of the relation between x and $f(x)$.

---

[2] The *extension principle* of Zadeh [1965,1975] provides a general method for extending or carrying over nonfuzzy mathematical notions or entities to the realm of fuzzy concepts, i.e., this principle enables us to fuzzify ordinary sets, numbers, etc. Specifically, for universes X and Y, if $f : X \to Y$ is a function and A is a fuzzy set in X, then A induces, via f, a fuzzy set B in Y with

$$\mu_B(y) = \begin{cases} \sup\limits_{\substack{x \\ y=f(x)}} \mu_A(x); \\ 0 \text{ if } f^{-1}(y) = \varnothing. \end{cases}$$

[Extension Principle]

As a practical matter, we need only replace crisp sets by fuzzy sets or, equivalently, replace crisp variable values by membership values that are associated with fuzzy constructs.

Fuzzy Numbers and Fuzzy Arithmetic [Mares 1994]

A *fuzzy quantity* a is any fuzzy subset of the real line R having membership function $\mu_a : R \rightarrow [0, 1]$ such that:

1. $\sup\{\mu_a(x), x \in R\} = 1$
2. There exist elements $x_1, x_2 \in a$, with $x_1 < x_2$, such that for all $x \in R$, $x \notin (x_1, x_2)$ implies $\mu_a(x) = 0$. (8.14)

(If the support set of a, $\{x \in R | \mu_a(x) > 0\}$, is bounded, then property 1 of Equation 8.14 becomes: there is a unique $x_0 \in R$ such that $\mu_a(x_0) = 1$.)

In this regard, a *fuzzy number* is a fuzzy quantity a that generalizes the notion of a real number $r \in R$. Specifically, $\mu_a(x)$ constitutes a measure of how well a approximates r (Figure 8.5), where $\mu_a(r) = 1$.

More formally, a fuzzy number is a fuzzy quantity a that satisfies the following:

1. $\mu_a(x) = 1$ for exactly one $x \in R$.
2. The support of a (supp(a)) is bounded and the α-cuts of a ($a\alpha$) are closed intervals in R. (8.15)
3. A fuzzy number is convex, i.e., for $x_1, x_2 \in R$ with $x_1 \leq x_2$ and any $x \in [x_1, x_2]$, $\mu_a(x) \geq \min\{\mu_a(x_1), \mu_a(x_2)\}$. (Here x can be represented as $x = \lambda x_1 + (1 - \lambda)x_2$, $\lambda \in [0, 1]$.)

By virtue of these properties, it follows that, if a is a fuzzy number with $\mu_a(r) = 1$, then a is nondecreasing on $(-\infty, r]$ and non-increasing on $[r + \infty)$. Here r is termed the *modal value* of fuzzy number a.

On the basis of this discussion, it is evident that the elements of R are fuzzy numbers. (Examples of fuzzy numbers abound: "a bit more than 9;" "around 5;" "more or less between 10 and 12," etc.)

To define arithmetic operations for fuzzy quantities, let us invoke the extension principles introduced previously. This device will enable us to extend operations on R to $F(R)$, the collection of fuzzy subsets of R. That is, if f is a function from universe X to universe Y, then f is extended to a function from $F(X)$ to $F(Y)$, where $F(X)$ (respectively, $F(Y)$) is the set of fuzzy subsets of X (respectively, of Y).

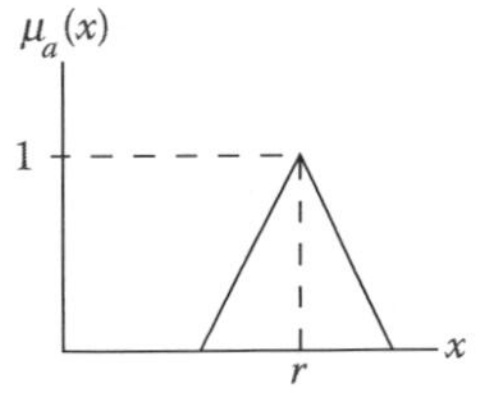

FIGURE 8.5
Fuzzy quantity a approximates r.

The function f is typically taken to be the sum, product, difference, or quotient of, say, fuzzy quantities a and b over $F(R)$. Hence, we can extend, via f, the arithmetic operations of addition, multiplication, subtraction, and division to fuzzy sets, thus developing a fuzzy arithmetic. What is needed is a generalized method for determining the membership function of fuzzy $c=f(a,b)$ from the membership functions of a and b.

To this end, our objective is to extend the *binary operation*[3] described by $f: R \times R \rightarrow R$ to the operation on $F(R)$. For $a, b \in F(R)$, let $c = f(a, b)$. Then the membership function for c is determined from the membership functions for a and b according to

$$\mu_c(z) = \sup_{z=f(x,y)} [\mu_a(x) \wedge \mu_b(y)] \tag{8.16}$$

for $x, y, z \in R$. Then for all $x, y, z \in R$, the aforementioned fundamental arithmetic operations are as follows:

Addition

$$1.\ \mu_{a+b}(z) = \max_{z=x+y} \left[\min\{\mu_a(x), \mu_b(y)\}\right];$$

Multiplication

$$2.\ \mu_{a.b}(z) = \max_{z=x.y} \left[\min\{\mu_a(x), \mu_b(y)\}\right]; \tag{8.17}$$

Subtraction

$$3.\ \mu_{a-b}(z) = \max_{z=x-y} \left[\min\{\mu_a(x), \mu_b(y)\}\right];$$

Division

$$4.\ \mu_{a\div b}(z) = \max_{\substack{z=x\div y \\ y\neq 0}} \left[\min\{\mu_a(x), \mu_b(y)\}\right].$$

Example 8.5

For fuzzy quantities $a,b \in R$ with

$$\mu_a(1) = \frac{1}{2},\ \mu_a(2) = \frac{3}{4},\ \mu_a(3) = 1, \text{ and } \mu_a(x) = 0 \text{ for } x \notin \{1,2,3\};$$

and

$$\mu_b(4) = \frac{1}{3},\ \mu_b(5) = \frac{2}{3},\ \mu_b(6) = 1, \text{ and } \mu_b(x) = 0 \text{ for } x \notin \{4,5,6\},$$

[3] In general, a binary operation on R is the function $f: R \times R \rightarrow R$, i.e., if f constitutes, say, "addition," then $(p, q) \in R \times R$ is mapped by f into $(p + q) \in R$.

let us determine $\mu_{a+b}(7)$. Because 7 = 1 + 6 or 3 + 4 or 2 + 5, we have

$$\begin{aligned}\mu_{a+b}(7) &= \max\left[\min\{\mu_a(1),\ \mu_b(6)\},\ \min\{\mu_a(2),\ \mu_b(5)\},\ \min\{\mu_a(3),\ \mu_b(4)\}\right] \\ &= \max\left[\min\left\{\frac{1}{2},1\right\},\ \min\left\{\frac{3}{4},\frac{2}{3}\right\},\ \min\left\{1,\frac{1}{3}\right\}\right] \\ &= \max\left[\frac{1}{2},\frac{2}{3},\frac{1}{3}\right] = \frac{2}{3}.\end{aligned}$$

Additionally, let us find $\mu_{a.b}(12)$. Because 12 = 2 × 6 or 3 × 4, we have

$$\begin{aligned}\mu_{a.b}(12) &= \max\left[\min\{\mu_a(2),\ \mu_b(6)\},\ \min\{\mu_a(3),\ \mu_b(4)\}\right] \\ &= \max\left[\min\left\{\frac{3}{4},1\right\},\ \min\left\{1,\frac{1}{3}\right\}\right] \\ &= \max\left[\frac{3}{4},\frac{1}{3}\right] = \frac{3}{4}.\end{aligned}$$

L-*R* and Triangular Representation of Fuzzy Numbers

At times, the use of the extension principle for fuzzy arithmetic calculations can be very tedious. To overcome this inefficiency, an assortment of stylized representations of fuzzy numbers have been developed. First, suppose that a fuzzy number M appears in the *L-R form* with membership function (Figure 8.6)

$$\mu_M(x) = \begin{cases} L\left(\dfrac{m-x}{\alpha}\right), & \alpha > 0, \text{ for all } x \le m; \\ R\left(\dfrac{x-m}{\beta}\right), & \beta > 0, \text{ for all } x \ge m, \end{cases} \tag{8.18}$$

where the function L (called the *left reference*) is such that $L(x) = L(-x)$; (2) $L(0) = 1$; and (3) L is decreasing over $[0, +\infty)$; and the function R (called the *right reference*) is such that

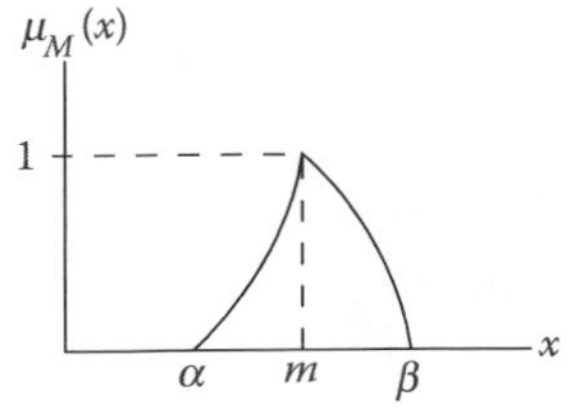

FIGURE 8.6
L-R representation of a fuzzy number M (= $(m, \alpha, \beta)_{LR}$)).

(1) $R(x) = R(-x)$; $R(0) = 1$; and (3) R is decreasing on $[0, +\infty)$. In Equation 8.18, m is the *mean value* of the fuzzy number M, α and β are the left and right spreads, respectively (so as the spreads increase, the fuzziness of M likewise increases), and, when $\alpha = \beta = 0$, the fuzzy number M becomes crisp.

For convenience, a fuzzy number M is written as the triple $M = (m_M, \alpha_M, \beta_M)$ or $M = (m, \alpha, \beta)_{LR}$. Then arithmetic operations involving L-R fuzzy numbers can be expressed in terms of their mean values and left and right spreads. Following Dubois and Prade [1980], suppose M and N are two fuzzy numbers in the L-R form or

$$M = (m, \alpha, \beta)_{LR}, N = (n, \gamma, \delta)_{LR}.$$

Then the *sum* of M and N is

$$M + N = (m + n, \alpha + \gamma, \beta + \delta)_{LR}. \tag{8.19}$$

Because the *opposite* of a fuzzy number can be written as

$$-M = (-m, \beta, \alpha)_{RL} \tag{8.20}$$

(note that the left and right references have been interchanged), it follows that the *difference* between M and N is

$$M - N = (m-n, \alpha+\delta, \beta+\gamma)_{LR}. \tag{8.21}$$

Next, if it is assumed that α and γ are small relative to m and n, then the product of M and N can be expressed (approximately) as

1. $(M, N > 0)$ $M \cdot N \approx (mn, m\gamma + n\alpha, m\delta + n\beta)_{LR}$.
2. $(M < 0, N > 0)$ $M \cdot N \approx (mn, n\alpha - m\delta, n\beta - m\gamma)_{RL}$.
3. $(M, N < 0)$ $M \cdot N \approx (mn, -n\beta - m\delta, -n\alpha - m\gamma)_{RL}$. (8.22)

For λ a real scalar and M a fuzzy number in the L-R form, *scalar multiplication* is expressed as follows:

1. $(\lambda > 0)$ $\lambda M = (\lambda m, \lambda\alpha, \lambda\beta)_{LR}$.
2. $(\lambda < 0)$ $\lambda M = (\lambda m, -\lambda\beta, -\lambda\alpha)_{RL}$. (8.23)

Suppose M is a positive L-R fuzzy number. Then the *inverse* of M near $1/m$ is (approximately)

$$M^{-1} = (m, \alpha, \beta)_{LR}^{-1} \approx \left(m^{-1}, \beta m^{-2}, \alpha m^{-2}\right)_{RL}. \tag{8.24}$$

For M negative, because $-(M^{-1}) = (-M^{-1})$, the inverse of M is (approximately)

$$M^{-1} \approx (-m^{-1}, \alpha m^{-2}, \beta m^{-2})_{LR}. \tag{8.25}$$

Finally, if M is a positive L-R fuzzy number and N is a positive R-L fuzzy number, then the quotient between M and N can be approximated as $M \div N = M \cdot N^{-1}$ or

$$\begin{aligned} M \div N &= (m,\alpha,\beta)_{LR} \div (n,\gamma,\delta)_{RL} \\ &\approx \left(\frac{m}{n}, \frac{\delta m + \alpha n}{n^2}, \frac{\gamma m + \beta n}{n^2}\right)_{LR}. \end{aligned} \tag{8.26}$$

We may further simplify the L-R representation of a fuzzy number by assuming that L and R are linear functions of x, i.e., we may formulate what are called *triangular* or *linear* fuzzy numbers. In this regard, if M is a triangular fuzzy number, then its membership function (Figure 8.7.A) appears as

$$\mu_M(x) = \begin{cases} \dfrac{x-\alpha}{m-\alpha}, & \alpha \le x \le m; \\ \dfrac{x-\beta}{m-\beta}, & m < x \le \beta; \\ 0 \text{ if } & x < \alpha \text{ or } x > \beta \end{cases} \tag{8.27}$$

for $\alpha \le m \le \beta$. Here the *modal value* of M is m. A triangular fuzzy number M is denoted as the triple $M = (\alpha, m, \beta)$ consisting of the crisp real values α, m, and β.

Hence, arithmetic operations on triangular fuzzy numbers are thus reduced to operations over triples of crisp real numbers, e.g., if $M = (\alpha, m, \beta)$ and $N = (\gamma, n, \delta)$, then

$$M + N = (\alpha + \gamma, m + n, \beta + \delta). \tag{8.19.1}$$

For a real scalar $\lambda > 0$,

$$\lambda M = (\lambda\alpha, \lambda m, \lambda\beta); \tag{8.23.1}$$

if $\lambda < 0$,

$$\lambda M = (\lambda\beta, \lambda m, \lambda\alpha). \tag{8.23.2}$$

In what follows, extensive use will be made of *symmetric triangular* fuzzy numbers. That is, in Equation 8.27, let us set $\alpha = m - c$ and $\beta = m + c$ so that this expression becomes

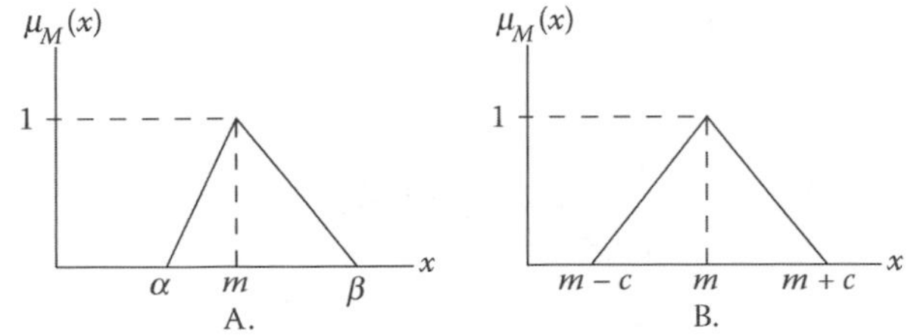

FIGURE 8.7
Triangular fuzzy number $M = (\alpha, m, \beta)$, and symmetric triangular fuzzy number $M = (m, c)$.

$$\mu_M(x) = \begin{cases} \dfrac{x-(m-c)}{m-(m-c)} & , \ m-c \le x \le m \ ; \\ \dfrac{x-(m+c)}{m-(m+c)} & , \ m < x \le m+c \ ; \\ 0 & \text{otherwise} \end{cases}$$

$$= \begin{cases} 1-\dfrac{(m-x)}{c} & , \ m-x > 0 \ ; \\ 1-\dfrac{(x-m)}{c} & , \ m-x < 0 \ ; \\ 0 & \text{otherwise} \end{cases} \tag{8.28}$$

$$= \begin{cases} 1-\dfrac{|x-m|}{c} & , \ m-c \le x \le m+c \ ; \\ 0 & \text{otherwise.} \end{cases}$$

A symmetric triangular fuzzy number will be denoted as $M = (m, c)$.

If a fuzzy number M is symmetric but not necessarily triangular, then M is denoted $M = (m, c)_L$ and defined as

$$\mu_M(x) = L\left(\left.(x-m)\middle/ c\right.\right), \ c > 0, \tag{8.29}$$

where the shape function $L(x)$ is such that $L(x) = L(-x)$; $L(0) = 1$; and L is strictly decreasing on $[0, +\infty)$.

Quite often, Equation 8.28 is arrived at via a different route. That is, let $M = (m, c)_L$ denote a symmetric fuzzy number defined by Equation 8.29. If $L(x) = \{0, 1-|x|\}$, then Equation 8.29 simplifies to Equation 8.28 (see Figure 8.7.B).

We next consider inclusion relations for symmetric fuzzy numbers $M = (m, c)_L$, where $\mu_M(x) = L\left(\left.(x-m)\middle/ c\right.\right)$. The inclusion of such fuzzy numbers with degree $h \in [0, 1]$ is denoted as $A \supset B$ and is defined as $[A]_h \supset [B]_h$, where $[A]_h = \{x | \mu_A(x) \ge h\}$ and $[B]_h = \{x | \mu_B(x) \ge h\}$ are h level sets. If $\mu_A(x) = L\left(\left.(x-m_1)\middle/ c_1\right.\right)$ and $\mu_B(x) = L\left(\left.(x-m_2)\middle/ c_2\right.\right)$, then it is readily demonstrated that $[A]_h \supset [B]_h$ is equivalent to

$$m_1 \le m_2 + |L^{-1}(h)|(c_1 - c_2)$$

$$m_1 \ge m_2 + |L^{-1}(h)|(c_1 - c_2). \tag{8.30}$$

Because $c_1 < c_2$ leads to a contradiction of these inequalities (in terms of m_1 and m_2) for $|L^{-1}(h)| > 0$, $c_1 \ne c_2$ and $m_1 \ne m_2$, we require that $c_1 \ge c_2$. Hence, $A \supset B$ for fuzzy A and B if and only if Equation 8.30 holds and $c_1 \ge c_2$.

Fuzzy Linear Regression Model (Nonfuzzy Data) [Kim, Moskowitz, and Koksalan 1996; Tanaka, Uejima, and Asai 1982]

Consider sets X and Y and a function $y = f(x; a_0, a_1)$, with parameters a_0 and a_1, which is a mapping from X to Y. Suppose the parameters are given by fuzzy sets A_0 and A_1. Then f is called a fuzzy function and defined as

$$f: X \to Y \subset F(y);\ Y = f(x; A_0, A_1), \tag{8.31}$$

where, for a fixed crisp X, fuzzy output Y is the image of fuzzy $A = A_0 \times A_1$ in X (A is defined as the product set of A_0, A_1) and $F(y)$ is the set of all fuzzy subsets of Y. Here the membership function of A is

$$\mu_A(a) = \inf_j \left\{\mu_{A_j}(a)\right\} = \mu_{A_o}(a_o) \wedge \mu_{A_1}(a_1) \tag{8.32}$$

and thus the membership function of Y is

$$\mu_Y(y) = \begin{cases} \sup\limits_{\{a|y=f(x;a_o,a_1)\}} \mu_A(a), & \{a|y = f(x;a_o,a_1)\} \neq \varnothing; \\ 0, & \{a|y = f(x;a_o,a_1)\} = \varnothing \end{cases} \tag{8.33}$$

If A_0 and A_1 are taken to be symmetric triangular fuzzy numbers ($A_j = (\alpha_j, c_j)$), then, from Equation 8.27,

$$\mu_{A_j}(a_j) = \begin{cases} 1 - \dfrac{|\alpha_j - a_j|}{c_j}, & \alpha_j - c_j \le a_j \le \alpha_j + c_j; \\ 0 & \text{otherwise,} \end{cases} \tag{8.34}$$

with $c_j > 0$, $j = 0, 1$. If we now set A = $\{(\alpha_0, \alpha_1), (c_0, c_1)\}$, then the fuzzy linear function $Y = A_0 + A_1 x$ has membership function[4]

$$\mu_Y(y) = \begin{cases} 1 - \dfrac{|y - \alpha_o - \alpha_1 x|}{c_0 + c_1|x|}, & c_o + c_1|x| > |y - \alpha_o - \alpha_1 x|; \\ 0, & c_o + c_1|x| \le |y - \alpha_o - \alpha_1 x|. \end{cases} \tag{8.35}$$

To actualize the fuzzy linear regression model, let us impose the requirement that the fuzzy output (Y) of the model should cover all of the actual observations y_i, $i = 1,\ldots, n$, to a certain (arbitrary) degree $h_i \in [0, 1]$. That is, we seek to determine fuzzy parameters A_j, $j = 0, 1$, such that the membership value of each observation y_i or $\mu_{Y_i}(y_i)$, $i = 1,\ldots, n$, is at least as great as the threshold value h_i or y_i is a member of the h_i-level set of Y_i,

[4] For the deviation of this expression see Tanaka, Uejima, and Asai [1982]. Also, for a more general version of Equation 8.35, see Appendix 8.A ("Membership Function for a Fuzzy Linear Function").

$$\left(Y_i\right)_{h_i} = \left\{y_i \middle| \mu_{Y_i}\left(y_i\right) \geq h_i\right\}. \tag{8.36}$$

This restriction is imposed because, under the assumptions of our model, fuzzy Y_i is the fitted or predicted y_i value given x_i and thus a measure of the degree of the fitting of the Y_i's to the y_i's is reflected by these h_i-level sets. Hence, for each y_i, we require that

$$\mu_{Y_i}(y_i) \geq h_i,\ i = 1, \ldots, n. \tag{8.36.1}$$

These restrictions will be termed *threshold conditioning inequalities.* As we shall now see, each of these restrictions produces a pair of inequalities for each data point (x_i, y_i), $i = 1, \ldots, n$. Then from Equations 8.35 and 8.36.1,

$$\mu_{Y_i}(y_i) = 1 - \frac{|y_i - \alpha_o - \alpha_1 x_i|}{c_o + c_1|x_i|} \geq h_i,\ i = 1, \ldots, n,$$

or

$$(1 - h_i)(c_0 + c_1|x_i|) - |y_0 - \alpha_0 - \alpha_1 x_i| \geq 0, i = 1, \ldots, n,$$

or

$$\left.\begin{aligned}(1-h_i)(c_o + c_1|x_i|) + \alpha_o + \alpha_1 x_i \geq y_i\\ (h_i - 1)(c_o + c_1|x_i|) + \alpha_o + \alpha_1 x_i \leq y_i\end{aligned}\right\} i = 1, \ldots, n.^5 \tag{8.37}$$

As Figure 8.8 reveals (see Kim, Moskowitz, and Koksalan1996), each level set $\left(Y_i\right)_{h_i}$ can be represented by the interval $[y', y''] = [(\alpha_0 + \alpha_1 x_i) + (h_i - 1)(c_0 + c_1|x_i|), (\alpha_0 + \alpha_1 x_i) + (1 - h_i)(c_0 + c_1|x_i|)]$, with any point y_i in this interval yielding a membership value of at least h_i in Y_i so that Equation 8.37 is satisfied.

Next, let us express the *simple* or *naïve vagueness criterion* as

$$J = c_0 + c_1. \tag{8.38}$$

Here J represents the degree of *total vagueness* of the fuzzy linear regression model and consists of the sum of the individual spreads of the fuzzy parameters. A reasonable decision rule for estimating the fuzzy parameters A_0 and A_1 is to choose A^* so as to minimize total vagueness subject to the requirement that the fuzzy output set Y_i^* contains y_i to at least degree h_i with $c_j \geq 0$, $j = 0, 1$. That is, the problem of finding the fuzzy regression parameters amounts to solving the linear program

$$\begin{aligned}&\min_{\alpha_j, c_j} J = c_o + c_1 \quad \text{s.t.}\\ &\left.\begin{aligned}(1-h_i)(c_o + c_1|x_i|) + \alpha_o + \alpha_1 x_i \geq y_i\\ (h_i - 1)(c_o + c_1|x_i|) + \alpha_o + \alpha_1 x_i \leq y_i\end{aligned}\right\} i = 1, \ldots, n,\\ &c_o, c_1 \geq 0;\ \ \alpha_j \text{ unrestricted}, j = 0, 1,\end{aligned} \tag{8.39}$$

[5] Note that $|z_i| = \begin{cases} z_i, & z_i > 0; \\ -z_i, & z_i < 0, \end{cases}$
where $|z_i| = |y_i - \alpha_0 + \alpha_1 x_i|$, $i = 1, \ldots, n$.

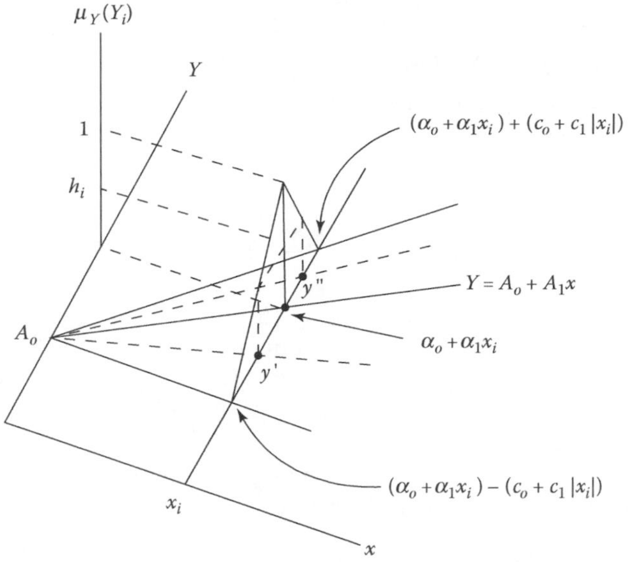

FIGURE 8.8
Given fuzzy $Y_i = A_o + A_1 x_i$, $\mu_{Y_i}(y_i) \geq h_i$ determines the degree h_i to which y_i is included in Y_i.

where "s.t." is short for "subject to."[6] (A generalization of Equation 8.39 is provided in Appendix 8.A [see "Membership Function for a Fuzzy Linear Function"].)

As indicated previously, for each data point (x_i, y_i), $i = 1,\ldots, n$, there are two structural constraints implied in Equation 8.39 or

$$(i=1)\begin{cases}(1-h_1)(c_o + c_1|x_1|) + \alpha_o + \alpha_1 x_1 \geq y_1\\(h_1 - 1)(c_o + c_1|x_1|) + \alpha_o + \alpha_1 x_1 \leq y_1\end{cases}$$

$$(i=2)\begin{cases}(1-h_2)(c_o + c_1|x_2|) + \alpha_o + \alpha_1 x_2 \geq y_2\\(h_2 - 1)(c_o + c_1|x_2|) + \alpha_o + \alpha_1 x_2 \leq y_2\end{cases}$$

$$\vdots$$

$$(i=n)\begin{cases}(1-h_n)(c_o + c_1|x_n|) + \alpha_o + \alpha_1 x_n \geq y_n\\(h_n - 1)(c_o + c_1|x_n|) + \alpha_o + \alpha_1 x_n \leq y_n\end{cases}$$

A modification of Equation 8.39 involves minimizing the sum of the fuzzy widths or spreads $J' = nc_o + c_1\sum_{i=1}^{n}|x_i|$ around the predicted Y_i values subject to the same structural constraints and non-negativity conditions.

[6] Readers not familiar with the salient features of linear programming are directed to Appendix 8.B.

Fuzzy Linear Regression Model (Fuzzy Output Data)

Following Tanaka, Uejima, and Asai [1982], let us consider a set of fuzzy output data values expressed as symmetric triangular fuzzy numbers $Y_i = (y_i, e_i)$, $i = 1,\ldots, n$, with membership function

$$\mu_{Y_i}(y) = 1 - \frac{|y_i - y|}{e_i}, \; e_i > 0. \tag{8.40}$$

We next specify the fuzzy linear regression model as

$$Y' = A_0 + A_1 x, \tag{8.41}$$

where the fuzzy parameters A_0, A_1 are defined by their membership functions in Equation 8.34 and, from Equation 8.35, the membership function of Y' is

$$\mu_{Y_i'}(y) = \begin{cases} 1 - \dfrac{|y_i - \alpha_o - \alpha_1 x_i|}{c_o + c_1|x_i|}, & c_o + c_1|x_i| > |y_i - \alpha_o - \alpha_1 x_i|; \\ 0, \quad c_o + c_1|x_i| \le |y_i - \alpha_o - \alpha_1 x_i|. \end{cases} \tag{8.42}$$

Given the h-level sets ($h \in [0, 1]$)

$$(Y_i)_h = \{y \,|\, \mu_{Y_i}(y) \ge h\}; \tag{8.43}$$

$$(Y_i')_h = \{y \,|\, \mu_{Y_i'}(y) \ge h\}, \tag{8.44}$$

the degree of fitting the fuzzy linear regression model in Equation 8.41 to the fuzzy output data $Y_i = (y_i, e_i)$, $i = 1,\ldots, n$, is measured by the index $\bar{h}_i$, which maximizes h subject to $(Y_i)_h \subset (Y_i')_h$. It can be shown that

$$\bar{h}_i = 1 - \frac{|y_i - \alpha_o - \alpha_1 x_i|}{(c_o + c_1|x_i|) - e_i}, \; i = 1,\ldots,n \tag{8.45}$$

(see Figure 8.9) [Tanaka, Uejima, and Asai 1982]. In addition, the degree of fitting Equation 8.41 to all of the data $Y_1,\ldots, Y_n$ is obtained by finding $\min(\bar{h}_i)$.

If again $J = c_0 + c_1$ defines the total vagueness of the fuzzy linear regression model, then our objective is to obtain fuzzy parameters $A_j^* = (\alpha_j, c_j)$, $j = 0, 1$, which minimizes J subject to $\bar{h}_i \ge H$ for all i and $c_j \ge 0$, $j = 0,1$, where H is arbitrarily chosen as the *degree of fitting* of the model. It is easily verified that, via Equation 8.45, the requirement $\bar{h}_i \ge H$ implies

$$\left.\begin{aligned} (1-H)(c_o + c_1|x_i|) + \alpha_o + \alpha_1 x_i \ge y_i + (1-H)e_i \\ (H-1)(c_o + c_1|x_i|) + \alpha_o + \alpha_1 x_i \le y_i + (H-1)e_i \end{aligned}\right\} i = 1,\ldots,n. \tag{8.46}$$

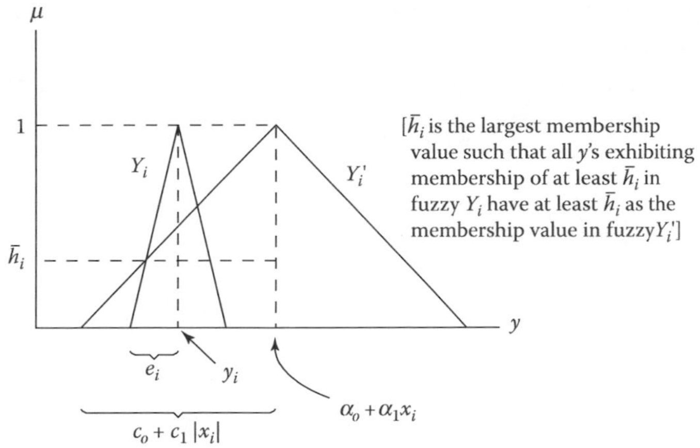

FIGURE 8.9
Degree of fitting fuzzy Y_i' to fuzzy Y_i.

Hence, we ultimately desire to choose fuzzy parameters A_j^* to

$$\min_{\alpha_j, c_j} J = c_o + c_1 \quad \text{s.t.}$$
$$\left.\begin{aligned}(1-H)(c_o + c_1|x_i|) + \alpha_o + \alpha_1 x_i &\geq y_i + (1+H)e_i\\ (H-1)(c_o + c_1|x_i|) + \alpha_o + \alpha_1 x_i &\leq y_i + (H-1)e_i\end{aligned}\right\} i = 1,\ldots,n, \tag{8.47}$$
$$c_o, c_1 \geq 0;\ \alpha_j \text{ unrestricted}, j = 0,1.$$

The modified objective function J' introduced at the end of the preceding section can also be used in Equation 8.47. (Wang and Tsaur [2000] note that, because the observed output data is fuzzy, such fuzziness should be accounted for by forming the objective function as

$$J'' = \sum_{j=0}^{1}\left(\sum_{j=1}^{n} c_i|x_i| - e_i\right).$$

Hence, their modified linear program is to minimize J'' subject to the constraints in Equation 8.47.)

Fuzzy Least-Squares Linear Regression [Savic and Pedrycz 1991]

It is evident that the α_j's are not arguments of the objective function J in Equation 8.36. Hence, only an increase in the spreads c_j will unfavorably affect the total vagueness objective. On the basis of this observation, Savic and Pedrycz [1991] offer a two-phase method for specifying a fuzzy linear regression model:

1. Phase I: Use classical OLS to fit a regression line through the set of observation (x_i, y_i), $i = 1,\ldots, n$. The estimated least-squares line has the form $\hat{y} = \hat{\alpha}_0 + \hat{\alpha}_1 x$, where

$\hat{\alpha}_0$ and $\hat{\alpha}_1$ are the OLS estimates of the centers or modal values of the symmetric triangular numbers $A_j = (\alpha_j, c_j)$, $j = 0, 1$.

2. Phase II: Substitute the α_j's determined in Phase I into Equation 8.39. Now the α_j's are no longer decision variables but simply predetermined variables. Hence, the only decision variables in Equation 8.39 are the c_j's $j = 0, 1$.

So for the Phase II program, we desire to

$$\begin{aligned} &\min\ J = c_0 + c_1 \quad \text{s.t.} \\ &\left.\begin{aligned}(1-h_i)(c_0 + c_1|x_i|) \geq y_i - \hat{\alpha}_0 - \hat{\alpha}_1 x_i \\ (h_i - 1)(c_0 + c_1|x_i|) \leq y_i - \hat{\alpha}_0 - \hat{\alpha}_1 x_i\end{aligned}\right\} i = 1,\ldots,n, \\ &c_0, c_1 \geq 0. \end{aligned} \tag{8.39.1}$$

How does fuzzy least-squares linear regression (FLSLR) (Equation 8.39.1) compare with what we shall call the *basic Tanaka model* (Equation 8.39)? We note briefly the following:

1. Because the FLSLR method uses OLS as its starting point, the spreads c_j obtained using this technique are not as small as would be the case if, say, the minimum vagueness (or basic Tanaka) criterion was exclusively used to solve Equation 8.39. Moreover, the FLSLR estimates of the c_j's are insensitive to the values of the modal or center points α_j of the observed fuzzy numbers.
2. In FLSLR, use of $\hat{\alpha}_j$ generally causes the centers of the Y_j' estimated values to be closer to the Y_j observed values, thus yielding higher membership values relative to Equation 8.39.
3. Unlike the model described by Equation 8.39, FLSLR produces a well-defined solution to the fuzzy linear regression program. That is, Equation 8.39 is scale dependent and not well defined (in that an infinite number of possible solutions can exist). In addition, the optimal basic feasible solution to the linear program in Equation 8.39 may be such that some of the fuzzy regression coefficients may become crisp. In this circumstance, if, say, A_1 turns crisp, then x_1 does not contribute to the overall fuzziness of the output variable Y.
4. The basic Tanaka model (Equation 8.39) assumes no underlying model by which the data are generated. In this regard, it is essentially a nonparametric method.

Example 8.6

Let us consider the dataset used by Tanaka, Uejima, and Asai (1982) and illustrated in Table 8.1. Here the x_i's are crisp and the Y_i's are symmetric triangular fuzzy numbers with centers y_i and spreads e_i, $i = 1,\ldots,5$. If we consider what we shall call the basic Tanaka model (Equation 8.39) (here both x_i and y_i are taken to be crisp, whereas the regression parameters A_0, A_1 are fuzzy sets), then, for $h_i = 0$ for all i, we obtain the linear program.

TABLE 8.1

Observations on Crisp x_i and Fuzzy Y_i

X_i	$Y_i = (y_i, e_i)$
1	(8.0, 1.8)
2	(6.4, 2.2)
3	(9.5, 2.6)
4	(13.5, 2.6)
5	(13.0, 2.4)

$$
\begin{aligned}
&\min J = c_o + c_1 \quad \text{s.t.} \\
&c_o + c_1 + \alpha_o' - \alpha_o'' + \alpha_1' - \alpha_1'' \geq 8.0 \\
&-c_o - c_1 + \alpha_o' - \alpha_o'' + \alpha_1' - \alpha_1'' \leq 8.0 \\
&c_o + 2c_1 + \alpha_o' - \alpha_o'' + 2\alpha_1' - 2\alpha_1'' \geq 6.4 \\
&-c_o - 2c_1 + \alpha_o' - \alpha_o'' + 2\alpha_1' - 2\alpha_1'' \leq 6.4 \\
&c_o + 3c_1 + \alpha_o' - \alpha_o'' + 3\alpha_1' - 3\alpha_1'' \geq 9.5 \\
&-c_o - 3c_1 + \alpha_o' - \alpha_o'' + 3\alpha_1' - 3\alpha_1'' \leq 9.5 \\
&c_o + 4c_1 + \alpha_o' - \alpha_o'' + 4\alpha_1' - 4\alpha_1'' \geq 13.5 \\
&-c_o - 4c_1 + \alpha_o' - \alpha_o'' + 4\alpha_1' - 4\alpha_1'' \leq 13.5 \\
&c_o + 5c_1 + \alpha_o' - \alpha_o'' + 5\alpha_1' - 5\alpha_1'' \geq 13.0 \\
&-c_o - 5c_1 + \alpha_o' - \alpha_o'' + 5\alpha_1' - 5\alpha_1'' \leq 13.0 \\
&\quad c_o,\ c_1,\ \alpha_o',\ \alpha_o'',\ \alpha_1',\ \alpha_1'' \geq 0.
\end{aligned}
$$

Note that, because α_0 and α_1 are unrestricted in sign, both α_0 and α_1 can be expressed as the difference between two nonnegative variables, i.e., $\alpha_o = \alpha_o' - \alpha_o''$; $\alpha_1 = \alpha_1' - \alpha_1''$. The solution[7] to this problem is $c_0 = 0$, $c_1 = 0.858$, $\alpha_0 = 6.166$, and $\alpha_1 = 0.975$. Hence, the estimated fuzzy linear regression equation appears as $\hat{Y} = \hat{A}_o + \hat{A}_1 x = (\alpha_o,\ c_o) + (\alpha_1,\ c_1)x = (6.166,\ 0) + (0.975,\ 0.858)x$.

Next, let us respecify this problem with $h_i = 0.7$ for all i. Hence, the revised linear program appears as

$$
\begin{aligned}
&\min J = c_o + c_1 \quad \text{s.t.} \\
&0.3c_o + 0.3c_1 + \alpha_o' - \alpha_o'' + \alpha_1' - \alpha_1'' \geq 8.0 \\
&-0.3c_o - 0.3c_1 + \alpha_o' - \alpha_o'' + \alpha_1' - \alpha_1'' \leq 8.0 \\
&0.3c_o + 0.6c_1 + \alpha_o' - \alpha_o'' + 2\alpha_1' - 2\alpha_1'' \geq 6.4 \\
&-0.3c_o - 0.6c_1 + \alpha_o' - \alpha_o'' + 2\alpha_1' - 2\alpha_1'' \leq 6.4 \\
&0.3c_o + 0.9c_1 + \alpha_o' - \alpha_o'' + 3\alpha_1' - 3\alpha_1'' \geq 9.5 \\
&-0.3c_o - 0.9c_1 + \alpha_o' - \alpha_o'' + 3\alpha_1' - 3\alpha_1'' \leq 9.5 \\
&0.3c_o + 1.2c_1 + \alpha_o' - \alpha_o'' + 4\alpha_1' - 4\alpha_1'' \geq 13.5 \\
&-0.3c_o - 1.2c_1 + \alpha_o' - \alpha_o'' + 4\alpha_1' - 4\alpha_1'' \leq 13.5 \\
&0.3c_o + 1.5c_1 + \alpha_o' - \alpha_o'' + 5\alpha_1' - 5\alpha_1'' \geq 13.0 \\
&-0.3c_o - 1.5c_1 + \alpha_o' - \alpha_o'' + 5\alpha_1' - 5\alpha_1'' \leq 13.0 \\
&\quad c_o,\ c_1,\ \alpha_o',\ \alpha_o'',\ \alpha_1',\ \alpha_1'' \geq 0.
\end{aligned}
$$

[7] The linear programming solutions that follow are all generated by the LINDO System.

The solution to this new problem is

$$c_0 = 0,\ c_1 = 2.861,\ \alpha_0 = 6.166, \text{ and } \alpha_1 = 0.975.$$

Hence, $\hat{Y} = \hat{A}_o + \hat{A}_1 x = (6.166,\ 0) + (0.975,\ 2.861)x$.

It is obvious that, for varying values of $h_i = h$ for all i, the fuzzy centers α_0 and α_1 remain unchanged. The higher the value of $h \in [0, 1]$, the fuzzier the (fuzzy) regression model. For $h = 0$, the (basic Tanaka) fuzzy regression model displays the minimum fuzzy spread among all allowed h's , i.e., $h = 0$ renders the narrowest fuzzy width.

Example 8.7

Let us estimate the fuzzy parameters of the fuzzy regression equation $Y = A_0 + A_1 x$, where x_i is crisp and $Y_i = (y_i, e_i)$ is symmetric triangular (Table 8.1). Then the basic Tanaka program (Equation 8.47) with $H = 0.5$ appears as

$$\begin{aligned}
&\min J = c_o + c_1 \quad \text{s.t.}\\
&0.5c_o + 0.5c_1 + \alpha_o' - \alpha_o'' + \alpha_1' - \alpha_1'' \ge 8.9\\
&-0.5c_o - 0.5c_1 + \alpha_o' - \alpha_o'' + \alpha_1' - \alpha_1'' \le 7.1\\
&0.5c_o + c_1 + \alpha_o' - \alpha_o'' + 2\alpha_1' - 2\alpha_1'' \ge 7.5\\
&-0.5c_o - c_1 + \alpha_o' - \alpha_o'' + 2\alpha_1' - 2\alpha_1'' \le 5.3\\
&0.5c_o + 1.5c_1 + \alpha_o' - \alpha_o'' + 3\alpha_1' - 3\alpha_1'' \ge 10.8\\
&-0.5c_o - 1.5c_1 + \alpha_o' - \alpha_o'' + 3\alpha_1' - 3\alpha_1'' \le 8.2\\
&0.5c_o + 2c_1 + \alpha_o' - \alpha_o'' + 4\alpha_1' - 4\alpha_1'' \ge 14.8\\
&-0.5c_o - 2c_1 + \alpha_o' - \alpha_o'' + 4\alpha_1' - 4\alpha_1'' \le 12.2\\
&0.5c_o + 2.5c_1 + \alpha_o' - \alpha_o'' + 5\alpha_1' - 5\alpha_1'' \ge 14.2\\
&-0.5c_o - 2.5c_1 + \alpha_o' - \alpha_o'' + 5\alpha_1' - 5\alpha_1'' \le 11.8\\
&c_o,\ c_1,\ \alpha_o',\ \alpha_o'',\ \alpha_1',\ \alpha_1'' \ge 0.
\end{aligned}$$

The solution to this linear program is $c_0 = 0$, $c_1 = 2.783$, $\alpha_0 = 6.93$, and $\alpha_1 = 0.575$. Hence, the estimated fuzzy regression equation is $\hat{Y} = \hat{A}_o + \hat{A}_1 x = (6.93,\ 0) + (0.575,\ 2.783)x$.

Example 8.8

Let us apply the Savic and Pedrycz [1991] two-phase FLSLR routine to the crisp x_i, y_i data values appearing in Table 8.1. Stage 1 has us determine the OLS estimates of α_0 and α_1 as $\hat{Y} = \hat{\alpha}_o + \hat{\alpha}_1 x = 4.95 + 1.71x$. Then Equation 8.39.1 becomes, for $h_i = 0$ for all i,

$$\begin{aligned}
&\min J = c_0 + c_1 \text{ s.t.}\\
&c_0 + c_1 \ge 1.34\\
&-c_0 - c_1 \le 1.34\\
&c_0 + 2c_1 \ge -1.97\\
&-c_0 - 2c_1 \le -1.97\\
&c_0 + 3c_1 \ge 0.58\\
&-c_0 - 3c_1 \le 0.58\\
&c_0 + 4c_1 \ge 1.71
\end{aligned}$$

$$-c_0 - 4c_1 \leq 1.71$$
$$c_0 + 5c_1 \geq -0.50$$
$$-c_0 - 5c_1 \leq -0.50$$
$$c_0,\ c_1 \geq 0.$$

(For instance, the first structural constraint is

$$c_o + c_1(1) \geq y_1 - \hat{\alpha}_o - \hat{\alpha}_1(1) = 8 - 4.95 - 1.71 = 1.34.)$$

The solution to this program is $c_0 = 0$, $c_1 = 1.34$. Hence, the estimated FLSLR equation is $\hat{Y} = \hat{A}_o + \hat{A}_1 x = (\hat{\alpha}_o,\ c_o) + (\hat{\alpha}_1,\ c_1)x = (4.95,\ 0) + (1.71,\ 1.34)x$.

Example 8.9

Suppose we apply the Savic and Pedrycz [1991] two-phase method to Equation 8.47 using $H = 0.5$ for x_i crisp and $Y_i = (y_i, e_i)$ symmetric triangular (see Table 8.1). Again using the OLS estimates $\hat{\alpha}_o = 4.95$ and $\hat{\alpha}_1 = 1.71$, the resulting linear program appears as

$$\min J = c_0 + c_1 \text{ s.t.}$$
$$0.5\ c_0 + 0.5c_1 \geq 2.24$$
$$-0.5c_0 - 0.5c_1 \leq 0.44$$
$$0.5c_0 + c_1 \geq -0.87$$
$$-0.5c_0 - c_1 \leq -3.07$$
$$0.5c_0 + 1.5c_1 \geq 1.88$$
$$-0.5c_0 - 1.5c_1 \leq -0.72$$
$$0.5c_0 + 2c_1 \geq 3.01$$
$$-0.5c_0 - 2c_1 \leq 1.41$$
$$0.5c_0 + 2.5c_1 \geq 0.70$$
$$-0.5c_0 - 2.5c_1 \leq -1.70$$
$$c_0,\ c_1 \geq 0.$$

(For example, the first structural constraint is determined as

$$(1 - 0.5)(c_o + c_1(1)) \geq y_1 - \hat{\alpha}_o - \hat{\alpha}_1(1) + (1 - 0.5)(1.8) = 2.24.)$$

The solution to this linear program is $c_0 = 0$ and $c_1 = 4.48$. Hence, $\hat{Y} = \hat{A}_o + \hat{A}_1 x = (\hat{\alpha}_o,\ c_o) + (\hat{\alpha}_1,\ c_1)x = (4.95,\ 0) + (1.71,\ 4.48)x$.

Fuzzy Least-Squares [Diamond 1988; Diamond and Tanaka 1996; Wang and Tsaur 2000]

Our next approach to determining fuzzy regression parameters from fuzzy data is provided by what is termed the fuzzy least-squares (FLS) technique. This method offers a fuzzy analogue to the usual OLS routine and can be used to compute fuzzy parameters for regression models involving (but not limited to) symmetric triangular fuzzy numbers.

Consider the fuzzy regression model $Y = A_0 + A_1 x$, where x_i is crisp and $A_0 = (\alpha_0, c_0)$, $A_1 = (\alpha_1, c_1)$, and $Y_i = (y_i, e_i)$, $i = 1,\ldots, n$, are all symmetric triangular fuzzy numbers. To mirror the OLS approach, let us adopt the decision criterion of minimizing the aggregate fuzziness between the estimated values of the fuzzy Y_i's and the observed fuzzy Y_i's. That is, we seek to minimize the total fuzzy distance between $Y = A_0 + A_1x$ and $Y_i = (y_i, e_i)$, $i = 1,\ldots, n$, which is defined by

$$D = \sum_i d(A_o + A_1x_i, Y_i)^2. \tag{8.48}$$

To this end, let

$$\begin{aligned} A_0 &= (\alpha_0 - c_0, \alpha_0, \alpha_0 + c_0), \\ A_1 &= (\alpha_1 - c_1, \alpha_1, \alpha_1 + c_1), \text{ and} \\ Y_i &= (y_i - e_i, y_i, y_i + e_i). \end{aligned} \tag{8.49}$$

Then

$$\begin{aligned} A_0 + A_1x_i &= (\alpha_0 + \alpha_1x_i, c_0 + c_1x_i) \\ &= (\alpha_0 + \alpha_1x_i - c_0 - c_1x_i, \alpha_0 + \alpha_1x_i, \alpha_0 + \alpha_1x_i + c_0 + c_1x_i). \end{aligned} \tag{8.50}$$

Following Diamond [1988], let us use the metric

$$d\,(u, v)^2 = (u^- - v^-)^2 + (u^m - v^m)^2 + (u^+ - v^+)^2, \tag{8.51}$$

where $u = (u^-, u^m, u+)$, $v = (v^-, v^m, v+)$, and u^m and v^m are the modal values of the triangular fuzzy numbers u and v, respectively. If we set $u = A_0 + A_1x_i$ and $v = Y_i$, where the former is determined from Equation 8.50 and the latter from Equation 8.49, then the metric in Equation 8.51 allows us to rewrite Equation 8.48 as

$$\begin{aligned} D &= \sum_i d(A_o + A_1x_i, Y_i)^2 \\ &= \sum_i \{(\alpha_o + \alpha_1x_i - c_o - c_1x_i - y_i + e_i)^2 \\ &\quad + (\alpha_o + \alpha_1x_i - y_i)^2 + (\alpha_o + \alpha_1x_i + c_o + c_1x_i - y_i - e_i)^2\}. \end{aligned} \tag{8.48.1}$$

Suppose the x_i's are nondegenerate, i.e., all x_i values are not the same. Then a set of equations similar to the least-squares normal equations may be obtained from Equation 8.48.1 by setting $\partial D/\partial x_o = \partial D/\partial x_1 = \partial D/\partial c_o = \partial D/\partial c_1 = 0$. The resulting equation system can then be solved simultaneously to obtain estimates of x_0, x_1, c_0, and c_1. However, because the signs of these fuzzy parameters need to be considered, the following special cases present themselves. Specifically

(a) $\alpha_0 > 0$, $\alpha_1 > 0$ and c_0 and c_1 unrestricted in sign. Then

$$\begin{aligned} \alpha_1 &= \frac{n\sum_i x_iy_i - \sum_i x_i \sum_i y_i}{n\sum_i x_i^2 - \left(\sum_i x_i\right)^2}, \quad \alpha_o = \bar{y} - \alpha_1\bar{x}; \\ c_1 &= \frac{n\sum_i x_ie_i - \sum_i x_i \sum_i e_i}{n\sum_i x_i^2 - \left(\sum_i x_i\right)^2}, \quad c_o = \bar{e} - c_1\bar{x}, \end{aligned} \tag{8.52}$$

where $\bar{x} = \sum_i x_i/n$, $\bar{y} = \sum_i y_i/n$, and $\bar{e} = \sum_i e_i/n$.

(b) $\alpha_0 < 0$, $\alpha_1 > 0$, $c_0 < 0$, and c_1 unrestricted in sign. Then,

$$\alpha_1 = \frac{n\sum_i x_i y_i - \sum_i x_i \sum_i y_i}{n\sum_i x_i^2 - \left(\sum_i x_i\right)^2},\ \alpha_o = \bar{y} - \alpha_1 \bar{x};$$
$$c_1 = \frac{n\sum_i x_i e_i - \sum_i x_i \sum_i e_i}{n\sum_i x_i^2 - \left(\sum_i x_i\right)^2},\ c_o = -(\bar{e} - c_1 \bar{x}). \tag{8.53}$$

(c) $\alpha_0 > 0$, $\alpha_1 < 0$, c_0 unrestricted in sign, and $c_1 < 0$. Then,

$$\alpha_1 = \frac{n\sum_i x_i y_i - \sum_i x_i \sum_i y_i}{n\sum_i x_i^2 - \left(\sum_i x_i\right)^2},\ \alpha_o = \bar{y} - \alpha_1 \bar{x};$$
$$c_1 = -\frac{n\sum_i x_i e_i - \sum_i x_i \sum_i e_i}{n\sum_i x_i^2 - \left(\sum_i x_i\right)^2},\ c_o = \bar{e} - c_1 \bar{x}. \tag{8.54}$$

(d) $\alpha_0 < 0$, $\alpha_1 < 0$, $c_0 < 0$, and $c_1 < 0$. Then,

$$\alpha_1 = \frac{n\sum_i x_i y_i - \sum_i x_i \sum_i y_i}{n\sum_i x_i^2 - \left(\sum_i x_i\right)^2},\ \alpha_o = \bar{y} - \alpha_1 \bar{x};$$
$$c_1 = -\frac{n\sum_i x_i e_i - \sum_i x_i \sum_i e_i}{n\sum_i x_i^2 - \left(\sum_i x_i\right)^2},\ c_o = -(\bar{e} - c_1 \bar{x}). \tag{8.55}$$

(e) $\alpha_0 < 0$, $\alpha_1 < 0$, $c_0 < 0$, and c_1 unrestricted in sign. Then, Equation 8.53 applies.

(f) $\alpha_0 < 0$, $\alpha_1 < 0$, c_0 unrestricted in sign, $c_1 < 0$. Equation 8.54 applies.

Example 8.10

For the dataset appearing in Table 8.1, determine the fuzzy regression parameters of $Y = A_0 + A_1 x$, where x_i is crisp, fuzzy Y_i is symmetric triangular, $A_0 = (\alpha_0, c_0)$, and $A_1 = (\alpha_1, c_1)$. Looking to the

TABLE 8.2

Calculations for Fuzzy Regression Parameter Estimation

x_i	y_i	e_i	x_iy_i	x_ie_i	x_i^2
1	8.0	1.8	8.0	1.8	1
2	6.4	2.2	12.8	4.4	4
3	9.5	2.6	28.5	7.8	9
4	13.5	2.6	54.0	10.4	16
5	13.0	2.4	65.0	12.0	25
15	50.4	11.6	168.3	36.4	55.0

sums in Table 8.2, we can readily determine that $\bar{x} = 3$, $\bar{y} = 10.08$, and $\bar{e} = 2.32$. Then, from Equation 8.52,

$$\alpha_1 = \frac{5(168.3)-(15)(50.4)}{5(55.0)-(15)^2} = 1.71, \ \alpha_o = 10.08-(1.71)(3) = 4.95;$$
$$c_1 = \frac{5(36.4)-(15)(11.6)}{5(55.0)-(15)^2} = 0.16, \ c_o = 2.32-(0.16)(3) = 1.84.$$

Hence, our estimated FLS regression equation is $\hat{Y} = (4.95, 1.84) + (1.71, 0.16)x$.

Appendix A: The General Fuzzy Linear Regression Model

Membership Function for a Fuzzy Linear Function

In general, given sets X and Y, suppose that $y = f(x_1,\ldots, x_m; a_1,\ldots, a_m) = f(x; a)$ is a mapping from X to Y, with $x = (x_1,\ldots, x_m)$ and $a = (a_1,\ldots, a_m)$. If the m parameters are given by fuzzy sets $A_1,\ldots, A_m$, then f is the fuzzy function

$$f: X \to Y \subset F(y);\ y = f(x; A_1,\ldots, A_m), \tag{8.A.1}$$

where, for a fixed x, fuzzy Y is the image of fuzzy $A = A_1 \times \ldots \times A_m$ in X and $F(y)$ is the set of all fuzzy subsets of Y. Then the membership function of A is

$$\begin{aligned}\mu_A(a) &= \inf_j\left[\mu_{A_j}(a_j)\right] \\ &= \mu_{A_1}(a_1) \wedge \ldots \wedge \mu_{A_m}(a_m)\end{aligned} \tag{8.A.2}$$

and the membership function of Y is, for $a = (a_1,\ldots, a_m)$ and $x = (x_1,\ldots, x_m)$,

$$\mu_Y(y) = \begin{cases} \sup\limits_{\{a|y=f(x;a)\}} \mu_A(a), & \{a|y = f(x;a)\} \neq \varnothing; \\ 0, & \{a|y = f(x;a)\} = \varnothing \end{cases} \tag{8.A.3}$$

When $(A_1, \ldots, A_m)$ are taken to be symmetric triangular fuzzy numbers $(A_j = (\alpha_j, c_j), j = 1, \ldots, m)$, the membership function of A_j is

$$\mu_{A_j}(a_j) = \begin{cases} 1 - \dfrac{|\alpha_j - a_j|}{c_j}, & \alpha_j - c_j \le a_j \le \alpha_j + c_j; \\ 0 & \text{otherwise}, \end{cases} \tag{8.A.4}$$

with $c_j > 0, j = 1, \ldots, m$. If we set $A = (\alpha, c)$, with $\alpha = (\alpha_1, \ldots, \alpha_m)$ and $c = (c_1, \ldots, c_m)$, then the fuzzy linear function $Y = A_1x_1 + \ldots + A_mx_m$ has membership function

$$\mu_Y(y) = \begin{cases} 1 - \dfrac{\left|y - \sum_{j=1}^{m} \alpha_j x_j\right|}{\sum_{j=1}^{m} c_j |x_j|}, & x \ne (0, \ldots, 0); \\ 1, \; x = (0, \ldots, 0), \; y = 0; \\ 0, \; x = (0, \ldots, 0), \; y \ne 0, \end{cases} \tag{8.A.5}$$

with $\mu_Y(y) = 0$ when $\sum_{j=1}^{m} c_j |x_j| \le \left| y - \sum_{j=1}^{m} \alpha_j x_j \right|$ (see Tanaka, Uejima, and Asai 1982).

If we impose the restriction that the fuzzy output (Y) of the model should cover all of the y_i's, $i = 1, \ldots, n$, to degree $h_i \in [0, 1]$, then $y_i \in (Y_i)_{h_i} = \{y_i \,|\, \mu_{Y_i}(y_i) \ge h_i\}$. Hence, under the *threshold conditioning inequalities* $\mu_{Y_i}(y_i) \ge h_i,\; i = 1, \ldots, n$, we obtain, via Equation 8.A.5,

$$\mu_{Y_i}(y_i) = 1 - \frac{\left|y_i - \sum_{j=1}^{m} \alpha_j x_{ij}\right|}{\sum_{j=1}^{m} c_j |x_{ij}|} \ge h_i, \; i = 1, \ldots, n,$$

or

$$\left.\begin{aligned} (1 - h_i)\left(\sum_{j=1}^{m} c_j |x_{ij}| + \sum_{j=1}^{m} \alpha_j x_{ij} \ge y_i\right) \\ (h_i - 1)\left(\sum_{j=1}^{m} c_j |x_{ij}| + \sum_{j=1}^{m} \alpha_j x_{ij} \le y_i\right) \end{aligned}\right\} i = 1, \ldots, n. \tag{8.A.6}$$

Next, if

$$J = \sum_{j=1}^{m} c_j \tag{8.A.7}$$

depicts the *naive vagueness criterion* (the total vagueness of the fuzzy model), then our objective is to choose A^* so as to minimize J subject to the requirement that the fuzzy output set Y_i^* contains y_i to at least degree h_i with $c_j \ge 0, j = 1, \ldots, m$. To this end, we must solve the linear programming problem

$$\begin{aligned} &\min_{\alpha_j, c_j} J = \sum_{i=1}^{m} c_j \quad \text{s.t.} \\ &\left.\begin{aligned} (1 - h_i)\left(\sum_{j=1}^{m} c_j |x_{ij}| + \sum_{j=1}^{m} \alpha_j x_{ij} \ge y_i\right) \\ (h_i - 1)\left(\sum_{j=1}^{m} c_j |x_{ij}| + \sum_{j=1}^{m} \alpha_j x_{ij} \le y_i\right) \end{aligned}\right\} i = 1, \ldots, n, \\ &c_j \ge 0; \; \alpha_j \text{ unrestricted}, j = 1, \ldots, m. \end{aligned} \tag{8.A.8}$$

An arrangement of the n nonfuzzy observations on the output variable y as well as the n nonfuzzy observations on the m (input) variables $x_1,\ldots, x_m$ is displayed in Table 8.A.1. Here x_{ij} depicts the ith observation on the jth input variable x_j, $i = 1,\ldots, n$; $j = 1,\ldots, m$. Note that, if $x_{i1} = 1$ for all i (the first input column associated with variable x_1 in Table 8.A.1 is simply a column of all 1's), then c_1 is a constant and thus is the intercept of the regression hyperplane $y = c_1 + c_2x_2 + \cdots + c_mx_m$.

Given the arrangement of data values appearing in Table 8.A.1, we may explicitly represent the structural constraints in Equation 8.A.8 as

$$(i=1)\begin{cases}(1-h_1)\sum_{j=1}^{m} c_j\left|x_{1j}\right| + \sum_{j=1}^{m}\alpha_j x_{1j} \geq y_1\\ (h_1-1)\sum_{j=1}^{m} c_j\left|x_{1j}\right| + \sum_{j=1}^{m}\alpha_j x_{1j} \leq y_1\end{cases}$$

$$(i=2)\begin{cases}(1-h_2)\sum_{j=1}^{m} c_j\left|x_{2j}\right| + \sum_{j=1}^{m}\alpha_j x_{2j} \geq y_2\\ (h_2-1)\sum_{j=1}^{m} c_j\left|x_{2j}\right| + \sum_{j=1}^{m}\alpha_j x_{2j} \leq y_2\end{cases}$$

.

.

.

$$(i=n)\begin{cases}(1-h_n)\sum_{j=1}^{m} c_j\left|x_{nj}\right| + \sum_{j=1}^{m}\alpha_j x_{nj} \geq y_n\\ (h_n-1)\sum_{j=1}^{m} c_j\left|x_{nj}\right| + \sum_{j=1}^{m}\alpha_j x_{nj} \leq y_n\end{cases}$$

The modified objective function J' introduced at the end of the section "Fuzzy Linear Regression Model" can also be used in Equation 8.A.8, i.e.,

$$\begin{aligned}J' &= nc_o + \sum_{i=1}^{n}\sum_{j=1}^{n} c_j\left|x_{ij}\right|\\ &= nc_o + c_1\sum_{i=1}^{n}\left|x_{i1}\right| + \cdots + c_m\sum_{i=1}^{n}\left|x_{im}\right|.\end{aligned}$$

TABLE 8.A.1

Arrangement of Data for Structural Constraint Construction

Output y	**Inputs** x_j, $j=1,\ldots,m$			
	x_1	x_2	...	x_m
y_1	x_{11}	x_{12}	...	x_{1m}
y_2	x_{21}	x_{22}	...	x_{2m}
.	.	.		.
.	.	.		.
.	.	.		.
y_n	x_{n1}	x_{n2}	...	x_{nm}

General Membership Function for a Fuzzy Linear Function

In general, if a symmetric fuzzy number (not necessarily triangular) appears as $A_i = (\alpha_i, c_i)_L$ (see Equation 8.29), then

$$\mu_{A_i}(a_i) = L\left(\left(\alpha_i - a_i\right) \big/ c_i \right) \tag{8.A.9}$$

and, for fuzzy linear function $Y = A_1 x_1 + \cdots + A_m x_m$,

$$\mu_Y(y) = \begin{cases} L\left(\left(y - \sum_{j=1}^{m} \alpha_j x_j \right) \Big/ \sum_{j=1}^{m} c_j |x_j| \right), & x \neq (0,\ldots,0); \\ 1, \; x = (0,\ldots,0), \; y = 0; \\ 0, \; x = (0,\ldots,0), \; y \neq 0. \end{cases} \tag{8.A.10}$$

If as above we require that the threshold conditioning inequalities are in effect, then, for $h_i \in [0,1]$, $\mu_{Y_i}(y_i) \geq h_i$, $i = 1,\ldots,n$, and

$$L\left(\left(y_i - \sum_{j=1}^{m} \alpha_j x_{ij} \right) \Big/ \sum_{j=1}^{m} c_j |x_{ij}| \right) \geq h_i, \; i = 1,\ldots,n,$$

or

$$\left. \begin{aligned} \sum_{j=1}^{m} \alpha_j x_{ij} + |L^{-1}(h_i)| \sum_{j=1}^{m} c_j |x_{ij}| \geq y_i \\ \sum_{j=1}^{m} \alpha_j x_{ij} - |L^{-1}(h_i)| \sum_{j=1}^{m} c_j |x_{ij}| \leq y_i \end{aligned} \right\} i = 1,\ldots,n. \tag{8.A.11}$$

Now our objective is to minimize total vagueness J subjective to Equation 8.A.11, $c_j \geq 0$, and α_j unrestricted, $j = 1,\ldots, m$.

Note that, if the shape function is of the form $L(x) = 1 - |x|$, then $|L^{-1}(h)| = 1 - h$.

Appendix B: Linear Programming Essentials [Panik 1996]

We may express the *basic structure* of a linear programming problem as

$$\max f = c_1 x_1 + c_2 x_2 \text{ s.t.} \tag{8.B.1}$$

$$\begin{aligned} a_{11}x_1 + a_{12}x_2 \leq b_1 \\ a_{21}x_1 + a_{22}x_2 \leq b_2 \end{aligned} \tag{8.B.2}$$

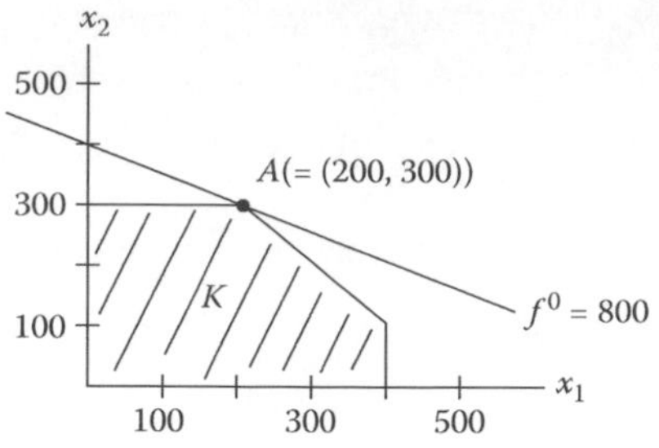

FIGURE 8.B.1
The feasible region.

$$x_1, x_2 \geq 0. \tag{8.B.3}$$

Here Equation 8.B.1 represents the *objective function hyperplane*, Equation 8.B.2 depicts the system of *structural constraint inequalities* (b_1, $b_2 \geq 0$ depict *structural constraint capacities*), and Equation 8.B.3 is a set of *non-negativity conditions*. Together, Equations 8.B.2 and 8.B.3 constitute the *constraint system*.

Each of the inequalities in Equations 8.B.2 and 8.B.3 is a closed half-plane or half-space. If we define the concept of a *solution set* as the set of points satisfying a linear inequality (or equality), then the *feasible region K* or region of admissible solutions to the linear programming problem is formed as the intersection of the solution sets of all the constraints; it is the intersection of a finite number of closed half-planes and thus represents a *convex polyhedron*. Moreover, if this convex polyhedron is bounded, then it constitutes a *convex polytope*: a strictly bounded closed convex set with a finite number of extreme points.[8]

Example 8.B.1

Let the feasible region K be determined by the system of linear inequalities

$$K \begin{cases} x_1 \leq 400 \\ x_2 \leq 300 \\ x_1 + x_2 \leq 500 \\ x_1, x_2 \geq 0. \end{cases} \tag{8.B.4}$$

Then K is illustrated in Figure 8.B.1.

Under what circumstances will the linear program (Equations 8.B.1 through 8.B.3) possess an optimal (maximal) solution? And where will it be found if such a solution exists? The next two theorems answer these questions:

1. Existence theorem: If the feasible region K is a convex polytope, then the linear program always has a finite optimal solution.
2. Extreme point theorem: The objective function assumes its optimum at an extreme point of the feasible region K. If it assumes its optimum at more than one extreme point, then it takes on that same value at each convex combination of those extreme points.

---

[8] A set is *bounded* if its diameter is finite. A set is *closed* if it contains all of its boundary points. Let us express the *convex combination* of points A and B as $C = \lambda A + (1 - \lambda)B$, $0 \leq \lambda \leq 1$ (λ a scalar). Then an *extreme point* is a point that cannot be written as a convex combination of any two other points of a set, i. e., an extreme point does not lie between any two other points of a set.

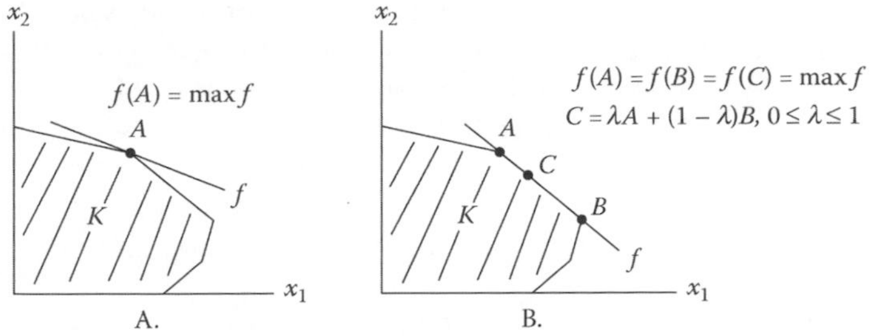

FIGURE 8.B.2
A unique optimal solution at extreme point *A*, and multiple optimal solutions all along edge *AB* (two extreme point solutions and an infinite number of nonextreme point solutions).

The two subcases of this extreme point theorem are illustrated in Figure 8.B.2

Suppose we want to maximize $f = x_1 + 2x_2$ subject to Equation 8.B.4. It is readily demonstrated that tangency to *K* (from above) occurs at extreme point *A* in Figure 8.B.1. How can the coordinates of *A* be determined? Because *A* is at the intersection of the second and third structural constraint equalities (when a constraint holds as a strict equality, it is said to be *binding*), we simply solve the relevant binding constraints simultaneously to obtain the coordinates of *A*, i.e., the simultaneous solution of

$$x_2 = 300$$

$$x_1 + x_2 = 500$$

yields $(x_1^o, x_2^o) = (200, 300)$. Then the optimal objective function value is $f^0 = 200 + 2(300) = 800$.

Note that, because $x_1^o = 200$ and the total capacity of the first structural constraint in Equation 8.B.4 is 400, we have 200 units of unused capacity or *slack* in this constraint. In this regard, any binding constraint exhibits zero slack (all of the right-hand side capacity is used), whereas any *nonbinding* constraint exhibits positive slack. To formalize these notions, let us explicitly introduce a set of non-negative *slack variables* x_3, x_4, and x_5 into Equation 8.B.4, i.e.,

$$\begin{aligned} x_1 \quad + x_3 \qquad &= 400 \\ x_2 \quad + x_4 \qquad &= 300 \\ x_1 + x_2 \qquad + x_5 &= 500 \\ x_1, \ldots, x_5 \geq 0. \end{aligned} \tag{8.B.5}$$

Hence, at point *A* in Figure 8.B.1, $x_3^0 = 200$ and $x_4^0 = x_5^0 = 0$. Clearly the role of a slack variable is to account for the excess of the right-hand side over the left-hand side of a structural constraint.

But what if a structural constraint appears as $3x_1 + 2x_2 \geq 10$? Now, this constraint can be rewritten as the equality $3x_1 + 2x_2 - x_3 = 10$, and the excess of the left-hand side over the right-hand side is thus accounted for by a non-negative *surplus variable* x_3. If this constraint is binding at a particular solution, then $x_3 = 0$; if it is not binding at the said solution, then $x_3 > 0$.

Let us refer to Equation 8.B.5 as the constraint system in *standard form*. Note that this system represents an *underdetermined system* of simultaneous linear equations, i.e., there are more variables than equations. Suppose that Equation 8.B.5 consists of *m* equations in *n* variables (original plus slack variables). Here, $m = 3$ and $n = 5$ (with $p = 2$ original variables). Let us divide our variables into two categories: *basic* and *nonbasic*. In general, the number of basic variables will be m. Hence, we are left with n-m nonbasic variables. If we set the nonbasic variables equal to zero, then we obtain a *basic solution* to Equation 8.B.5, i.e., if x_1 and x_2 are deemed nonbasic and equated to zero, then a basic solution to this system is $x_3 = 400$, $x_4 = 300$, and $x_5 = 500$. Because each basic variable is non-negative, this constitutes a *basic feasible solution* to Equation 8.B.5. In fact, we have obtained what is commonly termed an *initial basic feasible solution* to the linear programming problem. Interestingly enough, there exists a one-to-one correspondence between basic feasible solutions to Equation 8.B.5 and extreme point solutions of the feasible region K.

Up to this point, we have not admitted the objective function to our discussion. Clearly $f = 0$ at the initial basic feasible solution to Equation 8.B.5. Once the objective function is incorporated into Equation 8.B.5 and f is treated as a basic variable, then a procedure called *pivoting* can be undertaken to derive, iteratively, an optional basic feasible solution (see any standard text on linear programming solution routines).

Let us refer to Equation 8.B.1 as the *primal problem*. Then associated with the primal is another linear programming problem called its *dual*. That is, the *(symmetric) dual* of Equation 8.B.1 is

$$\begin{aligned} &\min g = b_1u_1 + b_2u_2 \text{ s.t.} \\ &a_{11}u_1 + a_{21}u_2 \geq c_1 \\ &a_{12}u_1 + a_{22}u_2 \geq c_2 \\ &u_1, u_2 \geq 0. \end{aligned} \tag{8.B.6}$$

[9]

The dual structural constraint inequalities can be transformed into a system of equalities by subtracting a set of nonnegative *dual surplus variables* from the left-hand side of Equation 8.B.6, i.e., if $u_3, u_4 \geq 0$ are dual surplus variables, the we obtain

$$\begin{aligned} a_{11}u_1 + a_{21}u_2 - u_3 &= c_1 \\ a_{12}u_1 + a_{22}u_2 - u_4 &= c_2. \end{aligned}$$

What is the relationship between the optimal solutions to the primal and dual programs? Specifically,

1. The primal maximum and dual minimum problems possess optimal solutions if and only if both have feasible solutions.
2. If at feasible solutions to the primal maximum and dual minimum problems we have $f^0 = g^0$, then these feasible solutions are also optimal.
3. Either both the primal maximum and dual minimum problems have optimal solutions with $f^0 = g^0$ or neither has an optimal solution.

---

[9] What are the essential features of the primal-dual pair of problems Equations 8.B.1 and 8.B.6? Specifically,

1. If the primal problem involves maximization, then the dual problem involves minimization and conversely. (It is immaterial which problem is actually called the dual because the dual of the dual is the primal problem itself.)
2. A new set of variables appears in the dual.
3. The primal objective function coefficients become the right-hand sides of the dual structural constraints and conversely.
4. The inequality signs in the structural constraints of the primal problem are reversed in the dual problem and conversely.
5. The coefficients on x_j in the primal problem become the coefficients of the j th dual structural constraint.
6. If the primal problem has p original variables and m slack variables, then the dual problem has m original variables and p surplus variables.

4. One (and thus both) of the primal maximum and dual minimum problems has an optimal solution if and only if either f or g is bounded on its (nonempty) feasible region.
5. Solutions to the primal maximum and dual minimum problems are optimal if and only if the *complementary slackness conditions* hold:

$$\begin{aligned} x_j^o u_{m+j}^o &= 0, \quad j = 1,\ldots,p; \\ u_i^o x_{p+i}^o &= 0, \quad i = 1,\ldots,m, \end{aligned} \tag{8.B.7}$$

where x_j (u_i) is the jth original primal (dual) variable and x_{p+i} (u_{m+j}) is the ith primal slack (dual surplus) variable. Hence, at least one of each pair in Equation 8.B.7 must equal zero, i.e.,

$$\text{if } x_j^o > 0, \text{ then } u_{m+j}^o = 0;$$
$$\text{if } u_{m+j}^o > 0, \text{ then } x_j^o = 0.$$

And

$$\text{if } u_j^o > 0, \text{ then } x_{p+i}^o = 0;$$
$$\text{if } x_{p+i}^o > 0, \text{ then } u_i^o = 0.$$

So if $u_i^o > 0$, its complementary primal structural constraint must be binding, and, if $x_j^o > 0$, its complementary dual structural constraint is binding.

Example 8.B.2

For the primal problem

$$\max f = x_1 + 2x_2 \text{ s.t.}$$
$$x_1 \le 400$$
$$x_2 \le 300$$
$$x_1 + x_2 \le 500$$
$$x_1, x_2 \ge 0,$$

the (symmetric) dual problem is

$$\min g = 400u_1 + 300u_2 + 500u_3 \text{ s.t.}$$
$$u_1 + u_3 \ge 1$$
$$u_2 + u_3 \ge 2$$
$$u_1, u_2, u_3 \ge 0.$$

In standard form, these problems are, respectively,

$$\max f = x_1 + 2x_2 \text{ s.t.}$$
$$x_1 + x_3 = 400$$
$$x_2 + x_4 = 300$$
$$x_1 + x_2 + x_5 = 500$$
$$x_1,\ldots, x_5 \ge 0,$$

and

$$\min g = 400u_1 + 300u_2 + 500u_3 \text{ s.t.}$$
$$u_1 + u_3 - u_4 = 1$$
$$u_2 + u_3 - u_5 = 2$$
$$u_1, \ldots, u_5 \geq 0.$$

Here x_3, x_4, and x_5 are nonnegative primal slack variables and u_4, u_5 are nonnegative dual surplus variables. We determined previously that an optimal (basic) feasible solution to the primal was $(x_1^o, x_2^o) = (200, 300)$ with $f^0 = 800$. Because the second and third primal structural constraints are binding, we have $x_4^o = x_5^o = 0$. From the first primal structural constraint, $x_3^o = 200$.Looking to the complementary slackness conditions, we see that, because $x_1^o > 0$ and $x_2^o > 0$, we must have $u_4^o = u_5^o = 0$, i.e., the dual structural constraints must be binding (no nonzero surplus). Hence,

$$u_1 + u_3 = 1$$
$$u_2 + u_3 = 2.$$

Because $x_3^o > 0$, we must have $u_1^o = 0$ via the complementary slackness conditions. Hence, the preceding system of equalities becomes

$$u_3 = 1$$
$$u_2 + u_3 = 2$$

or $u_2^o = 1 = u_3^o$. Then $g^0 = 300\ (1) + 500\ (1) = 800 = f^0$. Looking to the complete set of complementary slackness conditions, we have

$$x_1^o u_4^o = (200)(0) = 0$$
$$x_2^o u_5^o = (300)(0) = 0$$
$$u_1^o x_3^o = (0)(200) = 0$$
$$u_2^o x_4^o = (1)(0) = 0$$
$$u_3^o x_5^o = (1)(0) = 0.$$

Hence, the requisite complementarity between original primal variables and dual surplus variables holds. Likewise, we have a complementarity between original dual variables and primal slack variables.

Exercises

8-1. For crisp $A = \{x | 2 \leq x \leq 10\} \subset X = \{x | 0 \leq x \leq 15\}$, illustrate the indicator function for membership in A.

8-2. For $X = \{1, 2, \ldots, 10\}$, subjectively express fuzzy A as the set of small numbers and write fuzzy B as the set of large numbers. For your specification of A and B, does $A \cap B = \emptyset$?

TABLE E.8-1

Observations on Variables X and Y

X	2	4	6	8	10	12
$Y = (y_i, e_i)$	(14,1.7)	(10, 1.3)	(12, 2.0)	(6, 2.3)	(6, 1.8)	(4, 1.6)

TABLE E.8-2

Observations on Variables X and Y

X	2	3	4	5	6	7
$Y = (y_i, e_i)$	(2,2.1)	(4, 1.9)	(2, 1.8)	(5, 1.9)	(4, 1.9)	(6, 2.3)

8-3. Suppose X ={1,2,...,10} and subjectively,

A=0.007/3 + 0.5/4 + 1/5 + 1/6 + 0.5/7 + 0.07/8,

B=1/1 + 0.8/2 + 0.3/3 + 0.02/4 + 0.02/7 + 0.3/8 + 0.8/9 + 1/10.

Find

a. $A \cup B, A \cap B, \bar{A}$, and $\bar{B}$

b. $A_{0.7}$, $A_{0.1}$, and $B_{0.7}$

c. Supp(A) and Supp(B)

8-4. Let fuzzy quantities a, b, $\in R$ with$\mu_a(2) = \frac{2}{3}, \mu_a(3) = \frac{1}{4}, \mu_a(5) = \frac{1}{2}$, and $\mu_a(x) = 0$ for $x \notin$ {2, 3, 5}; and $\mu_b(4) = \frac{1}{5}$, $\mu_b(6) = \frac{1}{2}$, $\mu_b(7) = \frac{1}{3}$, and $\mu_b(x) = 0$ when $x \notin$ {4, 6, 7}. Determine (a) $\mu_{a+b}(9)$; and $\mu_{a-b}(12)$.

8-5. For fuzzy numbers $A = (6, 2, 3)_{LR}$ and $B = (8, 2, 1)_{LR}$, find: (a) $A + B$, (b) $A - B$, (c) $A \cdot B$, (d) λA for $\lambda = -2$, (e) B^{-1} (approximately), and (f) $A \div B$ (approximately).

8-6. Let A= (1,3,6) be a triangular fuzzy number. What is the form of its membership function? If triangular fuzzy B = (2,4,6), find $A + B$. If $\lambda = -$, find λA.

8-7. Let A = (8,3) be a symmetric triangular fuzzy number. What is the form of its memberships function?

8-8. Suppose $A = (4, 1)_L$ is a symmetric fuzzy number. Determine its membership function.

8-9. Given the Table E.8-1 dataset (the x_i's are crisp and the Y_i's are symmetric triangular fuzzy numbers), estimate the fuzzy linear regression equation using the basic Tanaka model (Equation 8.39) (here the x_i's and y_i's are considered crisp, whereas the regression parameters A_0, A_1 are fuzzy sets) with $h_i = 0$ for all i. Then reestimate the said regression equation with $h_i = 0.5$ for all i.

8-10. Using the Table E.8-1 dataset, use Tanaka's Equation 8.47 to estimate the fuzzy parameters of the fuzzy regression equation with x_i crisp and $Y_i = (y_i, e_i)$ symmetric triangular. Choose $H = 0.65$.

8-11. Apply the two-phase FLSLR method to the crisp x_i and y_i data values of Table E.8-1 using Equation 8.39.1 under $h_i = 0$ for all i.

8-12. Apply the two-phase FLSLR technique to Equation 8.47 with $H = 0.4$ for x_i crisp and $Y_i = (y_i, e_i)$ symmetric triangular (Table E.8-1).

8-13. Apply the FLS routine to the Table E.8-2 dataset to estimate the fuzzy regression parameters of $Y = A_0 + A_1 x$ with x_i crisp and fuzzy y_i symmetric triangular.

9

Random Coefficients Regression

Introduction

We previously specified the basic bivariate regression equation as

$$Y_i = \beta_0 + \beta_1 X_i + \varepsilon_i, \quad i = 1,\ldots,n, \tag{9.1}$$

where X_i is fixed and ε_i is a random error term satisfying the assumptions of the (weak) classical linear regression model. Under this structure, the population regression line is of the form

$$E\,(Y_i | X_i) = \beta_0 + \beta_1 X_i, \quad i = 1,\ldots,n, \tag{9.2}$$

i.e., the conditional mean of Y_i is a linear function of X_i, with $\beta_0 = E\,(Y_i|0)$ and β_1 is the average rate of change in Y per unit change in X.

At times, the constancy of β_0 and β_1 at successive data points can legitimately be questioned. For example, if β_1 depicts the response in regional sales of a certain product to regional advertising expenditure, it is evident that this sales response is influenced by, say, other factors such as regional economic conditions and the activity of competitors (if any). If these other factors can be held constant, then β_1 might be expected to be (approximately) constant as well. Furthermore, if these supplemental factors are observable, then they can be explicitly treated as explanatory variables in the model (we obviously have the case of multiple regression) so that β_1 can again be viewed as roughly constant because these other factors are being controlled for. However, if these additional factors are unobservable, then one can essentially view β_1 as the *average* of a random response rate in sales per unit change in advertising expenditure.

A Pure Random Coefficients Model [Hildreth and Houck 1968; Hsiao 1975; Johnston 1984]

Suppose we now assume that the regression coefficients follow a *random coefficients scheme* in that

$$\begin{aligned} Y_i &= \beta_{0i} + \beta_{1i} X_i \\ &= (\beta_0 + \varepsilon_{0i}) + (\beta_1 + \varepsilon_{1i}) X_i \\ &= \beta_0 + \beta_1 X_i + (\varepsilon_{0i} + \varepsilon_{1i} X_i) \qquad (9.3) \\ &= \beta_0 + \beta_1 X_i + \nu_i, \quad i = 1,\ldots,n, \end{aligned}$$

where $\nu_i = \varepsilon_{0i} + \varepsilon_{1i} X_i$. Here the vertical intercept is, on average, β_0, but there is random variation around β_0 that is described by the random variable ε_{0i}. Similarly, β_1 is the average response rate, but random variation around β_1 is picked up by the random variable ε_{1i}. Thus, β_0 and β_1 are unknown constants common to all observations, whereas ε_{0i} and ε_{1i} are random error terms that influence the values of the regression coefficients at the ith sample point. This is in contrast to Equation 9.1 wherein all of the uncertainty connected with the regression model is incorporated into the random disturbance ε_i and with β_0 and β_1 held constant at each sample point.

With respect to the behavior of the random variables ε_{0i} and ε_{1i}, suppose that,

$$\begin{aligned}
&E(\varepsilon_{ji}) = 0; j = 0,1, \quad i = 1,\ldots,n;\\
&E(\varepsilon_{ji}^2) = V(\varepsilon_{ji}) = \alpha_j, \quad j = 0,1, i = 1,\ldots,n;\\
&E(\varepsilon_{0i}\varepsilon_{1i}) = 0, \quad i = 1,\ldots,n; \text{ and}\\
&E(\varepsilon_{0r}\varepsilon_{1s}) = 0, \quad r \neq s, \text{ with } r,s = 1,\ldots,n.
\end{aligned} \tag{9.4}$$

Thus, the stochastic components of the random coefficients β_{0i} and β_{1i} have zero means, zero covariances at any given sample point, and are uncorrelated between sample points.

Given Equation 9.4 and $\nu_i = \varepsilon_{0i} + \varepsilon_{1i}X_i$, $i = 1,\ldots,n$, it follows that

$$\begin{aligned}
&E(\nu_i) = 0, \quad i = 1,\ldots,n;\\
&E(\nu_i^2) = V(\nu_i) = \sigma_i^2 = \alpha_0 + \alpha_1 X_i^2, \quad i = 1,\ldots,n, \text{ and}\\
&E(\nu_r \nu_s) = 0, \quad r \neq s, \text{ with } r,s = 1,\ldots,n.
\end{aligned} \tag{9.5}$$

A glance at σ_i^2 reveals that Equation 9.3 represents a regression model with a *heteroscedastic disturbance term*, in which the variance at each data point is expressible as a linear combination of the squares of the explanatory variables at that point.

Given Equations 9.1 and 9.5, it should be evident that the parameters of interest in this random coefficients model are β_0, β_1, α_0, and α_1, where, from Equation 9.4, $\alpha_0 = V(\varepsilon_{0i})$ and $\alpha_1 = V(\varepsilon_{1i})$. To estimate these parameters, let us consider the following stepwise procedure:[1]

1. Fit Equation 9.3 via OLS and square each residual e_i to obtain $e_1^2, e_2^2, \ldots, e_n^2$.
2. Regress the e_i^2 on the variables Z_1 and Z_2 (the columns of the array $\boldsymbol{Z} = \dot{\boldsymbol{M}}\dot{\boldsymbol{X}}$ that will be generated from the X_i values by SAS/IML) with no constant term to obtain $\hat{\alpha}_0, \hat{\alpha}_1$. (Discard any negative α's.)
3. Substitute $\hat{\alpha}_0, \hat{\alpha}_1$ into σ_i^2 (see Equation 9.5) to obtain estimates of $S_i^2 = \hat{\alpha}_0 + \hat{\alpha}_1 X_i^2$, $i = 1,\ldots,n$.
4. Using the S_i^2 values, determine the weights

$$w_i = S_i = (\hat{\alpha}_0 + \hat{\alpha}_1 X_i^2)^{1/2}, \quad i = 1,\ldots,n, \tag{9.6}$$

to execute the weighted least-square regression of Y_i/w_i on $1/w_i$ and X_i/w_i (with no constant term). The standard errors thus obtained for $\hat{\beta}_0$ and $\hat{\beta}_1$ are asymptotic standard errors.

---

[1] Details concerning this procedure are offered in Appendix 9.A. Readers not familiar with matrix notation might want to skip this discussion and proceed directly to the estimation steps themselves.

Example 9.1

Given the following set of n=8 observations on the variables X and Y (Table 9.1), let us apply the preceding random coefficients procedure using the model in Equation 9.3. We first generate the columns of the matrix $\boldsymbol{Z} = \dot{\boldsymbol{M}}\dot{\boldsymbol{X}}$ by invoking *proc IML* (Exhibit 9.1). (Readers not familiar with the *IML* procedure should consult Appendix 9.B.)

Exhibit 9.1 Calculation of the Z Matrix

```
proc iml;
start matrices;
X={1 1, 1 2, 1 3,
  1 4, 1 5, 1 6,
  1 7, 1 8}; ①
Xt=t(X); ②
p=Xt*X; ③
pinv=inv(p); ④
XpinvXt=X*pinv*Xt;⑤
in=i(8); ⑥
M=in-XpinvXt; ⑦
MSQ=M##2; ⑧
XSQ=X##2; ⑨
MXSQ=MSQ*XSQ; ⑩
print, MXSQ;
finish;
run matrices;
quit;
```

① We specify the rows of the (8×2) matrix $\boldsymbol{X}$.
② We take the transpose of $\boldsymbol{X}$ (denoted $\boldsymbol{X}'$).
③ We form the product $\boldsymbol{X}'\boldsymbol{X}$.
④ We determine the inverse of $\boldsymbol{X}'\boldsymbol{X}$ or $(\boldsymbol{X}'\boldsymbol{X})^{-1}$.
⑤ We form the product $\boldsymbol{X}(\boldsymbol{X}'\boldsymbol{X})^{-1}\boldsymbol{X}'$.
⑥ We request an identity matrix of order eight ($\boldsymbol{I}_8$).
⑦ We form $\boldsymbol{M} = \boldsymbol{I}_8 - \boldsymbol{X}(\boldsymbol{X}'\boldsymbol{X})^{-1}\boldsymbol{X}'$.
⑧ We square the elements in $\boldsymbol{M}$ to form $\dot{\boldsymbol{M}}$.
⑨ We square the elements in $\boldsymbol{X}$ to form $\dot{\boldsymbol{X}}$.
⑩ We form the product $\dot{\boldsymbol{M}}\dot{\boldsymbol{X}}$.

TABLE 9.1

Observations on the Variables X and Y

X	Y
1	1
2	3
3	1
4	5
5	2
6	8
7	4
8	9

TABLE 9.2

Observations on the Variables Z_1 and Z_2

Z_1	Z_2
0.5833333	4.0833333
0.7261905	3.7619048
0.8214286	7.1785714
0.8690476	14.047619
0.8690476	22.940476
0.8214286	31.285714
0.7261905	35.369048
0.5833333	30.333333

The output generated by this program appears in Table 9.2, with $\boldsymbol{Z} = \dot{\boldsymbol{M}}\dot{\boldsymbol{X}} = (Z_1, Z_2)$.

We next consider the following regression program (Exhibit 9.2) that uses the original X and Y variables along with Z_1, Z_2.

Exhibit 9.2 SAS Code for Weighted Regression

```
data random;
  input x, y, z1,z2;
xsq=x**2; ①
datalines;
1   1   0.5833   4.0833
2   3   0.7262   3.7619
3   1   0.8214   7.1786
4   5   0.8690   14.0476
5   2   0.8690   22.9404
6   8   0.8214   31.2857
7   4   0.7262   35.3690
8   9   0.5833   30.3333
;
proc reg data=random; ②
model y= x/r;
output out=new r=ry; ③
data new1; set new; ④
rysq = ry**2;
data randnew; ⑤
merge random new1;
proc print data =randnew;
proc reg data =randnew
  outest=param; ⑥
model rysq=z1 z2/noint; ⑦
proc print data=param;
data param1; set param;
  keep z1 z2; ⑧
```

① We calculate the squares of the values of X.

② We invoke the regression procedure and regress Y on X.

③ We store the regression residuals e_i as the variable *ry* in the output dataset *new*.

④ We store the squared residuals e_i^2 as the variable *rysq* in the dataset *new1*.

⑤ We place all of the variables X, Y, Z_1, Z_2, *xsq* and *rysq* in the dataset *randnew*.

⑥ We invoke the regression procedure using the dataset *randnew* and store the parameter estimates in the output dataset *param*.

⑦ We regress e_i^2 on Z_1 and Z_2 using the *no intercept* option.

⑧ We form dataset *param1*, which houses the estimated regression coefficients on the variables Z_1, Z_2.

⑨ we name the estimated coefficient on Z_1 *alpha0* and estimated coefficient on Z_2 *alpha1*. These alphas are placed in the dataset *param2*.

⑩ *param3* is a dataset that has eight observations each on the two variables

```
data param2; set param1;
  rename z1=alpha0
    z2=alpha1; ⑨
proc print data =param2;
  var alpha0 alpha1;
data param3; set param2;
  do _n_=1 to 8 by 1; ⑩
alpha0=alpha0;
  alpha1=alpha1; end;
data weight;
merge  randnew  param3; ⑪
w=(alpha0+alpha1*xsq)
  **0.5; ⑫
yw=y/w; invw=1/w; xw=x/w; ⑬
proc print data=weight;
proc reg data =weight;
model yw=invw xw/noint; ⑭
run;
```

alpha0 and *alpha1*, where each value of *alpha0* is *alpha0* and each value of *alpha1* is *alpha1*.

⑪ We merge datasets *randnew* and *param3* to form the dataset *weight* containing the variables *X*, *Y*, Z_1, Z_2, xsq, rysq, alpha0, and alpha1.

⑫ We form the weights according to Equation 9.6.

⑬, ⑭ We create the variables *Y*/*w*, 1/*w*, and *X*/*w* to perform the requisite weighted least-squares regression to obtain estimates of β_0 and β_1. Again, the *no intercept* option is used.

We now turn to the output generated from this program (Table 9.3):

TABLE 9.3
Output for Example 9.1

The SAS System
①
The REG Procedure
Model: MODEL1
Dependent Variable: *y*
Analysis of Variance

Source	DF	Sum of Squares	Mean Square	*F* Value	Pr > *F*
Model	1	37.14881	37.14881	8.04	0.0297
Error	6	27.72619	4.62103		
Corrected total	7	64.87500			

Root MSE	2.14966	*R*-Square	0.5726
Dependent mean	4.12500	Adj *R*-Sq	0.5014
Coeff Var	52.11293		

Parameter Estimates

Variable	DF	Parameter Estimate	Standard Error	*t* Value	Pr > |*t*|
Intercept	1	–0.10714	1.67500	–0.06	0.9511
x	1	0.94048	0.33170	2.84	0.0297

(continued)

TABLE 9.3 *(continued)*

The SAS System
The REG Procedure
Model: MODEL1
Dependent Variable: *y*
Output Statistics

Obs	Dep Var y	Predicted value	Std Error mean predict	(e_i) residual	Std Error residual	Student residual
1	1.0000	0.8333	1.3876	0.1667	1.642	0.102
2	3.0000	1.7738	1.1248	1.2262	1.832	0.669
3	1.0000	2.7143	0.9048	−1.7143	1.948	−0.880
4	5.0000	3.6548	0.7779	1.3452	2.004	0.671
5	2.0000	4.5952	0.7779	−2.5952	2.004	−1.295
6	8.0000	5.5357	0.9084	2.4643	1.948	1.265
7	4.0000	6.4762	1.1248	−2.4762	1.832	−1.352
8	9.0000	7.4167	1.3876	1.5833	1.642	0.964

Output Statistics

```
Obs   -2 -1  0  1  2    Cook's D
1      |     |     |      0.004
2      |     |*    |      0.084
3      |    *|     |      0.084
4      |     |*    |      0.034
5      |   **|     |      0.126
6      |     |**   |      0.174
7      |   **|     |      0.344
8      |     |*    |      0.332
```

Sum of residuals	0
Sum of squared residuals	27.72619
Predicted Residual SS (PRESS)	46.59659

The SAS System
②

Obs	x	y	z_1	z_2	xsq	(e_i) ry	(e_i^2) rysq
1	1	1	0.5833	4.0833	1	0.16667	0.02778
2	2	3	0.7262	3.7619	4	1.22619	1.50354
3	3	1	0.8214	7.1786	9	−1.71429	2.93878
4	4	5	0.8690	14.0476	16	1.34524	1.80967
5	5	2	0.8690	22.9404	25	−2.59524	6.73526
6	6	8	0.8214	31.2857	36	2.46429	6.07270
7	7	4	0.7262	35.3690	49	−2.47619	6.13152
8	8	9	0.5833	30.3333	64	1.58333	2.50694

TABLE 9.3 *(continued)*

The SAS System

③

The REG Procedure

Model: MODEL1

Dependent Variable: rysq

Note: No intercept in model. *R*-Square is redefined.

Analysis of Variance

Source	DF	Sum of Squares	Mean Square	*F* Value	Pr > *F*
Model	2	123.94387	61.97193	22.74	0.0016
Error	6	16.35061	2.72510		
Uncorrected total	8	140.29448			

Root MSE	1.65079	*R*-Square	0.8835
Dependent mean	3.46577	Adj *R*-Sq	0.8446
Coeff Var	47.63115		

Parameter Estimates

Variable	DF	Parameter Estimate	Standard Error	*t* Value	Pr > |*t*|
z_1	1	1.61850	1.38976	1.16	0.2884
z_2	1	0.12843	0.04737	2.71	0.0350

④

Obs	_MODEL_	_TYPE_	_DEPVAR_	_RMSE_	z_1	z_2	rysq
1	MODEL1	PARMS	rysq	1.65079	1.61850	0.12843	–1

⑤

Obs	alpha0	alpha1
1	1.61850	0.12843

⑥

Obs	*x*	*y*	z_1	z_2	xsq	ry	rysq	alpha0
1	1	1	0.5833	4.0833	1	0.16667	0.02778	1.61850
2	2	3	0.7262	3.7619	4	1.22619	1.50354	1.61850
3	3	1	0.8214	7.1786	9	–1.71429	2.93878	1.61850
4	4	5	0.8690	14.0476	16	1.34524	1.80967	1.61850
5	5	2	0.8690	22.9404	25	–2.59524	6.73526	1.61850
6	6	8	0.8214	31.2857	36	2.46429	6.07270	1.61850
7	7	4	0.7262	35.3690	49	–2.47619	6.13152	1.61850
8	8	9	0.5833	30.3333	64	1.58333	2.50694	1.61850

Obs	alpha1	w	yw	invw	xw
1	0.12843	1.32172	0.75659	0.75659	0.75659
2	0.12843	1.46021	2.05449	0.68483	1.36966
3	0.12843	1.66565	0.60037	0.60037	1.80110

(continued)

TABLE 9.3 *(continued)*

Obs	alpha1	w	yw	invw	xw
4	0.12843	1.91661	2.60877	0.52175	2.08702
5	0.12843	2.19756	0.91010	0.45505	2.27525
6	0.12843	2.49841	3.20204	0.40026	2.40153
7	0.12843	2.81276	1.42209	0.35552	2.48866
8	0.12843	3.13658	2.86937	0.31882	2.55055

⑦

The REG Procedure

Model: MODEL1

Dependent variable: *yw*

NOTE: No intercept in model. *R*-Square is redefined.

Analysis of Variance

Source	DF	Sum of Squares	Mean Square	*F* Value	Pr > *F*
Model	2	27.69366	13.84683	14.83	0.0048
Error	6	5.60280	0.93380		
Corrected total	8	33.29646			

Root MSE	0.96633	*R*-Square	0.8317
Dependent mean	1.80298	Adj *R*-Sq	0.7756
Coeff Var	53.59652		

Parameter Estimates

Variable	DF	Parameter Estimate	Standard Error	*t* Value	Pr > |*t*|
invw	1	0.20437	1.19407	0.17	0.8697
xw	1	0.86145	0.30991	2.78	0.0320

① For the initial regression of *Y* on *X*, we find that $\hat{\beta}_0 = -0.10714$ and $\hat{\beta}_1 = 0.94048$ with R^2=0.5726. Only $\hat{\beta}_1$ is statistically significant because its (two-tailed) *p*-value = 0.03 < 0.05. The section on *output statistics* lists the residuals e_i and the accompanying graph of the standardized residuals indicates a mild degree of heteroscedasticity for increasing values of *X*.

② The dataset *randnew* lists the values of all variables in *random* as well as the residuals (e_i) and their squares (e_i^2).

③ A regression of e_i^2 on Z_1 and Z_2 provides us with the estimates $\hat{\alpha}_0 = 1.61850$ and $\hat{\alpha}_1 = 0.12843$. Both alphas are positive, and the overall fit is good (R^2= 0.884, the *p*-value for the *F* statistic is 0.0016 < 0.05, and the *p*-value for the coefficient on Z_2 is 0.04 < 0.05).

④ We get a print of the contents of the *param* dataset.

⑤ We get a print of the contents of the *param2* dataset with the estimated coefficients on Z_1 and Z_2 named alpha0 and alpha1, respectively.

⑥ We print the contents of the *weight* dataset.

⑦ The weighted regression results yield new estimates for the original regression parameters β_0 and β_1, namely $\hat{\beta}_0 = 0.2044$ and $\hat{\beta}_1 = 0.8615$, respectively. Note that these results indicate a marked improvement over those displayed in Section ① above, i.e., for this weighted regression, the estimated *F* value increases, R^2 increases, $\sqrt{MSE}$ decreases, and the standard errors of the estimated coefficients decrease.

The Linear Random Coefficients Mixed Model

The traditional or conventional linear regression model generates only an estimated or predicted response in Y for subject i to a given level j of X or $Y_{ij} = \beta_0 + \beta_1 X_{ij} + \varepsilon_{ij}$, where ε_{ij} has constant variability and estimates of β_0 and β_1 and the estimated response $\hat{Y}_{ij}$ are simply "averages" taken over the entire population of subjects, with β_0 and β_1 serving as "fixed" population parameters. However, if a subject is "randomly assigned" to a specific level of X, then the subject-specific random response can be assessed.

For instance, if we again let Y_{ij} denote the response of subject i at level j of X, then we can write the *linear random coefficients mixed model* as

$$Y_{ij} = \beta_0 + a_i^* + (\beta_1 + b_i^*)X_{ij} + \varepsilon_{ij}, \quad i = 1,\ldots,t;\ j = 1,\ldots,n_i, \tag{9.7}$$

where again β_0 and β_1 are the fixed intercepts and slope parameters, respectively, with additional subject-specific random regression terms given by

$a_i^* = a_i - \beta_0$ (the random deviation of the ith subject's intercept from β_0); and

$b_i^* = b_i - \beta_1$ (the random deviation of the ith subject's slope from β_1).

(Note that, if a_i^* and b_i^* are substituted into Equation 9.7, then we obtain $Y_{ij} = a_i + b_i X_{ij} + \varepsilon_{ij}$ so that each subject has its own specific random intercept and slope terms.) In this regard, Equation 9.7 can be partitioned in terms of the model's mixed (fixed and random) effects as

$$Y_{ij} = \underbrace{\beta_0 + \beta_1 X_{ij}}_{\text{fixed effects}} + \underbrace{a_i^* + b_i^* X_{ij} + \varepsilon_{ij}}_{\text{random effects}}. \tag{9.7.1}$$

Given Equation 9.7, in addition to estimating the population average fixed effects $\beta_0 + \beta_1 X_{ij}$, we can also determine the specific response of a subject to a given level of the explanatory variable X or $\beta_0 + a_i^* + (\beta_1 + b_i^*)X_{ij}$. Here we assume, for each subject i, that ε_{ij} is independent and identically distributed $N(0,\sigma_\varepsilon^2)$, whereas each random intercept and slope pair $(a_i, b_i)'$ is independent and identically distributed $N((\beta_0,\beta_1)',\Psi)$(or $(a_i^*, b_i^*)'$ is independent and identically distributed $N((0,0)',\Psi)$ with variance-covariance matrix

$$\Psi = \begin{bmatrix} \sigma_a^2 & \sigma_{ab} \\ \sigma_{ba} & \sigma_b^2 \end{bmatrix}. \text{[2]} \tag{9.8}$$

The regression coefficients a_i^* and b_i^* within Equation 9.7 (or in Equation 9.7.1) are assumed to represent a random sample taken from some population distribution of possible coefficients. Clearly such coefficients will most assuredly influence the variability of the data. Random coefficient models apply whenever the data arise from independent subjects or clusters/categories and the regression model for each specific subject or cluster is viewed as a random deviation from some population regression equation so that the subject-specific regression equations are scattered about the overall population regression equation.

[2] For a discussion on the generalized linear mixed model, see Appendices 9.C and 9.D.

Whereas the average fixed effect is linear in terms of one set of parameters (the fixed-effects parameters), the variances and covariances are expressed in terms of an additional set of parameters (called *covariance parameters*). Hence, these two sets of parameters are needed to fully determine the probability distribution of the data. Given Equation 9.7 (or Equation 9.7.1), the variances and covariances of the random-effects parameters serve as the covariance parameters.

We may view the aforementioned notion of a *cluster* as consisting of the following:

a. A set of measures pertaining to some outcome or attribute for a specific subject
b. Test/inspection results in an educational/manufacturing setting
c. Observations on subjects in a social, geographical, political, or occupational setting
d. The daily amount of an effluent discharged from a refinery over a given time period (this is an example of a *longitudinal series* that results when a subject's characteristics are measured at designated points over some period of time)

to mention but a few possibilities.

In this regard, when we seek to make inferences about the entire population of clusters, the random coefficients approach enables us to examine the between-cluster differences, which are taken to represent the entire population of clusters. Moreover, the clusters actually observed are assumed to have been randomly chosen from an infinite population of clusters, with the elements found in a given cluster constituting a random sample drawn from the underlying population cluster.

To see how random coefficients models are constructed, let us suppose that three methods of preparation for the written portion of a high school drivers' education test were randomly selected from among the population of instructional techniques for drivers' education programs. Thirty high school seniors are to be enrolled in the program, and it is anticipated that their score on the written portion of their test might be associated with or influenced by their pre-enrollment grade-point average determined at the end of their junior year, e.g., those students who display high overall academic achievement might generally score higher on the written portion of the drivers' education test.

The three methods of instruction are each randomly assigned to 10 different students. Because the methods themselves were randomly selected, the resulting regression equation for each method is a random coefficients model from the population of instructional methods models. In this regard, the fixed effects of the model are the intercept β_0 and slope β_1 which are the expected values of the population distribution of intercepts and slopes of the various instructional methods, whereas the random effects correspond to random treatment effects. That is, for instructional method (treatment) i, substituting $a_i^* = a_i - \beta_0$ and $b_i^* = b_i - \beta_i$ into Equation 9.7 yields the method-specific random coefficients equation

$$Y_{ij} = a_i + b_i X_{ij} + \varepsilon_{ij}, \; i = 1,2,3; \; j = 1,\ldots,10.$$

For this model, each instructional method gives rise to a separate or method-specific cluster of 10 students. Because these clusters can be viewed as a random sample of clusters drawn from some population of clusters, the assessment of between-cluster differences makes sense and inferences concerning the entire population of clusters become relevant.

How do we know whether the effects of "instructional method" are fixed or random? To answer this question, we need only focus on how the data are generated. That is, if the instructional methods used can be considered to be a random sample from a population of such methods having a probability distribution, then the method effects are random. However, if the methods used were the only ones available, then their effects would be considered fixed. So when inferences about the population distribution of methods are to be made on the basis of a random sample taken from that distribution, then the effects must be random, and if inferences are limited to the effects within the model proper, then the effects are treated as fixed.

To see exactly how the preceding random sampling scenario involving information on drivers' education instructional methods (1, 2, or 3) and student performance is handled in a mixed (fixed- and random-effect) modeling environment, let us look to the following example problem.

Example 9.2

Given the dataset appearing in Table 9.4, let us first examine the SAS code provided in Exhibit 9.3.

Exhibit 9.3 SAS Code for PROC MIXED

```
data method;
  input id method score gpa;
datalines;
1  1  59  59.01
2  1  62  65.00
.  .  .     .
.  .  .     .
.  .  .     .
29  3  69  96.66
30  3  68  99.17
;
        ①
proc mixed data = method
  scoring =10 covtest;②,③
  class method;④
          ⑤           ⑥      ⑦  ⑧
  model score = gpa/covb cl s;
     ⑨       ⑩           ⑪
  random int  gpa/type = un
    sub = method ⑫
  ⑬ ⑭
  cl s;
run;
```

① We invoke the MIXED procedure to estimate a linear mixed model that contains both fixed- and random-effects parameters.

② We request that the likelihood-based *scoring* method be used for the first 10 iterations (for details on the scoring algorithm, see Appendix 18.D).

③ We request asymptotic standard errors and Wald Z tests for the covariance parameter estimates.

④ The *class* statement declares classification or qualitative variables that specify indicator variables in design matrices.

⑤ The *MODEL* statement specifies the dependent variable (test) *score* as a function of the fixed-effects variables (here only *gpa* appears).

⑥ The *covb* option produces the approximate variance-covariance matrix of the fixed-effects parameter estimates.

⑦ The *cl* option requests that *t*-type confidence limits be constructed for each of the fixed-effects parameters.

⑧ The *s* option requests the solution for the fixed-effects parameters.

⑨, ⑩ The *random* statement specifies the random effects. The keyword *int* (for intercept) is interpreted as a random effect with all 1's.

(PROC MIXED does not include the intercept by default as in, say, PROC REG.)

⑪ The *type*= option specifies the variance-covariance structure, e.g., *type* = *un* specifies an "unstructured" variance-covariance matrix parameterized directly in terms of variances and covariances. We obtain estimates of the different variances of the random intercepts and slopes and the covariances between them.

⑫ The *subject*= method option specifies that the intercept and slope of one method is distributed independently of the intercepts and slopes of the other methods, with the intercept and slope of each method being correlated so that an estimate of σ_{ab} is obtained.

⑬ The *cl* option requests that *t*-type confidence limits be constructed for each of the random-effects parameters.

TABLE 9.4

Method of Instruction and Student Performance

Identification Number (ID)	Method of Instruction (Method)	Score on Drivers' Ed Written Test (Score)	Scholastic Grade Point Average (GPA)
1	1	59	59.01
2	1	62	65.00
3	1	65	69.98
4	1	66	75.12
5	1	68	79.38
6	1	71	83.60
7	1	74	89.90
8	1	73	90.81
9	1	77	94.71
10	1	80	98.19
11	2	60	60.23
12	2	65	69.31
13	2	67	72.19
14	2	69	75.10
15	2	71	80.02
16	2	76	85.17
17	2	78	90.12
18	2	82	94.14
19	2	83	97.10
20	2	85	99.27
21	3	56	59.66
22	3	57	63.33
23	3	60	65.29
24	3	62	69.89
25	3	60	75.60
26	3	63	80.14
27	3	67	86.62
28	3	67	90.22
29	3	69	96.66
30	3	68	99.17

⑭ The *s* option requests the solution for the random-effects parameters. These estimates are the empirical best linear unbiased predictors (EBLUPs) and are used to compare the random effects from different experimental units.

The output generated by this SAS program appears in Table 9.5.

① Our estimate of the variance-covariance matrix Ψ (Equation 9.8) is

$$\hat{\Psi} = \begin{bmatrix} 66.6711 & -1.3088 \\ -1.3088 & 0.0255 \end{bmatrix}.$$

Because $\hat{\sigma}_{ab} < 0$, it is obvious that the random effects (coefficients) are negatively correlated. That is, if a method's intercept is larger than others, then its slope will tend to be smaller than others. (If $\hat{\sigma}_{ab} > 0$ and a method's intercept is larger relative to others, then its slope tends

TABLE 9.5

Output for Example 9.2 (Unstructured Model)

The SAS System

The Mixed Procedure

Model Information	
Data set	WORK.METHOD
Dependent variable	Score
Covariance structure	Unstructured
Subject effect	Method
Estimation method	REML
Residual variance method	Profile
Fixed effects SE method	Model-based
Degrees of freedom method	Containment

Class Level Information

Class	Levels	Values
Method	3	1 2 3

Dimensions	
Covariance parameters	4
Columns in X	2
Columns in Z per subject	2
Subjects	3
Max Obs per subject	10

Number of Observations	
Number of Observations Read	30
Number of Observations Used	30
Number of Observations Not Used	0

(continued)

TABLE 9.5. *(continued)*

Iteration History

Iteration	Evaluations	−2 Res log like	Criterion
0	1	173.45862286	
1	2	103.65572917	0.00189420
2	1	103.57828485	0.00045553
3	1	103.55891318	0.00015238
4	1	103.55228055	0.00006079
5	1	103.54959777	0.00002676
6	1	103.54840875	0.00001252
7	1	103.54785025	0.00000609
8	1	103.54757777	0.00000304
9	1	103.54744133	0.00000155
10	1	103.54737177	0.00000080
11	1	103.54733585	0.00000153
12	1	103.54729630	0.00000000

Convergence criteria met.

①

Covariance Parameter Estimates

Cov Parm	Subject	Estimate	Standard error	Z value	Pr Z
UN(1,1)	method	66.6711	71.7475	0.93	0.1764
UN(2,1)	method	−1.3088	1.3702	−0.96	0.3395
UN(2,2)	method	0.02550	0.02623	0.97	0.1655
Residual		1.1536	0.3551	3.25	0.0006

Fit statistics	
−2 Res log likelihood	103.5
AIC (smaller is better)	111.5
AICC (smaller is better)	113.3
BIC (smaller is better)	107.9

②

Null Model Likelihood Ratio Test

DF	Chi-Square	Pr > ChiSq
3	69.91	<0.0001

TABLE 9.5. *(continued)*

③

Solution for Fixed Effects

Effect	Estimate	Standard error	DF	*t* value	Pr > |*t*|	Alpha	Lower	Upper
Intercept	29.1015	4.8672	2	5.98	0.0269	0.05	8.1594	50.04
gpa	0.4892	0.09340	2	5.24	0.0346	0.05	0.08734	0.89

Covariance Matrix for Fixed Effects

Row	Effect	Col1	Col2
1	Intercept	23.6900	-0.4541
2	gpa	-0.4541	0.008723

④

Solution for Random Effects

Effect	Method	Estimate	Std Err Pred	DF	*t* Value	Pr > |*t*|	Alpha	Lower	Upper
Intercept	1	–5.0200	3.8202	24	–1.31	0.2012	0.05	–12.9046	2.8646
gpa	1	0.07059	0.08670	24	0.81	0.4235	0.05	–0.1083	0.2495
Intercept	2	–5.2103	4.0108	24	–1.30	0.2063	0.05	–13.4883	3.0676
gpa	2	0.1173	0.08794	24	1.33	0.1948	0.05	–0.06422	0.2988
Intercept	3	10.2304	3.5106	24	2.91	0.0076	0.05	2.9848	17.4760
gpa	3	–0.1879	0.08491	24	–2.21	0.0367	0.05	–0.3631	–0.01264

⑤

Type 3 Tests of Fixed Effects

Effect	Num DF	Den DF	*F* Value	Pr > *F*
gpa	1	2	27.43	0.0346

to be larger as well.) Because the *p*-value corresponding to the *Z* statistic associated with each element within $\hat{\Psi}$ exceeds 0.05, we can conclude that each of these covariances is not significantly different from zero. Also, because the *p*-value for the *Z* statistic associated with $\hat{\sigma}_\varepsilon = 1.1536$ lies far below 0.05, it follows that σ_ε^2 is significantly different from zero.

② The low *p*-value for the null model likelihood ratio test statistic indicates a statistically significant improvement over a (null) model consisting of no random effects and a homogenous random error term. Hence, the null model is rejected in favor of one incorporating random coefficients and a heterogeneous error variance.

③ The fixed-effects estimates are $\hat{\beta}_0 = 29.1015$ and $\hat{\beta}_1 = 0.4892$ (which represent the estimated means for the random intercept and slope terms, respectively). The low *p*-values for the reported *t* statistics indicate that the fixed-effects parameters β_0 and β_1 depart significantly from zero.

④ Random effects solutions represent the estimated deviations from the mean intercept and slope for each method, i.e.,

Method 1

$$\hat{a}_1^* = \hat{a}_1 - \hat{\beta}_0 = -5.0200$$
$$\hat{b}_1^* = \hat{b}_1 - \hat{\beta}_1 = 0.07059$$

Method 2

$$\hat{a}_2^* = \hat{a}_2 - \hat{\beta}_0 = -5.2103$$
$$\hat{b}_2^* = \hat{b}_2 - \hat{\beta}_1 = 0.1173$$

Method 3

$$\hat{a}_3^* = \hat{a}_3 - \hat{\beta}_0 = 10.2304$$
$$\hat{b}_3^* = \hat{b}_3 - \hat{\beta}_1 = -0.1879$$

Looking to the *p*-values associated with the accompanying *t* statistics reveals that, for Methods 1 and 2, none of the random deviations are statistically significant. However, for Method 3, both *p*-values are below 0.05, thus pointing to a statistically significant set of random deviations for the intercept and slope. Thus, the intercept for Method 3 lies significantly above $\hat{\beta}_0$, whereas the slope for this method falls significantly below $\hat{\beta}_1$. Moreover, given the $\hat{\beta}_0$ and $\hat{\beta}_1$ values, we can readily find the random effects:

Method 1	Method 2	Method 3
$\hat{a}_1^* = 24.0815$	$\hat{a}_2^* = 23.8912$	$\hat{a}_1^* = 39.3319$
$\hat{b}_1^* = 0.5598$	$\hat{b}_2^* = 0.6065$	$\hat{b}_1^* = 0.3013$

⑤ Here we test H_0: $\beta_1 = 0$ vs. $H_1 : \beta_1 \neq 0$. Given the low *p*-value, we reject H_0 in favor of H_1, adding *gpa* to a regression equation that already contains an intercept contributes significantly to the explanatory power of the model.

Next, suppose that the random statement in Exhibit 9.3 is replaced by

random int gpa/sub = method cl s;

By omitting *type=un*, we fit a random coefficients model under the restriction that $\sigma_{ab} = 0$. An abbreviated version of the resulting output appears in Table 9.6. Although the change in the estimates of the fixed effects is negligible, none of the estimates of the fixed effects is statistically significant. Moreover, in terms of the fit statistics for these two regression equations, –2 Res Log Like.=103.5 for the unstructured model, whereas –2 Res Log Like.=114.9 for the structured *or* restricted model ($\sigma_{ab} = 0$). Because these statistics exhibit a difference of 11.4, which corresponds asymptotically to the value of a χ^2 random variable with one degree of freedom and a *p*-value < 0.001, we can conclude that an unstructured random coefficients model has more explanatory power or fits the data better than one that restricts σ_{ab} to zero.

TABLE 9.6

Output for Example 9.2 (Structured Model)

The SAS System
The Mixed Procedure
Convergence criteria met.

Covariance Parameter Estimates

Cov Parm	Subject	Estimate	Standard Error	Z Value	Pr Z
Intercept	Method	67.0573	71.5827	0.94	0.1744
gpa	Method	0.02462	0.02527	0.97	0.1650
Residual		1.1318	0.3267	3.46	0.0003

Fit Statistics

−2 Res Log Likelihood	114.9
AIC (smaller is better)	120.9
AICC (smaller is better)	121.9
BIC (smaller is better)	118.2

Solution for Fixed Effects

Effect	Estimate	Standard Error	DF	*t* Value	Pr > |*t*|	Alpha	Lower	Upper
Intercept	29.1908	4.8886	2	5.97	0.0269	0.05	8.1569	50.22
gpa	0.4881	0.09185	2	5.31	0.0336	0.05	0.09287	0.88

Covariance Matrix for Fixed Effects

Row	Effect	Col1	Col2
1	Intercept	23.8982	−0.01871
2	gpa	−0.01871	0.008437

Solution for Random Effects

Effect	Method	Estimate	Std Err Pred	DF	*t* Value	Pr > |*t*|	Alpha	Lower	Upper
Intercept	1	−0.1434	5.0221	24	−0.03	0.9775	0.05	−10.5084	10.2216
gpa	1	0.01400	0.09393	24	0.15	0.8815	0.05	−0.1778	0.2058
Intercept	2	−7.8527	5.0304	24	−1.56	0.1316	0.05	−18.2349	2.5295
gpa	2	0.1474	0.09295	24	1.59	0.1260	0.05	−0.04448	0.3392
Intercept	3	7.9961	4.9992	24	1.60	0.1228	0.05	−2.3217	18.3138
gpa	3	−0.1614	0.09279	24	−1.74	0.0949	0.05	−0.3529	0.03015

(continued)

TABLE 9.6 *(continued)*

The SAS System

Type 3 Tests of Fixed Effects

Effect	Num DF	Den DF	*F* value	Pr > *F*
gpa	1	2	28.23	0.0336

Appendix A: A Random Coefficient Estimation Procedure

It was determined in Equation 9.5 that $\sigma_i^2 = \alpha_0 + \alpha_1 X_i^2$, $i = 1, \ldots, n$. If we set

$$\boldsymbol{X} = \begin{bmatrix} 1 & X_1 \\ \cdot & \cdot \\ \cdot & \cdot \\ \cdot & \cdot \\ 1 & X_n \end{bmatrix}, \dot{\boldsymbol{X}} = \begin{bmatrix} 1 & X_1^2 \\ \cdot & \cdot \\ \cdot & \cdot \\ \cdot & \cdot \\ 1 & X_n^2 \end{bmatrix}, \boldsymbol{\sigma}^2 = \begin{bmatrix} \sigma_1^2 \\ \cdot \\ \cdot \\ \cdot \\ \sigma_n^2 \end{bmatrix} \text{ and } \boldsymbol{\alpha} = \begin{bmatrix} \alpha_0 \\ \alpha_1 \end{bmatrix},$$

then

$$\boldsymbol{\sigma}^2 = \begin{bmatrix} \sigma_1^2 \\ \cdot \\ \cdot \\ \cdot \\ \sigma_n^2 \end{bmatrix} = \dot{\boldsymbol{X}}\boldsymbol{\alpha} = \begin{bmatrix} \alpha_0 + \alpha_1 X_1^2 \\ \vdots \\ \alpha_0 + \alpha_1 X_n^2 \end{bmatrix}. \tag{9.A.1}$$

(Note that the elements of $\dot{\boldsymbol{X}}$ are simply the squares of the corresponding elements in $\boldsymbol{X}$.)

If estimates S_i^2 of σ_i^2 were available, then a regression of these estimates on $\dot{\boldsymbol{X}}$ would yield an estimate of the components of $\boldsymbol{\alpha}$. Hence, we need to obtain the S_i^2 values. To this end, let us fit Equation 9.3 via OLS and determine the residuals e_i, $i = 1, \ldots, n$, and set

$$\boldsymbol{e} = \begin{bmatrix} e_1 \\ \cdot \\ \cdot \\ \cdot \\ e_n \end{bmatrix}, \dot{\boldsymbol{e}} = \begin{bmatrix} e_1^2 \\ \cdot \\ \cdot \\ \cdot \\ e_n^2 \end{bmatrix}.$$

Then it can be shown that

$$e = MV, \tag{9.A.2}$$

where

$$M = I_n - X(X'X)^{-1}X', \quad V = \begin{bmatrix} v_1 \\ \cdot \\ \cdot \\ \cdot \\ v_n \end{bmatrix}, \tag{9.A.3}$$

and I_n is an identity matrix of order n or

$$I_n = \begin{bmatrix} 1 & 0 & . & . & . & 0 \\ 0 & 1 & . & . & . & 0 \\ . & . & & & & . \\ . & . & & & & . \\ . & . & & & & . \\ 0 & 0 & . & . & . & 1 \end{bmatrix}.$$

Because

$$E(ee') = ME(VV')M,$$

we have, from Equations 9.5 and 9.A.1,

$$E(\dot{e}) = \begin{bmatrix} E(e_1^2) \\ \cdot \\ \cdot \\ \cdot \\ E(e_n^2) \end{bmatrix} = \dot{M}\sigma^2 = \dot{M}\dot{X}\alpha \tag{9.A.4}$$

where $\dot{M}$ is a matrix obtained from M by replacing each element in M by its square. Here Equation 9.A.4 can be used to estimate α_0 and α_1 by regressing $\dot{e}$ on $Z = \dot{M}\dot{X}$, i.e., we seek to fit

$$\dot{e} = Z\alpha + u,$$

where

$$u = \begin{bmatrix} u_1 \\ \cdot \\ \cdot \\ \cdot \\ u_n \end{bmatrix}$$

is a vector of random disturbance terms. Because Z does not contain a column of 1's, we need to suppress the intercept term.

Appendix B: The IML Procedure for Matrix (Vector) Operations

The IML procedure defines a set of vector or matrix operations, in which a *matrix* is a rectangular array of elements (typically real numbers) arranged in rows and columns. A *vector* can be viewed as a matrix having a single row or column. The size of a matrix is called its *order* and written as (*mxn*), in which *m* is the number of rows, "x" is read *by*, and *n* is the number of columns. That is, the matrices

$$\underset{(3\times 2)}{\boldsymbol{A}} = \begin{bmatrix} 1 & -1 \\ 0 & 2 \\ 3 & 1 \end{bmatrix}, \quad \underset{(2\times 4)}{\boldsymbol{B}} = \begin{bmatrix} 1 & 0 & -1 & 2 \\ 3 & 2 & 1 & 0 \end{bmatrix}$$

have orders "3 by 2" and " 2 by 4", respectively. In what follows, it is assumed that the reader has some familiarity with matrix algebra.

Braces "{.}" are used to define the rows of a matrix (vector), and a comma "," is used to separate rows. For example,

1. {3 1 4} indicates a single row of a (1x3) vector.
2. {3,1,4} indicates the rows of a (3x1) vector.
3. {3 1 2, 1 0 –1,0 1 5} indicates the rows of a (3x3) matrix.

We next consider the operations of addition, subtraction, and scalar multiplication. Given matrices

$$\boldsymbol{A} = \begin{bmatrix} 3 & -1 \\ 4 & 2 \end{bmatrix} \quad \text{and} \quad \boldsymbol{B} = \begin{bmatrix} -1 & 0 \\ 5 & -4 \end{bmatrix}:$$

1. The operation of addition "+" forms the sum

$$\boldsymbol{A} + \boldsymbol{B} = \boldsymbol{C} = \begin{bmatrix} 2 & -1 \\ 9 & -2 \end{bmatrix}$$

(we add corresponding elements).

2. The operation of subtraction "–" forms the difference

$$\boldsymbol{A} - \boldsymbol{B} = \boldsymbol{D} = \begin{bmatrix} 4 & -1 \\ -1 & 6 \end{bmatrix}$$

(we subtract corresponding elements).

Note that, for $\boldsymbol{A} + \boldsymbol{B}$ ($\boldsymbol{A} - \boldsymbol{B}$) to exist, $\boldsymbol{A}$ and $\boldsymbol{B}$ must be *conformable for addition (subtraction)*, i.e., they must be of the same order.

3. The operation of scaler multiplication "#" enables us to, for instance, multiply the matrix **A** above by the scaler (constant) λ or

$$\lambda \# \boldsymbol{A} = \boldsymbol{E} = \begin{bmatrix} \lambda(3) & \lambda(-1) \\ \lambda(4) & \lambda(2) \end{bmatrix}$$

(we multiply each element of **A** by λ). If λ = 3, then

$$3\# \boldsymbol{A} = \begin{bmatrix} 9 & -3 \\ 12 & 6 \end{bmatrix}.$$

To see exactly how these operations are performed, the reader should execute the following program for the indicated vectors:

```
proc iml;
start vectors;
a={3 1 6};
b={6 2 1};
c=a+b;
print, a, b, c;
d={1, 4, 2 };
e={1, 1, 6};
f=d-e;
print, d, e, f;
g=10#a;
h=5#e;
print, g, h;
finish;
run vectors;
quit;
```

Additional matrix (vector) operations that are useful include the following:

1. t(*A*) forms the *transpose* of **A** (denoted by **A'**).
   (**A'** is a new matrix derived from **A** by interchanging its rows and columns, i.e., row *i* of **A** becomes column i of **A'**.) If

$$\boldsymbol{A} = \begin{bmatrix} 2 & 1 \\ 3 & 0 \\ -1 & 4 \end{bmatrix}, \text{ then } \boldsymbol{A}' = \begin{bmatrix} 2 & 3 & -1 \\ 1 & 0 & 4 \end{bmatrix}.$$

   Note that, if **A** is of order (*mxn*), then **A'** is of order (*nxm*).

2. * indicates matrix multiplication. For **A** * **B**, **A** is called the *premultiplier matrix* and **B** is the *postmultiplier matrix*. The product **A** * **B** exists if **A** and **B** are *conformable for multiplication*, i.e., the number of columns in **A** must be same as the number of rows in **B**. (Note that matrix multiplication is *generally not commutative* or **A** * **B** ≠ **B** * **A**. In fact, if **A** * **B** exists, **B** * **A** may not exist.)

3. i(*n*) creates an *identity matrix of order n*, e.g.,

$$i(3) = \begin{bmatrix} 1 & 0 & 0 \\ 0 & 1 & 0 \\ 0 & 0 & 1 \end{bmatrix}$$

   is an identity matrix of order three. Thus, an identity matrix has ones along its *main diagonal* and zeros elsewhere. It is usually denoted as $\boldsymbol{I}_n$.
4. inv (***A***) creates the *inverse* of an *n*th order (square) matrix ***A***. (The inverse of a square matrix is itself a square matrix and is usually denoted as $\boldsymbol{A}^{-1}$.) If $\boldsymbol{A}^{-1}$ exists, then

$$\boldsymbol{A} * \boldsymbol{A}^{-1} = \boldsymbol{A}^{-1} * \boldsymbol{A} = \boldsymbol{I}_n.$$

5. det (***A***) calculates the value of the *determinant* of an *n*th order matrix ***A***. (A determinant is a scaler quantity associated with a square matrix ***A*** and is usually denoted as $|\boldsymbol{A}|$. If $|\boldsymbol{A}| \neq 0$, ***A*** is said to be *nonsingular.* Note that only nonsingular matrices have inverses as defined above.)
6. # denotes elementwise multiplication. That is, for

$$\boldsymbol{A} = \begin{bmatrix} 2 & 2 \\ 3 & 4 \end{bmatrix}, \boldsymbol{B} = \begin{bmatrix} 4 & 5 \\ 1 & 0 \end{bmatrix}$$

   we have

$$\boldsymbol{A} \# \boldsymbol{B} = \begin{bmatrix} 2(4) & 2(5) \\ 3(1) & 4(0) \end{bmatrix} = \begin{bmatrix} 8 & 10 \\ 3 & 0 \end{bmatrix}.$$

7. ## denotes an elementwise power operation, i.e., if

$$\boldsymbol{A} = \begin{bmatrix} 2 & 3 \\ 1 & 4 \end{bmatrix},$$

   then

$$\boldsymbol{A} \#\# 2 = \boldsymbol{B} = \begin{bmatrix} (2)^2 & (3)^2 \\ (1)^2 & (4)^2 \end{bmatrix} = \begin{bmatrix} 4 & 9 \\ 1 & 16 \end{bmatrix}.$$

8. J (a, b, c) creates an $(a \times b)$ matrix all of whose elements are *c*.
9. Echelon (***A***) transforms an $(n \times n + 1)$ matrix ***A*** to a *row-reduced echelon matrix* of the form

$$\begin{bmatrix} 1 & e_{12} & e_{13} & e_{14} \\ 0 & 1 & e_{23} & e_{24} \\ 0 & 0 & 1 & e_{34} \end{bmatrix}.$$

To see exactly how these operations work, the reader is encouraged to execute the following program on the indicated matrices:

```
proc iml;
start matrices;
a={2  3, -1  7};
b={8  1, 9  6};
c=a+b;
print, a,b,c;
d=2#a;
print, d;
e={2  3  -1, -1  7  0};
et=t(e);
print, e, et;
v={1, 2, 3};
vt=t(v);
print, v, vt;
f={6  1, -2  9, 3  4};
g=e*f;
print, e, f, g;
r=et*e;
s=e*et;
v1=vt*v;
v2=v*vt;
print, r, s, v1, v2;
imat=i(3);
print, imat;
if=imat*f;
print, if;
finish;
run matrices;
quit;
```

Appendix C: Generalized Least Squares [Johnston 1984; Kmenta 1986; Davidson and Mackinnon 1993; Greene 1993]

Generalizing the Error Variance

Let us express the *generalized linear regression model* as

> $\boldsymbol{Y} = \boldsymbol{X\beta} + \boldsymbol{\varepsilon}$, $\boldsymbol{X}$ is an $(n \times p)$ matrix with fixed or nonstochastic elements and $\rho(\boldsymbol{X}) = m + 1 = p < n$, $\boldsymbol{\beta}$ is a vector of unknown parameters, and $\boldsymbol{\varepsilon}$ is a random error term with $E(\boldsymbol{\varepsilon}) = \boldsymbol{0}$, $E(\boldsymbol{\varepsilon\varepsilon}') = \sigma_\varepsilon^2\boldsymbol{\Omega}$, and $\boldsymbol{\Omega}$ is a known $(n \times n)$ positive definite, symmetric, and nonstochastic matrix. (9.C.1)

Note that previously $\boldsymbol{\Omega} = \boldsymbol{I}_n$ so that the elements of $\boldsymbol{\varepsilon}$ were uncorrelated ($E(\varepsilon_i\varepsilon_j) = 0, i \neq j$) and had identical variances ($E(\varepsilon_i^2) = \sigma_\varepsilon^2$ for all i). Hence, replacing $\boldsymbol{I}_n$ by $\boldsymbol{\Omega}$ admits heteroscedastic errors (we can have differing elements on the main diagonal of $\boldsymbol{\Omega}$) as well as interdependence ($\boldsymbol{\Omega}$ can contain nonzero off-diagonal elements).

What happens if we apply classical OLS to estimate $\boldsymbol{\beta}$, mistakenly thinking that $E(\boldsymbol{\varepsilon\varepsilon}') = \sigma_\varepsilon^2\boldsymbol{I}_n$, when in fact the generalized least-squares (GLS) linear regression model (Equation 9.C.1) is appropriate? In this circumstance, our estimator is, as before,

$\hat{\boldsymbol{\beta}} = (\boldsymbol{X}'\boldsymbol{X})^{-1}\boldsymbol{X}'\boldsymbol{Y} = \boldsymbol{\beta} + (\boldsymbol{X}'\boldsymbol{X})^{-1}\boldsymbol{X}'$. Because $E(\hat{\boldsymbol{\beta}}) = \boldsymbol{\beta}$, the classical OLS estimator of $\boldsymbol{\beta}$ is unbiased And because we can write $\hat{\boldsymbol{\beta}} = \boldsymbol{\beta} + \left(\frac{1}{n}\boldsymbol{X}'\boldsymbol{X}\right)^{-1}\left(\frac{1}{n}\boldsymbol{X}'\boldsymbol{\varepsilon}\right)$, it is evident that the value of $\hat{\boldsymbol{\beta}}$ collapses onto $\boldsymbol{\beta}$ as $n \to \infty$ so that $\hat{\boldsymbol{\beta}}$ is also consistent for $\boldsymbol{\beta}$. Previously, under $E(\boldsymbol{\varepsilon}\boldsymbol{\varepsilon}') = \sigma_\varepsilon^2\boldsymbol{I}_n$, the variance-covariance matrix of $\hat{\boldsymbol{\beta}}$ was determined as $V(\hat{\boldsymbol{\beta}}) = \sigma_\varepsilon^2(\boldsymbol{X}'\boldsymbol{X})^{-1}$. Now, with $E(\boldsymbol{\varepsilon}\boldsymbol{\varepsilon}') = \sigma_\varepsilon^2\boldsymbol{\Omega}$,

$$
\begin{aligned}
V(\hat{\boldsymbol{\beta}}) &= E\left[(\hat{\boldsymbol{\beta}} - \boldsymbol{\beta})(\hat{\boldsymbol{\beta}} - \boldsymbol{\beta})'\right] = E\left[(\boldsymbol{X}'\boldsymbol{X})^{-1}\boldsymbol{X}'\boldsymbol{\varepsilon}\boldsymbol{\varepsilon}'\boldsymbol{X}(\boldsymbol{X}'\boldsymbol{X}')^{-1}\right] \\
&= \sigma_\varepsilon^2(\boldsymbol{X}'\boldsymbol{X})^{-1}\boldsymbol{X}'\boldsymbol{\Omega}\boldsymbol{X}(\boldsymbol{X}'\boldsymbol{X})^{-1}.
\end{aligned}
\tag{9.C.2}
$$

Also, the OLS estimator of σ_ε^2, $S_\varepsilon^2 = \boldsymbol{e}'\boldsymbol{e}/(n-p)$, is no longer unbiased because it is readily shown (via Appendix 12.D) that $E(\boldsymbol{e}'\boldsymbol{e}) = tr\left[\boldsymbol{M}E(\boldsymbol{\varepsilon}\boldsymbol{\varepsilon}')\right] = \sigma_\varepsilon^2 tr(\boldsymbol{M}\boldsymbol{\Omega}) = \sigma_\varepsilon^2\left\{tr(\boldsymbol{\Omega}) - tr\left[(\boldsymbol{X}'\boldsymbol{X})^{-1}(\boldsymbol{X}'\boldsymbol{\Omega}\boldsymbol{X})\right]\right\} \neq \sigma_\varepsilon^2(n-p)$. So when $\boldsymbol{\Omega} \neq \boldsymbol{I}_n$, $\hat{\boldsymbol{\beta}}$ is still linear and unbiased, but it is no longer "best" because $V(\hat{\boldsymbol{\beta}})$ exhibits two sources of bias: (1) $(\boldsymbol{X}'\boldsymbol{X})^{-1}\boldsymbol{X}'\boldsymbol{\Omega}\boldsymbol{X}(\boldsymbol{X}'\boldsymbol{X})^{-1} \neq (\boldsymbol{X}'\boldsymbol{X})^{-1}$; and (2) $E(S_\varepsilon^2) \neq \sigma_\varepsilon^2$.

Given that the classical OLS expression for the variance of $\hat{\boldsymbol{\beta}}$ is no longer efficient, the application of OLS to Equation 9.C.1 leads to misleading inferences; the biased variance estimators cause the usual test statistics to depart from their assumed distributions. In this regard, our objective is to now find a remedy for this situation.

A GLS Estimator for $\boldsymbol{\beta}$

Because Ω is positive definite, there exists a nonsingular matrix $\boldsymbol{P}$ such that $\boldsymbol{P}'\boldsymbol{P} = \boldsymbol{\Omega}^{-1}$ with $\boldsymbol{\Omega} = (\boldsymbol{P}'\boldsymbol{P})^{-1} = \boldsymbol{P}^{-1}(\boldsymbol{p}')^{-1}$ and $\boldsymbol{P}\Omega\boldsymbol{P}' = \boldsymbol{P}\boldsymbol{P}^{-1}(\boldsymbol{P}')^{-1}\boldsymbol{P}' = \boldsymbol{I}_n$.[3]

Let us consider the *transformed error term* $\boldsymbol{\varepsilon}_T = \boldsymbol{P}\boldsymbol{\varepsilon}$. Then $E(\boldsymbol{\varepsilon}_T) = \boldsymbol{P}E(\boldsymbol{\varepsilon}) = 0$, whereas $V(\boldsymbol{\varepsilon}_T) = E(\boldsymbol{\varepsilon}_T\boldsymbol{\varepsilon}_T') = E(\boldsymbol{P}\boldsymbol{\varepsilon}\boldsymbol{\varepsilon}'\boldsymbol{P}') = \sigma_\varepsilon^2\boldsymbol{P}\boldsymbol{\Omega}\boldsymbol{P}' = \sigma_\varepsilon^2\boldsymbol{I}_n$. Hence, the transformed random error vector $\boldsymbol{\varepsilon}_T$ satisfies the usual (weak) classical OLS assumptions; it has the desirable properties of homogeneity and independence.

In this regard, if we multiply the (Equation 9.C.1) regression equation by $\boldsymbol{P}$, then we obtain a *transformed regression model*

$$
\boldsymbol{PY} = \boldsymbol{PX\beta} + \boldsymbol{P\varepsilon} \text{ or } \boldsymbol{Y}_T = \boldsymbol{X}_T\boldsymbol{\beta} + \boldsymbol{\varepsilon}_T, \tag{9.C.3}
$$

where $\boldsymbol{Y}_T = \boldsymbol{PY}$, $\boldsymbol{X}_T = \boldsymbol{PX}$, and $\boldsymbol{\varepsilon}_T = \boldsymbol{P\varepsilon}$, which satisfies the usual OLS assumptions. Then applying OLS to the transformed variables $\boldsymbol{Y}_T = \boldsymbol{PY}$ and $\boldsymbol{X}_T = \boldsymbol{PX}$ yields estimates that possess all of the optimal properties of OLS. More specifically, we state

---

3 A matrix $\boldsymbol{C}$ is *orthogonal* if it satisfies $\boldsymbol{C}'\boldsymbol{C} = \boldsymbol{C}\boldsymbol{C}' = \boldsymbol{I}_n$ so clearly $\boldsymbol{C}' = \boldsymbol{C}^{-1}$. If Ω is symmetric positive definite, a nonsingular matrix $\boldsymbol{P}$ can be found such that $\boldsymbol{P}'\boldsymbol{P} = \Omega^{-1}$. To see this, let us first *diagonalize* Ω by writing $\boldsymbol{C}'\Omega\boldsymbol{C} = \Lambda$, where $\boldsymbol{C}$ is an orthogonal matrix with columns corresponding to the eigenvectors of Ω and the eigenvalues of Ω are arranged along the main diagonal of Λ, i.e., $\Lambda = \text{diag}(\lambda_1,\ldots,\lambda_n)$, where $\lambda_i > 0$ for all i because Ω is positive definite. Then Ω can be factored as $\Omega = \boldsymbol{C}\Lambda\boldsymbol{C}'$ while Λ can be factored as $\Lambda = \Lambda^{\frac{1}{2}}\Lambda^{\frac{1}{2}}$, where $\Lambda^{\frac{1}{2}} = \text{diag}(\lambda_1^{1/2},\ldots,\lambda_n^{1/2})$. Similarly, Ω^{-1} and Λ^{-1} can be factored as $\Omega^{-1} = \boldsymbol{C}\Lambda^{-1}\boldsymbol{C}'$ and $\Lambda^{-1} = \Lambda^{\frac{1}{2}}\Lambda^{\frac{1}{2}}$, respectively, so that

$$
\Omega^{-1} = \boldsymbol{C}\Lambda^{-\frac{1}{2}}\Lambda^{-\frac{1}{2}}\boldsymbol{C}' = \left(\Lambda^{-\frac{1}{2}}\boldsymbol{C}'\right)'\Lambda^{-\frac{1}{2}}\boldsymbol{C}' = \boldsymbol{P}'\boldsymbol{P},
$$

where obviously $\boldsymbol{P} = \Lambda^{-\frac{1}{2}}\boldsymbol{C}'$. (Note that, if $\mathbf{T} = \boldsymbol{C}\Lambda^{\frac{1}{2}}$, then $\Omega = \mathbf{T}\mathbf{T}'$.)

Aitken's Generalized Gauss-Markov Theorem [Aitken 1935]
For the generalized linear regression model
$\boldsymbol{Y} = \boldsymbol{X\beta} + \boldsymbol{\varepsilon}, E(\boldsymbol{\varepsilon}) = \boldsymbol{0}$, and $E(\boldsymbol{\varepsilon\varepsilon}') = \sigma_\varepsilon^2 \boldsymbol{\Omega}$,
the BLUE of $\boldsymbol{\beta}$ is the *generalized least squares estimator*

$$\boldsymbol{\beta}^* = (\boldsymbol{X}'\Omega^{-1}\mathbf{X})^{-1}\boldsymbol{X}'\Omega^{-1}\boldsymbol{Y} \tag{9.C.4}$$

with variance-covariance matrix

$$V(\boldsymbol{\beta}^*) = \sigma_\varepsilon^2(\boldsymbol{X}'\boldsymbol{\Omega}^{-1}\boldsymbol{X})^{-1}. \tag{9.C.5}$$

To rationalize this result, let us consider the transformed model (Equation 9.C.3) with $\boldsymbol{P}'\boldsymbol{P} = \boldsymbol{\Omega}^{-1}, \boldsymbol{P\Omega P}' = \boldsymbol{I}_n$, and $E(\boldsymbol{\varepsilon}_T\boldsymbol{\varepsilon}_T') = \sigma_\varepsilon^2\boldsymbol{P\Omega P}' = \sigma_\varepsilon^2\boldsymbol{I}_n$. Then the classical model applies and thus, from the Gauss-Markov Theorem, the BLUE of $\boldsymbol{\beta}$ from the transformed model is $\boldsymbol{\beta}^* = (\boldsymbol{X}_T'\boldsymbol{X}_T)^{-1}\boldsymbol{X}_T'\boldsymbol{Y} = (\boldsymbol{X}'\boldsymbol{P}'\boldsymbol{PX})^{-1}\boldsymbol{X}'\boldsymbol{P}'\boldsymbol{PY} = (\boldsymbol{X}'\boldsymbol{\Omega}^{-1}\boldsymbol{X})^{-1}\boldsymbol{X}'\boldsymbol{\Omega}^{-1}\boldsymbol{Y}$, whereas $V(\boldsymbol{\beta}^*) = \sigma_\varepsilon^2(\boldsymbol{X}_T'\boldsymbol{X}_T)^{-1} = \sigma_\varepsilon^2(\boldsymbol{X}'\boldsymbol{P}'\boldsymbol{PX})^{-1} = \sigma_\varepsilon^2(\boldsymbol{X}'\boldsymbol{\Omega}^{-1}\boldsymbol{X})^{-1}$.[4]

In addition, we can obtain an unbiased estimator for σ_ε^2 in the following manner. With $\boldsymbol{Y}^* = \boldsymbol{X\beta}^*$, the *GLS residual*

$$\begin{aligned}
\boldsymbol{e}^* &= \boldsymbol{Y} - \boldsymbol{Y}^* = \boldsymbol{X\beta} + \boldsymbol{\varepsilon} - \boldsymbol{X\beta}^* \\
&= \boldsymbol{X\beta} + \boldsymbol{\varepsilon} - \mathbf{X}[\boldsymbol{\beta} + (\boldsymbol{X}'\boldsymbol{\Omega}^{-1}\mathbf{X})^{-1}\boldsymbol{X}'\Omega^{-1}\boldsymbol{\varepsilon}] \\
&= \boldsymbol{\varepsilon} - \boldsymbol{X}(\boldsymbol{X}'\boldsymbol{\Omega}^{-1}\mathbf{X})^{-1}\boldsymbol{X}'\boldsymbol{\Omega}^{-1}\boldsymbol{\varepsilon} \\
&= [\boldsymbol{I}_n - \mathbf{X}(\boldsymbol{X}'\boldsymbol{\Omega}^{-1}\mathbf{X})^{-1}\boldsymbol{X}'\boldsymbol{\Omega}^{-1}]\,\boldsymbol{\varepsilon} = \boldsymbol{M}^*\boldsymbol{\varepsilon},
\end{aligned}$$

where $\boldsymbol{M}^* = \boldsymbol{I}_n - \boldsymbol{X}(\boldsymbol{X}'\Omega^{-1}\boldsymbol{X})^{-1}\boldsymbol{X}'\Omega^{-1}$ is idempotent (i.e., $(\boldsymbol{M}^*)' = \boldsymbol{M}^*$ and $(\boldsymbol{M}^*)^2 = \boldsymbol{M}^*$). Given that $(\boldsymbol{M}^*)'\,\Omega^{-1}\,\boldsymbol{M}^* = \Omega^{-1}\boldsymbol{M}^*$, it follows that

$$(\boldsymbol{e}^*)'\,\Omega^{-1}\boldsymbol{e}^* = \boldsymbol{\varepsilon}'(\boldsymbol{M}^*)'\,\Omega^{-1}\boldsymbol{M}^*\boldsymbol{\varepsilon} = \boldsymbol{\varepsilon}'\,\Omega^{-1}\boldsymbol{M}^*\,\boldsymbol{\varepsilon}$$

with

$$\begin{aligned}
E\left[(\boldsymbol{e}^*)'\boldsymbol{\Omega}^{-1}\boldsymbol{e}^*\right] &= E(\boldsymbol{\varepsilon}'\boldsymbol{\Omega}^{-1}\boldsymbol{M}^*\boldsymbol{\varepsilon}) = E\left[tr(\boldsymbol{\varepsilon}'\boldsymbol{\Omega}^{-1}\boldsymbol{M}^*\boldsymbol{\varepsilon})\right] \\
&= E\left[tr(\boldsymbol{M}^*\boldsymbol{\varepsilon\varepsilon}'\boldsymbol{\Omega}^{-1})\right] = tr\left[E(\boldsymbol{M}^*\boldsymbol{\varepsilon\varepsilon}'\boldsymbol{\Omega}^{-1})\right] \\
&= \sigma_\varepsilon^2 tr(\boldsymbol{M}^*) = \sigma_\varepsilon^2(n-p)
\end{aligned}$$

because $\boldsymbol{\varepsilon}'\Omega^{-1}\,\boldsymbol{M}^*\boldsymbol{\varepsilon}$ is a scalar, $tr(\boldsymbol{AB}) = tr(\boldsymbol{BA})$, trace is a linear operator, and $tr(\alpha\boldsymbol{A}) = \alpha tr(\boldsymbol{A})$, α a scalar.[5] Hence an unbiased estimator of σ_ε^2 is

$$(S^*)^2 = \frac{(\boldsymbol{e}^*)'\boldsymbol{\Omega}^{-1}\boldsymbol{e}^*}{n-p} \tag{9.C.6}$$

---

[4] That $\boldsymbol{\beta}^*$ (an arbitrary estimator for $\boldsymbol{\beta}$) minimizes the *generalized sum of squares*. $S^* = (\boldsymbol{e}^*)'\Omega^{-1}\boldsymbol{e}^* = (\boldsymbol{Y} - \boldsymbol{X\beta}^*)'\Omega^{-1}(\boldsymbol{Y} - \boldsymbol{X\beta}^*) = \boldsymbol{Y}'\Omega^{-1}\boldsymbol{Y} - 2\boldsymbol{\beta}^*\boldsymbol{X}'\Omega^{-1}\boldsymbol{Y} + (\boldsymbol{\beta}^*)'\boldsymbol{X}\,\Omega^{-1}\boldsymbol{X\beta}^*$ can be seen by setting $\partial S^*/\partial\boldsymbol{\beta}^* = \boldsymbol{0}$ and solving the *GLS normal equations* $\boldsymbol{X}'\,\Omega^{-1}\boldsymbol{X\beta}^* = \boldsymbol{X}'\,\Omega^{-1}\boldsymbol{Y}$ for $\boldsymbol{\beta}^*$.

[5] $tr(\boldsymbol{M}^*) = tr[\boldsymbol{I}_n - \mathbf{X}(\mathbf{X}'\,\Omega^{-1}\mathbf{X})^{-1}\mathbf{X}'\,\Omega^{-1}] = tr(\boldsymbol{I}_n) - tr[\mathbf{X}(\mathbf{X}'\,\Omega^{-1}\mathbf{X})^{-1}\mathbf{X}'\,\Omega^{-1}] = n - tr[\mathbf{X}'\,\Omega^{-1}\mathbf{X}(\mathbf{X}'\,\Omega^{-1}\mathbf{X})^{-1}] = n - tr(\boldsymbol{I}_p) = n - p$.

(because clearly $E\left[(S^*)^2\right] = \sigma_\varepsilon^2$), whereas an unbiased estimator of $V(\boldsymbol{\beta}^*)$ is

$$\boldsymbol{\Sigma}^* = (S_\varepsilon^*)^2 (\boldsymbol{X}'\boldsymbol{\Omega}^{-1}\boldsymbol{X})^{-1}. \tag{9.C.7}$$

Looking to the finite-sample properties of the GLS estimator $\boldsymbol{\beta}^*$, we know from Aitken's Theorem that $\boldsymbol{\beta}^*$ is BLUE. Moreover, for large samples and under some very mild restrictions on $\boldsymbol{PX}$, $\boldsymbol{\beta}^*$ is consistent, asymptotically efficient, and asymptotically normal.

The GLS Estimator as a Maximum-Likelihood Estimator

In the preceding section, no assumption about the distribution of $\boldsymbol{\varepsilon}$ was made. However, if we now posit that $\boldsymbol{\varepsilon} : N(\boldsymbol{0}, \sigma_\varepsilon^2\boldsymbol{\Omega})$, where again Ω is taken to be known and positive definite, then it is easily demonstrated that the maximum likelihood estimator and GLS estimator are one in the same. To this end, if we work with the transformed regression (Equation 9.C.3), then for $\boldsymbol{\varepsilon}_T = \boldsymbol{Y}_T - \boldsymbol{X}_T\boldsymbol{\beta}$ multivariate normal, the likelihood function for the sample of $\boldsymbol{Y}_T$ values is

$$\begin{aligned}\mathcal{L}(\boldsymbol{\beta}, \sigma_\varepsilon^2; \boldsymbol{Y}_T, n) &= \frac{1}{(2\pi)^{n/2}\sigma_\varepsilon^n}\exp\left(-\frac{1}{2\sigma_\varepsilon^2}\boldsymbol{\varepsilon}_T'\boldsymbol{\varepsilon}_T\right) \\ &= \frac{1}{(2\pi)^{n/2}\sigma_\varepsilon^n}\exp\left[-\frac{1}{2\sigma_\varepsilon^2}(\boldsymbol{Y}-\boldsymbol{X\beta})'\boldsymbol{\Omega}^{-1}(\boldsymbol{Y}-\boldsymbol{X\beta})\right]\end{aligned} \tag{9.C.8}$$

with $\boldsymbol{\varepsilon}_T'\boldsymbol{\varepsilon}_T = (\boldsymbol{PY}-\boldsymbol{PX\beta})'(\boldsymbol{PY}-\boldsymbol{PX\beta}) = (\boldsymbol{Y}-\boldsymbol{X\beta})'\Omega^{-1}(\boldsymbol{Y}-\boldsymbol{X\beta})$. Because Equation 9.C.8 is the likelihood function for the sample of $\boldsymbol{Y}_T$ values, it must be transformed to a likelihood function for a sample of $\boldsymbol{Y}$ values. After performing a *multivariate change of variable*,[6] Equation 9.8.C transforms to

$$\mathcal{L}(\boldsymbol{\beta}, \sigma_\varepsilon^2, \boldsymbol{\Omega}; \boldsymbol{Y}, n) = \frac{1}{(2\pi)^{n/2}\sigma_\varepsilon^n|\boldsymbol{\Omega}|^{1/2}}\exp\left[-\frac{1}{2\sigma_\varepsilon^2}(\boldsymbol{Y}-\boldsymbol{X\beta})'\boldsymbol{\Omega}^{-1}(\boldsymbol{Y}-\boldsymbol{X\beta})\right] \tag{9.C.8.1}$$

---

[6] Suppose the relationship between the sets of variables

$$\left.\begin{aligned} Y_i &= \gamma_i(X_1,\ldots,X_n) \\ X_i &= \phi_i(Y_i,\ldots,Y_n) \end{aligned}\right\} i = 1,\ldots,n,$$

is one-to-one with the Jacobian determinant $|\boldsymbol{J}| = |\partial(X_1,\ldots,X_n)/\partial(Y_1,\ldots,Y_n)| \neq 0$. To transform $f(X_1,\ldots,X_n)$ to a function of $Y_1,\ldots,Y_n$ we need only determine

$$g(Y_1,\ldots,Y_n) = f[\phi_1(Y_1,\ldots,Y_n),\ldots,\phi_n(Y_1,\ldots,Y_n)]\cdot|\boldsymbol{J}|.$$

In this regard, because the relationship between $\boldsymbol{Y}_T$ and $\boldsymbol{Y}$ is one-to-one and $|\boldsymbol{J}| = |\partial\boldsymbol{Y}_T/\partial\boldsymbol{Y}| = |\boldsymbol{P}| \neq 0$ (because $\boldsymbol{P}$ is nonsingular), let us write $|\boldsymbol{P}'\boldsymbol{P}| = |\boldsymbol{P}'|\cdot|\boldsymbol{P}| = |\boldsymbol{P}|^2 = |\Omega^{-1}|$ so that $|\boldsymbol{P}| = |\Omega^{-1}|^{1/2} = |\Omega|^{-1/2}$. Then we can transform the likelihood function in terms of $\boldsymbol{Y}_T$ to one involving $\boldsymbol{Y}$ by replacing $\mathcal{L}(\boldsymbol{\beta}, \sigma_\varepsilon^2; \boldsymbol{Y}_T, n)$ by $\mathcal{L}(\boldsymbol{\beta}, \sigma_\varepsilon^2; \boldsymbol{PY}, n)$ and multiplying the latter by $|\boldsymbol{J}| = |\Omega|^{-1/2}$ so as to obtain

$$\mathcal{L}(\boldsymbol{\beta}, \sigma_\varepsilon^2, \boldsymbol{\Omega}; \boldsymbol{Y}, n) = \mathcal{L}(\boldsymbol{\beta}, \sigma_\varepsilon^2; \boldsymbol{PY}, n)\cdot|\boldsymbol{\Omega}|^{-1/2}.$$

or, in log-likelihood form,

$$\ln \mathcal{L} = \frac{n}{2}\ln(2\pi) - \frac{n}{2}\ln\sigma_\varepsilon^2 - \frac{1}{2}\ln|\mathbf{\Omega}| - \frac{1}{2\sigma_\varepsilon^2}(Y - X\tilde{\boldsymbol{\beta}})'\mathbf{\Omega}^{-1}(Y - X\tilde{\boldsymbol{\beta}}), \tag{9.C.9}$$

where $\tilde{\boldsymbol{\beta}}$ is an arbitrary estimator for $\boldsymbol{\beta}$. To maximize $\ln \mathcal{L}$ with respect to $\tilde{\boldsymbol{\beta}}$, we need only minimize the *generalized sum of squares* $\tilde{S} = (\boldsymbol{Y} - \boldsymbol{X}\tilde{\boldsymbol{\beta}})'\mathbf{\Omega}^{-1}(\boldsymbol{Y} - \boldsymbol{X}\tilde{\boldsymbol{\beta}}) = \boldsymbol{Y}'\mathbf{\Omega}^{-1}\boldsymbol{Y} - 2\tilde{\boldsymbol{\beta}}\boldsymbol{X}'\mathbf{\Omega}^{-1}\boldsymbol{Y} + \tilde{\boldsymbol{\beta}}\boldsymbol{X}'\mathbf{\Omega}^{-1}\boldsymbol{X}\tilde{\boldsymbol{\beta}}$ with respect to the same so that setting $\partial \boldsymbol{S} / \partial \boldsymbol{\beta} = \mathbf{0}$ yields

$$\tilde{\boldsymbol{\beta}} = (\boldsymbol{X}'\mathbf{\Omega}^{-1}\boldsymbol{X})^{-1}\boldsymbol{X}'\mathbf{\Omega}^{-1}\boldsymbol{Y}. \tag{9.C.10}$$

Clearly this maximum-likelihood estimator for $\boldsymbol{\beta}$ is the same as the GLS estimator $\boldsymbol{\beta}^*$ (Equation 9.C.4) and thus is BLUE for $\boldsymbol{\beta}$. So with normally distributed disturbances, the GLS estimator for $\boldsymbol{\beta}$ is also the maximum-likelihood estimator.

If in Equation 9.C.9 we replace σ_ε^2 by the arbitrary estimator $\tilde{\sigma}_\varepsilon^2$, then setting

$$\frac{\partial \ln \mathcal{L}}{\partial \tilde{\sigma}_\varepsilon^2} = -\frac{n}{2}\frac{1}{\tilde{\sigma}_\varepsilon^2} + \frac{1}{\tilde{\sigma}_\varepsilon^4}(Y - \boldsymbol{X}\tilde{\boldsymbol{\beta}})'\mathbf{\Omega}^{-1}(Y - \boldsymbol{X}\tilde{\boldsymbol{\beta}}) = 0$$

renders

$$\tilde{\sigma}_\varepsilon^2 = (Y - \boldsymbol{X}\dot{\boldsymbol{\beta}})'\mathbf{\Omega}^{-1}(Y - \boldsymbol{X}\dot{\boldsymbol{\beta}}) / n \tag{9.C.11}$$

Thus, the maximum-likelihood estimator $\tilde{\sigma}_\varepsilon^2$ for σ_ε^2 is biased. However, from Equation 9.C.6, an unbiased estimator for σ_ε^2 is

$$(\tilde{S}_\varepsilon)^2 = \frac{(Y - X\tilde{\beta})'\mathbf{\Omega}^{-1}(Y - X\tilde{\beta})}{n - p}. \tag{9.C.12}$$

Feasible Generalized Least Squares [Zellner 1962; Kakwani 1967; Schmidt 1976]

It is important to note that Equations 9.C.4, 9.C.6, and 9.C.7 (as well as Equations 9.C.10 and 9.C.12) are directly applicable only if Ω is a "known" matrix. If Ω is unknown and must be estimated, then Aitken's Theorem ceases to apply. In fact, the GLS estimator $\boldsymbol{\beta}^*$ is no longer a feasible estimator of $\boldsymbol{\beta}$ (although it is still unbiased). When Ω is estimated by $\hat{\mathbf{\Omega}}$ (clearly $\hat{\mathbf{\Omega}}$ must be of full rank n so that $\hat{\mathbf{\Omega}}^{-1}$ exists), we obtain the *feasible generalized least squares* (FGLS) estimator

$$\boldsymbol{\beta}^{**} = (\boldsymbol{X}'\hat{\mathbf{\Omega}}_{-1}\boldsymbol{X})_{-1}\boldsymbol{X}'\hat{\mathbf{\Omega}}_{-1}\boldsymbol{Y}. \tag{9.C.13}$$

For this expression to be efficacious, we must assume that the elements of Ω are functions of a small number of unknown parameters $\boldsymbol{\theta}' = (\theta_1, \ldots, \theta_h)$ which must be independent of the $m + 1 = p$ parameters in $\boldsymbol{\beta}$. Then we require that $p + k < n$ so that $\Omega = \Omega(\boldsymbol{\theta})$ must be estimated by fewer than $n - p$ parameters. If $\hat{\boldsymbol{\theta}}$ is a consistent estimator for $\boldsymbol{\theta}$ (which means that each component of $\boldsymbol{\theta}$ is consistently estimated), the FGLS estimator of β is Equation 9.C.13 and is determined under the assumption that $\hat{\mathbf{\Omega}}^{-1} = \hat{\mathbf{\Omega}}^{-1}(\boldsymbol{\theta})$.

Given Equation 9.C.13, the varian ce-covariance matrix of $\boldsymbol{\beta}^{**}$ may be consistently estimated by

$$\boldsymbol{\Sigma}^{**} = (S_{\varepsilon}^{**})^2 (\boldsymbol{X}'\boldsymbol{\Omega}^{-1}\boldsymbol{X})^{-1}, \tag{9.C.14}$$

where again $\hat{\boldsymbol{\Omega}}^{-1} = \hat{\boldsymbol{\Omega}}^{-1}(\boldsymbol{\theta})$ and

$$(S_{\varepsilon}^{**})^2 = \left[(\boldsymbol{Y} - \boldsymbol{X}\boldsymbol{\beta}^{**})'\hat{\boldsymbol{\Omega}}^{-1}(\boldsymbol{Y} - \boldsymbol{X}\boldsymbol{\beta}^{**})\right]\Big/(n-p). \tag{9.C.15}$$

Although the small or finite sample properties of the GLS estimator $\boldsymbol{\beta}^{*}$ mirror those of the OLS estimator $\hat{\beta}$, this is not the case for the FGLS estimator $\boldsymbol{\beta}^{**}$ because ε and $\hat{\boldsymbol{\Omega}}$ are highly correlated. So although $\boldsymbol{\beta}^{**}$ tends to be unbiased for most standard estimators of Ω, it is neither linear (it is not a linear function of $\boldsymbol{Y}$ because Ω^{-1} depends on $\boldsymbol{Y}$) nor best (because its finite-sample variance-covariance matrix is not readily determined). Additionally, under some very general conditions, β^{**} has the same large-sample or asymptotic properties as $\boldsymbol{\beta}^{*}$: it too is consistent, asymptotically efficient, and asymptotically normal because $\boldsymbol{\beta}^{**}$ and $\boldsymbol{\beta}^{*}$ have the same limiting distribution if the parameters in Ω are consistently estimated.

FGLS and Maximum Likelihood [Magnus 1978]

Suppose $\Omega = \Omega(\boldsymbol{\theta})$ is known with arguments that are twice-differentiable functions of $\theta_1, \ldots, \theta_k$ and that the conditions underlying the transformed error term $\boldsymbol{\varepsilon}_T$ and Equations 9.C.4, 9.C.5, and 9.C.13 hold. Additionally, suppose $\boldsymbol{\varepsilon}$ follows a multivariate normal distribution and, for convenience, that $\sigma_{\varepsilon}^2 = 1$. Then $\tilde{\tilde{\boldsymbol{\beta}}}$ and $\tilde{\tilde{\boldsymbol{\theta}}}$ are maximum-likelihood estimators of $\boldsymbol{\beta}$ and $\boldsymbol{\Omega}$, respectively if:

$$\begin{aligned} &\text{a. } \tilde{\tilde{\boldsymbol{\beta}}} = (\boldsymbol{X}'\tilde{\tilde{\boldsymbol{\Omega}}}^{-1}\boldsymbol{X})^{-1}\boldsymbol{X}'\tilde{\tilde{\boldsymbol{\Omega}}}\boldsymbol{Y}, \\ &\text{b. } tr\left(\boldsymbol{\Omega}\frac{\partial \boldsymbol{\Omega}^{-1}}{\partial \theta_j}\right)_{\boldsymbol{\theta}=\tilde{\tilde{\boldsymbol{\theta}}}} = \tilde{\tilde{\boldsymbol{e}}}'\left(\frac{\partial \boldsymbol{\Omega}^{-1}}{\partial \theta_j}\right)_{\boldsymbol{\theta}=\tilde{\tilde{\boldsymbol{\theta}}}}\tilde{\tilde{\boldsymbol{e}}}, j = 1, \ldots, k, \end{aligned} \tag{9.C.16}$$

where $\tilde{\tilde{\boldsymbol{e}}} = \boldsymbol{Y} - \boldsymbol{X}\tilde{\tilde{\boldsymbol{\beta}}}$. (If $|\Omega|$ is independent of θ_j, then the jth equation in Equation 9.C.16.b simplifies to

$$\tilde{\tilde{\boldsymbol{e}}}'\left(\frac{\partial \boldsymbol{\Omega}^{-1}}{\partial \theta_j}\right)_{\boldsymbol{\theta}=\tilde{\tilde{\boldsymbol{\theta}}}}\tilde{\tilde{\boldsymbol{e}}} = 0.)$$

To see this, let the log-likelihood function for the sample again appear as Equation 9.C.9 with $\sigma_{\varepsilon}^2 = 1$ or

$$\begin{aligned} \ln \mathcal{L} &= -\frac{n}{2}\ln(2\pi) - \frac{1}{2}\ln|\boldsymbol{\Omega}| - \frac{1}{2}(\boldsymbol{Y} - \boldsymbol{X}\boldsymbol{\beta})'\boldsymbol{\Omega}^{-1}(\boldsymbol{Y} - \boldsymbol{X}\boldsymbol{\beta}) \\ &= -\frac{n}{2}\ln(2\pi) + \frac{1}{2}\ln|\boldsymbol{\Omega}^{-1}| - \frac{1}{2}(\boldsymbol{Y} - \boldsymbol{X}\ \boldsymbol{\beta})'\boldsymbol{\Omega}^{-1}(\boldsymbol{Y} - \boldsymbol{X}\ \boldsymbol{\beta}). \end{aligned} \tag{9.C.17}$$

Then the first-order conditions for a maximum of $\ln \mathcal{L}$ at $\boldsymbol{\beta} = \tilde{\tilde{\boldsymbol{\beta}}}$ and $\boldsymbol{\theta} = \tilde{\tilde{\boldsymbol{\theta}}}$ are:

$$\text{a.}\quad \frac{\partial \ln \mathcal{L}}{\partial \boldsymbol{\beta}} = 2\boldsymbol{X}'\boldsymbol{\Omega}^{-1}\boldsymbol{Y} - 2\boldsymbol{X}'\boldsymbol{\Omega}^{-1}\boldsymbol{X\beta} = \boldsymbol{0},$$

$$\text{b.}\quad \frac{\partial \ln \mathcal{L}}{\partial \theta_j} = \frac{1}{2} tr\left(\boldsymbol{\Omega}\frac{\partial \boldsymbol{\Omega}^{-1}}{\partial \theta_j}\right) - \frac{1}{2}(\boldsymbol{Y} - \boldsymbol{X\beta})'\frac{\partial \boldsymbol{\Omega}^{-1}}{\partial \theta_j}(\boldsymbol{Y} - \boldsymbol{X\beta}) \tag{9.C.18}$$

$$= \frac{1}{2} tr\left(\boldsymbol{\Omega}\frac{\partial \boldsymbol{\Omega}^{-1}}{\partial \theta_j}\right) - \frac{1}{2}\boldsymbol{e}'\frac{\partial \boldsymbol{\Omega}^{-1}}{\partial \theta_j}\boldsymbol{e},\ j = 1,\ldots,k,$$

or Equation 9.C.16.[7]

Because the system in Equation 9.C.16.b is highly nonlinear in the unknowns $\tilde{\tilde{\theta}}_k$, these stationary levels of θ_k can be approximated by using the following iterative procedure:

1. Step 1 (Initialize): Choose a feasible value of $\boldsymbol{\theta}$ (denote it $\boldsymbol{\theta}_0$).
2. Step 2: Find $\boldsymbol{\Omega}_0^{-1} = \boldsymbol{\Omega}^{-1}(\boldsymbol{\theta}_0), \boldsymbol{\beta}_0$ (from Equation 9.C.16.a), and $\boldsymbol{e}_0 = \boldsymbol{Y} - \boldsymbol{X\beta}_0$.
3. Step 3: Insert $\boldsymbol{e}_0$ into Equation 9.C.16.b and determine $\boldsymbol{\theta}_1$.
4. Step 4: find $\boldsymbol{\Omega}_1^{-1} = \boldsymbol{\Omega}^{-1}(\boldsymbol{\theta}_1), \boldsymbol{\beta}_1$ (again using Equation 9.C.16.a), and $\boldsymbol{e}_1 = \boldsymbol{Y} - \boldsymbol{X\beta}_1$.
5. Step 5: Substitute $\boldsymbol{e}_1$ into Equation 9.C.16.b and determine $\boldsymbol{\theta}_2$, and so on.
6. End: This process is continued until $||\boldsymbol{\beta}_i - \boldsymbol{\beta}_{i-1}|| < \eta_\beta$ and $||\boldsymbol{\theta}_i - \boldsymbol{\theta}_{i-1}|| < \eta_\theta$, 1,2,..., where η_β and η_θ are predetermined acceptable tolerance levels.

It can be demonstrated that a consistent estimate of the asymptotic variance-covariance matrix for $\tilde{\tilde{\boldsymbol{\beta}}}$ and $\tilde{\tilde{\boldsymbol{\theta}}}$ is

$$\tilde{\tilde{\boldsymbol{\Sigma}}} = \begin{bmatrix} (\boldsymbol{X}'\tilde{\tilde{\boldsymbol{\Omega}}}^{-1}\boldsymbol{X})^{-1} & \boldsymbol{0} \\ \boldsymbol{0} & 2\boldsymbol{\Gamma}^{-1} \end{bmatrix},$$

---

[7] If the elements of a nonsingular matrix $\boldsymbol{V}$ are functionally independent, then

$$\frac{\partial \log|\boldsymbol{V}|}{\partial \boldsymbol{V}} = \frac{1}{|\boldsymbol{V}|}\frac{\partial |\boldsymbol{V}|}{\partial \boldsymbol{V}} = (\boldsymbol{V}^{-1})'. \tag{9.C.19}$$

And for θ a scalar,

$$\frac{\partial \log|\boldsymbol{V}|}{\partial \theta} = \frac{1}{|\boldsymbol{V}|} tr\left[\left(\frac{\partial |\boldsymbol{V}|}{\partial \boldsymbol{V}}\right)'\frac{\partial \boldsymbol{V}}{\partial \theta}\right] \tag{9.C.20}$$

$$= tr\left[\left(\frac{1}{|\boldsymbol{V}|}\frac{\partial |\boldsymbol{V}|}{\partial \boldsymbol{V}}\right)'\frac{\partial \boldsymbol{V}}{\partial \theta}\right] = tr\left(\boldsymbol{V}^{-1}\frac{\partial \boldsymbol{V}}{\partial \theta}\right)$$

via (9.C.19). If $\boldsymbol{V} = \Omega^{-1}$, then the first term in the difference on the right-hand side of (9.C.18.b) immediately follows.

where Γ is a symmetric $(k \times k)$ matrix whose (i,j) th element is

$$\Gamma_{ij} = tr\left(\frac{\partial \boldsymbol{\Omega}^{-1}}{\partial \theta_i}\boldsymbol{\Omega}\frac{\partial \boldsymbol{\Omega}^{-1}}{\partial \theta_j}\boldsymbol{\Omega}\right), i, j = 1, \ldots, k,$$

with all of the unknown parameters in Ω and $\boldsymbol{\Gamma}$ evaluated at their maximum-likelihood estimates.

Finally, we note briefly that, under some mild restrictions, $\tilde{\tilde{\boldsymbol{\beta}}}$ and $\tilde{\tilde{\boldsymbol{\theta}}}$ are consistent, asymptotically efficient, and asymptotically normal. In fact, as mentioned above, the GLS estimator $\boldsymbol{\beta}^{*}$and the FGLS estimator $\boldsymbol{\beta}^{**}$ determined under consistent estimation of $\hat{\boldsymbol{\Omega}}$ have coincident asymptotic distributions and, under the normality of $\boldsymbol{\varepsilon}$, are both asymptotically efficient. So given the asymptotic efficiency of $\boldsymbol{\beta}^{**}$, the asymptotic distributions of $\tilde{\tilde{\boldsymbol{\beta}}}$ and $\boldsymbol{\beta}^{**}$ also coincide.

Appendix D: The Generalized Linear Mixed Model [Henderson 1950, 1973, 1984; Harville 1990; Robinson 1991; Wolfinger, Tobias, and Sall 1994]

Suppose we structure the *generalized linear mixed model* (GLMM) as

$$\boldsymbol{Y} = \boldsymbol{X\beta} + \boldsymbol{ZU} + \boldsymbol{\varepsilon} = (\boldsymbol{X}, \boldsymbol{Z})(\boldsymbol{\beta}, \boldsymbol{U})' + \boldsymbol{\varepsilon} \tag{9.D.1}$$

where $\boldsymbol{Y}$ is an $(n \times 1)$ vector of observable random variables, $\boldsymbol{X}$ and $\boldsymbol{Z}$ are, respectively, $(n \times p)$ and $(n \times q)$ known matrices, $\boldsymbol{\beta}$ is a $(p \times 1)$ vector of constants, and $\boldsymbol{U}$ and ε are $(q \times 1)$ and $(n \times 1)$ vectors of (unobservable) random variables. In particular, $\boldsymbol{\beta}$ is a vector of p fixed-effects parameters and $\boldsymbol{U}$ is a vector of q random-effects parameters. Under this representation, the components of $\boldsymbol{\varepsilon}$ are no longer required to be independent and homogeneous; we thus have greater flexibility in the specification of the covariance structure of $\boldsymbol{\varepsilon}$.

Let us assume that

$$E(\boldsymbol{U}) = E(\boldsymbol{\varepsilon}) = \boldsymbol{0},\ V\begin{pmatrix}\boldsymbol{U}\\ \boldsymbol{\varepsilon}\end{pmatrix} = \begin{bmatrix}V(\boldsymbol{U}) & \boldsymbol{0}\\ \boldsymbol{0} & V(\boldsymbol{\varepsilon})\end{bmatrix} = \begin{bmatrix}\boldsymbol{G} & \boldsymbol{0}\\ \boldsymbol{0} & \boldsymbol{R}\end{bmatrix}, \text{ and } COV(\boldsymbol{U}, \boldsymbol{\varepsilon}) = \boldsymbol{0},$$

where $\boldsymbol{G}$ and $\boldsymbol{R}$ are known positive definite matrices. Furthermore,

$E(\boldsymbol{Y}) = \boldsymbol{X\beta}$ and

$$\begin{aligned} \boldsymbol{V} = V(\boldsymbol{Y}) &= V(\boldsymbol{X\beta} + \boldsymbol{ZU} + \boldsymbol{\varepsilon}) \\ &= E[(\boldsymbol{X\beta} + \boldsymbol{ZU} + \boldsymbol{\varepsilon})(\boldsymbol{X\beta} + \boldsymbol{ZU} + \boldsymbol{\varepsilon})' \\ &= E[(\boldsymbol{ZU} + \boldsymbol{\varepsilon})(\boldsymbol{ZU} + \boldsymbol{\varepsilon})'] \\ &= E[(\boldsymbol{ZUU'Z'} + \boldsymbol{ZU\varepsilon'} + \boldsymbol{\varepsilon U'\varepsilon'} + \boldsymbol{\varepsilon\varepsilon'})] \\ &= \boldsymbol{Z}E(\boldsymbol{UU'})\boldsymbol{Z'} + E(\boldsymbol{\varepsilon\varepsilon'}) = \boldsymbol{ZGZ'} + \boldsymbol{R}.^{8} \end{aligned} \tag{9.D.2}$$

[8] Remember that, in general, for random $\boldsymbol{X}$, scalar vector $\boldsymbol{a}$, and scalar matrix $\boldsymbol{A}$, $E(\boldsymbol{a}'\boldsymbol{X}) = \boldsymbol{a}'E(\boldsymbol{X}), E(\boldsymbol{AX}) = \boldsymbol{A}E(\boldsymbol{X}), V(\boldsymbol{X}) = E[(\boldsymbol{X} - E(\boldsymbol{X}))(\boldsymbol{X} - E(\boldsymbol{X}))'] = \Sigma$, $V(\boldsymbol{a}'\boldsymbol{X}) = \boldsymbol{a}'\Sigma\boldsymbol{a}$, and $V(\boldsymbol{AX}) = \boldsymbol{A}\Sigma\boldsymbol{A}'$.

(Note that, when $\boldsymbol{R} = \sigma_\varepsilon^2 \boldsymbol{I}_n$ and $\boldsymbol{Z} = \boldsymbol{0}$, the mixed model Equation 9.D.1 reduces to the standard linear regression model $\boldsymbol{Y} = \boldsymbol{X\beta} + \boldsymbol{\varepsilon}$.) In what follows, we shall assume that $\boldsymbol{U}$ is multivariate $N(\boldsymbol{0}, \boldsymbol{G})$, $\boldsymbol{\varepsilon}$ is multivariate $N(\boldsymbol{0}, \boldsymbol{R})$, and $\boldsymbol{Y}$ is multivariate $N(\boldsymbol{X}\beta, \boldsymbol{ZGZ}' + \boldsymbol{R})$.

Given that $\boldsymbol{U}$ and $\boldsymbol{\varepsilon}$ are normally distributed, let us maximize the joint probability density function of $\boldsymbol{Y}$ and $\boldsymbol{U}$,

$$f = (2\pi\sigma^2)^{-\frac{1}{2}(n+q)\begin{vmatrix} G & 0 \\ 0 & R \end{vmatrix}^{-\frac{1}{2}}} \times \exp\left\{-\frac{1}{2\sigma^2}\begin{pmatrix} \hat{U} \\ Y - X\hat{\beta} - Z\hat{U} \end{pmatrix}' \begin{bmatrix} G & 0 \\ 0 & R \end{bmatrix}^{-1} \begin{pmatrix} \hat{U} \\ Y - X\hat{\beta} - Z\hat{U} \end{pmatrix}\right\}, \tag{9.D.3}$$

with respect to $\hat{\boldsymbol{\beta}}$ and $\hat{\boldsymbol{U}}$, where $\hat{\boldsymbol{\beta}}$ and $\hat{\boldsymbol{U}}$ are arbitrary estimators of $\boldsymbol{\beta}$ and $\boldsymbol{U}$, respectively. Transforming to logarithms and noting that

$$\begin{bmatrix} G & 0 \\ 0 & R \end{bmatrix}^{-1} = \begin{bmatrix} G^{-1} & 0 \\ 0 & R^{-1} \end{bmatrix},$$

it is readily seen that maximizing log$f$ with respect to $\hat{\boldsymbol{\beta}}$ and $\hat{\boldsymbol{U}}$ is equivalent to minimizing

$$\begin{aligned} h &= \begin{pmatrix} \hat{U} \\ Y - X\hat{\beta} - Z\hat{U} \end{pmatrix}' \begin{bmatrix} G^{-1} & 0 \\ 0 & R^{-1} \end{bmatrix} \begin{pmatrix} \hat{U} \\ Y - X\hat{\beta} - Z\hat{U} \end{pmatrix} \\ &= \hat{U}'G^{-1}\hat{U} + (Y - X\hat{\beta} - Z\hat{U})'R^{-1}(Y - X\hat{\beta} - Z\hat{U}) \\ &= \hat{U}'G^{-1}\hat{U} + Y'R^{-1}Y - 2Y'R^{-1}X\hat{\beta} - 2Y'R^{-1}Z\hat{U} \\ &\quad + 2\hat{\beta}'X'R^{-1}Z\hat{U} + \hat{\beta}'X'R^{-1}X\hat{\beta} + \hat{U}'Z'R^{-1}Z\hat{U} \end{aligned}$$

with respect to $\hat{\boldsymbol{\beta}}$ and $\hat{\boldsymbol{U}}$. To this end, let us set

$$\begin{aligned} \frac{\partial h}{\partial \hat{\beta}} &= -2X'R^{-1}Y + 2X'R^{-1}Z\hat{U} + 2X'R^{-1}X\hat{\beta} = 0 \\ \frac{\partial h}{\partial \hat{U}} &= 2G^{-1}\hat{U} - 2Z'R^{-1}Y + 2Z'R^{-1}X\hat{\beta} + 2Z'R^{-1}Z\hat{U} = 0, \end{aligned}$$

from which we obtain the system of *mixed model equations*

$$\begin{aligned} X'R^{-1}X\hat{\beta} + X'R^{-1}Z\hat{U} &= X'R^{-1}Y \\ Z'R^{-1}X\hat{\beta} + (Z'R^{-1}Z + G^{-1})\hat{U} &= Z'R^{-1}Y \end{aligned} \tag{9.D.4}$$

or, in matrix form,

$$\begin{bmatrix} X'R^{-1}X & X'R^{-1}Z \\ Z'R^{-1}X & Z'R^{-1}Z + G^{-1} \end{bmatrix} \begin{bmatrix} \hat{\beta} \\ \hat{U} \end{bmatrix} = \begin{bmatrix} X'R^{-1}Y \\ Z'R^{-1}Y \end{bmatrix}.$$

Then Equation 9.D.4 can be solved simultaneously for

$$\hat{\boldsymbol{\beta}} = (\boldsymbol{X}'\boldsymbol{V}^{-1}\boldsymbol{X})^{-1}\boldsymbol{X}'\boldsymbol{V}^{-1}\boldsymbol{Y}$$
$$\hat{\boldsymbol{U}} = \boldsymbol{G}\boldsymbol{Z}'\boldsymbol{V}^{-1}(\boldsymbol{Y} - \boldsymbol{X}\hat{\boldsymbol{\beta}}), \tag{9.D.5}$$

where again $\boldsymbol{V} = \boldsymbol{ZGZ}' + \boldsymbol{R}$. Additionally, for $\rho(\boldsymbol{X}) = p$, the variance-covariance matrix of $\hat{\boldsymbol{\beta}}$ and $\hat{\boldsymbol{U}}$ is

$$E\left[\begin{pmatrix}\hat{\boldsymbol{\beta}}-\boldsymbol{\beta}\\ \hat{\boldsymbol{U}}-\boldsymbol{U}\end{pmatrix}\begin{pmatrix}\hat{\boldsymbol{\beta}}-\boldsymbol{\beta}\\ \hat{\boldsymbol{U}}-\boldsymbol{U}\end{pmatrix}'\right] = \sigma^2\begin{bmatrix}\boldsymbol{X}'\boldsymbol{R}^{-1}\boldsymbol{X} & \boldsymbol{X}'\boldsymbol{R}^{-1}\boldsymbol{Z}\\ \boldsymbol{Z}'\boldsymbol{R}^{-1}\boldsymbol{X} & \boldsymbol{Z}'\boldsymbol{R}^{-1}\boldsymbol{Z}+\boldsymbol{G}^{-1}\end{bmatrix}^{-1}. \tag{9.D.6}$$

It should be apparent that the estimation of random effects should be an improvement over simply estimating a set of fixed effects in a regression model. When we estimate only fixed effects, we obtain $\hat{\boldsymbol{\beta}}$ as an "estimator" of $\boldsymbol{\beta}$ (in particular, $\hat{\boldsymbol{\beta}}$ is BLUE for $\boldsymbol{\beta}$), but, when we incorporate random effects in the model, we use the term "predictor" to highlight this distinction, i.e., $\hat{\boldsymbol{\beta}}$ and $\hat{\boldsymbol{U}}$ are the *best linear unbiased predictors* (BLUPs) of $\boldsymbol{\beta}$ and $\boldsymbol{U}$, respectively, where best is interpreted as minimum mean square error of prediction. Here the BLUP estimators in Equation 9.D.5 correspond to maximum-likelihood estimators. Moreover, Equation 9.D.5 renders a solution for which $\hat{\boldsymbol{\beta}}$ is identical to the GLS estimator for $\boldsymbol{\beta}$, and as $\boldsymbol{G}^{-1} \to \boldsymbol{0}$, Equation 9.D.5 tends to the GLS estimators for $\boldsymbol{\beta}$ and $\boldsymbol{U}$ when the components of $\boldsymbol{U}$ are considered to be fixed effects.

The preceding discussion has assumed that $\boldsymbol{V}$ is known (and thus $\boldsymbol{G}$ and $\boldsymbol{R}$ must be known matrices). However, if $\boldsymbol{G}$ and $\boldsymbol{R}$ are unknown, then still assuming that $\boldsymbol{U}$ and $\boldsymbol{\varepsilon}$ are multivariate normal, we can use a technique called *restricted maximum likelihood* (REML), which involves only the parameters in $\boldsymbol{G}$ and $\boldsymbol{R}$, i.e., for

$$\hat{\boldsymbol{r}} = \boldsymbol{Y} - \boldsymbol{X}\hat{\boldsymbol{\beta}} = \boldsymbol{Y} - \boldsymbol{X}(\boldsymbol{X}'\hat{\boldsymbol{V}}^{-1}\boldsymbol{X})^{-1}\boldsymbol{X}'\hat{\boldsymbol{V}}^{-1}\boldsymbol{Y},$$

the REML log-likelihood function is

$$l(\hat{\boldsymbol{G}},\hat{\boldsymbol{R}}) = -\frac{n-p}{2}\left[1+\log\left(2\pi/(n-p)\right)\right] - \frac{1}{2}\log\left|\hat{\boldsymbol{V}}\right|$$
$$-\frac{1}{2}\log\left|\boldsymbol{X}'\hat{\boldsymbol{V}}^{-1}\boldsymbol{X}\right| - \frac{n-p}{2}\log\hat{\boldsymbol{r}}\hat{\boldsymbol{V}}^{-1}\hat{\boldsymbol{r}}, \tag{9.D.7}$$

where $\hat{\boldsymbol{V}} = \boldsymbol{Z}\hat{\boldsymbol{G}}\boldsymbol{Z}' + \hat{\boldsymbol{R}}$ and $\hat{\boldsymbol{G}}$ and $\hat{\boldsymbol{R}}$ are arbitrary estimators for $\boldsymbol{G}$ and $\boldsymbol{R}$, respectively. After solving $\partial l/\partial\hat{\boldsymbol{G}} = \partial l/\partial\hat{\boldsymbol{R}} = \boldsymbol{0}$ for $\hat{\boldsymbol{G}}$ and $\hat{\boldsymbol{R}}$ (and thus for $\hat{\boldsymbol{V}}$), and provided that $\hat{\boldsymbol{G}}^{-1}$ exists, Equation 9.D.5 can be used to determine

$$\hat{\boldsymbol{\beta}} = (\boldsymbol{X}'\boldsymbol{V}^{-1}\boldsymbol{X})^{-1}\boldsymbol{X}'\hat{\boldsymbol{V}}^{-1}\boldsymbol{Y}$$
$$\hat{\boldsymbol{U}} = \hat{\boldsymbol{G}}\boldsymbol{Z}'\hat{\boldsymbol{V}}^{-1}(\boldsymbol{Y} - \boldsymbol{X}\hat{\boldsymbol{\beta}}). \tag{9.D.5.1}$$

Here $\hat{\boldsymbol{\beta}}$ and $\hat{\boldsymbol{U}}$ are termed the *estimated or empirical BLUPs* or EBLUPs for $\boldsymbol{\beta}$ and $\boldsymbol{U}$, respectively.

Exercises

Random Coefficients Estimation

9-1. Given the dataset appearing in Table E.9-1, use the stepwise procedure presented above (see "A Pure Random Coefficients Model") to estimate the parameters of the pure random coefficients regression model.

9-2. Suppose six varieties of rice are randomly selected from the population of rice varieties that thrive in East-Asian growing conditions. The experimental units are quarter-acre terraced plots of land on a 60 plot hillside. Each rice variety is randomly assigned to 10 plots of land. It is anticipated that remnants of a rice blight that attacked the last planting could adversely affect the ultimate yield of the new planting. Here the response variable "yield" (Y) is measured in bushels per plot and the measured number of blight spores/100 square centimeters ($\boldsymbol{X}$) is the explanatory variable (Table E.9-2). Given this information, estimate the parameters of the implied linear random coefficients mixed model. In particular,

a. Explain why this constitutes a random effects model.

b. Are the random effects positively or negatively correlated?

c. Is the null model appropriate?

d. What are the fixed and random effects?

e. Reestimate the model under the restriction that $\sigma_{ab} = 0$. What is your conclusion regarding this constraint?

TABLE E.9-1

Observations on the Variables X and Y

X	1	2	3	4	5	6	7	8	9	10
Y	8	6	8	4	7	2	6	1	5	0

TABLE E.9-2

Rice Yield (Y) versus Blight Spores (X)

Id #	Variety	Yield	Blight(00)
1	1	30	65
2	1	25	108
3	1	32	110
4	1	27	240
5	1	22	350
6	1	19	405
7	1	27	408
8	1	26	490

(continued)

TABLE E.9-2 *(continued)*

Id #	Variety	Yield	Blight(00)
9	1	18	600
10	1	19	690
11	2	44	60
12	2	42	120
13	2	39	160
14	2	39	250
15	2	31	380
16	2	30	445
17	2	28	505
18	2	26	510
19	2	29	570
20	2	22	610
21	3	59	70
22	3	50	100
23	3	48	185
24	3	35	203
25	3	31	298
26	3	32	303
27	3	31	320
28	3	21	390
29	3	17	450
30	3	10	505
31	4	63	40
32	4	53	50
33	4	52	150
34	4	50	250
35	4	45	340
36	4	40	450
37	4	38	550
38	4	40	610
39	4	31	700
40	4	28	750
41	5	50	50
42	5	47	150
43	5	51	203

TABLE E.9-2 *(continued)*

Id #	Variety	Yield	Blight(00)
44	5	50	300
45	5	46	395
46	5	47	500
47	5	46	590
48	5	45	650
49	5	40	690
50	5	42	727
51	6	70	200
52	6	60	260
53	6	58	310
54	6	55	370
55	6	42	405
56	6	39	500
57	6	32	570
58	6	25	640
59	6	15	645
60	6	22	715

9-3. Four types of packaging of a given product are randomly drawn from the population of packaging options. The experimental units are 40 grocery stores. Each packaging type is randomly assigned to 10 of the stores, and it is expected that the amount of available shelf space could influence sales. Let the response variable be "units sold" (Y) with the explanatory variable taken to be "shelf space measured in linear feet" (X) (Table E.9-3). Estimate the parameters of a linear random coefficients mixed model and fully explain your results. (Hint: let parts a through e of Exercise 9-2 guide your analysis of the regression results.)

TABLE E.9-3

Units Sold (Y) versus Shelf Space (X)

Id #	Type	Sales	Space
1	1	9	17
2	1	7	22
3	1	12	25
4	1	9	26
5	1	11	31
6	1	15	31
7	1	13	37
8	1	15	41

(continued)

TABLE E.9-3 *(continued)*

Id #	Type	Sales	Space
9	1	11	42
10	1	17	49
11	2	6	15
12	2	8	22
13	2	8	28
14	2	13	34
15	2	17	36
16	2	14	45
17	2	19	49
18	2	22	56
19	2	25	60
20	2	23	64
21	3	13	17
22	3	15	22
23	3	14	29
24	3	16	37
25	3	18	40
26	3	17	45
27	3	16	54
28	3	19	57
29	3	16	60
30	3	20	62
31	4	9	8
32	4	11	11
33	4	12	15
34	4	13	19
35	4	13	22
36	4	15	35
37	4	17	42
38	4	19	55
39	4	21	60
40	4	23	60

9-4. Three separate weight-loss programs are randomly selected from the population of such programs at a particular health clinic. Thirty new clients (the experimental units) have agreed to have one of the programs randomly assigned to them, and it is anticipated that their pre-program weight level could influence their final weight loss amount. Let the response variable be "weight loss in pounds after 6 months" (Y) and let the explanatory variable be "pre-program weight in pounds" (X) (Table E.9-4). Estimate the parameters of a linear random coefficients mixed regression model and fully explain your findings. (Hint: let parts a through e of Exercise 9-2 serve as your guide.)

TABLE E.9-4

Weight Loss (Y) versus Pre-Program Weight (X)

Id #	Program	Loss	Weight
1	1	21	140
2	1	23	170
3	1	25	225
4	1	24	265
5	1	28	288
6	1	33	295
7	1	28	305
8	1	33	310
9	1	32	330
10	1	36	330
11	2	15	135
12	2	16	151
13	2	17	170
14	2	20	200
15	2	18	210
16	2	15	230
17	2	20	249
18	2	20	275
19	2	25	305
20	2	23	341
21	3	9	141
22	3	12	169
23	3	18	172
24	3	13	199
25	3	8	200
26	3	20	220
27	3	13	233
28	3	17	262
29	3	25	290
30	3	20	330

IML Matrix Operations

9-5. Given matrices

$$A = \begin{bmatrix} 1 & 0 & 1 \\ 3 & -1 & 6 \\ 1 & 2 & 5 \end{bmatrix} \text{ and } B = \begin{bmatrix} 3 & 2 & 0 \\ 1 & 0 & 2 \\ 3 & -1 & 0 \end{bmatrix},$$

find:

a. $\boldsymbol{A} + \boldsymbol{B}$; $\boldsymbol{B} - \boldsymbol{A}$; $\boldsymbol{A}\ \boldsymbol{B}$; $\boldsymbol{A}'$; $\boldsymbol{B}' + \boldsymbol{A}$; and $\boldsymbol{A}^{-1}$.

b. $\lambda \boldsymbol{A}$ for $\lambda = 2$; and $\boldsymbol{A}\boldsymbol{A}^{-1}$.

c. A matrix $\boldsymbol{C}$ whose elements are the squares of those in $\boldsymbol{B}$.

d. Verify that $\boldsymbol{B}*i(3) = i(3)*\boldsymbol{B}$.

10

L_1 and q-Quantile Regression

Introduction

We noted in previous chapters that, if the random error terms ε_i in the population regression line $Y_i = \beta_o + \beta_1 X_i + \varepsilon_i$, $i = 1, \ldots, n$, are independent with ε_i: $N(0, \sigma_\varepsilon)$, then the least-squares estimators $\hat{\beta}_o$ and $\hat{\beta}_1$ are maximum-likelihood estimators of β_o and β_1, respectively. That is, if ε_i follows the probability density function

$$f(\varepsilon_i) = \frac{1}{\sqrt{2\pi}\,\sigma_\varepsilon} e^{-\varepsilon_i^2/2\sigma_\varepsilon^2}, \; i = 1, \ldots, n,$$

then the likelihood function for the sample random variables Y_i, $i = 1, \ldots, n$, is

$$\mathcal{L}(\beta_o, \beta_1, \sigma_\varepsilon; Y_1, \ldots, Y_n, n) = \frac{1}{(2\pi)^{n/2}\sigma_\varepsilon^n} e^{-\sum_{i=1}^n \varepsilon_i^2/2\sigma_\varepsilon^2},$$

or the log-likelihood function is

$$\log \mathcal{L} = -\frac{n}{2}\log(2\pi) - n\log(\sigma_\varepsilon) - \frac{1}{2\sigma_\varepsilon^2}\sum_{i=1}^n \varepsilon_i^2. \tag{10.1}$$

Then for σ_ε fixed, maximizing $\log \mathcal{L}$ is equivalent to minimizing $\sum_{i=1}^n \varepsilon_i^2$. So if the normality assumption concerning ε_i is tenable, then applying the *OLS criterion*, choose $\hat{\beta}_o$ and $\hat{\beta}_1$ so as to minimize $\sum_{i=1}^n e_i^2$ is appropriate, where $e_i = Y_i - \hat{Y}_i$ is the ith deviation or residual from the sample regression line.

However, if the preceding normality assumption is untenable, then OLS should not be used and a more robust regression procedure is called for. In this regard, suppose one feels that the tails of the error distribution are "much heavier" than those of the normal probability density function. Then we obviously need to apply a regression criterion that weights each observation unequally (least squares weights each observation equally) in that outliers should be downweighted in the estimation process.

One way to accomplish all this is to assume that the random errors ε_i, $i = 1, \ldots, n$, are independent and follow a *double-exponential (Laplace) distribution*

$$f(\varepsilon_i) = (2\sigma_\varepsilon)^{-1} e^{-|\varepsilon_i|/\sigma_\varepsilon}, \; i = 1, \ldots, n. \tag{10.2}$$

This distribution is symmetrical about zero, has a peak that is much sharper than that of a normal distribution, tails off to zero as $|\varepsilon_i| \to +\infty$, and has tails that are heavier than those of a normal distribution (because $e^{-\varepsilon_i^2}$ goes to zero faster than does $e^{-|\varepsilon_i|}$ when $\varepsilon_i \to \pm\infty$) (see Figure 10.1).

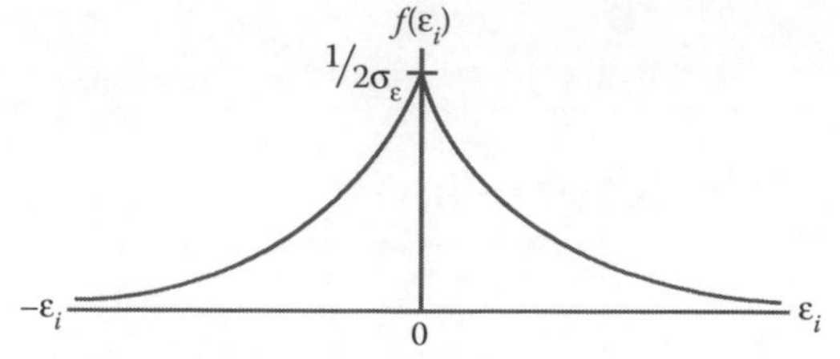

FIGURE 10.1
Double- exponential distribution.

Given Equation 10.2, the likelihood function appears as

$$\mathcal{L}(\beta_o,\ \beta_1,\ \sigma_\varepsilon;\ Y_1,\ldots,Y_n,n) = (2\sigma_\varepsilon)^{-n} e^{-\sum_{i=1}^{n}|\varepsilon_i|/\sigma_\varepsilon}$$

or

$$\log \mathcal{L} = -n\log(2\sigma_\varepsilon) - \frac{1}{\sigma_\varepsilon}\sum_{i=1}^{n}|\varepsilon_i|. \tag{10.3}$$

Hence, for a fixed σ_ε, maximizing $\log \mathcal{L}$ involves minimizing $\sum_{i=1}^{n}|\varepsilon_i|$.

Let us examine the formal mathematical structure of $\sum_{i=1}^{n}|\varepsilon_i|$. Suppose we have a *vector* **X** with components (or a point **X** with coordinates) $X_1,\ldots,X_n$. For p a real number with $1 \le p < +\infty$, the L_p *norm* of the n-tuple of scalars $\mathbf{X} = (X_1,\ldots,X_n)$ is defined by

$$\|\boldsymbol{X}\|_p = \left(\sum_{i=1}^{n}|X_i|^p\right)^{\frac{1}{p}}.\text{[1]} \tag{10.4}$$

If $p = 1$, Equation 10.4 simplifies to the L_1 *norm*

$$\|\boldsymbol{X}\|_1 = \sum_{i=1}^{n}|X_i|. \tag{10.6}$$

So for the ith residual $e_i = Y_i - \hat{\beta}_o - \hat{\beta}_1 X_i$, we may define an L_1 *regression estimator* or a *least absolute value regression estimator* as one for which $\hat{\beta}_o$ and $\hat{\beta}_1$ are chosen to minimize $\sum_{i=1}^{n}|e_i|$, i.e., the L_1 estimator is given by

$$\min_{\hat{\beta}_o,\hat{\beta}_1} \sum_{i=1}^{n}|e_i|. \tag{10.7}$$

(Note that the least-squares estimator is an L_2 *regression estimator*.) The motivation for using the L_1 regression estimator over the least-squares estimator is the observation that the

---

[1] In general, a *norm* on R^n is a function that assigns to each $\boldsymbol{X} \in R^n$ a number $\|\mathbf{X}\|$ such that

a. $\|\mathbf{X}\| \ge 0$ and $\|\mathbf{X}\| = 0$ if and only if $X_i = 0$ for all $i = 1,\ldots,n$.

b. $\|\mathbf{X} + \mathbf{Y}\| \le \|\mathbf{X}\| + \|\mathbf{Y}\|$ (the *triangle inequality*), where $\mathbf{X} + \mathbf{Y}$ denotes componentwise addition. (10.5)

c. $\|c\mathbf{X}\| = |c|\ \|\mathbf{X}\|$ (homogeneity), where c is a scalar and c**X** denotes componentwise multiplication.

d. $\|\mathbf{X}\cdot\mathbf{Y}\| \le \|\mathbf{X}\|\ \|\mathbf{Y}\|$ (*Cauchy-Schwarz inequality*), where $\boldsymbol{X}\cdot\boldsymbol{Y} = \sum_{i=1}^{n} X_i Y_i$ denotes the *scalar product* of the vectors **X**, **Y**.

value of $\hat{\theta}$ that minimizes the quantity $\sum_{i=1}^{n}\left|x_i-\hat{\theta}\right|$ is the median (that is robust against outliers). Hence, replacing e_i^2 by $|e_i|$ could possibly render an increase in the robustness of our regression parameter estimates. In fact, unlike the least-squares regression estimates, the L_1 regression estimates are robust with respect to outliers in the Y direction. However, L_1 regression is adversely affected by outlying X values that are influential leverage points, And, if any such X value is highly influential, the L_1 estimated line can actually pass through it, thus providing us with a sample regression equation of dubious worth. A least-squares fit, although vulnerable to influential leverage points, will not be compromised to the same extent. Moreover, the L_1 regression estimate may not be unique and its efficiency can decrease as the sample size n increases.

We next look to the definition of a q-quantile regression estimator. As pointed out by Mosteller and Tukey [1977] and, more recently, by Koenker [2005], we need to go beyond the simple OLS determination of the conditional mean of Y given X; such regression estimates need to be supplemented with an assortment of conditional quantile estimates of Y given X. Without doing so leaves us with an abridged or incomplete picture of the distribution of Y given X (just as examining only the mean of, say, a set of stock prices overlooks other important characteristics of a stock portfolio, such as riskiness, as evidenced by the standard deviation, etc.).

To examine the motivation underlying q-quantile regression, let us follow the very insightful discussion offered by Koenker [2005]. That is, q-quantiles are expressable as the solution to the following optimization problem. Specifically, let us first define the *indicator function* as

$$I(x<0)=\begin{cases}1, & x<0;\\ 0 & \text{otherwise.}\end{cases} \tag{10.8}$$

Next, a simple decision theoretic piecewise linear (asymmetric) loss function is specified as

$$\begin{aligned}\rho_q(x)&=x(q-I(x<0))\\ &=\begin{cases}xq, & x>0;\\ x(q-1), & x<0,\end{cases}\end{aligned} \tag{10.9}$$

for $q\in(0,1)$.

Given Equation 10.9 and a random variable X with cumulative distribution function $F(x)$, the problem is to find a point estimator $\hat{x}$ that minimizes expected loss.

To this end, let

$$\rho_q(X-\hat{x})=\begin{cases}q(X-\hat{x}), & X>\hat{x};\\ (q-1)(X-\hat{x}), & X<\hat{x}.\end{cases} \tag{10.10}$$

Then

$$\begin{aligned}E\,\rho_q(X-\hat{x})&=\int_{-\infty}^{+\infty}(x-\hat{x})f(x)dx\\ &=(q-1)\int_{-\infty}^{\hat{x}}(x-\hat{x})dF(x)+q\int_{\hat{x}}^{+\infty}(x-\hat{x})dF(x).^2\end{aligned} \tag{10.11}$$

2 For $g(x)$ a single-valued real function of a continuous random variable X, $E(g(x))=\int_{-\infty}^{+\infty}g(x)f(x)dx$, where $f(x)$ is the probability density function of X. In addition, at every point of continuity of f, $dF(x)/dx=f(x)$.

Differentiating Equation 10.11 with respect to $\hat{x}$ renders

$$0 = (1-q)\int_{-\infty}^{\hat{x}} dF(x) - q\int_{\hat{x}}^{+\infty} dF(x)$$
$$= \int_{-\infty}^{\hat{x}} dF(x) - q\int_{-\infty}^{+\infty} dF(x) = F(\hat{x}) - q$$

or $F(\hat{x}) = q$. With F monotone increasing in x, expected loss is minimized for $x = \hat{x}$, and, if $\hat{x}$ is uniquely determined, then $\hat{x} = F^{-1}(q)$ so that $\hat{x}$ is termed the qth quantile of X. Note that, if $q = 0.5$, then $\hat{x} = F^{-1}(0.5)$ determines the median of X. So given the asymmetric linear loss function Equation 10.10, the optimal point estimator $\hat{x}$ specifies the q-quantiles of X.

Within the context of the preceding discussion, if $X_1,\ldots,X_n$ denotes a set of sample random variables, then the problem of finding the qth sample quantile involves finding the solution $\hat{\theta}(q)$ to the optimization problem

$$\min_{\hat{\theta}\in R} \sum_{i=1}^{n} \rho_q(X_i - \hat{\theta}). \tag{10.12}$$

If we define the qth *conditional quantile function* as $Q_Y(q|X) = \beta_o(q) + \beta_1(q)X$, then our objective is to find estimators $\hat{\beta}_o(q)$ and $\hat{\beta}_1(q)$, which solve the optimization problem

$$\min_{\hat{\beta}_o,\hat{\beta}_1} \sum_{i=1}^{n} \rho_q(Y_i - \hat{\beta}_o - \hat{\beta}_1 X_i). \tag{10.13}$$

As we shall see below, both the L_1 regression problem (Equation 10.7) and the q-quantile regression problem (Equation 10.13) have a straightforward linear programming formulation.

L_1 Regression

Let us express the problem of L_1 *regression* or *least absolute value* (LAV) *regression* as

$$\min_{\hat{\beta}_o,\hat{\beta}_1} \sum_{i=1}^{n} \left|Y_i - \hat{\beta}_o - \hat{\beta}_1 X_i\right|. \tag{10.7.1}$$

Alternatively, Equation 10.7.1 may be rewritten as a linear program by introducing $2n$ new (slack) variables u_i and v_i, $i = 1, \ldots, n$, to account for the positive and negative values, respectively, of the residuals e_i (when $e_i > 0$, $Y_i = \hat{Y}_i + u_i$; and when $e_i < 0$, $Y_i = \hat{Y}_i - v_i$) or

$$\begin{aligned}
&\min_{\hat{\beta}_o,\hat{\beta}_1} \sum_{i=1}^{n} (u_i + v_i) \quad \text{s.t.}\\
&\hat{\beta}_o + \hat{\beta}_1 X_1 + u_1 - v_1 = Y_1\\
&\quad\vdots\\
&\hat{\beta}_o + \hat{\beta}_1 X_n + u_n - v_n = Y_n\\
&u_i,\ v_i \ge 0;\ i = 1,\ldots,n.
\end{aligned} \tag{10.14}$$

As the following example problem reveals, Equations 10.7.1 and 10.14 can be solved via the *SAS LAV* subroutine.

Example 10.1

Given the $n = 10$ data points appearing in Table 10.1, determine the values of $\hat{\beta}_o$ and $\hat{\beta}_1$, which solve the L_1-norm regression problem.

The SAS code for the LAV procedure appears in Exhibit 10.1.

Exhibit 10.1 SAS Code for LAV Procedure

```
proc iml; ①
a={1 1, 1 2, 1 3, 1 4, 1 5, 1 6, 1 7, 1 8, 1 9, 1 10}; ②
m=nrow(a);
b={2, 5, 4, 6, 5, 8, 6, 8, 10, 8};
opt={. 5 0 1}; ③
call lav(rc, xr, a, b,  ,opt); ④
quit;
```

① We invoke *proc iml*.

② The first column of the *a* matrix is a column of 1's, thus indicating that an intercept term is included in the regression model. The second column of *a* contains the observations on the variable *X*. Missing values are not permitted in the *a* (or *b*) argument.

③ *opt* is a vector used to specify options. *opt* [1] specifies the maximum number of iterations (here the default is chosen); *opt*[2] specifies the amount of printed output (*opt* [2] =5 requests that all output be printed); *opt* [3]=0 requests that the computation of the asymptotic standard errors and covariance matrix elements be based on the McKean-Schrader [1987] estimate of the variance of the median of the nonzero residuals; and *opt* [4]=1 calls for a convergence test for the optimality of the solution.

④ We call for the *LAV* subroutine in which *rc* is a scalar return code (e.g., $rc = 0$ when termination is successful, $rc = 1$ when termination is successful but approximate standard errors and elements within the covariance matrix cannot be determined, and $rc < 0$ if termination is unsuccessful). *xr* specifies a vector or matrix with *m* columns, where *m* is the number of

TABLE 10.1

Observations on Variables *X* and *Y*

X	*Y*
1	2
2	5
3	4
4	6
5	5
6	8
7	6
8	8
9	10
10	8

parameters to be estimated. a is an (nxm) matrix, b is an $(n \times 1)$ vector containing the values of the variable Y, and opt is defined as in ③.

The L_1 regression results determined from this program appear in Table 10.2.

① The starting point of the L_1 subroutine is the OLS solution. Here the OLS line of best fit is $\hat{Y} = \hat{\beta}_o + \hat{\beta}_1 X = 2.4666 + 0.6787X$ with $\sum_{i=1}^{10} e_i^2 = 11.5879$. Hence, an estimate of σ_ε is $s_\varepsilon = \left(\sum_{i=1}^{10} e_i^2 / (n-2)\right)^{1/2} = 1.2035$.

② Using the OLS estimates to initialize the LAV algorithm, the iteration history is provided along with an indication that termination is successful. The L_1 objective function value (Equation 10.7.1) is 9.6.

③ The estimated L_1 regression equation is $\hat{Y} = \hat{\beta}_o + \hat{\beta}_1 X = 1.6 + 0.8X$; the asymptotic standard errors of $\hat{\beta}_o$ and $\hat{\beta}_1$ are $s_{\hat{\beta}_o} = 1.3802$ and $s_{\hat{\beta}_1} = 0.2224$, respectively.

The role of the factor $\lambda = 2.02034$ in L_1 regression is similar to that of S_ε when OLS is performed: it gives us an indication of the "overall size" of the random errors. In fact, for L_1 regression, the approximate standard error of the slope is

$$s_{\hat{\beta}_1} \approx \lambda \Big/ \sqrt{\sum (X_i - \bar{X})^2} = 2.02034/9.083 = 0.2224.$$

TABLE 10.2

Output for Example 10.1

The SAS System

①

LS Solution		
Est	2.4666666667	0.6787878788

Sum-of-squares (LS) residual = 11.587878788

Estimate of largest eigenvalue = 392.36

Estimate of smallest eigenvalue = 2.0997697033

②

LAV (L1) Estimation

Start with LS Solution

Start Iter: gamma=1.4606060606 ActEqn=10

Iter	N Huber	Act Eqn	Rank	Gamma	L1(x)	F(Gamma)
1	1	3	2	0.8182	9.963636	6.245971
1	2	3	2	0.8182	10.021837	6.207457
2	3	2	2	0.4198	9.784647	7.865725
2	3	2	2	0.0000	9.600000	7.865725

Algorithm converged

Objective function L1(X) = 9.6

Number search directions = 6

Number refactorizations = 4

Total execution time = 0.0000

Necessary and sufficient optimality condition satisfied.

TABLE 10.2 *(continued)*

③

	L1 Solution with ASE	
Est	1.6	0.8
ASE	1.3801564794	0.222432357

Cov Matrix: McKean-Schrader

1.9048319076	–0.272118844
–0.272118844	0.0494761534

Factor lambda = 2.0203422134

④

Predicted Values and Residuals

N	Observed	Predicted	Residual
1	2.0000	2.4000	–0.400000
2	5.0000	3.2000	1.800000
3	4.0000	4.0000	–2.22045E-16
4	6.0000	4.8000	1.200000
5	5.0000	5.6000	–0.600000
6	8.0000	6.4000	1.600000
7	6.0000	7.2000	–1.200000
8	8.0000	8.0000	–4.44089E-16
9	10.0000	8.8000	1.200000
10	8.0000	9.6000	–1.600000

(For OLS, $S_{\hat{\beta}_1} = S_\varepsilon \Big/ \sqrt{\sum (X_i - \bar{X})^2}$.)

A key property of this L_1 regression equation is that it always passes through two of the actual sample points (in this instance, it passes through point 3 with coordinates (3,4) and point 8 with coordinates (8,8)). This is easily verified because $\hat{\beta}_1 = \frac{8-4}{8-3} = 0.8$ and $\hat{\beta}_o = 8 - 0.8(8) = 1.6$. Also, the residuals for these points (see④) are effectively zero.

Is the L_1 regression slope significantly different from zero? Here we test H_o: $\beta_1 = 0$ against H_1: $\beta_1 \neq 0$. Looking to the sample t statistic, we have $|t| = \left|\hat{\beta}_1 / s_{\hat{\beta}_1}\right| = |0.8/0.2224| = 3.5971$. Because $0.001 < p\text{-value} < 0.01$, we answer this question in the affirmative.

q-Quantile Regression

We noted above (Equation 10.13) that the *q-quantile regression model* is expressible as

$$\min_{\hat{\beta}_o,\, \hat{\beta}_1} \sum_{i=1}^{n} \rho_q (Y_i - \hat{\beta}_o - \hat{\beta}_1 X_i).$$

Alternatively, this problem can be reformulated as a linear program once $2n$ additional (slack) variables u_i and v_i, $i = 1,\ldots,n$ (which account for the positive and negative values, respectively, of the residuals e_i), are admitted so that we now seek to

$$\min_{\hat{\beta}_o,\hat{\beta}_1} q\sum_{i=1}^{n} u_i + (1-q)\sum_{i=1}^{n} v_i \quad \text{s.t.}$$
$$\hat{\beta}_o + \hat{\beta}_1 X_1 + u_1 - v_1 = Y_1$$
$$\vdots$$
$$\hat{\beta}_o + \hat{\beta}_1 X_n + u_n - v_n = Y_n$$
$$u_i,\ v_i \geq 0,\ i = 1,\ldots,n. \qquad (10.15)$$

Note that, if $q = 0.5$, Equation 10.15 simplifies to the L_1 regression problem specified previously (see Equation 10.14).

Example 10.2

Using the dataset presented in Table 10.1, let us use the *regression quantiles* (*RQ*) subroutine listed in *SAS IML* as developed by Koenker and Bassett [1978] and Bassett and Koenker [1982]. The *SAS* code for generating the *q*-quantile regression estimates via linear programming appears as Exhibit 10.2.

Exhibit 10.2 SAS Code for *q*-quantile Regression

```
proc iml; ①
start rq(yname, y, xname,x,b,predict,error,q); ②
bound=1.0e10;
coef=x`;
m=nrow(coef);
n=ncol(coef);
r=repeat(0, m+2, 1);                                          ⎫
L=repeat(q-1, 1, n)||repeat(0, 1, m) ||-bound ||-bound;       ⎬ ③
u=repeat (q, 1, n)|| repeat ( . , 1, m) ||{ . . };            ⎭
a=(y`|| repeat(0, 1, m)|| {-1 0})//                           ⎫
     (repeat (0, 1, n)|| repeat(-1, 1, m)|| {0-1})//          ⎪
      (coef|| I(m)|| repeat(0, m, 2)) ;                       ⎬ ④
basis=n+m+2 - (0:n+m+1);                                      ⎭
call lp(rc, p, d, a, r, , u, L, basis);⑤
L=repeat(q-1, 1, n)|| repeat(0, 1, m)||-bound||{0};           ⎫
u=repeat (q, 1, n)|| repeat (0, 1, m)||{ . 0 };               ⎬ ⑥
call  lp (rc, p, d, a, r, n+m+1, u, L, basis);                ⎭
```

```
variable=xname`;
b=d[3:m+2];
predict=X*b ;
error=y-predict ;
wsum=sum(choose(error<0, (q-1)*error, q*error)) ;
print ,,'Regression Quantile Estimation'   ,
         'Dependent Variable: '  yname  ,
                 'Regression Quantile:   ' q  ,           ⑦
         'Number  of  Observations: '  n  ,
                  'Sum of Weighted Absolute Errors: ' wsum,
                   variable b,
          x y predict error;

finish rq;
z={2 1 1, 5 1 2, 4 1 3, 6 1 4, 5 1 5, 8 1 6, 6 1 7, 8 1 8, 10 1 9,
8 1 10}; ⑧
y=z[, 1]; ⑨
x=z[, 2 : 3]; ⑩
run rq ('y', y, {'intercept' 'x'}, x, b1, pred, resid, .5); ⑪
quit;
```

① We invoke *proc iml*.
② We initiate the *RQ* subroutine (*q* denotes the *q*th quantile).
③ We construct the right-hand sides of the constraints and bounds.
④ We construct the coefficient matrix and basis matrix.
⑤ We determine a feasible solution.
⑥ We determine an optimal solution.
⑦ We report the *RQ* subroutine results.
⑧ The columns of the *Z* matrix correspond to the variables *Y*, intercept, and *X*, where the intercept values are all 1's, i.e.,

$$Z = \begin{bmatrix} Y_1 & 1 & X_1 \\ Y_2 & 1 & X_2 \\ \cdot & \cdot & \cdot \\ \cdot & \cdot & \cdot \\ \cdot & \cdot & \cdot \\ Y_n & 1 & X_n \end{bmatrix}.$$

⑨ The values of the variable *Y* appear in the first column of *Z*.
⑩ Columns two and three of *Z* represent the columns of the matrix *X*, where the first column of *X* is a set of 1's and the second column of *X* corresponds to the observations on the variable *X*.
⑪ We run *RQ* with $q = 0.5$.

The output forthcoming from this program appears in Table 10.3.a.

The estimated 0.5-quantile regression equation is $\hat{Y} = \hat{\beta}_o + \hat{\beta}_1 X = 1.6 + 0.8X$. Note that, as expected, we obtain the exact same regression equation as generated under L_1 estimation. As noted above, this regression line passes through sample points 3 and 8.

If we now set $q = 0.25$ (this can be done interactively in *SAS* using the *run* statement II by replacing 0.5 by 0.25), the new regression output appears in Table 10.3b.

The estimated 0.25-quantile regression equation is $\hat{Y} = \hat{\beta}_o + \hat{\beta}_1 X = 1.3333 + 0.6667X$. An examination of the residuals reveals that this regression line passes through sample points 1, 7, and 10. If we now (interactively) select $q = 0.75$, our final set of quantile regression results emerges in Table 10.3.c. Here the fitted 0.75-quantile regression equation is $\hat{Y} = \hat{\beta}_o + \hat{\beta}_1 X = 3.5714 + 0.7142X$; it passes through sample data points 2 and 9.

A graphical comparison of all three of these estimated q-quantile regressions is offered in Figure 10.2. (The estimated *OLS* regression equation has the form $\hat{Y} = 2.4667 + 0.6787\ X$.) In addition, Table 10.4 summarizes the information contained in Figure 10.2.

TABLE 10.3.A

Output for Example 10.2 ($q = 0.5$)

The SAS System

Regression Quantile Estimation

	YNAME
Dependent variable: *y*	
Regression quantile:	0.5
Number of observations:	10
Sum of weighted absolute errors:	4.8

VARIABLE	B
intercept	1.6
x	0.8

X		*Y*	**Predict**	**Error**
1	1	2	2.4	–0.4
1	2	5	3.2	1.8
1	3	4	4	0
1	4	6	4.8	1.2
1	5	5	5.6	–0.6
1	6	8	6.4	1.6
1	7	6	7.2	–1.2
1	8	8	8	0
1	9	10	8.8	1.2
1	10	8	9.6	–1.6

As this table reveals, the *OLS* slope of 0.6787 represents the conditional mean effect of X on Y; this response in Y is taken uniformly over the entire conditional distribution of Y. If only the intercept changes, we have a *pure location shift* that is also taken uniformly over the conditional distribution of Y, i.e., the estimated slope over each quantile corresponds to the mean effect of the *OLS* regression function. Obviously a modest location shift occurs as q increases from 0.25 to 0.5, but a sizeable location shift takes place when q increases from 0.5 to 0.75.

However, the first quartile of the conditional distribution of Y exhibits the lowest slope or rate of increase in Y per unit increase in X (0.6667). As we move to the second quartile of the conditional Y distribution, the slope increases (to 0.8), and, finally, when we reach the third quartile, the slope decreases a bit (to 0.7142) but still exceeds its estimated value in the first quartile.

In summary, both the estimated intercept and slope vary over the conditional distribution of Y. Clearly relying solely on the *OLS* estimates provides us with an incomplete picture of the conditional distribution of Y.

TABLE 10.3.B

Output for Example 10.2 ($q = 0.25$)

The SAS System

Regression Quantile Estimation

	YNAME
Dependent variable: y	
Regression quantile:	0.25
Number of observations:	10
Sum of weighted absolute errors:	3

VARIABLE	B
intercept	1.3333333
x	0.6666667

X		Y	**Predict**	**Error**
1	1	2	2	0
1	2	5	2.6666667	2.3333333
1	3	4	3.3333333	0.6666667
1	4	6	4	2
1	5	5	4.6666667	0.3333333
1	6	8	5.3333333	2.6666667
1	7	6	6	0
1	8	8	6.6666667	1.3333333
1	9	10	7.3333333	2.6666667
1	10	8	8	0

TABLE 10.3.C

Output for Example 10.2 ($q = 0.75$)

The SAS System

Regression Quantile Estimation

	YNAME
Dependent variable: y	
Regression quantile:	0.75
Number of observations:	10
Sum of weighted absolute errors:	3.3928571
VARIABLE	B
intercept	3.5714286
x	0.7142857

X		Y	Predict	Error
1	1	2	4.2857143	−2.285714
1	2	5	5	0
1	3	4	5.7142857	−1.714286
1	4	6	6.4285714	−0.428571
1	5	5	7.1428571	−2.142857
1	6	8	7.8571429	0.1428571
1	7	6	8.5714286	−2.571429
1	8	8	9.2857143	−1.285714
1	9	10	10	0
1	10	8	10.714286	−2.714286

TABLE 10.4

q-quantile versus OLS Estimates

	$q = 0.25$	$q = 0.5$	$q = 0.75$	*OLS*
intercept	1.3333	1.6	3.5714	2.4667
slope	0.6667	0.8	0.7142	0.6787

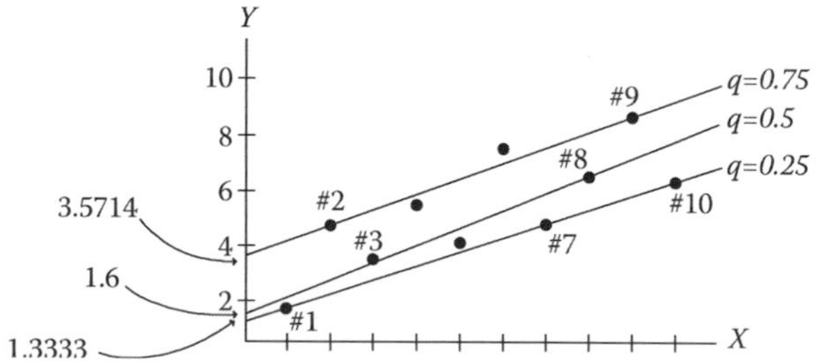

FIGURE 10.2
q-quantile regression lines.

Exercises

10-1. Given the Table E.10-1 dataset, determine the values of $\hat{\beta}_0$ and $\hat{\beta}_1$ via L_1-norm regression. (Hint: let Exhibit 10.1 guide your efforts.) In particular,

a. Determine which two sample points lie on the L_i norm regression equation.

b. Determine whether the L_1-norm regression slope is significantly different from zero.

c. Compare the L_1-norm regression results with those obtained using OLS.

10-2. Using the Table E.10-1 dataset, obtain q-quantile regression equations for $q = 0.25, 0.5$, and 0.75. In addition, determine the following:

a. Which sample points lie on these individual estimated regression lines?

b. What location shifts occur as q increases in value?

c. How does the regression slope vary as q increases in value?

d. Does the estimated OLS regression equation adequately represent the behavior of this dataset?

TABLE E.10-1

Observations on Variables X and Y

X	14	19	33	43	29	38	5	57	55	27	33	37	12	14
Y	41	44	28	34	44	28	24	16	16	31	17	33	40	38

11

Regression in a Spatial Domain

Introduction

In previous chapters, we modeled some explained variable Y as linear function of an explanatory variable X and a random error term ε or $Y = \beta_0 + \beta_1 X + \varepsilon$. Here the X and Y values are typically indexed by the passage of time or could be specified on a cross-sectional basis (e.g., taken over households, firms, countries, etc., at a particular point in time).

Let us now introduce a spatial dimension to the X and Y measurements. That is, we now have as set of *spatial data*: a set of region- or geographic-specific observations on the X and Y variables. In this regard, not only are we interested in the magnitudes of the components of some data point (X_k, Y_k), we are also interested in where this data point occurs.

So if our focus is on the analysis of data in a spatial context, then both an attribute feature as well as a spatial feature must be considered (*aspatial data* has only an attribute component). The attribute feature focuses on the character of the variables, whereas the spatial feature addresses or maps their location. For example, variables that have a spatial reference include the following: single-family dwellings indexed by zip code, median income referenced by census tract, maximum temperature recorded on a particular date over each state within the continental United States, and so on.

Types of Spatial Data [Cressie 1993; Schabenberger and Gotway 2005]

Spatial data may be classified according to its spatial domain. In this regard, let $\{Y(\boldsymbol{S}) : \boldsymbol{S} \in \mathcal{D} \subset R^2\}$ depict a *spatial process* in two dimensions, where Y denotes the realization of some attribute, $\boldsymbol{S} = (S_1, S_2)$ depicts, via a set of Cartesian coordinates S_1 and S_2, the location at which Y is observed, and $\mathcal{D}$ is the *spatial domain* of the process. What are the characteristics of the spatial domain $\mathcal{D}$?

a. If $\mathcal{D}$ is a continuous and fixed (nonrandom) set so that Y can be observed everywhere within $\mathcal{D}$, then the spatial process generates what will be termed a *geostatistical dataset*. It is important to note that it matters not whether Y itself is continuous or discrete or how the locations $\boldsymbol{S}$ within $\mathcal{D}$ are selected. What is important is that we have a spatial dataset that exhibits continuous variation. For instance, one can ostensibly get a reading of the ambient temperature Y at any point $\boldsymbol{S}$ within a fixed area $\mathcal{D}$ and map a continuous temperature surface $Y(\boldsymbol{S})$.

b. If $\mathcal{D}$ is a discrete and fixed set with Y observable at only a countable or enumerable set of locations within $\mathcal{D}$, then the spatial process produces *lattice or regional data.* Such data do not depict spatial locations by a point in space; spatial locations are areal regions or sites (which can be infinite in number). However, the convention is that one typically assigns a representative set of spatial coordinates $\mathbf{S}$ that serves as the locator for the entire site, e.g., attributes can be located by census tract, zip code, or area code, etc. In this regard, we can measure grain production by bushels per plot $Y(\mathbf{S}_i)$, where $\mathbf{S}_i$ depicts the coordinates of the centroid of the ith plot within $\mathcal{D}$. Because this type of measurement can be taken for each plot within $\mathcal{D}$, a key feature of lattice or regional data is complete or exhaustive observation or enumeration of all sites.

c. Suppose the spatial process is actually a random process, with each realization of this process producing a different set of outcome points falling within some area in $\mathcal{D}$. Let us group the resulting outcomes in a random set $\mathcal{D}^* \subset \mathcal{D}$. The number of points in the *random domain* $\mathcal{D}^*$ as well as their configuration is called a *point pattern dataset.* For instance, multiple lightening strikes over a particular area $\mathcal{D}$ occur randomly and collectively constitute a point pattern within some random domain $\mathcal{D}^*$. (Note that $\mathcal{D}^*$ excludes all points within $\mathcal{D}$ where a lightening strike did not occur.)

In what follows, we shall work exclusively with geostatistical or regional (lattice) data.

Spatial Nonstationarity

If we fit the aforementioned regression equation $Y = \beta_0 + \beta_1 X + \varepsilon$ to a set of spatial (geostatistical) data over some spatial domain $\mathcal{D}$, we obtain what may be termed a set of *global parameter estimates,* i.e., the set of regression results is assumed to apply uniformly across the entire domain. For this global fit, there emerges a single set of results that portrays the *average behavior* of the regressions relationship at all points over the whole study area.

A serious drawback of any *global analysis* of the conditional relationship between X and Y is that it provides us with no insight concerning the possibility that there may exist significant *spatial variation* in the said relationship. That is, global analysis renders only *general information* concerning the region under study (such information may not be indicative of the actual relationship prevailing in any portion of the region); it does not allow one to determine whether the regression relationship varies significantly across the study area. In fact, there may be substantial spatial variations in the regression estimates that are masked by the global results.

To overcome this shortcoming of global analysis (which assumes that the regression relationship is *stationary* over the entire study area), let us assume that the regression equation exhibits *spatial nonstationarity* over the spatial domain. That is, the regression fit obtained depends, *ceteris paribus,* on where the estimate is made. To this end, we shall engage in a process of *local analysis* or modeling, i.e., we can look for the presence of any *spatial variation* or differences across the study region by finding individual sets of *local parameter estimates* taken over suitably restricted areas. After all, it may be the case that different areas within the spatial domain exhibit inherently different characteristics, i.e., there can exist spatial variations in personal, cultural, and political factors that admit systematic spatial differences.

Based on these observations, it is evident that what is needed is an estimation technique that allows us to engage in the local modeling of the regression relationship between X and Y over some spatial domain. One such technique is termed *geographically weighted regression* (GWR).

Geographically Weighted Regression [Hastie and Tibshirani 1990; Loader 1999; Fotheringham, Brunsdon, and Charlton 2002]

The Local Regression Equation

Given a regression equation of the form

$$Y_i = \beta_o + \beta_1 X_i + \varepsilon_i,\ i = 1,\ldots,n, \tag{11.1}$$

the (OLS) global parameter estimates $\hat{\beta}_o$ and $\hat{\beta}_1$ are taken to be invariant across the entire spatial domain. However, suppose the regression coefficients β_o and β_1 are expressable as continuous functions of the variables u and v so that both $\beta_o = \beta_o(u, v)$ and $\beta_1 = \beta_1(u, v)$ define surfaces within R^3. For (u_i, v_i) a *regression point* in the spatial domain $\mathcal{D}$, the realizations of these intercept and slope functions can be denoted as $\beta_{oi} = \beta_o(u_i, v_i)$ and $\beta_{1i} = \beta_1(u_i, v_i)$, respectively. Hence, β_o and β_1 are specified as deterministic functions of spatial locations. In view of these parameter respecifications, the global expression in Equation 11.1 is readily transformed into the *local regression equation*

$$Y_i = \beta_o(u_i, v_i) + \beta_1(u_i, v_i)X_i + \varepsilon_i(u_i, v_i),\ i = 1,\ldots,n. \tag{11.2}$$

Clearly Equation 11.2 enables us to obtain localized parameter estimates for *any* point within the spatial domain $\mathcal{D}$ (we are not restricted to just the observed data points themselves). Note that, as we vary the regression point of application (u_i, v_i), we generate a surface map of local parameter estimates that indicates how these parameter estimates vary over the spatial domain. (See Figure 11.1 for the slope estimates map. Note that, if Equation 11.2 is spatially stationary, the implied surface is simply a three-dimensional plane whose height coincides with the OLS global estimate $\hat{\beta}_1$.) It is the behavior of these parameter surfaces that reveal any pattern of spatial variability or spatial nonstationarity in the regression relationship between X and Y. For the most part, the regression points will be the data points themselves because localized residuals can only be determined for points exhibiting an actual Y coordinate. However, the regression point (u_i, v_i) can be any point in the spatial domain. In addition, it is important to mention that X_i and Y_i need to be functions of the spatial coordinates u_i and v_i; they can be prices, income levels, attributes, etc.

The Spatial-Weighting Function

To reflect the local flavor of this analysis, it is assumed that data points located near the regression point (u_i, v_i) (which itself will be taken to be one of the data points) should exert a greater influence in the local estimation of β_o and β_1 than do data points that are located farther away from (u_i, v_i). Hence, a data point should be weighted according to its

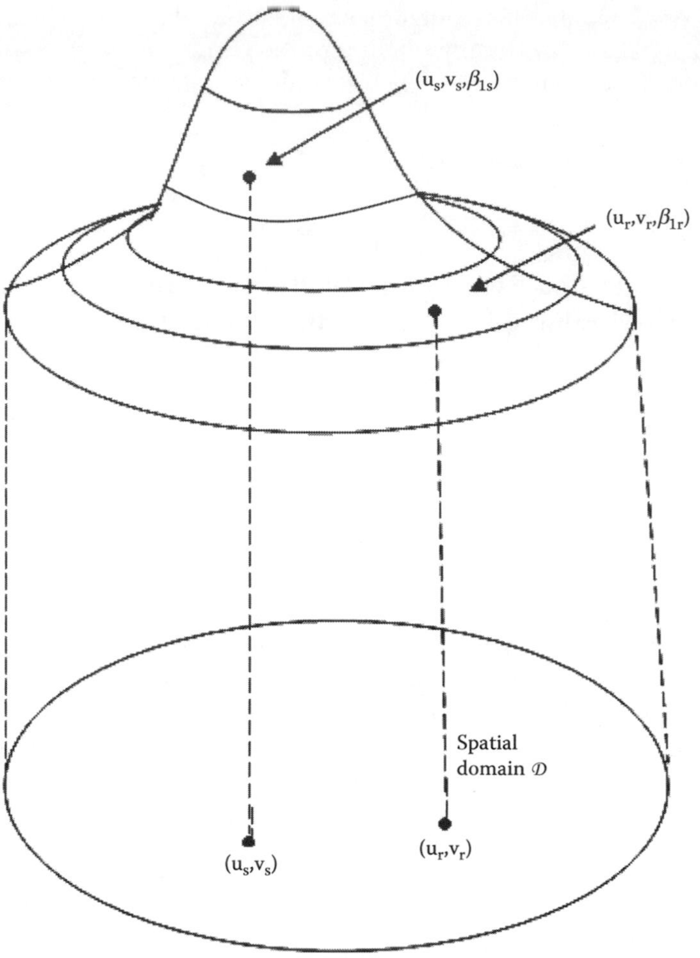

FIGURE 11.1
As we vary the regression point (u_i, v_i), we map out the surface of slope estimates $\beta_1 = \beta_1(u, v)$.

distance from the regression point. Obviously, data points close to the regression point are weighted more heavily than data points farther away. How should these weights be determined?

The weight-assignment scheme offered herein involves fitting a *fixed point kernel* over each regression point so that the observations around this point are weighted via a *spatial-weighting or distance-decay function*. That is, given a regression point (u_i, v_i), the distance between this point and any data point (X_j, Y_j) is given by the *Euclidean distance*

$$d_{ij} = [(u_i - X_j)^2 + (v_i - Y_j)^2]^{1/2}, \; i, j = 1,\ldots,n. \tag{11.3}$$

Because the distance-decay function must be chosen so that the influence of a data point diminishes with increasing distance from the regression point, let us use the *Gaussian spatial-weighting function*

$$w_{ij} = e^{-\frac{1}{2}\left(d_{ij}/b\right)^2}, \tag{11.4}$$

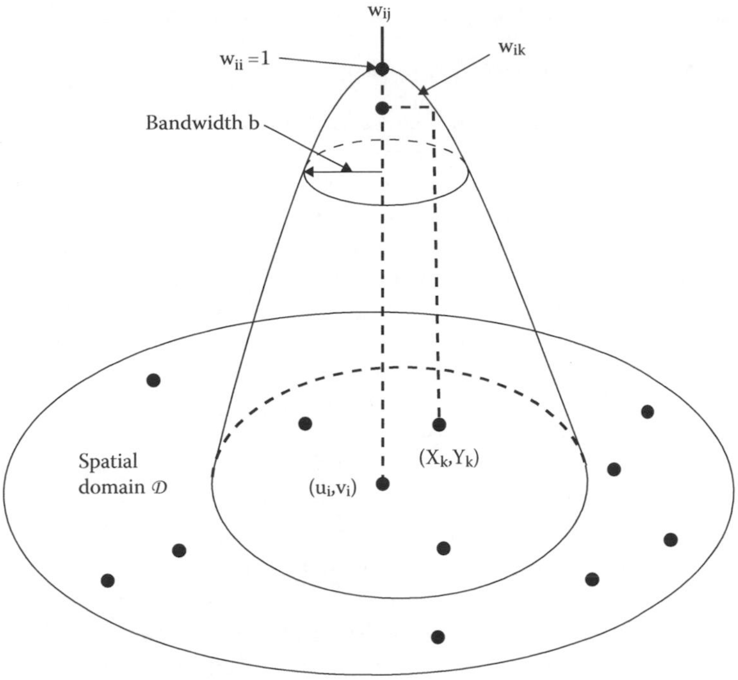

FIGURE 11.2
Fixed point kernel and Gaussian spatial weighting function.

where w_{ij} denotes the weight given to data point j in the tuning or calibration of the model around regression point i, d_{ij} is the distance between regression point i and data point j, b is the chosen *bandwidth*, and $e = \lim_{n \to \infty} \left(1 + \frac{1}{n}\right)^n$ (numerically, $e \approx 2.71828$ (see Figure 11.2).

As Figure 11.2 reveals, the bandwidth b controls the actual distance-decay rate in Equation 11.4; it impacts how sharp or flat the spatial kernel is at a given regression point. That is, a spatial kernel with a small bandwidth has a steeper weighting function and consequently a surface that is much sharper relative to a kernel constructed with a larger bandwidth. If $i = j$ in Equation 11.4, the regression point and the data point coincide and thus $w_{ii} = 1$ (the maximum of Equation 11.4). For $i \neq j$, the weight assigned to all other data points using Equation 11.4 decreases as distance from the regression point increases, i.e., $w_{ij} \to 0$ as $d_{ij} \to \infty$. In addition, as $b \to \infty$, $w_{ij} \to 1$ so that the GWR parameter estimates become equivalent to those obtained by OLS. Conversely, as $b \to 0$, $w_{ij} \to 0$ and the GWR parameter estimates tend to rely more heavily on those data points near the regression point.

It is obvious that the local tuning of the GWR model is achieved by perturbing the regression point across the spatial domain (typically to nearby data points), with each new regression point producing a set of weighted (according to Equation 11.4) least-squares estimates of the regression parameters.[1]

1 Appendix 11.A contains details pertaining to the structure of the WLS estimators for GWR.

Additional GWR Model Considerations

In the GWR model (Equations 11.3 and 11.4), it is implicitly assumed that the weighting scheme in Equation 11.4 is applied uniformly at each regression point over the spatial domain, i.e., the bandwidth b is the same at each regression point. However, it seems appropriate to tailor the selection of the bandwidth size to the density of the data points within the study region. That is, one can choose to use a *spatially adaptive distance-decay function* that exhibits a relatively large (small) bandwidth in regions where the data points tend to be sparse (dense). Examples of these *spatially varying kernels* are provided by Fotheringham, Brundson, and Charlton [2002, 57–59]. The advantage of spatially varying kernels is that they tend to enhance the efficiency of GWR parameter estimates.

Additionally, the issue of tuning or calibrating the distance-decay function by selecting a bandwidth that is optimal in some sense also needs to be considered. A discussion of *cross-validation* and *generalized cross-validation* approaches to addressing this problem are likewise offered in the study by Fotheringham, Brundson, and Charlton (2002, 59–62).

It is also important to mention that, in GWR, a tradeoff is made between coefficient bias and standard error reduction. Biased parameter estimates occur because, under spatial nonstationarity, local estimation at regression point i is undertaken using data points obtained from locations other than point i. So although the spatial tuning of the regression process helps to decrease standard errors, such tuning can actually precipitate biased parameter estimates if data points that are relatively far from regression point i (data points for which, ostensibly, a different set of regression estimates would obtain) are weighted too heavily.

Finally, an important issue concerning parameter estimation using spatial data is the notion of *spatial autocorrelation*. Specifically, let $Y(S)$ depict the realization of some attribute at location $\mathbf{S}$ in the spatial domain $\mathcal{D}$. Then spatial autocorrelation refers to the dependency or association between the attribute at two different locations $\mathbf{S}_1$ and $\mathbf{S}_2$, i.e., between $Y(\mathbf{S}_1)$ and $Y(\mathbf{S}_2)$. In this regard, *positive spatial autocorrelation* is said to exist if high values of Y tend to cluster together in some parts of $\mathcal{D}$, whereas low values of Y tend to cluster together in other parts of $\mathcal{D}$ (an object in a spatial domain tends to be more closely related to objects nearby than with objects farther away), and *negative spatial autocorrelation* exists if high values of Y lie in close proximity to low Y values and vice versa. However, in the absence of spatial autocorrelation, the fact that the two points $\mathbf{S}_1$ and $\mathbf{S}_2$ are found in close proximity has no bearing on the relationship between $Y(\mathbf{S}_1)$ and $Y(\mathbf{S}_2)$.

Let us now consider the notion of spatial dependency in the context of regression analysis. If one is interested in detecting the presence of spatial autocorrelation in a global regression model, then an analysis of the regression residuals is warranted. Furthermore, if, say, positive spatial autocorrelation is detected, it may be the result of fitting a global regression equation to nonstationary spatial data. Instead of having the effect of nonstationarity manifested through the error term in the global model, the problem of spatial nonstationarity can be addressed directly by using GWR.

The systematic estimation of a collection of local models rather than fitting a single global equation tends to mitigate the problem of spatial autocorrelation; spatial nonstationarity can be modeled through spatially varying parameters rather than through the random error term. Hence, it is expected that the degree of spatial autocorrelation can be reduced through the application of GWR techniques. Thus, GWR allows us to trade off, via local estimation, spatial autocorrelation in the error term of a global regression equation with spatial nonstationarity in the regression parameters.

TABLE 11.1

Observations and Cartesian Coordinates for Variables *X* and *Y*

X	Y	CX	CY
10	50	1	2
20	55	2	4
30	70	3	2
40	60	4	4
50	75	4	6
60	90	6	5
70	80	6	6
80	90	7	4
90	100	5	7
100	120	5	2
110	100	7	7
120	110	8	5
130	130	3	7
140	135	5	5
150	137	6	8

Example 11.1

Given the following set of $n = 15$ data points involving values of the variables *X* and *Y* along with their Cartesian coordinates *CX* and *CY*, respectively (Table 11.1), let us plot the locations of these *X* and *Y* combinations using their *CX*, *CY* coordinate values (Figure 11.3).

The accompanying *X* and *Y* values are listed as the points (*X*,*Y*) appearing directly beneath the (*CX*,*CY*) Cartesian points. Each of the (*CX*,*CY*) points illustrated in Figure 11.3 will serve as a

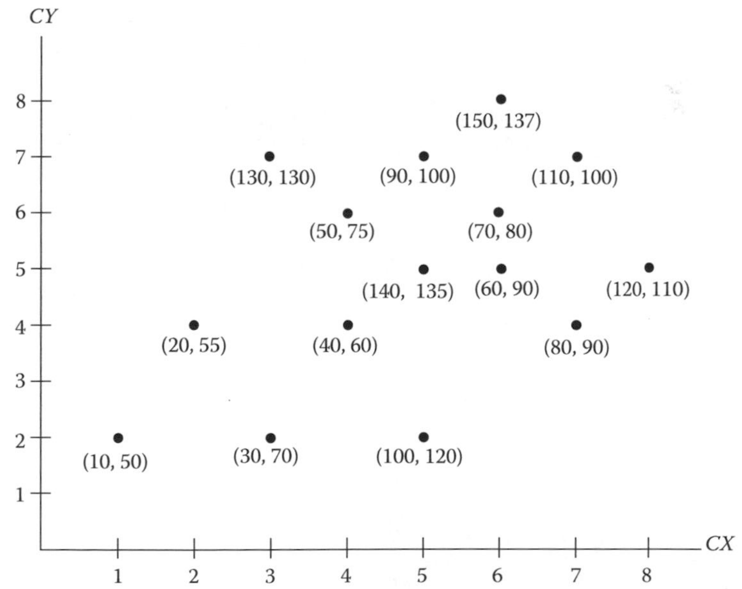

FIGURE 11.3
(*X*, *Y*) combinations plotted at the Cartesian coordinate locations (*CX*, *CY*).

regression point. In addition, the bandwidth is chosen to be fixed at $b = 1.2$. Exhibit 11.1 contains the requisite SAS code.[2]

Exhibit 11.1 SAS Code for Spatial Regression

```
data geog;
  input x y cx cy;
datalines;
 10   50  1  2
 20   55  2  4
 30   70  3  2
 40   60  4  4
 50   75  4  6
 60   90  6  5
 70   80  6  6
 80   90  7  4
 90  100  5  7
100  120  5  2
110  100  7  7
120  110  8  5
130  130  3  7
140  135  5  5
150  137  6  8
;
proc print data=geog;
run;
proc reg data=geog;
  model y=x;①
run;
%macro calcdist;
%Do j=1 %To 15;②
data geog; set geog;
idnum=symget('j') ;
if _n_=idnum then do;
call symput('xstart', cx);
call symput('ystart', cy);
end;
run;
data geog;
set geog;
d=sqrt((cx-&xstart)**2+
(cy-&ystart)**2);③
w=exp(-0.5*(d/1.2)**2);④
proc print data=geog;
run;
proc reg outest=est&j; weight w;⑤
 model y=x;
proc print data=est&j;⑥
proc append base=final data=est&j;⑦
```

① We request an OLS regression run involving the entire sample dataset against which to compare the GWR results.

② We define a macro variable *j* that determines the number of weighted regressions to run. Here *j* goes from 1 to 15(= *n*). The value of *n* can be changed appropriately for different datasets.

③ We define the set of Euclidean distances for each weighted regression run.

④ We define the Gaussian spatial-weighting function with bandwidth *b* = 1.2. Other bandwidths may be specified interactively.

⑤ We invoke the weighted regression procedure, with the weights determined in ④.

⑥ We request a print of the set of regression results determined at each regression point.

⑦, ⑧ We construct (and print) a table containing all *n*(= 15) estimated intercepts and slopes.

⑨, ⑩ We merge (and print) our initial dataset *geog* with the set of regression estimates *final*. The resulting dataset is named *merg*.

⑪ Using dataset *merg*, we request a three-dimensional surface plot of the regression slope at each coordinate (*CX*, *CY*).

[2] Many thanks to John Stewart and Mahmoud Wahab for their help with the SAS macro.

```
run;
%end;
%mend calcdist;
%calcdist;
proc print data=final;⑧
run;
data merg; merge geog final;⑨
proc print data=merg;⑩
run;
proc g3d data=merg;⑪
title 'surface plot of regression
slope';
scatter cx*cy=x/xticnum=5 yticnum=5
grid zmin=0.4 zmaxnum=0.9;
label cx='cx'
cy='cy'
x='beta1'
;
run;
```

Table 11.2 lists the OLS regression results when the fit is taken across the entire spatial domain. A glance at this table indicates that, on all accounts, we have a highly significant linear relationship between *X* and *Y*. Here the *global* regression equation is of the form $\hat{Y} = \hat{\beta}_o + \hat{\beta}_1 X = 43.6381 + 0.6228X$.

TABLE 11.2

Output for Example 11.1 (OLS Regression Results for Entire Domain)

The SAS System
The REG Procedure
Model : MODEL 1
Dependent Variable: y
Analysis of Variance

Source	**DF**	**Sum of Squares**	**Mean Square**	***F* Value**	**Pr > *F***
Model	1	10863	10863	190.55	<0.0001
Error	13	741.10476	57.00806		
Corrected total	14	11604			

Root MSE	7.55037	R-Square	0.9361
Dependent mean	93.46667	Adj R-Sq	0.9312
Coeff Var	8.07814		

Parameter Estimates

Variable	**DF**	**Parameter Estimates**	**Standard Error**	***t* Value**	**Pr > |*t*|**
Intercept	1	43.63810	4.10255	10.64	<0.0001
X	1	0.62286	0.04512	13.80	<0.0001

Next, Table 11.3 presents the GWR results when (*CX*, *CY*) = (1, 2) serves as the (first) regression point. (Although similar results are sequentially generated at each of the 14 remaining regression points, only the output pertaining to the first regression point is listed in detail.)

As this table reveals:

TABLE 11.3

Output for Example 11.1 (GWR Results at (CX, CY) = (1,2))

The SAS System

Obs	*X*	*y*	*CX*	*CY*	idnum	① *d*	② *w*
1	10	50	1	2	1	0.00000	1.00000
2	20	55	2	4	1	2.23607	0.17620
3	30	70	3	2	1	2.00000	0.24935
4	40	60	4	4	1	3.60555	0.01096
5	50	75	4	6	1	5.00000	0.00017
6	60	90	6	5	1	5.83095	0.00001
7	70	80	6	6	1	6.40312	0.00000
8	80	90	7	4	1	6.32456	0.00000
9	90	100	5	7	1	6.40312	0.00000
10	100	120	5	2	1	4.00000	0.00387
11	110	100	7	7	1	7.81025	0.00000
12	120	110	8	5	1	7.61577	0.00000
13	130	130	3	7	1	5.38516	0.00004
14	140	135	5	5	1	5.00000	0.00017
15	150	137	6	8	1	7.81025	0.00000

The SAS System

③

The REG Procedure

Model: MODEL 1

Dependent Variable: *y*

Weight: *w*

Analysis of Variance

Source	DF	Sum of Squares	Mean Square	*F* Value	Pr > *F*
Model	1	91.18141	91.18141	161.69	<0.0001
Error	13	7.33117	0.56394		
Corrected total	14	98.51257			

Root MSE	0.75096	R-Square	0.9256
Dependent mean	54.35232	Adj R-Sq	0.9199
Coeff Var	1.38165		

TABLE 11.3 *(continued)*

Parameter Estimates

Variable	DF	Parameter Estimates	Standard Error	t Value	Pr > |t|
Intercept	1	41.32069	1.20072	34.41	<0.0001
x	1	0.85859	0.06752	12.72	<0.0001

① Column d lists all of the Euclidean distances from the regression point (1,2).

② Column w displays the weights w_{ij} used in the WLS regression at regression point (1,2). (Note that w_{ii} = 1 with data points near the regression point receiving a higher weight than data points located farther away from the regression point.)

③ The WLS estimates of β_o and β_1 at the regression point (1,2) are

$$\hat{\beta}_o(1,2) = 41.3207 \text{ and } \hat{\beta}_1(1,2) = 0.8586.$$

Once each Cartesian point (CX, CY) has served as a regression point, the complete set of GWR intercept and slope estimates $\hat{\beta}_o(CX, CY)$ and $\hat{\beta}_1(CX, CY)$, respectively, for all 15(= n) regression points is summarized in Table 11.4.

Dataset *merg* (Table 11.5) is then used to obtain a surface plot of the regression slope $\hat{\beta}_1(CX, CY)$ at each Cartesian point (CX,CY) over the spatial domain (Figure 11.4).

TABLE 11.4

Output for Example 11.1 (Complete Set of GWR Estimates)

The SAS System

Obs	_MODEL_	_TYPE_	_DEPVAR_	_RMSE_	$(\hat{\beta}_0(CX,CY))$ Intercept	$(\hat{\beta}_1(CX,CY))$ x	y
1	MODEL1	PARMS	y	0.75096	41.3207	0.85859	−1
2	MODEL1	PARMS	y	1.56194	41.9674	0.65367	−1
3	MODEL1	PARMS	y	2.01650	44.3035	0.74211	−1
4	MODEL1	PARMS	y	2.88582	37.1571	0.71261	−1
5	MODEL1	PARMS	y	2.03879	39.1385	0.68427	−1
6	MODEL1	PARMS	y	3.91175	44.2273	0.61585	−1
7	MODEL1	PARMS	y	3.91384	43.4506	0.60596	−1
8	MODEL1	PARMS	y	2.79815	50.3704	0.52931	−1
9	MODEL1	PARMS	y	2.60202	40.7211	0.64519	−1
10	MODEL1	PARMS	y	2.10159	40.4082	0.77713	−1
11	MODEL1	PARMS	y	2.97048	38.6498	0.61657	−1
12	MODEL1	PARMS	y	1.83282	54.3429	0.43361	−1
13	MODEL1	PARMS	y	0.80047	39.5527	0.69089	−1
14	MODEL1	PARMS	y	3.34073	39.6543	0.67660	−1
15	MODEL1	PARMS	y	2.27823	36.0191	0.66289	−1

TABLE 11.5

Output for Example 11.1 (Initial Data and GWR Estimates)

The SAS System

Obs	$(\hat{\beta}_1(CX,CY))$ X	y	CX	CY	idnum	d	w	_MODEL_	_TYPE_	_DEPVAR_	_RMSE_	Intercept
1	0.85859	−1	1	2	15	7.81025	0.00000	MODEL1	PARMS	y	0.75096	41.3207
2	0.65367	−1	2	4	15	5.65685	0.00001	MODEL1	PARMS	y	1.56194	41.9674
3	0.74211	−1	3	2	15	6.70820	0.00000	MODEL1	PARMS	y	2.01650	44.3035
4	0.71261	−1	4	4	15	4.47214	0.00096	MODEL1	PARMS	y	2.88582	37.1571
5	0.68427	−1	4	6	15	2.82843	0.06218	MODEL1	PARMS	y	2.03879	39.1385
6	0.61585	−1	6	5	15	3.00000	0.04394	MODEL1	PARMS	y	3.91175	44.2273
7	0.60596	−1	6	6	15	2.00000	0.24935	MODEL1	PARMS	y	3.91384	43.4506
8	0.52931	−1	7	4	15	4.12311	0.00273	MODEL1	PARMS	y	2.79815	50.3704
9	0.64519	−1	5	7	15	1.41421	0.49935	MODEL1	PARMS	y	2.60202	40.7211
10	0.77713	−1	5	2	15	6.08276	0.00000	MODEL1	PARMS	y	2.10159	40.4082
11	0.61657	−1	7	7	15	1.41421	0.49935	MODEL1	PARMS	y	2.97048	38.6498
12	0.46361	−1	8	5	15	3.60555	0.01096	MODEL1	PARMS	y	1.83282	54.3429
13	0.69089	−1	3	7	15	3.16228	0.03105	MODEL1	PARMS	y	0.80047	39.5527
14	0.67660	−1	5	5	15	3.16228	0.03105	MODEL1	PARMS	y	3.34073	39.6543
15	0.66289	−1	6	8	15	0.00000	1.00000	MODEL1	PARMS	y	2.27823	36.0191

Appendix A: WLS Estimators for GWR

Equations 3.5 and 3.6 provided WLS estimators for β_o and β_1 using fixed weights v_i, $i = 1,\ldots,n$, applied to the entire sample. Now, each regression point (u_i, v_i), $i = 1,\ldots,n$, generates its own specific set of weights w_{ij} using Equation 11.4, where w_{ij} is the weight given to data point j, $j = 1,\ldots,n$, when estimates of β_o and β_1 are made at regression point i. So at regression point i, WLS estimators for β_o and β_1 are

$$\hat{\beta}_1(i) = \frac{\sum_j w_{ij}\left[X_j - \bar{X}(i)\right]\left[Y_j - \bar{Y}(i)\right]}{\sum_j w_{ij}\left[X_j - \bar{X}(i)\right]^2}, \tag{11.A.1}$$

$$\hat{\beta}_o(i) = \bar{Y}(i) - \hat{\beta}_1(i)\bar{X}(i), \tag{11.A.2}$$

respectively, where

$$\bar{X}(i) = \sum_j w_{ij}X_j \Big/ \sum_j w_{ij}, \tag{11.A.3}$$

$$\bar{Y}(i) = \sum_j w_{ij}Y_j \Big/ \sum_j w_{ij}. \tag{11.A.4}$$

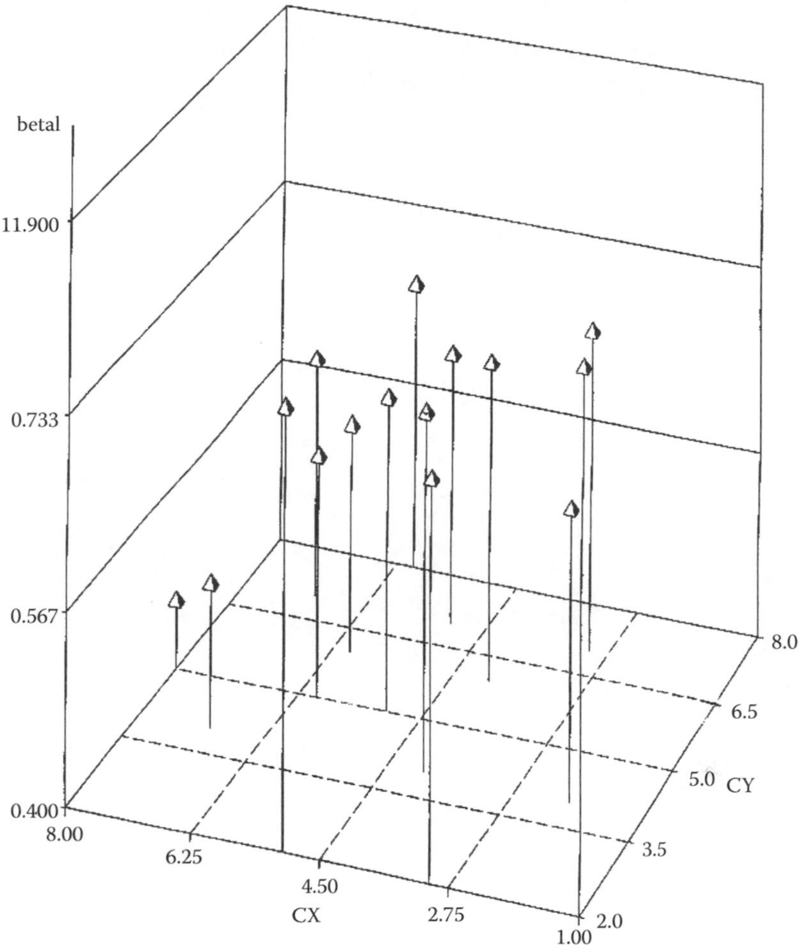

FIGURE 11.4
Surface plot of regression slope.

Here $\bar{X}(i)$ is the weighted mean of the X_j's at regression point i while $\bar{Y}(i)$ is the weighted mean of the Y_j's at regression point i. So as the regression point changes, the w_{ij}'s likewise change, thus precipitating changes in Equations 11.A.1 through 11.A.4.

Exercises

11-1. The objective is to study the relationship between housing prices (for simplicity, we ignore their structural characteristics) and location. Table E.11-1 provides data for the prices of residential units sold within the last six months and the location of the houses. Specifically, we have observations on the average prices of houses (in thousands of dollars) sold in the 25 towns of a particular county (Y), an index of population density (X), and normalized values of the latitude and longitude of the geographic centroids of the town boundaries ((CX, CY)). Using a fixed band width of $b = 1.3$, find the following:

TABLE E.11-1
Observations on Housing Prices (Y), Population Density (X), and Location ((CX,CY))

X	$Y(000)$	CX	CY
50	2250	80	100
49	1996	83	85
52	1052	90	99
50	1014	75	81
53	986	62	90
59	919	68	70
61	857	57	70
63	826	67	57
60	804	49	59
59	750	12	18
61	699	24	29
63	648	20	27
67	613	36	25
66	514	22	60
72	500	24	71
72	498	30	52
75	482	31	65
80	471	35	79
85	466	36	60
88	327	38	42
91	325	53	25
91	319	65	33
93	310	45	45
95	307	47	35
97	296	54	37

a. The global regression equation
b. The set of GWR results
c. A surface plot of the regression slope over the spatial domain

11-2. Appearing in Table E.11-2 is a set of observations involving the number of burglaries (Y), the population turnover rate as measured by the percentage of residents having moved within the last calendar year (X), and the normalized values of the latitude and longitude of the geographic centroids of the boundaries ((CX, CY)) of 20 towns in a given metropolitan area. Find the following:

TABLE E.11-2

Burglaries (Y), Population Turnover Rate (X), and Location ((CX,CY))

X (%)	*Y*	*CX*	*CY*
10.5	526	87	70
9.8	575	79	81
7.2	615	70	90
6.9	524	79	60
6.0	416	71	62
5.9	410	65	72
4.7	315	84	50
3.8	201	74	42
3.9	219	65	53
3.2	190	55	60
4.1	114	55	50
3.8	101	74	30
3.1	91	68	22
2.9	73	65	32
3.1	62	54	36
2.1	60	45	45
1.9	51	60	15
1.8	40	53	26
1.1	32	47	52
1.0	27	36	42

a. The global regression equation

b. The set of GWR results

c. A surface plot of the regression slope over the spatial domain

12

Multiple Regression

The Multiple Regression Model

The bivariate regression model introduced in Chapter 2 was of the form

$$Y = \beta_o + \beta_1 X + \varepsilon, \tag{12.1}$$

where $\beta_o + \beta_1 X$ was termed the *systematic* portion (systematic in the sense that it behaves in a predictable fashion) and ε was termed the *random error* component of the model. In addition, it was indicated that the role of ε was to account for the *net effect* of all excluded factors or variables. The time has now come to *pick apart* ε, i.e., to disentangle or unjumble the random errors by removing additional relevant explanatory variables form ε and explicitly incorporating them into Equation 12.1. In this way, the systematic component of the model becomes more inclusive and important (it can explain more of the variation in Y) and the random error portion becomes less important or influential (the unexplained variation in Y is diminished).

To this end, let us structure the *multivariate regression hyperplane* as

$$Y = \beta_o + \beta_1 X_1 + \beta_2 X_2 + \cdots + \beta_m X_m + \varepsilon, \tag{12.2}$$

where we now have m predetermined or nonrandom explanatory variables or *regressors* $X_1,\ldots,X_m$ with ε also denoting the (now less influential) random error term. Here Equation 12.2 depicts a *linear regression model* because it is linear in the parameters $\beta_o, \beta_1,\ldots, \beta_m$. In what follows, we shall assume that the assumptions underlying the strong classical linear regression model made in Chapter 2 hold here also, with three additional assumptions now offered:

A.1. Each of the X_j's, $j = 1,\ldots,m$, is nonstochastic with finite variance.

A.2. $n > p = m + 1$ so that we have a sufficient number of degrees of freedom for estimation and testing.

A.3. No exact linear relationship exists between the X_j's, $j = 1,\ldots,m$, i.e., none of the explanatory variables is perfectly correlated with any other explanatory variable or with any linear combination of the other explanatory variables.

Assumption A.3 posits the absence of *perfect* or *exact multicollinearity*. (If perfect multicollinearity exists, no solution to the multiple regression problem exists.) However, as will be indicated below, near-perfect multicollinearity can be quite troublesome in that its presence generally makes it difficult to determine the separate effects of the explanatory variables on the explained or *response variable* Y.

For fixed X_j's, $j = 1,\dots,m$, the *population regression hyperplane* is defined as the conditional mean of Y given the X_j's or

$$E(Y|X_1,\dots,X_m) = \beta_o + \beta_1 X_1 + \beta_2 X_2 + \cdots + \beta_m X_m \tag{12.3}$$

because $E(\varepsilon) = 0$. This population hyperplane with parameters $\beta_o, \beta_1,\dots, \beta_m$ is unobserved and must be estimated from sample data. Given Equation 12.3,

$$\beta_o = E(Y|X_1 = 0,\dots,X_m = 0), \tag{12.4}$$

i.e., the population regression intercept is the average value of Y given that all of the explanatory variables are set at zero. Next,

$$\beta_j = \partial E(Y|X_1,\dots,X_m)/\partial X_j, \quad j = 1,\dots,m, \tag{12.5}$$

is termed the jth *partial regression coefficient*, i.e., in the absence of multicollinearity, as X_j increases by one unit, the average value of Y changes by β_j units, given that all remaining explanatory variables are held constant (we control for the effect of all other factors or variables). Once the values of β_k, $k = 0,1,\dots,p = m + 1$, are estimated from sample information, we obtain the *sample regression hyperplane*

$$\hat{Y} = \hat{\beta}_o + \hat{\beta}_1 X_1 + \cdots + \hat{\beta}_m X_m \tag{12.6}$$

that serves as our proxy for the population regression hyperplane, where $\hat{Y}$ is the fitted or estimated value of Y and the $\hat{\beta}_k$'s, $k = 0,1,\dots,p$, represent the (arbitrary) estimates of the population parameters. Then $e_i = Y_i - \hat{Y}_i$, $i = 1,\dots,n$, denotes the ith *residual* or *deviation* from the sample regression hyperplane.

Estimating the Parameters of the Multiple Regression Model

Suppose $n(> p)$ observations are available on some response or explained variable Y and on m regressors or explanatory variables X_j, $j = 1,\dots,m$. If Y_i is the ith response level of Y and X_{ij} is the ith level of regressor X_j, $i = 1,\dots,n$, then our sample information consisting of n observations can be displayed in Table 12.1.

Then from Equation 12.2,

$$Y_i = \beta_o = \beta_1 X_{i1} + \beta_2 X_{i2} + \cdots + \beta_m X_{im} + \varepsilon_i, \quad i = 1,\dots,n. \tag{12.2.1}$$

One approach to estimating the parameters of Equation 12.2.1 is to apply the *principle of least squares*. That is, let us choose $\hat{\beta}_o, \hat{\beta}_1,\dots, \hat{\beta}_m$ (here $\hat{\beta}_o, \hat{\beta}_1,\dots, \hat{\beta}_m$ constitute an arbitrary set of estimators for $\beta_o, \beta_1,\dots, \beta_m$) so as to minimize the sum of the squared deviations from the sample regression hyperplane. To this end, we seek to

TABLE 12.1

Sample Observations for Multiple Linear Regression

Observation No.	Response Level	(m) Regressors		
i	Y	X_1	X_2 ...	X_m
1	Y_1	X_{11}	X_{12} ...	X_{1m}
2	Y_2	X_{21}	X_{22} ...	X_{2m}
.	.	.	.	.
.	.	.	.	.
.	.	.	.	.
n	Y_n	X_{n1}	X_{n2} ...	X_{nm}

$$\min\left\{\sum_{i=1}^{n} e_i^2 = \sum_{i=1}^{n}(Y_i - \hat{Y}_i)^2 = \sum_{i=1}^{n}\left(Y_i - \hat{\beta}_o - \sum_{j=1}^{m}\hat{\beta}_j X_{ij}\right)^2 = F\left(\hat{\beta}_o,\ \hat{\beta}_1, \ldots,\ \hat{\beta}_m\right)\right\}. \tag{12.7}$$

(For convenience, the operators $\sum_{i=1}^{n}$ and $\sum_{j=1}^{m}$ will often be simplified to Σ, where it is to be understood that we sum over all values of the i or j index.)

If we set

$$\frac{\partial F}{\partial \hat{\beta}_o} = \frac{\partial F}{\partial \hat{\beta}_1} = \cdot\ \cdot\ \cdot = \frac{\partial F}{\partial \hat{\beta}_m} = 0$$

(assume that the second-order conditions for a minimum hold), then the resulting simultaneous linear equation system can be expressed as

$$\begin{aligned} &\sum e_i = 0 \\ &\sum X_{i1} e_i = 0 \qquad \left[\text{system of least squares normal equations}\right] \\ &\cdot\ \cdot\ \cdot\ \cdot\ \cdot\ \cdot\ \cdot \\ &\sum X_{im} e_i = 0 \end{aligned} \tag{12.8}$$

or

$$\begin{aligned} &n\hat{\beta}_o + \hat{\beta}_1 \sum X_{i1} + \hat{\beta}_2 \sum X_{i2} + \cdots + \hat{\beta}_m \sum X_{im} = \sum Y_i \\ &\hat{\beta}_o \sum X_{i1} + \hat{\beta}_1 \sum X_{i1}^{\ 2} + \hat{\beta}_2 \sum X_{i1} X_{i2} + \cdots + \hat{\beta}_m \sum X_{i1} X_{im} = \sum X_{i1} Y_i \\ &\left[\text{system of least squares normal equations}\right] \\ &\cdot\ \cdot\ \cdot\ \cdot\ \cdot\ \cdot\ \cdot\ \cdot\ \cdot\ \cdot\ \cdot\ \cdot\ \cdot\ \cdot\ \cdot\ \cdot \\ &\hat{\beta}_o \sum X_{im} + \hat{\beta}_1 \sum X_{im} X_{i1} + \hat{\beta}_2 \sum X_{im} X_{i2} + \cdots + \hat{\beta}_m \sum X_{im}^{\ 2} = \sum X_{im} Y_i \end{aligned} \tag{12.8.1}$$

Here Equation 12.8.1 represents a simultaneous linear equation system involving $p(= m + 1)$ equations in p unknowns. Solving this system simultaneously yields the set of OLS parameter estimators $\hat{\beta}_o$, $\hat{\beta}_1, \ldots, \hat{\beta}_m$. According to the Gauss-Markov Theorem (for a proof of this theorem, see Appendix 12.A), these least-squares estimators are BLUE of the β_k's, $k = 0, 1, \ldots, m$.

The system in Equation 12.8.1 may be rewritten in terms of deviations from the means of the variables by expressing the first equation as

$$\hat{\beta}_o = \bar{Y} - \hat{\beta}_1 \bar{X}_1 - \hat{\beta}_2 \bar{X}_2 - \ldots - \hat{\beta}_m \bar{X}_m, \tag{12.9}$$

where $\bar{Y} = \sum Y_i / n$, $\bar{X}_j = \sum X_{ij} / n$, $j = 1, \ldots, m$. (Note that Equation 12.9 implies that the least-squares hyperplane passes through the point of means of the regression variables.)

Substituting Equation 12.9 into the remaining m least-squares normal equations and simplifying yields

$$\begin{aligned}
\hat{\beta}_1 M_{11} + \hat{\beta}_2 M_{12} + \cdots + \hat{\beta}_m M_{1m} &= M_{1y} \\
\hat{\beta}_1 M_{21} + \hat{\beta}_2 M_{22} + \cdots + \hat{\beta}_m M_{2m} &= M_{2y} \\
\cdots\cdots\cdots\cdots\cdots\cdots\cdots \\
\hat{\beta}_1 M_{m1} + \hat{\beta}_2 M_{m2} + \cdots + \hat{\beta}_m M_{mm} &= M_{my}
\end{aligned} \tag{12.10}$$

where $M_{jy} = \sum (X_{ij} - \bar{X}_j)(Y_i - \bar{Y}) = \sum X_{ij} Y_i - n\bar{X}_j \bar{Y}$, $j = 1, \ldots, m$; and $M_{jl} = \sum (X_{ij} - \bar{X}_j)(X_{il} - \bar{X}_l) = \sum X_{ij} X_{il} - n\bar{X}_j \bar{X}_l$, with both $j, l = 1, \ldots, m$. This simultaneous equation system can be solved for the $\hat{\beta}_j$'s, $j = 1, \ldots, m$; $\hat{\beta}_o$ is then obtained from Equation 12.9. (A parallel development of these results in matrix terms is offered in Appendix 12.B.)

An alternative to OLS estimation is maximum-likelihood estimation. In this regard, if the random errors ε_i are normally distributed, then the OLS estimators are also maximum-likelihood estimators of $\beta_o, \beta_1, \ldots, \beta_m$, with the maximum-likelihood estimators serving as minimum variance unbiased estimators of these regression parameters. So under the assumptions of the strong classical linear regression model, the ε_i are independent and normally distributed with the joint probability density function

$$\prod_{i=1}^{n} f(\varepsilon_i; 0, \sigma_\varepsilon^2) = (2\pi\sigma_\varepsilon^2)^{-\frac{n}{2}} e^{-\sum \varepsilon_i^2 / 2\sigma_\varepsilon^2}.$$

Because

$$\varepsilon_i = Y_i - \tilde{\beta}_o - \sum_{j=1}^{m} \tilde{\beta}_j X_{ij}$$

($\tilde{\beta}_o, \tilde{\beta}_1, \ldots, \tilde{\beta}_m$ are arbitrary estimators of $\beta_o, \beta_1, \ldots, \beta_m$) transforms linearly to Y_i, the likelihood function for the sample is

$$\mathcal{L}(\tilde{\beta}_o, \tilde{\beta}_1, \ldots, \tilde{\beta}_m, \sigma_\varepsilon^2; Y_1, \ldots, Y_n, n) = (2\pi\sigma_\varepsilon^2)^{-\frac{n}{2}} e^{-\frac{1}{2\sigma_\varepsilon^2} \sum_i \left(Y_i - \tilde{\beta}_o - \sum_j \tilde{\beta}_j X_{ij} \right)^2} \tag{12.11}$$

or, in log-likelihood form,

$$\log \mathcal{L} = -\frac{n}{2} \log(2\pi) - \frac{n}{2} \log \sigma_\varepsilon^2 - \frac{1}{2\sigma_\varepsilon^2} \sum_i \left(Y_i - \tilde{\beta}_o - \sum_j \tilde{\beta}_j X_{ij} \right)^2. \tag{12.11.1}$$

Now, for σ_ε^2 fixed, maximizing $\log \mathcal{L}$ with respect to $\tilde{\beta}_o, \tilde{\beta}_1, \ldots, \tilde{\beta}_m$ is equivalent to minimizing the expression $\sum_i \left(Y_i - \tilde{\beta}_o - \sum_j \tilde{\beta}_j X_{ij} \right)^2$ with respect to the same so that setting

$$\frac{\partial \log \mathcal{L}}{\partial \tilde{\beta}_o} = \frac{\partial \log \mathcal{L}}{\partial \tilde{\beta}_1} = \cdots = \frac{\partial \log \mathcal{L}}{\partial \tilde{\beta}_m} = 0$$

yields Equation 12.8.1 with $\hat{\beta}_k$ replaced by $\tilde{\beta}_k$, $k = 0,1,\ldots,m$, where the $\tilde{\beta}_k$'s represent maximum-likelihood estimators of the regression parameters $\beta_o, \beta_1,\ldots, \beta_m$. Clearly the maximum-likelihood estimators of these parameters are the same as the OLS estimators of $\beta_o, \beta_1,\ldots, \beta_m$.

Moreover, setting $\partial \log \mathcal{L}/\partial \sigma_\varepsilon^2 = 0$ yields the maximum-likelihood estimator of σ_ε^2 or

$$\tilde{\sigma}_\varepsilon^2 = \frac{1}{n}\sum e_i^2. \tag{12.12}$$

Thus, the maximum-likelihood estimator of σ_ε^2 is the sample variance of the OLS residuals e_i, $i = 1,\ldots,n$. As will be explained below (Appendix 12.D), $\tilde{\sigma}_\varepsilon^2$ is a biased estimator of σ_ε^2. (Appendix 12.C presents the derivation of maximum-likelihood estimators for $\beta_o, \beta_1,\ldots, \beta_m$ in matrix terms.)

Mean, Variance, and Sampling Distribution of the OLS Estimator $\hat{\boldsymbol{\beta}}$

Given the vector $\hat{\boldsymbol{\beta}}(= (\boldsymbol{X'X})^{-1}\boldsymbol{X'Y})$ of OLS estimators of $\boldsymbol{\beta}$, it is readily verified (Appendix 12.D) that $E(\hat{\boldsymbol{\beta}}) = \boldsymbol{\beta}$ and $V(\hat{\boldsymbol{\beta}}) = \sigma_\varepsilon^2(\boldsymbol{X'X})^{-1} = \sigma_\varepsilon^2\boldsymbol{C}$, where

$$\underset{(m+1\times m+1)}{\boldsymbol{C}} = (\boldsymbol{X'X})^{-1} = \begin{bmatrix} C_{00} & C_{01} & \cdots & C_{0m} \\ C_{10} & C_{11} & \cdots & C_{1m} \\ \cdot & \cdot & & \cdot \\ \cdot & \cdot & & \cdot \\ \cdot & \cdot & & \cdot \\ C_{m0} & C_{m1} & \cdots & C_{mm} \end{bmatrix}. \tag{12.13}$$

Hence,

$$V(\hat{\beta}_k) = \sigma_\varepsilon^2 C_{kk},\ k = 0,\ 1,\ldots,\ m;$$
$$COV(\hat{\beta}_l, \hat{\beta}_k) = \sigma_\varepsilon^2 C_{lk},\ \text{with } l,k = 0,\ 1,\ldots,\ m;$$

and σ_ε^2 is the variance of the unknown random error terms ε_i, $i = 1,\ldots,n$. In general, $V(\boldsymbol{\varepsilon}) = \sigma_\varepsilon^2\boldsymbol{I}_n$, where $\boldsymbol{I}_n$ denotes an $(n \times n)$ identity matrix. Let us use $S_\varepsilon^2 = \sum_{i=1}^n e_i^2/(n-m-1)$ as an estimator for σ_ε^2. Then $V(\hat{\beta}_k)$ is estimated by $S_{\hat{\beta}_k}^2 = S_\varepsilon^2 C_{kk}$ and the standard error of $\hat{\beta}_k$ is

$$S_{\hat{\beta}_k} = S_\varepsilon\sqrt{C_{kk}}. \tag{12.14}$$

Because $\hat{\boldsymbol{\beta}}$ is a linear function of the set of independent normal random variables $\boldsymbol{Y}$, it follows (Appendix 12.E.e) that $\hat{\boldsymbol{\beta}} : N(\boldsymbol{\beta}, \sigma_\varepsilon^2(\boldsymbol{X'X})^{-1})$ or $\hat{\beta}_k : N(\beta_k, \sigma_\varepsilon^2 C_{kk})$. Hence,

$$\frac{\hat{\beta}_k - \beta_k}{\sigma_\varepsilon\sqrt{C_{kk}}} : N(0,1).$$

If we replace σ_ε by S_ε, then

$$\frac{\hat{\beta}_k - \beta_k}{S_\varepsilon \sqrt{C_{kk}}} : t_{n-m-1}. \tag{12.15}$$

Precision of the OLS Estimators $\hat{\beta}_o, \hat{\beta}_1, \ldots, \hat{\beta}_m$: Confidence Intervals

To determine just how precisely the population parameters β_k, $k = 0, 1, \ldots, m$, have been estimated from the sample dataset, let us look to the following collection of individual $100(1-\alpha)\%$ confidence intervals for these regression parameters. Specifically, for confidence coefficient $1-\alpha$, let $-t_{\frac{\alpha}{2}, n-m-1}$ and $t_{\frac{\alpha}{2}, n-m-1}$ depict lower and upper percentage points, respectively, for the t distribution. Then a $100(1-\alpha)\%$ confidence interval for β_k is

$$\hat{\beta}_k \pm t_{\frac{\alpha}{2}, n-m-1} S_{\hat{\beta}_k}, \tag{12.16}$$

where $S_{\hat{\beta}_k}$ is determined from Equation 12.14.

Three equivalent ways of interpreting Equation 12.16 can be offered:

1. We may be $100(1-\alpha)\%$ confident that the population regression coefficient β_k lies between $\hat{\beta}_k - t_{\frac{\alpha}{2}, n-m-1} S_{\hat{\beta}_k}$ and $\hat{\beta}_k + t_{\frac{\alpha}{2}, n-m-1} S_{\hat{\beta}_k}, k = 0, 1, \ldots, m$.
2. Error bound interpretation: We may be $100(1-\alpha)\%$ confident that $\hat{\beta}_k$ will not differ from β_k by more than $\pm\, t_{\frac{\alpha}{2}, n-m-1} S_{\hat{\beta}_k}$, $k = 0, 1, \ldots, m$.
3. Precision interpretation: How precisely have we estimated β_k? We are within $\pm\, t_{\frac{\alpha}{2}, n-m-1} S_{\hat{\beta}_k}$ units of β_k with $100(1-\alpha)\%$ reliability.

(Remember that reliability is a long-run concept that emerges under repeated sampling from the same population. That is, if we took many samples of size n from the population and if an interval such as Equation 12.16 was computed for each of them, then, in the long run, $100(1-\alpha)\%$ of all the intervals would contain β_k and $100(\alpha)\%$ of them would not. Hence, reliability depicts the proportion of time the confidence interval will contain β_k.)

The $100(1-\alpha)\%$ confidence intervals (Equation 12.16) are individual or *statement confidence intervals* for the β_k's, i.e., they are constructed one at a time and interpreted as such. However, we can easily determine a set of joint confidence intervals for any $r(\leq p = m + 1)$ regression parameters that hold simultaneously with confidence coefficient $1 - \alpha$. To this end, let us form a set of *Bonferroni family confidence intervals* with *family confidence coefficient* of at least $1-\alpha$ for $\beta_o, \beta_1, \ldots, \beta_m$ (here $r = p = m + 1$) as

$$\hat{\beta}_k \pm t_{\alpha/2p,\, n-p} S_{\hat{\beta}_k}, k = 0, 1, \ldots, m, \tag{12.17}$$

where $t_{\alpha/2p,n-p}$ is chosen so that we are at least $100(1-\alpha)\%$ confident that all of these intervals are jointly correct. So if p parameters are to be estimated with a family confidence coefficient of $1-\alpha$, then each statement confidence coefficient must be $1-\alpha/p$. Hence, α/p must be divided between the two tails of the t distribution, thus requiring $t_{\alpha/2p,n-p}$.

Decomposition of the Sample Variation of Y

We now turn to the issue of assessing the adequacy of the linear regression model (Equation 12.2). Specifically, we seek to determine whether the regressors X_j, $j = 1,\ldots,m$, satisfactorily explain the variation in the response variable Y. That is, is there a statistically significant linear relationship between the response variable Y and the regressors, or can most of the variation in Y be attributed to the random errors ε_i, $i = 1,\ldots,n$? If it is determined that the regression model is statistically correct, which of the individual regressors seem to be important in explaining Y?

To answer these questions, let us first decompose the total variation in Y into two mutually exclusive components: a regression source or *explained variation* (the variation in Y explained by the linear influence of the regressors) and an error source or *unexplained variation* (variation in Y attributed to random factors or to ε). This decomposition is illustrated in the analysis of variance (ANOVA) or *partitioned sum of squares table* (Table 12.2), with the actual sums of squares appearing as

$$\begin{aligned} SSR &= \sum_{i=1}^{n}(\hat{Y}_i - \bar{Y})^2 = \hat{\boldsymbol{\beta}}'\boldsymbol{X}'\boldsymbol{Y} - n\bar{Y}^2, \\ SSE &= \sum_{i=1}^{n} e_i^2 = SST - SSR = \boldsymbol{Y}'\boldsymbol{Y} - \hat{\boldsymbol{\beta}}'\boldsymbol{X}'\boldsymbol{Y}, \text{ and} \\ SST &= \sum_{i=1}^{n}(Y_i - \bar{Y})^2 = \boldsymbol{Y}'\boldsymbol{Y} - n\bar{Y}^2 \end{aligned} \tag{12.18}$$

(details on the derivation of these expressions are provided in Appendix F).

Armed with these considerations, we now want to see whether the model matters, i.e., is there really a statistically significant linear relationship between Y and the regressors X_i, $i = 1,\ldots,m$, or did the observed relationship occur primarily because of chance or random factors. In this regard, a *global test* (we test significance of the regression equation as a whole) of model adequacy is provided by

$$H_o\text{: } \beta_1 = \beta_2 = \cdots = \beta_m = 0 \text{ (the model doesn't matter) vs.}$$
$$H_1\text{: at least one } \beta_j \neq 0, j = 1,\ldots,m \text{ (the model does matter).}$$

We thus have a joint test that all regression coefficients save for the constant term are zero. Our test statistic is provided by Table 12.2: under H_o,

$$F = \frac{MSR}{MSE} = \frac{SSR/m}{SSE/(n-m-1)} : F_{m,\, n-m-1} \tag{12.19}$$

TABLE 12.2

Summary of the Partitioned Sums of Squares

Source of Variation in Y	**Sum of Squares (SS)**	**Degrees of Freedom (d.f.)**	**Mean Square (MS = SS/d.f.)**
Regression (explained variation in Y)	SSR	m	$MSR = \frac{SSR}{m}$
Error (unexplained variation in Y)	Error sum of squares (SSE = SST − SSR)	$n - m - 1$	$MSE = \frac{SSE}{n-m-1}$
Total variation in Y	SST	$n - 1$	$F = MSR/MSE$, $R^2 = SSR/SST$

(the justification that F follows an $F_{m,n-m-1}$ distribution is offered in Appendix 12.G). For this test, conducted at the α level of significance, the region of rejection $\mathcal{R} = \left\{F \middle| F > F_{\alpha,\, m,\, n-m-1}\right\}$. So if we reject H_o at the α level, then we can conclude that the model matters: there exists a statistically significant linear relationship between the explained variable Y and the explanatory variables $X_1,\ldots,X_m$; that is, the linear influence of the regressors explains a significant portion of the variation in Y.

How much of the variation in Y can be attributed to the linear influence of the regressors? We can readily answer this question by introducing an overall summary statistic that reflects *goodness of fit* as evidenced by the proportion of the variation in Y explained or accounted for by the model, namely, the *coefficient of multiple determination*:

$$R^2 = \frac{SSR}{SST} = 1 - \frac{SSE}{SST},\quad 0 \le R^2 \le 1, \tag{12.20}$$

where $R^2 = 0$ when $\hat{\beta}_j = 0,\ j = 1,\ldots,m$; and $R^2 = 1$ when $\hat{Y}_i = Y_i,\ i = 1,\ldots,n$.

It is important to note that inserting additional regressors into the regression equation can only increase R^2 and can never reduce it (because SST is fixed and SSE can never increase as more explanatory variables are introduced into the model). Thus, R^2 can be made large simply by including a large number of regressors in the regression equation. To guard against *overfitting* the regression model (i.e., increasing the number of regressors although they display only scant explanatory power and are essentially unnecessary), a modification of R^2 is used that explicitly recognizes the number of explanatory variables in the model and assesses a penalty for adding regressors that do not reduce MSE = SSE/($n - m - 1$). This modification of R^2 is called the *adjusted coefficient of multiple determination:*

$$\bar{R}^2 = 1 - \frac{SSE/(n-m-1)}{SST/(n-1)}. \tag{12.21}$$

That is, if the reduction in SSE is more than offset by the reduction in degrees of freedom $n - m - 1$, then MSE increases and $\bar{R}^2$ concomitantly decreases. Hence, a penalty is imposed in terms of a tradeoff between a decrease in SSE and the loss of degrees of freedom as more explanatory variables are inserted into the regression equation. This safeguards against the frivolous inclusion of regressors into the model simply for the sake of having a high R^2 value.

Testing Hypotheses about the Regression Parameters β_k[1]

Tests on Individual Regression Parameters

Given a set of estimates $\hat{\beta}_j = 0,\ j = 1,\ldots,m$, of the regression slopes, a question that naturally arises is whether or not an individual regressor X_j makes a statistically significant contribution to explaining variation in Y given the presence of other regressors in the regression

[1] The global F test offered in "Decomposition of the Sample Variation of Y" along with the various hypothesis tests to be developed in this section are all special cases of a more comprehensive or unified approach to testing sets of linear restrictions (hypotheses) on the regression parameters $\beta_o, \beta_1,\ldots,\beta_m$. Such restrictions typically involve sets of linear combinations of the β_k's, $k = 0,1,\ldots,m$. For details, see Appendix 12.H, which covers OLS estimation under a linear restriction or constraint, and Appendix 12.I, which addresses the issue of testing sets of linear restrictions.

equation. (Hence, this is a partial test and not a global test.) That is, we seek to determine whether X_j and Y are truly linearly related given that the remaining regressors are held fixed.

The most common null hypothesis tested is that there is no linear relationship between X_j and Y, i.e., the conditional expectation of Y given the regressors does not depend linearly on X_j. Hence, the implied null hypothesis of no linear relationship between X_j and Y is H_0: $\beta_j = 0$. Then the appropriate alternative hypothesis is H_1: $\beta_j \neq 0$: there is a linear relationship between X_j and Y. From Equation 12.15, the test statistic is t distributed with $n-m-1$ degrees of freedom and appears under H_0 as

$$t = \frac{\hat{\beta}_j}{S_\varepsilon \sqrt{C_{jj}}}. \tag{12.22}$$

Then for a (two-tailed) test conducted at the α level, $\mathcal{R} = \left\{t \middle| \, |t| > t_{\alpha/2,n-m-1}\right\}$. So if H_0 is rejected, then we can legitimately conclude that, at the α level, there exists a statistically significant linear relationship between X_j and Y.

Additional alternative hypotheses are as follows:

a. $H_1 : \beta_j > 0$, so that $\mathcal{R} = \left\{t \middle| \, t > t_{\alpha,\, n\text{-}m\text{-}1}\right\}$;

or

b. $H_1 : \beta_j < 0$, with $\mathcal{R} = \left\{t \middle| \, t < -t_{\alpha,\, n\text{-}m\text{-}1}\right\}$.

So if $H_0 : \beta_j = 0$ is rejected under $H_1 : \beta_j > 0$ ($H_1 : \beta_j < 0$), we are consequently led to conclude at the α level that there exists a statistically significant positive (negative) linear relationship between X_j and Y.

It is important to mention that we can also test $H_0 : \beta_j = \beta_j^o (\neq 0)$ versus

a. $H_1 : \beta_j \neq \beta_j^o$, with $\mathcal{R} = \left\{t \middle| \, |t| > t_{\alpha/2,n-m-1}\right\}$;

or

b. $H_1 : \beta_j > \beta_j^o$, with $\mathcal{R} = \left\{t \middle| \, t > t_{\alpha,\, n-m-1}\right\}$;

or

c. $H_1 : \beta_j < \beta_j^o$, with $\mathcal{R} = \left\{t \middle| \, t < -t_{\alpha,\, n-m-1}\right\}$.

at the α level by using the test statistic (via Equation 12.15)

$$t = \frac{\beta_j - \beta_j^o}{S_\varepsilon \sqrt{C_{jj}}}. \tag{12.23}$$

Furthermore, hypothesis test concerning the intercept parameter β_0 can be conducted by using the test statistic

$$t = \frac{\beta_0 - \beta_0^o}{S_\varepsilon \sqrt{C_{00}}}, \tag{12.24}$$

where the null hypothesis can be written $H_0 : \beta_0 = \beta_0^o = 0$ or $H_0 : \beta_0 = \beta_0^o \neq 0$.

Tests on Subsets of Regression Parameters

Testing the Effect of Additional Regressors on the Conditional Mean of Y

The procedure that follows may generally be viewed as a test of the hypothesis that some of the regression parameters are zero. In this regard, it will be framed in terms of augmenting an existing regression equation by a set of additional explanatory variables and then assessing whether or not the additional variables, collectively, make a significant contribution to explained variation in Y.

Specifically, suppose we have obtained parameter estimates for some *basic* or *initial model*

$$Y = \beta_o + \beta_1 X_1 + \cdots + \beta_h X_h + \varepsilon \tag{12.25}$$

and we seek to determine whether $m - h$ additional regressors $X_{h+1}, X_{h+2},\ldots,X_m$ make a significant contribution to explaining variation in Y. Once these additional regressors are introduced into Equation 12.25, we obtain the *extended* or *augmented model*

$$Y = \beta_o + \beta_1 X_1 + \cdots + \beta_h X_h + \beta_{h+1} X_{h+1} + \beta_{h+2} X_{h+2} + \cdots + \beta_m X_m + \varepsilon. \tag{12.26}$$

Here we seek to test H_0: $\beta_{h+1} = \beta_{h+2} = \cdots = \beta_m = 0$, against H_1: at least one regression parameter in H_0 is different from zero (or simply H_0 is not true).

Fitting Equation 12.24 (assuming H_0 true) yields $SST = SSR_h + SSE_h$, whereas fitting Equation 12.25 renders $SST = SSR_m + SSE_m$. Let us form the *extra* or *incremental regression sum of squares* resulting from the introduction of the $m - h$ regressors $X_{h+1},\ldots,X_m$ into the regression equation given that $X_1,\ldots,X_h$ are already in that equation or

$$\begin{aligned} SSR(\beta_{h+1},\ldots,\beta_m \mid \beta_1,\ldots,\beta_h) &= \Delta SSR \\ &= SSR_m - SSR_h, \end{aligned}$$

where SSR_m has m degrees of freedom, SSR_h has h degrees of freedom, and ΔSSR has $m - h$ degrees of freedom. Now, if the mean square of ΔSSR relative to the mean square of SSE_m is "large," then we can conclude that the extra $m - h$ regressors $X_{h+1},\ldots,X_m$ are relevant, i.e., they significantly contribute to explaining variation in Y. So under H_0 above, our test statistic is

$$F = \frac{\Delta SSR/(m-h)}{SSE/(n-m-1)} : F_{m-h,\, n-m-1}. \tag{12.27}$$

(Technical details regarding the structure of Equation 12.27 are provided in Appendix 12.G.)

For a given level of significance α, the region of rejection is $\mathcal{R} = \{F \mid F > F_{\alpha,\, m-h,\, n-m-1}\}$. It is important to note that, under H_0, the statistic in Equation 12.27 is the basis for conducting *a partial F test*: we are measuring the contribution to explained variation in Y of some subset of regressors $X_{h+1},\ldots,X_m$ given that the regressors $X_1,\ldots,X_h$ are already in the model.

Testing for the Equality of Two Regression Parameters

At times, it may be necessary to test the hypothesis that two regression parameters, say β_j and β_k, are equal. That is, we seek to test $H_0: \beta_j = \beta_k,\ j \neq k$, against any of the possible alternative hypotheses : $\beta_j > \beta_k$ or $\beta_j < \beta_k$ or $\beta_j \neq \beta_k$. To construct this test, let us first consider a couple of key properties of the sampling distribution of the difference between the OLS estimators $\hat{\beta}_j$ and $\hat{\beta}_k$ or properties of the sampling distribution of $\hat{\beta}_j - \hat{\beta}_k$. Specifically, we may readily determine that

$$E(\hat{\beta}_j - \hat{\beta}_k) = \beta_j - \beta_k \text{; and}$$

$$V(\hat{\beta}_j - \hat{\beta}_k) = V(\hat{\beta}_j) + V(\hat{\beta}_k) - 2COV(\hat{\beta}_j, \hat{\beta}_k)$$
$$= \sigma_\varepsilon^2 C_{jj} + \sigma_\varepsilon^2 C_{kk} - 2\sigma_\varepsilon^2 C_{jk}.$$

Under $H_0: \beta_j - \beta_k = 0$, we know (via Appendix 12.E) that $\hat{\beta}_j - \hat{\beta}_k : N(0,\ V(\hat{\beta}_j - \hat{\beta}_k))$ so that

$$\frac{\hat{\beta}_j - \hat{\beta}_k}{\sqrt{\sigma_\varepsilon^2 (C_{jj} + C_{kk} - 2C_{jk})}} : N(0,1).$$

If S_ε^2 serves as an estimator for σ_ε^2, then our test statistic is the quantity

$$\frac{\hat{\beta}_j - \hat{\beta}_k}{S_\varepsilon \sqrt{(C_{jj} + C_{kk} - 2C_{jk})}} : t_{n-m-1}.$$

Standardized Regression Coefficients

We know from our previous discussion of the slope estimates $\hat{\beta}_j$, $j = 1,\ldots,m$, that each of these coefficients must be interpreted in terms of the stated units of measurement of their associated explanatory variable X_j. However, under some appropriate scaling of the response variable Y and the set of regressors, the estimated regression coefficients can be rendered independent of units. The resulting collection of dimensionless coefficients are referred to as *standardized regression coefficients*. Two equivalent methods for generating standardized regression coefficients now follow.

Transformation to Z-Scores

Given $Y_i = \beta_o + \beta_1 X_{i1} + \cdots + \beta_m X_{im} + \varepsilon_i$, $i = 1,\ldots,n$, and $\bar{Y} = \beta_o + \beta_1 \bar{X}_1 + \cdots + \beta_m \bar{X}_m + \bar{\varepsilon}$, we can readily form

$$\frac{Y_i - \bar{Y}}{\sigma_Y} = \beta_1 \frac{X_{i1} - \bar{X}_1}{\sigma_1} \frac{\sigma_1}{\sigma_Y} + \cdots + \beta_m \frac{X_{im} - \bar{X}_m}{\sigma_m} \frac{\sigma_m}{\sigma_Y} + \frac{\varepsilon_i - \bar{\varepsilon}}{\sigma_Y}$$

or

$$Z_{iY} = \beta_1 \left(\frac{\sigma_1}{\sigma_Y}\right) Z_{i1} + \cdots + \beta_m \left(\frac{\sigma_m}{\sigma_Y}\right) Z_{im} + u_i$$
$$= \beta_1^* Z_{i1} + \cdots + \beta_m^* Z_{im} + u_i,\ i = 1,\ldots,n, \tag{12.28}$$

where σ_y is the standard deviation of Y, σ_j, $j = 1,\ldots,m$, denotes the standard deviation of X_j, and u_i is a random error term. Here each transformed variable Z_y, Z_j, $j = 1,\ldots,m$, has been *centered to zero* by subtracting its mean, and each of these new variables has a standard deviation of unity. Note that the new intercept $\beta_0^* = 0$.

Applying OLS to Equation 12.28 yields

$$\hat{Z}_Y = \hat{\beta}_i^* Z_1 + \cdots + \hat{\beta}_m^* Z_m, \tag{12.29}$$

where the estimated quantity $\hat{\beta}_j^*$, $j = 1,\ldots,m$, is a standardized regression coefficient. Hence, $\hat{\beta}_j^*$ measures the average rate of change in Y per unit change in X_j in *standard deviation units*, i.e., when X_j increases by one standard deviation, Y changes by $\hat{\beta}_j^*$ standard deviations on the average. Obviously $\hat{\beta}_j^*$ is independent of units. Given $\hat{\beta}_j^*$, the usual regression slope can be recovered as $\hat{\beta}_j = \hat{\beta}_j^*(S_Y/S_j)$ with $\hat{\beta}_0 = \bar{Y} - \sum_j \hat{\beta}_j \bar{X}_j$.

In matrix terms, for

$$Z_{ij} = \frac{X_{ij} - \bar{X}_j}{S_j},\ i = 1,\ldots,n,\ j = 1,\ldots,m,\ Z_{iY} = \frac{Y_i - \bar{Y}}{S_Y},$$

the original $\boldsymbol{X}$ and $\boldsymbol{Y}$ matrices transform to

$$\underset{(n\times m)}{\boldsymbol{Z_X}} = \begin{bmatrix} Z_{11} & Z_{12} & \ldots & Z_{1m} \\ Z_{21} & Z_{22} & \ldots & Z_{2m} \\ \cdot & \cdot & & \cdot \\ \cdot & \cdot & & \cdot \\ \cdot & \cdot & & \cdot \\ Z_{n1} & Z_{n2} & \ldots & Z_{nm} \end{bmatrix},\quad \underset{(n\times 1)}{\boldsymbol{Z_Y}} = \begin{bmatrix} Z_{1Y} \\ \cdot \\ \cdot \\ \cdot \\ Z_{nY} \end{bmatrix}$$

respectively. Then $\hat{\boldsymbol{\beta}}^* = (\boldsymbol{Z}_X'\boldsymbol{Z}_X)^{-1}\boldsymbol{Z}_X'\boldsymbol{Z}_Y$.

Unit Length Scaling

Let us again start with $Y_i = \beta_o + \sum_j \beta_j X_{ij} + \varepsilon_i$, $i = 1,\ldots, n$, and $\bar{Y} = \beta_o + \sum_j \beta_j \bar{X}_j + \bar{\varepsilon}$ but this time form the expression

$$\frac{Y_i - \bar{Y}}{SST^{1/2}} = \beta_1 \frac{X_{i1} - \bar{X}_1}{M_{11}^{1/2}}\left(\frac{M_{11}}{SST}\right)^{\frac{1}{2}} + \cdots + \beta_m \frac{X_{im} - \bar{X}_m}{M_{mm}^{1/2}}\left(\frac{M_{mm}}{SST}\right)^{\frac{1}{2}} + \frac{\varepsilon_i - \bar{\varepsilon}}{SST^{1/2}}$$

or

$$\begin{aligned} Y_i^* &= \beta_1\left(\frac{M_{11}}{SST}\right)^{1/2} W_{i1} + \cdots + \beta_m\left(\frac{M_{mm}}{SST}\right)^{1/2} W_{im} + v_i \\ &= \beta_1^{**}W_{i1} + \cdots + \beta_m^{**}W_{im} + v_i,\ i = 1,\ldots,n, \end{aligned} \tag{12.30}$$

where $M_{jj} = \sum_i (X_{ij} - \bar{X}_j)^2$, $j = 1,\ldots,m$. Now, each transformed variable Y^* and W_j, $j = 1,\ldots,m$, has been centered to zero $(\bar{Y}^* = \bar{W}_j = 0)$ and, in particular, each new regressor W_j has unit length[2] because $\|W_j\| = \left[\sum_i |W_{ij} - \bar{W}_j|^2\right]^{1/2} = \left[\sum_i |W_{ij}|^2\right]^{1/2} = 1$.

[2] The distance between two points $\boldsymbol{X} = (x_1,\ldots,x_n)$ and $\boldsymbol{Y} = (y_1,\ldots,y_n)$ in n-dimensional space is given by the norm $\|\boldsymbol{X} - \boldsymbol{Y}\| = \left[\sum_i |x_i - y_i|^2\right]^{\frac{1}{2}}$, whereas the distance between, say, $\boldsymbol{X}$ and the origin $\boldsymbol{O} = (0,\ldots,0)$ is the *length of* $\boldsymbol{X}$ and written as the norm $\|\boldsymbol{X}\| = \left[\sum_i |x_i|^2\right]^{\frac{1}{2}}$.

If Equation 12.30 is fit via OLS, we obtain

$$\hat{Y}^* = \hat{\beta}_1^{**} W_1 + \cdots + \hat{\beta}_m^{**} W_m, \tag{12.31}$$

where $\hat{\beta}_j^{**}, j = 1,\ldots,m,$ is the estimated standardized (dimensionless) regression coefficient. Hence, increasing the standardized value of X_j, W_j, by one unit precipitates a $\hat{\beta}_j^{**}$ unit change in the standardized value of Y, Y^*, on average. Because

$$Z_{ij} = \frac{X_{ij} - \bar{X}_j}{S_j} = \sqrt{n-1}\,\frac{X_{ij} - \bar{X}_j}{M_{jj}^{1/2}} = \sqrt{n-1}\,W_{ij},$$

it follows that $\hat{\beta}_j^{**} = \hat{\beta}_j^{*}$. Once $\hat{\beta}_j^{**}$ is obtained, we can transform back to the ordinary regression slope according to $\hat{\beta}_j = \hat{\beta}_j^{**}(SST/M_{jj})^{1/2}$, $j = 1,\ldots,m,$ with $\hat{\beta}_0 = \bar{Y} - \sum_j \hat{\beta}_j \bar{X}_j$.

In matrix notation, with

$$W_{ij} = \frac{X_{ij} - \bar{X}_j}{M_{jj}^{1/2}},\ i = 1,\ldots,n,\ j = 1,\ldots,m,\ Y_i^* = \frac{Y_i - \bar{Y}}{SST^{1/2}},$$

the usual **X** and **Y** matrices become

$$\underset{(n\times m)}{\mathbf{W}} = \begin{bmatrix} W_{11} & W_{12} & \ldots & W_{1m} \\ W_{21} & W_{22} & \ldots & W_{2m} \\ \cdot & \cdot & & \cdot \\ \cdot & \cdot & & \cdot \\ \cdot & \cdot & & \cdot \\ W_{n1} & W_{n2} & \ldots & W_{nm} \end{bmatrix},\quad \underset{(n\times 1)}{\mathbf{Y}^*} = \begin{bmatrix} Y_1^* \\ Y_2^* \\ \cdot \\ \cdot \\ \cdot \\ Y_n^* \end{bmatrix}$$

respectively, so that $\hat{\boldsymbol{\beta}}^{**} = (\mathbf{W}'\mathbf{W})^{-1}\mathbf{W}'\mathbf{Y}^*$.

It is instructive to note that, under unit length scaling, both $W'W$ and $W'Y^*$ are actually correlation matrices. That is,

$$\underset{(m\times m)}{\mathbf{W}'\mathbf{W}} = \begin{bmatrix} 1 & R_{12} & R_{13} & \ldots & R_{1m} \\ R_{21} & 1 & R_{23} & \ldots & R_{2m} \\ R_{31} & R_{32} & 1 & \ldots & R_{3m} \\ \cdot & \cdot & \cdot & & \cdot \\ \cdot & \cdot & \cdot & & \cdot \\ \cdot & \cdot & \cdot & & \cdot \\ R_{m1} & R_{m2} & R_{m3} & \ldots & 1 \end{bmatrix},$$

where $R_{jl} = M_{jl}/(M_{jj}M_{ll})^{1/2}$, $j, l = 1,\ldots,m,$ is the Pearson or product-moment correlation coefficient between regressors X_j and X_l. Additionally,

$$W'Y^* = \begin{bmatrix} R_{iY} \\ R_{2Y} \\ \cdot \\ \cdot \\ \cdot \\ R_{mY} \end{bmatrix},$$

where $R_{jy} = M_{jy}/(M_{jj}SST)^{½}$ is the product-moment correlation coefficient between regressor X_j and the explained variable Y.

Decomposition of SSR into Incremental Sums of Squares

An *incremental sum of squares* depicts the marginal increase in SSR when one or more additional regressors are included in the regression equation given that other regressors are already in that equation. Equivalently, it is the marginal reduction in SSE when at least one additional regressor is included in the regression model given the presence of other regressors already in the model. Moreover, in what follows it must be remembered that because SST = SSR + SSE is fixed in value, introducing new regressors into an existing model usually increases SSR and decreases SSE so that any increase in SSR must be offset exactly by a decrease in SSE.

Given the model $Y = \beta_0 + \beta_1 X + \varepsilon$, estimation yields $SSR(X_1)$ and $SSE(X_1)$. If the model changes to $Y = \beta_0 + \beta_1 X_1 + \beta_2 X_2 + \varepsilon$, then estimation renders $SSR(X_1, X_2)$ and $SSE(X_1, X_2)$ with

$$SSR(X_1, X_2) > SSR(X_1) \text{ and } SSE(X_1, X_2) < SSE(X_1).$$

Let $SSR(X_2|X_1)$ represent the marginal effect of introducing X_2 into the regression equation given that X_1 is already in the same. Hence, this incremental sum of squares can be written as either

$$SSR(X_2|X_1) = SSR(X_1, X_2) - SSR(X_1) \tag{12.32}$$

[increase in *SSR* due to including X_2 in a model which already contains X_1]

or

$$SSR(X_2|X_1) = SSE(X_1) - SSE(X_1, X_2) \tag{12.33}$$

[reduction in *SSE* due to including X_2 in a model which already contains X_1]

If the model is now $Y = \beta_0 + \beta_1 X_1 + \beta_2 X_2 + \beta_3 X_3 + \varepsilon$, with $SSR(X_1, X_2, X_3)$ and $SSE(X_1, X_2, X_3)$ forthcoming under estimation, where

$$SSR(X_1, X_2, X_3) > SSR(X_1, X_2) \text{ and } SSE(X_1, X_2, X_3) < SSE(X_1, X_2),$$

then $SSR(X_3|X_1, X_2)$ depicts the marginal effect of introducing X_3 into a regression model given that both X_1 and X_2 are already included in that model. Hence, the incremental sums of squares are either

$$SSR(X_3|X_1, X_2) = SSR(X_1, X_2, X_3) - SSR(X_1, X_2) \quad (12.34)$$

[increase in *SSR* due to including X_3 in a model which already contains X_1 and X_2]

or

$$SSR(X_3|X_1, X_2) = SSE(X_1, X_2) - SSE(X_1, X_2, X_3) \quad (12.35)$$

[reduction in *SSE* due to including X_3 in a model which already contains X_1 and X_2].

Incremental sums of squares for the insertion of X_4, X_5, etc. into the regression model are defined and structured in a similar manner.

Suppose ($SST = SSR(X_1, X_2, X_3, X_4) + SSE(X_1, X_2, X_3, X_4)$. (Let us use the preceding relationships to decompose SSR into incremental sums of squares. To this end, we have

$$\begin{aligned} SST &= SSR(X_1, X_2, X_3, X_4) + SSE(X_1, X_2, X_3, X_4) \\ &= SSR(X_4|X_1, X_2, X_3) + SSR(X_1, X_2, X_3) + SSE(X_1, X_2, X_3, X_4) \\ &= SSR(X_4|X_1, X_2, X_3) + SSR(X_3|X_1, X_2) + SSR(X_1, X_2) + SSE(X_1, X_2, X_3, X_4) \\ &= SSR(X_4|X_1, X_2, X_3) + SSR(X_3|X_1, X_2) + SSR(X_1|X_2) + SSR(X_1) + SSE(X_1, X_2, X_3, X_4). \end{aligned} \quad (12.36)$$

Hence, $SSR(X_1, X_2, X_3, X_4)$ has been decomposed into the incremental or marginal components:

1. $SSR(X_1)$: the increase in *SSR* due to introducing X_1 into the model
2. $SSR(X_2|X_1)$: the increase in *SSR* due to introducing X_2 into the model given that X_1 is already in the model
3. $SSR(X_3|X_1, X_2)$: the increase in *SSR* due to introducing X_3 into the model given that X_1 and X_2 are already in the model
4. $SSR(X_4|X_1, X_2, X_3)$: the increase in *SSR* due to introducing X_4 into the model given that X_1, X_2, and X_3 are already in the model

(See Table 12.3.)

TABLE 12.3

Summary of Partitioned Sum of Squares and Decomposition of SSR

Source of Variation in *Y*	Sum of Squares	Degrees of Freedom (d.f.)	Mean Square (SS/d.f.)
Regression	$SSR(X_1, X_2, X_3, X_4)$	$m = 4$	$MSR(X_1, X_2, X_3, X_4)$
X_1	$SSR(X_1)$	1	$MSR(X_1)$
$X_2 \mid X_1$	$SSR(X_2 \mid X_1)$	1	$MSR(X_2 \mid X_1)$
$X_3 \mid X_1, X_2$	$SSR(X_3 \mid X_1, X_2)$	1	$MSR(X_3 \mid X_1, X_2)$
$X_4 \mid X_1, X_2, X_3$	$SSR(X_4 \mid X_1, X_2, X_3)$	1	$MSR(X_4 \mid X_1, X_2, X_3)$
Error	$SSE(X_1, X_2, X_3, X_4)$	$n - m\text{-}1 = n - 5$	$MSE(X_1, X_2, X_3, X_4)$
Total variation in *Y*	SST	$n - 1$	

In general,

$$SSR(X_1,X_2,\ldots, X_m) = SSR(X_1) + SSR(X_2|X_1) + SSR(X_3|X_1, X_2) + \cdots + SSR(X_m|X_1,\ldots, X_{m-1}), \tag{12.37}$$

where these incremental sums of squares are determined via

$$SSR(X_r|X_1,\ldots, X_{r-1}) = SSR(X_1,\ldots, X_r) - SSR(X_1,\ldots, X_{r-1}),\ r = 2,\ldots,m. \tag{12.38}$$

Example 12.1

A. ABC Corp. wants to examine the ability of certain factors to explain or predict the number of units sold of its product (Y) over the past quarter by its sales force. Its market research department extracts a random sample of $n = 35$ salesmen from its records and identifies the following $m = 4$ variables as possible regressors or predictors: number of "cold" calls (visits) made (X_1); number of follow-up calls (visits) made (X_2); average amount of time (in minutes) spent with a client (X_3); and the number of months of sales experience (X_4). These observations appear in Table 12.4. In what follows, the aforementioned variables will be coded as such: *units* (Y); *cold* (X_1); *foll* (X_2); *clt* (X_3); and *exp* (X_4). Exhibit 12.1 contains the SAS code for generating the regression results.

Exhibit 12.1 SAS Code for PROC REG and PROC CORR

```
data ABCsold;  ①
     input units cold foll clt exp;  ②
datalines;
14   62    10   20   12.00
21   120   22   45   27.25
16   100   20   27   24.25
.    .     .    .    .
.    .     .    .    .
.    .     .    .    .
17   91    21   39   19.25
;
proc print data = ABCsold;  ③
proc plot  data = ABCsold;  ④
     plot units * cold="*";
     plot units * foll ="*";
     plot units * clt ="*";
     plot units * exp ="*";
proc reg data = ABCsold;  ⑤
     model units = cold foll clt exp/all; ⑥
proc corr data = ABCsold;  ⑦
     var units cold foll clt exp;  ⑧
run;
```

① The dataset is named *ABCsold.*

② The *input statement* names all variables to be used in the regression study.

③ We request a print of the data by invoking the *print procedure* (the output of this procedure is not included below).

④ We request a plot of units (Y) against each regressor by invoking the plot *procedure* (the plots are not included below).

⑤ We request that the *regression procedure* be applied to our dataset.

⑥ We specify the *model* units $= \beta_0 + \beta_1$(cold) $+ \beta_2$(foll) $+ \beta_3$(clt) $+ \beta_4$(exp) $+ \varepsilon$. Additionally, we request that all relevant output be supplied. (We select *all* because we want to tie most

TABLE 12.4

ABC Corp. Sales Force Characteristics

Salesman	Units Sold (Y)	Number of Cold Calls (X_1)	Number of Follow-Up Calls (X_2)	Average Time Spent with Client (Mins.) (X_3)	Number of Months Experience (X_4)
1	14	62	10	20	12.00
2	21	120	22	45	27.25
3	16	100	20	27	24.25
4	13	50	8	25	10.50
5	16	85	19	30	20.75
6	20	130	16	46	24.50
7	20	115	17	50	24.25
8	15	80	14	28	15.00
9	22	125	20	62	30.75
10	14	59	10	21	15.75
11	13	53	15	25	16.50
12	13	50	14	25	17.50
13	13	60	10	20	21.00
14	19	111	17	55	23.25
15	16	89	10	30	20.00
16	20	126	22	58	30.50
17	15	77	15	29	18.25
18	14	60	9	26	20.25
19	21	119	18	45	31.75
20	22	110	15	51	36.25
21	16	79	17	30	22.50
22	12	60	8	20	10.25
23	13	60	7	22	9.00
24	17	85	18	31	25.75
25	15	71	17	33	19.75
26	17	90	19	29	22.25
27	18	92	19	35	21.25
28	20	109	25	60	30.25
29	21	119	21	63	30.75
30	20	97	27	61	19.50
31	15	63	17	19	10.25
32	13	66	10	22	15.50
33	17	80	20	36	18.50
34	16	75	17	31	25.75
35	17	91	21	39	19.25

of the output to the preceding textual materials. If *all* is not requested, then specific options in the *MODEL* statement can be selected from the partial list of options appearing in Part B of this example problem).

⑦ We request the computation of the set of Pearson product-moment correlation coefficients between all variables by invoking the *correlation procedure*.

⑧ We name the variables to be correlated in a *variables statement*.

B. Partial list of *MODEL* statement options

NOINT: fits a model without an intercept parameter.

XPX: produces a matrix that exhibits sums of squares and crossproduct terms associated with all variables in the *MODEL* statement. It is structured as

$$\begin{bmatrix} X'X & X'Y \\ Y'X & Y'Y \end{bmatrix}.$$

I: produces a matrix that displays the inverse of $X'X$ as well as $\hat{\beta}$ and SSE. It is structured as

$$\begin{bmatrix} (X'X)^{-1} & \hat{\beta} \\ \hat{\beta}' & SSE \end{bmatrix}.$$

CORRB: displays the correlation matrix of parameter estimates; the $(X'X)^{-1}$ matrix is scaled to a matrix having 1's on the main diagonal.

COVB: displays the variance-covariance matrix of the parameter estimates, i.e., it displays an estimate of $V(\hat{\beta}) = \sigma_\varepsilon^2(X'X)^{-1}$ or $S_\varepsilon^2(X'X)^{-1} = S_\varepsilon^2 C$.

SPEC: this option performs a model specification test of H_0: the errors are homoscedastic, independent of the regressors, and the first and second moments of the model are correctly specified. When the model is correct, rejection of H_0 provides evidence of heteroscedasticity.

ACOV: displays a consistent estimate of the (asymptotic) variance-covariance matrix of the parameter estimates in the presence of heteroscedasticity.

STB: displays standardized parameter estimates, i.e.,

$$\beta_j^* = \beta_j\left(\frac{S_j}{S_Y}\right), \; j = 1,\ldots,m,$$

where S_j is the sample standard deviation of X_j and S_y is the sample standard deviation of Y.

TOL: displays tolerance values for the parameter estimates.

VIF: displays variance-inflation factors.

COLLIN: provides an analysis of collinearity among the regressors.

COLLINOINT: same analysis as COLLIN but with the intercept suppressed.

INFLUENCE: provides an assortment of influence statistics.

DW: provides the Durbin-Watson Statistic.

CLB: provides $100(1 - \alpha)\%$ confidence limits for the parameter estimates (the default is $1 - \alpha = 0.95$).

CLI: provides $100(1 - \alpha)\%$ confidence limits for an individual predicted value of Y (the default is $1 - \alpha = 0.95$).

CLM: provides $100(1 - \alpha)\%$ confidence limits for the expected value Y (default is $1 - \alpha = 0.95$).

SS1: provides Type I (sequential) sums of squares.

SS2: provides Type II (partial) sums of squares.

P: provides predicted values of Y (not necessary if the *CLI* or *CLM* or *R* options are requested).

R: provides statistics for an analysis of the regression residuals.

C. Regression results for ABC sold

The output generated by the SAS code presented in Exhibit 12.1 appears as Table 12.5.

TABLE 12.5

Output for Example 12.1

The SAS System

The REG Procedure

Model : MODEL 1

①

Model Cross-Products X'X X'Y Y'Y

Variable	Intercept	cold	foll	clt	exp	units
Intercept	35	3018	564	1250	740.75	584
cold	3018	280770	51449	117900	68434.5	52749
foll	564	51449	9950	21862	12632.5	9789
clt	1250	117900	21862	51352	28887.75	22177
exp	740.75	68434.5	12632.5	28887.75	17232.0625	12946.5
units	584	52749	9789	22177	12946.5	10058

The REG Procedure

Model : MODEL 1

Dependent Variable: Units

②

X'X Inverse, Parameter Estimates, and SSE

Variable	Intercept	cold	foll	clt	exp	units
Intercept	0.5029767452	–0.0048282	–0.013210453	0.0069260371	–0.004373312	6.7182401281
cold	–0.0048282	0.0002464583	–0.000076349	–0.000236261	–0.000319186	0.0614773371
foll	–0.013210453	–0.000076349	0.0024437653	–0.000452292	–0.000162176	0.049695946
clt	0.0069260371	–0.000236261	–0.000452292	0.0006850125	–0.000176238	0.0646405485
exp	–0.004373312	–0.000319186	–0.000162176	–0.000176238	0.001927957	0.0735657007
units	6.7182401281	0.0614773371	0.049695946	0.0646405485	0.0735657007	19.254303645

③

Analysis of Variance

Source	DF	Sum of Squares	Mean Square	*F* Value	Pr > *F*
Model	4	294.28855	73.57214	114.63	<0.0001
Error	30	19.25430	0.64181		
Corrected total	34	313.54286			

Root MSE	0.80113	*R*-Square	0.9386
Dependent mean	16.68571	Adj *R*-Sq	0.9304
Coeff Var	4.80130		

(continued)

TABLE 12.5 *(continued)*

The REG Procedure
Model: MODEL 1
Dependent Variable: Units
④
Parameter Estimates

Variable	DF	Ⓐ Parameter Estimates	Ⓑ Standard Error	Ⓒ t Value	Pr > |t|	Ⓓ Type I SS	Ⓔ Type II SS
Intercept	1	6.71824	0.56817	11.82	<0.0001	9744.45714	89.73526
cold	1	0.06148	0.01258	4.89	<0.0001	278.55508	15.33510
foll	1	0.04970	0.03960	1.25	0.2192	5.31242	1.01061
clt	1	0.06464	0.02097	3.08	0.0044	7.61398	6.09974
exp	1	0.07357	0.03518	2.09	0.0451	2.80707	2.80707

Variable	DF	Ⓕ Standardized Estimate	Squared Semipartial Corr Type I	Squared Partial Corr Type I	Squared Semipartial Corr Type II	Squared Partial Corr Type II	Ⓖ Tolerance
Intercept	1	0					
cold	1	0.49749	0.88841	0.88841	0.04891	0.44335	0.19762
foll	1	0.08238	0.01694	0.15184	0.00322	0.04987	0.47497
clt	1	0.29901	0.02428	0.25658	0.01945	0.24058	0.21759
exp	1	0.16381	0.00895	0.12724	0.00895	0.12724	0.33364

Variable	DF	Ⓗ Variance Inflation	Ⓘ 95% Confidence Limits	
Intercept	1	0	5.55789	7.87860
cold	1	5.06032	0.03579	0.08716
foll	1	2.10541	−0.03119	0.13058
clt	1	4.59585	0.2182	0.10746
exp	1	2.99724	0.00173	0.14541

Ⓙ
Covariance of Estimates

Variable	Intercept	cold	foll	clt	exp
Intercept	0.322815566	−0.003098787	−0.008478602	0.0044452007	−0.002806836
cold	−0.003098787	0.0001581794	−0.000049002	−0.000151635	−0.000204857
foll	−0.008478602	−0.000049002	0.0015684333	−0.000290286	−0.000104086
clt	0.0044452007	−0.000151635	−0.000290286	0.0004396479	−0.000113111
exp	−0.002806836	−0.000204857	−0.000104086	−0.000113111	0.0012373823

TABLE 12.5 *(continued)*

The REG Procedure

Model: MODEL 1

Dependent Variable: Units

Correlation of Estimates

Variable	Intercept	cold	foll	clt	exp
Intercept	1.0000	−0.4337	−0.3768	0.3731	−0.1404
cold	−0.4337	1.0000	−0.0984	−0.5750	−0.4630
foll	−0.3768	−0.0984	1.0000	−0.3496	−0.0747
clt	0.3731	−0.5750	−0.3496	1.0000	−0.1534
exp	−0.1404	−0.4630	−0.0747	−0.1534	1.0000

Sequential Parameter Estimates

Intercept	cold	foll	clt	exp
16.685714	0	0	0	0
6.642117	0.116476	0	0	0
6.188825	0.101976	0.105721	0	0
6.885114	0.073657	0.055884	0.071365	0
6.718240	0.061477	0.049696	0.064641	0.073566

Ⓚ

Consistent Covariance of Estimates

Variable	Intercept	cold	foll	clt	exp
Intercept	0.2563908445	−0.002258589	−0.004403555	0.0037520188	−0.006248462
cold	−0.002258589	0.0001193572	0.0002202622	−0.000190016	−0.000216737
foll	−0.004403555	0.0002202622	0.00184541	−0.000539431	−0.001169871
clt	0.0037520188	−0.000190016	−0.000539431	0.0004886237	0.0001573339
exp	−0.006248462	−0.000216737	−0.001169871	0.0001573339	0.0018477435

Test of First and Second Moment Specification

DF	Chi-Square	Pr > ChiSq
14	13.29	0.5037

(continued)

TABLE 12.5 *(continued)*

The REG Procedure
Model: MODEL 1
Dependent Variable: Units
Ⓛ
Output Statistics

Obs	Dep Var Units	Predicted Value	Std Error Mean Predict	95% CL Mean		95% CL Predict		Residuals
1	14.0000	13.2024	0.2477	12.6966	13.7082	11.4899	14.9149	0.7976
2	21.0000	20.1023	0.2933	19.5032	20.7014	18.3600	21.8447	0.8977
3	16.0000	17.3892	0.3926	16.5873	18.1910	15.5671	19.2112	–1.3892
4	13.0000	12.5781	0.3482	11.8670	13.2893	10.7941	14.3621	0.4219
5	16.0000	16.3537	0.2313	15.8814	16.8261	14.6508	18.0567	–0.3537
6	20.0000	20.3459	0.4176	19.4931	21.1987	18.5009	22.1909	–0.3459
7	20.0000	19.6490	0.2657	19.1063	20.1916	17.9252	21.3727	0.3510
8	15.0000	15.2456	0.2239	14.7883	15.7029	13.5468	16.9444	–0.2456
9	22.0000	21.6667	0.3116	21.0303	22.3031	19.9111	23.4222	0.3333
10	14.0000	13.3585	0.2244	12.9003	13.8167	11.6594	15.0576	0.6415
11	13.0000	13.5518	0.2791	12.9818	14.1219	11.8192	15.2844	–0.5518
12	13.0000	13.3913	0.3111	12.7559	14.0266	11.6361	15.1464	–0.3913
13	13.0000	13.7415	0.3088	13.1109	14.3721	11.9881	15.4950	–0.7415
14	19.0000	19.6527	0.3129	19.0136	20.2918	17.8962	21.4092	–0.6527
15	16.0000	16.0972	0.2806	15.5242	16.6703	14.3636	17.8308	–0.0972
16	20.0000	21.5506	0.2656	21.0081	22.0931	19.8269	23.2743	–1.5506
17	15.0000	15.4146	0.1543	15.0994	15.7298	13.7484	17.0808	–0.4146
18	14.0000	14.0245	0.3165	13.3781	14.6709	12.2653	15.7837	–0.0245
19	21.0000	20.1731	0.3055	19.5492	20.7970	18.4221	21.9242	0.8269
20	22.0000	20.1896	0.4480	19.2747	21.1045	18.3151	22.0642	1.8104
21	16.0000	16.0142	0.2022	15.6013	16.4271	14.3268	17.7016	–0.0142
22	12.0000	12.8513	0.3033	12.2319	13.4707	11.1019	14.6008	–0.8513
23	13.0000	12.8389	0.3558	12.1123	13.5656	11.0487	14.6292	0.1611
24	17.0000	16.7365	0.2626	16.2001	17.2729	15.0147	18.4583	0.2635
25	15.0000	15.5140	0.2067	15.0919	15.9361	13.8243	17.2037	–0.5140
26	17.0000	16.7068	0.2685	16.1584	17.2552	14.9812	18.4324	0.2932
27	18.0000	17.1441	0.1936	16.7487	17.5394	15.4609	18.8273	0.8559
28	20.0000	20.7655	0.3483	20.0541	21.4768	18.9814	22.5495	–0.7655
29	21.0000	21.4122	0.3237	20.7511	22.0732	19.6475	23.1768	–0.4122
30	20.0000	19.4009	0.5233	18.3323	20.4696	17.4467	21.3552	0.5991
31	15.0000	13.4184	0.3534	12.6967	14.1401	11.6301	15.2066	1.5816
32	13.0000	13.8351	0.2178	13.3903	14.2798	12.1396	15.5306	–0.8351
33	17.0000	16.3184	0.2340	15.8404	16.7963	14.6139	18.0229	0.6816
34	16.0000	16.0720	0.2963	15.4669	16.6772	14.3276	17.8165	–0.0720
35	17.0000	17.2934	0.2437	16.7956	17.7912	15.5832	19.0036	–0.2934

TABLE 12.5 *(continued)*

The REG Procedure
Model: MODEL 1
Dependent Variable: Units
Output Statistics

Obs	Std Error Residual	Student Residual	−2 −1 0	0 1 2	Cook's D
1	0.762	1.047		* *	0.023
2	0.745	1.204		* *	0.045
3	0.698	−1.989	* * *		0.250
4	0.721	0.585		*	0.016
5	0.767	−0.461			0.004
6	0.684	−0.506	*		0.019
7	0.756	0.464			0.005
8	0.769	−0.319			0.002
9	0.738	0.452			0.007
10	0.769	0.834		*	0.012
11	0.751	−0.735	*		0.015
12	0.738	−0.530	*		0.010
13	0.739	−1.003	* *		0.035
14	0.737	−0.885	*		0.028
15	0.750	−0.130			0.000
16	0.756	−2.052	* * * *		0.104
17	0.786	−0.527	*		0.002
18	0.736	−0.0333			0.000
19	0.741	1.117		* *	0.042
20	0.664	2.726		* * * * *	0.676
21	0.775	−0.0184			0.000
22	0.742	−1.148	* *		0.044
23	0.718	0.224			0.002
24	0.757	0.348			0.003
25	0.774	−0.664	*		0.006
26	0.755	0.388			0.004
27	0.777	1.101		* *	0.015
28	0.721	−1.061	* *		0.052
29	0.733	−0.562	*		0.012
30	0.607	0.988		*	0.145
31	0.719	2.200		* * * *	0.234
32	0.771	−1.083	* *		0.019
33	0.766	0.890		*	0.015
34	0.744	−0.0968			0.000
35	0.763	−0.384			0.003

Sum of Residuals	0
Sum of Squared Residuals	19.25430
Predicted Residual SS (PRESS)	29.68694

(continued)

TABLE 12.5 *(continued)*

Ⓜ

The CORR Procedure

5 Variables: units cold foll clt exp

Simple Statistics

Variable	N	Mean	Std Dev	Sum	Minimum	Maximum
units	35	16.68571	3.03675	584.00000	12.00000	22.00000
cold	35	86.22857	24.57412	3018	50.00000	130.00000
foll	35	16.11429	5.03384	564.00000	7.00000	27.00000
clt	35	35.71429	14.04734	1250	19.00000	63.00000
exp	35	21.16429	6.76196	740.75000	9.00000	36.25000

Pearson correlation coefficients, $N = 35$

Prob > |r| under H0: Rho = 0

	units	cold	foll	clt	exp
units	1.00000	0.94256 <0.0001	0.72778 <0.0001	0.91001 <0.0001	0.84014 <0.0001
cold	0.94256 <0.0001	1.00000	0.66956 <0.0001	0.86176 <0.0001	0.80724 <0.0001
foll	0.72778 <0.0001	0.66956 <0.0001	1.00000	0.71506 <0.0001	0.60126 0.0001
clt	0.91001 <0.0001	0.86176 <0.0001	0.71506 <0.0001	1.00000	0.75316 <0.0001
exp	0.84014 <0.0001	0.80724 <0.0001	0.60126 0.0001	0.75316 <0.0001	1.00000

① This (symmetric) model crossproducts matrix displays the following structure:

$$\begin{bmatrix} \boldsymbol{X}'\boldsymbol{X} & \boldsymbol{X}'\boldsymbol{Y} \\ \boldsymbol{Y}'\boldsymbol{X} & \boldsymbol{Y}'\boldsymbol{Y} \end{bmatrix} = \begin{bmatrix} n & \sum X_{i1} & \sum X_{i2} & \sum X_{i3} & \sum X_{i4} & \sum Y_i \\ & \sum X_{i1}^2 & \sum X_{i1}X_{i2} & \sum X_{i1}X_{i3} & \sum X_{i1}X_{i4} & \sum X_{i1}Y_i \\ & & \sum X_{i2}^2 & \sum X_{i2}X_{i3} & \sum X_{i2}X_{i4} & \sum X_{i2}Y_i \\ & & & \sum X_{i3}^2 & \sum X_{i3}X_{i4} & \sum X_{i3}Y_i \\ & & & & \sum X_{i4}^2 & \sum X_{i4}Y_i \\ & & & & & \sum Y_i^2 \end{bmatrix}.$$

② This (symmetric) matrix has the following structure:

$$\begin{bmatrix} (\boldsymbol{X}'\boldsymbol{X})^{-1} & \hat{\boldsymbol{\beta}} \\ \hat{\boldsymbol{\beta}}' & SSE \end{bmatrix} = \begin{bmatrix} \boldsymbol{C} & \hat{\boldsymbol{\beta}} \\ \hat{\boldsymbol{\beta}}' & SSE \end{bmatrix}$$

$$= \begin{bmatrix} C_{00} & C_{01} & C_{02} & C_{03} & C_{04} & \hat{\beta}_0 \\ & C_{11} & C_{12} & C_{13} & C_{14} & \hat{\beta}_1 \\ & & C_{22} & C_{23} & C_{24} & \hat{\beta}_2 \\ & & & C_{33} & C_{34} & \hat{\beta}_3 \\ & & & & C_{44} & \hat{\beta}_4 \\ & & & & & SSE \end{bmatrix}.$$

③ This partitioned sum of squares or ANOVA table displays the information appearing in Table 12.2. Degrees of freedom for the model (SSR), error (SSE), and total (SST) are $m = 4$, $n - m - 1 = 30$, and $n - 1 = 34$, respectively. In addition

$$\begin{aligned}
SST &= \mathbf{Y}'\mathbf{Y} - n\bar{Y}^2 = 10058 - 35(584/35)^2 = 313.54286,\\
SSR &= \hat{\boldsymbol{\beta}}'\mathbf{X}'\mathbf{Y} - n\bar{Y}^2 = (6.71824)(584) + (0.06148)(52749)\\
&\quad + (0.04970)(9789) + (0.06464)(22177)\\
&\quad + (0.07357)(12946.5) - 35(584/38)^2\\
&= 294.28855,\\
SSE &= SST - SSR = 313.54286 - 294.28855 = 19.25430.
\end{aligned}$$

Then $MSR = SSR/m = 75.57214$, $MSE = SSE/(n-m-1) = 0.64181$, and the calculated or sample $F = MSR/MSE = 114.63$.

Under the null hypothesis $H_0 : \beta_1 = \beta_2 = \beta_3 = \beta_4 = 0$ (the model does not explain any of the variation in Y), MSR/MSE follows an F distribution with m and $n - m - 1$ degrees of freedom. This hypothesis is tested against the alternative hypothesis H_1: at least one of the β_j's in H_0 is different from zero (the model explains some of the variation in Y). Because the p-value is less than 0.05, we reject H_0 in favor of H_1; the model explains a statistically significant portion of the variation in Y.

To determine exactly how much of the variation in Y is attributed to the linear influence of the regressors, we need only look to the coefficient of multiple determinations $R^2 = SSR/SST = 0.9386$, i.e., the model explains almost 94% of the variation in Y; only 6% is left unexplained or attributed to random factors. Because the adjusted coefficient of the determination $\bar{R}^2$ is virtually identical to R^2 itself, we have not overfitted the model.

Finally, the standard error of estimate or $\sqrt{MSE} = 0.80113 = s_\varepsilon$ (our estimate of σ_ε). Remember that $S_\varepsilon^2 = SSE/(n-m-1) = MSE$.

④ A. Given the set of parameter estimates, our sample regression equation has the form

$$\begin{aligned}
\hat{Y} &= \hat{\beta}_0 + \hat{\beta}_1 X_1 + \hat{\beta}_2 X_2 + \hat{\beta}_3 X_3 + \hat{\beta}_4 X_4\\
&= 6.71824 + 0.06148X_1 + 0.04970X_2\\
&\quad + 0.06464X_3 + 0.07354X_4.
\end{aligned}$$

Does each estimated regression coefficient have the expected sign? (The answer is yes.)

Here $\hat{\beta}_0 = 6.71824$ is the average value of Y when $X_1 = \cdots = X_4 = 0$. Moreover, when X_1 increases by 1 unit, the average value of Y increases by $\hat{\beta}_1 = 0.06148$ units, given that the remaining regressors are held constant. (The other sample regression slopes may be interpreted in a similar fashion.)

B. From Equation 12.14, the estimated standard errors are

$$\begin{aligned}
&s_{\hat{\beta}_0} = s_\varepsilon\sqrt{C_{00}} = 0.80113\sqrt{0.50297}\\
&= 0.56817; s_{\hat{\beta}_1} = s_\varepsilon\sqrt{C_{11}} = 0.80113\sqrt{0.0002465} = 0.01258;\ s_{\hat{\beta}_2} = s_\varepsilon\sqrt{C_{22}}\\
&= 0.80113\sqrt{0.0024437} = 0.03960, \text{etc.}
\end{aligned}$$

C. From Equation 12.22, the estimated or calculated t values are all determined under the null hypothesis $H_0 : \beta_k = 0$, $k = 0,1,2,3,4$. For instance

(*i*) for $H_0 : \beta_1 = 0$ versus $H_0 : \beta_1 \neq 0$,

$$t = \frac{\hat{\beta}_1}{s_{\hat{\beta}_1}} = \frac{0.06148}{0.01258} = 4.89.$$

Because the two-tailed p-value < 0.0001, we reject H_0 in favor of H_1; β_1 is significantly different from zero.

$$(ii) \text{ for } H_0 : \beta_2 = 0 \text{ versus } H_0 : \beta_2 \neq 0,$$

$$t = \frac{\hat{\beta}_1}{s_{\hat{\beta}_1}} = \frac{0.04970}{0.03960} = 1.25.$$

Because the two-tailed p-value = 0.2192 > 0.05, we cannot reject H_0; β_2 is not significantly different from zero. A similar set of hypothesis tests involving β_3 and β_4 reveals that both of these regression slopes differ significantly from zero because each of their p-values is less than 0.05. Given that $\hat{\beta}_1$, $\hat{\beta}_3$ and $\hat{\beta}_4$ were found to be statistically significant, does this necessarily mean that X_1, X_3, and X_4 represent the "best subset" of regressors for explaining Y? For a discussion of this account, see Appendix 12.L. I. The 95% confidence intervals for β_k, $k=0,1,2,3,4$, are obtained by using Equation 12.16. For instance, given $t_{\frac{\alpha}{2}, n-m-1} = t_{0.025,\,30} = 2.042$, a 95% confidence interval for β_1 is $\beta_1 \pm t_{\frac{\alpha}{2}, n-m-1}\, s_{\hat{\beta}_1}$ or $0.06148 \pm 2.042\,(0.01258)$ or 0.06148 ± 0.0257. Here we may be 95% confident that β_1 assumes a value between 0.03579 and 0.08716. So how precisely have we estimated β_1? We are within ± 0.0257 units of β_1 with 95% reliability. Confidence intervals for the remaining regression parameters have been determined in a similar manner. Are these confidence interval estimates consistent with the hypothesis test results given in Part 4.C?

D. The Type I sum of squares (SS1) is a sequential sum of squares; it decomposes or partitions SSR into incremental sums of squares due to each regressor as that regressor is introduced sequentially into the model, with the actual sequence or ordering of the variables specified in the *MODEL* statement. Mirroring the structure of Table 12.3, we see from the ANOVA Table 3 that $SSR(X_1,X_2,X_3,X_4) = 294.28855$. It is decomposed into sums of squares as

$$\begin{array}{r} SSR(X_1, X_2, X_3, X_4) \\ \hline SSR(X_1) = 278.55508 \\ SSR(X_2|X_1) = \quad 5.31242 \\ SSR(X_3|X_1, X_2) = \quad 7.61398 \\ SSR(X_4|X_1, X_2, X_3) = \quad 2.80707 \\ \hline 294.28855 \end{array}$$

Note that

$$SSR(X_1,X_2) = SSR(X_2|X_1) + SSR(X_1) = 5.31242 + 278.5508 = 283.86322$$

and

$$\begin{aligned} SSR(X_1,X_2,X_3) &= SSR(X_3|X_1,X_2) + SSR(X_1,X_2) \\ &= 7.61398 + 283.86322 = 291.4772. \end{aligned}$$

E. The Type II sum of squares (SS2) for a particular variable is a partial sum of squares that is not order dependent; it is the increase in SSR attributable to introducing that variable into a regression model that already contains all other variables mentioned in the *MODEL* statement. Given the variables list

Variable	Type II Sum of Squares (SS2)
$cold(X_1)$	$SSR(X_1|X_2, X_3, X_4) = 15.33510$
$foll(X_2)$	$SSR(X_2|X_1, X_3, X_4) = 1.01061$
$clt(X_3)$	$SSR(X_3|X_1, X_2, X_4) = 6.09974$
$\exp(X_4)$	$SSR(X_4|X_1, X_2, X_3) = 2.80707$

Ⓙ The estimated variance-covariance matrix is $V(\hat{\boldsymbol{\beta}}) = S_\varepsilon^2(\boldsymbol{X}'\boldsymbol{X})^{-1} = S_\varepsilon^2\boldsymbol{C}$, where $V(\hat{\beta}_k) = S_\varepsilon^2 C_{kk}$, $COV(\hat{\beta}_l, \hat{\beta}_k) = S_\varepsilon^2 C_{lk}$, and $S_\varepsilon^2 = MSE$ is an estimator for σ_ε^2. For instance, because $s_\varepsilon = \sqrt{MSE} = 0.80113$, $s_{\hat{\beta}_2} = \sqrt{V(\hat{\beta}_2)} = s_\varepsilon\sqrt{C_{22}} = 0.80113\sqrt{0.00244} = 0.03960$ while $COV(\hat{\beta}_1, \hat{\beta}_2) = s_\varepsilon^2 C_{12} = 0.64181(-0.000076) = -0.000049$.

Ⓚ The elements within this matrix yield consistent estimates of the (asymptotic) variances and covariances of the parameter estimates in the presence of heteroscedasticity. That is, the adjusted or asymptotic value of $V(\hat{\beta}_1) = 0.000119$, whereas its unadjusted value (from Ⓙ) is 0.000158. Similarly, the asymptotic covariance $COV(\hat{\beta}_1, \hat{\beta}_3) = -0.00019$, whereas the unadjusted value (again from Ⓙ) is −0.00015. The model specification test (White's test) considers H_0: the errors are homoscedastic and independent of the regressors, versus H_1: not H_0. With the p-value = 0.5037, we cannot reject the null hypothesis. Hence, heteroscedasticity does not seem to pose the problem of adversely influencing the efficiency of the OLS parameter estimates (it does not bias the sample variances of these estimates).

Ⓛ For a discussion on *output statistics* (e.g., detection of outliers, leverage points, influence points, etc.) see Chapter 3 ("Detection of Outliers") and Appendix 3.D. The reader should consider an assessment of observation 20 in this context.

Ⓜ Simple or Pearson product-moment correlation coefficients,

$$r_{lk} = \frac{M_{lk}}{\sqrt{M_{ll}}\sqrt{M_{kk}}} = \frac{\sum(X_{il} - \bar{X}_l)(X_{ik} - \bar{X}_k)}{\sqrt{\sum(X_{il} - \bar{X}_l)^2}\sqrt{\sum(X_{ik} - \bar{X}_k)^2}}$$

and

$$r_{lY} = \frac{M_{lY}}{\sqrt{M_{ll}}\sqrt{M_{YY}}} = \frac{\sum(X_{il} - \bar{X}_l)(Y_i - \bar{Y})}{\sqrt{\sum(X_{il} - \bar{X}_l)^2}\sqrt{\sum(Y_i - \bar{Y})^2}},$$

l, k =1,2,3,4, along with their associated (two-tail) p-values, are calculated for all pairs of variables.

Here we test $H_0 : \rho_{lk} = 0$ against $H_1 : \rho_{lk} \neq 0$ and $H_0 : \rho_{ly} = 0$ against $H_1 : \rho_{ly} \neq 0$ for all l, k values. Because each p-value < 0.01, we reject H_0 in all cases and conclude that each of these population simple correlation coefficients differs significantly from zero.

F. The procedure for generating standardized regression coefficients has been discussed above (see "Standardized Regression Coefficients"). For instance, for the regressor *cold* (X_1), we have $\hat{\beta}_1^* = \hat{\beta}_2^{**} = 0.4975$. Using the z-transformation, we have (from the *simple statistics* section of the *proc corr* output)

$$\hat{\beta}_1^* = \hat{\beta}_1\left(\frac{s_1}{s_Y}\right) = 0.06148\left(\frac{24.57412}{3.03675}\right) = 0.4975;$$

and from unit length scaling,

$$\hat{\beta}_1^{**} = \hat{\beta}_1\left(\frac{M_{11}}{SST}\right)^{\frac{1}{2}} = 0.06148\left(\frac{20532.1714}{313.54286}\right)^{\frac{1}{2}} = 0.4975,$$

where the corrected sum of squares $M_{11} = \sum_i(X_{i1} - \bar{X}_1)^2 = \sum_i X_{i1}^2 - (\sum_i X_{i1})/n = 280770 - (3018)^2/35 = 20532.1714$. Note that the value M_{11} has been determined from information on the uncorrected sums of squares appearing within the $\boldsymbol{X'X}$ matrix.)

The following classes of hypothesis tests (Examples 12.2 through 12.5) provide a vehicle for assessing the contribution of individual regressors or groups of regressors to explaining variation in the dependent variable Y. These are the so-called added-last tests and added-in-order tests, which can be calculated on an individual or group basis.

Example 12.2

A *variable added-last test* compares the full or complete model with a reduced model obtained by deleting, say, the *r*th variable from the complete model. That is, it seeks to determine whether including the *r*th variable in a regression equation that already contains the $r-1$ remaining variables makes a significant contribution to explaining variation in the dependent variable Y. When the *r*th variable is deleted from the full model, the resulting or *reduced model* will be denoted *m_r*.

Let us apply this procedure to the Example 12.1 dataset. Because we have $m = 4$, there are four such tests that can be specified from a model pool containing five separate regression models (four reduced models *m_j*, $j = 1,2,3,4$, and the full model):

Model Pool

$$
\begin{aligned}
m\_1: Y &= \beta_0 + \beta_2X_2 + \beta_3X_3 + \beta_4X_4 + \varepsilon \\
m\_2: Y &= \beta_0 + \beta_1X_1 + \beta_3X_3 + \beta_4X_4 + \varepsilon \\
m\_3: Y &= \beta_0 + \beta_1X_1 + \beta_2X_2 + \beta_4X_4 + \varepsilon \\
m\_4: Y &= \beta_0 + \beta_1X_1 + \beta_2X_2 + \beta_3X_3 + \varepsilon \\
full: Y &= \beta_0 + \beta_1X_1 + \beta_2X_2 + \beta_3X_3 + \beta_4X_4 + \varepsilon
\end{aligned}
$$

For a variable added-last test we consider $H_0: \beta_r = 0$ vs. $H_0: \beta_r \neq 0$. Under H_0, the incremental regression sum of squares is $\Delta SSR = SSR(full) - SSR(m\_r)$ (or the incremental error sum of squares is $\Delta SSE = SSE(m\_r) - SSE(full)$). Note that, in all instances, we compare a reduced model with the full model. Then the added-last calculated or sample test statistic is

$$
F = \frac{\Delta SSR/(m+1-(m+1-1))}{SSE(full)/(n-m-1)} = \frac{\Delta SSR}{MSE(full)}, \tag{12.39}
$$

with critical region $\mathcal{R} = \{F \mid F > F_{\alpha,1,n-m-1}\}$.

The SAS code for the preceding battery of models appears in Exhibit 12.2.

Exhibit 12.2 SAS Code for Variable Added-Last Tests

```
proc reg data=ABCsold;
      m_1: model  units = foll clt exp;
      m_2: model  units = cold clt exp;
      m_3: model  units = cold foll exp;
      m_4: model  units = cold foll clt;
      full: model  units = cold foll clt exp;
run;
```

The ANOVA table for each of theses models appears in Table 12.6.

Given the regression results appearing in Table 12.6, we can conduct a series of variable added-last *F* tests using Equation 12.39 as follows:

m_1: $\Delta SSR = SSR(full) - SSR(m\_1) = 294.2855 - 278.95345 = 15.33510$. Then $F = 15.33510/0.64181 = 23.89091$, *p*-value <0.005. Hence, the regressor *cold* (X_1) contributes significantly to a model that already incorporates the regressors *foll*, *clt*, and *exp*.

TABLE 12.6

ANOVA Tables for Variable Added-Last Tests

Variable Deleted	Source	DF	Sum of Squares	Mean Square	*F* Value	Pr > *F*
m_1	Model	3	278.95345	92.98448	83.34	<0.0001
	Error	31	34.58949	1.11579		
	Corrected Total	34	313.54286			
m_2	Model	3	293.27795	97.75932	149.55	<0.0001
	Error	31	20.26491	0.65371		
	Corrected Total	34	313.54286			
m_3	Model	3	288.18881	96.16049	117.45	<0.0001
	Error	31	25.35405	0.81787		
	Corrected Total	34	313.54286			
m_4	Model	3	291.48148	97.57214	136.53	<0.0001
	Error	31	22.06137	0.71166		
	Corrected Total	34	313.54286			
full	Model	4	294.28855	73.57214	114.63	<0.0001
	Error	30	19.25430	0.64181		
	Corrected Total	34	313.54286			

m_2: $\Delta SSR = SSR(full) - SSR(m\_2) = 294.28855 - 293.27795 = 1.01076$. Here F=1.01076/0.64181= 1.574688, p-value >0.10. Thus, the regressor *foll* (X_2) does not make a significant contribution to a model that already involves the regressors *cold, clt*, and *exp*.

m_3: $\Delta SSR = SSR(full) - SSR(m\_3) = 6.09974$ with F=9.50293, p-value <0.005. Clearly the regressor *clt* (X_3) contributes significantly to a model that already contains the regressors *cold, foll*, and *exp*.

m_4: $\Delta SSR = SSR(full) - SSR(m\_4) = 2.80707$ with F=4.37320, 0.025 < p-value < 0.05. Hence, *exp* (X_4) contributes significantly to a model already incorporating *cold, foll*, and *clt*.

Are these variable added-last F tests equivalent to the individual t tests performed in Example 12.1? (See Section 4.C of Table 12.5.)

Example 12.3

A *variable added-in-order (sequential) test* compares a model with regressors $X_1,\ldots,X_r$ to a model with regressors $X_1,\ldots,X_{r-1}$. That is, we seek to determine whether including the regressor X_r in a regression equation that already contains the first $r - 1$ regressors makes a significant contribution to explaining variation in the dependent variable Y. When the rth variable X_r is included in a model involving $X_1,\ldots,X_{r-1}$, the resulting or augmented model will be denoted *mr*.

If we apply this process to the Example 12.1 dataset, then we will have, with $m\cdot = 4$, a model pool containing four augmented models *mj*, $j = 1,2,3,4$, where the full model is *m*4:

Model Pool

$$m0: Y = \beta_0 \qquad + \varepsilon$$
$$m1: Y = \beta_0 + \beta_1 X_1 \qquad + \varepsilon$$
$$m2: Y = \beta_0 + \beta_1 X_1 + \beta_2 X_2 \qquad + \varepsilon$$
$$m3: Y = \beta_0 + \beta_1 X_1 + \beta_2 X_2 + \beta_3 X_3 \qquad + \varepsilon$$
$$full: Y = \beta_0 + \beta_1 X_1 + \beta_2 X_2 + \beta_3 X_3 + \beta_4 X_4 + \varepsilon.$$

For the variable added-in-order test, we have $H_0: \beta_{r|1,\ldots,r-1} = 0$ vs. $H_1: H_0$ not true. Under H_0, the incremental regression sum of squares $\Delta SSR = SSR(mj) - SSR(m(j-1))$ (or the incremental error sum of squares is $\Delta SSE = SSE(m(j-1)) - SSE(mj)$). Note that, in all instances, these incremental sums of squares involve the comparison of two successive augmented models. The added-in-order calculated or sample test statistic is

$$F = \frac{\Delta SSR/(j-(j-1))}{SSE(full)/(n-m-1)} = \frac{\Delta SSR}{MSE(full)}, \tag{12.40}$$

with the critical region specified as $\mathcal{R} = \{F \mid F > F_{\alpha,1,n-m-1}\}$.

The *SAS* code for this series of added-in-order tests is in Exhibit 12.3.

Exhibit 12.3 SAS Code for Added-In-Order Tests

```
proc reg data=ABCsold;
      m0: model  units = ;
      m1: model  units = cold;
      m2: model  units = cold foll;
      m3: model  units = cold foll clt;
      full: model  units = cold foll clt exp;
run;
```

The ANOVA results for each of these models are summarized in Table 12.7.

Given these regression results, the set of sequential or added-in-order F tests can be conducted using Equation 12.40.

m1: $\Delta SSR = SSR(m1) - SSR(0) = 278.55508$. Then $F = 278.55508/0.64181 = 434.01486$, p-value <0.0001. Hence, including the regressor *cold* (X_1) significantly enhances the explanatory power of the model over and above that displayed by including only an intercept.

TABLE 12.7

ANOVA Tables for Variable Added-in-Order Tests

Augmented Model	Source	DF	Sum of Squares	Mean Square	*F* Value	Pr > *F*
mo	Model	0	0	-	-	-
	Error	34	313.54286	9.22184		
	Corrected Total	34	313.54286			
m1	Model	1	278.55508	278.55508	262.73	<0.0001
	Error	33	34.98778	1.06024		
	Corrected Total	34	313.54286			
m2	Model	2	283.86750	141.93375	153.05	<0.0001
	Error	32	29.67536	0.92735		
	Corrected Total	34	313.54286			
m3	Model	3	291.48148	97.16049	136.53	<0.0001
	Error	31	22.06137	0.71166		
	Corrected Total	34	313.54286			
full	Model	4	294.28855	73.57214	114.63	<0.0001
	Error	30	19.25430	0.64181		
	Corrected Total	34	313.54286			

m2: $\Delta SSR = SSR(m2) - SSR(m1) = 283.86750 - 278.55508 = 5.31242$. Then $F = 5.31242/0.64181 = 8.27725$, p-value <0.01. So given that an intercept and *cold* (X_1) are already in the model, also including *foll* (X_2) significantly increases the explanatory power of the model.

m3: $\Delta SSR = SSR(m3) - SSR(m2) = 291.48148 - 283.86750 = 7.61398$. Then $F = 7.61398/0.64181 = 11.86329$, p-value <0.01.Given that β_0 along with the regressors *cold* (X_1) and *foll* (X_2) are already in the model, augmenting the same with *clt* (X_3) significantly contributes to the model's explanatory power.

m4: $\Delta SSR = SSR(full) - SSR(m3) = 294.28855 - 291.48148 = 2.80707$. Thus, $F = 2.80707/0.64181 = 4.37368$, p-value <0.01. Adding *exp*(X_4) to a regression equation that currently admits an intercept and regressors *cold* (X_1), *foll* (X_2), and *clt* (X_3) significantly increases the explanatory power of that equation.

Note that the various ΔSSR values determined above are just the SSI entries provided in Section 4.D of Table 12.5 of Example 12.1. Note also that the order of the regressors appearing in the full *MODEL* statement must be preserved when using the SSI values to conduct these sequential tests; the ordering of the regressors affects the test results.

Example 12.4

A *group added-last test* (a generalization of the individual variable added-last test) compares the full or complete model with a reduced model obtained by deleting a subset of variables from the full model. Here the complete set of regressors is partitioned into two (or more) subsets of variables $A = \{X_1,\ldots,X_r\}$ and $B = \{X_{r+1},\ldots,X_m\}$, with r variables in subset A and $m - r$ variables in subset B. In this regard, we seek to determine whether including a particular subset of variables, say, B, in a regression equation that already contains all variables in A makes a significant contribution to explaining variation in Y. When the variables in subset A (respectively, B) are deleted from the full model, the reduced model will be denoted *m_A* (respectively, *m_B*).

Looking to the Example 12.1 dataset, let us form the two subsets of variables $A = \{cold(X_1), foll(X_2)\}$ *and* $B = \{clt(X_3), \exp(X_4)\}$. Hence, the model pool can be structured as

Model Pool

$$\begin{aligned} m\_B:&\ Y = \beta_0 + \beta_1 X_1 + \beta_2 X_2 + \varepsilon \\ m\_A:&\ Y = \beta_0 + \beta_3 X_3 + \beta_4 X_4 + \varepsilon \\ full:&\ Y = \beta_0 + \beta_1 X_1 + \beta_2 X_2 + \beta_3 X_3 + \beta_4 X_4 + \varepsilon. \end{aligned}$$

To conduct a group added-last test on subset A, we must compare the full model with model *m_A*. Hence, we test H_0: $\beta_1 = \beta_2 = 0$ vs. H_1 : H_0 not true. Under H_0, the incremental regression sum of squares is $\Delta SSR = SSR(full) - SSR(m\_A)$ (or the incremental error sum of squares is $\Delta SSE = SSE(m\_A) - SSE(full)$). Then the group added-last sample test statistic is

$$F = \frac{\Delta SSR/[m+1-(m+1-r)]}{SSE(full)/(n-m-1)} = \frac{\Delta SSR/r}{MSE(full)}, \tag{12.41}$$

with critical region $\mathcal{R} = \{F | F > F_{\alpha, r, n-m-1}\}$.

To conduct a group added-last test on subset B, we now compare the full model with model *m_B*, i.e., we test H_0: $\beta_3 = \beta_4 = 0$ against H_1 : H_0 not true. Under H_0, the incremental regression sum of squares is $\Delta SSR = SSR(full) - SSR(m\_B)$ (or the incremental error sum of squares is $\Delta SSE = SSE(m\_B) - SSE(full)$). In this instance, the group added-last sample test statistic is

$$F = \frac{\Delta SSR/[m+1-(m+1-(m-r))]}{SSE(full)/(n-m-1)} = \frac{\Delta SSR/(m-r)}{MSE(full)}, \tag{12.42}$$

where $\mathcal{R} = \left\{F \middle| F > F_{\alpha, m-r, n-m-1}\right\}$. Note that each of these group added-last tests involves comparing a reduced model with the full model.

The SAS code for this group added-last test is in Exhibit 12.4.

Exhibit 12.4 SAS Code for Group Added-Last Tests

```
proc reg data=ABCsold;
      m_B: model  units = cold foll;
      m_A: model  units = clt exp;
      full: model  units = cold foll clt exp;
run;
```

The ANOVA results for the preceding three models appear in Table 12.8.

In the light of these regression results, group added-last F tests can be conducted using Equations 12.41 and 12.42:

1. Test of Group A: Compare the full model with model *m_A*. Then $\Delta SSR = SSR(full) - SSR(m\_A) =$ 294.28855 − 277.0095 = 17.28760 and, from Equation 12.41, F = (17.28760/2)/0.64181 = 13.46785, p-value <0.005. Hence, including the subset A regressors in a regression equation that already includes the subset B regressors significantly enhances the explanatory power of the model.
2. Test of Group B: Compare the full model with model *m_B*. Here $\Delta SSR = SSR(full) - SSR(m\_B) =$ 294.28855 − 283.86750 = 10.42105 and, from Equation 12.42, F = (10.42105/2)/0.64181 = 8.11849, p-value <0.005. Thus, inserting the subset B regressors into a regression model that already houses the subset A regressors significantly increases the explanatory power of the model.

Example 12.5

Let us revisit the group added-last tests presented in Example 12.4. Instead of calculating the sample F values in Equations 12.41 and 12.42 directly, let us use the *SAS TEST* statement to automatically obtain the requisite F values and their accompanying p-values.

A *TEST* statement is used to test a (joint) hypothesis that a linear function of the parameters is equal to a specified constant, in which one or more *TEST* statements can accompany a *MODEL*

TABLE 12.8

ANOVA Tables for Group Added-Last Tests

Subset Deleted	Source	DF	Sum of Squares	Mean Square	F Value	Pr > F
m_B	Model	2	283.86750	141.93375	153.05	<0.0001
	Error	32	16.68571	0.92735		
	Corrected Total	34	313.54286			
m_A	Model	2	277.00095	138.50047	121.29	<0.0001
	Error	32	36.54191	1.14193		
	Corrected Total	34	313.54286			
full	Model	4	294.28855	73.57214	114.63	<0.0001
	Error	30	19.25430	0.64181		
	Corrected Total	34	313.54286			

statement in the *PROC REG* routine. Hence, we can test hypotheses about the regression parameters estimated in the *MODEL* statement, where the elements of the hypotheses are separated by commas within a TEST statement.

For a given *TEST* statement, *PROC REG* performs an *F* test for the joint hypothesis specified by that statement; it compares the full model presented in the *MODEL* statement with a reduced model arrived at by imposing the *TEST* statement restrictions on the full model.

For example, suppose the dependent variable *Y* is specified as a linear function of the regressors X1, X2, X3, and X4. Then we may conduct the following set of hypothetical hypothesis tests:

```
model Y = X1  X2  X3  X4;
        test1: X1=0, X2=0;
        test2: X1=0, X3=0;
        test3: INTERCEPT=0;
        test4: X1-X4=0;
        test5: X1+X2=1;
run;
```

Here test 1 compares the full model to a reduced model containing only X3, X4 $(H_0 : \beta_1 = \beta_2 = 0)$. Test 2 compares the full model to a reduced model containing only X2, X4 $(H_0 : \beta_1 = \beta_3 = 0)$. Test 3 compares the full model to a reduced model sans intercept $(H_0 : \beta_0 = 0)$. Test 4 tests the hypothesis that the regression coefficient on X1 equals the regression coefficient on X4 $(H_0 : \beta_1 - \beta_4 = 0)$. Test 5 tests the hypothesis that the sum of the regression coefficients on X1 and X2 equals unity $(H_0 : \beta_1 + \beta_2 = 1)$.

Using the *TEST* statement, the *SAS* code presented in Example 12.4 can be replaced by the code in Exhibit 12.5.

Exhibit 12.5 SAS Code for Group Added-Last Tests (TEST Statement)

```
proc reg data= ABCsold;
      model units= cold foll clt exp;
            test_A : test cold=0, foll=0;
            test_B : test clt=0, exp=0;
run;
```

The resulting ANOVA output is replicated in Table 12.9.

Note that, for test_A, $\Delta SSR/2$ = 8.64380, whereas for test_B, $\Delta SSR/2$ = 5.21053. For each of these tests, *MSE*(*full*) = 0.64181. Hence, the *F* values for these two tests are equal to their Example 12.4 counterparts so that the conclusions offered in that example apply here also. This example problem clearly demonstrates the usefulness of the *TEST* statement.

TABLE 12.9

ANOVA Tables for Group Added-Last Tests

		Test_A Results:		
Source	**DF**	**Mean Square**	***F* Value**	**Pr > *F***
Numerator	2	8.64380	13.47	<0.0001
Denominator	30	0.64181		
		Test_B Results:		
Source	**DF**	**Mean Square**	***F* Value**	**Pr > *F***
Numerator	2	5.21053	8.12	0.0015
Denominator	30	0.64181		

The next example problem considers the issue of testing linear restrictions on the regression parameters. Although the theoretical underpinnings for this process have been offered in Appendices 12.H and 12.I, the practical execution of these procedures can readily be demonstrated using the *SAS* system.

Example 12.6

One or more *RESTRICT* statements, which follow a *MODEL* statement, are used to place linear equality restrictions or constraints on the parameters of the variables in the *MODEL* statement. That is, each restriction is structured as a linear combination of the model parameters equated to a constant r_q, where the index q identifies the particular constraint.

As indicated in Appendix 12.H, in the presence of a set parameter restrictions, we seek to minimize the Lagrangian of $\sum e_i^2$ (formed as $\sum e_i^2$ plus a linear combination of the parameter constraints, where the coefficients attached to the constraints in the linear combination are unrestricted Lagrange multiplier parameters). It is important to note that we lose one degree of freedom for each restriction imposed on the model because each Lagrange multiplier parameter is estimated along with the original model parameters. Hence, the model degrees of freedom will decrease by 1 for each restriction imposed.

Estimates of both the model parameters and the multiplier parameters are displayed in the SAS output along with their test statistic realizations and *p*-values, with the estimates of the multiplier parameters housed under the heading *RESTRICT.* (Note that, if a multiplier is zero at the constrained minimum of $\sum e_i^2$, then its associated restriction is not binding, i.e., the restricted parameter estimates are equal to the unrestricted ones.)

For the *q*th restriction, the Lagrange multiplier parameter λ_q measures the incremental change in $\sum e_i^2$ per unit change in the constraint constant r_q, i.e., $\lambda_q = \partial \sum e_i^2 / \partial r_q$. So if this constraint constant changes from r_q to $r_q + \Delta r_q$, then $\sum e_i^2$ changes by $2\lambda_q \Delta r_q$ (see the structure of the Langrangian in Appendix H).

For example, suppose the dependent variable *Y* is deemed a linear function of the regressors X1, X2, X3, and X4. Then we may estimate the parameters of this model subject to the following collection of parameter restrictions:

```
model Y = X1  X2  X3  X4;
          restrict  X1 + X2 = 1;
          restrict  X3 - X4 = 0;
          restrict 0.5X2 + X3 = 0;
          restrict intercept = 3;
run;
```

If the indicated parameter restrictions are to be lifted, then a new (unrestricted) *MODEL* statement must be made.

To examine the impact of a RESTRICT statement on the regression results, let us return to the Example 12.1 dataset and introduce the following (somewhat artificial) parameter constraint. The requisite SAS code is in Exhibit 12.6.

Exhibit 12.6 SAS Code for Testing a Restriction

```
proc reg data = ABCsold;
     model units = cold foll clt exp;
          restrict cold + foll = 0.2;
run;
```

The resulting output is summarized in Table 12.10.

Ⓐ Although the overall ANOVA results are quite good, they are marginally less favorable relative to the unrestricted results presented in Table 12.5.

TABLE 12.10

Output for Example 12.6

The SAS System
The REG Procedure
Model : MODEL 1
Dependent Variable: Units

Note: Restrictions have been applied to parameter estimates.

Ⓐ
Analysis of Variance

Source	DF	Sum of squares	Mean square	*F* value	Pr > *F*
Model	3	291.17915	97.05972	134.54	<0.0001
Error	31	22.36370	0.72141		
Corrected total	34	313.54286			

Root MSE	0.84936	*R*-Square	0.9287
Dependent mean	16.68571	Adj *R*-Sq	0.9218
Coeff Var	5.09033		

Ⓑ
Parameter Estimates

Variable	DF	Parameter Estimate	Standard Error	*T* Value	Pr > |*t*|
Intercept	1	6.08679	0.51995	11.71	<0.0001
cold	1	0.06743	0.01302	5.18	<0.0001
foll	1	0.13257	0.01302	10.18	<0.0001
clt	1	0.04054	0.01896	2.14	0.0405
exp	1	0.05672	0.03640	1.56	0.1294
RESTRICT	–1	–35.00525	16.86110	–2.08	0.0356*

* Probability computed using beta distribution.

Ⓑ With the restriction binding (the multiplier parameter *RESTRICT* $\neq 0$), a reversal has occurred between the statistical significance or explanatory power of *foll* and *exp*, i.e., without the constraint, the *p*-value of *foll* was 0.2192 and the *p*-value of *exp* was 0.0451. Under the constraint, the *p*-value of *foll* is <0.0001, whereas the *p*-value of *exp* increases to 0.1294. Moreover, the change in degrees of freedom is as expected and the multiplier parameter *RESTRICT* has a (statistically significant) *p*-value = 0.0356.

Point and Interval Predictions

Prediction Interval for an Individual Value of *Y*

We noted in Chapter 2 that the primary purpose of regression analysis is prediction. In this regard, suppose we have estimated a regression equation as $\hat{Y} = \hat{\beta}_0 + \hat{\beta}_1 X_1 + \cdots + \hat{\beta}_m X_m$ and, given a particular set of regressor values $X_1^0, X_2^0, \ldots, X_m^0$, we seek to determine a new *predicted individual value of* Y at $X_1^0, \ldots, X_m^0$, denoted $\hat{Y}_0$. Clearly $\hat{Y}_0 = \hat{\beta}_0 + \hat{\beta}_1 X_1^0 + \cdots + \hat{\beta}_m X_m^0$ is the best *point predictor* of the individual response in Y at $X_1^0, \ldots, X_m^0$.

If the *actual* Y at $X_1^0, \ldots, X_m^0$ is $Y_0 = \beta_0 + \beta_1 X_1^0 + \cdots + \beta_m X_m^0 + \varepsilon_0$, with ε_0 serving as the actual value of the random error term ε at $X_1^0, \ldots, X_m^0$, then the *prediction error* is $e_0 = Y_0 - \hat{Y}_0$. It can be shown (Appendix 12.J) that $e_0 : N(0, V(e_0))$, where

$$V(e_0) = \sigma_\varepsilon^2 + \sigma_\varepsilon^2 \boldsymbol{X}_0 (\boldsymbol{X}'\boldsymbol{X})^{-1} \boldsymbol{X}_0'$$

and

$$\underset{(1\times m+1)}{\boldsymbol{X}_0} = \begin{bmatrix} 1 & X_1^0 & X_2^0 & \ldots & X_m^0 \end{bmatrix}.$$

Then with $S_\varepsilon^2 = \boldsymbol{e}'\boldsymbol{e}/(n-m-1)$ serving as an estimator for σ_ε^2, a 100(1 – α)% prediction interval for Y_0 (the actual Y at X_0) is

$$\boldsymbol{X}_0 \hat{\boldsymbol{\beta}} \pm t_{\frac{\alpha}{2}, n-m-1} S_\varepsilon \sqrt{1 + \boldsymbol{X}_0 (\boldsymbol{X}'\boldsymbol{X})^{-1} \boldsymbol{X}_0'}. \tag{12.43}$$

(The reader is again referred to Appendix 12.J for details regarding the derivation of Equation 12.43.) Here Equation 12.43 indicates how precisely Y_0 has been predicted from the sample regression equation.

Given Equation 12.13, the term $\boldsymbol{X}_0 (\boldsymbol{X}'\boldsymbol{X})^{-1} \boldsymbol{X}_0'$ in Equation 12.43 can be calculated as

$$\begin{aligned} \boldsymbol{X}_0 (\boldsymbol{X}'\boldsymbol{X})^{-1} \boldsymbol{X}_0' &= \boldsymbol{X}_0 \boldsymbol{C} \boldsymbol{X}_0' \\ &= C_{00} + 2\sum_{j=1}^m X_j^0 C_{0j} + \sum_{j=1}^m \left(X_j^0\right)^2 C_{jj} \\ &\quad + \sum_{\substack{l=1 \\ l<j}}^m \sum_{j=1}^m X_l^0 X_j^0 C_{lj}. \end{aligned} \tag{12.44}$$

Prediction Interval for the Average Value of *Y*

Given the actual Y at $X_1^0, \ldots, X_m^0$ or $Y_0 = \boldsymbol{X}_0\beta + \varepsilon_0$, the mean response at $X_1^0, \ldots, X_m^0$ is $E(Y_0) = X_0\beta$ since $E(\varepsilon_0) = 0$, whereas the estimated mean response at $X_1^0, \ldots, X_m^0$ is the point predictor $\hat{Y}_0 = \boldsymbol{X}_0 \hat{\boldsymbol{\beta}}$. Hence, the prediction error is $e_0^* = E(Y_0) - \hat{Y}_0$ with $e_0^* : N(0, V(e_0^*))$, where $V(e_0^*) = \sigma_\varepsilon^2 \boldsymbol{X}_0 (\boldsymbol{X}'\boldsymbol{X})^{-1} \boldsymbol{X}_0'$ (see Appendix 12.J). Once S_ε^2 is substituted for σ_ε^2 in $V(e_0^*)$, we can determine a 100(1 – α)% prediction interval for $E(Y_0)$ (the mean value of Y at X_0) as

$$\boldsymbol{X}_0 \hat{\boldsymbol{\beta}} \pm t_{\frac{\alpha}{2}, n-m-1} S_\varepsilon \sqrt{\boldsymbol{X}_0 (\boldsymbol{X}'\boldsymbol{X})^{-1} \boldsymbol{X}_0'}, \tag{12.45}$$

with s_ε serving as the sample realization of S_ε. Why is the interval in Equation 12.43 wider than that provided by Equation 12.45? (Hint: consider $V(e_0)$ vs. $V(e_0^*)$.)

Example 12.7

Given the regression results from Example 12.1, find 95% prediction intervals for $\hat{Y}_0$ (the individual value of Y at X_0) and $E(Y_0)$ (the mean response in Y at X_0), where X_0 = [1 135 30 64 37]. From Section ④ of Example 12.1 we first determine

$$X_0\hat{\boldsymbol{\beta}} = \begin{bmatrix} 1 & 135 & 30 & 64 & 37 \end{bmatrix} \begin{bmatrix} 6.71824 \\ 0.06148 \\ 0.04970 \\ 0.06464 \\ 0.07357 \end{bmatrix} = 23.36809.$$

Then from Section ② of Example 12.1,

$$C = (X'X)^{-1} = \begin{bmatrix} 0.50297 & -0.00482 & -0.01321 & 0.00692 & -0.00437 \\ -0.00482 & 0.00024 & -0.00007 & -0.00023 & -0.00032 \\ -0.01321 & -0.00007 & 0.00244 & -0.00045 & -0.00016 \\ 0.00692 & -0.00023 & -0.00045 & 0.00069 & -0.00017 \\ -0.00437 & -0.00032 & -0.00016 & -0.00017 & 0.00193 \end{bmatrix}$$

so that $X_0CX_0' = 0.38324$.

With $t_{0.025,30} = 2.042$ and $s_\varepsilon = \sqrt{MSE} = 0.80113$, Equation 12.43 yields $23.36809 \pm 2.042(0.80113)\sqrt{1.38324}$ or 23.36809 ± 1.92401. Hence, we may be 95% confident that the predicted value of Y at X_0 or $\hat{Y}_0$ lies between 21.44468 and 25.29210.

From Equation 12.45, a 95% prediction interval for the mean value of Y at X_0 or $E(Y_0)$ is $23.36809 \pm 2.042(0.80113)\sqrt{0.38324}$ or (22.35536, 24.38082).

Multicollinearity

When it comes to the application of multiple regression techniques to actual datasets, there often exists a set of underlying interrelationships or *collinearities* among variables that span the columns of the data matrix X. That is, certain subsets of regressors may be connected by unobserved *auxiliary regression equations* that must be identified and accounted for if one is to generate reliable estimates of the model regression coefficients $\beta_0, \beta_1, \ldots, \beta_m$. The basic data problem just described is classified under the heading or *multicollinearity.*

It must be emphasized, however, that some degree of multicollinearity is always present in most datasets (very few, if any, variables are truly independent of all other variables); what is important is not the presence or absence of multicollinearity per se but the *degree of multicollinearity* in a dataset. In this regard, we are primarily interested in the case in which the regressors are highly (although not perfectly) collinear; the case of what is termed *near multicollinearity.* (Although the following discussion on near multicollinearity is quite general in nature, a more technical treatment of multicollinearity is offered in Appendix 12.K.)

What are the consequences of near multicollinearity?

1. It may be very difficult to identify the separate effects of the individual regressors on the dependent variable Y, and, as the degree of multicollinearity increases, our ability to disentangle their separate effects diminishes.
2. The estimated regression coefficients lack precision; $V(\hat{\beta}_k)$, $k = 0,1,\ldots,m$, tends to be artificially inflated. Moreover, if regressors X_k and X_l are highly collinear, then $COV(\hat{\beta}_k, \hat{\beta}_l)$ also tends to be inordinately large.

3. Some of the estimated regression coefficients may be highly sensitive to the addition or deletion of a few data points or to the deletion of a regressor that lacks statistical significance.
4. A classical symptom of multicollinearity is that the overall model seems to display significant explanatory power (the regressors are jointly significant in that the R^2 and global F statistics are high) and yet, paradoxically, most (if not all) of the regressors are viewed as being individually unimportant in explaining variation in the dependent variable Y (the t value associated with a regressor's estimated coefficient is not statistically significant).
5. An individual regression coefficient may display the wrong sign or may appear in the estimated equation at an implausibly high (in absolute value) level.

An important multicollinearity diagnostic is the *variance inflation factor* (*VIF*) for an estimated regression coefficient. That is, for the jth regression coefficient $\hat{\beta}_j$,

$$VIF_j = \frac{1}{1-R_j^2}, \quad j=1,\ldots,m, \tag{12.46}$$

where R_j^2 is the coefficient of multiple determination obtained by regressing X_j on the remaining $m-1$ regressors. In this regard, VIF_j is a measure of how much $V(\hat{\beta}_j)$ is inflated by the relationship of X_j to the other columns of $\boldsymbol{X}$; it reflects the combined effect of the near dependencies among the regressors on $V(\hat{\beta}_j)$. Note that, if R_j^2 is near unity, VIF_j will be very large. In fact, the general consensus is that if any of the VIF_j's exceed 10, then there may be a serious problem with near multicollinearity: $\hat{\beta}_j$ may be imprecisely estimated. In the absence of any collinearity between X_j and the remaining columns of $\boldsymbol{X}$, $R_j^2 = 0$ and $VIF_j = 1$.

Another useful diagnostic device for gauging the degree of severity of multicollinearity in a dataset is the performance of an *eigenanalysis* (see Appendix 12.K, "Eigenanalysis"). Specifically, the magnitudes of the eigenvalues or characteristic roots of the $\boldsymbol{X'X}$ matrix serve as the focal point of this mode of analysis. That is, one or more very small eigenvalues is a symptom of the existence of near-linear dependencies among the columns of $\boldsymbol{X}$. In fact, the number of exact linear dependencies corresponds to the number of zero eigenvalues. More formally, certain statistics derived from the eigenvalues can be used to ultimately determine which of the estimated regression coefficients have the greatest potential of being affected by near collinearity.

Let us begin with the *condition number* of the $\boldsymbol{X'X}$ matrix

$$\kappa = \sqrt{\frac{\lambda_{\max}}{\lambda_{\min}}}, \tag{12.47}$$

where $\lambda_{\max}$ and $\lambda_{\min}$ are, respectively, the largest and smallest eigenvalues of the $\boldsymbol{X'X}$ matrix. If the eigenvalues are sorted by numerical size, beginning with the largest, then the condition number involves the first and last eigenvalues. As a general guideline, if the condition number ranges between 20 and 30, then moderate to serious near multicollinearity is indicated. A condition number in excess of 30 reflects severe or problematic near multicollinearity.

Next, we may determine the condition index for each eigenvalue of $\boldsymbol{X'X}$ as

$$\kappa_k = \sqrt{\frac{\lambda_{\max}}{\lambda_k}}, \quad k=0,1,\ldots,m, \tag{12.48}$$

where obviously the largest condition index is the condition number κ. A condition index of at least 30 indicates that near multicollinearity is a problem, and the number of condition indexes displaying a value of at least 30 is a good indication of the number of near collinearities in the dataset.

Finally, a critical feature of eigenanalysis is that it enables us to explore the actual structure of any near-linear dependencies in the dataset. In particular, we may obtain the *variance decomposition* of each estimated regression coefficient, i.e., for each $V(\hat{\beta}_k), k = 0,1,\ldots,m$, we can examine the *variance decomposition proportion* π_{ld} for the lth eigenvalue $\lambda_l, l = 0,1,\ldots,m$. Here π_{ld} is the proportion of the variance of $\hat{\beta}_k$ accounted for or contributed by the lth eigenvalue. In general, when the condition index for some eigenvalue exceeds 30 and, at the same time, a high proportion (typically greater than 0.5) of the variance of at least two regression coefficients is accounted for by that eigenvalue, then serious near dependencies among the columns of **X** are highly probable.

Example 12.8

Are there any serious near multicollinearities in the Example 12.1 dataset? To answer this question, we need to make a slight adjustment to the SAS code appearing in Exhibit 12.1. Specifically, we can request variance inflation factors and the results of an eigenanalysis by modifying the *MODEL* statement in *PROC REG* (see Exhibit 12.7).

Exhibit 12.7 SAS Code for Collinearity Diagnostics

```
data ABCsold;
       input units cold foll clt exp;
datalines;
 .
 .
 .
;
proc reg data = ABCsold;                        ①        ②       ③
       model units = cold foll clt exp/ collin   tol    vif;
run;
```

① We request the COLLIN option that provides an analysis of collinearity among the regressors. Here the $X'X$ matrix is scaled to have 1's on the main diagonal.

② The TOL option displays the tolerance values for the parameter estimates $\hat{\beta}_j$, $j = 1,\ldots,m$, where *tolerance* is $1 - R_j^2$; it serves as a measure of *uniqueness*. Obviously, large tolerance values (near 1) are desirable.

③ The VIF option requests variance inflation factors for each regressor $X_1,\ldots,X_m$.

The output generated by these options appears in Table 12.11.

① The lowest tolerance value is 0.19762, indicating that R_1^2 (the regressor X_1 is *cold*) is about 0.802. Hence, X_1 is moderately related to the other regressors. A similar conclusion holds for the regressor X_3 (*clt*) because $R_3^2 = 0.782$.

② None of the VIF's exceeds 10, whereas the largest VIF's are exhibited by X_1 (*cold*) and X_3 (*clt*). Examining the variance inflation factors does not lead us to suspect any troublesome multicollinearity.

③,④ The largest eigenvalue is λ_0 = 4.85894 (corresponding to the intercept) and the smallest eigenvalue is 0.01064. Hence, the condition number is $\kappa = 21.37$ Clearly κ is at the low end of the range for suspected near collinearities among the regressors. Note that this is also the value of the condition index κ_4 for regressor X_4 (*exp*). Thus, κ_4 is the only condition index that provides even a slight hint of multicollinearity.

⑤ Looking to the variance decomposition proportions, because $\kappa_4 = 21.37$, an examination of the last row of this subtable reveals that, under the $V(\hat{\beta}_1)$ and $V(\hat{\beta}_3)$ headings (corresponding to the regressors X_1 (*cold*) and X_3 (*clt*) , respectively), the variance decomposition proportions are $\pi_{14} = 0.95895$ and $\pi_{34} = 0.47914$, respectively. Although π_{34} does not exceed 0.5, we can still conclude that X_1 and X_3 are moderately collinear.

TABLE 12.11

Output for Example 12.8

The SAS System
The REG Procedure
Model : MODEL 1
Dependent Variable: Units

Analysis of Variance

Source	DF	Sum of Squares	Mean Square	*F* Value	Pr > *F*
Model	4	294.28855	73.57214	114.63	<.0001
Error	30	19.25430	0.64181		
Corrected total	34	313.54286			

Root MSE	0.80113	R- Square	0.9386
Dependent mean	16.68571	Adj R-Sq	0.9304
Coeff Var	4.80130		

Parameter Estimates

Variable	DF	Parameter Estimate	Standard Error	*t* Value	Pr > | *t* |	① Tolerance	② Variance Inflation
Intercept	1	6.71824	0.56817	11.82	<0.0001	.	0
cold	1	0.06148	0.01258	4.89	<0.0001	0.19762	5.06032
foll	1	0.04970	0.03960	1.25	0.2192	0.47497	2.10541
clt	1	0.06464	0.02097	3.08	0.0044	0.21759	4.59585
exp	1	0.07357	0.03518	2.09	0.0451	0.33364	2.99724

TABLE 12.11 *(continued)*

Collinearity Diagnostics

Number	③ Eigenvalue	④ Condition index
(intercept) 1	4.85894	1.00000
2	0.07100	8.27232
3	0.03669	11.50732
4	0.02272	14.62414
5	0.01604	21.36698

Collinearity Diagnostics

⑤

Proportion of Variation

Number	Intercept	cold	foll	clt	exp
(intercept)1	0.00229	0.00060546	0.00169	0.00117	0.00125
2	0.53236	0.00270	0.00000187	0.12086	0.00824
3	0.01420	0.01821	0.73757	0.00099123	0.23398
4	0.21331	0.01954	0.25289	0.39787	0.56475
5	0.23785	0.95895	0.00784	0.47912	0.19178

Example 12.9

Exhibit 12.8 contains the requisite SAS code for generating collinearity diagnostics for the dataset appearing in Malinvaud [1970, 19]. This dataset spans the period from 1947 to 1966 and presents observations on the variables: imports (I); gross domestic product (GDP); stock formation (SF); and consumption (C). We want to estimate the parameters of the regression equation

$$Y = \beta_0 + \beta_1 X_1 + \beta_2 X_2 + \beta_3 X_3 + \varepsilon$$

or

$$I = \beta_0 + \beta_1 GDP + \beta_2 SF + \beta_3 C + \varepsilon.$$

Table 12.12 contains the SAS output that displays the classic symptoms of multicollinearity: high F and R^2 values and insignificant t values for the regressors. In addition,

① The tolerances for $X_1(GDP)$ and $X_3(C)$ are quite low, indicating that R_1^2 and R_3^2 are very high in value (each is approximately 1).

② The variance inflation factors VIF_1(=469.74) and VIF_3(= 469.37) are extremely large, thus indicating that there is probably a serious problem with near multicollinearity: there is a near-linear dependency of the second (and also the fourth) column of **X** on the other columns of **X** (remember that the first column of **X** contains all 1's).

TABLE 12.12
Output for Example 12.9

The SAS System
The REG Procedure
Model: MODEL 1
Dependent Variable: I

Analysis of Variance

Source	DF	Sum of Squares	Mean Square	*F* Value	Pr > *F*
Model	3	2576.92075	858.97358	168.45	<0.0001
Error	14	71.39037	5.09931		
Corrected total	17	2648.31111			

Root MSE	2.25817	*R*-Square	0.9730
Dependent mean	30.07778	Adj *R*-Sq	0.9673
Coeff Var	7.50775		

Parameter Estimates

Variable	DF	Parameter Estimate	Standard Error	*t* Value	Pr > | *t* |	① Tolerance	② Variance Inflation
Intercept	1	−19.72511	4.12525	−4.78	0.0003	.	0
GDP	1	0.03220	0.18688	0.17	0.8656	0.00213	469.74214
SF	1	0.41420	0.32226	1.29	0.2195	0.95249	1.04988
C	1	0.24275	0.28536	0.85	0.4093	0.00213	469.37134

Collinearity Diagnostics

			⑤ Proportion of Variation			
Number	**③ Eigenvalue**	**④ Condition Index**	**Intercept**	**GDP**	**SF**	**C**
1	3.82821	1.00000	0.00110	0.00000900	0.01021	0.00000786
2	0.13287	5.36763	0.00424	0.00010959	0.94601	0.00009077
3	0.03886	9.92526	0.30445	0.00059564	0.04225	0.00032481
4	0.00006258	247.33056	0.69021	0.99929	0.00513	0.99958

Exhibit 12.8 SAS Code for Collinearity Diagnostics (Malinvaud data)

```
data malvd;
     input I  GDP  SF  C;
datalines;
15.9  149.3  4.2  108.1
16.4  161.2  4.1  114.8
19.0  171.5  3.1  123.2
```

```
19.1  175.5  3.1  126.9
18.8  180.8  1.1  132.1
20.4  190.7  2.2  137.7
22.7  202.1  2.1  146.0
26.5  212.4  5.6  154.1
28.1  226.1  5.0  162.3
27.6  231.9  5.1  164.3
26.3  239.0  0.7  167.6
31.1  258.0  5.6  176.8
33.3  269.8  3.9  186.6
37.0  288.4  3.1  199.7
43.3  304.5  4.6  213.9
49.0  323.4  7.0  223.8
50.3  336.8  1.2  232.0
56.6  353.9  4.5  242.9
;
proc reg data=malvd;
     model I=GDP  SF  C/collin  tol vif;
run;
```

③ The eigenvalue for $X_3(C)$ is $\lambda_3 = 0.000063$, an indication that a near collinearity exists among the columns of **X** and that **X'X** might be near singular (because$|\mathbf{X'X}| = \lambda_0 \cdot \lambda_1 \cdot \lambda_2 \cdot \lambda_3$).

④ The condition number $\kappa = 247.33$ while the largest condition index is $\kappa_3 = 247.33$. Hence, this rather sizeable condition index informs us that the variance decomposition proportions for λ_3 need to be examined.

⑤ The last vow of this submatrix contains the variance decomposition proportions π_{kl} of $V(\hat{\beta}_k), k = 0,1,2,3$, accounted for by λ_3. Three of these variance decomposition proportions exceed 0.5: $\pi_{03} = 0.69021$; $\pi_{13} = 0.99929$; and $\pi_{33} = 0.99958$. Hence, there is an almost exact linear dependency among the first, second, and fourth columns of **X**.

Let us estimate this implied relationship or near-linear dependency, i.e., let us fit $C = \alpha_0 + \alpha_1 GDP + \varepsilon$ (the regression output appears in Table 12.13). The resulting F value is extremely large while $R^2 = 0.998$ Hence, there is almost an exact linear relationship or dependency among the aforementioned columns of ***X***.

If multicollinearity is detected in a dataset, what are the remedies for dealing with it? If we remember that multicollinearity is basically a data problem, then one or more of the particulars on the following laundry list of remedies might apply:

1. Add additional observations that are less collinear then the original data values.
2. Obtain outside estimates of some parameters from an independent source or unrelated study, e.g., cross-section estimates of certain parameters can be inserted into a time-series model.
3. Use a *ridge estimator* instead of an OLS estimator for β (a discussion of ridge regression is offered in Chapter 14).
4. Judiciously drop a variable (or variables) from the regression equation. Theory should dictate which variable (or variables) to discard; otherwise, one runs the risk of introducing a specification bias into the regression model.
5. Transform the collinear variables to logarithms, first differences, ratios, etc.

TABLE 12.13

Output for Regression of C on GDP

The SAS System
The REG Procedure
Model : MODEL 1
Dependent Variable: C

Analysis of Variance

Source	DF	Sum of Squares	Mean Square	F Value	Pr > F
Model	1	29330	29330	7485.35	<0.0001
Error	16	62.69322	3.91833		
Corrected total	17	29393			

Root MSE	1.97948	R-Square	0.9979
Dependent mean	167.37778	Adj R-Sq	0.9977
Coeff Var	1.18264		

Parameter Estimates

Variable	DF	Parameter Estimate	Standard Error	t Value	Pr > |t|
Intercept	1	12.05408	1.85491	6.50	<0.0001
GDP	1	0.65395	0.00756	86.52	<0.0001

Appendix A: Gauss-Markov Theorem

Given the regression equation $\boldsymbol{Y} = \boldsymbol{X\beta} + \boldsymbol{\varepsilon}$, the OLS estimator $\hat{\boldsymbol{\beta}} = (\boldsymbol{X}'\boldsymbol{X})^{-1}\boldsymbol{X}'\boldsymbol{Y}$ is BLUE, i.e., $\hat{\boldsymbol{\beta}}$ is a best linear unbiased estimator for $\boldsymbol{\beta}$.

The verification of this assertion proceeds as follows:

1. Clearly $\hat{\boldsymbol{\beta}} = (\boldsymbol{X}'\boldsymbol{X})^{-1}\boldsymbol{X}'\boldsymbol{Y} = \boldsymbol{BY}$ is a linear function of $\boldsymbol{Y}$ for nonstochastic $\boldsymbol{X}$.
2. $\hat{\boldsymbol{\beta}}$ is an unbiased estimator for $\boldsymbol{\beta}$ because with $\hat{\boldsymbol{\beta}} = (\boldsymbol{X}'\boldsymbol{X})^{-1}\boldsymbol{X}'(\boldsymbol{X\beta} + \boldsymbol{\varepsilon}) = \boldsymbol{\beta} + (\boldsymbol{X}'\boldsymbol{X})^{-1}\boldsymbol{X}'\boldsymbol{\varepsilon}, E(\hat{\boldsymbol{\beta}}) = \boldsymbol{\beta}$ given that **X** is nonstochastic and $E(\boldsymbol{\varepsilon}) = 0$.
3. Suppose $\boldsymbol{A}$ is an arbitrary nonstochastic $(p \times n)$ matrix $(p = m + 1)$ and that $E(Y_i^2) < +\infty$ $i = 1, \ldots, n$. Then for $\tilde{\boldsymbol{\beta}} = \boldsymbol{AY}$ any unbiased estimator for $\boldsymbol{\beta}$, we seek to demonstrate that $\hat{\boldsymbol{\beta}}$ is best or $V(\hat{\beta}_k) \le V(\tilde{\beta}_k)$, $k = 0, 1, \ldots, m$.

Let $V(\hat{\boldsymbol{\beta}})$ represent the variance-covariance matrix of $\hat{\boldsymbol{\beta}}$. With $\hat{\boldsymbol{\beta}} - \boldsymbol{\beta} = (\boldsymbol{X}'\boldsymbol{X})^{-1}\boldsymbol{X}'\boldsymbol{\varepsilon}$,

$$V(\hat{\boldsymbol{\beta}}) = E\left[(\hat{\boldsymbol{\beta}} - \boldsymbol{\beta})(\hat{\boldsymbol{\beta}} - \boldsymbol{\beta})'\right] = E\left[(\boldsymbol{X}'\boldsymbol{X})^{-1}\boldsymbol{X}'\boldsymbol{\varepsilon\varepsilon}'\boldsymbol{X}(\boldsymbol{X}'\boldsymbol{X})^{-1}\right]$$
$$= (\boldsymbol{X}'\boldsymbol{X})^{-1}\boldsymbol{X}'\sigma_\varepsilon^2\boldsymbol{I}_n\boldsymbol{X}(\boldsymbol{X}'\boldsymbol{X})^{-1} = \sigma_\varepsilon^2(\boldsymbol{X}'\boldsymbol{X})^{-1}$$

(because $(\boldsymbol{X}'\boldsymbol{X})^{-1}$ is symmetric, $V(\varepsilon_i^2) = \sigma_\varepsilon^2$, and $COV(\varepsilon_i, \varepsilon_j) = 0$, $i \ne j$).

Because $E(Y_i^2)$ finite implies that $E(Y_i)$ is finite, $\boldsymbol{\beta} = E(\tilde{\boldsymbol{\beta}}) = \boldsymbol{A}E(\boldsymbol{Y}) = \boldsymbol{AX\beta}$ or $(\boldsymbol{I}_p - \boldsymbol{AX})\,\boldsymbol{\beta} = 0$ for $\boldsymbol{I}_p = \boldsymbol{AX}$. Let $\boldsymbol{U}$ represent an arbitrary nonstochastic $(p \times 1)$ vector.

Because

$$V(\boldsymbol{U}'\hat{\boldsymbol{\beta}}) = \sigma_\varepsilon^2 \boldsymbol{U}'(\boldsymbol{X}'\boldsymbol{X})^{-1}\boldsymbol{U}$$

and

$$\begin{aligned} V(\boldsymbol{U}'\tilde{\boldsymbol{\beta}}) &= V(\boldsymbol{U}'\boldsymbol{AY}) = V(\boldsymbol{U}'\boldsymbol{A}(\boldsymbol{X\beta}+\boldsymbol{\varepsilon})) \\ &= E(\boldsymbol{U}'\boldsymbol{A\varepsilon\varepsilon}'\boldsymbol{A}'\boldsymbol{U}) = \boldsymbol{U}'\boldsymbol{A}\sigma_\varepsilon^2\boldsymbol{I}_n\boldsymbol{A}'\boldsymbol{U} \\ &= \sigma_\varepsilon^2\boldsymbol{U}'\boldsymbol{AA}'\boldsymbol{U}, \end{aligned}$$

we see that

$$V(\boldsymbol{U}'\tilde{\boldsymbol{\beta}}) - V(\boldsymbol{U}'\hat{\boldsymbol{\beta}}) = \sigma_\varepsilon^2\boldsymbol{U}'\left[\boldsymbol{AA}' - (X'X)^{-1}\right]\boldsymbol{U} \geq 0 \tag{12.A.1}$$

if $\boldsymbol{AA}' - (\boldsymbol{X}'\boldsymbol{X})^{-1}$ is positive semidefinite (or nonnegative definite). That is, the quadratic form in Equation12.A.1 is positive semidefinite if its coefficient matrix $\boldsymbol{AA}' - (\boldsymbol{X}'\boldsymbol{X})^{-1}$ is positive semidefinite.

Let us rewrite $\boldsymbol{AA}' - (\boldsymbol{X}'\boldsymbol{X})^{-1}$ (remember that $\boldsymbol{I}_p = \boldsymbol{AX}$) as

$$\begin{aligned} \boldsymbol{AA}' - (\boldsymbol{X}'\boldsymbol{X})^{-1} &= \boldsymbol{AA}' - (\boldsymbol{X}'\boldsymbol{X})^{-1} - (\boldsymbol{X}'\boldsymbol{X})^{-1} + (\boldsymbol{X}'\boldsymbol{X})^{-1} \\ &= \boldsymbol{AA}' - \boldsymbol{I}_p(\boldsymbol{X}'\boldsymbol{X})^{-1} - \boldsymbol{I}_p(\boldsymbol{X}'\boldsymbol{X})^{-1} + \boldsymbol{I}_p(\boldsymbol{X}'\boldsymbol{X})^{-1} \\ &= \boldsymbol{AA}' - \boldsymbol{AX}(\boldsymbol{X}'\boldsymbol{X})^{-1} - \boldsymbol{AX}(\boldsymbol{X}'\boldsymbol{X})^{-1} + (\boldsymbol{X}'\boldsymbol{X})^{-1}(\boldsymbol{X}'\boldsymbol{X})(\boldsymbol{X}'\boldsymbol{X})^{-1} \\ &= \boldsymbol{AA}' - \boldsymbol{AX}(\boldsymbol{X}'\boldsymbol{X})^{-1} - (\boldsymbol{X}'\boldsymbol{X})^{-1}\boldsymbol{X}'\boldsymbol{A}' + (\boldsymbol{X}'\boldsymbol{X})^{-1}(\boldsymbol{X}'\boldsymbol{X})(\boldsymbol{X}'\boldsymbol{X})^{-1} \\ &= (\boldsymbol{A} - (\boldsymbol{X}'\boldsymbol{X})^{-1}\boldsymbol{X}')(\boldsymbol{A}' - \boldsymbol{X}(\boldsymbol{X}'\boldsymbol{X})^{-1}) \\ &= (\boldsymbol{A} - (\boldsymbol{X}'\boldsymbol{X})^{-1}\boldsymbol{X}')(\boldsymbol{A} - (\boldsymbol{X}'\boldsymbol{X})^{-1}\boldsymbol{X}')' = \boldsymbol{DD}'. \end{aligned}$$

If $\boldsymbol{W} = \boldsymbol{D}'\boldsymbol{U}$, then $\boldsymbol{U}'\boldsymbol{DD}'\boldsymbol{U} = \boldsymbol{W}'\boldsymbol{W} \geq 0$. Hence $\boldsymbol{DD}' = \boldsymbol{AA}' - (\boldsymbol{X}'\boldsymbol{X})^{-1}$ is positive semidefinite.

Appendix B: OLS Parameter Estimators in Matrix Notation

Let us set $\boldsymbol{Y} = \begin{bmatrix} Y_1 \\ Y_2 \\ \cdot \\ \cdot \\ \cdot \\ Y_n \end{bmatrix}$, $\hat{\boldsymbol{\beta}} = \begin{bmatrix} \hat{\beta}_0 \\ \hat{\beta}_1 \\ \cdot \\ \cdot \\ \cdot \\ \hat{\beta}_m \end{bmatrix}$, and $\boldsymbol{X} = \begin{bmatrix} 1 & X_{11} & X_{12} & \cdots & X_{1m} \\ 1 & X_{21} & X_{22} & \cdots & X_{2m} \\ \cdot & \cdot & \cdot & & \cdot \\ \cdot & \cdot & \cdot & & \cdot \\ \cdot & \cdot & \cdot & & \cdot \\ 1 & X_{n1} & X_{n2} & \cdots & X_{nm} \end{bmatrix}$

with

$$X'Y = \begin{bmatrix} \sum Y_i \\ \sum X_{i1}Y_i \\ \cdot \\ \cdot \\ \cdot \\ \sum X_{im}Y_i \end{bmatrix} \text{ and } X'X = \begin{bmatrix} n & \sum X_{i1} & \sum X_{i2} & \cdots & \sum X_{im} \\ \sum X_{i1} & \sum X_{i1}^2 & \sum X_{i1}X_{i2} & \cdots & \sum X_{i1}X_{im} \\ \cdot & \cdot & \cdot & & \cdot \\ \cdot & \cdot & \cdot & & \cdot \\ \cdot & \cdot & \cdot & & \cdot \\ \sum X_{im} & \sum X_{im}X_{i1} & \sum X_{im}X_{i2} & \cdots & \sum X_{im}^2 \end{bmatrix},$$

where a prime denotes the operation of matrix transposition. Then the OLS normal Equations 12.8.1 can be written as

$$(\boldsymbol{X}'\boldsymbol{X})\hat{\boldsymbol{\beta}} = \boldsymbol{X}'\boldsymbol{Y}.^{3} \tag{12.B.1}$$

From Assumption A.3 of "The Multiple Regression Model" above, we have $\rho(\boldsymbol{X}) = m + 1$ (the columns of $\boldsymbol{X}$ are linearly independent or $|\boldsymbol{X}'\boldsymbol{X}| \neq 0$ ($\boldsymbol{X}'\boldsymbol{X}$ is nonsingular)). Hence, $(\boldsymbol{X}'\boldsymbol{X})^{-1}$ exists and

$$\hat{\boldsymbol{\beta}} = (\boldsymbol{X}'\boldsymbol{X})^{-1}\boldsymbol{X}'\boldsymbol{Y}. \tag{12.B.2}$$

Alternatively, the system in Equation 12.10 can be written in matrix form as

$$(\underline{\boldsymbol{X}'\boldsymbol{X}})\underline{\hat{\boldsymbol{\beta}}} = \underline{\boldsymbol{X}'\boldsymbol{Y}} \tag{12.B.3}$$

where

$$\underline{\hat{\boldsymbol{\beta}}} = \begin{bmatrix} \hat{\beta}_1 \\ \hat{\beta}_2 \\ \cdot \\ \cdot \\ \cdot \\ \hat{\beta}_m \end{bmatrix}, \quad \underline{\boldsymbol{X}'\boldsymbol{Y}} = \begin{bmatrix} M_{1Y} \\ M_{2Y} \\ \cdot \\ \cdot \\ \cdot \\ M_{mY} \end{bmatrix}, \text{ and } (\underline{\boldsymbol{X}'\boldsymbol{X}}) = \begin{bmatrix} M_{11} & M_{12} & \cdots & M_{1m} \\ M_{21} & M_{22} & \cdots & M_{2m} \\ \cdot & \cdot & & \cdot \\ \cdot & \cdot & & \cdot \\ \cdot & \cdot & & \cdot \\ M_{m1} & M_{m2} & \cdots & M_{mm} \end{bmatrix}.$$

[3] These OLS normal equations can be derived as follows. Let the regression model be written as $\boldsymbol{Y} = \boldsymbol{X}\boldsymbol{\beta} + \varepsilon$. Then the vector of residuals is $\boldsymbol{e} = \boldsymbol{Y} - \boldsymbol{X}\hat{\boldsymbol{\beta}}$ ($\hat{\boldsymbol{\beta}}$ is a vector of arbitrary estimators for the components of $\boldsymbol{\beta}$) and

$$e'e = f(\hat{\boldsymbol{\beta}}) = Y'Y - 2\hat{\boldsymbol{\beta}}'X'Y + \hat{\boldsymbol{\beta}}'X'X\hat{\boldsymbol{\beta}}.$$

Then according to the principle of least squares, we must choose $\hat{\boldsymbol{\beta}}$ so as to minimize $\boldsymbol{e}'\boldsymbol{e} = \Sigma e_i^2$. Setting

$$\frac{\partial f(\hat{\boldsymbol{\beta}})}{\partial \hat{\boldsymbol{\beta}}} = -2X'Y + 2X'X\hat{\boldsymbol{\beta}} = \mathbf{0}$$

we immediately obtain Equation 12.B.1. (Note that the derivative of a linear form $\boldsymbol{a}'X$ with respect to X is $\boldsymbol{a}$; and the derivative of the quadratic form $X'\boldsymbol{A}X$ with respect to X is $2\boldsymbol{A}X$.)

Then, with $|\underline{X'X}| \neq 0$,

$$\underline{\hat{\beta}} = (\underline{X'X})^{-1}(\underline{X'Y}). \tag{12.B.4}$$

Given $\hat{\boldsymbol{\beta}}$, $\hat{\beta}_0$ is determined from Equation 12.9.

Appendix C: Maximum-Likelihood Parameter Estimators in Matrix Notation

Given $\boldsymbol{Y} = \boldsymbol{X\beta} + \boldsymbol{\varepsilon}$, the assumptions of the strong classical linear regression model, along with A.1 through A.3 of "The Multiple Regression Model" above, imply that $E(\boldsymbol{\varepsilon}) = \boldsymbol{0}$ and, because each ε_i has the same variance and the ε_i's are uncorrelated, the variance-covariance matrix of these random error terms is

$$V(\boldsymbol{\varepsilon}) = E(\boldsymbol{\varepsilon\varepsilon}') = \begin{bmatrix} V(\varepsilon_1) & COV(\varepsilon_1,\varepsilon_2) & \dots & COV(\varepsilon_1,\varepsilon_n) \\ COV(\varepsilon_2,\varepsilon_1) & V(\varepsilon_2) & \dots & COV(\varepsilon_2,\varepsilon_n) \\ \cdot & \cdot & & \cdot \\ \cdot & \cdot & & \cdot \\ \cdot & \cdot & & \cdot \\ COV(\varepsilon_n,\varepsilon_1) & COV(\varepsilon_n,\varepsilon_2) & \dots & V(\varepsilon_n) \end{bmatrix}$$

$$= \begin{bmatrix} \sigma_\varepsilon^2 & 0 & \dots & 0 \\ 0 & \sigma_\varepsilon^2 & \dots & 0 \\ \cdot & \cdot & & \cdot \\ \cdot & \cdot & & \cdot \\ \cdot & \cdot & & \cdot \\ 0 & 0 & \dots & \sigma_\varepsilon^2 \end{bmatrix} = \sigma_\varepsilon^2 \boldsymbol{I}_n, \tag{12.C.1}$$

where $\boldsymbol{I}_n$ denotes an identity matrix of order n.

In addition, because $\boldsymbol{\varepsilon}$ has a multivariate normal distribution, we may write $\boldsymbol{\varepsilon} : N(\boldsymbol{0},\ \sigma_\varepsilon^2 \boldsymbol{I}_n)$. Hence, the joint probability density function of the ε_i's is

$$f(\boldsymbol{\varepsilon};0,\ \sigma_\varepsilon^2) = (2\pi\sigma_\varepsilon^2)^{-\frac{n}{2}} e^{-\boldsymbol{\varepsilon}'\boldsymbol{\varepsilon}/2\sigma_\varepsilon^2}.$$

For $\boldsymbol{\varepsilon} = \boldsymbol{Y} - \boldsymbol{X}\tilde{\boldsymbol{\beta}}$ ($\tilde{\boldsymbol{\beta}}$ depicts an arbitrary set of estimators for $\boldsymbol{\beta}$), the likelihood function for the sample is

$$\mathcal{L}(\tilde{\boldsymbol{\beta}},\ \sigma_\varepsilon^2; Y_1,\dots,Y_n, n) = (2\pi\sigma_\varepsilon^2)^{-\frac{n}{2}} e^{-\frac{1}{2\sigma_\varepsilon^2}(\boldsymbol{Y}-\boldsymbol{X}\tilde{\boldsymbol{\beta}})'(\boldsymbol{Y}-\boldsymbol{X}\tilde{\boldsymbol{\beta}})}$$

or, in log-likelihood form,

$$\begin{aligned} \log \mathcal{L} = &-\frac{n}{2}\log(2\pi) - \frac{n}{2}\log\sigma_\varepsilon^2 \\ &-\frac{1}{2\sigma_\varepsilon^2}(\boldsymbol{Y}'\boldsymbol{Y} - 2\tilde{\boldsymbol{\beta}}'\boldsymbol{X}'\boldsymbol{Y} + \tilde{\boldsymbol{\beta}}'\boldsymbol{X}'\boldsymbol{X}\tilde{\boldsymbol{\beta}}). \end{aligned}$$

Then

$$\frac{\partial \log \mathcal{L}}{\partial \tilde{\boldsymbol{\beta}}} = -\frac{1}{2\sigma_\varepsilon^2}(-2\boldsymbol{X}'\boldsymbol{Y} + 2\boldsymbol{X}'\boldsymbol{X}\tilde{\boldsymbol{\beta}}) = \frac{1}{\sigma_\varepsilon^2}(\boldsymbol{X}'\boldsymbol{Y} - \boldsymbol{X}'\boldsymbol{X}\tilde{\boldsymbol{\beta}}) = \boldsymbol{0}$$

and thus,

$$\tilde{\boldsymbol{\beta}} = (\boldsymbol{X}'\boldsymbol{X})^{-1}\boldsymbol{X}'\boldsymbol{Y} \tag{12.C.2}$$

(or Equation 12.A.2). Hence, $\tilde{\boldsymbol{\beta}}$ is an unbiased estimator for $\boldsymbol{\beta}$. Again we see that the maximum-likelihood estimators are equivalent to the OLS estimators for $\boldsymbol{\beta}$. From

$$\frac{\partial \log \mathcal{L}}{\partial \sigma_\varepsilon^2} = -\frac{1}{2\tilde{\sigma}_\varepsilon^2} + \frac{1}{2\tilde{\sigma}_\varepsilon^4}(\boldsymbol{Y} - \boldsymbol{X}\tilde{\boldsymbol{\beta}})'(\boldsymbol{Y} - \boldsymbol{X}\tilde{\boldsymbol{\beta}}) = 0,$$

we obtain $\tilde{\sigma}_\varepsilon^2 = \boldsymbol{e}'\boldsymbol{e}/n$ as an estimator for σ_ε^2. This maximum-likelihood estimator for σ_ε^2 is biased. (An unbiased estimator for σ_ε^2 is derived in Appendix 12.D.)

Appendix D: (1) the Variance-Covariance Matrix of the OLS Estimators $\hat{\boldsymbol{\beta}}$ and (2) an Unbiased Estimator for σ_ε^2

1. Because $\hat{\boldsymbol{\beta}}$ is unbiased for $\boldsymbol{\beta}$, we may write

$$\begin{aligned}\hat{\boldsymbol{\beta}} - E(\hat{\boldsymbol{\beta}}) = \hat{\boldsymbol{\beta}} - \boldsymbol{\beta} &= (\boldsymbol{X}'\boldsymbol{X})^{-1}\boldsymbol{X}'\boldsymbol{Y} - \boldsymbol{\beta} \\ &= (\boldsymbol{X}'\boldsymbol{X})^{-1}\boldsymbol{X}'(\boldsymbol{X}\boldsymbol{\beta} + \boldsymbol{\varepsilon}) - \boldsymbol{\beta} \\ &= (\boldsymbol{X}'\boldsymbol{X})^{-1}\boldsymbol{X}'\boldsymbol{X}\boldsymbol{\beta} + (\boldsymbol{X}'\boldsymbol{X})^{-1}\boldsymbol{X}'\boldsymbol{\varepsilon} - \boldsymbol{\beta} \\ &= \boldsymbol{\beta} + (\boldsymbol{X}'\boldsymbol{X})^{-1}\boldsymbol{X}'\boldsymbol{\varepsilon} - \boldsymbol{\beta} = (\boldsymbol{X}'\boldsymbol{X})^{-1}\boldsymbol{X}'\boldsymbol{\varepsilon}.\end{aligned} \tag{12.D.1}$$

Then from Equation 12.D.1,

$$\begin{aligned}V(\hat{\boldsymbol{\beta}}) &= E\left[(\hat{\boldsymbol{\beta}} - \boldsymbol{\beta})(\hat{\boldsymbol{\beta}} - \boldsymbol{\beta})'\right] \\ &= E\left[(\boldsymbol{X}'\boldsymbol{X})^{-1}\boldsymbol{X}'\boldsymbol{\varepsilon}\boldsymbol{\varepsilon}'\boldsymbol{X}(\boldsymbol{X}'\boldsymbol{X})^{-1}\right] \\ &= (\boldsymbol{X}'\boldsymbol{X})^{-1}\boldsymbol{X}'\sigma_\varepsilon^2\boldsymbol{I}_n\boldsymbol{X}(\boldsymbol{X}'\boldsymbol{X})^{-1} = \sigma_\varepsilon^2(\boldsymbol{X}'\boldsymbol{X})^{-1}.\end{aligned} \tag{12.D.2}$$

Here $V(\hat{\boldsymbol{\beta}})$ is a $(p \times p)$ symmetric matrix whose kth diagonal element is the variance (of the sampling distribution) of $\hat{\boldsymbol{\beta}}_k$ and where the (l, k)th off-diagonal element is the covariance between $\hat{\boldsymbol{\beta}}_l$ and $\hat{\boldsymbol{\beta}}_k$. That is,

$$V(\hat{\boldsymbol{\beta}}) = \begin{bmatrix} V(\hat{\beta}_0) & \mathrm{COV}(\hat{\beta}_0, \hat{\beta}_1) & \mathrm{COV}(\hat{\beta}_0, \hat{\beta}_2) & \cdots & \mathrm{COV}(\hat{\beta}_0, \hat{\beta}_m) \\ \mathrm{COV}(\hat{\beta}_1, \hat{\beta}_0) & V(\hat{\beta}_1) & \mathrm{COV}(\hat{\beta}_1, \hat{\beta}_2) & \cdots & \mathrm{COV}(\hat{\beta}_1, \hat{\beta}_m) \\ \mathrm{COV}(\hat{\beta}_2, \hat{\beta}_0) & \mathrm{COV}(\hat{\beta}_2, \hat{\beta}_1) & V(\hat{\beta}_2) & \cdots & \mathrm{COV}(\hat{\beta}_2, \hat{\beta}_m) \\ \cdot & \cdot & \cdot & & \cdot \\ \cdot & \cdot & \cdot & & \cdot \\ \cdot & \cdot & \cdot & & \cdot \\ \mathrm{COV}(\hat{\beta}_m, \hat{\beta}_0) & \mathrm{COV}(\hat{\beta}_m, \hat{\beta}_1) & \mathrm{COV}(\hat{\beta}_m, \hat{\beta}_2) & \cdots & V(\hat{\beta}_m) \end{bmatrix}. \tag{12.D.2.1}$$

In this regard, if we set $\boldsymbol{C} = \boldsymbol{(X'X)}^{-1}$, then $V(\hat{\boldsymbol{\beta}}) = \sigma_\varepsilon^2 \boldsymbol{C}$, where

$$\boldsymbol{C} = \begin{bmatrix} C_{00} & C_{01} & \ldots & C_{0m} \\ C_{10} & C_{11} & \ldots & C_{1m} \\ \cdot & \cdot & & \cdot \\ \cdot & \cdot & & \cdot \\ \cdot & \cdot & & \cdot \\ C_{m0} & C_{m1} & \ldots & C_{mm} \end{bmatrix}.$$

Then the variance of $\hat{\beta}_k$ is $V(\hat{\beta}_k) = \sigma_\varepsilon^2 C_{kk}$; and the covariance between $\hat{\beta}_l$ and $\hat{\beta}_k$ is $COV(\hat{\beta}_l, \hat{\beta}_k) = \sigma_\varepsilon^2 C_{lk}$.

In terms of deviations from the means of the variables,

$$V(\underline{\hat{\boldsymbol{\beta}}}) = \sigma_\varepsilon^2 (\underline{\boldsymbol{X}'\boldsymbol{X}})^{-1} = \begin{bmatrix} V(\hat{\beta}_1) & COV(\hat{\beta}_1, \hat{\beta}_2) & \ldots & COV(\hat{\beta}_1, \hat{\beta}_m) \\ COV(\hat{\beta}_2, \hat{\beta}_1) & V(\hat{\beta}_2) & & COV(\hat{\beta}_2, \hat{\beta}_m) \\ \cdot & \cdot & & \cdot \\ \cdot & \cdot & & \cdot \\ \cdot & \cdot & & \cdot \\ COV(\hat{\beta}_m, \hat{\beta}_1) & COV(\hat{\beta}_m, \hat{\beta}_2) & \ldots & V(\hat{\beta}_m) \end{bmatrix}. \quad (12.D.3)$$

Because $\hat{\beta}_0$ is excluded from $\underline{\hat{\boldsymbol{\beta}}}$, it can be shown that

$$V(\hat{\beta}_0) = \frac{\sigma_\varepsilon^2}{n} + \sum_{j=1}^{m} \bar{X}_j V(\hat{\beta}_j) + 2 \sum_{l=1}^{m} \sum_{\substack{j=1 \\ l<j}}^{m} \bar{X}_l \bar{X}_j COV(\hat{\beta}_l, \hat{\beta}_j).$$

2. Because $\boldsymbol{Y} = \boldsymbol{X}\hat{\boldsymbol{\beta}} + \boldsymbol{e}$,

$$\begin{aligned} \boldsymbol{e} &= \boldsymbol{Y} - \boldsymbol{X}\hat{\boldsymbol{\beta}} = \boldsymbol{Y} - \boldsymbol{X}(\boldsymbol{X}'\boldsymbol{X})^{-1}\boldsymbol{X}'\boldsymbol{Y} \\ &= \left[\boldsymbol{I}_n - \boldsymbol{X}(\boldsymbol{X}'\boldsymbol{X})^{-1}\boldsymbol{X}'\right]\boldsymbol{Y} = \boldsymbol{MY}, \end{aligned} \quad (12.D.4)$$

where $\boldsymbol{M} = \boldsymbol{I}_n - \boldsymbol{X}(\boldsymbol{X'X})^{-1}\boldsymbol{X'}$ is symmetric ($\boldsymbol{M} = \boldsymbol{M'}$) and idempotent ($\boldsymbol{M} = \boldsymbol{M}^2$) with $\boldsymbol{MX} = \boldsymbol{0}$. From Equation 12.D.4,

$$\boldsymbol{e} = \boldsymbol{M}\,(\boldsymbol{X\beta} + \boldsymbol{\varepsilon}) = \boldsymbol{M\varepsilon}$$

so that

$$\boldsymbol{e}'\boldsymbol{e} = \boldsymbol{\varepsilon}'\boldsymbol{M\varepsilon} = \sum_{r=1}^{m} \sum_{s=1}^{m} \varepsilon_r \varepsilon_s M_{rs}. \quad (12.D.5)$$

Then

$$E(\boldsymbol{e}'\boldsymbol{e}) = E(\sum_{r=1}^{m}\sum_{s=1}^{m} M_{rs}) = \sigma_\varepsilon^2 \sum_i M_{ii}$$
$$= \sigma_\varepsilon^2 tr(\boldsymbol{M}) = \sigma_\varepsilon^2 \left[tr(\boldsymbol{I}_n) - tr(\boldsymbol{X}(\boldsymbol{X}'\boldsymbol{X})^{-1}\boldsymbol{X}') \right]$$
$$= \sigma_\varepsilon^2 (n - m - 1)$$

because $E(\varepsilon_i^2) = \sigma_\varepsilon^2$ and $E(\varepsilon_i\, \varepsilon_j) = COV(\varepsilon_i, \varepsilon_j) = 0$. If we set

$$S_\varepsilon^2 = \frac{e'e}{n-m-1} = \frac{\sum_i e_i^2}{n-m-1}, \tag{12.D.6}$$

then $E(S_\varepsilon^2) = \sigma_\varepsilon^2$, i.e., Equation 12.D.6 represents an unbiased estimator for σ_ε^2. Clearly the maximum-likelihood estimator (Equation 12.12) is a biased estimator for σ_ε^2. Also, $S_\varepsilon = \sqrt{S_\varepsilon^2} = \sqrt{MSE}$ (the standard error of estimate).

In view of this discussion, we may write the estimated variance of $\hat{\beta}_k$ as $S_{\hat{\beta}_k}^2 = S_\varepsilon^2 C_{kk}$ and thus the estimated standard error of $\hat{\beta}_k$ is

$$S_{\hat{\beta}_k} = S_\varepsilon \sqrt{C_{kk}}. \tag{12.D.7}$$

If we choose to work with the data in the form of deviations from the means of the variables, then

$$S_\varepsilon^2 = \frac{1}{n-m}(M_{YY} - \hat{\beta}_1 M_{Y1} - \hat{\beta}_2 M_{Y2} - \cdots - \hat{\beta}_m M_{Ym})$$
$$= \frac{1}{n-m}(\boldsymbol{Y}'\boldsymbol{Y} - \boldsymbol{Y}'\underline{\boldsymbol{X}}\underline{\hat{\boldsymbol{\beta}}}),$$

where $\underline{X}$ does not contain the columns of 1's corresponding to the intercept.

Appendix E: Foundations for Statistical Inference in the OLS Model

This appendix presents the essentials of a collection of theorems pertaining to distributions, linear and (idempotent) quadratic forms, and conditions for the independence of linear and quadratic forms. Only their highlights (offered without proof) are included.

Distributions

a. Let $X_1,\ldots,X_n$ constitute a set of independent $N(0, 1)$ random variables with $V = \sum_{i=1}^{n} X_i^2$. Then V follows a χ^2 *distribution* with n degrees of freedom or $V : \chi_n^2$.
b. Let $X_1,\ldots,X_n$ constitute a set of independent $N(0, \sigma^2)$ random variables with $W = \sum_{i=1}^{n} (X_i / \sigma)^2$. Then W follows a χ^2 *distribution* with n degrees of freedom or $W : \chi_n^2$.
c. Let $X : N(0,1)$, $V : \chi_k^2$, and X and V are independent. Then the ratio between X and $\sqrt{V/k}$ follows a *t distribution* with k degrees of freedom, i.e., $X/\sqrt{V/k} : t_k$.

d. Let $V_1 : \mathcal{X}_r^2, V_2 : \mathcal{X}_s^2$ and suppose V_1/r and V_2/s are independent. Then the ratio between V_1/r and V_2/s follows an *F distribution* with r and s degrees of freedom, i.e.,

$$\frac{V_1 / r}{V_2 / s} : F_{r,s}.$$

Linear and Quadratic Forms

a. If a matrix $\boldsymbol{B}$ is of order $(m \times n)$, $\boldsymbol{X}$ is $(n \times 1)$, and $\boldsymbol{Y}$ is $(m \times 1)$, then $\boldsymbol{Y} = \boldsymbol{BX}$ is termed *a linear form* in $\boldsymbol{X}$. Here the jth component of $\boldsymbol{Y}$ is expressable as $Y_j = \sum_{k=1}^{n} b_{jk} X_k$, $j = 1, \ldots, m$, where b_{jk} is the element in the jth row and kth column of $\boldsymbol{B}$. If $\boldsymbol{b}$ is of order $(n \times 1)$, then $\boldsymbol{Y} = \boldsymbol{b}'\boldsymbol{X}$ is also a linear form in $\boldsymbol{X}$.

b. If a symmetric matrix $\boldsymbol{A}$ is of order $(n \times n)$ ($\boldsymbol{A}$ is symmetric if $\boldsymbol{A} = \boldsymbol{A}'$) and $\boldsymbol{X}$ is $(n \times 1)$, then the scalar $Q = \boldsymbol{X}'\boldsymbol{AX} = \sum_{i=1}^{n} \sum_{j=1}^{n} a_{ij} X_i X_j = \sum_{i=1}^{n} a_{ii} X_i^2 + \sum_{i=1}^{n} \sum_{\substack{j=1 \\ i<j}}^{n} a_{ij} X_i X_j$ is called a *quadratic form* in X. Here Q is a homogenous second-degree polynomial in the variables $X_1, \ldots, X_n$. If $\rho(\boldsymbol{A}) = r$, then Q is said to be a *quadratic form of rank r.*

c. A symmetric $(n \times n)$ matrix $\boldsymbol{A}$ is said to be *positive definite* (*negative definite*) if and only if $\boldsymbol{X}'\boldsymbol{AX} > 0 (<0)$ for all $\boldsymbol{X} \neq \boldsymbol{0}$.

d. A symmetric $(n \times n)$ matrix $\boldsymbol{A}$ is termed *nonnegative definite* (*nonpositive definite*) if and only if $\boldsymbol{X}'\boldsymbol{AX} \geq 0 (\leq 0)$ for all $\boldsymbol{X}$. (Nonnegative definite is also known as *positive semidefinite*, and nonpositive definite is alternatively characterized as *negative semidefinite*.

Idempotent Quadratic Forms

a. Suppose $\boldsymbol{A}$ is an nth order idempotent matrix ($\boldsymbol{A}$ is idempotent if and only if $\boldsymbol{A} = \boldsymbol{A}'$, $\boldsymbol{A} = \boldsymbol{A}^2$, and $\rho(\boldsymbol{A}) = \text{tr}(\boldsymbol{A})$ with $\rho(\boldsymbol{A}) = r$. Then the quadratic form $Q = \boldsymbol{X}'\boldsymbol{AX}$ is said to be idempotent.

b. Let the $(n \times 1)$ vector $\mathbf{X}$ follow a $N(\boldsymbol{0}, \boldsymbol{I}_n)$ distribution. Then $\boldsymbol{X}'\boldsymbol{X}$ is distributed as $\mathcal{X}_n^2$. (Here $\boldsymbol{X}'\boldsymbol{X} = \sum_{i=1}^{n} X_i^2$. Because $X_1, \ldots, X_n$ constitute a set of independent $N(0, 1)$ variables, $\sum_{i=1}^{n} X_i^2 : \mathcal{X}_n^2$.)

c. Let the $(n \times 1)$ vector $\boldsymbol{X}$ follow a $N(\boldsymbol{0}, \boldsymbol{I}_n)$ distribution and suppose the $(n \times n)$ matrix $\boldsymbol{A}$ is idempotent with $\rho(\boldsymbol{A}) = r$. Then $\boldsymbol{X}'\boldsymbol{AX} : \mathcal{X}_r^2$. More generally,

d. Let the $(n \times 1)$ vector $\boldsymbol{X}$ follow a $N(\boldsymbol{0}, \sigma^2 \boldsymbol{I}_n)$ distribution and suppose the $(n \times n)$ matrix $\boldsymbol{A}$ is idempotent with $\rho(\boldsymbol{A}) = r$. Then $\boldsymbol{X}'\boldsymbol{AX} : \sigma^2 \mathcal{X}_r^2$.

Independence of Linear and Quadratic Forms

a. Let the $(n \times 1)$ vector $\boldsymbol{X}$ follow a $N(\boldsymbol{0}, \boldsymbol{I}_n)$ distribution. Suppose $\boldsymbol{A}$ is an $(n \times n)$ idempotent matrix with $\rho(\boldsymbol{A}) = r$ and let $\boldsymbol{B}$ be an $(m \times n)$ matrix with $\boldsymbol{BA} = \boldsymbol{0}$. Then the linear form $\boldsymbol{Y} = \boldsymbol{BX}$ is distributed independently of the quadratic form $Q = \boldsymbol{X}'\boldsymbol{AX}$. More generally,

b. Let the $(n \times 1)$ vector $\boldsymbol{X}$ follow a $N(\boldsymbol{0}, \sigma^2 \boldsymbol{I}_n)$ distribution. Suppose $\boldsymbol{A}$ is an $(n \times n)$ idempotent matrix with $\rho(\boldsymbol{A}) = r$ and let $\boldsymbol{B}$ be an $(m \times n)$ matrix with $\boldsymbol{BA} = \boldsymbol{0}$. Then the linear form $\boldsymbol{Y} = \boldsymbol{BX}$ is distributed independently of the quadratic form $Q = \boldsymbol{X}'\boldsymbol{AX}$.

Independence of Quadratic Forms

a. Let the $(n \times 1)$ vector $\boldsymbol{X}$ follow a $N(\mathbf{0}, \boldsymbol{I}_n)$ distribution. Suppose $\boldsymbol{A}$ is an $(n \times n)$ idempotent matrix with $\rho(\boldsymbol{A}) = r$, $\boldsymbol{B}$ is an $(n \times n)$ idempotent matrix with $\rho(\boldsymbol{B}) = s$, and $\boldsymbol{BA} = \mathbf{0}$. Then the quadratic form $Q_A = \boldsymbol{X'AX}$ is distributed independently of the quadratic form $Q_B = \boldsymbol{X'}\,\boldsymbol{BX}$. More generally,
b. Let the $(n \times 1)$ vector $\boldsymbol{X}$ follow a $N(\mathbf{0}, \sigma^2\boldsymbol{I}_n)$ distribution. Suppose $\boldsymbol{A}$ is an $(n \times n)$ idempotent matrix with $\rho(\boldsymbol{A}) = r$, $\boldsymbol{B}$ is an $(n \times n)$ idempotent matrix with $\rho(\boldsymbol{B}) = s$, and $\boldsymbol{BA} = \mathbf{0}$. Then the quadratics form $Q_A = \boldsymbol{X'}\,\boldsymbol{AX}$ is distributed independently of the quadratic form $Q_B = \boldsymbol{X'}\,\boldsymbol{BX}$.

Distributions Revisited

a. Suppose X_i is a random variable for which $E(X_i^2) < +\infty i = 1,\ldots,n$. Then $E(X_i) = \mu_i$ exists for all i. For $\boldsymbol{X}$ an $(n \times 1)$ random vector, let $E(\boldsymbol{X}) = \boldsymbol{\mu}$ with $V(\boldsymbol{X}) = E[(\boldsymbol{X} - \boldsymbol{\mu})(\boldsymbol{X} - \boldsymbol{\mu})'] = \Sigma$. Here

$$\boldsymbol{\Sigma} = \begin{bmatrix} \sigma_1^2 & \sigma_{12} & \cdots & \sigma_{1n} \\ \sigma_{21} & \sigma_2^2 & \cdots & \sigma_{2n} \\ \cdot & \cdot & & \cdot \\ \cdot & \cdot & & \cdot \\ \cdot & \cdot & & \cdot \\ \sigma_{n1} & \sigma_{n2} & \cdots & \sigma_n^2 \end{bmatrix}$$

represents the variance-covariance matrix of $\boldsymbol{X}$. In addition,
b. If the $(n \times 1)$ random vector $\boldsymbol{X}$ is $N(\boldsymbol{\mu}, \Sigma)$, then for $\boldsymbol{b}$ an $(n \times 1)$ nonstochastic vector and $\boldsymbol{A}$ an $(n \times n)$ nonstochastic matrix with $\rho(\boldsymbol{A}) = n$, if follows that $\boldsymbol{AX} + \boldsymbol{b}$: $N(\boldsymbol{A\mu} + \boldsymbol{b}, \boldsymbol{A\Sigma A'})$.
c. If the $(n \times 1)$ random vector $\boldsymbol{X}$ is $N(\mathbf{0}, \sigma_\varepsilon^2 \boldsymbol{I}_n)$ (the components of $\boldsymbol{X}$ are mutually independent normally distributed random variables with zero means and constant variances σ_ε^2), then $\boldsymbol{X}'\boldsymbol{X} / \sigma_\varepsilon^2 = \boldsymbol{X}'(\sigma_\varepsilon^2 \boldsymbol{I}_n)^{-1}\boldsymbol{X} : \mathcal{X}_n^2$. More generally, if $\boldsymbol{X}$ is $N(\mathbf{0}, \Sigma)$ and Σ is positive definite, then $\boldsymbol{X}'\boldsymbol{\Sigma}^{-1}\boldsymbol{X} : \mathcal{X}_n^2$.
d. If the random vector $\hat{\boldsymbol{\beta}} : N(\boldsymbol{\beta}, \sigma_\varepsilon^2(\boldsymbol{X}'\boldsymbol{X})^{-1})$, then $\hat{\boldsymbol{\beta}} - \boldsymbol{\beta} : N(\mathbf{0}, \sigma_\varepsilon^2(\boldsymbol{X}'\boldsymbol{X})^{-1})$. Additionally, the quadratic form $Q = (\hat{\boldsymbol{\beta}} - \boldsymbol{\beta})'\boldsymbol{X}'\boldsymbol{X}(\hat{\boldsymbol{\beta}} - \boldsymbol{\beta}) : \sigma_\varepsilon^2 \mathcal{X}_{m+1}^2$ (this latter result will be verified in Appendix 12.G).
    i. For $\boldsymbol{a}$ an $(n \times 1)$ nonstochastic vector, let $\boldsymbol{Y} = \boldsymbol{a}'\boldsymbol{X}$. Then

    $$E(\boldsymbol{Y}) = \boldsymbol{a}'\boldsymbol{\mu} \text{ and } V(\boldsymbol{Y}) = \boldsymbol{a}'\boldsymbol{\Sigma a} = \sum\nolimits_i \sum\nolimits_j a_i a_j \sigma_{ij}.$$

    ii. For $\boldsymbol{A}$ an $(n \times n)$ nonstochalistic matrix with $\rho(\boldsymbol{A}) = n$, let $\boldsymbol{Y} = \boldsymbol{AX}$. Then $E(\boldsymbol{Y}) = \boldsymbol{A\mu}$ and $V(\boldsymbol{Y}) = \boldsymbol{A\Sigma A'}$.

e. Let the $(n \times 1)$ random vector $\boldsymbol{X}$ follow a $N(\mathbf{0}, \boldsymbol{I}_n)$ distribution and let $\boldsymbol{C}$ be an $(n \times n)$ matrix such that $\boldsymbol{C'C} = \boldsymbol{I}_n$. Then $\boldsymbol{C'X} : N(\mathbf{0}, \boldsymbol{I}_n)$. (An nth order matrix $\boldsymbol{C}$ is said to be *orthogonal* if $\boldsymbol{C}' = \boldsymbol{C}^{-1}$ so that $\boldsymbol{C'C} = \boldsymbol{I}_n$.)

f. Under the assumptions of the strong classical linear regression model, it follows that the random error terms ε_i, $i = 1,\ldots,n$, constitute a set of mutually independent normally distributed random variables with zero means and constant variances σ_ε^2, i.e., the random vector $\boldsymbol{\varepsilon} : N(\mathbf{0}, \sigma_\varepsilon^2 \boldsymbol{I}_n)$ since $E(\boldsymbol{\varepsilon}) = \mathbf{0}$ and $E(\boldsymbol{\varepsilon\varepsilon}') = \sigma_\varepsilon^2 \boldsymbol{I}_n$ (see Appendix 12.C). In addition, from Appendix 12.D, we determined that $\hat{\boldsymbol{\beta}} = \boldsymbol{\beta} + (\boldsymbol{X}'\boldsymbol{X})^{-1}\boldsymbol{X}'\boldsymbol{\varepsilon}$ is a linear form in the normally distributed vector $\boldsymbol{\varepsilon}$ so that $\hat{\boldsymbol{\beta}} : N(\boldsymbol{\beta}, \sigma_\varepsilon^2(\boldsymbol{X}'\boldsymbol{X})^{-1})$ since $E(\hat{\boldsymbol{\beta}}) = \boldsymbol{\beta}$ and $V(\hat{\boldsymbol{\beta}}) = \sigma_\varepsilon^2(\boldsymbol{X}'\boldsymbol{X})^{-1}$. So componentwise, each element of $\hat{\boldsymbol{\beta}}$ is $N(\beta_k, V(\hat{\beta}_k))$ and thus

$$Z_k = \frac{\hat{\beta}_k - \beta_k}{\sigma_{\hat{\beta}_k}} = \frac{\hat{\beta}_k - \beta_k}{\sigma_\varepsilon \sqrt{C_{kk}}} : N(0,1),\ k = 0,1,\ldots,m, \tag{12.E.1}$$

where $\sigma_{\hat{\beta}_k} = \sqrt{V(\hat{\beta}_k)}$ and C_{kk} is the kth diagonal element of $\boldsymbol{C} = (\boldsymbol{X}'\boldsymbol{X})^{-1}$. Because σ_ε^2 is unknown, Z_k cannot be used for inferential purposes.

Now, we determined earlier (Appendix 12.D) that $\boldsymbol{e}'\boldsymbol{e} = \boldsymbol{\varepsilon}'\boldsymbol{M}\boldsymbol{\varepsilon}$, where $\boldsymbol{M} = \boldsymbol{I}_n - \boldsymbol{X}(\boldsymbol{X}'\boldsymbol{X})^{-1}\boldsymbol{X}'$ is idempotent with $\rho(\boldsymbol{M}) = n - m - 1$. So with $\boldsymbol{\varepsilon}'\boldsymbol{M}\boldsymbol{\varepsilon}$ an idempotent quadratic form of rank $n - m - 1$ in normal variables $\boldsymbol{\varepsilon}$, $\boldsymbol{e}'\boldsymbol{e} : \sigma_\varepsilon^2 \chi^2_{n-m-1}$ by Theorem 3.d of Appendix 12.E. With $S_\varepsilon^2 = \boldsymbol{e}'\boldsymbol{e}/(n-m-1)$, it follows that $S_\varepsilon^2/\sigma_\varepsilon^2 : \chi^2_{n-m-1}/(n-m-1)$.

Consider the quantity

$$\frac{\hat{\beta}_k - \beta_k}{S_{\hat{\beta}_k}} = \frac{(\hat{\beta}_k - \beta_k)/\sigma_{\hat{\beta}_k}}{S_{\hat{\beta}_k}/\sigma_{\hat{\beta}_k}} \tag{12.E.2}$$

Here Equation 12.E.2 is the ratio of a standard normal variable and $S_\varepsilon/\sigma_\varepsilon$, where the latter is the square root of $\chi^2_{n-m-1}/(n-m-1)$. Then according to Theorem 4.b of Appendix 12.E, because $(\boldsymbol{X}'\boldsymbol{X})^{-1}\boldsymbol{X}'\boldsymbol{M} = \mathbf{0}$ (since $\boldsymbol{MX} = \mathbf{0}$), the linear form $\hat{\boldsymbol{\beta}} - \boldsymbol{\beta} = (\boldsymbol{X}'\boldsymbol{X})^{-1}\boldsymbol{X}'\boldsymbol{\varepsilon}$ in $\boldsymbol{\varepsilon}$ is independent of the quadratic form $\boldsymbol{\varepsilon}'\boldsymbol{M}\boldsymbol{\varepsilon}$ in $\boldsymbol{\varepsilon}$ and thus $\hat{\beta}_k$ is distributed independently of S_ε^2. Then from Equation 12.E.2 (and from the definition of a t_{n-m-1} distribution provided in Appendix E, "Distributions," item c), the quantity

$$\frac{\hat{\beta}_k - \beta_k}{S_{\hat{\beta}_k}} = \frac{\hat{\beta}_k - \beta_k}{S_\varepsilon \sqrt{C_{kk}}} : t_{n-m-1}. \tag{12.E.3}$$

g. Suppose the $(n \times 1)$ random vector $\boldsymbol{X} : N(\mathbf{0}, \sigma^2 \boldsymbol{I}_n)$ and let the $(n \times n)$ matrix $\boldsymbol{A}$ be idempotent of rank r, the $(n \times n)$ matrix $\boldsymbol{B}$ be idempotent of rank s, and suppose $\boldsymbol{BA} = \mathbf{0}$. Then $Q_A = \boldsymbol{X}'\boldsymbol{AX} : \sigma^2\chi^2_r$, $Q_B = \boldsymbol{X}'\boldsymbol{BX} : \sigma^2\chi^2_s$, and Q_A and Q_B are independent. Then the ratio

$$\frac{(\boldsymbol{X}'\boldsymbol{AX}/\sigma)/r}{(\boldsymbol{X}'\boldsymbol{BX}/\sigma)/s} : F_{r,s}.$$

Appendix F: Decomposition of the Sample Variation of Y

We noted earlier that the OLS estimator of $\boldsymbol{\beta}$ is $\hat{\boldsymbol{\beta}} = (\boldsymbol{X}'\boldsymbol{X})^{-1}\boldsymbol{X}'\boldsymbol{Y}$. Then

$$\begin{aligned} \boldsymbol{X}'\boldsymbol{X}\hat{\boldsymbol{\beta}} - \boldsymbol{X}'\boldsymbol{Y} &= -\boldsymbol{X}'(\boldsymbol{Y} - \boldsymbol{X}\hat{\boldsymbol{\beta}}) = -\boldsymbol{X}'(\boldsymbol{Y} - \hat{\boldsymbol{Y}}) \\ &= -\boldsymbol{X}'\boldsymbol{e} = \boldsymbol{0} \end{aligned} \tag{12.F.1}$$

so that $\boldsymbol{X}$ and $\boldsymbol{e}$ are orthogonal vectors.

Let us now from

$$SST = \sum_{i=1}^{n} (Y_i - \bar{Y})^2 = \boldsymbol{Y}'\boldsymbol{Y} - n\bar{Y}^2,$$

where $\boldsymbol{Y} = \hat{\boldsymbol{Y}} + \boldsymbol{e}$. Then because $\hat{\boldsymbol{Y}} = \boldsymbol{X}\hat{\boldsymbol{\beta}}$,

$$\begin{aligned} \boldsymbol{Y}'\boldsymbol{Y} &= (\hat{\boldsymbol{Y}}' + \boldsymbol{e}')(\hat{\boldsymbol{Y}} + \boldsymbol{e}) = \hat{\boldsymbol{Y}}'\hat{\boldsymbol{Y}} + 2\hat{\boldsymbol{Y}}'\boldsymbol{e} + \boldsymbol{e}'\boldsymbol{e} \\ &= \hat{\boldsymbol{\beta}}'\boldsymbol{X}'\boldsymbol{X}\hat{\boldsymbol{\beta}} + 2\hat{\boldsymbol{\beta}}'\boldsymbol{X}'\boldsymbol{e} + \boldsymbol{e}'\boldsymbol{e} \\ &= \hat{\boldsymbol{\beta}}'\boldsymbol{X}'(\boldsymbol{Y} - \boldsymbol{e}) + 2\hat{\boldsymbol{\beta}}'\boldsymbol{X}'\boldsymbol{e} + \boldsymbol{e}'\boldsymbol{e} \\ &= \hat{\boldsymbol{\beta}}'\boldsymbol{X}'\boldsymbol{Y} + \boldsymbol{e}'\boldsymbol{e} \end{aligned} \tag{12.F.2}$$

via Equation 12.F.1. Then

$$\begin{aligned} SST &= \boldsymbol{Y}'\boldsymbol{Y} - n\bar{Y}^2 \\ &= \underbrace{\hat{\boldsymbol{\beta}}'\boldsymbol{X}'\boldsymbol{Y} - n\bar{Y}^2}_{SSR} + \underbrace{\boldsymbol{e}'\boldsymbol{e}}_{SSE}. \end{aligned} \tag{12.F.3}$$

Hence, SST partitions neatly into SSR and SSE.

Expression 12.F.3 can be rewritten in terms of the individual estimated regression coefficients and variable values as

$$\begin{aligned} SST &= \sum_{i=1}^{n} Y_i^2 - n\bar{Y}^2, \\ SSR &= \hat{\beta}_0 \sum_{i=1}^{n} Y_i + \sum_{j=1}^{m} \left(\hat{\beta}_j \sum_{i=1}^{n} X_{ij} Y_i\right) - n\bar{Y}^2 \\ &= \hat{\beta}_0 \sum_{i=1}^{n} Y_i + \hat{\beta}_1 \sum_{i=1}^{n} X_{i1} Y_i + \hat{\beta}_2 \sum_{i=1}^{n} X_{i2} Y_i \\ &\quad + \cdots + \hat{\beta}_m \sum_{i=1}^{n} X_{im} Y_i - n\bar{Y}^2, \text{ and} \\ SSE &= SST - SSR \\ &= \sum_{i=1}^{n} Y_i^2 - \hat{\beta}_0 \sum_{i=1}^{n} Y_i - \hat{\beta}_1 \sum_{i=1}^{n} X_{i1} Y_i - \hat{\beta}_2 \sum_{i=1}^{n} X_{i2} Y_i \\ &\quad - \cdots - \hat{\beta}_m \sum_{i=1}^{n} X_{im} Y_i. \end{aligned} \tag{12.F.4}$$

Equation 12.F.3 can also be expressed in terms of derivations from the means of the regression variables as

$$\begin{aligned}
SST &= \sum_{i=1}^{n}(Y_i - \bar{Y})^2,\\
SSR &= \sum_{i=1}^{n}(\hat{Y}_i - \bar{Y})^2 = \sum_{i=1}^{n}\Big(\hat{\beta}_0 + \sum_{j=1}^{m}\hat{\beta}_j X_{ij} - \bar{Y}\Big)^2\\
&= \sum_{i=1}^{n}\Big(\bar{Y} - \sum_{j=1}^{m}\hat{\beta}_j \bar{X}_j + \sum_{j=1}^{m}\hat{\beta}_j X_{ij} - \bar{Y}\Big)^2\\
&= \sum_{i=1}^{n}\Big(\sum_{j=1}^{m}\hat{\beta}_j (X_{ij} - \bar{X}_j)\Big)^2\\
&= \sum_{j=1}^{m}\hat{\beta}_j^2 M_{jj} + 2\sum_{l=1}^{m}\sum_{\substack{j=1\\ l<j}}^{m}\hat{\beta}_l\hat{\beta}_j M_{lj},
\end{aligned} \tag{12.F.5}$$

where $M_{jj} = \sum_{i=1}^{n}(X_{ij} - \bar{X}_j)^2$ and $M_{lj} = \sum_{i=1}^{n}(X_{il} - \bar{X}_l)(X_{ij} - \bar{X}_j)$, and $SSE = SST - SSR$.

Appendix G: The Distribution of *MSR*/*MSE*

We noted previously that the SST was decomposable into an SSR and an SSE as $SST = SSR + SSE$, where

$$\begin{aligned}
SST &= \sum_{i=1}^{n}(Y_i - \bar{Y})^2 = \boldsymbol{Y}'\boldsymbol{Y} - n\bar{Y}^2,\\
SSR &= \sum_{i=1}^{n}(\hat{Y}_i - \bar{Y}_i)^2 = \hat{\boldsymbol{\beta}}'\boldsymbol{X}'\boldsymbol{Y} - n\bar{Y}^2, \text{ and}\\
SSE &= \sum_{i=1}^{n}e_i^2 = \boldsymbol{Y}'\boldsymbol{Y} - \hat{\boldsymbol{\beta}}'\boldsymbol{X}'\boldsymbol{Y}.
\end{aligned}$$

To test the significance of the overall regression equation under $H_0: \beta_1 = \beta_2 = \cdots = \beta_m = 0$, we formed as our test statistic the ratio of the mean squares as

$$\frac{MSR}{MSE} = \frac{SSR/m}{SSE/(n-m-1)}. \tag{12.G.1}$$

We now want to demonstrate that this ratio is F distributed with m and $n - m - 1$ degrees of freedom.

Let us first consider the more general null hypothesis $H_0: \boldsymbol{\beta} = \boldsymbol{\beta}_0$. From Appendix 12.A, $\hat{\boldsymbol{\beta}} - \boldsymbol{\beta}_0 = (\boldsymbol{X}'\boldsymbol{X})^{-1}\boldsymbol{X}'\boldsymbol{\varepsilon}$ or $\boldsymbol{X}'\boldsymbol{X}(\hat{\boldsymbol{\beta}} - \boldsymbol{\beta}_0) = \boldsymbol{X}'\boldsymbol{\varepsilon}$. Hence,

$$\begin{aligned}
Q &= (\hat{\boldsymbol{\beta}} - \boldsymbol{\beta}_0)'\boldsymbol{X}'\boldsymbol{X}(\hat{\boldsymbol{\beta}} - \boldsymbol{\beta}_0)\\
&= \boldsymbol{\varepsilon}'\boldsymbol{X}(\boldsymbol{X}'\boldsymbol{X})^{-1}\boldsymbol{X}'\boldsymbol{\varepsilon} = \boldsymbol{\varepsilon}'(\boldsymbol{I}_n - \boldsymbol{M})\boldsymbol{\varepsilon}
\end{aligned} \tag{12.G.2}$$

is an idempotent quadratic form of rank $m + 1$ (because $\rho(\boldsymbol{I}_n - \boldsymbol{M}) = tr(\boldsymbol{I}_n - \boldsymbol{M}) = tr(\boldsymbol{I}_n) - tr(\boldsymbol{M}) = n - (n - m - 1) = (m + 1)$). So under $H_0: \boldsymbol{\beta} = \boldsymbol{\beta}_0$,

$$Q = \boldsymbol{\varepsilon}'(\boldsymbol{I}_n - \boldsymbol{M})\boldsymbol{\varepsilon} : \sigma_\varepsilon^2 \chi_{m+1}^2. \tag{12.G.3}$$

Because $\boldsymbol{X}(\boldsymbol{X}'\boldsymbol{X})^{-1}\boldsymbol{X}'\boldsymbol{M} = (\boldsymbol{I}_n - \boldsymbol{M})\boldsymbol{M} = \boldsymbol{0}$, the quadratic forms Q and $\boldsymbol{\varepsilon}'\boldsymbol{M}\boldsymbol{\varepsilon}$ are independent so that

$$\frac{(Q/\sigma_\varepsilon^2)/(m+1)}{(SSE/\sigma_\varepsilon^2)/(n-m-1)} : F_{m+1,\, n-m-1}. \tag{12.G.4}$$

We next consider a null hypothesis on a subset of parameters in $\boldsymbol{\beta}$, say $\beta_{h+1},\ldots,\beta_m$, with no hypothesis on $\beta_0,\beta_1,\ldots,\beta_h$. (Note that, if $h = 0$, we are back to $H_0: \beta_1 = \beta_2 = \cdots = \beta_m = 0$.) To this end, let us form the partitions $\boldsymbol{X} = [\boldsymbol{X}_1, \boldsymbol{X}_2]$ and $\boldsymbol{\beta}' = [\boldsymbol{\beta}'_1, \boldsymbol{\beta}'_2]$. Then $\boldsymbol{Y} = \boldsymbol{X}\boldsymbol{\beta} + \varepsilon$ is partitioned as

$$\boldsymbol{Y} = \boldsymbol{X}_1\boldsymbol{\beta}_1 + \boldsymbol{X}_2\boldsymbol{\beta}_2 + \boldsymbol{\varepsilon}, \tag{12.G.5}$$

where

$$\underset{(n\times h+1)}{\boldsymbol{X}_1} = \begin{bmatrix} 1 & X_{11} & \cdots & X_{1h} \\ 1 & X_{21} & \cdots & X_{2h} \\ \cdot & \cdot & & \cdot \\ \cdot & \cdot & & \cdot \\ \cdot & \cdot & & \cdot \\ 1 & X_{n1} & \cdots & X_{nh} \end{bmatrix}, \quad \underset{(n\times m-h)}{\boldsymbol{X}_2} = \begin{bmatrix} X_{1,\,h+1} & \cdots & X_{1m} \\ X_{2,\,h+1} & \cdots & X_{2m} \\ \cdot & & \cdot \\ \cdot & & \cdot \\ \cdot & & \cdot \\ X_{n,\,h+1} & \cdots & X_{nm} \end{bmatrix},$$

$$\underset{(h+1\times 1)}{\boldsymbol{\beta}_1} = \begin{bmatrix} \beta_0 \\ \beta_1 \\ \cdot \\ \cdot \\ \cdot \\ \beta_h \end{bmatrix}, \text{ and } \underset{(m-h\times 1)}{\boldsymbol{\beta}_2} = \begin{bmatrix} \beta_{h+1} \\ \beta_{h+2} \\ \cdot \\ \cdot \\ \cdot \\ \beta_m \end{bmatrix}.$$

By virtue of this partitioning, $\hat{\boldsymbol{\beta}} = (\boldsymbol{X}'\boldsymbol{X})^{-1}\boldsymbol{X}'\boldsymbol{Y}$ becomes

$$\begin{bmatrix} \hat{\boldsymbol{\beta}}_1 \\ \hat{\boldsymbol{\beta}}_2 \end{bmatrix} = \begin{bmatrix} \boldsymbol{X}_1'\boldsymbol{X}_1 & \boldsymbol{X}_1'\boldsymbol{X}_2 \\ \boldsymbol{X}_2'\boldsymbol{X}_1 & \boldsymbol{X}_2'\boldsymbol{X}_2 \end{bmatrix}^{-1} \begin{bmatrix} \boldsymbol{X}_1'\boldsymbol{Y} \\ \boldsymbol{X}_2'\boldsymbol{Y} \end{bmatrix}$$

so that, under partitioned inversion,

$$\begin{bmatrix} \hat{\boldsymbol{\beta}}_1 \\ \hat{\boldsymbol{\beta}}_2 \end{bmatrix} = \begin{bmatrix} (\boldsymbol{X}_1'\boldsymbol{X}_1)^{-1}\boldsymbol{X}_1'\boldsymbol{Y} - (\boldsymbol{X}_1'\boldsymbol{X}_1)^{-1}\boldsymbol{X}_1'\boldsymbol{X}_2\boldsymbol{D}^{-1}\boldsymbol{X}_2'\boldsymbol{M}_1\boldsymbol{Y} \\ \boldsymbol{D}^{-1}\boldsymbol{X}_2'\boldsymbol{M}_1\boldsymbol{Y} \end{bmatrix}, \tag{12.G.6}$$

where

$$\underset{(m-h\times m-h)}{\boldsymbol{D}} = \boldsymbol{X}_2'\boldsymbol{M}_1\boldsymbol{X}_2,$$

$$\underset{(n\times n)}{\boldsymbol{M}_1} = \boldsymbol{I}_n - \boldsymbol{X}_1(\boldsymbol{X}_1'\boldsymbol{X}_1)^{-1}\boldsymbol{X}_1'.$$

Let us set

$$\hat{\boldsymbol{\beta}}_2 = \boldsymbol{D}^{-1}\boldsymbol{X}_2'\boldsymbol{M}_1\boldsymbol{Y} = \boldsymbol{D}^{-1}\boldsymbol{X}_2'\boldsymbol{M}_1(\boldsymbol{X}_1\boldsymbol{\beta}_1 + \boldsymbol{X}_2\boldsymbol{\beta}_2 + \boldsymbol{\varepsilon})$$
$$= \boldsymbol{\beta}_2 + \boldsymbol{D}^{-1}\boldsymbol{X}_2'\boldsymbol{M}_1\boldsymbol{\varepsilon}$$

or $\hat{\boldsymbol{\beta}}_2 - \boldsymbol{\beta}_2 = \boldsymbol{D}^{-1}\boldsymbol{X}_2'\boldsymbol{M}_1\boldsymbol{\varepsilon}$. Then proceeding as above, under $H_0 : \boldsymbol{\beta}_2 = \boldsymbol{\beta}_2^0, \hat{\boldsymbol{\beta}}_2 - \boldsymbol{\beta}_2^0 = \boldsymbol{D}^{-1}\boldsymbol{X}_2'\boldsymbol{M}_1\boldsymbol{\varepsilon}$ or $\boldsymbol{D}(\hat{\boldsymbol{\beta}}_2 - \boldsymbol{\beta}_2^0) = \boldsymbol{X}_2'\boldsymbol{M}_1\boldsymbol{\varepsilon}$. Hence,

$$Q_2 = (\hat{\boldsymbol{\beta}}_2 - \boldsymbol{\beta}_2^0)'\boldsymbol{D}(\hat{\boldsymbol{\beta}}_2 - \boldsymbol{\beta}_2^0)$$
$$= \boldsymbol{\varepsilon}'\boldsymbol{M}_1\boldsymbol{X}_2\boldsymbol{D}^{-1}\boldsymbol{X}_2'\boldsymbol{M}_1\boldsymbol{\varepsilon} = \boldsymbol{\varepsilon}'\boldsymbol{E}\boldsymbol{\varepsilon}, \tag{12.G.7}$$

where $\boldsymbol{E} = \boldsymbol{M}_1\boldsymbol{X}_2\boldsymbol{D}^{-1}\boldsymbol{X}_2'\boldsymbol{M}_1$ is idempotent of rank $\rho(\boldsymbol{E}) = tr(\boldsymbol{E}) = tr(\boldsymbol{D}^{-1}\boldsymbol{X}_2'\boldsymbol{M}_1\boldsymbol{M}_1\boldsymbol{X}_2) = tr(\boldsymbol{D}^{-1}\boldsymbol{D})$ $= tr(\boldsymbol{I}_{m-h}) = m - h$. So for $H_0 : \boldsymbol{\beta}_2 = \boldsymbol{\beta}_2^0$,

$$Q_2 = (\hat{\boldsymbol{\beta}}_2 - \boldsymbol{\beta}_2^0)'\boldsymbol{D}(\hat{\boldsymbol{\beta}}_2 - \boldsymbol{\beta}_2^0) : \sigma_\varepsilon^2 \mathcal{X}_{m-h}^2. \tag{12.G.8}$$

Because $\boldsymbol{EM} = \boldsymbol{M}_1\boldsymbol{X}_2\boldsymbol{D}^{-1}\boldsymbol{X}_2'\boldsymbol{M}_1\boldsymbol{M} = \boldsymbol{M}_1\boldsymbol{X}_2\boldsymbol{D}^{-1}\boldsymbol{X}_2'\boldsymbol{M}_1(\boldsymbol{M}_1 - \boldsymbol{E}) = \boldsymbol{E} - \boldsymbol{E} = \boldsymbol{0}$, the quadratic forms Q_2 and $\boldsymbol{\varepsilon}'\boldsymbol{M}\boldsymbol{\varepsilon}$ are independent so that

$$\frac{(Q/\sigma_\varepsilon^2)/(m-h)}{(SSE/\sigma_\varepsilon^2)/(n-m-1)} : F_{m-h,\, n-m-1}. \tag{12.G.9}$$

Let us now consider a couple of special cases of Equation 12.G.8. First, suppose we choose to test $H_0 : \boldsymbol{\beta}_2 = \boldsymbol{\beta}_2^0 = \boldsymbol{0}$. Then via Equations 12.G.6 and 12.G.7,

$$Q_2' = \hat{\boldsymbol{\beta}}_2'\boldsymbol{D}\hat{\boldsymbol{\beta}}_2 = \boldsymbol{Y}'\boldsymbol{M}_1\boldsymbol{X}_2\boldsymbol{D}^{-1}\boldsymbol{X}_2'\boldsymbol{M}_1\boldsymbol{Y}$$

and thus

$$\frac{(Q_2'/\sigma_\varepsilon^2)/(m-h)}{(SSE/\sigma_\varepsilon^2)/(n-m-1)} : F_{m-h,\, n-m-1}. \tag{12.G.9.1}$$

Next, we know from above that

$$SSR + n\bar{Y}^2 = SSR + \boldsymbol{Y}'\left(\frac{1}{n}\boldsymbol{1}\boldsymbol{1}'\right)\boldsymbol{Y}$$
$$= \hat{\boldsymbol{\beta}}'\boldsymbol{X}'\boldsymbol{Y} = \boldsymbol{Y}'\boldsymbol{X}\hat{\boldsymbol{\beta}}$$
$$= \boldsymbol{Y}'(\boldsymbol{X}_1\hat{\boldsymbol{\beta}}_1 + \boldsymbol{X}_2\hat{\boldsymbol{\beta}}_2).$$

Then from Equation 12.G.6,

$$\begin{aligned} SSR + \boldsymbol{Y}'\left(\frac{1}{n}\boldsymbol{1}\boldsymbol{1}'\right)\boldsymbol{Y} &= \boldsymbol{Y}'\boldsymbol{X}_1(\boldsymbol{X}_1'\boldsymbol{X}_1)^{-1}\boldsymbol{X}_1'\boldsymbol{Y} - \boldsymbol{Y}'\boldsymbol{X}_1(\boldsymbol{X}_1'\boldsymbol{X}_1)\boldsymbol{X}_1'\boldsymbol{X}_2\boldsymbol{D}^{-1}\boldsymbol{X}_2'\boldsymbol{M}_1\boldsymbol{Y} \\ &\quad + \boldsymbol{Y}'\boldsymbol{X}_2\boldsymbol{D}^{-1}\boldsymbol{X}_2'\boldsymbol{M}_1\boldsymbol{Y} \\ &= \boldsymbol{Y}'\boldsymbol{X}_1(\boldsymbol{X}_1'\boldsymbol{X}_1)^{-1}\boldsymbol{X}_1'\boldsymbol{Y} - \boldsymbol{Y}'(\boldsymbol{I}_n - \boldsymbol{M}_1)\boldsymbol{X}_2\boldsymbol{D}^{-1}\boldsymbol{X}_2'\boldsymbol{M}_1\boldsymbol{Y} \\ &\quad + \boldsymbol{Y}'\boldsymbol{X}_2\boldsymbol{D}^{-1}\boldsymbol{X}_2'\boldsymbol{M}_1\boldsymbol{Y} \\ &= \boldsymbol{Y}'\boldsymbol{X}_1(\boldsymbol{X}_1'\boldsymbol{X}_1)^{-1}\boldsymbol{X}_1'\boldsymbol{Y} + \boldsymbol{Y}'\boldsymbol{M}_1\boldsymbol{X}_2\boldsymbol{D}^{-1}\boldsymbol{X}_2'\boldsymbol{M}_1\boldsymbol{Y}. \end{aligned} \tag{12.G.10}$$

Suppose $h = 0$ so that $\boldsymbol{X}_1 = \boldsymbol{1}$. Then

$$SSR + \boldsymbol{Y}'\left(\frac{1}{n}\boldsymbol{1}\boldsymbol{1}'\right)\boldsymbol{Y} = \boldsymbol{Y}'\boldsymbol{1}n^{-1}\boldsymbol{1}'\boldsymbol{Y} + \boldsymbol{Y}'\boldsymbol{M}_1\boldsymbol{X}_2\boldsymbol{D}^{-1}\boldsymbol{X}_2'\boldsymbol{M}_1\boldsymbol{Y}$$

or

$$SSR = \boldsymbol{Y}'\boldsymbol{M}_1\boldsymbol{X}_2\boldsymbol{D}^{-1}\boldsymbol{X}_2'\boldsymbol{M}_1\boldsymbol{Y} = Q_2.$$

Hence, under $H_0: \beta_1 = \beta_2 = \cdots = \beta_m = 0$,

$$\frac{(SSR/\sigma_\varepsilon^2)/m}{(SSE/\sigma_\varepsilon^2)/(n-m-1)} : F_{m,\, n-m-1}$$

or Equation 12.G.1.

For $h \neq 0$, suppose only $\boldsymbol{X}_1$ is an argument of the regression Equation 12.G.5, i.e., the conditional expectation of Y does not depend on $X_{h+1}, X_{h+2}, \ldots, X_m$ or, equivalently, $\beta_{h+1} = \beta_{h+2} = \ldots = \beta_m = 0$. Then our estimate of $\boldsymbol{\beta}_1$ can be written as $\hat{\boldsymbol{\beta}}_1 = (\boldsymbol{X}_1'\boldsymbol{X}_1)^{-1}\boldsymbol{X}_1'\boldsymbol{Y}$ and the corresponding regression sum of squares is $SSR_1 = \hat{\boldsymbol{\beta}}_1'\boldsymbol{X}_1'\boldsymbol{Y} - n\bar{Y}^2$ or

$$\begin{aligned} SSR_1 + \boldsymbol{Y}'\left(\frac{1}{n}\boldsymbol{1}\boldsymbol{1}'\right)\boldsymbol{Y} &= \hat{\boldsymbol{\beta}}_1'\boldsymbol{X}_1'\boldsymbol{Y} \\ &= \boldsymbol{Y}'\boldsymbol{X}_1(\boldsymbol{X}_1'\boldsymbol{X}_1)^{-1}\boldsymbol{X}_1'\boldsymbol{Y}. \end{aligned} \tag{12.G 11}$$

After substracting Equation 12.G.11 from Equation 12.G.10, we obtain

$$\Delta SSR = SSR - SSR_1 = \boldsymbol{Y}'\boldsymbol{M}_1\boldsymbol{X}_2\boldsymbol{D}^{-1}\boldsymbol{X}_2'\boldsymbol{M}_1\boldsymbol{Y}. \tag{12.G 12}$$

Let us refer to this expression as the *incremental regression sum of squares* resulting from the introduction of $\boldsymbol{X}_2$ into the regression equation given that $\boldsymbol{X}_1$ is already in that equation. Then by virtue of Equation 12.G.9.1, our statistic for testing $H_0 = \beta_{h+1} = \beta_{h+2} = \ldots = \beta_m = 0$ vs. H_1: at least one regression parameter in H_0 is different from zero, is

$$\frac{\Delta SSR/(m-h)}{SSE/(n-m-1)} : F_{m-h,\, n-m-1}. \tag{12.G.12}$$

Appendix H: Least-Squares Estimation under a Linear Restriction

Suppose we face a situation in which least-squares estimation of $\boldsymbol{\beta}$ involves sample information $\boldsymbol{Y} = \boldsymbol{X\beta} + \boldsymbol{\varepsilon}$ as well as nonsample information incorporated in a set of linear restrictions or constraints $\boldsymbol{R\beta} = \boldsymbol{r}$, where $\boldsymbol{R}$ is an $(r \times p)$ (with $p = m + 1$ and $r \leq p$) matrix of constants and $\boldsymbol{r}$ is an $(r \times 1)$ vector of constants. Let $\rho(\boldsymbol{R}) = r$ (no linear dependencies among the constraint rows of $\boldsymbol{R}$, i.e., no row of $\boldsymbol{R}$ can be expressed as a linear combination[4] of one or more other rows of $\boldsymbol{R}$). If the vector of residuals is $\boldsymbol{e} = \boldsymbol{Y} - \boldsymbol{X\beta}_*$ ($\boldsymbol{\beta}_*$ is a vector of arbitrary estimators for the components of $\boldsymbol{\beta}$), then our goal is to find a vector $\boldsymbol{\beta}_*$ that minimizes $\boldsymbol{e}'\boldsymbol{e} = \sum e_i^2$ subject to the side relation $\boldsymbol{r} - \boldsymbol{R\beta}_* = \boldsymbol{0}$.

Let us solve this problem using the technique of Lagrange. Specifically, let us form the Lagrangian function

$$\mathcal{L}(\boldsymbol{\beta}_*, \boldsymbol{\lambda}) = \boldsymbol{e}'\boldsymbol{e} + 2\boldsymbol{\lambda}'(\boldsymbol{r} - \boldsymbol{R\beta}_*),$$

where λ is an $(r \times 1)$ vector of undetermined Lagrange multipliers λ_h, $h = 1,\ldots,r$ (there is a separate Lagrange multiplier for each linear constraint in $\boldsymbol{R\beta}_* = \boldsymbol{r}$), and the "2" has been inserted for mathematical convenience.

Then

$$\mathcal{L}(\boldsymbol{\beta}_*, \boldsymbol{\lambda}) = \boldsymbol{Y}'\boldsymbol{Y} - 2\boldsymbol{\beta}_*'\boldsymbol{X}'\boldsymbol{Y} + \boldsymbol{\beta}_*'\boldsymbol{X}'\boldsymbol{X\beta}_* + 2\boldsymbol{\lambda}'(\boldsymbol{r} - \boldsymbol{R\beta}_*).$$

Setting

$$\begin{aligned} &\text{(a)}\quad \frac{\partial \mathcal{L}}{\partial \boldsymbol{\beta}_*} = -2\boldsymbol{X}'\boldsymbol{Y} + 2\boldsymbol{X}'\boldsymbol{X\beta}_* + 2\boldsymbol{\lambda}_*\boldsymbol{R}'\boldsymbol{\lambda}_* = \boldsymbol{0} \\ &\text{(b)}\quad \frac{\partial \mathcal{L}}{\partial \boldsymbol{\lambda}} = 2(\boldsymbol{r} - \boldsymbol{R\beta}_*) = \boldsymbol{0} \end{aligned} \tag{12.H.1}$$

(here $\lambda = \lambda_*$ is the vector of Lagrange multipliers that satisfies these first-order conditions) we obtain, from Equation 12.H.1.a

$$\begin{aligned} \boldsymbol{\beta}_* &= (\boldsymbol{X}'\boldsymbol{X})^{-1}\boldsymbol{X}'\boldsymbol{Y} + (\boldsymbol{X}'\boldsymbol{X})^{-1}\boldsymbol{R}'\boldsymbol{\lambda}_* \\ &= \hat{\boldsymbol{\beta}} + (\boldsymbol{X}'\boldsymbol{X})^{-1}\boldsymbol{R}'\boldsymbol{\lambda}_*, \end{aligned} \tag{12.H.2}$$

where $\hat{\boldsymbol{\beta}}$ is the OLS or unrestricted estimator of $\boldsymbol{\beta}$. Then from Equation 12.H.1.b,

$$\boldsymbol{R\beta}_* = \boldsymbol{R}\hat{\boldsymbol{\beta}} + \boldsymbol{R}(\boldsymbol{X}'\boldsymbol{X})^{-1}\boldsymbol{R}'\boldsymbol{\lambda}_* = \boldsymbol{r} \tag{12.H.3}$$

(the constraint is binding at $\boldsymbol{\beta}_*$). We assume that the second-order conditions also hold at $(\boldsymbol{\beta}_*, \boldsymbol{\lambda}_*)$.

[4] A vector $\boldsymbol{b}$ is a linear combination of the vectors $X_1,\ldots,X_s$ if there exist constants or scalars $\theta_1,\ldots,\theta_s$ such that $\boldsymbol{b} = \theta_1 X_1 + \cdots + \theta_s X_s$.

We know from our previous analysis that $(X'X)^{-1}$ is positive definite with rank $p = m + 1$. Hence, $R(X'X)^{-1}R'$ is positive definite with the same rank as $(X'X)^{-1}$ and thus $R(X'X)^{-1}R'$ is nonsingular (i.e., has an inverse) so that, from Equation 12.H.3,

$$\begin{aligned}\boldsymbol{\lambda}_* &= \left[\boldsymbol{R}(\boldsymbol{X}'\boldsymbol{X})^{-1}\boldsymbol{R}'\right]^{-1}(\boldsymbol{R}\boldsymbol{\beta}_* - \boldsymbol{R}\hat{\boldsymbol{\beta}}) \\ &= \left[\boldsymbol{R}(\boldsymbol{X}'\boldsymbol{X})^{-1}\boldsymbol{R}'\right]^{-1}(\boldsymbol{r} - \boldsymbol{R}\hat{\boldsymbol{\beta}}).\end{aligned} \tag{12.H.4}$$

After substituting Equation 12.H.4 into Equation 12.H.2, we obtain

$$\boldsymbol{\beta}_* = \hat{\boldsymbol{\beta}} + (\boldsymbol{X}'\boldsymbol{X})^{-1}\boldsymbol{R}'\left[\boldsymbol{R}(\boldsymbol{X}'\boldsymbol{X})^{-1}\boldsymbol{R}'\right]^{-1}(\boldsymbol{r} - \boldsymbol{R}\hat{\boldsymbol{\beta}}), \tag{12.H.2.1}$$

the *restricted least-squares estimator* of $\boldsymbol{\beta}$. Whereas $\hat{\boldsymbol{\beta}}$ uses sample data exclusively in the estimation of $\boldsymbol{\beta}$, $\boldsymbol{\beta}_*$ relies on both sample data and nonsample or extraneous information to estimate $\boldsymbol{\beta}$. Clearly $\boldsymbol{\beta}_*$ equals $\hat{\boldsymbol{\beta}}$ plus a linear function of the constraint relation $\boldsymbol{r} = \boldsymbol{R}\hat{\boldsymbol{\beta}}$ (with the latter also satisfied at $\hat{\boldsymbol{\beta}}$).

Looking to the properties of the sampling distribution of $\boldsymbol{\beta}_*$, it is easily demonstrated that, if the linear restriction is binding ($\boldsymbol{R\beta} = \boldsymbol{r}$), then $E(\boldsymbol{\beta}_*) = \boldsymbol{\beta}$ so that $\boldsymbol{\beta}_*$ is an unbiased estimator for $\boldsymbol{\beta}$ subject $\boldsymbol{r} - \boldsymbol{R\beta} = \boldsymbol{0}$. In addition, for $\boldsymbol{r} - \boldsymbol{R\beta} = \boldsymbol{0}$,

$$\begin{aligned}V(\boldsymbol{\beta}_*) &= E[(\boldsymbol{\beta}_* - E(\boldsymbol{\beta}_*))(\boldsymbol{\beta}_* - E(\boldsymbol{\beta}_*))'] \\ &= E[(\boldsymbol{\beta}_* - \boldsymbol{\beta})(\boldsymbol{\beta}_* - \boldsymbol{\beta})'].\end{aligned}$$

From Equation 12.H.2.1, let $\boldsymbol{\beta}_* - \boldsymbol{\beta} = \hat{\boldsymbol{\beta}} + A(\boldsymbol{r} - \boldsymbol{R}\hat{\boldsymbol{\beta}}) - \boldsymbol{\beta}$, where $A = (X'X)^{-1}R'[R(X'X)^{-1}R']^{-1}$.

Because $V(\hat{\boldsymbol{\beta}}) = E(\hat{\boldsymbol{\beta}}\hat{\boldsymbol{\beta}}') - \boldsymbol{\beta}\boldsymbol{\beta}' = \sigma_\varepsilon^2(\boldsymbol{X}'\boldsymbol{X})^{-1}$,

$$\begin{aligned}V(\boldsymbol{\beta}_*) &= \sigma_\varepsilon^2\left[(\boldsymbol{X}'\boldsymbol{X})^{-1} - (\boldsymbol{X}'\boldsymbol{X})^{-1}\boldsymbol{R}'\boldsymbol{A}' - \boldsymbol{AR}(\boldsymbol{X}'\boldsymbol{X})^{-1} + \boldsymbol{AR}(\boldsymbol{X}'\boldsymbol{X})^{-1}\boldsymbol{R}'\boldsymbol{A}'\right] \\ &= \sigma_\varepsilon^2(\boldsymbol{I}_n - \boldsymbol{AR})(\boldsymbol{X}'\boldsymbol{X})^{-1}(\boldsymbol{I}_n - \boldsymbol{R}'\boldsymbol{A}') \\ &= \sigma_\varepsilon^2\boldsymbol{M}_*(\boldsymbol{X}'\boldsymbol{X})^{-1}\boldsymbol{M}_*',\end{aligned}$$

where $\boldsymbol{M}_* = \boldsymbol{I}_n - \boldsymbol{AR}$. Because $\boldsymbol{M}_*$ is idempotent,

$$V(\boldsymbol{\beta}_*) = \sigma_\varepsilon^2\boldsymbol{M}_*(\boldsymbol{X}'\boldsymbol{X})^{-1}. \tag{12.H.5}$$

Let us express Equation 12.H.5 as

$$V(\boldsymbol{\beta}_*) = V(\hat{\boldsymbol{\beta}}) - \sigma_\varepsilon^2\boldsymbol{AR}(\boldsymbol{X}'\boldsymbol{X})^{-1}$$

or

$$V(\hat{\boldsymbol{\beta}}) - V(\boldsymbol{\beta}_*) = \boldsymbol{\Sigma}_{\hat{\beta}}\boldsymbol{\Sigma}_{\beta^*} = \sigma_\varepsilon^2\boldsymbol{AR}(\boldsymbol{X}'\boldsymbol{X})^{-1}. \tag{12.H.6}$$

Because the right-hand side of Equation 12.H.6 is positive definite, it follows that $V(\hat{\beta}_k) \geq V(\beta_k^*)$ for all $k = 0,1,\ldots, m$, where $\hat{\beta}_k$ is the OLS estimator of β_k and β_k^* is the restricted least-squares estimator for β_k. Hence, $\boldsymbol{\beta}_*$ is BLUE for $\boldsymbol{\beta}$ subject to $\boldsymbol{R\beta} = \boldsymbol{r}$.

Appendix I: Testing Sets of Linear Restrictions

In Appendix H, we introduced a set of linear restrictions on the regression parameters $\boldsymbol{\beta}$ of the form $\boldsymbol{R\beta} = \boldsymbol{r}$, where $\boldsymbol{R}$ is an $r \times p$(with $p = m + 1$ and $r \leq p$) matrix of constants with $\rho(\boldsymbol{R}) = r$ and $\boldsymbol{r}$ is an r-component vector of constants. Let us specify our null and alternative hypotheses as $H_0: \boldsymbol{R\beta} = \boldsymbol{r}$ and $H_1: \boldsymbol{R\beta} \neq \boldsymbol{r}$, respectively, with $\boldsymbol{R}\hat{\boldsymbol{\beta}}$ serving as an obvious estimator for $\boldsymbol{R\beta}$.

Looking to the sampling distribution of $\boldsymbol{R}\hat{\boldsymbol{\beta}}$, we have

$$\begin{aligned} E(\boldsymbol{R}\hat{\boldsymbol{\beta}}) &= \boldsymbol{R\beta} \text{ and} \\ V(\boldsymbol{R}\hat{\boldsymbol{\beta}}) &= E\left[(\boldsymbol{R}\hat{\boldsymbol{\beta}} - E(\boldsymbol{R}\hat{\boldsymbol{\beta}}))(\boldsymbol{R}\hat{\boldsymbol{\beta}} - E(\boldsymbol{R}\hat{\boldsymbol{\beta}}))'\right] \\ &= E\left[\boldsymbol{R}(\hat{\boldsymbol{\beta}} - \boldsymbol{\beta})(\boldsymbol{R}(\hat{\boldsymbol{\beta}} - \boldsymbol{\beta}))'\right] \\ &= E\left[\boldsymbol{R}(\hat{\boldsymbol{\beta}} - \boldsymbol{\beta})(\hat{\boldsymbol{\beta}} - \boldsymbol{\beta})'\boldsymbol{R}'\right] \\ &= \sigma_\varepsilon^2 \boldsymbol{R}(\boldsymbol{X}'\boldsymbol{X})^{-1}\boldsymbol{R}' \end{aligned} \tag{12.I.1}$$

via Appendix 12.A. Because $\hat{\boldsymbol{\beta}} : N(\boldsymbol{\beta}, \sigma_\varepsilon^2(\boldsymbol{X}'\boldsymbol{X})^{-1})$ it follows that $\boldsymbol{R}\hat{\boldsymbol{\beta}} : N(\boldsymbol{R\beta}, \sigma_\varepsilon^2 \boldsymbol{R}(\boldsymbol{X}'\boldsymbol{X})^{-1}\boldsymbol{R}')$ and thus

$$\boldsymbol{R}(\hat{\boldsymbol{\beta}} - \boldsymbol{\beta}) : N(\boldsymbol{0}, \sigma_\varepsilon^2 \boldsymbol{R}(\boldsymbol{X}'\boldsymbol{X})^{-1}\boldsymbol{R}'). \tag{12.I.2}$$

Under H_0, Equation 12.I.2 becomes

$$\boldsymbol{R}\hat{\boldsymbol{\beta}} - \boldsymbol{r} : N(\boldsymbol{0}, \sigma_\varepsilon^2 \boldsymbol{R}(\boldsymbol{X}'\boldsymbol{X})^{-1}\boldsymbol{R}'). \tag{12.I.3}$$

Because $\boldsymbol{R}(\boldsymbol{X}'\boldsymbol{X})^{-1}\boldsymbol{R}'$ is positive definite with the same rank as $(\boldsymbol{X}'\boldsymbol{X})^{-1}$, it follows that $[\boldsymbol{R}(\boldsymbol{X}'\boldsymbol{X})^{-1}\boldsymbol{R}']^{-1}$ exists so that, from Appendix 12.E, "Distributions Revisted" (item c),

$$(\boldsymbol{R}\hat{\boldsymbol{\beta}} - \boldsymbol{r})'\left[\sigma_\varepsilon^2 \boldsymbol{R}(\boldsymbol{X}'\boldsymbol{X})^{-1}\boldsymbol{R}'\right]^{-1}(\boldsymbol{R}\hat{\boldsymbol{\beta}} - \boldsymbol{r}) : \mathcal{X}_r^2. \tag{12.I.4}$$

We know from Theorem 3.d of Appendix 12.E that

$$\boldsymbol{e}'\boldsymbol{e} / \sigma_\varepsilon^2 : \mathcal{X}_{n-m-1}^2. \tag{12.I.5}$$

Moreover, with Equations 12.I.4 and 12.I.5 independently distributed via Appendix 12.G (because $\boldsymbol{e}'\boldsymbol{e}/\sigma_\varepsilon^2$ is independent of $\hat{\boldsymbol{\beta}}$ and thus independent of $\boldsymbol{R}\hat{\boldsymbol{\beta}}$), it follows from Appendix 12.E, "Distributions Revisited" (item g), that

$$F = \frac{(R\hat{\beta} - r)' \left[R(X'X)^{-1}R' \right]^{-1} (R\hat{\beta} - r) / r}{e'e/(n-m-1)} : F_{r,\, n-m-1}. \quad (12.I.6)$$

Here Equation 12.I.6 serves as our test statistic for handling sets of linear restrictions. What special cases involving the choices for R and r typically apply?

Testing a Single Regression Parameter

Here $r = 1$, $R = e_k'(e_k$ is a *unit column vector*; it has unity in its kth position, $k = 0,1,\ldots,m$, and zeros elsewhere, with $k = 0$ included to account for the regression intercept), and $r = 0$. Hence, H_0: $R\beta = r$ implies H_0: $\beta_k = 0$. (If $r = \beta_k^0$, then we are testing $H_0 : \beta_k = \beta_k^0$.) Then $R\hat{\beta} - r = \hat{\beta}_k$, $\left[R(X'X)^{-1}R' \right]^{-1} = C_{kk}^{-1}$, and thus Equation 12.I.6 becomes

$$F = \frac{\hat{\beta}_k^2}{S_\varepsilon^2 C_{kk}} : F_{1,\, n-m-1} = t_{n-m-1}^2 \quad (12.I.7)$$

(or $\hat{\beta}_k / S_\varepsilon \sqrt{C_{kk}} : t_{n-m-1}$).

Testing the Equality of Two Regression Parameters β_j, β_k

Set $r = 1$, $R = e_j' - e_k'$, and $r = 0$. Hence $H_0 : R\beta = r$ implies $H_0 : \beta_j - \beta_k = 0$. Then $R\hat{\beta} - r = \hat{\beta}_j - \hat{\beta}_k = 0, \left[R(X'X)^{-1}R' \right]^{-1} = (C_{jj} + C_{kk} - 2C_{jk})^{-1}$ and thus Equation 12.I.6 becomes

$$F = \frac{(\hat{\beta}_j - \hat{\beta}_k)^2}{S_\varepsilon^2 (C_{jj} + C_{kk} - 2C_{jk})} : F_{1,\, n-m-1} = t_{n-m-1}^2 \quad (12.I.8)$$

or

$$\frac{\hat{\beta}_j - \hat{\beta}_k}{S_\varepsilon \sqrt{C_{jj} + C_{kk} - 2C_{jk}}} : t_{n-m-1}.$$

A Global Test of the Regression Parameters

For $r = m$, let

$$\underset{(m \times m+1)}{R} = \begin{bmatrix} 0 & 1 & 0 & \ldots & 0 \\ 0 & 0 & 1 & \ldots & 0 \\ \cdot & \cdot & \cdot & & \cdot \\ \cdot & \cdot & \cdot & & \cdot \\ \cdot & \cdot & \cdot & & \cdot \\ 0 & 0 & 0 & \ldots & 1 \end{bmatrix} = [0, I_m] \text{ and } \underset{(m \times 1)}{r} = 0$$

so that $H_0: \boldsymbol{R\beta} = \boldsymbol{r}$ is equivalent to $H_0: \beta_1 = \beta_2 = \cdots = \beta_m = 0$. Then

$$\boldsymbol{R\hat{\beta}} - \boldsymbol{r} = \boldsymbol{\hat{\beta}}_2 = \begin{bmatrix} \hat{\beta}_1 \\ \cdot \\ \cdot \\ \cdot \\ \hat{\beta}_m \end{bmatrix}.$$

If we partition $\boldsymbol{X}$ as $\boldsymbol{X} = [1, \boldsymbol{X}_2]$, where $\boldsymbol{X}_2$ is of order $(n \times m)$, then

$$\boldsymbol{X}'\boldsymbol{X} = \begin{bmatrix} n & \boldsymbol{1}'\boldsymbol{X}_2 \\ \boldsymbol{X}_2'\boldsymbol{1} & \boldsymbol{X}_2'\boldsymbol{X}_2 \end{bmatrix}$$

and

$$(\boldsymbol{X}'\boldsymbol{X})^{-1} = \begin{bmatrix} \frac{1}{n} + \frac{1}{n}\boldsymbol{1}'\boldsymbol{X}_2\boldsymbol{A}^{-1}\boldsymbol{X}_2'\boldsymbol{1}\frac{1}{n} & -\frac{1}{n}\boldsymbol{1}'\boldsymbol{X}_2\boldsymbol{A}^{-1} \\ -\boldsymbol{A}^{-1}\boldsymbol{X}_2'\boldsymbol{1} & \boldsymbol{A}^{-1} \end{bmatrix},$$

where $\boldsymbol{A} = \boldsymbol{X}_2'\left(\boldsymbol{I}_n - \boldsymbol{1}\frac{1}{n}\boldsymbol{1}')\boldsymbol{X}_2\right)$ is a symmetric idempotent matrix of order $(m \times m)$. Then our test statistic Equation 12.I.6 becomes

$$F = \frac{\boldsymbol{\hat{\beta}}_2'\boldsymbol{X}_2'\left(\boldsymbol{I}_n - \boldsymbol{1}\frac{1}{n}\boldsymbol{1}'\right)\boldsymbol{X}_2\boldsymbol{\hat{\beta}}_2 / m}{\boldsymbol{e}'\boldsymbol{e}/(n-m-1)} : F_{m,\, n-m-1}. \tag{12.I.9}$$

Now, we know that $\boldsymbol{Y} = \boldsymbol{\hat{Y}} + \boldsymbol{e} = \boldsymbol{X\hat{\beta}} + \boldsymbol{e}$. Under the preceding partitioning scheme, we have

$$\boldsymbol{Y} = [\boldsymbol{1}, \boldsymbol{X}_2]\begin{bmatrix} \hat{\beta}_0 \\ \boldsymbol{\hat{\beta}}_2 \end{bmatrix} + \boldsymbol{e} = \boldsymbol{1}\hat{\beta}_0 + \boldsymbol{X}_2\boldsymbol{\hat{\beta}}_2 + \boldsymbol{e}.$$

Let $\boldsymbol{A} = \boldsymbol{X}_2'\boldsymbol{B}\boldsymbol{X}_2$, where $\boldsymbol{B} = \boldsymbol{I}_n - \boldsymbol{1}\frac{1}{n}\boldsymbol{1}'$ is symmetric idempotent. Then

$$\begin{aligned} \boldsymbol{BY} &= \boldsymbol{B1}\hat{\beta}_0 + \boldsymbol{BX}_2\boldsymbol{\hat{\beta}}_2 + \boldsymbol{Be} \\ &= \boldsymbol{BX}_2\boldsymbol{\hat{\beta}}_2 + \boldsymbol{e} \end{aligned}$$

because $\boldsymbol{B1} = \boldsymbol{0}$ and $\mathbf{Be} = \mathbf{e}$. Let us write

$$\boldsymbol{Y}' = \hat{\beta}_0\boldsymbol{1}' + \boldsymbol{\hat{\beta}}_2'\boldsymbol{X}_2' + \boldsymbol{e}'$$

and form

$$\boldsymbol{Y}'\boldsymbol{BY} = \boldsymbol{\hat{\beta}}_2'\boldsymbol{X}_2'\boldsymbol{BX}_2\boldsymbol{\hat{\beta}}_2 + \boldsymbol{e}'\boldsymbol{e} \tag{12.I.10}$$

because $\boldsymbol{1}'\boldsymbol{B} = \boldsymbol{0}'$, $\boldsymbol{1}'\boldsymbol{e} = 0$, $\boldsymbol{X}_2'\boldsymbol{e} = \boldsymbol{0}$ (via (12.F.1)), and $\boldsymbol{e}'\boldsymbol{B}' = \boldsymbol{e}'$.

With $\boldsymbol{BY} = \boldsymbol{Y} - \mathbf{1}\bar{Y}$, we have $\boldsymbol{Y}'\boldsymbol{BY} = \boldsymbol{Y}'\boldsymbol{Y} - n\bar{Y}^2 = SST$. Because $e'e = SSE$, it follows from Equation 12.I.10 that $\hat{\boldsymbol{\beta}}_2'\boldsymbol{X}_2'\boldsymbol{BX}_2\hat{\boldsymbol{\beta}}_2 = SSR$. Then Equation 12.I.9 can be written as

$$F = \frac{SSR/m}{\boldsymbol{e}'\boldsymbol{e}/(n-m-1)} : F_{m,\, n-m-1} \tag{12.I.11}$$

(which is just Equation 12.19 or 12.G.1).

A Partial Test on a Subset of Regression Parameters

For $r = m-h$, let

$$\underset{(m-h\times m+1)}{\boldsymbol{R}} = \begin{bmatrix} \underset{(m-h\times h+1)}{\mathbf{0}} & \boldsymbol{I}_{m-h} \end{bmatrix} \text{ and } \underset{(m-h\times 1)}{\boldsymbol{r}} = \mathbf{0}$$

with $H_0: \boldsymbol{R\beta} = \boldsymbol{r}$ equivalent to $H_0: \beta_{h+1} = \beta_{h+2} = \cdots = \beta_m = 0$. Then

$$\boldsymbol{R}\hat{\boldsymbol{\beta}} - \boldsymbol{r} = \underset{(m-h\times 1)}{\hat{\boldsymbol{\beta}}_2} = \begin{bmatrix} \hat{\beta}_{h+1} \\ \cdot \\ \cdot \\ \cdot \\ \hat{\beta}_m \end{bmatrix}.$$

If we partition $\boldsymbol{X}$ as $\boldsymbol{X} = [\boldsymbol{X}_1, \boldsymbol{X}_2]$, where $\boldsymbol{X}_1$ is of order $(n \times h + 1)$ and $\boldsymbol{X}_2$ is of order $(n \times m - h)$, then

$$\boldsymbol{X}'\boldsymbol{X} = \begin{bmatrix} \boldsymbol{X}_1'\boldsymbol{X}_1 & \boldsymbol{X}_1'\boldsymbol{X}_2 \\ \boldsymbol{X}_2'\boldsymbol{X}_1 & \boldsymbol{X}_2'\boldsymbol{X}_2 \end{bmatrix}$$

and

$$(\boldsymbol{X}'\boldsymbol{X})^{-1} = \begin{bmatrix} (\boldsymbol{X}_1'\boldsymbol{X}_1)^{-1} + (\boldsymbol{X}_1'\boldsymbol{X}_1)^{-1}\boldsymbol{X}_1'\boldsymbol{X}_2\boldsymbol{D}^{-1} & -(\boldsymbol{X}_1'\boldsymbol{X}_1)^{-1}\boldsymbol{X}_1'\boldsymbol{X}_2\boldsymbol{D}^{-1} \\ -\boldsymbol{D}^{-1}\boldsymbol{X}_2'\boldsymbol{X}_1(\boldsymbol{X}_1'\boldsymbol{X}_1)^{-1} & \boldsymbol{D}^{-1} \end{bmatrix},$$

where $\boldsymbol{D} = \boldsymbol{X}_2'\boldsymbol{X}_2 - \boldsymbol{X}_2'\boldsymbol{X}_1(\boldsymbol{X}_1'\boldsymbol{X}_1)^{-1}\boldsymbol{X}_1'\boldsymbol{X}_2 = \boldsymbol{X}_2'(\boldsymbol{I}_n - \boldsymbol{X}_1(\boldsymbol{X}_1'\boldsymbol{X}_1)^{-1}\boldsymbol{X}_1')\boldsymbol{X}_2 = \boldsymbol{X}_2'\boldsymbol{M}_1\boldsymbol{X}_2$ is of order $(m-h\times m-h)$ and $\boldsymbol{M}_1 = \boldsymbol{I}_n - \boldsymbol{X}_1(\boldsymbol{X}_1'\boldsymbol{X}_1)^{-1}\boldsymbol{X}_1'$ is of order $(n\times n)$. Then our test statistic Equation 12.I.6 can be expressed as

$$F = \frac{\hat{\boldsymbol{\beta}}_2'\boldsymbol{X}_2'\boldsymbol{M}_1\boldsymbol{X}_2\hat{\boldsymbol{\beta}}_2/(m-h)}{\boldsymbol{e}'\boldsymbol{e}/(n-m-1)} : F_{m-h,\, n-m-1}. \tag{12.I.12}$$

From Equation 12.G.12, $\Delta SSR = SSR - SSR_1 = \boldsymbol{Y}'\boldsymbol{M}_1\boldsymbol{X}_2\boldsymbol{D}^{-1}\boldsymbol{X}_2'\boldsymbol{M}_1\boldsymbol{Y}$. Because from Equation 12.G.6 we have $\hat{\boldsymbol{\beta}}_2 = \boldsymbol{D}^{-1}\boldsymbol{X}_2'\boldsymbol{M}_1\boldsymbol{Y}$, it follows that $\boldsymbol{D}\hat{\boldsymbol{\beta}}_2 = \boldsymbol{X}_2'\boldsymbol{M}_1\boldsymbol{Y}$ and thus $\hat{\boldsymbol{\beta}}_2'\boldsymbol{D} = \boldsymbol{Y}'\boldsymbol{M}_1\boldsymbol{X}_2$. Hence, $\Delta SSR = \hat{\boldsymbol{\beta}}_2'\boldsymbol{D}\hat{\boldsymbol{\beta}}_2 = \hat{\boldsymbol{\beta}}_2'\boldsymbol{X}_2'\boldsymbol{M}_1\boldsymbol{X}_2\hat{\boldsymbol{\beta}}_2$ and thus Equation 12.I.12 can be rewritten as

$$F = \frac{\Delta SSR/(m-h)}{\boldsymbol{e}'\boldsymbol{e}/(n-m-1)} : F_{m-h,\, n-m-1} \tag{12.I.12.1}$$

or Equation 12.G.12.

Other possibilities involving choices for $\boldsymbol{R}$ and $\boldsymbol{r}$ abound. For instance:

a. Suppose $r = 1$ and

$$\underset{(1\times m+1)}{\boldsymbol{R}} = [0 \;\; 1 \;\; 1 \;\; 1 \;\; 0 \;\; \ldots \;\; 0], \; \boldsymbol{r} = 1.$$

Then $H_0: \boldsymbol{R\beta} = \boldsymbol{r}$ is equivalent to $H_0: \beta_1 + \beta_2 + \beta_3 = 1$.

b. Let $r = 2$ and

$$\underset{(2\times m+1)}{\boldsymbol{R}} = \begin{bmatrix} 0 & 1 & 0 & -1 & 0 & 0 & \ldots & 0 \\ 0 & 0 & 1 & 0 & 1 & 0 & \ldots & 0 \end{bmatrix}, \; \boldsymbol{r} = \boldsymbol{0}.$$

Then $H_0: \boldsymbol{R\beta} = \boldsymbol{r}$ is equivalent to $H_0: \beta_1 - \beta_3 = 0, \beta_2 + \beta_4 = 0$.

c. For $r = 3$ and

$$\underset{(3\times m+1)}{\boldsymbol{R}} = \begin{bmatrix} 0 & 1 & 0 & 0 & 1 & 0 & 0 & \ldots & 0 \\ 0 & 1 & -4 & 0 & 0 & 0 & 0 & \ldots & 0 \\ 0 & 1 & 0 & 3 & 0 & -1 & 0 & \ldots & 0 \end{bmatrix}, \; \boldsymbol{r} = \begin{bmatrix} 1 \\ 0 \\ 0 \end{bmatrix},$$

$H_0: \boldsymbol{R\beta} = \boldsymbol{r}$ is equivalent to $H_0: \beta_1 + \beta_4 = 1, \beta_1 - 4\beta_2 = 0, \beta_1 + 3\beta_3 - \beta_5 = 0$.

Appendix J: Point and Interval Predictions

Prediction Interval for an Individual Value of *Y*

Given the estimated regression equation $\hat{Y} = \hat{\beta}_0 + \hat{\beta}_1 X_1 + \ldots + \hat{\beta}_m X_m$ and a specific set of regressor values $X_1^0, X_2^0, \ldots, X_m^0$, let us represent a new or predicted value of Y as $\hat{Y}_0 = \hat{\beta}_0 + \hat{\beta}_1 X_1^0 + \ldots + \hat{\beta}_m X_m^0$ or

$$\hat{Y}_0 = \boldsymbol{X}_0\hat{\boldsymbol{\beta}} = \begin{bmatrix} 1 & X_1^0 & \ldots & X_m^0 \end{bmatrix} \begin{bmatrix} \hat{\beta}_0 \\ \hat{\beta}_1 \\ \cdot \\ \cdot \\ \cdot \\ \hat{\beta}_m \end{bmatrix}. \tag{12.J.1}$$

Here Equation 12.J.1 serves as the best *point predictor* of the individual response in Y at $\boldsymbol{X}_0$.

To determine how precisely this individual response in Y has been predicted at $\boldsymbol{X}_0$, let us distinguish between *the predicted* Y at $\boldsymbol{X}_0$ or $\hat{Y}_0 = \boldsymbol{X}_0\hat{\boldsymbol{\beta}}$ (Equation 12.J.1) and the *actual* Y at $\boldsymbol{X}_0$ or

$$Y_0 = \mathbf{X}_0\boldsymbol{\beta} + \varepsilon_0, \tag{12.J.2}$$

where ε_0 is the actual value of the random disturbance ε when $Y = Y_0$. Moreover, given the assumptions of the strong classical linear regression model, $Y_0 : N(\boldsymbol{X}_0\boldsymbol{\beta}, \sigma_\varepsilon^2)$. Then from Equations 12.J.1 and 12.J.2 we may write the *prediction error* as

$$\begin{aligned} e_0 = Y_0 - \hat{Y}_0 &= \boldsymbol{X}_0\boldsymbol{\beta} + \varepsilon_0 - \boldsymbol{X}_0\hat{\boldsymbol{\beta}} \\ &= -\boldsymbol{X}_0(\hat{\boldsymbol{\beta}} - \boldsymbol{\beta}) + \varepsilon_0. \end{aligned} \tag{12.J.3}$$

What are the characteristics of the sampling distribution of e_0? First, $E(e_0) = 0$ because $E(\hat{\boldsymbol{\beta}}) - \boldsymbol{\beta} = 0$ and $E(\varepsilon_0) = 0$. Second,

$$\begin{aligned} V(e_0) &= E\left\{\left[-\boldsymbol{X}_0(\hat{\boldsymbol{\beta}} - \boldsymbol{\beta}) + \varepsilon_0\right]\left[-\boldsymbol{X}_0(\hat{\boldsymbol{\beta}} - \boldsymbol{\beta}) + \varepsilon_0\right]'\right\} \\ &= E\left\{\left[-\boldsymbol{X}_0(\hat{\boldsymbol{\beta}} - \boldsymbol{\beta}) + \varepsilon_0\right]\left[-(\hat{\boldsymbol{\beta}} - \boldsymbol{\beta})'\boldsymbol{X}_0' + \varepsilon_0\right]\right\} \\ &= E\left[\boldsymbol{X}_0(\hat{\boldsymbol{\beta}} - \boldsymbol{\beta})(\hat{\boldsymbol{\beta}} - \boldsymbol{\beta})'\boldsymbol{X}_0' - 2\boldsymbol{X}_0(\hat{\boldsymbol{\beta}} - \boldsymbol{\beta})\varepsilon_0 + \varepsilon_0^2\right] \\ &= \sigma_\varepsilon^2\boldsymbol{X}_0(\boldsymbol{X}'\boldsymbol{X})^{-1}\boldsymbol{X}_0' + \sigma_\varepsilon^2 \end{aligned} \tag{12.J.4}$$

from Equation 12.I.1 and the independence of $\hat{\boldsymbol{\beta}}$ and ε_0, i.e., $E\left[(\hat{\boldsymbol{\beta}} - \boldsymbol{\beta})\varepsilon_0\right] = E(\hat{\boldsymbol{\beta}} - \boldsymbol{\beta})E(\varepsilon_0) = 0$ because $E(\hat{\boldsymbol{\beta}}) = \boldsymbol{\beta}$ and $E(\varepsilon_0) = 0$. Thus $e_0 :$ N(0, $V(e_0)$) so that, for $e_0 = Y_0 - \hat{Y}_0 = Y_0 - \boldsymbol{X}_0\hat{\boldsymbol{\beta}}$,

$$\frac{Y_0 - \boldsymbol{X}_0\hat{\boldsymbol{\beta}}}{\sigma_\varepsilon\sqrt{1 + \boldsymbol{X}_0(\boldsymbol{X}'\boldsymbol{X})^{-1}\boldsymbol{X}_0'}} : N(0,\ 1).$$

Let us form (see Appendix 12.E)

$$\frac{\dfrac{Y_0 - \boldsymbol{X}_0\hat{\boldsymbol{\beta}}}{\sigma_\varepsilon\sqrt{1 + \boldsymbol{X}_0(\boldsymbol{X}'\boldsymbol{X})^{-1}\boldsymbol{X}_0'}}}{\sqrt{\dfrac{(n - m - 1)S_\varepsilon^2/\sigma_\varepsilon^2}{n - m - 1}}} = \frac{Y_0 - \boldsymbol{X}_0\hat{\boldsymbol{\beta}}}{S_\varepsilon\sqrt{1 + \boldsymbol{X}_0(\boldsymbol{X}'\boldsymbol{X})^{-1}\boldsymbol{X}_0'}} : t_{n-m-1} \tag{12.J.5}$$

and thus

$$P\left(-t_{\frac{\alpha}{2},\, n-m-1} \le \frac{Y_0 - \boldsymbol{X}_0\hat{\boldsymbol{\beta}}}{S_\varepsilon\sqrt{1+\boldsymbol{X}_0(\boldsymbol{X}'\boldsymbol{X})^{-1}\boldsymbol{X}_0'}} \le t_{\frac{\alpha}{2},\, n-m-1}\right) = 1-\alpha$$

or

$$P\left(\boldsymbol{X}_0\hat{\boldsymbol{\beta}} - t_{\frac{\alpha}{2},\, n-m-1} S_\varepsilon\sqrt{1+\boldsymbol{X}_0(\boldsymbol{X}'\boldsymbol{X})^{-1}\boldsymbol{X}_0'} \le Y_0 \le \boldsymbol{X}_0\hat{\boldsymbol{\beta}} + t_{\frac{\alpha}{2},\, n-m-1} S_\varepsilon\sqrt{1+\boldsymbol{X}_0(\boldsymbol{X}'\boldsymbol{X})^{-1}\boldsymbol{X}_0'}\right) = 1-\alpha.$$

Hence, a $100(1-\alpha)\%$ *prediction interval for* Y_0 (the actual Y at $\boldsymbol{X}_0$) is

$$\boldsymbol{X}_0\hat{\boldsymbol{\beta}} \pm t_{\frac{\alpha}{2},\, n-m-1}\, s_\varepsilon\sqrt{1+\boldsymbol{X}_0(\boldsymbol{X}'\boldsymbol{X})^{-1}\boldsymbol{X}_0'}, \tag{12.J.6}$$

where s_ε is the sample realization of $S_\varepsilon = \sqrt{\boldsymbol{e}'\boldsymbol{e}/(n-m-1)}$.

Prediction Interval for the Mean Value of *Y*

We next consider the construction of a prediction interval for the mean value of Y at $\boldsymbol{X}_0$, $E(Y_0)$. Given $Y_0 = \boldsymbol{X}_0\boldsymbol{\beta} + \varepsilon_0$, the *mean response* at $\boldsymbol{X}_0$ is $E(Y_0) = \boldsymbol{X}_0\boldsymbol{\beta}$ because $E(\varepsilon_0) = 0$, whereas the *estimated mean response at* $\boldsymbol{X}_0$ is the point predictor $\hat{Y}_0 = \boldsymbol{X}_0\hat{\boldsymbol{\beta}}$. Hence, the *prediction error* is $e_0^* = E(Y_0) - \hat{Y}_0 = \boldsymbol{X}_0\boldsymbol{\beta} - \boldsymbol{X}_0\hat{\boldsymbol{\beta}} = -\boldsymbol{X}_0(\hat{\boldsymbol{\beta}}-\boldsymbol{\beta})$ with $E(e_0^*) = 0$ (since $E(\hat{\boldsymbol{\beta}}) = \boldsymbol{\beta}$) and, proceeding as above,

$$\begin{aligned} V(e_0^*) &= E\left\{\left[-\boldsymbol{X}_0(\hat{\boldsymbol{\beta}}-\boldsymbol{\beta})\right]\left[-\boldsymbol{X}_0(\hat{\boldsymbol{\beta}}-\boldsymbol{\beta})\right]'\right\} \\ &= \sigma_\varepsilon^2\boldsymbol{X}_0(\boldsymbol{X}'\boldsymbol{X})^{-1}\boldsymbol{X}_0'. \end{aligned} \tag{12.J.7}$$

Hence, $e_0^* : N(0, V(e_0^*))$ and thus, for $e_0^* = E(Y_0) - \hat{Y}_0 = E(Y_0) - \boldsymbol{X}_0\hat{\boldsymbol{\beta}}$,

$$\frac{E(Y_0) - \boldsymbol{X}_0\hat{\boldsymbol{\beta}}}{\sigma_\varepsilon\sqrt{\boldsymbol{X}_0(\boldsymbol{X}'\boldsymbol{X})^{-1}\boldsymbol{X}_0'}} : N(0,\ 1).$$

Then using the same line of argumentation used in deriving Equations 12.J.5 and 12.J.6, we may conclude that a $100(1-\alpha)\%$ *prediction interval for* $E(Y_0)$ (the mean value of Y at $\boldsymbol{X}_0$) is

$$\boldsymbol{X}_0\hat{\boldsymbol{\beta}} \pm t_{\frac{\alpha}{2},\, n-m-1}\, s_\varepsilon\sqrt{\boldsymbol{X}_0(\boldsymbol{X}'\boldsymbol{X})^{-1}\boldsymbol{X}_0'}, \tag{12.J.8}$$

where again s_ε is the realized value of S_ε.

Appendix K: Multicollinearity

We noted in previous sections of this chapter that the OLS stimator of $\boldsymbol{\beta}$ is $\hat{\boldsymbol{\beta}} = (\boldsymbol{X}'\boldsymbol{X})^{-1}\boldsymbol{X}'\boldsymbol{Y}$ with $V(\hat{\boldsymbol{\beta}}) = \sigma_\varepsilon^2(\boldsymbol{X}'\boldsymbol{X})^{-1}$. A glance at the equation for $V(\hat{\boldsymbol{\beta}})$ reveals that this matrix depends on the sample values of the regressors $X_1,\ldots,X_m$ or on the columns of $\boldsymbol{X} = [\boldsymbol{X}_0, \boldsymbol{X}_1, \ldots, \boldsymbol{X}_m]$, where $\boldsymbol{X}_0$ is a column of 1's. We also noted above that the data problem of multicollinearity emerges because of certain collinearities or interrelations among the columns of $\boldsymbol{X}$. Obviously, these said collinearities will affect $V(\hat{\boldsymbol{\beta}})$. Our aim is to determine what sorts of interconnections involving the columns of $\boldsymbol{X}$ pose problems for estimaton purposes.

Orthogonal Regressors

If the regressors X_h and X_l are *orthogonal*, then $\boldsymbol{X}_h'\boldsymbol{X}_l = 0$ for all $h \neq l$, where $\boldsymbol{X}_h$ and $\boldsymbol{X}_l$ are the columns of $\boldsymbol{X}$ corresponding to the regressors X_h and X_l, respectively. In this instance, $COV(X_h, X_l) = 0$ for all $h \neq l$, and thus the columns of $\boldsymbol{X}$ are linearly independent[5] so that $\boldsymbol{X}$ has full column rank[6] $\rho(\boldsymbol{X}) = m + 1$. Hence, $\boldsymbol{X}'\boldsymbol{X}$ is nonsingular or $|\boldsymbol{X}'\boldsymbol{X}| \neq 0$ (actually $|\boldsymbol{X}'\boldsymbol{X}| = 1$) so that $(\boldsymbol{X}'\boldsymbol{X})^{-1}$ exists and thus the estimated regression coefficients $\hat{\beta}_0, \hat{\beta}_1, \ldots, \hat{\beta}_m$ are uniquely determined from the system of least-squares normal equations.

Exact Collinearity (or Perfect Multicollinearity)

If the regressors X_h and X_l are exactly collinear (e.g., $\boldsymbol{X}_h = \theta\boldsymbol{X}_l$), then $\rho(\boldsymbol{X}) < m + 1$. More generally, $\boldsymbol{X}$ is not of full column rank if at least one linear relation exists among the columns of $\boldsymbol{X}$, i.e., there is at least one linear relation of the form $\sum_{k=0}^{m+1} \lambda_k \boldsymbol{X}_k = \boldsymbol{0}$, with not all of the λ_k's equal to zero. If, say, $\lambda_{m+1} \neq 0$, then $\boldsymbol{X}_{m+1} = \sum_{k \neq m+1} (-\lambda_k / \lambda_{m+1})\boldsymbol{X}_k$ and thus $\boldsymbol{X}_{m+1}$ can be written as an *exact linear combination* of the remaining columns of $\boldsymbol{X}$, i.e., $\boldsymbol{X}_{m+1}$ is linearly dependent on the remaining columns of $\boldsymbol{X}$. So under the linear dependence of the columns of $\boldsymbol{X}$, $\rho(\boldsymbol{X}) < m + 1$. Hence, in this circumstance, $|\boldsymbol{X}'\boldsymbol{X}| = 0$ and $(\boldsymbol{X}'\boldsymbol{X})^{-1}$ does not exist. (It was mentioned above that $|\boldsymbol{X}'\boldsymbol{X}| = 1$ under the orthogonality of the regressors. However, as the degree of collinearity among the regressors increases, $|\boldsymbol{X}'\boldsymbol{X}|$ declines in value and equals zero when exact collinearity is attained.) In this regard, given that the columns of $\boldsymbol{X}$ are linearly dependent, the OLS normal equations $(\boldsymbol{X}'\boldsymbol{X})\hat{\boldsymbol{\beta}} = \boldsymbol{X}'\boldsymbol{Y}$ are con-

---

[5] A vector $\boldsymbol{X}$ is a *linear combination* of the vectors $\boldsymbol{X}_j$, $j = 1,\ldots,m$, if there exist scalars λ_j such that $\boldsymbol{X} = \sum_j \lambda_j \boldsymbol{X}_j$. A set of vector $\{\boldsymbol{X}_j\}$ is *linearly dependent* if there exist scalars λ_j not all zero such that $\sum_j \lambda_j \boldsymbol{X}_j = \boldsymbol{0}$ (the null vector is a linear combination of the vectors $\boldsymbol{X}_j$). If the only set of scalars for which $\Sigma j \lambda_j \boldsymbol{X}_j = \boldsymbol{0}$ is $\lambda_j = 0$, then the vectors $\boldsymbol{X}_j$ are *linearly independent* (the trivial combination is the only linear combination of the $\boldsymbol{X}_j$ which equals the null vector). If the set $\{\boldsymbol{X}_j\}$ is linearly dependent, then one of the vectors is a linear combination of the others.

[6] The *rank* of an $(m \times n)$ matrix $\boldsymbol{A}$, denoted $\rho(\boldsymbol{A})$, is the order of the largest nonvanishing determinant of $\boldsymbol{A}$ (or the order of the largest nonsingular submatrix of $\boldsymbol{A}$). The $(n \times m + 1)$ matrix $\boldsymbol{X}$, $n > m + 1$, is of full column rank if $\rho(\boldsymbol{X}) = m + 1$ (none of the columns of $\boldsymbol{X}$ are linearly dependent on the others). Alternatively, we may view the rank of $\boldsymbol{X}$ as the maximum number of linearly independent vectors $\boldsymbol{X}_k$, $k = 0,1,\ldots,m$, which span the columns of $\boldsymbol{X}$. With $\rho(\boldsymbol{X}) = m + 1$, it follows that $\rho(\boldsymbol{X}'\boldsymbol{X}) = m + 1$ or $|\boldsymbol{X}'\boldsymbol{X}| \neq 0$.

sistent but do not admit a unique solution. In fact, this system has an infinite number of particular solutions.[7]

Although the extreme cases of orthogonality and exact collinearity are quite rare, the instance of near collinearity is, unfortunately, very prevalent and, at times, highly troublesome. This observation thus takes us to the following discussion.

Near Multicollinearity

If $\rho(X) = m + 1$ (X has full column rank), then $(X'X)^{-1}$ exists despite the fact that $X'X$ is close to being singular or $|X'X| \approx 0$ and thus the OLS estimator $\hat{\boldsymbol{\beta}} = (\boldsymbol{X}'\boldsymbol{X})^{-1}\boldsymbol{X}'\boldsymbol{Y}$ is unique as well as BLUE. However, in the presence of near-linear dependencies among columns of $\boldsymbol{X}$, there exists at least one linear relation of the form

$$\sum_{k=0}^{m+1} \lambda_k \boldsymbol{X}_k \approx \boldsymbol{0}, \tag{12.K.1}$$

where "≈" means "almost or approximately equal to." Hence, the linear relationship among the columns of $\boldsymbol{X}$ is not an exact one: it " holds only approximately or nearly so." In the presence of an expression such as Equation 12.K.1, the variances of the $\hat{\beta}_k$'s, $k = 0, 1, \ldots, m$, can increase sharply.

To fully assess the impact of near multicollinearity on $V(\hat{\boldsymbol{\beta}})$, let us partition $\boldsymbol{X}$ as = $[\boldsymbol{X}_r, \boldsymbol{X}_{(-r)}]$, where $\boldsymbol{X}_r$ is the column of observations on regressor X_r and $\boldsymbol{X}_{(-r)}$ is the $(n \times m)$ submatrix containing all remaining columns of $\boldsymbol{X}$. Then

$$\boldsymbol{X}'\boldsymbol{X} = \begin{bmatrix} \boldsymbol{X}_r'\boldsymbol{X}_r & \boldsymbol{X}_r'\boldsymbol{X}_{(-r)} \\ \boldsymbol{X}_{(-r)}'\boldsymbol{X}_r & \boldsymbol{X}_{(-r)}'\boldsymbol{X}_{(-r)} \end{bmatrix}$$

and, under partitioned inversion,

$$(\boldsymbol{X}'\boldsymbol{X})^{-1} = \begin{bmatrix} (\boldsymbol{X}_r'\boldsymbol{M}_{(-r)}\boldsymbol{X}_r)^{-1} & -\boldsymbol{D}' \\ -\boldsymbol{D} & \boldsymbol{E} \end{bmatrix},$$

where

$$(\boldsymbol{X}_r'\boldsymbol{M}_{(-r)}\boldsymbol{X}_r)^{-1} = \boldsymbol{X}_r'\boldsymbol{X}_r - \boldsymbol{X}_r'\boldsymbol{X}_{(-r)}(\boldsymbol{X}_{(-r)}'\boldsymbol{X}_{(-r)})^{-1}(\boldsymbol{X}_{(-r)}'\boldsymbol{X}_r)$$

and

$$\boldsymbol{M}_{(-r)} = \boldsymbol{I}_n - \boldsymbol{X}_{(-r)}(\boldsymbol{X}_{(-r)}'\boldsymbol{X}_{(-r)})^{-1}\boldsymbol{X}_{(-r)}'.$$

---

[7] For a linear system $\boldsymbol{AX} = \boldsymbol{b}$, either $\boldsymbol{AX} = \boldsymbol{b}$ is inconsistent (no solution exists) or $\boldsymbol{AX} = \boldsymbol{b}$ consistent (meaning that exactly one solution exists or an infinity of solutions exists). Specifically, let $\boldsymbol{A}$ be of order $(m \times n)$. Then, if

a. $\rho[\boldsymbol{A}, \boldsymbol{b}] > \rho(\boldsymbol{A})$, the system is inconsistent.
b. $\rho[\boldsymbol{A}, \boldsymbol{b}] = \rho(\boldsymbol{A})$ = number of unknowns n, the system is consistent and possesses a unique solution.
c. $\rho[\boldsymbol{A}, \boldsymbol{b}] = \rho(\boldsymbol{A}) = k <$ number of unknowns n, the system is consistent and possesses an infinity of solutions, where arbitvary values may be assigned to $n - m$ of the variables.

Additionally,

$$D' = (X_r'M_{(-r)}X_r)^{-1}X_r'X_{(-r)}(X_{(-r)}'X_{(-r)})^{-1}$$

and

$$E = (X_{(-r)}'X_{(-r)})^{-1} + (X_{(-r)}'X_{(-r)})^{-1}X_{(-r)}'X_r(X_r'M_{(-r)}X_r)^{-1}X_r'X_r(X_{(-r)}'X_{(-r)})^{-1}.$$

Because from Equation 12.13 and Appendix 12.D we know that $V(\hat{\boldsymbol{\beta}}) = \sigma_\varepsilon^2(X'X)^{-1}$, it follows that

$$V(\hat{\beta}_r) = \frac{\sigma_\varepsilon^2}{X_r'M_{(-r)}X_r}. \tag{12.K.2}$$

How should we interpret the denominator of Equation 12.K.2? Suppose we use an auxiliary regression equation to regrees X_r on the remaining $m - 1$ explanatory variables (including the intercept). That is, given $\lambda_r \neq 0$, we may use Equation 12.K.1 to form the expression

$$X_r \approx \sum_{k \neq r}(-\lambda_k / \lambda_r)X_k$$

or, with $\beta_{kr} = -\lambda_k/\lambda_r$,

$$X_r \approx \sum_{k \neq r}\beta_{kr}X_k + \boldsymbol{\varepsilon}_r = X_{(-r)}\beta_r + \boldsymbol{\varepsilon}_r, \tag{12.K.3}$$

where $\boldsymbol{\varepsilon}_r$ is a random error term. Then from Equation 12.D.4, because $\hat{X}_r = X_{(-r)}\hat{\boldsymbol{\beta}}_r + e_r$,

$$e_r = \hat{X}_r - X_{(-r)}\hat{\boldsymbol{\beta}}_r = M_{(-r)}X_r$$

and thus

$$e_r'e_r = X_r'M_{(-r)}X_r = SSE_r \tag{12.K.4}$$

(because $M_{(-r)}$ is idempotent), i.e., $X_r'M_{(-r)}X_r$ is the error or residual sum of squares obtained from regressing X_r on the remaining $m - 1$ explanatory variables. (Note that the better the fit of Equation 12.K.3 the more exact Equation 12.K.1 becomes and thus the moves severe the multicollinearity among the columns of **X**. In the extreme, $SSE_r = 0$ and "≈" is replaced by "=" in Equation 12.K.1, the case of perfect or exact multicollinearity.) Hence, Equation 12.K.2 becomes

$$V(\hat{\beta}_r) = \frac{\sigma_\varepsilon^2}{SSE_r}. \tag{12.K.2.1}$$

Because SSE_r decreases with increasing collinearity between X_r and the remainig regressors, we see from the preceding expression that $V(\hat{\beta}_r)$ increases with increasing collinearity between X_r and the said regressors. However, one should be cognizant of the fact that not all regressors will exhibit the same SSE value when regressed on the other explanatory variables. Hence, $V(\hat{\beta}_r) = \sigma_\varepsilon^2 / SSE_r$ will obviously change according to the choice of *r*.

From Appendix 12.F, we know that $SSE_r = SST_r - SSR_r$ so that

$$\frac{SSE_r}{SST_r} = 1 - \frac{SSR_r}{SST_r} = 1 - R_r^2, \tag{12.K.5}$$

where R_r^2 is the auxiliary coefficient of determination resulting from the regression of X_r on the remaining explanatory variables. (Although the intercept is included in the auxiliary regression equation, it does not have an auxiliary coefficient of determination associated with it.) In view of Equation 12.K.5, Equation 12.K.2.1 can be expressed as

$$V(\hat{\beta}_r) = \frac{\sigma_\varepsilon^2}{SSE_r(1 - R_r^2)}. \tag{12.K.2.2}$$

Now, if the *m* regressors $X_1, \ldots, X_m$ are orthogonal, then $R_r^2 = 0$ for any choice of *r* and thus, if $\hat{\beta}_{r0}$ denotes the OLS estimate of β_r under orthogonality, then $V(\hat{\beta}_{r0}) = \sigma_\varepsilon^2 / SST_r$ and thus

$$V(\hat{\beta}_r) = \frac{V(\hat{\beta}_{r0})}{(1 - R_r^2)}$$

or

$$\frac{V(\hat{\beta}_r)}{V(\hat{\beta}_{r0})} = \frac{1}{(1 - R_r^2)}. \tag{12.K.6}$$

Here $(1 - R_r^2)^{-1}$ is termed the *variance inflation factor* for X_r. That is, for $V(\hat{\beta}_{r0})$ constant, Equation 12.K.6 measures the relative increase in $V(\hat{\beta}_r)$ (over and above that which would exist in the strictly orthogonal case) resulting from near multicollinearity among the regressors [Marquardt 1970].

Two additional implications of near multicollinearity merit our consideration. First, the estimated covariances between $\hat{\beta}_r$ and the other OLS slope estimates will be inflated (in absolute value), the greater the estimated impact of X_k on X_r (as indicated by the magnitude of $\hat{\beta}_{kr}$ when auxiliary regression Equation 12.K.3 is fit) and the better the overall fit to Equation 12.K.3 (as indicated by a small SSE_r term). That is, from $(\boldsymbol{X}'\boldsymbol{X})^{-1}$ above, $-\boldsymbol{D} = -(SSE_r)^{-1}\hat{\boldsymbol{\beta}}_r$, where $\hat{\boldsymbol{\beta}}_r = (\boldsymbol{X}_{(-r)}'\boldsymbol{X}_{(-r)})^{-1}\boldsymbol{X}_{(-r)}'\boldsymbol{X}_r$ is the vector of OLS estimates of the regression coefficients of the auxiliary regression Equation 12.K.3. Then the covariances between $\hat{\beta}_r$ and the remaining *m* OLS estimators are, from $V(\hat{\boldsymbol{\beta}}) = \sigma_\varepsilon^2(\boldsymbol{X}'\boldsymbol{X})^{-1}$,

$$\begin{bmatrix} COV(\hat{\beta}_0,\hat{\beta}_r) \\ \cdot \\ \cdot \\ \cdot \\ COV(\hat{\beta}_{r-1},\hat{\beta}_r) \\ COV(\hat{\beta}_{r+1},\hat{\beta}_r) \\ \cdot \\ \cdot \\ \cdot \\ COV(\hat{\beta}_m,\hat{\beta}_r) \end{bmatrix} = -\sigma_\varepsilon^2 \boldsymbol{D} = -\frac{\sigma_\varepsilon^2 \hat{\boldsymbol{\beta}}_r}{SSE_r}$$

or

$$COV(\hat{\beta}_k,\hat{\beta}_r) = \frac{-\sigma_\varepsilon^2 \hat{\beta}_{kr}}{SSE_r}, k \neq r \ ; \ k = 0,1,\ldots,m.$$

Second, it can be demonstrated that, in the presence of near multicollinearity, the estimated regression coefficients are a bit larger (in absolute value) than would be the case if any near-linear dependencies in $\boldsymbol{X}$ did not exist.

Detecting Multicollinearity with the Correlation Matrix: VIFs Revisited

It was determined above that VIF_r is a measure of the amount by which $V(\hat{\beta}_r)$ is inflated by the relationship of $\boldsymbol{X}_r$ to the remaining columns of $\boldsymbol{X}$ (including the intercept); it measures the combined effect of any near dependencies among the regressors on the estimated variance of $\hat{\beta}_r$.

Suppose we transform $\boldsymbol{X}$ to $\boldsymbol{W}$ via unit length scaling (see above, "Standardized Regression Coefficients") so that each transformed variable W_j has been centered to zero and has unit length ($\|\boldsymbol{W}_j\| = 1$) and the intercept is eliminated in the process. Then the $(m \times m)$ matrix

$$\boldsymbol{W}'\boldsymbol{W} = (\boldsymbol{X}'\boldsymbol{X})_s = \begin{bmatrix} 1 & r_{12} & \ldots & r_{1m} \\ r_{21} & 1 & \ldots & r_{2m} \\ \cdot & \cdot & & \cdot \\ \cdot & \cdot & & \cdot \\ \cdot & \cdot & & \cdot \\ r_{m1} & r_{m2} & \ldots & 1 \end{bmatrix}$$

is a correlation matrix containing Pearson product-moment correlation coefficients.

We noted above that, if $\boldsymbol{X}$ is of full (column) rank $m + 1$, then the variance-covariance matrix of $\hat{\boldsymbol{\beta}}$ is $V(\hat{\boldsymbol{\beta}}) = \sigma_\varepsilon^2(\boldsymbol{X}'\boldsymbol{X})^{-1}$. However, the process of centering and scaling the original variables X_j, $j = 1,\ldots,m$ (remember that this process eliminates the intercept) transforms $V(\hat{\boldsymbol{\beta}})$ to the $(m \times m)$ matrix

$$V(\hat{\boldsymbol{\beta}})_s = \sigma_\varepsilon^2(\boldsymbol{X}'\boldsymbol{X})_s^{-1} = \sigma_\varepsilon^2 \boldsymbol{C}^0, \tag{12.K.7}$$

where

$$\begin{aligned} V(\hat{\boldsymbol{\beta}}_r) &= \sigma_\varepsilon^2 C_{rr}^0 = \sigma_\varepsilon^2(1-R_r^2)^{-1} \\ &= \sigma_\varepsilon^2 VIF_r, \end{aligned} \tag{12.K.7.1}$$

where now R_r^2 represents the coefficient of multiple determination resulting from the regression of W_r on the remaining $m-1$ centered and scaled variables W_j, $j \neq r$.[8] Thus, the main diagonal elements of $(\boldsymbol{X}'\boldsymbol{X})_s^{-1}$, the inverse of the correlation matrix $(\boldsymbol{X}'\boldsymbol{X})_s$, are the variance inflation factors $C_{jj}^0 = VIF_j = (1-R_j^2)^{-1}, j = 1,\ldots,m$; C_{jj}^0 is the factor by which $V(\hat{\beta}_j)$ increases as a result of near multicollinearity among the regressors. That is, if W_r is nearly orthogonal to the remaining regressors W_j, $j \neq r$, then R_r^2 is small and $C_{rr}^0 = VIF_r$ is close to unity and thus precipitates a negligible increase in $V(\hat{\beta}_r)$. However, if R_r^2 is near unity (W_r exhibits a substantial degree of near collinearity with the other regressors W_j, $j \neq r$), then $C_{rr}^0 = VIF_r$ is large and consequently significantly inflates $V(\hat{\beta}_r)$, thus compromising the precision of our estimate of β_r.

Eigenanalysis

Performing an eigenanalysis of the $\boldsymbol{X}'\boldsymbol{X}$ matrix is an alternative and highly efficient way to expose near dependencies (- collinearities) among the set of regressors. To set the stage for an eigensystem analysis of $\boldsymbol{X}'\boldsymbol{X}$, let us develop some important terminology.

Eigenvalues and Eigenvectors

Let us pose the *eigenvalue* or *characteristic value problem* as finding values of a sclalar λ for which there exist vectors $\boldsymbol{X}(\neq \boldsymbol{0})$ such that

$$\boldsymbol{AX} = \lambda \boldsymbol{X}, \tag{12.K.8}$$

where $\boldsymbol{A}$ is an $(n \times n)$ matrix and $\boldsymbol{X}$ is an $(n \times 1)$ vector. So if nontrivial $\boldsymbol{X}$ satisfies Equation 12.K.8 for a given λ, then $\boldsymbol{AX}$ generates a vector that is a scalar multiple of $\boldsymbol{X}$. Suppose we rewrite Equation 12.K.8 as the homogeneous linear equation system in n unknowns

$$(\boldsymbol{A} - \lambda \boldsymbol{I}_n)\boldsymbol{X} = 0. \tag{12.K.8.1}$$

Then for a given λ, any $\boldsymbol{X}$ that satisfies Equation 12.K.8 must also satisfy Equation 12.K.8.1.

[8] With $(\boldsymbol{X}'\boldsymbol{X})_s$ a correlation matrix, it follows that $0 \leq |(\boldsymbol{X}'\boldsymbol{X})_s| \leq 1$. So if the regressors W_j, $j = 1,\ldots,m$, are orthogonal, then $|(\boldsymbol{X}'\boldsymbol{X})_s| = 1$ ($r_{jl} = 0$ for all $j, l = 1,\ldots,m$); and if exact collinearity holds among the regressors, then $|(\boldsymbol{X}'\boldsymbol{X})_s| = 0$ and $(\boldsymbol{X}'\boldsymbol{X})_s^{-1}$ does not exist.

The system in Equation 12.K.8.1 will admit a nontrivial solution ($\boldsymbol{X} \neq \boldsymbol{0}$) if and only if

$$f(\lambda) = |\boldsymbol{A} - \lambda \boldsymbol{I}_n| = 0 \tag{12.K.9}$$

or

$$f(\lambda) = \begin{vmatrix} a_{11}-\lambda & a_{12} & \cdots & a_{1n} \\ a_{21} & a_{22}-\lambda & \cdots & a_{2n} \\ \cdot & \cdot & & \cdot \\ \cdot & \cdot & & \cdot \\ \cdot & \cdot & & \cdot \\ a_{n1} & a_{n2} & \cdots & a_{nn}-\lambda \end{vmatrix} = 0.^{9} \tag{12.K.9.1}$$

Here Equation 12.K.9 (or Equation 12.K.9.1) is an nth degree polynominal in λ and is termed the *characteristic equation* for $\boldsymbol{A}$. This nth degree polynominal has n roots $\lambda_1, \ldots, \lambda_n$ that are termed the eigenvalues or characteristic values or *latent roots* of $\boldsymbol{A}$. Moreover, any vector ($\boldsymbol{X} \neq \boldsymbol{0}$) that satisfies Equation 12.K.8 (or Equation 12.K.8.1) is called an eigenvector or characteristic vector of $\boldsymbol{A}$, i.e., each eigenvalue λ_i, when substituted back into, say, Equation 12.K.8, yields the corresponding eigenvector $\boldsymbol{X}_i$.

Example 12.K.1

For instance, if

$$\boldsymbol{A} = \begin{bmatrix} a_{11} & a_{12} \\ a_{21} & a_{22} \end{bmatrix},$$

then the characteristic equation for $\boldsymbol{A}$ is

$$\begin{aligned} f(\lambda) &= \begin{vmatrix} a_{11}-\lambda & a_{12} \\ a_{21} & a_{22}-\lambda \end{vmatrix} = (a_{11}-\lambda)(a_{22}-\lambda) - a_{12}a_{21} \\ &= \lambda^2 - (a_{11}+a_{22})\lambda + (a_{11}a_{22} - a_{12}a_{21}) \\ &= \lambda^2 - (a_{11}+a_{22})\lambda + |\boldsymbol{A}|. \end{aligned} \tag{12.K.10}$$

The two roots or eigenvalues of this expression are (via the quadratic formula[10])

$$\lambda = \frac{1}{2}\left[(a_{11}+a_{22}) \pm \sqrt{(a_{11}+a_{22})^2 - 4|\boldsymbol{A}|}\right]. \tag{12.K.11}$$

---

[9] If $\boldsymbol{A} - \lambda \boldsymbol{I}_n$ is nonsingular (i.e., $|\boldsymbol{A} - \lambda \boldsymbol{I}_n| \neq 0$), then the only solution to Equation 12.K.8.1 is the trivial solution or $\boldsymbol{X} = \boldsymbol{0}$. For Equation 12.K.8.1 to admit a nontrivial solution, $\boldsymbol{A} - \lambda \boldsymbol{I}_n$ must be singular or Equation 12.K.9 must hold.

[10] To find the roots (or zeros) of the quadratic function $ax^2 + bx + c = 0$, $a \neq 0$, determine

$$x = \frac{-b \pm \sqrt{b^2 - 4ac}}{2a}.$$

These roots will be real if $\boldsymbol{A}$ is a real symmetric matrix. (In fact, in all that follows, it is assumed that $\boldsymbol{A}$ is a real symmetric matrix.) If we denote these eigenvalues as λ_1 and λ_2, then the characteristic equation for $\boldsymbol{A}$ can be written in the (factored) form

$$f(\lambda) = (\lambda_1 - \lambda)(\lambda_2 - \lambda) = \lambda^2 - (\lambda_1 + \lambda_2)\,\lambda + \lambda_1\lambda_2 = 0. \tag{12.K.12}$$

A comparison between Equations 12.K.10 and 12.K.12 reveals that

$$tr(\boldsymbol{A}) = \lambda_1 + \lambda_2 = a_{11} + a_{22};$$

$$|\boldsymbol{A}| = \lambda_1\lambda_2 = a_{11}a_{22} - a_{12}a_{21}.$$

(In general, for $\boldsymbol{A}$ an nth order matrix, $\rho(\boldsymbol{A})$ = number of nonzero eigenvalues, $tr(\boldsymbol{A}) = \sum_{i=1}^{n} \lambda_i$ and $|\boldsymbol{A}| = \prod_{i=1}^{n} \lambda_i$.) Finally, we may use Equation 12.K.8.1 to determine the eigenvectors $\boldsymbol{X}_1$ and $\boldsymbol{X}_2$ of $\boldsymbol{A}$ corresponding to λ_1 and λ_2 by solving the homogenous equation systems

$$(\boldsymbol{A} - \lambda_1 \boldsymbol{I}_n)\boldsymbol{X} = \boldsymbol{0}$$

$$(\boldsymbol{A} - \lambda_2 \boldsymbol{I}_n)\boldsymbol{X} = \boldsymbol{0}. \tag{12.K.13}$$

The resulting eigenvectors are real if λ_1 and λ_2 are real.

Some of the salient features of the set of eigenvectors corresponding to the eigenvalues of a real symmetric matrix $\boldsymbol{A}$ are as follows:

a. Eigenvectors corresponding to distinct eigenvalues are pairwise orthogonal, i.e., if $\boldsymbol{X}_h$ and $\boldsymbol{X}_l$ are the eigenvectors corresponding to the eigenvalues λ_h and λ_l respectively ($\lambda_h \neq \lambda_l$) then $\boldsymbol{X}_h'\boldsymbol{X}_l = 0$.

b. If an eigenvalue λ_t has multiplicity k (it is repeated k times), then there will be k orthogonal eigenvectors corresponding to this eigenvalue.

c. If an nth order matrix $\boldsymbol{A}$ has eigenvalues $\lambda_1,\ldots,\lambda_n$ (possibly not all of them distinct), then there exists a set of n pairwise orthogonal eigenvectors $\{\boldsymbol{X}_1,\ldots,\boldsymbol{X}_n\}$ such that

$$\boldsymbol{X}_i'\boldsymbol{X}_j = 0,\ i \neq j\,;\ i, j = 1,\ldots,n. \tag{12.K.14}$$

d. Suppose $\boldsymbol{X}_t$ ($\neq \boldsymbol{0}$) is an eigenvector of $\boldsymbol{A}$ corresponding to the eigenvalue λ_t. Then any scalar multiple of $\boldsymbol{X}_t$, $c\boldsymbol{X}_t$, is also an eignvalue of $\boldsymbol{A}$ for c a constant, i.e., if $\boldsymbol{A}\boldsymbol{X}_t = \lambda_t\boldsymbol{X}_t$, then $\boldsymbol{A}(c\boldsymbol{X}_t) = \lambda_t(c\boldsymbol{X}_t)$.

Orthogonal Matrices

When working with a set of eigenvectors of a symmetric matrix $\boldsymbol{A}$, it is customary to normalize them to length unity, i.e., $\boldsymbol{X}_i$ is transformed to $\boldsymbol{X}_i/\|\boldsymbol{X}_i\| = \boldsymbol{U}_i$, $i = 1,\ldots,n$, where the *length* or *norm* of $\boldsymbol{X}_i$ is $\|\boldsymbol{X}_i\| = (\boldsymbol{X}_i'\boldsymbol{X}_i)^{1/2} = \left(\sum_{i=1}^{n} |x_i|^2\right)^{1/2}$. Hence,

$$\|\boldsymbol{U}_i\| = 1 \text{ or } \boldsymbol{U}_i'\boldsymbol{U}_i = 1,\ i = 1,\ldots,n, \tag{12.K.15}$$

and, as required by Equation 12.K.14,

$$\boldsymbol{U}_i'\boldsymbol{U}_j = 0,\ i \neq j;\ i, j = 1,\ldots,n. \tag{12.K.16}$$

Conditions in Equations 12.K.15 and 12.K.16 are easily combined to yield

$$U_i'U_j = \delta_{ij} = \begin{cases} 0, & i \neq j; \\ 1, & i = j. \end{cases} \tag{12.K.17}$$

Any set of vectors satisfying this equation is termed an *orthonormal set of vectors.*

Let us rewrite Equation 12.K.17 in matrix form as

$$U'U = \left[U_i'U_j\right] = \left[\delta_{ij}\right] = I_n, \tag{12.K.18}$$

where $U = [U_1, U_2, \ldots, U_n]$ is a matrix whose columns constitute a set of pairwise orthogonal eigenvectors of unit length , i.e., the columns of U represent an orthogonal set of eigenvectors for A. Clearly, Equation 12.K.18 implies that

$$U' = U^{-1}; \tag{12.K.19}$$

and any matrix U whose inverse is its transpose is said to be an *orthogonal matrix*. Because U^{-1} is unique, it follows that $U'U = UU' = I_n$ so that both the rows and columns of U are orthogonal.

Similarity Transformations

Suppose A is an nth order symmetric matrix and let the eigenvector X correspond to the eigenvalue λ so that $AX = \lambda X$ as required by Equation 12.K.8. Additionally, let P be an $(n \times n)$ nonsingular matrix and form $PAX = \lambda PX = \lambda Y$. Clearly, $Y = PX$ is not an eigenvector of A. However, we may write $X = P^{-1}PX$ so that $PAP^{-1}PX = \lambda PX$ or $PAP^{-1}Y = \lambda Y$. Hence, Y is an eigenvector of PAP^{-1} corresponding to the eigenvalue λ.

Now, set $B = PAP^{-1}$ so that $BY = \lambda Y$ (thus λ is an eigenvalue of B). Then $P^{-1}BY = \lambda P^{-1}Y$. If we write $Y = PP^{-1}Y$, then $P^{-1}BPP^{-1}Y = \lambda P^{-1}Y$ or $AX = \lambda X$ because $A = P^{-1}BP$ and $X = P^{-1}Y$. Hence, X is an eigenvector of $P^{-1}BP$ corresponding to eigenvalue λ. In summary, if λ is an eigenvalue of A, then λ is also an eigenvalue of $B = PAP^{-1}$ for P an nth order nonsingular matrix; and if λ is an eigenvalue of B, then λ is also an eigenvalue of $A = P^{-1}BP$ In this regard, matrices having identical sets of eigenvalues are termed similar. More formally, if P is an nth-order nonsingular matrix such that $B = PAP^{-1}$, then the nth-order matrices A and B are *similar*, with B obtained from A by way of a *similarity transformation*.

Diagionalizing a Symmetric Matrix

For U_j corresponding to λ_j we have, from Equation 12.K.8, $AU_j = \lambda_j U_j$ and thus $U_i'AU_j = \lambda_j U_i'U_j = \lambda_j \delta_{ij}$. However, this is just the i, jth element of a diagonal matrix D. In this regard, given an orthogonal matrix of eigenvectors U,

$$\begin{aligned} U'AU &= \left[U_i'AU_j\right] = \left[\lambda_j U_i'U_j\right] = \left[\lambda_j \delta_{ij}\right] \\ &= D = \text{diag}(\lambda_1, \ldots, \lambda_n) \end{aligned} \tag{12.K.20}$$

is an nth-order diagonal matrix with the diagonal elements being the eigenvalues of A. Hence, the orthogonal matrix of eigenvectors U *diagionalizes* symmetric A or $U'AU = D$.

Note that, because U is orthogonal, $U'AU = U^{-1}AU$ in Equation 12.K.20 so that A is similar to D. Hence, to diagionalize a symmetric matrix A, we need only perform an orthogonal similarity transformation on A.

One final point is in order. We may represent the so-called *spectral or eigenvalue decomposition* of an nth-order symmetric matrix A as

$$A = U\text{diag}(\boldsymbol{\lambda})U', \tag{12.K.21}$$

where U is an orthogonal matrix, $\boldsymbol{\lambda}' = (\lambda_1,\ldots,\lambda_n)$ is a $(1 \times n)$ vector of eigenvalues of A, and diag($\boldsymbol{\lambda}$) denotes a diagonal matrix with the eigenvalues of A on the diagonal. By convention, the eigenvalues within $\boldsymbol{\lambda}$ are arranged from largest to smallest and their associated eigenvectors within U are arranged correspondingly.

Example 12.K.2

Given

$$A = \begin{bmatrix} 2 & 1 \\ 1 & 2 \end{bmatrix},$$

determine the spectral decomposition of A. From the characteristic equation

$$|A - \lambda I_2| = \begin{bmatrix} 2-\lambda & 1 \\ 1 & 2-\lambda \end{bmatrix} = \lambda^2 - 4\lambda + 3 = 0$$

we obtain the two roots or eigenvalues $\lambda_1 = 3$, $\lambda_2 = 1$. To determine the eigenvectors X_1 and X_2 corresponding to these eigenvalues, we must solve the homogenous equation systems $(A - \lambda_i I_2)X = 0$, $i = 1,2$.

a. For $\lambda_1 = 3$,

$$\begin{bmatrix} 2-\lambda_1 & 1 \\ 1 & 2-\lambda_1 \end{bmatrix}\begin{bmatrix} x_1 \\ x_2 \end{bmatrix} = \begin{bmatrix} 0 \\ 0 \end{bmatrix} \text{ or } \begin{matrix} -x_1 + x_2 = 0 \\ x_1 - x_2 = 0 \end{matrix}$$

and thus $x_1 = x_2$. Because the eigenvector X_1 must be normalized to unith length, set $U_1 = X_1/||X_1||$. With $u_1 = u_2$ (since $x_1 = x_2$), $\|U_1\|^2 = u_1^2 + u_2^2 = 2u_2^2 = 1$ or $u_2^2 = \frac{1}{2}$. Then $u_2 = 1/\sqrt{2} = u_1$ (we take the positive root). Hence, the eigenvector of unit length corresponding to $\lambda_1 = 3$ is $U_1' = (1/\sqrt{2}, 1/\sqrt{2})$.

b. For $\lambda_2 = 1$,

$$\begin{bmatrix} 2-\lambda_2 & 1 \\ 1 & 2-\lambda_2 \end{bmatrix}\begin{bmatrix} x_1 \\ x_2 \end{bmatrix} = \begin{bmatrix} 0 \\ 0 \end{bmatrix} \text{ or } \begin{matrix} x_1 + x_2 = 0 \\ x_1 + x_2 = 0 \end{matrix}$$

so that $x_1 = -x_2$. Then proceeding as in part a, with $u_1 = -u_2$, $\|U_2\|^2 = u_1^2 + u_2^2 = 2u_2^2 = 1$ or $u_2^2 = \frac{1}{2}$. Then $u_2 = 1/\sqrt{2}$ and $u_1 = -1/\sqrt{2}$. Thus, the eigenvector of unit length corresponding to $\lambda_2 = 1$ is $U_2' = (1/\sqrt{2}, -1/\sqrt{2})$. Given these results, the orthonormal set of vectors satisfying Equation 12K.17

is $\{U_1, U_2\}$, where $||U_1|| = ||U_2|| = 1$ and $U_1'U_2 = 0$. The orthogonal matrix $U = [U_1, U_2]$ diagonalizes A because, from Equation 12.K.20,

$$U'AU = \begin{bmatrix} 1/\sqrt{2} & 1/\sqrt{2} \\ 1/\sqrt{2} & -1/\sqrt{2} \end{bmatrix} \begin{bmatrix} 2 & 1 \\ 1 & 2 \end{bmatrix} \begin{bmatrix} 1/\sqrt{2} & 1/\sqrt{2} \\ 1/\sqrt{2} & -1/\sqrt{2} \end{bmatrix}$$

$$= \begin{bmatrix} 3 & 0 \\ 0 & 1 \end{bmatrix} = \begin{bmatrix} \lambda_1 & 0 \\ 0 & \lambda_2 \end{bmatrix} = D.$$

Finally, the *spectral decomposition* of A is

$$U\text{diag}(\lambda)U' = \begin{bmatrix} 1/\sqrt{2} & 1/\sqrt{2} \\ 1/\sqrt{2} & -1/\sqrt{2} \end{bmatrix} \begin{bmatrix} 3 & 0 \\ 0 & 1 \end{bmatrix} \begin{bmatrix} 1/\sqrt{2} & 1/\sqrt{2} \\ 1/\sqrt{2} & -1/\sqrt{2} \end{bmatrix}$$

$$= \begin{bmatrix} 2 & 1 \\ 1 & 2 \end{bmatrix} = A.$$

Given the preceding developments, we shall now demonstrate that eigenanalysis can identify the actual structure of the near-linear dependencies in the data. (The material that follows is based on the approach taken by Belsey, Kuh, and Welsch [1980] and Belsley [1991].) Let us apply the eigenvalue decomposition (Equation 12.K.21) to the $X'X$ matrix. If X has full column rank m +1 and the $(m + 1 \times m + 1)$ matrix U is orthogonal, then $X'X = U\text{diag}(\lambda)U'$ and

$$(X'X)^{-1} = U\text{diag}(\lambda^{-1})U'. \tag{12.K.22}$$

(Note that, if any eigenvalue of $X'X$ is zero, then $|X'X| = \lambda_0 \cdot \lambda_1 \ldots \lambda_m = 0$. Hence, $X'X$ is singular and thus $(X'X)^{-1}$ does not exist.)

Looking to the variance-covariance matrix of $\hat{\beta}$, we can write

$$V(\hat{\beta}) = \sigma_\varepsilon^2 (X'X)^{-1} = \sigma_\varepsilon^2 U \text{diag}(\lambda^{-1})U' \tag{12.K.23}$$

so that the kth diagonal element of this matrix is

$$V(\hat{\beta}_k) = \sigma_\varepsilon^2 \sum_{l=0}^{m} \frac{u_{kl}^2}{\lambda_l}, \; k = 0,1,\ldots,m, \tag{12.K.24}$$

where $u_{k0}, u_{k1},\ldots,u_{km}$ are the elements of the kth row of U. Here Equation 12.K.24 is termed the *variance decomposition* of $\hat{\beta}_k$. If u_{kl}^2/λ_l is replaced by its *variance decomposition proportion* π_{kl} of $\sigma_\varepsilon^{-2} V(\hat{\beta}_k)$, where $\sum_{l=0}^{m} \pi_{kl} = 1$, then these proportions can be arranged as in Table 12.K.1. Once this array is constructed, any near-linear dependencies in the columns of X can be detected.

For instance, the elements $\pi_{10}, \pi_{11},\ldots,\pi_{1m}$ under the $V(\hat{\beta}_1)$ heading are the proportions of the variance of $\hat{\beta}_1$ contributed by or accounted for by the lth eigenvalue, $l = 0,1,\ldots,m$. For a given row of this table, if a condition index is above 30 (or the accompanying eigenvalue is near zero) and the variance decomposition proportions for two or more variances of the regression coefficients are in excess of 0.5, then serious near-multicollinearity among the associated columns of X is indicated.

Suppose that k_3 exceeds 30 (λ_3 is close to zero) and π_{13}, π_{23}, and , π_{63} are all above 0.5. Hence, there exists a near-linear dependency among the regressors X_1, X_2, and X_6. What is the nature of this dependency? We know that U is an $(m + 1 \times m + 1)$ orthogonal matrix whose columns

TABLE 12.K.1

Variance-Decomposition Proportions for $V(\hat{\beta}_k)$

			Proportion of Variance			
Number	**Eigenvalue**	**Condition Index**	$V(\hat{\beta}_0)$	$V(\hat{\beta}_1)$	$\cdots$	$V(\hat{\beta}_m)$
(intercept)0	λ_0	κ_0	π_{00}	π_{10}	$\cdots$	π_{m0}
1	λ_1	κ_1	π_{01}	π_{11}	$\cdots$	π_{m1}
.	.	.	.	.		.
.	.	.	.	.		.
.	.	.	.	.		.
m	λ_m	κ_m	π_m	π_{1m}	$\cdots$	π_{mm}
Column Sums			1	1		1

$\boldsymbol{U}_k$, $k=0,1,\ldots,m$, are the eigenvectors of $\boldsymbol{X}'\boldsymbol{X}$. Hence, the components of the eigenvector $\boldsymbol{U}_3$ must reflect the form of this linear dependency. In particular, $u_{13}X_1+u_{23}X_2+u_{63}X_6=0$ from the definition of linear dependency. In fact, an auxiliary regression equation involving the variables X_1, X_2, and X_6 can be formed with, say, X_2 and X_6 serving as regressors. Then one can determine the proportion of the variation in X_1 explained by the linear influence of X_2 and X_6.

Appendix L: "Best Subset" Regression and the Variable Selection Problem

Given that one has chosen a particular regression technique, selecting the "best" regression model from a pool of models is not a trivial task. One must consider the tradeoff between model size and goodness of fit, i.e., in a multiple regression setting, one often looks for the "best subset" of regressors from among a large collection of explanatory variables. However, one must remember that a model that enables us to draw valid inferences (in terms of estimation and testing) may not generate the most reliable predictions.

In this regard, given a pool of candidate regressors, suppose theory or previous experience dictates which regressors are important in explaining the variation in the dependent variable Y. Then it is a fairly straightforward process to construct a suitable regression model. However, if theory or experience cannot dictate which explanatory variables should be included in the model, then we must address the so-called *variable selection problem*. This problem is concerned with identifying an appropriate subset of regressors (in terms of certain criteria) from among a pool of suspected influential regressors to determine which variables should constitute the final model. Solving the variable selection problem puts us on track for finding the best subset regression model. (Suppose the *full model* contains m regressors and we delete r ($<m$) of them. Then we are left with a *subset model* involving $m-r$ regressors.)

Our approach to addressing the variable selection problem and the attendant issue of best subset regression is a heuristic one, i.e., it uses a specific variable selection strategy and then iterates in a stepwise manner (by adding or deleting regressors one at a time) to find a terminal subset regression model. However, there is no icon-clad guarantee that the terminal model is the best subset model. In fact, there may be a few different subset models that are equally attractive in terms of the usual regression diagnostics.

What sorts of model or variable selection strategies can we apply? And what steps must be followed in the implementation of any particular strategy? Suggested guidelines to include the following:

1. Specify the full (base) model: the model containing all regressors under consideration.
2. Specify a model selection criterion: this enables us to identify the best subset model. To this end, suppose we delete r regressors from the full model so that the subset model contains $p = m - r$ regressors. Then a set of possible criteria for optimal subset model selection involves the following:
   a. Determine the maximum $\bar{R}_p^2$ value for the collection of subset regressors, where $\bar{R}_p^2 = SSR_p / SST$. This is actually equivalent to part b below.
   b. Finding the minimum MSE_p value for the subset regressors, where $MSE_p = SSE_p/(n - p - 1)$.
   c. Choosing a subset model that has a small $PRESS_p$ statistic. Given that $PRESS_p = \sum_{i=1}^{n}(Y_i - \hat{Y}_{(i)})^2$ (see Equations 3.55 and 3.58), it is evident that small differences between Y_i and $\hat{Y}_{(i)}$ reveal that the subset model predicts quite well.
   d. The p regressor subset model has virtually the same predictive power as the full model if the F statistic

$$F_p = \frac{[SSE_p - SSE(full)]/(m-p)}{MSE(full)} \tag{12.L.1}$$

   is not statistically significant, where $MSE(full) = SSE(full)/(n - m - 1)$ and $\mathcal{R} = \{F_p | F_p > F_{\alpha,m-p,n-m-1}\}$. With F_p insignificant, SSE_p must be near $SSE(full)$ so that the subset model has approximately the same predictive power as the full model. Hence, a p regressor subset model is desirable if F_p is not significant and p is as small as possible.
   e. One can also use *Mallow's C(p) statistics*, which is of the form

$$C(p) = \frac{SSE_p}{MSE(full)} - [n - 2(p+1)]. \tag{12.L.2}$$

To see exactly what this criterion measures, let us express the total mean square error of the *i th* estimated value of Y, $\hat{Y}_i$, as

$$\begin{aligned} MSE(\hat{Y}_i) &= E[\hat{Y}_i - E(Y_i)]^2 \\ &= [E(Y_i) - E(\hat{Y}_i)]^2 + V(\hat{Y}_i), \end{aligned} \tag{12.L.3}$$

where $E(\hat{Y}_i)$ is the mean response for the subset regression model and $E(Y_i)$ is the mean response for the (albeit unknown) tune or population model. Note that this equation has a *squared bias component* ($[E(Y_i) - E(\hat{Y}_i)]^2$) and a *random error component* represented by the variance of $\hat{Y}_i$ ($V(\hat{Y}_i)$). Then the total mean square error for all n estimated values of Y for the subset model is

$$MSE_P = \sum_{i=1}^{n}[E(Y_i) - E(\hat{Y}_i)]^2 + \sum_{i=1}^{n} V(\hat{Y}_i), \tag{12.L.4}$$

where the first term on the right-hand side of Equation 12.L.4 is the *total squared bias*.

To form the Mallow's Γ_p criterion, we must standardize the MSE for the subset model by σ_ε^2 or

$$\Gamma_p = \frac{MSE_p}{\sigma_\varepsilon^2}. \tag{12.L.5}$$

Hence, Γ_p small is a signal that MSE_p is small relative to σ_ε^2. Now, we know that

$$\sum_{i=1}^{n} V(\hat{Y}_i) = (p+1)\sigma_\varepsilon^2 \tag{12.L.6}$$

(there are p+1 parameters estimated, including the intercept), whereas

$$E(SSE_P) = \sum_{i=1}^{n}[E(Y_i) - E(\hat{Y}_i)]^2 + [n-(p+1)]\sigma_\varepsilon^2. \tag{12.L.7}$$

Hence, Equations 12.L.6 and 12.L.7 can be used to rewrite Equation 12.L.5 as

$$\begin{aligned}\Gamma_p &= \frac{1}{\sigma_\varepsilon^2}\left\{E(SSE_p) - [n-(p+1)]\sigma_\varepsilon^2 + (p+1)\sigma_\varepsilon^2\right\}\\ &= \frac{E(SSE_p)}{\sigma_\varepsilon^2} - [n-2(p+1)].\end{aligned} \tag{12.L.8}$$

If we replace $E(SSE_p)$ and σ_ε^2 by their sample estimates SSE_p and $MSE(full)$, respectively, then a good estimator for Γ_p is

$$C(p) = \frac{SSE_p}{MSE(full)} - [n-2(p+1)], \tag{12.L.9}$$

where again p is the number of regressors. Thus, $C(p)$ serves as an estimate of the total sum of squares of deviations (which includes bias error plus variance error) between the estimated Y values and the true ones.

If the p regressor subset model has a small or zero bias (remember that the regression model is unbiased if $E(\hat{Y}) - E(Y) = 0$), then $\sum_{i=1}^{n}[E(Y_i) - E(\hat{Y}_i)]^2 = 0$ and thus $E(SSE_p) = [n-(p+1)]\sigma_\varepsilon^2$ so that

$$\begin{aligned}&E[C(p) \mid E(Y_i) - E(\hat{Y}_i) = 0 \text{ for all } i]\\ &\qquad = \frac{[n-(p+1)]\sigma_\varepsilon^2}{\sigma_\varepsilon^2} - [n-2(p+1] = p+1.\end{aligned} \tag{12.L.10}$$

Because the magnitude of $C(p)$ reflects the size of the total mean square error of the n estimated values of the subset regression model, it follows that if there is no bias in a subset model with p regressors, then $C(p) \approx p+1$. Hence, subset models with substantial bias (they thus exhibit a poor fit to the data) will have $C(p)$ far in excess of $p + 1$.

As a practical matter, the $C(p)$ criterion selects as the best subset of regressors the model that displays a small $C(p)$ value that is approximately equal to $p + 1$. This latter requirement implies that the subset model is unbiased or that the bias is negligible. In reality, subset models will be somewhat biased because $r = m - p$ regressors have been deleted from the full model. However, as stated previously, the bias is expected to be small if $C(p) \approx p+1$. So if one is interested in minimizing total mean square error, then the subset model to be selected is the one with the smallest $C(p)$ value, provided one is confident that the bias is small.

Given items a through e, it is important to note that no single subset selection criterion is uniformly best in all circumstances. In fact, different criteria can lead to different subset model selections. For instance, if our objective is to make "good" out-of-sample predictions or to predict the conditional mean of Y, then the PRESS statistic is an appealing criterion to apply. This said, perhaps the subset selection criterion that is highly reliable as well as easy to use is Mallow's $C(p)$ criterion.

3. Specify a strategy for implementing the model selection criterion; we can run all possible regressions or use a model selection heuristic (that evaluates a limited number of subset regressions by either inserting or deleting regressors one at a time) such as the following:

   a. Forward selection: given a user-specified significance level, the process begins with the null model (no regressors in the equation save for the intercept) and then commences by identifying that single regressor out of all m regressors (call it X_1) that displays the largest significant F value in a bivariate regression equation with Y. The procedure subsequently admits to the regression equation that variable of the remaining $m - 1$ regressors (denote it as X_2) that displays the largest significant partial F value, i.e., X_2 is admitted to the regression equation if

$$F_2 = \frac{SSR(X_2|X_1)}{MSE(X_1,X_2)}$$

   is the largest significant partial F statistic. Now, two regressors are in the model. The procedure is continued until no remaining variables display a statistically significant partial F value. Thus, at each step, the regressor displaying the greatest contribution to the model is admitted. Note that, once a regressor has been chosen for inclusion in the model, it remains in the model.

   b. Backward elimination: given a user-specified significance level, this routine begins with the full model and deletes from that model the regressor whose coefficient displays the smallest insignificant partial F value. Here the partial F statistic is determined for each regressor as if it were the last one to enter the model. The resulting equation (with one fewer regressor than the base model) is estimated and again the regressor having the smallest insignificant partial F value is deleted, and so on. Thus, at each step the regressor displaying the smallest (insignificant) contribution to the model is deleted. The procedure is continued until all variables remaining in the model have coefficients that are statistically significant at the chosen level. Note that, once a regressor has been deleted, it never reenters the regression equation.

c. Stepwise selection: given a user-specified significance level, the procedure begins as in forward selection (a variable is added to the model on the basis of a significant partial F value), but, at any point after a set of regressors has been introduced into the model, the current equation is examined to see whether any variable presently in the model should be deleted on the grounds that its partial F value is not statistically significant. (This is because a regressor inserted at an earlier step may become redundant in the presence of some recently admitted variables.) The routine continues until no additions/deletions of variables are warranted. Clearly, variables already in the model may not necessarily remain in the same.

Given the preceding collection of techniques for generating an optional subset regression model, it is important to note that none of these variable selection procedures guarantees that the best subset regression model (in terms of generally accepted selection criteria) will be found. In fact, these methods may lead to quite different terminal regression models. Furthermore, there probably will be several good (or at least acceptable) models uncovered. All things considered, perhaps stepwise selection is the most widely applied variable selection technique.

The reader must keep in mind, however, that no matter what variable selection technique is chosen, an assessment of the validity of the basic regression assumptions (pertaining to the random error term ε, multicollinearity, etc.) for the full and terminal models must be made to determine which model is best. In addition, because the forward selection, backward elimination, and stepwise methods are heuristic or exploratory in nature and not purely theory driven, great care must be taken when it comes to interpreting the results in that the same set of regressors may not exhibit statistical significance if a larger or new sample is used, i.e., measurement (without theory) that is based on data-driven methods can be dangerous and can produce unstable results. It is important to note that none of these variable selection procedures guarantees that the best subset regression model (in terms of generally accepted selection criteria) will be found. In fact, these methods may lead to quite different terminal regression models. Furthermore, there probably will be several "good" (or at least acceptable) models uncovered. All things considered, perhaps stepwise selection is the most widely applied variable selection technique.

Example 12.L.1

Given the Example 12.1 dataset, find the best subset regression model using stepwise selection. The requisite SAS code appears in Exhibit 12.L.1.

Exhibit 12.L.1 SAS Code for Stepwise Selection

```
data ABCsold;
       input units cold foll clt exp;
datalines;
 14  62   10  20  12.00
 21  120  22  45  27.25
  .   .    .   .    .
```

```
 .    .    .    .    .
 .    .    .    .    .
 17  91   21   39  19.25
 ;
proc reg data=ABCsold;
                                        ①
      model units=cold foll clt exp/selection=stepwise tol vif; ②
run;
      ① We request the SELECTION=stepwise option in the MODEL statement.
        If desired, one can specify significance levels for stopping the
        insertion or deletion of variables:
            SLENTRY=0.xxx specifies the significance level for
            terminating the introduction of variables. If not specified,
            the default is 0. 15.
            SLSTAY=0.xxx specifies the significance level for
            terminating the elimination of variables. If not specified,
            the default is 0.15.
      ② We request exact collinearity (singularity) and
        multicollinearity diagnostics.
```

The output resulting from the stepwise routine appears in Table 12.L.1. As indicated above, stepwise search produces a sequence of subset regression models whereby, at each step, a partial F test is used to introduce or delete a regressor. (For notational convenience, let us set $cold=X_1$, $foll=X_2$, $clt=X_3$, and $exp=X_4$.)
Specifically,

TABLE 12.L.1

Output from Stepwise Regression

The SAS System
The REG Procedure
Model: MODEL 1

Dependent Variable: Units

Number of Observations Read	35
Number of Observations Used	35

①
Stepwise Selection: Step 1
(X_1)
Variable cold entered: $R^2 = 0.8884$ and $C(p) = 23.5142$

Analysis of Variance

Source	DF	Sum of Squares	Mean Square	F Value	Pr > F
Model	1	278.55508	278.55508	262.73	<0.0001
Error	33	34.98778	1.06024		
Corrected total	34	313.54286			

Variable	Parameter Estimate	Standard Error	Type II SS	F Value	Pr > F
Intercept	6.64212	0.64361	112.91854	106.5	<0.0001
cold	0.11648	0.00719	278.55508	262.73	<0.0001 ②
Bounds on condition number : 1, 1					

(continued)

TABLE 12.L.1 *(continued)*

The SAS System

③

Stepwise Selection: Step 2

(X_3)

Variable clt entered: $R^2 = 0.9255$ and $C(p) = 7.3760$

Analysis of Variance

Source	DF	Sum of Squares	Mean Square	*F* Value	Pr > *F*
Model	2	290.19635	145.09817	198.88	<0.0001
Error	32	23.34651	0.72958		
Corrected total	34	313.54286			

Variable	Parameter Estimate	Standard Error	Type II SS	*F* Value	Pr > *F*	
Intercept	7.19737	0.55170	124.16981	170.19	<0.0001	
cold	0.07603	0.01175	30.54725	41.87	<0.0001	} ④
Clt	0.08211	0.02056	11.64127	15.96	0.0004	

Bounds on condition number: 3.8854, 15.541

⑤

Stepwise Selection: Step 3

(X_4)

Variable exp entered: $R^2 = 0.9354$ and $C(p) = 4.5746$

Analysis of Variance

Source	DF	Sum of Squares	Mean Square	*F* Value	Pr > *F*
Model	3	293.27795	97.75932	149.55	<0.0001
Error	31	20.26491	0.65371		
Corrected total	34	313.54286			

Variable	Parameter Estimate	Standard Error	Type II SS	*F* Value	Pr > *F*	
Intercept	6.98689	0.53115	113.11548	173.04	<0.0001	
cold	0.06303	0.01263	16.27700	24.90	<0.0001	} ⑥
Clt	0.07384	0.01983	9.06714	13.87	0.0008	
exp	0.07686	0.03540	3.08160	4.71	0.0377	

Bounds on condition number: 5.0113, 36.078

All variables left in the model are significant at the 0.1500 level. No other variable met the 0.1500 significance level for entry into the model.

TABLE 12.L.1 *(continued)*

The SAS System

⑦

Summary of Stepwise Selection

Step	Variable Entered	Variable Removed	Number Vars In	Partial R^2	Model R^2	$C(p)$	F Value	Pr > F
1	cold		1	0.8884	0.8884	23.5142	262.73	<0.0001
2	clt		2	0.0371	0.9255	7.3760	15.96	0.0004
3	exp		3	0.0098	0.9354	4.5746	4.71	0.0377

⑧

Analysis of Variance

Source	DF	Sum of Squares	Mean Square	F Value	Pr > F
Model	3	293.27795	97.75932	149.55	<0.0001
Error	31	20.26491	0.65371		
Corrected total	34	313.54286			

Root MSE	0.80852	*R*-Square	0.9354
Dependent Mean	16.68571	Adj *R*-Sq	0.9291
Coeff Var	4.84559		

Parameter Estimates

Variable	DF	Parameter Estimate	Standard Error	t Value	Pr > |t|	Tolerance	Variance Inflation
Intercept	1	6.98689	0.53115	13.15	<.0001	.	0
cold	1	0.06303	0.01263	4.99	<.0001	0.19955	5.01135
clt	1	0.07384	0.01983	3.72	0.0008	0.24788	4.03422
exp	1	0.07686	0.03540	2.17	0.0377	0.33551	2.9805

①, ② Step 1.

A bivariate regression equation is estimated for each of the p regressors. For each such model the associated F statistic, $F_j = MSR_j/MSE_j$, is determined under $H_0 : \beta_j = 0$, $j = 1, 2, 3, 4$, versus H_1 not H_0. Given that MSR_j depicts the variation in Y explained by the regressor X_j, it follows that the regressor with the largest F value is a candidate for entry into the regression equation. Hence, $cold = X_1$ has the largest F value ($F_1 = 278.55508/1.06024 = 262.73$) and thus the greatest amount of explanatory power. Because F_1 is statistically significant (its p-value <0.001), *cold* is inserted into the model. (If none of the F_j's are significant, then the best regression model would be $\hat{Y} = \bar{Y}$, a model with no regressors.)

③, ④ Step 2.

Given that *cold* has been chosen for entry into the regression model, either *foll* or *clt* or *exp* is a candidate for the variable to next enter the regression equation. Hence, the stepwise routine now fits all two-regressor models, in which *cold* is a regressor in each of them. For this set of regression models, the partial F statistics are of the form

$$F_j = \frac{MSR(X_j|X_1)}{MSE(X_1,X_j)}, j = 2,3,4.$$

At this step, we are testing $H_0 : \beta_j = 0$ given that X_1 and X_j are in the model, $j = 2,3,4$, versus H_1 not H_0. Here too the regressor with the largest significant partial F value will be inserted into the regression equation (or provided that the highest partial F value exceeds a prespecified or default percentage point of the F distribution.) Because $clt{=}X_3$ has the largest significant partial F value ($F_3 = 11.64127/0.72958 = 15.96$, p-value = 0.0004), it follows that *clt* enters the regression model. Remember that $MSR(X_3/X_1) = 11.64129$ corresponds to a Type II (partial) sum of squares: it is the reduction in *SSE* attributable to inserting X_3 into a model that already contains X_1. Thus, X_3 provides the largest significant reduction in *SSE*.

It is at this point that the stepwise method determines whether any variable already in the model (here it is $cold{=}X_1$) should be deleted from the same. It makes this determination by calculating the partial F value

$$F_1^* = \frac{MSR(X_1|X_3)}{MSE(X_1,X_3)} = \frac{30.54725}{0.72958} = 41.87, \textit{p}\text{-value} < 0.0001.$$

Because F_1^* is statistically significant (F_1^* does not fall below a predetermined or default percentage point of the F distribution), X_1 is retained in the regression model. Otherwise X_1 would be dropped from the regression equation and only X_3 would remain.

⑤ ⑥ Step 3.

Given that X_1 and X_3 are currently in the model, the stepwise technique now determines whether either $foll{=}X_2$ or $exp{=}X_4$ should also be included in the regression model. To this end, the stepwise method now fits all three-variable regression equations, where the regressors *cold* and *clt* are in each of them. For this set of models, the partial F statistics appear as

$$F_j = \frac{MSR(X_j|X_1,X_3)}{MSE(X_1,X_3,X_j)}, j = 2,4.$$

Now we are testing $H_0 : \beta_j = 0$ given that X_1, X_3, and X_j are in the model, $j = 2,4$, versus H_1 : not H_0. As usual, the regressor with the largest significant partial F value will be inserted into the regression equation. Given that $exp{=}X_4$ displays the largest partial F value ($F_4 = 3.08160/0.65371 = 4.71$, p-value = 0.0377), we see that *exp* enters the regression equation; it provides us with the greatest reduction in *SSE* in the presence of X_1 and X_3.

Are any of the variables currently in the model redundant or superfluous in the presence of *exp*? To answer this question, let us look to the partial F values

$$F_1^{**} = \frac{MSR(X_1|X_3,X_4)}{MSE(X_1,X_3,X_4)} = \frac{16.27700}{0.65371} = 24.90,$$

p-value <0.0001;

$$F_3^{**} = \frac{MSR(X_3|X_1,X_4)}{MSE(X_1,X_3,X_4)} = \frac{9.06714}{0.65371} = 13.87,$$

p-value = 0.0008. Because each of these partial F's is statistically significant (neither F_1^{**} nor F_3^{**} falls short of the default percentage point of the F distribution), both X_1 and X_3 will remain in the regression equation.

⑦, ⑧ Summary of stepwise selection

Given that X_1, X_3, and X_4 are presently in the regression model, the stepwise method now tests to see whether *foll* = X_2 should also be entered into the regression equation. Specifically, the stepwise routine fits a four-regressor equation, where X_2 is added last. Then

$$F_2 = \frac{MSR(X_2|X_1,X_3,X_4)}{MSE(X_1,X_2,X_3,X_4)}$$

is calculated under $H_0 : \beta_2 = 0$ given that X_1, X_3 and X_4 are already in the model versus H_1 : not H_0. Because from the full model (see Example 12.1) we find that F_2 = 1.01061/0.64181 = 1.57, p-value >0.15, *foll* does not enter the regression model.

Given Steps 1 through 3, the stepwise method offers as the best subset regression equation the expression

$$\begin{aligned}\hat{Y} &= \hat{\beta}_0 + \hat{\beta}_1 X_1 + \hat{\beta}_3 X_3 + \hat{\beta}_4 X_4 \\ &= 6.98689 + 0.06303X_1 + 0.07384X_3 + 0.07686X_4.\end{aligned}$$

Note that the *tol* and *vif* values present no red flags as far as exact linear dependencies or near collinearities are concerned, i.e., given that the minimum $tol = 1 - R_j^2$ value (R_j^2 is the auxiliary coefficient of determination which results when X_j is regressed on the other independent variables in the model) is less than 0.20, it follows that the regressor *cold* has the largest R_j^2 value, which is only slightly in excess of 0.80. Hence, it appears that there are no near dependencies existing among the regressors in the model. Note also that, as expected, R^2 improves (as does $\bar{R}^2$) with each successive addition of a variable, and *MSE* does likewise. Additionally, Mallow's $C(p)$ decreases to 4.57 $\approx p + 1 = 5$. Because $C(p) < p + 1$, we see that the final subset model displays a small amount of bias.

As the preceding example problem has indicated, stepwise regression is an objective selection method that, for a sizeable number of regressors, involves a large number of regression hyperplane estimates and supporting tests. This being the case, a few cautions are in order:

1. The probability is high that we can make a Type I error (we include an unimportant regressor in the model) and a Type II error (we drop an important regressor from the model).
2. The best subset regression model can be made up of variables that are highly correlated. Hence, it is imperative that one examine the usual set of collinearity diagnostics.
3. Once the so-called best subset of regressors has been found, we should avoid the temptation to present the associated model as the " best model"; it is at this point that other important consideration pertaining to model building emerge, e.g., intuition, experience, etc. The point is that the best subset model is generally not the final model, especially if the purpose of the model-building exercise is prediction.

Exercises

12-1. Given the Table E.12-1 dataset,

a. Find the hyperplane of best fit.

b. Find the percentage of the variation in Y attributed to the linear influence of the regressors.

TABLE E.12-1

Observations on Variables Y, X, and X_2

Y	X_1	X_2
6	5	256
25	43	402
23	37	443
14	23	266
10	16	209
21	37	261
16	27	245
12	18	325
22	38	380
13	21	284
7	9	235
17	27	372
8	10	301
19	33	245
11	20	214
9	8	247
24	35	450
15	22	275

c. Conduct a test of H_0: the model does not explain any of the variation in Y, versus H_1: the model explains some of the variation in Y. Pick $\alpha = 0.05$. What is your conclusion based on the p-value for the F statistic?

d. Discuss whether or not the individual regression coefficients are statistically significant. Let $\alpha = 0.05$. What is your conclusion based on their p-values?

e. How precisely have the regression parameters been estimated?

f. Using unit length scaling, express the estimated regression slopes in standardized form. (Hint: find $\hat{\beta}_j^{**}, j = 1,2$.)

g. Discuss the results for the model specification (White's) test. What is H_0?

h. Are the simple correlation coefficients statistically significant?

12-2. Given the Exercise 12-1 regression results, use the set of output statistics to detect any outliers, influence points, etc.

12-3. For the Exercise 12-1 regression results, find 99% prediction intervals for $\hat{Y}_0$ (the individual value of Y at $\mathbf{X}_0$), and $E(Y_0)$ (the mean response in Y at $\mathbf{X}_0$), where $\mathbf{X}_0 = [1,27,246]$.

12-4. Given the dataset appearing in Table E.12-2,

a. Determine the hyperplane of best fit.

b. Find the portion of the variation in Y attributed to the linear influence of the regressors.

c. Conduct a test of H_0: the model does not explain any of the variation in Y, versus H_1: the model explains some of the variation in Y. Focus on the p-value of the F statistic.

TABLE E.12-2

Observations on Variables $Y, X_1, \ldots, X_4$

Y	X_1	X_2	X_3	X_4
79	7	25	17	13
74	2	30	20	8
104	11	55	8	21
87	10	31	14	19
96	8	52	5	25
109	12	56	9	29
103	3	70	17	19
73	1	31	22	17
93	4	54	17	22
116	21	47	4	35
84	5	39	24	25
113	12	66	9	29
109	11	70	8	36
84	8	27	17	21
91	16	40	20	23
102	17	40	15	31
94	18	30	20	21
86	9	47	25	24
77	10	50	22	19
112	24	63	8	28

d. Are the individual regression coefficients statistically significant? Focus on the p-values associated with the t statistics. Also, test $H_0: \beta_1 \geq 0.5$ versus $H_1: \beta_1 < 0.5$ and $H_0: \beta_3 \geq -0.6$ versus $H_1: \beta_3 < -0.6$ using $\alpha = 0.05$.

e. How precisely have the regression parameters been estimated?

f. Using the the Z-transform, write the estimated regression coefficients in terms of standard deviation units. Interpret your results. (Hint: find the $\hat{\beta}_j^*$'s, $j = 1,2,3,4$.)

g. Discuss the results for the model specification (White's) test. State H_0.

h. Are the simple correlation coefficients statistically significant?

12-5. Given the Exercise 12-4 regression output, discuss/interpret the values of the sequential sums of squares and the partial sums of squares.

12-6. Given the Exercise 12-4 regression results, use the set of output statistics to detect any outliers, influence points, etc.

12-7. Using the Exercise 12-4 regression output,

a. Conduct a series of variable added-last F tests. Interpret your results.

b. Conduct a series of added-in-order F tests. Interpret your results.

c. Consider the subsets of variables $A = \{X_1, X_3\}$ and $B = \{X_2, X_4\}$. Conduct a group added-last F test on subsets A and B.

d. For subsets A and B specified in part c, use a TEST statement to conduct a group added-last test on subsets A and B. Also, use TEST statements to test $H_0: X_1 - X_3 = 0$ and $H_0: X_1 - X_4 = 0$.

e. Use a RESTRICT statement to reestimate the parameters of the regression model subject to the restriction $X_2 - X_4 = 0.4$.

12-8. Given the Exercise 12-4 regression results, find 95% prediction intervals for $\hat{Y}_0$ (the individual value of Y at $\mathbf{X}_0$) and $E(Y_0)$ (the mean response in Y at $\mathbf{X}_0$), where $\mathbf{X}_0 = [1, 25, 72, 26, 38]$.

12-9. Given the Exercise 12-4 regression output, give an assessment of the various multicollinearity diagnostics to uncover any serious near multicollinearities, i.e., interpret:

a. Tolerance values.

b. Variance inflation factors.

c. The condition number of the $\mathbf{X'X}$ matrix.

d. The condition indexes for each eigenvalue of $\mathbf{X'X}$.

e. Variance decomposition proportions.

12-10. Given the dataset presented in Table E.12-3,

a. Perform a regression and correlation study.

b. Determine prediction intervals for $\hat{Y}_0$ and $E(Y_0)$ given $\mathbf{X}_0 = [1, 62, 27, 21]$.

c. Are there any near multicollinearities to worry about? If so, what are their implications for estimation?

TABLE E.12-3

Observations on Variables Y, X_1, X_2, and X_3

Y	X_1	X_2	X_3
10	10	10	10
11	0	20	12
9	5	6	7
6	5	7	6
9	10	10	7
10	10	10	9
9	10	15	8
12	15	10	14
9	20	5	9
10	20	5	11
16	25	25	16
16	25	25	17
10	30	5	12
9	30	5	10
13	35	5	14
13	35	5	15
14	40	15	11
14	40	10	12
15	45	15	16
17	45	15	18
19	50	20	20
17	60	20	20

TABLE E.12-4

Observations on Variables Y, X_1, and X_2

Y	X_1	X_2
16	15	25
25	43	40
23	37	44
14	23	26
10	16	21
21	37	26
16	27	25
12	18	33
22	38	38
13	21	28

12-11. Given the Table E.12-4 dataset, let us use PROC IML (introduced in Appendix 9.B) to generate some basic multiple regression results. Exhibit E.12-1 houses the SAS program that you must replicate and execute.

Exhibit E.12-1

```
proc iml;
start matrices;
x={ 1  15  25,
     1    43    40,
     1    37    44,
     1    23    26,
     1    16    21,
     1    37    26,
     1    27    25,
     1    18    33,
     1    38    38,
     1    21    28}; ①
y={16, 25, 23, 14, 10, 21,
16, 12, 22, 13}; ②
xt= t(x); ③
yt= t(y); ④
ixtx= inv(xt*x); ⑤
b= ixtx*xt*y; ⑥
bt=t(b);
yhat=x*b; ⑦
e= y-yhat; ⑧
et= t(e);
n= nrow(x); ⑨
p= ncol(x); ⑩
```

① We specify the **X** matrix with regressors X_1 and X_2 (the column of 1's corresponds to the intercept term).

② We specify the matrix **Y** containing values of the response variable Y.

③ We form the transpose of $\boldsymbol{X}$ or $\boldsymbol{X}'$.

④ We form the transpose of $\boldsymbol{Y}$ or $\boldsymbol{Y}'$.

⑤ We form the inverse of $\boldsymbol{X}'\boldsymbol{X}$.

⑥ We form the vector of parameter estimates $\hat{\beta} = (X'X)^{-1}X'Y$.

⑦ We calculate a vector of estimated Y values.

⑧ We calculate the set of residuals.

⑨ We specify the sample size n.

⑩ We specify the number of parameters $(m+1)$.

⑪ We determine the degrees of freedom for MSE.

⑫ We calculate the error sum of squares.

⑬ We form the error mean square.

```
df= n-p; ⑪
sse= et*e; ⑫
mse= sse/df; ⑬
covb= ixtx#mse; ⑭
ssy= yt*y; ⑮
rsquare= (ssy-sse)/ssy; ⑯
print, "Regression Results",
     sse  df  mse  covb  rsquare;
stdb= sqrt(vecdiag(ixtx)#mse); ⑰
t= b/stdb; ⑱
probt= 1-probf(t#t, 1, df); ⑲
print, "Parameter Estimates",
          b  stdb  t   probt;
print,    y, yhat, e;
finish;
run matrices;
quit;
```

⑭ We form the variance-covariance matrix of the estimated regression coefficients as $S_\varepsilon^2(X'X)^{-1}$.

⑮ We calculate the total sum of squares.

⑯ We form the coefficient of determination R^2 = SSR/SST.

⑰ We calculate the standard errors the estimated regression coefficients.

⑱ We find sample t statistics under $H_0 : \beta_i = 0$, i=1,…,p.

⑲ We calculate the p-values associated with the sample t statistics.

Once this program runs, it is prudent to check your results via PROCREG. (Remember that PROC REG automatically includes an intercept term.)

12-12. Given the Table E.12-5 dataset pertaining to the factors that might explain the presence or absence of respiratory ailments, obtain a logistic regression fit to the data and fully explain your results. The variables are described as follows:

Y: Presence/absence of a respiratory ailment, where Y_i = 1(presence) or 0 (absence)

X_1: age in years

X_2: smoker/nonsmoker, where X_{i2} = 1(Yes) or 0(No)

X_3: Income in the thousands of dollars

X_4: location of primary residence, where X_{i4} = 1(within city limits) or 0(outside city limits)

12-13. Given the Table E.12-2 dataset, determine the best subset regression hyperplane via the following:

a. Forward selection. (Hint: the *MODEL* statement may be written as

model $Y = X_1\ X_2\ X_3\ X_4$ /selection = forward tol vif;)

b. Backward elimination. (Hint: the *MODEL* statement may be written as

model $Y = X_1\ X_2\ X_3\ X_4$ /selection = backward tol vif;)

c. Stepwise selection. (Hint: the *MODEL* statement may be written as

model $Y = X_1\ X_2\ X_3\ X_4$ /selection = stepwise tol vif;)

TABLE E.12-5

Factors Involved in Respiratory Aliments

Y	X_1	X_2	X_3	X_4	Y	X_1	X_2	X_3	X_4
1	55	1	100	1	0	27	0	35	1
0	41	1	53	0	1	70	1	85	0
0	49	0	47	0	1	68	0	77	1
1	62	1	85	1	1	53	1	63	1
0	37	0	52	0	0	41	0	50	0
1	53	0	65	1	0	37	0	50	0
0	37	1	37	0	1	75	1	120	1
0	41	1	55	1	0	40	0	53	0
0	40	0	50	0	0	38	0	47	0
1	71	1	79	0	0	39	0	51	0
1	65	1	82	0	0	37	1	40	0
0	45	0	50	0	1	61	1	82	1
1	60	0	91	1	0	60	0	71	0
0	41	0	48	0	1	69	1	89	1
0	47	0	53	0	0	40	0	50	0
1	63	1	88	1	1	68	1	105	1
0	50	0	60	0	1	68	1	111	1
0	45	1	62	0	0	52	0	67	0
0	44	0	53	0	0	61	0	98	0
1	51	1	60	1	0	38	0	45	1

13

Normal Correlation Models

Regression and Correlation Models Compared

In the basic linear regression model, the regressors $X_1,\ldots,X_m$ are taken to be fixed constants, with our primary objective being to make inferences about or predict values of the explained variable Y. In this instance, only Y is random because the error term ε is random. Furthermore, under the assumptions of the strong classical linear regression model, $\varepsilon : N(0, \sigma_\varepsilon)$, σ_ε constant. In the case of a single fixed regressor X, the sample coefficient of correlation between X and Y, $R = \sum x_i y_i / \sqrt{\sum x_i^2}\sqrt{\sum y_i^2}$, is a *purely descriptive* measure of the linear association between X and Y. Here correlation only serves to measure (through $R^2 = SSR/SST$) the goodness of fit of the sample linear regression equation; it only reflects the proportion of the variation in Y explained by the linear influence of X.

In a correlation model, all variables are taken to be random, with no one variable designated as explained or dependent. Hence, the correlation model is used to study the mutual relationship, in terms of covariability, between the random variables. Moreover, the joint distribution or probability density function of these random variables must be *completely specified*. In this regard, it is assumed that the random variables are jointly normally distributed (for details on the multivariate normal distribution, see Appendix 13.A, "The Multivariate Normal Case"). If we again consider the two-variable case in which now both X and Y are random variables, then X and Y are assumed to follow a joint bivariate normal distribution (see Appendix 13.A, "The Bivariate Normal Case"). In this instance, R serves as an *estimator* of the population coefficient of correlation between the random variables X and Y, $\rho = \sigma_{XY}/\sigma_X\sigma_Y$, where $\sigma_{XY} = COV(X, Y)$ (if only Y is random, $COV(X, Y)$ does not exist). Additionally, as we shall now see, inferences on one random variable conditional on the other can be made via strong classical linear regression techniques.

Given that the random variables X and Y follow a joint bivariate normal probability distribution, it follows that the conditional distribution of Y given $X = x$ is univariate $N(E(Y|X = x), \sqrt{V(Y|X = x)})$, with $E(Y \mid X = x) = \mu_Y + \rho(\sigma_Y/\sigma_X)(x - \mu_x) = \beta_0 + \beta_1 x$ and $\sqrt{V(Y|X = x)} = \sigma_Y\sqrt{1-\rho^2}$; and the marginal distribution of X, $g(x) = (\sqrt{2\pi}\sigma_X)^{-1} \exp\left\{-\frac{1}{2}\left[(x-\mu_X)/\sigma_X\right]^2\right\}$, $-\infty < x < +\infty$, is devoid of any of the regression model parameters (on all this see, Appendix 13.A, "The Bivariate Normal Case"). So if we extract a random sample (X_i, Y_i), $i = 1,\ldots,n$, from a bivariate normal population and we choose to make inferences about Y given $X = x$, then the strong classical linear regression model applies because the Y_i's are independent and follow the preceding conditional univariate normal distribution. Hence, the conditional correlation model of Y given $X = x$ is equivalent to the strong classical linear regression model.

Now, what if in a regression model the regressors $X_1,\ldots,X_m$ are *all random variables* and not fixed constants? We noted in Chapter 2 that, for the bivariate case, if both X and Y are random variables and $\varepsilon : N(0, \sigma_\varepsilon)$ with X independent of ε, then the usual OLS results involving estimation and testing still apply, provided, of course, that the remaining assumptions of the strong classical linear regression model hold. (Also, as long as the probability distribution of X does not involve the regression parameters β_0, β_1, and σ_ε^2, the maximum-likelihood estimators will correspond to the OLS estimators for these parameters.)

Let us suppose further that the random variables X and Y are jointly distributed but the form of their joint probability distribution is unspecified. If the conditional distribution of Y given X is $N(E(Y|X = x), \sqrt{V(Y|X=x)})$, where $E(Y|X = x) = \beta_0 + \beta_1 x$, and the X values are independent random variables whose marginal distributions do not involve the regression parameters β_0, β_1, and $V(Y|X = x)$, then, again, the usual OLS results apply. Hence, we do not need to completely specify the joint probability density function connecting X and Y (as in the correlation model); only the conditional distribution of Y needs to be completely determined and a mild restriction placed on the marginal distribution of X. Although these requirements have been stated for the instance in which both X and Y are random variables, the same conclusion is applicable to the general case in which Y and all of the regressors $X_1,\ldots,X_m$ are random variables. Hence, the conditional multivariate correlation model is equivalent to the strong multivariate classical linear regression model.

Connections between Regression and Correlation: A Review of the Bivariate Cases

We determined the following in Chapter 2 :

1. For the strong classical linear regression model $Y = \beta_0 + \beta_1 X + \varepsilon$, where the regressor X is fixed and $\varepsilon : N(0, \sigma_\varepsilon)$, $\hat{\beta}_1 = \sum x_i y_i / \sum x_i^2$ and $R^2 = SSR/SST$ indicates the proportion of the variation in Y explained by the linear influence of X (it thus serves as a measure of goodness of fit or the strength of the linear relationship between X, Y). Stated alternatively, $R^2 = (SST - SSE)/SSE$ depicts the proportionate reduction in SST obtained by using OLS to estimate Y as a linear function of X. If R is treated as a purely descriptive measure of linear association or closeness of fit between X and Y, then

$$R = \frac{\sum x_i y_i}{\sqrt{\sum x_i^2}\sqrt{\sum y_i^2}} = \frac{\sum x_i y_i}{\sum x_i^2}\frac{\sqrt{\sum x_i^2}}{\sqrt{\sum y_i^2}} = \hat{\beta}_1\left({}^{S_X}\!/_{S_Y}\right).$$

2. For X and Y jointly distributed random variables, R serves as a measure of covariability between X and Y because now $COV(X, Y)$ exists.
3. If the random variables X and Y follow a joint bivariate normal distribution, then the conditional distribution of Y given $X = x$ is $N(E(Y|X = x), \sqrt{V(Y|X=x)})$, where $E(Y|X = x) = \beta_0 + \beta_1 x$ and $V(Y|X=x) = \sigma^2_{Y|X=x} = \sigma_Y^2(1-\rho^2)$ so that R^2 serves as an estimator for $\rho^2 = (\sigma_Y^2 - \sigma^2_{Y|X=x})/\sigma_Y^2$. Moreover, under bivariate normality, R is a maximum-likelihood estimator of ρ and the sampling distribution of R can be determined according to whether ρ is assumed to equal zero or not.

Correlations Involving Multiple Independent Variables or Regressors

In what follows, we shall assume that $Y, X_1,\ldots,X_m$ follow a joint multivariate normal distribution. We then consider *zero-order* (alternatively called simple or Pearson product-moment) *correlations* as well as a variety of *higher-order correlation* measures. In general, the *order* of a correlation coefficient is determined by the number of variables that are *controlled for*, i.e., in the higher-order case, we correlate two variables after we adjust one or both of them by removing the effect of at least one of the remaining variables.

Zero-Order Correlations

For instance, for variables $Y, X_1,\ldots,X_m$, zero-order correlations are denoted as

R_{y1} (we find the simple correlation between Y and X_1);

R_{24} (we determine the simple correlation between X_2 and X_4), etc.

Here neither variable within each indicated pair is controlled for the effect of any other variable. Moreover, all of the various zero-order correlations serve to measure the strength of the linear relationship between the selected pairs of variables.

The collection of all possible zero-order correlation coefficients constitutes the set of elements within the $(m + 1 \times m + 1)$ correlation matrix

$$\boldsymbol{R} = \begin{bmatrix} 1 & R_{Y1} & R_{Y2} & \ldots & R_{Ym} \\ R_{1Y} & 1 & R_{12} & \ldots & R_{1m} \\ \cdot & \cdot & \cdot & & \cdot \\ \cdot & \cdot & \cdot & & \cdot \\ \cdot & \cdot & \cdot & & \cdot \\ R_{mY} & R_{m1} & R_{m2} & \ldots & 1 \end{bmatrix}, \tag{13.1}$$

where

$$R_{Yj} = \frac{\sum y_i x_{ij}}{\sqrt{\sum y_i^2}\sqrt{\sum x_{ij}^2}}, \; j = 1,\ldots,m; \tag{13.2}$$

$$R_{lh} = \frac{\sum x_{il} x_{ih}}{\sqrt{\sum x_{il}^2}\sqrt{\sum x_{ih}^2}}, \; l,h = 1,\ldots,m. \tag{13.3}$$

Example 13.1

To generate the correlation matrix $\boldsymbol{R}$, we can invoke the correlation procedure (PROC CORR) in the SAS System. Looking to the dataset used in Example 12.1 (see Table 12.4), let us repeat the portion of the SAS code in Exhibit 12.1, which produces (among other items) the correlation matrix.

Exhibit 13.1 SAS Code for Correlation Matrix

```
proc corr data = ABCsold;
   var units cold foll clt exp;
run;
```

The complete output generated by this procedure was labeled as Section ⓂM of Table 12.5. The portion of that output involving the correlation matrix ***R*** is replicated here as Table 13.1. Note that beneath each of the estimated zero-order correlations is the associated *p*-value determined under the null hypothesis $H_0 : \rho_{rs} = 0$ (which is tested against the alternative $H_1 : \rho_{rs} \neq 0$). Because each reported *p*-value <0.05, it is evident that each of these zero-order correlation coefficients is statistically significant.

Coefficient of Multiple Determination

We noted above that the *sample coefficient of determination* $R^2 = (SST - SSE)/SST$ depicts the proportionate reduction in SST obtained by using OLS to estimate *Y* as a linear function of *X*. It serves as an estimator for the *population coefficient of determination* $\rho^2 = (\sigma_Y^2 - \sigma_{Y|X}^2)/\sigma_Y^2$, the proportionate reduction in the unconditional variation in *Y* attributable to conditioning on *X*. The *sample coefficient of multiple determination*, denoted $R^2_{Y|X_1,\ldots,X_m}$, measures the proportionate reduction in SST obtained by using OLS to estimate *Y* as a linear combination of $X_1,\ldots,X_m$; it measures the proportion of the variation in *Y* explained by the "full model." (In Example 12.1, the coefficient of multiple determination obtained from Section ③ of Table 12.5 is $R^2_{Y|X_1,X_2,X_3,X_4} = 0.9386$.) In this regard, $R^2_{Y|X_1,\ldots,X_m} = 1-(SSE/SST)$ serves as an estimator for the *population coefficient of multiple determination*

$$\rho^2_{Y|X_1,\ldots,X_m} = \frac{\sigma_Y^2 - \sigma^2_{Y|X_1,\ldots,X_m}}{\sigma_Y^2},$$

the proportionate reduction in the unconditional variation in *Y* attributable to conditioning on the full set of regressors $X_1,\ldots,X_m$.

If we take the square root of $R^2_{Y|X_1,\ldots,X_m}$, we obtain the *sample multiple correlation coefficient* $R_{Y|X_1,\ldots,X_m} \geq 0$, which measures the strength of the linear association between *Y* and the OLS determined linear combination of $X_1,\ldots,X_m$. Because the OLS linear combination of the

TABLE 13.1

Output for Example 13.1

The SAS System
Pearson Correlation Coefficients, $N = 35$
Prob > |r| under H_0 : Rho=0

	(Y) units	(X_1) cold	(X_2) foll	(X_3) clt	(X_4) exp
(Y) units	1.00000	0.94256	0.72778	0.91001	0.84014
		<0.0001	<0.0001	<0.0001	<0.0001
(X_1) cold	0.94256	1.00000	0.66956	0.86176	0.80724
	<0.0001		<0.0001	<0.0001	<0.0001
(X_2) foll	0.72778	0.66956	1.00000	0.71506	0.60126
	<0.0001	<0.0001		<0.0001	0.0001
(X_3) clt	0.91001	0.86176	0.71506	1.00000	0.75316
	<0.0001	<0.0001	<0.0001		<0.0001
(X_4) exp	0.84014	0.80724	0.60126	0.75316	1.00000
	<0.0001	<0.0001	0.0001	<0.0001	

full set of regressors is expressable as $\hat{Y} = \hat{\beta}_0 + \sum_{j=1}^{m} \hat{\beta}_j X_j$, we can also view the multiple correlation coefficient as the zero-order correlation coefficient between the observed and estimated values of Y or

$$R_{Y\hat{Y}} = \frac{COV(Y,\hat{Y})}{S_Y S_{\hat{Y}}} = \frac{\sum y_i \hat{y}_i}{\sqrt{\sum y_i^2}\sqrt{\sum \hat{y}_i^2}}, \tag{13.4}$$

where $y_i = Y_i - \bar{Y}$ and $\hat{y}_i = \hat{Y}_i - \bar{\hat{Y}}_i = \hat{Y}_i - \bar{Y}$.

Coefficients of Partial Determination and Partial Correlation

In general, a *population coefficient of partial determination* between Y and X_1, given that $X_2,\ldots,X_m$ are already in the regression model, measures the proportionate reduction in the conditional variance of Y given $X_2,\ldots,X_m$ attributable to admitting $X_1$to the conditioning set, i.e.,

$$\rho^2_{YX_1|X_2,\ldots,X_m} = \frac{\sigma^2_{Y|X_2,\ldots,X_m} - \sigma^2_{Y|X_1,X_2,\ldots,X_m}}{\sigma^2_{Y|X_2,\ldots,X_m}}. \tag{13.5}$$

The *sample coefficient of partial determination* can be obtained as

$$R^2_{YX_1|X_2,\ldots,X_m} = \frac{SSE(X_2,\ldots,X_m) - SSE(X_1X_2,\ldots,X_m)}{SSE(X_2,\ldots,X_m)}, \tag{13.6}$$

where $SSE(X_2,\ldots,X_m)$ measures the reduction in the variation in Y when $X_2,\ldots,X_m$ are in the regression model and $SSE(X_1, X_2,\ldots,X_m)$ is the reduction in Y variation when $X_1, X_2,\ldots,X_m$ appear in the model. Hence, Equation 13.6 represents the proportionate reduction in the error sum of squares attributable to introducing X_1 into a model that already contains $X_2,\ldots,X_m$ as regressors.

We may express Equation 13.6 alternatively as

$$R^2_{YX_1|X_2,\ldots,X_m} = \frac{SSR(X_1|X_2,\ldots,X_m)}{SSE(X_2,\ldots,X_m)}, \tag{13.6.1}$$

where $SSR(X_1|X_2,\ldots,X_m)$ depicts the increase in SSR when X_1 is added to a regression model that already contains regressors $X_2,\ldots,X_m$ (this increment in SSR must equal $SSE(X_2,\ldots,X_m) - SSE(X_1, X_2,\ldots,X_m)$, which is the marginal decrease in the variation in Y when X_1 is inserted into the model as an additional regressor).

Similarly, the sample coefficient of partial determination between Y and X_2, given that $X_1, X_3,\ldots,X_m$ are already in the regression model, appears as

$$\begin{aligned} R^2_{YX_2|X_1,X_3,\ldots,X_m} &= \frac{SSR(X_2|X_1,X_3,\ldots,X_m)}{SSE(X_1,X_3,\ldots,X_m)} \\ &= \frac{SSE(X_1,X_3,\ldots,X_m) - SSE(X_1,\ldots,X_m)}{SSE(X_1,X_3,\ldots,X_m)}. \end{aligned} \tag{13.7}$$

It measures the proportionate reduction in the error sum of squares attributable to introducing X_2 into a model that already contains $X_1, X_3,\ldots,X_m$ as regressors. Other sample coefficients of partial determination are defined in a similar manner.

More concretely, given the explained variable Y and the regressors X_1, X_2, and X_3:

$$R^2_{YX_1|X_2,X_3} = \frac{SSR(X_1|X_2,X_3)}{SSE(X_2,X_3)} = \frac{SSE(X_2,X_3) - SSE(X_1,X_2,X_3)}{SSE(X_2,X_3)};$$

[coefficient of partial determination between Y and X_1 given that X_2, X_3 are already in the regression model]

$$R^2_{YX_2|X_1,X_3} = \frac{SSR(X_2|X_1,X_3)}{SSE(X_1,X_3)} = \frac{SSE(X_1,X_3) - SSE(X_1,X_2,X_3)}{SSE(X_1,X_3)};$$

[coefficient of partial determination between Y and X_2 given that X_1, X_3 are already in the regression model]

$$R^2_{YX_3|X_1,X_2} = \frac{SSR(X_3|X_1,X_2)}{SSE(X_1,X_2)} = \frac{SSE(X_1,X_2) - SSE(X_1,X_2,X_3)}{SSE(X_1,X_2)}$$

[coefficient of partial determination between Y and X_3 given that X_1, X_2 are already in the regression model].

We now turn to the notion of partial correlation. Specifically, let the *population coefficient of partial correlation* between the random variables Y and X_1 be denoted as $\rho_{YX_1|X_2,\ldots,X_m}$ (the correlation between Y and X_1 in the conditional distribution of Y and X_1 given $X_2,\ldots,X_m$). It serves as a measure of the degree of linear association between Y and X_1 after controlling both of these variables for the linear influence of $X_2,\ldots,X_m$ That is, we seek to determine the extent to which Y and X_1 are linearly related after removing the influence of the remaining regressors from both Y and X_1.

The order of a partial correlation coefficient is specified by the number of variables being controlled for. Some examples include the following:

ρ_{YX_1}: a zero-order (partial) correlation coefficient between Y and X_1 (no control variables indicated);

$\rho_{YX_1|X_2}$: a first-order partial correlation coefficient (we control for the effect of X_2 on both Y and X_1);

$\rho_{YX_1|X_2X_3}$: a second-order partial correlation coefficient (we control for the effects of X_2 as well as X_3 on both Y and X_1), etc.

Obviously, $\rho_{YX_1|X_2,\ldots,X_m}$ is an $(m-1)$st-order partial correlation coefficient.

To estimate $\rho_{YX_1|X_2,\ldots,X_m}$, let us determine the *sample coefficient of partial correlation* between Y and X_1, denoted $R_{YX_1|X_2,\ldots,X_m}$, by calculating the simple or zero-order correlation coefficient between Y and X_1 after adjusting each of these variables for the linear effects of all remaining variables. Hence, both Y and X_1 are purged of any linear association with the regressors $X_2,\ldots,X_m$. This is accomplished by completely partialling out from both Y and X_1 their linear association with the remaining regressors. However, this then implies that we need to determine the simple correlation coefficient between two sets of residuals.

To see this, suppose, for the sake of simplicity, that we have only three random variables Y, X_1, and X_2 and that we choose to determine the first-order partial correlation coefficient $R_{YX_1|X_2}$, a measure of the strength of the linear association between Y and X_1 after controlling for or partialling out X_2. Estimates of the two regression equations

$$Y = \beta_0 + \beta_1 X_2 + \varepsilon,$$
$$X_1 = \beta_0' + \beta_1' X_2 + \varepsilon'$$

will be denoted, respectfully, as

$$\hat{Y} = \hat{\beta}_0 + \hat{\beta}_1 X_2,$$
$$\hat{X}_1 = \hat{\beta}_0' + \hat{\beta}_1' X_2.$$

Let us next form the two sets of residuals $e_i = Y_i - \hat{Y}_i$ and $e_i' = X_{i1} - \hat{X}_{i1}$, $i = 1,\ldots,n$. Here e_i (respectively, e_i') reflects the unexplained portion of Y_i (respectively X_{i1}) after the linear influence of X_2 has been removed. Hence, the first-order partial correlation between Y and X_1 after the effect of X_2 on each of these variables has been removed or partialled out is simply the zero-order correlation between the preceding two sets of residuals, i.e., $R_{YX_1|X_2} = R_{ee'}$ measures the strength of the linear association between the random e and e', and this in turn provides us with an index of the strength of the linear relationship between Y and X_1 after controlling for X_2.

Example 13.2

Using the dataset in Example 12.1, let us determine the first-order partial correlation coefficient between *units* (Y) and *cold* (X_1) while controlling for the effect of *exp* (X_4) on the preceding two variables, i.e., we want to find $R_{YX_1|X_4}$. We may obtain the value of this partial correlation coefficient in two distinct ways. The first uses the SAS System's PROC CORR and the second uses PROC REG.

First, the requisite SAS code is as follows in Exhibit 13.2.

Exhibit 13.2 SAS Code for Partial Correlation (PROC CORR)

```
proc corr data = ABCsold;
          var units cold;
          partial exp;
run;
```

Here we specify *units* and *cold* as the correlates and *exp* as the control variable. The output generated from this code appears as Table 13.2.

① We have requested the correlation between *units* and *cold* after partialling out from each of these variables the influence of *exp*.

②, ③ The ordinary or uncontrolled standard deviations of *units* and *cold* are 3.0368 and 24.5741, respectively, whereas, as expected, the standard deviations of these variables decrease after the effect of *exp* has been removed, i.e., the *partial standard deviations* of *units* and *cold* are now 1.6718 and 14.7225, respectively.

④ The sample realization of $R_{YX_1|X_4}$ is $r_{YX_1|X_4} = 0.8258$, thus indicating a fairly high degree of linear association between *units* and *cold* after controlling for the effect of *exp*. Moreover, because the *p*-value associated with this statistic is less than 0.0001, we can reject $H_0: \rho_{YX_1|X_4} = 0$ in favour of $H_1: \rho_{YX_1|X_4} \neq 0$.

Perhaps a more transparent way to interpret this outcome is to use $r^2_{YX_1|X_4} = 0.6820$, i.e., when *cold* is added to a model already containing *exp*, *SSE* is reduced by 68.2%. The proportionate reduction in SSE that results from adding *cold* to a model already containing *exp* can be written, from Equation 13.6, as

TABLE 13.2

Output for Example 13.2 (PROC CORR)

The SAS System
The CORR procedure
①

1	Partial variables:	exp
2	Variables:	units cold

Simple Statistics
②

Variable	N	Mean	Std Dev	Sum	Minimum	Maximum
exp	35	21.16429	6.76196	740.75000	9.00000	36.25000
units	35	16.68571	3.03675	584.00000	12.00000	22.00000
cold	35	86.22857	24.57412	3018	50.00000	130.00000

Simple Statistics
③

Variable	Partial Variance	Partial Std Dev
exp		
units	2.79500	1.67182
cold	216.75065	14.72245

④
Pearson Partial Correlation Coefficients, $N = 35$
Prob > |r| under H_0: Partial Rho = 0

	units	cold
units	1.00000	0.82582 <0.0001
cold	0.82582 <0.0001	1.00000

$$R^2_{YX_1|X_4} = \frac{SSR(X_4) - SSE(X_1, X_4)}{SSE(X_4)}.$$

Given the structure of this expression, it should be evident that, if we want to perform a test to determine whether $r_{YX_1|X_4}$ is significantly different from zero (i.e., to test $H_0 : \rho_{YX_1|X_4} = 0$, against $H_1 : \rho_{YX_1|X_4} \neq 0$), then we simply need to perform a variable added-last test. As Example 12.2 revealed, a variable added-last test seeks to determine whether including a particular variable in a regression equation that already contains other regressors makes a statistically significant contribution to explaining variation in some dependent variable Y. This test was structured as a partial or incremental F test and is formally equivalent to a test of significance of a partial correlation coefficient. In this regard, a test of $\rho_{YX_1|X_4} = 0$ via a variable added-last test now follows. In fact, this is the motivation for offering an alternative approach for determining the value of $R_{YX_1|X_4}$.

In this second approach, our strategy is to use PROC REG. Specifically, the requisite SAS code now follows in Exhibit 13.3.

Exhibit 13.3 SAS Code for Partial Correlation (PROC REG)

```
proc reg data = ABCsold;
    m1 : model units = exp;
```

```
    m2 : model units = exp cold / pcorr1 pcorr2;
run;
```

Here the *pcorr1* option determines the set of sequential or variables added-in-order (Type I) squared partial correlation coefficients, whereas the *pcorr2* option provides us with the set of variables-added-last (Type II) squared partial correlations. In addition, listing *cold* last in *m2* guarantees that the added-in-order squared partial correlation equals the added-last squared partial correlation because *cold* is included after *exp* is already in the regression equation, i.e., *units* and *cold* are correlated after controlling for *exp*.

The output produced by this SAS code is given in Table 13.3.

①, ② $SSR(m1) = 221.30794$ and $SSR(m2) = 284.21081$.
Hence, $\Delta SSR = SSR(m2) - SSR(m1) = 62.90287$ and, from Equation 12.39, $F = \frac{\Delta SSR}{MSE(m2)} = \frac{62.90287}{0.91663} = 68.62406$, p-value <0.0001. (Equivalently, $\Delta SSR = SSR(m1) - SSR(m2) = 92.23491 - 29.33205 = 62.90286$.) We thus reject $H_0: \beta_{cold} = 0$, in favor of

TABLE 13.3

Output for Example 13.2 (PROC REG)

The SAS System
The REG Procedure
Model: m1
Dependent Variable: Units
Analysis of Variance
①

Source	**DF**	**Sum of Squares**	**Mean Square**	***F* Value**	**Pr > *F***
Model	1	221.30794	221.30794	79.18	<0.0001
Error	33	92.23491	2.79500		
Corrected total	34	313.54286			

Root MSE	1.67182	*R*-Square	0.7058
Dependent mean	16.68571	Adj *R*-Sq	0.6969
Coeff Var	10.01950		

Parameter Estimates

Variable	**DF**	**Parameter Estimate**	**Standard Error**	***t* Value**	**Pr > |*t*|**
Intercept	1	8.70043	0.94084	9.25	<0.0001
exp	1	0.37730	0.04240	8.90	<0.0001

The REG Procedure
Model: m2
Dependent Variable: Units
Analysis of Variance
②

Source	**DF**	**Sum of Squares**	**Mean Square**	***F* Value**	**Pr > *F***
Model	2	284.21081	142.10540	155.03	<0.0001
Error	32	29.33205	0.91663		
Corrected total	34	313.54286			

Root MSE	0.95741	*R*-Square	0.9064
Dependent mean	16.68571	Adj *R*-Sq	0.9006
Coeff Var	5.73788		

(continued)

TABLE 13.3 *(continued)*

Parameter Estimates

Variable	DF	Parameter Estimate	Standard Error	t Value	Pr > |t|	③ Squared Partial Corr Type I	Squared Partial Corr Type II
Intercept	1	6.43663	0.60413	10.65	<0.0001	.	.
exp	1	0.10219	0.04114	2.48	0.0184	0.70583	0.16165
cold	1	0.09378	0.01132	8.28	<0.0001	0.68199	0.68199

$H_0: \beta_{cold} \neq 0$; the regressor *cold* contributes significantly to a model that already incorporates the regressor *exp*.

③ The preceding *F* test is equivalent to a test of $H_0: \rho_{YX_1|X_4} = 0$, against $H_1: \rho_{YX_1|X_4} \neq 0$. Because $r^2_{YX_1|X_4} = 0.68199$, $r_{YX_1|X_4} = 0.82583$. (Note that this partial correlation coefficient is given the same sign as the estimated coefficient on the variable *cold* in *m*2.)

Example 13.3

Let us use the Example 12.1 dataset to determine the second-order partial correlation coefficient between *units* (*Y*) and *cold* (X_1) while controlling for the effect of *clt* (X_3) and *exp* (X_4) on both *units* and *cold*. That is, we want to determine $R_{YX_1|X_3,X_4}$. Our first approach to finding $R_{YX_1|X_3,X_4}$ is to use PROC CORR. The SAS code follows in Exhibit 13.4.

Exhibit 13.4 SAS Code for Second-Order Partial Correlation (PROC CORR)

```
proc corr data = ABCsold;
      var units cold;
      partial clt exp;
run;
```

We thus request the correlation between *units* and *cold* in their conditional distribution after we specify both *clt* and *exp* as control variables. The output resulting from this SAS code appears in Table 13.4.

① We have asked for the partial correlation between *units* and *cold* after partialling out from each of these variables the influence of *clt* and *exp*.

②, ③ The uncontrolled standard deviations of *units* and *cold* are 3.0368 and 24.5741, respectively, whereas the standard deviations of these variables decrease to 1.06861 and 11.31528, respectively, after the effects of both *clt* and *exp* have been removed (their partial standard deviation levels).

④ The sample realization of $R_{YX_1|X_3,X_4}$ is $r_{YX_1|X_3,X_4} = 0.66741$, *p*-value <0.0001. Here we reject $H_0: \rho_{YX_1|X_3,X_4} = 0$ in favor of $H_1: \rho_{YX_1|X_3,X_4} \neq 0$: we have a statistically significant degree of linear association between *units* and *cold* after controlling for (essentially fixing the values of) both *clt* and *exp*.

Let us now form $r^2_{YX_1|X_3,X_4} = 0.44544$. So when *cold* is added to a model already containing *clt* and *exp*, SSE decreases by 44.54%.This proportionate reduction in SSE obtained by adding *cold* to a model incorporating *clt* and *exp* can be determined directly from the statistic

TABLE 13.4

Output for Example 13.3 (PROC CORR)

The SAS System
The CORR Procedure
①

2 Partial variables:	clt	exp
2 Variables:	units	cold

Simple Statistics
②

Variable	*N*	Mean	Std Dev	Sum	Minimum	Maximum
clt	35	35.71429	14.04734	1250	19.00000	63.00000
exp	35	21.16429	6.76196	740.75000	9.00000	36.25000
units	35	16.68571	3.03675	584.00000	12.00000	22.00000
cold	35	86.22857	24.57412	3018	50.00000	130.00000

Simple Statistics
③

Variable	Partial Variance	Partial Std Dev
clt		
exp		
units	1.14193	1.06861
cold	128.03548	11.31528

④
Pearson Partial Correlation Coefficients, *N* = 35
Prob > |r| under H_0: Partial Rho = 0

	units	cold
units	1.00000	0.66741 <0.0001
cold	0.66741 <0.0001	1.00000

$$R^2_{YX_1|X_3,X_4} = \frac{SSE(X_3,X_4) - SSE(X_1,X_3,X_4)}{SSE(X_3,X_4)}$$

(see Equation 13.6). The structure of this expression reveals that if we want to perform a test to determine whether $r_{YX_1|X_3,X_4}$ differs significantly from zero (i.e., to test $H_0: \rho_{YX_1|X_3,X_4} = 0$ versus $H_1: \rho_{YX_1|X_3,X_4} \neq 0$), then, as in the preceding example problem, we need only perform a variable-added-last test using *PROC REG* (the basis of our second approach to determining $R_{YX_1|X_3,X_4}$). To this end, the appropriate SAS code follows in Exhibit 13.5.

Exhibit 13.5 SAS Code for Second-Order Partial Correlation (PROC REG)

```
proc reg data = ABCsold;
      m1 : model units = exp;
      m2 : model units = clt exp cold / pcorr1 pcorr2;
run;
```

A discussion of the *pcorr1* and *pcorr2* options is given in Example 13.2. The output associated with this program is given in Table 13.5.

①, ② $\Delta SSR = SSR(m2) - SSR(m1) = 293.27795 - 277.00095 = 16.27700$ and thus, from Equation 12.39,

$$F = \frac{\Delta SSR}{MSE(m2)} = \frac{16.27700}{0.65371} = 24.89942,$$

TABLE 13.5

Output for Example 13.3 (PROC REG)

The SAS System
The REG Procedure
Model: m1
Dependent Variable: Units
①
Analysis of Variance

Source	DF	Sum of Squares	Mean Square	F Value	Pr > F
Model	2	277.00095	138.50047	121.29	<0.0001
Error	32	36.54191	1.14193		
Corrected total	34	313.54286			

Root MSE	1.06861	R-Square	0.8835
Dependent mean	16.68571	Adj R-Sq	0.8762
Coeff Var	6.40436		

Parameter Estimates

Variable	DF	Parameter Estimate	Standard Error	t Value	Pr > $\|t\|$
Intercept	1	8.34031	0.60358	13.82	<0.0001
clt	1	0.13850	0.01983	6.98	<0.0001
exp	1	0.16060	0.04120	3.90	0.0005

The REG Procedure
Model : m2
Dependent Variable: Units
Analysis of Variance
②

Source	DF	Sum of Squares	Mean Square	F Value	Pr > F
Model	3	293.27795	97.75932	149.55	<0.0001
Error	31	20.26491	0.65371		
Corrected total	34	313.54286			

TABLE 13.5 *(continued)*

Root MSE	0.80852	*R*-Square	0.9354
Dependent Mean	16.68571	Adj *R*-Sq	0.9291
Coeff Var	4.84559		

Parameter Estimates

Variable	DF	Parameter Estimate	Standard Error	*t* Value	Pr > |*t*|	③ Squared Partial Corr Type I	Squared Partial Corr Type II
Intercept	1	6.98689	0.53115	13.15	<.0001	.	.
clt	1	0.07384	0.01983	3.72	0.0008	0.82811	0.30912
exp	1	0.07686	0.03540	2.17	0.0377	0.32196	0.13199
cold	1	0.06303	0.01263	4.99	<.0001	0.44543	0.44543

p-value <0.0001. Hence, we reject $H_0: \beta_{cold} = 0$ in favor of $H_1: \beta_{cold} \neq 0$: the insertion of the variable *cold* into a regression model that already contains regressors *clt* and *exp* makes a statistically significant contribution to explaining the variability of *units*.

③ The preceding incremental *F* test is equivalent to testing $H_0: \rho_{YX_1|X_3,X_4} = 0$, against $H_1: \rho_{YX_1|X_3,X_4} \neq 0$. With $r^2_{YX_1|X_3,X_4} = 0.45543$, it follows that $r_{YX_1|X_3,X_4} = 0.66741$.

Example 13.4

Using the Example 12.1 dataset, let us now turn to an examination of the complete output generated by the *pcorr1* and *pcorr2* options in the *MODEL* statement when all four regressors are used:

Exhibit 13.6 SAS Code for Partial Correlations

```
prog reg data = ABCsold;
      model units = cold foll clt exp/ pcorr1 pcorr2;
run;
```

Remember that *pcorr1* requests the sequential or variables-added-in-order squared partial correlation coefficients while *pcorr2* requests the variables-added-last squared partial correlation coefficients.

Table 13.6 houses the output determined by the preceding lines of SAS code.

① The entries appearing in this column are

$$0.88841 = r^2_{\text{units, cold}};$$

$$0.15184 = r^2_{\text{units, foll}|\text{cold}};$$

$$0.25658 = r^2_{\text{units, clt}|\text{cold, foll}};$$

$$0.12724 = r^2_{\text{units, exp}|\text{cold, foll, clt}}.$$

Here we have a sequence of sample coefficients of partial determination between units and each individual regressor added in order and after controlling for the preceding regressors. The set of associated sample partial correlation coefficients are thus

TABLE 13.6

Output for Example 13.4

The SAS System
The REG Procedure
Model : MODEL1
Dependent Variable: Units
Analysis of Variance

Source	DF	Sum of Squares	Mean Square	F Value	Pr > F
Model	4	294.28855	73.57214	114.63	<0.0001
Error	30	19.25430	0.64181		
Corrected Total	34	313.54286			

Root MSE	0.80113	R-Square	0.9386
Dependent mean	16.68571	Adj R-Sq	0.9304
Coeff Var	4.80130		

Parameter Estimates

Variable	DF	Parameter Estimate	Standard Error	t Value	Pr > |t|	① Squared Partial Corr Type I	② Squared Partial Corr Type II
Intercept	1	6.71824	0.56817	11.82	<0.0001	.	.
cold	1	0.06148	0.01258	4.89	<0.0001	0.88841	0.44335
foll	1	0.04970	0.03960	1.25	0.2192	0.15184	0.04987
clt	1	0.06464	0.02097	3.08	0.0044	0.25658	0.24058
exp	1	0.07357	0.03518	2.09	0.0451	0.12724	0.12724

$r_{units,\ cold} = 0.94256;$
$r_{units,\ foll|cold} = 0.38967;$
$r_{units,\ clt|cold,\ foll} = 0.50654;$
$r_{units,\ exp|cold,\ foll,\ clt} = 0.35671,$

with the sign of each dictated by the sign of the accompanying estimated regression coefficient.

② The entries displayed in this column are

$$0.44335 = r^2_{\text{units, cold|foll, clt, exp}};$$

$$0.04987 = r^2_{\text{units, foll|cold, clt, exp}};$$

$$0.24058 = r^2_{\text{units, clt|cold, foll, exp}};$$

$$0.12724 = r^2_{\text{units, exp|cold, foll, clt}}.$$

Now each sample coefficient of partial determination between *units* and an individual regressor controls for all remaining regressors.

The collection of associated sample partial correlation coefficients appears as

$r_{units,\ cold|foll,\ clt,\ exp} = 0.66585;$
$r_{units,\ foll|cold,\ clt,\ exp} = 0.22332;$
$r_{units,\ cold|foll,\ clt,\ exp} = 0.49049;$
$r_{units,\ exp|cold,\ foll,\ clt} = 0.35671.$

Example 13.5

It is interesting to note that we may express a coefficient of partial determination in terms of an appropriate set of coefficients of multiple determination. That is,

$$R^2_{YX_j|X_1,\ldots,X_{j-1},X_{j+1},\ldots,X_m} = \frac{R^2_{Y|X_1,\ldots,X_m} - R^2_{Y|X_1,\ldots,X_{j-1},X_{j+1},\ldots,X_m}}{1-R^2_{Y|X_1,\ldots,X_{j-1},X_{j+1},\ldots,X_m}}.$$

For instance, using the Example 12.1 dataset, let us find

$$R^2_{YX_3|X_1,X_2,X_4} = \frac{R^2_{Y|X_1,X_2,X_3,X_4} - R^2_{Y|X_1,X_2,X_4}}{1-R^2_{Y|X_1,X_2,X_4}}.$$

The SAS code for this task follows in Exhibit 13.7.

Exhibit 13.7 SAS Code for Partial Determination Coefficient

```
proc reg data = ABCsold;
      m1 : model units = cold foll exp;
      m2 : model units = cold foll clt exp;
run;
```

The output is found in Table 13.7.

Because $r^2_{Y|X_1,X_2,X_3,X_4} = 0.9386$ and $r^2_{Y|X_1,X_2,X_4} = 0.9191$, it follows that

$$r^2_{YX_3|X_1,X_2,X_4} = \frac{0.9386-0.9191}{1-0.9191} = 0.2410$$

with $r_{YX_3|X_1,X_2,X_4} = 0.4910$.

It is interesting to note that partial correlation coefficients can be expressed in a variety of forms that are fundamentally equivalent to the structure of $R_{YX_1|X_2,\ldots,X_m}$ determined from Equation 13.6.

First, because any measure of correlation is dimensionless, let us convert the estimated OLS regression coefficient $\hat{\beta}_j$ to its standardized form $\hat{\beta}^*_j = \hat{\beta}_j(S_j / S_Y)$ (see Equation 12.29). Because $\hat{\beta}^*_j$ is independent of units, it can legitimately be compared across regressors, i.e., $\hat{\beta}^*_j$ depicts the average rate of change in Y per unit change in X_j in standard deviation units (when X_j increases by one standard deviation, Y changes by $\hat{\beta}^*_j$ standard deviations on the average).

Let $R_{Y\hat{Y}}$ represent the multiple correlation coefficient between the observed and estimated values of Y, where the $\hat{Y}_i$'s, $i = 1,\ldots,n$, are obtained by regressing Y on the $m-1$ regressors $X_1,\ldots,X_{j-1}$, $X_{j+1},\ldots,X_m$. Similarly, let $R_{X_j\hat{X}_j}$ depict the multiple correlation coefficient between the observed and estimated values of X_j, where the $\hat{X}_{ij}$'s, $i = 1,\ldots,n$, are obtained by regressing X_j on the same remaining $m-1$ regressors $X_1,\ldots,X_{j-1},X_{j+1},\ldots,X_m$. Then it can be demonstrated that the coefficient of partial correlation between Y and X_j after controlling for the effect of regressors $X_1,\ldots,X_{j-1}$, $X_{j+1},\ldots,X_m$ on both Y and X_j is

$$R_{YX_j|X_1,\ldots,X_{j-1},X_{j+1},\ldots,X_m} = \hat{\beta}_j \frac{S_j\sqrt{1-R^2_{X_j\hat{X}_j}}}{S_Y\sqrt{1-R^2_{Y\hat{Y}}}}, \tag{13.8}$$

TABLE 13.7

Output for Example 13.5

The SAS System
The REG Procedure
Model : m1
Dependent Variable: Units
Analysis of Variance

Source	DF	Sum of Squares	Mean Square	*F* Value	Pr > *F*
Model	3	288.18881	96.06394	117.45	<0.0001
Error	31	25.35405	0.81787		
Corrected Total	34	313.54286			

Root MSE	0.90436	*R*-Square	0.9191
Dependent mean	16.68571	Adj *R*-Sq	0.9113
Coeff Var	5.41998		

Parameter Estimates

Variable	DF	Parameter Estimate	Standard Error	*t* Value	Pr > |*t*|
Intercept	1	6.06467	0.59506	10.19	<0.0001
cold	1	0.08377	0.01162	7.21	<0.0001
foll	1	0.09238	0.04189	2.21	0.0350
exp	1	0.09020	0.03924	2.30	0.0284

The REG Procedure
Model : m2
Dependent Variable: Units
Analysis of Variance

Source	DF	Sum of Squares	Mean Square	*F* Value	Pr > *F*
Model	4	294.28855	73.57214	114.63	<0.0001
Error	30	19.25430	0.64181		
Corrected total	34	313.54286			

Root MSE	0.80113	*R*-Square	0.9386
Dependent mean	16.68571	Adj *R*-Sq	0.9304
Coeff Var	4.80130		

Parameter Estimates

Variable	DF	Parameter Estimate	Standard Error	*t* Value	Pr > |*t*|
Intercept	1	6.71824	0.56817	11.82	<0.0001
cold	1	0.06148	0.01258	4.89	<0.0001
foll	1	0.04970	0.03960	1.25	0.2192
clt	1	0.06464	0.02097	3.08	0.0044
exp	1	0.07357	0.03518	2.09	0.0451

where $S_j\sqrt{1-R^2_{X_j\hat{X}_j}}\Big/S_Y\sqrt{1-R^2_{Y\hat{Y}}}$ represents the standard deviation of the X_j residuals $e_i' = X_{ij} - \hat{X}_{ij}$ divided by the standard deviation of the Y residuals $e_i = Y_i - \hat{Y}_i, i = 1,\ldots,n$.

Next, an mth-order partial correlation coefficient can be expressed in terms of partial correlation coefficients of order $m - 1$. In this regard, the first-order partial correlation coefficient between Y and X_1 after controlling for X_2 can be written in terms of zero-order correlation coefficients as

$$R_{YX_1|X_2} = \frac{R_{YX_1} - R_{YX_2}R_{X_1X_2}}{\sqrt{1-R^2_{YX_2}}\sqrt{1-R^2_{X_1X_2}}}. \tag{13.9}$$

Here we modify the zero-order correlation R_{YX_1} to incorporate the effect of the control variable X_2. To define higher-order partial correlation coefficients in like manner, we simply use the relevant next lower-order partial correlation coefficient, e.g., the second-order partial correlation coefficient is stated in terms of appropriate first-order correlations. For instance, we modify the first-order partial correlation to reflect the application of an additional control variable. To this end,

$$R_{YX_1|X_2,X_3} = \frac{R_{YX_1|X_2} - R_{YX_3|X_2}R_{X_1X_3|X_2}}{\sqrt{1-R^2_{YX_3|X_2}}\sqrt{1-R^2_{X_1X_3|X_2}}} \tag{13.10}$$

[we control on X_2 in all first-order partial correlations].

Similarly, a third-order partial correlation coefficient is obtained by adjusting the second-order partial correlation to account for the effect of controlling for another variable. Thus,

$$R_{YX_1|X_2,X_3,X_4} = \frac{R_{YX_1|X_2,X_3} - R_{YX_4|X_2,X_3}R_{X_1X_4|X_2,X_3}}{\sqrt{1-R^2_{YX_4|X_2,X_3}}\sqrt{1-R^2_{X_1X_4|X_2,X_3}}}, \tag{13.11}$$

[we control on X_2 and X_3 in all second-order partial correlations]

and so on.

Coefficients of Multiple Partial Determination and Multiple Partial Correlation

Let us denote the *population coefficient of multiple partial determination* between Y and the random variables $X_1,\ldots,X_m$, given that the random variables $U_1,\ldots,U_p$ are already in the regression model, as $\rho^2_{Y(X_1,\ldots,X_m)|U_1,\ldots,U_p}$. This coefficient measures the proportionate reduction in the conditional variance of Y given $U_1,\ldots,U_p$ attributable to admitting $X_1,\ldots,X_m$ to the conditioning set or

$$\rho^2_{Y(X_1,\ldots,X_m)|U_1,\ldots,U_p} = \frac{\sigma^2_{Y|U_1,\ldots,U_p} - \sigma^2_{Y|X_1,\ldots,X_m,U_1,\ldots,U_p}}{\sigma^2_{Y|U_1,\ldots,U_p}}. \tag{13.12}$$

Once $X_1,\ldots,X_m$ are admitted to the conditioning set, the *complete* or *full regression model* is

$$Y = \beta_0 + \beta_1U_1 + \cdots + \beta_pU_p + \gamma_1X_1 + \cdots + \gamma_mX_m + \varepsilon. \tag{13.13}$$

The *sample coefficient of multiple partial determination* can be obtained as

$$R^2_{Y(X_1,\ldots,X_m)|U_1,\ldots,U_p} = \frac{SSE(U_1,\ldots,U_p) - SSE(X_1,\ldots,X_m,U_1,\ldots,U_p)}{SSE(U_1,\ldots,U_p)}. \tag{13.14}$$

Here $SSE(U_1,\ldots,U_p)$ depicts the reduction in the variation in Y when $U_1,\ldots,U_p$ are in the regression model, and $SSE(X_1,\ldots,X_m, U_1,\ldots,U_p)$ measures the reduction in Y variation when $X_1,\ldots,X_m, U_1,\ldots,U_p$ appear in the model. Hence, Equation 13.14 represents the proportionate reduction in SSE attributable to introducing $X_1,\ldots,X_m$ into a model that already has $U_1,\ldots,U_p$ as regressors.

We may express Equation 13.14 in an alternative manner as

$$R^2_{Y(X_1,\ldots,X_m)|U_1,\ldots,U_p} = \frac{SSR(X_1,\ldots,X_m|U_1,\ldots,U_p)}{SSE(U_1,\ldots,U_p)}, \tag{13.14.1}$$

where $SSR(X_1,\ldots,X_m|U_1,\ldots,U_p) = SSR(X_1,\ldots,X_m, U_1,\ldots,U_p) - SSR(U_1,\ldots,U_p)$ represents the increment in SSR when $X_1,\ldots,X_m$ are inserted into a regression model that already contains regressors $U_1,\ldots,U_p$; or as

$$R^2_{Y(X_1,\ldots,X_m)|U_1,\ldots,U_p} = \frac{R^2_{Y|X_1,\ldots,X_m,U_1,\ldots,U_p} - R^2_{Y|U_1,\ldots,U_p}}{1 - R^2_{Y|U_1,\ldots,U_p}}, \tag{13.14.2}$$

where the R^2 terms on the right-hand side of Equation 13.14.2 are coefficients of multiple determination between Y and the indicated conditioning variables.

We next denote the *population coefficient of multiple partial correlation* between the random variables Y and $X_1,\ldots,X_m$ after controlling for $U_1,\ldots,U_p$ as $\rho_{Y(X_1,\ldots,X_m)|U_1,\ldots,U_p}$ (which represents the correlation between Y and $X_1,\ldots,X_m$ in the conditional distribution of Y and $X_1,\ldots,X_m$ given $U_1,\ldots,U_p$). It measures the degree of linear association between Y and $X_1,\ldots,X_m$ after controlling Y and $X_1,\ldots,X_m$ for $U_1,\ldots,U_p$, i.e., we partial out the effects of $U_1,\ldots,U_p$ on Y and $X_1,\ldots,X_m$. This population coefficient is estimated by the *sample coefficient of multiple partial correlation* $R_{Y(X_1,\ldots,X_m)|U_1,\ldots,U_p} \geq 0$.

Example 13.6

Looking to the Example 12.1 dataset, let us find the multiple partial correlation coefficient between *units* (Y) and *clt* (X_3) and *exp* (X_4) while controlling for *cold* (X_1) and *foll* (X_2), i.e., we want to determine $R_{Y(X_3,X_4)|X_1,X_2}$. We may accomplish this task via the SAS regression procedure. Specifically, let us first specify groups A = {*cold*(X_1), *foll*(X_2)} and B = {*clt*(X_3), *exp*(X_4)} so that a group added-last regression experiment can be performed. Then we may determine, from Equation 13.14,

$$R^2_{Y(B)|A} = \frac{SSE(A) - SSE(A,B)}{SSE(A)}.$$

Fortunately, the heavy lifting for this calculation has already been performed for us in Example 12.4. If we examine the model pool therein, we see that we need to compare the full model with the reduced model *m_B* (which indicates that the variables in set B are deleted). Thus,

$$R^2_{Y(B)|A} = \frac{SSE(m\_B) - SSE(full)}{SSE(m\_B)}$$

so that, from Table 12.8, the realized value of $R^2_{Y(B)|A}$ is

$$r^2_{Y(B)|A} = \frac{29.67536 - 19.25430}{29.67536} = 0.3511$$

(when *clt* and *exp* are added to a model already containing *cold* and *foll*, SSE is reduced by 35.11%) and thus $r_{y(B)|A} = 0.5926$.

Next, to determine whether the value of $r_{Y(B)|A}$ is statistically significant, let us test ($H_0: \rho_{Y(B)|A} = 0$, against $H_1: \rho_{Y(B)|A} \neq 0$ (which is equivalent to testing $H_0: \gamma_1 = \gamma_2 = 0$, against $H_1: H_0$ not true). Then from Equation 12.42, the sample F statistic is

$$F = \frac{\left[SSR(A,B) - SSR(A)\right]/m}{MSE(A,B)} = \frac{\left[SSR(full) - SSR(m\_B)\right]/m}{MSE(full)}$$
$$= \frac{(294.28855 - 283.86750)/2}{0.64181} = 16.23697,$$

p-value <0.001. Here we reject $H_0: \gamma_1 = \gamma_2 = 0$ in favor of H_1 : not H_0 (hence, at least one of these γ_j's $\neq 0$); the insertion of the variables *clt* and *exp* into a regression model that already contains *cold* and *foll* makes a statistically significant contribution to explaining the variation of *units* (equivalently, we have a statistically significant degree of linear association between *units* and *clt* and *exp* after controlling for *cold* and *foll*).

Similarly, let us determine the multiple partial correlation coefficient between *units* (Y) and *cold*(X_1) and *foll* (X_2) while controlling for *clt* (X_3) and *exp* (X_4) or $R_{y(A)|B}$. Then again using Equation 13.14,

$$R^2_{Y(A)|B} = \frac{SSE(B) - SSE(A,B)}{SSE(B)}.$$

In the context of Example 12.4, we must compare the full model with the reduced model *m_A* (the variables in set A are deleted). Then

$$R^2_{Y(A)|B} = \frac{SSE(m\_A) - SSE(full)}{SSE(m\_A)}$$

and thus, from Table 12.8, the sample realization of $R^2_{Y(A)|B}$ is

$$r^2_{Y(A)|B} = \frac{36.54191 - 19.25430}{36.54191} = 0.4731$$

(when *cold* and *foll* are added to a model already containing *clt* and *exp*, SSE falls by 47.31%) and thus $r_{y(A)|B} = 0.6878$.

Let us test $H_0: \rho_{Y(A)|B} = 0$, against $H_1: \rho_{Y(A)|B} \neq 0$ (equivalently, we can test $H_0: \beta_1 = \beta_2 = 0$, against H_1 : not H_0). Again looking to Table 12.8, the sample F statistic is

$$F = \frac{\left[SSR(A,B) - SSR(B)\right]/p}{MSE(A,B)} = \frac{\left[SSR(full) - SSR(m\_A)\right]/p}{MSE(full)}$$
$$= \frac{(294.28855 - 277.00095)/2}{0.64181} = 26.93570,$$

p-value <0.001. Here we reject $H_0: \beta_1 = \beta_2 = 0$ (or $H_0: \rho_{Y(A)|B} = 0$) and conclude that there exists a statistically significant degree of linear association between *units* and *cold* and *foll* after controlling for the effect of *clt* and *exp*.

A Note on the Coefficient of Multiple Semipartial Correlation

We previously defined a sample coefficient of partial correlation between Y and X_1 given that $X_2,\ldots,X_m$ are already in the regression model (denoted $R_{YX_1|X_2,\ldots,X_m}$) as the simple correlation coefficient between Y and X_1 after adjusting each of these variables for the linear effects of all remaining variables, i.e., the regressors $X_2,\ldots,X_m$ are partialled out from both Y and X_1 and the residuals $e_i = Y_i - \hat{Y}_i$ and $e_i' = X_{i1} - \hat{X}_{i1}$ $i = 1,\ldots,n$, are then correlated to yield $R_{ee'}$.

However, instead of partialling out the regressors $X_2,\ldots,X_m$ from *both* Y and X_1, we can simply find the zero-order correlation coefficient between Y and X_1 with only X_1 controlled for $X_2,\ldots,X_m$ or the simple correlation coefficient between Y and X_1 with only Y controlled for $X_2,\ldots,X_m$. In each instance, we have determined a *sample coefficient of multiple semipartial correlation*. If only X_1 is adjusted for the linear effects of $X_2,\ldots,X_m$, then the resulting semipartial correlation coefficient is denoted as

$$R_{Y,(X_1|X_2,\ldots,X_m)} = R_{Ye'};$$

if only Y is adjusted for the same, then the implied semipartial correlation coefficient can be written as

$$R_{(Y|X_2,\ldots,X_m),X_1} = R_{eX_1}.$$

Example 13.7

Using the Example 12.1 dataset, let us find the multiple semipartial correlation coefficient between *units* (Y) and *clt* (X_3), controlling only *clt* for the linear effects of *cold* (X_1), *foll* (X_2), and exp (X_4), i.e., we want to determine $R_{Y,\,(X_3|X_1,X_2,X_4)}$.

The SAS code needed to accomplish this task appears as follows in Exhibit 13.8.

Exhibit 13.8 SAS Code for Multiple Semipartial Correlation (PROC CORR)

```
proc reg data = ABCsold;
     model clt = cold foll exp; ①
     output out = e residual = e_clt; ②
run;
proc corr data = e;
     var units e_clt; ③
run;
```

① We regress *clt* on *cold, foll,* and *exp.*
② We request the residuals $e\_clt = X_{i3} - \hat{X}_{i3}$, $i = 1,\ldots,n$, and store them in output dataset e.
③ We request the simple correlation coefficient between *units* and the residual e_*clt*.

The resulting output appears in Table 13.8.

The sample realization of $R_{Y,\,(X_3|X_1,X_2,X_4)} = R_{Ye\_clt}$ is $r_{ye\_clt} = 0.13948$, *p*-value = 0.4242. Given the high *p*-value, it is obvious that the value of r_{ye_clt} is not statistically significant, i.e., we cannot reject $H_0: \rho_{Y,\,(X_3|X_1,X_2,X_4)} = 0$, in favor of $H_1: \rho_{Y,\,(X_3|X_1,X_2,X_4)} \neq 0$.

TABLE 13.8

Output for Example 13.7 (PROC CORR)

The SAS System
The REG Procedure
Model: MODEL1
Dependent Variable: clt
Analysis of Variance

Source	**DF**	**Sum of Squares**	**Mean Square**	***F* Value**	**Pr > *F***
Model	3	5249.31544	1749.77181	37.16	<0.0001
Error	31	1459.82742	47.09121		
Corrected total	34	6709.14286			

Root MSE	6.86230	*R*-Square	0.7824
Dependent mean	35.71429	Adj *R*-Sq	0.7614
Coeff Var	19.21445		

Parameter Estimates

Variable	**DF**	**Parameter Estimate**	**Standard Error**	***t* Value**	**Pr > |*t*|**
Intercept	1	–10.11082	4.51532	–2.24	0.0325
cold	1	0.34490	0.08814	3.91	0.0005
foll	1	0.66027	0.31783	2.08	0.0461
exp	1	0.25728	0.29775	0.86	0.3942

The CORR Procedure
2 Variables: units e_clt
Simple Statistics

Variable	***N***	**Mean**	**Std Dev**	**Sum**	**Minimum**	**Maximum**	**Label**
units	35	16.68571	3.03675	584.00000	12.00000	22.00000	
e_clt	35	0	6.55256	0	–16.82355	14.81134	Residual

Pearson Correlation Coefficients, *N* = 35
Prob > |r| under H_0: Rho = 0

	units	**e_clt**
units	1.00000	0.13948 0.4242
e_clt Residual	0.13948 0.4242	1.00000

An alternative approach to calculating $R_{Y,(X_3|X_1,X_2,X_4)}$ is to examine the output generated by the *scorr1* and *scorr2* options in the *MODEL* statement: *scorr1* displays the squared semipartial correlation coefficients based on added-in-order (or Type I) sums of squares; *scorr2* gives the squared semipartial correlation coefficients using added-last (or Type II) sums of squares. The requisite SAS code is in Exhibit 13.9.

Exhibit 13.9 SAS Code for Multiple Semipartial Correlation (PROC REG)

```
proc reg data = ABCsold;
     model units = cold foll exp clt / scorr1 scorr 2;
run;
```

Note that lsiting *clt* last leads to an equality between the added-in-order sums of squares and the added-last sums of squares, with the corresponding squared semipartial correlations being equal once *clt* is corrected for *cold, foll,* and *exp*. Table 13.9 displays the resulting output.

① The entries appearing in this column are

$$0.88841 = r^2_{\text{units, cold}};$$
$$0.01694 = r^2_{\text{units, (foll|cold)}};$$
$$0.01378 = r^2_{\text{units, (exp|cold, foll)}};$$
$$0.01945 = r^2_{\text{units, (clt|cold, foll, exp)}}.$$

Hence,

$$r_{\text{units, (clt|cold, foll, exp)}} = 0.13946.$$

TABLE 13.9

Output for Example 13.7 (PROC REG)

The SAS System
The REG Procedure
Model: MODEL1
Dependent Variable: Units
Analysis of Variance

Source	DF	Sum of Squares	Mean Square	*F* Value	Pr > *F*
Model	4	294.28855	73.57214	114.63	<0.0001
Error	30	19.25430	0.64181		
Corrected total	34	313.54286			

Root MSE	0.80113	*R*-Square	0.9386
Dependent mean	16.68571	Adj *R*-Sq	0.9304
Coeff Var	4.80130		

Parameter Estimates

Variable	DF	Parameter Estimate	Standard Error	*t* value	Pr > |*t*|	① Squared Semipartial Corr Type I	② Squared Semipartial Corr Type II
Intercept	1	6.71824	0.56817	11.82	<0.0001	.	.
cold	1	0.06148	0.01258	4.89	<0.0001	0.88841	0.04891
foll	1	0.04970	0.03960	1.25	0.2192	0.01694	0.00322
exp	1	0.07357	0.03518	2.09	0.0451	0.01378	0.00895
clt	1	0.06464	0.02097	3.08	0.0044	0.01945	0.01945

② The entries displayed in this column are

$$0.04891 = r^2_{\text{units, (cold|foll, clt, exp)}};$$
$$0.00322 = r^2_{\text{units, (foll|cold, clt, exp)}};$$
$$0.00895 = r^2_{\text{units, (exp|cold, foll, clt)}};$$
$$0.01945 = r^2_{\text{units, (clt|cold, foll, exp)}}.$$

Taking the positive square root of the last entry in this column again yields. $r_{\text{units, (clt|cold, foll, exp)}} = 0.13946$.

Example 13.8

Using the Example 12.1 dataset, let us again determine $R_{Y,(X_3|X_1,X_2,X_4)}$. This time, however, we shall compare two separate regression models: the full model or model *m*2 that contains all four regressors (*cold, foll, clt,* and *exp*) and a reduced model or model *m*1 that contains all regressors save *clt*. The SAS code follows in Exhibit 13.10.

Exhibit 13.10 SAS Code for Multiple Semipartial Correlation (Full Model versus m2)

```
proc reg data = ABCsold;
     m1: model units = cold foll exp;
     m2: model units = cold foll clt exp;
run;
```

The resulting output appears as Table 13.10.

Let us express the coefficient of multiple determination for model *m*1 as $R^2_{Y|X_1,X_2,X_4}$, whereas the like coefficient for model *m*2 will be written as $R^2_{Y|X_1,X_2,X_3,X_4}$.

Then

$$r^2_{Y,(X_3|X_1,X_2,X_4)} = r^2_{Y|X_1,X_2,X_3,X_4} - r^2_{Y|X_1,X_2,X_4}$$
$$= 0.9386 - 0.9191 = 0.0195$$

and thus $r_{Y,(X_3|X_1,X_2,X_4)} = 0.1396$ (attributable to rounding).

This output also provides us with the requisite information needed to perform a significance test for $r_{Y,(X_3|X_1,X_2,X_4)}$. Specifically, we want to test $H_0 : \rho_{Y,(X_3|X_1,X_2,X_4)} = 0$, against $H_1 : \rho_{Y,(X_3|X_1,X_2,X_4)} \neq 0$. Our approach will be to use a variable-added-last test, the information for which is obtained from Table 13.10. To this end, our sample test statistic is, from Equation 12.39,

$$F = \frac{SSE(m1) - SSE(m2)}{MSE(m2)} = \frac{25.35405 - 19.25430}{0.64181} = 9.50,$$

p-value < 0.005. Hence, we reject H_0 in favor of H_1.

TABLE 13.10

Output for Example 13.8

The SAS System
The REG Procedure
Model: m1
Dependent Variable: Units

Analysis of Variance

Source	DF	Sum of Squares	Mean Square	*F* Value	Pr > *F*
Model	3	288.18881	96.06294	117.45	<0.0001
Error	31	25.35405	0.81787		
Corrected total	34	313.54286			

Root MSE	0.90436	*R*-Square	0.9191
Dependent mean	16.68571	Adj *R*-Sq	0.9113
Coeff Var	5.41998		

Parameter Estimates

Variable	DF	Parameter Estimate	Standard Error	*t* Value	Pr > |*t*|
Intercept	1	6.06467	0.59506	10.19	<0.0001
cold	1	0.08377	0.01162	7.21	<0.0001
foll	1	0.09238	0.04189	2.21	0.0350
exp	1	0.09020	0.03924	2.30	0.0284

The REG Procedure
Model: m2
Dependent Variable: Units

Analysis of Variance

Source	DF	Sum of Squares	Mean Square	*F* Value	Pr > *F*
Model	4	294.28855	73.57214	114.63	<0.0001
Error	30	19.25430	0.64181		
Corrected total	34	313.54286			

Root MSE	0.80113	*R*-Square	0.9386
Dependent mean	16.68571	Adj *R*-Sq	0.9304
Coeff Var	4.80130		

Parameter Estimates

Variable	DF	Parameter Estimate	Standard Error	*t* Value	Pr > |*t*|
Intercept	1	6.71824	0.56817	11.82	<0.0001
cold	1	0.06148	0.01258	4.89	<0.0001
foll	1	0.04970	0.03960	1.25	0.2192
clt	1	0.06464	0.02097	3.08	0.0044
exp	1	0.07357	0.03518	2.09	0.0451

Example 13.9

We now turn to a slightly different type of coefficient of multiple semipartial correlation. For example, instead of working with $R_{Y,(X_3|X_1,X_2,X_4)}$, let us increase the number of variables that are adjusted for the effects of other regressors, i.e., we can, for instance, determine $R_{Y,(X_1,X_3|X_2,X_4)}$; here we want the multiple semipartial correlation coefficient between Y and the group or subset of variables $A = \{X_1, X_3\}$ while controlling only subset A for the linear affects of the group $B = \{X_2, X_4\}$ variables. Using the Example 12.1 data, we can find the sample value of $R_{Y,(X_1,X_3|X_2,X_4)} = R_{Y,(A|B)}$, as well as test its statistical significance, by estimating the following group-added-last SAS model pool (Exhibit 13.11).

Exhibit 13.11 SAS Code for Multiple Semipartial Correlation (Group Added-Last Model)

```
proc reg data = ABCsold;
      m_B : model units = cold          clt;
      m_A : model units =          foll        exp;
       full: model units = cold foll  clt  exp;
run;
```

The output appears as Table 13.11.

From the estimated coefficients of multiple determination for models *m_A* and full we have

$$
\begin{aligned}
r^2_{Y,(X_1,X_3|X_2,X_4)} &= r^2_{full} - r^2_{m\_A} \\
&= r^2_{Y|X_1,X_2,X_3,X_4} - r^2_{Y|X_2,X_4} \\
&= 0.9386 - 0.7835 = 0.1551
\end{aligned}
$$

and thus $r_{Y,(X_1,X_3|X_2,X_4)} = 0.3938$. Next, to test $H_0 : \rho_{Y,(X_1,X_3|X_2,X_4)} = 0$, against $H_0 : \rho_{Y,(X_1,X_3|X_2,X_4)} \neq 0$, we need to conduct a group-added-last test on subset A; we must compare the full model with model *m_A* given $H_0 = \beta_1 = \beta_3 = 0$ vs. $H_1 : H_0$ not true. Under H_0, $\Delta SSR = SSR(full) - SSR(m\_A)$ and thus, from Equation 12.41, our calculated

$$F = \frac{(\Delta SSR / r)}{MSE(full)} = \frac{(294.28855 - 245.64985)/2}{0.64181} = 37.89,$$

p-value <0.0001. Hence, we may reject H_0 in favor of H_1; the realized value of $R_{Y,(X_1,X_3|X_2,X_4)}$ is statistically significant.

TABLE 13.11

Output for Example 13.9

The SAS System
The REG Procedure
Model : m_B
Dependent Variable: Units
Analysis of Variance

Source	DF	Sum of Squares	Mean Square	*F* Value	Pr > *F*
Model	2	290.19635	145.09817	198.88	<0.0001
Error	32	23.34651	0.72958		
Corrected total	34	313.54286			

Root MSE	0.85415	*R*-Square	0.9255
Dependent mean	16.68571	Adj *R*-Sq	0.9209
Coeff Var	5.11907		

(continued)

TABLE 13.11 *(continued)*

Parameter Estimates

Variable	DF	Parameter Estimate	Standard Error	*t* Value	Pr > |*t*|
Intercept	1	7.19737	0.55170	13.05	<0.0001
cold	1	0.07603	0.01175	6.47	<0.0001
clt	1	0.08211	0.02056	3.99	0.0004

The REG Procedure
Model : m_A
Dependent Variable: Units
Analysis of Variance

Source	DF	Sum of Squares	Mean Square	*F* Value	Pr > *F*
Model	2	245.64985	122.82492	57.89	<0.0001
Error	32	67.89301	2.12166		
Corrected total	34	313.54286			

Root MSE	1.45659	*R*-Square	0.7835
Dependent mean	16.68571	Adj *R*-Sq	0.7699
Coeff Var	8.72957		

Parameter Estimates

Variable	DF	Parameter Estimate	Standard Error	*t* Value	Pr > |*t*|
Intercept	1	7.30339	0.91763	7.96	<0.0001
foll	1	0.21036	0.06210	3.39	0.0019
exp	1	0.28314	0.04623	6.12	<0.0001

The REG Procedure
Model: full
Dependent Variable: Units
Analysis of Variance

Source	DF	Sum of Squares	Mean Square	*F* Value	Pr > *F*
Model	4	294.28855	73.57214	114.63	<0.0001
Error	30	19.25430	0.64181		
Corrected total	34	313.54286			

Root MSE	0.80113	*R*-Square	0.9386
Dependent mean	16.68571	Adj *R*-Sq	0.9304
Coeff Var	4.80130		

Parameter Estimates

Variable	DF	Parameter Estimate	Standard Error	*t* Value	Pr > |*t*|
Intercept	1	6.71824	0.56817	11.82	<0.0001
cold	1	0.06148	0.01258	4.89	<0.0001
foll	1	0.04970	0.03960	1.25	0.2192
clt	1	0.06464	0.02097	3.08	0.0044
exp	1	0.07357	0.03518	2.09	0.0451

Example 13.10

It is instructive to compute $R_{Y,(X_1,X_3|X_2,X_4)}$ from the Example 12.1 dataset in a slightly different manner. Specifically, because both X_1 and X_3 are corrected for the linear effect of X_2 and X_4, let us first regress X_1 and X_3 on X_2 and X_4 and find the resulting two sets of residuals {*el_cold*} and {*e2_clt*}. Then we can regress *units* (*Y*) on *el_cold* and *e2_clt* to find the coefficient of multiple determination $R^2_{Y|e1\_cold,e2\_clt} = R^2_{Y,(X_1,X_3|X_2,X_4)}$.

The SAS code is shown in Exhibit 13.12.

Exhibit 13.12 SAS Code for Multiple Semipartial Correlation (Residual Approach)

```
proc reg data = ABCsold;
      model cold = foll   exp;
            output out = e1 residual = e1_cold;
run;
      model clt = foll exp;
            output out = e2 residual = e2_clt;
run;
data data1;
      set e1;
data data2;
      set e2;
data data3;
      merge data1 data2 ABCsold;
proc reg data = data 3;
      model units = e1_cold    e2_clt;
run;
```

Table 13.12 houses the resulting output.

Clearly, $r^2_{Y|e1\_cold,\,e2\_clt} = 0.1551$ and thus $r_{Y|e1\_cold,\,e2\_clt} = 0.3938 = r_{Y,(X_1,X_3|X_2,X_4)}$.

A couple of additional thoughts on semipartial correlation are now offered. First, by a discussion similar to that given in support of Equation 13.8, it can be shown that the sample coefficient of multiple semipartial correlation between Y and X_j after controlling for the effects of the regressors $X_1,\ldots,X_{j-1}, X_{j+1},\ldots,X_m$ on X_j alone is

$$R_{Y,\,(X_j|X_1,\ldots,X_{j-1},X_{j+1},\ldots,X_m)} = \frac{\hat{\beta}_j S_j \sqrt{1-R^2_{X_j\hat{X}_j}}}{S_Y}. \tag{13.15}$$

Next, it can be demonstrated that, for variables X_1, Y, and a control variable X_2, a semipartial correlation coefficient can be written in terms of zero-order correlation coefficients as

$$\text{(a) } R_{Y,\,(X_1|X_2)} = \frac{R_{YX_1} - R_{YX_2}R_{X_1X_2}}{\sqrt{1-R^2_{X_1X_2}}}; \text{ and}$$

$$\text{(b) } R_{(Y|X_2),X_1} = \frac{R_{YX_1} - R_{YX_2}R_{X_1X_2}}{\sqrt{1-R^2_{YX_2}}}. \tag{13.16}$$

TABLE 13.12.
Output for Example 13.10

The SAS System
The REG Procedure
Model: MODEL1
Dependent Variable: Cold

Analysis of Variance

Source	DF	Sum of Squares	Mean Square	*F* Value	Pr > *F*
Model	2	14471	7235.26438	38.20	<0.0001
Error	32	6061.64268	189.42633		
Corrected total	34	20532			

Root MSE	13.76322	*R*-Square	0.7048
Dependent mean	86.22857	Adj *R*-Sq	0.6863
Coeff Var	15.96133		

Parameter Estimates

Variable	DF	Parameter Estimate	Standard Error	*t* Value	Pr > |*t*|
Intercept	1	14.78681	8.67059	1.71	0.0978
foll	1	1.40839	0.58682	2.40	0.0224
exp	1	2.30325	0.43685	5.27	<0.0001

The REG Procedure
Model: MODEL1
Dependent Variable: clt

Analysis of Variance

Source	DF	Sum of Squares	Mean Square	*F* Value	Pr > *F*
Model	2	4528.24518	2264.12259	33.22	<0.0001
Error	32	2180.89768	68.15305		
Corrected total	34	6709.14286			

Root MSE	8.25549	*R*-Square	0.6749
Dependent mean	35.711429	Adj *R*-Sq	0.6546
Coeff Var	23.11536		

Parameter Estimates

Variable	DF	Parameter Estimate	Standard Error	*t* Value	Pr > |*t*|
Intercept	1	−5.01084	5.20081	−0.96	0.3425
foll	1	1.14602	0.35199	3.26	0.0027
exp	1	1.05167	0.26203	4.01	0.0003

The REG Procedure
Model: MODEL1
Dependent Variable: Units

Analysis of Variance

Source	DF	Sum of Squares	Mean Square	*F* Value	Pr > *F*
Model	2	48.63871	24.31935	2.94	0.0674
Error	32	264.90415	8.27825		
Corrected total	34	313.54286			

TABLE 13.12 *(continued)*

Root MSE	2.87720	*R*-Square	0.1551
Dependent mean	16.68571	Adj *R*-Sq	0.1023
Coeff Var	17.24347		

Parameter Estimates

Variable	Label	DF	Parameter Estimate	Standard Error	*t* Value	Pr > |*t*|
Intercept	Intercept	1	16.68571	0.48633	34.31	<0.0001
e1_cold	Residual	1	0.06148	0.04517	1.36	0.1830
e2_clt	Residual	1	0.06464	0.07530	0.86	0.3971

Appendix A: The Multivariate Normal Probability Distribution

The Bivariate Normal Case

In this section, we seek to generalize the univariate normal distribution involving the random variable X with probability density function

$$f(x) = \frac{1}{\sigma\sqrt{2\pi}} e^{-\frac{1}{2}\left(\frac{x-\mu}{\sigma}\right)^2}, -\infty < x < +\infty,$$

to the bivariate case involving the random variables X and Y. Specifically, let the two-dimensional random variable (X, Y) have the joint probability density function

$$f(x,y) = \frac{1}{2\pi\sigma_X\sigma_Y(1-\rho^2)^{1/2}} e^{-\frac{1}{2}Q}, \quad -\infty < x, y < +\infty, \tag{13.A.1}$$

where

$$Q = \frac{1}{1-\rho^2}\left[\left(\frac{x-\mu_X}{\sigma_X}\right)^2 - 2\rho\left(\frac{x-\mu_X}{\sigma_X}\right)\left(\frac{y-\mu_Y}{\sigma_Y}\right) + \left(\frac{y-\mu_Y}{\sigma_Y}\right)^2\right], \tag{13.A.2}$$

and μ_x, μ_y, σ_x, σ_y and ρ are all parameters with, $-\infty < \mu_x, \mu_y < +\infty$, $\sigma_x > 0$, $\sigma_y > 0$, and $-1 < \rho < 1$ ($f(x, y)$ is undefined for $\rho = \pm 1$). Here $\mu_X = E(X)$, $\mu_Y = E(Y)$, $\sigma_X^2 = V(X)$, $\sigma_Y^2 = V(Y)$, and ρ is the coefficient of (linear) correlation between X and Y, i.e., it can be demonstrated that

$$\begin{aligned}
\rho_{XY} &= \frac{COV(X,Y)}{\sigma_X\sigma_Y} \\
&= E\left[\left(\frac{X-\mu_X}{\sigma_X}\right)\left(\frac{Y-\mu_Y}{\sigma_Y}\right)\right] \\
&= \int_{-\infty}^{+\infty}\int_{-\infty}^{+\infty}\left(\frac{x-\mu_X}{\sigma_X}\right)\left(\frac{y-\mu_Y}{\sigma_Y}\right) f(x,y)\, dy dx = \rho.
\end{aligned}$$

If Equations 13.A.1 and 13.A.2 hold, then X and Y are said to follow a *bivariate normal distribution* with joint probability density function in Equation 13.A.1. In this regard, any random variable (X, Y) having a probability density function given by Equation 13.A.1 is said to be bivariate $N(\mu_X, \mu_Y, \sigma_X, \sigma_Y, \rho)$.

Looking to the properties of Equation 13.A.1, $f(x, y)$ depicts a bell-shaped surface that is centered around the point (μ_X, μ_Y) in the x, y-plane (so that any plane perpendicular to the x, y-plane will intersect the surface in a curve that has the univariate normal form) with

a. $f(x, y) > 0$.
b. the probability that a point (X, Y) will lie within a region $\mathcal{A}$ of the x, y-plane is

$$P[(X,Y) \in \mathcal{A}] = \iint_{\mathcal{A}} f(x,y)\, dy\, dx.$$

and

c. as required of any legitimate joint probability density function,

$$\int_{-\infty}^{+\infty}\int_{-\infty}^{+\infty} f(x,y)\, dy\, dx = 1.$$

Note that if in Equation 13.A.1 we set

$$u = \frac{x-\mu_X}{\sigma_X} \text{ and } v = \frac{y-\mu_Y}{\sigma_Y}, \tag{13.A.3}$$

then $f(x, y)$ can be expressed in terms of a single parameter ρ.

Suppose that the random variable (X, Y) is bivariate $N(\mu_X, \mu_Y, \sigma_X, \sigma_Y, \rho)$. Then it can be shown that the marginal distribution of X is $N(\mu_X, \sigma_X)$, and the marginal distribution of Y is $N(\mu_Y, \sigma_Y)$, i.e., the individual marginal distributions of X and Y are univariate normal. Here the *marginal distribution of* X is obtained by integrating out y in Equation 13.A.1 or

$$g(x) = \int_{-\infty}^{+\infty} f(x,y)\, dy = \frac{1}{\sqrt{2\pi}\sigma_X} e^{-\frac{1}{2}\left(\frac{x-\mu_X}{\sigma_X}\right)^2}, -\infty < x < +\infty; \tag{13.A.4}$$

and the *marginal distribution of* Y is obtained by integrating out x in Equation 13.A.1 or

$$h(y) = \int_{-\infty}^{+\infty} f(x,y)\, dx = \frac{1}{\sqrt{2\pi}\sigma_Y} e^{-\frac{1}{2}\left(\frac{y-\mu_Y}{\sigma_Y}\right)^2}, -\infty < y < +\infty. \tag{13.A.5}$$

(It is important to note that, if the marginal distributions of the random variables X and Y are each univariate normal, this does not imply that (X, Y) is bivariate normal.)

We noted previously (Chapter 2) that, if the random variables X and Y are independent, then they must also be uncorrelated or $\rho_{XY} = 0$. However, the converse of this statement does not generally follow. Interestingly enough, it does apply if X and Y follow a bivariate normal distribution, as Theorem 13.A.1 attests:

Theorem 13.A.1: Let the random variables X and Y follow a bivariate normal distribution with probability density function given by Equation 13.A.1. Then X and Y are independent random variables if and only if $\rho_{XY} = \rho = 0$.

Hence, by virtue of this theorem, $\rho = 0$ is a necessary and sufficient condition for independence under the assumption of bivariate normality. Given that $\rho = 0$, Equation 13.A.1 factors as $f(x, y) = g(x)h(y)$, where $g(x)$ and $h(y)$ are the univariate normal marginal distributions appearing in Equations 13.A.4 and 13.A.5, respectively.

Next, if the random variable (X, Y) is bivariate $N(\mu_X, \mu_Y, \sigma_X, \sigma_Y, \rho)$, then the conditional distribution of X given $Y= y$ is univariate $N(\mu_X + \rho\frac{\sigma_X}{\sigma_Y}(y-\mu_Y), \sigma_X\sqrt{1-\rho^2})$. Here the conditional mean of X given $Y= y$ is $E(X\,|\,Y=y) = \mu_X + \rho\left(\frac{\sigma_X}{\sigma_Y}\right)(y-\mu_Y)$ and the conditional variance of X given $Y= y$ is $V(X\,|\,Y=y) = \sigma_X^2(1-\rho^2)$. (Note that $V(X|Y = y) \leq V(X)$. Also, as $\rho \to \pm 1$, the conditional variance $\sigma_X^2(1-\rho^2) \to 0$ while if $\rho = 0$, then $E(X|Y = y) = \mu_x$ and $V(X\,|\,Y=y) = \sigma_X^2$.) The conditional distribution of X given $Y = y$ is obtained from the joint and marginal distributions in Equations 13.A.1 and 13.A.5, respectively, for $h(y) \neq 0$, as

$$\begin{aligned} g(x\,|\,y) &= \frac{f(x,y)}{h(y)} = \frac{1}{\sqrt{2\pi}\sqrt{V(X\,|\,Y=y)}} e^{-\frac{(x-E(X/Y=y))^2}{2V(X/Y=y)}} \\ &= \frac{1}{\sqrt{2\pi}\sigma_X\sqrt{1-\rho^2}} e^{-\frac{(x-\mu_X-\rho(\sigma_X/\sigma_Y)(y-\mu_Y))^2}{2\sigma_X^2(1-\rho^2)}}, \quad -\infty < x < +\infty. \end{aligned} \tag{13.A.6}$$

Similarly, the conditional distribution of Y given $X = x$ is univariate $N\left(\mu_Y + \rho(\frac{\sigma_Y}{\sigma_X})(x-\mu_X), \sigma_Y\sqrt{1-\rho^2}\right)$. Thus the conditional mean of Y given $X = x$ is $E(Y\,|\,X=x) = \mu_Y + \rho\left(\frac{\sigma_Y}{\sigma_X}\right)(x-\mu_X)$ and the conditional variance of Y given $X = x$ is $V(Y\,|\,X=x) = \sigma_Y^2(1-\rho^2)$. Hence, the conditional distribution of Y given $X = x$ is obtained from the joint and marginal distributions in Equations 13.A.1 and 13.A.4, respectively, given that $g(x) \neq 0$, as

$$\begin{aligned} h(y\,|\,x) &= \frac{f(x,y)}{g(x)} = \frac{1}{\sqrt{2\pi}\sqrt{V(Y\,|\,X=x)}} e^{-\frac{(y-E(Y|X=x))^2}{2V(Y|X=x)}} \\ &= \frac{1}{\sqrt{2\pi}\sigma_Y\sqrt{1-\rho^2}} e^{-\frac{(y-\mu_Y-\rho(\sigma_Y/\sigma_X)(x-\mu_X))^2}{2\sigma_Y^2(1-\rho^2)}}, \quad -\infty < y < +\infty. \end{aligned} \tag{13.A.7}$$

It is well known that the mean of a random variable in a conditional distribution is termed a regression curve when expressed as a function of the fixed variable in the said distribution. For instance, the regression of Y on $X = x$ is the mean of Y in the conditional density of Y given $X = x$ and is written as a function $y = j(x)$, where

$$j(x) = E(Y\,|\,X=x) = \int_{-\infty}^{+\infty} yh(y\,|\,x)\,dy = \int_{-\infty}^{+\infty} y\frac{f(x,y)}{g(x)}\,dy. \tag{13.A.8}$$

The expression $y = j(x)$, the locus of the means of the conditional distribution $E(Y|X = x)$ when plotted in the x, y-plane, gives the regression curve of y on x for Y given $X = x$. That is to say, it is a curve that specifies the location of the means of Y for various values of X in the conditional distribution of Y given $X = x$. (The regression curve of x on y is defined in a similar manner.)

Although the regression function defined in Equation 13.A.8 need not be a straight line, it will be linear if the random variable (X, Y) is bivariate $N(\mu_X, \mu_Y, \sigma_X, \sigma_Y, \rho)$, i.e., the regression curve of y on x for X and Y jointly normally distributed has the linear form $y = j(x)$ or

$$y = E(Y \mid X=x) = \mu_Y + \rho\frac{\sigma_Y}{\sigma_X}(x-\mu_X)$$
$$= \left(\mu_Y - \rho\left(\frac{\sigma_Y}{\sigma_X}\right)\mu_X\right) + \rho\left(\frac{\sigma_Y}{\sigma_X}\right)x = \beta_0 + \beta_1 x, \tag{13.A.9}$$

where $\beta_0 = \mu_Y - \rho\left(\frac{\sigma_Y}{\sigma_X}\right)\mu_X$ is the vertical intercept and $\beta_1 = \rho\left(\frac{\sigma_Y}{\sigma_X}\right)$ is the slope of the linear regression curve of y on x. So if the random variable (X, Y) follows a bivariate $N(\mu_X, \mu_Y, \sigma_X, \sigma_Y, \rho)$ distribution, then a linear regression model (such as Equation 13.A.9) is the correct specification to be used when studying the behavior of Y conditioned on X.

The Multivariate Normal Case

Let the random variables $X_1,\ldots,X_n$ constitute the components of the $(n \times 1)$ vector $\boldsymbol{X}$ and let the means of $X_1,\ldots,X_n$, denoted $\mu_1,\ldots,\mu_n$ respectively, represent the components of the $(n \times 1)$ mean vector $\boldsymbol{\mu}$. Additionally, let $\boldsymbol{\Sigma}$ serve as the nth order variance-covariance matrix of $X_1,\ldots,X_n$, where

$$\boldsymbol{\Sigma} = \begin{bmatrix} \sigma_1^2 & \sigma_{12} & \cdots & \sigma_{1n} \\ \sigma_{21} & \sigma_2^2 & \cdots & \sigma_{2n} \\ \cdot & \cdot & & \cdot \\ \cdot & \cdot & & \cdot \\ \cdot & \cdot & & \cdot \\ \sigma_{n1} & \sigma_{n2} & \cdots & \sigma_n^2 \end{bmatrix} = \begin{bmatrix} \sigma_1^2 & \rho_{12}\sigma_1\sigma_2 & \cdots & \rho_{1n}\sigma_1\sigma_n \\ \rho_{21}\sigma_2\sigma_1 & \sigma_2^2 & \cdots & \rho_{2n}\sigma_2\sigma_n \\ \cdot & \cdot & & \cdot \\ \cdot & \cdot & & \cdot \\ \cdot & \cdot & & \cdot \\ \rho_{n1}\sigma_n\sigma_1 & \rho_{n2}\sigma_n\sigma_2 & \cdots & \sigma_n^2 \end{bmatrix} \tag{13.A.10}$$

is positive definite symmetric and $\rho_{kl} = \sigma_{kl}/\sigma_k\sigma_l$, $k, l = 1,\ldots,n$.

Then the joint probability density function for the *multivariate normal distribution* appears as

$$f(\boldsymbol{X};\boldsymbol{\mu},\boldsymbol{\Sigma}) = N(\boldsymbol{\mu},\boldsymbol{\Sigma})$$
$$= (2\pi)^{-n/2}\left|\boldsymbol{\Sigma}\right|^{-\frac{1}{2}} e^{-\frac{1}{2}(\boldsymbol{X}-\boldsymbol{\mu})'\boldsymbol{\Sigma}^{-1}(\boldsymbol{X}-\boldsymbol{\mu})}, -\infty < \mu_i < +\infty,\ i = 1,\ldots,n. \tag{13.A.11}$$

Given Equation 13.A.11, it can be demonstrated that

1. The marginal distribution of each X_i is $N(\mu_i, \sigma_i)$, $i = 1,\ldots,n$.
2. The conditional distribution of X_k given fixed values of the other variables is

$$N(E(X_k \mid X_1,\ldots,X_{k-1},X_{k+1},\ldots,X_n),\ \sigma^2_{X_k|X_1,\ldots,X_{k-1},X_{k+1},\ldots,X_n}),$$

where the conditional mean $E(X_k|X_1,\ldots,X_{k-1}, X_{k+1},\ldots,X_n)$ is a linear regression function and the conditional variance $\sigma^2_{X_k|X_1,\ldots,X_{k-1},X_{k+1},\ldots,X_n}$ is constant.

Example 13.A.1

Suppose $X_1,\ldots,X_4$ are jointly normally distributed random variables with joint probability density function given by Equation 13.A.11. Then we can generalize the discussion underlying Equations 13.A.6, and 13.A.7 and write the regression model $E(X_1 \mid X_2, X_3, X_4) = \beta_0 + \sum_{j=2}^{4} \beta_j X_j$ in terms of partial correlation coefficients and conditional variances. This can be accomplished by expressing the conditional distribution of X_1 given fixed values of X_2, X_3, and X_4 as $N(E(X_1|X_2,X_3,X_4),\sigma^2_{X_1|X_2,X_3,X_4})$, where

$$\begin{aligned}
E(X_1|X_2,X_3,X_4) &= \mu_1 + \rho_{X_1X_2|X_3X_4}\left(\frac{\sigma_{X_1|X_3X_4}}{\sigma_{X_2|X_3X_4}}\right)(X_2-\mu_2) \\
&\quad + \rho_{X_1X_3|X_2X_4}\left(\frac{\sigma_{X_1|X_2X_4}}{\sigma_{X_3|X_2X_4}}\right)(X_3-\mu_3) + \rho_{X_1X_4|X_2X_3}\left(\frac{\sigma_{X_1|X_2X_3}}{\sigma_{X_4|X_2X_3}}\right)(X_4-\mu_4) \\
&= \beta_{1|234} + \rho_{X_1X_2|X_3X_4}\left(\frac{\sigma_{X_1|X_3X_4}}{\sigma_{X_2|X_3X_4}}\right)X_2 \\
&\quad + \rho_{X_1X_3|X_2X_4}\left(\frac{\sigma_{X_1|X_2X_4}}{\sigma_{X_3|X_2X_4}}\right)X_3 + \rho_{X_1X_4|X_2X_3}\left(\frac{\sigma_{X_1|X_2X_3}}{\sigma_{X_4|X_2X_3}}\right)X_4 \\
&= \beta_{1|234} + \beta_{12|34}X_2 + \beta_{13|24}X_3 + \beta_{14|23}X_4,
\end{aligned}$$

where : $\beta_{1|234} = \mu_1 - \beta_{12|34}\mu_2 - \beta_{13|24}\mu_3 - \beta_{14|23}\mu_4; \beta_{12|34}$ is the (population) regression coefficient attached to X_2 when X_1 is regressed on X_2, X_3, and X_4, etc. and $\sigma^2_{X_1|X_2X_3X_4} = \sigma^2_{X_1|X_3X_4}(1-\rho^2_{X_1X_2|X_3X_4})$.

3. If R represents an nth-order correlation matrix with components ρ_{kl}, k, l = 1,…,n, then Equation 13.A.11 can be written as

$$f(\boldsymbol{U}) = (2\pi)^{-n/2}(\sigma_1\sigma_2\ldots\sigma_n)^{-1}|\boldsymbol{R}|^{-\frac{1}{2}} e^{-\frac{1}{2}\boldsymbol{U}'\boldsymbol{R}^{-1}\boldsymbol{U}}, \tag{13.A.12}$$

where $u_i = (X_i - \mu_i)/\sigma_i$, $i = 1,\ldots,n$. Furthermore, if $\rho_{kl} = 0$, $k \neq l$ (the jointly distributed normal variables are uncorrelated), then $\boldsymbol{R} = \boldsymbol{I}_n$ and

$$f(\boldsymbol{U}) = (2\pi)^{-n/2}(\sigma_1\sigma_2\ldots\sigma_n)^{-1} e^{-\frac{1}{2}\boldsymbol{U}'\boldsymbol{U}}. \tag{13.A.12.1}$$

If it is also true that $\mu_i = 0$ and $\sigma_i = \sigma$ for all i, then $u_i = X_i/\sigma$, $i = 1,\ldots,n$, and thus Equation 13.A.12.1 appears as

$$f(\boldsymbol{U}) = (2\pi)^{-n/2}\sigma^{-n} e^{-\frac{1}{2}\boldsymbol{U}'\boldsymbol{U}}. \tag{13.A.12.2}$$

If $\sigma = 1$, then $\boldsymbol{U}$ follows a *multivariate standard normal distribution* and thus

$$f(\boldsymbol{U}) = (2\pi)^{-n/2} e^{-\frac{1}{2}\boldsymbol{U}'\boldsymbol{U}}. \tag{13.A.12.3}$$

4. Given Equation 13.A.11, if the random variables $X_1,\ldots,X_n$ are mutually independent, then

$$f(\mathbf{X};\boldsymbol{\mu},\boldsymbol{\Sigma}) = \prod_{i=1}^{n} f(X_i;\mu_i,\sigma_i)$$
$$= \prod_{i=1}^{n} (2\pi)^{-\frac{1}{2}} \sigma_i^{-\frac{1}{2}} e^{-\frac{1}{2}\left(\frac{X_i-\mu_i}{\sigma_i}\right)^2}, \tag{13.A.13}$$

where $f(X_i; \mu_i, \sigma_i)$ is univariate normal. However, Equation 13.A.13 also obtains if $\boldsymbol{\Sigma}$ is a diagonal matrix ($\boldsymbol{\Sigma}$ is diagonal if the normally distributed random variables X_i, $i = 1,\ldots,n$, are uncorrelated or have zero covariance). In this regard, under Equation 13.A.11, the variables $X_1,\ldots,X_n$ are independent if and only if $\boldsymbol{\Sigma}$ is a diagonal matrix. Thus, zero covariance is equivalent to independence under multivariate normality.

Exercises

13-1. Given the dataset provided by Table E.12-1, determine the matrix of zero-order correlations and discuss the statistical significance of each. What are the null and alternative hypotheses?

13-2. Given the Table E.12-2 dataset, generate the collection of simple Pearson product-moment correlations and determine whether any of them are statistically significant.

13-3. Using the Table E.12-3 dataset, find the multiple correlation coefficient $R_{Y|X_1,X_2,X_3}$ by finding the zero-order correlation coefficient between Y and $\hat{Y}$. (Hint: use PROC REG to determine the $\hat{Y}$ values and then use PROC CORR to find $R_{Y\hat{Y}}$.)

13-4. Using the Table E.12-2 dataset, find $R_{Y|X_1,\ldots,X_4}$ by determining $R_{Y\hat{Y}}$.

13-5. Given the Table E.12-3 dataset: (a) find the coefficients of partial correlation $R_{YX_1|X_2}$, $R_{YX_3|X_2}$, $R_{YX_2|X_1}$, and $R_{YX_2|X_3}$ using PROC CORR; and (b) find $R_{YX_3|X_1}$ and $R_{YX_3|X_2}$ using PROC REG. (Hint: let Example 13.2 serve as your guide.) For each set of calculations, fully explain your results.

13-6. Given the dataset provided by Table E.12-2: (a) find $R_{YX_2|X_1}$, $R_{YX_2|X_3}$, and $R_{YX_2|X_4}$ using PROC CORR and fully explain your results; and (b) find $R_{YX_4|X_1}$, $R_{YX_4|X_2}$, and $R_{YX_4|X_3}$ via PROC REG and again fully explain your findings. Discuss the statistical significance of your findings.

13-7. For the Table E.12-3 dataset: (a) find the second-order partial correlations $R_{YX_2|X_1,X_3}$ and $R_{YX_3|X_1,X_2}$ using PROC CORR; and (b) find $R_{YX_1|X_2,X_3}$ using PROC REG and discuss the statistical significance of your results. (Hint: consider the programs introduced in Example 13.3.) For each set of calculations, fully explain your findings.

13-8. Given the Table E.12-2 set of observations: (a) find $R_{YX_1|X_2,X_4}$ via PROC CORR; and (b) determine $R_{YX_2|X_3,X_4}$ using PROC REG. Discuss the statistical significance of your results. (Hint: let Example 13.3 guide your efforts.)

13-9. Using the Table E.12-3 dataset, find the multiple partial correlation coefficient $R_{Y(X_1,X_2)|X_3}$. Is this correlation coefficient statistically significant? (Hint: specify groups A={X$_3$} and B={X$_1$, X$_2$} and then perform a group added-last regression experiment.)

13-10. Given the Table E.12-2 dataset, find $R_{Y(X_1,X_3)|X_2,X_4}$ and $R_{Y(X_2,X_4)|X_1,X_3}$. Are these correlation coefficients statistically significant? (Hint: let A={X_2, X_4} and B={X_1, X_3} and then perform a group added-last regression run.)

13-11. Using the Table E.12-3 data series, find the multiple semipartial correlation coefficient $R_{(Y|X_1,X_3),X_2} = R_{eX_2}$. Is it statistically significant? (Hint: see Examples 13.7, 13.8.)

13-12. For the Table E.12-2 dataset, determine $R_{Y,(X_2|X_1,X_3,X_4)} = R_{Ye'}$. Is it statistically significant? (Hint: follow the lead of Examples 13.7, 13.8.)

13-13. Given the Table E.12-3 dataset, find $R_{Y,(X_2,X_3|X_1)} = R_{Y,(A|B)}$ where A={ X_2, X_3} and B={X_1}. Determine whether it is statistically significant. (Hint: see Example 13.9.)

13-14. Using the Table E.12-2 dataset, find $R_{Y,(X_1,X_3|X_2,X_4)} = R_{Y,(A|B)}$, where A = {$X_1$, X_3} and B = {X_2, X_4}. Is it statistically significant? (Hint: see Example 13.9.)

13-15. Find $R_{Y,(X_2,X_4|X_1,X_3)}$ using the approach outlined in Example 13.10.

14

Ridge Regression

A Method for Dealing with Multicollinearity [Hoerl 1962; Hoerl and Kennard 1970a,b; Mayer and Willke 1973; Lowerre 1974; Theobald 1974; Hoerl, Kennard, and Baldwin 1975; Marquardt and Snee 1975; Hocking 1976; Obenchain 1977]

Ridge regression is an alternative to the technique of OLS that can be effectively used when the regressors are highly correlated, as in the case of near multicollinearity (the columns of $\mathbf{X}$ depart significantly from orthogonality and $\mathbf{X'X}$ is near singular). We noted in Chapter 12 that, in the presence of near-linear dependencies among the columns of $\mathbf{X}$, the OLS parameters tend to be too large (in absolute value) and unstable (the parameters can change appreciably from sample to sample), and the standard errors of the estimated regression coefficients are artificially inflated (as evidenced by the high values assumed by the VIFs associated with these coefficients).

Remember that the OLS estimators for the regression parameters are, via the Gauss-Markov Theorem, BLUE. So of the class of all unbiased linear estimators of the regression parameters, the least-squares estimators have *minimum variance*. However, this said minimum may itself be quite large. To circumvent this potential problem, we may want to consider an alternative estimator for $\boldsymbol{\beta}$. In this regard, we might look to the class of *biased estimators*. Perhaps among the elements of this class, we could find a linear estimator that has a variance that is smaller than the OLS estimator. This is what ridge regression attempts to do. It provides us with an answer to the question: "Can we find a linear estimator that has a small bias and a small variance relative to one that is unbiased but exhibits a large variance?" Hence, a ridge estimator trades off unbiasedness for increased stability and precision of the regression parameter estimates. In the next section, we examine the said trade-off graphically as well as analytically (using the mean square error concept).

The Mean Square Error Criterion

First, suppose for $m = 1$, we let $\hat{\beta}_1$ represent the OLS estimator for the regression slope $\boldsymbol{\beta}_1$, whereas $\tilde{\beta}_1$ is a slightly biased estimator for the same with a variance that is small relative to the variance of $\hat{\beta}_1$. The sampling distributions of these estimators appear in Figure 14.1. Clearly, $\tilde{\beta}_1$ has a small bias and is much more precise relative to the unbiased OLS estimator $\hat{\beta}_1$. Hence, $\tilde{\beta}_1$ is the preferred estimator because it has a higher probability of being close

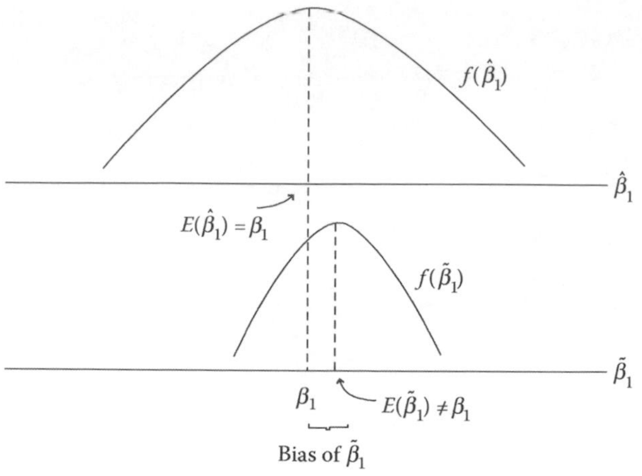

FIGURE 14.1
Biased versus unbiased estimators.

to $\boldsymbol{\beta}_1$ than does $\hat{\beta}_1$. The estimates obtained from the sampling distribution of $\tilde{\beta}_1$ are more stable relative to those obtained from the sampling distribution of $\hat{\beta}_1$.

All of the foregoing discussion can be incorporated into the notion of the MSE of $\tilde{\beta}_1$. Specifically, $MSE(\tilde{\beta}_1)$ is the expected squared distance between $\tilde{\beta}_1$ and $\boldsymbol{\beta}_1$ or

$$\begin{aligned} MSE(\tilde{\beta}_1) &= E\left[(\tilde{\beta}_1 - \beta_1)^2\right] = V(\tilde{\beta}_1) + \left[E(\tilde{\beta}_1) - \beta_1\right]^2 \\ &= V(\tilde{\beta}_1) + \left[bias(\tilde{\beta}_1)\right]^2 \end{aligned} \tag{14.1}$$

(see Appendix 14.A for a derivation of this result). Here $V(\tilde{\beta}_1)$ measures the precision of $\tilde{\beta}_1$, whereas the term $\left[bias(\tilde{\beta}_1)\right]^2$ measures its accuracy. So if we are willing to accept a small amount of bias, then perhaps we can find a biased estimator $\tilde{\beta}_1$ such that $V(\tilde{\beta}_1)$ is small enough to render $MSE(\tilde{\beta}_1) < V(\hat{\beta}_1) = MSE(\hat{\beta}_1)$ (because $\hat{\beta}_1$ is unbiased).

This then is how *ridge regression* or ridge estimation works. It addresses the multicollinearity issue by modifying the method of OLS to allow for the biased estimation of the regression parameters. Although our OLS estimator for the parameter vector $\boldsymbol{\beta}$ was given as $\hat{\boldsymbol{\beta}} = (\boldsymbol{X}'\boldsymbol{X})^{-1}\boldsymbol{X}'\boldsymbol{Y}$, its adjustment under ridge regression appears as

$$\boldsymbol{\beta}_R = (\boldsymbol{X}'\boldsymbol{X} + k\boldsymbol{I}_{m+1})^{-1}\boldsymbol{X}'\boldsymbol{Y}, \tag{14.2}$$

where $k(\geq 0)$ is termed the *biasing parameter* (discussion of the properties of $\boldsymbol{\beta}_R$ appears in Appendix 14.B). Note that, if $k = 0$, then we have $\boldsymbol{\beta}_R = \hat{\boldsymbol{\beta}}$, the OLS estimator for $\boldsymbol{\beta}$. Clearly, $\boldsymbol{\beta}_R$ is obtained by biasing $\hat{\boldsymbol{\beta}}$, with the bias introduced by uniformly increasing the diagonal elements of $\boldsymbol{X}'\boldsymbol{X}$ by an amount k while leaving the off-diagonal elements invariant, i.e., $\boldsymbol{X}'\boldsymbol{X}$ is replaced by $\boldsymbol{X}'\boldsymbol{X} + k\boldsymbol{I}_{m+1}$, $k \geq 0$.

If $\boldsymbol{\beta}_R$ is viewed as a random variable, then, in a sense, it purports to render a more accurate estimate of $\boldsymbol{\beta}$ than $\hat{\boldsymbol{\beta}}$; it is close to $\boldsymbol{\beta}$ in that the distance between $\boldsymbol{\beta}_R$ and $\boldsymbol{\beta}$ or $||\boldsymbol{\beta}_R - \boldsymbol{\beta}||$ tends to be small. This elevated accuracy of $\boldsymbol{\beta}_R$ is reflected by the fact that its MSE is small ($\boldsymbol{\beta}_R$ has a small variance and a small bias). In this regard, $\boldsymbol{\beta}_R$ is chosen over $\hat{\boldsymbol{\beta}}$

because, ostensibly, the increase in squared bias will not exceed the decrease in variance (Appendix 14.C).

What is the connection between ridge regression and OLS? It is readily observed that ridge regression attempts to improve the accuracy of $\hat{\boldsymbol{\beta}}$ by shrinking or decreasing its magnitude $\|\hat{\boldsymbol{\beta}}\|$. That is, given the OLS system of normal equations $\boldsymbol{X}'\boldsymbol{X}\hat{\boldsymbol{\beta}} = \boldsymbol{X}'\boldsymbol{Y}$, Equation 14.2 can be written in terms of $\hat{\boldsymbol{\beta}}$ as

$$\begin{aligned} \boldsymbol{\beta}_R &= (\boldsymbol{X}'\boldsymbol{X} + k\boldsymbol{I}_{m+1})^{-1}\boldsymbol{X}'\boldsymbol{X}\hat{\boldsymbol{\beta}} \\ &= \boldsymbol{D}\hat{\boldsymbol{\beta}}, \end{aligned} \tag{14.2.1}$$

where $\boldsymbol{D} = (\boldsymbol{X}'\boldsymbol{X} + k\boldsymbol{I}_{m+1})^{-1}\,\boldsymbol{X}'\boldsymbol{X}$ is a transformation matrix that expresses the ridge estimator $\boldsymbol{\beta}_R$ as a linear combination of the OLS estimator $\hat{\boldsymbol{\beta}}$, with $\|\boldsymbol{D}\hat{\boldsymbol{\beta}}\|^2 < \|\hat{\boldsymbol{\beta}}\|^2$. As mentioned above, to gauge the accuracy of $\boldsymbol{\beta}_R$, we need to examine its MSE. In addition, we must determine a nonstochastic k value which yields minimum MSE because $\boldsymbol{D}$ obviously depends on k. Although details on these issues are provided in Appendix 14.C, suffice it to say that Hoerl and Kennard [1970a] demonstrate that there is always a nonstochastic $k \geq 0$ for which $MSE(\boldsymbol{\beta}_R) < V(\hat{\boldsymbol{\beta}})$ for finite $\boldsymbol{\beta}'\boldsymbol{\beta}$.

Additional considerations regarding ridge estimation are as follows:

1. The ridge estimator attains an MSE advantage over the OLS estimator by shrinking the latter toward zero; for $k > 0$, $\boldsymbol{\beta}_R'\boldsymbol{\beta}_R < \hat{\boldsymbol{\beta}}'\hat{\boldsymbol{\beta}}$ and $\boldsymbol{\beta}_R'\,\boldsymbol{\beta}_R \rightarrow \boldsymbol{0}$ as $k \rightarrow 0$. In this regard, the biasing parameter k is alternatively termed a *shrinkage parameter.*
2. A particularly straightforward procedure for determining k (which minimizes the MSE of $\boldsymbol{\beta}_R$) is to set

$$k = (m+1)S_\varepsilon^2 \,/\, \hat{\boldsymbol{\beta}}'\hat{\boldsymbol{\beta}}, \tag{14.3}$$

   where S_ε^2 and $\hat{\boldsymbol{\beta}}$ are obtained from a standard OLS fit to the data (see Appendix 14.F) [Hoerl, Kennard, and Baldwin 1975]. Other methods for choosing k have been explored by the following: Dempster, Schatzoff, and Wermuth 1971; Lindley and Smith 1972; Mallows 1973; Goldstein and Smith 1974; Lawless and Wang 1974; Draper and Van Nostrand 1975; McDonald and Golarnean 1975; Obenchain 1975; Hoerl andKennard 1976; Van Nostrand 1980.
3. As a practical matter, if we can find an arbitrarily small value of k for which the magnitudes of the ridge estimates tend to become stable, then this k value can be chosen as our operational k. This can readily be accomplished by examining the so-called *ridge trace*, a plot of the components of $\boldsymbol{\beta}_R$ against increasing values of k, $0 \leq k \leq 1$.
4. Let us express the residual sum of squares from ridge estimation as

$$SSE_R + k\ \|\boldsymbol{\beta}_R\|^2 = SST - SSR_R$$

   (see Appendix 14.D). As k becomes larger, so does the left-hand side of this expression. However, this means that, with SST fixed in value, SSR_R must concomitantly decrease. Hence, ridge estimation may not provide us with the hyperplane that best fits the dataset; we sacrifice goodness of fit for an increase in the stability of the estimated regression coefficients.

5. From a computational viewpoint, all regressors are typically standardized before performing ridge regression. That is, $X_j \rightarrow Z_j = (X_j - \bar{X}_j)/S_j$, $j = 1,\ldots,m$. Hence, under standardization, $Y_i = \beta_0 + \beta_1 X_{i1} + \cdots + \beta_m X_{im} + \varepsilon_i$ is replaced by $Y_i = \beta_0^* + \beta_1^* Z_{i1} + \cdots + \beta_m^* Z_{im} + \varepsilon_i^*$ or $\boldsymbol{Y} = \boldsymbol{1}\beta_0^* + \boldsymbol{Z}\boldsymbol{\beta}^* + \boldsymbol{\varepsilon}^*$. Then the OLS estimators for β_0^* and $\boldsymbol{\beta}^*$ are, respectively, $\hat{\beta}_0^* = \bar{Y}$ and $\hat{\boldsymbol{\beta}}^* = (\boldsymbol{Z}'\boldsymbol{Z})^{-1}\boldsymbol{Z}'\boldsymbol{Y}$, whereas the ridge estimators for the same are $\hat{\beta}_{0R}^* = \bar{Y}$ and $\hat{\boldsymbol{\beta}}_R^* = (\boldsymbol{Z}'\boldsymbol{Z} + k\boldsymbol{I}_m)^{-1}\boldsymbol{Z}'\boldsymbol{Y}$, where $k = mS_\varepsilon^2/(\hat{\boldsymbol{\beta}}_R^*)'\hat{\boldsymbol{\beta}}_R^*$.

Moreover, standardization of the regressors removes nonessential ill-conditioning, thus reducing the impact of variance inflation of the parameter estimates [Marquardt and Snee 1975]. This is appropriate in the presence of a constant term in the regression equation because, in this circumstance, standardization removes the correlation between the constant term and all remaining (regressor) terms.

Biased Estimation Reduces Near Multicollinearity

We noted in Chapter 12 that near multicollinearity among the regressors means that at least one column of $\boldsymbol{X}$ is approximately a linear combination of some or all of the others, thus resulting in $\boldsymbol{X}'\boldsymbol{X}$ (or $\boldsymbol{Z}'\boldsymbol{Z}$ from item 4 above) being near singular. The upshot of this phenomenon is that the sampling variances of the OLS parameter estimates are inflated. Thus, $V(\hat{\beta}_j)$, $j = 1,\ldots,m$, is VIF_j times larger than would be the case if X_j were not correlated with some of the other regressors. So if the VIFs for the estimated OLS regression coefficients are large, then ridge regression can be used to generate stable regression coefficients (Appendix 14.E).

Ridge regression helps to overcome the problem of near multicollinearity by modifying $\boldsymbol{X}'\boldsymbol{X}$, i.e., $\boldsymbol{X}'\boldsymbol{X}$ is replaced by $\boldsymbol{X}'\boldsymbol{X} + k\boldsymbol{I}_{m+1}$ so as to move $\boldsymbol{X}'\boldsymbol{X}$ away from near singularity and closer to the state of orthogonality among the columns of $\boldsymbol{X}'\boldsymbol{X}$. Once the ridge estimation process commences, VIFs for the ridge regression coefficients are defined in a manner analogous to those for OLS regression coefficients: they measure the magnitudes of the sample variances of the ridge regression estimates relative to what they would be if the regressors were uncorrelated. As k varies and the ridge regression parameter estimates, say, decrease in value and eventually stabilize, then so do the ridge VIFs.

Example 14.1

In Example 12.9, we determined via OLS that the Malinvaud [1970, 19] dataset (named *malvd*) consisting of the variables imports (I), gross domestic product (GDP), stock formation (SF), and consumption (C) was highly collinear (as evidenced by $VIF_1 = 469.74$ and $VIF_3 = 469.37$). Let us now reestimate the regression model $I = \beta_0 + \beta_1 GDP + \beta_2 SF + \beta_3 C + \varepsilon$ using ridge regression. Exhibit 14.1 contains the requisite SAS code.

Exhibit 14.1 SAS Code for Ridge Regression

```
data malvd;
  input I GDP SF C;
datalines;
```

```
15.9   149.3  4.2    108.1
.        .      .      .
.        .      .      .
.        .      .      .
56.6   353.9  4.5    242.9
;
                                ①            ②
proc reg data = malvd outest = out noprint
          ③             ④
      outseb outvif ridge = 0 to 0.02 by 0.002;
      model I =GDP  SF  C;
      plot / ridgeplot ; ⑤
proc print data = out ; ⑥
run;
data out2 ; set out; Ⓐ
      if _ridge_=0.  or _ridge_=0.016; Ⓑ
proc print data = out2 ; Ⓒ
      var _type_  _rmse_  intercept  - - C; Ⓓ
run;
```

① The output dataset *out* will contain the ridge regression results.

② The *noprint* option suppresses the usual regression results (which, if needed, can be found in Table 12.12 of Example 12.9).

③ The *outseb* and *outvif* options request that the estimated regression coefficient standard errors and VIFs be included in the output dataset.

④ The *MODEL* option *ridge = list* generally requests that ridge regression be performed for values of the biasing constant k specified in the *list*. The *list* may consist of specific values, separated by commas, or it may be expressed in the form *ridge = a to b by c*. In this example, we let k go from 0.00 to 0.02 in increments of 0.002. (It is important to note that SAS performs ridge regression by first centering and scaling *all* of the variables in the regression model [for details on performing this process see Section 12.7.] This results in $\boldsymbol{\beta}_R = [\mathbf{R}_{XX}(\mathbf{I}_m + \mathbf{D}_k)]^{-1}\,\mathbf{R}_{XY}$, where $\mathbf{R}_{XX}$ is the (product-moment) correlation matrix of all the regressor variables, $\mathbf{R}_{XY}$ is a column vector of (product- moment) correlations between each regressor and the explanatory variable Y, and $\mathbf{D}_k$ is a diagonal matrix with all k's on the main diagonal.)

⑤ We request the ridge trace graph.

⑥ We request a print of the contents of the output dataset *out*.

The output generated by this SAS code appears in Table 14.1. Each observation (appearing under the leading *obs*) in the *out* dataset is a separate line of output, with the line or observation numbers ranging from 1 to 35. In this regard, observations 1 and 2 contain the OLS root mean square error ($\sqrt{MSE}$), regression parameter estimates, and coefficient standard errors. Observations 3, 4, and 5 contain the ridge estimation results for $k = 0.000$. That is, observation 3 is a print of the estimated regression coefficient VIFs, observation 4 gives $\sqrt{MSE}$ along with the estimated regression coefficient values, and observation 5 exhibits the estimated regression coefficient standard errors.

TABLE 14.1
Output for Example 14.1 ($k = 0$ to 0.02)

The SAS System

Obs	_MODEL_	_TYPE_	_DEPVAR_	(k) _RIDGE_	_PCOMIT_	_RMSE_	Intercept	GDP	SF	C	I
1	MODEL1	PARMS	I	.	.	2.25817	–19.7251	0.032	0.41420	0.243	–1
2	MODEL1	SEB	I	.	.	2.25817	4.1253	0.187	0.32226	0.285	–1
3	MODEL1	RIDGEVIF	I	0.000	.	.	.	469.742	1.04988	469.371	–1
4	MODEL1	RIDGE	I	0.000	.	2.25817	–19.7251	0.032	0.41420	0.243	–1
5	MODEL1	RIDGESEB	I	0.000	.	2.25817	4.1253	0.187	0.32226	0.285	–1
6	MODEL1	RIDGEVIF	I	0.002	.	.	.	56.974	1.04399	56.930	–1
7	MODEL1	RIDGE	I	0.002	.	2.26215	–18.9229	0.073	0.41213	0.179	–1
8	MODEL1	RIDGESEB	I	0.002	.	2.26215	2.5976	0.065	0.32192	0.100	–1
9	MODEL1	RIDGEVIF	I	0.004	.	.	.	21.031	1.03942	21.015	–1
10	MODEL1	RIDGE	I	0.004	.	2.26410	–18.7182	0.082	0.41224	0.166	–1
11	MODEL1	RIDGESEB	I	0.004	.	2.26410	2.4185	0.040	0.32149	0.061	–1
12	MODEL1	RIDGEVIF	I	0.006	.	.	.	10.936	1.03496	10.928	–1
13	MODEL1	RIDGE	I	0.006	.	2.26520	–18.6038	0.086	0.41269	0.160	–1
14	MODEL1	RIDGESEB	I	0.006	.	2.26520	2.3647	0.029	0.32096	0.044	–1
15	MODEL1	RIDGEVIF	I	0.008	.	.	.	6.744	1.03055	6.740	–1
16	MODEL1	RIDGE	I	0.008	.	2.26601	–18.5199	0.088	0.41323	0.157	–1
17	MODEL1	RIDGESEB	I	0.008	.	2.26601	2.3411	0.022	0.32039	0.034	–1
18	MODEL1	RIDGEVIF	I	0.010	.	.	.	4.611	1.02618	4.608	–1
19	MODEL1	RIDGE	I	0.010	.	2.26671	–18.4501	0.089	0.41382	0.155	–1
20	MODEL1	RIDGESEB	I	0.010	.	2.26671	2.3283	0.019	0.31981	0.028	–1
21	MODEL1	RIDGEVIF	I	0.012	.	.	.	3.380	1.02183	3.378	–1
22	MODEL1	RIDGE	I	0.012	.	2.26739	–18.3879	0.090	0.41444	0.153	–1
23	MODEL1	RIDGESEB	I	0.012	.	2.26739	2.3202	0.016	0.31923	0.024	–1
24	MODEL1	RIDGEVIF	I	0.014	.	.	.	2.606	1.01752	2.605	–1

25	MODEL1	RIDGE	I	0.014	.	2.26809	−18.3303	0.090	0.41506	0.152	−1
26	MODEL1	RIDGESEB	I	0.014	.	2.26809	2.3146	0.014	0.31865	0.021	−1
27	MODEL1	RIDGEVIF	I	0.016	.	.	.	2.087	1.01323	2.086	−1
28	MODEL1	RIDGE	I	0.016	.	2.26881	−18.2757	0.091	0.41569	0.151	−1
29	MODEL1	RIDGESEB	I	0.016	.	2.26881	2.3104	0.013	0.31808	0.019	−1
30	MODEL1	RIDGEVIF	I	0.018	.	.	.	1.723	1.00897	1.722	−1
31	MODEL1	RIDGE	I	0.018	.	2.26959	−18.2232	0.091	0.41632	0.150	−1
32	MODEL1	RIDGESEB	I	0.018	.	2.26959	2.3071	0.011	0.31752	0.017	−1
33	MODEL1	RIDGEVIF	I	0.020	.	.	.	1.457	1.00474	1.457	−1
34	MODEL1	RIDGE	I	0.020	.	2.27041	−18.1722	0.091	0.41695	0.149	−1
35	MODEL1	RIDGESEB	I	0.020	.	2.27041	2.3044	0.010	0.31697	0.016	−1

TABLE 14.2

Output for Example 14.1 ($k = 0.016$)

The SAS System

Obs		_TYPE_	_RMSE_	Intercept	(X_1) GDP	(X_2) SF	(X_3) C
OLS	1	PARMS	2.25817	−19.7251	0.03220	0.41420	0.24275
	2	SEB	2.25817	4.1253	0.18688	0.32226	0.28536
RIDGE	3	RIDGEVIF	.	.	2.08710	1.01323	2.08625
	4	RIDGE	2.26881	−18.2757	0.09083	0.41569	0.15086
	5	RIDGESEB	2.26881	2.3104	0.01252	0.31808	0.01911

(Note that, for $k = 0$, these results must be exactly the same as the preceding set of OLS results.) Observations 6, 7, and 8 present the ridge regression results when $k = 0.002$, etc. Notice how the successive VIFs on *GDP* and *C* monotonically decrease, whereas the estimated ridge regression parameter values begin to stabilize for $k > 0.006$.

The second part of the SAS code (appearing immediately after the first *run* statement) developed in Exhibit 14.1 was added once the first portion was run and the output was scrutinized. Because a glance at the ridge trace reveals that the estimated coefficients seem to become quite stable for $k \geq 0.016$, we requested a comparison of the OLS results with only the ridge regression output for $k = 0.016$. (Note that according to (14.3), $k = 4(5.09933) / 389.3113 = 0.0524$.) That is

Ⓐ We create a new output dataset *out* 2 that contains the same output as the *out* dataset.
Ⓑ dataset *out*2 contains the ridge regression results for $k = 0.000$ and $k = 0.016$.
Ⓒ We request a print of the *out*2 dataset.
Ⓓ We list the specific results wanted from *out*2.

Table 14.2 houses the output determined from the second portion of our SAS code. Observations 1 and 2 depict the OLS results, whereas observations 3, 4, and 5 represent the ridge regression output when $k = 0.016$. Note that the sample *t* values (under $H_0: \beta_j = 0, j = 1,2,3$) for the OLS regression coefficient estimates are, respectively $\hat{\beta}_1 / S_{\hat{\beta}_1} = 0.1723$, $\hat{\beta}_2 / S_{\hat{\beta}_2} = 1.2853$, and $\hat{\beta}_3 / S_{\hat{\beta}_3} = 0.8506$. Clearly, none of these estimated regression coefficients is statistically significant. However, under the same set of null hypotheses, the corresponding *t* values for the estimated regression coefficients obtained via ridge regression are, respectively $\hat{\beta}_{R1} / S_{\hat{\beta}_{R1}} = 7.2550$, $\hat{\beta}_{R2} / S_{\hat{\beta}_{R2}} = 1.3069$, and $\hat{\beta}_{R3} / S_{\hat{\beta}_{R3}} = 7.8943$. Hence, the coefficients on *GDP* and *C* are highly significant.

Appendix A: The Mean Square Error of the Estimator $\tilde{\beta}_1$

Suppose $\tilde{\beta}_1$ is a linear biased estimator for β_1 with $E(\tilde{\beta}_1) = \theta$ and $V(\tilde{\beta}_1) = E[(\tilde{\beta}_1 - \theta)^2]$. Then

$$\begin{aligned} MSE(\tilde{\beta}_1) &= E[(\tilde{\beta}_1 - \beta_1)^2] = E\left\{[(\tilde{\beta}_1 - \theta) + (\theta - \beta_1)]^2\right\} \\ &= E[(\tilde{\beta}_1 - \theta)^2 + (\theta - \beta_1)^2 + 2(\tilde{\beta}_1 - \theta)(\theta - \beta_1)] \\ &= E[(\tilde{\beta}_1 - \theta)^2] + E[(\theta - \beta_1)^2] + 2E[(\tilde{\beta}_1 - \theta)(\theta - \beta_1)] \\ &= V(\tilde{\beta}_1) + (\theta - \beta_1)^2 + 2(\theta - \beta_1)E[(\tilde{\beta}_1 - \theta)] \\ &= V(\tilde{\beta}_1) + (E(\tilde{\beta}_1) - \beta_1)^2 + 2(\theta - \beta_1)(0) \\ &= V(\tilde{\beta}_1) + [bias(\tilde{\beta}_1)]^2 \end{aligned} \tag{14.A.1}$$

because $E(\tilde{\beta}_1) - \theta = 0$.

Appendix B: Properties of the Ridge Estimator $\boldsymbol{\beta}_R$

The ridge estimator of the vector of regression parameters $\boldsymbol{\beta}$ is

$$\boldsymbol{\beta}_R = (\boldsymbol{X'X} + k\boldsymbol{I}_{m+1})^{-1}\boldsymbol{X'Y}, \tag{14.B.1}$$

a modification of the OLS estimator for $\boldsymbol{\beta}$, $\hat{\boldsymbol{\beta}} = (\boldsymbol{X'X})^{-1}\boldsymbol{X'Y}$. If we set $\boldsymbol{X'Y} = \boldsymbol{X'X}\hat{\boldsymbol{\beta}}$ (the OLS normal equation system), then we can rewrite Equation 14.B.1 as

$$\boldsymbol{\beta}_R = (\boldsymbol{X'X} + k\boldsymbol{I}_{m+1})^{-1}\boldsymbol{X'X}\hat{\boldsymbol{\beta}} = \boldsymbol{D}\hat{\boldsymbol{\beta}}. \tag{14.B.2}$$

For k nonstochastic, $E(\boldsymbol{\beta}_R) = E(\boldsymbol{D}\hat{\boldsymbol{\beta}}) = \boldsymbol{D}E(\hat{\boldsymbol{\beta}}) = \boldsymbol{D\beta}$, i.e., $\boldsymbol{\beta}_R$ is a biased estimator for $\boldsymbol{\beta}$.

Next, let us express Equation 14.B.1 as

$$\begin{aligned}\boldsymbol{\beta}_R &= (\boldsymbol{X'X} + k\boldsymbol{I}_{m+1})^{-1}\boldsymbol{X'}(\boldsymbol{X}\hat{\boldsymbol{\beta}} + \boldsymbol{\varepsilon})\\ &= \boldsymbol{D\beta} + (\boldsymbol{X'X} + k\boldsymbol{I}_{m+1})^{-1}\boldsymbol{X'\varepsilon}.\end{aligned} \tag{14.B.3}$$

This specification of $\boldsymbol{\beta}_R$ can be used to determine the variance-covariance matrix of $\boldsymbol{\beta}_R$ or

$$\begin{aligned}V(\boldsymbol{\beta}_R) &= E[(\boldsymbol{\beta}_R - E(\boldsymbol{\beta}_R))(\boldsymbol{\beta}_R - E(\boldsymbol{\beta}_R))']\\ &= E[(\boldsymbol{\beta}_R - \boldsymbol{D\beta})(\boldsymbol{\beta}_R - \boldsymbol{D\beta}')]\\ &= E[(\boldsymbol{X'X} + k\boldsymbol{I}_{m+1})^{-1}\boldsymbol{X'\varepsilon\varepsilon'X}(\boldsymbol{X'X} + k\boldsymbol{I}_{m+1})^{-1}]\\ &= \sigma_\varepsilon^2(\boldsymbol{X'X} + k\boldsymbol{I}_{m+1})^{-1}\boldsymbol{X'X}(\boldsymbol{X'X} + k\boldsymbol{I}_{m+1})^{-1}\end{aligned} \tag{14.B.4}$$

(again assuming that k is nonstochastic).

Finally, let us determine the MSE for $\boldsymbol{\beta}_R$. We know that, in general,

$$MSE(\boldsymbol{\beta}_R) = V(\boldsymbol{\beta}_R) + [bias(\boldsymbol{\beta}_R)]^2. \tag{14.B.5}$$

Although $V(\boldsymbol{\beta}_R)$ is given by Equation 14.B.4, we need to find an expression for the square of the bias of $\boldsymbol{\beta}_R$. To this end, let us write

$$\begin{aligned}bias(\boldsymbol{\beta}_R) &= E(\boldsymbol{\beta}_R) - \boldsymbol{\beta} = \boldsymbol{D\beta} - \boldsymbol{\beta}\\ &= (\boldsymbol{X'X} + k\boldsymbol{I}_{m+1})^{-1}\boldsymbol{X'X\beta} - \boldsymbol{\beta}\\ &= [(\boldsymbol{X'X} + k\boldsymbol{I}_{m+1})^{-1}\boldsymbol{X'X} - \boldsymbol{I}_{m+1}]\,\boldsymbol{\beta}.\end{aligned} \tag{14.B.6}$$

Now, because

$$(\boldsymbol{X'X} + k\boldsymbol{I}_{m+1})^{-1}(\boldsymbol{X'X} + k\boldsymbol{I}_{m+1}) = \boldsymbol{I}_{m+1},$$

it follows that

$$(\boldsymbol{X'X} + k\boldsymbol{I}_{m+1})^{-1}\boldsymbol{X'X} + (\boldsymbol{X'X} + k\boldsymbol{I}_{m+1})^{-1}k\boldsymbol{I}_{m+1} = \boldsymbol{I}_{m+1}$$

and thus

$$(\boldsymbol{X'X} + k\boldsymbol{I}_{m+1})^{-1}\boldsymbol{X'X} - \boldsymbol{I}_{m+1} = -k(\boldsymbol{X'X} + k\boldsymbol{I}_{m+1})^{-1}.$$

A substitution of this expression into Equation 14.B.6 yields

$$bias(\boldsymbol{\beta}_R) = -k(\boldsymbol{X}'\boldsymbol{X} + k\boldsymbol{I}_{m+1})^{-1}\boldsymbol{\beta}. \tag{14.B.6.1}$$

Then

$$\begin{aligned}[bias(\boldsymbol{\beta}_R)]^2 &= (bias(\boldsymbol{\beta}_R))'\,(bias(\boldsymbol{\beta}_R))\\ &= k^2\,\boldsymbol{\beta}'(\boldsymbol{X}'\boldsymbol{X} + k\boldsymbol{I}_{m+1})^{-2}\boldsymbol{\beta}\end{aligned} \tag{14.B.7}$$

and thus

$$MSE(\boldsymbol{\beta}_R) = V(\boldsymbol{\beta}_R) + k^2\boldsymbol{\beta}'(\boldsymbol{X}'\boldsymbol{X} + k\boldsymbol{I}_{m+1})^{-2}\boldsymbol{\beta}. \tag{14.B.5.1}$$

Appendix C: The Accuracy of the Ridge Estimator $\boldsymbol{\beta}_R$

To address the issue of the accuracy of $\boldsymbol{\beta}_R$, we need to first determine its (total) MSE. A useful scalar measure of $MSE\,(\boldsymbol{\beta}_R)$ is $tr(MSE(\boldsymbol{\beta}_R))$, the loss or risk incurred by using $\boldsymbol{\beta}_R$ to estimate $\boldsymbol{\beta}$. Then from Equations 14.B.5.1 and 14.B.4,

$$\begin{aligned}tr(MSE(\boldsymbol{\beta}_R)) &= tr(V(\boldsymbol{\beta}_R)) + k^2 tr[\boldsymbol{\beta}'(\boldsymbol{X}'\boldsymbol{X} + k\boldsymbol{I}_{m+1})^{-2}\boldsymbol{\beta}]\\ &= \sigma_\varepsilon^2 tr(\boldsymbol{X}'\boldsymbol{X} + k\boldsymbol{I}_{m+1})^{-2}\boldsymbol{X}'\boldsymbol{X}\\ &\quad + k^2 tr[\boldsymbol{\beta}'(\boldsymbol{X}'\boldsymbol{X} + k\boldsymbol{I}_{m+1})^{-2}\boldsymbol{\beta}].\end{aligned} \tag{14.C.1}$$

If $\mathbf{U}$ is the orthogonal matrix of characteristic vectors of $\boldsymbol{X}'\boldsymbol{X}$ then $\mathbf{U}'\boldsymbol{X}'\boldsymbol{X}\mathbf{U} = \boldsymbol{G} = diag(\lambda_0, \lambda_1, \ldots, \lambda_m)$, where $\lambda_0, \lambda_1, \ldots, \lambda_m$ are the eigenvalues of $\boldsymbol{X}'\boldsymbol{X}$ (see Appendix 12.K.D for a review of these concepts). Also, let $\mathbf{Z} = \boldsymbol{X}\mathbf{U}$ and $\alpha = \mathbf{U}'\boldsymbol{\beta}$.[1] Then Equation 14.C.1 becomes

$$\begin{aligned}tr(MSE(\boldsymbol{\beta}_R)) &= \sum_{j=0}^{m} MSE(\beta_{Rj})\\ &= \sigma_\varepsilon^2 tr(\boldsymbol{G} + k\boldsymbol{I}_{m+1})^{-2}\boldsymbol{G}\\ &\quad + k^2 tr[\boldsymbol{\alpha}'(\boldsymbol{G} + k\boldsymbol{I}_{m+1})^{-2}\boldsymbol{\alpha}]\\ &= \sigma_\varepsilon^2 \sum_{j=0}^{m} \frac{\lambda_j}{(\lambda_j + k)^2} + k^2 \sum_{j=0}^{m} \frac{\alpha_j^2}{(\lambda_j + k)^2}\end{aligned} \tag{14.C.2}$$

Here the first summation term (the sum of the variances of the parameter estimates) on the right-hand side of Equation 14.B.9 is a monotonically decreasing function of k, whereas the second summation term (the squared bias introduced when $\boldsymbol{\beta}_R$ is used to estimate $\boldsymbol{\beta}$) is a monotonically increasing function of k. In this regard, it is easily verified that the derivative of the first summation term with respect to nonstochalistic k is negative, whereas the derivative of the second summation term with respect to k is positive. Hence,

[1] Given $\boldsymbol{G}$, $\boldsymbol{Z}$, and $\boldsymbol{\alpha}$, the linear model $\boldsymbol{Y} = \boldsymbol{X\beta} + \boldsymbol{\varepsilon}$ can be written as $\boldsymbol{Y} = (\mathbf{ZU}')(\mathbf{U}\boldsymbol{\alpha}) + \boldsymbol{\varepsilon} = \boldsymbol{Z\alpha} + \boldsymbol{\varepsilon}$ (remember that $\mathbf{U}' = \mathbf{U}^{-1}$). Here the columns of $\mathbf{Z} = [\mathbf{Z}_0, \mathbf{Z}_1, \ldots, \mathbf{Z}_m]$ are said to define a set of orthogonal regressors called principal components. Then the OLS estimator for $\boldsymbol{\alpha}$ is the solution to $\mathbf{Z}'\mathbf{Z}\hat{\boldsymbol{\alpha}} = \mathbf{Z}'\boldsymbol{Y}$ or $\hat{\boldsymbol{\alpha}} = \mathbf{G}^{-1}\mathbf{Z}'\boldsymbol{Y}$. (Given $\hat{\boldsymbol{\alpha}}$, the original $\hat{\boldsymbol{\beta}} = \mathbf{U}\,\hat{\boldsymbol{\alpha}}$)

as k increases, the smaller is the variance of $\boldsymbol{\beta}_R$, whereas the larger is the squared bias of $\boldsymbol{\beta}_R$ (As $k \to +\infty$, the squared bias of $\boldsymbol{\beta}_R$ approaches $\boldsymbol{\beta}'\boldsymbol{\beta}$ as an upper limit.) If ridge regression is to be an attractive alternative to OLS, then the decrease in the variance of $\boldsymbol{\beta}_R$ must be greater than the increase in its squared bias so that $MSR(\boldsymbol{\beta}_R) < V(\hat{\boldsymbol{\beta}})$. In fact, Hoerl and Kennard [1970a] demonstrate that there always exists a $k > 0$ such that $MSE(\boldsymbol{\beta}_R) < V(\hat{\boldsymbol{\beta}})$, i.e., they show that $d[trMSE(\boldsymbol{\beta}_R)]/dk < 0$ for nonstochastic k restricted to the interval $(0, \sigma_\varepsilon^2 / \alpha^*)$, where $\alpha^* = \max(\alpha_j^2), j = 0, 1, \ldots, m$, and $\boldsymbol{\alpha} = \mathbf{U}'\boldsymbol{\beta} = (\alpha_0, \alpha_1, \ldots, \alpha_m)'$. Then at least for this range of k values, the ridge estimator $\boldsymbol{\beta}_R$ is better or more accurate than the OLS estimator $\hat{\boldsymbol{\beta}}$. Keep in mind, however, the ridge estimator $\boldsymbol{\beta}_R$ is an improvement over the OLS estimator $\hat{\boldsymbol{\beta}}$ only for a restricted range of k values, and, unfortunately, this range depends on the unknowns $\boldsymbol{\beta}$ and σ_ε^2.

Appendix D: Residual Sum of Squares for $\boldsymbol{\beta}_R$

Let $\boldsymbol{\theta}$ be any estimator for the parameter vector $\boldsymbol{\beta}$. Then the residual sum of squares for $\boldsymbol{\theta}$ is

$$\begin{aligned}\phi &= (\boldsymbol{Y} - \boldsymbol{X\theta})'(\boldsymbol{Y} - \boldsymbol{X\theta}) \\ &= \boldsymbol{Y}'\boldsymbol{Y} - \boldsymbol{Y}'\boldsymbol{X\theta} - \boldsymbol{\theta}'\boldsymbol{X}'\boldsymbol{Y} + \boldsymbol{\theta}'\boldsymbol{X}'\boldsymbol{X\theta}.\end{aligned} \tag{14.D.1}$$

Now, if $\boldsymbol{\theta} = \hat{\boldsymbol{\beta}}$ (the OLS estimator for $\boldsymbol{\beta}$), then

$$\begin{aligned}\phi_{OLS} &= \boldsymbol{Y}'\boldsymbol{Y} - \boldsymbol{Y}'\boldsymbol{X}\hat{\boldsymbol{\beta}} - \hat{\boldsymbol{\beta}}'\boldsymbol{X}'\boldsymbol{Y} + \hat{\boldsymbol{\beta}}'\boldsymbol{X}'\boldsymbol{X}\boldsymbol{\beta} \\ &= \boldsymbol{Y}'\boldsymbol{Y} - \hat{\boldsymbol{\beta}}'\boldsymbol{X}'\boldsymbol{X}\boldsymbol{\beta}\end{aligned}$$

and thus

$$\boldsymbol{Y}'\boldsymbol{Y} = \phi_{OLS} + \hat{\boldsymbol{\beta}}'\boldsymbol{X}'\boldsymbol{X}\boldsymbol{\beta}. \tag{14.D.2}$$

Substituting Equation 14.D.2 into Equation 14.D.1 yields

$$\phi = \phi_{OLS} + \hat{\boldsymbol{\beta}}'\boldsymbol{X}'\boldsymbol{X}\hat{\boldsymbol{\beta}} - \boldsymbol{Y}'\boldsymbol{X\theta} - \boldsymbol{\theta}'\boldsymbol{X}'\boldsymbol{Y} + \boldsymbol{\theta}'\boldsymbol{X}'\boldsymbol{X\theta}. \tag{14.D.1.1}$$

Next, we can write

$$\begin{aligned}\hat{\boldsymbol{\beta}}'\boldsymbol{X}'\boldsymbol{X}\hat{\boldsymbol{\beta}} &= (\boldsymbol{\theta} - \hat{\boldsymbol{\beta}})'\boldsymbol{X}'\boldsymbol{X}(\boldsymbol{\theta} - \hat{\boldsymbol{\beta}}) - \boldsymbol{\theta}'\boldsymbol{X}'\boldsymbol{X\theta} \\ &\quad + \boldsymbol{\theta}'\boldsymbol{X}'\boldsymbol{X}\hat{\boldsymbol{\beta}} + \hat{\boldsymbol{\beta}}'\boldsymbol{X}'\boldsymbol{X\theta}.\end{aligned} \tag{14.D.3}$$

Then substituting Equation 14.D.3 into Equation 14.D.1.1 renders

$$\phi = \phi_{OLS} = (\boldsymbol{\theta} - \hat{\boldsymbol{\beta}})'\boldsymbol{X}'\boldsymbol{X}(\boldsymbol{\theta} - \hat{\boldsymbol{\beta}}). \tag{14.D.1.2}$$

Note that, if $\boldsymbol{\theta} = \hat{\boldsymbol{\beta}}$, then $\phi = \phi_{OLS} = \min \phi$. Hence, Equation 14.D.1.2 tells us that ϕ equals its minimum value (ϕ_{OLS}) plus the value of a quadratic form in $(\boldsymbol{\theta} - \hat{\boldsymbol{\beta}})$.

When $\boldsymbol{\theta} = \boldsymbol{\beta}_R$, (14.D.1.2) becomes

$$\phi_R = \phi_{OLS} = (\boldsymbol{\beta}_R - \hat{\boldsymbol{\beta}})'\boldsymbol{X}'\boldsymbol{X}(\boldsymbol{\beta}_R - \hat{\boldsymbol{\beta}}). \tag{14.D.4}$$

Hence, the residual sum of squares for $\boldsymbol{\beta}_R$ exceeds the residual sum of squares for $\hat{\boldsymbol{\beta}}$.

Additionally, let us write ϕ_R as

$$\begin{aligned}\phi_R &= (\boldsymbol{Y} - \boldsymbol{X}\boldsymbol{\beta}_R)'(\boldsymbol{Y} - \boldsymbol{X}\boldsymbol{\beta}_R)\\ &= \boldsymbol{Y}'\boldsymbol{Y} - \boldsymbol{Y}'\boldsymbol{X}\boldsymbol{\beta}_R - \boldsymbol{\beta}_R'\boldsymbol{X}'\boldsymbol{Y} + \boldsymbol{\beta}_R'\boldsymbol{X}'\boldsymbol{X}\boldsymbol{\beta}_R.\end{aligned}$$

If we rewrite Equation 14.B.1 as $\boldsymbol{X}'\boldsymbol{X}\boldsymbol{\beta}_R = \boldsymbol{X}'\boldsymbol{Y} - k\boldsymbol{\beta}_R$, then the preceding expression becomes

$$\begin{aligned}\phi_R &= \boldsymbol{Y}'\boldsymbol{Y} - \boldsymbol{\beta}_R'\boldsymbol{X}'\boldsymbol{Y} - k\boldsymbol{\beta}_R'\boldsymbol{\beta}_R\\ &= SST - SSR_R - k\|\boldsymbol{\beta}_R\|^2.\end{aligned} \tag{14.D.5}$$

Hence, the residual sum of squares for $\boldsymbol{\beta}_R$ equals the total sum of squares less the regression sum of squares for $\boldsymbol{\beta}_R$ with an adjustment that depends on k and the squared length of $\boldsymbol{\beta}_R$.

Appendix E: Deriving the Ridge Estimator $\boldsymbol{\beta}_R$

Let us form, for $\boldsymbol{Y} = \mathbf{X}\boldsymbol{\beta} + \varepsilon$,

$$\begin{aligned}\hat{\boldsymbol{\beta}}'\hat{\boldsymbol{\beta}} &= [(\boldsymbol{X}'\boldsymbol{X})^{-1}\boldsymbol{X}'\boldsymbol{Y}]'(\boldsymbol{X}'\boldsymbol{X})^{-1}\boldsymbol{X}'\boldsymbol{Y}\\ &= (\boldsymbol{\beta}'\boldsymbol{X}' + \boldsymbol{\varepsilon}')\boldsymbol{X}(\boldsymbol{X}'\boldsymbol{X})^{-1}(\boldsymbol{X}'\boldsymbol{X})^{-1}\boldsymbol{X}'(\boldsymbol{X}\boldsymbol{\beta} + \boldsymbol{\varepsilon})\\ &= \boldsymbol{\beta}'\boldsymbol{\beta} + \boldsymbol{\beta}'(\boldsymbol{X}'\boldsymbol{X})^{-1}\boldsymbol{X}'\boldsymbol{\varepsilon} + \boldsymbol{\varepsilon}'\boldsymbol{X}(\boldsymbol{X}'\boldsymbol{X})^{-1}\boldsymbol{\beta}\\ &\quad + \boldsymbol{\varepsilon}'\boldsymbol{X}(\boldsymbol{X}'\boldsymbol{X})^{-1}(\boldsymbol{X}'\boldsymbol{X})^{-1}\boldsymbol{X}'\boldsymbol{\varepsilon}.\end{aligned}$$

Then

$$E(\hat{\boldsymbol{\beta}}'\hat{\boldsymbol{\beta}}) = \boldsymbol{\beta}'\boldsymbol{\beta} + E[\boldsymbol{\varepsilon}'\boldsymbol{X}(\boldsymbol{X}'\boldsymbol{X})^{-1}(\boldsymbol{X}'\boldsymbol{X})^{-1}\boldsymbol{X}'\boldsymbol{\varepsilon}].$$

Because the term in square brackets is a scalar,

$$\begin{aligned}E(\hat{\boldsymbol{\beta}}'\hat{\boldsymbol{\beta}}) &= \boldsymbol{\beta}'\boldsymbol{\beta} + E\{tr[\boldsymbol{\varepsilon}'\boldsymbol{X}(\boldsymbol{X}'\boldsymbol{X})^{-1}(\boldsymbol{X}'\boldsymbol{X})^{-1}\boldsymbol{X}'\boldsymbol{\varepsilon}]\}\\ &= \boldsymbol{\beta}'\boldsymbol{\beta} + E\{tr[(\boldsymbol{X}'\boldsymbol{X})^{-1}\boldsymbol{X}'\boldsymbol{\varepsilon}\boldsymbol{\varepsilon}'\boldsymbol{X}(\boldsymbol{X}'\boldsymbol{X})^{-1}]\}\\ &= \boldsymbol{\beta}'\boldsymbol{\beta} + \sigma_\varepsilon^2 tr(\boldsymbol{X}'\boldsymbol{X})^{-1} \qquad (14.E.1)\\ &= \boldsymbol{\beta}'\boldsymbol{\beta} + \sigma_\varepsilon^2\sum_{j=0}^{m}(1/\lambda_j) > \boldsymbol{\beta}'\boldsymbol{\beta} + \frac{\sigma_\varepsilon^2}{\lambda_{\min}},\end{aligned}$$

where λ_j is the jth eigenvalue of $\boldsymbol{X}'\boldsymbol{X}$ (note that the eigenvalues of $(\boldsymbol{X}'\boldsymbol{X})^{-1}$ are the reciprocals of the eigenvalues of $\boldsymbol{X}'\boldsymbol{X}$) and $\lambda_{\min}$ is the minimum eigenvalue of $\boldsymbol{X}'\boldsymbol{X}$ (see the discussion underlying the derivation of Equation 14.C.2). In the instance of ill-conditioned data or near multicollinearity, if $\lambda_{\min}$ is small, then $E(\hat{\boldsymbol{\beta}}'\hat{\boldsymbol{\beta}}) - \boldsymbol{\beta}'\boldsymbol{\beta} > 0$; and the smaller is $\lambda_{\min}$, the greater is the difference. Hence, ridge regression can be used to control coefficient instability and variance inflation.

The preceding discussion has indicated that, on average, the distance between $\hat{\boldsymbol{\beta}}$ and $\boldsymbol{\beta}$ will be sizeable if λ_{min} is very small. However, this means that the greater is the degree of ill-conditioning, the farther we can move away from $\hat{\boldsymbol{\beta}}$ without experiencing a considerable increase in the residual sum of squares ϕ (see Equation 14.D.1.2). So given Equation 14.E.1 and $\phi = \min\phi + (\boldsymbol{\theta} - \hat{\boldsymbol{\beta}})'\boldsymbol{X}'\boldsymbol{X}(\boldsymbol{\theta} - \hat{\boldsymbol{\beta}}) = \min\phi + \phi_0$, any feasible departure from min ϕ should be in the direction of shortening the length of $\boldsymbol{\theta}$ given that ϕ is held fixed. (The contour surface of constant ϕ is a hyperellipsoid centered at $\hat{\boldsymbol{\beta}}$.)

In this regard, our goal is to select a ϕ of minimum length given that $(\boldsymbol{\theta} - \hat{\boldsymbol{\beta}})'\boldsymbol{X}'\boldsymbol{X}(\boldsymbol{\theta} - \hat{\boldsymbol{\beta}}) = \phi_0 =$ constant. That is, we seek to

$$\min \boldsymbol{\theta}'\boldsymbol{\theta} \text{ s.t. } (\boldsymbol{\theta} - \hat{\boldsymbol{\beta}})'\boldsymbol{X}'\boldsymbol{X}(\boldsymbol{\theta} - \hat{\boldsymbol{\beta}}) = \phi_0.$$

The Lagrangian associated with this problem is

$$\mathcal{L}(\boldsymbol{\theta}, \lambda) = \boldsymbol{\theta}'\boldsymbol{\theta} + \lambda(\phi_0 - (\boldsymbol{\theta} - \hat{\boldsymbol{\beta}})'\boldsymbol{X}'\boldsymbol{X}(\boldsymbol{\theta} - \hat{\boldsymbol{\beta}}), \tag{14.E.2}$$

where λ is the Lagrange multiplier.

Setting

$$\begin{aligned} &(\text{a}) \partial\mathcal{L}/\partial\boldsymbol{\theta} = 2\boldsymbol{\theta} - 2\lambda\boldsymbol{X}'\boldsymbol{X}(\boldsymbol{\theta} - \hat{\boldsymbol{\beta}}) = 0 \\ &(\text{b}) \partial\mathcal{L}/\partial\lambda = \phi_0 - (\boldsymbol{\theta} - \hat{\boldsymbol{\beta}})'\boldsymbol{X}'\boldsymbol{X}(\boldsymbol{\theta} - \hat{\boldsymbol{\beta}}) = 0, \\ &\qquad \text{[the constraint is binding]} \end{aligned} \tag{14.E.3}$$

we can solve Equation 14.E.3.a for $\boldsymbol{\theta}$ as follows: write

$$\begin{aligned} \boldsymbol{\theta}/\lambda &= \boldsymbol{X}'\boldsymbol{X}\boldsymbol{\theta} - \boldsymbol{X}'\boldsymbol{X}\hat{\boldsymbol{\beta}} \\ &= \boldsymbol{X}'\boldsymbol{X}\boldsymbol{\theta} - \boldsymbol{X}'\boldsymbol{Y} \end{aligned}$$

or

$$\frac{\boldsymbol{\theta}}{\lambda} - \boldsymbol{X}'\boldsymbol{X}\boldsymbol{\theta} = -\boldsymbol{X}'\boldsymbol{Y}$$

and thus

$$\left(\boldsymbol{I}_{m+1}\frac{1}{\lambda} - \boldsymbol{X}'\boldsymbol{X}\right)\boldsymbol{\theta} = -\boldsymbol{X}'\boldsymbol{Y}.$$

Set $\lambda = -\frac{1}{k}$ (because with strict equality constraints λ is unrestricted in sign). Then

$$\boldsymbol{\theta} = (\boldsymbol{X}'\boldsymbol{X} - k\boldsymbol{I}_{m+1})^{-1}\boldsymbol{X}'\boldsymbol{Y} = \boldsymbol{\beta}_R.$$

Because the constraint is binding at $\boldsymbol{\theta} = \boldsymbol{\beta}_R$ (with $\lambda = -k^{-1}$), we can readily determined the value of ϕ_0. To this end, from Equation 14.E.3.a,

$$\boldsymbol{\theta} - \hat{\boldsymbol{\beta}} = (\boldsymbol{X}'\boldsymbol{X})^{-1}(\boldsymbol{\theta}/\lambda).$$

Then substituting this expression into Equation 14.E.3.b yields

$$\begin{aligned}\phi_0 &= (\boldsymbol{\theta}-\hat{\boldsymbol{\beta}})'\boldsymbol{X}'\boldsymbol{X}(\boldsymbol{\theta}-\hat{\boldsymbol{\beta}})\\ &= \left(\frac{\boldsymbol{\theta}'}{\lambda}\right)(\boldsymbol{X}'\boldsymbol{X})^{-1}\boldsymbol{X}'\boldsymbol{X}(\boldsymbol{X}'\boldsymbol{X})^{-1}\left(\frac{\boldsymbol{\theta}}{\lambda}\right)\\ &= k^2\boldsymbol{\theta}'(\boldsymbol{X}'\boldsymbol{X})^{-1}\boldsymbol{\theta} = k^2\boldsymbol{\beta}_R'(\boldsymbol{X}'\boldsymbol{X})^{-1}\boldsymbol{\beta}_R.\end{aligned}$$

Hence, the residual sum of squares for $\boldsymbol{\beta}_R$ is

$$\phi_R = \phi_{OLS} + k^2\boldsymbol{\beta}_R'(\boldsymbol{X}'\boldsymbol{X})^{-1}\boldsymbol{\beta}_R.$$

Appendix F: Determining a Practicable Value of the Shrinkage Parameter *k*

Before we proceed with our derivation of an easily applicable *k*, let us review some useful rules for differentiating a matrix with respect to a scalar.

We offer the following, without proof:

1. If $\boldsymbol{A}$ is a matrix that depends on a scalar θ, then

$$\frac{\partial}{\partial\theta} tr\boldsymbol{A} = tr\frac{\partial \boldsymbol{A}}{\partial\theta}. \tag{14.F.1}$$

2. If $\boldsymbol{A}$ and $\boldsymbol{B}$ are matrices that depend on a scalar θ, then, provided the product $\boldsymbol{AB}$ exists,

$$\frac{\partial}{\partial\theta}\boldsymbol{AB} = \frac{\partial \boldsymbol{A}}{\partial\theta}\boldsymbol{B} + \boldsymbol{A}\frac{\partial \boldsymbol{B}}{\partial\theta} \tag{14.F.2}$$

while

$$\frac{\partial}{\partial\theta} tr\boldsymbol{AB} = tr\left(\frac{\partial \boldsymbol{A}}{\partial\theta}\boldsymbol{B}\right) + tr\left(\boldsymbol{A}\frac{\partial \boldsymbol{B}}{\partial\theta}\right). \tag{14.F.3}$$

If a matrix $\boldsymbol{C}$ also depends on θ and the product $\boldsymbol{ABC}$ exists, then

$$\frac{\partial}{\partial\theta} tr\boldsymbol{ABC} = tr\left(\frac{\partial \boldsymbol{A}}{\partial\theta}\boldsymbol{BC}\right) + tr\left(\boldsymbol{A}\frac{\partial \boldsymbol{B}}{\partial\theta}\boldsymbol{C}\right) + tr\left(\boldsymbol{AB}\frac{\partial \boldsymbol{C}}{\partial\theta}\right). \tag{14.F.4}$$

3. If $y = f(\boldsymbol{X})$, the matrix $\boldsymbol{X}$ is of order $(m \times n)$, and $\boldsymbol{X}$ depends on a scalar θ, then

$$\frac{\partial y}{\partial\theta} = tr\frac{\partial f}{\partial \boldsymbol{X}}\left(\frac{\partial \boldsymbol{X}}{\partial\theta}\right)' \tag{14.F.5}$$

where

$$\underset{(m\times n)}{\frac{\partial f}{\partial \boldsymbol{X}}} = \begin{bmatrix} \partial f/\partial x_{11} & \dots & \partial f/\partial x_{1n} \\ \cdot & & \cdot \\ \cdot & & \cdot \\ \cdot & & \cdot \\ \partial f/\partial x_{m1} & \dots & \partial f/\partial x_{mn} \end{bmatrix}.$$

and $\partial \boldsymbol{X}/\partial\theta$ is a matrix whose (i, j)th element is $\partial x_{ij}/\partial\theta$.

4. We next look to the derivative of a matrix inverse $\boldsymbol{A}^{-1}$ with respect to a scalar θ. Because

$$\frac{\partial}{\partial\theta}\boldsymbol{A}\boldsymbol{A}^{-1} = \frac{\partial \boldsymbol{A}}{\partial\theta}\boldsymbol{A}^{-1} + \boldsymbol{A}\frac{\partial \boldsymbol{A}^{-1}}{\partial\theta},$$

it follows that with $\boldsymbol{A}\boldsymbol{A}^{-1}$ constant (it is the identity matrix),

$$\boldsymbol{A}\frac{\partial \boldsymbol{A}^{-1}}{\partial\theta} = -\frac{\partial \boldsymbol{A}}{\partial\theta}\boldsymbol{A}^{-1}$$

or

$$\frac{\partial \boldsymbol{A}^{-1}}{\partial\theta} = -\boldsymbol{A}^{-1}\frac{\partial \boldsymbol{A}}{\partial\theta}\boldsymbol{A}^{-1}. \tag{14.F.6}$$

5. Finally, on the basis of the foregoing discussion, it is readily verified that

$$\frac{\partial}{\partial\theta}\boldsymbol{B}\boldsymbol{A}^{-1} = \frac{\partial \boldsymbol{B}}{\partial\theta}\boldsymbol{A}^{-1} - \boldsymbol{B}\boldsymbol{A}^{-1}\frac{\partial \boldsymbol{A}}{\partial\theta}\boldsymbol{A}^{-1}. \tag{14.F.7}$$

Given this foundation, let us express, via Equation 14.B.6, the squared bias of $\boldsymbol{\beta}_R$ as

$$\begin{aligned}[bias(\boldsymbol{\beta}_R)]^2 &= \|E(\boldsymbol{\beta}_R) - \boldsymbol{\beta}\|^2 \\ &= \|\boldsymbol{D}\boldsymbol{\beta} - \boldsymbol{\beta}\|^2 = (\boldsymbol{D}\boldsymbol{\beta} - \boldsymbol{\beta})'(\boldsymbol{D}\boldsymbol{\beta} - \boldsymbol{\beta}),\end{aligned}$$

where, from Equation 14.B.2, $\boldsymbol{D} = (\boldsymbol{X}'\boldsymbol{X} + k\boldsymbol{I}_{m+1})^{-1}\boldsymbol{X}'\boldsymbol{X}$. Then

$$\begin{aligned}MSE(\boldsymbol{\beta}_R) &= V(\boldsymbol{\beta}_R) + \|\boldsymbol{D}\boldsymbol{\beta} - \boldsymbol{\beta}\|^2 \\ &= tr\left[\sigma_\varepsilon^2(\boldsymbol{X}'\boldsymbol{X} + k\boldsymbol{I}_{m+1})^{-1}\boldsymbol{X}'\boldsymbol{X}(\boldsymbol{X}'\boldsymbol{X} + k\boldsymbol{I}_{m+1})^{-1}\right] \\ &\quad + \boldsymbol{\beta}'\boldsymbol{D}'\boldsymbol{D}\boldsymbol{\beta} - \boldsymbol{\beta}'\boldsymbol{D}'\boldsymbol{\beta} - \boldsymbol{\beta}'\boldsymbol{D}\boldsymbol{\beta} - \boldsymbol{\beta}'\boldsymbol{\beta}.\end{aligned} \tag{14.F.8}$$

Using Equations 14.F.4 and 14.F.6, the derivative of the first term on the right-hand side of Equation 14.F.8 with respect to k is

$$\frac{d}{dk}tr[.] = -2\sigma_\varepsilon^2 tr(\boldsymbol{X'X} + k\boldsymbol{I}_{m+1})^{-3}\boldsymbol{X'X}. \tag{14.F.9}$$

Next, the derivative of the squared bias component of (14.F.8) with respect to k is

$$\begin{aligned}\frac{d}{dk}(\boldsymbol{D\beta} - \boldsymbol{\beta})'(\boldsymbol{D\beta} - \boldsymbol{\beta}) &= -2\boldsymbol{\beta}'\boldsymbol{X'X}(\boldsymbol{X'X} + k\boldsymbol{I}_{m+1})^{-3}\boldsymbol{X'X\beta} \\ &\quad + 2\boldsymbol{\beta}'(\boldsymbol{X'X} + k\boldsymbol{I}_{m+1})^{-2}\boldsymbol{X'X\beta}.\end{aligned} \tag{14.F.10}$$

Because $\boldsymbol{X'X} = (\boldsymbol{X'X} + k\boldsymbol{I}_{m+1}) - k\boldsymbol{I}_{m+1}$, Equation 14.F.10 becomes

$$\begin{aligned}\frac{d}{dk}(\boldsymbol{D\beta} - \boldsymbol{\beta})'(\boldsymbol{D\beta} - \boldsymbol{\beta}) &= -2\boldsymbol{\beta}'\boldsymbol{X}'\boldsymbol{X}(\boldsymbol{X}'\boldsymbol{X} + k\boldsymbol{I}_{m+1})^{-3} \\ &\quad \times [(\boldsymbol{X}'\boldsymbol{X} + k\boldsymbol{I}_{m+1}) - k\boldsymbol{I}_{m+1}]\boldsymbol{\beta} + 2\boldsymbol{\beta}'(\boldsymbol{X}'\boldsymbol{X} + k\boldsymbol{I}_{m+1})^{-2}\boldsymbol{X}'\boldsymbol{X\beta} \\ &= 2k\boldsymbol{\beta}'(\boldsymbol{X}'\boldsymbol{X} + k\boldsymbol{I}_{m+1})^{-3}\boldsymbol{X}'\boldsymbol{X\beta}.\end{aligned} \tag{14.F.10.1}$$

Then from Equations 14.F.9 and 14.F.10.1,

$$\begin{aligned}\frac{d}{dk}MSE(\boldsymbol{\beta}_R) &= \frac{d}{dk}tr[.] + \frac{d}{dk}(\boldsymbol{D\beta} - \boldsymbol{\beta})'(\boldsymbol{D\beta} - \boldsymbol{\beta}) \\ &= -2\sigma_\varepsilon^2 tr\boldsymbol{E} + 2k\boldsymbol{\beta}'\boldsymbol{E\beta},\end{aligned} \tag{14.F.11}$$

where $\boldsymbol{E} = (\boldsymbol{X'X} + k\boldsymbol{I}_{m+1})^{-3}\boldsymbol{X'X}$.

Because we seek a k which minimizes $MSE(\boldsymbol{\beta}_R)$, a necessary condition for a minimum is $dMSE(\boldsymbol{\beta}_R)/dk = 0$ or

$$k = \sigma_\varepsilon^2 tr\boldsymbol{E} / \boldsymbol{\beta}'\boldsymbol{E\beta} \tag{14.F.12}$$

Because $\boldsymbol{E}$ itself involves k our practicable choice for k is to set $\boldsymbol{E} = \boldsymbol{I}_{m+1}$. Then Equation 14.F.12 becomes

$$k' = (m+1)\sigma_\varepsilon^2 / \boldsymbol{\beta}'\boldsymbol{\beta}.$$

If σ_ε^2 and $\boldsymbol{\beta}$ are now replaced by their OLS estimators S_ε^2 and $\hat{\boldsymbol{\beta}}$, respectively, then we obtain Equation 14.3 or

$$k_{OLS} = (m+1)S_\varepsilon^2 / \hat{\boldsymbol{\beta}}'\hat{\boldsymbol{\beta}}.$$

Exercises

14-1. Given the Table E. 14-1 dataset: (a) determine the OLS hyperplane of best fit and check for the presence of near multicollinearity (with the *noprint* option deleted in Exhibit

14.1); and (b) reestimate the regression model using ridge regression. Compare the two sets of output.

14-2. For the Table E.14-2 data series: (a) estimate the OLS hyperplane that best fits the sample observations (do not use the *noprint* option in Exhibit 14.1); and (b) use ridge regression to refit the regression hyperplane and compare these two sets of results.

TABLE E.14-1

Observations on Variables Y, X_1, X_2, and X_3

Y	X_1	X_2	X_3
83	9	95	93
70	13	89	87
90	20	88	86
69	26	74	74
78	32	73	71
51	38	65	67
71	40	62	60
50	51	51	50
59	58	47	47
40	60	30	31
30	70	27	28
39	76	23	21
22	85	20	20
38	89	16	15
21	94	5	9

TABLE E.14-2

Observations on Variables Y, X_1,..., X_4

Y	X_1	X_2	X_3	X_4
77.9	7	25.9	6	59
74.1	1	29.0	14	50
103.7	10	55.4	9	21
87.0	12	31.2	8	47
96.2	7	51.4	6	30
110.1	11	56.7	8	21
102.0	4	71.0	16	6
74.1	2	31.2	23	44
93.3	1	53.8	18	23
116.1	19	46.6	4	26
83.4	1	41.0	25	33
112.9	10	65.2	9	11
110.3	11	68.1	7	11
72.5	1	31.0	22	44
93.1	2	54.0	18	22

15

Indicator Variables

Introduction

Up to this point in our treatment of regression analysis, we have been working with explanatory variables whose values are explicitly defined on some unambiguous scale of measurement, e.g., variables such as disposable income per year, retail sales figures per month, man hours worked per week, elapsed time in minutes, distance traveled in miles per day, etc. Note that all of the aforementioned variables are *quantitative*: they are measured on a numerical scale, with a definite positional ordering existing among the observed values of the variable.

However, it may be the case that a regressor is of the *qualitative* variety, i.e., it is categorical or classificatory in nature in that we can group the observations into mutually exclusive and collectively exhaustive categories. Here an ordinal scale is used, with differences between the categories of the variable possessing no relevant or useful numerical significance. For instance, certain phenomena can only be counted and not numerically measured; some characteristic is either present or absent; something either occurs or does not occur; a process is either operational or not; the speed of a particular brand of table fan is not continuous but can only be set at low, medium, or high; etc.

In what follows, it is the regression modeling with indicator variables along with the interpretation of the estimated coefficients on those variables that is important. All of the preceding regression methods are then applicable in a completely straightforward manner.

Regression Modeling with Indicator Variables as Regressors

A Single Conceptual Variable with Two Categories

Suppose our goal is to determine whether there is a statistically significant difference between the average yearly salaries received by females versus males. Let us formulate our regression model as

$$Y_i = \beta_0 + \beta_1 D_i + \varepsilon_i, \quad i = 1,\ldots, n, \tag{15.1}$$

where Y_i represents annual salary ($) and the random error term ε_i satisfies the assumptions of the (strong) classical linear regression model. Here D_i serves as an *indicator variable* (also termed a *dummy* or *binary variable*) that is categorical or classificatory in nature because it serves to index the two categories of sex: female and male.

Our convention for ascribing values to D_i is as follows: assign a unique numerical value to each category of an indicator variable and then code each sample observation accordingly. In this regard, if a sample observation corresponds to a female, let us record a zero (0); and if the data point is associated with a male, let us record a one (1). Hence,

$$D_i = \begin{cases} 1, \text{ male;} \\ 0, \text{ female,} \end{cases} \tag{15.2}$$

$i = 1,\ldots n$. That is, we have coded the *conceptual variable* sex (having two categories) with the numerical values 0, 1. The category coded "0" will always serve as our *reference* or *baseline* or *control category*, i.e., all comparisons will be made with or revert back to the reference category.

From Equations 15.1 and 15.2, two response or regression functions are implied:

$$\text{E}(Y_i|D_i = 0) = \beta_0 = \mu_F;$$

$$\text{E}(Y_i|D_i = 1) = \beta_0 + \beta_1 = \mu_M\,. \tag{15.3}$$

Hence, the mean of Y depends on whether some characteristic or attribute is present or not. Given $\beta_0 = \mu_F$, the intercept of the population regression line represents the mean salary of females; and $\beta_1 = \mu_M - \beta_0 = \mu_M - \mu_F$, the slope represents the difference between the mean salaries of males and females.

Suppose we estimate Equation 15.1 via OLS and obtain $\hat{Y} = \hat{\beta}_0 + \hat{\beta}_1 D$. We may then perform a t test of $H_0 : \beta_1 = 0$, against $H_1 : \beta_1 \neq 0$ (which is equivalent to testing $H_0 : \mu_M - \mu_F = 0$ versus $H_1 : \mu_M - \mu_F \neq 0$). If we reject H_0 at a given α level, then we may conclude that there exists a statistically significant difference between the average yearly salaries received by males versus females.

A Single Conceptual Variable with More Than Two Categories

Let us now consider the case in which average annual salary (Y) is to be explained by the conceptual variable "education." Now our classification scheme is structured to reflect three categories or levels of educational attainment: high school diploma, college degree, and graduate degree (obviously other legitimate classification schemes abound). Let us structure the regression model as

$$Y_i = \beta_0 + \beta_1 D_{i1} + \beta_2 D_{i2} + \varepsilon_i, \quad i = 1,\ldots, n, \tag{15.4}$$

where

$$D_{i1} = \begin{cases} 1, \text{ college degree;} \\ 0, \text{ otherwise,} \end{cases}$$
$$D_{i2} = \begin{cases} 1, \text{ graduate degree;} \\ 0, \text{ otherwise,} \end{cases} \tag{15.5}$$

where ε_i, $i = 1,\ldots, n$, is well behaved. Table 15.1 reveals the coding scheme for the sample observations.

TABLE 15.1

Categories of Educational Attainment

Educational Attainment	D_{i1}	D_{i2}
High school diploma	0	0
College degree	1	0
Graduate degree	0	1

Note that we only need two, and not three, dummy variables (we shall see why this is the case shortly) and the category "high school diploma" is coded with zero for each of these variables. That is, high school diploma is the reference or baseline category. Note also that, if $D_{i1} = 1$, then $D_{i2} = 0$, and conversely.

From Equations 15.4 and 15.5, the three response functions are

$$E(Y_i | D_{i1} = 0, D_{i2} = 0) = \beta_0 = \mu_H;$$

$$E(Y_i | D_{i1} = 1, D_{i2} = 0) = \beta_0 + \beta_1 = \mu_C;$$

$$E(Y_i | D_{i1} = 0, D_{i2} = 1) = \beta_0 + \beta_2 = \mu_G. \tag{15.6}$$

How may we interpret these expressions? As stated above, all intragroup comparisons are made relative to the reference group or category displaying $D_{i1} = D_{i2} = 0$ Thus,

$\beta_0 = \mu_A$: the intercept represents the mean salary of high school graduates;

$\beta_1 = \mu_C - \beta_0 = \mu_C - \mu_H$: the regression coefficient on D_1 represents the difference between mean salaries for college graduates and high school graduates;

$\beta_2 = \mu_G - \beta_0 = \mu_G - \mu_H$: the regression slope on D_2 represents the difference between mean salaries for holders of graduate degrees and high school graduates; and

$\beta_1 - \beta_2 = \mu_C - \mu_G$: represents the difference between mean salaries for college graduates and graduate degree holders.

Suppose we estimate Equation 15.4 using OLS and obtain $Y_i = \hat{\beta}_0 + \hat{\beta}_1 D_1 + \hat{\beta}_2 D_2$. To determine whether the individual intercepts β_1 and β_2 are statistically significant or significantly different from the reference intercept β_0 (i.e., to test $H_0 : \beta_1 = 0$ against $H_1 : \beta_1 \neq 0$ or to test $H_0 : \beta_2 = 0$ against $\beta_2 \neq 0$), we can simply use individual t tests on the estimated slopes. To test $H_0 : \beta_1 = \beta_2 = 0$ (no intercept effects) versus H_1 : not both $\beta_1 = 0$ and $\beta_2 = 0$, we can use the usual F test with $F = MSR/MSE : F_{2,n\text{-}3}$.

At this point in our discussion of categorical variables, the reader may be wondering why we had three distinct categories of educational attainment but only two binary variables D_1 and D_2 Why not three such variables, one for each group or classification? To see why using three binary variables poses a problem, suppose that, of a sample of size $n = 7$, two individuals hold high school diplomas, three are college graduates, and two have earned graduate degrees. Let

$$D_{i1} = \begin{cases} 1, & \text{high school diploma;} \\ 0, & \text{otherwise,} \end{cases}$$

$i = 1, 2$;

$$D_{i2} = \begin{cases} 1, & \text{college degree;} \\ 0, & \text{otherwise,} \end{cases}$$

$i = 3, 4, 5$; and

$$D_{i3} = \begin{cases} 1, & \text{graduate degree;} \\ 0, & \text{otherwise,} \end{cases}$$

$i = 6, 7$. Then the columns of the $\boldsymbol{X}$ matrix appear as

$$\begin{matrix} & \boldsymbol{X}_0 & \boldsymbol{X}_1 & \boldsymbol{X}_2 & \boldsymbol{X}_3 \end{matrix}$$
$$\boldsymbol{X} = \begin{bmatrix} 1 & 1 & 0 & 0 \\ 1 & 1 & 0 & 0 \\ 1 & 0 & 1 & 0 \\ 1 & 0 & 1 & 0 \\ 1 & 0 & 1 & 0 \\ 1 & 0 & 0 & 1 \\ 1 & 0 & 0 & 1 \end{bmatrix},$$

where $\boldsymbol{X}_0$ is a column of 1's (because an intercept term is present) and $\boldsymbol{X}_1$, $\boldsymbol{X}_2$, and $\boldsymbol{X}_3$ are vectors of observations pertaining to the indicator variables D_1, D_2, and D_3, respectively.

The difficulty posed by using three dummy variables should be clear once we scan the columns of $\boldsymbol{X}$: the first column can be written as the sum of the three other columns or

$$X_0 = \theta_1 X_1 + \theta_2 X_2 + \theta_3 X_3,$$

i.e., $\boldsymbol{X}_0$ is a linear combination of $\boldsymbol{X}_1$, $\boldsymbol{X}_2$, and $\boldsymbol{X}_3$ (with $\theta_j = 1$ for $j = 1, 2, 3$). However, this means that the columns of X are not linearly independent and thus $\boldsymbol{X'X}$ is singular or $|\boldsymbol{X'X}| = 0$: the case of perfect multicollinearity. In this circumstance, the OLS normal equations $\boldsymbol{X'X}\hat{\boldsymbol{\beta}} = \boldsymbol{X'Y}$ have no solution because $(\boldsymbol{X'X})^{-1}$ does not exist. To extricate ourselves from this so-called *dummy variable trap* (which is sprung when the number of dummy variables equals the number of categories so that the dummy variable columns sum to the unit column vector and consequently the intercept column of $\boldsymbol{X}$ is linearly dependent on the dummy variable columns of the same), all we need to do is use one fewer dummy variable than the number of categories. In general, we have the *indicator variable assignment rule*: (1) if the regression model contains a constant term, then to avoid an exact linear dependency among the columns of $\boldsymbol{X}$, a conceptual variable exhibiting k separate classes must be modeled with only $k - 1$ indicator variables $D_1, D_2, \ldots, D_{k-1}$ (the constant term serves as the intercept for the baseline or reference group and the regression coefficients on the dummy variables measure differences or shifts in intercepts from the baseline); however, (2) if the regression equation does not exhibit a constant term, then we can define an indicator variable for *each* of the k individual classes (here the regression coefficients attached to the dummy variables $D_1, \ldots, D_k$ represent the intercepts for the various classes).

Multiple Conceptual Variables

We now turn to the situation in which annual salary (Y) is explained by the *sex* of the individual and their *level of educational attainment*. Our regression model appears as

$$Y_i = \beta_0 + \beta_1 D_{i1} + \beta_2 D_{i2} + \beta_3 D_{i3} + \varepsilon_i, \quad i = 1, \ldots, n, \tag{15.7}$$

with

$$D_{i1} = \begin{cases} 1, \text{ college degree;} \\ 0, \text{ otherwise,} \end{cases}$$
$$D_{i2} = \begin{cases} 1, \text{ graduate degree;} \\ 0, \text{ otherwise,} \end{cases} \tag{15.8}$$
$$D_{i3} = \begin{cases} 1, \text{ male;} \\ 0, \text{ female,} \end{cases}$$

and ε_i satisfies the usual OLS assumptions. The dummy variable coding scheme is illustrated in Table 15.2. The reference category (the one coded with all zero-valued dummies) is obviously "females with a high school diploma."

From Equations 15.7 and 15.8, our set of response or regression functions appears as

$$E(Y_i | D_{i1} = 0, D_{i2} = 0, D_{i3} = 0) = \beta_0 = \mu_{H0};$$
$$E(Y_i | D_{i1} = 1, D_{i2} = 0, D_{i3} = 0) = \beta_0 + \beta_1 = \mu_{C0};$$
$$E(Y_i | D_{i1} = 0, D_{i2} = 1, D_{i3} = 0) = \beta_0 + \beta_2 = \mu_{G0}; \tag{15.9}$$
$$E(Y_i | D_{i1} = 0, D_{i2} = 0, D_{i3} = 1) = \beta_0 + \beta_3 = \mu_{H1};$$
$$E(Y_i | D_{i1} = 1, D_{i2} = 0, D_{i3} = 1) = \beta_0 + \beta_1 + \beta_3 = \mu_{C1};$$
$$E(Y_i | D_{i1} = 0, D_{i2} = 1, D_{i3} = 1) = \beta_0 + \beta_2 + \beta_3 = \mu_{G1}.$$

TABLE 15.2

Categories of Educational Attainment for Males and Females

Educational Attainment	D_{i1}	D_{i2}	D_{i3}
High school diploma	0	0	0
College degree	1	0	0
Graduate degree	0	1	0
High school diploma	0	0	1
College degree	1	0	1
Graduate degree	0	1	1

We can then directly determine that

$\beta_0 = \mu_{H0}$: the intercept represents the mean salary of female high school graduates.

$\beta_1 = \mu_{C0} - \beta_0 = \mu_{C0} - \mu_{H0}$: the regression coefficient on D_1 represents the difference between mean salaries for female college graduates and female high school graduates.

$\beta_2 = \mu_{G0} - \beta_0 = \mu_{G0} - \mu_{H0}$: the regression coefficient on D_2 represents the difference between mean salaries for female holders of graduate degrees and female high school graduates.

$\beta_3 = \mu_{H1} - \beta_0 = \mu_{H1} - \mu_{H0}$: the regression coefficient on D_3 represents the difference between mean salaries for male high school graduates and female high school graduates. (In addition, from the last two equations in Equation 15.9, we also have $\beta_3 = \mu_{C1} - \beta_0 - \beta_1 = \mu_{C1} - \mu_{C0}$ and $\beta_3 = \mu_{G1} - \beta_0 - \beta_2 = \mu_{G1} - \mu_{G0}$.)

$\beta_1 - \beta_2 = \mu_{C0} - \mu_{G0}$: represents the difference between mean salaries for females with a college degree and females with a graduate degree.

$\beta_1 - \beta_3 = \mu_{C0} - \mu_{H1}$: represents the difference between mean salaries for females with a college degree and males with a high school diploma.

$\beta_2 - \beta_3 = \mu_{G0} - \mu_{H1}$: represents the difference between mean salaries for females with a graduate degree and males with a high school diploma.

Suppose we estimate Equation 15.7 via OLS and obtain $\hat{Y} = \hat{\beta}_0 + \hat{\beta}_1 D_1 + \hat{\beta}_2 D_2 + \hat{\beta}_3 D_3$. Here any statistically significant intercept shifts can be explained by differences in the sex of the individual and/or by differences in the level of educational attainment. In the former case, if under a t test, we reject $H_0 : \beta_3 = 0$ in favor of $H_1 : \beta_3 \neq 0$ at a given α level, then we can conclude that differences in the sex of the individual contribute significantly to explaining differences in average annual income. For the latter case, either individual t tests involving $H_0 : \beta_1 = 0$ against $H_1 : \beta_1 \neq 0$ or $H_0 : \beta_2 = 0$ against $H_1 : \beta_2 \neq 0$ can be conducted or, to test for the absence of any effects of educational attainment on average annual income, we can test $H_0 : \beta_1 = \beta_2 = 0$, versus not both $\beta_1 = 0$ and $\beta_2 = 0$. This can be accomplished, for instance, by conducting either a partial F test using Equation 12.27, where $SSR(\beta_1, \beta_2 | \beta_3) = SSR(\beta_1, \beta_2, \beta_3) - SSR(\beta_3)$ with $m = 3$ and $k = 1$, or, equivalently, a group added-last test. Specifically, let us define set $A = \{D_1, D_2\}$. Then to specify a group added-last test on set A, we need to compare the full model with m_A. That is, for the model pool

$$m\_A: Y = \beta_3 D_3 + \varepsilon$$

$$full: Y = \beta_0 + \beta_1 D_1 + \beta_2 D_2 + \beta_3 D_3 + \varepsilon,$$

if H_0 is assumed true, then $\Delta SSR = SSR(full) - SSR(m\_A)$. Then the group added-last test statistic is, via Equation 12.41, $F = (\Delta SSR / 2) / MSE(full)$. If $F > F_{\alpha,2,n-m-1}$, then we reject H_0 at the α level: differences in educational attainment contribute significantly to explaining differences in average annual income.

Analysis of Covariance

The preceding section considered regression models that exclusively involved indicator variables as regressors. Such models fall into the domain of what is termed analysis of variance. This section involves regression models having both indicator or qualitative variables and quantitative or numerical variables as regressors. Such models come under the heading of analysis of covariance.

Modeling an Intercept Shift

Suppose we posit a naive investment model involving net plant and equipment investment at time $t(I_t)$ as a linear function of autonomous investment at time $t(\beta_0)$ and the change in aggregate output $(X_t) = \Delta Y_t = Y_t - Y_{t-1})$ or

$$I_t = \beta_0 + \beta_1 X_t + \varepsilon_t, \quad t = 1,\ldots, n, \tag{15.10}$$

where ε_t is a well-behaved random error term. To model an intercept shift, let us introduce the indicator variable

$$D_t = \begin{cases} 1, \text{ an event occurs at time } t; \\ 0, \text{ an event does not occur at time } t \end{cases} \tag{15.11}$$

into Equation 5.10 as

$$I_t = \beta_0 + \beta_1 X_t + \beta_2 D_t + \varepsilon_t, \quad t = 1,\ldots, n. \tag{15.10.1}$$

In this regard, although a dummy variable was used previously to indicate classification or membership in a group, we now see that it can also be used to represent a particular time period. For example, we may want to differentiate between years of *peacetime* ($D_t = 0$) versus *wartime* ($D_t = 1$), or between a period of *low inflation* ($D_t = 0$) versus *high inflation* ($D_t = 1$), or between a period of *low unemployment* ($D_t = 0$) versus *high unemployment* ($D_t = 1$), or possibly between years with *a Democrat as president* ($D_t = 0$) versus years with a *Republican as president* ($D_t = 1$).

From Equations 15.10.1 and 15.11, the two implied regression response functions are

$$E(I_t | X_t, D_t = 0) = \beta_0 + \beta_1 X_t;$$

$$E(I_t | X_t, D_t = 1) = (\beta_0 + \beta_2) + \beta_1 X_t \tag{15.12}$$

(see Figure 15.1.A for $\beta_2 > 0$). Hence, the intercept changes from β_0 when $D_t = 0$ to $\beta_0 + \beta_2$, $\beta_2 > 0$, when $D_t = 1$. Testing $H_0 : \beta_2 = 0$, against $H_1 : \beta_2 \neq 0$ using the typical t test determines whether the parallel shift in the regression equation is statistically significant.

Modeling a Change in the Slope of a Quantitative Regressor

Given Equations 15.10 and 15.11, let us model a change in the regression coefficient on X_t by inserting D_t into Equation 15.10 as

$$I_t = \beta_0 + \beta_1 X_t + \beta_2 D_t X_t + \varepsilon_t, \quad t = 1,\ldots, n, \tag{15.10.2}$$

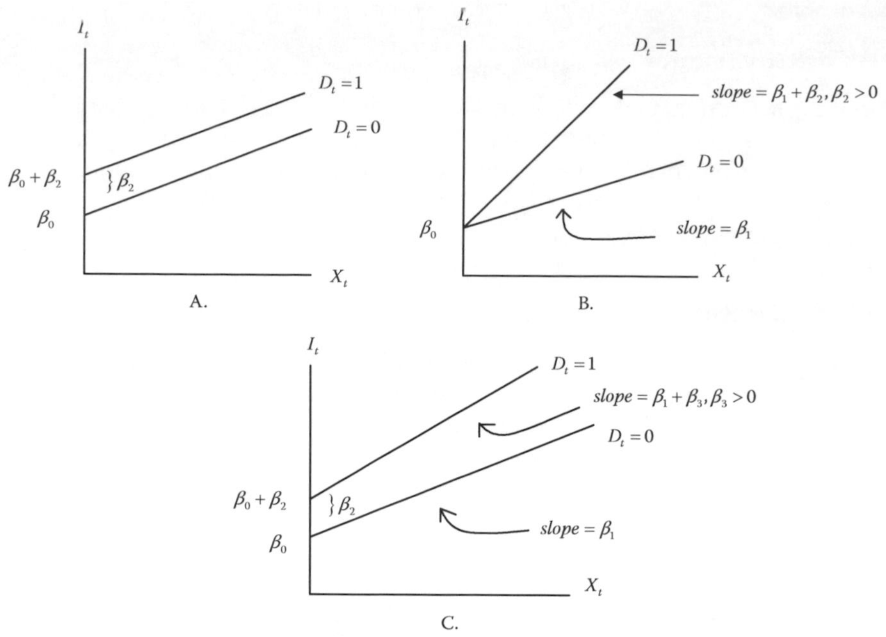

FIGURE 15.1
Modeling changes in regression slopes and intercepts.

where ε_t is again assumed to be well behaved with D_tX_t termed an *interaction regressor.* Now the response functions appears as

$$E(I_t|X_t, D_t = 0) = \beta_0 + \beta_1 X_t;$$
$$E(I_t|X_t, D_t = 1) = \beta_0 + (\beta_1 + \beta_2)X_t \tag{15.13}$$

(see Figure 15.1.B for $\beta_2 > 0$). Clearly, the regression slope increases from β_1 when $D_t = 0$ to $\beta_1 + \beta_2$, $\beta_2 > 0$, when $D_t = 1$. A test of $\beta_2 = 0$, versus $H_1 : \beta_2 \neq 0$ using the t statistic can readily determine whether the change in the slope is statistically significant.

Modeling a Simultaneous Change in the Intercept and in the Slope of a Quantitative Regressor

Given that both the intercept and slope of Equation 15.10 can be affected by the temporal occurrence of some event, as indexed by Equation 15.11, let us again use this indicator to modify Equation 15.10 as

$$I_t = \beta_0 + \beta_1 X_t + \beta_2 D_t + \beta_3 D_t X_t + \varepsilon_t, \quad t = 1,\ldots, n, \tag{15.10.3}$$

where ε_t is specified as a well-behaved random error term. Now the regression responses are

$$E(I_t|X_t, D_t = 0) = \beta_0 + \beta_1 X_t;$$
$$E(I_t|X_t, D_t = 1) = (\beta_0 + \beta_2) + (\beta_1 + \beta_3)X_t \tag{15.14}$$

(see Figure 15.1.C for $\beta_2 > 0$ and $\beta_3 > 0$). Again individual t tests on the regression coefficients in Equation 15.14 can be used to determine whether statistically significant changes in the intercept and/or slope occur. (This particular case will be addressed more fully in the next section.)

Using Indicator Variables to Compare Two Regression Equations

Chapter 2 (in the section on "Comparing Two Linear Regression Equations") addressed the issue of comparing two population regression lines with respect to the similarity of their intercepts or slopes. Specifically, we used two independent datasets (X_{Ai}, Y_{Ai}), $i = 1,\ldots, n_A$, and (X_{Bi}, Y_{Bi}), $i = 1,\ldots, n_B$, to obtain, via OLS, the individual sample regression equations

$$\hat{Y}_A = \hat{\beta}_{0A} + \hat{\beta}_{1A}X_A,$$
$$\hat{Y}_B = \hat{\beta}_{0B} + \hat{\beta}_{1B}X_B$$

as estimates of the population regression lines or responses

$$E(Y_A|X_A) = \beta_{0A} + \beta_{1A}X_A,$$
$$E(Y_B|X_B) = \beta_{0B} + \beta_{1B}X_B,$$

respectively. We then conducted separate hypothesis tests involving $H_0 : \beta_{0A} = \beta_{0B}$ versus $H_1 : \beta_{0A} \neq \beta_{0B}$ and $H_0 : \beta_{1A} = \beta_{1B}$ versus $H_1 : \beta_{1A} \neq \beta_{1B}$.

In this section, our aim is to see how multiple regression techniques and the application of indicator variables allows us to take a more comprehensive and efficient approach to comparing two regression lines. Specifically, instead of fitting two separate regression equations, we need only estimate a single regression equation and then test it to see whether the two associated or implied population regression lines have the same slope, share the same intercept, or actually coincide (they simultaneously have the same intercept and slope).

The omnibus or overall regression model just alluded to will be written as

$$Y_i = \beta_0 + \beta_1X_i + \beta_2D_i + \beta_3D_iX_i + \varepsilon_i, \quad i = 1,\ldots, n, \tag{15.15}$$

where the random error term ε_i is taken to be well behaved, $n = n_A + n_B$, and

$$D_i = \begin{cases} 1, & \text{the observation is from data set } A; \\ 0, & \text{the observation is from data set } B. \end{cases} \tag{15.16}$$

Then from Equations 15.15 and 15.16, the two regression response functions are

$$[\text{Group } A]\ E(Y_i|X_i, D_i = 1) = (\beta_0 + \beta_2) + (\beta_1 + \beta_3)X_i;$$

$$[\text{Group } B]\ E(Y_i|X_i, D_i = 0) = \beta_0 + \beta_1 X_i. \tag{15.17}$$

As we shall now see, estimating Equation 15.15 is formally equivalent to estimating two separate regression equations. The advantage afforded us by undertaking single-equation estimation is that the process of hypothesis testing is streamlined: we can test a variety of hypotheses by directly applying the extra sum-of squares methodologies developed in Chapter 12.

Test for Equal Slopes (Parallelism)

The test for the equality of slopes (allowing for unequal intercepts) or a test that the two regression lines are parallel is executed by simply testing $H_0 : \beta_3 = 0$, against, $H_1 : \beta_3 \neq 0$. Clearly if $\beta_3 = 0$, then $\beta_1 + \beta_3$, the slope for the full or Group A model

$$(\text{Full Model})\ Y_i = \beta_0 + \beta_1 X_i + \beta_2 D_i + \beta_3 D_i X_i + \varepsilon_i, \quad i = 1,\ldots, n,$$

reduces to β_1, the slope of the reduced or Group B model

$$(\text{Reduced Model})\ Y_i = \beta_0 + \beta_1 X_i + \beta_2 D_i + \varepsilon_i, \quad i = 1,\ldots, n.$$

To test $H_0 : \beta_3 = 0$, versus $H_1 : \beta_3 \neq 0$, let us use a variable added-last test. The model pool appears as

$$m\_1\text{: } Y = \beta_0 + \beta_1 X + \beta_2 D + \varepsilon$$

$$full\text{: } Y = \beta_0 + \beta_1 X + \beta_2 D + \beta_3 DX + \varepsilon.$$

Under H_0, $\Delta SSR = SSR(full) - SSR(m\_1)$. Then the variable added-last test statistic is, from Equation 12.39, $F = \Delta SSR / MSE(full)$. If $F > F_{\alpha,1,n-m-1}$, then we reject H_0 at the α level and conclude that the slopes of the two regression lines are significantly different.

Test for Equality of Intercepts

A test for the equality of intercepts (allowing for unequal slopes) is conducted by testing $H_0 : \beta_2 = 0$, against $H_1 : \beta_2 \neq 0$. If $\beta_2 = 0$, then $\beta_0 + \beta_2$, the intercept for the full or Group A model

$$(\text{Full Model})\ Y_i = \beta_0 + \beta_1 X_i + \beta_2 D_i + \beta_3 D_i X_i + \varepsilon_i, \quad i = 1,\ldots, n,$$

reduces to β_0, the intercept of the reduced or Group B model

$$(\text{Reduced Model})\ Y_i = \beta_0 + \beta_1 X_i + \beta_3 D_i X_i + \varepsilon_i, \quad i = 1,\ldots, n.$$

Here too a variable added-last test will be used to test $H_0: \beta_2 = 0$, vs. $H_1: \beta_2 \neq 0$. The requisite model pool is

$$m\_2: Y = \beta_0 + \beta_1 X + \beta_3 DX + \varepsilon$$

$$full: Y = \beta_0 + \beta_1 X + \beta_2 D + \beta_3 DX + \varepsilon.$$

If H_0 is assumed true, then $\Delta SSR = SSR(full) - SSR(m\_2)$ and again $F = \Delta SSR / MSE(full)$. If H_0 is rejected for a chosen α, then we may conclude that the intercepts of the two regression equations differ significantly.

Test of Coincidence

A test to determine whether the two regression lines are identical or coincide is executed by testing the joint null hypothesis $H_0: \beta_2 = \beta_3 = 0$, against $H_1:$ not both $\beta_2 = 0$ and $\beta_3 = 0$. If $\beta_2 = \beta_3 = 0$, then both the intercept and slope for the full or Group *A* model ($\beta_0 + \beta_2$ and $\beta_1 + \beta_3$ respectively)

$$\text{(Full Model) } Y_i = \beta_0 + \beta_1 X_i + \beta_2 D_i + \beta_3 D_i X_i + \varepsilon_i, \quad i = 1, \ldots, n,$$

reduce to β_0 and β_1 the respective intercept and slope of the reduced model

$$\text{(Reduced Model) } Y_i = \beta_0 + \beta_1 X_i + \varepsilon_i, \quad i = 1, \ldots, n.$$

To test $H_0: \beta_2 = \beta_3 = 0$, versus $H_1:$ not both $\beta_2 = 0$ and $\beta_3 = 0$, we need to perform a group added-last test (see Chapter 12). To this end, let us first define set $R = \{D, DX\}$. Then, to conduct a group added-last test on set R, we need to compare the full model with model m_R. Thus, the model pool is

$$m\_R: Y = \beta_0 + \beta_1 X + \varepsilon$$

$$full: Y = \beta_0 + \beta_1 X + \beta_2 D + \beta_3 DX + \varepsilon.$$

Under H_0, $\Delta SSR = SSR(full) - SSR(m\_R)$. Then the group added-last test statistic is, from Equation 12.41, $F = (\Delta SSR / 2) / MSE(full)$. If $F > F_{\alpha,2,n-m-1}$, then we reject H_0 at the α level: the two regression lines are not identical or coincident.

Example 15.1

The Department of Human Resource Development at ABC Corp. is looking into the effect of employee age (X) on absenteeism (Y). The director takes a random sample of 34 employees and checks their employment records to determine their pattern of absenteeism (number of days absent during the first quarter of 2007) and their age. The director then fits the regression equation $Y = \beta_0 + \beta_1 X + \varepsilon$ to the sample dataset and obtains

$$\hat{Y} = \hat{\beta}_0 + \hat{\beta}_1 X = \underset{(-0.75)}{-2.1791} + \underset{(3.62)}{0.2112}, \; R^2 = 0.2906$$

(numbers in parentheses beneath the estimated coefficients are t values).

Encouraged by the discovery of a statistically significant linear relationship between absenteeism and age, and, based on the director's previous experience elsewhere, she feels that smoking may play an important role in further explaining the degree of absenteeism. The employees are then contacted and asked whether they are smokers (naturally each of them cheerfully responds to the query). Given this supplemental information, the director now chooses to fit

$$Y = \beta_0 + \beta_1 X + \beta_2 D + \beta_3 DX + \varepsilon,$$

where

$$D = \begin{cases} 1, & \text{the employee is a smoker;} \\ 0, & \text{the employee is a nonsmoker} \end{cases}$$

and ε is taken to be a well-behaved random error term.

The director ultimately wants to determine whether all of the employees can be considered as having been drawn from the same population, i.e., she wants to determine whether the relationship between absenteeism and age is the same for both smokers and nonsmokers. Looked at in an alternative manner, are the regression equations for each separate category, as indexed by D, identical or coincident? (For the sake of completeness, we shall also test for equality of slopes and equality of intercepts. Exhibit 15.1 contains the complete sample dataset and requisite SAS code.

Exhibit 15.1 SAS Code for Absenteeism Study

```
 data smk;
  input Y X D;

DX=D*X;   ①
datalines;
10    40    1
 5    40    0
 1    45    0
 2    46    0
 9    47    1
 8    47    0
 7    49    0
 7    51    0
 2    30    1
 2    31    0
 8    33    1
 5    33    0
 8    35    1
 5    35    0
15    52    1
15    55    1
 6    55    0
 3    55    0
14    57    1
 5    25    1
 2    25    0
```

① We form the interaction variable *DX* as *D* times *X*.

② m_1 versus the *full* model is used to test for equality of slopes.

③ m_2 versus the *full* model is used to test for equality of intercepts.

④ *m_R* versus the *full* model is used to test for coincidence, where *R* = {*D*, *DX*}.

```
 5    27   1
 3    27   0
19    60   1
 3    60   0
14    61   1
 6    61   0
16    62   1
 6    62   0
20    64   1
18    64   1
 7    65   0
 9    65   0
 5    65   0
;
proc reg data=smk;
  m_1: model Y=X D ; ②
  m_2: model Y=X DX; ③
  m_R: model Y=X ; ④
  full: model Y=X D DX;
run;
```

To test for the coincidence between the regression equations for smokers versus nonsmokers, let us test $H_0 : \beta_2 = \beta_3 = 0$, against H_1 : not both $\beta_2 = 0$ and $\beta_3 = 0$. Under H_0, the incremental regression sum of squares for the group added-last test is, from parts Ⓐ and Ⓑ of Table 15.3, $\Delta SSR = SSR(full) - SSR(m\_R) = 823.2656 - 273.7133 = 549.5523$. Then $F = (\Delta SSR / 2) / MSE(full) = 274.7762/3.9539 = 69.495$, p-value < 0.0001. Hence, we reject H_0 in favor of H_1: the smoker versus nonsmoker regression lines are not coincident and thus can be thought of as being associated with different employee populations.

TABLE 15.3

Output for Example 15.1

The SAS System

Ⓐ

The REG Procedure

Model: full

Dependent Variable: Y

Analysis of Variance

Source	**DF**	**Sum of Squares**	**Mean Square**	***F* Value**	**Pr > *F***
Model	3	823.26561	274.42187	69.41	<0.0001
Error	30	118.61674	3.95389		
Corrected total	33	941.88235			

Root MSE	1.98844	*R*-Square	0.8741
Dependent mean	7.94118	Adj *R*-Sq	0.8615
Coeff Var	25.03961		

(continued)

TABLE 15.3 (*continued*)

Parameter Estimates

Variable	DF	Parameter Estimate	Standard Error	*t* Value	Pr > |t|
Intercept	1	0.58015	1.75327	0.33	0.7430
X	1	0.08831	0.03508	2.52	0.0174
D	1	−5.95238	2.52841	−2.35	0.0253
DX	1	0.27487	0.05086	5.40	<0.0001

Ⓑ
The REG Procedure
Model: m_R
Dependent Variable: *Y*

Analysis of Variance

Source	DF	Sum of Squares	Mean Square	*F* Value	Pr > *F*
Model	1	273.71330	273.71330	13.11	0.0010
Error	32	668.16906	20.88028		
Corrected total	33	941.88235			

Root MSE	4.569949	*R*-Square	0.2906
Dependent mean	7.94118	Adj *R*-Sq	0.2684
Coeff Var	57.54179		

Parameter Estimates

Variable	DF	Parameter Estimate	Standard Error	*t* Value	Pr > |t|
Intercept	1	−2.17910	2.90297	−0.75	0.4584
X	1	0.21123	0.05834	3.62	0.0010

Ⓒ
The REG Procedure
Model: m_1
Dependent Variable: *Y*

Analysis of Variance

Source	DF	Sum of Squares	Mean Square	*F* Value	Pr > *F*
Model	2	707.76232	353.88116	46.86	<0.0001
Error	31	234.12003	7.55226		
Corrected total	33	941.88235			

TABLE 15.3 (*continued*)

Ⓒ

The REG Procedure

Root MSE	2.74814	*R*-Square	0.7514
Dependent mean	7.94118	Adj *R*-Sq	0.7354
Coeff Var	34.60617		

Parameter Estimates

Variable	DF	Parameter Estimate	Standard Error	*t* Value	*Pr* > |*t*|
Intercept	1	−5.73052	1.80763	−3.17	0.0034
X	1	0.21906	0.03510	6.24	<0.0001
D	1	7.19904	0.94961	7.58	<0.0001

Ⓓ

The REG Procedure

Model: m_2

Dependent Variable: *Y*

Analysis of Variance

Source	DF	Sum of Squares	Mean Square	*F* Value	Pr > *F*
Model	2	801.35207	400.67603	88.39	<0.0001
Error	31	140.53028	4.53323		
Corrected total	33	941.88235			

Root MSE	2.12914	*R*-Square	0.8508
Dependent mean	7.94118	Adj *R*-Sq	0.8412
Coeff Var	26.81139		

Parameter Estimates

Variable	DF	Parameter Estimate	Standard Error	*t* Value	*Pr* > |*t*|
Intercept	1	−2.28200	1.35266	−1.69	0.1016
X	1	0.14360	0.02790	5.15	<0.0001
DX	1	0.15965	0.01480	10.79	<0.0001

Given that these regression equations are not coincident, we may next determine whether they are parallel or have the same slope. Here we test $H_0 : \beta_3 = 0$, against $H_1 : \beta_3 \neq 0$. Under H_0, the incremental regression sum of squares for the variable added-last test is, from parts Ⓐ and Ⓒ of Table 15.3, $\Delta SSR = SSR(full) - SSR(m\_1) = 823.2356 - 707.7623 = 115.5033/3.9539 = 29.2125$, p-value <0.001. Clearly, we may reject H_0 in favor of H_1: the smoker versus nonsmoker regression lines are not parallel.

Do these regression equations exhibit different intercepts? To answer this question, we test $H_0 : \beta_2 = 0$, against $H_1 : \beta_2 \neq 0$. Given H_0, the incremental regression sum of squares for this variable added-last test is, from parts Ⓐ and Ⓓ of Table 15.3, $\Delta SSR = SSR(full) - SSR(m\_2) = 823.2656 -$

801.3521 = 21.9135. Then $F = \Delta SSR / MSE(full) = 21.9135/3.9539 = 5.5423$, with $0.025 < p\text{-value} < 0.05$. Hence, we may reject H_0 in favor of H_1: the smoker versus nonsmoker regression equations also have significantly different intercepts.

Exercises

15-1. Given the Table E.15-1 dataset involving individual (cash) contributions to political parties (Y) and political activism (D), where

$$D = \begin{cases} 1, \text{the individual is politically active;} \\ 0, \text{the individual is not politically active,} \end{cases}$$

determine whether there is a statistically significant difference between the annual cash contributions given by those considered politically active versus those not so. What is the reference category? What are the two regression response functions? Using regression model Equation 15.1, how are β_0 and β_1 interpreted?

TABLE E.15-1

Contributions (Y) and Political Activism (X)

Y	D
325	0
425	1
50	0
170	1
220	1
60	1
20	0
35	1
150	0
300	1
500	1
100	0
70	1
50	1
200	0
100	0

15-2. Suppose the variable Y represents the value of regional furniture sales and it is hypothesized that such sales are affected by aggregate regional income (X) and seasonal (quarterly) factors (see Tables E.15-2 and E.15-3). What is the reference quarter? For the regression model

$$Y_t = \beta_0 + \beta_1 X_t + \beta_2 D_{t2} + \beta_3 D_{t3} + \beta_4 D_{t4} + \varepsilon_i, \quad t = 1,\ldots, n,$$

determine the set of regression response functions. Are the seasonal effects statistically significant? (Hint: test $H_0: \beta_2 = \beta_3 = \beta_4 = 0$, versus H_1: not H_0, via a group added-last test.)

TABLE E.15-2

Seasonal Factors

Quarter	D_2	D_3	D_4
Spring	0	0	0
Summer	1	0	0
Fall	0	1	0
Winter	0	0	1

TABLE E.15-3

Regional Sales (Y) versus Income (X)

Quarter	Y_t(000)	X_t(0,000)
I.	180	56
II.	200	103
III.	170	170
IV.	165	200
I.	205	205
II.	210	280
III.	203	300
IV.	200	400
I.	300	420
II.	310	470
III.	290	510
IV.	280	595
I.	310	605
II.	380	675
III.	350	746
IV.	310	798
I.	380	850
II.	410	890
III.	400	975
IV.	380	1000

(continued)

TABLE E.15-3 (*continued*)

I.	410	1025
II.	490	1100
III.	450	1153
IV.	435	1189
I.	520	1200
II.	545	1225
III.	517	1300
IV.	520	1390
I.	600	1427
II.	630	1450
III.	605	1580
IV.	550	1600

15-3. Given the Table E.15-4 data values, suppose we model the number of new housing starts (Y) as a linear function of the unemployment rate (X, defined as the percentage of the civilian labor force unemployed) and conjecture that during a period of high unemployment ($X \geq 6\%$), both the intercept and slope are subject to changes.
Let

$$D = \begin{cases} 1, \text{ high unemployment;} \\ 0, \text{ low unemployment.} \end{cases}$$

What is the reference category? What are the regression response functions? Are the expected changes statistically significant? (Hint: follow the test procedures offered in support of Equations 15.10.1 through 15.10.3.)

15-4. For the Table E.15-5 dataset, construct separate tests to determine whether the Group A and Group B regression equations (a) are parallel, (b) have unequal intercepts, and (c) are coincident, where

$$D = \begin{cases} 1, \text{ the observation belongs to Group A;} \\ 0, \text{ the observation belongs to Group B.} \end{cases}$$

What are the appropriate regression response functions for parts a, b, and c? (Hint: let Example 15.1 serve as your guide.)

TABLE E.15-4

New Housing Starts (Y) versus Unemployment Rate (X)

Y(00)	X(%)	D
448	4.0	0
400	4.5	0
550	3.0	0
500	3.7	0
450	4.0	0
350	5.0	0
360	4.9	0
325	5.5	0
159	6.1	1
157	6.3	1
165	6.0	1
158	6.1	1
155	6.4	1
160	6.0	1
318	5.9	0
320	5.8	0
330	5.5	0
390	4.7	0
349	5.0	0
380	4.8	0

TABLE E.15-5

Observations on Variables Y and X

Group A		Group B	
Y	X	Y	X
76	11	39	4
59	14	30	8
69	16	33	15
42	17	20	19
51	18	21	24
28	21	13	30
38	22	15	34
10	23	6	39
18	25	7	47
3	26	2	53

16

Polynomial Model Estimation

Polynomial Models in the Variables X_j, $j = 1,\dots,m$

The Univariate Case

We have been working with linear (in the parameters) regression equations in which the regressors involved only *first-degree terms*, i.e., for either the univariate ($Y = \beta_0 + \beta_1 X + \varepsilon$) or multivariate ($Y = \beta_0 + \beta_1 X_1 + \cdots + \beta_m X_m + \varepsilon$) cases, each regressor appears only to the first power. In general, as we shall soon see, the *degree* of a polynomial model is the maximum power of any term in the model, whereas the *order* of the model is the maximum sum of powers for any individual term. The name given to a polynomial model reflects the highest degree (or largest power) of any single regressor. Thus, for univariate polynomial models, their degree always equals their order.

Although the fitting of first-order models is quite common, it may be the case that *nonlinearities in the variables* need to be introduced to adequately model some phenomenon. For instance, suppose we have a set of n data points (X_i, Y_i), $i = 1,\dots,n$, and the scatter of sample points is illustrated in Figure 16.1.A. Clearly, the visual pattern of these data points indicates that a *first-order model* of the form

$$Y = \beta_0 + \beta_1 X + \varepsilon \tag{16.1}$$

is appropriate. However, what if the scatter of data points displays a pattern such as in Figure 16.1.B? To pick up the nonlinearity, we need to introduce into Equation 16.1 the second-degree term X^2. Hence, we should fit a *second-order polynomial* in X or

$$Y = \beta_0 + \beta_1 X + \beta_2 X^2 + \varepsilon. \tag{16.2}$$

If the set of sample points displays a *point of inflection* (a point at which a curve crosses over its tangent line and changes the direction of its concavity from upward to downward or vice versa), then, to model any such point (Figure 16.1.C), we need to introduce into Equation 16.2 the third-degree term X^3. We now need to fit a *third-order polynomial* in X or

$$Y = \beta_0 + \beta_1 X + \beta_2 X^2 + \beta_3 X^3 + \varepsilon. \tag{16.3}$$

Polynomial expressions such as Equations 16.2 and 16.3 are termed *natural polynomials*. As evidenced by the form of Equations 16.1 through 16.3, the degree of a polynomial in X is the degree of its term of highest degree.

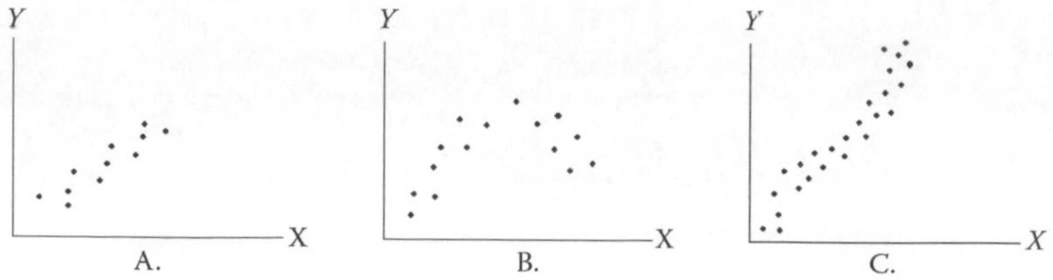

FIGURE 16.1
Linear versus nonlinear scatter plots.

For the latter two cases (Equations 16.2 and 16.3), in spite of the inclusion of second- and third-degree terms in X, each is characterized as a linear regression equation because each is still linear in the parameters. Moreover, if in the mth order polynomial

$$Y = \beta_0 + \sum_{j=1}^{m} \beta_j X^j + \varepsilon \tag{16.4}$$

we set $W_j = X^j$, $j = 1,\ldots m$, then we obtain the more familiar first-order structure $Y = \beta_0 + \sum_{j=1}^{m} W_j + \varepsilon$, and thus essentially all of the preceding estimation techniques apply.

The Multivariate Case

A first-order polynomial model involving two first-degree regressors X_1 and X_2,

$$Y = \beta_0 + \beta_1 X_1 + \beta_2 X_2 + \varepsilon, \tag{16.5}$$

can be used to estimate the parameters of a regression hyperplane H (Figure 16.2.A). Hence, no nonlinearities need to be accounted for. However, to estimate the parameters of a curved surface (called a *response surface*) S (Figure 16.2.B), we would need to specify a second-order polynomial model (arrived at by inserting the second-order terms X_1^2, X_1X_2, and X_2^2 into Equation 16.5) of the form

$$Y = \beta_0 + \beta_1 X_1 + \beta_2 X_2 + \beta_{11} X_1^2 + \beta_{12} X_1 X_2 + \beta_{22} X_2^2 + \varepsilon, \tag{16.6}$$

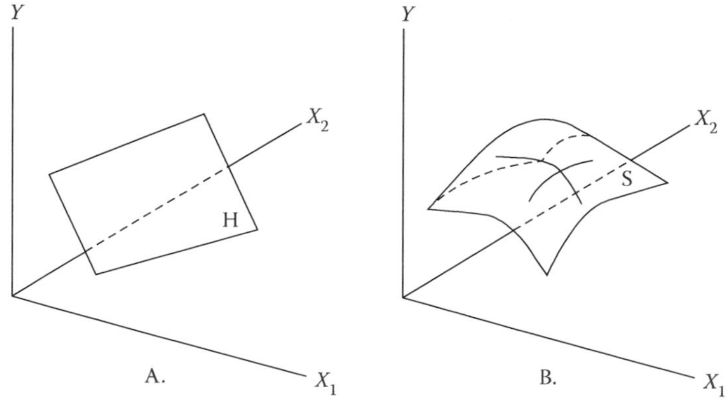

FIGURE 16.2
Regression hyperplane versus response surface.

where X_1X_2 is called an *interaction term*. (Notice how the regression coefficients on the nonlinear terms in Equation 16.6 are indexed, i.e., β_{11} is the coefficient on $X_1X_1 = X_1^2$, β_{12} is attached to X_1X_2, and β_{22} is associated with $X_2X_2 = X_2^2$.) Although Equation 16.6 is a second-degree (as well as second-order) regression model, a *fully saturated second-degree regression model* appears as

$$\begin{aligned} Y = \beta_0 + \beta_1X_1 + \beta_2X_2 + \beta_{11}X_1^2 + \beta_{12}X_1X_2 + \beta_{22}X_2^2 \\ + \beta_{112}X_1^2X_2 + \beta_{122}X_1X_2^2 + \beta_{1122}X_1^2X_2^2 + \varepsilon \end{aligned} \tag{16.7}$$

and is of order four.

In a similar vein, a second-order polynomial model involving the three regressors X_1, X_2, and X_3 is of the form

$$\begin{aligned} Y = \beta_0 + \beta_1X_1 + \beta_2X_2 + \beta_3X_3 + \beta_{11}X_1^2 + \beta_{22}X_2^2 \\ + \beta_{33}X_3^2 + \beta_{12}X_1X_2 + \beta_{13}X_1X_3 + \beta_{23}X_2X_3 + \varepsilon \end{aligned} \tag{16.8}$$

whereas a third-order polynomial model in the regressors X_1 and X_2 appears as

$$\begin{aligned} Y = \beta_0 + \beta_1X_1 + \beta_2X_2 + \beta_{11}X_1^2 + \beta_{12}X_1X_2 + \beta_{22}X_2^2 \\ + \beta_{111}X_1^3 + \beta_{112}X_1^2X_2 + \beta_{122}X_1X_2^2 + \beta_{222}X_2^3 + \varepsilon. \end{aligned} \tag{16.9}$$

(Here β_{111} is the coefficient on $X_1X_1X_1 = X_1^3$, β_{112} is attached to $X_1X_1X_2 = X_1^2X_2$, etc.) Obviously, other modeling configurations involving variations in the number of regressors and/or in the order of the polynomial can be readily constructed.

Looking to the practical aspects of polynomial regression, the following should be remembered when fitting polynomial equations in X. (1) The order of the polynomial regression equation should be kept as low as possible yet consistent with the characteristics of the phenomenon being modeled; failure to do so wastes precious degrees of freedom, increases the odds that the $\mathbf{X'X}$ matrix will be ill-conditioned, and can produce highly misleading predictions. (2) If the range of the X values is highly restrictive or narrow and/or the order of the estimating polynomial is high, then near-multicollinearity can occur between the columns associated with X, X^2, X^3,..., thus causing $\mathbf{X'X}$ to be ill-conditioned. A possible remedy for such ill-conditioning (although by no means a failsafe) is to center the X values or to correct X for its average (i.e., the origin of X is changed from zero to $\bar{X}$). Hence, any ill-conditioning induced by the arbitrary choice of origin can be removed. For example, the centered (to $\bar{X}$) versions of Equations 16.2 and 16.3 are, respectively,

$$Y = \beta_0 + \beta_1(X - \bar{X}) + \beta_2(X - \bar{X})^2 + \varepsilon \tag{16.2.1}$$

and

$$Y = \beta_0 + \beta_1(X - \bar{X}) + \beta_2(X - \bar{X})^2 + \beta_3(X - \bar{X})^3 + \varepsilon. \tag{16.3.1}$$

It must be remembered, however, that these two polynomial models are valid or defined only over the original range of X values; for Equations 16.2.1 and 16.3.1, the range of the regressor X is centered at $\bar{X}$. Finally, (3) a recommended estimation strategy for polynomial regression is to use a *backward selection process* that is based on added-in-order sums of squares. That is, we start with a sufficiently high-order model (possibly Equation 16.3) and then successively estimate smaller-order models, with model reduction stopping at the term of highest degree displaying statistical significance, where, say, $\alpha = 0.05$ is used throughout. For instance, if a test of the term $X^3|X, X^2$ is significant, we stop; our polynomial model of choice is Equation 16.3.

Example 16.1

Table 16.1 presents $n = 27$ observations on the variables X and Y. Does the cubic regression Equation 16.3 serve as an adequate model for estimating the relationship between X and Y? The required SAS code is outlined in Exhibit 16.1.

Exhibit 16.1 SAS Code for Cubic Polynomial Estimation

```
data cube;
   input  x  y;
zx = x;  ①
datalines;
 1      7
 5      8
 .      .
 .      .
 .      .
29     31
30     33
```

① We create a new variable *ZX* that has the same set of values as *X*.

② PROC STANDARD is used to center *X* via *ZX* (we stipulate that *ZX* must have a mean of zero); the variable representing the centered values of *X* is labeled *X_C*, which is housed in the output dataset *cube2*.

③, ④ Dataset *cube3* contains all variables found in *cube* and *cube2* along with *X_C2* and *X_C3*, the variables representing, respectively, the squared and cubed values of *X_C*.

TABLE 16.1

Observations on Variables X and Y

Observation Number	X	Y	Observation Number	X	Y
1	1	7	15	15	26
2	5	8	16	16	20
3	5	9	17	16	27
4	8	8	18	17	23
5	9	10	19	17	28
6	11	10	20	19	27
7	13	12	21	19	30
8	13	14	22	23	30
9	14	16	23	24	29
10	14	17	24	27	31
11	15	16	25	28	32
12	15	21	26	29	31
13	15	22	27	30	33
14	15	25			

```
;
proc standard data = cube
   out = cube2 (rename =(zx=x_c))
   m = 0;
   var  zx;
run;
data cube3; merge cube cube2; ③
   x_c2=x_c**2;   ④
   x_c3=x_c**3;
run;
proc reg data=cube3;
```

⑤ We specify a cubic polynomial model in *X_C* and request tolerance (*tol*) and variance inflation ② factors (*vif*) to check for the presence of near multi-collinearity. We also request sequential or added-in-order (Type I) sums of squares (*ss1*), sequential parameter estimates (*seqb*), and sequential or added-in-order partial correlation coefficients (*pcorr1*).

```
   model y=x_c x_c2 x_c3/tol vif ss1 seqb  pcorr1; ⑤
run;
proc reg data=cube3;
   model y=x_c;
      output out=l p=pl r=rl;
   model y=x_c x_c2;
      output out=q  p=pq  r=rq;⑥
   model y=x_c  x_c2  x_c3;
      output  out=s  p=ps  r=rs;
```

⑥ We fit first-order, second-order, and third-order polynomials in *X_C* and obtain the predicted as well as residual values for each.

```
data all; merge l  q  s;   ⑦
run;
symbol1  v=star  I=none  c=black;
symbol2  v='1'   l=1  I=spline  c=black;
symbol3  v='2'   l=1  I=spline  c=black;   ⑧
symbol4  v='3'   l=1  I=spline  c=black;
proc gplot  data=all;  ⑨
    plot y*x_c=1 pl*x_c=2  pq*x_c=3  ps*x_c=4/overlay;   ⑩
run;
proc gplot data=s;
    plot rs*x_c/vref=0;⑪
run;
```

⑦ Output datasets *l*, *q*, and *s* from ⑥ are merged into the dataset *all*.

⑧ A set of plot statements is specified:
   a. the (*v*=) *star* symbol is used to indicate the (centered) data points ($(X\_C)_i$, Y_i), $i = 1, \ldots, 27$; the I=NONE option requests that the points not be connected by a smooth line.
   b. the order of a polynomial is used as the plotting symbol (*v='1' or '2' or '3'*) for the estimated polynomial equation; the I=SPLINE option requests that a smooth line join all points; and the *l=1* option requests a solid line plot.

⑨ PROC GPLOT is SAS/GRAPH procedure that produces high-resolution graphic/plots.

⑩ We request that the plots of the predicted values for each estimated polynomial be superimposed on the plot of *Y* against *X_C*.

⑪ We request a residual plot for the cubic polynomial model; the VREF=0 option requests that a horizontal line be drawn at the zero value on the vertical-axis.

The output generated from the preceding SAS program appears in Table 16.2.

① To assess the statistical significance of the full or third-order model, let us test $H_0: \beta_1 = \beta_2 = \beta_3 = 0$, against H_1: not all βj's = 0, j = 1, 2, 3. Because the p-value for the sample F statistic is less than 0.0001, we can safely reject H_0 in favor of H_1 and conclude that a third-order polynomial model in *X_C* explains a statistically significant portion of the variation in *Y*. In fact, $R^2 = 0.8891$ reveals that the full model explains about 89% of said variation. With $\sqrt{MSE} = 3.09$, we see that a third-order polynomial fits the data reasonably well.

②, ③ The estimated third-order polynomial equation appears as

$$\hat{Y} = \hat{\beta}_0 + \hat{\beta}_1 X\_C + \hat{\beta}_2 X\_C2 + \hat{\beta}_3 X\_C3$$

$$= 22.0814 + 1.7651X\_C - 0.0188X\_C2 - 0.0048X\_C3.$$

The p-value associated with each estimated coefficient's t statistic is less than $\alpha = 0.05$, thus indicating that the intercept as well as each regressor contributes significantly to explaining variation in *Y*, i.e., we reject $H_0 : \beta_k = 0$, in favor of $H_1 : \beta_k \neq 0$, $k = 0, 1, 2, 3$.

④, ⑤, ⑦ remember that Type I or sequential sums of squares represent a partitioning of SSR into individual component sums of squares attributed to each regressor as it is introduced sequentially into the model (in the exact order specified in the *MODEL* statement); it provides the contribution to SSR of each regressor as it is inserted, in turn, into the regression equation. Hence, the Type I sum of squares is used to determine what order of polynomial is actually needed.

The various *TYPE* 1 sums of squares are as follows:

SSR(X_C|intercept) = 1615.172 : the regression sum of squares for a model containing only X_C and an intercept;

SSR(X_C2|intercept, *X_C*) = 23.859 : the increment in the regression sum of square attributable to the addition of *X_C2* to a model already containing an intercept and *X_C;*

SSR(X_C3|intercept, *X_C*, *X_C2*) = 116.035 : the increase in the regression sum of squares attributable to the addition of *X_C3* to a model already containing an intercept, *X_C*, and *X_C2*.

Dividing each Type I sum of squares value by *MSE(full)* = 9.5221 (see ①) provides a set of *F* statistics that are used to determine whether each additional regressor makes a statistically significant contribution to SSR and thus to explaining variation in *Y*. The sequential *F* values are

$$F = \frac{SSR(X\_C|\text{ intercept})}{MSE(full)} = \frac{1615.172}{9.522} = 169.625,$$

p-value <0.0001. A first-degree term is warranted. Next,

$$F = \frac{SSR(X\_C2|\text{ intercept}, X\_C)}{MSE(full)} = \frac{23.859}{9.522} = 2.505,$$

p-value <0.10. This result is not statistically significant; a second-degree term may not be appropriate. Finally,

$$F = \frac{SSR(X\_C3|\text{ intercept}, X\_C, X\_C2)}{MSE(full)} = \frac{116.035}{9.522} = 12.185,$$

TABLE 16.2

Output for Example 16.1

The SAS System
The REG Procedure
Model : MODEL1
Dependent Variable: y
①

Analysis of Variance

Source	**DF**	**Sum of Squares**	**Mean Square**	**F Value**	**Pr > F**
Model	3	1755.06565	585.02188	61.44	<.0001
Error	23	219.00843	9.52211		
Corrected total	26	1974.07407			

Root MSE	3.08579	*R*-Square	0.8891
Dependent Mean	20.81481	Adj *R*-Sq	0.8746
Coeff Var	14.82497		

Parameter Estimates

Variable	**DF**	② **Parameter Estimate**	**Standard Error**	③ **t Value**	**Pr > |t|**	④ **Type I SS**	⑤ **Squared Partial Corr Type I**
Intercept	1	22.08137	0.76616	28.82	<.0001	11698.	.
x_c	1	1.76516	0.21023	8.40	<.0001	1615.171921	0.81819
x_c2	1	−0.01877	0.00896	−2.10	0.0473	23.85920	0.06648
x_c3	1	−0.00477	0.00137	−3.49	0.0020	116.03453	0.34633

Parameter Estimates

Variable	**DF**	**Tolerance**	⑥ **Variance Inflation**
Intercept	1	.	0
x_c	1	0.15401	6.49326
x_c2	1	0.96118	1.04039
x_c3	1	0.15603	6.40911

(continued)

TABLE 16.2 *(continued)*

⑦

Sequential Parameter Estimates

Intercept	x_c	x_c2	x_c3
20.814815	0	0	0
20.814815	1.074500	0	0
21.540826	1.091162	−0.014012	0
22.081365	1.765164	−0.018771	−0.004771

⑧

The REG Procedure
Model : MODEL1
Dependent Variable: y

Analysis of Variance

Source	DF	Sum of Squares	Mean Square	F Value	Pr > F
Model	1	1615.17192	1615.17192	112.51	<.0001
Error	25	358.90215	14.35609		
Corrected total	26	1974.07407			

Root MSE	3.78894	*R*-Square	0.8182
Dependent Mean	20.81481	Adj *R*-Sq	0.8109
Coeff Var	18.20310		

Parameter Estimates

Variable	DF	Parameter Estimate	Standard Error	t Value	Pr > |t|
Intercept	1	20.81481	0.72918	28.55	<.0001
x_c	1	1.07450	0.10130	10.61	<.0001

⑨

The REG Procedure
Model : MODEL2
Dependent Variable: y

Analysis of Variance

Source	DF	Sum of Squares	Mean Square	F Value	Pr > F
Model	2	1639.03112	819.51556	58.70	<.0001
Error	24	335.04295	13.96012		
Corrected total	26	1974.07407			

Root MSE	3.73632	*R*-Square	0.8303
Dependent mean	20.81481	Adj *R*-Sq	0.8161
Coeff Var	17.95031		

TABLE 16.2 *(continued)*

Parameter Estimates

Variable	DF	Parameter Estimate	Standard Error	t Value	Pr > |t|
Intercept	1	21.54083	0.90854	23.71	<.0001
x_c	1	1.09116	0.10070	10.84	<.0001
x_c2	1	–0.01401	0.01072	–1.31	0.2035

⑩

The REG Procedure

Model : MODEL3

Dependent Variable: y

Analysis of Variance

Source	DF	Sum of Squares	Mean Square	F Value	Pr > F
Model	3	1755.06565	585.02188	61.44	<.0001
Error	23	219.00843	9.52211		
Corrected total	26	1974.07407			

Root MSE	3.08579	*R*-Square	0.8891
Dependent Mean	20.81481	Adj *R*-Sq	0.8746
Coeff Var	14.82497		

Parameter Estimates

Variable	DF	Parameter Estimate	Standard Error	t Value	Pr > |t|
Intercept	1	22.08137	0.76616	28.82	<.0001
x_c	1	1.76516	0.21023	8.40	<.0001
x_c2	1	–0.01877	0.00896	–2.10	0.0473
x_c3	1	–0.00477	0.00137	–3.49	0.0020

p-value <0.001. A third-degree term is justified. (Because the third-order term is significant, all lower-order terms within the polynomial will be retained.)

As explained in Chapter 14, *pcorr1* determines the set of sequential or variables added-in-order squared partial correlation coefficients based on added-in-order Type I sums of squares:

$r^2_{YX\_C} = 0.818$: when *X_C* is added to a regression model that only displays an intercept, SSE decreases by 81.82%;

$r^2_{YX\_C2|X\_C} = 0.0664$: when *X_C2* is added to a regression model already containing *X_C*, SSE decreases by 6.64%; and

$r^2_{YX\_C3|X\_C,X\_2C} = 0.3463$: when *X_C3* is introduced into a regression model that already contains *X_C* and *X_C2*, SSE decreases by 34.63%.

The *SEQB* option prints the coefficients of the successive polynomials fitted by adding regressors in the order specified in the *MODEL* statement. If the preceding sequential *F* test revealed that a polynomial of degree lower than three was appropriate, then estimates of the coefficients of lower-order polynomials are found in the various rows of ⑦:

row 1: coefficients of the zero-order polynomial $\hat{Y} = \hat{\beta}_0 = 20.8148$;

row2: coefficients of the first-order polynomial
$\hat{Y} = \hat{\beta}_0 + \hat{\beta}_1 X\_C = 20.8148 + 1.0745X\_C$;

row 3: coefficients of the second-order polynomial
$\hat{Y} = \hat{\beta}_0 + \hat{\beta}_1 X\_C + \hat{\beta}_2 X\_C2 = 21.5408 + 1.0911X\_C - 0.0140X\_C2$; and

row 4: coefficients of the full or third-order polynomial model (see ②).

⑥ each *vif* is less than 10, thus indicating that near-multicollinearity is not a serious problem. Moreover, the tolerances associated with regressors *X_C*, *X_C2*, and *X_C3* are 0.1540, 0.9611, and 0.1560, respectively.
Because tolerance $= 1 - R_j^2$, $j = 1,2,3$, it follows that $R_1^2 = 0.8460$, $R_2^2 = 0.0389$, and $R_3^2 = 0.8440$. Thus, *X_C* and *X_C3* are moderately related to the other regressors.

⑧, ⑨, ⑩ First-, second-, and third-order polynomials in *X_C* are estimated so that predicted and residual values (not printed) can be determined. Figure 16.3 houses a plot of the sample

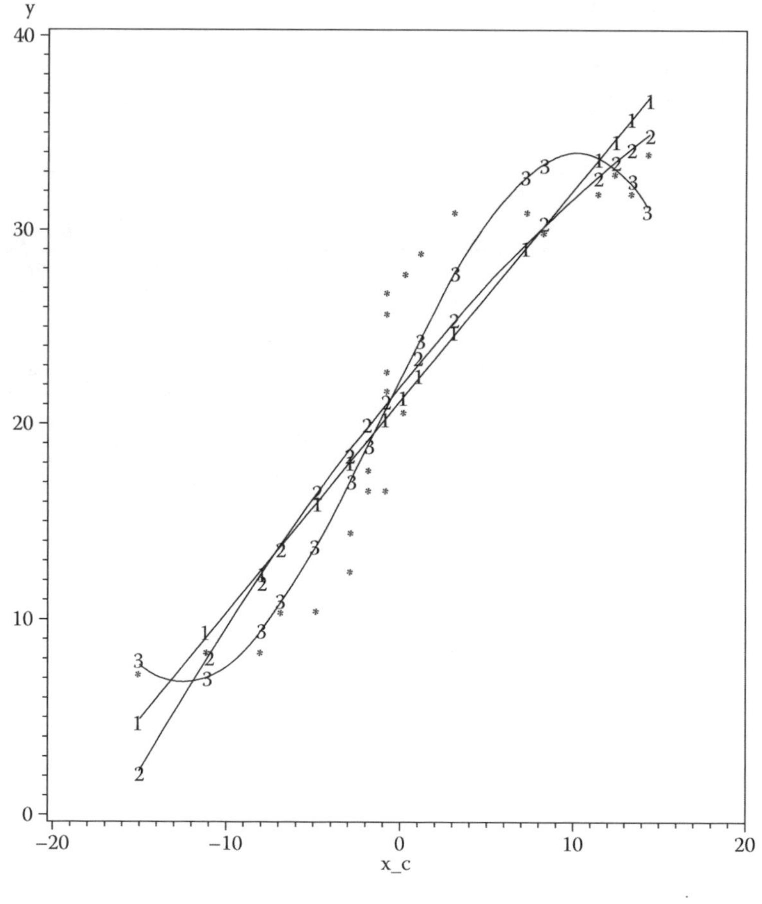

FIGURE 16.3
Polynomial plots.

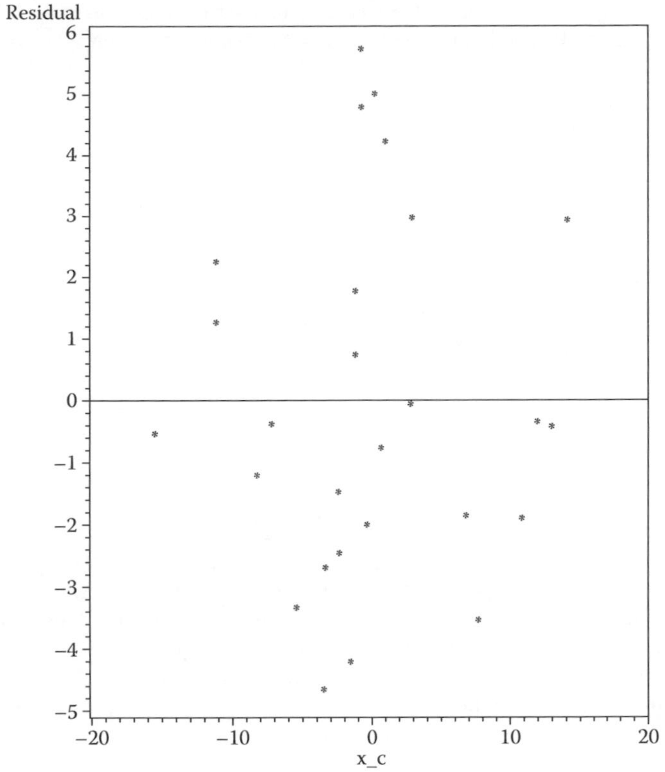

FIGURE 16.4
Residual plot for third-order polynomial.

data points along with the set of polynomial plots. Notice how the fit of the estimated polynomial equation to the pattern of data points improves as successively higher-order terms are included in the model. Finally, Figure 16.4 plots the residuals obtained from the third-order model against *X_C*. The third-order model seems appropriate because no troublesome pattern to the indicated points emerges.

Example 16.2

We noted previously that one possible strategy for fitting a polynomial model in *X* was that of *backward selection/elimination*. This approach basically starts with all regressors in the model and then systematically deletes them, one at a time, according to a prespecified elimination criterion. More specifically, after fitting the entire *m*th-order polynomial regression model, we delete from the same the variable displaying the "smallest" estimated *F* value.[1] The model stated in terms of the surviving variables (there is now one fewer regressor) is subsequently estimated, and the variable that again has the "smallest" *F* value is a candidate for deletion, and so on. The process stops when all variables remaining in the regression equation are statistically significant at some chosen level. (It is important to note that, once a variable has been eliminated, it never reenters into the regression equation.)

[1] At each step, the regressor displaying the smallest incremental contribution to the model is a candidate for elimination. This is determined by a regressor's Type II sum of squares value (given that all other regressors are currently in the model). Then the value of the added-last *F* statistic associated with the regressor is compared with some critical *F* value to determine whether elimination of the variable is warranted.

This said, let us use the Example 16.1 dataset and reestimate the third-order polynomial model using backward elimination. To this end, our modified MODEL statement appears as in Exhibit 16.2.

Exhibit 16.2 SAS Code for Cubic Polynomial Estimation (Backward Elimination)

```
proc reg data = cube3;
     model y = x_c  x_c2  x_c3/ selection =
         backward tol vif;
run;
```

The SAS output generated by this modified *MODEL* statement appears in Table 16.3.

① Backward elimination starts with the full model (Step 0).

② Mallows' [1973] $C(p) = (SSE_p/MSE(full)) - [n - 2(p + 1)]$ statistic is a useful criterion for *subset model selection*, in which SSE_p is the error sum of squares for a subset model containing $p(\leq m)$ regressors (and an intercept) selected from the full model, and *MSE(full)* is the mean square for the full model. $C(p)$serves as a measure of total squared error for the subset model containing p regressors, in which the total squared error consists of the error variance plus the bias incurred by not including important variables in the model. It is thus a useful indicator of when backward selection is deleting too many variables, e.g., for $C(p) > p + 1$, too many relevant variables have been dropped from the full model and thus parameter estimates tend to be biased; if $C(p) < p + 1$, the full model contains too many variables (it is overspecified); and if $C(p) \approx p + 1$, the correct model has been fit and the parameter estimates are unbiased. Here we find that, for $p = 3$, $C(p) = p + 1 = 4$.

③ The set of added-last F values are determined from the Type II sum of squares entries as follows:

$$\frac{SSR(X\_C|X\_C2,\ X\_C3)}{MSE(full)} = \frac{671.29522}{9.52211} = 70.50;$$

$$\frac{SSR(X\_C2|X\_C,\ X\_C3)}{MSE(full)} = \frac{41.82787}{9.52211} = 4.39; \text{ and}$$

$$\frac{SSR(X\_C3|X\_C,\ X\_C2)}{MSE(full)} = \frac{116.03453}{9.52211} = 12.19.$$

Based on the magnitudes of these sample F values, none of the regressors will be deleted from the model and consequently the backward elimination process ends with the full model itself. No additional steps are executed.

④ The default level of significance for each added-last F test (or for stopping the backward elimination routine) is 0.10; once the process ends, each variable remaining in the model does so at the $\alpha \leq 0.10$ level. If 10% significance is not acceptable, then the option *slstay* = level can be inserted into the *MODEL* statement. For example, if $\alpha = 0.01$ is deemed an appropriate level of significance, then the preceding *MODEL* statement is amended to read as in Exhibit 16.3.

TABLE 16.3

Output for Example 16.2

The SAS System
The REG Procedure
Model : MODEL1
Dependent Variable: y

①

Backward Elimination: Step 0

②

All Variables Entered: $R^2 = 0.8891$ and C(p) = 4.0000

Analysis of Variance

Source	**DF**	**Sum of Squares**	**Mean Square**	**F Value**	**Pr > F**
Model	3	1755.06565	585.02188	61.44	<.0001
Error	23	219.00843	9.52211		
Corrected total	26	1974.07407			

Variable	**Parameter Estimate**	**Standard Error**	**Type II SS**	③ **F Value**	**Pr > F**
Intercept	22.08137	0.76616	7909.34743	830.63	<.0001
x_c	1.76516	0.21023	671.29522	70.50	<.0001
x_c2	−0.01877	0.00896	41.82787	4.39	0.0473
x_c3	−0.00477	0.00137	116.03453	12.19	0.0020

Bounds on condition number: 6.4933, 41.828

④ All variables left in the model are significant at the 0.1000 level.

The REG Procedure
Model : MODEL1
Dependent Variable: y

Analysis of Variance

Source	**DF**	**Sum of Squares**	**Mean Square**	**F Value**	**Pr > F**
Model	3	1755.06565	585.02188	61.44	<.0001
Error	23	219.00843	9.52211		
Corrected total	26	1974.07407			

(continued)

TABLE 16.3 (continued)

Root MSE	3.08579	*R*-Square	0.8891
Dependent Mean	20.81481	Adj *R*-Sq	0.8746
Coeff Var	14.82497		

Parameter Estimates

Variable	DF	Parameter Estimate	Standard Error	t Value	Pr > |t|	Tolerance	Variance Inflation
Intercept	1	22.08137	0.76616	28.82	<.0001	.	0
x_c	1	1.76516	0.21023	8.40	<.0001	0.15401	6.49326
x_c2	1	−0.01877	0.00896	−2.10	0.0473	0.96118	1.04039
x_c3	1	−0.00477	0.00137	−3.49	0.0020	0.15603	6.40911

Exhibit 16.3 SAS Code for Cubic Polynomial Estimation (slstay=0.01)

```
model y = x_c x_c2 x_c3/ selection=
    backward slstay = 0.01 tol vif;
run;
```

If *slstay* = 0.01 is actually used, then the resulting output appears as Table 16.4. Note that, with the regressor *X_C2* removed in Step 1, the overall R^2 value along with the added-last *F* values for *X_C* and *X_C3* decrease a bit, whereas Mallows' *C*(*p*) increases from 4.0 to 6.39, thus indicating that some relevant regressor has been eliminated from the model.

TABLE 16.4

Output for Example 16.2 (slstay=0.01)

The SAS System

The REG Procedure

Model : MODEL1

Dependent Variable: y

Backward Elimination: Step 0

All Variables Entered: $R^2 = 0.8891$ and C(p) = 4.0000

Analysis of Variance

Source	DF	Sum of Squares	Mean Square	F Value	Pr > F
Model	3	1755.06565	585.02188	61.44	<.0001
Error	23	219.00843	9.52211		
Corrected total	26	1974.07407			

TABLE 16.4 *(continued)*

Variable	Parameter Estimate	Standard Error	Type II SS	F Value	Pr > F
Intercept	22.08137	0.76616	7909.34743	830.63	<.0001
x_c	1.76516	0.21023	671.29522	70.50	<.0001
x_c2	–0.01877	0.00896	41.82787	4.39	0.0473
x_c3	–0.00477	0.00137	116.03453	12.19	0.0020

Bounds on condition number: 6.4933, 41.828

Backward Elimination: Step 1
variable x_c2 Removed: R^2 = 0.8679 and C(p) = 6.3927

Analysis of Variance

Source	DF	Sum of Squares	Mean Square	F Value	Pr > F
Model	2	1713.23778	856.61889	78.82	<.0001
Error	24	260.83630	10.86818		
Corrected total	26	1974.07407			

Backward Elimination: Step 1

Variable	Parameter Estimate	Standard Error	Type II SS	F Value	Pr > F
Intercept	21.08189	0.64065	11769	1082.88	<.0001
x_c	1.68176	0.22054	632.00092	58.15	<.0001
x_c3	–0.00433	0.00144	98.06585	9.02	0.0061

Bounds condition number: 6.2606, 25.042

All variables left in the model are significant at the 0.0100 level.

Summary of Backward Elimination

Step	Variable Removed	Number Vars In	Partial R^2	Model R^2	C(p)	F Value	Pr > F
1	x_c2	2	0.0212	0.8679	6.3927	4.39.	0.0473

(continued)

TABLE 16.4 *(continued)*

The REG Procedure
Model : MODEL1
Dependent Variable: y

Analysis of Variance

Source	DF	Sum of Squares	Mean Square	F Value	Pr > F
Model	2	1713.23778	856.61889	78.82	<.0001
Error	24	260.83630	10.86818		
Corrected total	26	1974.07407			

Root MSE	3.29669	*R*-Square	0.8679
Dependent mean	20.81481	Adj *R*-Sq	0.8569
Coeff Var	15.83820		

Parameter Estimates

Variable	DF	Parameter Estimate	Standard Error	t Value	Pr > |t|	Tolerance	Variance Inflation
Intercept	1	21.08189	0.64065	32.91	<.0001	.	0
x_c	1	1.68176	0.22054	7.63	<.0001	0.15973	6.26059
x_c3	1	−0.00433	0.00144	−3.00	0.0061	0.15973	6.26059

Orthogonal Polynomials

Although centering the variable X to its mean tends to diminish the degree of near-multicollinearity or ill-conditioning that can occur when using the regressors 1, X, X^2, X^3,…, to fit a polynomial in X, it by no means eliminates the problem if the polynomial is of sufficiently high degree. An alternative approach to polynomial fitting (one designed to avoid the possibility of near-multicollinearity at the outset) is to use *orthogonal polynomials.* Such polynomials are used to estimate a polynomial model of essentially any order in X (the maximum possible order is $n - 1$ if n data points (X_i, Y_i), $i = 1,\ldots,n$, are available), where the X values are typically taken to be equally spaced.

To estimate the mth degree polynomial model

$$Y_i = \beta_0 + \beta_1 X_i + \beta_2 X_i^2 + \cdots + \beta_m X_i^m + \varepsilon_i,\ i = 1,\cdots,n, \tag{16.10}$$

using orthogonal polynomials, we replace the nonorthogonal columns of $\boldsymbol{X}$ corresponding to the regressors 1, X, X^2, X^3,…,X^m by a set of substitute columns that are generated by polynomials of varying degrees that are designed to be orthogonal. These stylized polynomials take the form

$$\begin{aligned}
&\phi_0(X_i) = 1 && \text{(zero-order)}\\
&\phi_1(X_i) = P_{01} + P_{11}X_i && \text{(first-order)}\\
&\phi_2(X_i) = P_{02} + P_{12}X_i + P_{22}X_i^2 && \text{(second-order)}\\
&\quad\bullet\\
&\quad\bullet\\
&\quad\bullet\\
&\phi_m(X_i) = P_{0m} + \cdots + P_{m-1,m}X_i^{m-1} + P_{mm}X_i^m, && (m\text{th order})
\end{aligned} \tag{16.11}$$

$i = 1,\ldots,n$, where the mutual orthogonality of the polynomials in Equation 16.11 is guaranteed by the *orthogonality requirement*

$$\begin{aligned}
&<\phi_j,\phi_l> = \textstyle\sum_{i=1}^{n}\phi_j(X_i)\phi_l(X_i) = 0,\ j \neq l,\\
&\text{with } j,l = 0,1,\ldots,m, \text{ and } \phi_0(X_i) \equiv 1.\\
&\left[\text{Orthogonal Polynomials}\right]
\end{aligned} \tag{16.12}$$

Given Equations 16.11 and 16.12, we may rewrite Equation 16.10 as

$$Y_i = \alpha_0\phi_0(X_i) + \alpha_1\phi_1(X_i) + \alpha_2\phi_2(X_i) + \cdots + \alpha_m\phi_m(X_i) + \varepsilon_i,\ i = 1,\ldots, m,$$

or

$\boldsymbol{Y} = \alpha\boldsymbol{X} + \boldsymbol{\varepsilon}$, where

$$\underset{(n\times m+1)}{\boldsymbol{X}} = \begin{bmatrix}
\phi_0(X_1) & \phi_1(X_1) & \ldots & \phi_m(X_1)\\
\phi_0(X_2) & \phi_1(X_2) & \ldots & \phi_m(X_2)\\
\cdot & \cdot & & \cdot\\
\cdot & \cdot & & \cdot\\
\cdot & \cdot & & \cdot\\
\phi_0(X_n) & \phi_1(X_n) & \ldots & \phi_m(X_n)
\end{bmatrix}.$$

Because the columns of $\boldsymbol{X}$ are orthogonal,

$$\boldsymbol{X}'\boldsymbol{X} = \begin{bmatrix} \sum_{i=1}^{n} \phi_0^2(X_i) & & & & \mathbf{0} \\ & \sum_{i=1}^{n} \phi_1^2(X_i) & & & \\ & & \cdot & & \\ & & & \cdot & \\ & & & & \cdot \\ \mathbf{0} & & & & \sum_{i=1}^{n} \phi_m^2(X_i) \end{bmatrix}.$$

And because $\boldsymbol{X'X}$ is a diagonal matrix,

$$(\boldsymbol{X}'\boldsymbol{X})^{-1} = \begin{bmatrix} 1\Big/\sum_{i=1}^{n} \phi_0^2(X_i) & & & & \mathbf{0} \\ & 1\Big/\sum_{i=1}^{n} \phi_1^2(X_i) & & & \\ & & \cdot & & \\ & & & \cdot & \\ & & & & \cdot \\ \mathbf{0} & & & & 1\Big/\sum_{i=1}^{n} \phi_m^2(X_i) \end{bmatrix}.$$

With

$$\boldsymbol{X}'\boldsymbol{Y} = \begin{bmatrix} \sum_{i=1}^{n} \phi_0(X_i)Y_i \\ \sum_{i=1}^{n} \phi_1(X_i)Y_i \\ \cdot \\ \cdot \\ \cdot \\ \sum_{i=1}^{n} \phi_m(X_i)Y_i \end{bmatrix},$$

the OLS estimator for α is $\hat{\boldsymbol{\alpha}} = (\boldsymbol{X}'\boldsymbol{X})^{-1}\boldsymbol{X}'\boldsymbol{Y}$ or

$$\begin{bmatrix} \hat{\alpha}_0 \\ \hat{\alpha}_1 \\ \cdot \\ \cdot \\ \cdot \\ \hat{\alpha}_m \end{bmatrix} = \begin{bmatrix} \sum_{i=1}^{n} \phi_0(X_i)Y_i \Big/ \sum_{i=1}^{n} \phi_0^2(X_i) \\ \sum_{i=1}^{n} \phi_1(X_i)Y_i \Big/ \sum_{i=1}^{n} \phi_1^2(X_i) \\ \cdot \\ \cdot \\ \cdot \\ \sum_{i=1}^{n} \phi_m(X_i)Y_i \Big/ \sum_{i=1}^{n} \phi_m^2(X_i) \end{bmatrix}. \tag{16.13}$$

In general,

$$\hat{\alpha}_k = \sum_{i=1}^{n} \phi_k(X_i)Y_i \Big/ \sum_{i=1}^{n} \phi_k^2(X_i), \quad k = 0,1,\ldots,m, \tag{16.13.1}$$

with $\hat{\alpha}_0 = \bar{Y}$ (since $\phi_0(X_i) \equiv 1$).

We know from Equations 12.13 and 12.14 (see also Appendix 12.D) that $V(\hat{\boldsymbol{\beta}}) = \sigma_\varepsilon^2(\boldsymbol{X}'\boldsymbol{X})^{-1}$. With σ_ε^2 estimated by S_ε^2, it follows that

$$S_{\hat{\alpha}_k} = S_\varepsilon \Big/ \sqrt{\sum_{i=1}^{n} \phi_k^2(X_i)}, \quad k = 0,1,\ldots,m. \tag{16.14}$$

Next, from Equation 12.18, we may write

$$\begin{aligned} SSE &= SST - SSR = SST - \sum_{j=1}^{m} \hat{\alpha}_j \left[\sum_{i=1}^{n} \phi_j(X_i)Y_i \right] \\ &= SST - \sum_{j=1}^{m} SSR(\hat{\alpha}_j). \end{aligned} \tag{16.15}$$

As this expression reveals, the sum of squares attributable to $\hat{\alpha}_j$ is

$$\begin{aligned} SSR(\hat{\alpha}_j) &= \hat{\alpha}_j \left[\sum_{i=1}^{n} \phi_j(X_i)Y_i \right] \\ &= \frac{\left(\sum_{i=1}^{n} \phi_j(X_i)Y_i \right)^2}{\sum_{i=1}^{n} \phi_j^2(X_i)}, \quad j = 1,\ldots,m, \end{aligned} \tag{16.16}$$

and is independent of the sum of squares attributable to any other estimated parameter.

Given Equations 16.15 and 16.16, we may now construct the partitioned sum of squares (ANOVA) table (Table 16.5).

To test H_0: $\alpha_k = 0$, against H_1: $\alpha_k \neq 0$, we may use the F statistic

$$F_k = \frac{MSR(\hat{\alpha}_k)}{MSE}, \quad k = 0,1,\ldots,m. \tag{16.17}$$

Then if $F_k > F_{\alpha,1,n-m-1}$, we reject H_0 in favor of H_1 at the α level of significance.

A few additional points are in order:

1. Suppose we fit a fourth-order polynomial model in orthogonal polynomials and we determine, via Equation 16.17, that H_0: $\alpha_4 = 0$ can safely be rejected (the

TABLE 16.5

Summary Table for the Partitioned Sums of Squares

Source of Variation in Y	Sum of Squares (SS)	Degrees of Freedom (d.f.)	Mean Square (MS=SS/d.f.)
Regression	$SSR = \sum_{j=1}^{m} SSR(\hat{\alpha}_j)$	m	$MSR = SSR/m$
$\hat{\alpha}_1$	$SSR(\hat{\alpha}_1)$	1	$MSR(\hat{\alpha}_1) = \hat{\alpha}_1 \sum_{i=1}^{n} \phi_1(X_i)Y_i$
.	.	.	.
.	.	.	.
.	.	.	.
$\hat{\alpha}_m$	$SSR(\hat{\alpha}_m)$	1	$MSR(\hat{\alpha}_m) = \hat{\alpha}_m \sum_{i=1}^{n} \phi_m(X_i)Y_i$
Error	$SSE = SST - \sum_{j=1}^{m} SSR(\hat{\alpha}_j)$	$n-m-1$	$MSE = SSE/(n-m-1)$
Total	$SST = \sum_{j=1}^{n} Y_i^2$	$n-1$	

associated p-value is, say, less than 0.05). Then the form of our estimated orthogonal polynomial model is

$$\hat{Y} = \hat{\alpha}_0 + \hat{\alpha}_1\phi_1(X) + \hat{\alpha}_2\phi_2(X) + \hat{\alpha}_3\phi_3(X) + \hat{\alpha}_4\phi_4(X), \tag{16.18}$$

with all lower-order terms retained.

2. If we choose to fit a higher-order model in orthogonal polynomials (the order changes from m to, say, $m + 2$), then all we need to do is include the additional terms $\alpha_{m+1}\phi_{m+1}(X_i) + \alpha_{m+2}\phi_{m+2}(X_i)$ in Equation 16.10.1. However, because the polynomials $\phi_0, \phi_1, \phi_2, \ldots, \phi_m, \phi_{m+1}$, and ϕ_{m+2} are mutually orthogonal, only $\hat{\alpha}_{m+1}$ and $\hat{\alpha}_{m+2}$ need to be computed; the original coefficients $\hat{\alpha}_k$, $k = 0, 1, \ldots, m$, do not change.
3. When the levels of X are equally spaced, the orthogonal polynomials $\phi_k(X_i)$ can readily be determined for any X_i's, $i = 1,\ldots,n$. Tables of values of $\phi_k(X_i)$ and $\sum_{i=1}^{n} \phi_k^2(X_i)$ for $k = 1,\ldots,6$ and $n \le 52$ are provided by Pearson and Hartley [1966]; for more extensive tables, see DeLury [1960] and Wilkie [1965]. A short table of orthogonal polynomial coefficients is provided in the Appendix at the end of the book (Table A.9). Orthogonal polynomial models that accommodate the unequal spacing of the X values can also be constructed [Wishart and Metakides 1953; Robson 1959; Seber 1977].

4. Let d (=constant) represent the distance between successive (equally spaced) values of X with λ_i (=constant) chosen so that the orthogonal polynomials have integer values. Then for $k = 0,1,\ldots 6$, the orthogonal polynomials for any n are

$$
\begin{aligned}
\phi_0(X) &\equiv 1,\\
\phi_1(X) &= \lambda_1\left(\frac{X-\bar{X}}{d}\right),\\
\phi_2(X) &= \lambda_2\left[\left(\frac{X-\bar{X}}{d}\right)^2 - \frac{1}{12}(n^2-1)\right],\\
\phi_3(X) &= \lambda_3\left[\left(\frac{X-\bar{X}}{d}\right)^3 - \frac{1}{20}(3n^2-7)\left(\frac{X-\bar{X}}{d}\right)\right],\\
\phi_4(X) &= \lambda_4\left[\left(\frac{X-\bar{X}}{d}\right)^4 - \frac{1}{14}(3n^2-13)\left(\frac{X-\bar{X}}{d}\right)^2 + \frac{3}{560}(n^2-1)(n^2-9)\right],\\
\phi_5(X) &= \lambda_5\left[\left(\frac{X-\bar{X}}{d}\right)^5 - \frac{5}{18}(n^2-7)\left(\frac{X-\bar{X}}{d}\right)^3\right.\\
&\qquad \left. + \frac{1}{1008}(15n^4-230n^2+407)\left(\frac{X-\bar{X}}{d}\right)\right], \text{ and}\\
\phi_6(X) &= \lambda_6\left[\left(\frac{X-\bar{X}}{d}\right)^6 - \frac{5}{44}(3n^2-31)\left(\frac{X-\bar{X}}{d}\right)^4\right.\\
&\qquad - \frac{1}{176}(5n^4-110n^2+329)\left(\frac{X-\bar{X}}{d}\right)^2\\
&\qquad \left. - \frac{5}{14784}(n^2-1)(n^2-9)(n^2-25)\right].
\end{aligned}
\tag{16.19}
$$

Example 16.3

Given the $n = 10$ data points displayed in Table 16.6, determine the equation of the third-order polynomial that best fits this dataset using orthogonal polynomials. Our goal is to fit the polynomial equation

$$Y_i = \alpha_0\phi_0(X_i) + \alpha_1\phi_1(X_i) + \alpha_2\phi_2(X_i) + \alpha_3\phi_3(X_i) + \varepsilon_i, \quad i = 1,\ldots,10.$$

For convenience, the coefficients of the orthogonal polynomials ϕ_0, ϕ_1, ϕ_2, and ϕ_3 appearing in Appendix Table A.9 are replicated as Table 16.7. Then:

TABLE 16.6
Observations on the Variables X and Y

X	Y
5	5
10	30
15	35
20	39
25	40
30	39
35	35
40	34
45	40
50	65

$$\sum_{i=1}^{10}\phi_0^2(X_i)=10=n,\ \sum_{i=1}^{10}\phi_1^2(X_i)=330,\ \sum_{i=1}^{10}\phi_2^2(X_i)=132,\ \sum_{i=1}^{10}\phi_3^2(X_i)=8580,$$

$$\sum_{i=1}^{10}\phi_0(X_i)Y_i=\sum Y_i=5+30+\cdots+65=362,$$

$$\sum_{i=1}^{10}\phi_1(X_i)Y_i=(-9)(5)+(-7)(30)+\cdots+(9)(65)=592,$$

$$\sum_{i=1}^{10}\phi_2(X_i)Y_i=(6)(5)+(2)(30)+\cdots+(6)(65)=-47,$$

$$\sum_{i=1}^{10}\phi_3(X_i)Y_i=(-42)(5)+(14)(30)+\cdots+(42)(65)=2551.$$

Then, from Equation 16.13,

$$\begin{bmatrix}\hat{\alpha}_0\\ \hat{\alpha}_1\\ \hat{\alpha}_2\\ \hat{\alpha}_3\end{bmatrix}=\begin{bmatrix}362/10\\ 592/330\\ -47/132\\ 2551/8580\end{bmatrix}=\begin{bmatrix}36.200\\ 1.794\\ -0.356\\ 0.297\end{bmatrix}$$

and thus the estimated third-order orthogonal polynomial model is

$$\hat{Y}=36.200+1.794\phi_1(X)+-0.356\phi_2(X)+0.297\phi_3(X).$$

With

$$SST=\sum_{i=1}^{10}Y_i^2-n\bar{Y}^2=(5)^2+\cdots+(65)^2-13104.40=1893.6,$$

$$SSR=\sum_{j=1}^{3}\hat{\alpha}_j\left(\sum_{i=1}^{10}\phi_j(X_i)Y_i\right)$$

$$=1.794(592)+(-0.356)(-47)+0.297(2551)$$

$$=1062.05+16.73+757.65=1836.43$$

and

$$SSE = SST - SSR = 57.17,$$

Table 16.8 provides us with the complete partitioned sums of squares or ANOVA results.

TABLE 16.7

Orthogonal Polynomial Coefficients

i	ϕ_0	ϕ_1	ϕ_2	ϕ_3
1	1	−9	6	−42
2	1	−7	2	14
3	1	−5	−1	35
4	1	−3	−3	31
5	1	−1	−4	12
6	1	1	−4	−12
7	1	3	−3	−31
8	1	5	−1	−35
9	1	7	2	−14
10	1	9	6	42
$\Sigma_{i=1}^{10}\phi_j^2(X_i)$	10	330	132	8580
λ_k	—	$\lambda_1 = 2$	$\lambda_2 = \frac{1}{2}$	$\lambda_3 = \frac{5}{3}$

TABLE 16.8

ANOVA Results for the Third-Order Orthogonal Polynomial Model

Source of Variation in Y	Sum of Squares (SS)	Degrees of Freedom (d.f.)	Mean Square (MS=SS/d.f.)
Regression	$SSR = 1836.43$	3	$MSR = 612.14$
$\hat{\alpha}_1$	$SSR(\hat{\alpha}_1) = 1062.05$	1	$MSR(\hat{\alpha}_1) = 1062.05$
$\hat{\alpha}_2$	$SSR(\hat{\alpha}_2) = 16.73$	1	$MSR(\hat{\alpha}_2) = 16.73$
$\hat{\alpha}_3$	$SSR(\hat{\alpha}_3) = 757.65$	1	$MSR(\hat{\alpha}_3) = 757.65$
Error	$SSE = SST - SSR$ $= 1893.60 - 1836.43$ $= 57.17$	$10 - 3 - 1 = 6$	$MSE = 57.15/6 = 9.53$
Total	$SST = 1893.60$	9	

Given the information presented in Table 16.8, we can readily test $H_0 : \alpha_1 = \alpha_2 = \alpha_3 = 0$, versus H_1 : not all α_j's equal to zero, $j = 1, 2, 3$. To this end, we have $F = MSR/MSE = 64.23$, p-value <0.0001. Hence, the cubic model in orthogonal polynomials explains a statistically significant portion of the variation in Y. In fact, $R^2 = 0.97$. Next, from Equation 16.17,

$$F_1 = \frac{MSR(\hat{\alpha}_1)}{MSE} = 111.44, \; p-\text{value} < 0.0001;$$

$$F_2 = \frac{MSR(\hat{\alpha}_2)}{MSE} = 1.76, \; p-\text{value} > 0.10; \text{ and}$$

$$F_3 = \frac{MSR(\hat{\alpha}_3)}{MSE} = 79.50, \; p-\text{value} < 0.0001.$$

Because F_3 is statistically significant, we see that this third-order model explains most of the variation in Y and thus serves as our model of choice.

Finally, let us use Equation 16.19 to express our estimated polynomial in terms of X as

$$\begin{aligned}\hat{Y} &= 36.200 + 1.794\phi_1(X) - 0.356\phi_2(X) + 0.297\phi_3(X) \\ &= 36.200 + 1.794(2)\left(\frac{X-27.5}{5}\right) - 0.356\left(\frac{1}{2}\right)\left[\left(\frac{X-27.5}{5}\right)^2 - \frac{99}{12}\right] \\ &\quad + 0.297\left(\frac{5}{3}\right)\left[\left(\frac{X-27.5}{5}\right)^3 - \frac{293}{20}\left(\frac{X-27.5}{5}\right)\right] \\ &= 37.669 - 0.733(X-27.5) - 0.007(X-27.5)^2 \\ &\quad + 0.004(X-27.5)^3.\end{aligned}$$

Example 16.4

Let us use the set of n =10 data points from the preceding example problem and use the SAS system to fit a third-order polynomial model using orthogonal polynomials. The SAS code appears in Exhibit 16.4.

Exhibit 16.4 SAS Code for Orthogonal Polynomial Regression

```
data cubeor;
   input  x  y;
datalines;       ① We invoke the IML procedure.
 5     5         ② We import values of X from dataset cubeor into a vector X1.
10     30        ③ The orpol function is used to obtain orthogonal
15     35          polynomial scores for  the X  elements of vector X1.
20     39          The resulting scores are placed in the matrix poly.
25     40        ④ mtx is a matrix that has the same number of rows
30     39          and orthogonal polynomial scores for φ1, φ2, and φ3
35     35          as does poly. (":" is the index operator: 2:4 selects
40     34          columns 2, 3, and 4 of poly.)
```

```
45      40        ⑤ A new dataset cube1 is created from the
50      65          columns of mtx. Variable names are specified using the
 ;                  previously defined vector X1.
proc iml;   ①
     use cubeor;
read all var "x" into x1;   ②
poly=orpol (x1,3) ;   ③
mtx=poly[ ,2:4] ;   ④
create cube1 var {x1_or1  x1_or2  x1_or3} ;   ⑤
     append  from  mtx;   ⑥
     close cube1 ;
run;
proc  print data=cube1 ;
run;
data cube2; merge cubeor cube1 ;   ⑦
proc reg data=cube2;
     model y=x1_or1  x1_or2  x1_or3/tol  vif  ss1  pcorr1;   ⑧
run;
```

⑥ The *APPEND* statement writes each row of *mtx* as an observation to the dataset *cube1*.

⑦ Our final dataset *cube2* results from the merger of *cubeor* and *cube1*.

⑧ Using dataset *cube2*, we regress *Y* on the orthogonal polynomial scores generated in ③ and named in ⑤.

The results of the preceding orthogonal regression program appear in Table 16.9. Specifically,

① We request a print of the orthogonal polynomial scores for ϕ_1, ϕ_2, and ϕ_3 (remember that $\phi_0 = 1$ for all *i*).

②, ③ The overall *F* value for testing the significance of the entire third-order orthogonal polynomial model is $F = 65.16$, *p*-value <0.0001. Hence, we reject $H_0: \alpha_j = 0$, $j = 1, 2, 3$, in favor of H_1 : not H_0. Also, because $R^2 = 0.97$, the model displays a high degree of explanatory power.

④ Our estimated third-order polynomial in orthogonal polynomials is the form

$$\hat{Y} = \hat{\alpha}_0\phi_0 + \hat{\alpha}_1\phi_1 + \hat{\alpha}_2\phi_2 + \hat{\alpha}_3\phi_3$$

$$= 36.200 + 32.589\phi_1 - 4.091\phi_2 + 27.540\phi_3.$$

⑤ The *t* values for the intercept and the estimated coefficients on regressors X1_OR1, X1_OR3 are highly significant; their accompanying *p*-values ≤ 0.0001. If the *t* values are squared, then we get (approximately) the set of F_j's, $j = 1, 2, 3$, calculated in the preceding example problem.

⑥ For the Type I sums of squares associated with the regressors, we obtain $SSR(\hat{\alpha}_1) = 1062.012$, $SSR(\hat{\alpha}_2) = 16.734$, and $SSR(\hat{\alpha}_3) = 758.461$. (If each of these regression sums of squares is divided by MSE, then we obtain the sample F_j's, $j = 1, 2, 3$.) Remember also that these Type I sums of squares are sequential sums of squares: $SSR(X1\_or1|\text{intercept}) = 1062.012$ is the regression sum of squares for a model containing only *X1_OR1* and an intercept, etc.

TABLE 16.9
Output for Example 16.4

The SAS System
①

Obs	X1_OR1	X1_OR2	X1_OR3
1	−0.49543	0.52223	−0.45343
2	−0.38534	0.17408	0.15114
3	−0.27524	−0.08704	0.37785
4	−0.16514	−0.26112	0.33467
5	−0.05505	−0.34816	0.12955
6	0.05505	−0.34816	−0.12955
7	0.16514	−0.26112	−0.33467
8	0.27524	−0.08704	−0.37785
9	0.38534	0.17408	−0.15114
10	0.49543	0.52223	0.45343

The SAS System
The REG Procedure
Model : MODEL1
Dependent Variable: y

Analysis of Variance

Source	DF	Sum of Squares	Mean Square	② F Value	Pr > F
Model	3	1837.20862	612.40287	65.16	<.0001
Error	6	56.39138	9.39856		
Corrected total	9	1893.60000			

③

Root MSE	3.06571	*R*-Square	0.9702
Dependent Mean	36.20000	Adj *R*-Sq	0.9553
Coeff Var	8.46881		

TABLE 16.9 (*continued*)

Parameter Estimates

Variable	DF	④ Parameter Estimate	Standard Error	⑤ t Value	Pr > |t|	⑥ Type I SS	⑦ Squared Partial Corr Type I
Intercept	1	36.20000	0.96946	37.34	<.0001	13104	.
X1_OR1	1	32.58853	3.06571	10.63	<.0001	1062.01212	0.56084
X1_OR2	1	–4.09082	3.06571	–1.33	0.2305	16.73485	0.02012
X1_OR3	1	27.54018	3.06571	8.98	0.0001	758.46166	0.93080

Parameter Estimates

Variable	DF	Tolerance	⑧ Variance Inflation
Intercept	1	.	0
X1_OR1	1	1.00000	1.00000
X1_OR2	1	1.00000	1.00000
X1_OR3	1	1.00000	1.00000

⑦ Here $r^2_{YX1\_OR1} = 0.5608$: when *X1_OR1* is added to a regression model that only displays an intercept, SSE decreases by 56.08%, etc.

⑧ As constructed, the regressors are orthogonal, as evidenced by the tolerance and variance inflation factors.

Spline Functions

It may be the case that a response variable Y behaves differently in different parts or segments of the range of some regressor X, e.g., if X = time in, say, years, then the entire time span may cover p distinct subperiods, with each subperiod corresponding to the years in which an incumbent president was either a Democrat or Republican, or if X = age (also in years), then the beginning of each of the p subperiods may correspond to some milestone event in a person's life. Whatever the nature of X happens to be, it should be clear that, throughout each subperiod or segment, the intercept as well as the slope of the regression equation can change, thus causing jumps or discontinuities in a regression function. To model such changes, spline functions are frequently used.

Spline functions are piecewise polynomials of order k that meet at threshold or join points called *knots*. Hence, the spline function is continuous at the knots. In this regard, not only are the function values equal at the knots, it is generally required that the values of the first $k - 1$ derivatives agree at the knots.

As a practical matter, some important considerations regarding the fitting of spline functions are (1) determining the requisite number and location of the knots and (2) determining the order of the polynomial that is appropriate for each subperiod or segment. General guidelines are (1) use as few knots as possible, (2) position the knots so that each subperiod or segment has at least five data points, (3) there should be only one maximum or minimum point within each subperiod or segment (with any such point centered in the segment), and (4) there should be no more than one point of inflection within a segment (and it should be located near a knot).

Linear Splines

Single Known Knot

Suppose the range of X is divided into two segments, denoted I and II (Figure 16.5) and that the regression model is structured as

$$\text{(I) } Y = \beta_{0\text{I}} + \beta_{1\text{I}}X + \varepsilon,\ X < a;$$
$$\text{(II) } Y = \beta_{0\text{II}} + \beta_{1\text{II}}X + \varepsilon,\ X \geq a. \tag{16.20}$$

Let us rewrite Equation 16.20 in what we shall call its estimable form. If D is a dummy variable with

$$D = \begin{cases} 0, & X < a; \\ 1, & X \geq a, \end{cases}$$

then the dashed lines in Figure 16.5 can be represented by the equation

$$Y = \beta_0 + \beta_1 X + \gamma D + \delta DX + \varepsilon, \tag{16.21}$$

i.e., for $D = 0$,

$$\text{(I) } Y = \beta_0 + \beta_1 X + \varepsilon;$$

and for $D = 1$

$$\text{(II) } Y = (\beta_0 + \gamma) + (\beta_1 + \delta)X + \varepsilon.$$

To make Equation 16.21 continuous at the knot $X = a$, our continuity restriction requires that the line segments join at point A of Figure 16.5. That is, their function values are the same at $X = a$ so that

$$\beta_0 + \beta_1 a = (\beta_0 + \gamma) + (\beta_1 + \delta)a$$

or $0 = \gamma + \delta a$ and thus $\gamma = -\delta a$. If this latter expression is substituted into Equation 16.21, we obtain the spline function

$$Y = \beta_0 + \beta_1 X + \delta D(X - a) + \varepsilon. \tag{16.22}$$

In this regard, if $X < a$ then $D = 0$ so that Equation 16.22 becomes

$$Y = \beta_0 + \beta_1 X + \varepsilon;$$

and if $X \geq a$, then $D = 1$ so that Equation 16.22 appears as

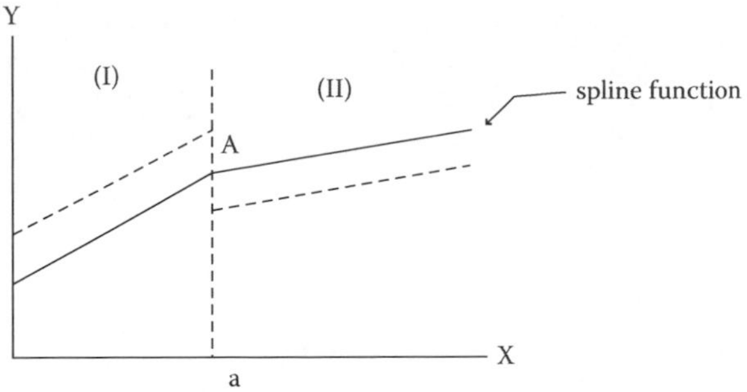

FIGURE 16.5
Spline function with single known knot.

$$Y = \beta_0 + \beta_1 X + \delta(X - a) + \varepsilon.$$

For notational convenience, let us rewrite the spline function (Equation 16.22) as

$$Y = \beta_0 + \beta_1 X + \delta(X - a)_+ + \varepsilon,$$

$$(X - a)_+ = \begin{cases} 0, & X < a; \\ X - a, & X \geq a. \end{cases} \qquad (16.22.1)$$

Example 16.5

Given the dataset presented in Table 16.10, let us estimate the first-order spline model (Equation 16.22.1), in which the knot occurs at $X = a = 6$ (Figure 16.6). The SAS code appears in Exhibit 16.5.

TABLE 16.10
Observations on Variables X and Y

X	*Y*
1	2
2	3
3	4.5
4	6
5	6
6	7
7	7.5
8	8.25
9	8
10	8.25
11	8.5
12	9
13	9.25
14	9.5
15	9

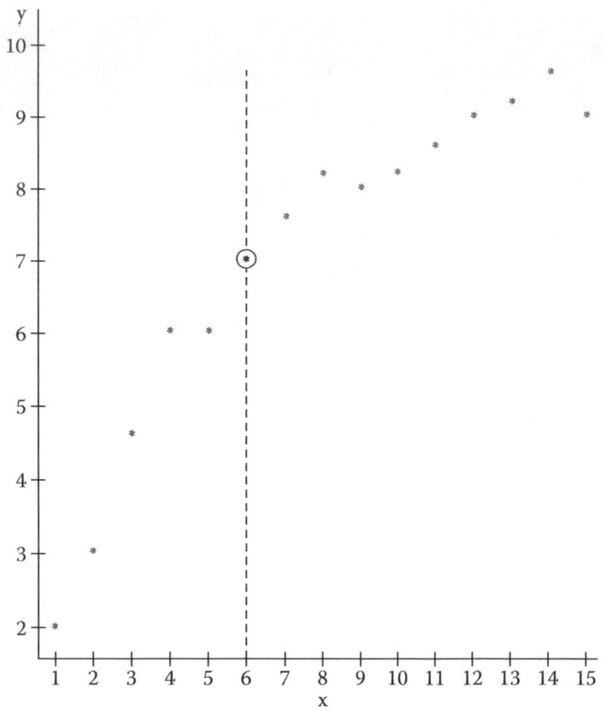

FIGURE 16.6
Plot of first-order spline.

Exhibit 16.5 SAS Code for First-Order Spline Regression (One Known Knot)

```
data spline1;
   input x y ;

datalines ;

 1      2

 2      3

 .      .
 .      .
 .      .
15      9
;       ①
data spline ; set spline1;
     xplus6=max (x-6,0);②
run;
proc reg data=spline;
     model y=x  xplus6/ss1 pcorr1;     ③
run;
```

① We create a new dataset *spline* that houses X, Y, and the variable

$$(X-6)_+ = \begin{cases} 0, & X < 6; \\ X-6, & X \ge 6. \end{cases}$$

② $(X-6)_+$ is created using the *max* function

$$\max(X - X^*, 0) = \begin{cases} 0, & X < X^*; \\ X - X^*, & X \ge X^*, \end{cases}$$

i.e., this function reports the maximum of the two values specified in the arguments. Hence, we define the variable *Xplus6* = max(X −6, 0) so that, for $X < 6$, *Xplus6* = 0; and for $X \ge 6$, *Xplus6* = $X - 6$.

③ The spline model is $Y = \beta_0 + \beta_1 X + \beta_2 Xplus6 + \varepsilon$.

```
proc plot data=spline1;
    plot y*x='x';
run;
```

The output generated by this SAS program appears in Table 16.11.

① Given that the overall *F* statistic is $F = 300.02$, p – value < 0.0001, we can readily reject $H_0 : \beta_1 = \beta_2 = 0$, in favor of H_1 : not H_0, and conclude that our first-order spline model explains a statistically significant portion of the variation in *Y*. In fact, the model explains 98.04% of said variation, thus indicating that the fit is quite good.

② The estimated regression equation is

$$\hat{Y} = 1.1320 + 1.0434X - 0.8098Xplus6.$$

Note that each estimated coefficient is highly significant because its associated *p*-value is less than or equal to 0.003 (Table 16.12). So for $X < 6$, $Xplus6 = 0$ and thus

TABLE 16.11

Output for Example 16.5

The SAS System

The REG Procedure

Model : MODEL1

Dependent Variable: y

①

Analysis of Variance

Source	DF	Sum of Squares	Mean Square	F Value	Pr > F
Model	2	73.67657	36.83829	300.02	<.0001
Error	12	1.47343	0.12279		
Corrected total	14	75.15000			

Root MSE	0.35041	*R*-Square	0.9804
Dependent Mean	7.05000	Adj *R*-Sq	0.9771
Coeff Var	4.97032		

Parameter Estimates

Variable	DF	② Parameter Estimate	Standard Error	t Value	Pr > |t|	③ Type I SS	④ Squared Partial Corr Type I
Intercept	1	1.13208	0.30568	3.70	0.0030	745.53750	.
x	1	1.04340	0.06807	15.33	<.0001	64.36808	0.85653
xplus6	1	–0.80975	0.09300	–8.71	<.0001	9.30849	0.86334

$$\hat{Y} = 1.1320 + 1.0434X;$$

and for $X \geq 6$, $Xplus6 = X - 6$ so that

$$\hat{Y} = 5.9908 + 0.2336X.$$

③ The Type I or sequential sums of squares are as follows:

$SSR(X|\text{intercept}) = 64.3680$: the regression sum of squares for a model containing only X and an intercept; and

$SSR(Xplus6|\text{intercept}, X) = 9.3085$: the increment in the regression sum of squares attributable to the addition of $Xplus6$ to a model already containing an intercept and X.

(If we divide each sequential sum of squares by $MSE = 0.1228$, we obtain a set of F statistics that can be used to determine whether each additional regressor contributes significantly to SSR.)

④ The sequential or variables added-in-order squared partial correlation coefficients are as follows:

$r^2_{YX} = 0.8595$: when X is added to a regression model that only displays an intercept, SSE decreases by 85.86%; and

$r^2_{YXplus6|X} = 0.8633$: when $Xplus6$ is added to a regression model that already contains X, SSE decreases by 86.33%.

Multiple Known Knots

Let us assume that the range of X is divided into three segments, denoted as I, II, and III (Figure 16.7). Then the regression model is structured as

$$\begin{aligned} &\text{(I) } Y = \beta_{0\text{I}} + \beta_{1\text{I}}X + \varepsilon, X < a; \\ &\text{(II) } Y = \beta_{0\text{II}} + \beta_{1\text{II}}X + \varepsilon, a \leq X < b; \\ &\text{(III) } Y = \beta_{0\text{III}} + \beta_{1\text{III}}X + \varepsilon, X \geq b. \end{aligned} \tag{16.23}$$

Let us specify two dummy variables D_1 and D_2 as

$$D_1 = \begin{cases} 0, & X < a; \\ 1, & X \geq a, \end{cases} \qquad D_2 = \begin{cases} 0, & X < b; \\ 1, & X \geq b. \end{cases}$$

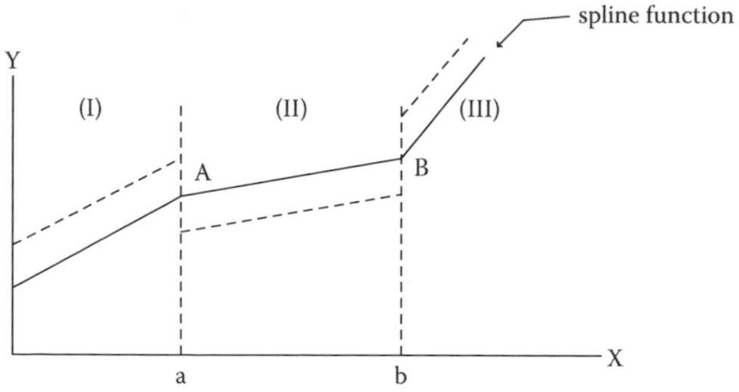

FIGURE 16.7
Spline function with multiple known knots.

Then Equation 16.23 can be rewritten as

$$Y = \beta_0 + \beta_1 X + \gamma_1 D_1 + \delta_1 D_1 X + \gamma_2 D_2 + \delta_2 D_2 X + \varepsilon \tag{16.24}$$

and used to represent the dashed lines in Figure 16.7, i.e., for $D_1 = D_2 = 0$,

$$\text{(I) } Y = \beta_0 + \beta_1 X + \varepsilon;$$

for $D_1 = 1$ and $D_2 = 0$,

$$\text{(II) } Y = (\beta_0 + \gamma_1) + (\beta_1 + \delta_1)X + \varepsilon;$$

and for $D_1 = D_2 = 1$,

$$\text{(III) } Y = (\beta_0 + \gamma_1 + \gamma_2) + (\beta_1 + \delta_1 + \delta_2)X + \varepsilon.$$

To make Equation 16.24 continuous, the line segments must join at the knots or at points A and B of Figure 16.7. To this end, we require that

$$\beta_0 + \beta_1 a = (\beta_0 + \gamma_1) + (\beta_1 + \delta_1)a$$

or $\gamma_1 = -\delta_1 a$. In addition, we must have

$$(\beta_0 + \gamma_1) + (\beta_1 + \delta_1)b = (\beta_0 + \gamma_1 + \gamma_2) + (\beta_1 + \delta_1 + \delta_2)b$$

or $\gamma_2 = -\delta_2 b$. Substituting these expressions for γ_1 and γ_2 into Equation 16.24 yields the spline function

$$Y = \beta_0 + \beta_1 X + \delta_1 D_1 (X - a) + \delta_2 D_2 (X - b) + \varepsilon. \tag{16.25}$$

In this regard, if $X < D_1$, then $D_1 = D_2 = 0$ and Equation 16.25 simplifies to

$$Y = \beta_0 + \beta_1 X + \varepsilon;$$

if $a \leq X < b$, then $D_1 = 1$ and $D_2 = 0$ so that Equation 16.25 appears as

$$Y = \beta_0 + \beta_1 X + \delta_1 (X - a) + \varepsilon;$$

and if $X \geq b$, then $D_1 = D_2 = 1$ and thus Equation 16.25 becomes

$$Y = \beta_0 + \beta_1 X + \delta_1 (X - a) + \delta_2 (X - b) + \varepsilon.$$

If we adopt the notational convention used in Equation 16.22.1, we can rewrite Equation 16.25 as

$$Y = \beta_0 + \beta_1 X + \delta_1 (X - a)_+ + \delta_2 (X - b)_+ + \varepsilon, \tag{16.25.1}$$

$$(X-a)_+ = \begin{cases} 0, & X < a; \\ X-a, & X \ge a, \end{cases}$$

$$(X-b)_+ = \begin{cases} 0, & X < b; \\ X-b, & X \ge b. \end{cases}$$

In general, if the range of X displays r known knots, $t_1 < t_2 < \ldots < t_r$, and it is required that the spline is continuous at the r knots, then the first-order spline function has the form

$$Y = \beta_0 + \beta_1 X + \sum_{j=1}^{r} \delta_j D_j (X - t_j) + \varepsilon,$$
$$D_j = \begin{cases} 0, & X < t_j; \\ X - t_j, & X \ge t_j, \; j = 1, \ldots, r, \end{cases} \tag{16.26}$$

or

$$Y = \beta_0 + \beta_1 X + \sum_{j=1}^{r} \delta_j (X - t_j)_+ + \varepsilon,$$
$$(X - t_j)_+ = \begin{cases} 0, & X < t_j; \\ X - t_j, & X \ge t_j, \; j = 1, \ldots, r. \end{cases} \tag{16.26.1}$$

Example 16.6

Let us use the dataset specified in Table 16.12 to estimate the spline function (Equation 16.26.1), where knots are found at $X = 6$ and $X = 16$ (Figure 16.8). Exhibit 16.6 displays the SAS code.

Exhibit 16.6 SAS Code for First-Order Spline Regression (Two Known Knots)

```
data spline2;
   input  x  y ;
datalines ;
 1    2
 2    3
 .    .
 .    .
 .  . .
```

① We create a new dataset *spline* that contains *X*, *Y*, and the variables

$$(X-6)_+ = \begin{cases} 0, & X < 6; \\ X-6, & X \ge 6. \end{cases}$$

$$(X-16)_+ = \begin{cases} 0, & X < 16; \\ X-16, & X \ge 16. \end{cases}$$

② $(X - 6)_+$ is created using the *max* function

$$\max(X-6, 0) = \begin{cases} 0, & X < 6; \\ X-6, & X \ge 6, \end{cases}$$

and $(X - 16)_+$ is created similarly as

$$\max(X-16, 0) = \begin{cases} 0, & X < 16; \\ X-16, & X \ge 16. \end{cases}$$

```
22      18
;   ①
data spline ; set spline2;
   xplus6 = max (x-6,0); ②
   xplus16 = max (x-16,0);
run;
proc reg data=spline;
   model y=x xplus6 xplus16/ss1 pcorr1; ③
run;
proc plot data=spline2;
   plot y*x='x';
run;
```

We then define the variables *Xplus6* = max(*X* − 6, 0), *Xplus16* = max(*X* − 16, 0).

③ The spline model is $Y = \beta_0 + \beta_1 X + \beta_2 Xplus6 + \beta_3 Xplus16 + \varepsilon$.

The output resulting from this SAS code appears in Table 16.13.

The reader can easily check that the overall fit is good (as evidenced by the realizations of the *F* statistic and R^2) and the *p*-values associated with the individual estimated regression coefficients

TABLE 16.12

Observations on Variables X and Y

X	*Y*
1	2
2	3
3	4.5
4	6
5	6
6	7
7	6
8	6.25
9	6
10	7
11	7.25
12	7.5
13	8
14	8.25
15	8.5
16	7.5
17	8
18	10
19	9
20	11
21	12
22	13

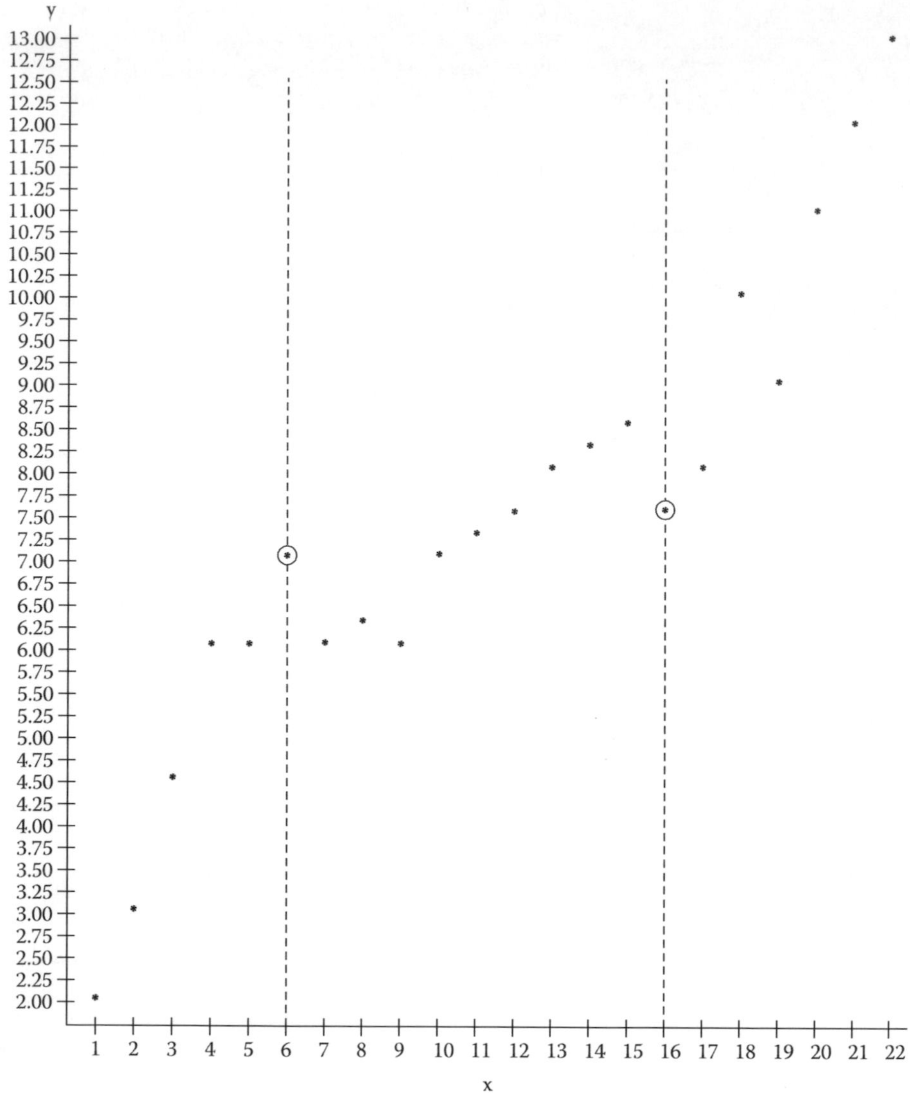

FIGURE 16.8
Plot of spline function with multiple known knots.

are all less than or equal to 0.0082. Clearly, the application of the first-order spline model produces some highly significant results. The estimated spline equation is

$$\hat{Y} = 1.6588 + 0.8176X - 0.6866\mathit{Xplus}6 + 0.6563\mathit{Xplus}16.$$

For $X < 6$, $\mathit{Xplus}6 = \mathit{Xplus}16 = 0$, thus

$$\hat{Y} = 1.6588 + 0.8176X;$$

for $6 \le X < 16$, $\mathit{Xplus}6 = X - 6$ and $\mathit{Xplus}16 = 0$ so that

TABLE 16.13

Output for Example 16.6

The SAS System

The REG Procedure

Model : MODEL1

Dependent Variable: y

Analysis of Variance

Source	DF	Sum of Squares	Mean Square	F Value	Pr > F
Model	3	136.38245	45.46082	109.34	<.0001
Error	18	7.48402	0.41578		
Corrected Total	21	143.86648			

Root MSE	0.64481	*R*-Square	0.9480
Dependent Mean	7.44318	Adj *R*-Sq	0.9393
Coeff Var	8.66309		

Parameter Estimates

Variable	DF	Parameter Estimate	Standard Error	t Value	Pr > |t|	Type I SS	Squared Partial Corr Type I
Intercept	1	1.65886	0.55865	2.97	0.0082	1218.82102	.
x	1	0.81763	0.12205	6.70	<.0001	124.57032	0.86587
xplus6	1	−0.68668	0.15693	−4.38	0.0004	1.84604	0.09567
xplus16	1	0.65633	0.13406	4.90	0.0001	9.96610	0.57112

$$\hat{Y} = 1.6588 + 0.8176X - 0.6866(X - 6)$$

$$= 5.7784 + 0.1310X;$$

and for $X \geq 16$, $Xplus6 = X - 6$ and $Xplus16 = X - 16$ so that

$$\hat{Y} = 1.6588 + 0.8176X - 0.6866(X - 6) + 0.6563(X - 16)$$
$$= -4.7224 + 0.7873X.$$

Quadratic Splines with Known Knots

It is readily apparent that at point (knot) *A* in Figure 16.5, there occurs an abrupt change in the slope of the spline function (the first derivative of the spline is discontinuous there). Given this observation, it can be argued that the actual change should be smooth (i.e., the estimated function should possess a continuous first derivative at all points within the range of *X* values). This requirement is easily addressed if we fit *quadratic spline functions*: piecewise second-order polynomials (parabolas) that meet at join points or knots.

For example, suppose the range of X is divided into two segments, denoted I and II, and that the regression model is specified as

$$\begin{aligned}&(\text{I})\ Y = \beta_{0\text{I}} + \beta_{1\text{I}}X + \beta_{2\text{I}}X^2 + \varepsilon,\ X < a;\\&(\text{II})\ Y = \beta_{0\text{II}} + \beta_{1\text{II}}X + \beta_{2\text{II}}X^2 + \varepsilon,\ X \geq a,\end{aligned} \tag{16.27}$$

where a single knot occurs at $X = a$. Given the dummy variable

$$D = \begin{cases} 0, & X < a; \\ 1, & X \geq a, \end{cases}$$

we can rewrite Equation 16.27 as

$$Y = \beta_0 + \beta_1 X + \beta_2 X^2 + \gamma_1 D + \delta_1 DX + \theta_1 DX^2 + \varepsilon. \tag{16.28}$$

Then for $D = 0$,

$$(\text{I})\ Y = \beta_0 + \beta_1 X + \beta_2 X^2 + \varepsilon; \tag{16.29.I}$$

and for $D = 1$,

$$(\text{II})\ Y = (\beta_0 + \gamma_1) + (\beta_1 + \delta_1)X + (\beta_2 + \theta_1)X^2 + \varepsilon. \tag{16.29.II}$$

We noted at the outset of our discussion of spline functions that the piecewise polynomials of order k are required to be continuous at the knots and that theirs first $k - 1$ derivatives also agree at these points. In this regard, if Equation 16.28 is to satisfy the continuity restriction at $X = a$, then, via Equations 16.29.I and 16.29.II), we require that

$$\beta_0 + \beta_1 a + \beta_2 a^2 = (\beta_0 + \gamma_1) + (\beta_1 + \delta_1)\text{a} + (\beta_2 + \theta_1)a^2$$

or $\gamma_1 = -\ \delta_1 a - \theta_1 a^2$. If this latter expression is substituted into Equation 16.28, we obtain

$$Y = \beta_0 + \beta_1 X + \beta_2 X^2 + \delta_1 D(X - a) - \theta_1 a^2 D + \theta_1 DX^2. \tag{16.28.1}$$

Next, the derivatives of Equations 16.29.I and 16.29.II with respect to X are, respectively,

$$\begin{aligned}&(\text{I})\ Y' = \beta_1 + 2\beta_2 X\\&(\text{II})\ \text{Y}' = \beta_1 + \delta_1 + 2(\beta_2 + \theta_1)X.\end{aligned}$$

If the values of these derivatives must also agree at $X = a$, then we require that

$$\beta_1 + 2\beta_2 a = \beta_1 + \delta_1 + 2(\beta_2 + \theta_1)a$$

or $\delta_1 = -2\theta_1 a$. If this expression for δ_1 is substituted into Equation 16.28.1, we ultimately obtain

$$Y = \beta_0 + \beta_1 X + \beta_2 X^2 + \theta_1 D(X - a)^2 + \varepsilon \tag{16.30}$$

(note that the term $(X - a)$ disappears) or

$$Y = \beta_0 + \beta_1 X + \beta_2 X^2 + \theta_1 (X - a)_+^2 + \varepsilon,$$
$$(X - a)_+ = \begin{cases} 0, & X < a; \\ X - a, & X \geq a. \end{cases} \quad (16.31)$$

So for $X < a$,

$$Y = \beta_0 + \beta_1 X + \beta_2 X^2 + \varepsilon;$$

and for $X \geq a$,

$$Y = \beta_0 + \beta_1 X + \beta_2 X^2 + \theta_1 (X - a)^2 + \varepsilon.$$

The parameters of an expression such as Equation 16.31 can be estimated by using the *SAS* code in Exhibit 16.7.

Exhibit 16.7 SAS Code for Quadratic Spline Estimation (Known Knot)

```
data spline3;
   input x  y;
datalines;
 .
 . }input values
 .
 ;   ①
data spline ; set spline3;
      x2=x**2;
      xplusa=max(x-a, 0);     ②
      xplusa2=xplusa**2;
run;
proc reg data = spline ;
      model y= x  x2  xplusa   xplusa2  / ss1  pcorr1; ③
      restrict  Xplusa = 0; ④
run;
```

Looking to the salient features of this program, the following applies:

①, ② We create a new dataset *spline* that houses *X*,*Y*, and the variables *X2* (the square of *X*), *Xplusa*, and *Xplusa2* (the square of *Xplusa*), where *Xplusa* is created via the *max* function.

③, ④ All variables are included in the *MODEL* statement, i.e., we seek to estimate the parameters of

$$E(Y) = \beta_0 + \beta_1 X + \beta_2 X2 + \beta_3 Xplusa + \beta_4 Xplusa2.$$

Because our smoothness restriction on the first derivative of the quadratic spline requires that $\beta_3 = 0$ (the term *Xplusa* or $X - a$ does not appear in Equation 16.31), we estimate this model subject to the linear restriction that $\beta_3 = 0$. This is performed by using the RESTRICT statement, i.e., we stipulate that the coefficient on *Xplusa* equals zero by inserting the statement

restrict xplusa = 0;

The RESTRICT statement offers the advantage of providing us with a test of the impact of the restriction that $\beta_3 = 0$ on the regression results, i.e., if the *p*-value associated with the variable RESTRICT is less than some chosen α, then we can reject $H_0 : \beta_3 = 0$ in favor of $H_1 : \beta_3 \neq 0$ and thus conclude that the quadratic spline model with the smoothness condition on the derivative is not appropriate.

If the range of the X values exhibits r known knots, $t_1 < t_2 < \ldots < t_r$, and it is required that the first derivative of the spline is continuous at the r knots, then the second-order or quadratic spline function has the form

$$Y = \beta_0 + \beta_1 X + \beta_2 X^2 + \sum\nolimits_{j=1}^{r} \theta_j (X - t_j)_+^2 + \varepsilon,$$
$$(X - t_j)_+ = \begin{cases} 0 \quad , X < t_j; \\ X - t_j, \; X \geq t_j, \; j = 1, \ldots, r. \end{cases} \tag{16.32}$$

Cubic Spline with Known Knots

As this topical heading suggests, we are interested in developing a methodology for fitting *cubic spline functions*: piecewise third-order polynomials that meet at join points or knots. To this end, let us assume that the range of the X values is divided into two segments (again denoted I and II), there is a single knot at $X = a$, and the regression model is formulated as

$$\text{(I)}\; Y = \beta_{0\text{I}} + \beta_{1\text{I}} X + \beta_{2\text{I}} X^2 + \beta_{3\text{I}} X^3, \; X < a;$$
$$\text{(II)}\; Y = \beta_{0\text{II}} + \beta_{1\text{II}} X + \beta_{2\text{II}} X^2 + \beta_{3\text{II}} X^3, \; X \geq a. \tag{16.33}$$

Introducing the dummy variable

$$D = \begin{cases} 0, \; X < a; \\ 1, \; X \geq a \end{cases}$$

enables us to rewrite Equation 16.33 as

$$Y = \beta_0 + \beta_1 X + \beta_2 X^2 + \beta_3 X^3 + \gamma_1 D + \delta_1 DX + \theta_1 DX^2 + \tau_1 DX^3 + \varepsilon. \tag{16.34}$$

So for $D = 0$,

$$\text{(I)}\; Y = \beta_0 + \beta_1 X + \beta_2 X^2 + \beta_3 X^3 + \varepsilon; \tag{16.35.I}$$

and for $D = 1$,

$$\text{(II) } Y = (\beta_0 + \gamma_1) + (\beta_1 + \delta_1)X + (\beta_2 + \theta_1)X^2 + (\beta_3 + \tau_1)X^3 + \varepsilon. \tag{16.35.II}$$

Let us now impose the requisite set of continuity restrictions on the spline function and on its first two derivatives at $X = a$. First, from Equations 16.35.I and 16.35.II, we must have

$$\beta_0 + \beta_1 a + \beta_2 a^2 + \beta_3 a^3 = (\beta_0 + \gamma_1) + (\beta_1 + \delta_1)a + (\beta_2 + \theta_1)a^2 + (\beta_3 + \tau_1)a^3$$

or

$$\gamma_1 = -\delta_1 a - \theta_1 a^2 - \tau_1 a^3. \tag{16.36}$$

Next, the derivatives of Equations 16.35.I and 16.35.II with respect to X are, respectively,

$$\text{(I) } Y' = \beta_1 + 2\beta_2 X + 3\beta_3 X^2,$$
$$\text{(II) } Y' = \beta_1 + \delta_1 + 2(\beta_2 + \theta_1)X + 3(\beta_3 + \tau_1)X^2.$$

Because the values of these derivatives must be equal at $X = a$, we require that

$$\beta_1 + 2\beta_2 a + 3\beta_3 a^2 = \beta_1 + \delta_1 + 2(\beta_2 + \theta_1)a + 3(\beta_3 + \tau_1)a^2$$

or

$$\delta_1 = -2\theta_1 a - 3\tau_1 a^2. \tag{16.37}$$

Finally, because the second derivatives of Equations 16.35.I and 16.35.II with respect to X must also agree at $X = a$, let us first obtain

$$\text{(I) } Y'' = 2\beta_2 + 6\beta_3 X,$$
$$\text{(II) } Y'' = 2(\beta_2 + \theta_1) + 6(\beta_3 + \tau_1)X.$$

Then the values of these second derivatives will be equal at $X = a$ if

$$2\beta_2 + 6\beta_3 a = 2(\beta_2 + \theta_1) + 6(\beta_3 + \tau_1)a$$

or

$$\theta_1 = -3\tau_1 a. \tag{16.38}$$

If Equations 16.36 through 16.38 are substituted sequentially into Equation 16.33, then we obtain

$$Y = \beta_0 + \beta_1 X + \beta_2 X^2 + \beta_3 X^3 + \tau_1 D(X - a)^3 + \varepsilon \tag{16.39}$$

(note that the $(X-a)$ and $(X-a)^2$ terms disappear) or

$$Y = \beta_0 + \beta_1 X + \beta_2 X^2 + \beta_3 X^3 + \tau_1 (X-a)_+^3 + \varepsilon,$$
$$(X-a)_+ = \begin{cases} 0 & , X < a; \\ X-a, & X \geq a. \end{cases} \tag{16.40}$$

So for $X < a$,

$$Y = \beta_0 + \beta_1 X + \beta_2 X^2 + \beta_3 X^3 + \varepsilon;$$

and for $X \geq a$,

$$Y = \beta_0 + \beta_1 X + \beta_2 X^2 + \beta_3 X^3 + \tau_1 (X-a)^3 + \varepsilon.$$

The SAS code leading to the estimation of Equation 16.40 is presented in Exhibit 16.8.

Exhibit 16.8 SAS Code for Cubic Spline Estimation (Known Knot)

```
data spline4;
     input x  y;
datalines;
 .
 .}input values
 .
  ;         ①
data spline ; set spline4;
       x2=x**2;
       x3=x**3;
       xplusa=max(x-a, 0);      ②
       xplusa2=xplusa**2;
       xplusa3=xplusa**3;
run;
proc reg data = spline ;
        model y= x  x2  x3  xplusa   xplusa2
                     xplusa3  / ss1  pcorr1; ③
       test1 : test xplusa = 0, xplusa2 = 0; ④
run;
```

The key details of this program are as follows:

①, ② We create a new dataset *spline* that contains *X*,*Y*, and the variables *X2* (the square of *X*), *X3* (the cube of *X*), *Xplusa*, *Xplusa2* (the square of *Xplusa*), and *Xplusa3* (the cube of *Xplusa*). *Xplusa* is again created by using the *max* function.

③, ④ The model statement includes all of the variables so that we seek to fit the equation

$$E(Y) = \beta_0 + \beta_1 X + \beta_2 X2 + \beta_3 X3 + \beta_4 Xplusa$$
$$+ \beta_5 Xplusa2 + \beta_6 Xplusa3.$$

Because the smoothness restriction on the first and second derivatives of the cubic spline dictates that $\beta_4 = \beta_5 = 0$ (the terms $(X - a)$ and $(X - a)^2$ drop out of Equation 16.40), we estimate this model under the linear restriction $\beta_4 = 0$ and $\beta_5 = 0$ or under the null hypothesis $H_0 : \beta_4 = \beta_5 = 0$. This can be accomplished by using the *TEST* statement, which follows the *MODEL* statement with a restricted model in which some subset of coefficients is constrained to have all zero values. The restricted model at hand is specified by the *TEST* statement

test1: test xplusa = 0, xplusa2 = 0;

We thus test the (null) hypothesis that the coefficients on both *Xplusa* and *Xplusa2* are zero; we compare the full model with the model containing only the variables *X*, *X2*, *X3*, and *Xplusa3*. If the *p*-value associated with the partial *F* statistic (taken as the ratio between a mean square determined from a Type II sum of squares and the error mean square for the full model) is less than some chosen α, then we can reject $H_0 : \beta_4 = \beta_5 = 0$ in favor of H_1: not H_0, and conclude that the cubic spline model incorporating the smoothness conditions on the first and second derivatives is not appropriate.

If the range of the set of *X*'s exhibits *r* known knots, $t_1 < t_2 < \ldots < t_r$ and we stipulate that the first two derivatives of the spline are continuous at the *r* knots, then the cubic spline function has the form

$$Y = \beta_0 + \beta_1 X + \beta_2 X^2 + \beta_3 X^3 + \sum_{j=1}^{r} \tau_1 (X - t_j)_+^3 + \varepsilon,$$

$$(X - t_j)_+ = \begin{cases} 0 & , X < t_j; \\ X - t_j, & X \geq t_j,\ j = 1, \ldots, r. \end{cases} \tag{16.41}$$

A Linear Spline Function with Unknown Knot

At times, it may not be apparent where the exact position of a knot is located (or where multiple knots may occur). In this instance, the knot is actually a parameter to be estimated along with an intercept and slope. For this more complicated estimation problem, linear regression cannot be used. However, we have at our disposal a nonlinear estimation procedure called PROC NLIN that uses, as a set of initial values, the PROC REG results obtained from estimation with a known knot. To see how this process works, we now turn to Example 16.7.

Example 16.7

Using the dataset used in Example 16.5 (see Table 16.10), let us reestimate the spline function Equation 16.22.1. Previously, the known knot was set at $X = a = 6$ and the estimates obtained were $\hat{\beta}_0 = 1.1320$, $\hat{\beta}_1 = 1.04304$, and $\hat{\beta}_2 = -0.80975$. Given that the placement of the knot is now assumed unknown, we can fit Equation 16.22.1 using the following SAS code (Exhibit 16.9).

Exhibit 16.9 SAS Code for First-Order Spline Regression (One Unknown Knot)

```
data  spline1;          ① We invoke PROC REG to get initial OLS estimates
    input  x  y ;         of β0, β1, and β2 = δ.
```

```
 datalines ;
 1    2
 2    3
 3    4.5
 4    6
 5    6
 6    7
 7    7.5
 8    8.25
 9    8
10    8.25
11    8.5
12    9
13    9.25
14    9.5
15    9
;
data spline; set spline1;
   xplus6=max(x-6,0);
run;
        ①                 ②
proc reg data=spline outest=ests;
    model y=x xplus6;
run;
proc print data=ests;   ③
    run;   ④   ⑤
    data ests1; set ests;

            call symput ( intercept);            }
            call symput (  trim (left(x)));      }  ⑥
            call symput (  trim(left(xplus6)));  }

run;   ⑦   ⑧
proc nlin data=spline
    parms intercept=&init_int beta1=&init_x beta2=&init_xplus6 ⑨
            knot=6;
    xpl=max(x-knot,0);   ⑩
    model  y=intercept+beta1*x+beta2*xpl;   ⑪
run;
```

② The results of PROC REG are written to the temporary file *ests* by the OUTEST option.

③ We request a print of the temporary file *ests*.

④, ⑤ We specify the SAS data file *ests1*, the contents of which are the entries in the temporary file *ests*.

⑥ We use a macro to indentify the initial OLS estimates of β_0, β_1, and β_2 as determined by PROC REG.

⑦ We invoke PROC NLIN to conduct the nonlinear estimation process.

⑧ The SAS dataset to be used in PROC NLIN is *spline*.

⑨ *Parms* receives, via the macro in ⑥, the initial OLS estimates appearing in the temporary file *ests* ($\hat{\beta}_0$, $\hat{\beta}_1$, and $\hat{\beta}_2$) along with the initial value of the parameter *knot* (=6).

⑩ We create the variable *Xpl* using the *max* function stated in terms of the parameter *knot*.

⑪ The *MODEL* statement now expresses *y* in terms of *X* and *Xpl*.

We now turn to the output generated by this set of SAS code. The PROC NLIN results are presented in Table 16.14.

①, ② We obviously get as our set of initial OLS estimates the regression results obtained previously (Table 16.11) using the known (fixed) knot located at $X = a = 6$.

③ We obtain a print of the contents of the temporary file *ests* or the single observation in the SAS dataset *ests*.

④ PROC NLIN uses the (default) Gauss-Newton method to perform two iteration. The convergence criterion *RPC(knot)* = 0.0032 indicates that the parameter *knot* changed by 0.0032 relative to its value at the last iteration.

TABLE 16.14

Output for Example 16.7

The SAS System
The REG Procedure
Model : MODEL1
Dependent Variable: y

①

Analysis of Variance

Source	**DF**	**Sum of Squares**	**Mean Square**	**F Value**	**Pr > F**
Model	2	73.67657	36.83829	300.02	<.0001
Error	12	1.47343	0.12279		
Corrected total	14	75.15000			

Root MSE	0.35041	*R*-Square	0.9804
Dependent Mean	7.05000	Adj *R*-Sq	0.9771
Coeff Var	4.97032		

②

Parameter Estimates

Variable	**DF**	**Parameter Estimate**	**Standard Error**	**t Value**	**Pr > |t|**
Intercept	1	1.13208	0.30568	3.70	0.0030
x	1	1.04340	0.06807	15.33	<.0001
xplus6	1	–0.80975	0.09300	–8.71	<.0001

③

Obs	**_MODEL_**	**_TYPE_**	**_DEPVAR_**	**_RMSE_**	**Intercept**	**x**	**xplus6**	**y**
1	MODEL1	PARMS	Y	0.35041	1.13208	1.04340	–0.80975	–1

④

The NLIN Procedure
Dependent Variable y
Method: Gauss-Newton
Iterative Phase

Iter	**Intercept**	**beta1**	**beta2**	**knot**	**Sum of Squares**
0	1.3121	1.0434	–0.8097	6.0000	1.4734
1	1.0000	1.1000	–0.8561	5.6632	1.4244
2	1.0000	1.1000	–0.8561	5.6814	1.4220

Note: Convergence criterion met.

(continued)

TABLE 16.14 *(continued)*

Estimation Summary

Method	Gauss-Newton
Iterations	2
R	0
PPC	0
RPC(knot)	0.003217
Object	0.001708
Objective	1.42197
Observations read	15
Observations used	15
Observations missing	0

⑤

Source	DF	Sum of Squares	Mean Square	F Value	Pr > F
Regression	4	819.3	204.8	190.11	<.0001
Residual	11	1.4220	0.1293		
Uncorrected total	15	820.7			
Corrected total	14	75.1500			

Parameter	Estimate	Approx Std Error	Approximate 95% Confidence Limits	
intercept	1.0000	0.3771	1.1700	1.8300
beta1	1.1000	0.1137	0.8498	1.3502
beta2	−0.8561	0.1204	−1.1210	−0.5911
knot	5.6814	0.4789	4.6273	6.7356

The NLIN Procedure

Approximate Correlation Matrix

	intercept	beta1	beta2	knot
intercept	1.0000000	−0.9045340	0.8542422	0.5053602
beta1	−0.9045340	1.0000000	−0.9444003	−0.7435656
beta2	0.8542422	−0.9444003	1.0000000	0.5492639
knot	0.5053602	−0.7435656	0.5492639	1.0000000

⑤ The estimated first-order spline function with unknown knot is

$$\hat{Y} = 1.0000 + 1.1000X - 0.8561Xpl,$$

where the estimate of the parameter *knot* is knot = 5.6814. In this regard, for $X < 5.6814$,

$$Y = 1.0000 + 1.1000\ X;$$

and for $X \geq 5.6814$,

$$\begin{aligned}\hat{Y} &= 1.0000 + 1.1000X - 0.8561(X - 5.6814)\\ &= 5.8382 + 0.2484X.\end{aligned}$$

Note that zero is not a member of any of the reported 95% confidence intervals for the unknown parameters β_0, β_1, β_2, and *knot*. Hence, we may conclude that each of these parameters differs significantly from zero at the 5% level. Note also that the overall fit is quite good ($F = 190.11$ with p-value <0.0001, $R^2 = SSR/SST = 0.9983$, and $MSE = 0.1293$) and is comparable with our previous results obtained with a predetermined knot at $X = a = 6$.

Estimating the Parameters of a Quadratic Response Surface

Suppose we want to fit the model depicted by Equation 16.6 or

$$\begin{aligned}Y &= f(X_1, X_2)\\ &= \beta_0 + \beta_1 X_1 + \beta_2 X_2 + \beta_{11} X_1^2 + \beta_{12} X_1 X_2 + \beta_{22} X_2^2 + \varepsilon\end{aligned}$$

to a set of data points (X_{i1}, X_{i2}, Y_i), $i = 1,\ldots,n$, and obtain a plot of the predicted values $\hat{Y}_i$, $i = 1,\ldots,n$, above the X_1, X_2 –plane. The resulting three-dimensional figure is called a *quadratic response surface.*

Example 16.8

To see exactly how the process of generating a quadratic response surface is performed, let us use the sample dataset ($n = 39$) provided in Table 16.15. The requisite SAS code is given in Exhibit 16.10.

Exhibit 16.10 SAS Code for Quadratic Response Surface

```
data quad;
  input  x1  x2  y;
datalines;
23  24  30
26  25  25
```

① To execute a response surface plot, we first produce a grid of values of the regressors *X1* and *X2*, where the grid should consist of at least 30 rows and 30 columns. The grid of values is housed in the *q1* dataset.

```
.   .   .             ② We integrate the grid dataset q1 with the original
.   .   .                dataset quad to produce dataset q2.
.   .   .             ③ We run PROC RSREG using dataset q2. The
27  42  10               option out = p1 names the output dataset to
;                        contain the predicted values.
data q1;
       do x1=8  to 40  by 0.5;  ①
       do x2=5  to 43  by 0.5;
output;  end;  end;
run;                             ④ The MODEL statement indicates that we
data q2; set quad q1; ②            regress Y on X1 and X2. Invoking the
run;                                option PREDICT requests that the output
proc rsreg data=q2 out=p1; ③       dataset p1 contains the predicted values.
     model  y=x1  x2/predict;  ④
run;
data plot; set  p1;  ⑤
run;
proc  gcontour  data=plot;
        plot  x1*x2=y/llevels=5 7 12 18 25 33 40 clevels=black  ⑥
        black  black  black  black  black  black;
run;
proc g3d  data=plot;  ⑦
     plot  x1*x2=y;
run;
```

⑤ Dataset *plot* is grid dataset *p1*.

⑥ We call PROC GCONTOUR to produce a contour or level curve plot of the estimated response surface, with the PLOT statement specifying *X1* as the row variable and *X2* as the column variable. Seven levels (the default number of levels) of the estimated response $\hat{Y}$ are plotted according to the different types of black lines specified by the options CLEVELS and LLEVELS.

⑦ A three-dimensional plot of the response surface is produced using PROC G3D.

Looking to the output generated by the SAS code (Table 16.16), we find the following:

① The F statistic for the full model is 28.77, p-value <0.0001, thus indicating that the model (Equation 16.6) explains a statistically significant portion of the variation in Y. In fact, because the full model $R^2 = 0.8134$, we find that 81.34% of the variation in Y is explained by the model.

② Neither linear terms (involving X_1 and X_2) nor the cross-product term $(X_1 \cdot X_2)$ is needed in the model because their associated F statistics indicate a lack of statistical significance, i.e., for the linear portion, $F = 1.01$ (p-value = 0.3746); for the cross-product term, $F = 0.32$ (p-value = 0.5728). Clearly, most of the variation in Y is explained by the quadratic portion of the model ($F = 70.75$, p-value <0.0001). That is, because Type I (sequential) sums of squares are used:

$$SSR(X_1^2, X_2^2 | \text{intercept}, X_1, X_2) = 927.3307.$$

TABLE 16.15

Observations on Variables X_1, X_2, and Y

Observation Number	X_1	X_2	Y	Observation Number	X_1	X_2	Y
1	23	24	30	21	15	28	15
2	26	25	25	22	20	33	15
3	22	25	25	23	33	26	15
4	24	23	25	24	29	28	15
5	21	21	25	25	26	33	15
6	30	22	20	26	40	22	10
7	25	12	20	27	36	15	10
8	21	12	20	28	32	10	10
9	18	21	20	29	28	6	10
10	18	25	20	30	10	10	10
11	21	26	20	31	8	20	10
12	30	25	20	32	8	26	10
13	25	29	20	33	10	30	10
14	33	22	15	34	14	40	10
15	30	11	15	35	20	43	10
16	26	10	15	36	39	26	10
17	20	7	15	37	37	28	10
18	16	11	15	38	30	37	10
19	12	22	15	39	27	42	10
20	13	26	15				

With $MSE(full) = 6.5532$,

$$F = \frac{927.3307/2}{6.5532} = 70.75.$$

③ The individual contributions of X_1 and X_2 in explaining variation in Y are now ascertained. Here we address the following question: can either X_1 or X_2 be deleted from the model without compromising the fit? To answer this question, let us first consider all terms involving X_1 and test $H_0 : \beta_1 = \beta_{11} = \beta_{12} = 0$, against H_1 : not H_0. Because $F = 35.13$ (p-value <0.0001) leads us to reject H_0, we conclude that X_1 cannot be omitted from the model. Similarly, if we next consider all terms involving X_2, then a test of $H_0 : \beta_2 = \beta_{22} = \beta_{12} = 0$ versus H_1 : not H_0 again prompts us to reject H_0 because now $F = 25.23$, p-value <0.0001. Thus, X_2 cannot be dropped from the model.

TABLE 16.16
Output for Example 16.8

The SAS System
The RSREG Procedure
Coding Coefficients for the Independent Variables

Factor	Subtracted off	Divided by
x1	24.000000	16.000000
x2	24.500000	18.500000

Response Surface for Variable y

Response mean	15.641026
Root MSE	2.559938
R^2	0.8134
Coefficient of variation	16.3668

Type 1 Sum

	Regression	DF	① Type I Sum of Squares	R^2	F Value	Pr > F
②	Linear	2	13.259634	0.0114	1.01	0.3746
②	Quadratic	2	927.330719	0.8001	70.75	<.0001
②	Crossproduct	1	2.125700	0.0018	0.32	0.5728
①	Total model	5	942.716054	0.8134	28.77	<.0001

Residual	DF	Sum of Squares	② Mean Square
Total error	33	216.258305	6.553282

④ Parameter	DF	Estimate	Standard Error	t Value	Pr > |t|	Parameter Estimate from Coded Data
Intercept	1	−20.679243	5.281315	−3.92	0.0004	22.194087
x1	1	2.391069	0.292466	8.18	<.0001	−1.822012
x2	1	1.403635	0.255448	5.49	<.0001	−2.488105
x1*x1	1	−0.054133	0.005295	−10.22	<.0001	−13.858081
x2*x1	1	0.003814	0.006697	0.57	0.5728	1.128969
x2*x2	1	−0.033258	0.003933	−8.46	<.0001	−11.382716

TABLE 16.16 *(continued)*

	Factor	DF	Sum of Squares	Mean Square	F Value	Pr > F
③	x1	3	690.665701	230.221900	35.13	<.0001
	x2	3	495.957863	165.319288	25.23	<.0001

The RSREG Procedure

Canonical Analysis of Response Surface Based on Coded Data

	⑤ Critical Value	
Factor	**Coded**	**Uncoded**
x1	–0.070332	22.874685
x2	–0.112781	22.413552

Predicted value at stationary point: 22.398466

The RSREG Procedure

Canonical Analysis of Response Surface Based on Coded Data

⑥	Eigenvectors	
Eigenvalues	**x1**	**x2**
–11.260067	0.212321	0.977200
–13.980730	0.977200	–0.212321

Stationary point is a maximum.

④, ⑤, ⑥ Our estimated quadratic (response) function is

$$\hat{Y} = f(X_1, X_2) = -20.6792 + 2.3910X_1 + 1.4036X_2$$
$$-0.0541X_1^2 + 0.0038X_1X_2 - 0.0333X_2^2.$$

Solving

$$f_1 = 2.3910 - 0.1082X_1 + 0.0038X_2 = 0$$

$$f_2 = 1.4036 + 0.0038X_1 - 0.0666X_2 = 0$$

simultaneously yields the critical values of X_1 and X_2 or $(X_1^0, X_2^0) = (22.8746,\ 22.4136)$. The predicted value of f at this point is $\hat{Y}^0 = f(X_1^0,\ X_2^0) = 22.3984$. Hence, the response surface has a stationary point at $(X_1^0,\ X_2^0,\ \hat{Y}^0)$. Because the Hessian matrix of f,

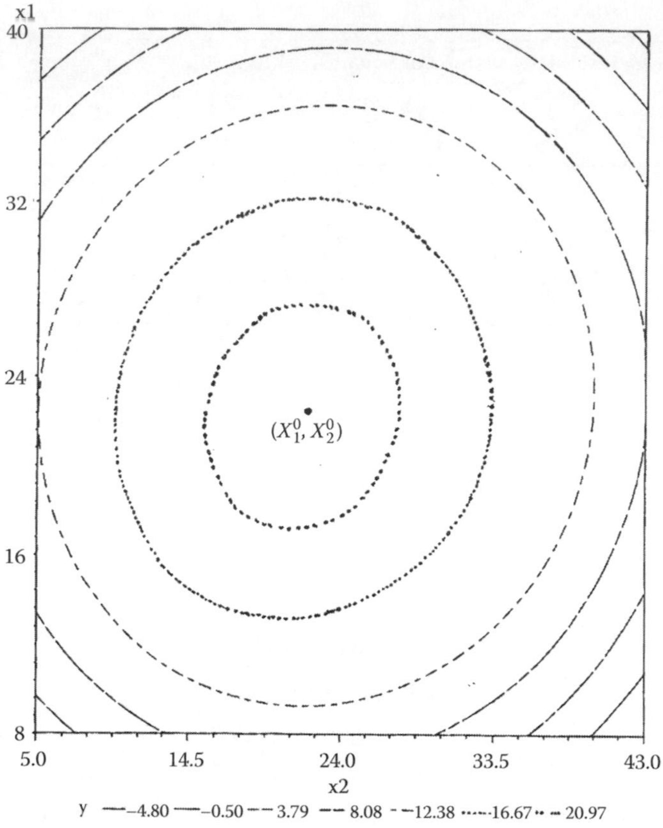

FIGURE 16.9
Contour map of quadratic response surface.

$$\boldsymbol{H} = \begin{bmatrix} f_{11}^0 & f_{12}^0 \\ f_{21}^0 & f_{22}^0 \end{bmatrix} = \begin{bmatrix} -0.1082 & 0.0038 \\ 0.0038 & -0.0666 \end{bmatrix},$$

is negative definite at (X_1^0, X_2^0) or $f_{11}^0 = -0.1082 < 0$ and $|\boldsymbol{H}| = 0.0072 > 0$, we may conclude that this stationary point corresponds to a strong global maximum of *f*.

Alternatively, we noted in Appendix K of Chapter 12 that the eigenvalues of a symmetric (Hessian) matrix are the solutions of the characteristic equation

$$\begin{bmatrix} f_{11}^0 - \lambda & f_{12}^0 \\ f_{21}^0 & f_{22}^0 - \lambda \end{bmatrix} = \begin{bmatrix} -0.1082 - \lambda & 0.0038 \\ 0.0038 & -0.0666 - \lambda \end{bmatrix} = 0.$$

These eigenvalues are –11.2600 and –13.9807. Because a (Hessian) matrix is negative definite if and only if all of its eigenvalues are negative, it again follows that *f* has a stationary point that represents a strong global maximum. (If both eigenvalues are positive, then *f* attains a strong global minimum at (X_1^0, X_2^0); and if the eigenvalues are of opposite sign, then *f* has a saddle point at (X_1^0, X_2^0).)

⑦ Figure 16.9 illustrates the *contour map* of the quadratic response surface. For example, along the contour $f = 20.97$, we have the locus or set of combinations of X_1 and X_2 that yield a function value of 20.97, etc.

⑧ Figure 16.10 depicts the graph of the entire response surface over its domain.

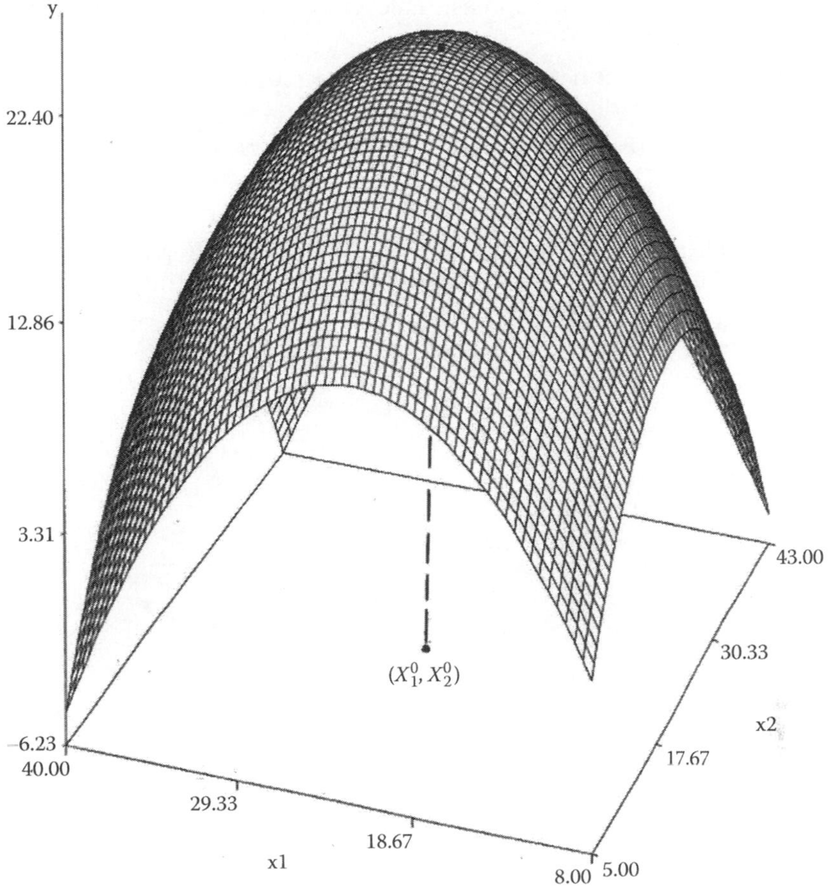

FIGURE 16.10
Quadratic response surface.

Exercises

16-1. How is the degree of a polynomial model defined? How is its order defined? Specify the degree and order of following polynomial models:

a. $Y = 10 - 3X + 6X^2$.

b. $Y = 5 + 2X + 4X^3$.

c. $Y = 3X + 2X^2 + X^5$.

d. $Y = 2 + 4X_1 + 6X_2 + X_1^2 + X_1X_2^2$.

16-2. Given the Table E.16-1 data values, does a cubic regression model adequately fit this set of observations? (Hint: see Example 16.1.)

16-3. Given the Table E.16-2 dataset, determine whether a cubic regression model is appropriate. (Hint: follow the lead of Example 16.1.)

16-4. Given the Exercise 16-2 dataset, reestimate a third-order polynomial model using backward elimination.

16-5. Using the Exercise 16-3 data values, reestimate a third-order polynomial model via backward elimination.

16-6. Given the dataset presented in Table E.16-3, determine the equation of the third-order polynomial of best fit using orthogonal polynomials via Table A.9. Discuss its statistical significance. Find R^2. Express the estimated polynomial in terms of X.

16-7. Using the Table E.16-3 data values, use the SAS system to fit a third-order polynomial using orthogonal polynomials. Fully explain your findings.

16-8. For the dataset appearing in Table E.16-4, fit a first-order spline model with known knot at $X = 7$. Analyze your results. (Hint: see Example 16.4.)

TABLE E.16-1

Observations on Variables X and Y

X	Y	X	Y
4	78	35	40
10	70	40	24
13	68	42	27
20	58	45	17
21	69	54	20
25	60	55	12
28	49	60	13
29	40	65	12
30	53	70	20
33	28	76	20

TABLE E.16-2

Observations on Variables X and Y

X	9	12	21	25	35	40	50	61	70	80
Y	11	21	22	38	41	40	51	48	52	42

TABLE E.16-3

Observations on Variables X and Y

X	3	6	9	12	15	18	21	24	27	30
Y	27	23	21	23	25	29	27	24	19	10

TABLE E.16-4

Observations on Variables X and Y

X	1	2	3	4	5	6	7	8	9	10	11	12	13	14
Y	1.0	1.2	1.8	1.7	2.0	1.9	4.0	5.0	6.1	7.0	8.1	8.5	11.0	10.9

16-9. Given the Table E.16-5 dataset, fit a spline function with known knots occurring at $X = 7$ and $X = 15$. Discuss your findings. (Hint: see Example 16.5.)

16-10. Using the dataset presented in Table E.16-6, fit a quadratic spline function with a known knot at $X = 8$. Interpret your results. (Hint: see Equation 16.31 and the accompanying SAS code.)

TABLE E.16-5

Observations on Variables *X* and *Y*

X	Y	X	Y
1	1.5	11	8.2
2	1.2	12	8.5
3	1.8	13	11.1
4	1.7	14	11.0
5	1.9	15	7.0
6	2.0	16	7.1
7	4.0	17	7.9
8	5.0	18	7.7
9	6.1	19	8.0
10	7.0	20	8.2

TABLE E.16-6

Observations on Variables *X* and *Y*

X	0.5	1.0	2.0	4.0	5.9	6.1	7.9	8.0	10.5	12.0	14.0	15.8	16.2
Y	33.0	38.0	42.0	45.0	46.0	46.1	45.0	40.0	39.0	35.0	31.0	28.0	24.0

TABLE E.16-7

Observations on Variables *X* and *Y*

X	Y	X	Y
0.4	32.0	14.0	31.0
1.0	38.0	15.8	28.0
2.0	42.0	16.2	24.0
4.0	45.0	16.5	33.0
5.9	46.0	18.0	29.0
6.1	46.1	20.0	26.0
7.9	45.0	21.6	25.0
8.0	40.0	24.1	25.5
10.5	39.0	28.0	26.0
12.0	35.0	29.2	27.3

16-11. For the Table E.16-7 dataset, estimate a cubic spline model with known knots at $X = 8$ and $X = 16.5$. Discuss your findings. (Hint: refer to Equations 16.40 and 16.41 and the supporting SAS code.)

16-12. For the dataset provided by Table E.16-8, fit a first-order spline model with unknown knot. Interpret your results. (Hint: follow the details provided in Example 16.7.)

16-13. Given the Table E.16-9 data values, estimate a quadratic response surface. Discuss your findings. (Hint: see Example 16.8.)

TABLE E.16-8

Observations on Variables X and Y

X	1	2	3	4	5	6	7	8	9	10	11	12	13	14
Y	0.9	1.0	1.2	1.8	1.7	2.0	1.9	3.9	5.0	6.1	7.0	8.1	8.5	11.0

TABLE E.16-9

Observations on Variables X_1, X_2, and Y

X_1	X_2	Y	X_1	X_2	Y
24	40	5	35	18	10
15	38	5	34	25	10
10	33	5	33	28	10
7	23	5	28	29	10
6	22	5	23	26	15
12	7	5	20	24	15
18	6	5	17	21	15
27	5	5	21	19	15
36	8	5	25	18	15
41	23	5	29	19	15
43	23	5	32	23	15
37	32	5	29	24	15
33	37	5	17	23	20
22	30	10	23	21	20
17	28	10	27	22	20
12	24	10	29	22	20
13	16	10	28	23	20
17	14	10	24	22	20
25	15	10	26	22	25
32	16	10			

17

Semiparametric Regression

Introduction

We have been assuming, for the most part, that our multiple regression model relates a response variable Y to a set of regressors $X_1,\ldots,X_m$ in a linear manner or

$$Y = \beta_0 + \sum\nolimits_{j=1}^{m} \beta_j X_j + \varepsilon, \tag{17.1}$$

where ε is random disturbance term that, according to the strong classical assumptions, is $N(0, \sigma_\varepsilon)$, with σ_ε constant. Here the first two components involving the parameters $\beta_0, \beta_1,\ldots,\beta_m$ appearing on the right side of Equation 17.1 constitute the parametric portion of the *parametric specification* of the regression model.

Obviously, linear parametric models tend to be somewhat restrictive. Occasionally, the need arises for greater flexibility in the specification of a regression equation, e.g., nonlinear relationships need to be modeled in a manner such that the shape of the regression function is not predetermined but can be adjusted to reflect the features of the data or characteristics of the modeling environment. This type of flexible nonlinear regression has come to be known as *nonparametric regression*. Specifically, let $Y = f(Z_1,\ldots,Z_p) + \varepsilon$, $E(\varepsilon) = 0$. Then $E(Y|Z_1,\ldots,Z_p) = f(Z_1,\ldots,Z_p)$ is taken to be an unspecified smooth function that is to be estimated from the data points $(Z_{i1},\ldots,Z_{ip}, Y_i)$, $i = 1,\ldots,n$. Clearly, nonparametric regression involves a model-free basis for estimating the response level of f over the dataset.

Suppose, however, that we now formulate a combined model made up of a parametric as well as a nonparametric segment, which we shall call a *semiparamteric regression model*, e.g.,

$$Y = \sum\nolimits_{j=1}^{m} \beta_j X_j + f(Z_i,\ldots,Z_p) + \varepsilon, \tag{17.2}$$

where $E(Y|X_1,\ldots,X_m, Z_1,\ldots,Z_p) = \sum \beta_j X_j + f(Z_1,\ldots,Z_p)$, $V(Y|X_1,\ldots,X_m, Z_1,\ldots,Z_p) = \sigma_\varepsilon^2 =$ constant, and f (taken only to be smooth) is not known to reside in any specific parametric class of functions. Here the particular type of semiparametric regression model displayed in Equation 17.2 is termed the *partial linear model*. Note that we have simply introduced a nonparametric component into a parametric regression equation.

In what follows, two approaches will be offered to generate a solution to Equation 17.2: (1) a *modular approach* involving the two operations of differencing and smoothing (in that order); and (2) a *penalized spline* smoothing routine. The smoothing portion of the first approach uses LOWESS (for LOcally WEighted Scatter plot Smoothing), whereas the second approach uses penalized least squares.

The Modular Approach [Rice 1984; Yatchew 1988, 1997, 1998, 2003; Powell 1994; Fan and Huang 2001]

Differencing [Yatchew 2003]

Suppose we have, for simplicity, a pure bivariate nonparametric model $Y = f(Z) + \varepsilon$, where f is smooth, the ε_i, $i = 1,\ldots,n$, are independent and identically distributed random variables with $E(\varepsilon_i) = 0$ and $V(\varepsilon_i) = \sigma_\varepsilon^2$ = constant for all i, and the Z_i's are *everywhere dense* (between any two Z_i's lies another Z value) in the domain of f. The process of differencing is based on the fundamental contention that the closeness of the Z_i's, with f smooth, results in corresponding function values $f(Z_i)$ that are also close. Hence, differencing can be effectively used to remove the nonparametric effect from the partial linear model (Equation 17.2).

To see this, let us rewrite the partial linear model in simplified form as

$$Y = \beta X + f(Z) + \varepsilon, \tag{17.2.1}$$

where β is a scalar, $E(\varepsilon \mid X, Z) = \sigma_\varepsilon^2$. Suppose Z is bounded on the domain of f and that the Z_i's are rearranged in increasing order or $Z_1 \le Z_2 \le \ldots \le Z_n$. Furthermore, suppose that Z effects X indirectly through $X = g(Z) + \eta$, η a well-behaved random error term. If g is also smooth, with g' bounded over the domain of g, $E(X|Z) = g(Z)$, and $V(X|Z) = \sigma_\eta^2$ = constant, then taking first differences of

$$Y = \beta(g(X) + \eta) + f(Z) + \varepsilon \tag{17.2.2}$$

yields

$$\begin{aligned} Y_i - Y_{i-1} &= \beta(X_i - X_{i-1}) + [f(Z_i) - f(Z_{i-1})] + \varepsilon_i - \varepsilon_{i-1} \\ &= \beta[g(Z_i) - g(Z_{i-1})] + \beta(\eta_i - \eta_{i-1}) + [f(Z_i) - f(Z_{i-1})] + \varepsilon_i - \varepsilon_{i-1} \\ &\approx \beta(\eta_i - \eta_{i-1}) + \varepsilon_i - \varepsilon_{i-1}. \end{aligned} \tag{17.3}$$

Thus, the direct effect $f(Z)$ of nonparametric Z and the indirect effect $g(Z)$ of Z, which occurs through X, are both removed via differencing. Hence, an estimate of β can be obtained by applying OLS to the differenced parametric component of Equation 17.2.1 or to

$$Y_i - Y_{i-1} \approx \beta(X_i - X_{i-1}) + \varepsilon_i - \varepsilon_{i-1}.$$

This yields

$$\hat{\beta}_{diff} = \frac{\sum_{i=1}^{n}(X_i - X_{i-1})(Y_i - Y_{i-1})}{\sum_{i=1}^{n}(X_i - X_{i-1})^2}. \tag{17.4}$$

If we substitute the approximations $X_i - X_{i-1} \approx \eta_i - \eta_{i-1}$ and $Y_i - Y_{i-1} \approx \beta(\eta_i - \eta_{i-1}) + \varepsilon_i - \varepsilon_{i-1}$ into Equation 17.4, then $\hat{\beta}_{diff}$ can be shown (under convergence in distribution) to be approximately normal, i.e.,

$$\hat{\beta}_{diff} \xrightarrow{d} N\left(\beta, \frac{1.5}{n}\frac{\sigma_\varepsilon^2}{\sigma_\eta^2}\right) \tag{17.5}$$

or

$$n^{1/2}(\hat{\beta}_{diff} - \beta) \xrightarrow{d} N\left(0, \frac{1.5\sigma_\varepsilon^2}{\sigma_\eta^2}\right), \tag{17.5.1}$$

where consistent estimators for σ_ε^2 and σ_η^2 are (under convergence in probability)

$$S_\varepsilon^2 \approx \frac{1}{2n}\sum_{i=2}^{n}(\varepsilon_i - \varepsilon_{i-1})^2 \xrightarrow{p} \sigma_\varepsilon^2 \tag{17.6}$$

and

$$S_\eta^2 \approx \frac{1}{2n}\sum_{i=2}^{n}(\eta_i - \eta_{i-1})^2 \xrightarrow{p} \sigma_\eta^2 \tag{17.7}$$

respectively.

How should S_ε^2 in Equation 17.6 be operationalized? To obtain an estimator for the residual variance σ_ε^2, let us consider

$$S_{diff}^2 = \frac{1}{2n}\sum_{i=2}^{n}(Y_i - Y_{i-1})^2. \tag{17.8}$$

This estimator is consistent because, for the Z_i's in increasing order and close to each other, differencing (as demonstrated above) tends to remove the nonparametric effect, i.e., $Y_i - Y_{i-1} = f(Z_i) - f(Z_{i-1}) + \varepsilon_i - \varepsilon_{i-1} \approx \varepsilon_i - \varepsilon_{i-1}$ and thus Equation 17.8 can be written as

$$S_{diff}^2 \approx \frac{1}{2n}\sum_{i=2}^{n}(\varepsilon_i - \varepsilon_{i-1})^2 \tag{17.8.1}$$

(which is simply Equation 17.6 again). Then, from this expression, it can be demonstrated that

$$n^{1/2}(S_{diff}^2 - \sigma_\varepsilon^2) \xrightarrow{d} N(0,\ E(\varepsilon^4)). \tag{17.9}$$

In summary, taking first differences of the regression Equation 17.2.1 (with the Z values in increasing order) serves to eliminate the nonparametric effect f, thus enabling us to ignore f, obtain $\hat{\beta}_{diff}$ by OLS, and thus perform the usual inferences on β, inferences that can be made as if we had a fully parametric specification of the regression model (such as Equation 17.1) at the start.

Having estimated β by $\hat{\beta}_{diff}$, we can now remove the estimated parametric component of Equation 17.2.1 and then turn to an analysis of f by using nonparametric smoothing techniques; we can now operate as if β were actually known. (Why smoothing? Remember that the Z's are arranged in an increasing sequence and are everywhere dense on the domain of f. By smoothing the Y's for which the associated X's are close or within some suitably restricted neighborhood of a given X, we can obtain a reasonable estimate of the nonparametric regression effect at that X value.)

To this end, let us form the collection of ordered pairs $(Z_i, Y_i - \hat{\beta}_{diff}X_i), i = 1, \ldots, n$, and fit the nonparametric regression equation

$$\begin{aligned} Y_i - \hat{\beta}_{diff}X_i &= (\beta - \hat{\beta}_{diff})X_i + f(Z_i) + \varepsilon_i \\ &\approx f(Z_i) + \varepsilon_i,\ i = 1, \ldots, n. \end{aligned} \tag{17.10}$$

(Note that, by using $Y_i - \hat{\beta}_{diff} X_i$ as the dependent variable in Equation 17.10, we remove the parametric effect from the original dependent variable level Y_i.)

The nonparametric smoothing technique to be applied to Equation 17.10 is LOWESS. As demonstrated by Yatchew [1997, 1998, 2003], consistency, optimal rate of convergence results, and the construction of confidence limits for f all hold because $\hat{\beta}_{diff}$ converges sufficiently quickly to β so that the approximation in Equation 17.10 leaves asymptotic or large sample properties unaffected.

Smoothing (Using LOWESS)

LOWESS is a nonparametric technique used to estimate a regression function of unspecified (unknown parametric) form. Suppose the ith observation Y_i on the response variable Y is related to the p observations $Z_{i1},\ldots,Z_{ip}$ on the regressors $Z_1,\ldots,Z_p$, respectively, via a generalized regression function f according to

$$Y_i = f(Z_{i1},\ldots,Z_{ip}) + \varepsilon_i, \quad i = 1,\ldots,n,$$

where the ε_i constitute a set of independent random error terms, each with a zero mean and a constant variance σ_ε^2.

As was the case with our previous development of geographically weighted regression, the population regression function (surface) $E(Y|Z_1,\ldots,Z_p) = f(Z_1,\ldots,Z_p)$ can be approximated locally by the response generated by a low-order polynomial, i.e., we fit, using weighted least squares, either a linear or quadratic function of the regressors $Z_1,\ldots,Z_p$ to a set of data points restricted to some chosen neighborhood of the point $(Z_{i1}^0,\ldots,Z_{ip}^0)$. LOWESS obtains a predicted response value at a given point of application $(Z_{i1}^0,\ldots,Z_{ip}^0)$ by using an OLS fit incorporating all of the observations that are in close proximity to the said point.

The radius of each neighborhood is chosen so that the neighborhood contains a specified fraction of the data points closest to its center $(Z_{i1}^0,\ldots,Z_{ip}^0)$. That is, the percentage of the data used to form neighborhoods, as well as the number of neighborhoods, is dictated by the value of the *smoothing parameter* s (typically <1). (If, for instance, $s = 0.40$, then the closest 40% of the total number of data points is used to form the neighborhood.) In fact, the number of neighborhoods controls the smoothness of the estimated response function, i.e., the smaller the number of neighborhoods, the smoother (albeit less well-fitting) is the response function. If $s = 1$, LOWESS estimates a linear or quadratic function using the entire dataset and thus renders the smoothest but poorest fitting response function. The final response function is an amalgam of individual response functions.

In general, given $s < 1$, let q_s represent the number of points in a local neighborhood of $(Z_{i1}^0,\ldots,Z_{ip}^0)$, with d_t, $t = 1,\ldots,q_s$, denoting the distance of each of these q_s points from the neighborhood center $(Z_{i1}^0,\ldots,Z_{ip}^0)$, with $d^* = \max\{d_t, t = 1,\ldots,q_s\}$. The point at distance d_t from the center receives weight w_t in the local OLS regression, where w_t is determined (typically) from a *tri-cube weight function* as

$$w_t(Z_{i1}^0,\ldots,Z_{ip}^0) = \frac{32}{5}\left(1-(d_t/d^*)^3\right)^3, \quad 0 \le d_t/d^* < 1. \tag{17.11}$$

Note that w_t decreases as the distance between the center $(Z_{i1}^0,\ldots,Z_{ip}^0)$ and the neighborhood point increases. (If $s > 1$, then d^* is replaced by $d^* s^{1/p}$ in Equation 17.11, where p is the number of regressors.)

In the special case for which $p = 1$, LOWESS has us fit either a first- or second-degree polynomial in a neighborhood of Z^0 so as to find coefficients $\{\alpha_j(Z^0)\}_{j=0}^{r}$ that minimize

$$\frac{1}{n}\sum_{Z_t \in N(Z^0)} w_t(Z_0)\left(Y_t - \sum_{j=0}^{r} \alpha_j(Z_0)Z_t^j\right)^2,$$

where Z_t is restricted to a neighborhood of Z^0, $w_t(Z^0)$ is determined from (17.11), and $r = 1$ or 2. Having discussed the details of this two-phase or modular approach (via differencing and smoothing) to estimate the partial linear regression model, let us now turn to an SAS program for accomplishing the same.

Example 17.1

Table 17.1 displays a set of $n = 19$ observations on the variables X_1, X_2 (which enter into the parametric portion of the partial linear model Equation 17.2) and Z (which is the argument in f, the nonparametric part of Equation 17.2). Hence, our objective is to estimate

$$Y = \beta_1 X_1 + \beta_2 X_2 + f(Z) + \varepsilon. \qquad (17.2.3)$$

The SAS code for executing the modular approach to the estimation of Equation 17.2.3 appears in Exhibit 17.1.

TABLE 17.1

Observations on Variables X_1, X_2, Z, and Y

X_1	X_2	Z	Y
10	3	5	10
15	6	7	13
25	11	10	19
20	8	7	15
28	13	11	21
35	19	12	30
50	31	19	48
42	29	17	42
38	24	14	35
54	35	21	50
60	38	22	55
67	41	23	59
85	58	28	76
80	52	25	69
83	55	26	73
73	46	24	63
88	62	31	79
89	64	30	81
91	67	32	82

Exhibit 17.1 SAS Code for Modular Estimation

```
data npls;
   input x1 x2 z y;
datalines;
10   3    5    10
15   6    7    13
.    .    .    .
.    .    .    .
.    .    .    .
.    .    .    .
91   67   32   82
;
proc sort data=npls out=snpls; ①
   by z;
proc print data=snpls;
run;
data data2; set snpls; ②
dx1=dif(x1);]
dx2=dif(x2);} ③
dy=dif(y);  ]
proc reg data=data2 outest=ests;
   model dy=dx1 dx2/noint;        ④
run;
proc print data=ests;
run;
data ests1; set ests; ⑤
call symput('init_dx1', trim(left(dx1)));} ⑥
call symput('init_dx2', trim(left(dx2)));}
run;
data data3; merge data2 ests1; ⑦
y1=y-&init_dx1*x1-&init_dx2*x2; ⑧
data data4; set data3; ⑨
proc print data=data4;
run;
proc loess data=data4; ⑩
   model y1=z/smooth=0.2 to 0.5 by 0.1
   residual; ⑪
ods output outputstatistics=a; ⑫
proc print data=a;
run;
symbol1 c=black v=dot;                                     ]
symbol2 c=black v=point i=join l=1;                        ]
goptions nodisplay;                                        ]
proc gplot data=a;                                         ]
      by smoothingparameter;                               ]
      plot depvar*z=1 pred*z=2/overlay name='fit';         } ⑬
run;                                                       ]
goptions display;                                          ]
proc greplay nofs tc=sashelp.templt template=12r2;         ]
      igout gseg;                                          ]
treplay 1:fit 2:fit2 3:fit1 4:fit3;                        ]
```

① We invoke PROC SORT to arrange the dataset *npls* by increasing values of Z; the resulting output dataset is called *snpls*.

② Our working dataset is *data2*.

③ We request the first differences of the variables *X1*, *X2*, and *Y*.

④ We invoke PROC REG to estimate a parametric regression model, without an intercept, in the differenced variables. The resulting parameter estimates $\hat{\beta}_{1,dif}$ and $\hat{\beta}_{2,diff}$ are written to the temporary file *ests* by the OUTEST option.

⑤ We specify the SAS data file *ests*1, the contents of which are the entries in the temporary file *ests*.

⑥ We use a macro to identify the OLS estimates $\hat{\beta}_{1,diff}$ and $\hat{\beta}_{2,diff}$ determined in ④.

⑦ We specify the SAS data file *data*3 that contains all variables from *data*2 along with the OLS parameter estimates from *ests*1.

⑧ Using the macro variables from ⑥, we create a new dependent variable $Y1 = Y - \hat{\beta}_{1,diff} * X1 - \hat{\beta}_{2,diff} * X2$ by removing the parametric effects of *X1* and *X2* from *Y*.

⑨ We specify the SAS data file *data*4 that contains all variables from *data*3 along with *Y1*.

```
run;
symbol1 c=black v=dot;
goptions nodisplay;
proc gplot data=a;
       by smoothingparameter;
       plot residual*z=1/vref=0 name='resids';
run;                                                      ⑭
goptions display;
proc greplay nofs tc=sashelp.templt template 12r2;
       igout gseg;
treplay 1:resids 2:resids2 3:resids1 4:resdis3;
run;
```

⑩, ⑪ We invoke PROC LOESS to smooth *Y1* as a function of *Z*. The *MODEL* option SMOOTH= specifies that the model will be fit sequentially with smoothing parameters 0.2, 0.3, 0.4, and 0.5. The RESIDUAL option in the *MODEL* statement is used.

⑫ The ODS OUTPUT statement specifies the various datasets to be created. The keyword OUTPUTSTATISTICS requests that output dataset *a* be created. Dataset *a* houses the residuals requested in ⑪ as well as predicted values, values of the dependent variable, and *Z*.

⑬ We request a 2 × 2 plot grid of the fitted LOWESS response curves for the range of the smoothing parameter.

⑭ We request a 2 × 2 grid of the fitted residuals against the *Z* values for the range of the smoothing parameter.

The results generated by the preceding SAS program are offered in Table 17.2.

Ⓐ The overall fit to the estimated regression equation in first differences,

$$\begin{aligned} dY &= \hat{\beta}_{1,diff} * dX1 + \hat{\beta}_{2,diff} * dX2 \\ &= 0.42302dX1 + 0.61152dX2, \end{aligned}$$

is quite good (R^2 is high, the p-value for the calculated F statistic is very low, and the p-values associated with the estimated regression coefficients are also both very low).

Ⓑ The results presented in this section are of limited applicability/use.

Ⓒ The estimation results for each of the smoothing parameters are listed in turn. At each *Z*, we are provided with the value of the dependent variable *Y1*, its predicted value, and the residual.

Figure 17.1 plots the smoothed LOWESS fit for each value of the smoothing parameter *s*, and Figure 17.2 plots the residuals against the predicted values for each *s*. From these plots, it is evident that a smoothing parameter of 0.3 appears to yield a good fit (a smoothing parameter of 0.2 overfits the data; it almost exactly interpolates the observations on *Y1*).

Let us now focus on the estimation results generated by the smoothing parameter value of 0.3. In particular, suppose we want to obtain 95% confidence limits for the predicted values. One way to obtain this extra output is to augment the SAS code in Exhibit 17.1 by the additional lines of code in Exhibit 17.2.

TABLE 17.2

Output for Example 17.1

The SAS System

Ⓐ

The REG Procedure
Model : MODEL1
Dependent Variable: dy

Note: No intercept in model. *R*-Square is redefined.

Analysis of Variance

Source	DF	Sum of Squares	Mean Square	F Value	Pr > *F*
Model	2	356.52672	178.26336	103.82	<.0001
Error	16	27.47328	1.71708		
Uncorrected total	18	384.00000			

Root MSE	1.31037	*R*-Square	0.9285
Dependent Mean	4.00000	Adj *R*-Sq	0.9195
Coeff Var	32.75935		

Parameter Estimates

Variable	DF	($\hat{\beta}_{diff}$) Paramater Estimate	Standard Error	*t* Value	Pr > |*t*|
dx1	1	($\beta_{1,diff}$) 0.42602	0.13492	3.14	0.0064
dx2	1	($\beta_{2,diff}$) 0.61152	0.16636	3.68	0.0020

The SAS System

Ⓑ

The LOESS procedure
Independent Variable Scaling
Scaling applied: None

Statistic	z
Minimum Value	5.00000
Maximum Value	32.00000

The LOESS Procedure
Smoothing Parameter: 0.2
Dependent Variable: y1
Fit Summary

Fit method	kd tree
Blending	Linear
Number of observations	19
Number of fitting points	18
kd tree bucket size	1
Degree of local polynomials	1
Smoothing parameter	0.20000
Points in local neighborhood	3
Residual sum of squares	0.89531

TABLE 17.2 *(continued)*

The LOESS Procedure
Smoothing Parameter: 0.3
Dependent Variable: y1
Fit Summary

Fit method	kd tree
Blending	Linear
Number of observations	19
Number of fitting points	18
kd tree bucket size	1
Degree of local polynomials	1
Smoothing parameter	0.30000
Points in local neighborhood	5
Residual sum of squares	5.16720

The LOESS Procedure
Smoothing Parameter: 0.4
Dependent Variable: y1
Fit Summary

Fit method	kd tree
Blending	Linear
Number of observations	19
Number of fitting points	18
kd tree bucket size	1
Degree of local polynomials	1
Smoothing parameter	0.40000
Points in local neighborhood	7
Residual sum of squares	8.43170

The LOESS Procedure
Smoothing Parameter: 0.5
Dependent Variable: y1
Fit Summary

Fit method	kd tree
Blending	Linear
Number of observations	19
Number of fitting points	18
Kd tree bucket size	1
Degree of local polynomials	1
Smoothing parameter	0.50000
Points in local neighborhood	9
Residual sum of squares	13.48940

(continued)

TABLE 17.2 (*continued*)

The SAS System
©

Obs	Smoothing Parameter	Obs	z	(Y1) DepVar	Pred	Residual
1	0.2	1	5.00000	3.93524	3.93524	0
2	0.2	2	7.00000	2.98558	2.31651	0.66907
3	0.2	3	7.00000	1.64744	2.31651	−0.66907
4	0.2	4	10.00000	1.69777	1.69777	−2.2204E−16
5	0.2	5	11.00000	1.20567	1.20567	0
6	0.2	6	12.00000	3.57539	3.57539	8.88178E−16
7	0.2	7	14.00000	4.24872	4.24872	0
8	0.2	8	17.00000	6.49903	6.49903	8.88178E−16
9	0.2	9	19.00000	7.89183	7.89183	0
10	0.2	10	21.00000	5.75366	5.75366	0
11	0.2	11	22.00000	6.38098	6.38098	0
12	0.2	12	23.00000	5.58527	5.58527	0
13	0.2	13	24.00000	3.98954	3.98954	0
14	0.2	14	25.00000	3.35927	3.35927	0
15	0.2	15	26.00000	4.25564	4.25564	0
16	0.2	16	28.00000	4.57503	4.57503	0
17	0.2	17	30.00000	4.21382	4.21382	0
18	0.2	18	31.00000	3.85988	3.85988	0
19	0.2	19	32.00000	2.53321	2.53321	0
20	0.3	1	5.00000	3.93524	3.79332	0.14193
21	0.3	2	7.00000	2.98558	2.51050	0.47508
22	0.3	3	7.00000	1.64744	2.51050	−0.86306
23	0.3	4	10.00000	1.69777	1.35500	0.34277
24	0.3	5	11.00000	1.20567	2.12055	−0.91488
25	0.3	6	12.00000	3.57539	2.87068	0.70471
26	0.3	7	14.00000	4.24872	4.39587	−0.14715
27	0.3	8	17.00000	6.49903	6.33463	0.16440
28	0.3	9	19.00000	7.89183	6.88435	1.00748
29	0.3	10	21.00000	5.75366	6.38599	−0.63233
30	0.3	11	22.00000	6.38098	5.80888	0.57210
31	0.3	12	23.00000	5.58527	5.35622	0.22906
32	0.3	13	24.00000	3.98954	4.26596	−0.27642
33	0.3	14	25.00000	3.35927	3.98006	−0.62079
34	0.3	15	26.00000	4.25564	4.04974	0.20591
35	0.3	16	28.00000	4.57503	4.43527	0.13977
36	0.3	17	30.00000	4.21382	3.97342	0.24040
37	0.3	18	31.00000	3.85988	3.46668	0.39320
38	0.3	19	32.00000	2.53321	2.91524	−0.38203
39	0.4	1	5.00000	3.93524	3.33968	0.59556
40	0.4	2	7.00000	2.98558	2.71165	0.27392
41	0.4	3	7.00000	1.64744	2.71165	−1.06422
42	0.4	4	10.00000	1.69777	2.19509	−0.49731
43	0.4	5	11.00000	1.20567	2.55976	−1.35410
44	0.4	6	12.00000	3.57539	2.87068	0.70471

TABLE 17.2 *(continued)*

The SAS System
©

Obs	Smoothing Parameter	Obs	z	(Y1) DepVar	Pred	Residual
45	0.4	7	14.00000	4.24872	4.32239	−0.07367
46	0.4	8	17.00000	6.49903	6.10243	0.39660
47	0.4	9	19.00000	7.89183	6.80077	1.09106
48	0.4	10	21.00000	5.75366	6.45681	−0.70315
49	0.4	11	22.00000	6.38098	5.77803	0.60295
50	0.4	12	23.00000	5.58527	5.13210	0.45318
51	0.4	13	24.00000	3.98954	4.59278	−0.60324
52	0.4	14	25.00000	3.35927	4.19825	−0.83898
53	0.4	15	26.00000	4.25564	4.13140	0.12424
54	0.4	16	28.00000	4.57503	4.27094	0.30410
55	0.4	17	30.00000	4.21382	3.78940	0.42442
56	0.4	18	31.00000	3.85988	3.49091	0.36897
57	0.4	19	32.00000	2.53321	3.15247	−0.61926
58	0.5	1	5.00000	3.93524	2.69390	1.24134
59	0.5	2	7.00000	2.98558	2.63645	0.34913
60	0.5	3	7.00000	1.64744	2.63645	−0.98901
61	0.5	4	10.00000	1.69777	2.74447	−1.04670
62	0.5	5	11.00000	1.20567	2.85439	−1.64873
63	0.5	6	12.00000	3.57539	3.07745	0.49795
64	0.5	7	14.00000	4.24872	4.32239	−0.07367
65	0.5	8	17.00000	6.49903	5.84426	0.65477
66	0.5	9	19.00000	7.89183	6.34405	1.54778
67	0.5	10	21.00000	5.75366	6.07702	−0.32336
68	0.5	11	22.00000	6.38098	5.73582	0.64516
69	0.5	12	23.00000	5.58527	5.20242	0.38286
70	0.5	13	24.00000	3.98954	4.77411	−0.78457
71	0.5	14	25.00000	3.35927	4.50146	−1.14219
72	0.5	15	26.00000	4.25564	4.22569	0.02995
73	0.5	16	28.00000	4.57503	4.01130	0.56374
74	0.5	17	30.00000	4.21382	3.70586	0.50794
75	0.5	18	31.00000	3.85988	3.54636	0.31352
76	0.5	19	32.00000	2.53321	3.36644	−0.83323

Exhibit 17.2 SAS Code for Modular Estimation (s = 0.3)

```
proc loess data=data4;
      model y1=z/smooth=0.3 residual clm;
ods output outputstatistics=a;
proc print data=a;
run;
symbol3  c=black  i=join  v=none;
symbol4  c=black  i=join  v=none;
```

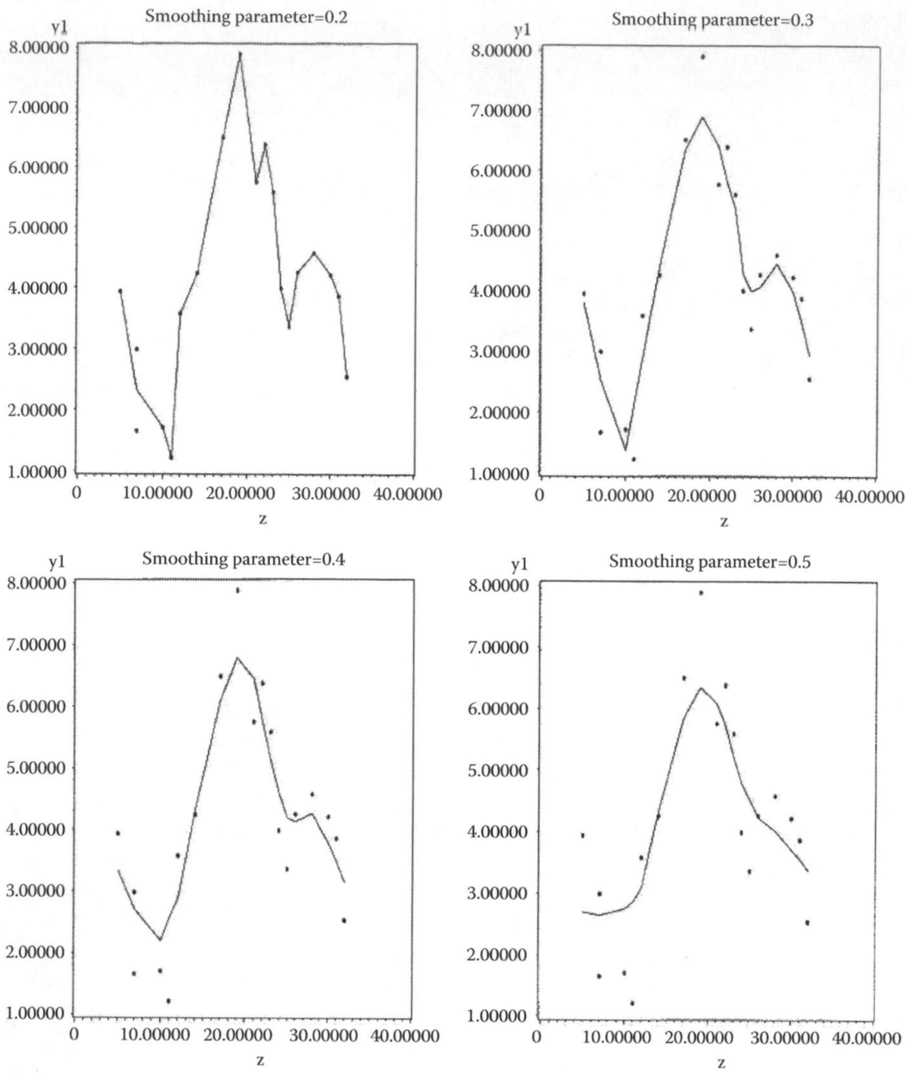

FIGURE 17.1
Plot of *Y1* versus *Z*.

```
proc gplot data=a;
      plot depvar*z  pred*z  lowercl*z  uppercl*z/overlay;
run;
```

The resulting output appears in Table 17.3, and Figure 17.3 displays the LOWESS fit for $s = 0.3$ along with the requested confidence limits.

An assortment of additional points merits our consideration. First, given the prominent peaks and troughs displayed by the smoothed LOWESS plot for $s = 0.3$ in Figure 17.1, we may want to conduct our smoothing process by using a second-degree polynomial rather than simply relying on the (default) first-degree polynomial for local approximations. The requisite adjustment to the model statement in Exhibit 17.2 is

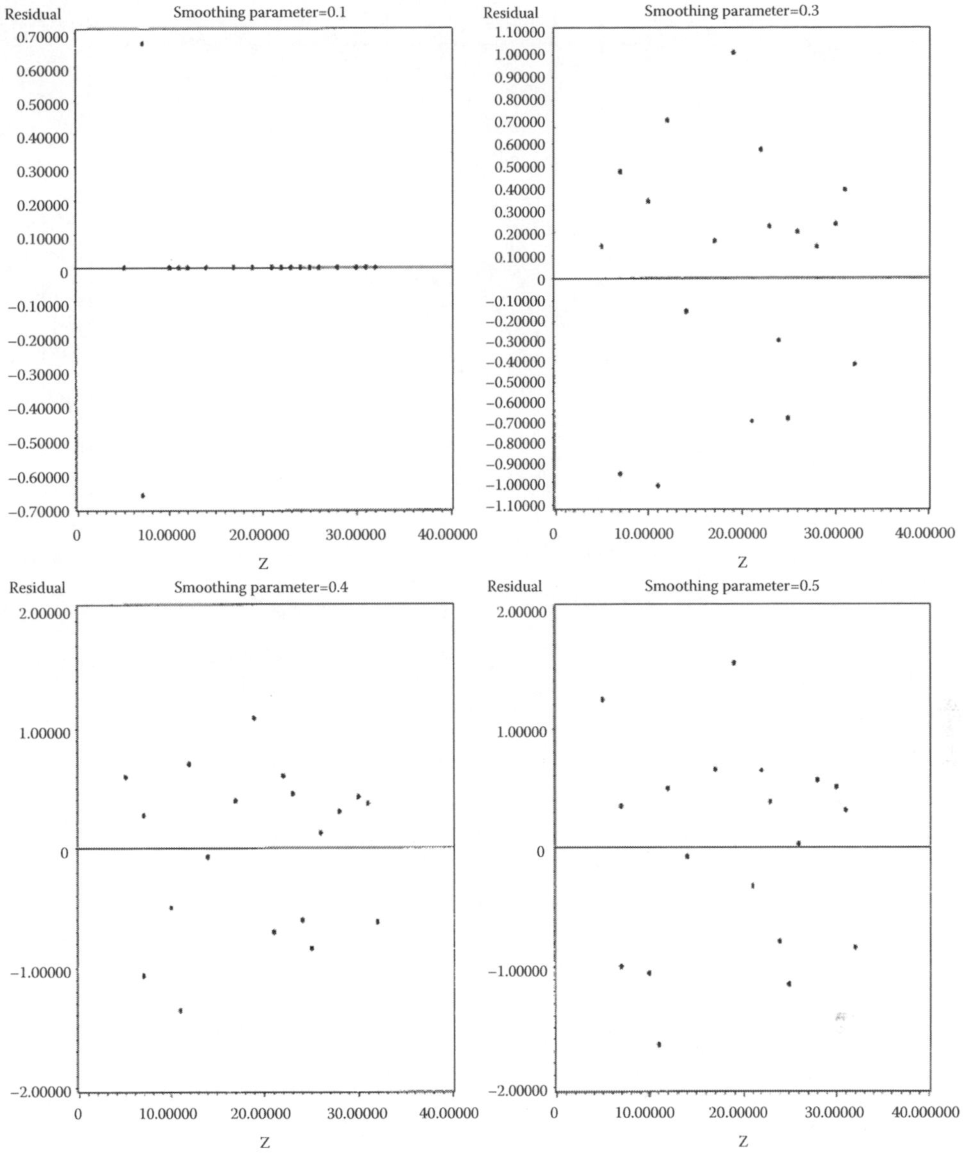

FIGURE 17.2
Plot of residual versus Z.

```
model y1=z/ degree=2 smooth=0.3 residual clm;
```

Second, PROC LOESS uses, by default, approximating local least-squares polynomial fits at a *representative sample of points* and generates estimated values at all other points by interpolation. If the dataset is relatively small, one can request that a local polynomial be fit at every data point by selecting the DIRECT option in the *MODEL* statement. In this regard, the preceding *MODEL* statement would read

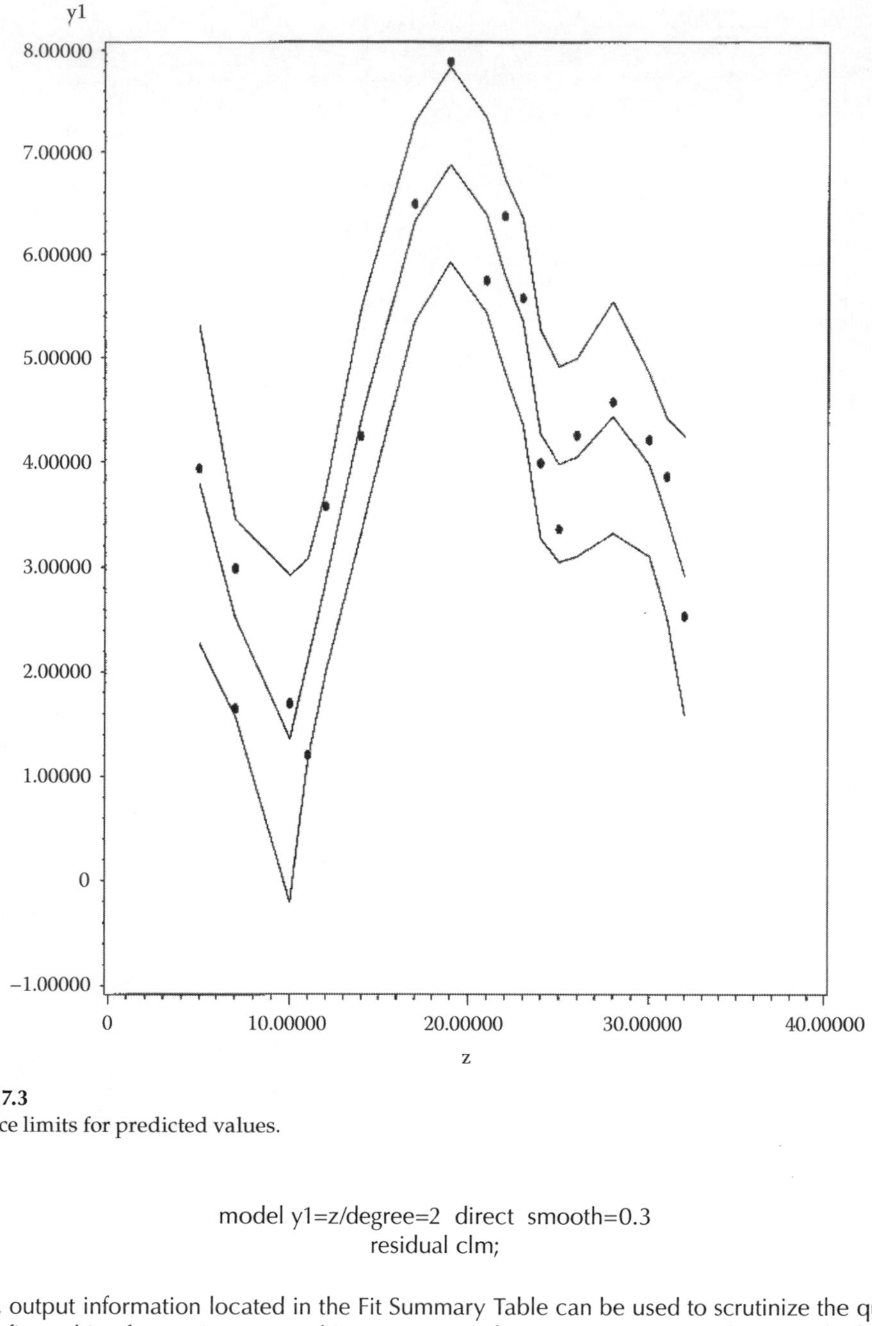

FIGURE 17.3
Confidence limits for predicted values.

```
model y1=z/degree=2 direct smooth=0.3
        residual clm;
```

Third, output information located in the Fit Summary Table can be used to scrutinize the quality of the fit resulting from using a smoothing parameter of, say, $s = 0.3$. We noted previously that the poorest fitting response function obtains when $s = 1$. In this regard, if our null model is taken to be the one displaying lack of fit, then it may be compared with an alternative model with smoothing parameter equal to 0.3 to assess, via an F test, the significance of the alternative model fit. To this end, we use the statistic

$$F = \frac{(SSE^0 - SSE)/(\delta_1^0 - \delta_1)}{SSE/\delta_1} : F_{\alpha,\nu_1,\nu_2}, \tag{17.12}$$

TABLE 17.3.
Output for Example 17.1 (s=0.3)

The SAS System
The LOESS procedure
Smoothing Parameter: 0.3
Dependent Variable: y1
Fit Summary

Fit method	kd tree
Blending	Linear
Number of observations	19
Number of fitting points	18
kd tree bucket size	1
Degree of local polynomials	1
Smoothing parameter	0.30000
Points in local neighborhood	5
Residual sum of squares	5.16720
Trace[L]	8.76141
GCV	0.04929
AICC	2.06757
AICC1	45.91683
Delta1	8.91296
Delta2	8.04057
Equivalent number of parameters	7.43579
Lookup degrees of freedom	9.88001
Residual standard error	0.76141

Obs	Smoothing Parameter	Obs	Z	DepVar	Pred	Residual	LowerCL	UpperCL
1	0.3	1	5.00000	3.93524	3.79332	0.14193	2.26970	5.31693
2	0.3	2	7.00000	2.98558	2.51050	0.47508	1.57164	3.44935
3	0.3	3	7.00000	1.64744	2.51050	−0.86306	1.57164	3.44935
4	0.3	4	10.00000	1.69777	1.35500	0.34277	−0.20849	2.91849
5	0.3	5	11.00000	1.20567	2.12055	−0.91488	1.16276	3.07833
6	0.3	6	12.00000	3.57539	2.87068	0.70471	2.00546	3.73590
7	0.3	7	14.00000	4.24872	4.39587	−0.14715	3.31949	5.47225
8	0.3	8	17.00000	6.49903	6.33463	0.16440	5.36173	7.30752
9	0.3	9	19.00000	7.89183	6.88435	1.00748	5.9147	7.83722
10	0.3	10	21.00000	5.75366	6.38599	−0.63233	5.43676	7.33521
11	0.3	11	22.00000	6.38098	5.80888	0.57210	4.87072	6.74704
12	0.3	12	23.00000	5.58527	5.35622	0.22906	4.35578	6.35665
13	0.3	13	24.00000	3.98954	4.26596	−0.27642	3.26553	5.26640
14	0.3	14	25.00000	3.35927	3.98006	−0.62079	3.04190	4.91822
15	0.3	15	26.00000	4.25564	4.04974	0.20591	3.10051	4.99896
16	0.3	16	28.00000	4.57503	4.43527	0.13977	3.31887	5.55167
17	0.3	17	30.00000	4.21382	3.97342	0.24040	3.09757	4.84927
18	0.3	18	31.00000	3.85988	3.46668	0.39320	2.51326	4.42010
19	0.3	19	32.00000	2.53321	2.91524	−0.38203	1.58191	4.24856

where

$$v_1 = \frac{(\delta_1^0 - \delta_1)^2}{\delta_2^0 - \delta_2},$$

$$v_2 = \text{lookup degrees of freedom} = \delta_1^2 / \delta_2,$$

and the superscript "0" refers to the null model. For example, suppose we add to the SAS code in Exhibit 17.1 the additional lines (Exhibit 17.3) as follows:

Exhibit 17.3 SAS Code for Modular Estimation (Second-Degree Smoothing)

```
proc loess data=data4;
        model y1=z/degree=2 direct smooth=0.3 1.0 residual clm;
ods output outputstatistics=b;
proc print data=b;
run;
```

The output that follows constitutes Table 17.4. Looking to selections Ⓐ and Ⓑ of this table, we can easily determine that $F = 3.34$, $v_1 = 10.19$, and $v_2 = 5.27$. Also, $F_{\alpha, v_1, v_2} = F_{0.05,\,10,\,5} = 4.74$. Because the sample F value falls short of 4.74, we cannot reject the null model. Hence, we must conclude, at the $\alpha = 0.05$ level, that the LOWESS model with second-degree local smoothing and $s = 0.3$ does not fit significantly better that the null model (with the latter using second-degree local smoothing and $s = 1$).

Next, up to this point in our discussion of PROC LOESS, we have been selecting the smoothing parameter on an ad hoc basis, i.e., by simply examining plots such as those represented in Figures 17.1 and 17.2. However, a macro yielding an automatic method for choosing a smoothing parameter is supplied by SAS [see SAS Institute Inc. 1999d, 1895–1899].

Finally, the time has come to perform some tests pertaining to the nonparametric portion of the partial linear model, i.e., on $Y1 \approx f(Z) + \varepsilon$. The questions that we shall now consider are as follows:

1. Does the LOWESS model with first degree smoothing and $s = 0.3$ fit significantly better than the null model (with first-degree smoothing and $s = 1$)?
2. Can the nonparametric portion of the partial linear model be adequately represented by a linear, quadratic, or cubic function of Z?

To answer these questions, let us insert the following lines of SAS code into Exhibit 17.1 (Exhibit 17.4):

Exhibit 17.4 SAS Code for Modular Estimation (PROC LOESS versus PROC REG)

```
data npls;
       input x1 x2 z y;
sz=z**2;⎫
        ⎬sz and qz denote Z² and Z³, respectively
qz=z**3;⎭
```

TABLE 17.4
Output for Example 17.1 (PROC LOESS, Second Degree Smoothing)

The SAS System

Ⓐ

The LOESS Procedure
Smoothing Parameter: 0.3
Dependent Variable: y1

Fit Summary

Fit method	Direct
Number of observations	19
Degree of local polynomials	2
Smoothing parameter	0.30000
Points in local neighborhood	5
Residual sum of squares	2.59636(= SSE)
Trace[L]	14.43054
GCV	0.12435
AICC	11.02038
AICC1	209.01815
Delta1	4.08715(= δ_1)
Delta2	3.16953(= δ_2)
Equivalent number of parameters	13.94823
Lookup degrees of freedom	5.27044(= v_2)
Residual standard error	0.79703

Ⓑ

The LOESS Procedure
Smoothing Parameter: 1
Dependent Variable: y1

Fit Summary

Fit method	Direct
Number of observations	19
Degree of local polynomials	2
Smoothing parameter	1.00000
Points in local neighborhood	19
Residual sum of squares	25.72592(= SSE^0)
Trace[L]	3.78667
GCV	0.11115
AICC	2.02758
AICC1	38.70758
Delta1	14.98964 (= δ_1^0)
Delta2	14.83151 (=δ_2^0)
Equivalent number of parameters	3.56297
Lookup degrees of freedom	15.14945
Residual standard error	1.31006

```
datalines;
 .
 .
 .
;
proc loess data=data4;
     model y1=z/smooth=0.3 1.0 residual clm;
ods output outputstatistics=c;
proc print data=c;
run;
proc reg data=npls;
      model y=x1 x2 z sz;
      model y=x1 x2 z sz qz;
run;
```

The output generated by this code appears in Table 17.5 (PROC LOESS results) and Table 17.6 (PROC REG results).

Addressing Question 1, let us use the information contained in Sections Ⓐ, Ⓑ of Table 17.5 and use Equation 17.12 to calculate $F = 6.85$ with $v_1 = 6.73$ and $v_2 = 9.88$. Because the value of this sample F statistic exceeds $F_{\alpha,v_1,v_2} = F_{0.05,\,6,\,9} = 3.37$, we can readily conclude, at the $\alpha = 0.05$ level, that a LOWESS model with first-degree local smoothing and $s = 0.3$ fits significantly better than the null model with first-degree local smoothing and $s = 1$.

The answer to Question 2 is based on the outcome of the following *specification test* [Yatchew 2003]. Suppose we want to test the null hypothesis that the f component of the partial linear model is a parametric function with argument Z (f may be hypothesized as, say, linear or quadratic or cubic in Z), i.e., we test the parametric null $h(Z;\gamma_1,\gamma_2,\ldots)$ (the γ's are unknown parameters) against a nonparametric alternative. In this regard, we may formulate the null hypothesis as H_0: $f(Z) = h(Z;\gamma_1,\gamma_2,\ldots)$ is true, with h a known function of Z.

Consider the test statistic

$$V = \frac{n^{1/2}(S_\varepsilon^2 - S_{diff}^2)}{S_{diff}^2}, \tag{17.13}$$

where

$$s_\varepsilon^2 = \frac{1}{n}\sum_{i=1}^{n}(Y_i - h(Z;\gamma_1,\gamma_2,\ldots))^2$$

is the usual OLS estimator of the residual variance obtained from a pure parametric specification for f, and S_{diff}^2 is determined from Equation 17.8.1 or from estimating the parametric portion of the partial linear model in differenced variables $dY = \sum_{j=1}^{m}\beta_{j,diff}dX_j + \varepsilon_i - \varepsilon_{i-1}$. Furthermore, it can be demonstrated that V is asymptotically standard normal under the null hypothesis and thus this hypothesis is rejected for V large.

From Section Ⓐ of Table 17.2, we see that $MSE/2 = 1.71708/2 = 0.85854 = S_{diff}^2$. (We divide MSE by 2 so that the variance of the residuals from the differenced equation corresponds to the residual variance of the undifferenced form $Y = \beta_1 X1 + \beta_2 X2 + f(Z) + \varepsilon$.) Let us test the hypothesis that f can be modeled by a quadratic function in Z or that H_0: $f(Z) = \gamma_0 + \gamma_1 Z + \gamma_2 Z^2$ is true. From Section Ⓐ of Table 17.6, $S_\varepsilon^2 = 1.71583$. Thus, from Equation 17.13, $V = (19)^{1/2}(1.77189 - 0.85854)/0.85854 = 4.64$, $p-$ value < 0.0001. Hence, we reject the null hypothesis that a quadratic specification for f is adequate. Next, from Section Ⓑ of Table 17.6, we find that $S_\varepsilon^2 = 0.86755$. This enables us to test the hypothesis that f can be modeled by a cubic expression in Z or that H_0: $f(Z) = \gamma_0 + \gamma_1 Z + \gamma_2 Z^2 + \gamma_3 Z^3$ is true. With $V = (19)^{1/2}(0.86755 - 0.85854)/0.85854 = 0.05$, we cannot reject H_0; a cubic specification for f seems appropriate.

TABLE 17.5

Output for Example 17.1 (PROC LOESS)

The SAS System

Ⓐ

The LOESS Procedure

Smoothing Parameter: 0.3

Dependent Variable: y1

Fit Summary

Fit method	kd tree
Blending	Linear
Number of observations	19
Number of fitting points	18
kd tree bucket size	1
Degree of local polynomials	1
Smoothing parameter	0.30000
Points in local neighborhood	5
Residual sum of squares	5.16720(= *SSE*)
Trace[L]	8.76141
GCV	0.04929
AICC	2.06757
AICC1	45.91683
Delta1	8.91296(= δ_1)
Delta2	8.04057(= δ_2)
Equivalent number of parameters	7.43579
Lookup degrees of freedom	9.88001(= v_2)
Residual standard error	0.76141

Ⓑ

The LOESS Procedure

Smoothing Parameter: 1

Dependent Variable: y1

Fit Summary

Fit method	kd tree
Blending	Linear
Number of observations	19
Number of fitting points	9
kd tree bucket size	3
Degree of local polynomials	1
Smoothing parameter	1.00000
Points in local neighborhood	19
Residual sum of squares	34.39592(= SSE^0)
Trace[L]	2.51111
GCV	0.12651
AICC	2.07816
AICC1	39.57901
Delta1	16.27240 (= δ_1^0)
Delta2	16.08475 (=δ_2^0)
Equivalent number of parameters	2.29462
Lookup degrees of freedom	16.46224
Residual standard error	1.45388

TABLE 17.6

Output for Example 17.1 (PROC REG)

The SAS System

Ⓐ

The REG Procedure
Model: MODEL1
Dependent Variable: y

Analysis of Variance

Source	DF	Sum of Squares	Mean Square	*F* Value	Pr > *F*
Model	4	11284	2820.95628	1592.06	<.0001
Error	14	24.80645	1.77189		
Corrected total	18	11309			

Root MSE	1.33112	*R*-Square	0.9978
Dependent Mean	48.42105	Adj *R*-Sq	0.9972
Coeff Var	2.74906		

Parameter Estimates

Variable	DF	Parameter Estimate	Standard Error	*t* Value	Pr > |*t*|
Intercept	1	−1.31954	1.82044	−0.72	0.4805
x1	1	0.17855	0.12560	1.42	0.1770
x2	1	0.80106	0.21949	3.65	0.0026
z	1	1.05846	0.31824	3.33	0.0050
sz	1	−0.01859	0.00647	−2.87	0.0123

Ⓑ

The REG Procedure
Model: MODEL2
Dependent Variable: y

Analysis of Variance

Source	DF	Sum of Squares	Mean Square	*F* Value	Pr > *F*
Model	5	11297	2259.47068	2604.42	<.0001
Error	13	11.27818	0.86755		
Corrected Total	18	11309			

Root MSE	0.93142	*R*-Square	0.9990
Dependent Mean	48.42105	Adj *R*-Sq	0.9986
Coeff Var	1.92360		

Parameter Estimates

Variable	DF	Parameter Estimate	Standard Error	t Value	Pr > |*t*|
Intercept	1	7.95447	2.67173	2.98	0.0107
x1	1	0.02361	0.09625	0.25	0.8100
x2	1	0.93833	0.15747	5.96	<.0001
z	1	−0.71792	0.50194	−1.43	0.1762
sz	1	0.09983	0.03033	3.29	0.0058
qz	1	−0.00213	0.00053949	−3.95	0.0017

Penalized Spline Smoothing

The second stage of the above modular approach to estimating the partial linear model (Equation 17.2.1) used a smoothing technique called LOWESS. Here the objective was to determine the underlying trend in the scatter plot of points $(Z_i, \tilde{Y}_i)$, where $\tilde{Y}_i = Y_i - \hat{\beta}_{diff} X_i = f(Z_i) + \varepsilon_i$, $i = 1, \ldots, n$, f was taken to be an unspecified smooth function, and no particular probability model was assumed.

The scatter plot smoothing method to be pursued now involves penalized splines. Remember that splines (Chapter 16) are piecewise polynomial functions of order k connected at knots $k_1, \ldots, k_t$. Although we previously modeled splines with a fixed (and known) number of knots, our current approach will admit considerable flexibility in adding knots. However, adding too many knots provides us with a choppy or saw-toothed pattern of fluctuations in the data. To dampen the roughness of the fit, an approach called *penalized spline regression*, which retains all of the knots but constraints their impact, will be used.

To this end, let us structure (via Equation 7.2.1) the *penalized least-squares problem* as finding a function estimate $\hat{f}$ so as to

$$\min_f \left\{ \frac{1}{n} \sum_{i=1}^{n} (Y_i - \beta X_i - f(Z_i))^2 + \lambda P(f) \right\}, \tag{17.14}$$

where $\lambda (> 0)$ is a smoothing parameter and $P(f)$ is a penalty on the roughness of $\hat{f}$. Here the first term on the right side of Equation 17.14 measures goodness of fit, whereas the second term measures the smoothness of $\hat{f}$. In this regard, the objective function is designed to tradeoff goodness of fit to the data against the smoothness of $\hat{f}$; a penalty is incurred if $\hat{f}$ fits the data extremely well but, at the same time, is very rough. Hence, the larger is λ, the smoother is the function f selected.

A common measure of smoothness is the integral of the square of the second derivative of f. Hence, our objective is to find a function estimate $\hat{f}$ that solves the penalized least-squares problem

$$\min_f \left\{ \frac{1}{n} \sum_{i=1}^{n} (Y_i - \beta X_i - f(Z_i))^2 + \lambda \int_{-\infty}^{+\infty} (d^2 f / dZ^2)^2 dZ \right\}. \tag{17.14.1}$$

Clearly, large values of λ heavily penalize estimates with large second derivatives so that, as $\lambda \to \infty$, the estimate becomes progressively smoother (the second derivative must concomitantly become progressively smaller) and, in the limit, the estimate becomes linear. As $\lambda \to 0$, the estimate fits the data progressively better and, in the limit, fits the data perfectly.

How should the appropriate amount of smoothing be determined? Our choice of the "optimal" value of the smoothing parameter λ will be made by minimizing a global error criterion function called the *cross-validation function*:

$$V(\lambda) = \frac{(1/n) \| \boldsymbol{I} - \boldsymbol{H}(\lambda) \boldsymbol{Y} \|^2}{[(1/n) tr(\boldsymbol{I} - \boldsymbol{H}(\lambda))]^2},$$

where $\boldsymbol{H}(\lambda)$ is the hat matrix for which $\hat{\boldsymbol{Y}} = \boldsymbol{H}(\lambda)\boldsymbol{Y}$.

Example 17.2

Using the dataset presented in Table 17.1, let us use penalized spline smoothing to find a function $\hat{f}$ that fits the data reasonably well, is sufficiently smooth, and enables us to obtain good predicted values of Y. The penalized least-squares estimates of Y will be generated by the SAS procedure TPSPLINE. The requisite SAS code appears in Exhibit 17.5.

Exhibit 17.5 SAS Code for Penalized Spline Smoothing

```
data tpspl;
   input x1 x2 z y;
datalines;
10  3  5  10
15  6  7  13
.   .  .   .
.   .  .   .
.   .  .   .
91 67 32  82
;
proc sort data=tpspl
   out=stpspl; ①
   by z;
proc print data=stpspl;

run;
data data2; set stpspl;

proc tpspline data=data2; ②
   model y=x1 x2 (z)/lognlambda=(-1 to 1 by 0.1); ③
output out=estimate pred resid uclm lclm; ④
run;
proc print data=estimate;
run;
data data3; merge data2 estimate;
proc print data=data3;
run;
proc plot data=data3; ⑤
   plot P_Y*z='*';
run;
proc tpspline data=data3;
   model y=x1 x2 (z)/lognlambda=(-1 to 1 by 0.1);
ods output GCVFunction=gcv; ⑥
   symbol1 i=join v=point l=1;
proc gplot data=gcv; ⑦
   plot gcv*lognlambda=1;
run;
```

① The dataset *tpspl* is sorted by increasing values of *Z*. The resulting output dataset is *stpspl*.

② We invoke PROC TPSPLINE using the dataset *data2*.

③ The *MODEL* statement specifies a linear relationship involving *X1* and *X2* and an unknown functional relationship in *Z*. The parentheses separate the regression variables *X1* and *X2* from the smoothing variable *Z*. (Obviously, there can be more than one smoothing variable placed in parentheses, e.g., we could fit a model such as *Y* = *X1 X2* (*Z1 Z2*).) The LOGNLAMBDA=option returns a list of generalized cross-validation (GCV) values with $\log_{10}(n\lambda)$ ranging from −1 to 1 by 0.1.

④ The OUTPUT statement returns the output dataset *estimate* that contains predicted values of Y(P_Y), residuals (R_Y), and 95% lower and upper confidence limits (LCLM_Y and UCLM_Y), respectively.

⑤ We request a plot of the predicted values of *Y* against *Z*.

⑥ To obtain a plot of the (generalized) cross-validation function, we invoke PROC TPSPLINE using dataset *data3*. We use the output delivery system (ODS) to create the output dataset *gcv* from the output table *GCVFunction* that houses the GCV function values for LOGNLAMBDA in the range from −1 to 1. Hence, the ODS OUTPUT statement transforms an output table into an output dataset named *gcv*.

⑦ `We invoke PROC GPLOT using dataset` *`gcv`* `and plot the GCV function values against` $\log_{10}(n\lambda)$ `values.`

The (abbreviated) output resulting from this SAS code appears in Table 17.7.

TABLE 17.7

Output for Example 17.2

Ⓐ

The TPSPLIN Procedure

Dependent Variable: y

Summary of Input Data SET

Number of Non-Missing Observations	19
Number of Missing Observations	0
Unique Smoothing Design Points	18

Summary of Final Model

Number of Regression Variables	2
Number of Smoothing Variables	1
Order of Derivative in the Penalty	2
Dimension of Polynomial Space	4

GCV Function

Log10(n*Lambda)	**GCV**
−1.000000	2.059753
−0.900000	1.966380
−0.800000	1.866016
−0.700000	1.762035
−0.600000	1.657854
−0.500000	1.556649
−0.400000	1.461168
−0.300000	1.373646
−0.200000	1.295793
−0.100000	1.228832
0	1.173550
0.100000	1.130324
0.200000	1.099134
0.300000	1.079558
0.400000	1.070761*
0.500000	1.071516
0.600000	1.080260
0.700000	1.095216
0.800000	1.114557
0.900000	1.136612
1.000000	1.160048

Note: * indicates minimum GCV Value.

(continued)

TABLE 17.7 *(continued)*

Summary Statistics of Final Estimation	
Log10(n*Lambda)	0.4408
Smoothing Penalty	0.4061
Residual SS	5.5699

The SAS System
The TPSPLINE Procedure
Dependent Variable: y

Summary Statistics of Final Estimation	
Tr(I-A)	9.9451
Model DF	9.0549
Standard Deviation	0.7484(= S_ε)

Ⓑ

Obs	x1	x2	Z	Y	P_y	R_y	LCLM_y	UCLM_y
1	10	3	5	10	9.7891	0.21093	8.4742	11.1039
2	15	6	7	13	12.5712	0.42878	11.5972	13.5452
3	20	8	7	15	15.4798	−0.47979	14.4529	16.5067
4	25	11	10	19	19.1250	−0.12502	18.1798	20.0702
5	28	13	11	21	21.9909	−0.99086	21.0617	22.9200
6	35	19	12	30	29.0391	0.96094	28.0602	30.0179
7	38	24	14	35	35.1261	−0.12610	34.0519	36.2003
8	42	29	17	42	42.2286	−0.22863	41.0668	43.3904
9	50	31	19	48	47.0769	0.92315	45.9779	48.1758
10	54	35	21	50	50.8537	−0.85370	49.8702	51.8372
11	60	38	22	55	54.5600	0.43998	53.7305	55.3895
12	67	41	23	59	58.5374	0.46260	57.6478	59.4270
13	73	46	24	63	63.5202	−0.52023	62.7319	64.3085
14	80	52	25	69	69.6562	−0.65616	68.7570	70.5553
15	83	55	26	73	72.7521	0.24790	71.8380	73.6662
16	85	58	28	76	75.7517	0.24826	74.7438	76.7597
17	89	64	30	81	80.7715	0.22848	79.7841	81.7590
18	88	62	31	79	78.7369	0.26313	77.6767	79.7971
19	91	67	32	82	82.4337	−0.43366	81.2107	83.6566

Ⓐ We have $n = 19$ observations with 18 unique design points (equivalent to the number of unknown parameters to be estimated). The value of $\log_{10}(n\lambda)$ that minimizes the GCV function and that is used to obtain the final smoothing spline estimate is 0.4408 (Figure 17.4). The standard error of estimate is

$$S_\varepsilon = \sqrt{\frac{SSE}{d.f.}} = \sqrt{\frac{SSE}{tr(\boldsymbol{I}-\boldsymbol{A})}}$$

$$= \sqrt{\frac{5.5699}{9.9451}} = 0.7484,$$

where $\boldsymbol{A}$ denotes the "hat matrix."

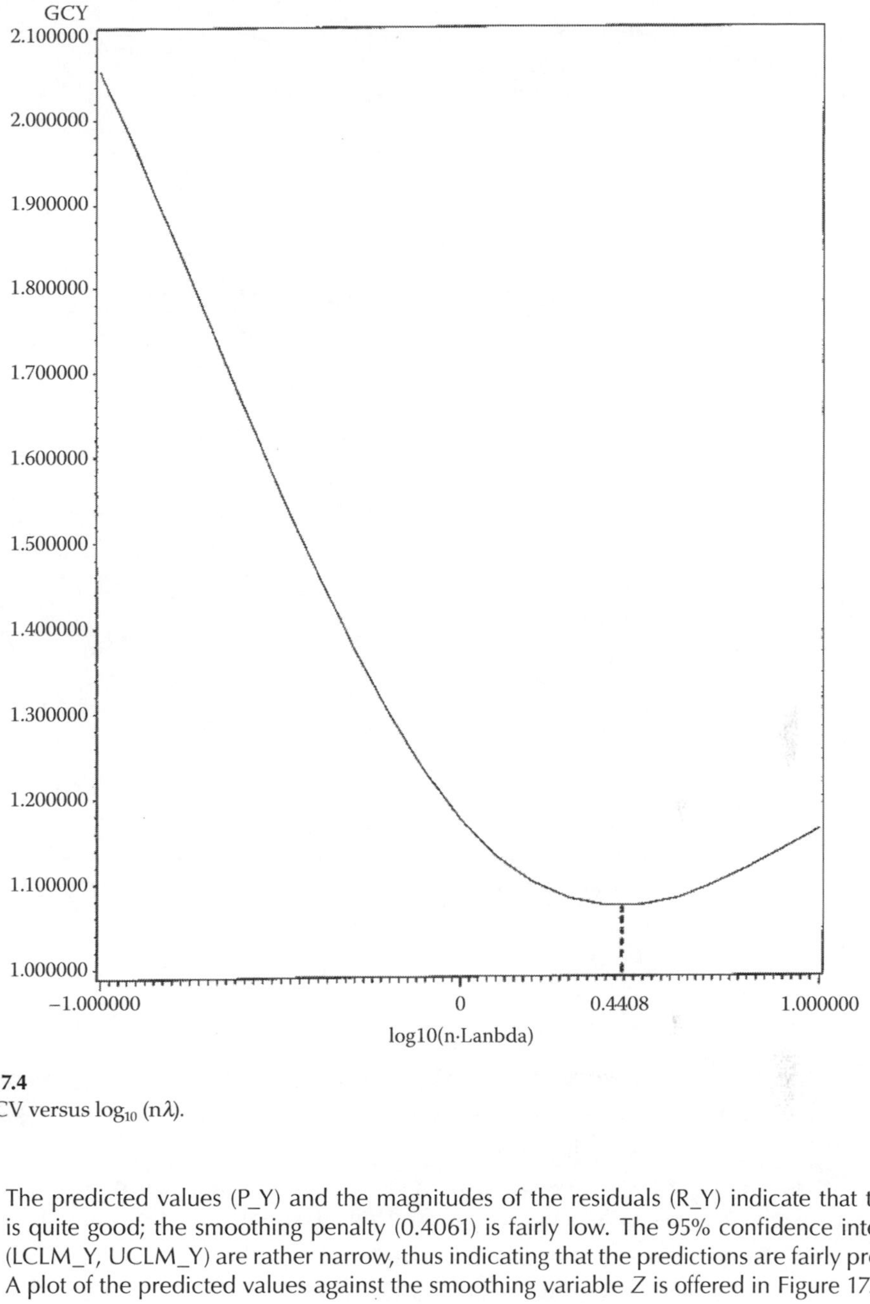

FIGURE 17.4
Plot of GCV versus $\log_{10}(n\lambda)$.

Ⓑ The predicted values (P_Y) and the magnitudes of the residuals (R_Y) indicate that the fit is quite good; the smoothing penalty (0.4061) is fairly low. The 95% confidence intervals (LCLM_Y, UCLM_Y) are rather narrow, thus indicating that the predictions are fairly precise. A plot of the predicted values against the smoothing variable Z is offered in Figure 17.5.

Exercises

17-1. Given the Table E.17-1 dataset, use the modular method to estimate the partial linear regression model. Fully explain your findings. (Hint: follow the lead of Example 17.1.)

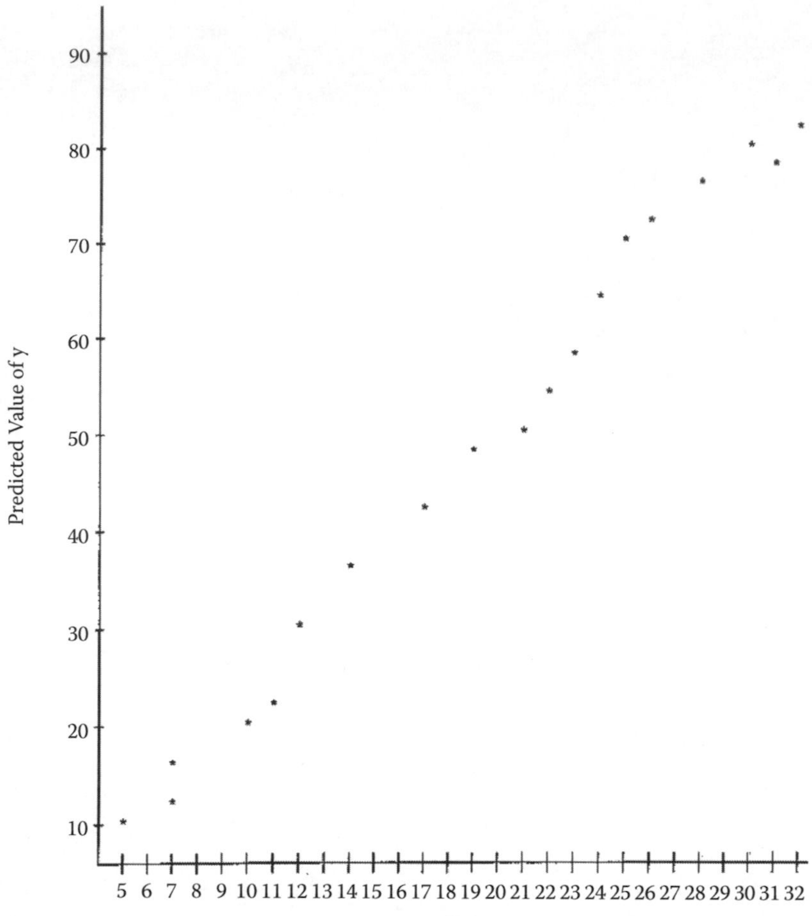

FIGURE 17.5
Plot of predicted *Y* versus *Z*.

17-2. Using the Table E.17-1 set of observations, use the penalized spline smoothing method to find a sufficiently smooth function estimate $\hat{f}$ that provides a good fit to the data. (Hint: see Example 17.2.)

17-3. For the Table E.17-2 data values, use the modular technique to estimate the partial linear regression model. Interpret your findings. (Hint: see Example 17.1.)

17-4. Using the Table E.17-2 dataset, use penalized spline smoothing to find a function estimate $\hat{f}$ that renders a good (smooth) fit to the observations. (Hint: consider Example 17.2.)

TABLE E.17-1

Observations on Variables Y, X_1, X_2, and Z

Y	X_1	X_2	Z
4.68	2.09	3.14	1.10
4.39	2.20	2.64	0.19
4.29	1.41	3.64	4.97
4.05	0.69	4.57	15.05
4.20	1.79	2.40	0.37
4.58	1.77	3.76	3.96
4.41	1.09	4.53	11.83
4.61	1.79	3.89	4.41
4.70	2.09	3.58	2.22
4.78	2.10	3.76	2.76
4.55	1.41	4.11	7.29
4.73	2.09	3.43	1.80
4.17	1.09	4.04	8.70
4.92	1.79	4.57	7.72
4.45	1.41	4.53	9.73
4.03	0.69	4.27	12.81
4.40	1.09	4.11	9.12
4.18	1.09	4.58	12.18
5.01	2.20	4.48	5.19
4.18	1.09	3.21	4.49
3.58	0.09	4.40	18.57
4.04	1.41	2.40	0.98
3.90	0.69	4.15	11.97
3.76	1.09	2.30	1.46
4.80	1.78	4.26	6.15
4.68	2.09	3.14	1.10
4.39	2.20	2.64	0.19
4.29	1.41	3.60	4.79

TABLE E.17-2

Observations on Variables Y, X_1, X_2, and Z

Y	X_1	X_2	Z
0.62	−2.81	−0.77	2.16
0.85	−1.96	−1.10	2.15
1.50	−1.35	−0.21	0.28
−1.66	−3.91	−2.66	10.40
1.45	−1.83	−0.44	0.81
0.96	−1.56	−1.14	1.77
1.34	−0.84	−0.56	0.47
−0.73	−3.21	−2.30	7.38
1.87	−2.04	−0.70	1.42
1.39	−1.83	−0.94	1.72
−0.44	−3.50	−2.04	7.14
−1.02	−3.91	−2.53	9.89
1.17	−1.96	−0.61	1.20
0.33	−2.41	−1.47	3.54
1.97	−0.27	−0.15	0.04
1.65	−1.35	−0.40	0.54
0.99	−2.04	−1.10	2.24
0.71	−1.90	−0.89	1.69
1.49	−0.80	−0.34	0.27
0.97	−1.90	−0.87	1.65
0.53	−2.66	−1.05	2.79
0.72	−2.53	−1.02	2.58
1.26	−1.77	−0.71	1.25
0.41	−2.30	−1.31	3.01
0.90	−1.90	−0.94	1.78
0.62	−2.81	−0.77	2.16

18

Nonlinear Regression

Intrinsic Linearity/Nonlinearity

In the (strong) classical linear regression model, the multivariate regression equation was specified as

$$Y_i = \beta_0 + \sum_{j=1}^{m} \beta_j X_{ij} + \varepsilon_i,\ i = 1,\ldots,n. \tag{18.1}$$

As mentioned previously, its functional form is said to be linear because it is *linear in the parameters*. Hence, linearity does not necessarily refer to the relationship between Y and the X_j's or between the X_j's themselves, $j = 1,\ldots, m$; but it does refer to how the parameters appear in the regression equation. Because Equation 18.1 is linear in both the parameters and variables, it will be termed *fully* or *completely linear*.

However, to address the issue of any possible nonlinearity in the variables, suppose $W_1,\ldots, W_p$ is a set of p independent regressors: $\phi^1(W_1,\ldots, W_p) = \phi^1(.),\ldots, \phi^m(W_1,\ldots, W_p) = \phi^m(.)$ constitutes a set of m independent functions of the W_k's, $k = 1,\ldots, p$; and $g(y)$ is a function of the explained variable Y. The regression equation

$$\begin{aligned} g(y) &= \beta_1\phi^1(.) + \beta_2\phi^2(.) + \cdots + \beta_m\phi^m(.) + \varepsilon \\ &= \beta_1 X_1 + \beta_2 X_2 + \cdots + \beta_m X_m + \varepsilon \end{aligned} \tag{18.2}$$

is *linear in the variables* $X_1 = \phi^1(.)$, $X_2 = \phi^2(.),\ldots, X_m = \phi^m(.)$.[1] In this regard, if the regression equation is linear in the parameters and nonlinear in the regressors but can be converted to a linear form by a suitable transformation or respecification of the regressors (such as in Equation 18.2), then the regression equation is said to be *intrinsically linear in the variables*. For instance, $Y = \theta_1 + \theta_2 W^{-1} + \varepsilon$ is intrinsically linear in the variables because it is linear in the parameters, nonlinear in W, and can be transformed to a linear form by setting $X = \phi(W) = W^{-1}$. Hence, $g(Y) = Y = \theta_1 + \theta_2 X + \varepsilon$ is now fully linear in the parameters and in the variables X and Y.

The preceding discussion of intrinsic linearity focused on nonlinearities in the variables and their respecification to achieve full or complete linearity. Let us now consider intrinsic linearity as it relates to the transformation or respecification of the parameters themselves.

[1] The expressions $X_j = \phi^j(W_1,\ldots,W_p) = \phi^j(.)$, $j = 1,\ldots,m$, $m \le p$, are *functionally independent* if every mth-order Jacobian determinant of these m functions with respect to m of the p variables does not identically vanish.

In this regard, for a regression equation that is linear in the regressors, suppose $\theta_1,\ldots,\theta_r$ is a set of *elemental* or *restricted parameters* (that is, parameters for which certain nonlinear relations among the regression coefficients must hold, e.g., one regression coefficient is the product or ratio of two other regression coefficients), $\beta_1,\ldots,\beta_r$ is a set of *unrestricted parameters*, and each unrestricted parameter β_1 is expressible as a one-to-one mapping or function[2] of the restricted parameters $\theta_1,\ldots,\theta_r$ or $\beta_l = \gamma^l(\theta_1,\ldots,\theta_r) = \gamma^l(.)$, $l = 1,\ldots,r$. Then the regression model

$$\begin{aligned} g(Y) &= \gamma^1(.)W_1 + \gamma^2(.)W_2 + \cdots + \gamma^r(.)W_r + \varepsilon \\ &= \beta_1 W_1 + \beta_2 W_2 + \cdots + \beta_r X_r + \varepsilon \end{aligned} \tag{18.3}$$

is *intrinsically linear in the restricted parameters* because it is fully linear in the unrestricted parameters.

If a regression equation is intrinsically linear in both the restricted parameters and regressors, then it is unambiguously said to be intrinsically linear. For example, $Y = \theta_1 W^{\theta_2}\eta$ is termed log-linear because it can be written as linear in the logs of the variables or $\log Y = \log\theta_1 + \theta_2 \log W + \log\eta$. However, this logarithmic expression is not linear in the parameters because $\log\theta_1$ is not linear. If we set $g(y) = \log Y$, $X = \phi(W) = \log W$, $\beta_1 = \gamma^1(\theta_1) = \log\theta_1$, $\beta_2 = \gamma^2(\theta_2) = \theta_2$ (clearly, γ^1 and γ^2 are one-to-one as required), and $\varepsilon = \log\eta$, then $g(Y) = \beta_1 + \beta_2 X + \varepsilon$ is fully linear in the parameters β_1 and β_2 and in the variables $g(Y)$ and X. Hence, the original expression involving Y and W is intrinsically linear.

Regression models that are nonlinear in the parameters as well as the variables are termed *intrinsically nonlinear*. For this class of functions, there is no transformation of variables that would induce linearity in the parameters. For example,

a. $Y = \theta_1 + (\theta_2 + W)^{-1} + \varepsilon$ is nonlinear in the parameters and with respect to W. No possible relabeling or respecification of the variables can lead to a fully linear regression model. Hence, this model is intrinsically nonlinear.
b. $Y = \theta_1 W_1^{\theta_2} W_2^{\theta_3} + \varepsilon$ is intrinsically nonlinear; no transformation of variables produces a regression equation that is fully linear in the parameters and variables. (How about if the specification is $Y = \theta_1 W_1^{\theta_2} W_2^{\theta_3} e^{\varepsilon}$?)

Intrinsic Linearity in the Restricted Parameters and Identification

We stated previously that the regression Equation 18.3 is intrinsically linear in the restricted parameters θ_l if the functions $\beta_l = \gamma^l(.)$, $l = 1,\ldots, r$, are one-to-one, where the β_l's represent a set of unrestricted parameters. That γ^l is one-to-one constitutes what is called an *identification requirement* or *condition*, i.e., if the γ^l's, $l = 1,\ldots,r$, are one-to-one, then the elemental or restricted parameters of the regression model, $\theta_1,\ldots,\theta_r$, are exactly identified in terms of the unrestricted parameters, $\beta_1,\ldots,\beta_r$, of the fully linear model in Equation 18.3. So if exact

[2] The mapping γ^l is one-to-one or injective if no β_l is the image of more than one $\theta = (\theta_1,\ldots,\theta_r)$. Hence, $\boldsymbol{\theta}_1 = (\theta_1^1,\ldots,\theta_r^1) \neq \boldsymbol{\theta}_2 = (\theta_1^2,\ldots,\theta_r^2)$ implies $\gamma^l(\boldsymbol{\theta}_1) \neq \gamma^l(\boldsymbol{\theta}_2)$.

identification holds, the model can be estimated in terms of the unrestricted β_l's and the restricted $\hat{\theta}_l$'s can be obtained from the $\hat{\beta}_l$'s, $l = 1,\ldots,r$; exact identification leads to a unique solution for the $\hat{\theta}_l$'s in terms of the $\hat{\beta}_l$'s, $l = 1,\ldots,r$.

Example 18.1

Suppose we want to obtain estimates of the fundamental or restricted parameters θ_1, θ_2, θ_3, and θ_4 appearing in the log-linear regression equation

$$g(Y) = \log Y = \log\theta_1 + \theta_2\theta_3\log W_1 + \theta_2(1-\theta_3)\log W_2 + \theta_2\theta_3(1-\theta_3)\theta_4\log W_3 + \varepsilon, \qquad (18.4)$$

where the ε_i are independent random variables with $\varepsilon_i : N(0,\sigma_\varepsilon^2)$ for $i = 1,\ldots,n$. Then via the preceding discussion, we can rewrite Equation 18.4 as

$$g(Y) = \beta_1 + \beta_2X_1 + \beta_3X_2 + \beta_4X_3 + \varepsilon, \qquad (18.4.1)$$

where obviously

$$\beta_1 = \gamma^1(.) = \log\theta_1 \qquad X_1 = \phi^1(.) = \log W_1$$

$$\beta_2 = \gamma^2(.) = \theta_2\theta_3 \qquad X_2 = \phi^2(.) = \log W_2$$

$$\beta_3 = \gamma^3(.) = \theta_2(1-\theta_3) \qquad X_3 = \phi^3(.) = \log W_3$$

$$\beta_4 = \gamma^4(.) = \theta_2\theta_3(1-\theta_3)\theta_4$$

with $X_j = \phi^j\,(W_1, W_2, W_3)$, $j = 1, 2, 3$, $(= m = p)$; and $\beta_l = \gamma^l(\theta_1, \theta_2, \theta_3, \theta_4)$, $l = 1,\ldots,4$ $(= r)$. Then Equation 18.4.1 can be estimated by OLS, with the result that

$$\hat{\theta}_1 = e^{\hat{\beta}_1}$$
$$\hat{\theta}_2 = \hat{\beta}_2 + \hat{\beta}_3$$
$$\hat{\theta}_3 = \frac{\hat{\beta}_2}{\hat{\beta}_2 + \hat{\beta}_3}$$
$$\hat{\theta}_4 = \frac{\hat{\beta}_4(\hat{\beta}_2 + \hat{\beta}_3)}{\hat{\beta}_2\hat{\beta}_3}.$$

Note that this result obtains because the θ^l's are exactly identified in terms of the β_l's, and this follows from the fact that $\beta_l = \gamma^l(.)$ is one-to-one. Clearly, the regression model in Equation 18.4 is intrinsically linear.

The reader may have noticed that, for the restricted parameters to be exactly identified in terms of the unrestricted parameters, the number of restricted parameters must be the same as the number of unrestricted ones. What happens when this is not the case? If the number of restricted parameters (the θs) exceeds the number of unrestricted parameters (the βs), then we cannot obtain a solution for the restricted parameters in terms of the unrestricted parameters. In this instance, we have *underidentification of the restricted parameters.* For example, suppose $Y_i = \theta_1 + \theta_2W_i + \varepsilon_i$ but that W_i is unobserved. However, let $W_i = \theta_3W_{i-1} + \eta_i$, where W_{i-1} is observed. Then

$$Y_i = \theta_1 + \theta_2\theta_3W_{i-1}\ \theta_2\eta_i + \varepsilon_i$$

$$= \beta_1 + \beta_2X_i + u_i, \qquad (18.5)$$

where $X_i = W_{i-1}$ and the random error term in Equation 18.5 is $u_i = \theta_2\eta_i + \varepsilon_i$ Here $\beta_1 = \theta_1$ while $\beta_2 = \theta_2\theta_3$. Applying OLS to Equation 18.5 yields $\hat{\beta}_1 = \hat{\theta}_1$ and $\hat{\beta}_2 = \widehat{\theta_2\theta_3}$. Thus, there is no solution for either $\hat{\theta}_2$ or $\hat{\theta}_3$: all we get is an estimate of the product of θ_2 and θ_3. This situation emerges because θ_1, θ_2, and θ_3 are underidentified in terms of β_1 and β_2.

Next, if there are fewer restricted parameters than unrestricted ones, then there is no unique solution for the restricted parameters in terms of the unrestricted parameters. This is the case of *overidentification of the restricted parameters*. (Note that, with overidentification of the restricted parameters, there is a solution but it is *not unique*; with underidentification, there is no solution at all.) For instance, let $Y = \theta_1(1-\theta_2) + \theta_3 W_1 - \theta_2\theta_3 W_2 + \theta_2 W_3 + \varepsilon$. The unrestricted version of this equation is

$$Y = \beta_1 + \beta_2 X_1 + \beta_3 X_2 + \beta_4 X_3 + \varepsilon, \tag{18.6}$$

with $X_k = W_k$, $k = 1, 2, 3$, and

$$\beta_1 = \theta_1(1-\theta_2)$$

$$\beta_2 = \theta_3$$

$$\beta_3 = -\theta_2\theta_3$$

$$\beta_4 = \theta_2.$$

Then an application of OLS to Equation 18.6 yields

$$\hat{\theta}_1 = \frac{\hat{\beta}_1}{1-\hat{\theta}_2} = \frac{\hat{\beta}_1}{1-\hat{\beta}_4}$$
$$\hat{\theta}_2 = \hat{\beta}_4 \text{ or } \hat{\theta}_2 = -\hat{\beta}_3/\hat{\theta}_3 = -\hat{\beta}_3/\hat{\beta}_2$$
$$\hat{\theta}_3 = \hat{\beta}_2 \text{ or } \hat{\theta}_3 = -\hat{\beta}_3/\hat{\theta}_2 = -\hat{\beta}_3/\hat{\beta}_4.$$

Thus, there is *no unique solution* for the $\hat{\theta}$'s in terms of the $\hat{\beta}$'s and thus the restricted parameters are overidentified.

Estimation of Intrinsically Linear and Intrinsically Nonlinear Regression Models

Intrinsic Linearity

If a regression equation is intrinsically linear and the elemental or restricted parameters are exactly identified in terms of the unrestricted parameters, then OLS can be used to fit the regression equation in unrestricted and fully linear form. Under exact identification, the restricted estimates (the $\hat{\theta}$'s) are unique (nonlinear) functions of the unrestricted OLS estimators (the $\hat{\beta}$'s). Moreover, the desirable asymptotic or large sample (but not small sample) properties (e.g., consistency) of these unrestricted estimators carry over to the restricted estimators. Although the variances of the $\hat{\beta}$'s are readily obtainable from the variance-covariance matrix of these OLS estimators, asymptotic or large sample variances for the $\hat{\theta}$'s can be approximated in terms of the variances and covariances of the $\hat{\beta}$'s (see Appendix 18.A).

Intrinsic Nonlinearity

When a regression equation is intrinsically nonlinear, estimates of the restricted parameters can be obtained by using either (1) nonlinear least squares (NLS) (we minimize $\sum_{i=1}^{n} e_i^2$, which is a nonlinear function of the restricted parameters) or (2) the method of maximum likelihood (ML) (we maximize the likelihood function of the sample with respect to the restricted parameters).

If NLS is used, then the estimates of the restricted parameters are the same as those obtained from OLS estimation applied to an unrestricted regression equation under exact identification. If the assumption that the ε_i's are independent and identically distributed $N(0,\sigma_\varepsilon^2)$ random variables is tenable, then ML estimation may be undertaken. In fact, if the maximization of the likelihood function is executed by minimizing $\sum_{i=1}^{n} e_i^2$, then the resulting estimates of the $\hat{\theta}$'s are the same as those obtained via NLS.

If the normality of the random error terms is not assumed, then the ML method is not applicable and thus NLS is the preferred method of estimating the $\hat{\theta}$'s. Interestingly enough, even in the absence of the normality of ε, the asymptotic distribution of an NLS estimator is itself normal and has the same mean and variance as the ML estimator that would have obtained had normality of ε been presumed at the start.

In what follows, we shall examine the salient features of both NLS and the ML routine. As we shall soon see, for each of these approaches to nonlinear estimation, the derived sets of normal equations tend to be highly nonlinear functions of the parameters and rather difficult to solve; indeed, the normal equations typically do not admit an explicit solution for the parameter estimates (the $\hat{\theta}$'s). Hence, the normal equations can only be solved for an approximation to the optimal set of parameter estimates using an iterative technique that starts from an initial best guess to the levels of those estimates.

Nonlinear Least Squares

Let us start with a nonlinear regression model involving a single parameter θ. For regressors $X_1,\ldots, X_m$, let $\boldsymbol{X}_i = (X_{i1}, X_{i2},\ldots, X_{im})$ and let $Y_i = f(\boldsymbol{X}_i, \theta) + \varepsilon_i$, $i = 1,\ldots, n$, where the ε_i's are independent and identically distributed random variables with a zero mean and constant variance σ_ε^2. The NLS estimator for θ is then the arbitrary estimator $\hat{\theta}$ that minimizes the residual sum of squares described by the least-squares function $S(\hat{\theta}) = \sum_{i=1}^{n} e_i^2$, i.e., we seek to

$$\min\left\{\sum_{i=1}^{n} e_i^2 = \sum_{i=1}^{n}(Y_i - f(\boldsymbol{X}_i,\hat{\theta}))^2 = S(\hat{\theta})\right\}. \tag{18.7}$$

To this end, our first-order condition for a minimum is

$$\frac{dS(\hat{\theta})}{d\hat{\theta}} = -2\sum_{i=1}^{n}(Y_i - f(\boldsymbol{X}_i,\hat{\theta}))\left(\frac{df(\boldsymbol{X}_i,\hat{\theta})}{d\hat{\theta}}\right) = 0$$

or

$$\sum_{i=1}^{n}(Y_i - f(\boldsymbol{X}_i,\hat{\theta}))\left(\frac{df(\boldsymbol{X}_i,\hat{\theta})}{d\hat{\theta}}\right) = 0. \tag{18.8}$$

So if we can find an estimator $\hat{\theta}$ that satisfies this equation, then $\hat{\theta}$ is termed a *NLS estimator of θ*; $\hat{\theta}$ thus yields the absolute smallest value (the global minimum) of $S(\theta)$. (Keep in mind, however, that multiple optimal solutions to Equation 18.8, both local and global, can exist.) As mentioned above, this equation may be highly nonlinear and thus it may only be possible to obtain an approximation to $\hat{\theta}$ via a numerical technique. (See Appendix 18.B for details on the Gauss-Newton iterative method.)

Example 18.2

Suppose $Y_i = f(X_i, \theta) + \varepsilon_i = \theta X_{i1} + \theta^{-1}X_{i2} + \varepsilon_i$, $i = 1,\ldots,n$. Then the NLS estimator for θ is the value $\hat{\theta}$ which, from Equation 18.7, minimizes

$$S(\hat{\theta}) = \sum_{i=1}^{n} e_i^2 = \sum_{i=1}^{n} (Y_i - \hat{\theta}X_{i1} - \hat{\theta}^{-1}X_{i2})^2.$$

Setting

$$\frac{dS(\hat{\theta})}{d\hat{\theta}} = 2\sum_{i=1}^{n} (Y_i - \hat{\theta}X_{i1} - \hat{\theta}^{-1}X_{i2})(-X_{i1} + \hat{\theta}^{-2}X_{i2}) = 0$$

and simplifying yields

$$-\hat{\theta}^{-3}\sum X_{i2}^2 + \hat{\theta}^{-2}\sum X_{i2}Y_i + \hat{\theta}\sum X_{i2}^2 - \sum X_{i1}Y_i = 0.$$

Obviously a numerical routine needs to be used to solve this expression for $\hat{\theta}$.

We next consider the case in which the nonlinear regression model contains r parameters $\theta_1,\ldots,\theta_r$ that are to be estimated using NLS. Proceeding as above, for regressors $X_1,\ldots,X_m$, again let $\boldsymbol{X}_i = (X_{i1}, X_{i2},\ldots, X_{im})$, $i = 1,\ldots,n$, but now $\boldsymbol{\theta} = (\theta_1,\ldots,\theta_r)$. Our regression equation is thus $Y_i = f(\boldsymbol{X}_i, \boldsymbol{\theta}) + \varepsilon_i$, $i = 1,\ldots,n$, where the ε_i constitute a set of independent and identically distributed random variables with a zero mean and constant variance σ_ε^2.

The least-squares function is

$$S(\hat{\boldsymbol{\theta}}) = \sum_{i=1}^{n} e_i^2 = \sum_{i=1}^{n} (Y_i - f(\boldsymbol{X}_i, \hat{\boldsymbol{\theta}}))^2.$$

Again applying the principle of least squares, we choose the arbitrary estimators $\hat{\boldsymbol{\theta}} = (\hat{\theta}_1,\ldots,\hat{\theta}_r)$ so as to minimize $\sum_{i=1}^{n} e_i^2$, i.e., the NLS estimator for $\boldsymbol{\theta}$ is the $\hat{\boldsymbol{\theta}}$ for which

$$\min\left\{\sum_{i=1}^{n} e_i^2 = \sum_{i=1}^{n} (Y_i - f(\boldsymbol{X}_i, \hat{\boldsymbol{\theta}}))^2 = S(\hat{\boldsymbol{\theta}})\right\}. \tag{18.9}$$

After setting

$$\frac{\partial S(\hat{\boldsymbol{\theta}})}{\partial \hat{\theta}_l} = 0, \; l = 1,\ldots,r$$

(the first-order condition for a minimum), we get the system of *NLS normal equations*

$$\sum_{i=1}^{n} (Y_i - f(\boldsymbol{X}_i, \hat{\boldsymbol{\theta}}))\left(\frac{\partial f(\boldsymbol{X}_i, \hat{\boldsymbol{\theta}})}{\partial \hat{\theta}_l}\right) = 0, \; l = 1,\ldots,r. \tag{18.10}$$

Then the estimator $\hat{\boldsymbol{\theta}}$ that satisfies this equation system is the NLS estimator for $\boldsymbol{\theta}$: $S(\hat{\boldsymbol{\theta}})$ is at its minimum (although multiple optimal solutions can occur).[3] (Appendix 18.C considers the Gauss-Newton method for the r parameter case.)

Example 18.3

Determine the set of NLS normal equations when

$$Y_i = f(\boldsymbol{X}_i,\boldsymbol{\theta}) + \varepsilon_i = \theta_1 + \theta_2 X_i^{\theta_3} + \varepsilon_i, \quad i = 1,\ldots,n.$$

---

[3] In matrix terms, let

$$\underset{(n\times 1)}{\boldsymbol{Y}} = \begin{bmatrix} Y_1 \\ \cdot \\ \cdot \\ \cdot \\ Y_n \end{bmatrix}, \quad \underset{(n\times m)}{\boldsymbol{X}} = \begin{bmatrix} \boldsymbol{X}_1 \\ \cdot \\ \cdot \\ \cdot \\ \boldsymbol{X}_n \end{bmatrix}, \quad \underset{(1\times m)}{\boldsymbol{X}_i} = (X_{i1}, X_{i2}, \ldots, X_{im}),$$

$$\underset{(n\times 1)}{\boldsymbol{f}(\boldsymbol{X},\boldsymbol{\theta})} = \begin{bmatrix} f(\boldsymbol{X}_1,\boldsymbol{\theta}) \\ \cdot \\ \cdot \\ \cdot \\ f(\boldsymbol{X}_n,\boldsymbol{\theta}) \end{bmatrix}$$

with $\mathbf{Y} = f(\mathbf{X}, \boldsymbol{\theta}) + \boldsymbol{\varepsilon}$, $\boldsymbol{\varepsilon}' = (\varepsilon_1,\ldots,\varepsilon_n)$, $E(\boldsymbol{\varepsilon}) = \mathbf{0}$ ($\mathbf{0}$ is the *null vector* containing all zeros), and $E(\boldsymbol{\varepsilon}\boldsymbol{\varepsilon}') = \sigma_\varepsilon^2 \boldsymbol{I}$. Set

$$\underset{(n\times 1)}{\boldsymbol{e}} = \left[\boldsymbol{Y} - \boldsymbol{f}(\boldsymbol{X},\hat{\boldsymbol{\theta}})\right] = \begin{bmatrix} Y_1 - f(\boldsymbol{X}_1,\hat{\boldsymbol{\theta}}) \\ \cdot \\ \cdot \\ \cdot \\ Y_n - f(\boldsymbol{X}_n,\hat{\boldsymbol{\theta}}) \end{bmatrix}$$

so that $S(\hat{\boldsymbol{\theta}}) = \boldsymbol{e}'\boldsymbol{e} = \left[\boldsymbol{Y} - \boldsymbol{f}(\boldsymbol{X},\hat{\boldsymbol{\theta}})\right]'\left[\boldsymbol{Y} - \boldsymbol{f}(\boldsymbol{X},\hat{\boldsymbol{\theta}})\right] = \sum_{i=1}^n \left[Y - f(\boldsymbol{X}_i,\hat{\boldsymbol{\theta}})\right]^2$. For

$$\underset{(n\times r)}{\frac{\partial \boldsymbol{f}(\boldsymbol{X},\hat{\boldsymbol{\theta}})}{\partial \hat{\boldsymbol{\theta}}}} = \frac{\partial\left(f(\boldsymbol{X}_1,\hat{\boldsymbol{\theta}}),\ldots,f(\boldsymbol{X}_n,\hat{\boldsymbol{\theta}})\right)}{\partial(\hat{\theta}_1,\ldots,\hat{\theta}_r)}$$

$$= \begin{bmatrix} \partial f(\boldsymbol{X}_1,\hat{\boldsymbol{\theta}})/\partial\hat{\theta}_1 & \ldots & \partial f(\boldsymbol{X}_1,\hat{\boldsymbol{\theta}})/\partial\hat{\theta}_r \\ \cdot & & \cdot \\ \cdot & & \cdot \\ \cdot & & \cdot \\ \partial f(\boldsymbol{X}_n,\hat{\boldsymbol{\theta}})/\partial\hat{\theta}_1 & \ldots & \partial f(\boldsymbol{X}_n,\hat{\boldsymbol{\theta}})/\partial\hat{\theta}_r \end{bmatrix}$$

the $(n \times r)$ *Jacobian matrix* of $\boldsymbol{f}(\boldsymbol{X},\hat{\boldsymbol{\theta}})$ with respect to $\hat{\boldsymbol{\theta}}$, the NLS normal equations are

$$\frac{\partial S(\hat{\boldsymbol{\theta}})}{\partial \hat{\boldsymbol{\theta}}} = \left(\frac{\partial \boldsymbol{f}(\boldsymbol{X},\hat{\boldsymbol{\theta}})}{\partial \hat{\boldsymbol{\theta}}}\right)' \boldsymbol{e} = \mathbf{0}, \tag{18.10.1}$$

where a prime denotes matrix transposition.

(Remember that $\frac{d}{dx}(a^x) = a^x \log_e a$.) Here

$$S(\hat{\boldsymbol{\theta}}) = \sum_{i=1}^{n} (Y_i - \hat{\theta}_1 - \hat{\theta}_2 X_i^{\hat{\theta}_3})^2$$

so that, with

$$\frac{\partial S(\hat{\boldsymbol{\theta}})}{\partial \hat{\theta}_1} = 2\sum_{i=1}^{n} (Y_i - \hat{\theta}_1 - \hat{\theta}_2 X_i^{\hat{\theta}_3})(-1) = 0$$

$$\frac{\partial S(\hat{\boldsymbol{\theta}})}{\partial \hat{\theta}_2} = 2\sum_{i=1}^{n} (Y_i - \hat{\theta}_1 - \hat{\theta}_2 X_i^{\hat{\theta}_3})(-X_i^{\hat{\theta}_3}) = 0$$

$$\frac{\partial S(\hat{\boldsymbol{\theta}})}{\partial \hat{\theta}_3} = 2\sum_{i=1}^{n} (Y_i - \hat{\theta}_1 - \hat{\theta}_2 X_i^{\hat{\theta}_3})(-\hat{\theta}_2 X_i^{\hat{\theta}_3} \log X_i) = 0,$$

the system of NLS normal equations is

$$\begin{aligned}
&\sum_{i=1}^{n} (Y_i - \hat{\theta}_1 - \hat{\theta}_2 X_i^{\hat{\theta}_3}) = 0 \\
&\sum_{i=1}^{n} (Y_i X_i^{\hat{\theta}_3} - \hat{\theta}_1 X_i^{\hat{\theta}_3} - \hat{\theta}_2 X_i^{2\hat{\theta}_3}) = 0 \\
&\sum_{i=1}^{n} (\hat{\theta}_2 Y_i X_i^{\hat{\theta}_3} \log X_i - \hat{\theta}_1 \hat{\theta}_2 X_i^{\hat{\theta}_3} \log X_i - \hat{\theta}_2 X_i^{2\hat{\theta}_3} \log X_i) = 0.
\end{aligned} \tag{18.11}$$

In terms of Equation 18.10.1,

$$\frac{\partial \boldsymbol{f}(\boldsymbol{X},\hat{\boldsymbol{\theta}})}{\partial \hat{\boldsymbol{\theta}}} = \begin{bmatrix} \partial f(\boldsymbol{X}_1,\hat{\boldsymbol{\theta}})/\partial \hat{\theta}_1 & \partial f(\boldsymbol{X}_1,\hat{\boldsymbol{\theta}})/\partial \hat{\theta}_2 & \partial f(\boldsymbol{X}_1,\hat{\boldsymbol{\theta}})/\partial \hat{\theta}_3 \\ \cdot & \cdot & \cdot \\ \cdot & \cdot & \cdot \\ \cdot & \cdot & \cdot \\ \partial f(\boldsymbol{X}_n,\hat{\boldsymbol{\theta}})/\partial \hat{\theta}_1 & \partial f(\boldsymbol{X}_n,\hat{\boldsymbol{\theta}})/\partial \hat{\theta}_2 & \partial f(\boldsymbol{X}_n,\hat{\boldsymbol{\theta}})/\partial \hat{\theta}_3 \end{bmatrix}$$

$$= \begin{bmatrix} 1 & X_1^{\hat{\theta}_3} & \hat{\theta}_2 X_1^{\hat{\theta}_3} \log X_1 \\ \cdot & \cdot & \cdot \\ \cdot & \cdot & \cdot \\ \cdot & \cdot & \cdot \\ 1 & X_n^{\hat{\theta}_3} & \hat{\theta}_2 X_n^{\hat{\theta}_3} \log X_n \end{bmatrix}.$$

Then the NLS normal equations in matrix form are

$$\begin{bmatrix} 1 & \cdots & 1 \\ X_1^{\hat{\theta}_3} & \cdots & X_n^{\hat{\theta}_3} \\ \hat{\theta}_2 X_1^{\hat{\theta}_3} \log X_1 & \cdots & \hat{\theta}_2 X_n^{\hat{\theta}_3} \log X_n \end{bmatrix} \begin{bmatrix} Y_1 - \hat{\theta}_1 - \hat{\theta}_2 X_1^{\hat{\theta}_3} \\ \cdot \\ \cdot \\ \cdot \\ Y_n - \hat{\theta}_1 - \hat{\theta}_2 X_n^{\hat{\theta}_3} \end{bmatrix} = \mathbf{0}$$

or Equation 18.11.

Maximum Likelihood

For the nonlinear regression model $Y_i = f(X_i, \theta) + \varepsilon_i$, $i = 1,\ldots, n$, suppose the ε_i's are taken to be independent with $\varepsilon_i : N(0,\sigma_\varepsilon^2), i=1,\ldots,n$. Then the joint probability density function, when taken as a function of the arbitrary estimators $\tilde{\theta}$ and $\tilde{\sigma}_\varepsilon^2$ for θ and σ_ε^2, respectively, is termed the *likelihood function* for the sample and written

$$\begin{aligned}\mathcal{L}(\tilde{\theta},\ \tilde{\sigma}_\varepsilon^2; Y, X, n) &= (2\pi\tilde{\sigma}_\varepsilon^2)^{-n/2} e^{-\frac{1}{2\tilde{\sigma}_\varepsilon^2}\sum_{i=1}^{n}(Y_i - f(X_i,\tilde{\theta}))^2} \\ &= (2\pi\tilde{\sigma}_\varepsilon^2)^{-n/2} e^{-\frac{1}{2\tilde{\sigma}_\varepsilon^2}S(\tilde{\theta})},\end{aligned} \tag{18.12}$$

where $Y' = (Y_1,\ldots, Y_n)$, $X' = (X_1,\ldots, X_n)$, and $S(\tilde{\theta})$ is the least-squares function. For convenience, we shall work with the log-likelihood function

$$\log \mathcal{L} = -\frac{n}{2}\ln 2\pi - \frac{n}{2}\ln \tilde{\sigma}_\varepsilon^2 - \frac{1}{2\tilde{\sigma}_\varepsilon^2}S(\tilde{\theta}). \tag{18.12.1}$$

Given Equation 18.12.1, the ML approach has us choose the estimator $\tilde{\theta}$ (the *maximum likelihood estimator* of θ) for which the probability of obtaining the observed sample is greatest. In this regard, the first-order condition for log $\mathcal{L}$ to attain a global maximum at $\theta = \tilde{\theta}$ is $\partial \log \mathcal{L}/\partial\theta\big|_{\theta=\tilde{\theta}} = 0$. Because the *likelihood equation* $\partial \log \mathcal{L}/\partial\theta$ will typically be highly nonlinear and not readily admit an explicit solution for $\tilde{\theta}$, an iterative method must be used to approximate the value of $\tilde{\theta}$ (see Appendix 18.D).

It is important to note that (1) the ML method is a large-sample estimation method, and (2) if in the regression equation $Y_i = f(X_i, \theta) + \varepsilon_i$ it is assumed that $\varepsilon_i : N(0,\sigma_\varepsilon^2)$, $i=1,\ldots,n$, then the ML estimator $\tilde{\theta}$ is identical to the NLS estimator $\hat{\theta}$ if the former is obtained by minimizing the least-squares function $S(\theta)$.

To address this first point, let us determine, using Equation 18.12.1, the ML estimator of σ_ε^2, $\tilde{\sigma}_\varepsilon^2$,by solving $\partial \log \mathcal{L}/\partial\tilde{\sigma}_\varepsilon^2 = 0$ for $\tilde{\sigma}_\varepsilon^2 = S(\tilde{\theta})/n$. So for large n, $\tilde{\sigma}_\varepsilon^2 \approx SSE/(n-1)$.

Next, substituting $\tilde{\sigma}_\varepsilon^2 = S(\tilde{\theta})/n$ into Equation 18.12.1 results in the *condensed log-likelihood function*

$$\begin{aligned}\mathcal{L}^* &= \log \mathcal{L}(\tilde{\theta},\ \tilde{\sigma}_\varepsilon^2; Y, X, n) \\ &= -\frac{n}{2}\ln 2\pi - \frac{n}{2}\ln(S(\tilde{\theta})/n) - \frac{n}{2}\end{aligned}$$

which expresses log $\mathcal{L}$ in terms of the argument $\tilde{\theta}$. Then setting $\partial\mathcal{L}^*/\partial\tilde{\theta} = 0$ implies $\partial S/\partial\tilde{\theta} = 0$, where this latter derivative condition is simply the first-order condition for minimizing $S(\theta)$ under NLS estimation. So in the normal error case, maximizing $\mathcal{L}^*$ is equivalent to minimizing the error sum of squares or least-squares function $S(\theta)$.

For the r parameter case, let the regression equation appear as $Y_i = f(\boldsymbol{X}_i, \boldsymbol{\theta}) + \varepsilon_i$, $i = 1,\ldots,n$, $\boldsymbol{\theta}' = (\theta_1,\ldots, \theta_r)$, or $\boldsymbol{Y} = \boldsymbol{f}(\boldsymbol{X}, \boldsymbol{\theta}) + \varepsilon$, and let us write the least-squares function as $S(\tilde{\boldsymbol{\theta}}) = \boldsymbol{e}'\boldsymbol{e} = (\boldsymbol{Y} - \boldsymbol{f}(\boldsymbol{X},\tilde{\boldsymbol{\theta}}))'(\boldsymbol{Y} - \boldsymbol{f}(\boldsymbol{X},\tilde{\boldsymbol{\theta}}))$, where $\tilde{\boldsymbol{\theta}}$ is an arbitrary estimator for $\boldsymbol{\theta}$. For $\boldsymbol{\varepsilon} : N(\boldsymbol{0},\sigma_\varepsilon^2\boldsymbol{I})$, the likelihood function is

$$\begin{aligned}\mathcal{L}(\tilde{\boldsymbol{\theta}},\ \tilde{\sigma}_\varepsilon^2; \boldsymbol{Y}, \boldsymbol{X}, n) &= (2\pi\tilde{\sigma}_\varepsilon^2)^{-\frac{n}{2}} e^{-\left[(\boldsymbol{Y}-\boldsymbol{f}(\boldsymbol{X},\tilde{\boldsymbol{\theta}}))'(\boldsymbol{Y}-\boldsymbol{f}(\boldsymbol{X},\tilde{\boldsymbol{\theta}}))/2\tilde{\sigma}_\varepsilon^2\right]} \\ &= (2\pi\tilde{\sigma}_\varepsilon^2)^{-\frac{n}{2}} e^{-S(\tilde{\boldsymbol{\theta}})/2\tilde{\sigma}_\varepsilon^2}\end{aligned} \tag{18.13}$$

with $\tilde{\sigma}_\varepsilon^2$ serving as an arbitrary estimator of σ_ε^2. Then the log-likelihood function is written as

$$\log \mathcal{L} = -\frac{n}{2}\ln 2\pi - \frac{n}{2}\ln \tilde{\sigma}_\varepsilon^2 - \frac{1}{2\tilde{\sigma}_\varepsilon^2} S(\tilde{\boldsymbol{\theta}}). \tag{18.13.1}$$

As indicated above, the ML method is concerned with finding an estimator of $\boldsymbol{\theta}, \hat{\boldsymbol{\theta}}$, for which log $\mathcal{L}$ attains a global maximum (we maximize the probability of obtaining the observed sample), and this maximum will occur where $\partial \log \mathcal{L}/\partial \boldsymbol{\theta}\big|_{\boldsymbol{\theta}=\tilde{\boldsymbol{\theta}}} = \mathbf{0}$. Here, too, the likelihood equation $\partial \log \mathcal{L}/\partial \tilde{\boldsymbol{\theta}}$ will be highly nonlinear and thus require an iterative method for approximating the components of $\tilde{\boldsymbol{\theta}}$ (see Appendix 18.D).

Testing Nonlinear Parameter Restrictions [Neyman and Pearson 1928; Wald 1943; Aitchison and Silvey 1958, 1960; Silvey 1959; Breusch and Pagan 1980; Buse 1982; Engle 1982, 1984; Rao 1984]

The ML estimation technique provides us with a convenient vehicle for testing nonlinear restrictions on the parameters $\theta_1,\ldots,\theta_r$. The three tests to follow (the likelihood ratio (LR) test, Wald (W) test, and Lagrange multiplier (LM) test) are large-sample tests that are asymptotically equivalent and can be used when a suitable finite or small-sample exact test statistic is not at hand. Although their associated test statistics have unknown small-sample distributions, each is asymptotically distributed as $\mathcal{X}_v^2$, where v is the number of restrictions being tested.

Suppose we want to use the ML method to estimate the parameters $\boldsymbol{\theta}' = (\theta_1,\ldots,\theta_r)$ of the regression model $\mathbf{Y} = f(\mathbf{X}, \boldsymbol{\theta}) + \boldsymbol{\varepsilon}, \boldsymbol{\varepsilon} : N(\mathbf{0}, \sigma_\varepsilon^2 \mathbf{I})$. If these parameters are unrestricted, then we will estimate $\boldsymbol{\theta}$ freely and denote the unrestricted ML estimator of $\boldsymbol{\theta}$ as $\tilde{\boldsymbol{\theta}}' = (\tilde{\theta}_1,\ldots,\tilde{\theta}_r)$, with the maximum value of the associated likelihood $\mathcal{L}(\boldsymbol{\theta})$ appearing as $\mathcal{L}(\tilde{\boldsymbol{\theta}})$. However, if $\boldsymbol{\theta}$ is to be estimated subject to the nonlinear restriction or side relation $H_0 : g(\boldsymbol{\theta}) = 0$ on the parameters, then the restricted ML estimator of $\boldsymbol{\theta}$ will be written $\boldsymbol{\theta}_*' = (\theta_1^*,\ldots,\theta_r^*)$ and the restricted maximum of $\mathcal{L}(\boldsymbol{\theta})$ will thus appear as $\mathcal{L}(\boldsymbol{\theta}_*)$.

The Likelihood Ratio Test

In general, the LR test procedure compares the maximum value of the log-likelihood function when $H_0 : g(\boldsymbol{\theta}) = 0$ is assumed true to the maximum value of the log-likelihood function in the absence of any restriction(s) on $\boldsymbol{\theta}$. More specifically, the LR test is based on the supposition that, if $H_0 : g(\boldsymbol{\theta}) = 0$ is true, then $\mathcal{L}(\boldsymbol{\theta}_*)$ cannot differ substantially from $\mathcal{L}(\tilde{\boldsymbol{\theta}})$, with $\mathcal{L}(\boldsymbol{\theta}_*) \leq \mathcal{L}(\tilde{\boldsymbol{\theta}})$ (the restricted maximum of $\mathcal{L}$ cannot exceed its unrestricted maximum) or $0 < \mathcal{L}(\boldsymbol{\theta}_*)/\mathcal{L}(\tilde{\boldsymbol{\theta}}) \leq 1$. Let us set $\lambda = \mathcal{L}(\boldsymbol{\theta}_*)/\mathcal{L}(\tilde{\boldsymbol{\theta}})$. Then, if the null hypothesis or restriction $g(\boldsymbol{\theta}) = 0$ is valid, λ will be close to unity, and, if this restriction is not valid, then λ will be appreciably less than unity. Under $H_0 : g(\boldsymbol{\theta}) = 0$, it can be shown that the *LR test statistic*

$$LR = -2\log\lambda = 2(\log \mathcal{L}(\tilde{\boldsymbol{\theta}}) - \log \mathcal{L}(\boldsymbol{\theta}_*)) \tag{18.14}$$

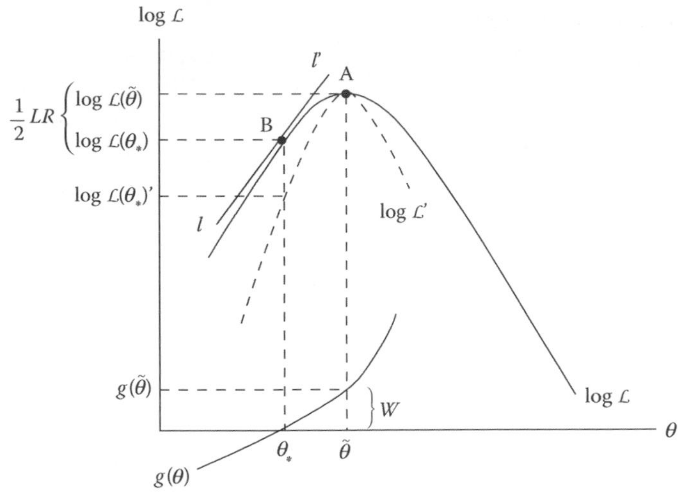

FIGURE 18.1
Log-likelihood function.

is asymptotically distributed as $\mathcal{X}_1^2$.[4] In this regard, H_0 will be rejected if $LR > \mathcal{X}_{\alpha,1}^2$, where $\mathcal{X}_{\alpha,1}^2$ is the upper α percentile of the $\mathcal{X}_1^2$ distribution.

As indicated in Figure 18.1, the log-likelihood function $\log \mathcal{L}$ has an unconstrained maximum at $\tilde{\theta}$ (point A), whereas its constrained maximum occurs at θ_* (point B) with $g(\theta_*) = 0$. Then from Equation 18.14, $\frac{1}{2}LR = \log \mathcal{L}(\tilde{\theta}) - \log \mathcal{L}(\theta_*)$. Hence, a large disparity between $\log \mathcal{L}(\tilde{\theta})$ and $\log \mathcal{L}(\theta_*)$ favors the terms on which we would reject H_0.

A glance back at Figure 18.1 reveals that, if at θ_* the curvature of the log-likelihood function is increased (suppose we replace the solid $\log \mathcal{L}$ function by its dashed counterpart $\log \mathcal{L}'$), then obviously $\log \mathcal{L}(\tilde{\theta}) - \log \mathcal{L}(\theta_*) < \log \mathcal{L}(\tilde{\theta}) - \log \mathcal{L}(\theta_*)'$ and thus $\frac{1}{2}LR$ is directly related to the curvature of $\log \mathcal{L}$ at $\tilde{\theta}$. The point being made is that the quantity $\frac{1}{2}LR$ depends on the difference $\tilde{\theta} - \theta_*$ as well as on the curvature of $\log \mathcal{L}$ at $\tilde{\theta}$, where the latter is measured by the negative of its second derivative at $\tilde{\theta}$ or by $-(d^2 \log \mathcal{L}/d\theta^2)_{\theta=\tilde{\theta}}$. A test statistic that incorporates both of these pieces of information in its construction is the Wald statistic, to which we now turn.

The Wald Test

The W test is based on the extent to which the constraint $g(\theta) = 0$ is violated at the unconstrained ML estimator $\tilde{\theta}$ or the extent to which $g(\tilde{\theta})$ departs from zero ($g(\theta) = 0$ is obviously satisfied at the constrained ML estimator θ_* or $g(\theta_*) = 0$) as well as on the curvature of the log-likelihood function at the unrestricted ML estimator $\tilde{\theta}$, i.e., we weight the squared distance between $\tilde{\theta}$ and θ_* by a measure or estimate of the curvature of the log-likelihood function at $\tilde{\theta}$ so as to obtain the Wald test statistic

$$W = (\tilde{\theta} - \theta_*)^2 \left(-\frac{d^2 \log \mathcal{L}}{d\theta^2} \right)_{\theta=\tilde{\theta}}. \tag{18.15}$$

[4] If $\mathbf{g}(\boldsymbol{\theta})' = g^1(\boldsymbol{\theta}),\ldots,g^v(\boldsymbol{\theta}))$ is a $(1 \times v)$ vector-valued function displaying v nonlinear restrictions on $\boldsymbol{\theta}$, then the null hypothesis appears as $H_0 : \mathbf{g}(\boldsymbol{\theta}) = \mathbf{0}$, $\mathbf{0}' = (0,\ldots, 0)$, and LR is asymptotically distributed as $\mathcal{X}_v^2$.

Under $H_0 : g(\theta) = 0$, it can be shown that W is χ_1^2. Here too we reject H_0 if $W > \chi^2_{\alpha,1}$, i.e., $\tilde{\theta}$ violates the restriction on θ by a significant amount or $g(\tilde{\theta})$ departs significantly from zero.

Let us express Equation 18.15 in a slightly different form. The quantity $I(\theta) = E(-\partial^2 \log \mathcal{L}/\partial\theta^2)$ represents the information on θ in the sample and is called the *information number for* θ. In view of this expression, we may respecify Equation 18.15 by weighting the squared distance between $\tilde{\theta}$ and θ_* by the expected curvature of the log-likelihood function at $\tilde{\theta}$ or now

$$W = (\tilde{\theta} - \theta_*)^2 I(\tilde{\theta}). \tag{18.15.1}$$

In general, for the r parameter case, with $\boldsymbol{\theta}' = (\theta_1, \ldots, \theta_r)$, the W test compares the value of the constraint function at the unrestricted ML estimator $\tilde{\boldsymbol{\theta}}$, $g(\tilde{\boldsymbol{\theta}})$, with its hypothesized value of zero. If H_0 is true, then $g(\tilde{\boldsymbol{\theta}})$ will be close to zero, where the proximity of $g(\tilde{\boldsymbol{\theta}})$ to zero is assessed in terms of the estimated variance-covariance matrix of $g(\tilde{\boldsymbol{\theta}})$ or by

$$V(g(\tilde{\boldsymbol{\theta}})) = \left(\frac{\partial g}{\partial \boldsymbol{\theta}}\right)'_{\boldsymbol{\theta}=\tilde{\boldsymbol{\theta}}} V(\tilde{\boldsymbol{\theta}}) \left(\frac{\partial g}{\partial \boldsymbol{\theta}}\right)_{\boldsymbol{\theta}=\tilde{\boldsymbol{\theta}}} \tag{18.16}$$

(see Equation 12.I.1 and Appendix 12.A), where the lth component of $(\partial g/\partial\boldsymbol{\theta})_{\boldsymbol{\theta}=\tilde{\boldsymbol{\theta}}}$ is $(\partial g/\partial\theta_l)_{\boldsymbol{\theta}=\tilde{\boldsymbol{\theta}}}$, $l = 1, \ldots, r$. To obtain a representation for $V(\tilde{\boldsymbol{\theta}})$, let us use the *information matrix for* $\boldsymbol{\theta}$ (a generalization of the information number for θ)

$$\boldsymbol{I}(\tilde{\boldsymbol{\theta}}) = E\left(-\frac{\partial^2 \log \mathcal{L}}{\partial\boldsymbol{\theta}\partial\boldsymbol{\theta}'}\right)_{\boldsymbol{\theta}=\tilde{\boldsymbol{\theta}}}, \tag{18.17}$$

where

$$\frac{\partial^2 \log \mathcal{L}}{\partial\boldsymbol{\theta}\partial\boldsymbol{\theta}'} = \begin{bmatrix} \left.\dfrac{\partial^2 \log \mathcal{L}}{\partial\theta_1^2}\right|_{\tilde{\boldsymbol{\theta}}} & \cdots & \left.\dfrac{\partial^2 \log \mathcal{L}}{\partial\theta_1\partial\theta_r}\right|_{\tilde{\boldsymbol{\theta}}} \\ \vdots & & \vdots \\ \left.\dfrac{\partial^2 \log \mathcal{L}}{\partial\theta_r\partial\theta_1}\right|_{\tilde{\boldsymbol{\theta}}} & \cdots & \left.\dfrac{\partial^2 \log \mathcal{L}}{\partial\theta_r^2}\right|_{\tilde{\boldsymbol{\theta}}} \end{bmatrix}.$$

Here $\boldsymbol{I}(\tilde{\boldsymbol{\theta}})$ serves as an estimate of the curvature of $\log \mathcal{L}(\boldsymbol{\theta})$ at $\tilde{\boldsymbol{\theta}}$. Then, subject to some very general regularity conditions on $\log \mathcal{L}$, the inverse of $\boldsymbol{I}(\tilde{\boldsymbol{\theta}})$ serves as variance-covariance matrix for $\tilde{\boldsymbol{\theta}}$, $V(\tilde{\boldsymbol{\theta}})$, and thus Equation 18.16 can be rewritten as

$$V(g(\tilde{\boldsymbol{\theta}})) = \left(\frac{\partial g}{\partial \boldsymbol{\theta}}\right)'_{\boldsymbol{\theta}=\tilde{\boldsymbol{\theta}}} \boldsymbol{I}(\tilde{\boldsymbol{\theta}})^{-1} \left(\frac{\partial g}{\partial \boldsymbol{\theta}}\right)_{\boldsymbol{\theta}=\tilde{\boldsymbol{\theta}}} \tag{18.16.1}$$

so that the r parameter version of the Wald statistic is (from Equation 12.I.4 and Appendix 12.E, "Distributions Revisited")

$$W = g(\tilde{\boldsymbol{\theta}})' V(g(\tilde{\boldsymbol{\theta}}))^{-1} g(\tilde{\boldsymbol{\theta}})$$

$$= g(\tilde{\boldsymbol{\theta}})' \left[\left(\frac{\partial g}{\partial \boldsymbol{\theta}} \right)'_{\boldsymbol{\theta}=\tilde{\boldsymbol{\theta}}} \boldsymbol{I}(\tilde{\boldsymbol{\theta}})^{-1} \left(\frac{\partial g}{\partial \boldsymbol{\theta}} \right)_{\boldsymbol{\theta}=\tilde{\boldsymbol{\theta}}} \right]^{-1} g(\tilde{\boldsymbol{\theta}}) \tag{18.18}$$

$$= \frac{g(\tilde{\boldsymbol{\theta}})' g(\tilde{\boldsymbol{\theta}})}{\left[\left(\frac{\partial g}{\partial \boldsymbol{\theta}} \right)'_{\boldsymbol{\theta}=\tilde{\boldsymbol{\theta}}} \boldsymbol{I}(\tilde{\boldsymbol{\theta}})^{-1} \left(\frac{\partial g}{\partial \boldsymbol{\theta}} \right)_{\boldsymbol{\theta}=\tilde{\boldsymbol{\theta}}} \right]}.$$

Under $H_0 : g(\boldsymbol{\theta}) = 0$, W is asymptotically distributed as $\mathcal{X}_1^2$ so that, for $W > \mathcal{X}_{\alpha,1}^2$, we are prompted to reject H_0 at the α level of significance and conclude that $g(\tilde{\boldsymbol{\theta}})$ is significantly different from zero.

The Lagrange Multiplier Test

The LM test appears in two equivalent forms: its *score form*, (in which the *score for* θ in the sample is $G(\theta) = \partial \log \mathcal{L}/\partial\theta$) and its *multiplier form* (which explicitly incorporates the optimal value of the Lagrange multiplier resulting when the log-likelihood function is maximized using the technique of Lagrange).

Looking to the score form, Figure 18.1 reveals that, if the log-likelihood function slope $G(\theta)$ is evaluated at the unrestricted ML estimator $\tilde{\theta}$ then $G(\tilde{\theta}) = 0$ (point A), but if $G(\theta)$ is evaluated at the restricted ML estimator θ_*, then $G(\theta_*) \neq 0$ (point B). Hence, the score version of the LM test is based on the extent to which $G(\theta)$ evaluated at θ_* differs from zero, i.e., it focuses on the slope of $\log \mathcal{L}$ at θ_* (as depicted by the slope of the tangent line ll' to $\log \mathcal{L}$ at point B). So if the restriction $H_0 : g(\theta) = 0$ is valid, the value of $\partial \log \mathcal{L}/\partial\theta$ determined at the restricted estimate θ_* should not be significantly different from zero.

To structure the LM test statistic in terms of the departure of $G(\theta_*)$ from zero, let us weight the squared score or squared slope of the log-likelihood function $\log \mathcal{L}$ at θ_* by the inverse of the curvature of $\log \mathcal{L}$ at this point or

$$LM = \frac{(G(\theta_*))^2}{-(\partial^2 \log \mathcal{L}/\partial\theta^2)_{\theta=\theta_*}} \tag{18.19}$$

(the reciprocal of the curvature is used because the value of LM will then be small when θ_* is close to $\tilde{\theta}$). Alternatively, this test statistic can be respecified by weighting the squared score at θ_* by the inverse of the expected curvature of the log-likelihood function at θ_* or

$$LM = (G(\theta_*))^2 I(\theta_*)^{-1}, \tag{18.19.1}$$

where $I(\theta_*)$ is the information number for θ_* in the sample. Under $H_0 : g(\theta) = 0$, LM is asymptotically $\mathcal{X}_1^2$; and H_0 is rejected for large LM values or for $LM > \mathcal{X}_{\alpha,1}^2$ (the upper α percentile of the $\mathcal{X}_1^2$ distribution).

For the r parameter case, with $\boldsymbol{\theta}_*' = (\theta_1^*, \ldots, \theta_r^*)$ the restricted ML estimator of $\boldsymbol{\theta}$,

$$LM = G(\boldsymbol{\theta}_*)' \boldsymbol{I}(\boldsymbol{\theta}_*)^{-1} G(\boldsymbol{\theta}_*), \tag{18.20}$$

where $G(\boldsymbol{\theta}_*) = (\partial \log \mathcal{L}/\partial \boldsymbol{\theta})_{\boldsymbol{\theta}=\boldsymbol{\theta}_*}$ and $\boldsymbol{I}(\boldsymbol{\theta}_*) = E(-\partial^2 \log \mathcal{L}/\partial \boldsymbol{\theta} \partial \boldsymbol{\theta}')_{\boldsymbol{\theta}=\boldsymbol{\theta}_*}$ is the information matrix for $\boldsymbol{\theta}_*$. If $H_0 : g(\boldsymbol{\theta}) = 0$ is true, then LM is asymptotically distributed as $\mathcal{X}_1^2$. Here, too, H_0 is rejected when $LM > \mathcal{X}^2_{\alpha,1}$.

Let us now focus on the multiplier version of the LM test. Because H_0: $g(\theta) = 0$ acts as a constraint on the regression model, we cannot maximize the log-likelihood function $\log\mathcal{L}(\theta)$ freely; it must be maximized subject to the restriction $g(\theta) = 0$. To this end, we need to find a global maximum of the *Lagrangian of* $\log \mathcal{L}$ or of the function $L(\theta, \lambda) = \log \mathcal{L}(\theta) - \lambda g(\theta)$. Setting $\partial L/\partial \theta = \partial L/\partial \lambda = 0$ and solving yields the value of the restricted ML estimator θ_* and the associated Lagrange multiplier λ_*, with $g(\theta_*) = 0$.

The magnitude of the Lagrange multiplier indicates how tightly the constraint function $g(\theta) = 0$ is binding at θ_*, i.e., the smaller is λ_*, the less tightly the constraint is binding at θ_* and vice versa. In this regard, the greater the degree of agreement between the sample dataset and $H_0 : g(\theta) = 0$, the closer θ_* is to $\tilde{\theta}$ and thus the smaller the value of λ_*. Hence, a test of the restriction $H_0 : g(\theta) = 0$ is tantamount to testing whether or not the Lagrange multiplier is significantly different from zero.

This said, in its multiplier form, the LM statistic is

$$LM = \left(\lambda_* \left.\frac{dg}{d\theta}\right|_{\theta=\theta_*} \right)^2 I(\theta_*)^{-1}, \tag{18.21}$$

where λ_* is the Lagrange multiplier determined at the restricted or constrained maximum of $\log \mathcal{L}$ and $(dg/d\theta)_{\theta=\theta_*}$ is the derivative of the constraint function evaluated at θ_*.[5] For r parameters, again let $\boldsymbol{\theta}_*' = (\theta_1^*,\ldots,\theta_r^*)$ so that now the first-order condition $\partial L/\partial \boldsymbol{\theta} = \mathbf{0}$ yields $G(\boldsymbol{\theta}_*) = \lambda_*(\partial g/\partial \boldsymbol{\theta})_{\boldsymbol{\theta}=\boldsymbol{\theta}_*}$ and thus Equation 18.20 transforms to

$$LM = \left(\lambda_* \left.\frac{\partial g}{\partial \boldsymbol{\theta}}\right|_{\boldsymbol{\theta}=\boldsymbol{\theta}_*} \right)' \boldsymbol{I}(\boldsymbol{\theta}_*)^{-1} \left(\lambda_* \left.\frac{\partial g}{\partial \boldsymbol{\theta}}\right|_{\boldsymbol{\theta}=\boldsymbol{\theta}_*} \right), \tag{18.21.1}$$

where $(\partial g/\partial \boldsymbol{\theta})'_{\boldsymbol{\theta}_*} = \left((\partial g/\partial \theta_1)_{\boldsymbol{\theta}_*},\ldots,(\partial g/\partial \theta_r)_{\boldsymbol{\theta}_*}\right)$. If the restriction incorporated in H_0 is true, then this alternative version of the LM test statistic is asymptotically distributed $\mathcal{X}_1^2$, with H_0 rejected if $LM > \mathcal{X}^2_{\alpha,1}$.

At this point in our discussion of testing nonlinear restrictions using the LR, W, or LM tests, it is important to note that, although these tests are asymptotically equivalent, they have different informational or computational requirements. That is,

1. The LR test uses both the *unrestricted and restricted ML estimates* of θ.
2. The W test incorporates only the unrestricted ML estimates of θ.
3. The LM test depends on only the restricted ML estimates of θ.

So although asymptotic equivalence holds, the test statistics LR, W, and LM typically yield different values for their sample realizations. Given that $g(\theta) = 0$ is nonlinear, it is generally true that $LR \geq LM$ (because W involves only unrestricted estimation). However, if $g(\theta) = 0$ is

[5] Given the Lagrangian $L(\theta, \lambda) = \log \mathcal{L} - \lambda g$, the first-order conditions for a constrained global maximum are $\partial L/\partial \theta = \partial \log \mathcal{L}/\partial \theta - \lambda(dg/d\theta) = 0$ and $\partial L/\partial \lambda = g = 0$ at (θ_*, λ_*). Then from $\partial L/\partial \theta = 0$, we obtained (in score terms) $G(\theta_*) = \lambda_*(dg/d\theta)_{\theta*}$. A substitution of this expression into Equation 18.19.1 renders Equation 18.21. Clearly, the score and multiplier versions of the LM test are equivalent.

linear in θ, then generally $W \geq LR \geq LM$. Hence, it is possible to reject $H_0 : g(\theta) = 0$ by the W test but not by the LR or LM tests. In the presence of a linear constraint function, the usual convention is to use the exact F test because it is applicable for small samples, although, asymptotically, it is equivalent to the preceding three tests.

Example 18.4

Given the dataset appearing in Table 18.1, (1) estimate the regression model $Y = \theta_0 + \theta_1 X_1 + \theta_2 X_2 + \varepsilon$, $\boldsymbol{\varepsilon} : N(\mathbf{0}, \sigma_\varepsilon \boldsymbol{I})$; and (2) test the restriction $\theta_0 - (\theta_1/\theta_2) = 0$ using the Wald test for $\alpha = 0.05$. Exhibit 18.1 displays the SAS code.

Exhibit 18.1 SAS Code for Conducting Wald Test

```
data nlg;
  input y x1 x2;
datalines;
160    270    145
122    184    124
134    200    148
220    380    180
 65     91    105
171    260    278
 81     98    101
192    330    245
220    198    264
 51     50    105
250    428    280
234    374    244
147    239    166
100    154    109
213    369    261
;
proc reg data=nlg; ①
  model y=x1  x2/covb; ②
run;
```

① We request an OLS fit to the dataset via PROC REG.

② We request a print of the (3 × 3) variance-covariance matrix (*covb*) for the estimated coefficients $\hat{\theta}_0, \hat{\theta}_1$, and $\hat{\theta}_2$.

From the output generated in Table 18.2, we immediately see the following.

① The set of estimated regression coefficients appears as

$$\hat{\theta}_0 = 8.1418,\ \hat{\theta}_1 = 0.3393,\ \text{and}\ \hat{\theta}_2 = 0.3658.$$

② The estimated variance-covariance matrix is

$$V(\hat{\boldsymbol{\theta}}) = \begin{bmatrix} 205.6628 & -0.0401 & -0.9321 \\ -0.0401 & 0.00468 & -0.0059 \\ -0.9321 & -0.0059 & 0.0129 \end{bmatrix}.$$

TABLE 18.1

Observations on the Variables Y, X_1, and X_2

Y	X	X_2
160	270	145
122	184	124
134	200	148
220	380	180
65	91	105
171	260	278
81	98	101
192	330	245
220	198	264
51	50	105
250	428	280
234	374	244
171	239	166
100	154	109
213	369	261

Given $H_0 : g(\boldsymbol{\theta}) = \theta_0 - (\theta_1/\theta_2) = 0$, it follows that $\partial g / \partial \theta_0 = 1$, $\partial g / \partial \theta_1 = -\theta_2^{-1}$, and $\partial g / \partial \theta_2 = \theta_1 \theta_2^{-2}$. Then $g(\hat{\boldsymbol{\theta}}) = 7.2143$,

$$\left(\frac{\partial g}{\partial \boldsymbol{\theta}}\right)_{\boldsymbol{\theta}=\hat{\boldsymbol{\theta}}} = \begin{bmatrix} 1 \\ -2.7337 \\ 2.5357 \end{bmatrix},$$

and thus

$$W = \frac{g(\hat{\boldsymbol{\theta}})^2}{\left(\frac{\partial g}{\partial \boldsymbol{\theta}}\right)'_{\hat{\boldsymbol{\theta}}} V(\hat{\boldsymbol{\theta}}) \left(\frac{\partial g}{\partial \boldsymbol{\theta}}\right)_{\hat{\boldsymbol{\theta}}}}$$
$$= \frac{52.0461}{201.3529} = 0.2584.$$

Because $\chi^2_{0.05,\,1} = 3.841$, we cannot reject H_0 at the 5% level.

TABLE 18.2

Output for Example 18.4

The SAS System
The REG Procedure
Model : MODEL1
Dependent Variable: y
Analysis of Variance

Source	DF	Sum of Squares	Mean Square	F Value	Pr > F
Model	2	53199	26599	71.56	<.0001
Error	12	4460.59018	371.71585		
Corrected Total	14	57659			

Root MSE	19.27993	*R*-Square	0.9226
Dependent Mean	157.33333	Adj *R*-Sq	0.9097
Coeff Var	12.25420		

Parameter Estimates

①

Variable	DF	Parameter Estimate	Standard Error	*t* Value	Pr > | *t* |
Intercept	1	8.14179	14.34110	0.57	0.5807
x1	1	0.33932	0.06842	4.96	0.0003
x2	1	0.36583	0.11355	3.22	0.0073

②

Covariance of Estimates

Variable	Intercept	x1	x2
Intercept	205.66727703	–0.040134108	–0.932053418
x1	–0.040134108	0.0046817984	–0.005941745
x2	–0.932053418	–0.005941745	0.0128927866

Example 18.5

Using the dataset appearing in Table 18.3, estimate the parameters of the following nonlinear regression model using the SAS procedure *NLIN*:

$$Y = \theta_1\left(e^{\theta_2 X_1} + e^{\theta_3 X_2}\right) + \varepsilon,$$

where $\boldsymbol{\varepsilon} \sim N(\mathbf{0}, \sigma_\varepsilon^2 \boldsymbol{I})$. Exhibit 18.2 houses the SAS code.

Exhibit 18.2 SAS Code for Nonlinear Estimation

```
data nlinreg;
  input y x1 x2;
datalines;
6    3    3
7    5    6
```

① We invoke PROC NLIN, a procedure for estimating regression equations that are nonlinear in the parameters. The default solution is the Gauss-Newton iterative routine.

TABLE 18.3
Observations on the Variables Y, X_1, and X_2

Y	X_1	X_2
6	3	3
7	5	6
8	7	7
10	8	11
13	10	13
15	12	14
17	18	18
20	15	20
25	20	25
27	27	27
31	28	34
32	30	36
40	32	41
50	35	42
62	37	45
77	41	48
90	49	49
120	53	55
148	56	58
171	60	62

```
.   .   .
.   .   .
.   .   .
148 56  58
171 60  62
;   ①
proc nlin data=nlinreg hougaard; ②
   parms theta1=0.1 theta2=0.1 ③
      theta3=0.1;
   model y=theta1*(exp(theta2*x1)+exp (theta3*x2); ④
run;
```

② We request the *Hougaard* option to obtain a skewness measure for the estimated parameters. Hougaard's measure of skewness, g_{11}, $l = 1,\ldots,r$, is used to determine whether a parameter estimate is close to linear (i.e., has desirable statistical properties similar to the estimates obtained by fitting a linear regression model) or whether it exhibits marked non-linearity. In this regard: (a) if $|g_{11}| < 0.1$, the estimator $\hat{\theta}_l$ is very close-to-linear in its character; (b) if $0.1 < |g_{11}| < 0.25$, the estimator $\hat{\theta}_l$ is fairly close-to-linear; (c) if $|g_{11}| > 0.25$, evidence of moderate skewness to $\hat{\theta}_l$ emerges; and (d) if $|g_{11}| > 1$, there is a considerable departure in $\hat{\theta}_l$ from the linear behavior of unbiasedness, minimum variance, etc.

③ The *PARMS* statement is used to specify the names of the parameters and to set their starting values to initiate the iterative estimation process.

④ The *MODEL* statement explicitly specifies the full model in terms of the named parameters and variables.

The output generated by this SAS code appears in Table 18.4.

① Starting from the set of initial values specified in the *PARMS* statement, the process converges after 10 iterations to a stable set of parameter estimates $\hat{\theta}_1 = 3.0693$, $\hat{\theta}_2 = 0.0559$, and $\hat{\theta}_3 = 0.0542$.

② $R = 2.77E{-}6$ is a convergence measure that approaches zero as the value of the objective function approaches its optimum. *PPC* is a prospective change measure that indicates the maximum relative change in the parameter estimates when the parameter-change vector is computed for the next iteration. Hence, *PPC*(*theta*3) = 9.846*E*–7 indicates that *theta*3 (it has the largest *PPC* value) would change by this relative amount if PROC NLIN were to perform an additional iteration. *RPC* is a retrospective change measure that indicates the maximum relative change in the parameters from the previous iteration step. Thus, *RPC*(*theta*3) = 0.000024 reveals that *theta*3 (it displays the largest *RPC* value) changed by this amount relative to its value at the last iteration. *Object* measures the relative change in the objective function value between successive iterations. Hence, *Object* = 1.692*E*–8 indicates that the objective function value changed by this relative amount from the last iteration.

③ The large *F* value with *p*-value <0.0001 indicates that the overall fit is quite good.

④,⑤ The estimates of the parameters θ_1, θ_2, and θ_3 are, respectively, $\hat{\theta}_1 = 3.0693$, $\hat{\theta}_2 = 0.0559$, and $\hat{\theta}_3 = 0.0542$. Because their associated 95% confidence limits do not contain zero, we can conclude that each of these parameters is significantly different from zero at the 5% level.

⑥ Looking to the Hougaard measure of skewness, we find that,

1. With $|g_{11}| = 0.49$, $\hat{\theta}_1$ is moderately skewed away from linear behavior;
2. Because $|g_{12}| = 1.24$, $\hat{\theta}_2$ is far removed from linear behavior; and
3. With $|g_{13}| = 0.57$, $\hat{\theta}_3$ is moderately skewed from linear behavior.

Example 18.6

Let us reestimate the regression function appearing in Example 18.5 using the Table 18.3 dataset. This time we shall use a nonlinear regression routine from the SAS ETS program library. The requisite code is found in Exhibit 18.3.

Exhibit 18.3 SAS Code for Nonlinear Estimation (ETS)

```
data mlreg;
  input  y  x1  x2;
datalines;
 6    3    3
 7    5    6
 8    7    7
10    8   11
13   10   13
15   12   14
17   18   18
20   15   20
25   20   25
27   27   27
31   28   34
32   30   36
40   32   41
50   35   42
62   37   45
```

① The *MODEL* procedure is used to perform nonlinear regression analysis and is a much more general procedure than *NLIN*, which was used in the preceding example problem.

② We write out explicitly the complete regression equation (including all parameters and arithmetic operations) to be estimated save for the random error term.

③ We use a *FIT* statement to run our fully specified regression against the *mlreg* dataset. Starting values for the unknown parameters are also supplied.

④ We request that the full-information maximum-likelihood *FIML* estimation technique be used.

⑤ The *FIT* statement options used herein are as follows:

out = *a* (specifies the output dataset *a* to have the residual and predicted values);

TABLE 18.4

Output for Example 18.5

The SAS System
The NLIN Procedure
Dependent Variable: y
Method: Gauss-Newton

①
Iterative Phase

Iter	theta1	theta2	theta3	Sum of Squares
0	0.1000	0.1000	0.1000	36171.7
1	0.1292	0.0951	0.0961	35388.4
2	0.1679	0.0903	0.0921	34444.1
3	0.2679	0.0812	0.0839	34394.0
4	0.3899	0.0752	0.0777	32624.5
5	0.6682	0.0668	0.0681	30405.8
6	1.6733	0.0532	0.0508	26698.5
7	2.8306	0.0582	0.0565	462.3
8	3.0562	0.0559	0.0542	240.4
9	3.0694	0.0559	0.0542	240.0
10	3.0693	0.0559	0.0542	240.0

Note: Convergence criterion met.

②
Estimation Summary

Method	Gauss-Newton
Iterations	10
Subiterations	11
Average subiterations	1.1
R	2.77E-6
PPC(theta3)	9.846E-7
RPC(theta3)	0.000024
Object	1.692E-8
Objective	240.0302
Observations read	20
Observations used	20
Observations missing	0

Note: An intercept was not specified for this model.

Source	DF	Sum of Squares	Mean Square	③ F Value	Approx Pr > F
Regression	3	91949.0	30649.7	2170.74	<.0001
Residual	17	240.0	14.1194		
Uncorrected total	20	92189.0			
Corrected total	19	45241.0			

TABLE 18.4 *(continued)*

The NLIN Procedure

Parameter	④ Estimate	Approx Std Error	⑤ Approximate 95% Confidence Limits		⑥ Skewness
theta1	3.0693	0.2791	2.4804	3.6582	0.4898
theta2	0.0559	0.00336	0.0488	0.0630	–1.2446
theta3	0.0542	0.00499	0.0436	0.0647	–0.5723

Approximate Correlation Matrix

	theta1	theta2	theta3
theta1	1.0000000	0.2616343	–0.7999940
theta2	0.2616343	1.0000000	–0.7821554
theta3	–0.7999940	–0.7821554	1.0000000

```
77    41    48          outresid (writes the residual values to the
90    49    49            output dataset a);
120   53    55          outpredict (writes the predicted values to the
148   56    58            output dataset a);
171   60    62          covb (we request a print of the variance-
;     ①                   covariance matrix of the estimated parameters);
proc model data=mlreg;
  y=theta1* (exp(theta2*x1)+exp(theta3*x2));②
③fit y start= (theta1 0.1 theta2 0.1 theta3 0.1)/fiml ④
    out=a outresid outpredict covb prl=both collin normal;⑤
    test theta1 - (theta2/theta3) = 1;
    test theta2 - theta3 = 0;⑥
proc print data=a;
run;
```

prl = both (we request both Wald and likelihood ratio (LR) 95% profile likelihood confidence intervals on the estimated parameters. The default is 95% LR confidence intervals.);

collin (we request a set of collinearity diagnostics);

normal (we request a test(s) of the regression residuals).

⑥ The *TEST* option enables us to perform tests of nonlinear (as well as linear) hypotheses on the regression parameters estimated by the immediately preceding *FIT* statement. Any number of individual *TEST* statements can be specified. The following options can also be used in any *TEST* statement: *WALD* (the Wald test, the default); *LM* (the Lagrange multiplier test); *LR* (the likelihood-ratio test); or *ALL* (all three test are requested).

Although not incorporated into the current SAS program, some additional options that can be used within the *FIT* statement are as follows:

outest = (writes the set of parameter estimates to a named dataset);

Breusch (performs the Breusch-Pagan test for heteroscedasticity);

corr (prints correlation matrices);

White (performs White's test for heteroscedasticity);

printall (provides all printing options).

Given the SAS code listed in Exhibit 18.3, the associated output generated appears in Table 18.5.

① The Gauss-Newton iteration routine is the default solution method.

② See the interpretation of these convergence criteria offered in Example 18.5.

③ It was mentioned previously in our discussion of near-perfect multicollinearity that a condition index with a value of at least 30 indicates serious multicollinearity. The third condition index with a value of 24 reveals that a nontrivial amount of multicollinearity is present; there is one (modest) near-linear dependency in the data.

④,⑤ The R^2 value of 0.9947 indicates that the regression model has considerable explanatory power, whereas the t value for each estimated regression coefficient has a p-value <0.0001. Hence, the parameter estimates $\hat{\theta}_1 = 3.0693$, $\hat{\theta}_2 = 0.0558$, and $\hat{\theta}_3 = 0.0541$ are highly significant.

⑥ We first test H_0: $\theta_1 - (\theta_2/\theta_3) = 1$. Given that the Wald test statistic $W = 18.26$ is highly significant (p-value <0.0001), we reject H_0 in favor of H_0 : not H_0. Next, we test $H_0 : \theta_2 - \theta_3 = 0$. Now the Wald statistic $W = 0.04$ is not significant (p-value = 0.8512) so that we cannot reject H_0 -- θ_2 and θ_3 can be thought of as being (approximately) equal in value.

⑦ Both the Wald and LR 95% confidence intervals verify the results of the above t tests for the estimated regression parameters; zero is not a member of any of these intervals. According to this set of results, each of our parameter estimates is significantly different from zero with 95% reliability. Note that a greater degree of precision is attained with the set of 95% LR confidence intervals relative to the 95% Wald confidence intervals (i.e., the LR intervals are narrower than the Wald intervals).

⑧ Maximum-likelihood estimation requires that the sample size be large and the random errors ε_i, $i = 1,\ldots, n$, be normally distributed with a mean of zero and a constant variance. Hence, it is important to test this normality assumption concerning the ε_i's. Let us focus on the Shapiro-Wilk (S-W) test. Here we test H_0 : the random errors are normal versus H_1 : not H_0. If the p-value associated with the S-W statistic is less than 0.05, then we reject H_0 in favor of H_1 and conclude that the observations have not been taken from a normal population, and, if the p-value exceeds 0.05, we cannot reject H_0. Because for this test the S-W statistic equals 0.97 and the associated p-value is 0.828, we cannot reject the null hypothesis of normality.

An important feature of PROC MODEL is the ability to incorporate nonlinear (as well as linear) restrictions on the model parameters into the estimation process associated with the current *FIT* statement. This is accomplished with the *RESTRICT* option, which can be written as

restrict ________ = (or ≤ , ≥, <, >)________;

Any number of individual *RESTRICT* statements can be specified. For instance, we can reestimate the current model by inserting the following *RESTRICT* option under the preceding *FIT* statement:

```
restrict theta1* theta2 + theta3=0.1;
```

The abbreviated SAS output results appear in Table 18.6.

① The Marquardt-Lerenberg method is a modification of the basic linearization (Taylor's expansion) approach; it is a ridge-like procedure that is designed to reduce the residual sum of squares at each iteration step and is used in combination with the Gauss-Newton technique [see Marquardt 1963; Bates and Watts 1988; Seber and Wild 1989].

② The parameter estimate Restrict0 = 4.9463 pertains to the optimal value of the Lagrange multiplier associated with the restriction or constraint imposed on the model parameters. (If the constraint is binding or holds as an equality at the minimum of the error sum of squares, then Restrict0 > 0, and, if the constraint is not binding at the said minimum, then Restrict0 = 0 and the constrained minimum of the likelihood function is the same as its unconstrained minimum.) Here the parameter restriction has a reasonably significant impact on the estimation results because the associated p-value =0.0590.

TABLE 18.5

Output for Example 18.6

The SAS System
The MODEL Procedure

Model Summary

Model Variables	1
Parameters	3
Equations	1
Number of statements	5

Model Variables	y		
Parameters (Value)	theta1(0.1)	theta2(0.1)	theta3(0.1)
Equations	y		

The Equation to estimate is

y = F(theta1, theta 2, theta3)

Note: At FIML Iteration 0 CONVERGE=0.001 Criteria Met.

The SAS System
The MODEL Procedure

FIML Estimation Summary

Data set options	
DATA=	MLREG
OUT=	A

Minimization Summary

Parameters estimated	3
Method	Gauss ①
Hessian	GLS
Covariance estimator	Cross
Iterations	0

②

Final Convergence Criteria

R	0.0008
PPC (theta3)	0.000024
RPC	.
Object	.
Trace(S)	12.00151
Gradient norm	1.748203
Log likelihood	−53.2291

Observations Processed

Read	20
Solved	20

(continued)

TABLE 18.5 *(continued)*

Collinearity Diagnostics

		③			
		Condition	Proportion of Variation		
Number	Eigenvalue	Number	theta1	theta2	theta3
1	2.968713	1.0000	0.0021	0.0022	0.0009
2	0.026143	10.6563	0.3460	0.3926	0.0001
3	0.005144	24.0237	0.6519	0.6051	0.9990

Nonlinear FIML Summary of Residual Errors

Equation	DF Model	DF Error	SSE	MSE	Root MSE	④ R^2	Adj R^2
y	3	17	240.0	14.1194	3.7576	0.9947	0.9941

⑤

Nonlinear FIML Parameter Estimates

Parameter	Estimate	Approx Std Err	t value	Approx Pr > $\|t\|$
theta1	3.06936	0.3794	8.09	<.0001
theta2	0.055884	0.00305	18.30	<.0001
theta3	0.054165	0.00639	8.48	<.0001

⑥

Test Results

Test	Type	Statistic	Pr > χ^2	Label
Test0	Wald	18.26	<.0001	theta1 - (theta2/theta3) = 1
Test1	Wald	0.04	0.8512	theta2 – theta 3 = 0

⑦

Parameter Wald 95% Confidence Intervals

Parameter	Value	Lower	Upper
theta1	3.0694	2.3257	3.8130
theta2	0.0559	0.0499	0.0619
theta3	0.0542	0.0416	0.0667

Parameter Likelihood Ratio 95% Confidence Intervals

Parameter	Value	Lower	Upper
theta1	3.0694	2.6060	3.6980
theta2	0.0559	0.0463	0.0605
theta3	0.0542	0.0430	0.0618

TABLE 18.5 *(continued)*

Number of Observations		Statistics for System	
Used	20	Log Likelihood	–53.2291
Missing	0		

⑧

Normality Test

Equation	Test Statistic	Value	Prob
y	Shapiro-Wilk W	0.97	0.8280
System	Mardia Skewness	0.01	0.9131
	Mardia Kurtosis	–0.41	0.6850
	Henze-Zirkler T	–1.93	0.0542

Covariance of Parameter Estimates

	theta1	theta2	theta3
theta1	0.1439591	0.0007535	–.0022845
theta2	0.0007535	0.0000093	–.0000169
theta3	–.0022845	–.0000169	0.0000408

The SAS System

Obs	_ESTYPE_	_TYPE_	_WEIGHT_	y	x1	x2
1	FIML	PREDICT	1	7.241	3	3
2	FIML	RESIDUAL	1	–1.241	3	3
3	FIML	PREDICT	1	8.037	5	6
4	FIML	RESIDUAL	1	–1.307	5	6
5	FIML	PREDICT	1	9.023	7	7
6	FIML	RESIDUAL	1	–1.203	7	7
7	FIML	PREDICT	1	10.369	8	11
8	FIML	RESIDUAL	1	–0.369	8	11
9	FIML	PREDICT	1	11.574	10	13
10	FIML	RESIDUAL	1	1.426	10	13
11	FIML	PREDICT	1	12.554	12	14
12	FIML	RESIDUAL	1	2.446	12	14
13	FIML	PREDICT	1	16.530	18	18
14	FIML	RESIDUAL	1	0.470	18	18
15	FIML	PREDICT	1	16.166	15	20
16	FIML	RESIDUAL	1	3.834	15	20
17	FIML	PREDICT	1	21.274	20	25
18	FIML	RESIDUAL	1	3.726	20	25
19	FIML	PREDICT	1	27.127	27	27
20	FIML	RESIDUAL	1	–0.127	27	27
21	FIML	PREDICT	1	34.034	28	34
22	FIML	RESIDUAL	1	–3.034	28	34

(continued)

TABLE 18.5 *(continued)*

The SAS System

Obs	_ESTYPE_	_TYPE_	_WEIGHT_	y	x1	x2
23	FIML	PREDICT	1	37.984	30	36
24	FIML	RESIDUAL	1	–5.984	30	36
25	FIML	PREDICT	1	46.634	32	41
26	FIML	RESIDUAL	1	–6.634	32	41
27	FIML	PREDICT	1	51.558	35	42
28	FIML	RESIDUAL	1	–1.558	35	42
29	FIML	PREDICT	1	59.393	37	45
30	FIML	RESIDUAL	1	2.607	37	45
31	FIML	PREDICT	1	71.669	41	48
32	FIML	RESIDUAL	1	5.331	41	48
33	FIML	PREDICT	1	91.076	49	49
34	FIML	RESIDUAL	1	–1.076	49	49
35	FIML	PREDICT	1	119.715	53	55
36	FIML	RESIDUAL	1	0.285	53	55
37	FIML	PREDICT	1	141.198	56	58
38	FIML	RESIDUAL	1	6.802	56	58
39	FIML	PREDICT	1	175.959	60	62
40	FIML	RESIDUAL	1	–4.959	60	62

TABLE 18.6

Output for Example 18.6 (RESTRICT Option)

The SAS System
The MODEL Procedure
Model Summary

Model variables	1
Parameters	3
Equations	1
Number of statements	3

Model Variables	y		
Parameters (Value)	theta1(0.1)	theta2(0.1)	theta3(0.1)
Equations	y		

The equation to estimate is

y = F(theta1, theta 2, theta3)

Note: Gauss method failed. Switching to Marquardt's method ①
Note: At FIML Iteration 0 CONVERGE=0.001 Criteria Met.

Nonlinear FIML Summary of Residual Errors

Equation	DF Model	DF Error	SSE	MSE	Root MSE	④ R^2	Adj R^2
y	2	18	509.2	28.2877	5.3186	0.9887	0.9881

TABLE 18.6 *(continued)*

Nonlinear FIML Parameter Estimates

Parameter	Estimate	Approx Std Err	t value	Approx Pr > |t|	Label
theta1	7.824571	0.6739	11.61	<.0001	
theta2	0.051747	0.00163	31.81	<.0001	
theta3	−0.3049	0.0225	−13.54	<.0001	
Restrict0 ②	4.9463	2.6463	1.87	0.0590	theta1*theta2 + theta3 =0.1

Parameter Wald 95% Confidence Intervals

Parameter	Value	Lower	Upper
theta1	7.8246	6.5038	9.1454
theta2	0.0517	0.0615	0.0618
theta3	−0.3049	−0.3470	−0.2617

Parameter Likelihood Ratio 95% Confidence Intervals

Parameter	Value	Lower	Upper
theta1	7.8246	6.5246	9.2570
theta2	0.0517	0.0486	0.0552
theta3	−0.3049	−0.3504	−0.2594

Number of Observations		Statistics for System	
Used	20	Log Likelihood	−60.7494
Missing	0		

Appendix A: Approximation of Asymptotic Variances of Restricted Estimators

Suppose the restricted estimator for θ, $\hat{\theta}$, is a function of the k unrestricted estimators $\hat{\beta}_1,\dots,\hat{\beta}_k$ or $\hat{\theta}=f(\hat{\beta}_1,\dots,\hat{\beta}_k)$. Let us develop a large sample approximation formula for the variance of $\hat{\theta}$ in terms of the variances and covariances of the $\hat{\beta}$'s.

In general,

$$V(\hat{\theta})\approx\sum_l\left(\frac{\partial f}{\partial\hat{\beta}_l}\right)^2V(\hat{\beta}_l)+2\sum_{j<l}\left(\frac{\partial f}{\partial\hat{\beta}_j}\right)\left(\frac{\partial f}{\partial\hat{\beta}_l}\right)\mathrm{COV}(\hat{\beta}_j,\hat{\beta}_l),\ j,\ l=1,\dots,k. \tag{18.A.1}$$

To make the rationalization of this expression as transparent or tractable as possible, let us assume, for convenience, that $\hat{\theta}=f(\hat{\beta}_1,\hat{\beta}_2)$. Moreover, we know that the OLS estimators are unbiased so that $E(\hat{\theta}) = E(f(\hat{\beta}_1, \hat{\beta}_2)) = f(\beta_1, \beta_2) = \theta$. In addition, $V(\hat{\theta}) = E(\hat{\theta} - E(\hat{\theta}))^2$ and $\text{COV}(\hat{\beta}_1, \hat{\beta}_2) = E\left[(\hat{\beta}_1 - \beta_1)(\hat{\beta}_2 - \beta_2)\right]$. Let us write the first-order Taylor expansion (or linearization) of f near the point (β_1, β_2) as

$$f(\hat{\beta}_1, \hat{\beta}_2) \approx f(\beta_1, \beta_2) + \left.\frac{\partial f}{\partial \hat{\beta}_1}\right|_{\hat{\beta}_1=\beta_1} (\hat{\beta}_1 - \beta_1) + \left.\frac{\partial f}{\partial \hat{\beta}_2}\right|_{\hat{\beta}_2=\beta_2} (\hat{\beta}_2 - \beta_2)$$

or

$$\hat{\theta} \approx E(\hat{\theta}) + \left.\frac{\partial f}{\partial \hat{\beta}_1}\right|_{\hat{\beta}_1=\beta_1} (\hat{\beta}_1 - \beta_1) + \left.\frac{\partial f}{\partial \hat{\beta}_2}\right|_{\hat{\beta}_2=\beta_2} (\hat{\beta}_2 - \beta_2).$$

Then

$$\begin{aligned}
V(\hat{\theta}) = E(\hat{\theta} - E(\hat{\theta}))^2 &\approx E\left[\left(\left.\frac{\partial f}{\partial \hat{\beta}_1}\right|_{\hat{\beta}_1=\beta_1}\right)^2 (\hat{\beta}_1 - \beta_1)^2\right. \\
&\quad \left. + \left(\left.\frac{\partial f}{\partial \hat{\beta}_2}\right|_{\hat{\beta}_2=\beta_2}\right)^2 (\hat{\beta}_2 - \beta_2)^2 + 2\left(\left.\frac{\partial f}{\partial \hat{\beta}_1}\right|_{\hat{\beta}_1=\beta_1}\right)\left(\left.\frac{\partial f}{\partial \hat{\beta}_2}\right|_{\hat{\beta}_2=\beta_2}\right)(\hat{\beta}_1 - \beta_1)(\hat{\beta}_2 - \beta_2)\right] \\
&= \left(\left.\frac{\partial f}{\partial \hat{\beta}_1}\right|_{\hat{\beta}_1=\beta_1}\right)^2 E(\hat{\beta}_1 - \beta_1)^2 + \left(\left.\frac{\partial f}{\partial \hat{\beta}_2}\right|_{\hat{\beta}_2=\beta_2}\right)^2 E(\hat{\beta}_2 - \beta_2)^2 \\
&\quad + \left(\left.\frac{\partial f}{\partial \hat{\beta}_1}\right|_{\hat{\beta}_1=\beta_1}\right)\left(\left.\frac{\partial f}{\partial \hat{\beta}_2}\right|_{\hat{\beta}_2=\beta_2}\right) E\left[(\hat{\beta}_1 - \beta_1)(\hat{\beta}_2 - \beta_2)\right] \\
&= \left(\left.\frac{\partial f}{\partial \hat{\beta}_1}\right|_{\hat{\beta}_1=\beta_1}\right)^2 V(\hat{\beta}_1) + \left(\left.\frac{\partial f}{\partial \hat{\beta}_2}\right|_{\hat{\beta}_2=\beta_2}\right)^2 V(\hat{\beta}_2) \\
&\quad + \left(\left.\frac{\partial f}{\partial \hat{\beta}_1}\right|_{\hat{\beta}_1=\beta_1}\right)\left(\left.\frac{\partial f}{\partial \hat{\beta}_2}\right|_{\hat{\beta}_2=\beta_2}\right) \text{COV}(\hat{\beta}_1, \hat{\beta}_2).
\end{aligned}$$

Example 18.A.1

We found in Example 18.1 that

$$\hat{\theta}_1 = e^{\hat{\beta}_1};$$
$$\hat{\theta}_2 = \hat{\beta}_2 + \hat{\beta}_3;$$
$$\hat{\theta}_3 = \frac{\hat{\beta}_2}{\hat{\beta}_2 + \hat{\beta}_3};$$
$$\hat{\theta}_4 = \frac{\hat{\beta}_4(\hat{\beta}_2 + \hat{\beta}_3)}{\hat{\beta}_2\hat{\beta}_3}.$$

Given these expressions involving the functional specification of the restricted parameters $\hat{\theta}_1,\ldots,\hat{\theta}_4$ in terms of the unrestricted parameters $\hat{\beta}_1,\ldots,\hat{\beta}_4$, use Equation 18.A.1 to find $V(\hat{\theta}_1),\ldots,V(\hat{\theta}_4)$. To this end we have

$$V(\hat{\theta}_1) = \left(\hat{\beta}_1 e^{\hat{\beta}_1}\right)^2 V(\hat{\beta}_1);$$
$$V(\hat{\theta}_2) = V(\hat{\beta}_2) + V(\hat{\beta}_3) + 2\mathrm{COV}(\hat{\beta}_2,\hat{\beta}_3);$$
$$\begin{aligned} V(\hat{\theta}_3) = {} & \left[\frac{1}{\hat{\beta}_2+\hat{\beta}_3} - \frac{\hat{\beta}_2}{(\hat{\beta}_2+\hat{\beta}_3)^2}\right]^2 V(\hat{\beta}_2) + \left[\frac{-\hat{\beta}_2}{(\hat{\beta}_2+\hat{\beta}_3)^2}\right]^2 V(\hat{\beta}_3) \\ & + 2\left[\frac{1}{\hat{\beta}_2+\hat{\beta}_3} - \frac{\hat{\beta}_2}{(\hat{\beta}_2+\hat{\beta}_3)^2}\right]\left[\frac{-\hat{\beta}_2}{(\hat{\beta}_2+\hat{\beta}_3)^2}\right]\mathrm{COV}(\hat{\beta}_2,\hat{\beta}_3); \end{aligned}$$
$$\begin{aligned} V(\hat{\theta}_4) = {} & \left(\frac{-\hat{\beta}_4}{\hat{\beta}_2^2}\right)^2 V(\hat{\beta}_2) + \left(\frac{-\hat{\beta}_4}{\hat{\beta}_3^2}\right)^2 V(\hat{\beta}_3) + (\hat{\beta}_3^{-1} + \hat{\beta}_2^{-1})^2 V(\hat{\beta}_4) \\ & + 2\left(\frac{-\hat{\beta}_4}{\hat{\beta}_2^2}\right)\left(\frac{-\hat{\beta}_4}{\hat{\beta}_3^2}\right)\mathrm{COV}(\hat{\beta}_2,\hat{\beta}_3) \\ & + 2\left(\frac{-\hat{\beta}_4}{\hat{\beta}_2^2}\right)(\hat{\beta}_3^{-1} + \hat{\beta}_2^{-1})\mathrm{COV}(\hat{\beta}_2,\hat{\beta}_4) \\ & + 2\left(\frac{-\hat{\beta}_4}{\hat{\beta}_3^2}\right)(\hat{\beta}_3^{-1} + \hat{\beta}_2^{-1})\mathrm{COV}(\hat{\beta}_3,\hat{\beta}_4). \end{aligned}$$

Appendix B: Gauss-Newton Iteration Scheme—The Single-Parameter Case

Given the regression equation $Y_i = f(\mathbf{X}_i, \theta) + \varepsilon_i$, $i = 1,\ldots,n$, and least-squares function, $S(\theta) = \sum_{i=1}^{n}(Y_i - f(\mathbf{X}_i,\theta))^2$, our objective is to find a $\theta = \hat{\theta}$ value that satisfies Equation 18.7. To this end, let us linearize or approximate $f(\mathbf{X}_i, \theta)$ at $\theta = \theta_0$ (our initial guess at $\hat{\theta}$) by the *first-order Taylor expansion*

$$\begin{aligned} f(\mathbf{X}_i,\theta) &\approx f(\mathbf{X}_i,\theta_0) + \left.\frac{df(\mathbf{X}_i,\theta)}{d\theta}\right|_{\theta=\theta_0}(\theta-\theta_0) \\ &= f(\mathbf{X}_i,\theta_0) + z_i(\theta_0)(\theta-\theta_0), \end{aligned} \tag{18.B.1}$$

where $z_i(\theta_0) = (df(\boldsymbol{X}_i,\theta)/d\theta)_{\theta=\theta_0}$. Then the *linearized regression model* is

$$\begin{aligned} Y_i &= f(\boldsymbol{X}_i, \theta) + \varepsilon_i \\ &= f(\boldsymbol{X}_i, \theta_0) + z_i(\theta_0)(\theta - \theta_0) + \varepsilon_i,\ i = 1,\ldots,n, \end{aligned}$$

or

$$Y_i^0 = Y_i - f(\boldsymbol{X}_i,\theta_0) + z_i(\theta_0)\theta_0 = z_i(\theta_0)\theta + \varepsilon_i,\ i = 1,\ldots,n.$$

Hence, we need to find an OLS estimate for the slope θ of the linearized model

$$Y_i^0 = z_i(\theta_0)\theta + \varepsilon_i,\ i = 1,\ldots,n, \tag{18.B.2}$$

which satisfies

$$\min\left\{\sum_{i=1}^{n}(Y_i^0 - z_i(\theta_0)\theta)^2 = S(\theta)\right\}. \tag{18.B.3}$$

Let us denote the θ for which this minimum holds as

$$\theta_1 = (\mathbf{Z}(\theta_0)'\,\mathbf{Z}(\theta_0))^{-1}\mathbf{Z}(\theta_0)'\mathbf{Y}_0, \tag{18.B.4}$$

where $\mathbf{Z}(\theta_0)' = (z_1(\theta_0),\ldots, z_n(\theta_0))$ and $(\boldsymbol{Y}_0)' = (Y_1^0,\ldots,Y_n^0)$. In terms of the components of Equation 18.B.4,

$$\theta_1 = \frac{\sum_{i=1}^{n} Y_i^0 z_i(\theta_0)}{\sum_{i=1}^{n} z_i(\theta_0)^2}. \tag{18.B.4.1}$$

Given this new (and ostensibly improved) $\theta = \theta_1$ value, we can now form a second linear approximation to $f(\boldsymbol{X}_i, \theta_0)$ (this time at θ_1), apply OLS so that $S(\theta)$ is minimized, and consequently obtain

$$\theta_2 = (\mathbf{Z}(\theta_1)'\,\mathbf{Z}(\theta_1))^{-1}\mathbf{Z}(\theta_1)'\mathbf{Y}_1, \tag{18.B.5}$$

where $\boldsymbol{Y}_1' = (Y_1',\ldots,Y_n')$ and $Y_i' = Y_i - f(\boldsymbol{X}_i,\theta_1) + z_i(\theta_1)\theta_1,\ i = 1,\ldots,n$. We continue this process and obtain, at the $(j + 1)$st linear approximation,

$$\begin{aligned} \theta_{j+1} &= (\boldsymbol{Z}(\theta_j)'\boldsymbol{Z}(\theta_j))^{-1}\boldsymbol{Z}(\theta_j)'\boldsymbol{Y}_j \\ &= (\boldsymbol{Z}(\theta_j)'\boldsymbol{Z}(\theta_j))^{-1}\boldsymbol{Z}(\theta_j)'\left[\boldsymbol{Y} - \boldsymbol{f}(\boldsymbol{X},\theta_j) + \boldsymbol{Z}(\theta_j)\theta_j\right] \\ &= \theta_j + (\boldsymbol{Z}(\theta_j)'\boldsymbol{Z}(\theta_j))^{-1}\boldsymbol{Z}(\theta_j)'(\boldsymbol{Y} - \boldsymbol{f}(\boldsymbol{X},\theta_j)),\ j = 0,1,\ldots, \end{aligned} \tag{18.B.6}$$

where

$$\underset{(n\times 1)}{\boldsymbol{f}(\boldsymbol{X},\theta_j)} = \begin{bmatrix} f(\boldsymbol{X}_1,\theta_j) \\ \cdot \\ \cdot \\ \cdot \\ f(\boldsymbol{X}_n,\theta_j) \end{bmatrix}.$$

For successive values of j, this procedure produces the sequence of OLS estimates θ_0, θ_1, θ_2,....

Note that, if the process in Equation 18.B.6 converges, then $\theta_{j+1} - \theta_j = 0$ and thus $\boldsymbol{Z}(\theta_j)'(\boldsymbol{Y} - \boldsymbol{f}(\boldsymbol{X}, \theta)) = 0$ or Equation 18.8 holds (our first-order condition for a minimum of $S(\theta)$). If $\theta_{j+1} - \theta_j \neq 0$, this iterative process is continued until $|(\theta_{j+1} - \theta_j)/\theta_j| < \delta$, $j = 0, 1, 2, \ldots$, where δ is a very small prespecified constant. It must be remembered that the process described by Equation 18.B.6 may not converge to $\hat{\theta}$ at all, and, if convergence to $\hat{\theta}$ does occur, it may be quite slow, with successive solutions oscillating wildly. If convergence to a minimum obtains, this minimum may only be local in character and not correspond to the global minimum of $S(\theta)$. Hence, it is prudent to repeat the process using a few different initial values for θ. Finally, we note briefly that, if the ε_i's are independent and identically distributed with zero mean and constant variance σ_ε^2, then the *Gauss-Newton NLS estimator* θ_{j+1} is consistent and can be taken to have an approximate normal distribution with mean θ and variance $\hat{\sigma}_\varepsilon^2$ consistently estimated by

$$S_\varepsilon^2(\boldsymbol{Z}(\theta_{j+1})'\boldsymbol{Z}(\theta_{j+1}))^{-1}, \text{ where } S_\varepsilon^2 = S(\theta_{j+1})/(n-1).$$

Example 18.B.1

Given the regression equation $Y_i = \theta X_{i1} + \theta^{-1}X_{i2} + \varepsilon_i = f(\boldsymbol{X}_i, \theta) + \varepsilon_i$, $\boldsymbol{X}_i = (X_{i1}, X_{i2})$, $i = 1,\ldots,n$, let us use the Gauss-Newton iteration routine specified by Equation 18.B.6. Let

$$z_i(\theta) = \frac{df(\boldsymbol{X}_i,\theta)}{d\theta} = X_{i1} - \theta^{-2}X_{i2},\ i = 1,\ldots,n.$$

Then

$$\boldsymbol{Z}(\theta) = \begin{bmatrix} z_1(\theta) \\ \cdot \\ \cdot \\ \cdot \\ z_n(\theta) \end{bmatrix} = \begin{bmatrix} X_{11} - \theta^{-2}X_{12} \\ \cdot \\ \cdot \\ \cdot \\ X_{n1} - \theta^{-2}X_{n2} \end{bmatrix},$$

$$\boldsymbol{Z}(\theta)'\boldsymbol{Z}(\theta) = \sum\nolimits_{i=1}^{n}(X_{i1} - \theta^{-2}X_{i2})^2, \text{ and}$$

$$\boldsymbol{Z}(\theta)'(\boldsymbol{Y} - \boldsymbol{f}(\boldsymbol{X},\theta)) = \sum\nolimits_{i=1}^{n}(X_{i1} - \theta^{-2}X_{i2})(Y_i - \theta X_{i1} - \theta^{-1}X_{i2}),$$

where

$$\boldsymbol{Y} - \boldsymbol{f}(\boldsymbol{X},\theta) = \begin{bmatrix} Y_1 - f(\boldsymbol{X}_1,\theta) \\ \cdot \\ \cdot \\ \cdot \\ Y_n - f(\boldsymbol{X}_n,\theta) \end{bmatrix} = \begin{bmatrix} Y_1 - \theta X_{11} - \theta^{-1} X_{12} \\ \cdot \\ \cdot \\ \cdot \\ Y_n - \theta X_{n1} - \theta^{-1} X_{n2} \end{bmatrix}.$$

From Equation 18.B.6,

$$\theta_{j+1} = \theta_j + \frac{\sum_{i=1}^{n} (X_{i1} - \theta_j^{-2} X_{i2})(Y_i - \theta_j X_{i1} - \theta_j^{-1} X_{i2})}{\sum_{i=1}^{n} (X_{i1} - \theta_j^{-2} X_{i2})^2}, \; j = 0,1,2,\ldots,$$

with

$$\hat{\sigma}_\varepsilon = S(\theta_{j+1})/(n-1) = \frac{\sum_{i=1}^{n} (Y_i - \theta_{j+1} X_{i1} - \theta_{j+1}^{-1} X_{i2})^2}{n-1}.$$

Appendix C: Gauss-Newton Iteration Scheme—The *r* Parameter Case

For the regression equation $Y_i = f(\boldsymbol{X}_i, \boldsymbol{\theta}) + \varepsilon_i$, $i = 1,\ldots, n$, $\boldsymbol{\theta}' = (\theta_1,\ldots, \theta_r)$, or $\boldsymbol{Y} = \boldsymbol{f}(\boldsymbol{X}, \boldsymbol{\theta}) + \boldsymbol{\varepsilon}$, let the least-squares function appear as $S(\boldsymbol{\theta}) = \boldsymbol{e}'\boldsymbol{e} = (\boldsymbol{Y} - \boldsymbol{f}(\boldsymbol{X}, \boldsymbol{\theta}))'(\boldsymbol{Y} - \boldsymbol{f}(\boldsymbol{X}, \boldsymbol{\theta}))$. Our goal is to determine a vector $\boldsymbol{\theta} = \hat{\boldsymbol{\theta}}$ for which Equation 18.9 is satisfied. We start by linearizing or approximating $\boldsymbol{f}(\boldsymbol{X}, \boldsymbol{\theta})$ at $\boldsymbol{\theta} = \boldsymbol{\theta}_0$ (an initial guess at $\hat{\boldsymbol{\theta}}$) via the first-order Taylor expansion

$$\boldsymbol{f}(\boldsymbol{X}, \boldsymbol{\theta}) \approx \boldsymbol{f}(\boldsymbol{X}, \boldsymbol{\theta}_0) + \boldsymbol{Z}(\boldsymbol{\theta}_0)(\boldsymbol{\theta} - \boldsymbol{\theta}_0), \tag{18.C.1}$$

where the Jacobian matrix of $\boldsymbol{f}(\boldsymbol{X}, \boldsymbol{\theta})$ with respect to $\boldsymbol{\theta}$ at $\boldsymbol{\theta}_0$ is

$$\underset{(n\times r)}{\boldsymbol{Z}(\boldsymbol{\theta}_0)} = \left.\frac{\partial \boldsymbol{f}(\boldsymbol{X},\boldsymbol{\theta})}{\partial \boldsymbol{\theta}}\right|_{\boldsymbol{\theta}=\boldsymbol{\theta}_0} = \begin{bmatrix} \left.\frac{\partial f(\boldsymbol{X}_1,\boldsymbol{\theta})}{\partial \boldsymbol{\theta}_1}\right|_{\boldsymbol{\theta}_0} & \cdots & \left.\frac{\partial f(\boldsymbol{X}_1,\boldsymbol{\theta})}{\partial \boldsymbol{\theta}_r}\right|_{\boldsymbol{\theta}_0} \\ \cdot & & \cdot \\ \cdot & & \cdot \\ \cdot & & \cdot \\ \left.\frac{\partial f(\boldsymbol{X}_n,\boldsymbol{\theta})}{\partial \boldsymbol{\theta}_1}\right|_{\boldsymbol{\theta}_0} & \cdots & \left.\frac{\partial f(\boldsymbol{X}_n,\boldsymbol{\theta})}{\partial \boldsymbol{\theta}_r}\right|_{\boldsymbol{\theta}_0} \end{bmatrix}.$$

Then the linearized regression model is

$$\begin{aligned} Y &= f(X, \boldsymbol{\theta}) + \varepsilon \\ &= f(X, \boldsymbol{\theta}_0) + \boldsymbol{Z}(\boldsymbol{\theta}_0)(\boldsymbol{\theta} - \boldsymbol{\theta}_0) + \boldsymbol{\varepsilon} \end{aligned}$$

or

$$Y_0 = Y - f(X, \boldsymbol{\theta}_0) + \boldsymbol{Z}(\boldsymbol{\theta}_0)\boldsymbol{\theta}_0 = \boldsymbol{Z}(\boldsymbol{\theta}_0)\boldsymbol{\theta} + \boldsymbol{\varepsilon}.$$

We thus need to determine the OLS estimate of θ for the linearized regression model

$$Y_0 = \boldsymbol{Z}(\boldsymbol{\theta}_0)\boldsymbol{\theta} + \boldsymbol{\varepsilon} \tag{18.C.2}$$

which satisfies

$$\min\{(Y_0 - \boldsymbol{Z}(\boldsymbol{\theta}_0)\boldsymbol{\theta})'(Y_0 - \boldsymbol{Z}(\boldsymbol{\theta}_0)\boldsymbol{\theta}) = \mathrm{S}(\boldsymbol{\theta})\}. \tag{18.C.3}$$

Let us denote the $\boldsymbol{\theta}$ for which this minimum is attained as

$$\boldsymbol{\theta}_1 = (\boldsymbol{Z}(\boldsymbol{\theta}_0)'\boldsymbol{Z}(\boldsymbol{\theta}_0))^{-1}\boldsymbol{Z}(\boldsymbol{\theta}_0)'Y_0. \tag{18.C.4}$$

In general, we continue this process and, for the $(j + 1)$st linear approximation of $f(X, \boldsymbol{\theta})$ at $\boldsymbol{\theta}_j$, the OLS estimate of $\boldsymbol{\theta}_{j+1}$ is

$$\begin{aligned} \boldsymbol{\theta}_{j+1} &= (\boldsymbol{Z}(\boldsymbol{\theta}_j)'\boldsymbol{Z}(\boldsymbol{\theta}_j))^{-1}\boldsymbol{Z}(\boldsymbol{\theta}_j)'Y_j \\ &= (\boldsymbol{Z}(\boldsymbol{\theta}_j)'\boldsymbol{Z}(\boldsymbol{\theta}_j))^{-1}\boldsymbol{Z}(\boldsymbol{\theta}_j)'(Y - f(X, \boldsymbol{\theta}_j) + \boldsymbol{Z}(\boldsymbol{\theta}_j)\boldsymbol{\theta}_j) \\ &= \boldsymbol{\theta}_j + (\boldsymbol{Z}(\boldsymbol{\theta}_j)'\boldsymbol{Z}(\boldsymbol{\theta}_j))^{-1}\boldsymbol{Z}(\boldsymbol{\theta}_j)'(Y - f(X, \boldsymbol{\theta}_j)),\ j = 0, 1, \ldots \end{aligned} \tag{18.C.5}$$

So for successive values of j, this procedure generates the sequence of OLS estimators $\boldsymbol{\theta}_0, \boldsymbol{\theta}_1, \boldsymbol{\theta}_2, \ldots$ When this process converges (or when $|(\boldsymbol{\theta}_{l,j+1} - \boldsymbol{\theta}_{l,j})/\boldsymbol{\theta}_{l,j}| < \delta$, δ small, $l = 1, \ldots, r$) $\boldsymbol{\theta}_{j+1} - \boldsymbol{\theta}_j = \mathbf{0}$ and thus $\boldsymbol{Z}(\boldsymbol{\theta}_j)'(Y - f(X, \boldsymbol{\theta}_j)) = 0$ or Equation 18.10 holds (the first-order condition for a minimum of $\mathrm{S}(\boldsymbol{\theta})$).

Because convergence to a global minimum may not occur, one should always try a few different starting points or initial values. In addition, if the ε_i are independent and identically distributed with a zero mean and constant variance σ_ε^2, then this Gauss-Newton NLS estimator $\boldsymbol{\theta}_{j+1}$ is consistent and can be taken to be approximately normally distributed with mean $\boldsymbol{\theta}$ and a variance-covariance matrix that is consistently estimated by $S_\varepsilon^2(\boldsymbol{Z}(\boldsymbol{\theta}_{j+1})'\boldsymbol{Z}(\boldsymbol{\theta}_{j+1}))^{-1}$, where $S_\varepsilon^2 = S(\boldsymbol{\theta}_{j+1})/(n-r)$.

Example 18.C.1

Given the regression equation

$$Y_i = \theta_1^2 X_{i1} + \theta_2^2 X_{i2} + \varepsilon_i = f(\boldsymbol{X}_i, \boldsymbol{\theta}) + \varepsilon_i,\ \boldsymbol{X}_i = (X_{i1}, X_{i2}),\ \boldsymbol{\theta}' = (\theta_1, \theta_2),\ i = 1, \ldots, n,$$

let us determine the Gauss-Newton iteration scheme represented by Equation 18.C.5. Given

$$\underset{(n\times 2)}{\boldsymbol{Z}(\boldsymbol{\theta})} = \begin{bmatrix} \frac{\partial f(\boldsymbol{X}_1,\boldsymbol{\theta})}{\partial\theta_1} & \frac{\partial f(\boldsymbol{X}_1,\boldsymbol{\theta})}{\partial\theta_2} \\ \frac{\partial f(\boldsymbol{X}_2,\boldsymbol{\theta})}{\partial\theta_1} & \frac{\partial f(\boldsymbol{X}_2,\boldsymbol{\theta})}{\partial\theta_2} \\ \cdot & \cdot \\ \cdot & \cdot \\ \cdot & \cdot \\ \frac{\partial f(\boldsymbol{X}_n,\boldsymbol{\theta})}{\partial\theta_1} & \frac{\partial f(\boldsymbol{X}_n,\boldsymbol{\theta})}{\partial\theta_2} \end{bmatrix} = \begin{bmatrix} 2\theta_1 X_{11} & 2\theta_2 X_{12} \\ 2\theta_1 X_{21} & 2\theta_2 X_{22} \\ \cdot & \cdot \\ \cdot & \cdot \\ \cdot & \cdot \\ 2\theta_1 X_{n1} & 2\theta_2 X_{n2} \end{bmatrix},$$

it follows that

$$\underset{(2\times 2)}{\boldsymbol{Z}(\boldsymbol{\theta})'\boldsymbol{Z}(\boldsymbol{\theta})} = \begin{bmatrix} 4\theta_1^2 \sum_{i=1}^n X_{i1}^2 & 2\theta_1\theta_2 \sum_{i=1}^n X_{i1}X_{i2} \\ 2\theta_2\theta_1 \sum_{i=1}^n X_{i2}X_{i1} & 4\theta_2^2 \sum_{i=1}^n X_{i2}^2 \end{bmatrix}$$

and

$$\underset{(2\times 1)}{\boldsymbol{Z}(\boldsymbol{\theta})'(\boldsymbol{Y}-\boldsymbol{f}(\boldsymbol{X},\boldsymbol{\theta}))} = \begin{bmatrix} 2\theta_1 X_{11} & 2\theta_1 X_{21} & \ldots & 2\theta_1 X_{n1} \\ 2\theta_2 X_{12} & 2\theta_2 X_{22} & \ldots & 2\theta_2 X_{n2} \end{bmatrix} \begin{bmatrix} Y_1 - \theta_1^2 X_{11} - \theta_2^2 X_{12} \\ Y_2 - \theta_1^2 X_{21} - \theta_2^2 X_{22} \\ \cdot \\ \cdot \\ \cdot \\ Y_n - \theta_1^2 X_{n1} - \theta_2^2 X_{n2} \end{bmatrix}$$

$$= \begin{bmatrix} 2\theta_1 \sum_{i=1}^n X_{i1}(Y_i - \theta_1^2 X_{i1} - \theta_2^2 X_{i2}) \\ 2\theta_2 \sum_{i=1}^n X_{i2}(Y_i - \theta_1^2 X_{i1} - \theta_2^2 X_{i2}) \end{bmatrix},$$

where

$$\boldsymbol{Y}-\boldsymbol{f}(\boldsymbol{X},\boldsymbol{\theta}) = \begin{bmatrix} Y_1 - f(\boldsymbol{X}_1,\boldsymbol{\theta}) \\ Y_2 - f(\boldsymbol{X}_2,\boldsymbol{\theta}) \\ \cdot \\ \cdot \\ \cdot \\ Y_n - f(\boldsymbol{X}_n,\boldsymbol{\theta}) \end{bmatrix} = \begin{bmatrix} Y_1 - \theta_1^2 X_{11} - \theta_2^2 X_{12} \\ Y_2 - \theta_1^2 X_{21} - \theta_2^2 X_{22} \\ \cdot \\ \cdot \\ \cdot \\ Y_n - \theta_1^2 X_{n1} - \theta_2^2 X_{n2} \end{bmatrix}.$$

Then from Equation 18.C.5,

$$\boldsymbol{\theta}_{j+1} = \boldsymbol{\theta}_j + \begin{bmatrix} 4\theta_1^2 \sum_{i=1}^n X_{i1}^2 & 2\theta_1\theta_2 \sum_{i=1}^n X_{i1}X_{i2} \\ 2\theta_2\theta_1 \sum_{i=1}^n X_{i2}X_{i1} & 4\theta_2^2 \sum_{i=1}^n X_{i2}^2 \end{bmatrix}^{-1} \times \begin{bmatrix} 2\theta_1 \sum_{i=1}^n X_{i1}(Y_i - \theta_1^2 X_{i1} - \theta_2^2 X_{i2}) \\ 2\theta_2 \sum_{i=1}^n X_{i2}(Y_i - \theta_1^2 X_{i1} - \theta_2^2 X_{i2}) \end{bmatrix}.$$

Appendix D: The Newton-Raphson and Scoring Methods

Given the regression equation $Y_i = f(X_i, \theta) + \varepsilon_i$, $i = 1,\ldots,n$, log-likelihood function $\log \mathcal{L} = -(n/2)\ln 2\pi - (n/2)\ln \sigma_\varepsilon^2 - (1/2\sigma_\varepsilon^2)S(\theta)$, and least-squares function $S(\theta) = \sum_{i=1}^{n}(Y_i - f(\boldsymbol{X}_i,\theta))^2$, our objective is to determine the value of θ, $\tilde{\theta}$, for which $(\partial \log \mathcal{L}/\partial\theta)_{\theta=\tilde{\theta}} = 0$. For $\theta = \theta_0$ an initial (best guess) estimate of $\tilde{\theta}$, let us construct a first-order Taylor expansion or linearization of $\partial \log \mathcal{L}/\partial\theta$ at θ_0 as

$$\frac{\partial \log \mathcal{L}}{\partial\theta} \approx \left.\frac{\partial \log \mathcal{L}}{\partial\theta}\right|_{\theta=\theta_0} + \left.\frac{\partial^2 \log \mathcal{L}}{\partial\theta^2}\right|_{\theta=\theta_0}(\theta-\theta_0). \tag{18.D.1}$$

At the global maximum of $\log \mathcal{L}$, $\partial \log \mathcal{L}/\partial\theta = 0$ and thus, from Equation 18.D.1,

$$(\theta =)\theta_1 = \theta_0 - \left(\frac{\partial^2 \log \mathcal{L}}{\partial\theta^2}\right)^{-1}_{\theta_0}\left(\frac{\partial \log \mathcal{L}}{\partial\theta}\right)_{\theta_0}. \tag{18.D.2}$$

In general, the $(j + 1)$st iteration produces

$$\theta_{j+1} = \theta_j - \left(\frac{\partial^2 \log \mathcal{L}}{\partial\theta^2}\right)^{-1}_{\theta_j}\left(\frac{\partial \log \mathcal{L}}{\partial\theta}\right)_{\theta_j}, \; j=0,1,2,\ldots. \tag{18.D.3}$$

This iteration technique is known as the *Newton-Raphson method*. Let us term the quantity

$$G(\theta_j) = \left(\frac{\partial \log \mathcal{L}}{\partial\theta}\right)_{\theta_j} = \frac{-1}{2\sigma_\varepsilon^2}\left(\frac{\partial S}{\partial\theta}\right)_{\theta_j} \tag{18.D.4}$$

the score for θ_j in the sample, whereas

$$I(\theta_j) = -E\left[\left(\frac{\partial^2 \log \mathcal{L}}{\partial\theta^2}\right)_{\theta_j}\right] = \frac{1}{\sigma_\varepsilon^2}\left((\boldsymbol{Z}(\theta_j)'\boldsymbol{Z}(\theta_j)\right) \tag{18.D.5}$$

is called the *information on* θ_j in the sample, where $\boldsymbol{Z}(\theta_j)' = (z_1(\theta_j),\ldots,z_n(\theta_j))$ and $z_i(\theta_j) = (df(\boldsymbol{X}_i,\theta)/d\theta)_{\theta=\theta_j}$. In this regard, for large n, the *method of scoring* has us replace $(\partial^2 \log \mathcal{L}/\partial\theta^2)^{-1}_{\theta_j}$ by $-I(\theta_j)^{-1}$ and $(\partial \log \mathcal{L}/\partial\theta)_{\theta_j}$ by $G(\theta_j)$, to wit Equation 18.D.3 becomes

$$\theta_{j+1} = \theta_j - \frac{1}{2}\left(\boldsymbol{Z}(\theta_j)'\boldsymbol{Z}(\theta_j)\right)^{-1}\left(\frac{\partial S}{\partial\theta}\right)_{\theta_j}, \; j=0,1,2,\ldots. \tag{18.D.6}$$

This process is continued until $\theta_{j+1} - \theta_j = 0$ or until $|(\theta_{j+1} - \theta_j)/\theta_j| < \delta$, $j = 0, 1, 2,\ldots$, δ a small predetermined constant.

Given that ε_i, $i = 1,\ldots, n$, are independent with $\varepsilon_i : N(0,\sigma_\varepsilon^2)$ for all i, the ML estimator $\tilde{\theta}$ is consistent and asymptotically normally distributed with the variance of $\tilde{\theta}$ consistently estimated by $I(\theta_{j+1})^{-1} = \sigma_\varepsilon^2\left(\boldsymbol{Z}(\theta_{j+1})'\boldsymbol{Z}(\theta_{j+1})\right)^{-1}$, with $\tilde{\sigma}_\varepsilon^2 = S(\theta_{j+1})/n$.

We may extend Equation 18.D.3 to the r parameter case by writing $\boldsymbol{\theta}' = (\theta_1,\ldots,\theta_r)$,

$$\boldsymbol{\theta}_{j+1} = \boldsymbol{\theta}_j - \left(\frac{\partial^2 \log \mathcal{L}}{\partial\boldsymbol{\theta}\partial\boldsymbol{\theta}'}\right)^{-1}_{\boldsymbol{\theta}_j}\left(\frac{\partial \log \mathcal{L}}{\partial\boldsymbol{\theta}}\right)_{\boldsymbol{\theta}_j}, \; j = 0,1,2,\ldots. \tag{18.D.7}$$

This expression is known as the *generalized Newton-Raphson method.*

Let

$$G(\boldsymbol{\theta}_j) = \left(\frac{\partial \log \mathcal{L}}{\partial\boldsymbol{\theta}}\right)_{\boldsymbol{\theta}_j} = \frac{-1}{2\sigma_\varepsilon^2}\left(\frac{\partial S}{\partial\boldsymbol{\theta}}\right)_{\boldsymbol{\theta}_j} \tag{18.D.8}$$

represent the score for $\boldsymbol{\theta}_j$ in the sample, whereas

$$I(\boldsymbol{\theta}_j) = -E\left[\left(\frac{\partial^2 \log \mathcal{L}}{\partial\boldsymbol{\theta}\partial\boldsymbol{\theta}'}\right)_{\boldsymbol{\theta}_j}\right] = \frac{1}{\sigma_\varepsilon^2}\left(\boldsymbol{Z}(\boldsymbol{\theta}_j)'\boldsymbol{Z}(\boldsymbol{\theta}_j)\right) \tag{18.D.9}$$

depicts the information on $\boldsymbol{\theta}_j$ in the sample. Then proceeding as above, for large n, the *generalized method of scoring* has us substitute $-I(\boldsymbol{\theta}_{j+1})^{-1}$ and $G(\boldsymbol{\theta}_j)$ into Equation 18.D.7 so as to obtain

$$\boldsymbol{\theta}_{j+1} = \boldsymbol{\theta}_j - \frac{1}{2}\left[\boldsymbol{Z}(\boldsymbol{\theta}_j)'\boldsymbol{Z}(\boldsymbol{\theta}_j)\right]^{-1}\left(\frac{\partial S}{\partial\boldsymbol{\theta}}\right)_{\boldsymbol{\theta}_j}, \; j = 0,1,2,\ldots, \tag{18.D.10}$$

where

$$\underset{(n\times r)}{\boldsymbol{Z}(\boldsymbol{\theta}_j)} = \left.\frac{\partial f(\boldsymbol{X},\boldsymbol{\theta})}{\partial\boldsymbol{\theta}}\right|_{\theta=\theta_j} = \begin{bmatrix} \left.\frac{\partial f(\boldsymbol{X}_1,\boldsymbol{\theta})}{\partial\theta_1}\right|_{\boldsymbol{\theta}_j} & \cdots & \left.\frac{\partial f(\boldsymbol{X}_1,\boldsymbol{\theta})}{\partial\theta_r}\right|_{\boldsymbol{\theta}_j} \\ \vdots & & \vdots \\ \left.\frac{\partial f(\boldsymbol{X}_n,\boldsymbol{\theta})}{\partial\theta_1}\right|_{\boldsymbol{\theta}_j} & \cdots & \left.\frac{\partial f(\boldsymbol{X}_n,\boldsymbol{\theta})}{\partial\theta_r}\right|_{\boldsymbol{\theta}_j} \end{bmatrix}$$

and $(\partial S/\partial\boldsymbol{\theta})' = (\partial S/\partial\theta_1,\ldots,\partial S/\partial\theta_r)_{\boldsymbol{\theta}_j}$. The iterations are continued until $\boldsymbol{\theta}_{j+1} - \boldsymbol{\theta}_j = \boldsymbol{0}$ or when $|(\theta_{l,j+1} - \theta_{l,j})/\theta_{lj}| < \delta$, δ small, $l = 1,\ldots, r$.

For $\boldsymbol{\varepsilon}: N(\boldsymbol{0},\sigma_\varepsilon^2\boldsymbol{I})$ and with the components of $\boldsymbol{\varepsilon}$ mutually independent, the ML estimator $\tilde{\boldsymbol{\theta}}$ is consistent and asymptotically normal with the variance-covariance matrix of $\tilde{\boldsymbol{\theta}}$ consistently estimated by $I(\boldsymbol{\theta}_{j+1})^{-1} = \tilde{\sigma}_\varepsilon^2\left[\boldsymbol{Z}(\boldsymbol{\theta}_{j+1})'\boldsymbol{Z}(\boldsymbol{\theta}_{j+1})\right]$, where $\tilde{\sigma}_\varepsilon^2 = S(\boldsymbol{\theta}_{j+1})/n$.

Exercises

18-1. Determine which of the following regression model specifications is intrinsically nonlinear:

a. $Y = \beta_0 + \beta_1 e^X + \varepsilon.$
b. $Y = \beta_0 + \beta_1 lnX + \varepsilon.$
c. $Y = \beta_0 + (\beta_1 + X)^{-1} + \varepsilon.$
d. $Y = \beta_0 + \beta_1 e^{\beta_2 X} + \varepsilon.$
e. $Y = \beta_2(1 + e^{\beta_0 + \beta_1 X})^{-1} + \varepsilon,\ \beta_2 > 0,\ \beta_1 < 0.$

18-2. Is the regression model

$$\begin{aligned} g(Y) = \log Y &= \theta_1 + \frac{\theta_2}{\theta_3}\log W_1 + \frac{\theta_3}{1-\theta_3}\log W_2 + \frac{\theta_3}{\theta_2}\frac{\theta_4}{1-\theta_2}\log W_3 + \varepsilon \\ &= \beta_1 + \beta_2 X_1 + \beta_3 X_2 + \beta_4 X_3 + \varepsilon \end{aligned}$$

intrinsically linear? That is, are the restricted parameters (the θ's) exactly identified in terms of the unrestricted parameters (the β's)? (Hint: see Example 18.1.)

18-3. Given the regression model

$$g(Y) = \ln Y = \log \theta_1 + \theta_2 \log W_1 + \frac{\theta_3 \theta_4}{1-\theta_2}\log W_2 + \varepsilon = \beta_1 + \beta_2 X_1 + \beta_3 X_2 + \varepsilon,$$

are the restricted parameters (theθ's) underidentified in terms of the unrestricted parameters (the β's)? Explain your answer.

18-4. Given the Exercise 18-2 regression model, find $V(\hat{\theta}_1),\ldots,V(\hat{\theta}_4)$. (Hint: see Equation 18.A.1 of Appendix 18.A.)

18-5. Given the regression model $Y = 5(X+\theta)^{-1} + \varepsilon, \varepsilon : N(0,\sigma_\varepsilon^2)$, determine the first-order condition for the least squares function $S(\hat{\theta})$ to attain a minimum.

18-6. For the regression function $Y = (X+\theta)^{-1} + \varepsilon, \varepsilon : N(0,\sigma_\varepsilon^2)$, determine the first-order condition for the log-likelihood function to attain a maximum at $\tilde{\theta}$.

18-7. Given the regression model $Y = \ln\theta_1 + X^{-\theta_2} + \varepsilon, \varepsilon : N(0,\sigma_\varepsilon^2)$, determine the first-order conditions (the NLS normal equations) for the least-squares function $S(\hat{\theta})$ to attain a minimum.

18-8. For the regression equation $Y = \theta_1 + \theta_2(X+\theta_3)^{-1} + \varepsilon, \varepsilon : N(0,\sigma_\varepsilon^2)$, determine the first-order conditions for the log-likelihood function to attain a maximum at $\tilde{\theta}$.

18-9. Given the Exercise 18.6 regression equation, use the Gauss-Newton iteration scheme to find θ_{j+1}. (Hint: see Example 18.B.1.)

18-10. For the Exercise 18.7 regression model, use the Gauss-Newton iteration routine to find θ_{j+1}. (Hint: see Example 18.C.1.)

18-11. Use the Gauss-Newton iteration method to determine θ_{j+1} for the Exercise 18.8 regression model. (Hint: follow Example 18.C.1.)

18-12. For the Exercise 18.6 regression function, use the Newton-Raphson method to find θ_{j+1}. (Hint: use Equation 18.D.3.)

18-13. Given the Exercise 18.6 regression equation, use the method of scoring to determineθ_{j+1}. (Hint: see Equation 18.D.6.)

18-14. For the Exercise 18.7 regression model, use the generalized Newton-Raphson technique to find θ_{j+1}. (Hint: use Equation 18.D.7.)

18-15. Using the Exercise 18.8 regression equation, use the generalized method of scoring to find θ_{j+1}. (Hint: see Equation 18.D.10.)

18-16. Discuss the difference between the Gauss-Newton and Newton-Raphson iteration techniques.

18-17. Given the Example 18.4 regression output (Table 18.2), test the parameter restriction $g(\theta)=\theta_0^{1/2}+\theta_2/\theta_1=0$ via the Wald test.

18-18. Using the Table E.18-1 dataset, estimate the parameters of the nonlinear regression model $Y=\theta_1+X_1^{\theta_2}+X_2^{\theta_3}+\varepsilon, \boldsymbol{\varepsilon}\sim N(\mathbf{0},\sigma_\varepsilon^2\mathbf{I})$, using PROC NLIN. Let the starting values of the parameters be θ_1 = 0.3, θ_2 = 0.3, and θ_3 = 0.5. Compare the results from PROC NLIN with an OLS fit. (Hint: run $\ln Y=\ln\theta_1+\theta_2\ln X_1+\theta_3\ln X_3+\eta, \boldsymbol{\eta}\sim N(\mathbf{0},\sigma_\eta^2\mathbf{I})$.)

18-19. Using the SAS ETS MODEL procedure and the Table E.18-1 dataset, estimate the regression equation

$$Y=\theta_1(\theta_2X_1^{-\theta_3}+\theta_4X_2^{-\theta_3})^{-1/\theta_3}+\varepsilon, \boldsymbol{\varepsilon}\sim N(\mathbf{0},\sigma_\varepsilon^2\mathbf{I}),$$

and test the following parameter restrictions using the LR, W, and LM test statistics:

a. $\theta_1-\frac{1}{2}\theta_3=0.$

b. $\theta_2-\theta_4=0.$

c. $\theta_2+\theta_4=1.$

Let the starting values for the parameters be θ_l = 0.5, l = 1, 2, 3, 4. (Hint: see Example 18.6.)

TABLE E.18-1

Observations on Variables Y, X_1, and X_2

Y	X_1	X_2	Y	X_1	X_2
33.0	131.6	27.3	48.2	187.8	35.7
36.3	136.7	28.4	47.9	192.5	36.3
36.8	142.3	29.6	48.5	198.0	37.3
38.8	147.7	30.5	51.3	204.3	37.9
38.1	152.2	30.4	49.8	207.3	37.5
40.7	157.7	31.8	53.7	210.3	37.7
45.4	164.5	33.1	58.8	215.7	40.1
46.7	171.3	33.8	59.0	220.2	41.5
42.6	175.3	33.1	55.4	223.9	44.0
47.8	182.1	34.8	61.1	229.4	42.3

18-20. Use the SAS ETS MODEL procedure and the Table E.18-1 data to estimate the regression model $Y=[\theta_1 X_1+\theta_2 X_2^{\theta_3}]^{\theta_4}+\varepsilon, \varepsilon: N(\mathbf{0},\sigma_\varepsilon^2\mathbf{I})$, and test the following restrictions:

a. $\theta_1 + \theta_2 + \theta_3 = 1$.
b. $\theta_3 - \theta_4 = 0$.

Let the starting values for the parameters be $\theta_1 = \theta_2 = 0.4$, $\theta_3 = 0.3$, and $\theta_4 = 0.5$.

18-21. For the regression equation

$$Y = \theta_1 X_1^{\theta_2} X_2^{\theta_3} e^{\theta_4 X_1 + \theta_5 X_2} + \varepsilon, \varepsilon : N(\mathbf{0}, \sigma_\varepsilon^2 \mathbf{I}),$$

estimate θ_l, $l = 1,\ldots,5$, using the Table E.18-1 dataset and the SAS ETS MODEL procedure. Let the starting parameter values be $\theta_1 = 0.3$, $\theta_2 = \theta_3 = 0.5$, and $\theta_4 = \theta_5 = 0.4$.

18-22. A Linear versus Log-Linear Specification Test [Davidson and Mackinnon 1981; Mackinnon, White, and Davidson 1983]:

An important modeling consideration involves the determination of whether a linear or log-linear specification of a regression equation is appropriate. In this regard, let the null and alternative hypotheses appear, respectively, as

$$\begin{aligned} &H_0 : Y = \beta_0 + \sum\nolimits_{j=1}^{m} \beta_j X_j + \varepsilon \quad \text{[linearity]} \\ &H_1 : \ln Y = \ln \gamma_0 + \sum\nolimits_{j=1}^{m} \gamma_j \ln X_j + \varepsilon \quad \text{[log-linearity].} \end{aligned} \tag{E.18.1}$$

In addition, let $\hat{Y}$ denote the estimated Y under H_0 and let $\widehat{\ln Y}$ represent the estimated value of lnY under H_1. Then the test is executed by first testing the significance (via an ordinary t test) of the estimated coefficient $\hat{\alpha}_1$ in the *artificial regression model*

$$Y=\beta_0+\sum_{j=1}^{m}\beta_j X_j+\alpha_1(\widehat{\ln Y}-\ln\hat{Y})+\varepsilon. \tag{E.18.2}$$

Next, let us reverse the status of the linear and log-linear models in (Equation E.18.1 and write

$$\begin{aligned} &H_0:\ln Y=\ln\gamma_0+\sum_{j=1}^{m}\gamma_j\ln X_j+\varepsilon \quad \text{[log-linearity]}\\ &H_1:Y=\beta_0+\sum_{j=1}^{m}\beta_j X_j+\varepsilon \quad \text{[linearity].}\end{aligned} \tag{E.18.3}$$

Now the artificial regression model has the form

$$\ln Y=\ln\gamma_0+\sum_{j=1}^{m}\gamma_j\ln X_j+\alpha_2(\hat{Y}-e^{\widehat{\ln Y}})+\varepsilon. \tag{E.18.4}$$

Again we are interested in testing the significance of the estimated coefficient $\hat{\alpha}_2$.

If only one of $\hat{\alpha}_1$ and $\hat{\alpha}_2$ is statistically significant, then the appropriate model is immediately revealed, i.e., if $\hat{\alpha}_1$ is significant, then we reject the null hypothesis of linearity and select the log-linear model as the appropriate model; if $\hat{\alpha}_2$ is significant, then the null of log linearity is rejected and we go with the alternative hypothesis of linearity. If both $\hat{\alpha}_1$ and $\hat{\alpha}_2$ depart significantly from zero, then both model specifications can be rejected, and if neither $\hat{\alpha}_1$ nor $\hat{\alpha}_2$ is significant, then these tests offer no hint as to which specification (linear or log-linear) is preferred.

This said, use the Table E.18-2 data values to determine whether a linear or log-linear specification is warranted. Exhibit E.18-1 houses the requisite SAS code.

TABLE E.18-2

Observations on Variables Y, X_1, and X_2

Y	X1	X2	Y	X1	X2
33.0	136.6	27.3	71.2	182.8	35.1
35.3	136.7	28.4	77.2	192.5	36.3
37.8	140.3	28.6	82.5	198.0	37.3
40.2	146.7	29.5	95.3	203.3	37.9
43.1	151.2	30.4	102.0	207.3	38.5
47.7	155.7	31.8	110.7	210.2	39.7
52.4	160.5	32.1	115.8	215.7	41.1
56.7	161.3	32.8	125.0	220.2	44.5
61.0	167.3	33.1	140.4	223.9	50.0
66.8	172.1	33.8	150.1	226.4	52.4

Exhibit E.18-1

```
data spec;
  input y x1 x2;
lny = log(y);    ]
lnx1 = log(x1);  }  ①
lnx2 = log(x2);  ]
datalines;
33.0    136.6   27.3
35.3    136.7   28.4
.       .       .
.       .       .
.       .       .
150.1   226.4   52.4
;
proc reg data=spec;
  model y=x1 x2; output out=out0 p=p0;②
  data spec0; set out0;③
run;
proc reg data=spec;
  model lny=lnx1 lnx2; output out=out1 p=p1;④
  data spec1; set out1;⑤
run;
data spec2; merge spec spec0 spec1;⑥
lnp0=log(p0);  ]
dif0=p1-lnp0;  }  ⑦
ep1=exp(p1);   }
dif1=p0-ep1;   ]
proc reg data=spec2;
  model y=x1 x2 dif0;⑧
  model lny=lnx1 lnx2 dif1;⑨
run;
```

① We transform all variables to natural logarithms.

② We estimate the linear equation $Y = \beta_0 + \beta_1 X_1 + \beta_2 X_2 + \varepsilon$ and house the predicted values of *Y*, *p0*, in the output dataset *out0*.

③ We create the SAS dataset *spec0* to store the predicted values *p0*.

④ We estimate the log-linear equation $\ln Y = \ln\beta_0 + \beta_1 \ln X_1 + \beta_2 \ln X_2 + \varepsilon$ and store the predicted values of ln*Y*, *p1*, in the output dataset *out1*.

⑤ We create the SAS dataset *spec1* to house the predicted values *p1*.

⑥ We merge datasets *spec*, *spec0*, and *spec1*.

⑦ We form the differences dif0= $\widehat{\ln Y} - \ln\hat{Y}$ and dif1 = p0 − ep1.

⑧ We estimate the artificial regression Equation E.18.2 given Equation E.18.1.

⑨ We estimate the reversed artificial regression Equation E.18.4 under Equation E.18.3.

19

Issues in Time Series Modeling and Estimation

Introduction

This chapter addresses regression estimation problems that typically arise when using time series data. (Readers not familiar with time series analysis should read the appendices to this chapter, which cover topics such as autoregressive and moving-average processes, autoregressive distributed lag (ADL) processes, unit root processes, integrated processes, cointegrated processes, etc.) In particular, we shall examine the detection of so-called unit roots, testing for cointegration among integrated series, and the use of an error-correcting term as a regressor.

Consideration of these issues is important because time series regression using trended data is potentially problematic, i.e., time series OLS regression results are reliable only if the data are trend free or stationary. Interpreting regression results involving (two) deterministically trended variables whose time paths diverge can be tricky, whereas spurious or nonsense regression results can emerge when the model variables are subject to stochastic trends. Although deterministic trends can be modeled before estimation begins or explicitly included in the regression equation, stochastic trends need considerably more attention. In what follows then, we shall focus on the detection and handling of stochastic trends in a regression setting.

The issue of stochastic trends in nonstationary (integrated) series is most directly addressed by differencing the data series involved so that stationarity is attained. However, for *difference-stationary data*, long-run information that is typically garnered in *levels of variables* is lost. We thus need to use techniques that combine both long-and short-run characteristics of the data series but ensure the stationarity of all variables involved. In this way, we can recapture information about levels of variables that is lost through differencing.

Given the linear regression model

$$Y_t = \beta_0 + \beta_1 X_t + \varepsilon_t, \tag{19.1}$$

the classical method of parameter estimation (OLS) assumes that $\{X_t\}$ and $\{Y_t\}$ are stationary and the regression errors $\{\varepsilon_t\}$ have a zero mean and a constant variance. Under stationarity, the means and variances of X_t and Y_t are constant and independent of time. But, if these variables are nonstationary or integrated, then using OLS to estimate Equation 19.1 can prompt one to make some highly misleading inferences: the *spurious regression problem*. Given that the means and variances of nonstationary variables are time dependent, all of the diagnostic statistics that use these quantities must also change over time and, consequently, fail to converge to their true values as the sample size n increases. The upshot is that, for a spurious regression, the OLS estimates are inconsistent, R^2 tends to be high, the t statistics appear to be significant at some nominal test level, the DW statistic is very low (in fact, a low DW is usually an indication that the regression variables are nonstationary), and a test of, say, $H_0 : \beta_1 = 0$ versus $H_1 : \beta_1 \neq 0$ is seriously biased toward rejecting this null hypothesis. So although the regression output "looks good," the results are not at all meaningful. Having touched on some of the difficulties connected with nonstationary data, additional issues with and insights into spurious regressions are offered in the next section.

The Spurious Regression Problem [Hooker 1901; Yule 1926a; Granger and Newbold 1974; Hendry 1980, 1986; Phillips 1986]

We mentioned above that nonsense regressions will occur when classical techniques are applied to integrated or nonstationary data. This conclusion emerges, in large part, from the work of Granger and Newbold [1974]. Specifically, suppose that Y_t and X_t represent two integrated but unrelated time series with data-generating (unit root) processes given by

$$\begin{aligned} Y_t &= Y_{t-1} + \varepsilon_{Yt} \\ X_t &= X_{t-1} + \varepsilon_{Xt} \end{aligned} \tag{19.2}$$

respectively, where the ε_{Yt}'s (respectively ε_{Xt}'s) are independent and identically distributed random variables with zero means and constant variances, and $\mathrm{COV}(\varepsilon_{Yt}, \varepsilon_{Xt}) = 0$ with $E(\varepsilon_{Yt}, \varepsilon_{Y,\, t-k}) = E(\varepsilon_{Xt}, \varepsilon_{X,\, t-k}) = 0,\ k \neq 0$. In this regard, Y_t and X_t can be characterized as uncorrelated random walks.

Given that X_t and Y_t are nonstationary random processes, with X_t neither affecting nor affected by Y_t, one would not expect the regression of Y_t on X_t to yield any statistically significant results. After all, because $\{X_t\}$ and $\{Y_t\}$ are independent processes, Equation 19.1 is necessarily meaningless; any relationship between these two variables must be spurious. Granger and Newbold generated many samples from Equation 19.2 and, for each of them, Equation 19.1 was estimated via OLS. Surprisingly, at the 5% level of significance, these researchers were able to reject $H_0 : \beta_1 = 0$ approximately 75% of the time. The reported regressions typically had a high R^2 and the residuals exhibited a low value of the DW statistic (thus signaling the presence of a high degree of autocorrelation).

To rationalize the Granger-Newbold results, we note first that Equation 19.1 is without foundation if the error process $\{\varepsilon_t\}$ is nonstationary. If $\{\varepsilon_t\}$ has a stochastic trend, then the error from period t never disappears; its effect is permanent. If we omit the intercept term in Equation 19.1, then we may use Equation 19.2 to express the error term as

$$\varepsilon_t = \sum\nolimits_{i=1}^{t} \varepsilon_{Yt} - \beta_1 \sum\nolimits_{i=1}^{t} \varepsilon_{Xt}, \tag{19.3}$$

$X_0 = Y_0 = 0$ (see Equations 19.A.30 and 19.A.37). Because the variance of this error term becomes infinitely large as t increases (see point two after Equation 19.A.30), the assumptions supporting the usual hypothesis tests are violated; t ratios, the F ratio, and R^2 are all unreliable and standard test procedures do not apply, i.e., the conventional t statistic for $\hat{\beta}_1$ (as well as for $\hat{\beta}_0$) is not t distributed and does not possess a limiting normal distribution as $n \to \infty$.

For example, under $H_0 : \beta_1 = 0$, Equation 19.1 becomes $Y_t = \beta_0 + \varepsilon_t$. Then given that $Y_t : I(1)$ (by virtue of Equation 19.2), it follows from Equation 19.3 that $\varepsilon_t = \sum_{i=1}^{t} \varepsilon_{Yt} : I(1)$. Hence, the notion that the error term is a *unit root process* is consistent with the distribution theory underlying classical regression techniques. In fact, the larger is n, the more likely it is that one falsely concludes that $\beta_1 \neq 0$ [Phillips 1986], and this is directly attributable to the nonstationarity of X_t and Y_t as determined by Equation 19.2.

Remember that regression methods detect correlations, and, with nonstationary time series variables, *spurious correlations* may emerge and may endure even in large samples and despite the fact that X_t and Y_t are not related. In fact, because X_t and Y_t are uncorrelated, one would expect that $\hat{\beta}_1$ and R^2 as determined from Equation 19.1 should converge in probability to zero. The fundamental issue is that an observed correlation between integrated time series variables must not be interpreted as if it occurred between stationary variables. As indicated previously, the estimated residuals from a spurious regression exhibit a high degree of autocorrelation. (In fact, the correlation coefficient between ε_t, ε_{t+1} goes to unity as t increases without bound.) Hence, it is the DW statistic that provides us with a device for discriminating between spurious and legitimate regressions because, for the latter, the DW statistic will converge to a nonzero value. As a rule of thumb, if we have a regression for which $R^2 > DW$, then this regression is likely to be spurious [Granger and Newbold 1974].

Testing for Unit Roots [Dickey and Fuller 1979, 1982]

The Structure of the Dickey-Fuller Test Equation

Based on our discussions in the preceding section, it should be evident that the nonstationarity of time series data precludes us from obtaining reliable regression parameter estimates. More specifically, nonstationary or unit root variables have means and variances that are time dependent so that any inferences based on OLS regression results are misleading.

To overcome this difficulty, we shall apply unit root tests to determine whether our regression variables are stationary or nonstationary. Remember that the series $\{Y_t\}$ may be nonstationary because it has a deterministic trend or a stochastic trend. In this regard, to examine the issue of a variable's nonstationarity, we must address the following question: should one detrend or difference a time series? The answer depends on whether the time series is a trend-stationary or difference-stationary process. As explained in Appendix 19.A, a (deterministic) trend-stationary process has the form

$$Y_t = \gamma_0 + \gamma_1 t + \varepsilon_t; \tag{19.4}$$

whereas a difference stationary process is structured as

$$Y_t = \alpha_0 + Y_{t-1} + \varepsilon_t. \tag{19.5}$$

If ε_t is an independent and identically distributed sequence of random variables, then Equation 19.5 constitutes a random walk with drift.

With both cases (Equations 19.4 and 19.5) subsumed under

$$\begin{aligned} Y_t &= \gamma_0 + \gamma_1 t + u_t \\ u_t &= \rho u_{t-1} + \varepsilon_t, \end{aligned} \qquad (19.6)$$

where ε_t is taken to be a stationary random variable (it has a zero mean and a constant variance σ_ε^2), we can rewrite Equation 19.6 as

$$Y_t = \gamma_0 + \gamma_1 t + \rho u_{t-1} + \varepsilon_t. \qquad (19.6.1)$$

Then substituting $Y_{t-1} = \gamma_0 + \gamma_1(t-1) + u_{t-1}$ into Equation 19.6.1 yields

$$Y_t = \gamma_0 + \gamma_1 t + \rho[Y_{t-1} - \gamma_0 - \gamma_1(t-1)] + \varepsilon_t. \qquad (19.6.2)$$

If we now subtract Y_{t-1} from both sides of Equation 19.6.2, our final result is

$$\begin{aligned} \Delta Y_t &= \gamma_0 - \rho\gamma_0 + \rho\gamma_1 + \gamma_1 t - \rho\gamma_1 t + (\rho - 1)Y_{t-1} + \varepsilon_t \\ &= \beta_0 + \beta_1 t + (\rho - 1)Y_{t-1} + \varepsilon_t, \end{aligned} \qquad (19.6.3)$$

where $\beta_0 = \gamma_0(1-\rho) + \rho\gamma_1$ and $\beta_1 = \gamma_1(1-\rho)$. In this regard, if $|\rho| < 1$, then Y_t is trend stationary and if $\beta_1 = 0$ and $|\rho| = 1$ (the unit root case), then Y_t is difference stationary.

Let us rewrite Equation 19.6.3 as

$$\Delta Y_t = \beta_0 + \beta_1 t + \phi Y_{t-1} + \varepsilon_t, \qquad (19.7)$$

where $\phi = \rho - 1$. Then to conduct a unit root test, set $H_0 : \phi = \rho - 1 = 0$ and $H_1 : \phi = \rho - 1 < 0$. However, because under the null hypothesis $Y_t : I(1)$, Equation 19.7 represents the regression of an $I(0)$ variable on an $I(1)$ variable so that the conventional t statistic does not have an asymptotic normal distribution and thus we cannot use the usual t distribution to test the null.

Nonstandard Critical Values

To see exactly what sort of modified distribution should be used, we know that for observed

$$Y_t = \rho Y_{t-1} + \varepsilon_t, \ |\rho| < 1, \ t = 1,2,\ldots,$$

the OLS estimator for ρ is

$$\hat{\rho} = \frac{\sum_{t=2}^{n} Y_t Y_{t-1}}{\sum_{t=2}^{n} Y_{t-1}^2}$$

with

$$\left[n/(1-\rho)^2\right]^{\frac{1}{2}}(\hat{\rho} - \rho) \xrightarrow{d} N(0,1), \qquad (19.8)$$

i.e., the quantity $\left[n/(1-\rho)^2\right]^{\frac{1}{2}}(\hat{\rho} - \rho)$ converges in distribution to a standard normal distribution function. However, for $H_0 : \rho = 1$ (versus $H_0 : \rho < 1$), Dickey and Fuller [1979] examine the distribution of $\hat{\rho}$ above under this null hypothesis and demonstrate that the statistic

$$n(\hat{\rho} - \rho) = n(\hat{\rho} - 1) \tag{19.9}$$

has a nonstandard limiting distribution (it is not normal or even symmetrical[1]) and thus the usual t ratio is not t distributed so that critical values from conventional t tables are not applicable for hypothesis testing. In fact, tables of critical values for the limiting distribution of Equation 19.9 were computed by Dickey and Fuller under various assumptions (concerning the presence of drift, a deterministic trend, and the behavior of ε_t). These new critical values will be termed *t-like critical values*.

To ascertain how Dickey-Fuller (unit root) tests are executed, it is readily seen that three separate regression models are nested in Equation 19.7:

$$\begin{aligned}&1.\ \Delta Y_t = \phi Y_{t-1} + \varepsilon_t.\\&2.\ \Delta Y_t = \beta_0 + \phi Y_{t-1} + \varepsilon_t.\\&3.\ \Delta Y_t = \beta_0 + \beta_1 t + \phi Y_{t-1} + \varepsilon_t.\end{aligned} \tag{19.10}$$

where $\phi = \rho - 1$. (Remember that if $\rho = 1$, then $\phi = 0$; and if $\rho < 1$, then $\phi < 0$.) In this regard, looking to Equation 19.10.1, to conduct a unit root test, set $H_0 : \phi = 0$ (Y_t is nonstationary) versus $H_1 : \phi < 0$ (Y_t is stationary) and let the conventional t statistic obtained from an OLS fit of Equation 19.10.1 be denoted as τ. However, τ does not follow the usual t distribution; the simulated asymptotic critical values of τ obtained under the null hypothesis that $\phi = 0$ are provided in Table A.6 [Fuller 1976]. It is important to note that these critical values are derived under the assumption that there is no drift ($\beta_0 = 0$) and no time trend ($\beta_1 = 0$). Hence, the limiting distribution of the test statistic τ and the associated critical values are incorrect if these assumptions do not hold.

Dickey and Fuller actually derive a separate limiting distribution, and thus separate critical values, for the OLS t statistic under the null hypothesis that $\phi = 0$ for each of the two remaining regressions in Equation 19.10 (but under the assumption that Equation 19.10.1 generates the data). When Equation 19.10.2 is fit, the OLS t statistic for $\phi = 0$ is denoted τ_μ; and when Equation 19.10.3 is run, the OLS statistic for $\phi = 0$ is denoted as τ_τ. Selected critical values for these statistics are also presented in Table A.6. Because in each case $H_1 : \phi < 0$, a calculated value smaller than the (negative) critical or tabular value would lead us to reject H_0 (a unit root) in favor of H_1 (stationarity).

In reality, we do not know the appropriate values for β_0 and/or β_1. Hence, we must test jointly for the presence of drift and/or a deterministic trend as well as for the existence of a unit root. To perform joint tests, Dickey and Fuller [1981] tabulated a collection of *F-like critical values* (denoted Φ_1, Φ_2, and Φ_3) for assessing the significance of various statistics determined via OLS and that are structured as conventional F statistics (ratios) that compare restricted and unrestricted residual sums of squares. So in addition to determining t-like critical values, Dickey and Fuller offer F-like critical values (Table A.7).

---

1 It has a negative mean, a standard deviation in excess of unity, and its peak is sharper than that of a normal distribution. Specifically, based on the distribution of ratios of quadratic forms, Dickey and Fuller [1979] demonstrate that

$$n(\hat{\rho} - 1) \xrightarrow{d} t,$$

where $t = \frac{1}{2}(u^2 - 1)/v$ is simulated and $u = \sum_{k=1}^{\infty} \sqrt{2}\gamma_k Z_k$, $\gamma_k = 2(-1)^{k+1}/(2k-1)\pi$, $v = \sum_{k=1}^{\infty} \gamma_k^2 Z_k^2$, and the Z_k's are independent and identically distributed $N(0, 1)$ variables.

For instance,

1. To test the joint null hypothesis $H_0 : \beta_0 = \phi = 0$ (a unit root and no drift), against the alternative H_1 : not H_0, we use the Dickey-Fuller F-like critical value Φ_1. Stated alternatively, H_0 : (19.10.1) is the true model, H_1 : (19.10.2) is the true model. Hence,

$$\text{Restricted model: } \Delta Y_t = \phi Y_{t-1} + \varepsilon_t;$$
$$\text{Unrestricted model: } \Delta Y_t = \beta_0 + \phi Y_{t-1} + \varepsilon_t.$$

Let SSE_R and SSE_U denote, respectively, the restricted and unrestricted error sums of squares. Then the sample F statistic is constructed as

$$F_{\Phi_1} = \frac{(SSE_R - SSE_U)/r}{SSE_U/(n-k)}, \tag{19.12}$$

where r is the number of parameter restrictions and $n - k$ corresponds to the number of observations less the number of parameters estimated in the unrestricted model. (Hence, $r = 2$ and $n - k = n - 2$.)

In this regard, if the restriction is not binding, then SSE_R should be close in value to SSE_U and thus we can expect F_{Φ_1} to be small. In this instance, if $F_{\Phi_1} < \Phi_1$, then we cannot reject H_0 in favor of H_1; we can, however, infer that $\beta_0 = 0$ and/or $\phi = 0$ (the model has a unit root). Furthermore, if the restriction is binding, then SSE_R should be larger than SSE_U and thus we can expect F_{Φ_1} to be large. If $F_{\Phi_1} > \Phi_1$, then we reject H_0 and conclude that either $\beta_0 \neq 0$ and/or $\phi \neq 0$ (no unit root).

2. To test the joint null hypothesis $H_0 : \beta_0 = \beta_1 = \phi = 0$ (a unit root but no drift or deterministic trend) against the alternative H_1 : not H_0, we will use the F-like critical value Φ_2. Equivalently, H_0 : (19.10.1) is the true model, H_1 : (19.10.3) is the true model. Now,

$$\text{Restricted model: } \Delta Y_t = \phi Y_{t-1} + \varepsilon_t;$$
$$\text{Unrestricted model: } \Delta Y_t = \beta_0 + \beta_1 t + \phi Y_{t-1} + \varepsilon_t.$$

Then the sample F statistic is

$$F_{\Phi_2} = \frac{(SSE_R - SSE_U)/r}{SSE_U/(n-k)}, \tag{19.13}$$

where $r = 3$ and $n - k = n - 3$.

Mirroring the decision rule in Part 1 immediately above, if the restriction under H_0 does not hold and $F_{\Phi_2} < \Phi_2$, then we will not reject H_0 -- $\beta_0 = 0$ and/or $\beta_1 = 0$ and/or $\phi = 0$ (the model has a unit root). If the restriction is binding and $F_{\Phi_2} > \Phi_2$, we are prompted to reject H_0 -- either $\beta_0 \neq 0$ and/or $\beta_1 \neq 0$ and/or $\phi \neq 0$ (no unit root).

3. To test the joint null hypothesis $H_0 : \beta_1 = \phi = 0$ (a unit root and no deterministic trend) against H_1 : not H_0, we use the F-like critical value Φ_3. Alternatively, we have H_0 : (19.10.2) is the true model, H_1 : (19.10.3) is the true model. Thus,

$$\text{Restricted model: } \Delta Y_t = \beta_0 + \phi Y_{t-1} + \varepsilon_t;$$
$$\text{Unrestricted model: } \Delta Y_t = \beta_0 + \beta_1 t + \phi Y_{t-1} + \varepsilon_t.$$

Now the sample F statistic appears as

$$F_{\Phi_3} = \frac{(SSE_R - SSE_U)/r}{SSE_U/(n-k)} \tag{19.14}$$

where $r = 2$ and $n - k = n - 3$.

So if the restriction under H_0 is not in force and $F_{\Phi_3} < \Phi_3$, then we do not reject $H_0 : \beta_1 = 0$ and/or $\phi = 0$ (the model has a unit root). If the restriction is in effect and $F_{\Phi_3} > \Phi_3$, then we will reject H_0, to wit either $\beta_0 \neq 0$ and/or $\phi \neq 0$ (no unit root).

Interestingly enough, this last test is useful in determining whether a time series $\{Y_t\}$ is difference stationary or trend stationary. That is, if we do not reject the null hypothesis ($\beta_1 = 0$ and $\phi = 0$ or $\rho = 1$), then $\{Y_t\}$ is difference stationary, and if we reject the null, then $|\rho| < 1$ so that $\{Y_t\}$ is trend stationary.

The Augmented Dickey-Fuller Test Equation

Thus far, we have taken the ε_t's to be independent and identically distributed random variables. If this assumption is not warranted, then the limiting distributions and critical values obtained by Dickey and Fuller cannot be assumed to hold. However, these limiting distributions and critical values are tenable even when ε_t is autoregressive if what is called the *augmented Dickey-Fuller (ADF) regression* is run. That is, suppose the data-generating process is Equation 19.10.1 with $\phi = 0$ or $\rho = 1$ and that ε_t is a stationary autoregressive process of order p or

$$\begin{aligned}\varepsilon_t &= \theta_1\varepsilon_{t-1} + \theta_2\varepsilon_{t-2} + \cdots + \theta_p\varepsilon_{t-p} + \eta_t \\ &= \sum\nolimits_{i=1}^{p} \theta_i\varepsilon_{t-i} + \eta_t,\end{aligned} \tag{19.15}$$

where $\{\eta_t\}$ constitutes a set of random variables that forms an independent and identically distributed process.

Consider the expression in Equation 19.10.3, where $H_0 : \phi = 0$ is to be tested against $H_1 : \phi < 0$. Then via Equation 19.15,

$$\Delta Y_t = \beta_0 + \beta_1 t + \phi Y_{t-1} + \sum\nolimits_{i=1}^{p} \theta_i\varepsilon_{t-1} + \eta_t, \tag{19.16}$$

which can be rewritten (because Equation 19.10.1 with $\rho = 1$ gives $\varepsilon_t = Y_t - Y_{t-1}$) as

$$\begin{aligned}\Delta Y_t &= \beta_0 + \beta_1 t + \phi Y_{t-1} + \theta_1(Y_{t-1} - Y_{t-2}) \\ &\quad + \theta_2(Y_{t-2} - Y_{t-3}) + \cdots + \theta_p(Y_{t-p} - Y_{t-p-1}) + \eta_t \\ &= \beta_0 + \beta_1 t + \phi Y_{t-1} + \sum\nolimits_{i=1}^{p} \theta_i \Delta Y_{t-i} + \eta_t.\end{aligned} \tag{19.17}$$

Dickey and Fuller demonstrate that, for this ADF regression, the t statistic under $H_0 : \phi = 0$ has the same nonstandard limiting distribution as does the τ_τ statistic. Thus, the critical values for a significance test based on the ADF regression are identical to those associated with the statistic τ_τ.

What value should p assume in Equation 19.17? One can generally start with an a priori chosen upper level for p and then drop the last lagged difference term if it is not statistically significant (based on the ordinary t test[2]). This process is repeated until the last lagged difference term is significant or until the residuals become white noise. Alternatively, a Lagrange multiplier test can be conducted on the residuals to check for a lack of serial correlation and thus for the adequacy of the chosen p.

Given that the ADF equation (Equation 19.17) is deemed appropriate for conducting a unit root test, we can easily modify, say, Equation 19.13 to read

$$F_{\Phi_2} = \frac{(SSE_R - SSE_U)/r}{SSE_U/(n-k-m)}, \tag{19.18}$$

with $r = 3$ and $n - k - m = n - 3 - m$, where m is the number of terms retained in the autoregression expression $\sum_{i=1}^{m} \theta_i \Delta Y_{t-i}$. Equations 19.12 and 19.14 are modified accordingly.

An important area of concern that emerges when testing for a unit root in a series is the issue of the *power of the Dickey-Fuller test* (i.e., the test's ability to detect a *false* null hypothesis). Specifically, the power of the Dickey-Fuller test is considered to be low. That is, because we are trying to ascertain whether a series is $I(1)$ (the null hypothesis) versus $I(0)$ (the alternative hypothesis), $I(0)$ alternatives in which the series may be *close to being* $I(1)$ but not exactly $I(1)$ are highly plausible because the true form of the data-generating process is unknown. Moreover, as evidenced by the Dickey-Fuller tables, the (simulated) critical values are sensitive to the specification of the test equation.

As a practical matter, how does one go about determining whether a time series $\{Y_t\}$ has a unit root? To answer this question, we now offer what can be called a systematic *testing sequence for a unit root*. It consists of the following steps:

1. Step 1: Estimate the ADF Equation 19.17 and, via the results of a conventional F test, retain only the significant augmentation terms $\theta_i \Delta Y_{t-1}$, $i = 1,\ldots,m$. That portion of the ADF equation that remains after the insignificant augmentation terms are eliminated will be termed the *adjusted ADF equation*.
2. Step 2: Estimate the adjusted ADF equation and
   (a) Test $H_0 : \phi = 0$ vs. $H_1 : \phi < 0$ by comparing the t value associated with $\hat{\phi}$ to the critical value τ_τ listed in the Dickey-Fuller tables. If H_0 is rejected, stop; there is no unit root. If H_0 is not rejected, proceed to Part b and test for the presence of a trend.
   (b) Test $H_0 : \beta_1 = \phi = 0$ (i.e., $\beta_1 = 0$ given $\phi = 0$) vs. H_1: not H_0. Here we compare the calculated F statistic (a modification of Equation 19.14)

$$F_{\Phi_3} = \frac{(SSE_R - SSE_u)/r}{SSE_u/(n-k-m)},$$

   where m is the number of augmentation terms retained in the autoregression expression $\sum_{i=1}^{m} \theta_i \Delta Y_{t-1}$, with the F-like critical value Φ_3 determined by Dickey and Fuller [1981]. If H_0 is rejected, proceed to Part c and test for a unit root by using the standard normal distribution. If H_0 is not rejected, proceed to Step 3.

[2] The regression coefficients on the lagged differences have limiting normal distributions. Hence, individual t statistics can be used to determine whether any particular lagged difference can be dropped from the ADF regression, whereas, for large samples, a standard F test can be conducted to see whether any subset of these lagged differences can be omitted from the same.

(c) Test $H_0 : \phi = 0$ against $H_1 : \phi < 0$ by comparing the t value for $\hat{\phi}$ with the appropriate standard normal critical value. Then,

  (i) If H_0 is rejected, stop; there is no unit root. However, we must determine whether or not the Y_t series has a linear trend. To this end, test $H_0 : \beta_1 = 0$ vs. $H_0 : \beta_1 \neq 0$ by comparing the t statistic for $\hat{\beta}_1$ with a standard normal critical value. If H_0 is not rejected, then the Y_t series is stationary with no linear trend. In either instance, conventional hypothesis tests apply to β_0.

  (ii) If H_0 is not rejected, we conclude that $\beta_1 \neq 0$ and there is a unit root.

3. Step 3: Reestimate the adjusted ADF equation without the trend term.

   (a) Test $H_0 : \phi = 0$ vs. $H_1 : \phi < 0$ by comparing the t value corresponding to $\hat{\phi}$ with the critical value τ_μ found in the Dickey-Fuller tables. If H_0 is rejected, stop; there is no unit root. If H_0 is not rejected, proceed to Part b and test for the presence of drift.

   (b) Test $H_0 : \beta_0 = \phi = 0$ (i.e., $\beta_0 = 0$ given $\phi = 0$) versus H_1 : not H_0. We now compare the calculated F value (a modification of Equation 19.12)

$$F_{\Phi_1} = \frac{(SSE_R SSE_u)/r}{SSE_u/(n-k-m)}$$

   to the Dickey-Fuller F-like critical value Φ_1. If H_0 is rejected, proceed to Part c and test for a unit root using the standard normal distribution. If H_0 is not rejected, proceed to Step 4.

   (c) Test $H_0 : \phi = 0$ against $H_1 : \phi < 0$ by comparing the t value for $\hat{\phi}$ with the requisite standard normal critical value. If H_0 is rejected, stop; there is no unit root (the Y_t series is stationary about a nonzero mean). If H_0 is not rejected, we conclude that $\beta_0 \neq 0$ and there is a unit root (the Y_t series is a random walk with drift).

4. Step 4: Reestimate the adjusted ADF equation without an intercept *and* a trend term. Test $H_0 : \phi = 0$ versus $H_1 : \phi < 0$ by comparing the t value for $\hat{\phi}$ with the value τ found in the Dickey-Fuller tables. If H_0 is rejected, stop; there is no unit root (the Y_t series is stationary about a zero mean). If H_0 is not rejected, we conclude that there is a unit root (the Y_t series is a random walk without drift).

5. End.

As an adjunct to this process, suppose that the null hypothesis of a unit root $H_0 : \phi = \rho - 1 = 0$) cannot be rejected. In this circumstance, it is possible that the order of integration of Y_t exceeds zero (or perhaps Y_t is not integrated at all). Hence, our next step is to see if possibly $Y_t : I(1)$. Now, we know that if $Y_t : I(1)$, then $\Delta Y_t : I(0)$. Hence, we must apply the Dickey-Fuller test to ΔY_t instead of to Y_t itself. To this end, our second round test equation is

$$\Delta(\Delta Y_t) = \Delta^2 Y_t = \phi \Delta Y_{t-1} + \varepsilon_t. \tag{19.19}$$

Again we are interested in the negativity of ϕ so that, under $H_0 : \phi = 0$ (ΔY_t has a unit root) versus $H_1 : \phi < 0$ (ΔY_t is stationary), if we reject H_0, then ΔY_t is stationary or $\Delta Y_t : I(0)$ and thus $Y_t : I(1)$.

If the null hypothesis is that ΔY_t has a unit root cannot be rejected, then we must conduct a test to determine whether possibly $Y_t : I(2)$. For $Y_t : I(2)$, $\Delta Y_t : I(1)$ so that we must apply the Dickey-Fuller test to $\Delta^2 Y_t$. Thus, the third round test equation appears as

$$\Delta(\Delta^2 Y_t) = \Delta^3 Y_t = \phi\, \Delta^2 Y_{t-1} + \varepsilon_t.$$

We repeat this process until we determine the order of integration for Y_t or we conclude that $\{Y_t\}$ is not a difference stationary process.

The approach to unit root testing just described is termed a *bottom-up strategy*. Two obvious drawbacks associated with this testing methodology are as follows: (1) it is possible that the Y_t series cannot eventually be transformed (by differencing) to stationarity; and (2) *over differencing* may occur, in which case one reaches the erroneous conclusion that an order of integration higher that the true one exists. This latter outcome is detected by the emergence of high positive values of the Dickey-Fuller test statistic.

To overcome these difficulties, it has been suggested [Dickey and Pantula 1987] that one might pursue a *general-to-specific testing strategy*: one starts from the highest reasonable or informed (suspected) order of integration (possibly $I(2)$) and then proceeds by *testing downward* until stationarity obtains. For more details on this approach, see Appendix 19.D.

Example 19.1

Given the time series dataset presented in Table 19.1 and illustrated in Figure 19.1, test the Y_t variable for the presence of a unit root. (To follow the details of the solution strategy, the reader should review the four-step procedure for finding a unit root outlined above.)

One approach to solving this problem is to use PROC REG and to follow the four step procedure outlined above. To this end, the requisite SAS code appears in Exhibit 19.1.

Exhibit 19.1 SAS Code for Unit Root Test

```
data df1;
   input year y;
datalines;
1900   52.0
1901   53.0
 .      .
 .      .
 .      .
2000   105.0
2001   108.0
;
proc plot data = df1;
   plot y * year='*';
run;
data df1; set df1;
ly=lag(y); ①
dy=dif(y); ②
dly=dif(lag(y));  ③
ddy=dif(dif(y));  ④
t =_n_;⑤
ldy=lag(dy);⑥
l2dy = lag2(dy); ⎫
l3dy = lag3(dy); ⎬ ⑦
l4dy = lag4(dy); ⎭
```

① Returns the lagged value of Y_t or Y_{t-1}.

② Returns the first difference of Y_t or $\Delta Y_t = Y_t - Y_{t-1}$.

③ Returns the first difference of Y_{t-1} or $\Delta Y_{t-1} = Y_{t-1} - Y_{t-2}$.

④ Returns the second difference of Y_t or $\Delta^2 Y_t = \Delta(\Delta Y_t)$.

⑤ Assigns the order of the realizations of Y_t (used to index the passage of time).

⑥ Returns the lagged value of ΔY_t or $\Delta Y_{t-1} = Y_{t-1} - Y_{t-2}$.

⑦ lagn(dy) returns the lagged value of ΔY_t n periods ago, where the lagging period is greater than unity, i.e., lag2(dy) $= \Delta Y_{t-2} = Y_{t-2} - Y_{t-3}$, etc.

⑧ Returns the first difference of ΔY_{t-1} or $\Delta(\Delta Y_{t-1}) = Y_{t-1} - Y_{t-2} - (Y_{t-2} - Y_{t-3})$.

⑨ We invoke the regression procedure.

⑩ We estimate via OLS the ADF regression equation (Equation 19.17)

$$\Delta Y_t = \beta_0 + \phi Y_{t-1} + \beta_1 t + \sum_{i=1}^{4} \theta_i \Delta Y_{t-i} + \eta_t$$

and we conduct a joint (F) test of the significance of the last three augmentation terms ΔY_{t-2}, ΔY_{t-3}, and ΔY_{t-4}.

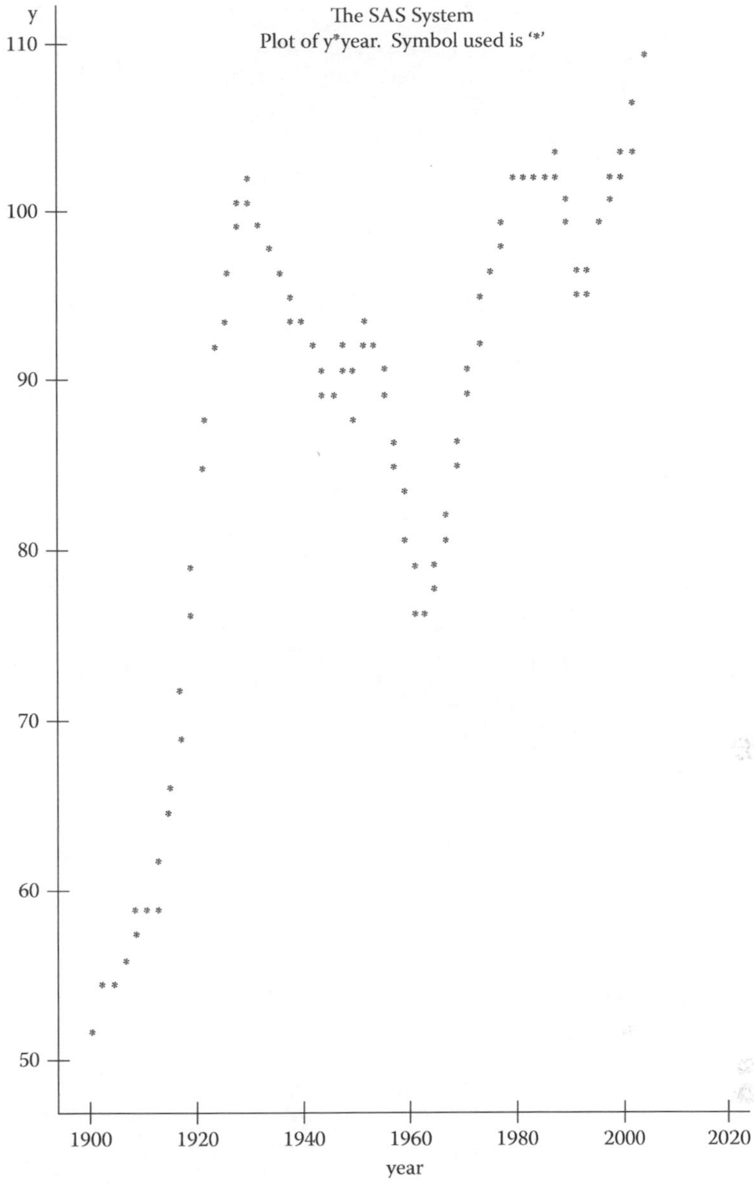

FIGURE 19.1
Plot of Y_t over time.

```
dldy=dif(lag(dy)); ⑧
proc reg data=df1; ⑨
    model dy=ly t ldy
    l2dy l3dy l4dy;
      test1: test l2dy=0,
      l3dy=0, l4dy=0; ⑩
  model dy=ly t ldy;⑪
  model dy=ly ldy;⑫
```

⑪ We estimate via OLS the adjusted ADF equation

$$\Delta Y_t = \beta_0 + \phi Y_{t-1} + \beta_1 t + \theta_1 \Delta Y_{t-1} + \eta_t .$$

⑫ We estimate via OLS the adjusted ADF equation without the deterministic trend.

⑬ We estimate via OLS the adjusted ADF equation without drift and a deterministic trend.

TABLE 19.1

Time Series Observations on Y; Y_t Values, 1900–2001 (n = 102)

Year	Y_t
1900	52.0
1901	53.0
1902	53.8
1903	54.0
1904	55.5
1905	55.0
1906	55.0
1907	56.5
1908	58.0
1909	58.1
1910	58.3
1911	58.1
1912	61.0
1913	64.2
1914	66.0
1915	69.0
1916	72.0
191 7	75.0
1918	78.0
1919	84.0
1920	87.0
1921	91.0
1922	92.1
1923	93.3
1924	95.1
1925	98.1
1926	100.4
1927	101.0
1928	100.0
1929	99.0
1930	98.0
1931	97.0
1932	97.6
1933	96.0
1934	95.5
1935	94.0
1936	93.0
1937	92.7
1938	92.5
1939	92.1
1940	91.2
1941	90.0
1942	89.0
1943	88.2
1944	88.0

(continued)

TABLE 19.1 (*continued*)

Year	Y_t
1945	90.0
1946	91.1
1947	87.8
1948	89.5
1949	92.0
1950	92.2
1951	92.0
1952	91.0
1953	90.0
1954	88.0
1955	86.0
1956	84.5
1957	82.7
1958	80.0
1959	78.0
1960	76.0
1961	75.9
1962	76.0
1963	76.8
1964	78.2
1965	80.0
1966	82.0
1967	84.0
1968	86.0
1969	88.0
1970	90.0
1971	92.0
1972	94.0
1973	95.0
1974	96.0
1975	97.3
1976	99.0
1977	101.0
1978	102.0
1979	101.0
1980	101.0
1981	101.5
1982	101.6
1983	101.5
1984	101.6
1985	102.0
1986	102.5
1987	100.0
1988	98.0
1989	86.0
1990	94.0
1991	94.2

(continued)

TABLE 19.1 (*continued*)

Year	Y_t
1992	96.0
1993	98.0
1994	99.0
1995	100.0
1996	101.0
1997	102.0
1998	102.5
1999	103.0
2000	105.0
2001	108.0

```
   model dy=ly ldy/noint;⑬
   model ddy=dly dldy;⑭
   model ddy=dly;⑮
run;
```

⑭ We estimate via OLS the regression equation

$$\Delta^2 Y_t = \beta_0 + \phi\Delta Y_{t-1} + \theta_1\Delta^2 Y_{t-1} + \eta_t.$$

⑮ We estimate via OLS the regression equation

$$\Delta^2 Y_t = \beta_0 + \phi\Delta Y_{t-1} + \eta_t.$$

The output generated by this SAS code appears in Table 19.2.

Ⓐ① We test $H_0 : \phi = 0$ (the unit root hypothesis), against $H_1 : \phi < 0$. The conventional t statistic associated with $\hat{\phi}$ is –2.83. From the Dickey-Fuller table (Table A.6), $\tau_\tau = -3.45$ for a significance level of 0.05. Clearly, we cannot reject H_0.

② We next test $H_0 : \theta_2 = \theta_3 = \theta_4 = 0$, against H_1 : not H_0. Here the calculated F value is 1.78 (p-value =0.1557). We thus cannot reject H_0 so that the last three augmentation terms will be omitted from the ADF equation. We consequently look to the estimation of the adjusted ADF equation.

Ⓑ① A test of $H_0 : \phi = 0$ versus $H_1 : \phi < 0$ reveals that we cannot reject the null hypothesis of a unit root because the sample t value of –2.22 is not less than the Dickey-Fuller critical value $\tau_\tau = -3.45$ at the 0.05 level of significance. We next test for the presence of a (linear) deterministic trend. Hence, we must compare the

$$\text{Restricted model: } \Delta Y_t = \beta_0 + \phi Y_{t-1} + \theta_1 \Delta Y_{t-1} + \eta_t$$

to the

$$\text{Unrestricted model: } \Delta Y_t = \beta_0 + \phi Y_{t-1} + \beta_1 t + \theta_1 \Delta Y_{t-1} + \eta_t.$$

Here we specify $H_0 : \beta_1 = \phi = 0$ vs. H_1 : not H_0. The regression output for the restricted model appears in Part Ⓒ of Table 19.2; the output for the unrestricted model is presented in Part Ⓑ of the same. From Equation 19.14,

$$F_{\Phi_3} = \frac{(SSE_R - SSE_U)/2}{SSE_U/(n-3-1)} = \frac{(134.035 - 131.766)/2}{131.766/98}$$
$$= 0.8435.$$

Because the Dickey-Fuller critical value of $\Phi_3 = 6.49$ (Table A.7) exceeds F_{Φ_3}, we cannot reject H_0 at the 0.05 level of significance (thus, $\beta_1 = 0$ and/or $\phi = 0$). Hence, we conclude that $\{Y_t\}$ is difference stationary. We now concentrate on the adjusted ADF equation without the trend term. The associated regression output was labeled Part Ⓒ.

TABLE 19.2

Output for Example 19.1

The SAS System

Ⓐ

The REG Procedure
Model : MODEL1
Dependent Variable: dy

Analysis of Variance

Source	**DF**	**Sum of Squares**	**Mean Square**	***F* Value**	**Pr > *F***
Model	6	152.38803	25.39800	18.38	<0.0001
Error	90	124.36702	1.38186		
Corrected total	96	276.75505			

Root MSE	1.17552	*R*-Square	0.5506
Dependent mean	0.54124	Adj *R*-Sq	0.5207
Coeff Var	217.19199		

Parameter Estimates

Variable	**DF**	**Parameter Estimate**	**Standard Error**	***t* Value**	**Pr > |*t*|**
Intercept	1	2.62818	0.92391	2.84	0.0055
ly	1	$-0.03553 = \hat{\phi}$	0.01254	−2.83①	0.0057
t	1	0.01147	0.00597	1.92	0.0578
ldy	1	$0.58834 = \hat{\theta}_1$	0.10294	5.72	<0.0001
l2dy	1	$0.01136 = \hat{\theta}_2$	0.11945	0.10	0.9244
l3dy	1	$0.10692 = \hat{\theta}_3$	0.11969	0.89	0.3741
l4dy	1	$0.11545 = \hat{\theta}_4$	0.10465	1.10	0.2729

Test test1 Results for Dependent Variable dy

Source	**DF**	**Mean Square**	***F* Value**	**Pr > *F***
Numerator	3	2.46645	1.78②	0.1557
Denominator	90	1.38186		

Ⓑ

The REG Procedure
Model: MODEL2
Dependent Variable: dy

Analysis of Variance

Source	**DF**	**Sum of Squares**	**Mean Square**	***F* Value**	**Pr > *F***
Model	3	144.98867	48.32956	34.11	<.0001
Error	93	131.76638	1.41684		
Corrected total	96	276.75505			

(continued)

TABLE 19.2 *(continued)*

Root MSE	1.19031	*R*-Square	0.5239
Dependent mean	0.54124	Adj *R*-Sq	0.5085
Coeff Var	219.92432		

Parameter Estimates

Variable	DF	Parameter Estimate	Standard Error	*t* Value	Pr > |*t*|
Intercept	1	2.14277	0.90870	2.36	0.0205
ly	1	$-0.02679 = \hat{\phi}$	0.01207	−2.22j	0.0288
t	1	0.00727	0.00574	1.27	0.2089
ldy	1	0.70534	0.07347	9.60	<0.0001

©
The REG Procedure
Model: MODEL3
Dependent Variable: dy
Analysis of Variance

Source	DF	Sum of Squares	Mean Square	*F* Value	Pr > *F*
Model	2	142.72036	71.36018	50.05	<0.0001
Error	94	134.03470	1.42590		
Corrected total	96	276.75505			

Root MSE	1.19411	*R*-Square 0.5157
Dependent mean	0.54124	Adj *R*-Sq 0.5054
Coeff Var	220.62622	

Parameter Estimates

Variable	DF	Parameter Estimate	Standard Error	*t* Value	Pr > |*t*|
Intercept	1	1.67216	0.83175	2.01	0.0473
ly	1	$-0.01695 = \hat{\phi}$	0.00925	−1.83①	0.0702
ldy	1	0.69408	0.07316	9.49	<0.0001

Ⓓ
The REG Procedure
Model: MODEL4
Dependent Variable: dy

Note: No intercept in model. R-Square is redefined.

Analysis of Variance

Source	DF	Sum of Squares	Mean Square	*F* Value	Pr > *F*
Model	2	165.37210	82.68605	56.19	< 0.0001
Error	95	139.79790	1.47156		
Corrected total	97	305.17000			

TABLE 19.2 (*continued*)

Root MSE	1.21308	*R*-Square	0.5419
Dependent mean	0.54124	Adj *R*-Sq	0.5323
Coeff Var	224.13050		

Parameter Estimates

Variable	DF	Parameter Estimate	Standard Error	*t* Value	Pr > |*t*|
ly	1	$0.00144 = \hat{\phi}$	0.00144	1.00	0.3202
ldy	1	$0.72088 = \hat{\theta}_1$	0.07308	9.86 ①	<0.0001

Ⓔ
The REG Procedure
Model: MODEL5
Dependent Variable: ddy
Analysis of Variance

Source	DF	Sum of Squares	Mean Square	*F* Value	Pr > *F*
Model	2	23.28959	11.64479	7.93	0.0007
Error	94	137.95722	1.46763		
Corrected total	96	161.24680			

Root MSE	1.21146	*R*-Square	0.1444
Dependent mean	0.01546	Adj *R*-Sq	0.1262
Coeff Var	7834.09365		

Parameter Estimates

Variable	DF	Parameter Estimate	Standard Error	*t* Value	Pr > |*t*|
Intercept	1	0.15544	0.12975	1.20	0.2339
dly	1	$-0.26343 = \hat{\phi}$	0.07987	−3.30①	0.0014
dldy	1	$-0.07925 = \hat{\theta}_1$	0.10344	−0.77	0.4455

Ⓕ
The REG Procedure
Model : MODEL6
Dependent Variable: ddy
Analysis of Variance

Source	DF	Sum of Squares	Mean Square	*F* Value	Pr > *F*
Model	1	22.42809	22.42809	15.35	0.0002
Error	95	138.81872	1.46125		
Corrected total	96	161.24680			

Root MSE	1.20882	*R*-Square	0.1391
Dependent mean	0.01546	Adj *R*-Sq	0.1300
Coeff Var	7817.04633		

(continued)

TABLE 19.2 *(continued)*

Parameter Estimates					
Variable	**DF**	**Parameter Estimate**	**Standard Error**	***t* Value**	**Pr > |*t*|**
Intercept	1	0.16655	0.12865	1.29	0.1986
dly	1	$-0.28736 = \hat{\phi}$	0.07335	−3.921	0.0002

ⒸⒸ① We test $H_0 : \phi = 0$ against $H_1 : \phi < 0$. The sample t value for $\hat{\phi}$ is –1.87. Given the Dickey-Fuller critical value of $\tau_\mu = -\ 2.89$ (Table A.6), we conclude that we cannot reject the null hypothesis at the 0.05 significance level. We next test for the presence of drift. To this end, we must compare the

$$\text{Restricted model: } \Delta Y_t = \phi Y_{t-1} + \theta_1 \Delta Y_{t-1} + \eta_t$$

to the

$$\text{Unrestricted model: } \Delta Y_t = \beta_0 + \phi Y_{t-1} + \theta_1 \Delta Y_{t-1} + \eta_t$$

by testing $H_0 : \beta_0 = \phi = 0$, vs. H_1 : not H_0. The requisite output for the restricted model appears in Part Ⓓ of Table 19.2, whereas the output for the unrestricted model is again provided by Part Ⓒ. From Equation 19.12,

$$F_{\Phi_1} = \frac{(SSE_R - SSE_U)/2}{SSE_U/(n-2-1)} = \frac{(139.798 - 134.035)/2}{134.035/99}$$
$$= 2.130.$$

Because the Dickey-Fuller critical value of $\Phi_1 = 4.71$ (Table A.7) exceeds F_{Φ_1}, we are not able to reject H_0 at the 0.05 significance level (thus, $\beta_1 = 0$ and/or $\phi = 0$).

Ⓓ① The Part Ⓓ regression output results from the estimation of the adjusted ADF equation without an intercept and a deterministic time trend term. Note that, in this instance $\hat{\phi}$(= 0.00144), is positive (although insignificant) rather than negative, whereas $\hat{\theta}_1$ is highly significant (its p-value <0.0001). Because the augmentation term ΔY_{t-1} provides virtually all of the explanatory power for this regression equation, it is clear that the null hypothesis of a unit root for $\{Y_t\}$ cannot be rejected. Our next step is to see whether $Y_t : I(1)$ by applying the Dickey-Fuller test to ΔY_t (and not to Y_t) by fitting, via OLS, the regression equation $\Delta(\Delta Y_t) = \beta_0 + \phi \Delta Y_{t-1} + \theta_1 \Delta(\Delta Y_{t-1}) + \eta_t$ (see Equation 19.19). The regression is depicted by Part Ⓔ of Table 19.2.

Ⓔ① We test $H_0 : \phi = 0$ (ΔY_t has a unit root), versus $H_1 : \phi < 0$(ΔY_t is stationary). The calculated t value for $\hat{\phi}$ is –3.30. Because the Dickey-Fuller critical value at the 0.05 significance level is $\tau_\mu = -\ 2.89$, we can now reject the null hypothesis $\Delta Y_t : I(0)$ and thus $Y_t : I(1)$, *i.e.*, Y_t has a unit root and thus is difference stationary. Because the augmentation term is not statistically significant at any reasonable probability level (the p-value for $\hat{\theta}_1$ is 0.4455), one may want to take this analysis a step further and fit the preceding regression equation without $\Delta^2 Y_{t-1}$ or simply fit $\Delta(\Delta Y_t) = \beta_0 + \phi \Delta Y_{t-1} + \eta_t$. As is readily seen in Part ⒻF of Table 19.2, our conclusion regarding Y_t being difference stationary is strengthened a bit; now the t value associated with $\hat{\phi}$ is –3.92.

Example 19.2

The four-step methodology used in Example 19.1 to test a time series for the presence of a unit root was essentially one of "brute force" (notwithstanding the fact that it was quite transparent and indicative of the type of reasoning involved in uncovering unit roots). This and the next example problem uses an "automatic" method for determining whether a series is difference stationary. The SAS code, which incorporates the Example 19.1 dataset, appears in Exhibit 19.2.

Exhibit 19.2 SAS Code for %DFTEST

```
data df1;
      input year y;
datalines;
1900   52.0
1901   53.0
1902   53.8
  .     .
  .     .
  .     .
2000   105.0
2001   108.0
;
data df1; set df1;
%dftest (df1, y, trend=2, outstat=dfstat1);  ①
%put p=&dftest;  ②
proc print data=dfstat1;
run;
%dftest (df1, y, dif=(1), ar=1, outstat=dfstat2);  ③
%put p=&dftest;
proc print data=dfstat2;
run;
```

① We invoke the *%DFTEST* macro. The two required arguments are (1) the name of the SAS dataset containing the time series variable to be tested for a unit root; and (2) the name of the time series variable itself. The basic form of this macro is thus

%dftest (df1, y [, *options*]);

where the various options, which must be inserted after the two required arguments and separated by commas, are as follows:

DIF= (differencing-list): specifies the degree of differencing to be applied to Y_t, e.g.,
DIF = (1): requests ΔY_t, i.e., the Y_t series is to be differenced once at lag 1.
DIF = (1,1): requests $\Delta^2 Y_t$, i.e., the Y_t series is differenced once at lag 1 and then differenced again at lag 1.
DLAG = |1|2|4|12: specifies the lag to be tested for a unit root (default is DLAG=1).
AR = *n*: specifies the order of the autoregressive model to be fit after any differencing is requested by the DIF = and DLAG= options (default is AR = 3).
OUT = SAS–data–set: writes residuals to an output dataset.
OUTSTAT = SAS–data–set: writes the test statistic, parameter estimates, and other statistics to an output dataset.
TREND = 0|1|2: specifies the status of a deterministic linear time trend included in the model:
TREND = 0: assumes a zero mean and no linear time trend.
TREND = 1: includes an intercept term.
TREND = 2: includes an intercept and a linear time trend.
The default is TREND = 1.

a. This first test is applied to Y_t. The test equation includes an intercept, a linear trend term, and three autoregressive terms.

b. OUTSTAT = dfstat1 names the output dataset.

② %PUT *p*= *&DFTEST* is the macro variable that defines the computed *p*-value for the Dickey-Fuller-type critical value.

③ The second test is applied to ΔY_t. The test equation includes an intercept term along with one autoregressive term.

Only an abridged version of the SAS output is presented below. Its salient features are as follows:

TEST 1

$H_0 : Y_t$ has a unit root; H_1 : not H_0.

TAU	_PVALUE_
–2.7974	0.20178

We cannot reject H_0 - - Y_t is nonstationary.

TEST 2

$H_0 : \Delta Y_t$ has a unit root; H_1 : not H_0.

TAU	_PVALUE_
–3.32740	0.01627

We reject H_0 - - ΔY_t is stationary or $\Delta Y_t : I(0)$ and thus $Y_t : I(1)$; Y_t has a unit root and is difference stationary.

Example 19.3

Our next approach for unit root detection is to use the STATIONARITY option in the *ARIMA* procedure. Again the Example 19.1 data series will be used. The requisite SAS code is provided in Exhibit 19.3.

Exhibit 19.3. SAS Code for STATIONARITY Test

```
 data df1;
       input year y;
datalines;
1900 52.0
1901 53.0
1902 53.8
  .    .
  .    .
  .    .
2000 105.0
2001 108.0
;
proc arima data=df1;  ①
         ②       ③                     ④
       identify var = y stationarity = (adf = (1));
run;
proc arima data=df1;
```

① We invoke the *ARIMA* procedure.

② The *IDENTIFY* statement reads the time series to be modeled, differences the series if necessary, and computes a variety of test statistic needed to help identify models to be estimated.

③ The *VAR=* option specifies the variable to be modeled. For the first test, *VAR* = *Y*.

```
                    ⑤
      identify var = y(1) stationarity = (adf = (1));
run;
```

④ The *STATIONARITY* = option performs unit root tests. We choose the augmented Dickey-Fuller test with one autoregressive term (*ADF* = (1)). (If, say, four autoregressive terms were warranted, then we would write ADF = (1,2,3,4).)

⑤ We next test ΔY_t for a unit root by specifying *VAR* = Y(1), which renders the first difference of the Y_t series. (If $\Delta^2 Y_t$ were to be tested for a unit root, then we would write *VAR* = Y(1,1), which responds with the second difference of Y_t.)

The (condensed) output generated by these two stationarity tests appears in Table 19.3.

Ⓐ① The residuals determined up to a specific lag are checked to see if they constitute a white noise process. Here we test H_0 : all of the autocorrelations of the residual series up to a given lag are zero, against H_1 : not H_0. If we do not reject H_0 for all lags, then an ARIMA model is not appropriate for the Y_t series. (The *NLAG* = option controls the number of lags for which autocorrelations are calculated. The default is *NLAG* = 24.) Because each of the *p*-values for the sample χ^2 statistics is quite small, we reject H_0 in favor of H_1 and conclude that the Y_t series is not stationary. Hence, the current models do not adequately explain the Y_t data series.

TABLE 19.3

Output for Example 19.3

The SAS System

Ⓐ

The ARIMA Procedure

Name of variable = y

Mean of working series	87.07549
Standard deviation	14.7346
Number of observations	102

The SAS System

The ARIMA Procedure

①

Autocorrelation Check for White Noise

To lag	χ^2	DF	Pr > χ^2	Autocorrelations					
6	432.17	6	<0.0001	0.955	0.905	0.850	0.789	0.724	0.654
12	550.58	12	<0.0001	0.580	0.506	0.432	0.360	0.291	0.222
18	556.66	18	<0.0001	0.157	0.096	0.039	−0.015	−0.062	−0.102
24	582.81	24	<0.0001	−0.137	−0.162	−0.181	−0.192	−0.200	−0.207

②

Augmented Dickey-Fuller Unit Root Tests

Type	Lags	Rho	Pr < Rho	Tau	Pr < Tau	*F*	Pr > *F*
Zero mean	1	0.5349	0.8127	1.06	0.9235		
Single mean	1	−4.8923	0.4375	−1.80	0.3800	2.59	0.4192
Trend	1	−8.4694	0.5308	−2.19	0.4869	2.45	0.6903

(continued)

TABLE 19.3 *(continued)*

Ⓑ

The ARIMA Procedure

Name of variable = y(1)

Period(s) of differencing	1
Mean of working series	0.554455
Standard deviation	1.659209
Number of observations	101
Observation(s) eliminated by differencing	1

①

Autocorrelation Check for White Noise

To lag	χ^2	DF	Pr > χ^2	Autocorrelations					
6	141.97	6	<0.0001	0.695	0.526	0.463	0.409	0.344	0.289
12	154.29	12	<0.0001	0.194	0.098	–0.035	–0.143	–0.115	–0.163
18	182.75	18	<0.0001	–0.179	–0.236	–0.226	–0.214	–0.162	–0.155
24	190.18	24	<0.0001	–0.098	–0.112	–0.138	–0.115	–0.044	0.033

②

Augmented Dickey-Fuller Unit Root Tests

Type	Lags	Rho	Pr < Rho	Tau	Pr < Tau	F	Pr > F
Zero mean	1	–20.9199	0.0009	–3.07	0.0024		
Single mean	1	–24.1880	0.0027	–3.33	0.0163	5.56	0.0254
Trend	1	–24.5744	0.0190	–3.29	0.0744	5.51	0.0983

② Three separate augmented Dickey-Fuller tests are performed using the following specifications: a zero mean; a single mean (or the presence of drift); and the presence of drift along with a deterministic linear time trend. In each case, the p-value associated with the appropriate τ statistic is too large to reject the null hypothesis of a unit root. Thus, the Y_t series is nonstationary.

Ⓑ① Given that each of the computed χ^2 values has a very small p-value, we reject the null hypothesis that the residuals of the ΔY_t process are white noise.

② The p-values associated with the τ statistics for the zero mean and single mean cases are both below 0.05, indicating that the rejection of the null hypothesis of a unit root for ΔY_t is warranted. Hence, we conclude that $\Delta Y_t : I(0)$ (ΔY is stationary) and thus $Y_t : I(1)$ - -Y_t has a unit root and is difference stationary.

The next example problem addresses the issue of testing to determine whether two time series variables X_t and Y_t are cointegrated. Although the properties of cointegrated processes were discussed in detail in Appendix 19.C, a brief review of some of the salient features of cointegration is now offered.

Given a set of nonstationary variables that are integrated of the same order, these variables are said to be cointegrated if their linear combination is stationary or $I(0)$. Hence, cointegrated variables have the same stochastic trend [Stock and Watson 1988] and are connected by a long-run equilibrium relationship that precludes their moving independently of each other. Because the time paths of cointegrated variables are linked, these time paths are related to the current deviation from the equilibrium equation; we should expect a connection between a change in the variables X_t and Y_t and the deviation ε_t from their long-run equilibrium track. If the said equilibrium is actual, then $\{\varepsilon_t\}$ must be stationary and thus any deviation from this equilibrium is only temporary. So if, for instance, X_t and Y_t are each $I(1)$ and $\varepsilon_t = Y_t - \beta_0 - \beta_1 X_t : I(0)$, then $X_t, Y_t : CI(1,1)$.

How do we actually execute a test of whether or not X_t and Y_t are cointegrated? Following Engle and Granger [1987], the test procedure involves two steps:

(1) Determine whether the variables X_t and Y_t are individually integrated of the same order (typically both X_t and Y_t are $I(1)$).

(2) Test to determine whether the residuals e_t of the *cointegrating regression* $Y_t = \hat{\beta}_o + \hat{\beta}_1 X_t + e_t$ possess a unit root. Here H_0: $\{e_t\}$ has a unit root (or X_t and Y_t are not cointegrated) is tested against H_1 : not H_0.

If H_0 is rejected, then the residuals are stationary and thus X_t and Y_t are cointegrated; they share a stochastic trend.[3]

More specifically, if it is determined that X_t and Y_t are $I(1)$ with $e_t : I(0)$, then $X_t, Y_t : CI(1, 1)$ with cointegrating vector $(1, -\beta_1)$. However, this is not the complete story. We know that the short-run behavior of the model is influenced by the magnitude of the departure from long-run equilibrium as represented by an *error-correction mechanism* (ECM) that incorporates the previous period's deviation from equilibrium. Hence, short-run dynamic or disequilibrium performance can be modeled by using long-run parameter estimates to structure an ECM that relates movements in X_t and Y_t to the previous period's departure from long-run equilibrium. In this regard, the ECM has the form $e_{t-1} = Y_{t-1} - \hat{Y}_{t-1}$ and is included in the set of regression equations

$$\begin{aligned}&\text{(a) } \Delta Y_t = \alpha_0 + \alpha_1 e_{t-1} + lagged(\Delta Y_t, \Delta X_t) + \eta_Y \\ &\text{(b) } \Delta X_t = \gamma_0 + \gamma_1 e_{t-1} + lagged(\Delta Y_t, \Delta X_t) + \eta_X\end{aligned} \tag{19.20}$$

or

$$\begin{aligned}&\text{(a) } \Delta Y_t = \alpha_0 + \alpha_1 e_{t-1} + \sum_{j=1}^{r} \alpha_{1j} \Delta Y_{t-j} + \sum_{j=1}^{s} \alpha_{2j} \Delta X_{t-j} + \eta_{Yt} \\ &\text{(b) } \Delta X_t = \gamma_0 + \gamma_1 e_{t-1} + \sum_{j=1}^{r} \gamma_{1j} \Delta Y_{t-j} + \sum_{j=1}^{s} \gamma_{2j} \Delta X_{t-j} + \eta_{Xt}\end{aligned} \tag{19.20.1}$$

where η_{Yt} and η_{Xt} are taken to be white noise. Here changes in, say, Y_t are explained by its own past behavior, lagged changes in X_t, and the previous period's deviation from the long-run equilibrium path. Hence, cointegrated series have an ECM representation, and, conversely, for an ECM to hold in $I(1)$ variables, these variables must be cointegrated or ECMs generate cointegrated series (the *Engle-Granger Representation Theorem*).

Note that the general ECM specification involves two expressions because, typically, each of the cointegrated variables will respond or adjust to a deviation from the long-run equilibrium relation. Here we expect $\alpha_1 < 0$ and/or $\gamma_1 < 0$ because otherwise the system would diverge from its long-run equilibrium path. In addition, for a given e_{t-1}, the larger is the speed of adjustment $\alpha_1(\gamma_1)$, the greater is the response of $Y_t(X_t)$ to the previous departure from long-run equilibrium. Now, it is possible that either α_1 or γ_1 (but not both) is zero. If, say, $\gamma_1 = 0$ then X_t does not respond to the previous period's departure from long-run equilibrium and thus Y_t does all of the adjustment. In this instance, X_t is said to be *weakly exogenous*. Hence, the ECM can be adequately modeled by using only Equation 19.20.a (or Equation 19.20.1.a).

---

[3] If the long-run components of the X_t and Y_t series are modeled as stochastic trends and if they move together, then these series should be cointegrated. In this regard, the Engle-Granger procedure can be restated as

1. Testing to see whether X_t and Y_t have a stochastic trend (via an ADF unit root test).
2. Testing to see whether the stochastic trends are related by determining whether the residuals e_t from the long-run relationship between X_t and Y_t (the cointegrating regression) have a stochastic trend (also via an ADF unit root test). If $e_t : I(0)$ given that X_t and Y_t are $I(1)$, then X_t and Y_t are cointegrated.

This last observation brings to the fore the concept of *Granger causality*, i.e., in a cointegrated system, Y_t does not Granger cause X_t if the lagged terms ΔY_{t-j} do not enter the ΔX_t equation and ΔX_t does not respond to the previous period's departure from long-run equilibrium (i.e., $\gamma_1 = 0$ and all γ_{1j}'s = 0). Clearly, X_t must be weakly exogenous. Looked at in another manner, if the presence of Y_t does not enhance our ability to forecast X_t, then Y_t does not Granger cause X_t. So if $X_t, Y_t : CI(1,1)$, Granger causality must exist in at least one direction; at least one variable must help forecast the other.

Remember that a stationary error ε_t in a regression model relating integrated variables involves a long-run equilibrium relationship between them, and, conversely, in the absence of any such relationship, there is no vehicle for joining the X_t and Y_t series in an estimated linear equation so that the residuals (which measure equilibrium error) would be nonstationary. In summary, we start by estimating the long-run relation between X_t and Y_t and then insert the lagged deviations e_{t-1} as an ECM in the short-run regression Equations 19.20 (or Equation 19.20.1), provided, of course, that $\alpha_1 < 0$ and/or $\gamma_1 < 0$.

So if X_t and Y_t are $I(1)$ and cointegrated, then regressing Y_t on X_t in levels makes no sense. What does make sense are the regressions specified by Equation 19.20 (or Equation 19.20.1)) because now all variables are $I(0)$ (which means that $\eta_{Yt} : I(0)$ and $\eta_{Xt} : I(0)$) and no loss of long-run information occurs. (Such information would be lost if our model involved only differences of variables.)

Example 19.4

Given the X_t and Y_t series provided in Table 19.4, determine whether these variables are $I(1)$ and cointegrated. If cointegration holds, estimate the parameters of Equation 19.20.1 and test for Granger causality. Exhibit 19.4 houses the SAS code.

Exhibit 19.4 SAS Code for Testing if X and Y Are I(1) and Cointegrated

```
data df2;
      input year y x;
dx = dif(x);
dy = dif(y);①
ldy=lag(dy);  ②
l2dy=lag2(dy);
l3dy=lag3(dy);  ③
l4dy=lag4(dy);
ldx=lag(dx);  ④
l2dx=lag2(dx);
l3dx=lag3(dx);  ⑤
l4dx=lag4(dx);
datalines;
1900  52.0  50.0
1901  53.0  53.1
1902  53.8  56.0
  .     .     .
  .     .     .
  .     .     .
2000 105.0 100.3
2001 108.0 105.0
;
```

① We create the first differences ΔX_t, ΔY_t.

② Returns the lagged value of ΔY_t

③ lagn(dY) returns the lagged value of ΔY_t n periods ago, where n>1.

④ Returns the lagged value of ΔX_t.

⑤ lagn(dX) returns the lagged value of ΔX_t n periods ago, where n>1.

⑥ We invoke the *ARIMA* procedure.

⑦ The *IDENTIFY* statement reads the time series to be modeled, differences the series if required, and computes an assortment of test statistics used to identify models to be estimated.

⑧ The *VAR*= option specifies the variable to be modeled.
For the first test, *VAR*=Y.

⑨ The *STATIONARITY*= option performs unit root tests. We choose the ADF test with four autoregressive terms (*ADF*=(1,2,3,4)).

TABLE 19.4

Time Series Observations on X and Y; X_t, Y_t Values, 1900–2001 (n = 102)

Year	Y_t	X_t
1900	52.0	50.0
1901	53.0	53.1
1902	53.8	56.0
1903	54.0	56.7
1904	55.5	58.0
1905	55.0	59.0
1906	55.0	59.0
1907	56.5	60.5
1908	58.0	60.0
1909	58.1	61.0
1910	58.3	61.1
1911	58.1	62.0
1912	61.0	61.8
1913	64.2	62.0
1914	66.0	63.0
1915	69.0	65.0
1916	72.0	66.0
1917	75.0	68.1
1918	78.0	70.0
1919	84.0	72.0
1920	87.0	74.1
1921	91.0	75.0
1922	92.1	80.3
1923	93.3	81.0
1924	95.1	82.1
1925	98.1	86.0
1926	100.4	87.0
1927	101.0	85.5
1928	100.0	87.8
1929	99.0	86.0
1930	98.0	87.0
1931	97.0	83.1
1932	97.6	84.0
1933	96.0	83.2
1934	95.5	84.0
1935	94.0	82.7
1936	93.0	82.3
1937	92.7	84.1
1938	92.5	80.0
1939	92.1	79.0
1940	91.2	83.7
1941	90.0	79.5
1942	89.0	81.1
1943	88.2	79.0
1944	88.0	77.0

(continued)

TABLE 19.4 *(continued)*

Year	Y_t	X_t
1945	90.0	78.0
1946	91.1	81.1
1947	87.8	79.0
1948	89.5	81.5
1949	92.0	83.0
1950	92.2	83.9
1951	92.0	84.0
1952	91.0	82.5
1953	90.0	83.3
1954	88.0	84.1
1955	86.0	80.0
1956	84.5	78.5
1957	82.7	79.9
1958	80.0	75.6
1959	78.0	76.0
1960	76.0	77.0
1961	75.9	74.1
1962	76.0	72.0
1963	76.8	76.1
1964	78.2	78.0
1965	80.0	80.0
1966	82.0	83.0
1967	84.0	86.0
1968	86.0	86.5
1969	88.0	87.0
1970	90.0	88.0
1971	92.0	90.1
1972	94.0	88.3
1973	95.0	89.5
1974	96.0	89.7
1975	97.3	91.0
1976	99.0	92.4
1977	101.0	93.0
1978	102.0	91.3
1979	101.0	91.0
1980	101.0	88.0
1981	101.5	87.2
1982	101.6	89.0
1983	101.5	90.1
1984	101.6	90.5
1985	102.0	91.0
1986	102.5	92.0
1987	100.0	91.0
1988	98.0	90.0
1989	86.0	86.0
1990	94.0	87.2

TABLE 19.4 *(continued)*

Year	Y_t	X_t
1991	94.2	87.5
1992	96.0	88.0
1993	98.0	86.0
1994	99.0	90.0
1995	100.0	91.0
1996	101.0	93.3
1997	102.0	94.0
1998	102.5	96.5
1999	103.0	99.0
2000	105.0	100.3
2001	108.0	105.0

```
proc arima data=df2; ⑥
      ⑦         ⑧                  ⑨
   identify var=y stationarity=(adf=(1,2,3,4));
run;
proc arima data=df2;
                ⑩
   identify var=y(1) stationarity=(adf=(1));
run;
proc arima data=df2;
                ⑪
   identify var=x stationarity=(adf=(1,2,3,4));
run;
proc arima data=df2;
                ⑫
   identify var=x(1) stationarity=(adf=(1));
run;
proc reg data=df2;
   model y=x;      ⑬
   output out=out1 r=r;  ⑭
data rdf2; set out1;  ⑮
lr=lag(r);  ⑯
run;
proc arima data=rdf2;
                ⑰
   identify var=r stationarity=(adf=(1,2,3,4));
run;
proc arima data=rdf2;
                ⑱
   identify var=r(1) stationarity=(adf =(1));
run;
data mdf2; merge df2 rdf2;  ⑲
ecm=lr;      ⑳
proc reg data=mdf2;
   model dy= ecm ldy l2dy l3dy l4dy ldx l2dx l3dx l4dx;㉑
   test1: test ldx=0, l2dx=0, l3dx=0, l4dx=0;㉒
   model dx= ecm ldy l2dy l3dy l4dy ldx l2dx l3dx l4dx;㉓
   test2: test ldy=0, l2dy=0, l3dy=0, l4dy=0;㉔
run;
```

⑩ We next test ΔY_t for a unit root by specifying *VAR*=Y(1), which produces the first difference in the Y_t series.

⑪ We test *VAR*=X for a unit root using four autoregressive terms.

⑫ We test ΔX_t for a unit root by specifying *VAR*=X(1), which produces the first difference of the X_t series.

⑬ We use PROC REG to estimate the parameters of the long-run equilibrium relationship between X_t and Y_t.

⑭ The output dataset housing the regression output is named *OUT*1; it contains the set of regression residuals *r* along with other information.

⑮ We create the *SAS* dataset *rdf*2 that contains all of the information appearing in *OUT*1.

⑯ We form the series of lagged residuals r_{t-1}.

⑰ We use PROC ARIMA to test the residuals r_t for a unit root by specifying *VAR*=r; we choose the ADF test with four autoregressive terms.
⑱ We next test Δr_t for a unit root by specifying *VAR*=r(1), which produces the first differences of the r_t series.
⑲ We merge *SAS* datasets *df2* and *rdf2* to obtain the *SAS* dataset *mdf2*.
⑳ We specify the values of the error-correction variable (*ecm*) to be the set of lagged residuals.
㉑ PROC REG uses the *mdf2* dataset to estimate Equation 19.20.1.a.
㉒ We test $H_0{:}\alpha_{2j} = 0$. $j = 1.2,3,4$, against H_1: not H_0, i.e., we conduct an *F* test to see whether Y_t is weakly exogenous.
㉓ PROC REG uses the *mdf2* dataset to estimate Equation 19.20.1.b.
㉔ We test $H_0{:}\gamma_{1j} = 0$, $j = 1, 2, 3, 4$, against H_1: not H_0, i.e., an *F* test is used to determine whether X_t is weakly exogenous.

The output generated by the preceding set of SAS code statements appears in Table 19.5.

Ⓐ①,② For the Y_t series, the *p*-values associated with the appropriate ADF τ statistics are much too large to warrant rejecting the null hypothesis of a unit root; for the ΔY_t series, the *p*-values associated with the τ statistics for the zero mean and single mean cases are both below 0.05. Hence, we can reject the null hypothesis of a unit root for ΔY_t so that $\Delta Y_t : I(0)(\Delta Y$ is stationary) and thus $Y_t : I(1)$.

Ⓑ①,② For the X_t series, the *p*-values for the various τ statistics are all too large to call for the rejection of the null hypothesis of a unit root; for the ΔX_t series, the *p*-value for each of the ADF τ statistics is far below 0.05, thus prompting us to reject the null hypothesis of a unit root for ΔX_t. Thus, $\Delta X_t : I(0)$ so that $X_t : I(1)$.

Ⓒ An estimate of the long-run equilibrium relationship between X_t and Y_t is $\hat{Y} = \hat{\beta}_0 + \hat{\beta}_1 X = -13.2452 + 1.2470X$. A glance at the *F*, R^2, and *p*-values for the regression coefficients indicates that the overall fit is quite good.

Ⓓ①,② For the r_t series, the *p*-values associated with the printed τ statistics are too large for us to reject the null hypothesis of a unit root; looking to the Δr_t series, the *p*-value for each of the ADF τ's is well below 0.05 so that we may reject the null hypothesis of a unit root for Δr_t. Hence, $\Delta r_t : I(0)$ and thus $r_t : I(1)$.[4]

Ⓔ① The estimated speed of adjustment $\hat{\alpha}_1 = -0.0623$ is statistically significant (*p*-value = 0.0494) and has the correct sign, thus indicating that the error of the previous period is corrected by a factor of approximately 0.0623 through an adjustment in Y_t.

② Given the large *p*-value (=0.8176) for the reported *F* statistic, we cannot reject the null hypothesis that Y_t is weakly exogenous: the ΔX_{t-j} terms do not significantly enter into the ΔY_t Equation 19.20.1.a. Thus, X_t does not Granger cause Y_t; current and past values of X_t do not help us forecast future values of Y_t.

Ⓕ① The estimated speed of adjustment $\hat{\gamma}_1 = 0.0218$ is not statistically significant (*p*-value =0.6432) and has the wrong sign. Hence, all the "heavy lifting" with respect to adjusting to departures from long-run equilibrium is undertaken by Y_t alone: X_t is unresponsive in this regard.

② Because the *p*-value (=0.0073) for the realized *F* statistic is very small, we reject the null hypothesis that X_t is weakly exogenous: ΔY_{t-j} terms enter significantly into the ΔX_t Equation 19.20.1.b. Hence, Y_t can be said to Granger cause X_t; knowledge of current and past Y_t values improves our ability to forecast future X_t's.

---

[4] Strictly speaking, because the residuals r_t are determined from a regression equation (the true errors ε_t are unknown), we cannot legitimately use the Dickey-Fuller critical values. Instead, we must use so-called critical values for (Engle-Granger) cointegration tests appearing in Table A.8 (see Engle and Yoo 1987; Phillips and Ouliaris 1990; Mackinnon 1991). For this particular two-variable case with $n = 102$, the appropriate critical value is -3.398 at the 0.05 level of significance. Given that the resulting τ for the ADF test with a single mean is –8.77, our basic conclusion involving the rejection of the null hypothesis of a unit root still stands.

TABLE 19.5

Output for Example 19.4

The SAS System

Ⓐ

The ARIMA Procedure

Name of variable = y

Mean of working series	87.07549
Standard deviation	14.7346
Number of observations	102

Autocorrelation Check for White Noise

To lag	χ^2	DF	Pr > χ^2	Autocorrelations					
6	432.17	6	<.0001	0.955	0.905	0.850	0.789	0.724	0.654
12	550.58	12	<.0001	0.580	0.506	0.432	0.360	0.291	0.222
18	556.66	18	<.0001	0.157	0.096	0.039	–0.015	–0.062	–0.102
24	582.81	24	<.0001	–0.137	–0.162	–0.181	–0.192	–0.200	–0.207

①

Augmented Dickey-Fuller Unit Root Tests

Type	Lags	Rho	Pr < Rho	Tau	Pr < Tau	*F*	Pr > *F*
Zero mean	1	0.5349	0.8127	1.06	0.9235		
	2	0.5227	0.8104	0.94	0.9066		
	3	0.5189	0.8087	0.79	0.8824		
	4	0.4672	0.7956	0.68	0.8602		
Single mean	1	–4.8923	0.4375	–1.80	0.3800	2.59	0.4192
	2	–5.7717	0.3569	–1.89	0.3361	2.61	0.4146
	3	–7.8226	0.2166	–2.17	0.2169	6.05	0.3028
	4	–8.2665	0.1938	–2.05	0.2648	2.62	0.4103
Trend	1	–8.4694	0.5308	–2.19	0.4869	2.45	0.6903
	2	–10.6543	0.3719	–2.38	0.3863	2.86	0.6101
	3	–15.7115	0.1407	–2.80	0.2018	3.93	0.4004
	4	–19.3688	0.0637	–2.83	0.1895	4.01	0.3853

The ARIMA Procedure

Name of variable = $y(1)$

Period (s) of differencing	1
Mean of working series	0.554455
Standard Deviation	1.659209
Number of Observations	101
Observation(s) eliminated by differencing	1

Autocorrelation Check for White Noise

To lag	χ^2	DF	Pr > χ^2	Autocorrelations					
6	141.97	6	<0.0001	0.695	0.526	0.463	0.409	0.344	0.289
12	154.29	12	<0.0001	0.194	0.098	–0.035	–0.143	–0.115	–0.163
18	182.75	18	<0.0001	–0.179	–0.236	–0.226	–0.214	–0.162	–0.155
24	190.18	24	<0.0001	–0.098	–0.112	–0.138	–0.115	–0.044	0.033

(continued)

TABLE 19.5 *(continued)*

The SAS System

②

Augmented Dickey-Fuller Unit Root Tests

Type	Lags	Rho	Pr < Rho	Tau	Pr < Tau	F	Pr > F
Zero mean	1	−20.9199	0.0009	−3.07	0.0024		
Single mean	1	−24.1880	0.0027	−3.33	0.0163	5.56	0.0254
Trend	1	−24.5744	0.0190	−3.29	0.0744	5.51	0.0983

Ⓑ

The ARIMA Procedure

Name of variable = x

Mean of working series	80.44902
Standard deviation	11.25186
Number of observations	102

Autocorrelation Check for White Noise

To lag	χ^2	DF	Pr > χ^2	Autocorrelations					
6	389.94	6	<0.0001	0.924	0.862	0.803	0.739	0.679	0.623
12	532.47	12	<0.0001	0.566	0.516	0.470	0.425	0.381	0.340
18	565.00	18	<0.0001	0.302	0.260	0.217	0.175	0.142	0.108
24	566.38	24	<0.0001	0.078	0.055	0.036	0.017	0.010	−0.002

①

Augmented Dickey-Fuller Unit Root Tests

Type	Lags	Rho	Pr < Rho	Tau	Pr < Tau	F	Pr > F
Zero mean	1	0.5896	0.8262	2.27	0.9944		
	2	0.5592	0.8187	1.84	0.9837		
	3	0.5531	0.8172	1.27	0.9477		
	4	0.5555	0.8177	1.09	0.9270		
Single mean	1	−2.5321	0.7100	−1.35	0.6049	4.04	0.0896
	2	−2.2730	0.7422	−1.02	0.7428	2.52	0.4366
	3	−3.1743	0.6301	−0.99	0.7555	1.49	0.6938
	4	−4.0667	0.5246	−1.05	0.7339	1.31	0.7397
Trend	1	−6.5408	0.6903	−1.95	0.6191	1.95	0.7876
	2	−7.5449	0.6059	−1.87	0.6651	1.74	0.8291
	3	−14.7701	0.1707	−2.32	0.4170	2.78	0.6261
	4	−24.0980	0.0212	−2.66	0.2537	3.70	0.4463

The ARIMA Procedure

Name of variable = x(1)

Period(s) of differencing	1
Mean of working series	0.554455
Standard deviation	2.008175
Number of observations	101
Observation(s) eliminated by differencing	1

TABLE 19.5 *(continued)*

The SAS System

Autocorrelation Check for White Noise

To lag	χ^2	DF	Pr > χ^2	Autocorrelations					
6	14.61	6	0.0236	−0.001	0.148	0.286	0.106	−0.086	0.122
12	17.57	12	0.1294	−0.083	0.026	−0.086	−0.085	−0.032	−0.055
18	28.35	18	0.0569	−0.183	0.033	−0.036	−0.159	0.166	0.015
24	35.61	24	0.0599	−0.063	0.116	−0.035	−0.153	0.114	−0.012

②

Augmented Dickey-Fuller Unit Root Tests

Type	Lags	Rho	Pr < Rho	Tau	Pr < Tau	*F*	Pr > *F*
Zero mean	1	−63.0667	<0.0001	−5.51	<0.0001		
Single mean	1	−73.9022	0.0009	−5.94	<0.0001	17.67	0.0010
Trend	1	−73.8039	0.0003	−5.87	<0.0001	17.48	0.0010

©

The REG Procedure

Model: MODEL1

Dependent Variable: y

Analysis of Variance

Source	DF	Sum of Squares	Mean Square	*F* value	Pr > *F*
Model	1	20081	20081	972.99	<0.0001
Error	100	2063.86272	20.63863		
Corrected total	101	22145			

Root MSE	4.54298	*R*-Square	0.9068
Dependent mean	87.07549	Adj *R*-Sq	0.9059
Coeff Var	5.21728		

Parameter Estimates

Variable	DF	Parameter Estimate	Standard Error	*t* value	Pr > |*t*|
Intercept	1	−13.24528	3.24746	−4.08	<0.0001
x	1	1.24701	0.03998	31.19	<0.0001

Ⓓ

The ARIMA Procedure

Name of variable = r

Mean of working series	−144E-16
Standard deviation	4.498216
Number of observations	102

(continued)

TABLE 19.5 *(continued)*

Autocorrelation Check for White Noise

To lag	χ^2	DF	Pr > χ^2	Autocorrelations					
6	290.12	6	<.0001	0.848	0.767	0.703	0.619	0.537	0.488
12	335.71	12	<.0001	0.409	0.327	0.250	0.192	0.144	0.092
18	343.80	18	<.0001	0.031	−0.030	−0.096	−0.132	−0.117	−0.152
24	367.31	24	<.0001	−0.168	−0.171	−0.197	−0.191	−0.154	−0.150

①

Augmented Dickey-Fuller Unit Root Tests

Type	Lags	Rho	Pr < Rho	Tau	Pr < Tau	F	Pr > F
Zero mean	1	−7.1909	0.0617	−1.62	0.1000		
	2	−5.4354	0.1063	−1.33	0.1698		
	3	−7.2947	0.0597	−1.54	0.1152		
	4	−10.3216	0.0236	−1.78	0.0714		
Single mean	1	−7.1148	0.2582	−1.59	0.4819	1.41	0.7139
	2	−5.3531	0.3936	−1.30	0.6255	0.95	0.8296
	3	−7.2051	0.2524	−1.52	0.5200	1.21	0.7643
	4	−10.2149	0.1183	−1.76	0.3997	1.59	0.6700
Trend	1	−8.5303	0.5260	−1.85	0.6718	2.06	0.7669
	2	−6.8811	0.6616	−1.64	0.7716	2.04	0.7712
	3	−9.1270	0.4794	−1.90	0.6490	2.67	0.6471
	4	−12.7763	0.2523	−2.18	0.4955	3.31	0.5234

The ARIMA Procedure
Name of variable = r(1)

Period(s) of differencing	1
Mean of working series	−0.12461
Standard deviation	2.279039
Number of observations	101
Observation(s) eliminated by differencing	1

Autocorrelation Check for White Noise

To lag	χ^2	DF	Pr > χ^2	Autocorrelations					
6	11.35	6	0.0781	−0.213	−0.059	0.127	0.047	−0.125	0.154
12	15.54	12	0.2131	0.020	0.044	−0.149	−0.065	0.030	0.084
18	27.52	18	0.0698	−0.075	0.032	−0.098	−0.197	0.195	−0.068
24	39.61	24	0.0235	−0.047	0.124	−0.167	−0.165	0.139	−0.011

②

Augmented Dickey-Fuller Unit Root Tests

Type	Lags	Rho	Pr < Rho	Tau	Pr < Tau	F	Pr > F
Zero mean	1	−153.122	0.0001	−8.80	<.0001		
Single mean	1	−153.993	0.0001	−8.77	<.0001	38.48	0.0010
Trend	1	−156.487	0.0001	−8.85	<.0001	39.32	0.0010

TABLE 19.5 *(continued)*

Ⓔ
The REG Procedure
Model : MODEL1
Dependent Variable: dy
Analysis of Variance

Source	DF	Sum of Squares	Mean Square	F Value	Pr > F
Model	9	149.98534	16.66504	11.44	<.0001
Error	87	126.76972	1.45712		
Corrected total	96	276.75505			

Root MSE	1.20711	*R*-Square	0.5419
Dependent Mean	0.54124	Adj *R*-Sq	0.4946
Coeff Var	223.02860		

①
Parameter Estimates

Variable	DF	Parameter Estimate	Standard Error	*t* Value	Pr > |*t*|
Intercept	1	0.14053	0.13351	1.05	0.2954
ecm	1	−0.06232	0.03127	−1.99	0.0494
ldy	1	0.63744	0.10919	5.84	<0.0001
l2dy	1	0.00714	0.12855	0.06	0.9558
l3dy	1	0.09234	0.12843	0.72	0.4740
l4dy	1	0.09705	0.11682	0.83	0.4084
ldx	1	−0.02361	0.08134	−0.29	0.7723
l2dx	1	0.04123	0.08287	0.50	0.6201
l3dx	1	−0.01158	0.08066	−0.14	0.8861
l4dx	1	−0.06467	0.07472	−0.87	0.3891

②
The REG Procedure
Model: MODEL1
Test test1 Results for Dependent Variable dy

Source	DF	Mean Square	F Value	Pr > F
Numerator	4	0.56354	0.39	0.8176
Denominator	87	1.45712		

Ⓕ
The REG Procedure
Model: MODEL2
Dependent Variable: dx
Analysis of Variance

Source	DF	Sum of Squares	Mean Square	*F* Value	Pr > *F*
Model	9	107.12307	11.90256	3.61	0.0007
Error	87	287.16374	3.30073		
Corrected total	96	394.28680			

Root MSE	1.81679	*R*-Square	0.2717
Dependent mean	0.48454	Adj *R*-Sq	0.1963
Coeff Var	374.95491		

(continued)

TABLE 19.5 *(continued)*

①
Parameter Estimates

Variable	DF	Parameter Estimate	Standard Error	*t* Value	Pr > |*t*|
Intercept	1	0.20378	0.20094	1.01	0.3133
ecm	1	0.02188	0.04706	0.46	0.6432
1dy	1	0.50472	0.16434	3.07	0.0028
12dy	1	0.05043	0.19348	0.26	0.7950
13dy	1	0.17729	0.19329	0.92	0.3616
14dy	1	−0.17757	0.17582	−1.01	0.3153
1dx	1	−0.27011	0.12242	−2.21	0.0300
12dx	1	−0.02710	0.12472	−0.22	0.8285
13dx	1	0.18046	0.12140	1.49	0.1408
14dx	1	0.08954	0.11246	0.80	0.4281

②
The REG Procedure
Model : MODEL2
Test test2 Results for Dependent Variable dx

Source	DF	Mean Square	*F* Value	Pr > F
Numerator	4	12.36770	3.75	0.0073
Denominator	87	3.30073		

Appendix A: Rudiments of Time Series Analysis [Box and Jenkins 1970; Anderson 1976; Fuller 1996]

Stochastic Processes and Time Series

A *stochastic process* $\{X_t\}$ is a collection of real-valued random variables ordered in (or indexed by) time, with the joint probability density function of these random variables denoted as $f(X_1, X_2,\ldots,X_t)$. A (discrete) *time series* $\{Y_t\}$ can be thought of as a single realization of the stochastic process $\{X_t\}$ represented by $f(X_1, X_2,\ldots,X_t)$ i.e., the elements of the time series $Y_1, Y_2,\ldots,Y_t$ can be viewed as depicting a specific outcome $f(X_1 = Y_1, X_2 = Y_2,\ldots,X_t = Y_t)$ of the joint probability density function f, where Y_i is the realized value of X_i in period i. Hence, any time series $\{Y_t\}$ can be thought of as having been generated by a set of jointly distributed random variables $\{X_t\}$; it is this joint probability distribution that links the realizations $Y_1,\ldots,Y_t$ across time.

For a fixed t, the first and second moments of the probability density function of X_t, $f(X_t)$, are assumed to exist and are specified as

$$\begin{aligned}&(a)\ E(X_t)=\mu_t,\\&(b)\ V(X_t)=\sigma_t^2,\\&(c)\ COV(X_t,X_{t-\tau})=\gamma_\tau.\end{aligned}\tag{19.A.1}$$

Note that each of these parameters is a function of time. Because Equation 19.A.1 holds for each value of t, each of the realizations $\{Y_t\}$ of $\{X_t\}$ has its own specific set of parameters in Equation 19.A.1. We are obviously faced with dealing with an unwieldy number of parameters.

To reduce that number, we need to impose some restrictions on the structure of the random process $\{X_t\}$. Specifically, we shall assume that (1) the process $\{X_t\}$, and thus the time series $\{Y_t\}$, is *stationary*, and (2) the process $\{X_t\}$, and thus $\{Y_t\}$, has *finite memory*.

In this regard, the time series $\{Y_t\}$ is said to be *strictly stationary* if the underlying stochastic process that generates the series is invariant with respect to time. If the characteristics of the process are fixed in time, then the joint distribution of $X_t, \ldots, X_{t+k}$ must be the same as the joint distribution of $X_{t+k}, \ldots, X_{t+k+\tau}$, i.e., the joint distribution of a strictly stationary process is unchanged under an arbitrary displacement in time. (If the characteristics of the process are not time invariant, then the process is termed *nonstationary*.) Because the form of the joint probability density function f is generally unknown, let us relax the notion of strict stationarity and frame our operational discussion of stationarity in terms of the first and second moments of the realization $\{Y_t\}$ of $\{X_t\}$.

Specifically, the time series $\{Y_t\}$ is said to be *covariance (weakly) stationary* if

$$\begin{aligned}&(a)\ E(Y_t) = E(Y_{t+m}) = \mu = \text{constant},\\ &(b)\ V(Y_t) = V(Y_{t+m}) = \sigma_Y^2 = \text{constant},\\ &(c)\ COV(Y_t, Y_{t-\tau}) = COV(Y_{t+m}, Y_{t+m-\tau}) = \gamma_\tau, \tau = 1, 2, \ldots.\end{aligned} \tag{19.A.2}$$

(Remember that $V(Y_t) = E[(Y_t - \mu)^2]$, whereas $COV(Y_t, Y_{t-\tau}) = E[(Y_t - \mu)(Y_{t-\tau} - \mu)]$.) Hence, a time series is covariance stationary if its mean, variance, and autocovariances are independent of time.[5] Note that the autocovariance γ_τ depends only on the gap τ between the two time points Y_t amd $Y_{t+\tau}$ and not on the actual point in time when the autocovariance is determined. In addition, for $\tau = 0$, γ_0 is equivalent to $V(Y_t)$. In what follows, we shall use the terms stationary and covariance stationary interchangeably.

For a stationary series $\{Y_t\}$, we next define the *autocorrelation function* between Y_t and $Y_{t+\tau}$ (or with lag τ) as

$$\begin{aligned}\rho_\tau &= CORR(Y_t, Y_{t-\tau}) = \frac{COV(Y_t, Y_{t-\tau})}{\sqrt{V(Y_t)}\sqrt{V(Y_{t-\tau})}}\\ &= \frac{COV(Y_t, Y_{t-\tau})}{V(Y_t)} = \gamma_\tau / \gamma_0, \tau = 1, 2, \ldots,\end{aligned} \tag{19.A.3}$$

because $V(Y_t) = V(Y_{t-\tau})$. Here the autocorrelation function indicates how much association or interdependence there is between neighboring data points in the $\{Y_t\}$ series. Because γ_τ and γ_0 are time independent, it follows that ρ_τ must be time independent so that the autocorrelation between, say, Y_t and Y_{t+1} must be the same as Y_{t+m} and Y_{t+m+1}. In addition, $\rho_0 = 1$.

A cornerstone of stochastic time series modeling is the notion of a *purely random* or *white noise process*. Specifically, the realizations $\{Y_t\}$ of the (discrete) process $\{X_t\}$ are white noise if they are independent and identically distributed with zero mean, constant variance, and zero autocovariance, i.e., for each value of t:

$$\begin{aligned}&(a)\ E(Y_t) = 0,\\ &(b)\ V(Y_t) = E(Y_t^2) = \sigma_Y^2,\\ &(c)\ COV(Y_t, Y_{t+\tau}) = E(Y_t Y_{t+\tau}) = 0 \text{ for any } \tau,\end{aligned} \tag{19.A.4}$$

and

---

[5] If $\{X_t\}$ follows a multivariate normal distribution, strict stationarity and covariance stationarity are equivalent.

$$\gamma_\tau = \begin{cases} \sigma_\varepsilon^2, & \tau = 0; \\ 0, & \tau \geq 1 \end{cases}$$

so that

$$\rho_\tau = \begin{cases} 1, & \tau = 0; \\ 0, & \tau \neq 0. \end{cases}$$

Given Equation 19.A.4.c, a white noise process is said to have *no memory*.

The Lag Operator L and the Difference Operator Δ

The linear operator[6] L^j is termed a *jth order lag operator* (or *backshift operator*) if for any realization Y_t and any j, $L^jY_t = Y_{t-j}$. Hence, L^j operating on Y_t lags Y_t by j periods. Thus, $L^0Y_t = Y_t$, $LY_t = Y_{t-1}$, $L^2Y_t = Y_{t-2}$, etc. The key properties of the lag operator are as follows:

1. For any constant c, $Lc = c$.
2. L satisfies the distributive law
   $(L^j + L^k)Y_t = L^jY_t + L^kY_t = Y_{t-j} + Y_{t-k}$.
3. L satisfies the *associative law (for multiplication)*
   $L^jL^kY_t = L^{j+k}Y_t = Y_{t-j-k}$; stated alternatively, $L^j(L^kY_t) = L^jY_{t-k} = Y_{t-k-j}$.
4. L raised to a negative power is termed a *lead operator*, i.e., $L^{-k}Y_t = Y_{t+k}$.
5. For $|a| < 1$,

$$\left[1 + aL + (aL)^2 + (aL)^3 + \cdots\right]Y_t = \frac{Y_t}{1 - aL}, \tag{19.A.5}$$

   where [.] is an infinite sum.[7]
6. For $|a| > 1$,

$$\left[1 + (aL)^{-1} + (aL)^{-2} + (aL)^{-3} + \cdots\right]Y_t = \frac{-aLY_t}{1 - aL}, \tag{19.A.6}$$

   where [.] is an infinite sum.

The linear operator Δ is termed a *first-order difference operator* if for the realization Y_t, $\Delta Y_t = Y_t - Y_{t-1}$. Additionally,

1. For any constant c, $\Delta c = 0$.
2. $\Delta(cY_t) = c\Delta Y_t$, c a constant.
3. Δ satisfies the *distributive law* $\Delta(Y_t \pm Z_t) = \Delta Y_t \pm \Delta Z_t$.

---

[6] A linear function f between two vector spaces X and Y is termed a *linear operator* if for elements $x \in X$ and $y \in Y$ and scalars α and β, $f(\alpha x + \beta y) = \alpha f(x) + \beta f(y)$.

[7] To see this, let us define the function $f(aL) = (1 - aL)^{-1}$ and consider the Taylor expansion of f near $aL = 0$ Then $f' = (1-aL)^{-2}, f'' = 2(1-aL)^{-3}, f''' = 6(1-aL)^{-4}$, etc. With $f(0) = 1, f'(0) = 1, f''(0) = 2, f'''(0) = 6$, etc., it follows that

$$(1 - aL)^{-1} = 1 + aL + (aL)^2 + (aL)^3 + \cdots = \sum_{i=0}^{\infty} a^i L^i$$

This sum converges if $|a| < 1$.

The operators L and Δ are connected by the relation $\Delta = 1 - L$. The *second-order difference operator* is $\Delta^2 Y_t = \Delta(\Delta Y_t) = \Delta(Y_t - Y_{t-1}) = \Delta Y_t - \Delta Y_{t-1} = Y_t - 2Y_{t-1} + Y_{t-2}$. Clearly, $\Delta^2 = (1 - L)^2$ In general, the *jth order difference operator* is $\Delta^j Y_t = \Delta(\Delta^{j-1} Y_t) = \Delta^{j-1}(\Delta Y_t) = \Delta^{j-1} Y_t - \Delta^{j-1} Y_{t-1}$. Stated alternatively, $\Delta^j Y_t = (1 - L)^j Y_t$ or

$$\Delta^j Y_t = \sum_{r=0}^{j} (-1)^r \binom{j}{r} Y_{t-r},$$

where

$$\binom{j}{r} = \frac{j!}{r!(j-r)!}.$$

Moving Average Process of Order *q*

Suppose $\{\varepsilon_t\}$ is a white noise process. Then the time series $\{Y_t\}$ specified as

$$\begin{aligned} Y_t &= \beta_0 \varepsilon_t + \beta_1 \varepsilon_{t-1} + \beta_2 \varepsilon_{t-2} + \cdots + \beta_q \varepsilon_{t-q} \\ &= \sum_{i=0}^{q} \beta_i \varepsilon_{t-i} \end{aligned} \tag{19.A.6}$$

is termed a *moving average process of order q* and denoted MA(q). (For convenience, the ε's can be scaled so that $\beta_0 = 1$.) Given that $E(\varepsilon_t) = 0$ for all t and the ε's are independent ($E(\varepsilon_t \varepsilon_{t-\tau}) = 0$, $\tau \neq 0$) with a common variance $V(\varepsilon_t) = \sigma_\varepsilon^2$, we can write, from Equation 19.A.6,

$$E(Y_t) = \sum_{i=0}^{q} \beta_i E(\varepsilon_{t-i}) = 0 \tag{19.A.7}$$

and

$$\begin{aligned} V(Y_t) &= E\left[(\varepsilon_t + \beta_1 \varepsilon_{t-1} + \beta_2 \varepsilon_{t-2} + \cdots + \beta_q \varepsilon_{t-q})^2\right. \\ &= \sum_{i=0}^{q} \beta_i^2 V(\varepsilon_{t-i}) = \sigma_\varepsilon^2 \sum_{i=0}^{q} \beta_i^2 \end{aligned} \tag{19.A.8}$$

(so that $V(Y_t)$ is finite if $\sum_{i=0}^{q} \beta_i^2$ is finite). In addition, the *autocovariance function for a τ-lag displacement* is

$$\begin{aligned} \gamma_\tau &= COV(Y_t, Y_{t-\tau}) = E(Y_t Y_{t-\tau}) \\ &= E\big[(\varepsilon_t + \beta_1 \varepsilon_{t-1} + \beta_2 \varepsilon_{t-2} + \cdots + \beta_q \varepsilon_{t-q}) \\ &\quad (\varepsilon_{t-\tau} + \beta_1 \varepsilon_{t-\tau-1} + \beta_2 \varepsilon_{t-\tau-2} + \cdots + \beta_q \varepsilon_{t-\tau-q})\big] \\ &= \sigma_\varepsilon^2 (\beta_\tau + \beta_1 \beta_{\tau+1} + \beta_2 \beta_{\tau+2} + \cdots + \beta_q \beta_{\tau+q}) \\ &= \begin{cases} \sigma_\varepsilon^2 \sum_{i=0}^{q-\tau} \beta_i \beta_{i+\tau}, & \tau = 0, 1, \cdots, q; \\ 0, \ \tau > q \end{cases} \end{aligned} \tag{19.A.9}$$

(here γ_τ is finite if $\sum_{i=0}^{q-\tau} \beta_i \beta_{i+\tau}$ is finite), whereas the *autocorrelation function*

$$\rho_\tau = \begin{cases} \gamma_\tau / \gamma_0, \tau = 0, 1, ..., q; \\ 0, \tau > q. \end{cases} \tag{19.A.10}$$

Note that, for $\tau = 0$, $\gamma_0 = V(Y_t)$ and $\rho_0 = 1$.

Because γ_τ is independent of t, the MA(q) process in Equation 19.A.6 is covariance stationary. In fact, a moving average process with finite q is always stationary because $\sum_{i=0}^{q} \beta_i^2$ is finite. An $MA(\infty)$ process will be stationary if both $\sum_{i=0}^{\infty} \beta_i^2$ and $\sum_{i=0}^{\infty} \beta_i \beta_{i+\tau}$ converge.

Given $\beta_0 = 1$, we can write the MA(q) process in terms of the lag operator L as

$$\begin{aligned} Y_t &= (1 + \beta_1 L + \beta_2 L^2 + \cdots + \beta_q L^q)\varepsilon_t \\ &= \left(\sum_{i=0}^{q} \beta_i L^i\right)\varepsilon_t = \beta(L)\varepsilon_t. \end{aligned} \tag{19.A.11}$$

Because the *MA(q) operator* or *lag polynomial* $\beta(L)$ has q roots $r_1,\ldots,r_q$, we can rewrite Equation 19.A.11 as

$$Y_t = (1 - r_1 L)(1 - r_2 L)\ldots(1 - r_q L)\varepsilon_t. \tag{19.A.12}$$

Hence, $r_1,\ldots,r_q$ are the *characteristic roots* of the *characteristic equation*

$$Z^q + \beta_1 Z^{q-1} + \beta_2 Z^{q-2} + \cdots + \beta_q = 0. \tag{19.A.3}$$

For $Y_t = \beta(L)\varepsilon_t$, under what conditions is this MA(q) process invertible? That is, under what conditions can we express $\varepsilon_t = \beta(L)^{-1} Y_t$? Suppose

$$\beta(L) = (1 - r_1 L)(1 - r_2 L)...(1 - r_q L) = \prod_{j=1}^{q} (1 - r_j L)$$

has q distinct roots $r_1,\ldots r_q$. Then expanding in partial fractions,

$$\beta(L)^{-1} = \left[\prod_{j=1}^{q} (1 - r_j L)\right]^{-1} = \sum_{j=1}^{q} \frac{m_j}{1 - r_j L},$$

where the m_j's constitute a set of undetermined coefficients. Hence, our *invertibility condition* is that $\beta(L)^{-1}$ converges inside the unit circle, and this will occur if $|r_j| < 1$, $j = 1,\ldots,q$. Stated alternatively, because the roots or zeros of the lag operator polynomial $\beta(L) = 0$ are $r_j^{-1}, j = 1, ..., q$, the MA(q) process is invertible if the inverses of the roots of $\beta(L) = 1 + \beta_1 L + \beta_2 L^2 + \cdots + \beta_q L^q = 0$ are all less than one in absolute value (or that the roots r_j^{-1} all lie outside the unit circle). Note that the same condition for invertibility still holds even if the zeros of $\beta(L) = 0$ are not all distinct.

It is instructive to examine two important cases of the MA(q) process: the instances in which the order is taken to be one and then two. First, for the MA(1) process

$$Y_t = \varepsilon_t + \beta_1 \varepsilon_{t-1} \text{ (with } \beta_0 = 1)$$

and $\{\varepsilon_t\}$ is white noise. Then

$$
\begin{aligned}
E(Y_t) &= E(\varepsilon_t) + \beta_1 E(\varepsilon_{t-1}) = 0,\\
V(Y_t) &= E(Y_t^2) = E\left[(\varepsilon_t + \beta_1\varepsilon_{t-1})^2\right]\\
&= E(\varepsilon_t^2 + 2\beta_1\varepsilon_t\varepsilon_{t-1} + \beta_1^2\varepsilon_{t-1}^2)\\
&= E(\varepsilon_t^2) + 2\beta_1 E(\varepsilon_t\varepsilon_{t-1}) + \beta_1^2 E(\varepsilon_{t-1}^2)\\
&= \sigma_\varepsilon^2(1+\beta_1^2),\\
\gamma_1 &= COV(Y_t, Y_{t-1}) = E(Y_tY_{t-1})\\
&= E\left[(\varepsilon_t + \beta_1\varepsilon_{t-1})(\varepsilon_{t-1} + \beta_1\varepsilon_{t-2})\right]\\
&= E(\varepsilon_t\varepsilon_{t-1} + \beta_1\varepsilon_t\varepsilon_{t-2} + \beta_1\varepsilon_{t-1}^2 + \beta_1^2\varepsilon_{t-1}\varepsilon_{t-2})\\
&= \beta_1\sigma_\varepsilon^2,
\end{aligned}
$$

with $\gamma_0 = E(Y_t^2) = V(Y_t)$. In general,

$$\gamma_\tau = \begin{cases}\beta_1\sigma_\varepsilon^2, \tau = 1;\\ 0, \tau > 1,\end{cases}$$

i.e., the autocovariance of a MA(1) process is zero when the displacement τ is more than one time period. In this regard, the MS(1) process is said to have only a *one period memory* in that, for any t, Y_t is correlated with Y_{t-1}, but not with any other value of $\{Y_t\}$. Hence, an MA(1) process "forgets" what transpired more than one time period in the past. Additionally,

$$\rho_1 = \frac{\gamma_1}{\gamma_0} = \frac{\beta_1}{1+\beta_1^2},$$

with $\rho_0 = 1$. In general,

$$\rho_\tau = \frac{\gamma_\tau}{\gamma_0} = \begin{cases}\dfrac{\beta_1}{1+\beta_1^2}, & \tau = 1\ ;\\ 0, & \tau > 1.\end{cases}$$

Finally, $Y_t = (1 + \beta_1 L)\varepsilon_t = \beta(L)\varepsilon_t$ is invertible if $|\beta_1| < 1$, i.e., for $|\beta_1| < 1$, $(1+\beta_1 L)^{-1} = \sum_{j=0}^{\infty}\beta_1^j L^j$ converges within the unit circle (see Equation 19.A.5.5). This is equivalent to saying that the inverse of the root of the lag operator polynomial $\beta(L) = 1 + \beta_1 L = 0$, $L = -\beta_1^{-1}$, must be less than one in absolute value (or $L = -\beta_1^{-1}$ itself lies outside the unit circle).

Turning to the MA(2) process

$$Y_t = \varepsilon_t + \beta_1\varepsilon_{t-1} + \beta_2\,\varepsilon_{t-2} \text{ (with } \beta_0 = 1),$$

where $\{\varepsilon_t\}$ is white noise. Then

$$
\begin{aligned}
E(Y_t) &= E(\varepsilon_t) + \beta_1 E(\varepsilon_{t-1}) + \beta_2 E(\varepsilon_{t-2}) = 0,\\
V(Y_t) &= E(Y_t^2) = E\left[(\varepsilon_t + \beta_1\varepsilon_{t-1} + \beta_2\varepsilon_{t-2})^2\right]\\
&= E(\varepsilon_t^2 + 2\beta_1\varepsilon_t\varepsilon_{t-1} + 2\beta_2\varepsilon_t\varepsilon_{t-2} + \beta_1^2\varepsilon_{t-1}^2\\
&\quad + 2\beta_1\beta_2\varepsilon_{t-1}\varepsilon_{t-2} + \beta_2^2\varepsilon_{t-2}^2)\\
&= \sigma_\varepsilon^2(1 + \beta_1^2 + \beta_2^2),
\end{aligned}
$$

$$\begin{aligned}
\gamma_1 &= COV(Y_t, Y_{t-1}) = E(Y_t Y_{t-1}) \\
&= E\left[(\varepsilon_t + \beta_1\varepsilon_{t-1} + \beta_2\varepsilon_{t-2})(\varepsilon_{t-1} + \beta_1\varepsilon_{t-2} + \beta_2\varepsilon_{t-3})\right] \\
&= E(\varepsilon_t\varepsilon_{t-1} + \beta_1\varepsilon_t\varepsilon_{t-2} + \beta_2\varepsilon_t\varepsilon_{t-3} + \beta_1\varepsilon_{t-1}^2 \\
&\qquad + \beta_1^2\varepsilon_{t-1}\varepsilon_{t-2} + \beta_1\beta_2\varepsilon_{t-1}\varepsilon_{t-3} + \beta_2\varepsilon_{t-1}\varepsilon_{t-2} \\
&\qquad + \beta_1\beta_2\varepsilon_{t-2}^2 + \beta_2^2\varepsilon_{t-2}\varepsilon_{t-3}) \\
&= \sigma_\varepsilon^2(\beta_1 + \beta_1\beta_2), \\
\gamma_2 &= COV(Y_t, Y_{t-2}) = E(Y_t Y_{t-2}) \\
&= E\left[(\varepsilon_t + \beta_1\varepsilon_{t-1} + \beta_2\varepsilon_{t-2})(\varepsilon_{t-2} + \beta_1\varepsilon_{t-3} + \beta_2\varepsilon_{t-4})\right] \\
&= \beta_2\sigma_\varepsilon^2,
\end{aligned}$$

with $\gamma_0 = E(Y_t^2) = V(Y_t)$. In general

$$\gamma_\tau = \begin{cases} \sigma_\varepsilon^2 \sum_{i=0}^{2-\tau} \beta_i\beta_{i+\tau}, \tau = 0,1,2; \\ 0, \tau > 2, \end{cases}$$

i.e., the autocovariance of a MA(2) process is zero when the displacement τ is more than two time periods. Hence, the MA(2) process is said to be have only a *two period memory* in that, for any t, Y_t is correlated with Y_{t-1} and Y_{t-2}, but not with any other value of $\{Y_t\}$; the MA(2) process "forgets" what transpired more than two periods in the past. Additionally,

$$\rho_1 = \frac{\gamma_1}{\gamma_0} = \frac{\beta_1 + \beta_1\beta_2}{1 + \beta_1^2 + \beta_2^2},$$

$$\rho_2 = \frac{\gamma_2}{\gamma_0} = \frac{\beta_2}{1 + \beta_1^2 + \beta_2^2},$$

with $\rho_0 = 1$. In general,

$$\rho_\tau = \frac{\gamma_\tau}{\gamma_0} = \begin{cases} \sum_{i=0}^{2-\tau} \beta_i\beta_{i+\tau} \Big/ \sum_{i=0}^{2} \beta_i^2, \ 0 \le \tau \le 2; \\ 0, \ \tau > 2. \end{cases}$$

Finally, $Y_t = (1 + \beta_1 L + \beta_2 L^2)\varepsilon_t = \beta(L)\varepsilon_t$ is invertible only if the roots of the lag operator polynomial $\beta(L) = 1 + \beta_1 L + \beta_2 L^2 = 0$ lie outside the unit circle or, stated alternatively, the roots r_1, r_2 of the second-order characteristic equation $Z^2 + \beta_1 Z + \beta_2 = 0$ are less than one in absolute value. Hence, we must have, for this quadratic case,

$$\left| \frac{-\beta_1 \pm \sqrt{\beta_1^2 - 4\beta_2}}{2\beta_2} \right| < 1,$$

i.e., the parameters β_1, β_2 must lie inside the triangular region

$$\begin{aligned}
\beta_1 + \beta_2 &> -1 \\
\beta_2 - \beta_1 &> -1 \\
-1 < \beta_2 &< 1.
\end{aligned}$$

Autoregressive Process of Order *p*

Let $\{\varepsilon_t\}$ constitute a white noise process. Then the time series $\{Y_t\}$ specified as

$$\begin{aligned} Y_t &= \alpha_1 Y_{t-1} + \alpha_2 Y_{t-2} + \cdots + \alpha_p Y_{t-p} + \varepsilon_t \\ &= \sum_{j=1}^{p} \alpha_j Y_{t-j} + \varepsilon_t,\ t = 1, 2, \ldots, \end{aligned} \tag{19.A.14}$$

is termed an *autogregressive process of order p* and denoted AR(p) (it is "autoregressive" in the sense that Y_t is regressed on its own past values). Given Equation 19.A.14, we may write the AR(p) process in terms of the lag operator L as

$$(1 - \alpha_1 L - \alpha_2 L^2 - \cdots - \alpha_p L^p) Y_t = \varepsilon_t \text{ or } \alpha(L) Y_t = \varepsilon_t, \tag{19.A.15}$$

where

$$\alpha(L) = 1 - \alpha_1 L - \alpha_2 L^2 - \cdots - \alpha_p L^p = 1 - \sum_{j=1}^{p} \alpha_i L^j$$

is called the *lag polynomial.*

For the AR(p) process to be stationary, it is necessary that $V(Y_t)$ is finite, and this occurs if all the zeros or roots of $\alpha(L)$ lie outside the unit circle. To see this, let us first determine, for

$$Y_t = \sum_{j=1}^{p} \alpha_j Y_{t-j} + \varepsilon_t,$$
$$E(Y_t) = \sum_{j=1}^{p} \alpha_j E(Y_{t-j}) + E(\varepsilon_t)$$

or $\mu = \mu \sum_{j=1}^{p} \alpha_j$ and thus $\mu(1 - \sum_{j=1}^{p} \alpha_j) = 0$. If $\sum_{j=1}^{p} \alpha_j < 0$, then we must have $\mu = 0$ so that $E(Y_t)$ is finite and independent of t.

Next, let us write

$$\alpha(L) = (1 - s_1 L)(1 - s_2 L) \ldots (1 - s_p L) = \prod_{j=1}^{p} (1 - s_j L),$$

where $s_1, \ldots s_p$ are the p *characteristic roots* (assumed distinct) of the *characteristic equation*

$$Z^p - \alpha_1 Z^{p-1} - \alpha_2 Z^{p-2} - \cdots - \alpha_p = 0.$$

Then

$$Y_t = \alpha(L)^{-1} \varepsilon_t = \left[\prod_{j=1}^{p} (1 - s_j L)\right]^{-1} \varepsilon_t \tag{19.A.16}$$

and, expanding in partial fractions, we obtain, from Equation 19.A.5,

$$\begin{aligned} Y_t &= \left(\sum_{j=1}^{p} \frac{u_j}{1 - s_j L}\right) \varepsilon_t = \sum_{j=1}^{p} \left(u_j \sum_{k=0}^{\infty} s_j^k \varepsilon_{t-k}\right) \\ &= \sum_{k=0}^{\infty} \left(\sum_{j=1}^{p} u_j s_j^k\right) \varepsilon_{t-k} \end{aligned}$$

(here the u_j's represent a set of undetermined coefficients) so that

$$V(Y_t) = E(Y_t^2) = \sum_{k=0}^{\infty}\left(\sum_{j=1}^{p} u_j s_j^k\right)^2 \sigma_\varepsilon^2$$

given that the ε_t's are independent and $E(\varepsilon_t^2) = \sigma_\varepsilon^2$ for all t. Because σ_ε^2 is finite, $V(Y_t)$ will be finite for $|s_j| < 1, j = 1,\ldots,p$ (the condition for the convergence of $\alpha(L)^{-1}$). In this regard, because the roots of $\alpha(L) = 0$ are $s_j^{-1}, j = 1,\ldots, p$, the AR($p$) process is stationary if the inverses of the roots of $\alpha(L) = 0$ are all less than one in absolute value (or that the roots s_j^{-1} all lie outside the unit circle). This same necessary condition for stationarity will hold even if the zeros of $\alpha(L) = 0$ are not all distinct.

To derive an operational version of $V(Y_t)$, let us first focus on obtaining the *autocovariance function* γ_τ. From this, we can readily get the *autocorrelation function* ρ_τ and then, as a special case of γ_τ, $\gamma_0 = V(Y_t)$. To this end, let us multiply both sides of Equation 19.A.14 by $Y_{t-\tau}$, $\tau > 0$, so as to obtain

$$Y_{t-\tau}Y_t = \alpha_1 Y_{t-\tau}Y_{t-1} + \alpha_2 Y_{t-\tau}Y_{t-2} + \cdots + \alpha_p Y_{t-\tau}Y_{t-p} + Y_{t-\tau}\varepsilon_t. \qquad (19.A.17)$$

Taking expectations of this expression yields

$$\gamma_\tau = \alpha_1\gamma_{\tau-1} + \alpha_2\gamma_{\tau-2} + \cdots + \alpha_p\gamma_{\tau-p}.\text{[8]} \qquad (19.A.18)$$

Then dividing both sides of Equation 19.A.18 by γ_0 renders

$$\rho_\tau = \alpha_1\rho_{\tau-1} + \alpha_2\rho_{\tau-2} + \cdots + \alpha_p\rho_{\tau-p}. \qquad (19.A.19)$$

It is important to note that we may use Equation 19.A.19 as a *recursive relation* to obtain ρ_τ, $\tau \geq 2$, i.e., we start with $\rho_0 = 1$ and solve for ρ_1, and so on (more on this later).

For $\tau = 1, 2,\ldots,p$ in Equation 19.A.19, we generate the *system of Yule-Walker equations* [Yule, 1926; Walker, 1931] (remember that $\gamma_{-\tau} = \gamma_\tau$ for any τ):

$$\begin{aligned}
(\tau = 1)\ \rho_1 &= \alpha_1 + \alpha_2\rho_1 + \cdots + \alpha_p\rho_{p-1} \\
(\tau = 2)\ \rho_2 &= \alpha_1\rho_1 + \alpha_2 + \cdots + \alpha_p\rho_{p-2} \\
&\ \ \vdots \\
(\tau = p)\ \rho_p &= \alpha_1\rho_{p-1} + \alpha_2\rho_{p-2} + \cdots + \alpha_p
\end{aligned} \qquad (19.A.20)$$

or, in matrix terms,

$$\boldsymbol{\rho} = \mathbf{P}\boldsymbol{\alpha}, \qquad (19.A.20.1)$$

[8] Remember that $E(\varepsilon_t Y_t) = E(\varepsilon_t^2) = \sigma_\varepsilon^2$, $E(\varepsilon_t Y_{t-\tau}) = 0$, $E(Y_t Y_{t-\tau}) = \gamma_\tau = E(Y_{t-j}Y_{t-j-\tau})$ and $E(Y_t Y_t) = E(Y_{t-j}Y_{t-j}) = \gamma_0$.

where

$$\underset{(p\times 1)}{\boldsymbol{\rho}} = \begin{bmatrix} \rho_1 \\ \rho_2 \\ \cdot \\ \cdot \\ \cdot \\ \rho_p \end{bmatrix}, \quad \underset{(p\times p)}{\boldsymbol{P}} = \begin{bmatrix} 1 & \rho_1 & \cdots & \rho_{p-1} \\ \rho_1 & 1 & \cdots & \rho_{p-2} \\ \cdot & \cdot & & \cdot \\ \cdot & \cdot & & \cdot \\ \cdot & \cdot & & \cdot \\ \rho_{p-1} & \rho_{p-2} & \cdots & 1 \end{bmatrix}, \quad \underset{(p\times 1)}{\boldsymbol{\alpha}} = \begin{bmatrix} \alpha_1 \\ \alpha_2 \\ \cdot \\ \cdot \\ \cdot \\ \alpha_p \end{bmatrix}.$$

Then from Equation 19.A.20.1, $\boldsymbol{X} = \boldsymbol{P}^{-1}\boldsymbol{\rho}$.

Let us now obtain $V(Y_t)$. Set $\tau = 0$ in Equation 19.A.17 so that $E(Y_{t-\tau}\,\varepsilon_t)$ is $E(\varepsilon_t^2) = \sigma_\varepsilon^2$. Hence, in this instance, Equation 19.A.18 becomes

$$\gamma_0 = \alpha_1\gamma_{-1} + \alpha_2\gamma_{-2} + \cdots + \alpha_p\gamma_{-p} + \sigma_\varepsilon^2.$$

Dividing both sides of this equation by $\gamma_0(= V(Y_t))$ (with $\gamma_{-\tau} = \gamma_\tau$) yields

$$1 = \alpha_1\rho_1 + \alpha_2\rho_2 + \cdots + \alpha_p\rho_p + \frac{\sigma_\varepsilon^2}{\gamma_0}$$

or

$$V(Y_t) = \gamma_0 = \frac{\sigma_\varepsilon^2}{1 - \alpha_1\rho_1 - \alpha_2\rho_2 - \cdots - \alpha_p\rho_p}. \quad (19.A.21)$$

What is the relationship between MA(q) and AR(p)? If $\beta(L)$ is invertible, then the MA(q) process $Y_t = \beta(L)\varepsilon_t$ can be written as $\beta(L)^{-1}Y_t = \pi(L)Y_t = \varepsilon_t$ so that this moving-average process is equivalent to an AR(∞) process. In a similar vein, a stationary AR(p) process $\alpha(L)Y_t = \varepsilon_t$ is equivalent to a MA(∞) process, i.e., we can write $Y_t = \alpha(L)^{-1}\,\varepsilon_t = \Psi(L)\varepsilon_t$ because $\alpha(L)$ is always invertible (i.e., $\alpha(L) = 1 - \alpha_1 L - \alpha_2 L^2 - \cdots - \alpha_p L^P$ is finite so no restrictions on the α_j's are needed).

Let us now focus on two important cases of the AR(p) process. In particular, we present the instances in which the order of the process is set at one and then two. First, for the AR(1) process

$$Y_t = \alpha_1 Y_{t-1} + \varepsilon_t$$

and $\{\varepsilon_t\}$ is white noise. Then

$$E(Y_t) = \alpha_1 E(Y_{t-1}) + E(\varepsilon_t)$$

or $\mu = \alpha_1\mu$. Because $\alpha_1 \neq 0$, it follows that $\mu = 0$. Also,

$$\begin{aligned} V(Y_t) &= E(Y_t^2) = E\left[(\alpha_1 Y_{t-1} + \varepsilon_t)^2\right] \\ &= E(\alpha_1^2 Y_{t-1}^2 + 2\alpha_1\varepsilon_t Y_{t-1} + \varepsilon_t^2) \\ &= \alpha_1^2 V(Y_t) + \sigma_\varepsilon^2 \end{aligned}$$

or

$$V(Y_t) = \frac{\sigma_\varepsilon^2}{1-\alpha_1^2}.$$

For $V(Y_t)$ to be finite and nonnegative, $-1 < \alpha_1 < 1$. Looking to autocovariances, we first form

$$Y_{t-\tau}Y_t = \alpha_1 Y_{t-\tau}Y_{t-1} + Y_{t-\tau}\varepsilon_t$$

and thus

$$E(Y_{t-\tau}Y_t) = \alpha_1 E(Y_{t-\tau}Y_{t-1}) + E(Y_{t-\tau}\varepsilon_t).$$

For $\tau \geq 1$,

$$\gamma_\tau = \alpha_1 \gamma_{\tau-1}.$$

This recursive relation indicates that, for any $\tau \geq 1$, knowing $\gamma_{\tau-1}$ enables us to obtain γ_τ. For instance, given $\gamma_0 = V(Y_t) = \sigma_\varepsilon^2/(1-\alpha_1^2)$:

$$\gamma_1 = \alpha_1\gamma_0 = \frac{\alpha_1\sigma_\varepsilon^2}{1-\alpha_1^2},$$

$$\gamma_2 = \alpha_1\gamma_1 = \frac{\alpha_1^2\sigma_\varepsilon^2}{1-\alpha_1^2}, \text{ etc.}$$

In general,

$$\gamma_\tau = \frac{\alpha_1^\tau\sigma_\varepsilon^2}{1-\alpha_1^2}, \quad \tau = 0,1,2,\ldots.$$

Also,

$$\rho_\tau = \frac{\gamma_\tau}{\gamma_0} = \alpha_1^\tau, \quad \tau = 0,1,2,\ldots.$$

(Note that, from Equation 19.A.21, $V(Y_t) = \sigma_\varepsilon^2/(1-\alpha_1\rho_1) = \sigma_\varepsilon^2/(1-\alpha_1^2)$.)

Given $Y_t = \alpha_1 Y_{t-1} + \varepsilon_t$ or $(1 - \alpha_1 L)Y_t = \varepsilon_t$, it follows (from Equation 19.A.5) that $Y_t = (1-\alpha_1 L)^{-1}\varepsilon_t = \sum_{i=0}^{\infty}\alpha_1^i\varepsilon_{t-i}$ because $(1-\alpha_1 L)^{-1} = \sum_{i=0}^{\infty}\alpha_1^i L^i$. For the stationarity of this AR(1) process, $(1 - \alpha_1 L)^{-1}$ must converge or α_1 must satisfy $|\alpha_1| < 1$. Because the root of $\alpha(L) = 1 - \alpha_1 L = 0$ is $L = \alpha_1^{-1}$, it follows that this condition is equivalent to requiring that the root of $\alpha(L) = 0$ must lie outside the unit circle.

Let us briefly explore the connection between the MA(1) and AR(1) processes. For MA(1): $Y_t = \varepsilon_t + \beta_1\varepsilon_{t-1}$, $\beta_0 = 1$ and $|\beta_1| < 1$, let us form the sequence of innovations

$$\begin{aligned}\varepsilon_t &= Y_t - \beta_1\varepsilon_{t-1}\\ \varepsilon_{t-1} &= Y_{t-1} - \beta_1\varepsilon_{t-2}\\ \varepsilon_{t-2} &= Y_{t-2} - \beta_1\varepsilon_{t-3}\\ &\text{etc.}\end{aligned}$$

Then, by back-substituting these innovations into the above MA(1) process, we obtain

$$\begin{aligned} Y_t &= \varepsilon_t + \beta_1(Y_{t-1} - \beta_1\varepsilon_{t-2}) \\ &= \varepsilon_t + \beta_1 Y_{t-1} - \beta_1^2\varepsilon_{t-2} \\ &= \varepsilon_t + \beta_1 Y_{t-1} - \beta_1^2(Y_{t-2} - \beta_1\varepsilon_{t-3}) \\ &= \varepsilon_t + \beta_1 Y_{t-1} - \beta_1^2 Y_{t-2} + \beta_1^3\varepsilon_{t-3} - \ldots. \end{aligned}$$

Hence, the autoregressive representation of the MA(1) process involves expressing Y_t in terms of the current shock or disturbance and lagged observed values of the series $\{Y_t\}$ (rather than writing it in terms of the current shock and lagged unobserved shocks).

For AR(1): $Y_t = \alpha_1 Y_{t-1} + \varepsilon_t$, set

$$\begin{gathered} Y_{t-1} = \alpha_1 Y_{t-2} + \varepsilon_{t-1} \\ Y_{t-2} = \alpha_1 Y_{t-3} + \varepsilon_{t-2} \\ \text{etc.} \end{gathered}$$

Then, via back-substitution,

$$\begin{aligned} Y_t &= \alpha_1(\alpha_1 Y_{t-2} + \varepsilon_{t-1}) + \varepsilon_t = \alpha_1^2 Y_{t-2} + \alpha_1\varepsilon_{t-1} + \varepsilon_t \\ &= \alpha_1^2(\alpha_1 Y_{t-3} + \varepsilon_{t-2}) + \alpha_1\varepsilon_{t-1} + \varepsilon_t \\ &= \alpha_1^3 Y_{t-3} + \alpha_1^2\varepsilon_{t-2} + \alpha_1\varepsilon_{t-1} + \varepsilon_t + \ldots. \end{aligned}$$

This moving-average representation of Y_t is convergent if $|\alpha_1| < 1$, in which case, Y_t can be expressed as an infinite moving average of the ε_t's or $Y_t = \sum_{i=0}^{\infty}\alpha_1^i\varepsilon_{t-i}$.

Turning to the AR(2) process,

$$Y_t = \alpha_1 Y_{t-1} + \alpha_2 Y_{t-2} + \varepsilon_t,$$

where $\{\varepsilon_t\}$ is white noise. Then

$$E(Y_t) = \alpha_1 E(Y_{t-1}) + \alpha_2 E(Y_{t-2}) + E(\varepsilon_t) \text{ or } \mu(1 - \alpha_1 - \alpha_2) = 0. \text{ For } \alpha_1 + \alpha_2 < 1,$$

we have $\mu = 0$ (for stationarity). To calculate autocovariances, let us form

$$Y_{t-\tau}Y_t = \alpha_1 Y_{t-\tau}Y_{t-1} + \alpha_2 Y_{t-\tau}Y_{t-2} + Y_{t-\tau}\varepsilon_t \quad (19.A.22)$$

with

$$E(Y_{t-\tau}Y_t) = \alpha_1 E(Y_{t-\tau}Y_{t-1}) + \alpha_2 E(Y_{t-\tau}Y_{t-2}) + E(Y_{t-\tau}\varepsilon_t).$$

For $\tau \geq 1$,

$$\gamma_\tau = \alpha_1\gamma_{\tau-1} + \alpha_2\gamma_{\tau-2} \quad (19.A.23)$$

or

$$\rho_\tau = \frac{\gamma_\tau}{\gamma_0} = \alpha_1\rho_{\tau-1} + \alpha_2\rho_{\tau-2}, \; \tau \geq 1 \quad (19.A.24)$$

$$\big[\text{Yule-Walker equations}\big]$$

(remember that $\rho_{-\tau} = \rho_\tau$), a recursive relation for obtaining ρ_τ, $\tau > 2$. That is, for $\rho_0 = 1$, $\rho_1 = \alpha_1\rho_0 + \alpha_2\rho_{-1} = \alpha_1 + \alpha_2\rho_1$ or

$$\rho_1 = \frac{\alpha_1}{1-\alpha_2}.$$

Then for $\tau = 2$, Equation 19.A.24 yields $\rho_2 = \alpha_1\rho_1 + \alpha_2\rho_0 = \alpha_1\rho_1 + \alpha_2$ or

$$\rho_2 = \alpha_1\left(\frac{\alpha_1}{1-\alpha_2}\right) + \alpha_2 = \frac{\alpha_1^2}{1-\alpha_2} + \alpha_2.$$

For $\tau = 3$ in Equation 19.A.24, $\rho_3 = \alpha_1\rho_2 + \alpha_2\rho_1$ or

$$\begin{aligned}\rho_3 &= \alpha_1\left(\frac{\alpha_1^2}{1-\alpha_2} + \alpha_2\right) + \alpha_1\left(\frac{\alpha_1}{1-\alpha_2}\right) \\ &= \frac{\alpha_1^3 + 2\alpha_1\alpha_2 - \alpha_1\alpha_2^2}{1-\alpha_2}, \text{ etc.}\end{aligned}$$

For $\tau = 0$ in Equation 19.A.22 so that

$$E(Y_t^2) = \alpha_1 E(Y_t Y_{t-1}) + \alpha_2 E(Y_t Y_{t-2}) + E(Y_t \varepsilon_t),$$

Equation 19.A.23 is replaced by

$$\gamma_0 = \alpha_1\gamma_{-1} + \alpha_2\gamma_{-2} + \sigma_\varepsilon^2. \tag{19.A.23.1}$$

Dividing both sides of this expression by $\gamma_0 (= V(Y_t))$ yields

$$1 = \alpha_1\rho_1 + \alpha_2\rho_2 + \frac{\sigma_\varepsilon^2}{\gamma_0}$$

or

$$V(Y_t) = \frac{\sigma_\varepsilon^2}{1 - \alpha_1\rho_1 - \alpha_2\rho_2}. \tag{19.A.25}$$

Because

$$\begin{aligned}1 - \alpha_1\rho_1 - \alpha_2\rho_2 &= 1 - \frac{\alpha_1^2}{1-\alpha_2} - \frac{\alpha_1^2\alpha_2}{1-\alpha_2} - \alpha_2^2 \\ &= \frac{(1+\alpha_2)(1-\alpha_1-\alpha_2)(1+\alpha_1-\alpha_2)}{1-\alpha_2},\end{aligned}$$

it follows that Equation 19.A.25 is expressible as

$$V(Y_t) = \frac{(1-\alpha_2)\sigma_\varepsilon^2}{(1+\alpha_2)(1-\alpha_1-\alpha_2)(1+\alpha_1-\alpha_2)}. \tag{19.A.25.1}$$

For AR(2) to be stationary, it is necessary that $V(Y_t)$ be finite and nonnegative. Hence, each factor in the denominator of Equation 19.A.25.1 must be positive with $1 - \alpha_2 > 0$. The resulting set of inequalities yields the triangular region of stationarity

$$\begin{aligned} &\alpha_1 + \alpha_2 < 1 \\ &-\alpha_1 + \alpha_2 < 1 \\ &-1 < \alpha_2 < 1. \end{aligned} \tag{19.A.26}$$

Looked at in another manner, $\alpha(L)Y_t = (1 - \alpha_1 L - \alpha_2 L^2)Y_t = \varepsilon_t$ is stationary if the inverses of the roots of the lag operator polynomial $\alpha(L) = 1 - \alpha_1 L - \alpha_2 L^2 = 0$ are all less than one in absolute value or the roots of $\alpha(L) = 0$ lie outside the unit circle. Equivalently, the roots s_1, s_2 of the characteristic polynomial $Z^2 - \alpha_1 Z - \alpha_2 = 0$ are less than one in absolute value or

$$\left| \frac{\alpha_1 \pm \sqrt{\alpha_1^2 + 4\alpha_2}}{2} \right| < 1.$$

Hence, the parameters α_1, α_2 must satisfy Equation 19.A.26.

Once ρ_1, ρ_2 are calculated from Equation 19.A.24, the parameters α_1, α_2 can be determined using Equation 19.A.20.1, i.e., $\boldsymbol{\alpha} = \boldsymbol{P}^{-1}\boldsymbol{\rho}$ or

$$\begin{aligned} \begin{bmatrix} \alpha_1 \\ \alpha_2 \end{bmatrix} &= \begin{bmatrix} 1 & \rho_1 \\ \rho_1 & 1 \end{bmatrix}^{-1} \begin{bmatrix} \rho_1 \\ \rho_2 \end{bmatrix} \\ &= \frac{1}{1-\rho_1^2} \begin{bmatrix} 1 & -\rho_1 \\ -\rho_1 & 1 \end{bmatrix} \begin{bmatrix} \rho_1 \\ \rho_2 \end{bmatrix} \\ &= \begin{bmatrix} \rho_1(1-\rho_2)/(1-\rho_1^2) \\ (\rho_2 - \rho_1^2)/(1-\rho_1^2) \end{bmatrix}. \end{aligned}$$

Mixed Autoregressive-Moving Average Process

A logical extension of the AR(p) and MA(q) processes is the *mixed autoregressive-moving average process of order* (p, q), which is denoted as ARMA(p, q) and written

$$\begin{aligned} Y_t &= \alpha_1 Y_{t-1} + \cdots + \alpha_p Y_{t-p} + \varepsilon_t + \beta_1 \varepsilon_{t-1} \\ &\quad + \cdots + \beta_q \varepsilon_{t-q} \\ &= \sum\nolimits_{j=1}^{p} \alpha_j Y_{t-j} + \sum\nolimits_{i=o}^{q} \beta_i \varepsilon_{t-i} \end{aligned} \tag{19.A.27}$$

with $\beta_0 = 1$. Alternatively, this expression may be rewritten in terms of the lag operator L as

$$(1 - \alpha_1 L - \alpha_2 L^2 - \cdots - \alpha_p L^p)Y_t = (1 + \beta_1 L + \beta_2 L^2 + \cdots + \beta_q L_q)\varepsilon_t$$

or

$$\alpha(L)Y_t = \beta(L)\varepsilon_t, \tag{19.A.28}$$

where $\alpha(L)$ and $\beta(L)$ are, respectively, lag polynomial functions of order p and q in L and ε_t is a white noise process.

As indicated previously, $\alpha(L)Y_t = \beta(L)\varepsilon_t$ is stationary provided that the lag polynomial $\alpha(L) = 0$ has all of its roots lying outside the unit circle, whereas the roots of $\beta(L) = 0$ must lie outside the same for invertibility. In this regard and under these restrictions, the ARMA (p, q) process can be expressed in either MA form as

$$Y_t = \psi(L)\varepsilon_t = \alpha^{-1}(L)\beta(L)\varepsilon_t,$$

where $\psi(L) = \alpha^{-1}(L)\beta(L)$; or in AR form as

$$\pi(L)Y_t = \beta^{-1}(L)\alpha(L)Y_t = \varepsilon_t,$$

where $\pi(L) = \beta^{-1}(L)\alpha(L)$.

From Equation 19.A.27, $E(Y_t) = 0$ because $\alpha(1) \neq 0$ (or $1 - \sum_{j=1}^{p} \alpha_j > 0$: a necessary condition for stationarity) and, for any integer τ, multiplying both sides of Equation 19.A.27 by $Y_{t-\tau}$ and passing to expectations yields the *autocovariance function*

$$\gamma_\tau = \alpha_1 \gamma_{\tau-1} + \cdots + \alpha_p\gamma_{\tau-p} + \gamma_{Y\varepsilon}(\tau) + \beta_1\gamma_{Y\varepsilon}(\tau - 1) + \cdots + \beta_q\gamma_{Y\varepsilon}(\tau - q), \qquad (19.A.29)$$

where $\gamma_{Y\varepsilon}(\tau) = COV(Y_{t-\tau}, \varepsilon_t)$ is the *cross covariance between* Y *and* ε *at lag* τ. Because $E(\varepsilon_t) = E(Y_{t-\tau}) = 0$, it follows that $\gamma_{Y\varepsilon}(\tau) = E(Y_{t-\tau}\varepsilon_t)$. Also, because $Y_{t-\tau}$ depends only on random shocks that have occurred up to time $t - \tau$ (ε_t is independent of any previous Y value), it also follows that $\gamma_{Y\varepsilon}(\tau) = 0$ for $\tau > 0$; and $\gamma_{Y\varepsilon}(\tau) \neq 0$ for $\tau \leq 0$.

For $\tau > q$, Equation 19.A.29 reduces to

$$\gamma_\tau = \alpha_1\gamma_{\tau-1} + \alpha_1\gamma_{\tau-1} + \alpha_2\gamma_{\tau-2} + \cdots + \alpha_p\gamma_{\tau-p} \qquad (19.A.30)$$

(or Equation 19.A.18) so that the *autocorrelation function*

$$\rho_\tau = \alpha_1\rho_{\tau-1} + \alpha_2\rho_{\tau-2} + \cdots + \alpha_p\rho_{\tau-p}$$

or

$$\alpha(L)\rho_\tau = 0 \qquad (19.A.31)$$

(or Equation 19.A.19). Note that, *after lag* q, neither Equation 19.A.30 nor Equation 19.A.31 involves the MA(q) parameters β_i, $i = 0, 1,\ldots,q$. Thus, after lag q, the autocovariances and autocorrelations behave exclusively as those of an AR(p) process. So for the ARMA(p, q) process in Equation 19.A.27, early ρ_τ's will also depend on the MA(q) parameters, i.e., for the ARMA(p, q) process, the values of the autocorrelations ρ_τ, $\tau = 1,\ldots,q$, will depend on the MA(q) parameters $\beta_1,\ldots\beta_q$ as well as on the AR(p) parameters $\alpha_1,\ldots,\alpha_p$; and when $\tau \geq q + 1$, ρ_τ depends only on $\alpha_1,\ldots,\alpha_p$.

When $\tau = 0$, the variance of Y_t is

$$\begin{aligned} V(Y_t) = \gamma_0 &= \alpha_1\gamma_1 + \ldots + \alpha_p\gamma_p + \sigma_\varepsilon^2 \\ &\quad + \beta_1\gamma_{Y\varepsilon}(-1) + \ldots + \beta_q\gamma_{Y\varepsilon}(-q) \end{aligned} \qquad (19.A.32)$$

which has to be solved along with the p equations in Equation 19.A.29, as $\tau = 1, 2,\ldots,p$, so as to obtain $\gamma_0, \gamma_1,\ldots,\gamma_p$.

As a special case of Equation 19.A.27, let us consider the ARMA(1, 1) process

$$Y_t - \alpha_1 Y_{t-1} = \varepsilon_t + \beta_1\varepsilon_{t-1} \qquad (19.A.33)$$

or

$$(1 - \alpha_1 L)Y_t = (1 + \beta_1 L)\varepsilon_t,$$

where ε_t is white noise. For stationarity and invertibility, we require, respectively, $|\alpha_1| < 1$ and $|\beta_1| < 1$. Because $\alpha_1 \neq 1$, $E(Y_t) = 0$.

From Equation 19.A.29 we obtain

$$\gamma_\tau = \alpha_1\gamma_{\tau-1} + \gamma_{Y\varepsilon}(\tau) + \beta_1\gamma_{Y\varepsilon}(\tau - 1)$$

so that, for $\tau = 0, 1$,

$$\begin{aligned} \gamma_0 &= \alpha_1\gamma_1 + \sigma_\varepsilon^2 + \beta_1\gamma_{Y\varepsilon}(-1), \\ \gamma_1 &= \alpha_1\gamma_0 + \beta_1\sigma_\varepsilon^2; \end{aligned} \tag{19.A.34}$$

and, for $\tau > 1$,

$$\gamma_\tau = \alpha_1\gamma_{\tau-1}. \tag{19.A.35}$$

Multiplying both sides of Equation 19.A.33 by ε_{t-1} and taking expectations renders

$$\gamma_{Y\varepsilon}(-1) - \alpha_1\sigma_\varepsilon^2 = \beta_1\sigma_\varepsilon^2. \tag{19.A.36}$$

Then using Equation 19.A.36 to eliminate $\gamma_{Y\varepsilon}(-1)$ from Equation 19.A.35 enables us to write

$$\begin{aligned} \gamma_0 &= \left[\frac{1 + 2\alpha_1\beta_1 + \beta_1^2}{1 - \alpha_1^2}\right]\sigma_\varepsilon^2, \\ \gamma_1 &= \left[\frac{(1 + \alpha_1\beta_1)(\alpha_1 + \beta_1)}{1 - \alpha_1^2}\right]\sigma_\varepsilon^2. \end{aligned}$$

Because $\rho_\tau = \gamma_\tau/\gamma_0$, $\tau = 0, 1, 2, \ldots$, it follows from the preceding two expressions and Equation 19.A.35 that

$$\begin{aligned} \rho_1 &= \frac{(1 + \alpha_1\beta_1)(\alpha_1 + \beta_1)}{1 + 2\alpha_1\beta_1 + \beta_1^2}, \\ \rho_2 &= \alpha_1\rho_1. \end{aligned}$$

Clearly, the autocorrelation function decays exponentially from its starting value ρ_1, with the said decay being smooth if $\alpha_1 > 0$ and oscillatory if $\alpha_1 < 0$. Moreover, the sign of ρ_1 is dictated by the sign of $\alpha_1 + \beta_1$, which thus determines from which side of zero the exponential decay commences.

Suppose that, for the ARMA(p, q) process $\alpha(L)Y_t = \beta(L)\varepsilon_t$: (1) all of the autoregressive roots $\alpha_1, \ldots, \alpha_p$ lie on or outside of the unit circle, with at most one of these α's found on the unit circle; and (2) all of the moving average roots $\beta_1, \ldots, \beta_q$ lie outside the unit circle. In this circumstance, Y_t is said to have a *unit autoregressive root* (or *unit root* for short) if one of the p roots of $\alpha(L) = 0$ is unity, in which case we can factor α(L) as $\alpha(L) = \alpha^*(L)(1 - L)$, where $\alpha^*(L)$ is a lag polynomial in L of degree p –1. Thus, Y_t is ARMA(p – 1, q) in first differences because $\alpha^*(L)(1-L)Y_t = \beta(L)\varepsilon_t$ is $\alpha^*(L)\Delta Y_t = \beta(L)\varepsilon_t$. Clearly, Y_t itself is not stationary because a root of the characteristic polynomial $\alpha(L) = 0$ is on the unit circle; however, ΔY_t constitutes a stationary and invertible ARMA($p-1$, q) process.

If within the ARMA(p, q) process the series Y_t is replaced by $\Delta^d Y_t = (1-L)^d Y_t$, then Y_t is said to be characterized as an *integrated autoregressive-moving average process of order* (p, d, q) and denoted ARIMA(p, d, q), where d denotes the *order of integration* and represents the number of times the ARIMA(p, d, q) process has to be differenced to attain stationarity (integrated processes are discussed in greater detail in Appendix 19.B). For instance, an ARIMA(p, 1, q) process is structured as

$$\alpha(L)(1-L)Y_t = \beta(L)\varepsilon_t$$

or

$$(1-L)Y_t = \alpha^{-1}(L)\beta(L)\varepsilon_t \tag{19.A.37}$$

with all roots of $\alpha(L) = 0$ and $\beta(L) = 0$ lying outside of the unit circle. Thus, ARIMA(p, 1, q) is a stationary and invertible ARMA(p, q) process in first differences and consequently is said to be integrated of order one (denoted $I(1)$). In general, an ARIMA(p, d, q) process is expressable as

$$\alpha(L)(1-L)^d Y_t = \beta(L)\varepsilon_t$$

or

$$(1-L)^d Y_t = \alpha(L)^{-1}\,\beta(L)\varepsilon_t, \tag{19.A.38}$$

with all of the roots of $\alpha(L) = 0$ and $\beta(L) = 0$ located outside of the unit circle. Hence, ARIMA(p, d, q) is a stationary and invertible ARMA(p, q) process after differencing d times and thus is termed *integrated of order d* (denoted $I(d)$). In summary, the ARIMA(p, d, q) process that is integrated of order d has d unit roots and thus the dth difference of any such process is stationary ARMA(p, q).

ADL Process

In a typical *auto regressive distributed lag (ADL) model*, the current value of an explained variable Y_t is specified as a function of its own lagged values (the autoregressive portion), the current and lagged values of all explanatory variables (the distributed lag part), and a random error term ε_t, where $\{\varepsilon_t\}$ is taken to be white noise. Hence, the ADL model appears as

$$Y_t = \alpha_1 Y_{t-1} + \alpha_2 Y_{t-2} + \cdots + \alpha_p Y_{t-p} + \beta_0 X_t + \beta_1 X_{t-1} + \beta_2 X_{t-2} + \cdots + \beta_k X_{t-k} + \varepsilon_t \tag{19.A.39}$$

or, in terms of the lag operator L,

$$\alpha(L)Y_t = \beta(L)X_t + \varepsilon_t, \tag{19.A.39.1}$$

where the lag polynomials

$$\alpha(L) = 1 - \alpha_1 L - \alpha_2 L^2 - \cdots - \alpha_p L^p = \sum\nolimits_{j=0}^{p} \alpha_j L^j,\ \alpha_0 = 1;$$
$$\beta(L) = \beta_0 + \beta_1 L + \beta_2 L^2 + \cdots + \beta_k L^k = \sum\nolimits_{s=0}^{k} \beta_s L^s.$$

As indicated in the preceding section, the $\{Y_t\}$ process will be stationary if the roots of $\alpha(L) = 0$ lie outside the unit circle, and because the autoregressive part of Equation 19.A.39 contains p lags and the distributed lag portion of the same exhibits k lags, this autoregressive distributed lag model will be denoted as ADL(p, k). If $p = k$, then it will simply be written as ADL(p).

Random Walk Processes

Let us consider the AR(1) process

$$Y_t = \alpha_1 Y_{t-1} + \varepsilon_t, \tag{19.A.40}$$

where $|\alpha_1| < 1$ ($\{Y_t\}$ is stationary) and $\{\varepsilon_t\}$ is white noise. A *random walk process* is a special case of Equation 19.A.40 in which $\alpha_1 = 1$, i.e., for convenience, let us modify our notation a bit and denote this particular process as

$$X_t = X_{t-1} + u_t, \tag{19.A.41}$$

where $\{u_t\}$ is white noise. Because in this instance $\alpha(L)X_t = (1 - L)X_t = u_t$ so that the root of $\alpha(L) = 1 - L = 0$ is unity, Equation 19.A.41 is also referred to as a unit root process. It is instructive to compare the properties of Equations 19.A.40 and 19.A.41. As we shall now see, Y_t in Equation 19.A.40 is stationary, whereas the unit root case depicted by X_t in Equation 19.A.41 is nonstationary.

If each of these autoregressive models is transformed into a moving-average model (by successive back-substitutions) and the initial observation of each is assumed to be zero (e.g., $X_0 = Y_0 = 0$), then

$$\begin{aligned} &\text{a. } Y_t = \sum\nolimits_{i=0}^{t-1} \alpha_1^i \varepsilon_{t-i}, \\ &\text{b. } X_t = \sum\nolimits_{i=0}^{t-1} u_{t-i} \end{aligned} \tag{19.A.42}$$

(note that each random shock u_{t-i} has a permanent or nondecaying effect on $\{X_t\}$ because its coefficient is unity). Looking to the statistical properties of these two time series specifications we have

$$\begin{aligned} &1.\ E(Y_t) = \sum\nolimits_{i=0}^{t-1} \alpha_1^i E(\varepsilon_{t-i}) = 0, \\ &\quad E(X_t) = \sum\nolimits_{i=0}^{t-1} E(u_{t-i}) = 0. \\ &2.\ V(Y_t) = \sum\nolimits_{i=0}^{t-1} \alpha_1^{2i} E(\varepsilon_{t-i}^2) = \frac{\sigma_\varepsilon^2}{1-\alpha_1^2}, \\ &\quad V(X_t) = \sum\nolimits_{i=0}^{t-1} E(u_{t-i}^2) = t\sigma_u^2 \to \infty \text{as } t \to \infty. \end{aligned}$$

So although the means of Y_t, X_t are each zero, the variance of Y_t converges to a constant asymptotically, whereas the variance of the random walk or unit root process increases as t increases. Moreover,

$$\begin{aligned} 3.\ COV(Y_{t-\tau}, Y_t) &= E(Y_{t-\tau} Y_t) \\ &= E\left[\left(\sum\nolimits_{i=0}^{t-\tau-1} \alpha_1^i \varepsilon_{t-\tau-i}\right)\left(\sum\nolimits_{i=0}^{t-1} \alpha_1^i \varepsilon_{t-i}\right)\right] \end{aligned}$$

$$= \gamma_\tau^Y = \frac{\alpha_1^\tau \sigma_\varepsilon^2}{1-\alpha_1^2},$$

$$COV(X_{t-\tau}, X_t) = E(X_{t-\tau} X_t)$$

$$= E\left[\left(\sum_{i=0}^{t-\tau-1} u_{t-\tau-i}\right)\left(\sum_{i=0}^{t-1} u_{t-i}\right)\right]$$

$$= \gamma_\tau^X = (t-\tau)\sigma_u^2.$$

Here, too, the autocovariance of Y_t converges asymptotically to a constant, whereas the autocovariance of X_t is time dependent. Finally, looking to autocorrelations,

4. $\rho_\tau^Y = \gamma_\tau^Y / \gamma_0^Y = \alpha_1^\tau \to 0$ as $\tau \to \infty$
(with $|\alpha_1| < 1$),

$$\rho_\tau^X = \gamma_\tau^X / \gamma_\tau^0 = \frac{t-\tau}{\tau}.$$

Hence, the autocorrelation function of X_t approaches zero only for very large values of τ.

The upshot of this discussion is that an AR(1) process with a unit root (a random walk process) is nonstationary. As Figure 19.A.1 reveals, the time path of stationary Y_t intersects its mean level at least once; Y_t converges toward its mean (it is said to be *mean-reverting*) while fluctuating randomly about the same. Contrary to the behavior of Y_t, X_t increases (decreases) systematically and is not mean-reverting as t increases. Interestingly enough, the random walk or unit root process (Equation 19.A.41) can be made stationary by differencing, i.e., the variable $\Delta X_t = X_t - X_{t-1} = u_t$ is stationary about zero because $\{u_t\}$ is white noise. We shall elaborate on this point later.

Before closing this section, let us consider two important modifications of Equations 19.A.40 (a stationary AR(1) process) and 19.A.41 (a nonstationary AR(1) process with a unit root). First, suppose we have the specification

$$Y_t = \mu + \alpha_1 Y_{t-1} + \varepsilon_t, \tag{19.A.43}$$

where $|\alpha_1| < 1$ and $\{\varepsilon_t\}$ is white noise. Here Equation 19.A.43 is a stationary AR(1) process with mean $E(Y_t) = \mu/(1 - \alpha_1)$. However, when $\alpha_1 = 1$ in Equation 19.A.43, we have (again modifying our notation a bit)

$$X_t = \mu + X_{t-1} + u_t, \tag{19.A.44}$$

where $\{u_t\}$ is white noise. This type of expression is termed a *random walk with nonzero* ($\mu \neq 0$) *drift*; it also constitutes a nonstationary AR(1) process with a unit root (Figure 19.A.2). Next, let

$$Y_t = \mu + \lambda t + \alpha_1 Y_{t-1} + \varepsilon_t, \tag{19.A.45}$$

where again $|\alpha_1| < 1$ with $\{\varepsilon_t\}$ white noise. Now $\{Y_t\}$ is a stationary AR(1) process about a linear trend (assuming $\lambda \neq 0$). However, if $\alpha_1 = 1$, the appropriate modification of Equation 19.A.45 is

$$X_t = \mu + \lambda t + X_{t-1} + u_t, \tag{19.A.46}$$

with $\{u_t\}$ taken to be white noise. Here $\{X_t\}$ is a *random walk process about a time trend* ($\lambda \neq 0$); it too is nonstationary AR(1) with a unit root (Figure 19.A.2).

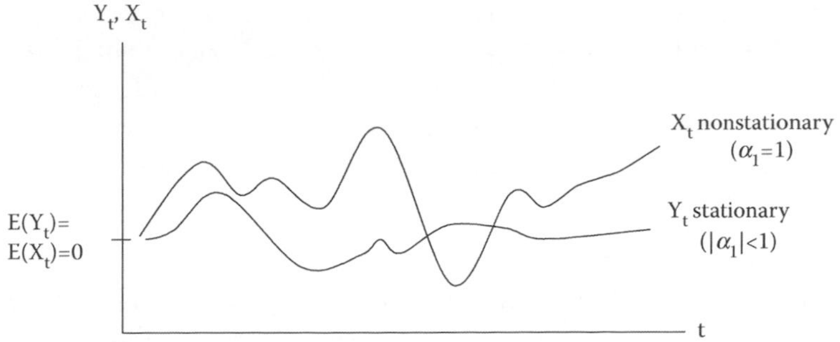

FIGURE 19.A.1
Stationary versus nonstationary processes.

If from Equation 19.A.44 we form the difference $\Delta X_t = X_t - X_{t-1} = \mu + u_t$, then ΔX_t is stationary about μ because $\{u_t\}$ is white noise. Similarly, from Equation 19.A.46, the difference $\Delta X_t = X_t - X_{t-1} = \mu + \lambda t + u_t$ is stationary about a linear trend.

Let us elaborate briefly on the notion of *trend*. Specifically, a trend component of a time series $\{Y_t\}$ can be characterized as *deterministic* or *stochastic*. To see this, suppose a time series $\{Y_t\}$ always changes by a constant amount λ from period to period or $\Delta Y_t = \lambda$. Then the time path of Y_t can be specified as

$$Y_t = Y_0 + \lambda t, \tag{19.A.47}$$

where Y_0 is the initial value of Y_t (for t = 0). In this instance, Y_t is said to have a nonstationary *deterministic linear time trend* component λt. If we now insert into Equation 19.A.47 a stationary moving-average scheme $\beta(L)\varepsilon_t = \varepsilon_t + \beta_1\varepsilon_{t-1}$, $|\beta_1| < 1$, so that

$$Y_t = Y_0 + \lambda t + \varepsilon_t + \beta_1\varepsilon_{t-1}, \tag{19.A.48}$$

then $\beta(L)\varepsilon_t$ is stationary and Y_t only temporarily derivates from its trend by the amount $\beta(L)\varepsilon_t$. Hence, Equation 19.A.48 is termed a *trend-stationary model*; no permanent departures from trend are exhibited.

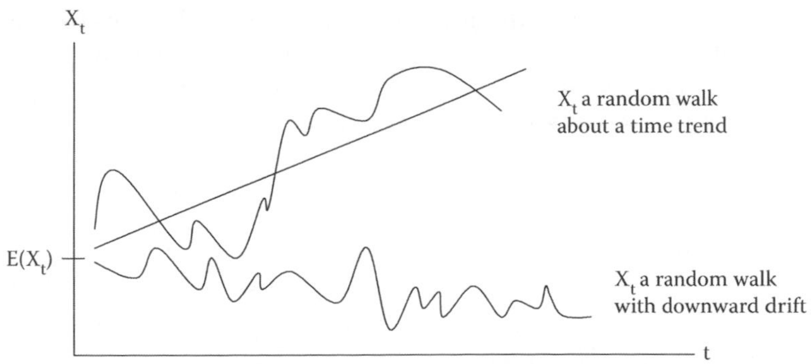

FIGURE 19.A.2
Random walk processes.

What if the inter-period change in Y_t is respecified as $\Delta Y_t = \lambda + \varepsilon_t$, where ε_t is white noise? Then the period-to-period change in Y_t is, on average, λ ($E(Y_t) = \lambda$ since $E(\varepsilon_t) = 0$). Now Y_t follows the time path

$$Y_t = Y_0 + \lambda t + \sum_{i=1}^{t} \varepsilon_i, \tag{19.A.49}$$

where again Y_0 is the starting value of Y_t. In addition to the deterministic trend term λt, the model in Equation 19.A.49 entails a *stochastic trend* component $\sum_{i=1}^{t} \varepsilon_i$ consisting of cumulated shocks. Because each coefficient on ε_i is unity, each of these random shocks or disturbances has a permanent effect on Y_t. Although Equation 19.A.49 is nonstationary (e.g., its mean is $E(Y_t) = Y_0 = \lambda t$), its first difference $\Delta Y_t = Y_t - Y_{t-1} = \lambda + \varepsilon_t$ is stationary.

As we shall see below, a nonstationary random walk or unit root process can be made stationary by differencing and will thus be termed a *difference stationary process*. We note briefly that, if a nonstationary process exhibiting a deterministic trend can be made stationary by subtracting λt from Y_t, then that process is said to be *trend stationary*.

Appendix B: Integrated Processes/Series [Engle and Granger 1991; Banerjee et al. 1993; Hargreaves 1994; Rao 1994; Charemza and Deadman 1997; Madalla and Kim 1998; McAleer and Oxley 1999; Enders 2004]

We indicated above that a time series $\{Y_t\}$ is (weakly) stationary if Equation 19.A.2 holds. Implicit in this definition is the notion that a stationary series cannot grow over time. However, the changes $\Delta Y_t = Y_t - Y_{t-1}$ in the series might be stationary. In this regard, the nonstationary variable Y_t is said to be *integrated of order one*, denoted $Y_t : I(1)$, if the changes in Y_t, ΔY_t, form a stationary series, in which case ΔY_t is said to be *integrated of order zero* and written $\Delta Y_t : I(0)$. Thus, an *integrated series* with order of integration greater than zero is nonstationary. So if $Y_t : I(0)$, then $\{Y_t\}$ is a stationary series.

Let us briefly consider the principal differences between time series that are integrated of order zero versus integrated of order one. If $Y_t : I(0)$, then Y_t has a fixed mean and a constant (or bounded) variance. This type of series has a tendency to frequently return to, and intersect, its mean value. Moreover, a value of Y_t far from its mean has a tendency to be followed by values nearer to the mean. In this regard, the mean of the series is termed its *attractor*.

As far as $Y_t : I(1)$ is concerned, in the absence of drift, any such series has no attractor but has a variance that increases with time. Hence, in this instance, Y_t rarely returns to any specific level and tends to wander erratically. An $I(1)$ series will be relatively smooth when its time profile is compared to that of any $I(0)$ series and displays fluctuations that are less pronounced than those of an $I(0)$ series.

In general, a nonstationary series $\{Y_t\}$ that can be transformed to a stationary series by differencing d times in succession is said to be integrated of order d and denoted $Y_t : I(d)$. Hence, $Y_t : I(d)$ is difference stationary if $\Delta^d Y_t : I(0)$. For example, if difference stationary $Y_t : I(2)$, then $\Delta^2 Y_t = \Delta(\Delta Y_t) : I(0)$ or $\Delta^2 Y_t$ is stationary. And, if $Y_t : I(0)$, then Y_t is stationary and no differencing is needed. Moreover, if $Y_t : I(d)$, then $\Delta^b Y_t : I(d - b)$, $d > b > 0$. It is also true that, if $Y_t : I(d)$, then $\alpha + \beta Y_t : I(d)$, where α and β are constants.

Looked at in another light, we can state that an *integrated series* accumulates past shocks or innovations, i.e., an integrated series is expressible as a sum. For example, we noted in

Appendix 19.A that the random walk $Y_t = Y_{t-1} + \varepsilon_t$, $\{\varepsilon_t\}$ white noise, can be "integrated" or written as the sum of all previous errors or $Y_t = \sum_{j=0}^{\infty} \varepsilon_{t-j}$, with $Y_0 = 0$. Hence, $Y_t : I(1)$ because old and new shocks have equal weight in determining Y_t so that Y_t obviously displays *permanent* or *long memory*. Hence, this process was deemed nonstationary because its future time path depends on all past innovations; the said path is not wedded to some mean level to which it eventually reverts.

However, $\Delta Y_t = Y_t - Y_{t-1} = \sum_{j=0}^{\infty} \varepsilon_{t-j} - \sum_{j=0}^{\infty} \varepsilon_{t-j-1} = \varepsilon_t : I(0)$, i.e., ΔY_t is stationary because it is white noise and thus exhibits *temporary* or *short memory*. So with a long-memory series, an "old" shock to the series has a noticeable impact on the current value of the series; for a short-memory series, an "old" shock to the series has little or no effect on the current level of the series.

In contrast to the random walk case, suppose $Y_t = \alpha Y_{t-1} + \varepsilon_t$, $\{\varepsilon_t\}$ white noise. After integrating or summing, $Y_t = \sum_{j=0}^{\infty} \alpha_1^j \varepsilon_{t-j}$, $Y_0 = 0$. If $|\alpha_1| < 1$ (the stationary condition), $\alpha_1^j \to 0$ as $j \to \infty$ so that $Y_t : I(0)$. Hence, the effect of or level of memory associated with past shocks declines as time passes. However, if $\alpha_1 = 1$, then $Y_t = \sum_{j=0}^{\infty} \varepsilon_{t-j} : I(1)$ since Y_t is now the sum of all past innovations.

Appendix C: Cointegrated Processes/Series [See the References for Appendix B]

In time series analysis, one is often looking to estimate some *long-run relationship* between a set of variables, say, X and Y. Such a relationship essentially holds, on the average, over time. (This is in contrast to a *short-run relationship* that involves an association that does not persist or endure over time.)

It may be the case that a set of variables will enter structurally into some *stable equilibrium state*, a state to which the variables are attracted and from which there is no inherent tendency to emerge or change. If this state corresponds to a *long-run equilibrium*, then we have a "precise relationship" to which the variables systematically gravitate or converge over time; a state in which the variables are devoid of any time index or subscript.

Given the variables X and Y, a long-run equilibrium relationship $Y = BX$ holds between these variables if the error or discrepancy $\varepsilon_t = Y_t - \beta X_t$ between the actual observations and the equilibrium relationship is a stationary process, i.e., $\{\varepsilon_t\} : I(0)$ so that $Y = BX$ serves as an attractor for the X_t, Y_t variables. Hence, the short-run discrepancy ε_t has no systematic tendency to grow over time. Note that, if both X_t, $Y_t : I(0)$ to begin with, then $\varepsilon_t = Y_t - \beta X_t : I(0)$ for "any" β.

In a long-run equilibrium setting, the variables X_t, Y_t are linked by the relation $Y = BX$ and thus their time paths should not steadily diverge. Although these variables may drift apart in the short term, divergence from a stable long-run equilibrium state must eventually come to a halt. The statistical representation of the concept of a sustainable equilibrium relation connecting X_t and Y_t is incorporated in the notion of *cointegration*. It is cointegration that enables us to depict the structure of an equilibrium or stationary relationship between variables that are not themselves stationary. In this regard, although X_t and Y_t are not stationary, under certain circumstances, these variables, in combination, may be stationary. Specifically, although X_t and Y_t are, say, integrated of order one, deviations from an equilibrium relation linking these variables may be stationary or integrated of order zero. So despite the fact that X_t and Y_t are integrated processes whose variances increase over time, the linear combination of X_t and Y_t, which defines the equilibrium structure joining these variables, may have a constant or finite variance and thus exhibit stationarity.

It is generally true that, if both $X_t, Y_t : I(d)$, then there exist constants θ_1, θ_2 such that the linear combination $\theta_1 X_t + \theta_2 Y_t : I(d)$, although it is possible that $\theta_1 X_t + \theta_2 Y_t : I(d - b)$, $b > 0$. In this regard, the variables X_t and Y_t are said to be *cointegrated of order d, b* ($d \geq b \geq 0$), denoted $X_t, Y_t : CI(d, b)$, if

1. Both $X_t, Y_t : I(d)$, and
2. There exists a linear combination $\theta_1 X_t + \theta_2 Y_t : I(d - b)$,

where $\boldsymbol{\theta}' = (\theta_1, \theta_2) \neq \boldsymbol{0}'$ is the *cointegrating vector.* (Note that X_t and Y_t are required to be integrated of the same order.) We may view b as the amount by which the order of integration is reduced by taking X_t and Y_t in combination. Hence, a linear relation exists between these variables that has an order of integration $(d - b)$ lower than those of either X_t and Y_t taken individually. If $d - b > 0$, then the variables X_t and Y_t may deviate from the linear relation by an ever-increasing amount. Hence, the said linear relation is not one which constitutes a long-run equilibrium.

Let us now turn to an important special case of the preceding definition of cointegrated series. It is generally not true that there exists a cointegrating vector that produces a linear combination of X_t and Y_t that is stationary. However, suppose $d = b = 1$ above and the components of the cointegrating vector are taken to coincide with the parameters of some long-run equilibrium relation $Y = BX$ between X_t and Y_t. That is, for the cointegrating vector $(1, -\beta)$, the discrepancy $\varepsilon_t = Y_t - \beta X_t$ can be considered to be equilibrium error: ε_t measures the short-run departure of X_t and Y_t from their long-run equilibrium relation.

In this regard, let both $X_t, Y_t : I(1)$. Then $X_t, Y_t : CI(1, 1)$ if there exists a cointegrating vector $(1, -\beta)$ such that the linear combination of these variables is stationary or $\varepsilon_t = Y_t - \beta X_t : I(0)$. Note that, if β exists, then it must be unique. Hence, X_t and Y_t share a common stochastic trend and thus do not drift apart without bound, and keeping track of the equilibrium error ε_t enables us to examine information concerning long-run behavior in an $I(0)$ setting. Because the discrepancy $\varepsilon_t : I(0)$, it is clear that ε_t will not drift far from its mean and will often cross its mean level so that equilibrium will sporadically be restored.

To amplify the preceding discussion, we may note that X_t and Y_t will be cointegrated (i.e., $X_t, Y_t : I(1)$ and $\varepsilon_t = Y_t - \beta X_t : I(0)$) if they exhibit a structural decomposition of the form

$$\begin{aligned} Y_t &= \beta W_t + \eta_{Yt} \\ X_t &= W_t + \eta_{Xt}, \end{aligned} \tag{19.C.1}$$

where it is assumed that $W_t : I(1)$, and both η_{Xt} and η_{Yt} are white noise. Given that $W_t : I(1)$, it follows that both $X_t, Y_t : I(1)$ (thus the $I(1)$ property of the X_t and Y_t series is attributed to the single $I(1)$ *common factor* W_t) with

$$\begin{aligned} \varepsilon_t &= Y_t - \beta X_t = \beta W_t + \eta_{Yt} - \beta(W_t + \eta_{Xt}) \\ &= \eta_{Yt} - \beta\eta_{Xt} : I(0). \end{aligned}$$

Hence, ε_t is a linear combination of the stochastic components η_{Xt} and η_{Yt} so that $X_t, Y_t : CI(1, 1)$. Conversely, if $X_t, Y_t : CI(1, 1)$, then they will display a common factor structure as in Equation 19.C.1.

Looked at it in another manner, if $\varepsilon_t : I(0)$, then the linear regression $Y_t = \beta X_t + \varepsilon_t$ makes sense; with X_t and Y_t cointegrated, these variables do not drift too far apart over time so that there exists a long-run equilibrium relation between them, i.e., the regression of Y_t on X_t is not *spurious*. Moreover, for $\varepsilon_t = Y_t - \beta X_t$:

a. If $Y_t : I(1)$ and $X_t : I(0)$, then $\varepsilon_t : I(1)$ and X_t and Y_t are not cointegrated. Moreover, we can expect that $\beta = 0$.
b. If both $X_t, Y_t : I(0)$, then $\varepsilon_t : I(0)$ and the notion of cointegration is trivial.
c. If $Y_t : I(0)$ and $X_t : I(1)$ then $\varepsilon_t : I(1)$ and X_t and Y_t are not cointegrated; here too, the only plausible value for β is zero.
d. If both $X_t, Y_t : I(1)$ and $(1, -\beta)$ is a cointegrating vector, then $\varepsilon_t : I(0)$.

To generalize the above definition of cointegrated series to the m variable case, suppose $\boldsymbol{X}_t$ constitutes a $(m \times 1)$ vector of time series variables $X_{1t}, X_{2t},\ldots,X_{mt}$. Then these variables are said to be *cointegrated of order d, b*, denoted $X_{1t},\ldots,X_{mt} : CI(d, b)$, if

1. Each $X_{jt} : I(d), j = 1,\ldots,m$; and
2. There exists a linear combination $\boldsymbol{\theta}'\boldsymbol{X}_t = \sum_{j=1}^{m} \theta_j X_{jt} : I(d-b)$,

where $\boldsymbol{\theta}' = (\theta_1,\ldots,\theta_m) \neq \boldsymbol{0}'$ is the *cointegrating vector.*

Given that the equilibrium relation $Y = BX$ functions as an attractor, it follows that ε_t serves as a measure of the extent to which the model is out of equilibrium and will thus be termed *equilibrium error.* In this regard, we can expect that the value of the next change in Y_t will be influenced by the sign and magnitude of the preceding equilibrium error ε_{t-1}. In this sense then, we have a dynamic disequilibrium adjustment process that is error correcting: it guides the model variables to the long-run equilibrium relation $Y = BX$.

As we shall soon see, this ECM encompasses a strategy for modeling simultaneously levels and differences; it involves the dynamics of both long-run levels and short-run changes. The focal point of the ECM is the term ε_{t-1}; it is introduced into a regression equation as an explanatory variable that provides additional information on the extent of an adjustment to a deviation from equilibrium in that, as constituted herein, a proportion of the disequilibrium from one period is corrected in the next period.

To set the stage for the development of a regression model incorporating an ECM, let us note first that we require the X_t and Y_t variables to be cointegrated and nonstationary. Remember that cointegration implies the presence of an adjustment mechanism that prevents the equilibrium errors $\varepsilon_t = Y_t - \beta X_t$ from becoming increasingly disparate because $\varepsilon_t : I(0)$. The connection between the long-run notion of a set of cointegrated variables and the short-run dynamics of an ECM is captured by the *Engle-Granger Representation Theorem* [Engle and Granger 1987, 1991]: Let both $X_t, Y_t : I(1)$. If $X_t, Y_t : CI(1, 1)$, with cointegrating vector $(1, -\beta)$, then X_t and Y_t may be taken to be generated by an ECM. Conversely, if an ECM incorporates the $I(1)$ variables X_t and Y_t, then these variables must be cointegrated. Hence, the ECM models adjustments in Y_t that depend not on the level of X_t but on the extent to which Y_t deviates from its equilibrium relation with X_t. So if the long-run equilibrium relation is $Y = BX$, then the error-correction term is $\varepsilon_{t-1} = Y_{t-1} - \beta X_{t-1}$. Our starting point for constructing an ECM model is the dynamic linear regression equation represented by the ADL(1,1) model. In its most basic form,

$$Y_t = \alpha_0 + \alpha_1 Y_{t-1} + \beta_0 X_t + \beta_1 X_{t-1} + \varepsilon_t, \tag{19.C.2}$$

where $|\alpha_1| < 1$ and the ε_t's are independent and identically distributed with mean zero and constant variance σ_ε^2. Let us rewrite Equation 19.C.2 as

$$Y_t - Y_{t-1} = \alpha_0 + (\alpha_1 - 1)Y_{t-1} + \beta_0 X_t + \beta_1 X_{t-1} + \beta_0 X_{t-1} - \beta_0 X_{t-1} + \varepsilon_t$$

or

$$\Delta Y_t = \alpha_0 + (\alpha_1 - 1)Y_{t-1} + \beta_0 \Delta X_t + (\beta_0 + \beta_1)X_{t-1} + \varepsilon_t. \tag{19.C.3}$$

Suppose the long-run equilibrium values of X_t and Y_t are given by $E(X_t) = E(X_{t-1}) = X_*$ and $E(Y_t) = E(Y_{t-1}) = Y_*$, respectively, with $E(\varepsilon_t) = 0$ for all t. Then taking the expectation of both sides of Equation 19.C.2 yields

$$Y_* = \alpha_0 + \alpha_1 Y_* + \beta_0 X_* + \beta_1 X_*$$

or

$$Y_* = \frac{\alpha_0}{1-\alpha_1} + \frac{\beta_0 + \beta_1}{1-\alpha_1} X_* = k_0 + k_1 X_*,$$

the *long-run equilibrium relation*. Because $k_1 = (\beta_0 + \beta_1)/(1 - \alpha_1)$, it follows that $\beta_0 + \beta_1 = -(\alpha_1 - 1)k_1$. A substitution of $\beta_0 + \beta$ into Equation 19.C.3 gives us

$$\Delta Y_t = \alpha_0 + (\alpha_1 - 1)(Y_{t-1} - k_1 X_{t-1}) + \beta_0 \Delta X_t + \varepsilon_t, \tag{19.C.4}$$

where $Y_{t-1} - k_1 X_{t-1} = \varepsilon_{t-1}$ represents the error-correction component. In this regard, we may view the term $\alpha_1 - 1 < 0$ as the short-run *speed of adjustment* in Y_t to deviations from the long-run equilibrium relation $Y_* = k_0 + k_1 X_*$ in the previous period.

In general, how does the ECM operate? Given the long-term attractor $Y = BX$, we know that the previous period's departure from this equilibrium error is $\varepsilon_{t-1} = Y_{t-1} - \beta X_{t-1}$. Specifically, suppose that Y_t is currently not on its long-run equilibrium path and that $\varepsilon_{t-1} > 0$ or $Y_{t-1} > \beta X_{t-1}$. Then, from Equation 19.C.4, $\Delta Y_t < 0$ when $\alpha_1 - 1 < 0$, i.e., the ECM ensures that $Y_t < Y_{t-1}$ when Y_{t-1} is above its long-run equilibrium value. For $\varepsilon_{t-1} < 0$ or $Y_{t-1} < \beta X_{t-1}$, Equation 19.C.4 guarantees that $\Delta Y_t > 0$ when $\alpha_1 - 1 < 0$ so that $Y_1 > Y_{t-1}$ when Y_{t-1} is below its long-run equilibrium level. So given $\alpha_1 - 1 < 0$, Y_t will be attracted to its long-run equilibrium path $Y = BX$, from either above or below, and the adjustment will occur as a proportion of the preceding period's error term ε_{t-1}.

To summarize, if both X_t, Y_t : $I(1)$ and cointegrated, with cointegrating vector $(1, -\beta)$, then the deviations $\varepsilon_t = Y_t - \beta X_t$ of Y_t from its long-run equilibrium trend $Y = BX$ are $I(0)$. Hence, a regression model in first differences and an error-correction term ε_{t-1} can be developed wherein ΔX_t, ΔY_t, and ε_{t-1} are all $I(0)$ and thus standard regression techniques apply, i.e., no stochastic or deterministic trends in the data can render a spurious regression equation. Moreover, provided $\alpha_1 - 1 < 0$ in Equation 19.C.4, the ECM enables us to incorporate long-run equilibrium factors into a dynamic regression model.

Appendix D: Testing Downward for Unit Roots [See the References for Appendix B]

Consider the AR(2) model

$$Y_t = \alpha_1 Y_{t-1} + \alpha_2 Y_{t-2} + \varepsilon_t \tag{19.D.1}$$

or $\alpha(L)Y_t = \varepsilon_t$, where the lag polynomial $\alpha(L) = 1 - \alpha_1 L - \alpha_2 L^2$ and ε_t is white noise. Let us rewrite $\alpha(L)$ as $\alpha(L) = (1 - s_1 L)(1 - s_2 L)$, where s_1 and s_2 are the two characteristic roots of the characteristic equation $Z^2 - \alpha_1 Z - \alpha_2 = 0$. We know from Appendix 19.A (see "Autoregressive Process of Order p") that the stationarity of Equation 19.D.1 requires that $|s_j| < 1, j = 1, 2$, or the roots $s_j^{-1}, j = 1, 2$, of $\alpha(L) = 0$ lie outside the unit circle.

Suppose that the AR(2) process in Equation 19.D.1 is nonstationary because $\alpha(L)$ has a single unit root, e.g., suppose, for instance, that $s_2 = 1$ and $s_1^{-1} > 1$. Then in this case,

$$\begin{aligned}\alpha(L)Y_t &= (1 - s_1 L)(1 - s_2 L) = (1 - s_1 L)(1 - L)\\ &= (1 - s_1 L)\Delta Y_t = \Delta Y_t - s_1 \Delta Y_{t-1} = \varepsilon_t.\end{aligned}$$

Hence, the nonstationary AR(2) process that describes Y_t implies that ΔY_t is generated by a stationary AR(1) process so that $Y_t : I(1)$ and $\Delta Y_t : I(0)$.

If $\alpha(L)$ has two unit roots, then both $s_1 = s_2 = 1$

and

$$\begin{aligned}\alpha(L)Y_t &= (1 - L)(1 - L)Y_t = (1 - L)^2 Y_t = \Delta^2 Y_t = \Delta(\Delta Y_t)\\ &= \Delta(Y_t - Y_{t-1}) = \Delta Y_t - \Delta Y_{t-1} = \varepsilon_t.\end{aligned} \tag{19.D.2}$$

In this instance, $\Delta Y_t : I(1)$ and $\Delta^2 Y_t : I(0)$. Hence, the Y_t series must be differenced twice in order for a stationary series to obtain, i.e., $Y_t : I(2)$.

Give the above discussion, for test purposes, suppose we start with Equation 19.D.2 and proceed in the reverse or downward direction. Because $\alpha(L)Y_t = \varepsilon_t = (1 - s_1 L)(1 - s_2 L)Y_t$, it follows that

$$(1 - s_1 L - s_2 L + s_1 s_2 L^2)Y_t = \varepsilon_t$$

or

$$Y_t = (s_1 + s_2)Y_{t-1} - s_1 s_2 Y_{t-2} + \varepsilon_t. \tag{19.D.3}$$

Then substituting Equation 19.D.3 into Equation 19.D.2 yields

$$\begin{aligned}\Delta^2 Y_t &= (s_1 s_2 - 1)\Delta Y_{t-1} - (1 - s_1 - s_2 + s_1 s_2)Y_{t-1} + \varepsilon_t\\ &= (s_1 s_2 - 1)\Delta Y_{t-1} - (1 - s_1)(1 - s_2)Y_{t-1} + \varepsilon_t\\ &= \delta_1 \Delta Y_{t-1} + \delta_2 Y_{t-1} + \varepsilon_t,\end{aligned} \tag{19.D.4}$$

where $\delta_1 = s_1 s_2 - 1$ and $\delta_2 = -(1 - s_1)(1 - s_2)$.

The process of testing downward for a unit root thus consists of the following steps:

1. Step 1: Test $H_0 : Y_t : I(2)$ vs. $H_1 : Y_t : I(1)$. Under the null hypothesis of two unit roots, $s_1 = s_2 = 1$ so that $\delta_1 = \delta_2 = 0$; under the alternative hypothesis, one of the characteristic roots, s_1 or s_2, is unity and, as required, the other root in absolute value is taken to exceed unity. Hence we may take, under the alternative hypothesis, $\delta_1 < 0$ and $\delta_2 = 0$. Clearly $\delta_2 = 0$ satisfies both the null and alternative hypotheses. Hence, a Dickey-Fuller test of the null hypothesis is conducted by the OLS estimation of Equation 19.D.4 subject to $\delta_2 = 0$ or by estimating

$$\Delta^2 Y_t = \delta_1 \Delta Y_{t-1} + \varepsilon_t \tag{19.D.5}$$

and determining whether $\hat{\delta}_1$ is significantly negative by comparing the t statistic for $\hat{\delta}_1$ with the τ critical value taken from the Dickey-Fuller tables. If H_0 is rejected, proceed to Step 2.

2. Step 2: Test $H_0 : Y_t : I(1)$ vs. $H_1 : Y_t : I(0)$. Under this null hypothesis of a single unit root, we take $\delta_1 < 0$ and $\delta_2 = 0$; and under the alternative hypothesis, we must have $\delta_1 < 0$ and $\delta_2 < 0$ (because we require $|s_j| < 1, j = 1, 2$, for the stationarity of Y_t). Equation 19.D.4 is then estimated via OLS and the negativity of $\hat{\delta}_2$ is tested by comparing its t value to the Dickey-Fuller critical value τ. If $\hat{\delta}_2$ is not significantly negative, then the null hypothesis cannot be rejected. However, if the null hypothesis is rejected, then we can conclude that $Y_t : I(0)$, i.e., Y_t has no unit root and thus is stationary.

3. End.

As a final point, we mention briefly that augmentation terms can be included in Equation 19.D.4 to address the issue of potential autocorrelation, whereas a constant as well as a deterministic trend term can also be inserted into Equation 19.D.4 if warranted.

Exercises

19-1. For the MA(1) process $Y_t = \varepsilon_t + 0.7\varepsilon_{t-1}$, find the following:

a. $E(Y_t)$.

b. $V(Y_t)$.

c. γ_1.

d. ρ_0 and ρ_1.

e. Is Y_t invertible?

19-2. Given the MA(2) process $Y_t = \varepsilon_t + 0.5\varepsilon_{t-1} - 0.4\,\varepsilon_{t-2}$, determine the following:

a. $E(Y_t)$.

b. $V(Y_t)$.

c. γ_1 and γ_2.

d. ρ_0, ρ_1, and ρ_2.

e. Is Y_t invertible?

19-3. For the AR(1) process $Y_t = 0.8Y_{t-1} + \varepsilon_t$, find the following:

a. $E(Y_t)$.

b. $V(Y_t)$.

c. γ_0, γ_1, γ_2, and γ_3.

d. ρ_0, ρ_1, ρ_2, and ρ_3.

e. Is Y_t stationary?

19-4. Given the AR(2) process $Y_t = 0.8Y_{t-1} - 0.6Y_{t-2} + \varepsilon_t$, determine the following:

a. $E(Y_t)$.

b. $V(Y_t)$.

c. ρ_0, ρ_1, ρ_2, and ρ_3.

d. Is the Y_t process stationary?

19-5. For the ARMA(1, 1) process $Y_t = 0.7Y_{t-1} + \varepsilon_t - 0.\ 8\varepsilon_{t-1}$, find the following:

a. $E(Y_t)$.

b. Is Y_t stationary and invertable?

c. γ_0, γ_1, and γ_2.

d. ρ_0, ρ_1, and ρ_2.

19-6. Test the time series variable Y_t (use Table E.19-1) for the presence of a unit root via the following:

a. The Example 19.1 four-step methodology.

b. The *%DFTEST* macro of Example 19.2.

c. The *STATIONARITY* option in PROC ARIMA (Example 19.3).

19-7. Does the time series variable Y_t (Table E.19-2) possess a unit root? Base your answer on the following:

a. The four-step procedure of Example 19.1.

b. The *%DFTEST* macro of Example 19.2.

c. The *STATIONARITY* option in PROC ARIMA (Example 19.3).

19-8. Using the Table E.19-3 dataset, determine whether the X_t and Y_t variables are $I(1)$ and cointequated via the Engle-Granger two-step procedure. If they are cointegrated, estimate the parameters of Equation 19.20.1 (remember that ECM terms are included) and test for Granger causality.

TABLE E.19-1

Time Series Observations on Variable Y

Year	Y_t
1970	534.8
1971	579.4
1972	600.8
1973	623.6
1974	616.1
1975	657.5
1976	671.6
1977	683.8
1978	680.9
1979	721.7
1980	737.2
1981	756.6
1982	800.3
1983	832.5
1984	876.4
1985	929.3
1986	984.6
1987	1011.4
1988	1058.1
1989	1087.6
1990	1231.6
1991	1298.2
1992	1369.7
1993	1438.6
1994	1479.4
1995	1475.0
1996	1512.2
1997	1480.0
1998	1534.7
1999	1639.0

TABLE E.19-2

Time Series Observations on Variable Y

Year	Y_t
1975	337.3
1976	341.6
1977	350.1
1978	363.4
1979	370.0
1980	394.1
1981	405.4
1982	413.8
1983	418.0
1984	440.4
1985	452.0
1986	461.4
1987	482.0
1988	500.5
1989	528.0
1990	557.5
1991	585.7
1992	602.7
1993	634.4
1994	657.9
1995	672.1
1996	696.8
1997	737.1
1998	767.9
1999	762.8
2000	779.4
2001	823.1
2002	864.3
2003	903.2
2004	927.6

TABLE E.19-3
Time Series Observations on Variables X, Y

Year	X_t	Y_t
1974	398.2	362.3
1975	402.9	369.3
1976	427.0	394.1
1977	446.5	405.4
1978	455.2	413.8
1979	461.0	418.0
1980	479.0	440.4
1981	489.6	452.0
1982	503.9	461.4
1983	524.8	482.0
1984	542.7	500.5
1985	580.5	528.0
1986	616.3	557.5
1987	647.0	585.7
1988	673.1	602.7
1989	701.4	634.4
1990	712.5	646.5
1991	722.7	657.9
1992	751.7	672.1
1993	779.1	698.8
1994	810.3	737.1
1995	865.2	767.9
1996	857.7	762.8
1997	874.8	779.4
1998	906.9	823.1
1999	943.3	864.3
2000	988.6	903.2
2001	1015.5	927.6
2002	1021.7	931.8
2003	1049.7	950.5
2004	1058.5	963.3
2005	1098.5	1009.2
2006	1170.3	1063.1
2007	1179.7	1072.3

Appendix A

TABLE A.1

Standard Normal Areas (Z is N(0, 1))

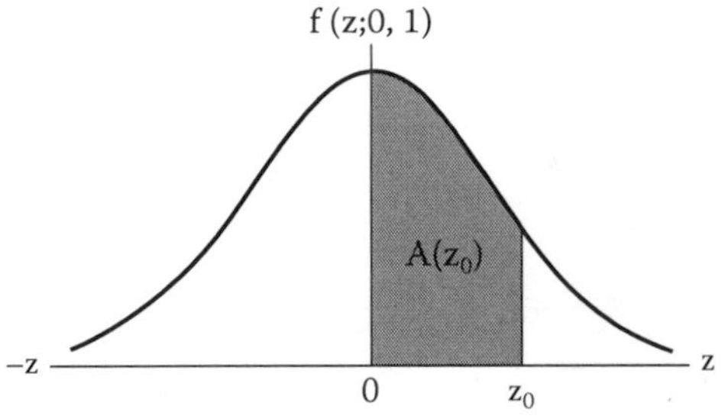

FIGURE A.1

$$A(z_0) = \frac{1}{\sqrt{2\pi}} \int_0^{z_0} e^{-z^2/2} dz.$$

$A(z_0)$ gives the total area under the standard normal distribution between 0 and any point z_0 on the positive z-axis (e.g., for $z_0 = 1.96$, $A(z_0) = 0.475$).

z	0.00	0.01	0.02	0.03	0.04	0.05	0.06	0.07	0.08	0.09
0.0	0.0000	0.0040	0.0080	0.0120	0.0150	0.0199	0.0239	0.0279	0.0319	0.0359
0.1	0.0398	0.0438	0.0478	0.0517	0.0557	0.0596	0.0636	0.0675	0.0714	0.0754
0.2	0.0793	0.0832	0.0871	0.0910	0.0948	0.0987	0.1026	0.1064	0.1103	0.1141
0.3	0.1179	0.1217	0.1253	0.1293	0.1331	0.1368	0.1406	0.1443	0.1480	0.1517
0.4	0.1554	0.1591	0.1628	0.1664	0.1700	0.1736	0.1772	0.1808	0.1844	0.1879
0.5	0.1915	0.1950	0.1985	0.2019	0.2054	0.2088	0.2123	0.2157	0.2190	0.2224
0.6	0.2258	0.2291	0.2324	0.2357	0.2389	0.2422	0.2454	0.2486	0.2518	0.2549
0.7	0.2580	0.2612	0.2642	0.2673	0.2704	0.2734	0.2764	0.2794	0.2823	0.2852
0.8	0.2881	0.2910	0.2939	0.2967	0.2996	0.3023	0.3051	0.3078	0.3106	0.3133
0.9	0.3159	0.3186	0.3212	0.3288	0.3264	0.3289	0.3315	0.3340	0.3365	0.3389

1.0	0.3413	0.3438	0.3461	0.3485	0.3508	0.3531	0.3554	0.3557	0.3559	0.3621
1.1	0.3642	0.3665	0.3686	0.3708	0.3729	0.3749	0.3770	0.3790	0.3810	0.3830
1.2	0.3849	0.3869	0.3888	0.3907	0.3925	0.3944	0.3962	0.3980	0.3997	0.4015
1.3	0.4032	0.4049	0.4066	0.4082	0.4099	0.4115	0.4131	0.4147	0.4162	0.4177
1.4	0.4192	0.4207	0.4222	0.4236	0.4251	0.4265	0.4279	0.4292	0.4306	0.4319
1.5	0.4332	0.4345	0.4357	0.4370	0.4382	0.4394	0.4406	0.4418	0.4429	0.4441
1.6	0.4452	0.4463	0.4474	0.4484	0.4495	0.4505	0.4515	0.4525	0.4535	0.4545
1.7	0.4554	0.4564	0.4573	0.4582	0.4591	0.4599	0.4608	0.4616	0.4625	0.4633
1.8	0.4641	0.4649	0.4656	0.4664	0.4671	0.4678	0.4686	0.4693	0.4699	0.4706
1.9	0.4713	0.4719	0.4726	0.4732	0.4738	0.4744	0.4750	0.4756	0.4761	0.4767
2.0	0.4772	0.4778	0.4783	0.4788	0.4793	0.4798	0.4803	0.4808	0.4812	0.4817
2.1	0.4821	0.4826	0.4830	0.4834	0.4838	0.4842	0.4846	0.4850	0.4854	0.4857
2.2	0.4861	0.4864	0.4868	0.4871	0.4875	0.4878	0.4881	0.4884	0.4887	0.4890
2.3	0.4893	0.4896	0.4898	0.4901	0.4904	0.4906	0.4909	0.4911	0.4913	0.4916
2.4	0.4918	0.4920	0.4922	0.4925	0.4927	0.4929	0.4931	0.4932	0.4934	0.4936
2.5	0.4938	0.4940	0.4941	0.4943	0.4945	0.4946	0.4948	0.4949	0.4951	0.4952
2.6	0.4953	0.4955	0.4956	0.4957	0.4959	0.4960	0.4961	0.4962	0.4963	0.4964
2.7	0.4965	0.4966	0.4967	0.4968	0.4969	0.4970	0.4971	0.4972	0.4973	0.4974
2.8	0.4974	0.4975	0.4976	0.4977	0.4977	0.4978	0.4979	0.4979	0.4980	0.4981
2.9	0.4981	0.4982	0.4982	0.4983	0.4984	0.4984	0.4985	0.4985	0.4986	0.4986
3.0	0.4987	0.4987	0.4987	0.4988	0.4988	0.4989	0.4989	0.4989	0.4990	0.4990

TABLE A.2

Quantiles of Student's t Distribution (T is t_v)

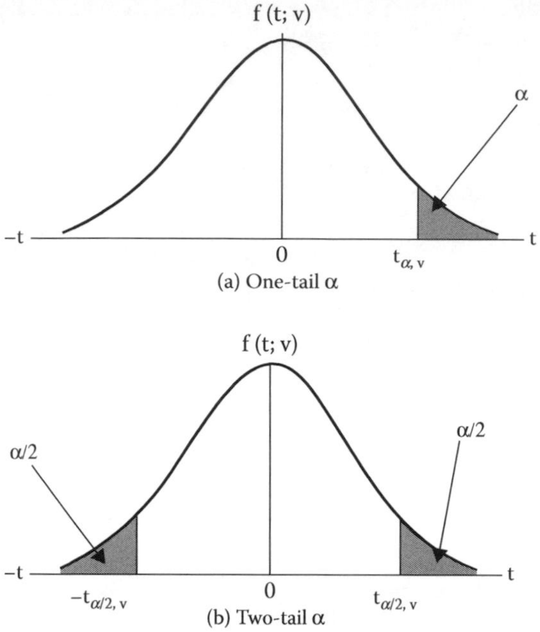

Given degrees of freedom v, the table gives either (1) the one-tail $t_{\alpha,v}$ value such that $P(T \geq t_{\alpha,v}) = \alpha$, or (2) the two-tail $\pm t_{\alpha/2,v}$ values for which $P(T \leq -t_{\alpha/2,v}) + P(T \geq t_{\alpha/2,v}) = \alpha/2 + \alpha/2 = \alpha$ (e.g., for $v = 15$ and $\alpha = 0.05$, $t_{0.05,15} = 1.753$, whereas $t_{0.025,15} = 2.131$).

FIGURE A.2

			One-Tail α			
	0.10	**0.05**	**0.025**	**0.01**	**0.005**	**0.001**
			Two-Tail α			
v	**0.20**	**0.10**	**0.05**	**0.02**	**0.01**	**0.002**
1	3.078	6.314	12.706	31.821	63.657	318.309
2	1.886	2.920	4.303	6.965	9.925	22.327
3	1.638	2.353	3.182	4.541	5.841	10.215
4	1.533	2.132	2.776	3.747	4.604	7.173
5	1.476	2.015	2.571	3.365	4.032	5.893
6	1.440	1.943	2.447	3.143	3.707	5.208
7	1.415	1.895	2.365	2.998	3.499	4.785
8	1.397	1.860	2.306	2.896	3.355	4.501
9	1.383	1.833	2.262	2.821	3.250	4.297
10	1.372	1.812	2.228	2.764	3.169	4.144
11	1.363	1.796	2.201	2.718	3.106	4.025
12	1.356	1.782	2.179	2.681	3.055	3.930
13	1.350	1.771	2.160	2.650	3.012	3.852
14	1.345	1.761	2.145	2.624	2.977	3.787
15	1.341	1.753	2.131	2.602	2.947	3.733
16	1.337	1.746	2.120	2.583	2.921	3.686
17	1.333	1.740	2.110	2.567	2.898	3.646
18	1.330	1.734	2.101	2.552	2.878	3.610
19	1.328	1.729	2.093	2.539	2.861	3.579

TABLE A.2 *(continued)*

	One-Tail α					
	0.10	**0.05**	**0.025**	**0.01**	**0.005**	**0.001**
	Two-Tail α					
ν	**0.20**	**0.10**	**0.05**	**0.02**	**0.01**	**0.002**
20	1.325	1.725	2.086	2.528	2.845	3.552
21	1.323	1.721	2.080	2.518	2.831	3.527
22	1.321	1.717	2.074	2.508	2.819	3.505
23	1.319	1.714	2.069	2.500	2.807	3.485
24	1.318	1.711	2.064	2.492	2.797	3.467
25	1.316	1.708	2.060	2.485	2.787	3.450
29	1.311	1.699	2.045	2.462	2.756	3.396
30	1.310	1.697	2.042	2.457	2.750	3.385
40	1.303	1.684	2.021	2.423	2.704	3.307
60	1.296	1.671	2.000	2.390	2.660	3.232
80	1.292	1.664	1.990	2.374	2.639	3.195
100	1.290	1.660	1.984	2.364	2.626	3.174
∞	1.282	1.645	1.960	2.326	2.576	3.090

TABLE A.3

Quantiles of the χ^2 Distribution (X is χ^2_v)

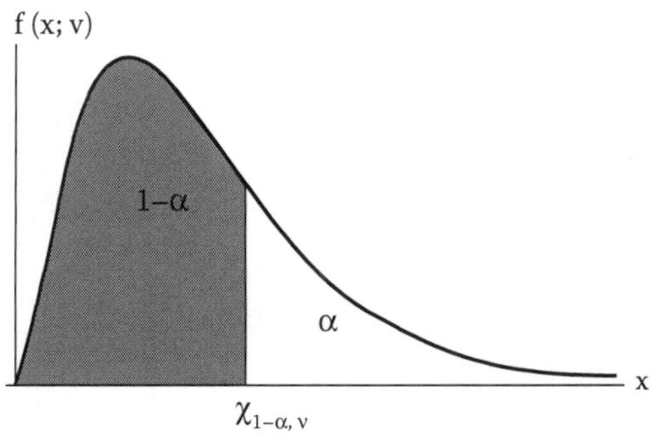

FIGURE A.3

For the cumulative probability 1− α and degrees of freedom v, the quantile $\chi_{1-\alpha,v}$ satisfies $F(\chi_{1-\alpha}; v) = P(X \le \chi_{1-\alpha,v}) = 1 - \alpha$ or, alternatively, $P(X > \chi_{1-\alpha,v}) = 1 - P(X \le \chi_{1-\alpha,v}) = \alpha$ (e.g., for $v = 10$ and $\alpha = 0.05$, $1 - \alpha = 0.95$ and thus $\chi_{0.95,10} = 18.31$).

	1 − α			1 − α			
v	**0.75**	**0.90**	**0.95**	**0.975**	**0.99**	**0.995**	**0.999**
1	1.3233	2.7100	3.8400	5.0200	6.6300	7.8800	10.8280
2	2.7726	4.6100	5.9900	7.3800	9.2100	10.6000	13.8160
3	4.1084	6.2500	7.8100	9.3500	11.3400	12.8400	16.2660
4	5.3853	7.7800	9.4900	11.1400	13.2800	14.8600	18.4670
5	6.6257	9.2400	11.0700	12.8300	15.0900	16.7500	20.5150
6	7.8408	10.6400	12.5900	14.4500	16.8100	18.5500	22.4580
7	9.0372	12.0200	14.0700	16.0100	18.4800	20.2800	24.3220
8	10.2188	13.3600	15.5100	17.5300	20.0900	21.9600	26.1250
9	11.3887	14.6800	16.9200	19.0200	21.6700	23.5900	27.8770
10	12.5489	15.9900	18.3100	20.4800	23.2100	25.1900	29.5880
11	13.7007	17.2800	19.6800	21.9200	24.7300	26.7600	31.2640
12	14.8454	18.5500	21.0300	23.3400	26.2200	28.3000	32.9090
13	15.9839	19.8100	22.3600	24.7400	27.6900	29.8200	34.5280
14	17.1170	21.0600	23.6800	26.1200	29.1400	31.3200	36.1230
15	18.2451	22.3100	25.0000	27.4900	30.5800	32.8000	37.6970

16	19.3688	23.5400	26.3000	28.8500	32.0000	34.2700	39.2520
17	20.4887	24.7690	27.5871	30.1910	33.4087	35.7185	40.7900
18	21.6049	25.9900	28.8700	31.5300	34.8100	37.1600	42.3120
19	22.7178	27.2036	30.1435	32.8523	36.1908	38.5822	43.8200
20	23.8277	28.4100	31.4100	34.1700	37.5700	40.0000	45.3150
21	24.9348	29.6151	32.6705	35.4789	38.9321	41.4010	46.7970
22	26.0393	30.8133	33.9244	36.7807	40.2894	42.7956	48.2680
23	27.1413	32.0069	35.1725	38.0757	41.6384	44.1813	49.7280
24	28.2412	33.1963	36.4151	39.3641	42.9798	45.5585	51.1790
25	29.3389	34.3816	37.6525	40.6465	44.3141	46.9278	52.6200
26	30.4345	35.5631	38.8852	41.9232	45.6417	48.2899	54.0520
27	31.5284	36.7412	40.1133	43.1944	46.9630	49.6449	55.4760
28	32.6205	37.9159	41.3372	44.4607	48.2782	50.9933	56.8920
29	33.7109	39.0875	42.5569	45.7222	49.5879	52.3356	58.3020
30	34.7998	40.2560	43.7729	46.9792	50.8922	53.6720	59.7030
40	45.6160	51.8050	55.7585	59.3417	63.6907	66.7659	73.4020
50	56.3336	63.1671	67.5048	71.4202	76.1539	79.4900	86.6610
60	66.9814	74.4000	79.0800	83.3000	88.3800	91.9500	99.6070
70	77.5766	85.5271	90.5312	95.0231	100.4250	104.2150	112.3170
80	88.1303	96.5782	101.8790	106.6290	112.3290	116.3210	124.8390
90	98.6499	107.5650	113.1450	118.1360	124.1160	128.2990	137.2080
100	109.1410	118.4980	124.3420	129.5610	135.8070	140.1690	149.4490

Source: Pearson, E.S., Hartley, H.O. (1976). *Biometrika Tables for Statisticians*, Vol.1, 3rd ed. London: Cambridge University Press, Table 8. With permission of the trustees of *Biometrika.*

TABLE A.4

Quantiles of Snedecor's $\boldsymbol{F}$ Distribution ($\boldsymbol{F}$ is F_{v_1,v_2})

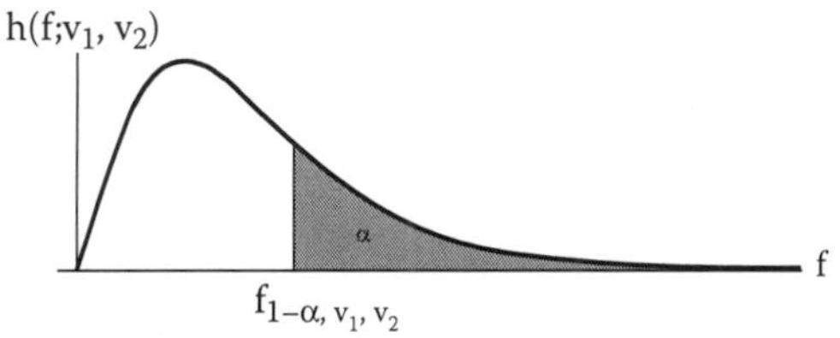

FIGURE A.4

Given the cumulative proportion $1 - \alpha$ and numerator and denominator degrees of freedom v_1 and v_2, respectively, the table gives the upper α quantile $f_{1-\alpha,v_1,v_2}$ such that $P(F \geq f_{1-\alpha,v_1,v_2}) = \alpha$ (e.g., for $1 - \alpha = 0.95$, $v_1 = 6$, and $v_2 = 10$, $f_{0.95,6,10} = 5.39$).

α = 0.10 (Upper 10% Fractile)

v_2 \ v_1	1	2	3	4	5	6	7	8	9	10	12	15	20	24	30	40	60	120	∞
1	39.86	49.50	53.59	55.83	57.24	58.20	58.91	59.44	59.86	60.19	60.71	61.22	61.74	62.00	62.26	62.53	62.79	63.06	63.33
2	8.53	9.00	9.16	9.24	9.29	9.33	9.35	9.37	9.38	9.39	9.41	9.42	9.44	9.45	9.46	9.47	9.47	9.48	9.49
3	5.54	5.46	5.39	5.34	5.31	5.28	5.27	5.25	5.24	5.23	5.22	5.20	5.18	5.18	5.17	5.16	5.15	5.14	5.13
4	4.54	4.32	4.19	4.11	4.05	4.01	3.98	3.95	3.94	3.92	3.90	3.87	3.84	3.83	3.82	3.80	3.79	3.78	3.76
5	4.06	3.78	3.62	3.52	3.45	3.40	3.37	3.34	3.32	3.30	3.27	3.24	3.21	3.19	3.17	3.16	3.14	3.12	3.10
6	3.78	3.46	3.29	3.18	3.11	3.05	3.01	2.98	2.96	2.94	2.90	2.87	2.84	2.82	2.80	2.78	2.76	2.74	2.72
7	3.59	3.26	3.07	2.96	2.88	2.83	2.78	2.75	2.72	2.70	2.67	2.63	2.59	2.58	2.56	2.54	2.51	2.49	2.47
8	3.46	3.11	2.92	2.81	2.73	2.67	2.62	2.59	2.56	2.54	2.50	2.46	2.42	2.40	2.38	2.36	2.34	2.32	2.29
9	3.36	3.01	2.81	2.69	2.61	2.55	2.51	2.47	2.44	2.42	2.38	2.34	2.30	2.28	2.25	2.23	2.21	2.18	2.16
10	3.29	2.92	2.73	2.61	2.52	2.46	2.41	2.38	2.35	2.32	2.28	2.24	2.20	2.18	2.16	2.13	2.11	2.08	2.06
11	3.23	2.86	2.66	2.54	2.45	2.39	2.34	2.30	2.27	2.25	2.21	2.17	2.12	2.10	2.08	2.05	2.03	2.00	1.97
12	3.18	2.81	2.61	2.48	2.39	2.33	2.28	2.24	2.21	2.19	2.15	2.10	2.06	2.04	2.01	1.99	1.96	1.93	1.90
13	3.14	2.76	2.56	2.43	2.35	2.28	2.23	2.20	2.16	2.14	2.10	2.05	2.01	1.98	1.96	1.93	1.90	1.88	1.85
14	3.10	2.73	2.52	2.39	2.31	2.24	2.19	2.15	2.12	2.10	2.05	2.01	1.96	1.94	1.91	1.89	1.86	1.83	1.80

15	3.07	2.70	2.49	2.36	2.27	2.21	2.16	2.12	2.09	2.06	2.02	1.97	1.92	1.90	1.87	1.85	1.82	1.79	1.76
16	3.05	2.67	2.46	2.33	2.24	2.18	2.13	2.09	2.06	2.03	1.99	1.94	1.89	1.87	1.84	1.81	1.78	1.75	1.72
17	3.03	2.64	2.44	2.31	2.22	2.15	2.10	2.06	2.03	2.00	1.96	1.91	1.86	1.84	1.81	1.78	1.75	1.72	1.69
18	3.01	2.62	2.42	2.29	2.20	2.13	2.08	2.04	2.00	1.98	1.93	1.89	1.84	1.81	1.78	1.75	1.72	1.69	1.66
19	2.99	2.61	2.40	2.27	2.18	2.11	2.06	2.02	1.98	1.96	1.91	1.86	1.81	1.79	1.76	1.73	1.70	1.67	1.63
20	2.97	2.59	2.38	2.25	2.16	2.09	2.04	2.00	1.96	1.94	1.89	1.84	1.79	1.77	1.74	1.71	1.68	1.64	1.61
21	2.96	2.57	2.36	2.23	2.14	2.08	2.02	1.98	1.95	1.92	1.87	1.83	1.78	1.75	1.72	1.69	1.66	1.62	1.59
22	2.95	2.56	2.35	2.22	2.13	2.06	2.01	1.97	1.93	1.90	1.86	1.81	1.76	1.73	1.70	1.67	1.64	1.60	1.57
23	2.94	2.55	2.34	2.21	2.11	2.05	1.99	1.95	1.92	1.89	1.84	1.80	1.74	1.72	1.69	1.66	1.62	1.59	1.55
24	2.93	2.54	2.33	2.19	2.10	2.04	1.98	1.94	1.91	1.88	1.83	1.78	1.73	1.70	1.67	1.64	1.61	1.57	1.53
25	2.92	2.53	2.32	2.18	2.09	2.02	1.97	1.93	1.89	1.87	1.82	1.77	1.72	1.69	1.66	1.63	1.59	1.56	1.52
26	2.91	2.52	2.31	2.17	2.08	2.01	1.96	1.92	1.88	1.86	1.81	1.76	1.71	1.68	1.65	1.61	1.58	1.54	1.50
27	2.90	2.51	2.30	2.17	2.07	2.00	1.95	1.91	1.87	1.85	1.80	1.75	1.70	1.67	1.64	1.60	1.57	1.53	1.49
28	2.89	2.50	2.29	2.16	2.06	2.00	1.94	1.90	1.87	1.84	1.79	1.74	1.69	1.66	1.63	1.59	1.56	1.52	1.48
29	2.89	2.50	2.28	2.15	2.06	1.99	1.93	1.89	1.86	1.83	1.78	1.73	1.68	1.65	1.62	1.58	1.55	1.51	1.47
30	2.88	2.49	2.28	2.14	2.05	1.98	1.93	1.88	1.85	1.82	1.77	1.72	1.67	1.64	1.61	1.57	1.54	1.50	1.46
40	2.84	2.44	2.23	2.09	2.00	1.93	1.87	1.83	1.79	1.76	1.71	1.66	1.61	1.57	1.54	1.51	1.47	1.42	1.38
60	2.79	2.39	2.18	2.04	1.95	1.87	1.82	1.77	1.74	1.71	1.66	1.60	1.54	1.51	1.48	1.44	1.40	1.35	1.29
120	2.75	2.35	2.13	1.99	1.90	1.82	1.77	1.72	1.68	1.65	1.60	1.55	1.48	1.45	1.41	1.37	1.32	1.26	1.19
∞	2.71	2.30	2.08	1.94	1.85	1.77	1.72	1.67	1.63	1.60	1.55	1.49	1.42	1.38	1.34	1.30	1.24	1.17	1.00

(continued)

TABLE A.4 *(continued)*

$\alpha = 0.05$ (Upper 5% Fractile)

v_2 \ v_1	1	2	3	4	5	6	7	8	9	10	12	15	20	24	30	40	60	120	∞
1	161.4	199.5	215.7	224.6	230.2	234	236.8	238.9	240.5	241.9	243.9	245.9	248	249.1	250.1	251.1	252.2	253.3	254.3
2	18.51	19	19.16	19.25	19.3	19.33	19.35	19.37	19.38	19.4	19.41	19.43	19.45	19.45	19.46	19.47	19.48	19.49	19.5
3	10.13	9.55	9.28	9.12	9.01	8.94	8.89	8.85	8.81	8.79	8.74	8.7	8.66	8.64	8.62	8.59	8.57	8.55	8.53
4	7.71	6.94	6.59	6.39	6.26	6.16	6.09	6.04	6	5.96	5.91	5.86	5.8	5.77	5.75	5.72	5.69	5.66	5.63
5	6.61	5.79	5.41	5.19	5.05	4.95	4.88	4.82	4.77	4.74	4.68	4.62	4.56	4.53	4.5	4.46	4.43	4.4	4.35
6	5.99	5.14	4.76	4.53	4.39	4.28	4.21	4.15	4.1	4.06	4	3.94	3.87	3.84	3.81	3.77	3.74	3.7	3.67
7	5.59	4.74	4.35	4.12	3.97	3.87	3.79	3.73	3.68	3.64	3.57	3.51	3.44	3.41	3.38	3.34	3.3	3.27	3.23
8	5.32	4.46	4.07	3.84	3.69	3.58	3.5	3.44	3.39	3.35	3.28	3.22	3.15	3.12	3.08	3.04	3.01	2.97	2.93
9	5.12	4.26	3.86	3.63	3.48	3.37	3.29	3.23	3.18	3.14	3.07	3.01	2.94	2.9	2.86	2.83	2.79	2.75	2.71
10	4.96	4.1	3.71	3.48	3.33	3.22	3.14	3.07	3.02	2.98	2.91	2.85	2.77	2.74	2.7	2.66	2.62	2.58	2.54
11	4.84	3.98	3.59	3.36	3.2	3.09	3.01	2.95	2.9	2.85	2.79	2.72	2.65	2.61	2.57	2.53	2.49	2.45	2.4
12	4.75	3.89	3.49	3.26	3.11	3	2.91	2.85	2.8	2.75	2.69	2.62	2.54	2.51	2.47	2.43	2.38	2.34	2.3
13	4.67	3.81	3.41	3.18	3.03	2.92	2.83	2.77	2.71	2.67	2.6	2.53	2.46	2.42	2.38	2.34	2.3	2.25	2.21
14	4.6	3.74	3.34	3.11	2.96	2.85	2.76	2.7	2.65	2.6	2.53	2.46	2.39	2.35	2.31	2.27	2.22	2.18	2.13
15	4.54	3.68	3.29	3.06	2.9	2.79	2.71	2.64	2.59	2.54	2.48	2.4	2.33	2.29	2.25	2.2	2.16	2.11	2.07
16	4.49	3.63	3.24	3.01	2.85	2.74	2.66	2.59	2.54	2.49	2.42	2.35	2.28	2.24	2.19	2.15	2.11	2.06	2.01
17	4.45	3.59	3.2	2.96	2.81	2.7	2.61	2.55	2.49	2.45	2.38	2.31	2.23	2.19	2.15	2.1	2.06	2.01	1.96
18	4.41	3.55	3.16	2.93	2.77	2.66	2.58	2.51	2.46	2.41	2.34	2.27	2.19	2.15	2.11	2.06	2.02	1.97	1.92
19	4.38	3.52	3.13	2.9	2.74	2.63	2.54	2.48	2.42	2.38	2.31	2.23	2.16	2.11	2.07	2.03	1.98	1.93	1.88
20	4.35	3.49	3.1	2.87	2.71	2.6	2.51	2.45	2.39	2.35	2.28	2.2	2.12	2.08	2.04	1.99	1.95	1.9	1.84
21	4.32	3.47	3.07	2.84	2.68	2.57	2.49	2.42	2.37	2.32	2.25	2.18	2.1	2.05	2.01	1.96	1.92	1.87	1.81
22	4.3	3.44	3.05	2.82	2.66	2.55	2.46	2.4	2.34	2.3	2.23	2.15	2.07	2.03	1.98	1.94	1.89	1.84	1.78
23	4.28	3.42	3.03	2.8	2.64	2.53	2.44	2.37	2.32	2.27	2.2	2.13	2.05	2.01	1.96	1.91	1.86	1.81	1.76

24	4.26	3.4	3.01	2.78	2.62	2.51	2.42	2.36	2.3	2.25	2.18	2.11	2.03	1.98	1.94	1.89	1.84	1.79	1.73
25	4.24	3.39	2.99	2.76	2.6	2.49	2.4	2.34	2.28	2.24	2.16	2.09	2.01	1.96	1.92	1.87	1.82	1.77	1.71
26	4.23	3.37	2.98	2.74	2.59	2.47	2.39	2.32	2.27	2.22	2.15	2.07	1.99	1.95	1.9	1.85	1.8	1.75	1.69
27	4.21	3.35	2.96	2.73	2.57	2.46	2.37	2.31	2.25	2.2	2.13	2.06	1.97	1.93	1.88	1.84	1.79	1.73	1.67
28	4.2	3.34	2.95	2.71	2.56	2.45	2.36	2.29	2.24	2.19	2.12	2.04	1.96	1.91	1.87	1.82	1.77	1.71	1.65
29	4.18	3.33	2.93	2.7	2.55	2.43	2.35	2.28	2.22	2.18	2.1	2.03	1.94	1.9	1.85	1.81	1.75	1.7	1.64
30	4.17	3.32	2.92	2.69	2.53	2.42	2.33	2.27	2.21	2.16	2.09	2.01	1.93	1.89	1.84	1.79	1.74	1.68	1.62
40	4.08	3.23	2.84	2.61	2.45	2.34	2.25	2.18	2.12	2.08	2	1.92	1.84	1.79	1.74	1.69	1.64	1.58	1.51
60	4	3.15	2.76	2.53	2.37	2.25	2.17	2.1	2.04	1.99	1.92	1.84	1.75	1.7	1.65	1.59	1.53	1.47	1.39
120	3.92	3.07	2.68	2.45	2.29	2.17	2.09	2.02	1.96	1.91	1.83	1.75	1.66	1.61	1.55	1.5	1.43	1.35	1.25
∞	3.84	3	2.6	2.37	2.21	2.1	2.01	1.94	1.88	1.83	1.75	1.67	1.57	1.52	1.46	1.39	1.32	1.22	1.00

(continued)

TABLE A.4 *(continued)*

α = 0.01 (Upper 1% Fractile)

v_2 \ v_1	1	2	3	4	5	6	7	8	9	10	12	15	20	24	30	40	60	120	∞
1	4052	4999.5	5403	5625	5764	5859	5928	5982	6022	6056	6106	6157	6209	6235	6261	6287	6313	6339	6366
2	98.50	99.00	99.17	99.25	99.30	99.33	99.36	99.37	99.39	99.40	99.42	99.43	99.45	99.46	99.47	99.47	99.48	99.49	99.50
3	34.12	30.82	29.46	28.71	28.24	27.91	27.67	27.49	27.35	27.23	27.05	26.87	26.69	26.60	26.50	26.41	26.32	26.22	26.13
4	21.20	18.00	16.69	15.98	15.52	15.21	14.98	14.80	14.66	14.55	14.37	14.20	14.02	13.93	13.84	13.75	13.65	13.56	13.46
5	16.26	13.27	12.06	11.39	10.97	10.67	10.46	10.29	10.16	10.05	9.89	9.72	9.55	9.47	9.38	9.29	9.20	9.11	9.02
6	13.75	10.92	9.78	9.15	8.75	8.47	8.26	8.10	7.98	7.87	7.72	7.56	7.40	7.31	7.23	7.14	7.06	6.97	6.88
7	12.25	9.55	8.45	7.85	7.46	7.19	6.99	6.84	6.72	6.62	6.47	6.31	6.16	6.07	5.99	5.91	5.82	5.74	5.65
8	11.26	8.65	7.59	7.01	6.63	6.37	6.18	6.03	5.91	5.81	5.67	5.52	5.36	5.28	5.20	5.12	5.03	4.95	4.85
9	10.56	8.02	6.99	6.42	6.06	5.80	5.61	5.47	5.35	5.26	5.11	4.96	4.81	4.73	4.65	4.57	4.48	4.40	4.31
10	10.04	7.56	6.55	5.99	5.64	5.39	5.20	5.06	4.94	4.85	4.71	4.56	4.41	4.33	4.25	4.17	4.08	4.00	3.91
11	9.65	7.21	6.22	5.67	5.32	5.07	4.89	4.74	4.63	4.54	4.40	4.25	4.10	4.02	3.94	3.86	3.78	3.69	3.60
12	9.33	6.93	5.95	5.41	5.06	4.82	4.64	4.50	4.39	4.30	4.16	4.01	3.86	3.78	3.70	3.62	3.54	3.45	3.36
13	9.07	6.70	5.74	5.21	4.86	4.62	4.44	4.30	4.19	4.10	3.96	3.82	3.66	3.59	3.51	3.43	3.34	3.25	3.17
14	8.86	6.51	5.56	5.04	4.69	4.46	4.28	4.14	4.03	3.94	3.80	3.66	3.51	3.43	3.35	3.27	3.18	3.09	3.00
15	8.68	6.36	5.42	4.89	4.56	4.32	4.14	4.00	3.89	3.80	3.67	3.52	3.37	3.29	3.21	3.13	3.05	2.96	2.87
16	8.53	6.23	5.29	4.77	4.44	4.20	4.03	3.89	3.78	3.69	3.55	3.41	3.26	3.18	3.10	3.02	2.93	2.84	2.75
17	8.40	6.11	5.18	4.67	4.34	4.10	3.93	3.79	3.68	3.59	3.46	3.31	3.16	3.08	3.00	2.92	2.83	2.75	2.65
18	8.29	6.01	5.09	4.58	4.25	4.01	3.84	3.71	3.60	3.51	3.37	3.23	3.08	3.00	2.92	2.84	2.75	2.66	2.57
19	8.18	5.93	5.01	4.50	4.17	3.94	3.77	3.63	3.52	3.43	3.30	3.15	3.00	2.92	2.84	2.76	2.67	2.58	2.49
20	8.10	5.85	4.94	4.43	4.10	3.87	3.70	3.56	3.46	3.37	3.23	3.09	2.94	2.86	2.78	2.69	2.61	2.52	2.42
21	8.02	5.78	4.87	4.37	4.04	3.81	3.64	3.51	3.40	3.31	3.17	3.03	2.88	2.80	2.72	2.64	2.55	2.46	2.36
22	7.95	5.72	7.82	4.31	3.99	3.76	3.59	3.45	3.35	3.26	3.12	2.98	2.83	2.75	2.67	2.58	2.50	2.40	2.31
23	7.88	5.66	4.76	4.26	3.94	3.71	3.54	3.41	3.30	3.21	3.07	2.93	2.78	2.70	2.62	2.54	2.45	2.35	2.26

24	7.82	5.61	7.72	4.22	3.90	3.67	3.50	3.36	3.26	3.17	3.03	2.89	2.74	2.66	2.58	2.49	2.40	2.31	2.21
25	7.77	5.57	4.68	4.18	3.85	3.63	3.46	3.32	3.22	3.13	2.99	2.85	2.70	2.62	2.54	2.45	2.36	2.27	2.17
26	7.72	5.53	4.64	4.14	3.82	3.59	3.42	3.29	3.18	3.09	2.96	2.81	2.66	2.58	2.50	2.42	2.33	2.23	2.13
27	7.68	5.49	4.60	4.11	3.78	3.56	3.39	3.26	3.15	3.06	2.93	2.78	2.63	2.55	2.47	2.38	2.29	2.20	2.10
28	7.64	5.45	4.57	4.07	3.75	3.53	3.36	3.23	3.12	3.03	2.90	2.75	2.60	2.52	2.44	2.35	2.26	2.17	2.06
29	7.60	5.42	4.54	4.04	3.73	3.50	3.33	3.20	3.09	3.00	2.87	2.73	2.57	2.49	2.41	2.33	2.23	2.14	2.03
30	7.56	5.39	4.51	4.02	3.70	3.47	3.30	3.17	3.07	2.98	2.84	2.70	2.55	2.47	2.39	2.30	2.21	2.11	2.01
40	7.31	5.18	4.31	3.83	3.51	3.29	3.12	2.99	2.89	2.80	2.66	2.52	2.37	2.29	2.20	2.11	2.02	1.92	1.80
60	7.08	4.98	4.13	3.65	3.34	3.12	2.95	2.82	2.72	2.63	2.50	2.35	2.20	2.12	2.03	1.94	1.84	1.73	1.60
120	6.85	4.79	3.95	3.48	3.17	2.96	2.79	2.66	2.56	2.47	2.34	2.19	2.03	1.95	1.86	1.76	2.66	1.53	1.38
∞	6.63	4.61	3.78	3.32	3.02	2.80	2.64	2.51	2.41	2.32	2.18	2.04	1.88	1.79	1.70	1.59	1.47	1.32	1.00

Source: Pearson, E.S., Hartley, H.O. (1976). *Biometrika Tables for Statisticians*, Vol.1, 3rd ed. London: Cambridge University Press, Table 18. With permission of the trustees of *Biometrika*.

TABLE A.5

Durbin-Watson DW Statistic: 5% Significance Points d_L and d_U (n is the sample size and k' is the number of regressors excluding the intercept)

	$k'=1$		$k'=2$		$k'=3$		$k'=4$		$k'=5$		$k'=6$		$k'=7$		$k'=8$		$k'=9$		$k'=10$	
n	d_L	d_U	d_L	d_U	d_L	d_U	d_L	d_U	d_L	d_U	d_L	d_U	d_L	d_U	d_L	d_U	d_L	d_U	d_L	d_U
6	0.610	1.400																		
7	0.700	1.356	0.467	1.896																
8	0.763	1.332	0.559	1.777	0.368	2.287														
9	0.824	1.320	0.629	1.699	0.455	2.128	0.296	2.588												
10	0.879	1.320	0.697	1.641	0.525	2.016	0.376	2.414	0.243	2.822										
11	0.927	1.324	0.758	1.604	0.595	1.928	0.444	2.283	0.316	2.645	0.203	3.005								
12	0.971	1.331	0.812	1.579	0.658	1.864	0.512	2.177	0.379	2.506	0.268	2.832	0.171	3.149						
13	1.010	1.340	0.861	1.562	0.715	1.816	0.574	2.094	0.445	2.390	0.328	2.692	0.230	2.985	0.147	3.266				
14	1.045	1.350	0.905	1.551	0.767	1.779	0.632	2.030	0.505	2.296	0.389	2.572	0.286	2.848	0.200	3.111	0.127	3.360		
15	1.077	1.361	0.946	1.543	0.814	1.750	0.685	1.977	0.562	2.220	0.447	2.472	0.343	2.727	0.251	2.979	0.175	3.216	0.111	3.438
16	1.106	1.371	0.982	1.539	0.157	1.728	0.734	1.935	0.615	2.157	0.502	2.388	0.398	2.624	0.304	2.860	0.222	3.090	0.155	3.304
17	1.133	1.381	1.015	1.536	0.897	1.710	0.779	1.900	0.664	2.104	0.554	2.318	0.451	2.537	0.356	2.757	0.272	2.975	0.198	3.184
18	1.158	1.391	1.046	1.535	0.933	1.696	0.820	1.872	0.710	2.060	0.603	2.257	0.502	2.461	0.407	2.667	0.321	2.873	0.244	3.073
19	1.180	1.401	1.074	1.336	0.967	1.685	0.859	1.848	0.752	2.023	0.649	2.206	0.549	2.396	0.456	2.589	0.369	2.783	0.290	2.974
20	1.201	1.411	1.100	1.537	0.998	1.676	0.894	1.828	0.792	1.991	0.692	2.162	0.595	2.339	0.502	2.521	0.416	2.704	0.336	2.885
21	1.221	1.420	1.125	1.538	1.026	1.669	0.927	1.812	0.829	1.964	0.732	2.124	0.637	2.290	0.547	2.460	0.461	2.633	0.380	2.806
22	1.239	1.429	1.147	1.541	1.053	1.664	0.958	1.797	0.863	1.940	0.769	2.090	0.677	2.246	0.588	2.407	0.504	2.511	0.424	2.734
23	1.257	1.437	1.168	1.543	1.078	1.660	0.986	1.785	0.895	1.920	0.804	2.061	0.715	2.208	0.628	2.360	0.545	2.514	0.465	2.670
24	1.273	1.446	1.188	1.546	1.101	1.656	1.013	1.775	0.925	1.902	0.837	2.035	0.751	2.174	0.666	2.318	0.584	2.464	0.506	2.613
25	1.288	1.454	1.206	1.550	1.123	1.654	1.038	1.767	0.953	1.886	0.868	2.012	0.784	2.144	0.702	2.280	0.621	2.419	0.544	2.560
26	1.302	1.461	1.224	1.553	1.143	1.652	1.062	1.759	0.979	1.873	0.897	1.992	0.816	2.117	0.735	2.246	0.657	2.379	0.581	2.513
27	1.316	1.469	1.240	1.556	1.162	1.651	1.084	1.753	1.004	1.861	0.925	1.974	0.845	2.093	0.767	2.216	0.691	2.342	0.616	2.470
28	1.328	1.476	1.255	1.560	1.181	1.650	1.104	1.747	1.028	1.850	0.951	1.958	0.874	2.071	0.798	2.188	0.723	2.309	0.650	2.431
29	1.341	1.483	1.270	1.563	1.198	1.650	1.124	1.743	1.050	1.841	0.975	1.944	0.900	2.052	0.826	2.164	0.753	2.278	0.682	2.396

30	1.352	1.489	1.284	1.567	1.214	1.650	1.143	1.739	1.071	1.833	0.998	1.931	0.926	2.034	0.854	2.141	0.782	2.251	0.712	2.363
31	1.363	1.496	1.297	1.570	1.229	1.650	1.160	1.735	1.090	1.825	1.020	1.920	0.950	2.018	0.879	2.120	0.810	2.226	0.741	2.333
32	1.373	1.502	1.309	1.574	1.244	1.650	1.177	1.732	1.109	1.819	1.041	1.909	0.972	2.004	0.904	2.102	0.836	2.203	0.769	2.306
33	1.383	1.508	1.321	1.577	1.258	1.651	1.193	1.730	1.127	1.813	1.061	1.900	0.994	1.991	0.927	2.085	0.861	2.181	0.795	2.281
34	1.393	1.514	1.333	1.580	1.271	1.652	1.208	1.728	1.144	1.808	1.080	1.891	1.015	1.979	0.950	2.069	0.885	2.162	0.821	2.257
35	1.402	1.519	1.343	1.584	1.283	1.653	1.222	1.726	1.160	1.803	1.097	1.884	1.034	1.967	0.971	2.054	0.908	2.144	0.845	2.236
36	1.411	1.525	1.354	1.587	1.295	1.654	1.236	1.724	1.175	1.799	1.114	1.877	1.053	1.957	0.991	2.041	0.930	2.127	0.868	2.216
37	1.419	1.530	1.364	1.590	1.307	1.655	1.249	1.723	1.190	1.795	1.131	1.870	1.071	1.948	1.011	2.029	0.951	2.112	0.891	2.198
38	1.427	1.535	1.373	1.594	1.318	1.656	1.261	1.722	1.204	1.792	1.146	1.864	1.088	1.939	1.029	2.017	0.970	2.098	0.912	2.180
39	1.435	1.540	1.382	1.597	1.328	1.658	1.273	1.722	1.218	1.789	1.161	1.859	1.104	1.932	1.047	2.007	0.990	2.085	0.932	2.164
40	1.442	1.544	1.391	1.600	1.338	1.659	1.285	1.721	1.230	1.786	1.175	1.854	1.120	1.924	1.064	1.997	1.008	2.072	0.952	2.149
45	1.475	1.566	1.430	1.615	1.383	1.666	1.336	1.720	1.287	1.776	1.238	1.835	1.189	1.895	1.139	1.958	1.089	2.022	1.038	2.088
50	1.503	1.585	1.462	1.628	1.421	1.674	1.378	1.721	1.335	1.771	1.291	1.822	1.246	1.875	1.201	1.930	1.156	1.986	1.110	2.044
55	1.528	1.601	1.490	1.641	1.452	1.681	1.414	1.724	1.374	1.768	1.334	1.814	1.294	1.861	1.253	1.909	1.212	1.959	1.170	2.010
60	1.549	1.616	1.514	1.652	1.480	1.689	1.444	1.727	1.408	1.767	1.372	1.808	1.335	1.850	1.298	1.894	1.260	1.939	1.222	1.984
65	1.567	1.629	1.536	1.662	1.503	1.696	1.471	1.731	1.438	1.767	1.404	1.805	1.370	1.843	1.336	1.882	1.301	1.923	1.266	1.964
70	1.583	1.641	1.554	1.672	1.525	1.703	1.494	1.735	1.464	1.768	1.433	1.802	1.401	1.837	1.369	1.873	1.337	1.910	1.305	1.948
75	1.598	1.652	1.571	1.680	1.543	1.709	1.515	1.739	1.487	1.770	1.458	1.801	1.428	1.834	1.399	1.867	1.369	1.901	1.339	1.935
80	1.611	1.662	1.586	1.688	1.560	1.715	1.534	1.743	1.507	1.772	1.480	1.801	1.453	1.831	1.425	1.861	1.397	1.893	1.369	1.925
85	1.624	1.671	1.600	1.696	1.575	1.721	1.550	1.747	1.525	1.774	1.500	1.801	1.474	1.829	1.448	1.857	1.422	1.886	1.396	1.916
90	1.635	1.679	1.612	1.703	1.589	1.726	1.566	1.751	1.542	1.776	1.518	1.801	1.494	1.827	1.469	1.854	1.445	1.881	1.420	1.909
95	1.645	1.687	1.623	1.709	1.602	1.732	1.579	1.755	1.557	1.778	1.535	1.802	1.512	1.827	1.489	1.852	1.465	1.877	1.442	1.903
100	1.654	1.694	1.634	1.715	1.613	1.736	1.592	1.758	1.571	1.780	1.550	1.803	1.528	1.826	1.506	1.850	1.484	1.874	1.462	1.898
150	1.720	1.746	1.706	1.760	1.693	1.774	1.679	1.788	1.665	1.802	1.651	1.817	1.637	1.132	1.622	1.847	1.608	1.862	1.594	1.877
200	1.758	1.778	1.748	1.789	1.738	1.799	1.728	1.810	1.718	1.820	1.707	1.831	1.697	1.841	1.686	1.852	1.675	1.863	1.665	1.874

(continued)

TABLE A.5 *(continued)*

	k′ = 11		k′ = 12		k′ = 13		k′ = 14		k′ = 15		k′ = 16		k′ = 17		k′ = 18		k′ = 19		k′ = 20	
n	d_L	d_U	d_L	d_U	d_L	d_U	d_L	d_U	d_L	d_U	d_L	d_U	d_L	d_U	d_L	d_U	d_L	d_U	d_L	d_U
16	0.098	3.503																		
17	0.138	3.378	0.087	3.557																
18	0.177	3.265	0.123	3.441	0.078	3.603														
19	0.220	3.159	0.160	3.335	0.111	3.496	0.070	3.642												
20	0.263	3.063	0.200	3.234	0.145	3.395	0.100	3.542	0.063	3.616										
21	0.307	2.976	0.240	3.141	0.182	3.300	0.132	3.448	0.091	3.583	0.058	3.105								
22	0.349	2.897	0.281	3.057	0.220	3.211	0.166	3.358	0.120	3.495	0.083	3.619	0.052	3.731						
23	0.391	2.826	0.322	2.979	0.259	3.128	0.202	3.272	0.153	3.409	0.110	3.535	0.076	3.650	0.048	3.753				
24	0.431	2.761	0.362	2.908	0.297	3.053	0.239	3.193	0.186	3.327	0.141	3.454	0.101	3.572	0.070	3.678	0.044	3.773		
25	0.470	2.702	0.400	2.844	0.335	2.983	0.275	3.119	0.221	3.251	0.172	3.376	0.130	3.494	0.094	3.604	0.065	3.702	0.041	3.790
26	0.508	2.649	0.438	2.784	0.373	2.919	0.312	3.051	0.256	3.179	0.205	3.303	0.160	3.420	0.120	3.531	0.087	3.632	0.060	3.724
27	0.544	2.600	0.475	2.730	0.409	2.859	0.348	2.987	0.291	3.112	0.238	3.233	0.191	3.349	0.149	3.460	0.112	3.563	0.081	3.658
28	0.578	2.555	0.510	2.680	0.445	2.805	0.383	2.928	0.325	3.050	0.271	3.168	0.222	3.283	0.178	3.392	0.138	3.495	0.104	3.592
29	0.612	2.515	0.544	2.634	0.479	2.755	0.418	2.874	0.359	2.992	0.305	3.107	0.254	3.219	0.208	3.327	0.166	3.431	0.129	3.528
30	0.643	2.477	0.577	2.592	0.512	2.708	0.451	2.823	0.392	2.937	0.337	3.050	0.286	3.160	0.238	3.266	0.195	3.368	0.156	3.465
31	0.614	2.443	0.608	2.553	0.545	2.665	0.484	2.776	0.425	2.887	0.370	2.996	0.317	3.103	0.269	3.208	0.224	3.309	0.183	3.406
32	0.703	2.411	0.638	2.517	0.576	2.625	0.515	2.733	0.457	2.840	0.401	2.946	0.349	3.050	0.299	3.153	0.253	3.252	0.211	3.343
33	0.731	2.382	0.668	2.484	0.606	2.588	0.546	2.692	0.488	2.796	0.432	2.899	0.379	3.000	0.329	3.100	0.283	3.198	0.239	3.293

34	0.758	2.355	0.695	2.454	0.634	2.554	0.575	2.654	0.518	2.754	0.462	2.854	0.409	2.954	0.359	3.051	0.312	3.147	0.267	3.240
35	0.783	2.330	0.722	2.425	0.662	2.521	0.604	2.619	0.547	2.716	0.492	2.813	0.439	2.910	0.388	3.005	0.340	3.099	0.295	3.190
36	0.808	2.306	0.748	2.398	0.689	2.492	0.631	2.586	0.575	2.680	0.520	2.774	0.467	2.868	0.417	2.961	0.369	3.053	0.323	3.142
37	0.831	2.285	0.772	2.374	0.714	2.464	0.657	2.555	0.602	2.646	0.548	2.738	0.495	2.829	0.445	2.920	0.397	3.009	0.351	3.091
38	0.854	2.265	0.796	2.351	0.739	2.438	0.683	2.526	0.628	2.614	0.575	2.703	0.522	2.792	0.472	2.880	0.424	2.968	0.318	3.054
39	0.815	2.246	0.819	2.329	0.763	2.413	0.707	2.499	0.653	2.585	0.600	2.671	0.549	2.757	0.499	2.843	0.451	2.929	0.404	3.013
40	0.896	2.228	0.840	2.309	0.785	2.391	0.731	2.473	0.678	2.557	0.626	2.641	0.575	2.724	0.525	2.808	0.477	2.892	0.430	2.974
45	0.988	2.156	0.938	2.225	0.887	2.296	0.838	2.367	0.788	2.439	0.740	2.512	0.692	2.586	0.644	2.659	0.598	2.733	0.553	2.801
50	1.064	2.103	1.019	2.163	0.973	2.225	0.927	2.287	0.882	2.350	0.836	2.414	0.792	2.479	0.747	2.544	0.703	2.610	0.660	2.675
55	1.129	2.062	1.087	2.116	1.045	2.170	1.003	2.225	0.961	2.281	0.919	2.338	0.877	2.396	0.836	2.454	0.795	2.512	0.754	2.571
60	1.184	2.031	1.145	2.079	1.106	2.127	1.068	2.177	1.029	2.227	0.990	2.278	0.951	2.330	0.913	2.382	0.874	2.434	0.836	2.487
65	1.231	2.006	1.195	2.049	1.160	2.093	1.124	2.138	1.088	2.183	1.052	2.229	1.016	2.276	0.980	2.323	0.944	2.371	0.908	2.419
70	1.272	1.986	1.239	2.026	1.206	2.066	1.172	2.106	1.139	2.148	1.105	2.189	1.072	2.232	1.038	2.275	1.005	2.318	0.971	2.362
75	1.308	1.970	1.277	2.006	1.247	2.043	1.215	2.080	1.184	2.118	1.153	2.156	1.121	2.195	1.090	2.235	1.058	2.275	1.021	2.315
80	1.340	1.957	1.311	1.991	1.283	2.024	1.253	2.059	1.224	2.093	1.195	2.129	1.165	2.165	1.136	2.201	1.106	2.238	1.076	2.275
85	1.369	1.946	1.342	1.977	1.315	2.009	1.287	2.040	1.260	2.073	1.232	2.105	1.205	2.139	1.177	2.172	1.149	2.206	1.121	2.241
90	1.395	1.937	1.369	1.966	1.344	1.995	1.318	2.025	1.292	2.055	1.266	2.085	1.240	2.116	1.213	2.148	1.187	2.179	1.160	2.211
95	1.418	1.929	1.394	1.956	1.370	1.984	1.345	2.012	1.321	2.040	1.296	2.068	1.271	2.097	1.247	2.126	1.222	2.156	1.197	2.186
100	1.439	1.923	1.416	1.948	1.393	1.974	1.371	2.000	1.347	2.026	1.324	2.053	1.301	2.080	1.277	2.108	1.253	2.135	1.229	2.164
150	1.579	1.892	1.564	1.908	1.550	1.924	1.535	1.940	1.519	1.956	1.504	1.972	1.489	1.989	1.474	2.006	1.458	2.023	1.443	2.040
200	1.654	1.885	1.643	1.896	1.632	1.908	1.621	1.919	1.610	1.931	1.599	1.943	1.588	1.955	1.576	1.967	1.565	1.979	1.554	1.991

Source: Savin N.E., White, K.J. (1977). The Durbin-Watson Test for Serial Correlation with Extreme Sample Sizes or Many Regressors. *Econometricia* 45:1989–1996. Corrections: Farebrother, R.W. (1980). Econometricia 48:1554. With permission of the Econometric Society.

TABLE A.6

Empirical Cumulative Distribution of τ for $\rho = 1$

Sample Size n	Significance Level **0.01**	**0.025**	**0.05**	**0.10**
The τ statistic: no constant or time trend ($\beta_0 = \beta_1 = 0$) (Equation 19.10.1)				
25	−2.66	−2.26	−1.95	−1.60
50	−2.62	−2.25	−1.95	−1.61
100	−2.60	−2.24	−1.95	−1.61
250	−2.58	−2.23	−1.95	−1.62
300	−2.58	−2.23	−1.95	−1.62
∞	−2.58	−2.23	−1.95	−1.62
The τ_μ statistic: constant but no time trend ($\beta_1 = 0$) (Equation 19.10.2)				
25	−3.75	−3.33	−3.00	−2.62
50	−3.58	−3.22	−2.93	−2.60
100	−3.51	−3.17	−2.89	−2.58
250	−3.46	−3.14	−2.88	−2.57
300	−3.44	−3.13	−2.87	−2.57
∞	−3.43	−3.12	−2.86	−2.57
The τ_τ statistic: constant + time trend (Equation 19.10.3)				
25	−4.38	−3.95	−3.60	−3.24
50	−4.15	−3.80	−3.50	−3.18
100	−4.04	−3.73	−3.45	−3.15
250	−3.99	−3.69	−3.43	−3.13
300	−3.98	−3.68	−3.42	−3.13
∞	−3.96	−3.66	−3.41	−3.12

Source: Adapted from Fuller, W.A. (1996). *Introduction to Statistical Time Series*, 2nd ed. New York: John Wiley & Sons, Inc. With permission of John Wiley & Sons, Inc.

TABLE A.7

Empirical Distribution of Φ

Sample Size	Significance Level			
n	**0.01**	**0.05**	**0.025**	**0.01**
			Φ_1 (H_0: $\beta_0 = \varphi = 0$)	
25	4.12	5.18	3.60	7.88
50	3.94	4.86	5.80	7.06
100	3.86	4.71	5.57	6.70
250	3.81	4.63	5.45	6.52
300	3.79	4.61	5.41	6.47
∞	3.78	4.59	5.38	6.43
			Φ_2 (H_0: $\beta_0 = \beta_1 = \varphi = 0$)	
25	4.67	5.68	6.75	8.21
50	4.31	5.13	5.94	7.02
100	4.16	4.88	5.59	6.50
250	4.07	4.75	5.40	6.22
300	4.05	4.71	5.35	6.15
∞	4.03	4.68	5.31	6.09
			Φ_3 (H_0: $\beta_1 = \varphi = 0$)	
25	5.91	7.24	8.65	10.61
50	5.61	6.73	7.81	9.31
100	5.47	6.49	7.44	8.73
250	5.39	6.34	7.25	8.43
300	5.36	6.30	7.20	8.34
∞	5.34	6.25	7.16	8.27

Source: Dickey, D.A., Fuller, W.A. (1981). Likelihood Ratio Statistics for Autoregressive Time Series with a Unit Root. *Econometrica* 49:1057–1072. With permission of the Econometric Society.

TABLE A.8

Critical Values for the Cointegration Test

Number of Variables k	Sample Size n	Significance Level 1%	5%	10%
1[a]	50	2.62	1.95	1.61
	100	2.60	1.95	1.61
	250	2.58	1.95	1.62
	500	2.58	1.95	1.62
	∞	2.58	1.95	1.62
1[b]	50	3.58	2.93	2.60
	100	3.51	2.89	2.58
	250	3.46	2.88	2.57
	500	3.44	2.87	2.57
	∞	3.43	2.86	2.57
2	50	4.32	3.67	3.28
	100	4.07	3.37	3.03
	200	4.00	3.37	3.02
3	50	4.84	4.11	3.73
	100	4.45	3.93	3.59
	200	4.35	3.78	3.47
4	50	4.94	4.35	4.02
	100	4.75	4.22	3.89
	200	4.70	4.18	3.89
5	50	5.41	4.76	4.42
	100	5.18	4.58	4.26
	200	5.02	4.48	4.18

[a] Critical values of $\hat{\tau}$.

[b] Critical values of $\hat{\tau}_\mu$.

Source: Fuller, W.A. (1996). *Introduction to Statistical Time Series*, 2nd ed. New York: John Wiley & Sons, Inc., p. 373; and Engle, R.F., Yoo, B.S. (1987). Forecasting and Testing in Co-integrated Systems. *J Econom* 35:143–159. With permission of *The Journal of Econometrics*.

TABLE A.9

Orthogonal Polynomial Coefficients

	n = 3		n = 4			n = 5				n = 6				
i	ϕ_1	ϕ_2	ϕ_1	ϕ_2	ϕ_3	ϕ_1	ϕ_2	ϕ_3	ϕ_4	ϕ_1	ϕ_2	ϕ_3	ϕ_4	ϕ_5
1	−1	1	−3	1	−1	2	2	−1	1	−5	5	−5	1	−1
2	0	−2	−1	−1	3	−1	−1	2	−4	−3	−1	7	−3	5
3	1	1	1	−1	−3	0	−2	0	6	−1	−4	4	2	−10
4			3	1	1	1	−1	−2	−4	1	−4	−4	2	10
5						2	2	1	1	3	−1	−7	−3	−5
6										5	5	5	1	1
$\sum_{i=1}^{n}\phi_j^2(X_i)$	2	6	20	4	20	10	14	10	70	70	84	180	28	252
λ_k	1	3	2	1	$\frac{10}{3}$	1	1	$\frac{5}{6}$	$\frac{35}{12}$	2	$\frac{3}{2}$	$\frac{5}{3}$	$\frac{7}{12}$	$\frac{21}{10}$

	n = 7						n = 8					
i	ϕ_1	ϕ_2	ϕ_3	ϕ_4	ϕ_5	ϕ_6	ϕ_1	ϕ_2	ϕ_3	ϕ_4	ϕ_5	ϕ_6
1	−3	5	−1	3	−1	1	−7	7	−7	7	−7	1
2	−2	0	1	−7	4	−6	−5	1	5	−13	23	−5
3	−1	−3	1	1	−5	15	−3	−3	7	−3	−17	9
4	0	−4	0	6	0	−20	−1	−5	3	9	−15	−5
5	1	−3	−1	1	5	15	1	−5	−3	9	15	−5
6	2	0	−1	−7	−4	−6	3	−3	−7	−3	17	9
7	3	5	1	3	1	1	5	1	−5	−13	−23	−5
8							7	7	7	7	7	1
$\sum_{i=1}^{n}\phi_j^2(X_i)$	28	84	6	154	84	924	168	168	264	616	2184	264
λ_k	1	1	$\frac{1}{6}$	$\frac{7}{12}$	$\frac{7}{20}$	$\frac{77}{60}$	2	1	$\frac{2}{3}$	$\frac{7}{12}$	$\frac{7}{10}$	$\frac{11}{60}$

(continued)

TABLE A.9 *(continued)*

	$n=9$						$n=10$					
i	ϕ_1	ϕ_2	ϕ_3	ϕ_4	ϕ_5	ϕ_6	ϕ_1	ϕ_2	ϕ_3	ϕ_4	ϕ_5	ϕ_6
1	−4	28	−14	14	−4	4	−9	6	−42	18	−6	3
2	−3	7	7	−21	11	−17	−7	2	14	−22	14	−11
3	−2	−8	13	−11	−4	22	−5	−1	35	−17	−1	10
4	−1	−17	9	9	−9	1	−3	−3	31	3	−11	6
5	0	−20	0	18	0	20	−1	−4	12	18	−6	−8
6	1	−17	−9	9	9	1	1	−4	−12	18	6	−8
7	2	−8	−13	−11	4	22	3	−3	−31	3	11	6
8	3	7	−7	−21	−11	−17	5	−1	−35	−17	1	10
9	4	28	14	14	4	4	7	2	−14	−22	−14	−11
10							9	6	42	18	6	3
$\sum_{i=1}^{n}\phi_j^2(X_i)$	60	2772	990	2002	468	1980	330	132	8580	2860	780	660
λ_k	1	3	$\frac{5}{6}$	$\frac{7}{12}$	$\frac{3}{20}$	$\frac{11}{60}$	2	$\frac{1}{2}$	$\frac{5}{3}$	$\frac{5}{12}$	$\frac{1}{10}$	$\frac{11}{240}$

Source: Pearson, E.S., Hartley, H.O. (1976). *Biometrika Tables for Statisticians*, Vol. 1, 3rd ed. London: Cambridge University Press. With permission of Oxford University Press.

TABLE A.10

Upper-Tail Probabilities for the Null Distribution of *D*

Given n, the entry in the body of the table for the critical value $d_{\alpha/2}$ is $P(D \geq d_{\alpha/2}) = \alpha/2$ $(=P(D \leq -d_{\alpha/2}))$. For example, if $n = 8$ and $d_{\alpha/2} = d_{0.031} = 16$, then $\alpha = 0.062$ and $1 - \alpha = 0.938$.

	n								
$d_{\alpha/2}$	**4**	**5**	**8**	**9**	**12**	**13**	**16**	**17**	**20**
0	.625	.592	.548	.540	.527	.524	.518	.516	.513
2	.375	.408	.452	.460	.473	.476	.482	.484	.487
4	.167	.242	.360	.381	.420	.429	.447	.452	.462
6	.042	.117	.274	.306	.369	.383	.412	.420	.436
8		.042	.199	.238	.319	.338	.378	.388	.411
10		.008	.138	.179	.273	.295	.345	.358	.387
12			.089	.130	.230	.255	.313	.328	.362
14			.054	.090	.190	.218	.282	.299	.339
16			.031	.060	.155	.184	.253	.271	.315
18			.016	.038	.125	.153	.225	.245	.293
20			.007	.022	.098	.126	.199	.220	.271
22			.002	.012	.076	.102	.175	.196	.250
24			.001	.006	.058	.082	.153	.174	.230
26			.000	.003	.043	.064	.133	.154	.211
28				.001	.031	.050	.114	.135	.193
30				.000	.022	.038	.097	.118	.176
32					.016	.029	.083	.102	.159
34					.010	.021	.070	.088	.144
36					.007	.015	.058	.076	.130
38					.004	.011	.048	.064	.117
40					.003	.007	.039	.054	.104
42					.002	.005	.032	.046	.093
44					.001	.003	.026	.038	.082
46					.000	.002	.021	.032	.073
48						.001	.016	.026	.064
50						.001	.013	.021	.056
52						.000	.010	.017	.049
54							.008	.014	.043
56							.006	.011	.037
58							.004	.009	.032
60							.003	.007	.027
62							.002	.005	.023
64							.002	.004	.020
66							.001	.003	.017
68							.001	.002	.014
70							.001	.002	.012
72							.000	.001	.010
74								.001	.008
76								.001	.007
78								.000	.006

(continued)

TABLE A.10 *(continued)*

	n								
$d_{\alpha/2}$	**4**	**5**	**8**	**9**	**12**	**13**	**16**	**17**	**20**
80									.005
82									.004
84									.003
86									.002
88									.002
90									.002
92									.001
94									.001
96									.001
98									.001
100									.000

TABLE A.10 *(continued)*

	n							
$d_{\alpha/2}$	6	7	10	11	14	15	18	19
1	.500	.500	.500	.500	.500	.500	.500	.500
3	.360	.386	.431	.440	.457	.461	.470	.473
5	.235	.281	.364	.381	.415	.423	.441	.445
7	.136	.191	.300	.324	.374	.385	.411	.418
9	.068	.119	.242	.271	.334	.349	.383	.391
11	.028	.068	.190	.223	.295	.313	.354	.365
13	.008	.035	.146	.179	.259	.279	.327	.339
15	.001	.015	.108	.141	.225	.248	.300	.314
17		.005	.078	.109	.194	.218	.275	.290
19		.001	.054	.082	.165	.190	.250	.267
21		.000	.036	.060	.140	.164	.227	.245
23			.023	.043	.117	.141	.205	.223
25			.014	.030	.096	.120	.184	.203
27			.008	.020	.079	.101	.165	.184
29			.005	.013	.063	.084	.147	.166
31			.002	.008	.050	.070	.130	.149
33			.001	.005	.040	.057	.115	.133
35			.000	.003	.031	.046	.100	.119
37				.002	.024	.037	.088	.105
39				.001	.018	.029	.076	.093
41				.000	.013	.023	.066	.082
43					.010	.018	.056	.072
45					.007	.014	.048	.062
47					.005	.010	.041	.054
49					.003	.008	.034	.047
51					.002	.006	.029	.040
53					.002	.004	.024	.034
55					.001	.003	.020	.029
57					.001	.002	.016	.025
59					.000	.001	.013	.021
61						.001	.011	.017
63						.001	.009	.014
65						.000	.007	.012
67							.005	.010
69							.004	.008
71							.003	.006
73							.003	.005
75							.002	.004
77							.001	.003
79							.001	.003

(continued)

TABLE A.10 *(continued)*

	n							
$d_{\alpha/2}$	**6**	**7**	**10**	**11**	**14**	**15**	**18**	**19**
81							.001	.002
83							.001	.002
85							.000	.001
87								.001
89								.001
91								.001
93								.000

Source: Adapted from Kaarsemaker L., van Wijngaarden, A. (1953). Tables for Use in Rank Correlation. *Stat Neerl* 7:41–54. With permission of *Stat Neerl* and Blackwell Publishers.

TABLE A.11

Fisher's $\hat{\rho}(=r)$ to ξ Transformation

The table gives values of $\xi(\hat{\rho}) = \frac{1}{2}\log_e\left(\frac{1+\hat{\rho}}{1-\hat{\rho}}\right)$, where $\xi(-\hat{\rho}) = -\xi(\hat{\rho})$ (e.g., for $\hat{\rho}$ = 0.76, $\xi(0.76)$ = 0.996; for $\hat{\rho}$ = −0.76, ξ (−0.76) = −0.996).

$\hat{\rho}$	$\hat{\rho}$ (3rd Decimal)					$\hat{\rho}$	$\hat{\rho}$ (3rd Decimal)				
	0.000	**0.002**	**0.004**	**0.006**	**0.008**		**0.000**	**0.002**	**0.004**	**0.006**	**0.008**
0.00	0.0000	0.0020	0.0040	0.0060	0.0080	**0.35**	0.3654	0.3677	0.3700	0.3723	0.3746
1	0.0100	0.0120	0.0140	0.0160	0.0180	6	0.3769	0.3792	0.3815	0.3838	0.3861
2	0.0200	0.0220	0.0240	0.0260	0.0280	7	0.3884	0.3907	0.3931	0.3954	0.3977
3	0.0300	0.0320	0.0340	0.0360	0.0380	8	0.4001	0.4024	0.4047	0.4071	0.4094
4	0.0400	0.0420	0.0440	0.0460	0.0480	9	0.4118	0.4142	0.4165	0.4189	0.4213
0.05	0.0500	0.0520	0.0541	0.0561	0.0581	**0.40**	0.4236	0.4260	0.4284	0.4308	0.4332
6	0.0601	0.0621	0.0641	0.0661	0.0681	1	0.4356	0.4380	0.4404	0.4428	0.4453
7	0.0701	0.0721	0.0741	0.0761	0.0782	2	0.4477	0.4501	0.4526	0.4550	0.4574
8	0.0802	0.0822	0.0842	0.0862	0.0882	3	0.4599	0.4624	0.4648	0.4673	0.4698
9	0.0902	0.0923	0.0943	0.0963	0.0983	4	0.4722	0.4747	0.4772	0.4797	0.4822
0.10	0.1003	0.1024	0.1044	0.1064	0.1084	**0.45**	0.4847	0.4872	0.4897	0.4922	0.4948
1	0.1104	0.1125	0.1145	0.1165	0.1186	6	0.4973	0.4999	0.5024	0.5049	0.4075
2	0.1206	0.1226	0.1246	0.1267	0.1287	7	0.5101	0.5126	0.5152	0.5178	0.5204
3	0.1307	0.1328	0.1348	0.1368	0.1389	8	0.5230	0.5256	0.5282	0.5308	0.5334
4	0.1409	0.1430	0.1450	0.1471	0.1491	9	0.5361	0.5387	0.5413	0.5440	0.5466
0.15	0.1511	0.1532	0.1552	0.1573	0.1593	**0.50**	0.5493	0.5520	0.5547	0.5573	0.5600
6	0.1614	0.1634	0.1655	0.1676	0.1696	1	0.5627	0.5654	0.5682	0.5709	0.5736
7	0.1717	0.1737	0.1758	0.1779	0.1799	2	0.5763	0.5791	0.5818	0.5846	0.5874
8	0.1820	0.1841	0.1861	0.1882	0.1903	3	0.5901	0.5929	0.5957	0.5985	0.6013
9	0.1923	0.1944	0.1965	0.1986	0.2007	4	0.6042	0.6070	0.6098	0.6127	0.6155
0.20	0.2027	0.2048	0.2069	0.2090	0.2111	**0.55**	0.6194	0.6213	0.6241	0.6270	0.6299
1	0.2132	0.2153	0.2174	0.2195	0.2216	6	0.6328	0.6358	0.6387	0.6416	0.6446
2	0.2237	0.2258	0.2279	0.2300	0.2321	7	0.6475	0.6505	0.6535	0.6565	0.6595
3	0.2342	0.2363	0.2384	0.2405	0.2427	8	0.6625	0.6655	0.6685	0.6716	0.6746
4	0.2448	0.2469	0.2490	0.2512	0.2533	9	0.6777	0.6807	0.6838	0.6869	0.6900
0.25	0.2554	0.2575	0.2597	0.2618	0.2640	**0.60**	0.6931	0.6963	0.6994	0.7026	0.7057
6	0.2661	0.2683	0.2704	0.2726	0.2747	1	0.7089	0.7121	0.7153	0.7185	0.7218
7	0.2769	0.2790	0.2812	0.2833	0.2855	2	0.7250	0.7283	0.7315	0.7348	0.7381
8	0.2877	0.2899	0.2920	0.2942	0.2964	3	0.7414	0.7447	0.7481	0.7514	0.7548
9	0.2986	0.3008	0.3029	0.3051	0.3073	4	0.7582	0.7616	0.7650	0.7684	0.7718
0.30	0.3095	0.3117	0.3139	0.3161	0.3183	**0.65**	0.7753	0.7788	0.7823	0.7858	0.7893
1	0.3205	0.3228	0.3250	0.3272	0.3294	6	0.7928	0.7964	0.7999	0.8035	0.8071
2	0.3316	0.3339	0.3361	0.3383	0.3406	7	0.8107	0.8144	0.8180	0.8217	0.8254
3	0.3428	0.3451	0.3473	0.3496	0.3518	8	0.8291	0.8328	0.8366	0.8404	0.8441
4	0.3541	0.3564	0.3586	0.3609	0.3632	9	0.8480	0.8518	0.8556	0.8595	0.8634

(continued)

TABLE A.11 *(continued)*

$\hat{\rho}$	$\hat{\rho}$ (3rd Decimal)					$\hat{\rho}$	$\hat{\rho}$ (3rd Decimal)				
	0.000	**0.002**	**0.004**	**0.006**	**0.008**		**0.000**	**0.002**	**0.004**	**0.006**	**0.008**
0.70	0.8673	0.8712	0.8752	0.8792	0.8832	**0.85**	1.2560	1.2630	1.2710	1.2780	1.2860
1	0.8872	0.8912	0.8953	0.8994	0.9035	**6**	1.2930	1.3010	1.3090	1.3170	1.3250
2	0.9076	0.9118	0.9160	0.9202	0.9245	**7**	1.3330	1.3410	1.3500	1.3580	1.3670
3	0.9287	0.9330	0.9373	0.9417	0.9462	**8**	1.3760	1.3850	1.3940	1.4030	1.4120
4	0.9505	0.9549	0.9549	0.9639	0.9684	**9**	1.4220	1.4320	1.4420	1.4520	1.4620
0.75	0.9730	0.9780	0.9820	0.9870	0.9910	**0.90**	1.4720	1.4830	1.4940	1.5050	1.5160
6	0.9960	1.0010	1.0060	1.0110	1.0150	**1**	1.5280	1.5390	1.5510	1.5640	1.5760
7	1.0200	1.0250	1.0300	1.0350	1.0400	**2**	1.5890	1.6020	1.6160	1.6300	1.6440
8	1.0450	1.0500	1.0560	1.0610	1.0660	**3**	1.6580	1.6730	1.6890	1.7050	1.7210
9	1.0710	1.0770	1.0820	1.0880	1.0930	**4**	1.7380	1.7560	1.7740	1.7920	1.8120
0.80	1.0990	1.1040	1.1100	1.1160	1.1210	**0.95**	1.8320	1.8530	1.8740	1.8970	1.9210
1	1.1270	1.1330	1.1390	1.1450	1.1510	**6**	1.9460	1.9720	2.0000	2.0290	2.0600
2	1.1570	1.1630	1.1690	1.1750	1.1820	**7**	2.0920	2.1270	2.1650	2.2050	2.2490
3	1.1880	1.1950	1.2010	1.2080	1.2140	**8**	2.2980	2.3510	2.4100	2.4770	2.5550
4	1.2210	1.2280	1.2350	1.2420	1.2490	**9**	2.6470	2.7590	2.9030	3.1060	3.4530

Source: Abridged from Pearson, E.S., Hartley, H.O. (1976). *Biometrika Tables for Statisticians*, Vol.1, 3rd ed. London: Cambridge University Press, Table 14. With permission of the trustees of *Biometrika*.

References

Aitken, J.C. (1935). On Least Squares and Linear Combinations of Observations. *Proc R Soc Edinburgh* 55:42–48.

Anderson, O.D. (1976). *Time Series Analysis and Forecasting: The Box-Jenkins Approach*. London: Butterworths & Co.

Banerjee, A., Dolado, J.J., Galbraith, J.W., Hendry, D.F. (1993). *Cointegration, Error Correction, and the Econometric Analysis of Nonstationary Data*. Oxford: Oxford University Press.

Bassett, G., Koenker, R. (1982). Tests of Linear Hypotheses and λ_1 Estimation. *Econometrica* 50:1577–1584.

Beaton, A.E., Tukey, J.W. (1974). The Fitting of Power Series, Meaning Polynomials, Illustrated on Band-Spectroscopic Data. *Technometrics* 16:147–185.

Belsley, D. (1991). *Conditioning Diagnostics, Collinearity and Weak Data in Regression*. New York: John Wiley & Sons, Inc.

Belsley, D., Kuh, E., Welsh, R.E. (1980). *Regression Diagnostics*. New York: John Wiley & Sons, Inc.

Bollerslev, T. (1986). Generalized Autoregressive Conditional Heteroskedasticity. *J Econom* 31:307–327.

Bolstad, W.M. (2004). *Introduction to Bayesian Statistics*. New York: John Wiley & Sons, Inc.

Box, G.E.P., Jenkins, G.M. (1970). *Time Series Analysis: Forecasting and Control*. San Francisco: Holden-Day.

Box, G.E.P., Tiao, G.C. (1992). *Bayesian Inference in Statistical Analysis*. Reading, MA: Addison-Wesley Publishing Co.

Charemza, W.W., Deadman, D.F. (1997). *New Directions in Econometric Practice*. Cheltenham: Edward Elgar.

Chow, G.C. (1960). Tests of the Equality Between Sets of Coefficients in Two Linear Regressions. *Econometrica* 28:591–605.

Cochrane, D., Orcutt, G.H. (1949). Application of Least Squares Regressions to Relationships Containing Autocorrelated Error Terms. *JASA* 44:32–61.

Collett, D. (1993). *Modelling Binary Data*. London: Chapman and Hall.

Cook, R.D. (1977). Detection of Influential Observations in Linear Regression. *Technometrics* 19:15–18.

Cook, R.D. (1979). Influential Observations in Linear Regression. *JASA* 74:169–174.

Cressie, N.A.C. (1993). *Statistics for Spatial Data*, revised ed. New York: John Wiley & Sons, Inc.

Davidson, R., Mackinnon, J.G. (1981). Several Tests for Model Specification in the Presence of Alternative Hypotheses. *Econometrica* 49:781–793.

Davidson, R., Mackinnon, J.G. (1993). *Estimation and Inference in Econometrics*. New York: Oxford University Press.

DeLury, D.B. (1960). *Values and Integrals of the Orthogonal Polynomials up to n=26*. Toronto: University of Toronto Press.

Dempster, A.P., Schatzoff, M., Wermuth, N. (1977). A Simulation Study of Alternatives to Ordinary Least Squares. *JASA* 72:77–90.

Diamond, P. (1988). Fuzzy Least Squares. *Inf Sci (Ny)* 46:141–157.

Diamond, P., Tanaka, H. (1996). *Fuzzy Regression Analysis. Fuzzy Sets in Decision Analysis, Operations Research, and Statistics*. Norwell, MA: Kluwer Academic Publishers.

Dickey, D.A, Fuller, W.A. (1979). Distributions of the Estimators for Autoregressive Time Series with a Unit Root. *JASA* 74:427–431.

Dickey, D.A., Fuller, W.A. (1981). Likelihood Ratio Statistics for Autoregressive Time Series with a Unit Root. *Econometrica* 49:1057–1072.

Dickey, D.A., Pantula, S.G. (1987). Determining the Order of Differencing in Autoregressive Processes. *J Business and Econ Statist* 5:459–461.
Draper, N.R., Van Nostrand, R.C. (1975). Ridge Regression and James-Stein Estimation: Review and Comments. *Technometrics* 21:451–466.
Dubois, D., Prade, H. (1980). *Fuzzy Sets and Systems*. New York: Academic Press.
Durbin, J., (1970). Testing for Serial Correlation in Least Squares Regression When Some of the Regressors are Lagged Dependent Variables. *Econometrica* 38:410–421.
Durbin, J., Watson, G.S. (1950). Testing for Serial Correlation in Least Squares Regression. *Biometrika* 37:409–428.
Durbin, J., Watson, G.S. (1951). Testing for Serial Correlation in Least Squares Regression. *Biometrika* 38:159–178.
Enders, W. (2004). *Applied Econometric Time Series*, 2nd ed. New York: John Wiley & Sons, Inc.
Engle, R.F. (1982). Autoregressive Conditional Heteroskedasticity with Estimates of the Variance of United Kingdom Inflation. *Econometrica* 50:987–1006.
Engle, R.F., Granger, C.W.J. (1987). Co-integration and Error Correction Representation, Estimation and Testing. *Econometrica* 55:251–276.
Engle, R.F., Granger, C.W.J., eds. (1991). *Long Run Economic Relations: Readings in Cointegration*. Oxford: Oxford University Press.
Engle, R.F., Yoo, B.S. (1987). Forecasting and Testing in Co-integrated Systems. *J Econom* 35:143–159.
Fan, J., Huang, L. (2001). Goodness-of-Fit Tests for Parametric Regression Models. *JASA* 96:640–652.
Fotheringham, A.S., Brundson, C., Charlton, M. (2002). *Quantitative Geography*. London: Sage Publications.
Fuller, W.A. (1996). *Introduction to Statistical Time Series*, 2nd ed. New York: John Wiley & Sons, Inc.
Glejser, H.(1969). A New Test for Heteroskedasticity. *JASA* 64:316–323.
Goldfeld, S.M., Quandt. R.E. (1965). Some Tests for Heteroskedasticity. *JASA* 60:539–547.
Goldstein, M., Smith, A.F.M. (1974). Ridge-Type Estimators for Regression Analysis. *J R Stat Soc B* 36:284–291.
Granger, C.W.J., Newbold, P. (1974). Spurious Regressions in Econometrics. *J Econom* 35:143–159.
Greene, W.H. (1993). *Econometric Analysis*, 2nd ed. New York: Macmillan Publishing Co.
Hargreaves, C.P. (1994). *Nonstationary Time Series and Cointegration*. Oxford: Oxford University Press.
Harville, D.A. (1990). BLUP (Best Linear Unbiased Prediction) and Beyond. In: *Advances in Statistical Methods for Genetic Improvement of Livestock* (Gianola D., Hammond K., eds.). New York: Springer.
Hastie, T., Tibshirani, R. (1990). *Generalized Additive Models*. New York: Chapman and Hall.
Henderson, C.R. (1950). Estimation of Genetic Parameters (abstract). *Ann Math Statist* 21:309–310.
Henderson, C.R. (1973). Sire Evaluation and Genetic Trends. In: *Proceedings of the Animal Breeding and Genetics Symposium in Honor of Dr. Jay L. Lush*. Champaign, IL: American Society of Animal Science and American Diary Science Association, 10–41.
Henderson, C.R. (1984). Applications of Linear Models in Animal Breeding. Guelph, ON, Canada: University of Guelph.
Hendry, D.F. (1980). Econometrics: Alchemy or Science? *Economica* 47:387–406.
Hendry, D.F. (1986). Econometric Methodology: A Personal Perspective. In: *Advances in Econometrics, Vol. 2* (Bewley, T.F., ed.). Cambridge: Cambridge University Press.
Hettmansperger, T.P. (1984). *Statistical Inference Based on Ranks*. New York: John Wiley & Sons, Inc.
Hettmansperger, T.P., McKean, J.W. (1977). A Robust Alternative Based on Ranks in the Linear Model. *JASA* 78:885–893.
Hildreth, C., Houck, J.P. (1968). Some Estimates for a Linear Model with Random Coefficients. *JASA* 63:584–595.
Hocking, R.R. (1976). *Methods and Applications of Linear Models*. New York: John Wiley & Sons, Inc.
Hoerl, A.E. (1962). Application of Ridge Analysis to Regression Problems. *Chem Eng Prog* 58:54–59.
Hoerl, A.E., Kennard, R.W. (1970a). Ridge Regression: Biased Estimation for Nonorthogonal Problems. *Technometrics* 12:55–67.

Hoerl, A.E., Kennard, R.W. (1970b). Ridge Regression: Applications to Nonorthogonol Problems. *Technometrics* 12:69–82.
Hoerl, A.E., Kennard, R.W. (1976). A Note on a Power Generalization of Ridge Regression. *Technometrics* 17:269.
Hoerl, A.E., Kennard, R.W., Baldwin, K.R. (1975). Ridge Regression: Some Simulations. *Commun Stat A* 4:105–123.
Holland, P.W., Walsh, R.E. (1977). Robust Regression Using Iteratively Reweighted Least Squares. *Comm Stat A* 6:813–827.
Hooker, R.H. (1901). Correlation of the Marriage-Rate with Trade. *J R Stat Soc* 64:485–492.
Hosmer, D.W. Jr., Lemeshow, S. (1989). *Applied Logistic Regression*. New York: John Wiley & Sons, Inc.
Hougaard, P. (1985). The Appropriateness of the Asymptotic Distribution in a Nonlinear Regression Model in Relation to Curvature. *J R Stat Soc Ser B* 47:103–114.
Hsiao, C. (1975). Some Estimation Methods for a Random Coefficient Model. *Econometrica* 43:305–325.
Huber, P.J. (1964). Robust Estimation of a Location Parameter. *Ann Math Statist* 35:73–101.
Huber, P.J. (1996). *Robust Statistical Procedures*, 2nd ed. Philadelphia: SIAM.
Huber, P.J. (2004). *Robust Statistics*. New York: John Wiley & Sons, Inc.
Jaeckel, L.A. (1972). Estimating Regression Coefficients by Minimizing the Dispersion of the Residuals. *Ann Math Statist* 43:1449–1458.
Johnston, J. (1984). *Econometric Methods*, 3rd ed. New York: McGraw-Hill.
Kaarsemaker L., van Wijngaarden A. (1953). Tables for Use in Rank Correlation. *Stat Neerl* 7:41–54.
Kakwani, N.C. (1967). The Unbiasedness of Zellner's Seemingly Unrelated Regression Estimators. *JASA* 82:141–142.
Kim, K.J., Moskowitz, H., Koksalan, M.(1996). Fuzzy Versus Statistical Linear Regression. *Eur J Oper Res* 92:417–434.
Kmenta, J. (1986). *Elements of Econometrics*, 2nd ed. New York: Macmillan Publishing Co.
Koenker, R. (2005). *Quantile Regression*. Cambridge: Cambridge University Press.
Koenker, R., Bassett, G.J. (1978). Regression Quantiles. *Econometrica* 46:33–50.
Lancaster, T. (2004). An Introduction to Modern Bayesian Econometrics. Malden, MA: Blackwell Publishing.
Lawless, J.F., Wang, P. (1974). A Simulation Study of Ridge and Other Regression Estimators. *Commun Stat* A5:307–323.
Lee, P.M. (2004). *Bayesian Statistics: An Introduction*. New York: Oxford University Press.
Li, H., Yen, V. (1995). *Fuzzy Sets and Fuzzy Decision-Making*. Boca Raton: CRC Press.
Lindley, D.V., Smith, A.F.M. (1972). Bayes Estimates for the Linear Model. *J R Stat Soc Series B* 34:1–41.
Loader, C. (1999). *Local Regression and Likelihood*. New York: Springer-Verlag.
Lowerre, J.M. (1974). On the Mean Square Error of Parameter Estimates for Some Biased Estimators. *Technometrics* 16:461–464.
Mackinnon, J.G., White, H., Davidson, R. (1983). Tests for Model Specification in the Presence of Alternative Hypotheses: Some Further Results. *J Econom* 21:53–70.
Maddala, G.S., Kim, I.M. (1998). *Unit Roots, Cointegration, and Structural Change*. Cambridge: Cambridge University Press.
Magnus, J. R. (1978). Maximum Likelihood Estimation of the GLS Model with Unknown Parameters in the Disturbance Covariance Matrix. *J Econom* 7:281–312.
Malinvaud, E. (1970). *Statictical Methods of Econometrics*, 2nd ed. Amsterdam: North-Holland Publishing Co.
Mallows, C.L. (1973). Some Comments on C_p. *Technometrics* 15:661–675.
Mares, M. (1994). *Computation Over Fuzzy Quantities*. Boca Raton: CRC Press. Marquardt, D.W. (1970). Generalized Inverses, Ridge Regression, Biased Linear Estimation, and Nonlinear Estimation. *Technometrics* 12:591–612.
Marquardt, D. W., Snee, R.D. (1975). Ridge Regression in Practice. *Am Stat* 29:3–19.
Mayer, L.S., Willke, T.A. (1973). On Biased Estimation in Linear Models. *Technometrics* 16:494–508.

McAleer, M., Oxley, L. (1999). *Practical Issues in Cointegration Analysis.* Malden, MA: Blackwell Publishers, Inc.

McDonald, G.C., Galarnean, D. I. (1975). A Monte Carlo Evaluation of Some Ridge-Type Estimators. *JASA* 70:407–416.

Mosteller, F., Tukey, J.W. (1977). *Data Analysis and Regression.* Reading, MA: Addison-Wesley Publishing Co.

Nguyen, H., Walker, E. (2000). *A First Course in Fuzzy Logic,* 2nd ed. Boca Raton: Chapman & Hall/CRC.

Obenchain, R.L. (1975). Good and Optimal Ridge Estimators. *Ann Stat* 6:1111–1121.

Obenchain, R.L. (1977). Classical F-Tests and Confidence Intervals for Ridge Regression. *Technometrics* 19:429–439.

Panik, M. (1996). *Linear Programming: Mathematics, Theory, and Algorithms.* Boston: Kluwer Academic Publishers.

Pearson, E.S., Hartley, H.O. (1976). *Biometrika Tables for Statisticians,* Vol. 1, 3rd ed. London: Cambridge University Press.

Phillips, P.C.B. (1986). Understanding Spurious Regressions in Econometrics. *J Econom* 33:311–340.

Powell, J. (1994). Estimation of Semiparametric Models. *The Handbook of Econometrics,* Vol. IV (Engle, R., McFadden, D., eds.). Amsterdam: North Holland.

Prais, S.J., Winsten, C.B. (1954). *Trend Estimators and Serial Correlation.* Discussion Paper 383. Chicago: Cowles Commission.

Pregibon, D. (1981). Logistic Regression Diagnostics. *Ann Statist* 9:705–724.

Puri, M.L., Sen, P.K. (1985). *Nonparametric Methods in General Linear Models.* New York: John Wiley & Sons, Inc.

Rao, B.B. (1994). *Cointegration for the Applied Economist.* Basingstoke: Macmillan Press Ltd.

Ratakowsky, D.A. (1990). *Handbook of Nonlinear Regression Models.* New York: Marcel Dekker.

Rice, J. (1984). Bandwidth Choice for Nonparametric Regression. *Ann Statist* 12:1215–1230.

Robinson, G.K. (1991). That BLUP is a Good Thing: The Estimation of Random Effects. *Stat Sci* 6:15–32.

Robson, D.S. (1959). A Simple Method for Constructing Orthogonal Polynomials When the Independent Variable is Unequally Spaced. *Biometrics* 15:187–191.

Ross, T., Booker, J., Parkinson, W. (2002). *Fuzzy Logic and Probability Applications: Bridging the Gap.* Philadelphia: SIAM.

Rousseeuw, P.J. (1984). Least Median of Squares Regression. *JASA* 79:871–880.

Rousseeuw, P.J., Hubert, M. (1996). Regression Depth. *JASA* 94:388–433.

Rousseeuw, P.J., Leroy, A.M. (1987). *A Robust Scale Estimator Based on the Shortest Half.* Technical Report. Delft, The Netherlands: Delft University of Technology.

Rousseeuw, P.J., Leroy, A.M. (2003). *Robust Regression and Outlier Detection.* New York: John Wiley & Sons, Inc.

Rousseeuw, P.J., Yohai, V. (1984). Robust Regression by Means of S-Estimators. In: *Robust and Nonlinear Time Series Analysis* (Franke, J., Härdle, W., Martin, R.D., eds.). New York: Springer-Verlag.

SAS Institute Inc. (1999a). *SAS/ETS User's Guide,* Version 8. Cary, NC: SAS Institute, Inc.

SAS Institute Inc. (1999b). *SAS/IML User's Guide,* Version 8. Cary, NC: SAS Institute, Inc.

SAS Institute Inc. (1999c). *SAS Procedures Guide,* Version 8. Cary, NC: SAS Institute, Inc.

SAS Institute Inc. (1999d). *SAS/STAT User's Guide,* Version 8. Cary, NC: SAS Institute, Inc.

Savic, D.A., Pedrycz, W. (1991). Fuzzy Linear Regression Models: Construction and Evaluation. In: *Fuzzy Regression Analysis* (Kacprzyk, J., Fedrizzi, M., eds.). Heidelberg: Physica-Verlag.

Savin N.E., White, K.J. (1977). The Durbin-Watson Test for Serial Correlation with Extreme Sample Sizes or Many Regressors. *Econometricia* 45:1989–1996. Corrections: Farebrother, R.W. (1980). *Econometricia* 48:1554.

Schabenberger, O., Gotway, C.A. (2005). Statistical Methods for Spatial Data Analysis. *JASA* 101:389–340.

Schmidt, P. (1976). *Econometrics.* New York: Marcel Dekker.

Seber, G.A.F. (1977). *Linear Regression Analysis.* New York: John Wiley & Sons, Inc.

Stock, J.H., Watson, M.W. (1988). Testing for Common Trends. *JASA* 83:1097–1107.

Tanaka, H., Uejima, S., Asai, K. (1982). Linear Regression Analysis with Fuzzy Model. *IEEE Trans Systems Man Cybern SMC* 12:903–907.

Theil, H. (1950a). A Rank-Invariant Method of Linear and Polynomial Regression Analysis, I. *Proc Koninklijke Nederlandse Akademie Wetenschappen* 53:386–392.

Theil, H. (1950b). A Rank-Invariant Method of Linear and Polynomial Regression Analysis, II. *Proc Koninklijke Nederlandse Akademie Wetenschappen* 53:521–525.

Theil, H. (1950c). A Rank-Invariant Method of Linear and Polynomial Regression Analysis, III. *Proc Koninklijke Nederlandse Akademie Wetenschappen* 53:1397–1412.

Theil, H., Nagar, A.L. (1961). Testing the Independence of Regression Disturbances. *JASA* 56:793–806.

Theobald, C.M. (1974). Generalizations of Mean Square Error Applied to Ridge Regression. *J R Stat Soc Series B* 36:103–106.

Van Nostrand, R.C. (1980). Comment on "A Critique of Some Ridge Regression Methods" by G. Smith and F. Campbell. *JASA* 75:92–94.

Walker, G. (1931). On Periodicity in Series of Related Terms. *Proc R Soc* A131:518–526.

Wang, H.F, Tsaur, R.C. (2000). Insight of a Fuzzy Regression Model. *Fuzzy Sets Syst* 112:355–369.

White, H. (1980). A Heteroscedasticity-Consistent Covariance Matrix Estimator and a Direct Test of Heteroscedasticity. *Econometrica* 48:817–838.

Wilkie, D. (1965). Complete Set of Leading Coefficients, λ (r, n), for Orthogonal Polynomials up to $n = 26$. *Technometrics* 7:644–648.

Wishart, J., Metakides, T. (1953). Orthogonal Polynomial Fitting. *Biometrika* 76:141–148.

Wolfinger, R., Tobias, R., Sall, J. (1994). Computing Gaussian Likelihoods and Their Derivatives for General Linear Mixed Models. *SIAM J Sci Comp* 15:1294–1310.

Yatchew, A. (1988). Some Tests of Nonparametric Regression Models. *Dynamic Econometric Modeling* (Barnett, W., Berndt, E., White, H., eds.). Cambridge: Cambridge University Press.

Yatchew, A. (1997). An Elementary Estimator of the Partial Linear Model. *Econ Lett* 57:135–143.

Yatchew, A. (1998). Nonparametric Regression Techniques in Economics. *J Econ Lit* 36:669–721.

Yatchew, A. (2003). *Semiparametric Regression for the Applied Econometrician*. Cambridge: Cambridge University Press.

Yule, G.U. (1926a). Why Do We Sometimes Get Nonsense-Correlations Between Time Series? A Study in Sampling and the Nature of Time Series. *J R Stat Soc* 89:1–64

Yule, G.U. (1926b). On a Method of Investigating Periodicities in Disturbed Series with Special Reference to Wolfer's Sunspot Numbers. *Philos Trans R Soc Lond A* 226:267–298.

Zadeh, L.A. (1965). Fuzzy Sets. *Inform Control* 8:338–353.

Zadeh, L.A., Fu, K.S., Tanaka, K., Shimura, M., eds. (1975). *Fuzzy Sets and their Applications to Cognitive and Decision Processes.* New York: Academic Press.

Zellner, A. (1962). An Efficient Method for Estimating Seemingly Unrelated Regressions and Tests for Aggregation Bias. *JASA* 57:348–368.

Zellner, A. (1971). *An Introduction to Bayesian Inference in Econometrics*. New York: John Wiley & Sons, Inc.

Zellner, A. (1975). Bayesian Analysis of Regression Error Terms. *JASA* 70:138–144.

Index